Radiation Oncology – Principles, Precepts and Practice

Anusheel Munshi
Tharmarnadar Ganesh
Biplab Sarkar • Atul Sharma
Indranil Mallick
Manur Gururajachar Janaki
Bidhu K. Mohanti
Editors

Radiation Oncology – Principles, Precepts and Practice

Volume I – Technical Aspects

Second Edition

Springer

BYWORD BOOKS™

Editors
Anusheel Munshi
Director, Department of Radiation Oncology
IOCI
Fortis Noida, Uttar Pradesh, India

Biplab Sarkar
Chief Medical Physicist, Department of Radiation Oncology
Apollo Multispeciality Hospitals
Kolkata, West Bengal, India

Indranil Mallick
Senior Consultant, Department of Radiation Oncology
Tata Medical Center
Kolkata, West Bengal, India

Bidhu K. Mohanti
Director-Academic and Head of Radiation Oncology
Bagchi Sri Shankara Cancer Centre & Research Institute
Bhubaneswar, Odisha, India

Tharmarnadar Ganesh
Manipal Hospital
New Delhi, Delhi, India

Atul Sharma
Vice Chairman, Department of Medical Oncology
Max Institute of Cancer Care, Max Super Specialty Hospital
New Delhi, Delhi, India

Manur Gururajachar Janaki
Professor, Department of Radiation Oncology
Ramaiah Medical College
Bengaluru, Karnataka, India

ISBN 978-981-97-8388-5 ISBN 978-981-97-8389-2 (eBook)
https://doi.org/10.1007/978-981-97-8389-2

Jointly published with Byword Books

1st edition: © Elsevier 2007
2nd edition: © The Editor(s) (if applicable) and The Author(s) 2025, corrected publication 2025

Co-publishing partnership between Byword Books Private Limited and Springer Nature Singapore Pte Ltd.
The registered company address is: 152 Beach Road, #21-01/04 Gateway East, Singapore 189721, Singapore

Preface

We are proud to introduce *Radiation Oncology: Principles, Precepts, and Practice*, a comprehensive textbook, that aims to provide both oncology students and practicing clinicians with a robust understanding of the principles, technologies, and practices that define the dynamic field of radiation oncology. This book is designed to serve as both a foundational resource for students entering the field and a practical reference for experienced practitioners looking to refine their expertise.

Radiation oncology, as a cancer-directed therapy (CDT), is a critical component of modern cancer treatment, responsible for treating a vast array of malignancies using ionizing radiation. Over the last few decades, advancements in technology, radiobiology, and treatment protocols have significantly enhanced the precision, effectiveness, and safety of radiation therapy. However, with these advancements also come challenges, such as the integration of new treatment modalities, the complexity of treatment planning, and the need for a nuanced understanding of cancer biology. To address these challenges, our goal in writing this textbook was not only to provide an up-to-date account of current practices and technologies but also to present a balanced view that incorporates evidence from both the developed and developing world.

To create a textbook that serves the global community of oncology students and practitioners, we have made a conscious effort to ensure that the content is applicable to a diverse range of healthcare settings. While much of the literature in radiation oncology has focused on the clinical practices of developed countries, we recognize that the challenges and opportunities in radiation therapy are vastly different in resource-limited settings of low- and middle-income countries (LMIC). By including evidence, case studies, and clinical insights from both high-resource and low-resource countries, this book provides a more inclusive perspective, highlighting the innovative approaches that clinicians in developing countries have adopted despite challenges like limited access to advanced technology, financial constraints, and a lack of specialized personnel.

The structure of this book is designed to serve as both a learning tool and a practical guide for everyday clinical practice. *Volume I* (*technical aspects*) of the book, in your hand/for your eyes, provides a solid foundation in the fundamental principles of radiation oncology, including an introduction to the cancer world, the physics of radiation oncology, radiobiology, and the molecular mechanisms underlying cancer development and progression.

These include detailed discussions on external beam radiation therapy (EBRT), stereotactic radiosurgery (SRS), brachytherapy, and emerging therapies, such as proton therapy and heavy ion therapy. These sections lay the groundwork for understanding the biological effects of radiation, treatment planning, and the role of radiation therapy in the broader context of cancer care.

Following this foundational knowledge, *Volume II* of the book has chapters that delve into the clinical application of radiation oncology for various body sites. In this volume, we have included treatment protocols and decision-making algorithms that are drawn from real-world clinical scenarios, ensuring that students and practitioners can connect theory to practice.

One of the key features of this textbook is its focus on multidisciplinary care. Radiation therapy does not exist in a vacuum, and the treatment of cancer requires the collaboration of a wide range of specialists, including surgeons, medical oncologists, pathologists, cancer researchers and epidemiologists, and palliative care teams. Each chapter of the book reflects this integrated approach, highlighting the importance of communication and cooperation among healthcare providers to deliver optimal care to cancer patients.

Other important aspects covered in this book include patient safety and quality assurance in the department. Radiation therapy often comes with challenges, such as skin reactions, fatigue, and organ-specific toxicities. The book provides evidence-based strategies for managing these side effects, enhancing the quality of life for patients undergoing radiation therapy. Furthermore, the psychological and emotional aspects of cancer treatment are explored, offering guidance on supporting patients through the emotional burdens of a cancer diagnosis and the rigors of treatment courses.

To ensure the textbook is of value to a global audience, we have included examples from both high-income countries (HIC), where advanced technology and resources are more readily available, and under-developed LMIC settings, where resource constraints often necessitate creative and adaptive approaches to treatment. By doing so, we aim to make this textbook relevant to oncology professionals around the world, whether they are working in well-equipped cancer centers or in environments with limited resources. Finally, the quality of multidisciplinary care provided to cancer patients across diverse settings should become a sensitive component of cancer care.

We would like to express our gratitude to the many experts who have contributed their insights and knowledge by authoring the chapters for this book. Their expertise has added much value to the content, ensuring that this textbook is both evidence-based, and practice oriented for real-world cancer care. The idea of this book was conceived in 2021, hence for the biomedical world, it is the creation out of the COVID-19 pandemic.

We acknowledge with gratitude the support and guidance given to us by the publishers Springer Nature and Byword, particularly the editors Jagjeet Kaur Saini, Naren Aggarwal, and Dinesh Sinha, as well as the production editors Ejaz Ahmed and Hemalatha Kothandaraman for bringing this project to fruition. We would like to acknowledge the support from Dr. Vikas Pandey and Mr. Rahul Mehra in everyday help and support in the making of this textbook.

And finally, we thank each one of you, the readers of this book. We eagerly wait to hear your comments and reviews.

Fortis Noida, Uttar Pradesh, India	Anusheel Munshi
New Delhi, Delhi, India	Tharmarnadar Ganesh
Kolkata, West Bengal, India	Biplab Sarkar
New Delhi, Delhi, India	Atul Sharma
Kolkata, West Bengal, India	Indranil Mallick
Bengaluru, Karnataka, India	Manur Gururajachar Janaki
Bhubaneswar, Odisha, India	Bidhu K. Mohanti
17th January 2025	

Contents

Introduction to Cancer World 1

Bidhu K. Mohanti, Prashant Mathur, Umesh Jayarajah, Biswa Mohan Biswal, and Shankar Prinja

1.1 A Preamble

Cancer might seem like our modern-day illness, but people have been getting cancer for thousands of years. We know this because people in ancient history wrote about cancer. The word 'cancer' was first used by Hippocrates, the Greek physician, considered as the 'Father of Medicine'. He used the Greek word 'carcinos' *when he wrote about tumours. This Greek word means a 'crab' because Hippocrates thought the insides of the tumours looked like crabs (crayfish).*

B. K. Mohanti (✉)
Academic and Head of Radiation Oncology, Bagchi Sri Shankara Cancer Centre & Research Institute, Bhubaneswar, Odisha, India

P. Mathur
National Centre for Disease Informatics and Research, Indian Council of Medical Research (Department of Health Research, Ministry of Health and Family Welfare, Govt. of India), Bengaluru, India
e-mail: director-ncdir@icmr.gov.in

U. Jayarajah
National Hospital of Sri Lanka, Colombo, Western Province, Sri Lanka

Postgraduate Institute of Medicine, University of Colombo, Colombo, Sri Lanka

B. M. Biswal
Department of Radiotherapy and Oncology, KPJ Ipoh Specialist Hospital, Ipoh, Perak, Malaysia

S. Prinja
Professor in Health Economics, Department Community Medicine and School of Public Health, Post Graduate Institute Medical Education and Research (PGIMER), Chandigarh, India

Formerly Executive Director (HP&QA), National Health Authority Ayushman Bharat PM-JAY Government of India, New Delhi, India

This chapter delineates the context of cancer in the world. The human habitation has recorded cancer since 4000 years. The history and evolution of cancer are important for the knowledge bank of the scientists and clinicians. During the last 100 years, cancer incidence is growing all over the world. This introductory chapter describes the following sub-sections with emphasis on India and low- and middle-income countries.

History of cancer in humans.
Evolution of cancer: nature versus nurture.
Cancer Epidemiology and Registration.
Cancer Burden in the World.
Cancer Burden in India and LMIC.
Cancer Care in India and LMIC.
Cancer Management in Developed versus Developing Countries.

1.2 Introduction

The origin and antiquity of cancer in humans should be of interest to scientists, students, researchers and clinicians working in different

A. Munshi et al. (eds.), *Radiation Oncology – Principles, Precepts and Practice*,
https://doi.org/10.1007/978-981-97-8389-2_1

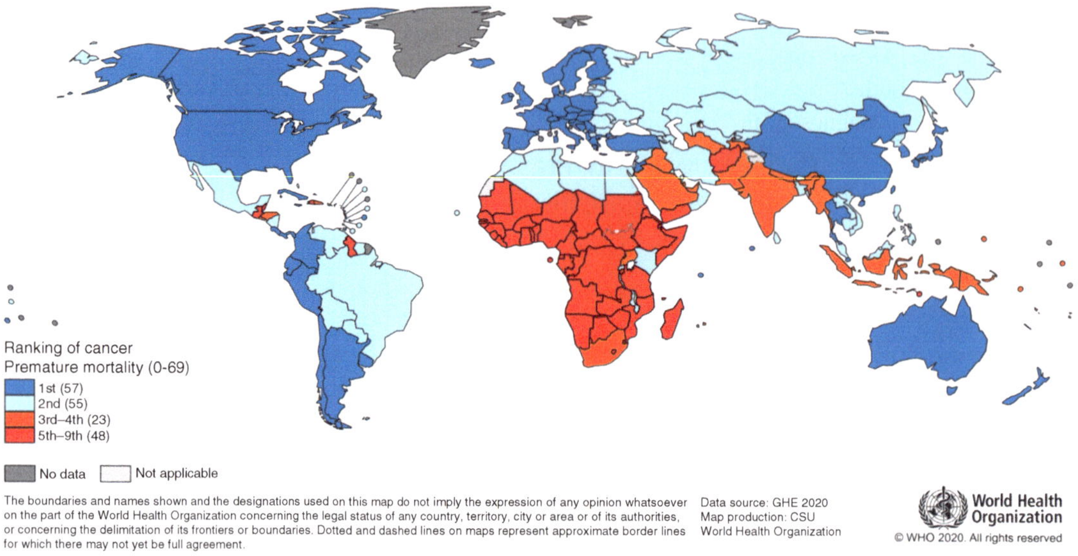

Fig. 1.1 Ranking of cancer as cause for pre-mature mortality (0–69 years of age) (Courtesy WHO map 2020)

specialties related to oncology. Although the terms such as cancer, neoplasm, malignancy and oncology are used interchangeably, the biomedical world has recognised that the progress in different areas and specialties of oncology and the management of cancer are relatively recent in the last one hundred years. The authentic records on human cancers are available for the last 4000 years, mainly from the palaeopathological studies [1]. Cancer was a rare disease in pre-human, ancient mankind and vertebrate animals. The evolution in terms of human habitation, environment and genetic factors has given rise to the increasing incidences. The ancient medical systems of India, namely Ayurveda and Siddha, mentioned about cancer-like diseases more than 2500 years back. The allopathic and western medicine reached India in the sixteenth century, and the published records on human cancers were presented since the last 400 years. The first population-based cancer registry (PBCR) in India was created in Bombay (now Mumbai) in 1963 by DJ Jussawalla, an eminent cancer surgeon of the time, with funding from the Indian Cancer Society and the National Cancer Institute of the USA. This PBCR provided reliable population-based data on the cancer incidence in Bombay [2].

The staggering increase in cancer burden globally, including the low- and middle-income countries (LMICs), is a new phenomenon in the human population over the last one century. For many countries, the cancer prevalence itself contributes to limit the life expectancy [1]. The mapping of diseases around the world gives significant importance to cancer. According to estimates from the World Health Organisation (WHO) in 2020, cancer is the first or second leading cause of death before the age of 70 years in 112 of 183 countries and ranks third or fourth in a further 23 countries (Fig. 1.1). Cancer ranks as a leading cause of death and an important barrier to increasing life expectancy in every country of the world [3]. Taking into account the accessibility to cancer care modalities, radiotherapy facilities remain the backbone of comprehensive cancer management and yet remain sparse or absent in many countries of the LMIC. In the coming decades, the development of radiation oncology specialty in terms of infrastructure, equipment and skilled manpower should be the focus.

1.3 History of Cancer in Humans

Cancer in human being is not a modern disease. The earliest documented case of osteosarcoma of the metatarsal bone was documented in an ancient human ancestor perhaps *Homo ergaster* or

Tranthropus robustus. This case was discovered in Swartkrans, a fossil-rich cave site near Johannesburg, South Africa. By 3D imaging, the researchers have diagnosed this tumour origin between 1.6 and 1.8 million years ago [4] (Fig. 1.2).

According to famous Egyptologists, the earliest written record generally regarded as describing human cancer appeared in ancient Egyptian manuscripts discovered in the nineteenth century, especially the Edwin Smith and George Ebers papyri that describe surgical, pharmacological, and magical treatments. They were written between 1500 and 1600 BC possibly based on material from thousands of years earlier [5]. Subsequently Greek physician Hippocrates (460–370 BC) coined the word *Karkinos* (carcinos) representing a crab. Thereupon, Italian surgeon-scientist Aulus Cornelius Celsus (25 BC–50 AD) translated *Karkinos* to cancer as Latin word for crab. In the second century AD, famous follower of Hippocrates, Claudius Galen (129–216 AD), used words *Oncos* (Greek word for swelling) and *Oma* (Greek word for growth) as cancerous lesion. Then onwards from Galen, the modern word oncology is assigned for cancer science. Galen popularised 'humoral theory' (that the human body consisted of four basic humours—black bile, yellow bile, blood and phlegm as the causes of cancer) and recommended treatment using diet, bleeding and laxatives to treat cancer [6]. Similarly in the eastern world, from the limited and ill-defined evidence of Mohenjo Daro & Harappan civilisation (2300 BC to 1750 BC) at Indian-Pakistan border region, the research documentation on neoplasms was rare but not unknown. Ancient documents of Indian traditional Ayurveda and Siddha systems of medicine (second century BCE) documents the human cancers and their treatment. The cause of cancer was based on *Tridosa theory* of Ayurveda. The name *Arbuda* was given for tumours, *Granthi* as minor tumours and *Asadhya Vrana* as malignant tumours documented in Atharva Veda as 'APACIT' [7]. Subsequent history of cancer in Asia and Indian subcontinent was evident from the nineteenth century onwards in cancer epidemiology such as Kangri cancer native to Kashmir and Dhoti cancer in Gujarat, India.

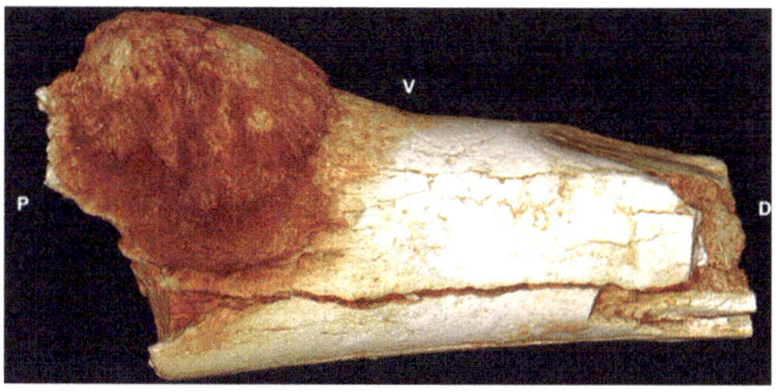

Fig. 1.2 Osteosarcoma in a hominin fifth metatarsal, exhibit a hemi-spherical bony mass located on the proximo-ventral aspect of the shaft, abutting the cortical bone surface (P, proximal; D, distal; V, ventral) [Reproduced from open source: Odes EJ et al. S Afr J Sci 2016;112(7/8):100–104]

1.4 Evolution of Cancer: Nature Versus Nurture

Cancer encompasses over 200 diseases; which are unified by uncontrolled cell growth and metastatic potential. Yet, during the last 100 years, the studies to identify the aetiological factors have revealed that there are wide variations. Obviously, the question arises, 'what causes cancer?' Broadly, cancer is a result of changes in the DNA of the cell, leading to somatic mutations, loss of homeostatic control and uncontrolled proliferation of cells [8]. This cascading biology of cancer is often called the 'hallmarks of cancer', which can be observed at multiple levels such as mutation, cell proliferation, immortality of cell, genome instability, angiogenesis, evasion of apoptosis, invasion and metastasis.

Majority of cancers arise through a complex series of interactions between genes ('nature') and the environment ('nurture'). These interactions vary between individuals, genders and ethnicities. Environmental and lifestyle related carcinogens can lead to sporadic mutations, which accumulate over the years inside the human body. On the other side, inherited genetic changes, often known as hereditary syndrome, would lead to cancer formation. As an estimate, it is counted that 95% of all cancers occur due to accumulated somatic mutations (*nurture*) during lifetime, whereas approximately 5% can be attributed to the inherited genetic changes or defects (*hereditary*). In terms of cancer causation, the development and progression of cancer are seen as a multistep process called carcinogenesis. Although, it is difficult to pinpoint a starting time of a tumour, the cellular and genetic injuries due to various endogenous (excess hormone or oxidative process inside the body) or exogenous (tobacco, alcohol, viral infection) factors lead to the initiation and promotion of cancer [9]. At the turn of the century, in 2000, Hanahan and Weinberg collected and analysed the cancer research carried out since the 1950s and set forth the origin of cancer or tumorigenesis in humans as a multistep process which occurs during the lifetime [10].

The environmental and lifestyle factors such as a chemical (nicotine) or biological (human papilloma virus) agent can trigger injury and change in the genome. Similarly, ageing causes senescence in the human cells. Hence, for a vast majority of cancers, such a multi-step process makes cancer a disease caused by accumulation of injuries and alterations at cellular, molecular and genetic levels. Hanahan and Weinberg in 2000 [10] suggested that most cancers acquire a set of functional capabilities; broadly, this set consists of six (points 1–6) alterations. In subsequent accumulation of evidences from cancer research in the last 20 years, four more cellular and humoral processes have been added (points 7–10) making a total of ten functional abnormalities acquired by human cells, named together as Hallmarks of Cancer [11]:

1. self-sufficiency in growth signals,
2. insensitivity to growth-inhibitory (anti-growth) signals,
3. evasion of programmed cell death (apoptosis),
4. limitless replicative potential,
5. sustained angiogenesis,
6. tissue invasion and metastasis,
7. deregulating cellular metabolism,
8. avoiding immune destruction,
9. tumour-promoting inflammation,
10. genome instability and mutation.

However, majority of the cancer types can be attributed to causative factors, and cancer prevention can reduce the life time cancer risks by giving attention to avoid or reduce exposure to the environmental and lifestyle agents. A global estimate showed that 2.4 million (35%) out of seven million deaths from cancer were attributable to nine potentially modifiable risk factors. These were enumerated as (1) overweight and obesity, (2) low fruit and vegetable intake, (3) physical inactivity, (4) smoking, (5) alcohol, (6) sexual transmission of HPV, (7) urban air pollution, (8) indoor smoke from fuel and (9) contaminated injection in healthcare setting [12].

It is still not completely clear why certain individuals are more susceptible to develop cancer. However, one possible explanation may be genetic polymorphism. Genetic polymorphisms are involved in the regulation of several vital biological processes in cell proliferation, such as DNA synthesis, cell cycle regulation, protein synthesis and apoptosis. Variations in such genetic polymorphisms, between one individual and another, are postulated to affect the degree of influence by environmental factors in carcinogenesis [13]. The complex interplay between genetics and the environment in the causation of cancer, and then its prognosis and response to different treatment modality different individually. Therefore, many studies are being conducted in the field of personalised medicine and have become a crucial futuristic perspective in cancer care and cancer research.

For our understanding of nature versus nurture interplay in the human cancers, another interesting area is the human migration. Epidemiological studies among immigrant populations who are likely to have the same genetic factors have highlighted the prominent role of environmental and life-style related factors in the causation of cancer [14]. Some studies from population-based twin cohort studies have shown that greater than 50% of cancer incidence was due to environmental factors [15]. This is crucial in the prevention of cancers because a large proportion of cancer cases could be prevented by modifying the environmental factors despite similar genetic susceptibility.

1.5 Cancer Epidemiology and Registration

From the above description of the evolution of cancer, it is evident that the study of human population is vital for the understanding of cancer. Epidemiology is defined as the study of the occurrence and distribution of diseases or health-related events in a specified population and its determining factors and applying this knowledge for disease prevention and control [16]. Cancer epidemiology aims to quantify the magnitude of cancer, understand the causative factors, evaluate disease outcomes and use the information to define disease control measures. The evolution of cancer epidemiology dates back to 1713 when Bernardino Ramazzini observed that a low cervical cancer incidence could be associated with celibacy [17]. There have been major breakthroughs in identifying the association of modifiable risk factors such as tobacco and alcohol use, diet, and physical activity with cancer.

Cancer registration is the systematic collection, analysis, interpretation and dissemination of cancer data to facilitate prevention and control programs [18]. They provide a significant basis for descriptive and analytical epidemiology, allowing for information on cancer incidence, distribution, and patterns. Regional variations give rise to the need to search for likely aetiology and formulation of suitable hypotheses. Research studies that look into cancer's crucial causes or risk factors and prevalence of exposure to such elements can be undertaken in the population covered by the cancer registry. The functioning of a cancer registry is shown in Fig. 1.3. The information could be obtained through an active collection (actual visits to the facility) or passive reporting (through notification forms and discharge reports).

There are two types of cancer registries: Population-Based Cancer Registries (PBCRs) and Hospital-Based Cancer Registries (HBCRs). The functioning of PBCRs and HBCRs complements and supplements each other. Population-based cancer registry data is used to describe the magnitude of cancer burden in a well-defined population and help identify and prioritise problems of public health concern. The data could be used as a source for conducting epidemiologic studies on causation and association. They help in monitoring and assessing the effectiveness of cancer control activ-

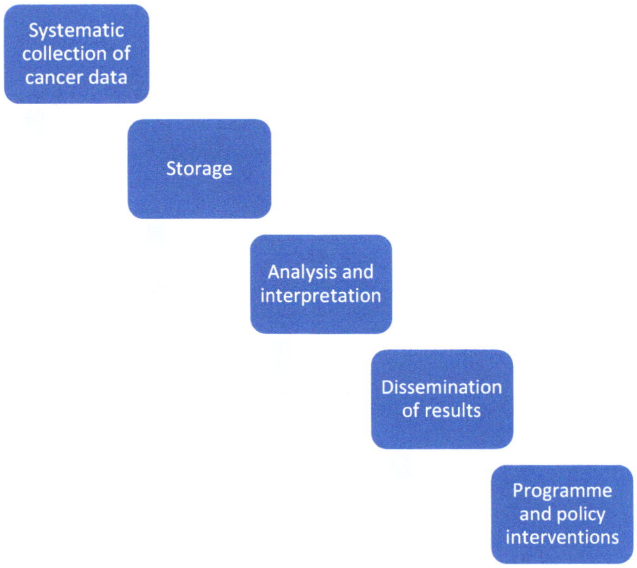

Fig. 1.3 Functioning of a cancer registry

ities. HBCRs' generate essential information on diagnosis, staging, treatment modalities, and outcomes in patients who avail of care at a specific hospital in any part of the country [19].

The primary sources of information of a population-based registry include (1) Cancer treating facilities, which include hospitals as well as private clinics (2) Diagnostic centres, which include laboratories and imaging centres (3) Death certificates from the civil registration system. The type of data collected by a cancer registry is shown in Table 1.1. Depending upon the cancer registry and its location, the data variables could be categorised as 'mandatory' or 'recommended/optional' based up on the resources and type of data available. The coding for tumour site, topography and morphology are done according to the tenth revision of the International Classification of Diseases (ICD-10) [20], and the third revision of the ICD for Oncology (ICD-O-3) [21]. The quality of data in a cancer registry is evaluated by its completeness and validity. The data and findings from a cancer registry are made available as reports and peer reviewed journal publications.

Table 1.1 Type of data collected in a cancer registry programme

Domain	Data variables
Sociodemographic details	– Patient details: Name, age, sex – Other sociodemographic details such as marital status, level of education, religion etc. – Incidence registration number – Hospital name and registration
Clinical details (tumour site and pathology, diagnosis, treatment and outcomes)	– First diagnosis date, method of diagnosis and complete pathological diagnosis – Topography and morphology of primary site according to ICD-O-3 – Site of tumour according to ICD-10 – Secondary site – Morphology of metastasis – Sequence – Anatomical site of specimen/smear – Laterality – Clinical extent of disease before treatment – Staging system followed and TNM staging – Intention to treat – Cancer-directed treatment – Disease status/performance status/death

Cancer surveillance is a continuous process and is defined as the systematic collection, analysis, interpretation and dissemination of cancer data to facilitate cancer prevention and control programs [22]. Therefore, cancer registries are an integral part of cancer surveillance and play a vital role in collecting and managing surveillance data for a region, country and overall for the world.

1.6 Cancer Burden in the World

Cancer ranks as a leading cause of death. According to estimates from the WHO in 2020, cancer is the first or second leading cause of death before the age of 70 years (Fig. 1.1) in 112 of 183 countries and ranks third or fourth in a further 23 countries [23]. The rising burden of cancer in terms of incidence and mortality could be attributed to the growing population, life expectancy, socioeconomic development and decline in infectious diseases, resulting in the emergence and changing profile of cancer risk factors.

1.6.1 Global Cancer Statistics in 2020 [24]

The total number of cancer cases was estimated to be 19.3 million in 2020, with a reported number of ten million deaths. The number of cases was higher in males (10.06 million) than females (9.22 million). The number of incident cases in different regions is shown in Fig. 1.4. The age-standardised incidence rate was an estimated as 222 per 100,000 for males and 186 per 100,000 for females in 2020. The highest age-standardised incidence rate was recorded in Australia and New Zealand (494.2 per 100,000 in males and 405.2

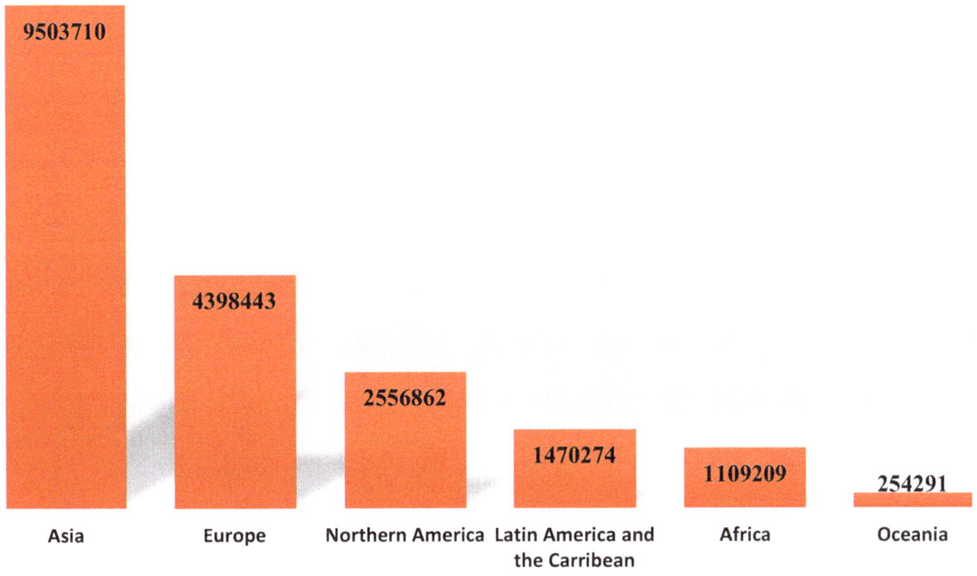

Fig. 1.4 Incident number of cancer cases among both sexes by worldwide regions, 2020. *Source*: Ref. [24]

per 100,000 in females). Western Africa reported the lowest age-standardised incidence rate in both genders (100.6 per 100,000 males and 123.2 per 100,000 females). As for incidence, Australia and New Zealand reported the highest age-standardised mortality rate of 85.8 per 100,000, while the mortality rate was 67 per 100,000 in South Central Asia, lower than other regions. Cancer incidence is higher than the world's average in North America, Europe, Polynesia, Eastern Asia and Southern Africa. Due to the geographic size and population, close to half (49.3%) of the incident cases are in Asia. The age-standardized rate (ASR) in various world regions is shown in Table 1.2.

Similarly, over half of the deaths (58.3%) were in Asia. Breast cancer accounted for 11.7% of the incident cases, followed by lung cancer (11.4%) in both sexes. As many as 18% of all cancer deaths world-wide were ascribed to lung cancer.

Table 1.2 Age-standardised incidence rates in both genders according to region

Age-standardised incidence rates (per 100,000)		
Region	Males	Females
Australia and New Zealand	494.2	405.2
Northern America	397.9	332.5
Western Europe	365.3	294.3
Northern Europe	343.6	296.5
Southern Europe	317.7	249.8
Central and Eastern Europe	293.8	220.9
Polynesia	248.3	215.8
Eastern Asia	243.5	196.7
Southern Africa	232.7	188.9
World	**222.0**	**186.0**
South America	217.1	192.1
Caribbean	213.8	174.4
Micronesia	202.6	165.8
Western Asia	198.3	162.3
Melanesia	192.6	202.5
South Eastern Asia	159.2	149.3
Northern Africa	145.7	140.1
Central America	140.9	141.1
Eastern Africa	112.9	148.0
Middle Africa	109.6	115.8
South Central Asia	103.2	102.5
Western Africa	100.6	123.2

Source: Ref. [24]

Table 1.3 Age-standardised incidence rate for all cancers and leading sites in both sexes

	Males			Females		
Sl. no.	Cancer site	Number	ASR	Cancer site	Number	ASR
1.	All sites excluding non-melanoma of skin	9,342,957	206.9	All sites excluding non-melanoma of skin	8,751,759	178.1
2.	Lung	1,435,943	31.5	Breast	2,261,419	47.8
3.	Prostate	1,414,259	30.7	Lung	770,828	14.6
4.	Non-melanoma of skin	722,348	15.1	Cervix uteri	604,127	13.3
5.	Stomach	719,523	15.8	Colon	547,619	10.0
6.	Liver	632,320	14.1	Non-melanoma of skin	475,725	7.9
7.	Colon	600,896	13.1	Thyroid	448,915	10.1
8.	Rectum	443,358	9.8	Corpus uteri	417,367	8.7
9.	Bladder	440,864	9.5	Stomach	369,580	7.0
10.	Oesophagus	418,350	9.3	Ovary	313,959	6.6

Source: Ref. [24]

1.6.2 Global Burden of Cancers in Specific Sites

The top ten leading cancer sites that account for incidence in both sexes are described in Table 1.3. The burden of a few specific cancers is described below:

1.6.2.1 Breast Cancer

As many as 2.3 million women were diagnosed with breast cancer in 2020, with a total of 685,000 reported deaths, and it is leading cancer in 159 countries [24]. Compared to any other cancer, breast cancer contributes to the highest number of Disability-Adjusted Life Years (DALYs). Australia and New Zealand, North and Western Europe, and North America recorded the highest incidence rate (>80 per 100,000. The lowest incidence rates were seen in Central America, South Central Asia, and Middle and Eastern Africa (<40 per 100,000). Since the early 2000s, breast cancer incidence rates appear to have stabilised in the western world, which could be due to increased participation in screening or decreased use of hormone replacement therapy (HRT) for menopause [25]. Breast cancer incidence rates seem to be rapidly rising in Asian and African countries, primarily due to changes in the social and economic environment, resulting in a risk factor profile similar to the Western countries [26]. Breast cancer mortality is highest in Sub-Saharan Africa, reflecting on the deficiencies in the health system functionality and performance. Compared to a 5-year age-standardised survival rate of 85% to 90% among breast cancer patients from 2010 to 2014 in high-income countries (HICs), the survival rate was as low as 66% for breast cancer patients diagnosed between 2008 and 2015 in 12 countries in sub-Saharan Africa [26].

1.6.2.2 Lung Cancer

Lung cancer comprised the second most commonly diagnosed cancer and was the leading cause of death among all cancers in 2020 [25]. It is the most frequently diagnosed cancer among males in 36 countries. The highest country-level incidence rates were recorded for males in Turkey and for females in Hungary. Region-wise, the highest lung cancer incidence in both genders was recorded in Micronesia/Polynesia. In Eastern Asia, high lung cancer incidence rates were observed in women, resulting from increased exposure to indoor and outdoor air pollution, mainly due to the use of solid fuels in households. Over time, trends in lung cancer incidence indicate a decline among younger male birth cohorts, while in females, a fall or stabilisation appears to be less pronounced [27]. The longitudinal time trends in the USA has shown that the relative risk of association between smoked tobacco use and lung cancer mortality among women in the USA has increased from the 1960s (2.7) to the 2000s (25.7) [28]. Such a trend is a likely risk in future for the LMICs.

1.6.2.3 Cervical Cancer

An estimated 604,000 new cases of cervical cancer and 342,000 deaths occurred in 2020, ranking fourth as the leading cause of cancer-related deaths in women [24]. Sub-Saharan Africa recorded the highest incidence and mortality rate, whereas the rates are 7–10 times lower in North America, Europe and the Middle East [24]. The decline in cervical cancer incidence in most parts of the world could be associated with improved socioeconomic levels leading to enhanced personal hygiene, delayed marriage and reduced parity [29].

The incidence, mortality, and distribution of leading cancer sites worldwide are very diverse and associated with various sociodemographic, socioeconomic, cultural, behavioural and health system-related factors. Nevertheless, cancer is a significant cause of morbidity and mortality and calls for strengthening cancer control measures globally through widespread international collaborations.

The incidence, mortality, and distribution of leading cancer sites worldwide are very diverse and associated with various sociodemographic, socioeconomic, cultural, behavioural and health system-related factors. For example, leading cause of cancer death is cervical cancer in Kenya, and lung cancer in Sri Lanka [23, 24]. Hence, cancer is a significant cause of morbidity and mortality and calls for strengthening cancer control measures globally through widespread international collaborations.

1.7 Cancer Burden in Low- and Middle-Income Countries (LMICs)

The cancer burden, both incident cases and deaths due to cancer, in LMICs is becoming a major part of health system approaches towards non-communicable diseases since the 1990s. Although LMICs have reported a lower cancer incidence than that reported from HICs, nearly 70% of all deaths due to cancer occur in the LMICs [30]. As mentioned in the introduction, it is a major concern that the premature mortality due to cancer in persons less than 65 years is higher in LMICs' than HICs [31]. Tables 1.4 and 1.5 show the age-standardised rates (ASR) of cancer incidence and mortality among men and women in countries with higher Human Development Index (HDI) versus lower HDI countries in 2020. Among both genders, cancer incidence in all sites except for lip and oral cavity cancer is more in countries with a high or very high HDI. The incidence of cervical cancer is higher in countries with a low or medium HDI. Between 84% and 90% of cervical cancers globally are reported in LMICs such as India, South Africa, Brazil and China [32, 33].

The rising burden of cancer in LMICs' could be due to a higher proportion of infection-related cancers attributable to the human papilloma virus, *Helicobacter pylori* and hepatitis B virus, which are associated with cancers of the cervix, stomach, oesophagus and liver. The proportion of cancers attributable to infection is as high as 50% in sub-Saharan Africa compared to 5% in the USA and Canada [34]. In addition, the rapid rise of Kaposi's sarcoma associated with HIV-infection was notably high in Africa.

Over the recent years, besides tobacco use, the high rates of obesity, low physical activity, and dietary factors such as low intake of fruits and vegetables are the rapidly emerging cancer risk factors of public health concern in the LMICs'. The prevalence of smoking was high among men in the LMICs in Asia in the 1980s. Although the prevalence of smoking decreased in the 1990s, the reduction in the prevalence of cancer has been relatively modest [35]. The use of smokeless tobacco accounts for a considerable magnitude of disease burden, including oral cancer in the LMICs [36]. The consumption of pickled and high salt content in traditional foods is associated with an increased risk of stomach cancer in many parts of LMICs.

The incidence of breast cancer has increased in many LMICs, resulting from socioeconomic

Table 1.4 Age-standardised rate per 100,000 males in countries according to HDI

Site	Incidence		Mortality	
	Very high/high HDI	Low/medium HDI	Very high/high HDI	Low/medium HDI
Lung	39.0	10.3	31.6	9.4
Prostate	37.5	11.3	8.1	5.9
Colorectum	29.0	7.4	13.1	4.7
Stomach	18.9	7.2	12.6	6.4
Liver	16.8	6.9	15.1	6.7
Oesophagus	10.3	6.4	8.9	6.0
Bladder	11.7	3.1	3.7	1.8
Lip, oral cavity	3.8	10.2	1.4	5.7
Non-Hodgkin's lymphoma	7.9	4.0	3.4	2.5
Leukaemia	7.6	3.8	4.4	3.0
Kidney	7.8	1.8	3.0	1.1
Pancreas	7.2	1.6	6.7	1.5
Larynx	3.4	3.9	1.7	2.5
Brain & nervous system	4.7	2.4	3.6	2.1
Melanoma of skin	5.2	0.4	0.9	0.2

Source: Ref. [33]

Table 1.5 Age-standardised rate per 100,000 females in countries according to HDI

Site	Incidence		Mortality	
	Very high/high HDI	Low/medium HDI	Very high/high HDI	Low/medium HDI
Breast	55.9	29.7	12.8	15.0
Cervix uteri	11.3	18.8	5.2	12.4
Colorectum	20.0	5.4	8.4	3.3
Lung	18.2	4.2	13.7	3.8
Thyroid	14.3	2.6	0.5	0.5
Corpus uteri	11.1	3.0	2.1	1.0
Ovary	7.1	5.8	4.1	4.2
Stomach	8.0	4.0	5.4	3.5
Liver	5.8	3.2	5.3	3.1
Leukaemia	5.4	2.9	2.9	2.2
Non-Hodgkin's lymphoma	5.5	2.8	2.1	1.7
Oesophagus	3.5	3.7	3.0	3.4
Pancreas	5.0	1.0	4.6	1.0
Brain & nervous system	3.8	1.7	2.8	1.4
Lip & oral cavity	1.7	3.6	0.6	2.2

Source: Ref. [33]

development leading to changes in the reproductive pattern in terms of delayed childbirth and reduced parity. Coupled with a rising incidence, the survival rate of breast cancer tends to be much lower in LMICs. Between 2005 and 2009, the 5-year survival rate of breast cancer patients was over 80% in North America, Northern and Western European countries, Australia, Israel and Brazil but ≤60% in many LMICs, such as South Africa, Algeria, India and Mongolia [37]. The higher burden of cervical cancer in LMICs could be explained by the presence of sustaining factors which include sexually transmitted infections, including HIV, traditional practices which favour early marriage and childbirth, and poor genital hygiene.

Most cancers tend to be diagnosed later (i.e. in advanced stages) in LMICs compared to HICs. This reflects on a weak health system and a lack of awareness about the disease leading to poor access to health services and treatment-seeking behaviour.

1.8 Cancer Burden in South and South East Asia

1.8.1 Cancer Incidence, DALYs' and Cumulative Risk and Mortality

South and South East Asia is a diverse population with vast differences in the culture and socioeconomic conditions. Out of 19 countries, there are 14 low or LMIC in South and South East Asia according to the WHO report in 2021. The age-standardised incidence rates (ASR) ranged from 80.9 to 165.8 per 100,000 populations with a cumulative risk of 81.9–162 per 100,000 populations according to the 2020 estimates of GLOBOCAN data as shown in Fig. 1.5. The lowest ASR was seen in Nepal and Bhutan, whereas the highest ASR was seen in Laos and Philippines. The age-standardised mortality rates (ASMR) ranged from 54.8 to 117.8 per 100,000 populations with a cumulative risk ranging from 10.97 to 27.33 per 100,000 populations as Fig. 1.6. The lowest ASMR were seen in Nepal and Sri Lanka and the highest ASMR in Laos and Vietnam. The 5-year prevalence of all cancers ranged from 99.0 to 363.5 per 100,000 populations (Fig. 1.7). The lowest prevalence was seen in Afghanistan and Timor-Leste and the highest in Sri Lanka and Vietnam. However, reporting biases should be considered when interpreting these findings. Interestingly, Sri Lanka was among the highest in prevalence but the lowest in the estimated mortality rates. The disability-adjusted life years (DALYs), according to the 2019 Global Burden of Diseases (GBD) estimates, ranged from 1754.7 to 3301.92 per 100,000 population years. The lowest DALYs were seen in Bhutan and Timor-Leste and the highest in Vietnam and Myanmar (Fig. 1.8).

It is worthwhile to reiterate that most countries in the South and South East Asia have a developing economy and belong to the LMIC group. Singapore and Brunei are classified as HIC and Thailand, Malaysia and Maldives are upper middle-income countries (UMIC). Compared to the HICs and UMICs, the LMICs have a lower incidence of malignancies. Similarly, the mortality and prevalence were also lower in LMICs of South and South-East Asia (Figs. 1.5 and 1.6). DALY's related to malignancies were

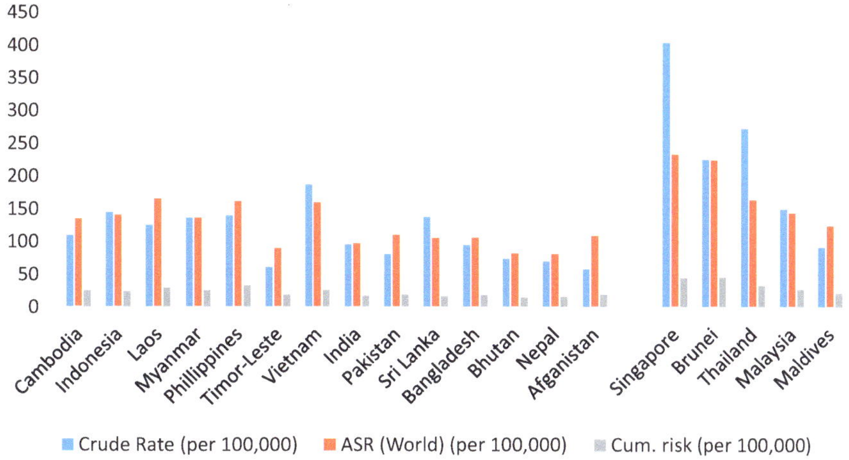

Fig. 1.5 Crude incidence rates, age-standardised incidence rates and cumulative incidence risk per 100,000 in the LMIC in South and South East Asia. *Source*: Ref. [33], Globocan (2020)

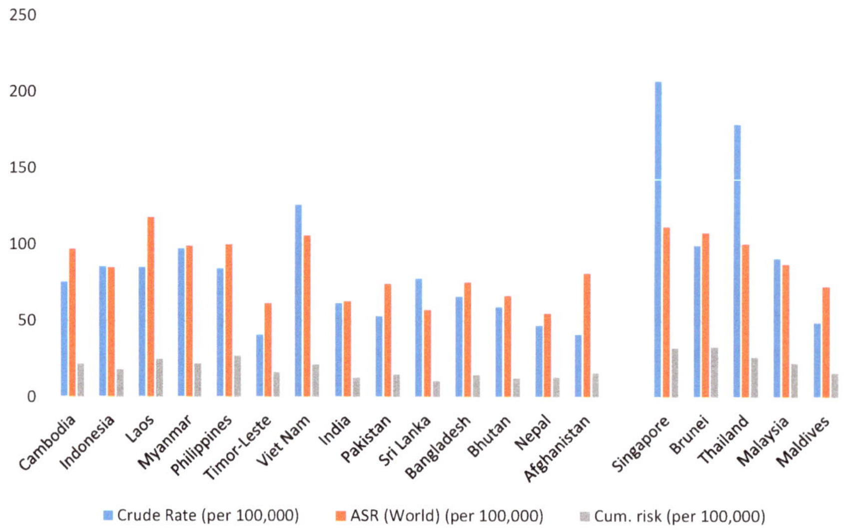

Fig. 1.6 Crude mortality rates, age-standardised mortality rates and cumulative mortality risk per 100,000 in South and South East Asia (LMIC vs. high and UMIC). *Source*: Ref. [33], Globocan (2020)

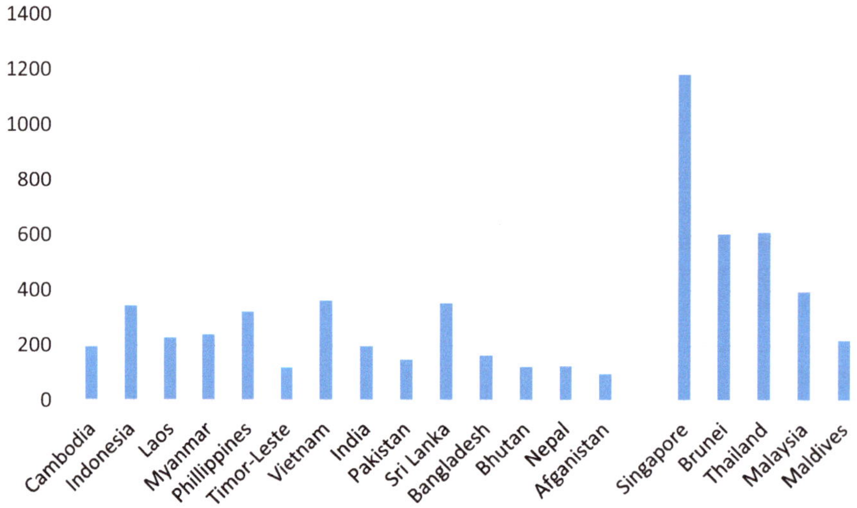

Fig. 1.7 The 5-year prevalence per 100,000 population in South and South East Asia (LMIC vs. high and UMIC). *Source*: Ref. [33], Globocan (2020)

generally better in high and UMIC in South and South East Asia [38].

The pronounced variations in social norms, ethnicity, cultural practices, economic development, dietary practices and habits are reflected in the different patterns and prevalence of malignancies in different parts of South and South East Asia. In South Asia, cancers of the oral cavity, pharynx, lung, stomach and oesophagus are the commonest in men whereas in South East Asia, lung, liver, stomach, colon and rectum and prostate are the commonest cancers. This variation can be explained by the prevailing life style related risk factors. In South Asia, smokeless tobacco use is very prevalent explaining the higher prevalence of oral and pharyngeal cancers, whereas in South East Asia, tobacco smoking is more prevalent. In

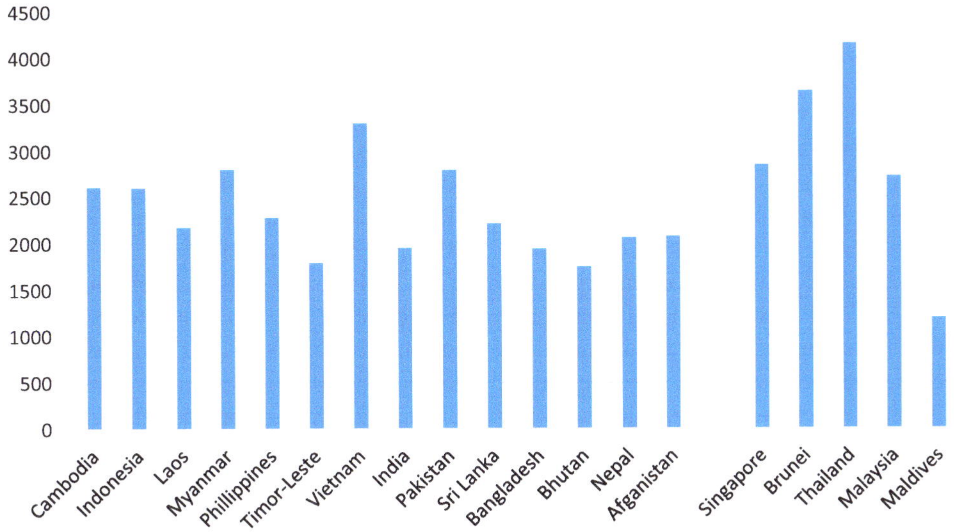

Fig. 1.8 The DALYs per 100,000 in the LMIC in South and South East Asia. *Source*: Ref. [38], GBD (2019)

females, cervix, breast, oral cavity, ovary and oesophagus are the commonest in South Asia, whereas in South East Asia, breast, cervix, colon and rectum, lung and liver are the top 5 commonest cancers [39].

1.9 Trends in the Burden of Cancer in Specific Countries in South Asia

Despite the limited studies conducted on the burden of cancer in developing countries, considerable differences are noted in the cancer patterns. Breast cancer is the commonest cancer in Pakistan and accounts for the highest cancer-related deaths followed by oral cancers [40]. The incidence and mortality related to breast cancer is rising at an alarming rate similar to the other regional countries. The high incidence of oral cancer in India, Pakistan and Sri Lanka can be explained by the high prevalence of smokeless tobacco usage [35, 36].

In Sri Lanka, cancer is currently the second commonest cause of hospital deaths thereby accounting for 14% of all hospital mortalities [41]. Based on the GLOBOCAN estimates, the incidence and mortality of cancer has been predicted to increase by 23% per annum till 2030 [42]. Breast cancer has become the commonest cancer in Sri Lanka surpassing oropharyngeal cancers. Furthermore, a significant increasing trend in the incidence of cancers was also noted in relation to the thyroid, stomach, oesophagus and colon and rectum. Over the last 25 years, the overall incidence of malignancies in Sri Lanka has doubled with a concomitant rise in cancer-related deaths [41]. However, quality data regarding the trends in mortality patterns are lacking due to poor reporting.

Bangladesh is one of the most populous countries in the world and faces increasing burden due to cancer. For a total population of 163 million in 2019, the country reported 150,781 new cancer cases, 108,137 cancer deaths; with cancer mortality contributing to 26% of all premature deaths due to NCDs [43]. The incidence and the burden of breast and cervical cancers in females and lung cancer in males are rising. Among males, cancers of the lung, oropharynx, oesophagus and stomach are common; whereas cancers of the breast, cervix and uterus are common in females in addition to the cancers of the oropharynx, lung and oesophagus [38]. Among children (less than 15 years), retinoblastoma and leukaemia were the commonest, and among adolescents (15–19 years), bone tumours (38%), gonadal and germ cell tumours (17%) and epithelial tumours

(16%) were the commonest. Cancer control in Bangladesh is gaining attention with focus on registration and developments of cancer centres [44, 45].

In Nepal, cancer contributes to around 10% of the total mortality and 5.6% of total DALYs. Similar to the regional countries, cancers of the breast, lung, uterine cervix, stomach and oropharyngeal cancers are the common cancers [46]. Similar to Sri Lanka, the estimated number of new cancers and cancer-related deaths has doubled over the last 2–3 decades. A published study of HBCR data from seven hospitals in Nepal has analysed a total of 55,931 cancer cases with known age registered from 2003 to 2012. Lung (incl. Trachea and bronchus) cancer (19.08%) was the major cancer in males followed by stomach cancer (7.86%) and pharynx cancer (5.4%). At the same time period, cervix cancer (21.9%) was the most common cancer in females followed by breast (15.48%) and lung (incl. Trachea and bronchus) cancer (10.47%) over the 10 years [47].

In Afghanistan, which has the sociogeographical and resource-constraint health system issues, a study by Oncology Department of Jamhuriyat Hospital, Kabul, Afghanistan, has analysed 1025 patients, diagnosed and treated from October 2015 to December 2017. Results showed that the most frequent cancers among Afghans were breast and oesophagus. The most common cancer in men was oesophagus and stomach at the age range of 50–70 years while in women, breast and oesophagus cancers were common and within the age range of 25–65 years old [48].

Maldives and Bhutan have a low population and low number of cancers. However, the burden of cancer has been gradually increasing over the last few decades. The Globocan 2020 has recorded 572 new patients, 456 cancer deaths and 961 prevalent cases for a population of 0.77 million in Bhutan [49]. The top five common cancers diagnosed in Bhutan are stomach, cervix, mouth & pharynx, and thyroid [50]. The number of diagnosed cancers and cancer related deaths are predicted to double by 2040. Cancer control, in terms of registration and comprehensive cancer management, is getting focus within the public health system in Bhutan, which is known to be a small country with strong social welfare measures [51]. The Bhutan Ministry of Health has promoted efforts to integrate cancer-related services, as evident in its National Cancer Control Strategy (2019–2025) [50].

In Maldives, a policy report in 2018 in the absence of NCCP has estimated total prevalence rate in the country at approximately 1% of the population (0.84% in males and 1.01% in females; 920 cases for 100,000 population) for all malignancies in 2013–2014 [52]. This high prevalence rate is a matter of concern for the health system. Therefore, population-based cancer registry and proper reporting would be vital for correct assessment of the burden of different types of cancer. In this connection, the published literature on challenges faced by small island nations of the Pacific region for cancer control is relevant. Pacific island countries and territories (PICTs) are diverse but face common challenges; unable to provide comprehensive cancer services, with some patients receiving cancer care in other countries where resources allow [53]. Small island countries of the oceanic regions in the world have gained attention in the last 30 years. In 1992, the UN formally recognised a group of small island nations that faced 'specific social, economic and environmental vulnerabilities'. Although no specific criteria exist in the classification of a small island nation, more than 50 countries (not all of them islands) are recognised as such. Small island nations are located across three broad geographical areas: the Pacific; the Caribbean; and the Atlantic and Indian Oceans, the Mediterranean Sea, and the South China Sea (AIMS). Cancer is a leading cause of death in small island nations and is forecast to increase substantially over the coming years [54].

Overall, LMICs generally have inadequate resources for health care, specifically for cancer care, and hence, mortality rates are higher than HICs. Patterns of cancer vary across countries of different income levels. From a policy perspective, it is important to identify the priorities for investment to make maximum health gains with the available budgetary resources, by using cost-effectiveness analysis at each level, such as: cancer registration, public awareness and advocacy, cancer prevention and diagnosis, anti-cancer treatment facilities, treatment outcomes, cancer follow-up and survivorship, and country-specific targets [55].

1.10 Cancer Burden in India

1.10.1 Cancer Incidence, DALYs' and Cumulative Risk and Mortality

Cancer contributed to 8.3% of deaths and 5.0% of DALYs in India in 2016, double its contribution in 1990 [37]. Among these, stomach cancer contributed to the highest proportion of DALYs (9%), followed by breast cancer (8.2%) and lung cancer (7.5%). India reported an estimated 1.39 million new cancer cases in 2020 [56]. The distribution of cancer cases in terms of crude incidence (CR) and age-standardised rate (ASR) for the leading cancer sites, gender-wise, is shown in Fig. 1.9. Cancers of the lip and oral cavity accounted for the highest CR and ASR in males, while the highest CR and ASR were observed for breast cancer among females.

India is a diverse country, where cancer incidence varies from one geographic region to another. The age-adjusted incidence rate (AAR) of cancer in the Population Based Cancer Registry covered areas is seen in Fig. 1.10. Aizawl district of Mizoram state had the highest AAR (269.4 per 100,000) in males, and Papum Pare district of Arunachal Pradesh recorded the most elevated AAR (219.8 per 100,000) in females for all cancer sites. Likewise, the crude mortality rate of cancer in all sites in males varied across the geographical areas ranging from 9.7 per 100,000 in Aurangabad to 115.0 per 100,000 in Aizawl district. Among females, the crude mortality rate varied from 6.8 per 100,000 in Nagaland PBCR to 69.6 per 100,000 in Aizawl district. The cumulative risk is the chance that a person will develop any cancer in their lifetime (0–74 years), which was observed to be the highest, about 1 in 4, for Papum Pare district in both the genders.

1.10.2 Trends in the Burden of Cancer in Specific Sites Over Time

There has been a significant increase in the AAR of breast cancer in many PBCR-covered geographical areas, as shown in Fig. 1.11. Breast cancer comprised either the first or second leading cause of cancer deaths among females in 28 Indian states in 2016 [37]. India has one of the highest rates of triple-negative breast cancer (TNBC) [57]. Also, breast cancer is observed to occur in a younger population compared to Western cohorts.

Cervical cancer constitutes the second leading cause of cancer deaths among females in 12 Indian states [37]. Most of the PBCRs observed a decline in the AAR of cervical cancer over time, except for Mizoram, Aurangabad and Sikkim (Fig. 1.11). Regarding head and neck cancers, a significant increase in incidence rates was observed among males in Aurangabad, Delhi, Chennai and Bhopal PBCRs, and among females, in Nagpur PBCR. As many as 33.2%, 29.8% and 20.9% of the lip and oral cavity cancer DALYs' have been attributed to smokeless tobacco, alcohol use and smoking, while 37.9% and 17.2% of the laryngeal cancer DALYs are reportedly attributable to smoking and alcohol use [56].

1.10.3 Projected Future Estimates of Cancer Incidence

By 2025, India will be expected to have 1.56 million new cases, among which cancers in sites associated with tobacco use would account for 27.2% of all cancers. The distribution of the projected number of new cases according to the anatomic site is displayed in Table 1.6.

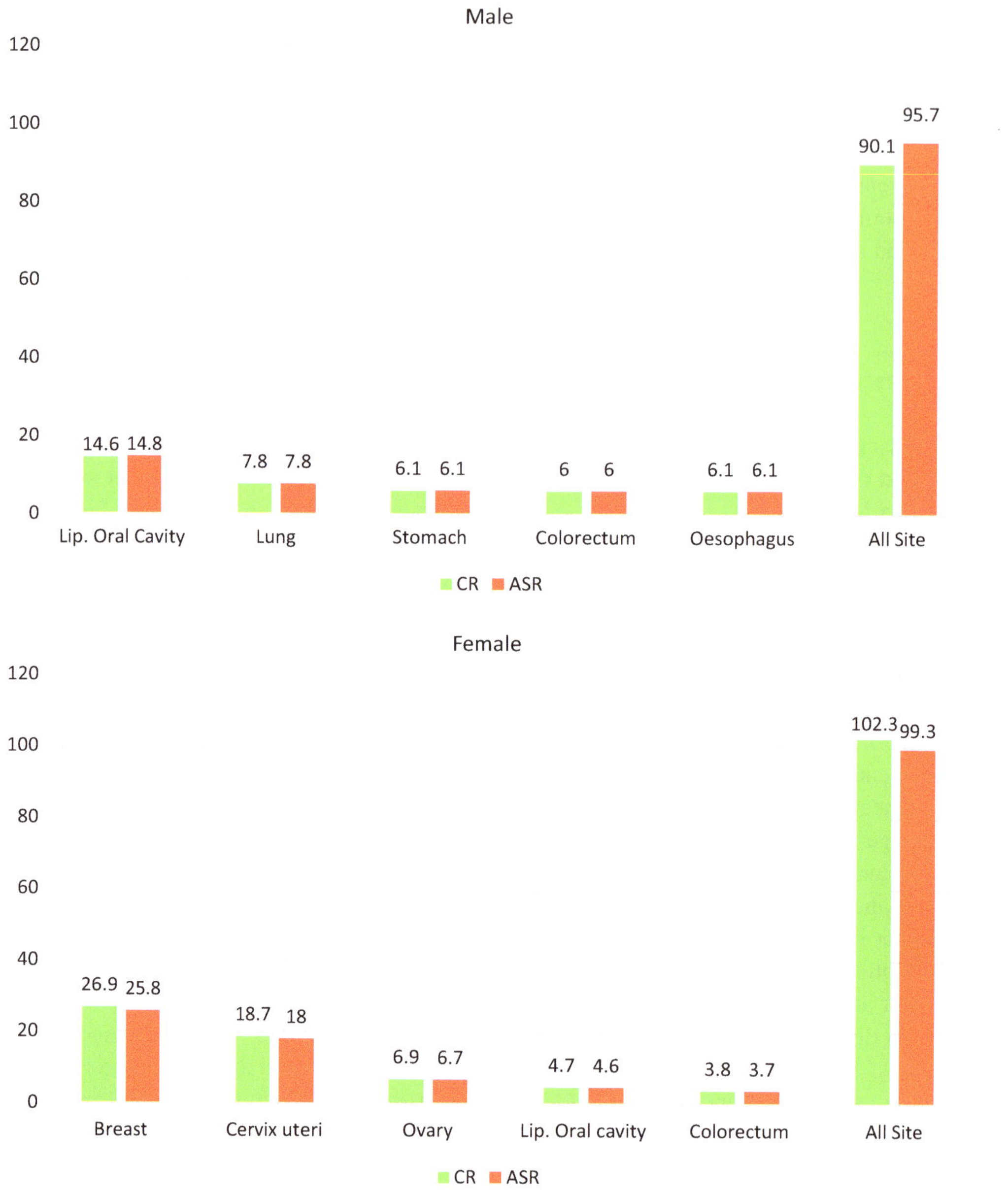

Fig. 1.9 Sex-wise distribution of incidence rate of leading cancers in India. *Source*: Ref. [24, 33]

1.11 Cancer Care in India and LMIC

The diagnosis and treatment of cancer in LMICs require special considerations, due to several prevailing factors, a few of which are common to these countries, including India.

1. Cancer registration and diagnostic facilities are not uniformly distributed.
2. Cancer treatments and Palliative Care are delivered in fragmented manner and usually lack comprehensive cancer management for majority of the cancer patients.
3. Affordable cancer care through government or non-profit hospital facilities are few in

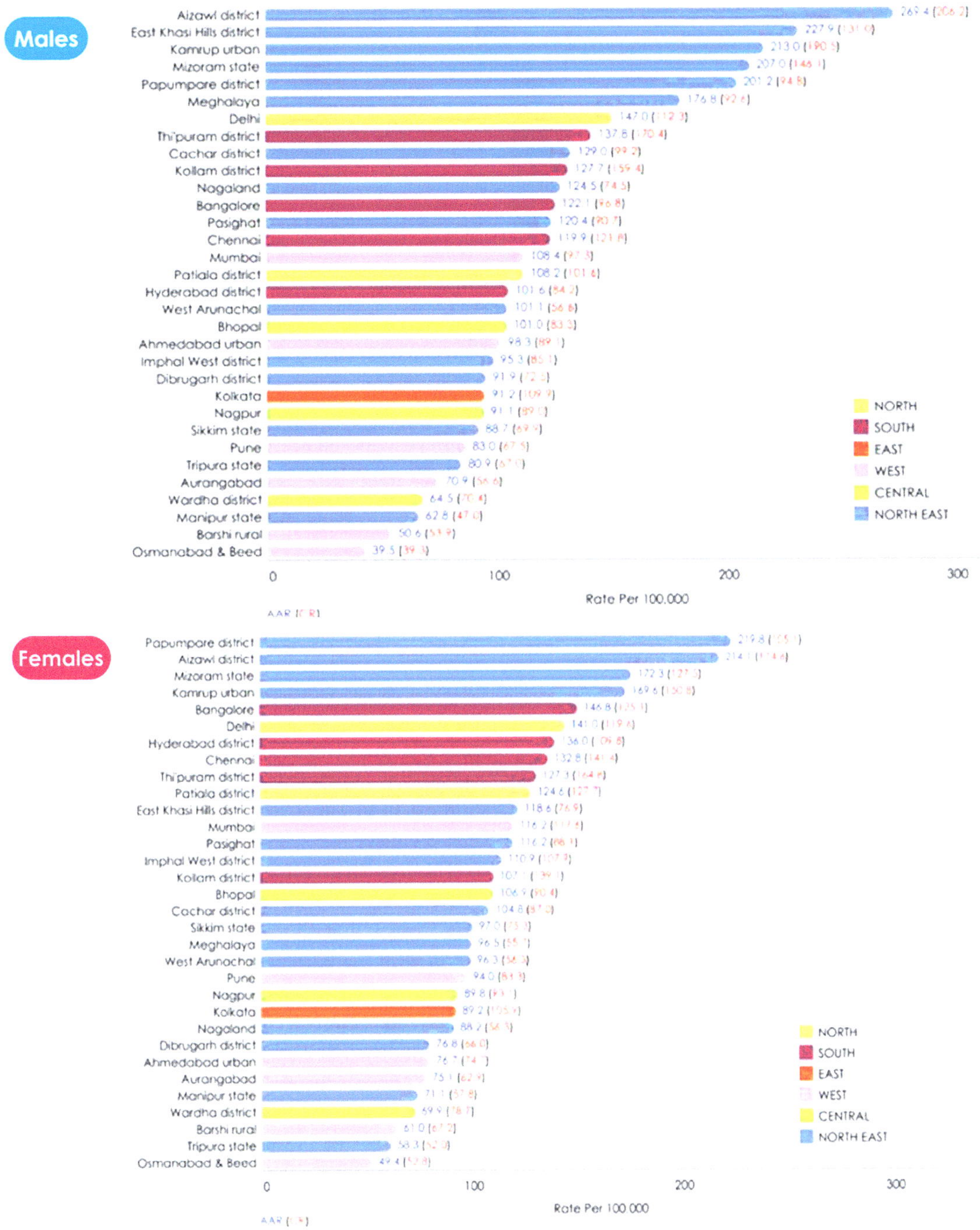

Fig. 1.10 Age-adjusted incidence rate of cancer in all sites in both sexes; from 28 PBCRs, India, 2020. *Source*: Ref. [56]

comparison to the expensive for-profit private centres.

4. Cancer research directed at tumours prevalent in the region or country, and appropriate analysis of treatment outcomes and survival are weak links in the cancer care.
5. Health workforce trained and devoted to cancer care remains a major challenge.

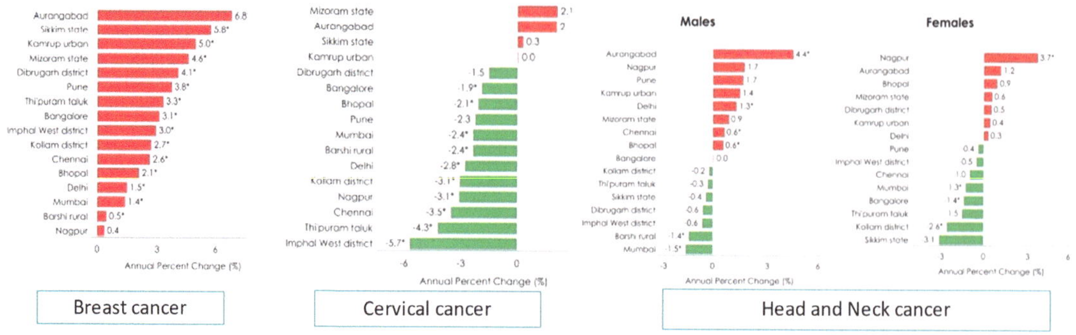

Fig. 1.11 Trends in the AAR of breast, cervical and head and neck cancers over time. *Source*: Ref. [56]

Table 1.6 Projected number of new cases according to broad anatomic site by 2025

Broad anatomical sites of cancer	2025	
	Number of cases	Percentage
All sites	**1,569,793**	**100.0**
Tobacco-related cancers	427,273	27.2
Gastrointestinal tract	310,142	19.8
Cervix uteri	85,241	5.4
Breast	232,832	14.8
Corpus uteri and ovary	79,765	5.1
Lymphoid & haematopoietic malignancies	138,592	8.8
Prostate	47,068	3.0
Central nervous system	36,258	2.3

Source: Ref. [56]

Since the beginning of the twenty-first century, it has been consistently recorded that the cancer burden and mortality are increasing in LMICs, whereas there is a decline in incident cases and improvement in cancer survival in the regions of HICs. In 2020, 19.3 million people were diagnosed with cancer and 9.9 million cancer-related deaths occurred world-wide. Out of this global cancer status, up to 60% of incident cases and 70% of cancer-related deaths were reported in LMICs [58]. There is increasing awareness within the medical community and at the level of health policy makers in many countries to improve the facilities and access for the cancer patients in LMICs, without which the global cancer status would not improve.

National Cancer Control Programme (NCCP) in India was one of the earliest in the global health systems, started in 1975, to coordinate the country-wide planning for cancer prevention and management [[59], see the link]. Coincidentally, in the USA, the National Cancer Act of 1971 was enacted [[60], see the link and Fig. 1.12].

Cancer registration is vital to gather information on incidence, mortality and survival; alongside records related to age, sex, risk factors and treatment types. All these combined together can drive cancer epidemiological research, health policy management and health care funding towards equity within a country or region. In spite of our knowledge that increase in burden is highest in low-resource settings, the population-based cancer registry (PBCR) coverage currently is for less than 5% of the population in Asia, Africa, and Latin America. WHO has emphasised this aspect that the planning of cancer control starts with the collection and analysis of data on incidence and mortality; whereas only 35% of the countries, mostly in HICs, collect and report high-quality PBCR data [61].

In spite of the early beginning in the 1970s, the cancer registration in India has a coverage of 15% and 1% of the urban and rural population, respectively [19]. Prior to the NCCP, Mumbai (earlier called Bombay) had established in 1963 a cancer registry to cover its urban population, which has become a part of the national registry since 1982. According to cancer registry operation expenditures for Mumbai and its satellite registries, the cost per each case of cancer patient registered would be 226 Indian rupees (approximately 4 US dollars) in 2014 [62]. This proves that cancer registration in India and LMIC can become a very cost-effective health system fund-

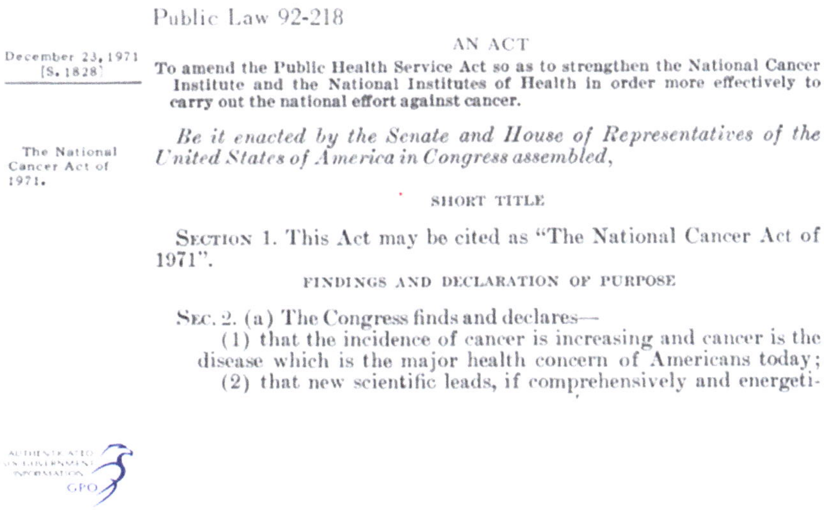

Public Law 92-218

December 23, 1971
[S. 1828]

AN ACT

To amend the Public Health Service Act so as to strengthen the National Cancer Institute and the National Institutes of Health in order more effectively to carry out the national effort against cancer.

The National Cancer Act of 1971.

Be it enacted by the Senate and House of Representatives of the United States of America in Congress assembled,

SHORT TITLE

SECTION 1. This Act may be cited as "The National Cancer Act of 1971".

FINDINGS AND DECLARATION OF PURPOSE

SEC. 2. (a) The Congress finds and declares—
(1) that the incidence of cancer is increasing and cancer is the disease which is the major health concern of Americans today;
(2) that new scientific leads, if comprehensively and energeti-

Fig. 1.12 The National Cancer Act, USA, 1971

ing. While there is widespread recognition of the need for cancer registration and the data they generate, the main hurdles in LMICs are pertaining to (a) trained manpower, (b) financial allocation and (c) policy implementation within a region or a country. A good example is Nigeria National System of Cancer Registries (NSCR) which began coordination of the country-wide activities from 2009. This NSCR is now in a position to collect and disseminate data related to cancer incidence and treatment outcomes online for the cancer clinicians, cancer researchers and health policy makers [63]. Within a decade, this has paved the way for harmonisation of cancer care in the Sub-Saharan Africa.

In the last 50 years, the HICs have invested on cancer registration, diagnostics and therapeutics including developments of technology and drugs. For example, although the modern radiotherapy is delivered by linear accelerator (LINAC) to millions of cancer patients in Asia and Africa, the equipment is not manufactured indigenously or originally from any country of LMICs. Hence, a LINAC is prohibitively expensive for the GDP of these countries, thus making the health policy makers unwilling for the growth of the radiation oncology in a region or country. For example, the cost of one linear accelerator in India, through import, would be in range from 70 million rupees (0.8 million $) for low energy equipment to 220 million rupees (2.6 million $) for state-of-art (multiple energies) machine. As a consequence, out of all patients in the need of RT only 10% of patients in low-income and 40% in middle-income countries have the access. A comparative availability can be seen. Africa has approximately 380 linacs for a population of 1.2 billion, while the USA has almost 4000 linacs for a population of 331 million [[64] see the link & [65]]. Similarly, the chemotherapy delivery in LMICs faces the hurdles related to the access and affordability for a vast majority of anti-cancer medicines [66]. The prices of cancer medicines are expensive in LMICs, often leading to abandonment of the treatment course due to low affordability as per individual patient's income. The protection from financial hardship and from non-compliance to cancer chemotherapy can be achieved by country-specific rationalisation of drug pricing and implementation of a national essential list of cancer medicines under NCCP. In order to improve the situation, a national essential cancer medicines list, with treatment guidelines, is currently available in 33% of the LMICs. For the remaining two-third of LMICs, the cancer chemotherapy access and drug affordability would require health policy implementation [66]. Likewise, a surgical procedure is needed by

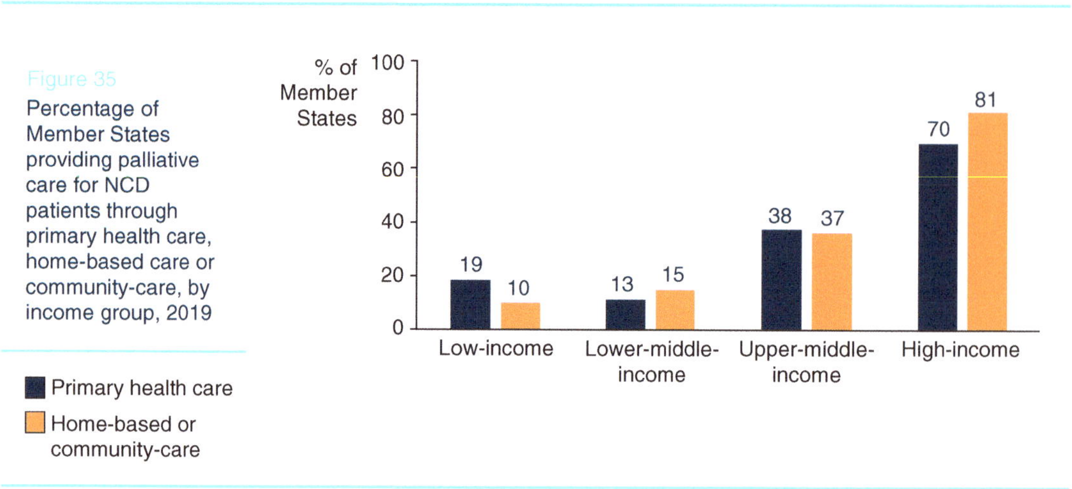

Fig. 1.13 Percentages of LMICs providing palliative care are shown in this figure (Ref. Global Atlas of Palliative Care. 2nd Edition, 2020. www.thewhpca.org)

nearly 80% of individuals with cancer. Yet for a vast majority of these patients, cancer surgery practices for biopsy, organ preservation, curative surgery, palliation, rehabilitation etc. are done by surgical specialists of different disciplines with inadequate knowledge and skill in oncologic concepts. An international prospective study, covering 82 countries, on primary surgery for breast, colorectal, or stomach cancers showed considerable disparities between HICs and LMICs [67]. The relevant key findings which require attention and improvements in LMICs: (1) a high proportion of advanced stage presentation, (2) the 30-day mortality was much higher than HICs for colorectal and gastric cancers, but was not different for breast cancer, (3) strengthening of perioperative care systems can improve cancer surgery outcomes [67]. Besides radiation oncology, medical oncology and surgical oncology, the continuum of cancer management requires Palliative Care (PC) to become integrated with anti-cancer therapy, and not only for the advanced stage or end-of-life care. WHO has emphasised the importance of PC for cancer patients since the 1990s, yet its availability within the health system is absent or sub-optimal in nearly 125 out of 198 countries of the world, accounting for a global population of 53.5%. For example, according to this world mapping of palliative care services, most countries of South-East Asia, India and its neighbouring countries like Bangladesh, Indonesia, Myanmar, Nepal and Sri Lanka, continue to have isolated or patchy access to PC for the needy patients [[68] and Fig. 1.13].

All these above features of deficiencies in cancer care services in LMICs are further compounded by

(a) Lack of awareness about cancer in the community [69].
(b) Delays and barriers to cancer diagnosis and treatment facilities [70].
(c) A low number of available manpower trained and skilled in cancer care [58].
(d) Paucity in national or regional funding for cancer research directly related to common cancers which are specific to the country and geographic regions [71].

Cancers are a major contributor to disease burden worldwide. The United Nations (UN) Sustainable Development Goals (SDGs) recognise the need for reducing cancer burden as part of its target goals by 2030. Low and low-middle sociodemographic index (SDI) countries show a higher rate of growth in the number of cases and deaths than high SDI locations during the last decade [72]. Establishment of NCCPs in LMICs, with policy framework, financial sanction and suitable mechanisms to attain the goals, can reduce the disparities between HICs and LMICs. The NCCP in India, presently a division of the National Centre for Disease Informatics and Research (NCDIR) under the Indian Council of Medical Research (ICMR), has contributed significantly towards cancer care in the country since 1982 [73].

Formed in 2012 following discussions at the World Cancer Congress in Montreal, the International Cancer Control Partnership (ICCP) is one of the key measures being undertaken to bridge the gaps in cancer control planning. The ICCP is a collective of key international organisations united in their efforts to promote and support cancer control. In 2013, 91 NCCPs from 42 countries were available on the ICCP Portal; this number has increased to 115 from 50 countries in 2015 and 224 from 93 countries in 2018. From such global analysis, it is now recognised that as countries move towards universal health coverage, greater emphasis is needed on developing NCCPs that are evidence-based, financed, and implemented to ensure translation into action [74].

1.11.1 What Makes a Good National Cancer Control Plan?

The WHO defines NCCPs as 'public health programmes designed to reduce cancer incidence and mortality and improve the quality of life of cancer patients, through the systematic and equitable implementation of evidence-based strategies for prevention, early detection, diagnosis, treatment and palliation, making the best use of available resources'.

A number of elements are central to developing a comprehensive NCCP that meets the current and future health needs of a country's cancer burden [[75] see the link]:

1. Consider the whole continuum of care: prevention, early detection, diagnosis, treatment, rehabilitation, palliation and research.
2. Adopt a health systems approach that sets out governance arrangements, resource allocation and financing around measurable goals to support a country's cancer control efforts.
3. Provide a sustainable strategic plan for cancer control, based on the country's cancer burden, cancer risk factor prevalence and the resources available to implement the plan.
4. Take into account the socioeconomic environment and healthcare system in the country.
5. Ensure the involvement of multisector stakeholders to set realistic objectives that respond to the population's cancer needs.
6. Develop an evidence-based approach—the availability of reliable cancer surveillance data, including cancer incidence, stage at diagnosis and mortality, generated by population-based cancer registries, is vital for developing targeted and effective NCCPs and for evaluating the impact of national programmes.

1.12 Cancer Management in Developed Versus Developing Countries

As per the mandate given by WHO, and stated earlier in this chapter, there are four pillars to be part of cancer management or cancer control policy for any region or country, namely, prevention, early detection, diagnosis and treatment, and palliative care. Yet, several differences in the cancer management have been observed between developed versus developing countries mainly because of the differences in sociocultural norms and economic and resource constraints as shown in Table 1.7.

1.12.1 Primary Prevention and Screening

Lifestyle and environment-related risk factors determine the geographical patterns of cancer. The developing nations account for the majority of the world's population, have lower incidence of malignancies than the developed countries but accounts for the majority of the global cancer burden. In the Western world, the main risk fac-

Table 1.7 Comparison of developed and developing countries of South Asia in relation to health system capacity and work force based on the WHO cancer country profile 2020 (per 10,000 cancer patients)

Country		USA	UK	India	Pakistan	Sri Lanka	Bangladesh	Nepal	Bhutan	Maldives	Afghanistan
Cost per year US$ per capita		High income	High income	1.5–2	1.5–2	1.5–2	1.5–2	0–0.5	1.5–2	1.5–2	0–0.5
Health system capacity	PBCR	High quality	High quality	High quality	Registration activity	Available	Registration activity	Available	Available	No information	Available
	Mortality registration	High	High	Very low	No coverage	No coverage	No coverage	No coverage	No coverage	Low	No coverage
Workforce (per 10,000 cancer patients)	Available staff for cancer care in the MOH	Yes	Yes	Yes	Yes	Yes	Yes	Yes	No	Yes	Yes
	Number of radiation oncologists	N/A	16	3.1	1.8	19.1	1.7	N/A	N/A	N/A	N/A
	Number of medical physicist	41.6	38	15.6	N/A	11	21.9	5	0	N/A	N/A
	Number of surgeons	538.6	485.2	272.7	128.7	50.1	173.4	97	102.6	142.5	N/A
	Number of radiologists	182.6	81.8	345.6	57.5	80.7	33.2	N/A	51.5	N/A	N/A
	Number of nuclear medicine physicians	18.3	2	3.5	14.4	1.3	6.6	0.8	0	N/A	N/A
	Number of medical and pathology lab scientists	2593.9	N/A	N/A	N/A	N/A	N/A	N/A	1076.9	237.5	N/A

tors of cancer are tobacco smoking, metabolic risk factors such as obesity, and life style related factors such as physical inactivity and unhealthy dietary practices. Although the public are educated about these risk factors, such risk factors are difficult to prevent due to practical issues. Therefore, pre-emptive prevention and early detection of highly prevalent cancers such as breast and bowel cancers are given sufficient priority in the developed countries [76].

Smokeless tobacco and betel chewing with areca nut are additional risk factors in developing countries in South Asia. Furthermore, due to the demographic trends and economic progress, there is an increasing prevalence of metabolic and life-style related risk factors similar to the developed countries leading to a rapid increase in the incidence of cancers, as reported from Sri Lanka [41].

A considerable emphasis is given for reducing exposure to carcinogens; however, much greater attention is vital for the control of metabolic risk factors such as physical inactivity, obesity, and unhealthy diet. Yet, such implementation would be challenging due to the westernisation of life-style in the developing countries. Therefore, the adoption of research from Western countries can benefit the LMICs. A good example is the vaccination programmes for HPV and Hepatitis-B currently underway in several developing South Asian countries. These vaccines, when they can cover the target population, are likely to decrease the incidence rates of cervical cancer and liver cancer respectively in future [41, 77, 78].

In countries such as the UK, routine screening is offered for breast cancer by screening mammogram; for colon cancer by colonoscopy and faecal testing; and cervical cancer by cervical screening. Opportunistic screening is offered for other cancers such as prostate and upper gastrointestinal cancers [79]. Due to economic constraints and limited resources, screening is not offered nationwide in developing countries and persons who are subjected to screening are limited. Therefore, the diagnosis is delayed and patients present at an advanced stage precluding curative treatment [80]. Table 1.8 describes the differences between developed countries (UK and USA) and South Asian countries in terms of the cancer management services, as per the four pillars of WHO mentioned earlier in this chapter. Although, nationwide screening programs are offered for breast cancer in developing countries such as Sri Lanka and Bangladesh, mammography machines and personnel are insufficient for the nationwide program of mammographic screening [41]. Therefore, clinical breast examination is promoted while mammographic screening is performed only in few centres. In certain developing countries in South Asia, Pap smear and oral examination are offered to screen for cervical and oral cancer respectively. However, such facilities are poorly utilised by the general public due to lack of awareness and stigma. There are no routine screening programmes for gastrointestinal malignancies and prostate cancer due to the relatively low prevalence and resource constraints. However, opportunistic screening is provided for these cancers in selected patients [41].

1.12.2 Diagnosis and Treatment

In most developing countries, cancer care is provided by the government sector with additional contribution from the private sector. However, some countries such as India have significant contribution from the private sector. In developing countries, specialised inpatient cancer centres are concentrated to a few major cities and only a limited proportion of patients are able to afford cancer care due to high cost and low accessibility [41, 81].

During the past few decades in South Asia, many tertiary care hospitals with state-of-the-art facilities for medical, radiation and surgical oncology have been established to manage all types of cancer [81, 82]. However, countries such as Afghanistan need further improvement in health care facilities to cater for the rising incidence of cancer. Furthermore, in Sri Lanka, certain facilities such as radiation oncology are only available in limited provinces in the government sector and at few centres in the private sector [41]. Similarly in India, a considerable proportion of patients need to travel long distances to

Table 1.8 Comparison of developed nations and developing countries of South Asia in relation to formulating response based on the WHO cancer country profile 2020

Country	USA	UK	India	Pakistan	Sri Lanka	Bangladesh	Nepal	Bhutan	Maldives	Afghanistan
Integrated NCD plan	Operational	Operational	Operational	N/A	Operational	Operational	Operational	Operational	Operational	Operational
NCCP	Operational	Operational	Operational	N/A	Operational	Operational	Operational	Under development	N/A	N/A
Guidelines	Yes	Yes	Yes	Yes	Yes	Yes	No	Yes	Yes	No
Number of treatment services	3	3	3	3	3	2	0	3	2	0
Breast cancer screening program	Yes	Yes	Yes	No	Yes	Yes	No	Yes	Yes	No
Public cancer centres per 100,000 cancer patients	N/A	0.4	2.6	N/A	3.8	0.3	1.1	17.1	47.5	0.5
Early detection program for breast, cervix, colon, childhood cancers	3	4	3	0	3	3	2	2	4	0
Pathology services	Generally available	Generally available	Generally available	Generally available	Generally available	Generally not available	Generally not available	Generally available	Generally available	Generally not available
Bone marrow transplant capacity	Generally available	Generally available	Generally not available	Generally not available	Generally not available	Generally not available	Generally not available	Generally not available	Generally not available	Generally not available
Community based palliative care	Generally available	Generally available	Generally not available	Generally not available	Generally not available	Generally not available	Generally not available	Generally available	Generally available	Generally not available
Availability[a] of opioids for pain management	32,394	7723	19	2	44	9	21	15	37	3
HPV vaccination program coverage	46%	80.6%	N/A	N/A	66.9%	N/A	N/A	89.6	N/A	N/A
Cervical cancer screening	Yes	Yes	Yes	No	Yes	Yes	Yes	Yes	Yes	No
Participation rates	<10%	>70%	>70%	N/A	50–70%	<10%	10–50%	50–70%	10–50%	N/A

[a]Defined daily doses for statistical purposes (S-DDD) per million inhabitants per day

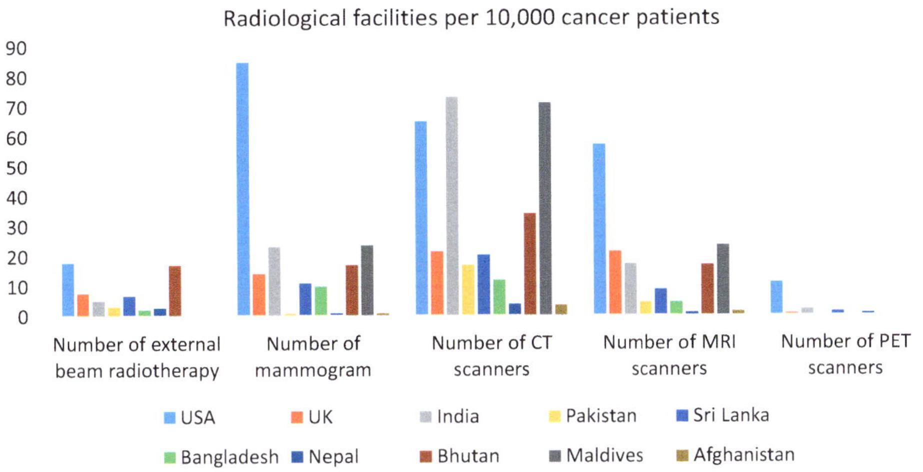

Fig. 1.14 Radiology facilities per 10,000 cancer patients based on the WHO cancer country profile 2020

receive optimal care resulting in considerable financial toxicities and higher non-compliance rates [81]. Another concern, radiological services are generally lacking in developing countries (Fig. 1.14). Compared to developed countries, health care personnel such as radiation oncologists, radiologists and nuclear medicine physicians are fewer (Table 1.7). Histopathological services are widely available in most centres in developing countries. However, novel marker and molecular genetic assessment for precision therapy are available only in few state-of-the-art centres in South Asia. Lack of awareness among the general public, insufficient infrastructure, inefficient allocation of health care personnel and resources, lack of cost-effectiveness and suboptimal access to cancer care and reduced affordability of therapy are important limitations that should be overcome with respect to cancer management in developing countries.

A considerable proportion of patients in developing South Asian countries with varying sociocultural beliefs and practices, seek alternative medical practitioners and alternative therapy for cancer care. Certain native or indigenous treatment may help to palliate adverse effects of cancer therapy. Collaboration with institutes of indigenous medicine is necessary to harness the positive aspects while regulating the native healers who proclaim cures for all cancers through alternative therapy [41].

1.12.3 Cancer Registry and Research

Establishment of cancer registries are vital to understand the cancer burden and optimal resource allocation. Developed countries have comprehensive and meticulous coverage of all diagnosed malignancies with mortality data leading to regular dissemination of key epidemiological findings. Countries such as India and Sri Lanka have well established cancer registries. However, the coverage has potential for improvement [19, 73]. Reporting of mortality data has been disappointingly low in the developing countries (Table 1.7). Most of the developing countries rely on Western guidelines for cancer care. Timely research and audits are required for updated policies and moreover, publications and research outputs in most of the developing countries are somewhat limited. Due to the sociocultural differences and variations in the patterns of cancer, individual countries should conduct their own studies and publish their guidelines suiting their setting. Country-specific basic and epidemiological studies, and clinical trials, combined with local guidelines can improve the resource-appropriate cancer management [71].

1.12.4 Palliative Care

Although the national authorities have concentrated substantially on curative aspects in recent years, palliative care services require major improvement in developing countries. Steps have been taken to allocate resources to tertiary care and community-based palliative care services with adequate resources. However, such facilities are not available widely especially in rural settings of developing countries (Table 1.8).

Summary Box

Cancer in human being has been recorded since 4000 years in palaeoanthropological studies. The correct documentation of cancer incidences has developed in the last 100 years. Cancer is a leading cause of death and reduces the life expectancy in low- and middle-income countries (LMIC). This is because of low or non-existent financial and manpower investment for comprehensive cancer care in most of the countries of LMIC. Radiotherapy is the key specialty for cancer control and cancer-directed therapy for more than 50% of all diagnosed cancer patients. Radiotherapy is recognised as a cost-effective modality, yet health policy makers and national health system in many parts of LMIC continue to neglect the establishment of adequate radiotherapy facilities. Less than 10% of cancer patients in low-income countries and not more than 40% of cancer patients in most middle-income countries in the need for radiotherapy during their cancer management would get the access to the facility. The global public health experts have identified certain key areas for cancer control in LMIC, such as (1) lack of awareness about cancer in the community, (2) delays and barriers to cancer diagnosis and treatment facilities, (3) a low number of available manpower trained and skilled in cancer care and (4) poor resource and budgetary allocation towards implementation of cancer prevention and cancer management practices. The International Cancer Control Partnership (ICCP) is one of the key measures being undertaken to bridge the gaps in cancer control planning between high-income countries (HIC) and LMIC.

Multiple Choice Questions

1. Cancer in human being is known to be recorded by palaeopathological method since:
   (a) 100 years
   (b) 500 years
   (c) 4000 years
   (d) Not recorded
2. There are two types of cancer registry process:
   (a) population-based cancer registry and hospital-based cancer registry
   (b) national cancer registry and world cancer registry
   (c) local cancer registry and regional cancer registry
   (d) All of the above
3. According to worldwide cancer data, which of following statement is correct at present:
   (a) 60% or more cancer cases occur in low- and middle-income countries (LMICs)
   (b) Less than 20% of all cancers in the world occur in LMIC
   (c) Cancer deaths are lower in LMIC than in developed countries
   (d) Any of above
4. Cancer management in LMIC has to improve in the following area(s):
   (a) Cancer treatments and palliative care should be done in a comprehensive manner
   (b) Cancer registration and diagnostic facilities should be available

(c) Treatments for common cancers and the results of survival require to be analysed
(d) All of above

5. Radiotherapy by linear accelerator (LINAC) should be available for modern cancer management. As per current estimate, LINAC accessibility in LMIC is noted as:
   (a) Is not a problem in many countries including India
   (b) Is less than 50% of the required machines in all countries of LMIC including India
   (c) Is similar to the available number of machines in the USA
   (d) All of above

Answers

1. (c)
2. (a)
3. (a)
4. (d)
5. (b)

References

1. Capasso LL. Antiquity of cancer. Int J Cancer. 2005;113(1):2–13.
2. Smith RD, Mallath MK. History of the Growing Burden of Cancer in India: From Antiquity to the 21st Century. J Glob Oncol. 2019;5:1–15.
3. Sung H, Ferlay J, Siegel RL, Laversanne M, Soerjomataram I, Jemal A, Bray F. Global Cancer Statistics 2020: GLOBOCAN Estimates of Incidence and Mortality Worldwide for 36 Cancers in 185 Countries. CA Cancer J Clin. 2021;71(3):209–49. https://doi.org/10.3322/caac.21660. Epub 2021 Feb 4.
4. Odes EJ, Randolph-Quinney PS, Steyn M, Throckmorton Z, et al. Earliest hominin cancer: 1.7 million-year-old osteosarcoma from Swartkrans cave, South Africa. S Afr J Sci. 2016;112(7/8) (internet) https://www.sajs.co.za/article/view/3566. Accessed 01/01/2022.
5. Faguet GB. A brief history of cancer: age-old milestones underlying our current knowledge database. Int J Cancer. 2015;136(9):2022–36. https://doi.org/10.1002/ijc.29134.
6. Karpozilos A, Pavlidis N. The treatment of cancer in Greek antiquity. Eur J Cancer. 2004;40(14):2033–40.
7. Prasad GC, Sahu M, Deshpande PJ. Concept of cancer in ayurveda. Anc Sci Life. 1982;1(3):172–6. PMID: 22556485; PMCID: PMC3336678.
8. Hyndman IJ. Review: the Contribution of both Nature and Nurture to Carcinogenesis and Progression in Solid Tumours. Cancer Microenviron. 2016;9(1):63–9.
9. Weinberg RA. The biology of cancer. 2nd ed. New York: Garland Science; 2014.
10. Hanahan D, Weinberg RA. The hallmarks of cancer. Cell. 2000;100(1):57–70.
11. Hanahan D. Hallmarks of Cancer: New Dimensions. Cancer Discov. 2022;12(1):31–46.
12. Danaei G, Vander Hoorn S, Lopez AD, Murray CJ, Ezzati M. Comparative Risk Assessment collaborating group (Cancers). Causes of cancer in the world: comparative risk assessment of nine behavioural and environmental risk factors. Lancet. 2005;366(9499):1784–93.
13. Ferlay J, Soerjomataram I, Dikshit R, Eser S, Mathers C, Rebelo M, Parkin DM, Forman D, Bray F. Cancer incidence and mortality worldwide: Sources, methods and major patterns in GLOBOCAN. Int J Cancer. 2015;136:E359–86. https://doi.org/10.1002/ijc.29210.
14. Mbemi A, Khanna S, Njiki S, Yedjou CG, Tchounwou PB. Impact of gene-environment interactions on cancer development. Int J Environ Res Public Health. 2020;17(21):8089. https://doi.org/10.3390/ijerph17218089. PMID: 33153024; PMCID: PMC7662361.
15. Hollman AL, Tchounwou PB, Huang H-C. The association between gene-environment interactions and diseases involving the human gst superfamily with SNP variants. Int J Environ Res Public Health. 2016;13:379. https://doi.org/10.3390/ijerph13040379.
16. Last JM, International Epidemiological Association. A dictionary of epidemiology. New York: Oxford University Press; 2001.
17. American Cancer Society. History of Cancer Epidemiology: 18th century to present. Available on: https://www.cancer.org/cancer/cancer-basics/history-of-cancer/cancer-epidemiology.html.
18. National Cancer Institute. Surveillance, epidemiology and end results programme. Available on: https://seer.cancer.gov/registries/cancer_registry/index.html
19. Bhatia A, Victora CG, Beckfield J, Budukh A, Krieger N. "Registries are not only a tool for data collection, they are for action": cancer registration and gaps in data for health equity in six population-based registries in India. Int J Cancer. 2021;148(9):2171–83. https://doi.org/10.1002/ijc.33391.
20. World Health Organization. International statistical classification of diseases and related health problems, 10th revision, Fifth edition, 2016. World Health Organization; 2015. https://apps.who.int/iris/handle/10665/246208
21. International classification of diseases for oncology (ICD-O)—3rd edition, 1st revision. World Health Organization 2013.
22. Brawley OW. The cancer registry as a cancer-control tool. Cancer. 2016;122(9):1343–5. https://doi.org/10.1002/cncr.29968. Epub 2016 Mar 9.

23. World Health Organization (WHO). Global health estimates 2020: deaths by cause, age, sex, by country and by region, 2000–2019.WHO; 2020.
24. Global cancer observatory. Available on: https://gco.iarc.fr/today/data/factsheets/cancers/39-All-cancers-fact-sheet.pdf
25. Breen N, Gentleman JF, Schiller JS. Update on mammography trends: comparisons of rates in 2000, 2005, and 2008. Cancer. 2011;117:2209–18.
26. Allemani C, Matsuda T, Di Carlo V, et al. Global surveillance of trends in cancer survival 2000-14 (CONCORD-3): analysis of individual records for 37 513 025 patients diagnosed with one of 18 cancers from 322 population-based registries in 71 countries. Lancet. 2018;391:1023–75.
27. Lortet-Tieulent J, Renteria E, Sharp L, Weiderpass E, et al. Convergence of decreasing male and increasing female incidence rates in major tobacco-related cancers in Europe in 1988-2010. Eur J Cancer. 2015;51(9):1144–63.
28. Thun MJ, Carter BD, Feskanich D, et al. 50-year trends in smoking-related mortality in the United States. N Engl J Med. 2013;368:351–64.
29. International Agency for Research on Cancer (IARC). IARC handbooks of cancer prevention. Volume 10. Cervix Cancer Screening. IARC Press; 2005. Accessed November 23, 2020. https://publications.iarc.fr/Book-And-Report-Series/Iarc-Handbooks-Of-Cancer-Prevention/Cervix-Cancer-Screening-2005.
30. Cancer [Internet]. Who.int. 2021 Available from: https://www.who.int/news-room/fact-sheets/detail/cancer
31. Torre LA, Siegel RL, Ward EM, et al. Global cancer incidence and mortality rates and trends—an update. Cancer Epidemiol Biomarkers Prev. 2016;25:16–27.
32. Hull R, Mbele M, Makhafola T, Hicks C, Wang SM, Reis RM, Mehrotra R, Mkhize-Kwitshana Z, Kibiki G, Bates DO, Bates DO, et al. Cervical cancer in low and middle-income countries (Review). Oncol Lett. 2020:2058–74.
33. GLOBOCAN International Agency for Research in Cancer 2020. Available from: https://gco.iarc.fr/.
34. Plummer M, de Martel C, Vignat J, et al. Global burden of cancers attributable to infections in 2012: A synthetic analysis. Lancet Glob Health. 2016;4:e609–16.
35. Islami F, Torre LA, Jemal A. Global trends of lung cancer mortality and smoking prevalence. Translational Health Lung. Cancer Res. 2015;4 https://doi.org/10.3978/j.issn.2218-6751.2015.08.04.
36. National Cancer Institute and Centers for Disease Control and Prevention. Smokeless tobacco and public health: a global perspective Bethesda, MD: U.S. Department of Health and Human Services, Centers for Disease Control and Prevention and National Institutes of Health, National Cancer Institute; NIH Publication No14–7983, 2014.
37. Dhillon P, Mathur P, Nandakumar A, Fitzmaurice C, Kumar G, Mehrotra R, et al. The burden of cancers and their variations across the states of India: the Global Burden of Disease Study 1990–2016. Lancet Oncol. 2018;19:1289–306.
38. Global Burden of Disease Cancer Collaboration, Fitzmaurice C, Abate D, Abbasi N, Abbastabar H, et al. Global, regional, and national cancer incidence, mortality, years of life lost, years lived with disability, and disability-adjusted life-years for 29 cancer groups, 1990 to 2017: a systematic analysis for the global burden of disease study. JAMA Oncol. 2019;5(12):1749–68. https://doi.org/10.1001/jamaoncol.2019.2996.
39. Sankaranarayanan R, Ramadas K, Qiao Y-L. Managing the changing burden of cancer in Asia. BMC Med. 2014;12:3–3.
40. Cancer Country Profile: Pakistan. https://www.who.int/publications/m/item/cancer-pak-2020
41. Jayarajah U, Abeygunasekera AM. Cancer services in Sri Lanka: current status and future directions. J Egypt Natl Canc Inst. 2021;33(1):1–7.
42. Jayarajah U, Varothayan S, Jayasinghe R, Seneviratne S. Present status of cancer burden in Sri Lanka based on GLOBOCAN estimates. South Asian J Cancer. 2021:108. https://doi.org/10.1093/bjs/znab309.043
43. Cancer country profile: Bangladesh. https://www.who.int/publications/m/item/cancer-bgd-2020
44. Hussain SA, Sullivan R. Cancer control in Bangladesh. Jpn J Clin Oncol. 2013;43(12):1159–69. https://doi.org/10.1093/jjco/hyt140.
45. Rahman MM, Opo FADM, Asiri AM. Comprehensive studies of different cancer diseases among less-developed countries. Healthcare (Basel). 2022;10(3):424. https://doi.org/10.3390/healthcare10030424.
46. Shrestha G, Thakur RK, Singh R, et al. Cancer burden in Nepal, 1990-2017: an analysis of the Global Burden of Disease study. PLoS One. 2021;16(8):e0255499. https://doi.org/10.1371/journal.pone.0255499.
47. Neupane PR, Poudel KK, Huang ZB, Steel R, Poudel JK. Distribution of cancer by sex and site in Nepal. Asian Pac J Cancer Prev. 2017;18(6):1611–5. https://doi.org/10.22034/APJCP.2017.18.6.1611.
48. Joya M, Stanikzai Z, Akbarzadeh I, Babaloui S, Bradley DA, Jafari SM. Prevalence of cancers diagnosed in Jamhuriyat Hospital, Kabul, Afghanistan. Heliyon. 2020;6(3):e03603. https://doi.org/10.1016/j.heliyon.2020.e03603.
49. https://gco.iarc.fr/today/data/factsheets/populations/64-bhutan-fact-sheets.pdf
50. https://www.moh.gov.bt/wp-content/uploads/ict-files/2021/02/Bhutan-Cancer-Control-Strategy.pdf
51. https://ascopost.com/issues/october-25-2021/cancer-on-the-global-stage-incidence-and-cancer-related-mortality-in-bhutan/
52. Saeed J: Policy report: introducing an effective national cancer control plan in the Maldives. In: American Society of Clinical Oncology; 2018: 152–152.
53. Sarfati D, Dyer R, Sam FA, et al. Cancer control in the Pacific: big challenges facing small Island states. Lancet Oncol. 2019;20(9):e475–92.

54. Sarfati D, Dyer R, Vivili P, et al. Cancer control in small Island nations: from local challenges to global action. Lancet Oncol. 2019;20(9):e535–48.
55. Gelband H, Jha P, Sankaranarayanan R, et al., editors. Cancer: disease control priorities, third edition (Volume 3). Washington (DC): The International Bank for Reconstruction and Development / The World Bank; 2015.
56. Report of National Cancer Registry Programme (ICMR-NCDIR), Bengaluru, India 2020.
57. Sandhu GS, Erqou S, Patterson H, Mathew A. Prevalence of triple-negative breast cancer in India: systematic review and meta-analysis. J Glob Oncol. 2016;2(6):412–21. https://doi.org/10.1200/JGO.2016.005397. PMID: 28717728; PMCID: PMC5493252.
58. Trapani D, Murthy SS, Boniol M, Booth C, et al. Distribution of the workforce involved in cancer care: a systematic review of the literature. ESMO Open. 2021;6:100292. https://doi.org/10.1016/j.esmoop.2021.100292.
59. https://main.mohfw.gov.in/Organisation/Departments-of-Health-and-Family-Welfare/national-cancer-control-programme
60. https://www.cancer.gov/about-nci/overview/history/national-cancer-act-1971
61. WHO report on cancer: setting priorities, investing wisely and providing care for all. Geneva: World Health Organization; 2020. Licence: CC BY-NC-SA 3.0 IGO.
62. Koyande S, Subramanian S, Edwards P, et al. Economic evaluation of Mumbai and its satellite cancer registries: Implications for expansion of data collection. Cancer Epidemiol. 2016;45(Suppl 1):S43–9. https://doi.org/10.1016/j.canep.2016.10.002.
63. Jedy-Agba EE, Oga EA, Odutola M, et al. Developing national cancer registration in developing countries—case study of the Nigerian National System of Cancer Registries. Front Public Health. 2015;3:186. https://doi.org/10.3389/fpubh.2015.00186. PMID: 26284233; PMCID: PMC4519655.
64. https://indico.cern.ch/event/1246707/attachments/2590978/4471255/Linacs%20to%20narrow%20radiotherapy%20gap%20%E2%80%93%20CERN%20Courier.pdf
65. Ige TA, Jenkins A, Burt G, Angal-Kalinin D, et al. Surveying the challenges to improve linear accelerator-based radiation therapy in Africa: a unique collaborative platform of all 28 African countries offering such treatment. Clin Oncol (R Coll Radiol). 2021;33(12):e521–9.
66. Ocran Mattila P, Ahmad R, Hasan SS, Babar ZU. Availability, affordability, access, and pricing of anti-cancer medicines in low- and middle-income countries: a systematic review of literature. Front Public Health. 2021;9:628744.
67. GlobalSurg Collaborative and National Institute for Health Research Global Health Research Unit on Global Surgery. Global variation in postoperative mortality and complications after cancer surgery: a multicentre, prospective cohort study in 82 countries. Lancet. 2021;397(10272):387–97. https://doi.org/10.1016/S0140-6736(21)00001-5.
68. Global atlas of palliative care. 2nd Edition, 2020. www.thewhpca.org
69. Schliemann D, Ismail R, Donnelly M, Cardwell CR, Su TT. Cancer symptom and risk factor awareness in Malaysia: findings from a nationwide cross-sectional study. BMC Public Health. 2020;20(1):464. https://doi.org/10.1186/s12889-020-08581-0.
70. Brand NR, Qu LG, Chao A, Ilbawi AM. Delays and barriers to cancer care in low- and middle-income countries: a systematic review. Oncologist. 2019;24(12):e1371–80. https://doi.org/10.1634/theoncologist.2019-0057.
71. Pramesh CS, Badwe RA, Bhoo-Pathy N, et al. Priorities for cancer research in low- and middle-income countries: a global perspective. Nat Med. 2022;28(4):649–57. https://doi.org/10.1038/s41591--022-01738-x. PMCID: PMC9108683.
72. Global Burden of Disease 2019 Cancer Collaboration, Kocarnik JM, Compton K, Dean FE, et al. Cancer incidence, mortality, years of life lost, years lived with disability, and disability-adjusted life years for 29 cancer groups from 2010 to 2019: a systematic analysis for the global burden of disease study 2019. JAMA Oncol. 2022;8(3):420–44. https://doi.org/10.1001/jamaoncol.2021.6987.
73. Sarin R. Indian National Cancer Control Programme: setting sight on shifting targets. J Cancer Res Ther. 2005;1(4):240–8. https://doi.org/10.4103/0973-1482.19603.
74. Romero Y, Trapani D, Johnson S, et al. National cancer control plans: a global analysis. Lancet Oncol. 2018;19(10):e546–55. https://doi.org/10.1016/S1470-2045(18)30681-8.
75. https://www.iccp-portal.org/developing-nccp
76. Magrath I, Litvak J. Cancer in developing countries: opportunity and challenge. J Natl Cancer Inst. 1993;85(11):862–74.
77. Sankaranarayanan R, Basu P, Kaur P, Bhaskar R, Singh GB, Denzongpa P, Grover RK, Sebastian P, Saikia T, Oswal K, et al. Current status of human papillomavirus vaccination in India's cervical cancer prevention efforts. Lancet Oncol. 2019;20(11):e637–44.
78. Lahariya C, Subramanya BP, Sosler S. An assessment of hepatitis B vaccine introduction in India: Lessons for roll out and scale up of new vaccines in immunization programs. Indian J Public Health. 2013;57(1):8–14.
79. NHS Screening, https://www.nhs.uk/conditions/nhs-screening/
80. Shah SC, Kayamba V, Peek RM Jr, Heimburger D. Cancer control in low- and middle-income countries: is it time to consider screening? J Glob Oncol. 2019;5:1–8.
81. Boby JM, Rajappa S, Mathew A. Financial toxicity in cancer care in India: a systematic review. Lancet Oncol. 2021;22(12):e541–9.

82. Munshi A, Ganesh T, Mohanti BK. Radiotherapy in India: history, current scenario and proposed solutions. Indian J Cancer. 2019;56(4):359–63. https://doi.org/10.4103/ijc.IJC_82_19.

Further Reading

Global Burden of Disease Cancer Collaboration. doi: https://doi.org/10.1001/jamaoncol.2019.2996.

Global Cancer Observatory: https://gco.iarc.fr/

History of cancer epidemiology: https://www.cancer.org/treatment/understanding-your-diagnosis/history-of-cancer/cancer-epidemiology.html

https://cerncourier.com/a/linacs-to-narrow-radiotherapy-gap/

https://www.iccp-portal.org/developing-nccp

International statistical classification of diseases and related health problems, tenth revision, Fifth edition, 2016. World Health Organization. https://apps.who.int/iris/handle/10665/246208

Report of National Cancer Registry Programme (ICMR-NCDIR), Bengaluru, India 2020. https://ncdirindia.org/All_Reports/Report_2020/default.aspx

Radiation Oncology Infrastructure and Manpower

2

Monidipa Mondal, Suman Mallick, and Biplab Sarkar

Radiotherapy is an integral part of cancer treatment. It is estimated that 50% of cancer patients require radiotherapy during their course of illness [1]. Knowledge of existing radiotherapy infrastructure and manpower, accessibility to such facilities, the actual requirement, problems in accessibility, and the existing deficit in infrastructure is, therefore, necessary to deliver good quality radiotherapy and cancer care and also for planning of various future policies for cancer control. In this chapter, the first half will deal with an overview of the existing infrastructure and manpower (worldwide and Indian scenario), radiotherapy infrastructure requirements, the gap between the available and required infrastructure, and policies to be adopted to decrease the gap. The second half of the chapter will deal with setting up radiotherapy infrastructure.

2.1 Quantification of Radiotherapy Infrastructure Requirement: How Much Is the Need?

Several guidelines have been proposed by different governing agencies regarding infrastructure and manpower requirements. These guidelines provide a generalized idea about the requirement of the different radiotherapy equipment and manpower and have to be curated according to the need of a particular country. Therefore, for a particular country, we also required to know the types and geographical distribution of cancer, stage of disease at presentation, time trends, incidence, and mortality rates of different types of cancer, the current radiotherapy utilization rate (RUR), and the existing infrastructure.

The International Atomic Energy Agency (IAEA) has proposed guidelines regarding the quantification of infrastructure and human resources requirements in radiotherapy [2]. It gives a crude general guideline regarding the need for linear accelerator (LINAC)/radiation oncologist/physicist per year. The number of teletherapy machines per million of the population is an essential indicator of the ability to deliver radiotherapy services. **IAEA recommends one machine per 250,000 people** [3]. Besides this, IAEA has suggested other parameters: the number of radiotherapy centers, the number of treatment machines (operational cobalt/Linac units); the number of orthovoltage machines; teletherapy machines per thousand new cases of cancer per year; the number of remote control after-loading brachytherapy systems; a number of CT simulators; the number of computerized treatment planning systems (TPSs); the number of centers implementing advanced radiotherapy techniques; the number of centers implementing particle ther-

M. Mondal (✉) · S. Mallick
Department of Radiotherapy, Narayana Superspeciality Hospital, Howrah, Shibpur, Howrah, West Bengal, India

B. Sarkar
Department of Radiation Oncology, Apollo Multispeciality Hospitals, Kolkata, West Bengal, India

A. Munshi et al. (eds.), *Radiation Oncology – Principles, Precepts and Practice*,
https://doi.org/10.1007/978-981-97-8389-2_2

apy (protons, carbon ions; the number of radiation oncologists per million population; the number of radiation oncologists per 1000 new cases of cancer per year; the number of medical physicists; and the number of full-time equivalent radiotherapy technologists (FTE RTTs).

IAEA guidelines regarding radiotherapy staff requirements are given in Table 2.1.

The **European Society for Radiotherapy and Oncology** has provided its recommendation in the form of **ESTRO-QUARTS** (QUAntification of Radiation Therapy Infrastructure And Staffing Needs) guideline, which is summarized in Table 2.2 [4]. For linear accelerators, QUARTS guideline recommends an average of 5.5 MV units/million in the high, 3.5 MV units/million in the medium, and 2.0 MV units/million in the low-resource countries. For radiation oncologists, the variation was between 1 per 150–400 patients, with an average of 250, and there was no clear difference between high-, medium-, and low-resource countries. For physicists, the average was 1 per 530 patients for high and 1 per 610 for medium or low-resource countries [4].

The **WHO** recommends one teletherapy machine for a population of one million [5].

2.1.1 Radiotherapy Utilization Rate (RUR)

It gives a practical estimate of the demand for radiotherapy under ideal conditions. It is defined as the proportion of cancer patients that receive at least one course of radiotherapy during their lifetime [2].

Table 2.1 IAEA guideline regarding radiotherapy staff requirement [1]

Category	Staffing
Radiation oncologist-in-chief	One per program
Staff radiation oncologist	One additional for 200–250 patients treated annually
Radiation physicist	One per center up to 400 patients annually
Dosimetrist or physics assistant	One per 300 patients treated annually
Radiotherapy technologist (RTT)	Two per megavoltage units up to 25 patients treated daily per unit, 4 per MV units up to 50 patients
RTT-Sim	Two for every 500 patients simulated annually

Table 2.2 Summary of ESTRO-QUARTS guidelines [3]

Category	Per patient guideline
Linear accelerators	General 1 per 450 patients/year, with increasing complexity 1 per 400–450 patients/year
Radiation oncologists	General 1 per 250 patients/year, with increasing complexity 1 per 200–250 patients/year
Physicists	General 1 per 450–500 patients/year

$$\text{RUR} = \text{Patient treated with radiotherapy first time} / \text{Total new cases}$$

- The **optimum RUR** is the proportion of all cancers with an indication for radiotherapy at some point during their illness. Therefore, it represents the ideal scenario where all patients who need radiotherapy (RT) have access to it. It is calculated based on the published evidence-based guidelines of different cancers stating the indications of RT for a particular type and stage of cancer. The optimal number of radiotherapy fractions per cancer patient and per treatment course, along with RUR, also predicts the radiotherapy workload and hence the manpower needed for that workload and therefore gives a benchmark for the planning of RT infrastructure in a particular country.
- The **actual RUR** is the proportion of cancer patients that actually receives radiation therapy. The proportion of cancer patients who need RT (as per guidelines) and who actually receive it gives the best estimate of the accessibility of radiotherapy in a country.
- The **difference** between the two (the optimum and actual RUR) gives an estimate of the **unmet need** for the radiotherapy infrastructure (percentage of patients who, despite having an indication for radiotherapy, do not have access to it). Knowledge of this unmet requirement will help to plan future policy-making to strengthen the infrastructure and accessibility of radiotherapy.

In developed countries, radiotherapy is indicated in more than 50% of cancer patients [6]. In developed countries, actual RUR is about 70–100% of the optimum RUR, meaning that 70–100% of the cancer patients requiring RT receive it [2]. The optimum RUR is slightly more in low-middle-income countries (LMICs) because cancer patients usually present with more advanced diseases. This fact, coupled with limited access to oncology surgery, results in higher demand for radiotherapy compared with high-income countries [7]. In developing countries such as India, the radiotherapy utilization rate is >55% and may reach 70–80% [8].

The IAEA conducted a project [7] to establish the optimal and actual RUR in 9 middle-income countries and concluded that though the optimal RUR did not differ significantly from that of the high-income countries, the actual RUR was consistently lower, in particular in countries with limited resources and a large population. Lower actual RUR reflects problems with access to RT, which is most likely due to the limited number of RT centers and/or RT machines availability relative to a large population, affordability issues, lack of awareness of patients and doctors, cultural beliefs, etc.

Therefore, in low-middle-income countries like India, having a large population growth, the difference between the optimum and actual RUR increases, resulting in a greater unmet need.

2.1.2 Assessment of Existing Facilities

Since 1959, the IAEA has maintained a computerized database of radiotherapy centers worldwide called DIRAC (DIrectory of RAdiotherapy Centres) [9].

2.1.3 Assessment of Cancer Incidence, Mortality, Geographical Distribution, and Time Trends

This information can be obtained from GLOBOCON 2020 data [10] and population-based cancer registry data for the particular country.

2.1.4 Radiotherapy Cost Assessment

- In 2013, the Global Task Force on Radiotherapy for Cancer Control (GTFRCC) was commissioned to quantify the investment needed and operational costs for radiotherapy across the globe. Analysis on the topic has been published by the Union for International Cancer Control's Global Task Force on Radiotherapy for Cancer Control (GTFRCC) [11]. Aiming to calculate the operational cost of radiotherapy treatment, resources required, total operational costs, and cost per course, GTFRCC has used a time-driven activity-based (TD ABC) costing model [12]. The total cost is calculated as the summated cost of the care process activities (such as treatment planning, delivery verification, or dose delivery) and related resources involved in generating that product, including the time these resources are committed to specific activities (e.g., minutes of RTT time needed to deliver one radiotherapy fraction) [13].

The Health Economics in Radiation Oncology (HERO) project, conducted by the European Society for Radiotherapy and Oncology, has provided a comprehensive overview of radiotherapy equipment and departments across European countries [14] and also reported the gap between the evidence-based optimal and actual radiotherapy treatments delivered across Europe [15].

2.2 Existing Infrastructure

2.2.1 World Scenario

Teletherapy According to the DIrectory of RAdiotherapy Centres (DIRAC) of IAEA, out of 214 countries, 156 countries have radiotherapy facilities. Worldwide, there are 7646 RT centers, with 15,095 RT machines (14,986 MV therapy machines and 109 light ion therapy machines), as shown in Fig. 2.1. The distribution of radiotherapy machines in different countries is summarized in Table 2.3 [9]. Out of these 15,095 RT machines, 62.2% (9397) are located in high-income (HIC) countries, 27.2% (4110) in upper-middle-income countries (UMIC), 10.3% (1547)

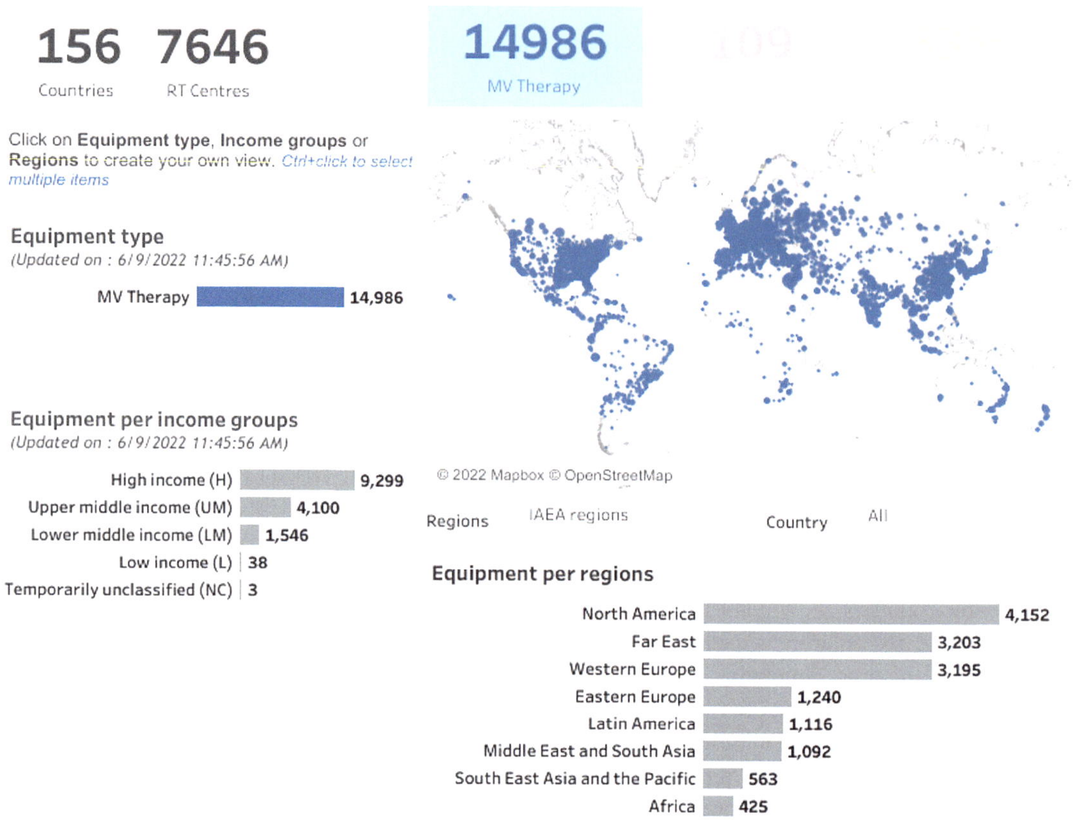

Fig. 2.1 Distribution of MV therapy machines across the world in 2022 (as per DIrectory of RAdiotherapy Centres)

Table 2.3 Radiotherapy infrastructure and manpower in South Asian countries

Countries	India	Pakistan	Bangladesh	Sri Lanka	Nepal
Radiotherapy centers [9]	432	32	21	9	6
MV machines [9]	689	63	35	25	9
kV therapy machines [9]	5	1	0	0	0
Brachytherapy units [18]	317	14	8	3	3
Light ion therapy machines [9]	1	0	0	0	0
MV machines/linear accelerators per 10,00,000 population [18]	0.5	0.3	0.22	0.95	0.3
Number of radiation/clinical oncologists [18]	3743	100–200	26	23	15
Number of radiation/clinical oncologists per 10,00,000 population [18]	2.7	0.46	0.8	3.0	1.4

in lower-middle-income countries (LMIC), and 0.2% (38) in low-income countries (LIC). There is a clear disparity in the distribution of RT machines and the population of the countries, which is reflected in the number of radiotherapy machines per million population. Most of the radiotherapy machines are concentrated in the HICs, which account for only 15.6% of the world's population, whereas 60% of incident cases and 70% of cancer-related deaths were reported in low- and middle-income countries [16]. Region-wise, the distribution of RT centers is as follows: North America has approximately 2600 centers, with the United States accounting for nearly 90% of these. Europe has around 1600 centers, while South America has 650 centers. Africa has 250 centers, and Asia has nearly 3000 RT centers.

Over the past decades, the number of linear accelerators has increased compared to telecobalt machines. In 2020, the number of linear accelerators was 12,411 (86.1%), whereas the number of telecobalt units was 2004 (13.9%) [17].

Among the South Asian countries, India has the highest number of radiotherapy centers/MV machines/brachytherapy units, followed by Pakistan, Bangladesh, Sri Lanka, and Nepal, as shown in Table 2.3 [9]. The number of MV machines/linear accelerators per 1,000,000 population is highest in Sri Lanka (0.95), followed by India (0.5) [18].

Brachytherapy Worldwide, in 131 countries, there are 2509 brachytherapy centers having 3350 brachytherapy machines [9]. Brachytherapy centers are mostly concentrated in Western Europe (836) and North America (824), as shown in Fig. 2.2.

Light ion therapy machines are also concentrated in North America (37) and Western Europe

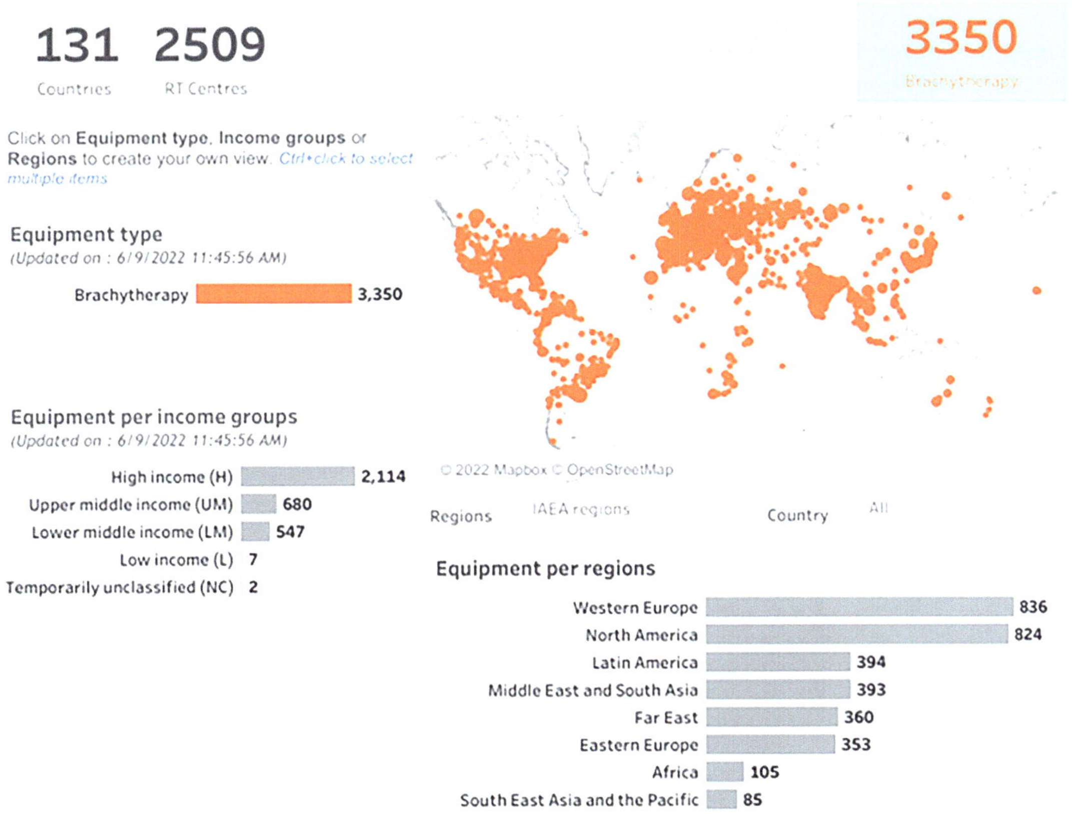

Fig. 2.2 Distribution of the brachytherapy facilities across the world in 2022 (as per DIrectory of RAdiotherapy Centres)

(32), with the United States (37) and Japan (23) having the highest number of facilities [17].

Simulators In European countries, the ESTRO-HERO project reported a median of 0.5 simulators per MV unit (range, 0.3–1.5) and 1.4 simulators per department (range, 0.4–4.4), whereas there are less than 100 simulators in the entire African subcontinent [19].

Manpower A systematic review of the workforce involved in cancer care showed a median of 0.28 radiation oncologists per 100,000 population and 0.15 per 100 cancer patients worldwide [20]. The distribution of the cancer care workforce in different countries grouped together as per their income is shown in Tables 2.4 and 2.5. There is a stark disparity in the distribution of manpower, with severe shortages of manpower in LMICs and LICs, particularly among healthcare professionals, where the cancer burden and mortality are high. Among the South Asian countries, the number of radiation/clinical oncologists per 1,000,000 population is highest in Sri Lanka (3), followed by India (2.7), Nepal (1.4), Bangladesh (0.8), and lowest in Pakistan (0.46) [18].

Cost of Radiotherapy Treatment Van Dyk et al. in 2017 reported that the base case cost per course was US$5368 in HICs and US$2028 in LICs; the annual operating cost was US$4,595,000 and US$1,736,000, respectively. This analysis also showed that the cost per course decreases as department size increases; reducing operating hours below eight hours per day has a significant impact on the cost per course, and the use of IMRT increases the cost per course by 22% [21].

2.2.2 Indian Scenario

Teletherapy According to the DIrectory of RAdiotherapy Centres (DIRAC) of IAEA, India has 432 radiotherapy centers, 688 MV therapy machines (telecobalt+linear accelerator), one particle therapy machines, 5 VK therapy machines, and 317 brachytherapy machines (updated in 2022) [9]. Distribution is shown in Fig. 2.3. The details of the radiotherapy equipment are summarized in Table 2.6. The number of MV machines per million populations is less than one (0.5).

Brachytherapy Munshi et al. reported 250 HDR units and 6 LDR units in 2019 [22]. The

Table 2.4 The distribution of the radiotherapy infrastructure in different countries grouped together as per their income [9]

Income group	Countries	Countries with RT	RT centers	MV therapy	Light-ion therapy	RT machines	Brachytherapy machines	Million population	RT machines per million populations
High income	74	60	4550	9299	98	9397 (62.2%)	2114	1215 (15.6%)	8
Upper middle income	54	44	2245	4100	10	4110 (27.2%)	680	2584 (33.1%)	2
Lower middle income	54	37	886	1546	1	1547 (10.2%)	547	3324 (42.7%)	0
Lower income	28	14	22	38	0	38 (0.25%)	7	666 (8.5%)	0
Temporarily unclassified	4	1	1	3	0	3 (0.01%)	2	1 (0.1%)	5
Total	**214**	**156**	**7704**	**14,986**	**109**	**15,095 (100%)**	**3350**	**7789 (100%)**	**2**

Countries are grouped into income groups as per World Bank gross national income (GNI) per capita classifications

Table 2.5 Median ratios of the radiotherapy manpower per 1,00,000 populations and per 1000 cancer patients in different countries grouped together on the basis of per capita income [20]

Countries	Median ratios	Clinical Oncologist	Radiation Oncologist	Medical Physicist	Radiotherapy Technologist
HIC	Per 1,00,000 population	0.88	1.24	0.78	0.44
	Per 1000 cancer patients	0.86–2.9	0.25	0.47	1.2
UMIC	Per 1,00,000 population	0.2	0.4	0.3	0.28
	Per 1000 cancer patients	0.37	0.19	0.28	0.7
LMIC	Per 1,00,000 population	0.03	0.04	0.05	0.06
	Per 1000 cancer patients	0.12	0.076	–	–
LIC	Per 1,00,000 population	0.006	–	0.1	–
	Per 1000 cancer patients	0.01	–	–	–
Global	Per 1,00,000 population	**0.028**	**0.28**	**0.23**	**0.19**
	Per 1000 cancer patients	**0.27**	**0.15**	**NA**	**NA**

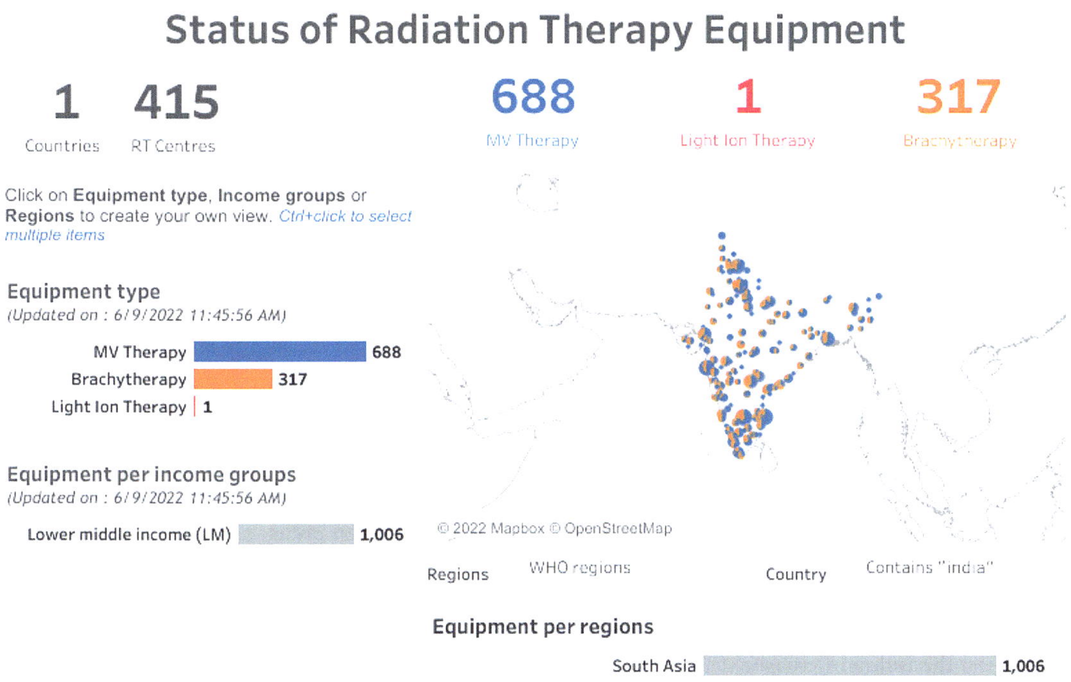

Fig. 2.3 Distribution of radiation therapy equipment across India in 2022 (as per DIrectory of RAdiotherapy Centres)

Atomic Energy Regulatory Board (AERB) of India reported 305 remote after-loading and 49 manual after-loading equipment in 2018 [23].

Manpower The number of radiation/clinical oncologists per 1,000,000 population is 2.7 [18].

Cost of Treatment The current average prices for Indian patients treated in hospitals are as follows: local patients in public hospital, \$0.6 ± 03 thousand; outstation patients in public hospitals, \$0.7 ± 0.3 thousand; local patients in private hospitals, \$7.05 ± 5.1 thousand; outstation patients in private hospitals, \$7.8 ± 5.1 thousand [24, 25].

Table 2.6 Summary of radiotherapy infrastructure in India

Equipment	Numbers	Last updated
Radiotherapy centers	432	2022 [9]
MV therapy machines (telecobalt+LINACs)	688	2022 [9]
Particle therapy machine	1	2022 [9]
Simulators	40	2019 [22]
CT simulators	50	2019 [22]
Tomotherapy machines	8	2019 [22]
CyberKnife	7	2019 [22]
Gamma Knife	7	2019 [22]
Brachytherapy machines	317	2022 [9]
KV therapy machines	5	2022 [9]
Facilities using I-125 Prostate Implant Seed Sources	47	2019 [22]
Facilities using Sr-90, Ru-106, I-125 Ophthalmic sources	2	2019 [22]
Number of machines per million population	<1 (0.5)	2022 [9]

2.3 Where Lies the Problem?

India has a higher requirement for radiotherapy infrastructure and manpower compared to high-income countries (HICs). Rosenblatt et al. study estimates the radiotherapy utilization rate in India is >55% and may reach 70–80%, whereas it is 50% in developed countries [7]. The higher requirement of radiotherapy infrastructure is because of the following reasons:

- Radiotherapy is the definitive treatment mode for head, neck, and cervical cancers. The incidence of this cancer is higher in India. According to the GLOBOCON 2020 data, 17.7% of cases of newly diagnosed cancer were head and neck cancer, and 9.4% were cervical cancer [26]. Therefore, the optimal radiotherapy utilization rate and consequently infrastructure requirement are higher compared to those in HICs. According to the GTFRCC report, optimum radiotherapy utilization rates for head and neck and cervical cancer are 74% and 71%, respectively [11]. In India, the majority of cancer patients are diagnosed in the more advanced stages of the disease. According to a report from the National Cancer Registry Program (2020), 66.6% of head and neck cancers, 60% of cervical cancers, and 57% of breast cancers are diagnosed at an advanced stage [27]. Advanced-stage presentation coupled with limited access to oncology surgery increases the radiotherapy requirement further.
- The high population growth rate in LMICs, like India, leads to increased cancer incidence and mortality rates compared to HICs, which further widens the gap between the availability and requirement of radiotherapy facilities. By 2030, LICs and LMICs are likely to have more than 70% of the global cancer burden [11]. Although many new radiotherapy facilities are opening up and the actual radiotherapy utilization rate is increasing promisingly, it remains disproportionate to the rising cancer burden in a given year.

The proportion of patients who actually receive radiation therapy (actual RTU rates) is lower due to the presence of various barriers to accessing radiotherapy facilities. They are:

The majority of radiotherapy centers are concentrated in urban areas, especially metropolitan cities (40%) [28], whereas most cancer patients come from rural or semi-urban areas. Another study from 2018 reported that eight metro cities in India, constituting only 10.9% of the national population, benefit from 38% of the total RT capacity of India [29]. In contrast, about 60% of teletherapy machines in India are located in Southern and Western India [22]. The costs of traveling to these centers and accommodation during a prolonged course of radiotherapy treatment also create a hindrance to the optimal utilization of radiotherapy facilities.

Also, 50–85% of these facilities are under non-governmental ownership where the cost of treatment is high, thus further reducing the accessibility of radiotherapy facilities to the lower income group patients [27]. In the financial year 2021–2022 (pandemic year), expenditure on health sector was 2.1% of GDP, and in 2019–20,220, it was 1.3%. The union budget allocation for the year 2022–2023 was 0.23% higher than the previous year [30]. However, the cancer burden accounted for as much as 0.36% of the GDP lost for India [31]. Consequently, providing equitable and affordable cancer care poses a major challenge.

2.4 The Unmet Need: Deficit

2.4.1 World Scenario

Table 2.7 summarizes the actual radiotherapy facilities available versus the total need in different regions of the world [12].

2.4.2 Indian Scenario

As per the Globocan 2020 data, India witnesses approximately 1,324,413 new cancer cases every year, with 2,720,251 prevalent cases and 851,678 cancer deaths per year. The recorded incidence of cancer in India is 97 per 100,000 people.

A Deficit in Teletherapy: The estimated shortfall in India for teletherapy machines is around 1200, as per the WHO estimate [32]. If one were to attempt to realize the ambitious goal of a global average of about four teletherapy machines for one million populations, the estimated deficit in India is nearly 5000 machines.

A Deficit in Brachytherapy: Chopra et al. [33] suggested that one brachytherapy machine is needed for every 100 new cervical cancer patients diagnosed. By this estimate, India had a shortfall of 127 brachytherapy machines.

Manpower: In India, there are 5000 radiation oncologists for an approximate annual incidence of 1.4 million new cases. The estimated number of radiation oncologists is less than 200 for an annual estimated cancer incidence of about one million [34].

Every year about 40 units are added, and 15 units are decommissioned. Therefore, 25 new units are added effectively per year. These 25 new units are insufficient to take care of India's annual population growth, which is 25 million per year [22].

2.5 Future Direction: What Needs to be Done

Several measures have been suggested in the literature to address these unmet needs.

- Investment in and Adoption of Indigenous Technology: The huge cost of a new MV radiotherapy machine is a significant deterrent to establishing new RT centers [17]. Indigenous technology could be more eco-

Table 2.7 Summary of the actual radiotherapy facilities available versus the total need in different region of the world [12]

Regions	Africa	Asia Pacific	Europe	Latin America	North America
Radiotherapy centers					
Actual	140	2585	1431	620	2787
Needed	407	3503	1449	624	1200
Radiotherapy MV machines					
Actual	277	3894	3751	968	4243
Needed	813	6406	4098	1106	2175
Actual coverage of the needs	**34%**	**61%**	**92%**	**88%**	**195%**
Cost per radiotherapy course in million US$					
Actual cost	1226	2423	3428	1939	6581
Total need	1306	2126	3487	2079	7084

nomical. Bhabhatron and Bhabhatron II telecobalt machines, developed by the Bhabha Atomic Research Centre (BARC) and Panacea Medical in India, and the Karknidon-I HDR brachytherapy machine are such examples.

- Encouraging public–private partnerships to provide affordable health care to all.
- Governmental Action: Recognition of the need for radiotherapy requirement in the National Cancer Control Program [35] by making a radiation oncology department mandatory in medical colleges by the National Medical Council and relaxing customs duty on radiotherapy equipment [36].
- Collaboration for Resource Exchange Within the Country and Cross-Country: It also facilitates the establishment of telecommunication networks to standardize cancer management, launch education initiatives, develop a skilled radiotherapy workforce, and collaborate on national-level cancer research. The National Cancer Grid is such an initiative.
- Strengthening referral system and telemedicine services.
- Training and research.

2.6 Setting Up of Radiotherapy Infrastructure

2.6.1 Guidelines and Regulatory Requirements for New Radiotherapy Facility

1. The first step for the concerned facility is to obtain requisite regulatory consent from the country's regulatory board. In India, the regulatory board is known as Atomic Energy Regulatory Board (AERB). As per the AERB Safety Guide, ensure compliance with regulatory requirements as mentioned in safety codes. It is essential to verify that the Unit (LINAC, Telecobalt, etc.) is type-approved and possesses a "No Objection Certificate" [37].
2. To submit the application form for obtaining requisite regulatory clearances, the employer of the institute must register their institute on the eLORA web portal [38].
3. **Shielding Calculation**.

   Primary Barrier: A primary barrier is an attenuating structure, such as a wall, ceiling, or floor, that directly intercepts radiation (Fig. 2.4).

   Secondary Barrier: A secondary barrier is a wall, ceiling, floor, or other structure that will intercept the secondary radiation, which consists of leakage radiation from the source housing and scattered radiation from the patient.

   Calculations:

   Primary Barrier: The number (n) of TVs required is given by n = log (reduction factor) = log(WUT/Pd^2).

   The thickness of the primary wall = $n \times$ TVL.

   W = Workload, U = Use factor, T = Occupancy factor.

   Workload = Workload (W) for radiotherapy is the time integral of the absorbed dose rate at a depth of the maximum absorbed dose at 1 m from the source, considering the maximum possible number of the patient treated and physics work. The most common period of time over which W is specified is 1 week. W is expressed in cGy/wk. at 1 m.

Width of the Primary Barrier

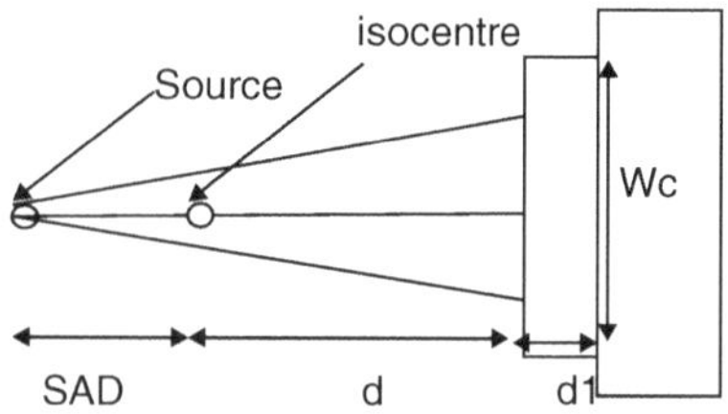

Width of the primary barrier (Wc) = $\frac{(SAD + d + d1) \times F}{SAD}$ + 2x30 cm

Fig. 2.4 Calculation of the width of primary barrier for radiotherapy bunker

$$W = \frac{\text{Dose}}{\text{pt.}} \times \text{cGy} \times \frac{70\text{pt.}}{d} \times \frac{5d}{\text{wk}} \times \frac{1}{PDD(10\text{cm depth})} = \_\_ \text{cGy / wk at 1m}$$

Use Factor (*U*).

The use factor (*U*) is the fraction of a primary beam workload that is directed toward a given primary barrier. The value of "*U*" will depend on the type of installation and protective barrier. For example,

For accelerators and Cobalt-60 units, the following values of factors (*U*) are generally used:

1: For gantry pointing down

¼: For each rotational direction

1: For scatter and leakage radiation and brachytherapy.

Occupancy Factor (*T*).

The occupancy factor (*T*) is the fraction of beam "ON" time that the exposed individual is present while the beam is "ON." The value of "*T*" will depend on the type of occupancy. For example,

$$T = 1 \text{ for full occupancy}$$

$$T = 1/4 \text{ for partial occupancy}$$

$$T = 1/16 \text{ for occasional occupancy}$$

Secondary Barrier Thickness Calculation.

Shielding Due to Head Leakage: The reduction Factor (RF) required to achieve the design goal (*P*) against head leakage radiation can be given by

$$\text{RF} = \frac{\alpha_0 WUT}{P(ds)^2} = \frac{0.001 \times WUT}{P(ds)^2}$$

where α_0 is the leakage from the head (0.1% of primary).

P is the design dose limit in CSV/wk.

d_s is the distance from the isocenter to the point of interest in m.

W is the workload in cGy/wk.

T is the occupancy factor.

U is the use factor equal to 1 for secondary barriers.

Thickness due to head leakage (t_{leak}) = No. of TVs × TVL value.

Shielding Due to Patient Scatter: The reduction factor (RF) against radiation scattered by the patient to achieve the design dose limit (P) is given by

$$\text{RF} = \frac{\alpha WUT \frac{F}{400}}{P(d_{sca} \times d_S)^2}$$

where

P, *W*, and *T* have their usual meaning.

α is the patient scatter fraction defined at d_{sca}.

(The scatter primary ratio (α) is dependent on the energy of the X or gamma photons and the scattering angle. Values of α are available per 400 cm^2 irradiated field area for all clinical beams.)

F is the max field area incident on the patient in cm^2.

d_{sca} is the distance from the radiation source to the patient in m.

d_s is the distance from the patient to the point of interest, in m.

Thickness due to patent scatters (t_{sc}) = No. of TVs × TVL value for patient scatter at 90 degrees.

- If the difference between thicknesses t_{leak} and t_{sc} is greater than one TVL, then. Thickness of secondary barrier = Max{$t_{leak,}$ t_{sc}}.
- If the difference between thicknesses t_{leak} and t_{sc} is lesser than one TVL, then. The thickness of the secondary barrier = Max{$t_{leak,}$ t_{sc}} + 1 HVT for leakage.

4. **Guidance to Prepare Site and Layout Plan Drawings** [37].

- The first step in establishing a radiotherapy facility is to submit the layout and site plan of the radiotherapy installation and get it approved from AERB, a typical layout shown in Fig. 2.5.

- Prepare the site layout plan drawing for the radiotherapy installation, with the name and address of the institution. The recommended scale is 1:200.
- The location of the radiotherapy installation should be away from unrelated facilities and close to related facilities such as the simulator room, mold room, radiation oncologist(s) room, medical physicist(s) room, etc. All such connected facilities should be shown in the site layout plan.
- All facilities within 20 m of the external walls of the radiotherapy room should be shown in the site layout plan with proper names, legends, and indicators, as the floor layout plan drawing for the proposed radiotherapy room, with the institute name and address.
- The floor layout plan drawing for the proposed radiotherapy room, with the institute name and address, should be prepared.
- The following should be well indicated in plan:

Photon energies, beam axis, thickness and wall length, length and width of maze, door with interlock, conduit, gamma zone monitors, CCTVS, LMO, warning lights, control console room, etc.

5. **Appointment of Radiotherapy Professionals**.

An Adequate number of full-time Radiation Oncologists, Medical Physicists, and Radiation Therapy Technologists must be engaged based on their qualification and experience. The relevant IAEA guidelines are already summarized in Table 2.1.

6. Other Requirements.
   - Procurements of TLD badges from AERB-accredited agencies.
   - Procurements of measuring and monitoring instruments.
   - Other associated instruments, such as the CT simulator, required site, and layout approvals prior to installation.
   - Before energizing equipment for radiation, a commissioning must be obtained from AERB. It is necessary to have requisite measuring, monitoring, and QA tools in

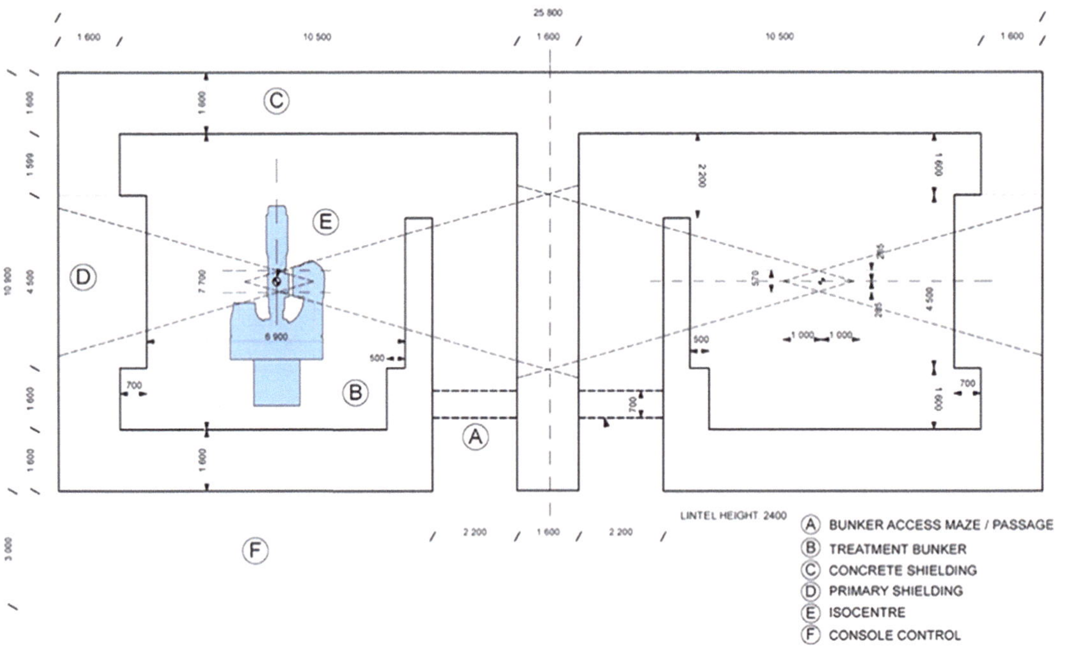

Fig. 2.5 A typical plan layout

place for obtaining commissioning approval.
- Commissioning approval is followed by survey approval and the QA tests.
- After completion of all mechanical, electrical, and radiation-related QA tests, the application is submitted for license. Patient treatment can only be started after obtaining a license for operation.

2.7 Summary

- It is estimated that 50% of cancer patients require radiotherapy during their course of illness.
- The International Atomic Energy Agency (IAEA) recommends one machine per 250,000 people. The WHO recommends one teletherapy machine for a population of one million.
- In developing countries such as India, the radiotherapy utilization rate is >55% and may reach 70–80%.
- The IAEA maintains a computerized database of radiotherapy centers worldwide called DIRAC (DIrectory of RAdiotherapy Centres).
- Worldwide, most radiotherapy machines are concentrated in the HICs, accounting for only 15.6% of the world's population, whereas 60% of incident cases and 70% of cancer-related deaths were reported in low- and middle-income countries.
- Among the South Asian countries, India has the highest number of radiotherapy centers/MV machines/brachytherapy units, followed by Pakistan, Bangladesh, Sri Lanka, and Nepal. According to the DIrectory of RAdiotherapy Centres (DIRAC) of IAEA, India has 432 radiotherapy centers, 688 MV therapy machines (telecobalt+linear accelerator), one particle therapy machine, 5 VK therapy machines, and 317 brachytherapy machines (updated in 2022). The number of MV machines per million population is less than one (0.5). The number of radiation/clinical oncologists per 10,00000 population is 2.7.
- In India, the requirement for radiotherapy infrastructure is more due to the high population growth rate, advanced-stage presentation, limited access to oncology surgery, and different barriers to access radiotherapy facilities.
- The estimated shortfall in India for teletherapy machines is around 1200, as per the WHO estimate. India had a shortfall of 127 brachytherapy machines. The estimated number of radiation oncologists is less than 200 for an annual estimated cancer incidence of about one million.
- Several measures are suggested for addressing the unmet need: Investment in and adoption of indigenous technology, encouraging public–private partnership, governmental action, collaboration for resource exchange within country and cross-country, strengthening referral system and telemedicine services, training, and research.
- For setting up a radiotherapy infrastructure, the first thing is to obtain requisite regulatory consent from AERB followed by registration of institute in eLORA web portal. Shielding calculation has to be done; the layout and site plan of the radiotherapy installation have to be approved by AERB, followed by the appointment of radiotherapy professionals, commissioning approval from AERB, survey approval, and the QA tests. Patient treatment can only be started after obtaining a license for operation.

Multiple Choice Questions

1. Increase in the annual patient load of a center will
   (a) Increase the RUR
   (b) Decrease the RUR
   (c) Will keep the RUR same
2. As per IAEA, one radiotherapy machine is recommended for how many people?
   (a) 250,000
   (b) 25,000
   (c) 300,000

3. As per WHO estimate, what is the shortfall in the number of teletherapy machines?
   (a) 1000
   (b) 1200
   (c) 2000
4. According to the GLOBOCON 2020 data, what percentage of head and neck and cervical cancer cases were reported?
   (a) 17.7% and 9.4%
   (b) 20.3% and 10.7%
   (c) 12.3% and 19.6%
5. What is DIRAC?
   (a) DIrectory of RAdiotherapy Centres of India
   (b) DIrectory of RAdiotherapy Centres
   (c) Directory of radiotherapy machines
6. What is the number of radiotherapy centers in Europe, Africa, and Asia?
   (a) 1600, 250, 3000
   (b) 3000, 1600, 250
   (c) 250, 3000, 1600
7. What is the radiotherapy utilization rate in India and developed countries?
   (a) 55% and 50%
   (b) 50% and 55%
   (c) 75% and 50%.
8. By 2030, LICs and LMICs are likely to have more than XX % of the global cancer burden, xx =
   (a) 70%
   (b) 50%
   (c) 30%.
9. What is "full occupancy" in radiotherapy shielding calculation?
   (a) $T = 1$
   (b) $T = 1/4$
   (c) $T = 1/16$
10. Value of "use factor" for secondary barrier =
   (a) 1
   (b) ¼
   (c) 1/16

Answers

1. (a)
2. (a)
3. (b)
4. (a)
5. (b)
6. (a)
7. (a)
8. (a)
9. (a)
10. (a)

References

1. Rosenblatt E, Zubizarreta E, editors. Radiotherapy in cancer care: facing the global challenge. Vienna: International Atomic Energy Agency; 2017.
2. International Atomic Energy Agency. IAEA human health series no. 14. Planning national radiotherapy services: a practical tool. Vienna, Austria: International Atomic Energy Agency; 2010. Available at: http://www-pub.iaea.org/MTCD/Publications/PDF/Pub1462_web.pdf. Accessed 16nth September 2022.
3. Grover S, et al. A systematic review of radiotherapy capacity in low- and middle-income countries. Front Oncol. 2014;4:380.
4. Slotman BJ, Cottier B, Bentzen SM, Heeren G, Lievens Y, Van den Bogaert W. Overview of national guidelines for infrastructure and staffing of radiotherapy. ESTRO-QUARTS: work package 1. Radiother Oncol. 2005;75(3):349–E1.
5. Indicator Metadata Registry Details. https://www.who.int/data/gho/indicator-metadata-registry/imr-details
6. Delaney G, Jacob S, Featherstone C, Barton M. The role of radiotherapy in cancer treatment: estimating optimal utilization from a review of evidence-based clinical guidelines. Cancer. 2005;104(6):1129–37.
7. Rosenblatt E, Fidarova E, Zubizarreta E, Barton MB, MacKillop W, Jones GW, et al. Radiation therapy utilization in middle-income countries. Int J Radiat Oncol Biol Phys. 2016;96:S37.
8. Barik SK, Behera BK, Majumdar SK, Parida DK. How prepared are the radiotherapy centers to deal with the COVID-19 pandemic? A nationwide survey from 46 cancer centers across India. J Cancer Res Ther. 2022;18(1):245.
9. https://dirac.iaea.org/. Accessed 16 Sept 2022.
10. https://www.uicc.org/news/globocan-2020-new-global-cancer-data. Accessed 16 Sept 2022.
11. Atun R, Jaffray DA, Barton MB, et al. Expanding global access to radiotherapy. Lancet Oncol. 2015;16(10):1153e1186.
12. Zubizarreta EV, Van Dyk J, Lievens Y. Analysis of global radiotherapy needs and costs by geo-

graphic region and income level. Clin Oncol. 2017;29(2):84–92.
13. Kaplan RS, Anderson SR. Time-driven activity-based costing. Harv Bus Rev. 2004;82:150.
14. Grau C, Defourny N, Malicki J, Dunscombe P, Borras JM, Coffey M, Slotman B, Bogusz M, Gasparotto C, Lievens Y, Kokobobo A. Radiotherapy equipment and departments in the European countries: final results from the ESTRO-HERO survey. Radiother Oncol. 2014;112(2):155–64.
15. Borras JM, Lievens Y, Dunscombe P, et al. The optimal utilization proportion of external beam radiotherapy in European countries: an ESTRO-HERO analysis. Radiother Oncol. 2015;116(1):38e44.
16. Allemani C, Matsuda T, Di Carlo V, Harewood R, Matz M, Nikšić M, Bonaventure A, Valkov M, Johnson CJ, Estève J, Ogunbiyi OJ. Global surveillance of trends in cancer survival 2000–14 (CONCORD-3): analysis of individual records for 37 513 025 patients diagnosed with one of 18 cancers from 322 population-based registries in 71 countries. Lancet. 2018;391(10125):1023–75.
17. Laskar SG, Sinha S, Krishnatry R, Grau C, Mehta M, Agarwal JP. Access to radiation therapy: from local to global and equality to equity. JCO Glob Oncol. 2022;8:e2100358.
18. Sarkar B, Munshi A, Ganesh T, Manikandan A, Bhattacharjee B, Pun R, Imbulgoda N, Yasmin T, Pradhan A. Growth characteristics of female radiation/clinical oncologists in South Asia: assessment of gender neutrality and leadership position. Clin Oncol. 2022;34(10):e437-45
19. Grau C, Defourny N, Malicki J, et al. Radiotherapy equipment and departments in the European countries: final results from the ESTRO-HERO survey. Radiother Oncol. 2014;112:155–64.
20. Trapani D, Murthy SS, Boniol M, et al. Distribution of the workforce involved in cancer care: a systematic review of the literature. ESMO Open. 2021;6:100292.
21. Van Dyk J, Zubizarreta E, Lievens Y. Cost evaluation to optimise radiation therapy implementation in different income settings: a time-driven activity-based analysis. Radiother Oncol J. 2017;125:178–85.
22. Munshi A, Ganesh T, Mohanti BK. Radiotherapy in India: history, current scenario and proposed solutions. Indian J Cancer. 2019;56:359–63.
23. Atomic Energy Regulatory Board Annual Report 2018, Government of India. Available at https://www.aerb.gov.in/images/PDF/Annual_report/ar2018/AERB-Annual-Report-2018-1.pdf. Accessed 4 Sept 2022.
24. Mohanti BK, Mukhopadhyay A, Das S, Sharma K, Dash S. The economic burden of cancer. Econ Polit Wkly. 2011;22:112–7.
25. Sarkar B, Munshi A, Rastogi K, Ganesh T, Bansal K, Manikandan A, Mohanti BK, Tyagi B, Vaishya S, Ghosh B, Bandyopadhyay S. Cancer care medical tourism in the national capital region of India-challenges for overseas patients treated in two private hospitals. Health Policy Technol. 2022;11(3):100659. Supplementary material
26. https://gco.iarc.fr/today/data/factsheets/populations/356-india-fact-sheets.pdf. Accessed 4 Sept 2022.
27. Mathur P, Sathishkumar K, Chaturvedi M, Das P, Sudarshan KL, Santhappan S, Nallasamy V, John A, Narasimhan S, Roselind FS, ICMR-NCDIR-NCRP Investigator Group. Cancer statistics, 2020: report from national cancer registry programme, India. JCO Glob Oncol. 2020;6:1063–75.
28. Singh AG, Chaukar D, Gupta S, et al. A prospective study to determine the cost of illness for oral cancer in India. Ecancermedicalscience. 2021;15:1252.
29. Chaudhuri S, Goel A, Awasthi A, et al. Disparity between metro-centric cancer care and rural outreach in India: situational analysis and future trends in context of developing countries. J Glob Oncol. 2018;4(suppl 2):59s.
30. https://www.indiabudget.gov.in/economicsurvey/ebook_es2022/files/basic-html/page377.html Accessed 4 Sept 2022.
31. Pearce A, Sharp L, Hanly P, et al. Productivity losses due to premature mortality from cancer in Brazil, Russia, India, China, and South Africa (BRICS): a population-based comparison. Cancer Epidemiol. 2018;53:27–34.
32. Lievens Y, Borras JM, Grau C. Provision and use of radiotherapy in Europe. Mol Oncol. 2020;14:1461–9.
33. Chopra S, Shukla R, Budukh A, et al. External radiation and brachytherapy resource deficit for cervical cancer in India: call to action for treatment of all. J Glob Oncol. 2019;5:1–5.
34. Elmore SNC, Prajogi GB, Rubio JAP, et al. The global radiation oncology workforce in 2030: estimating physician training needs and proposing solutions to scale up capacity in low- and middle-income countries. Appl Radiat Oncol. 2019;8:10–6.
35. Gupta M, Singh K. Steps to improve cancer treatment facilities especially radiotherapy infrastructure in India. J Cancer Res Ther. 2022;18:1–4.
36. Maitre P, Krishnatry R, Chopra S, Gondhowiardjo S, Likonda BM, Hussain QM, Zubizarreta EH, Agarwal JP. Modern radiotherapy technology: obstacles and opportunities to access in low-and middle-income countries. JCO Glob Oncol. 2022;8:e2100376.
37. AERB safety code: radiation therapy sources, equipment and installations: Available from https://www.aerb.gov.in/images/PDF/CodesGuides/RadiationFacility/MedicalApplicationsInvolvingRadiation/4.pdf. Accessed 26 Nov 2022.
38. Government of India: Atomic Energy Regulatory Board: e-Licensing of Radiation Applications (eLORA) System. Available from https://elora.aerb.gov.in/ELORA/populateLoginAction.htm. Accessed 26 Nov 2022.

Further Reading

http://www-pub.iaea.org/MTCD/Publications/PDF/Pub1462_web.pdf. Accessed 16 Sept 2022.

https://www.aerb.gov.in/images/PDF/Annual_report/ar2020/AERB-Annual-Report-2020-english.pdf

https://www.youtube.com/watch?v=Fknim8cYNNc

https://www.youtube.com/watch?v=qCCl5sy7ixU

https://www-pub.iaea.org/MTCD/Publications/PDF/pub1296_web.pdf

Cancer Biology

3

Ann Christy Saju, Sangeeta Kakoti, Godhanjali Chekuri, and Jayant S. Goda

3.1 Origin of Cancer

One of the key differences that are noted between a normal cell and a malignant cell is the limitless replicative potential of tumors, while normal cells follow tissue homeostasis. The subsequent sections elaborate and highlight the distinction between the two (Fig. 3.1).

3.1.1 The Clonal Theory

The clonal theory of tumorigenesis states that tumors arise from the mitotic division of a single cell. This is made possible when a single cell acquiring one or more somatic mutations undergo cell division. A quintessential example of this being the Philadelphia chromosome in chronic myeloid leukemia. Understanding the clonality of human tumors provides the basis of tumor development, growth, and differentiation potential [1, 2].

3.1.2 Hallmarks of Cancer

There are more than one hundred types of cancer with multiple subtypes within. Do these cells share a common trait that differentiates them from a normal cell have been extensively explored. Hanahan et al. in the year 2000 attempted to assemble six major hallmarks of cancer that postulate a malignant growth. A decade later, with advancement in technology and expanding knowledge on the tumor microenvironment, two more emerging hallmarks were added. Although it is true that the future may invite a lot more observations, these hallmarks have endured the test of time (Table 3.1 and Fig. 3.2) [3].

3.1.3 Self-Sufficiency in Growth Signals

Normal cells require a mitogenic growth signal to divide and proliferate. The extracellular growth factors bind to transmembrane receptors, which in turn activate various intracellular circuits to achieve proliferation. In contrast, tumor cells develop autonomy toward these growth signals through various strategies. The three main tactics

A. C. Saju · S. Kakoti · J. S. Goda (✉)
Department of Radiation Oncology, Tata Memorial Centre, Parel, Mumbai, Maharashtra, India

Homi Bhaba National Institute, Trombay, Mumbai, Maharashtra, India
e-mail: jgoda@actrec.gov.in

G. Chekuri
Homi Bhaba National Institute, Trombay, Mumbai, Maharashtra, India

Clinical Biology Lab, Department of Radiation Oncology, Tata Memorial Centre, Kharghar, Navi Mumbai, Maharashtra, India

A. Munshi et al. (eds.), *Radiation Oncology – Principles, Precepts and Practice*,
https://doi.org/10.1007/978-981-97-8389-2_3

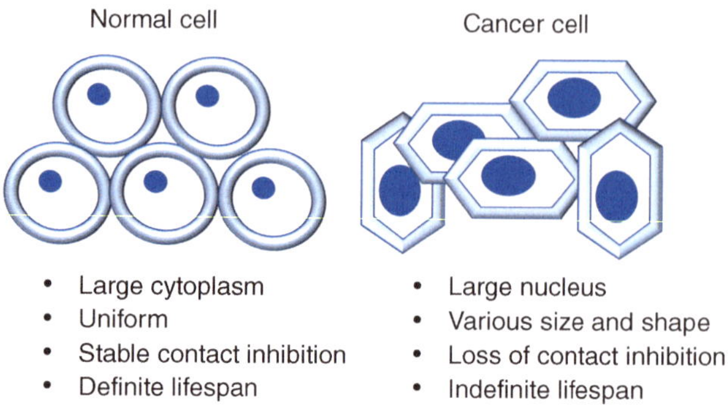

Fig. 3.1 Difference between normal cell and cancer cell

Fig. 3.2 Hallmarks of cancer

Table 3.1 Hallmarks of cancer

Time-tested hallmarks of cancer:	**Next-generation hallmarks**:
1. Self-sufficiency in growth signals 2. Insensitivity to growth-inhibitory signals 3. Evading apoptosis 4. Limitless replicative potential 5. Sustained angiogenesis 6. Tissue invasion and metastasis.	Reprogramming of energy metabolism Evading immune destruction **Enabling characteristics** Genomic instability and mutation Tumor-promoting inflammation

include alteration in the growth factors, alteration at transcellular receptors, or alteration at the level of intracellular circuits.

3.1.4 Insensitivity to Growth-Inhibitory Signals

Growth-inhibitory signals either force the cells to a quiescent state (G0) or induce cells to a postmitotic state associated with specific differentiation. The transition of cells through the cell cycle is governed by several proto-oncogenes and tumor suppressor genes like p53 and Rb gene at various cell cycle checkpoints. Mutations that cause its inactivation can lead to uninhibited cell proliferation.

3.1.5 Evading Apoptosis

Programmed cell death plays a salient role in the maintenance of normal tissue homeostasis, the balance of proliferation, and cell attrition. Acquired resistance to this is a hallmark is perhaps present in all tumors.

3.1.6 Limitless Replicative Potential

Normal cells possess an intrinsic program that limits the multiplication of cells after a certain number of divisions termed as senescence. Tumor cells escape senescence probably by multiple mechanisms including loss of tumor suppressor proteins and maintenance of telomeres. Telomeres are Chromatin segments at the chromosome ends that protect the terminal end from degradation. They become progressively shorter during cell division leading to the senescence of normal cells. Cancer cells express increased telomerase activity and subsequently repair telomeres leading to a lack of senescence.

3.1.7 Sustained Angiogenesis

Tumors cells appear to have an ability to sustain oxygen and nutrition supply by altering the angiogenic switch. The proangiogenic factors like vascular endothelial growth factor (VEGF) and platelet-derived growth factor (PDGF) outbalance the antiangiogenic factors like angiostatin and interferon-alpha. This unbalanced stimulus results in precocious capillary sprouting, excessive vessel branching, distorted enlarged vessels, and erratic blood flow.

3.1.8 Tissue Invasion and Metastasis

Tumor cells detach from the primary tumor and spread locally by invading into adjacent tissues, followed by intravasation into nearby blood and lymphatic vessels. After a transit through the lymphatic and hematological system, they adhere to endothelial cells and extravasate into the parenchyma of distant organs. Numerous altered cell–matrix interactions play a pivotal role in this process. The affected proteins include cell–cell adhesion molecules like immunoglobulins, cadherins (E-cadherin), and integrins. E-cadherin's function seems to be lost in many of the epithelial tumors, thus disturbing the cell–cell homeostasis [3, 4].

3.1.9 Reprograming of Energy Metabolism

Normal cells under aerobic conditions convert glucose to pyruvate and use glycolysis only in an anaerobic environment. Tumor cells reprogram their metabolism to favor glycolysis and this anomalous characteristic named as "Warburg effect" or "aerobic glycolysis" fuel the cancer cells.

3.1.10 Evading Immune Destruction

Immune surveillance is responsible for averting a vast amount of incipient cancer cells. Defective immunological monitoring by cytotoxic T lymphocytes, helper T cells, or natural killer cells, secretion of immunosuppressive agents by tumor cells, or evading surveillance by surface expression of altered antigens rendering tumor cells invisible could all be possible mechanisms whereby tumor cells evade their destruction.

3.1.11 Genomic Instability and Mutation

Mutations and epigenetic transformations like DNA methylation and histone modifications can trigger clonal expansions. These vary between tumor types but are certainly enabling characteristics associated with the hallmark features.

3.1.12 Tumor-Promoting Inflammation

Histological specimens of tumors show dense infiltration by inflammatory cells which were conventionally thought to be an attempt to eradicate tumor cells. Evolving evidence shows a paradoxical effect with innate immunity showing a propensity in promoting neoplastic growth by providing growth factors, bio-active molecules, and reactive oxygen species which are in turn mutagenic and enabling tumor progression [5].

3.1.13 Cancer Etiology

The International Agency for Research in Cancer (IARC) has identified and classified agents that are carcinogenic to humans into four main groups (Table 3.2) [6].

The etiologies of human cancers can be mainly categorized into chemical, biological, physical, dietary, and endogenous causes.

Tobacco has been identified as a carcinogen since the nineteenth century, and at least 72 measurable carcinogens have been found in cigarette smoke. The notable ones among them are the tobacco-specific nitrosamines, polycyclic amino hydrocarbons, e.g., benzo (a) pyrene and aromatic amines. There is sufficient evidence to establish that tobacco has a causal relationship in 10 or more cancers including but not limited to lung, oral cavity, pharynx, esophagus, larynx, pancreas, stomach, cervix, bladder, colon, and rectal cancers. The major mechanism of carcinogenesis by tobacco is through the formation of DNA adducts by its metabolic intermediaries.

Biological causes comprise viruses, bacteria, and fungal infections. Human papilloma virus (HPV) plays a crucial role in the development of cervical cancer. It has also been identified to have a causal relationship in the oropharynx, penile, anal, and bladder cancers. Other common viruses causing cancers comprise hepatitis B and C causing hepatocellular cancer, Epstein–Barr virus (EBV) causing Burkitt's lymphoma, nasopharyngeal carcinoma and other non-Hodgkin lymphomas, HHV-8 (human herpes virus-8) causing Kaposi sarcoma, human T-cell leukemia virus causing adult T-cell leukemia and lymphoma, and Merkel cell polyomavirus causing Merkel cell carcinoma.

Table 3.2 IARC classification of carcinogens [6]

Carcinogen groups		Agents identified
Group 1	Carcinogenic to humans	121 agents
Group 2A	Probably carcinogenic to humans	90 agents
Group 2B	Possibly carcinogenic to humans	322 agents
Group 3	Possibly carcinogenic to humans	498 agents

IARC identifies and provides an extensive list of chemical and environmental carcinogens like nickel, cadmium, benzene, nitrosamines, asbestos, etc.

Much of the evidence regarding ionizing radiation causing cancer comes from the victims of the atomic bomb explosion in Japan and the Chernobyl disaster. Epidemiological studies identified an increase in thyroid cancer and other solid tumors like breast, ovary, colon, and bladder cancers in these victims. Ionizing radiation has been theorized to show its effects through two main mechanisms. In direct mechanism, the secondary electrons produced by absorption of radiation cause double-stranded DNA breaks, whereas in indirect mechanism, secondary electron interactions produce free radicals which in turn cause DNA damage. UV (Ultraviolet) rays are also known to be a causative factor for skin cancers like melanoma.

Clustering of tumors in families led to the identification of many cancer susceptibility syndromes. The comprehension of these syndromes is ever-evolving and common ones include the Hereditary Breast and Ovarian Cancer (HBOC) syndrome with germline BRCA 1 and BRCA 2 mutations, Familial Adenomatous Polyposis (FAP) caused by germline APC (adenomatous polyposis coli) mutations causing colorectal cancers, and the Lynch or hereditary nonpolyposis colon cancer syndrome (HNPCC) also known to cause colorectal cancers.

Expanding knowledge regarding these etiologies provides a pathway for interventions aimed at preventing these tumors [6, 7].

3.1.14 Carcinogenesis

Carcinogenesis is a multistep process, comprising tumor initiation, tumor promotion, malignant conversion, and tumor progression (Fig. 3.3).

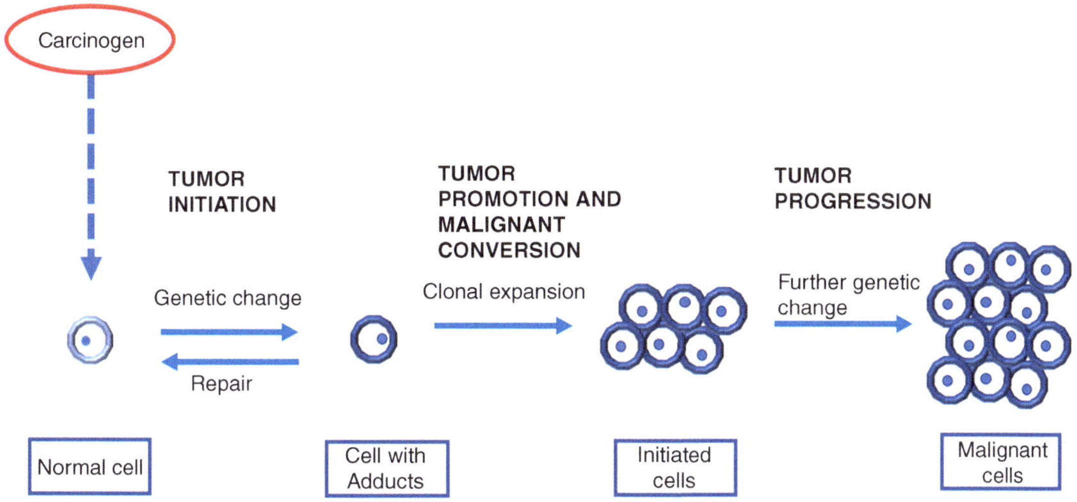

Fig. 3.3 Mechanism of action of carcinogenesis. Carcinogenesis is a multistage process which can be divided into three stages: initiation, promotion, and progression. Upon the carcinogen administration, there is an alteration or mutation in the genes which causes tumor initiation. During promotion stage, these carcinogen-initiated cells actively proliferate and form clusters. In progression stage, the cells undergo further mutations, and with the help of tumor-promoting substances, the cell proliferation rate increases, converting preneoplastic cells to cancer

3.1.15 Tumor Initiation

A carcinogen (chemical/biological/physical carcinogen) for example can form DNA adducts and result in accumulation of mutations causing either activation of a proto-oncogene or inactivation of a tumor suppressor gene resulting in the initiation of tumor growth.

3.1.16 Tumor Promotion

This constitutes a clonal expansion of initiated cells. Tumor promoters do not cause mutations on their own but mediate the biological effects of carcinogens. They are cocarcinogens that enhance the tumorigenicity of the carcinogen. For example, nicotine is not carcinogenic but acts as a tumor promoter in cigarette smoke.

3.1.17 Malignant Conversion

Preneoplastic cell conversion into malignant phenotype is further made possible by the accumulation of additional genetic.

3.1.18 Tumor Progression

Here the neoplastic cells acquire further aggressive characteristics which allow invasion and metastasis beyond the primary tumor site [8, 9].

3.2 Growth and Antigrowth Signals

Normal cells have proto-oncogenes which regulate biological tasks and function as growth factors, signal transducers, or nuclear transcription factors. Activating mutations of proto-oncogenes results in oncogenes which initiate tumor formation. The activating mutations could be point mutations (e.g., K-Ras in pancreatic, colon, and bladder cancers), amplifications (N-myc in neuroblastoma, HER2/neu in breast), or chromosome translocations (e.g., BCR-ABL in leukemias, C-Myc). Mutations in proto-oncogenes usually arise from somatic cells and are not inherited and mutation of a single allele is sufficient for its activation.

Tumor suppressor genes play a key role again in normal cell growth, their differentiation, and block tumorigenesis. Unlike proto-oncogenes, their mutations can be inherited, and both its alleles must be affected to result in tumor growth.

The archetypical examples of tumor suppressor genes are Rb (retinoblastoma) gene and the TP53 gene. The Rb gene acts as a cell cycle gatekeeper and decides cell cycle progression. Dephosphorylated Rb binds to EF2 (elongation factor 2) and prevents G1/S transition of the cell cycle. TP53, also known as the master regulator of the genome and regulator of the cell cycle, induces p21 to cause cell cycle arrest (G1 arrest) and repair DNA, cause apoptosis of damaged cells, and therefore maintain cellular and genomic stability. Mutation of p53 (Li–Fraumeni syndrome) is prevalent in various neoplasms including leukemias, breast carcinoma, soft tissue sarcomas, and brain tumors. Few other tumor suppressor genes include the BRCA gene (in breast and ovarian cancer), NF1 (neurofibromas), and APC (adenomatous polyposis coli causing colorectal cancers) [2, 9, 10].

3.3 Cell Cycle and Cancer

The continuation of human life is through a highly systematic and complex process of cell division. A hallmark attribute of a cancer cell is its potential for limitless replication. The reasons for this ability can be partly and largely decoded by anomalies in the cell cycle.

The longest phase of the mammalian cell cycle is the interphase comprising of G1, S, and G2 phases. The G1 phase (Gap 1 phase) is where the cell prepares for DNA replication. If the mitogenic stimulus is absent at this phase, the cell enters the G0 phase (quiescent phase). DNA replication occurs in the S phase, and in the G2 phase, the cell prepares for mitosis. In the M phase (Mitosis), the chromatin condenses to chromosomes and cells divide the genetic material and cell division occurs.

The progression through these phases is largely governed by cell cycle regulators which include the cyclins, the cyclin-dependent kinases (CDKs), and the cell cycle checkpoints. CDKs are crucial for cell cycle progression. They are a member of the Serine/Threonine kinases family and are activated after binding to specific proteins called cyclins. 20 or more CDKs have been recognized with CDK1,2,4 and 6 playing principal roles. Similarly, cyclins A to H has been described in mammals with each cyclin produced in distinct phases of the cell cycle. A specific combination of cyclins and CDKs are essential for cell cycle progression (Fig. 3.4).

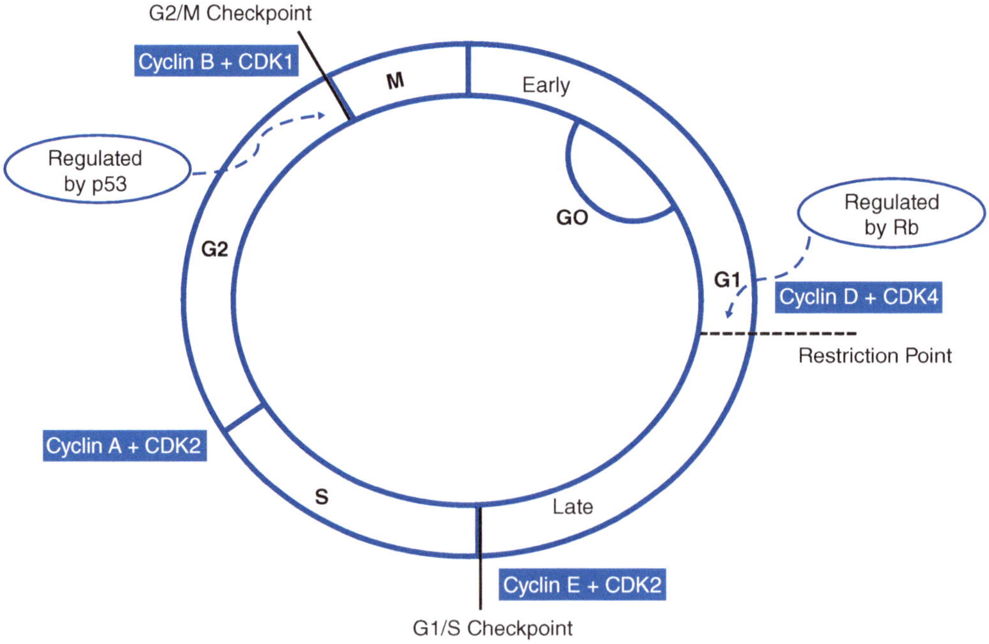

Fig. 3.4 Cell cycle. Interphase of cell cycle consists of G1 (gap phase 1), S (synthesis phase), G2 (gap phase 2), and G0 phase where cells enter upon mitogen stimulation. The cell cycle is controlled by cyclin-dependent kinases (CDKs), cyclins, p53, and Rb, and cells must pass through these checkpoints to enter into next phase

CDKs are inhibited by a specific group of proteins called CDK inhibitors (CKI). They belong to two classes, the Cip/Kip family components (p21, p27, p57) which are nonspecific, and the specific inhibitors are INK4 proteins (inhibitor of kinase 4/alternative reading frame) which include p16INK4a (blocks CDK4 and causes cell cycle arrest at late G1) and p14ARF (prevents feedback inhibition of p53).

Other vital components involved in cell cycle regulation are the checkpoints. They ensure the integrity of cell cycle events, including arresting the cell cycle at these points in case of damaged DNA. The principal ones are G1/S, S, and G2/M checkpoints (Fig. 3.4). A Few important regulators of these checkpoints include the p53 gene and Rb (retinoblastoma) gene. Hyperphosphorylated Rb releases E2F (Elongation factor 2) resulting in transcription of target genes necessary for S phase progression. Also, P53 is an important regulator of both the G1/S and G2/M phases.

Gain of function mutations in D-type cyclins (amplifications/translocations) and loss of function mutations in the p53 or Rb gene are commonly noted alterations in cell cycle metabolism seen in cancer cells. Several CDK-directed targeted therapies are evolving with a better understanding of these cell cycle regulators [11–13].

3.4 Apoptosis

Billions of cells produced daily in a human body should be counter-balanced with an equal amount of cellular death. *Apoptosis*, literally meaning falling off of leaves, is a process of programmed cell death, aiming at the same. Physiological processes happening as a result of apoptosis include amphibian metamorphosis, menstruation, etc. [14].

There are two different cellular pathways leading to apoptosis (Fig. 3.5) [15]:

1. Extrinsic pathway: stimulated when death ligands bind to a death receptor, e.g., TNF receptor 1, Fas (CD95), etc. This leads to formation of the death-inducing signaling complex (DISC), assembly and activation of procaspase-8 and finally cleavage and activation of downstream effector caspases.
2. Intrinsic pathway: activated by intracellular factors like nutrient deficiency, high cytosolic Ca^{2+}, DNA damaging agents, reactive oxygen species (ROS), some inflammatory cytokines, hypoxemia, misfolded proteins, etc. Consequently, there is increased mitochondrial permeability (via modulation of Permeability transition pores), release of cytochrome c, formation of *apoptosome* and finally activation of caspase-9.

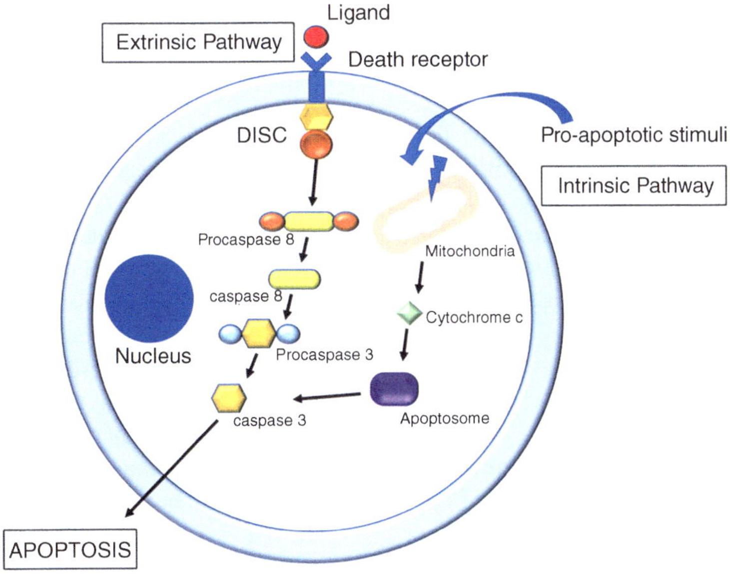

Fig. 3.5 Apoptosis is facilitated by intrinsic and extrinsic pathways. Intrinsic pathway is regulated by proapoptotic proteins (e.g., Bax, Bak) and mitochondria while the extrinsic pathways are activated by proapoptotic ligands (e.g., TRAIL) binding to the death receptors (e.g. TRAILRs). Both these pathways induce apoptosis through caspase cascades

Table 3.3 Anticancer drugs targeting apoptosis

Class of molecules	Drugs	Indications	Status of clinical use
Death receptor agonists [20]	Conatumumab (AMG 655), Dulanermin (rhApo2L/TRAIL), Mapatumumab, Lexatumumab, PRO95780, ABBV-621	Advanced solid tumors, AML and lymphoma	Phase I and II clinical trials
Selective BCL-2 inhibitors [21]	Venetoclax, S55746, Lisaftoclax (APG-2575)	CLL, AML, NHL	Phase I–III clinical trials
BH-3 mimetics [22]	Navitoclax, palcitoclax (APG-1252), Mirzotamab clezutoclax (ABBV-155), AMG-176	CLL, advanced solid tumors, lymphoma	Phase I–II clinical trials
Inhibitor of Apoptosis Proteins (IAP) antagonists [23]	LCL161, Birinapant	Myelofibrosis, advanced solid tumors, NHL	Phase I–II clinical trials

The morphological changes involved in apoptosis include shrinkage and separation from adjacent cells, condensation of the nuclear material (*pyknosis*), nuclear fragmentation (*karryorrhexis*), membrane blebbing, movement of the membrane lipid phosphatidylserine from the inner to the outer side of the plasma membrane (for recognition by phagocytes), and finally phagocytosis by a histiocyte or lysosome [16]. These changes are visible as a Ladder pattern in electrophoretic gel (a biochemical hallmark of apoptosis).

Apoptosis can be detected in vitro by mainly the following experiments [17]: TUNEL (terminal dUTP nick end labeling staining), Annexin V staining, DNA laddering, caspase activation, electron microscopy, or flow cytometry.

Factors regulating apoptosis mainly include [18]:

1. Proapoptotic proteins (e.g., Bax, Bak, Bad, Bcl-Xs, Bid, Bik, Bim, and Hrk).
2. Antiapoptotic proteins (e.g., Bcl-2, Bcl-XL, Bcl-W, Bfl-1, and Mcl-1).
3. Growth regulators, e.g., p53, Myc.
4. Miscellaneous other regulators (e.g., MCL-1, micro-RNAs namely miR15-a and miR-16-1).

3.4.1 Clinical Significance of Apoptosis [19]

Increased apoptosis leads to dysregulated cell death and is associated with various neurodegenerative disorders, hemolytic diseases, graft versus host rejections, and disease progression. On the other hand, suppression of apoptosis (usually by the inhibition of the intrinsic pathway) leads to uncontrolled cellular proliferation and finally development of a variety of malignant neoplasms. Tumors commonly arising as a result of perturbed apoptosis include follicular lymphomas, Wilm's tumor, etc. Apoptosis is also an important factor responsible for cell loss spontaneously or following cytotoxic chemotherapy for malignant tumors (Table 3.3).

3.5 Immortality and Senescence in Cancer

Cancer cells, unlike the traditional cells in our bodies, can grow forever. Neoplastic cell immortality ends up to massive tumors, metastatic spread, and potentially re-emergence. The ends of the chromosomes, which are called telomeres, keep track of cellular age. During each mitotic cycle, a little of telomere is lost; eventually, they become too short to guard the chromosomes and then the cell dies [24]. Cancers become immortal by reversing the traditional telomere shortening process and instead lengthen their telomeres. Telomerase activity is also known to be linked to cellular proliferation, and its activation seems to be a compulsory step in carcinogenesis; therefore, this cell immortalization is led by maintaining telomerase expression. The protein component of telomerase is a human telomerase reverse transcriptase, known as hTERT, which is lacking in somatic cells but expressed in immortalized cells [25].

Senescence of the cell could be a stable state of cell cycle arrest that is triggered in proliferative cells by multiple varieties of damage and stress. Senescent cells are known to usually acquire some structural changes as a result of stress-induced signaling mechanisms, including flattened, enlarged and aberrant morphologies with modified cytoplasmic compositions [26]. However, it's now recognized that senescence may be a dynamic multistep process. The senescence growth arrest is usually triggered by persistent DNA damage response (DDR) caused by either intrinsic or external stimulus. Cellular senescence is considered by implementation of senescence phenotypes referred to as senescence-associated secretory phenotype (SASP). The SASP consists of diverse proinflammatory cytokines, chemokines, and growth factors and might function in autocrine or paracrine manner exerting specific physiological or pathological effects. DDR induction is an extensively studied mechanism that triggers cellular senescence, once the DNA damage is identified, Ataxia-Telangiectasia mutated (ATM) and ATM- and Rad3-related (ATR) are recruited and activated which in turn activates downstream checkpoint kinase (CHK1 and CHK2) and effector protein p53 by arresting the cell in G1 or G2/M phase. Cellular senescence may be a defining feature of multiple forms of precancerous abrasions and responses to cancer therapies; thus, there is a need of more specific, evident, and targetable molecular biomarkers of cellular senescence [27].

3.6 Biological Principles of Cancer Metastasis and Invasion

The growth of secondary tumors in part of the body that is distant from the original primary cancer is termed as metastasis. The important threat and the reason for most cancer-related mortality is not the primary neoplasia but secondary tumors, the metastases. The event of metastases requires cancer cells to leave their primary location, circulate in the bloodstream, adapt to new cellular surroundings in a secondary location, and escape deadly conflict with immune cells. Therefore, activating invasion and metastasis is known to be a hallmark of cancer (Fig. 3.6) [28].

The invasion–metastasis cascade involves:

Step 1: Procurement of local invasiveness.
Step 2: Invasion of the cell into blood vessels (intravasation).
Step 3: Transport through the blood vessels to distant tissue locations.
Step 4: Escape of the cancer cells from circulation (extravasation).
Step 5: Ability to acclimatize to the local tissue environment and to proliferate.

3.6.1 Intravasation and EMT

Dissemination process is the initial step of invasion–metastasis cascade. During dissemination, cancer cells acquire ability to depart the primary tumor location to invade nearby tissues and travel to secondary tumor locations. The cascade is the consequence of chromosomal instability that is caused by continuous errors in chromosome segregation during mitosis. At this stage, epithelial–mesenchymal transition (EMT), a process during the detachment of cancer cells from the epithelial stratum by losing of their epithelial markers and gaining motility, has an important role on the cancer progression.

EMT involves the loss of cell–cell adhesion and therefore the polarized epithelial morphology through the characteristic loss of epithelial cell junctional proteins such as E-cadherin, claudins, and a subsequent increase in mesenchymal markers such as N-cadherin, vimentin, fibronectin, and cytoskeletal reorganization. Tumor cells, which have migratory and invasive feature, are more resistant to chemo- or radiotherapy than other cancer cells [29].

After escaping from the primary tumor, EMT, migration, and invasion through the basement membrane, tumor cells can directly enter by degrading the extracellular matrix by secretion of proteases like matrix metalloproteinase (MMP)

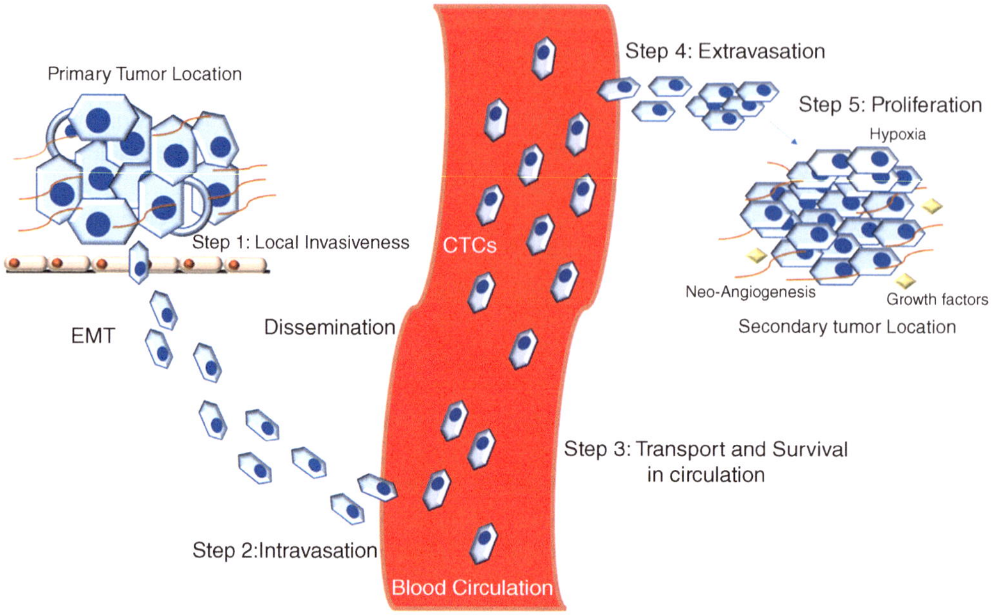

Fig. 3.6 The invasion–metastasis cascade. For the primary tumor cells to get motile and travel to a secondary location to grow, it needs to go through series of steps. The primary cells first are induced by EMT and escape the primary site and then intravasate into blood vessels. After they survive in the circulation, the CTCs extravasate and establish their cell growth in a distant location

and activation of the proteolytic urokinase type plasminogen activator system. During intravasation, tumor cells pass through tissue and reach the endothelial vessel. Intravasation is mediated actively or passively. This depends on the tumor type, microenvironment, and vasculature.

3.6.2 Circulation

When the cancer cells have achieved to intravasate into the blood vessels, they travel in the venous and arterial circulation, known as circulating tumor cells (CTCs). Less than 1% of CTCs flowing in the blood survive and have an opportunity to produce distant metastasis. Tumor cells protect themselves from physical stress and immune surveillance by secreting substances like thrombin, cathepsin B, cancer procoagulant, and MMP-2/-14 with the consequence of platelets aggregating around the tumor cells, a process called tumor cell-induced platelet aggregation [30].

3.6.3 Extravasation and Proliferation

For metastasis, CTCs leave the blood stream and adhere to endothelial cells (extravasation). Endothelial cells can either allow or block the adhesion of tumor cells, also possible transmigration. Therefore, endothelial cells are essential due to their role in the determination of secondary tumor location and regulation of metastatic formation [31].

Tumors are often intensely proliferating tissues and are, therefore, dependent on a sufficient connection to the vascular system. The term neoangiogenesis describes the process of tumor vascularization that is the formation of small capillaries. If a tumor does not possess sufficient vascularization, hypoxia occurs and induces the expression of several proangiogenic factors. The balance between pro and antiangiogenic factors such as VEGF, PDGFs, Glut-1, MMPs, ADAM12, and CCL2 regulates tumor vascularization. With the existence of advantageous premetastatic niche and angiogenic signals, clinically detectable metastases are the final result of highly complex invasion–metastasis cascade [32].

3.7 Evolving Therapeutics Arising out of Biological Basis of Cancer

3.7.1 Cancer Therapeutic Targets

Cancer cells display multiple layers of metabolic alterations affecting cell proliferation and its survival. Most types of targeted therapy help treat cancer by interfering with specific proteins or enzymes that contributes to its growth. Targeted drugs can block or turn off signals that make cancer cells grow, or can signal the cancer cells to destroy themselves. Because of their targeted action, these drugs have an effect on the cancer cells and mostly leave normal, healthy cells alone [33].

Most targeted therapies are either small-molecule drugs or monoclonal antibodies. Small molecule drugs are small enough to enter cells easily, so they are used for targets that are inside cells. Monoclonal antibodies, also known as therapeutic antibodies, are proteins produced in the lab. These proteins are designed to attach to specific targets found on cancer cells [34].

3.7.2 Cancer Immunology and Immunotherapy

Immune cells play an integral role in tumor biology, with defects in immunity contributing to carcinogenesis, cancer progression, and response to therapy. Harnessing the immune system using immunotherapy has demonstrated antitumor activity across a wide range of cancers. The immune system plays a major role in the surveillance against tumors. In order to avoid an attack by the immune system, tumor cells use a range of strategies, such as suppression of expression of tumor-associated antigens or of MHC class 1 molecules, or even counterattack against immune cells [35].

Currently, cancer immunotherapies are classified as active and passive treatments. Active treatments include vaccines designed to induce tumor cell recognition. Passive treatments, on the other hand, imply direct administration of antibodies and T cells, to the patient. Immune checkpoint inhibitors and adoptive T-cell therapy are among the most innovative approaches. The development of future treatments depends on finding effective immune-based biomarkers that can help to predict responses to treatment.

3.7.3 Novel Therapeutics Arising from Cancer Cell Biology Insights

Currently, a majority of cancer treatment strategies are based on the removal of tumor mass mainly by surgery. Chemical and physical treatments such as chemo- and radiotherapies have also made a major contribution in inhibiting rapid growth of malignant cells. Novel approaches such as ligand or receptor-based targeting, triggered release, intracellular drug targeting, gene delivery, cancer stem cell therapy, magnetic drug targeting, and ultrasound-mediated drug delivery have added new modalities for cancer treatment.

Tumor targeting ligands such as antibody, aptamer, siRNA, and peptides are being explored to target metastatic cells and block migration and invasion. In comparison to conventional chemotherapy which can cause high toxicity due to the lack of specificity, gene therapy provides a unique and powerful approach to combat cancer. Researchers have also used viruses to deliver therapeutic genes into cell nuclei because of high transfection efficiency, ability to penetrate, express and replicate in host cells. Cancer immunotherapy is an alternative treatment intended to activate the immune system to induce disease stabilization (Table 3.4) [36].

Table 3.4 Active cancer immunotherapies in present day

Cancer immunotherapy	Drugs (FDA-approved)	Mechanism	Indication
Immune checkpoint inhibitors	Nivolumab, Pembrolizumab, and Cemiplimab	PD-1 inhibitors	NSCLC, metastatic melanoma, HNSCC
	Atezolimumab, Durvalumab, and Avelumab	PDL-1 inhibitors	Bladder, MCC
	Ipilimumab	CTLA-4 inhibitor	Melanoma
Adoptive cell therapy (CAR-T cell therapies)	Kymriah (Tisagenlecleucel), Yescarta (Axicabtagene ciloleucel), Tecartus (Brexucabtagene auroleucel)	CD19 targetting	B-cell ALL, relapsed B cell lymphoma
Vaccines	Cervarix, Gardasil-4 and HEPLISAN-B (HBV)	Targets TSAs, TAAs, and viral related antigens	HPV-related cervical, vaginal, anal, head and neck cancers HBV-related hepatocellular carcinoma
Oncolytic viruses	T-VEC (Imlygic®)	Modified herpes simplex virus (HSV) that infects tumor cells	Melanoma lesion in skin and lymph nodes

3.8 Approaches to Biological Methods in Cancer

3.8.1 Histopathology and Immunohistochemistry (IHC)

Histopathology and cytopathology have been the main tools utilized in the diagnosis of cancer. These techniques have evolved from an era of diagnosis based on hematoxylin and eosin (H&E)-stained slides to the current regular evaluation of tumors by immunocytochemistry (IHC) to confirm tumor histogenesis and subtype.

In histology image analysis for cancer diagnosis, histopathologists visually examine the regularities of cell shapes and tissue distributions, decide whether tissue regions are cancerous, and determine the malignancy level. Common treatments employ a series of reagents to process the fixed tissue and embed it in a stable medium such as paraffin wax, plastic or resin. Such treatments include the main steps of dehydration, clearing, infiltration, and embedding [37].

Immunohistochemistry (IHC) is a method for detecting antigens or haptens in cells of a tissue section by exploiting the principle of antibodies binding specifically to antigens in biological tissues. The antibody–antigen binding can be visualized in different manners. Enzymes, such as horseradish peroxidase (HRP) or alkaline phosphatase (AP), are commonly used to catalyze a color-producing reaction [38].

3.8.2 Liquid Biopsy

In the recent years, research in oncology has focused on liquid biopsies, which rely on the detection of cancer-derived components, which include circulating tumor cells (CTCs), circulating tumor DNA (ctDNA), RNA, extracellular vesicles (EVs), and tumor educated platelets (TEPs), in the biofluids of patients, providing genomic, epigenetic, transcriptomic, and proteomic information about tumors and metastatic sites (Fig. 3.7) [39].

3.8.3 Flow Cytometry

Flow cytometry (FCM) is a multiparameter cell analysis technique used for detecting and measuring the physical and chemical properties of a particular cell population, including cell count,

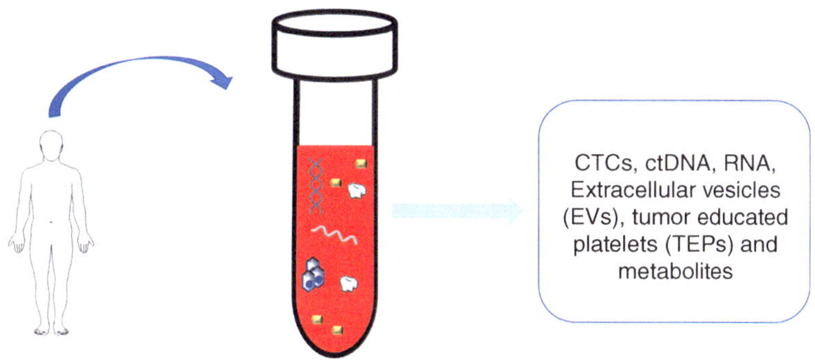

Fig. 3.7 Liquid biopsy can be used to acquire information of the components from primary tumor cells and also circulating tumor cells (CTCs). Liquid biopsy can assist in early stage disease screening, is a minimal invasive procedure, can assist in recurrence monitoring, and is a cheap method

cell size, and cell cycle. While the method has been routinely used for basic cell biological research, it also plays a vital role in the diagnosis, treatment, and management of certain bone and soft tissue tumors and blood cancers such as leukemia, lymphoma, and myeloma.

The presence of tumor markers, such as antigens, on the surface of the cells is also measured. The cells are stained with a light-sensitive dye, placed in a fluid, and then passed one at a time through a beam of light. The measurements are based on how the stained cells react to the beam of light [40].

3.8.4 Fluorescence In Situ Hybridization (FISH)

Fluorescence in situ hybridization is a technique useful in detecting various abnormalities, such as fusion of genes and genomic gains and deletions. In FISH testing, pieces of single-strand DNA (called DNA probes) are sent to find corresponding stretches of DNA from tumor cells sampled from the body. The probes, marked with a fluorescent dye, attach to those corresponding cells—that process is called hybridization. Once the probes attach to the chromosomes in the tumor cells, they glow when hit with fluorescent light, clearly showing where the probes match to the cancer cells, and where any chromosomal differences are [41].

3.8.5 Next-Generation Sequencing (NGS)

Next-generation sequencing (NGS) is one of the most significant technological advances in the biological sciences of the last 30 years. NGS has been widely implemented for de novo genome sequencing, DNA resequencing, transcriptome sequencing, and epigenomics. As cancer is a genetic disease driven by heritable or somatic mutations, new DNA sequencing technologies will have a significant impact on the detection, management, and treatment of disease. NGS method can be used for genetic testing within genes linked to familial cancer syndromes or for detection of mutations in genes of therapeutic importance within cancer cells or tissues [42].

3.9 Future of Cancer Biology

3.9.1 Challenges and Opportunities

In cancer, early tumor detection with noninvasive imaging such as CT, MRI, and PET scans known as radiomics is proving increasingly promising for discriminating between indolent and aggressive tumors. Challenges that need to be concentrated are improving access to resources for researchers, developing more predictive preclinical models of cancer, and using genomics and systems biology more effectively. Cancer biology must continue to move from the discovery of essential biological mechanisms that are needed to understand cancer to the translation of these results to the clinic [43].

3.9.2 Integrating Technology with Biology

Technologies and innovations like CRISPR, artificial intelligence, telehealth, the Infinium Assay, cryo-electron microscopy, and robotic surgery are helping accelerate progress against cancer. Cancer systems biology has the potential to usher in an era in which the effects of molecular aberrations and interactions within networks are integrated with molecular knowledge and pharmacogenomics. Systems biology can help to increase therapy efficacy or bypass the emergence of resistance, thus converting the current (often short term) effects of targeted therapies into durable responses, ultimately to improve quality of life and provide a cure [44].

3.10 Summary of the Chapter

- Cancer is a spectrum of disease that arises from a single somatic cell mutation leading to unlimited cell replication.
- With expanding knowledge and technology, the hallmarks of cancer are evolving over time.
- Carcinogenesis is multifactorial, and the etiologies of human cancers can be mainly categorized into chemical, biological, physical, dietary, and endogenous causes.
- Carcinogenesis is a multistage process comprising of tumor initiation, tumor promotion, malignant conversion, and tumor progression.
- Cancer growth is due to acquisition of deleterious mutations in proto-oncogenes, tumor suppressor genes, and DNA-repair genes.
- The progression of cell through each phase of cell cycle is governed by certain proteins called checkpoints which help in regulating the integrity of the cell. These checkpoints are disabled in cancer cells.
- Apoptosis is a process of programmed cell death which follows an extrinsic and intrinsic pathway. Suppression of apoptosis can lead to cellular proliferation and formation of malignant neoplasms.
- Normal cells have a finite number of cell division because of the presence of telomeres which shorten with every division, while cancer cells maintain the length of telomeres allowing them to replicate unlimited times making them immortal.
- For the cancer cell to metastasize, it goes through a series of steps which include local invasion, intravasation into blood vessels, cellular transport and survival in blood circulation, escape from the circulation, accumulation, and proliferation at the secondary site.
- Immune cells play an integral role in tumor biology, with defects in immunity contributing to carcinogenesis, cancer progression, and also response to antineoplastic therapy.
- Biological methods for detection of cancer involve liquid biopsy, histopathology, flow cytometry, next-generation sequencing, etc.

Multiple Choice Questions

1. Which of the following is not a hallmark of cancer?
   (a) Sustaining proliferative signaling
   (b) Evading growth suppressors
   (c) Sensitive stress response
   (d) Inducing VEGF and PDGF
2. Which of the following is NOT true for cancer cells?
   (a) Large amount of telomeres
   (b) Sensitive to external growth signals
   (c) Division in presence of checkpoint signals
   (d) Induction of proto-oncogene
3. Oncogenes are cancer causing genes in the cells but they do not express usually because of
   (a) Tumor suppressor genes
   (b) Proto oncogenes
   (c) Tumor promoting genes
   (d) Transposons

4. Level of cyclin A is highest during which cell cycle phase?
   (a) Early G1 phase
   (b) S phase
   (c) Late G2 phase
   (d) Late S phase
5. Function of Rb gene in cancer?
   (a) Differentiation
   (b) Apoptosis
   (c) Promotes cell cycle
   (d) All of the above
6. Cancer cells can produce immunosuppressive cytokines such as.
   (a) IL-4 and IL-5
   (b) IL-5 and IL-13
   (c) IL-10 and IL-12
   (d) IL-10 and TGF-β
7. An example of an oncolytic virus is:
   (a) Epstein–Barr virus
   (b) HIV-2
   (c) Herpes simplex virus
   (d) HBV
8. CD44 (Cancer stem cell) is a molecule involved in:
   (a) Cell adhesion
   (b) Antigen recognition
   (c) Metastasis
   (d) Apoptosis
9. Which of the following is a noninvasive diagnostic tool of cancer?
   (a) Liquid biopsy
   (b) Fine needle aspiration biopsy
   (c) Image-guided biopsy
   (d) Vacuum-assisted biopsy
10. Which of the following is correct sequence of invasion–metastatic cascade?
    A. Extravasation
    B. Proliferation
    C. EMT
    D. Intravasation
   (a) A-B-C-D
   (b) C-D-B-A
   (c) C-D-A-B
   (d) A-C-D-B

Answers

1. (c)
2. (d)
3. (a)
4. (d)
5. (d)
6. (d)
7. (c)
8. (c)
9. (a)
10. (c)

References

1. Greaves M, Maley CC. Clonal evolution in cancer. Nature. 2012;481(7381):306–13. https://doi.org/10.1038/nature10762.
2. Kontomanolis EN, Koutras A, Syllaios A, et al. Role of oncogenes and tumor-suppressor genes in carcinogenesis: a review. Anticancer Res. 2020;40(11):6009–15. https://doi.org/10.21873/anticanres.14622.
3. Hanahan D, Weinberg RA, Francisco S. The hallmarks of cancer. Cell. 2000;100:57–70.
4. Fouad YA, Aanei C. Revisiting the hallmarks of cancer. Am J Cancer Res. 2017;7(5):1016–36.
5. Hanahan D, Weinberg RA. Review hallmarks of cancer: the next generation. Cell. 2011;144(5):646–74. https://doi.org/10.1016/j.cell.2011.02.013.
6. IARC Monographs on the identification of carcinogenic hazards to humans. https://monographs.iarc.who.int/agents-classified-by-the-iarc/. Accessed 20 Jan 2012.
7. Cogliano VJ, Baan R, Straif K, et al. Preventable exposures associated with human cancers. J Natl Cancer Inst. 2010;103(24):1827–39. https://doi.org/10.1093/jnci/djr483.
8. Hennings H, Glick AB, Yuspa H. Critical aspects of initiation, promotion, and progression in multistage epidermal carcinogenesis. PSEBM. 1993;3:1–8.
9. Stanley LA. Molecular aspects of chemical carcinogenesis: the roles of oncogenes and tumour suppressor genes. Toxicology. 1995;96(3):173–94. https://doi.org/10.1016/0300-483X(94)02991-3.
10. Lehman TA, Reddel R, Pfeifer AMA, et al. Oncogenes and tumor-suppressor genes. Environ Health Perspect. 2014;93:133–44.
11. Garrett MD. Cell cycle control and cancer. Curr Sci. 2001;81(5):515–22. https://doi.org/10.1007/978-1-4899-3111-5_7.

12. Malumbres M, Barbacid M. Cell cycle, CDKs and cancer: a changing paradigm. Nat Rev Cancer. 2009;9(3):153–66. https://doi.org/10.1038/nrc2602.
13. Barnum KJ, O'Connell MJ. Cell cycle regulation by checkpoints. In: Cell Cycle Control: Mechanisms and Protocols, Methods in Molecular Biology. 2014; 1170:477–499. doi:https://doi.org/10.1007/978-1-4939-0888-2.
14. Wyllie AH, Kerr JFR, Currie AR. Cell death: the significance of apoptosis. Int Rev Cytol. 1980;68(C):251–306. https://doi.org/10.1016/S0074-7696(08)62312-8.
15. Xu G, Shi Y. Apoptosis signaling pathways and lymphocyte homeostasis. Cell Res. 2007;17(9):759–71. https://doi.org/10.1038/cr.2007.52.
16. Kerr JF, Wyllie AH, Currie AR. Apoptosis: a basic biological phenomenon with wide-ranging implications in tissue kinetics. Br J Cancer. 1972;26(4):239–57.
17. Bastian AM, Yogesh TL, Kumaraswamy KL. Various methods available for detection of apoptotic cells—a review. Indian J Cancer. 2013;50(3):274. https://doi.org/10.4103/0019-509X.118720.
18. Apoptosis Regulators | Request PDF. https://www.researchgate.net/publication/8886370_Apoptosis_Regulators. Accessed 26 Sept 2022.
19. Zangemeister-Wittke U, Simon HU. Apoptosis—regulation and clinical implications. Cell Death Differ. 2001;8(5):537–44. https://doi.org/10.1038/SJ.CDD.4400844.
20. Wiezorek J, Holland P, Graves J. Death receptor agonists as a targeted therapy for cancer. Clin Cancer Res. 2010;16(6):1701–8. https://doi.org/10.1158/1078-0432.CCR-09-1692/347575/P/DEATH-RECEPTOR-AGONISTS-AS-A-TARGETED-THERAPY-FOR.
21. Roberts AW. Therapeutic development and current uses of BCL-2 inhibition. Hematol Am Soc Hematol Educ Program. 2020;2020(1):1. https://doi.org/10.1182/HEMATOLOGY.2020000154.
22. Diepstraten ST, Anderson MA, Czabotar PE, Lessene G, Strasser A, Kelly GL. The manipulation of apoptosis for cancer therapy using BH3-mimetic drugs. Nat Rev Cancer. 2021;22(1):45–64. https://doi.org/10.1038/s41568-021-00407-4.
23. Infante JR, Dees EC, Olszanski AJ, et al. Phase i dose-escalation study of LCL161, an oral inhibitorof apoptosis proteins inhibitor, in patients with advanced solid tumors. J Clin Oncol. 2014;32(28):3103–10. https://doi.org/10.1200/JCO.2013.52.3993.
24. How do cancer cells achieve immortality? https://www.jax.org/news-and-insights/2018/june/how-do-cancer-cells-achieve-immortality. Accessed 24 Sept 2022.
25. Zipursky A. Telomerase, immortality, and cancer. Pediatr Res. 2000;47(2):174. https://doi.org/10.1203/00006450-200002000-00002.
26. Ou HL, Hoffmann R, Gonz Alez-Lopez C, Doherty GJ, Korkola JE, Munoz-Espin D. Cellular senescence in cancer: from mechanisms to detection. doi:https://doi.org/10.1002/1878-0261.12807.
27. Herranz N, Gil J. Mechanisms and functions of cellular senescence. J Clin Invest. 2018;128(4):1238–46. https://doi.org/10.1172/JCI95148.
28. Calibasi-Kocal G, Basbinar Y. Introductory chapter: cancer metastasis. Cancer Metastasis . Published online November. 2018;6 https://doi.org/10.5772/INTECHOPEN.81914.
29. Leber MF, Efferth T. Molecular principles of cancer invasion and metastasis (review). Int J Oncol. 2009;34(4):881–95. https://doi.org/10.3892/ijo_00000214.
30. Jiang WG, Sanders AJ, Katoh M, et al. Tissue invasion and metastasis: molecular, biological and clinical perspectives. Semin Cancer Biol. 2015;35:S244–75. https://doi.org/10.1016/J.SEMCANCER.2015.03.008.
31. Fares J, Fares MY, Khachfe HH, Salhab HA, Fares Y. Molecular principles of metastasis: a hallmark of cancer revisited. doi:https://doi.org/10.1038/s41392-020-0134-x.
32. Pachmayr E, Treese C, Stein U. Underlying mechanisms for distant metastasis—molecular biology. Visc Med. 2017;33(1):11–20. https://doi.org/10.1159/000454696.
33. Cheong H, Lu C, Lindsten T, Thompson CB. Therapeutic targets in cancer cell metabolism and autophagy. Nat Biotechnol. 2012;30(7):671–8. https://doi.org/10.1038/NBT.2285.
34. Targeted Therapy for Cancer—NCI. https://www.cancer.gov/about-cancer/treatment/types/targeted-therapies. Accessed 25 Sept 2022.
35. Ren X. Cancer immunology and immunotherapy. Cancer Biol Med. 2021;18(4):931. https://doi.org/10.20892/J.ISSN.2095-3941.2021.0036.
36. Mitra AK, Agrahari V, Mandal A, et al. Novel delivery approaches for cancer therapeutics. J Control Release. 2015;219:248. https://doi.org/10.1016/J.JCONREL.2015.09.067.
37. He L, Long LR, Antani S, Thoma GR. Histology image analysis for carcinoma detection and grading. https://doi.org/10.1016/j.cmpb.
38. Immunohistochemistry Principle. How IHC staining Works. https://www.bosterbio.com/protocol-and-troubleshooting/immunohistochemistry-ihc-principle. Accessed 25 Sept 2022.
39. Martins I, Ribeiro IP, Jorge J, et al. Liquid biopsies: applications for cancer diagnosis and monitoring. Published online 2021. doi:https://doi.org/10.3390/genes12030349.
40. Definition of flow cytometry—NCI Dictionary of Cancer Terms—NCI. https://www.cancer.gov/publications/dictionaries/cancer-terms/def/flow-cytometry. Accessed 25 Sept 2022.
41. Chrzanowska NM, Kowalewski J, Lewandowska MA. Use of fluorescence in situ hybridization (FISH) in diagnosis and tailored therapies in solid tumors. Molecules. 2020;25:8. https://doi.org/10.3390/MOLECULES25081864.

42. Meldrum C, Doyle MA, Tothill RW. Next-generation sequencing for cancer diagnostics: a practical perspective. Clin Biochem Rev. 2011;32(4):177. Accessed 25 Sept 2022 /pmc/articles/PMC3219767/
43. Cagan R, Meyer P. Rethinking cancer: current challenges and opportunities in cancer research. Dis Model Mech. 2017;10(4):349. https://doi.org/10.1242/DMM.030007.
44. Werner HMJ, Mills GB, Ram PT. Cancer systems biology: a peak into the future of patient care? Nat Rev Clin Oncol. 2014;11(3):167. https://doi.org/10.1038/NRCLINONC.2014.6.

Further Reading

https://www.jove.com/science-education/10987/cancer
https://www.youtube.com/watch?v=rrMq8uA_6iA
Oxford Textbook of Cancer Biology. (2019). *Oxford Textbook of Cancer Biology*. https://doi.org/10.1093/MED/9780198779452.001.0001
Schirmer EC, de las Heras JI (Eds.). Cancer biology and the nuclear envelope. 2014;773. https://doi.org/10.1007/978-1-4899-8032-8
Wong RS. Apoptosis in cancer: from pathogenesis to treatment. J Exp Clin Cancer Res. 2011;30:87. https://doi.org/10.1186/1756-9966-30-87.

Radiation Biology

4

Manur Gururajachar Janaki, Anusheel Munshi, Soumyajit Roy, Nikhil Joshi, and Bidhu K. Mohanti

Preamble

Radiation physics and radiobiology are the two pillars on which the success of radiation therapy stands. While radiation physics deals with differential delivery of dose to the tumor as opposed to normal tissues, radiobiology helps us to understand how a tumor is different from normal tissues in terms of cellular biology, response to radiation, repair of the damage, and so on. In this chapter, we will understand various concepts of radiobiology with their clinical application. The reader is suggested to use the online links for further in-depth understanding.

M. G. Janaki (✉)
Department of Radiation Oncology, Ramaiah Medical College, Bengaluru, Karnataka, India
e-mail: janakimg.rmc@msruas.ac.in

A. Munshi
Department of Radiation Oncology, IOCI, Fortis Noida, Uttar Pradesh, India

S. Roy
Department of Radiation Oncology, UH Seidman Cancer Center, Case Western Reserve University, Cleveland, OH, USA

N. Joshi
Department of Radiation Oncology, Rush University Medical Center, Chicago, IL, USA

B. K. Mohanti
Academic and Head of Radiation Oncology, Bagchi Sri Shankara Cancer Centre & Research Institute, Bhubaneswar, Odisha, India

4.1 Introduction

Radiobiology is the branch of science that deals with the study of the action of ionizing radiation on living things. The main goal of a radiation oncologist is to deliver the maximum dose to the tumor while minimizing the dose to the surrounding normal structures, which has been achieved with advances in technology and the use of varying fractionation schedules. Once the radiation is absorbed in the tissues, a series of events happen, resulting in either tumor response or radiation sequelae which is understood by various concepts. With advances in radiobiology research, many earlier concepts of radiobiology have become obsolete. We will try to understand the earlier as well as the newer concepts and how they can be implemented in clinical practice for better patient care.

4.2 History of Radiobiology

Three decades ago, Hall and Denekemp wrote, "to attain the objective of educating physicians to be skillful in the practice of radiation oncology, a solid grounding in radiation biology and cancer biology is essential. Without the inclusion of the basic sciences, the clinical practice of radiation oncology would be sterile and could not move forward" [1]. Biologic research is as old as the clinical practice of radiation oncology. A few

A. Munshi et al. (eds.), *Radiation Oncology – Principles, Precepts and Practice*,
https://doi.org/10.1007/978-981-97-8389-2_4

years after the discovery of X-rays by Roentgen in 1895, radiation dose delivered to human body and to the animals (ram testicle) was studied to show cellular radiosensitivity. Thus, the study of radiobiology and its contributions to clinical practice has several milestones for the last one hundred years and more. In his characteristic altruism, Fowler described his own journey, "When I came into radiotherapy in 1950, I was puzzled that some patients were treated to 3000 rads (cGy) in 3 weeks but others received 4000 in 5 or 6000 in 6 weeks. When I asked why, there were no convincing answers given" [2].

The DNA damage of mitotic cells as the basis of tumor cell-kill vis-à-vis normal tissue tolerance and recovery emerged as the scientific clinical foundation soon after the discovery of DNA in 1953. This has led to our key understanding of clinical radiobiology from the 1970s onward: 4 Rs/5 Rs; dose-fraction schedules; modeling and calculation of dose prescription (BED, TDF, L-Q, etc.); fractional equivalent dose for tumor radiosensitivity and normal tissue damage (FED, EUD, EQD2, etc.). Such information has become vital for improved precision in our daily practice of radiotherapy, be it conventional fractionation, hypofractionation, stereotactic radiotherapy (SRS, SRT, SBRT, etc.). For example, the dose per fraction in external beam radiotherapy and brachytherapy is now prescribed taking into consideration several parameters: (a) Organ or anatomic site (brain, head and neck, lung, breast, stomach, uterus, urinary bladder, rectum, limb, etc.); (b) adjacent normal structures (brain stem, lens, parotid, heart, kidney, spine, gonads, joint-space, etc.); (c) Tumor type (astrocytoma, carcinoma, sarcoma, lymphoma, etc.) [3]. Even though the progress and contribution of radiobiology are stellar in many ways, newer techniques of radiotherapy and the progress in cancer biology such as cancer genomics, molecular pathology, tumor microenvironment, carcinogenesis would require the repurposing of radiobiology curriculum. The enhanced approach to cancer biology can solve some of the challenges facing the radiation oncology community. The teaching and learning of radiobiology need to be integrated with cancer biology as shown in Fig. 4.1 [4]. More importantly, as cancer has evolved from an acute to a chronic dis-

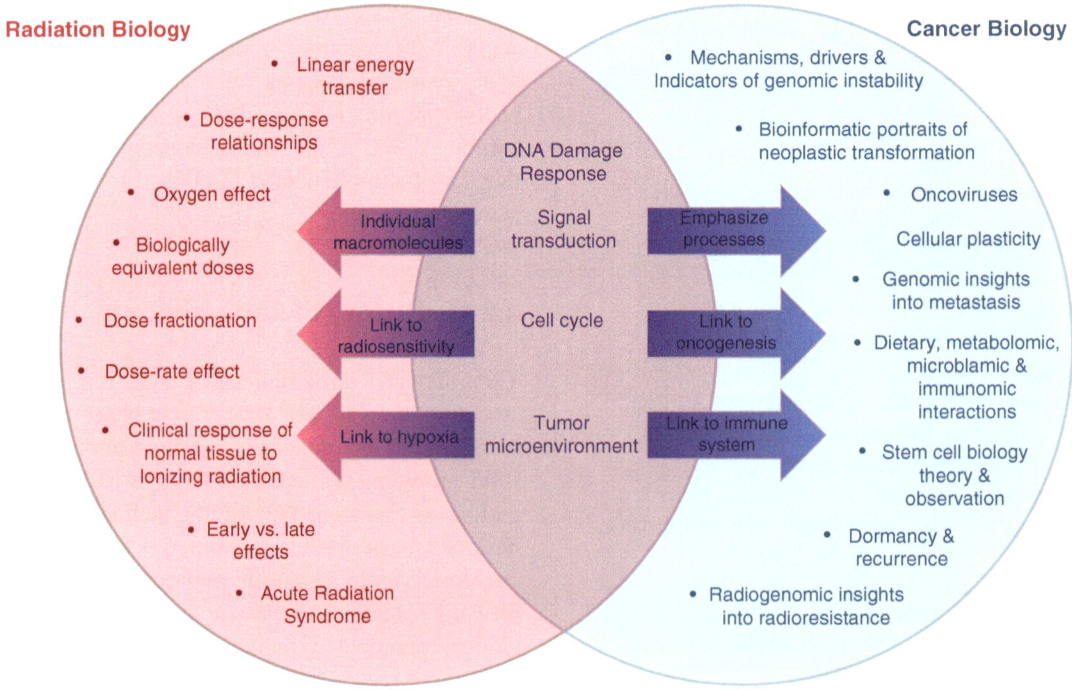

Fig. 4.1 Traditional radiobiology topics integrated with cancer biology topics [Ref. Vijayakumar S et al. doi: 10.7759/cureus.20662. open access journal]

ease (with longer survival for many cancers), radiation oncology is expected to maintain its role as a noninvasive, highly effective treatment modality that may be focused and strengthened in combination with surgery, systemic chemotherapy and immunotherapy. The horizon and engagement of radiobiology demands that residents, trainees, researchers and clinicians keep pace with all basic and translational progress.

4.2.1 Radiation-Induced Cellular Effects

Radiation can act by direct or indirect means. Direct ionization happens when charged particles (e.g., electrons, protons, and alpha particles) interact with cellular atoms or molecules to create free radicals. The process is known as direct because the interaction occurs directly between a particle and a cellular component. There is no intermediary step. Indirect ionization occurs when noncharged particles (e.g., photons) interact with water in the cell. The energy absorbed by an H_2O molecule forms ion pairs and reactive oxygen metabolites (hydroxyl radicals). These free radicals then interact with cellular atoms and molecules and damage cellular proteins in the process. This process is indirect because there is an intermediate step of H_2O-based free radical formation, as shown in Fig. 4.2. Once the cell has been irradiated, there is a chain of physical, chemical, and biological processes that are set into motion, as shown in Fig. 4.3.

Possible futures of any irradiated cell are that either there could be no effect at all or any one of the following can happen. Delay in division, apoptosis, reproductive failure, genomic instability, mutation, transformation, bystander effects, and adaptive responses. Cell death of static (nonprolif-

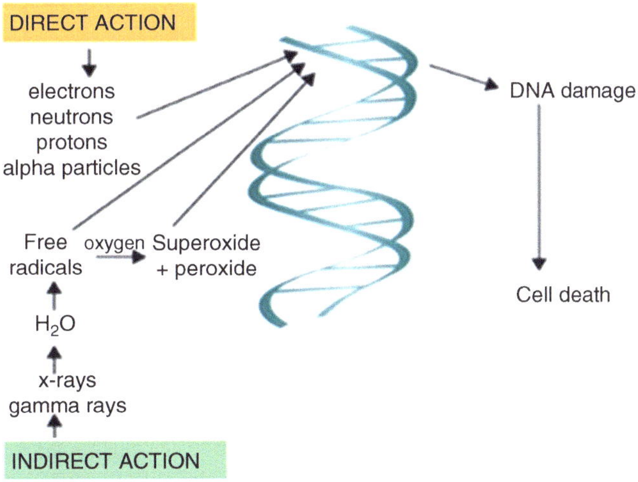

Fig. 4.2 Direct and indirect effects of radiation

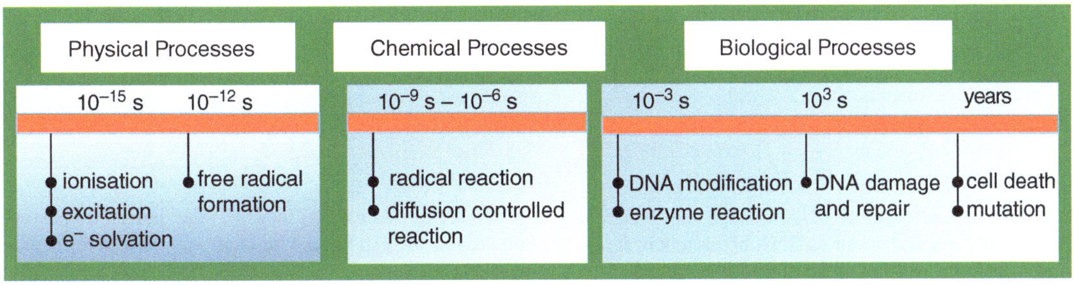

Fig. 4.3 Chain of events representing the interaction of ionizing radiation with tissue

erating) cells is defined as the loss of a specific function of the cell. In contrast, for stem cells and other cells capable of many divisions, it is defined as the loss of replicative power or reproductive integrity. The critical aim of treating malignancies with radiation therapy is to deprive tumor cells of their reproductive potential. The significant pathways by which radiation achieves this goal are apoptosis, mitotic catastrophe, and senescence. Induction of apoptosis alone is insufficient to account for the therapeutic effect of radiotherapy. The proliferative capacity of malignant cells following irradiation, especially with solid tumors, can occur via alternative cell death modalities or permanent cell cycle arrests, i.e., senescence.

4.2.1.1 Clinical Implications

1. In the clinically used X-ray/gamma rays range, two-third of DNA damage is due to indirect effects, and one-third is due to direct effects.
2. Neutrons/protons/alpha particles are densely ionizing, causing events close to each other and hence cause more direct DNA damage.

4.2.2 Radiation Effects on DNA

DNA is a double helical structure with two strands, each having a series of deoxynucleotides. When exposed to radiation, three types of breaks in the strands are observed: base damage, single-strand breaks (SSB), and double-strand breaks (DSB). Following these types of DNA damages, the single-strand and double-strand breaks are well-separated in their respective repair processes generally, as shown in Fig. 4.4. The double-strand break occurring close

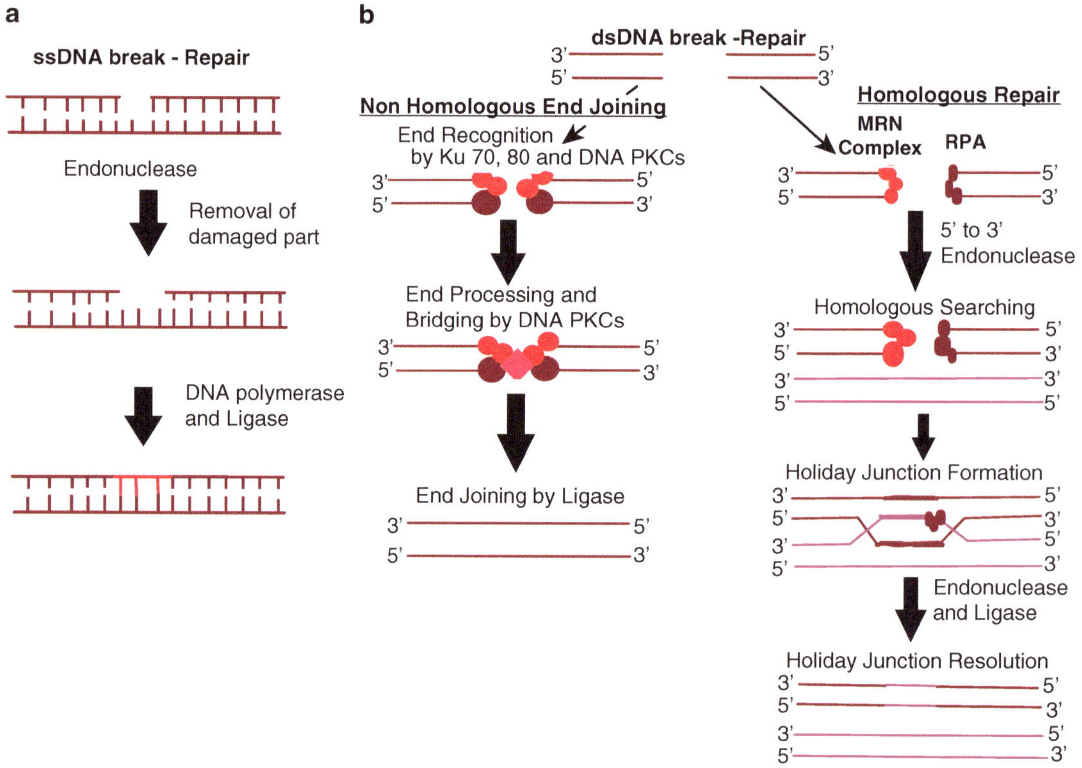

Fig. 4.4 Various types of DNA damage and their repair. (**a**) showing mechanism of ss DNA break repair which is simple and involves the processing of the damaged ends, synthesis of new strand using the undamaged strand as template and ligation of newly formed strand with remaining DNA. (**b**) showing two mechanisms of ds DNA break repair. Nonhomologous end joining (NHEJ) is an error-prone repair mechanism in which the damaged ends are processed, the gap filled with DNA and the ligase enzyme then binds the two broken ends. Homologous repair (HR) is an error free mechanism which takes a longer duration of about 6 h or more, and the main feature is the use of an undamaged sister chromatid sequence (Pink) as template for repair

to each other is lethal to the cell. Approximately 1–2 Gy of radiation will cause more than 1000 base damages, 1000 SSB, and 40 DSB.

Radiation damage is of three types.

1. Lethal damage (LD), which is irreversible and irreparable. It invariably leads to cell death.
2. Sublethal damage (SLD), which can be repaired in hours. However, if additional sublethal damage is added, it will eventually lead to lethal damage.
3. Potentially lethal damage (PLD) can be altered by modifying the environment when the cells are allowed to remain in a nondividing state.

4.3 Cell Kill and Cell Survival

A cell survival curve describes the relationship between the surviving fraction of cells and the absorbed dose. The surviving fraction (SF) is the fraction of irradiated cells that maintain their reproductive integrity. Cell survival is graphically represented by plotting the SF on a logarithmic scale on the ordinate against dose on a linear scale on the abscissa, as shown in Fig. 4.5. The type of radiation used significantly influences the shape of the cell survival curve. Densely ionizing radiations exhibit a cell survival curve that is a near-exponential function of dose. This is represented by an almost straight line on the log-linear plot. In contrast, for sparsely ionizing radiation, the curves show an initial slope followed by a shoulder region. Subsequently, the line becomes nearly straight at higher doses, as seen in Fig. 4.5. With every fraction, the shoulder of the cell survival curve repeats representing the repair of the sublethal damage from the initial fraction and the repair is more in acute responding compared to late responding tissues as shown in Fig. 4.5.

In the present times, the linear-quadratic model is most often used to describe the cell survival curve. This model assumes that there are two components to cell kill by radiation where S(D) is the fraction of cells surviving a dose D; α is a constant that describes the initial slope of the cell survival curve, depicting the linear kill, and β is a constant that describes the final slope representing the quadratic component of cell killing.

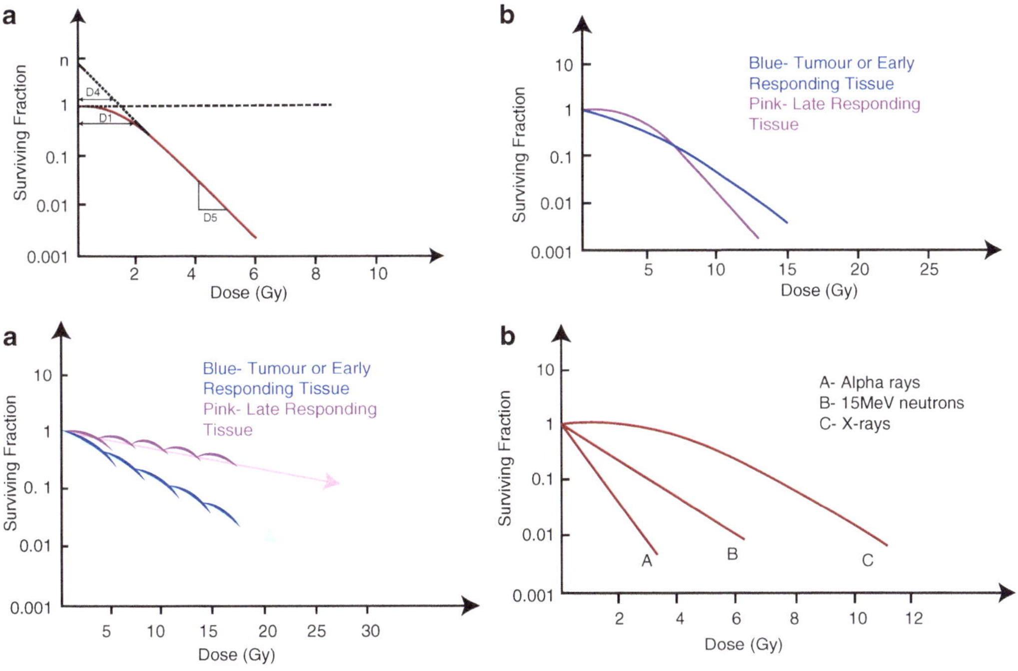

Fig. 4.5 The parts of a cell survival curve (**a**), cell survival curve for different types of tissue (**b**), effect of fractionation (**c**), and for different radiations (**d**)

The classical curve for X-rays/γ Rays has an initial broad portion called the shoulder at a lower dose and a later steep part at higher doses. The shoulder region is primarily due to a single event and is proportional to the dose. In contrast, multiple events cause the steep portion of the curve at higher doses which is proportional to the square of the dose. This is called the linear-quadratic model, and α /Β ratio is the dose at which both the components are equal as shown in Fig. 4.6.

The target in a cell is mostly the nucleus, and hence, the changes are confined to the cell. However, the neighboring non irradiated tissues show some changes, which are likely to be due to the release of certain substances and the presence of gap junctions between the cells, causing a "bystander effect" and is observed with clinically used X-rays and gamma rays. Another way of classifying tissues based on radiation response is early responding (skin, mucosa, and intestinal epithelium) and late responding (spinal cord, kidney, and heart).

The cells are in various phases of the cell cycle, and hence, the tumor is made of asynchronous cells. Studies done by radiation exposure to synchronized cell lines have shown cells in the G2/M phase are susceptible to radiation and cells are most resistant during the S Phase (Fig. 4.7). The cells in the G0 are not proliferating; hence, radiation is ineffective.

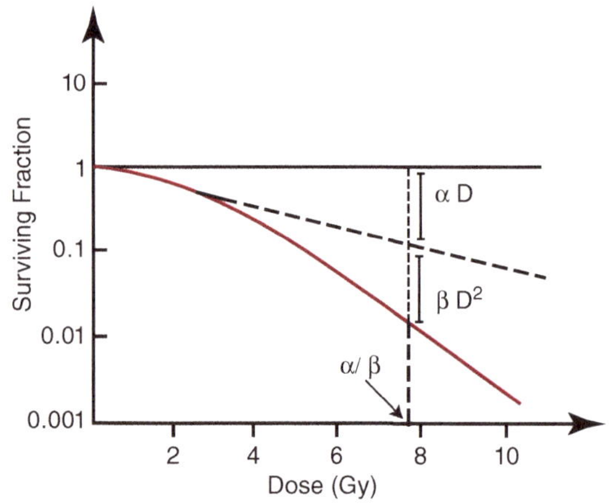

Fig. 4.6 The LQ model

4.3.1 Clinical Implications

1. Fractionation with a time gap between radiotherapy fractions helps the cells to move forward from the resistant to sensitive phase in the cell cycle and get synchronized; hence, fractionation is more effective, and this process is called reassortment.
2. Radiation acts on cells in G2/M phase. In combination with certain systemic chemotherapeutic agents, which work on other phases of the cell cycle, radiotherapy is more synergistic and effective.

4.4 Radiation Dose–Response Relationship Concepts

The goal of a radiation oncologist is to achieve 100% tumor control with no complications or at least maximum tumor control with minimal com-

Fig. 4.7 Cell survival curves in different phases of cell cycle

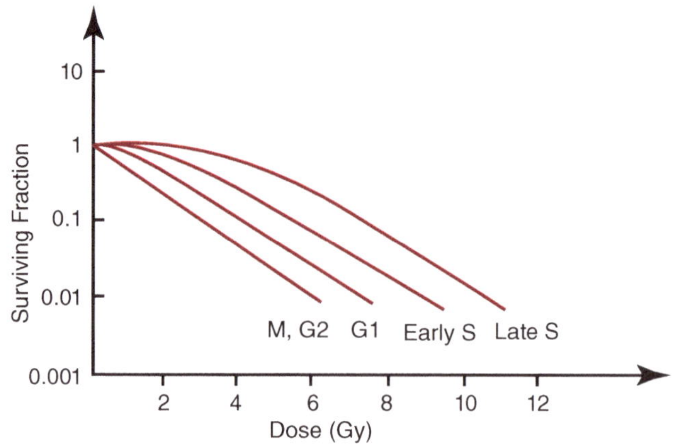

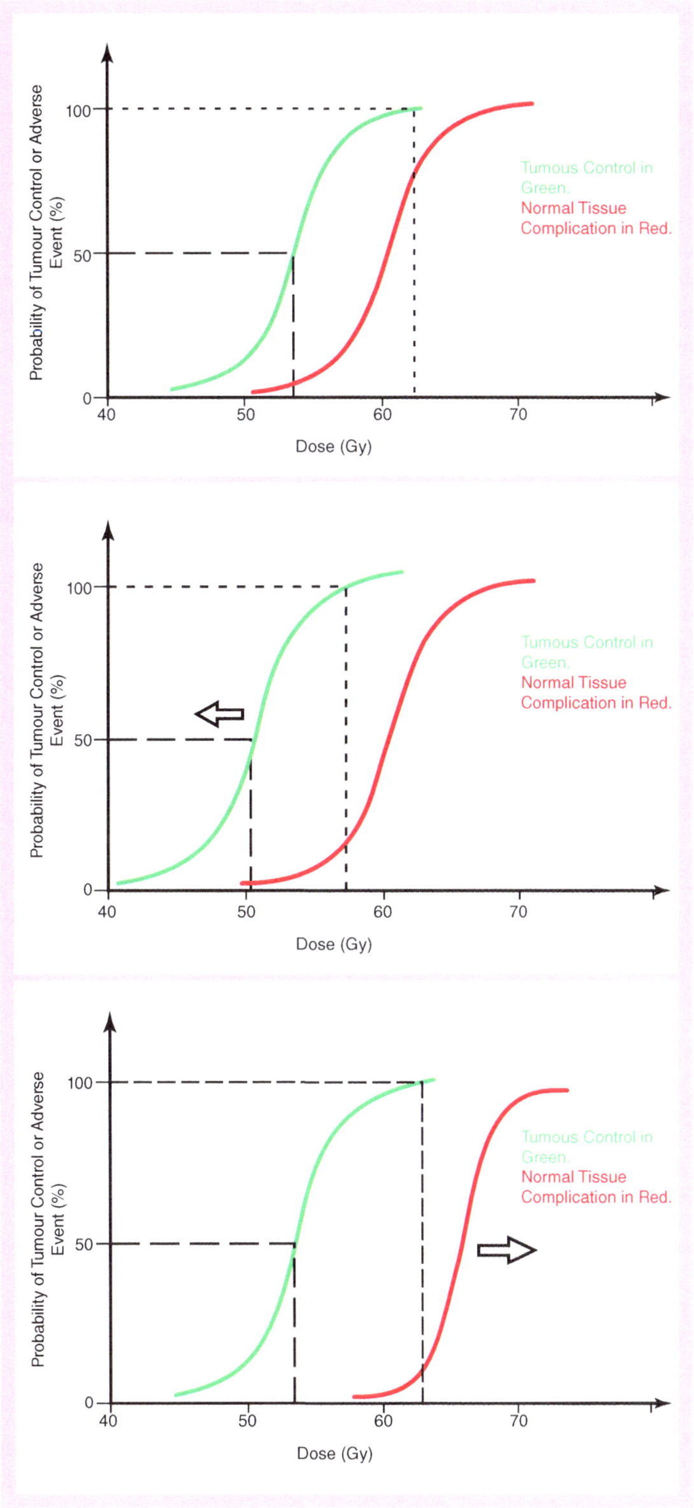

Fig. 4.8 Holthusen's curve depicting TCP and NTCP (**a**), increased therapeutic gain by shifting TCP to left (**b**), and NTCP to the right (**c**)

plications. We are familiar with the tumoricidal doses for different tumors and also the limiting doses for various normal structures. The relationship of dose to endpoints was initially described by Holthusen in 1936 as in the graph in Fig. 4.8. There are two curves which are both sigmoidal in shape but are separated from each other. The two curves represent the probability of tumor control (tumor control probability, TCP) and the probability of complications in the normal tissue (normal tissue complication probability, NTCP). The ratio of these two is called the therapeutic ratio or therapeutic index. From the graph, it is very clear that higher doses are required for better tumor control, but at the same time, the probability of complications also increases with higher doses. The space between the two curves could be called the therapeutic window, which could be widened using various strategies and would result in a better therapeutic gain, as shown in Fig. 4.8.

4.4.1 Clinical Implications

1. The steepness of TCP and NTCP curves varies with each endpoint.
2. There are certain tumors like the prostate where a dose escalation benefit is observed, while esophageal cancers do not show the same.
3. This graph represents conventional treatment, and the only dose is taken into consideration. The present radiotherapy technology is image-based, and the treatment is focused on the target (tumor region), including fewer and fewer normal tissues in the high-dose tumor region; hence, the therapeutic window, which refers to the gap between the two curves, would be wider, resulting in a better therapeutic gain.
4. Clinically, the side effects follow a predicted pattern in relation to the dose delivered, whereas tumor response is mostly unpredictable.
5. For specific sites, a higher dose could be delivered without increasing the side effects, as in prostate cancer. It is important to remember that the high-dose region gets not only a higher total dose but also a higher dose for every fraction (described as double trouble).
6. Optimization in brachytherapy helps in reducing the dose to critical normal structures resulting in a wider therapeutic window.
7. The use of radiation sensitizers and radio protectors also results in therapeutic gain leading to higher tumor control probability and lesser normal tissue complication probability.
8. Risk stratification, as in most pediatric tumors, and tailored reduction in dose, as in lymphoma, also shifts the TCP curve to the left and improves the therapeutic gain.

4.5 Radiobiological Parameters

4.5.1 Linear Energy Transfer

The deposition of energy within the tissue depends on the penetrating power of the incoming ionizing radiation. Electrons are lightweight, scatter and hence deposit their entire energy over a short distance, while protons are of heavy weight and they travel for a distance before depositing their entire energy over a short distance at the end of their track. The x-rays of the clinical range are sparsely ionizing, travel over a greater distance, and deposit their energy over the entire length. We need to understand various radiations in terms of their quality and absorbed dose, and this is done using LET, RBE, and OER.

Linear energy transfer (LET) is the energy transferred per unit length of the track and is expressed as keV/ μm. It is a measure of interaction between a particle and a medium. It is calculated in two ways, energy average and track average. Energy average is where the path of the radiation is divided into increments of equal energies, and the average length is calculated, while in track average, the path is divided into equal lengths, and the average energy is calculated.

The LET value ranges from 0.2 for Cobalt up to 200 for alpha particles, as shown in the Table 4.1. An optimum value of LET for maximum cell kill is 100 keV/μm at which the distance between two energy depositions is 20 nm which is the diameter of the DNA helix and leads to double-strand breaks which are lethal to the cell.

Table 4.1 Values of LET, RBE, OER, and radiation weighting factors for various ionizing radiations

Energy	LET (KeV/μm)	RB	OER	Radiation weighting factor
250 kV X-rays	2.0	1	2.7	1
Cobalt-60	0.2	0.8–0.9	2.5–3	1
6–15 MV photons	0.3	≈0.8	2.5–3	1
150 MeV proton (therapy energies)	0.5	Up to 10	2.6–2.8	2
Clinically used electrons range	0.1–3	1.0–1.5	–	1
Alpha particles	50–200	Up to 20	1	20
Neutrons	0.5–100	Up to 5	1.6	5–20
Carbon ions (spread out Bragg peak)	40–90	Up to 20	2.5–2.7	20

4.5.2 Relative Biological Effectiveness

Once we understand the quality of radiation, the next step is to understand the quantity of radiation absorbed in the tissue, which is called the absorbed dose and is expressed as Gy. There are different ways of deposition of energy in the tissues with various ionizing radiations and so results in varying effectiveness; hence, there is a need to compare them, which is done using the concept of Relative Biological Effectiveness (RBE).

RBE is defined as the ratio of the dose required by the reference radiation (250 kVp X-rays) to the dose required by the test radiation to achieve the same biological effect. The RBE values range from 0.8 for Cobalt to 10 for alpha particles as shown in Table 4.1. The RBE depends on the LET of the radiation (Optimal at 100 keV/μm), total dose (higher the dose, lesser the RBE), number of fractions and dose rate (greater the better), and the endpoint in question.

4.5.3 Oxygen Enhancement Ratio

The process of DNA damage repair starts as soon as energy is deposited in the tissue. Oxygen in the tissues fixes the damage. Tumors have varying degrees of hypoxia and so if we want the same cell kill, we have to use different doses of various ionizing radiations, and this is calculated using Oxygen Enhancement Ratio (OER) as illustrated in Fig. 4.9. It is expressed as a ratio of the dose of radiation required in hypoxic conditions to the dose required in the presence of oxygen to achieve the same endpoint. The OER values range from 1 for alpha particles to 2.5 for X-rays and Cobalt, as shown in Table 4.1.

The relationship between LET, RBE, and OER is shown in Fig. 4.10. An ideal therapeutic radiation should have a LET of 100 keV/μm and RBE of, and an OER of 1 so that it is effective even in hypoxic conditions.

4.5.4 Radiation Weighting Factor

We have now understood how energy deposition and absorption happen in the tissues leading to a biological endpoint. We measure the dose received by the radiation worker and consider them for radiation protection. However, the resulting biological effect will be different for different radiations (Alpha particles versus X-rays). This is overcome by using the Radiation weighting factor, which, when multiplied by the absorbed dose, can be used to compare different radiations. It is different for different radiations, as shown in Table 4.1, and is useful with respect to stochastic effects of radiation.

4.5.5 Clinical Implications

1. LET is more useful for a clinician to understand, and it explains the quality of radiation.
2. Photons such as X-rays and gamma rays have low LET because of their zero mass and

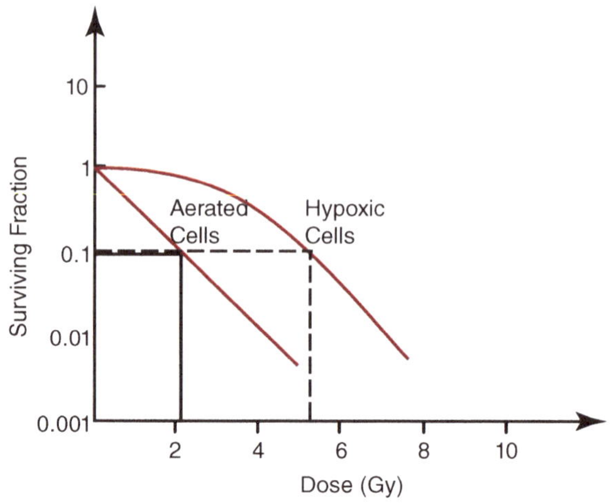

Fig. 4.9 The cell survival curve in the presence and absence of oxygen

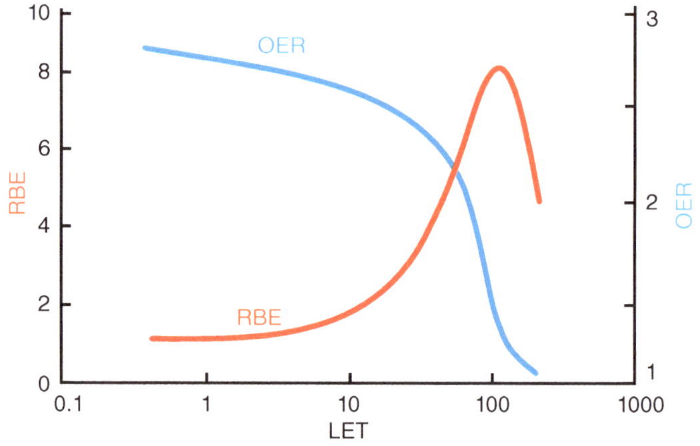

Fig. 4.10 The relationship of LET, RBE, and OER

greater penetration and so are sparsely ionizing, causing more single-strand breaks.

3. Protons, alpha particles, and neutrons have high LET because of their high mass and deposit their entire energy within the short range and so are densely ionizing, leading to clustered DNA damage, i.e. both direct and double-strand breaks.
4. In **proton** radiation therapy, a constant relative biological effectiveness (**RBE**) of 1.1 is usually assumed and overall prescribed doses are therefore lower as compared to photon therapy.
5. The knowledge of radiation weighting factors is applicable in radiation oncology as well as nuclear medicine.

4.6 5 Rs of Radiation Therapy

Withers in 1975 proposed the four Rs of radiobiology in clinical radiotherapy, which form the basis of fractionation. They are 1. Repair, 2. Redistribution, 3. Repopulation, and 4. Reoxygenation. In addition, there are fifth R (Radiosensitivity) and sixth R (Reactivation) which are called as newly added inherent components of radiobiology. The cells *per se* have varying degrees of sensitivity to radiation which has been established on various cell lines as shown in Fig. 4.11.

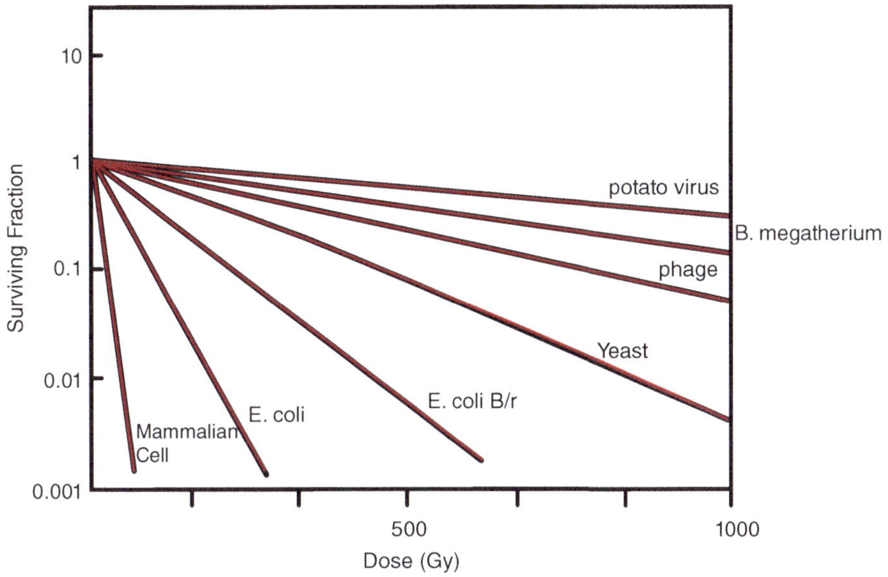

Fig. 4.11 Mainly animal cell lines inherent sensitivity

4.6.1 Repair (of Sublethal Damage)

Three categories of radiation damage are observed after irradiation. They are lethal damage (irreversible, irreparable, and leads to cell death), sublethal damage (repairs in hours if additional fractions are not given), and potentially lethal damage (that can be modified by the post-irradiation environment). There is an inherent capacity to repair during the interval between fractions which varies with slow repair seen in tissues with a low rate of proliferation, like late-responding tissues. It is also lesser in tumor tissues compared to normal tissues due to the tumor environment and inefficient enzymatic process.

4.6.2 Redistribution

When an asynchronous population of cells present in different phases of the cell cycle is exposed to a dose of radiation, more cells are killed in the sensitive (G2/M phase) than in the resistant (S Phase) phases of the cell cycle. The surviving population of cells progresses around the cell cycle and gets partly synchronized before the next fractionation when they are likely to be in the G2-M (sensitive) phase rather than S (resistant) phase. This repeats with every fraction. Thus, reassortment of cell population carries on between fractions and maximum tumor kill can be expected at the end of the radiation schedule.

4.6.3 Repopulation

It is the increase in the rate of cell division that is seen in normal and malignant cells at some point after radiation is started. Early responding tissues begin Repopulation at about 4 weeks, while late responding tissue begins Repopulation after a conventional course of radiation has been completed, and therefore repopulation has minimal effect on these tissues.

4.6.4 Reoxygenation

The presence of oxygen is very important, and it is lesser in tumor tissues. The free radical produced during ionization in the tissue reacts with DNA to form DNA radicals, which causes damage. This DNA radical is short-lived, and the presence of oxygen prevents its restoration, thereby fixing the damage, which is called the

"Oxygen fixation hypothesis." Let us understand why hypoxia occurs. As the tumor grows, there is an increase in the number of cells, and many cells are far away from the blood vessels. Cells that are away from the blood vessel will get less diffused oxygen. This is called chronic hypoxia or diffusion-limited hypoxia. The second one is called acute hypoxia or perfusion-limited hypoxia. As the tumor grows, new vessels form, which is abnormally thin and leaky with altered perfusion. When a growing tumor mass occludes these vessels, the blood supply is cut off temporarily, leading to varying concentrations of oxygen.

With every fraction of radiation, the tumor mass shrinks, opening the blood vessels, and perfusion improves, which addresses the acute hypoxia usually in 24 h. Chronic hypoxia is diffusion dependent and hence takes a longer time to recover.

To summarize 4Rs, reassortment, repair, and reoxygenation are the reasons for the success of clinical radiotherapy fractionation, while accelerated repopulation of tumor cells (during radiotherapy course) affects the outcome negatively. Repair occurs in normal cells as well as in tumor cells. Reassortment occurs in proliferating cells, mostly in tumors but as well in some normal cells. Reoxygenation occurs only in tumor cells. Repopulation occurs in the tumor cells. In recent years, 5th R which which explains the intrinsic sensitivity of tumor types (e.g. lymphoma is more sensitive than high-grade gliomas), and 6th R in radiobiology which shows that irradiation causes changes in the tumor microenvironment(TME) leading to reactivation of the immune response as local and systemic(abscopal) effects; both these are considered as important in radiobiology.

4.6.5 Clinical Implications

1. Tumors vary in sensitivity due to their inherent nature, like lymphoma being more sensitive compared to sarcoma.
2. The viable tumor cells are at a distance of up to 100 μm from a blood vessel. The hypoxic component is observed over 100–180 μm and beyond 180 anoxic cells are present.
3. Oxygen should be present at or immediately after exposure to radiation.
4. As the oxygen concentration increases, the tissue becomes more sensitive to radiation.
5. Any tumor which is bigger than 200 μm will definitely have a necrotic component. This means by the time a tumor is detected clinically or by imaging; hypoxia would have already set in.
6. The hypoxic cells neither respond to radiation nor die. They continue their metabolic function and, over time, develop into recurrent tumors.
7. An increase in partial pressure of O_2 from 0 to 30 mm of Hg enhances radiosensitivity to 50%.
8. There are various methods of measuring tissue oxygenation, such as the use of oxygen probes which is invasive, or the use of PET / SPECT with agents such as F18-MISO. Once the hypoxic mapping is done, they can be given a differential higher dose with IMRT anticipating better control.

4.7 Clinical Radiation Biology: Tumor Versus Normal Tissue Responses

Radiation cannot differentiate tumors from normal tissues and hence affects all the tissues in its path. However, based on various factors, the cells could be classified to understand the final endpoint in terms of tumor control or toxicity.

1. Broadly, there are two populations of cells, A. the first one being early-reacting tissues and tumors, which are mostly proliferating cells, and B. the second being the late-responding tissues, which are mostly differentiated cells with slow cell cycle. After a dose of radiation, the proliferating normal cells are killed; the classic examples are the stem cells in the basal layer of skin and intestinal

mucosa. However, the top layers are more differentiated and less responsive to radiation, and hence, no effects are seen till a couple of weeks when there are no stem cells to replace the top layers, and mucositis and enteritis are observed. Secondary to the death of stem cells, various inflammatory markers are released into the stroma, which may cause fever and fatigue during the first few days of radiation and lead to collagen deposition and fibrosis in the long run. Occasionally, the stem cells are completely depleted, and so the damage persists as a late effect which is termed as a consequential late effect, an example being vaginal stenosis.

In organs such as the liver, kidney, heart, and CNS, the effects of radiation are seen much later as they are generally made of slowly growing cells. The effects are also contributed by vascular damage in addition to parenchymal changes.

2. Normal tissues are also divided into parallel and serial organs based on the functional subunits (FSU). The brain stem, spinal cord, skin, and mucosa are examples of serial organs in which the FSU work in continuity; as opposed to the liver, kidney, and lungs which are considered as parallel organs, where the FSUs are distinct from each other. A small focus of a high dose in one FSU is enough to be detrimental to the entire organ in the case of serial organs, while many FSUs are to be damaged for parallel organs before the effect becomes evident. During repair, the stem cells migrate from the surrounding areas of FSU in the case of serial organs, which is not possible with parallel organs as they are discrete; hence, the tolerance dose is lesser for organs like liver, lungs, and kidneys.
3. The response to radiation also depends on the intrinsic radiosensitivity of the cell. The law of Bergonie and Tribondeau states that if the cells are less differentiated, they have a greater proliferation capacity and so are more sensitive to radiation. The only exception to this rule is lymphocytes which are well differentiated but most sensitive to radiation. Later, Casarett divided the cells into a classification, and their importance for a clinician is depicted in Table 4.2.

4.7.1 Clinical Implications

1. The exponential fall in tumor cell number does not allow for the complete elimination of tumor cells from the body. Every effort is made toward reducing them to a few at the end of treatment so that the immune mechanism of the body takes care of them. This knowledge also forms the basis for metronomic therapy and maintenance therapy.
2. The advanced technology involving image-guided planning and execution along with focused radiation delivery allows minimum normal tissue inclusion in the field of radiation and hence leads to an improved therapeutic ratio.
3. The onset of acute effects of radiation coincides with the turnaround time taken by the stem cells to replenish the functioning cells and is different for different tissues.
4. Enteritis during pelvic irradiation and mucositis during head and neck radiation are observed from the beginning of the third week and increase in severity till completion of treatment and then completely recovers.
5. When bone marrow is irradiated, the first ones to reduce in number are WBCs, followed by platelets, and anemia is the last to be seen as RBCs are replenished once in 120 days. Accordingly, the symptomatology follows.
6. A particular organ may show both early and late effects as the rectum, during radiation for cervical cancer. Acute radiation proctitis due to denudation of mucosa and late radiation proctitis due to telangiectasia and thinning of rectal mucosa.
7. In the case of serial organs, the dose constraint for a particular endpoint is always for a very small volume (representing point dose), such as 1 cc or 0.1 cc (as for brain stem), while for parallel organs, it's with respect to a greater volume (as for lung).
8. Total body irradiation kills the blast cells in the marrow achieving zero cells in the circula-

Table 4.2 Casarett's classification of cells and their importance (Arranged from Class I most radio-sensitive to Class IV least radio-sensitive)

Type of cell	Class	Nature	Sensitivity to radiation	Examples
Vegetative intermitotic (VIM),	I	Continuously divide	Highly sensitive	Basal layers of skin and intestines, blast cells in the marrow
Differentiating intermitotic (DIM)	II	Do divide and also differentiate	Varying sensitivity	Myelocyte
Reverting Postmitotic (RPM)	III	Occasionally divided and more differentiated	Less sensitive	Liver cells
Fixed Postmitotic (FPM)	IV	Do not divide and highly differentiated	Least sensitive	Nerve cells

tion within a couple of days and facilitating bone marrow transplantation.

4.8 Clinical Radiotherapy Fractionation Schedules

Conventional fractionation involves delivering 1.8–2 Gy per fraction given five fractions per week, over 6–7 weeks, to a total dose of 60–70 Gy. In this case, the fraction size is 1.8–2 Gy, and the overall treatment time (OTT) is 42–49 days. Altered fractionation is any schedule where the fraction size, fractions per day, and OTT are different, and each of them is based on various radiobiological concepts and has advantages and limitations [5].

An ideal situation is when good locoregional control is achieved with minimum acute/late effects. The probability of occurrence of late effects is directly proportional to the fraction size, and hence, reduction in a dose per fraction with a slightly increased total dose and giving two fractions per day is beneficial in H&N cancers, glioma, and lung cancer. This is called hypofractionation. Another approach is accelerated fractionation, typically used for proliferating tumors such as aggressive H&N tumors to overcome accelerated Repopulation that happens in the latter half of the treatment. It has many schedules; one is a modest decrease in OTT, such as giving an extra fraction on Mondays/Saturdays, giving a weekend boost on Saturdays, a concomitant boost (where the second fraction/day is delivered during the last 2 weeks), and the other is an accelerated hyper fractionation where the entire dose is delivered over 12 days giving three fractions/day with a gap of 6 h between fractions. There are other various schedules used, and the details are summarized in Table 4.3. Another altered fractionation schedule is split course radiation, where a gap of 3 weeks is given at third week of starting radiation with an OTT of almost 10 weeks.

4.8.1 Clinical Implications

1. Fraction size determines late effects, while fraction size and OTT affect early reactions.
2. Altered fractionation schedules were tried to overcome the shortcomings of conventional radiation and are useful in the era of conventional radiation. The idea was to spare the normal structures, which are achieved presently with image-guided planning.
3. When we change fractionation to a higher dose per fraction as compared to conventional fractionation, severe late effects are expected. If we keep the total dose to match acute effects and give two fractions per day, keeping the same overall time, late effects are less severe.
4. Ideal fractionation would be to give a high dose per fraction in shorter OTT as in stereotactic radiation therapy, but it can be done only in selected patients.

5. Radiosensitization with the addition of chemotherapy is proved to be beneficial, and hence, other fractionation schedules are not presently used in the present era.
6. Another concept is to give a high dose per fraction to the target with a higher tumor burden which can be achieved with a simultaneous integrated boost with a dose per fraction ranging from 1.6 to 2.2 Gy/fractions for various parts of the target.
7. The logistics of giving multiple fractions every day is very challenging.
8. Hypofractionation is typically used for prostate and breast cancer where the estimate of α/β ratio is around 2.5, similar to that of late-reacting tissues, and hence, good control is achieved sparing the normal tissues.
9. SRT/SBRT, where varying doses are delivered in fewer fractions and are possible only because of advanced technology which spares normal tissues to the maximum.
10. Hypofractionation can also be used in re-irradiation for selected cases which would spare the normal structures to a greater extent.

4.9 Fractionation: L-Q, Time, and beyond

A clinician uses fractions with different doses per fraction, total dose, and OTT based on various tumor and host-related factors along with the availability of technology. It's a challenge to understand the biological endpoints that could be achieved with various schedules, and we need a model on which these can be compared. Over the last century, efforts have been made to overcome the challenges associated with each of these models.

Bioeffect models were based on animal experiments, and when iso-effect lines were plotted, it was observed that keeping the same total dose, late effects were higher when the dose per fraction was increased. The next one is the **target theory** which says there are many targets in a cell, and a specified number of them needs to be inactivated for the cell kill. There are two concepts, the first one being the cell is killed following a single target hit, so the curve is straight. The second concept is that there are multiple targets, all of which need to be hit to kill the cell, and so there is an initial shoulder followed by an exponential component in the cell survival curve. This led to the most commonly used **Linear Quadratic** (LQ) model, which says that at lower doses, the cell kill is due to a single hit resulting in a shallow part of the curve, while at higher doses, the cell kill is due to multiple hits, and hence, the curve is steeper.

The conventional fractionation that is time tested was initially based mostly on logistics rather than a scientific one. When we look at the cell survival curve, we can see that at lower doses, survival is higher for late-responding normal tissues as compared to tumors, while the reverse is observed at the higher dose. This means as we increase the dose per fraction, the late effects are likely to be higher, meaning late-responding tissues are more sensitive to altered fractionation than early-responding tissues, which behave like tumor tissues. In addition, if we can use image guidance to reduce the amount of normal tissues getting radiation, the late effects could be further lesser, and this forms the basis for hypofractionation, SBRT, and brachytherapy, where a very high dose is delivered.

In conventional two-dimensional (2-D) radiation technique, normal tissues and tumors get a similar dose while in image-based conformal therapy, there will be non-uniform dose distribution, and we are always keeping the dose to normal critical structures less than that of the tumor. Hence, it is expected to notice differential biological effects.

The α/β ratio is similar for tumors as well as for early responding normal tissues. Late-reacting normal tissues have a sparing effect when dose/fraction is lesser, while at higher fractions size the sparing is lesser. It is assumed that α/β ratio is around 10 for tumors as well as early-reacting tissues while it is around 2.5–3 for late-reacting tissues. The equivalent uniform dose (EUD) is a concept useful in conformal therapy (the dose across the target is less homogenous). EUD is the dose that, if uniformly delivered in the tissue,

Table 4.3 Various fractionation schedules

Sl No	Name of the fractionation	The total dose in Gy/no of fractions	Dose per fraction in Gy	No. of fractions per week	OTT in days	Basis	Advantages	Disadvantages	Used for	References
1	Conventional	70 Gy/35 Fr	1.8–2	5	49	4 Rs	Time-tested and adequate local control with acceptable side effects	Relatively longer duration of treatment and so less patient-friendly	Most sites	
2	Hyperfractionation	80.5 Gy/70 Fr Marginally increased	~1.15	10	49	Reduced late effects	Reduced late effects with up to 8% better in local control	The patient has to come double the time for radiation	Advanced H&N cancer	[6]
3	Accelerated with weekend boost	68 Gy/34 Fr	1.8–2	6	39	Overcome accelerated repopulation	Higher response and better DFS	Increased mucositis but transient	Advanced H&N cancer	[7]
		68 Gy/34 Fr	1.8–2	7	35	Overcome accelerated repopulation	Increased early tumor response	Early mucositis lasting for a more extended time	Advanced H&N cancer	[8]
4	Accelerated with concomitant boost	72 Gy/42 Fr	1.8–2	5 (10 during the last 2 weeks)	32	Overcome accelerated repopulation	Better OS and DFS with no increased toxicity	Logistics for the patient	Advanced H&N cancer	[9]
5	CHART	54 Gy/36 Fr	1.8–2	18	12	Complete radiation before accelerated repopulation begins	Good initial response that persists for a long time	Severe acute effects requiring admission	Advanced H&N cancer	[10]
6	Hypofractionation	40 Gy/15 Fr (Breast)	2.66	5	21	α/β ratio is 2.5–3	Shorter OTT and no higher toxicity	Strict planning and execution are required	Adjuvant breast/chest wall irradiation	[11]
		60 Gy/20 Fr (prostate)	3	5	28	α/β ratio is 2.5–3	Shorter OTT and no higher toxicity	Strict planning and execution are required	Radical	[12]

	Extreme hypofractionation	26 Gy/5 Fr	5.2	5	5	α/β ratio is 3.5–3.7	Similar control and cosmesis	Strict planning and execution are required	Early breast post BCS	[13]
7	Simultaneous integrated boost	66–59.40–54 Gy /30 Fr	2/1.8/1.6	5	42	Tumor is heterogeneous and different areas need different doses	A modest decrease in OTT with no higher acute toxicity	Nothing in particular	Head and neck cancer	[14]
8	Flash RT	15 Gy	~40 Gy/s	Once	1	Ultra HDR induces hypoxia in normal tissues	Differentially saves normal tissues	Yet to be studied Equipment to produce a HDR would be costly	Few humans treated	[15]

produces the same endpoint as a non-uniform dose would result in. Simply put, with emergence of image-based radiotherapy technique, such as intensity-modulated radiotherapy (IMRT), and other improved techniques, the EUD provides a comparison between different treatment plans of non-homogenous dose distributions.

4.9.1 Clinical Implications

1. The various models will only help us to understand, and the clinician's judgment plays an equal role regarding various fractionation schedules.
2. There is no single fractionation schedule that's appropriate for all cancers.
3. Whenever higher doses per fraction are used, one has to be extremely careful at every level of planning and execution.
4. The L-Q model does not take into consideration the non-homogeneous dose distribution with various techniques, which is overcome by the concept of EUD.

4.10 Modified and New Fractionation Regimens in Practice

Traditionally, radiation time-dose-fractionation (TDF) regimens are classified into conventional fractionation, hyperfractionation, accelerated hyperfractionation, and hypofractionation [16]. The choice of 2 Gy by convention evolved empirically in the early twentieth century, based largely on the intensity of early skin reactions and local control rates in squamous cell carcinomas of the head and neck (HNSCC) and uterine cervix [16]. The 4 Rs of radiation oncology—repair, reoxygenation, repopulation, and reassortment—are pivotal to the understanding of the rationale of these altered fractionation regimens [17]. In addition to these, one should also consider intrinsic radiosensitivity (the 5th R), which is defined as the responsiveness of tumors to radiotherapy, and this radiosensitivity correlates with the intrinsic radiosensitivity of the cells in vitro [18].

In the past decade, there has been a paradigm shift toward the use of hypofractionated radiation regimens. Hypofractionated radiation has been somewhat arbitrarily subdivided into two distinct approaches: moderate hypofractionation, defined here as 2.4–3.4 Gy per fraction; and ultra-hypofractionation or stereotactic body radiation therapy (SBRT), defined as greater than 5 Gy per fraction. In either form, it has the obvious added advantage of reducing the number of treatment fractions and overall treatment time [19], although paradoxically, these regimens challenge the fundamental concepts related to the benefits of fractionated radiation therapy. Safe and effective delivery of these hypofractionated regimens could be attributed to major advancements in image guidance and treatment delivery techniques that have enabled us to deliver large doses to tumors with reduced margins with high-dose gradients minimizing doses to relatively large volumes of surrounding normal tissue [20]. Similarly, radiobiological modeling has witnessed major strides of advancement in the recent past. Macroscopic tumor control probability (TCP) and normal tissue complication probability (NTCP) modeling can now account for heterogeneous dose distributions in the tumor and critical normal tissues as well as their volume effects and are far more effective in predicting normal tissue toxicity [21]. We can now plausibly model the anticipated changes in the clinical outcome and the therapeutic ratio with changes in the number of fractions [22].

4.10.1 α/β Ratio and Hypofractionation

Based on the standard Linear Quadratic (LQ) model, the surviving fraction (S) after n fractions, with each one delivering a radiation dose of d, is characterized by the following equation—$S = \alpha d - \beta d^2)]^n$.

Therefore, α/β is defined as the ratio when the linear and quadratic components are equal. As the tumor α/β gradually reduces, the relative advantage of a higher number of fractions diminishes. In cases where α/β ratio of the tumor is equal to that of the adjacent critical normal tissue (e.g., ~3 Gy), the dependence on fraction size disappears the concept of which is illustrated in Fig. 4.12. Finally, for tumor α/β of 1.5 Gy, i.e., lower than that for nearby critical structures, the TCP increases as the number of fractions are reduced, with the maximum TCP being achieved with a single fraction [22, 23].

There is now sufficient evidence to show that the α/β of breast and prostate cancer falls in the range of 1.5–3 Gy, and therefore, hypofractionation is favored in these tumors to optimize the therapeutic ratio [24–27]. However, the validity of the LQ model at high dose per fraction is controversial. Some studies demonstrate that the predictions from LQ model tend to deviate from that derived from Repair–Misrepair (RMR) and Lethal–Potentially Lethal (LPL) the model above 5 Gy [28], others show that the LQ model can fit experimental survival data well up to 10 Gy or maybe even for 15 to 20 Gy delivered in a single fraction [29, 30]. Based on data from large doses per fraction, it has been shown that LQ formalism underestimates the sparing effect of small doses per fraction provided optimal interfraction time is allowed for the repair of sublethal damage. The magnitudes of errors were 6–19% for 2–3 fraction schedules in V79 and EMT6 single cells, 18–30% for 2–5 fraction schedules in V79 spheroids, and 21 to 42% for 2–5 fraction schedules in EMT6 tumors. The large discrepancy between single cells and in vivo tumors was largely attributed to reoxygenation during interfraction intervals in the hypo-fractionated groups [31, 32]. While alternative models such as Linear-Quadratic-Linear (LQL) or Universal Survival Curve (USC) models can provide superior visual fit to in-vitro studies, there are certain demerits of these models, including an increase in the number of variable model parameters (3 in LQL and 5 in USC) which adds to the complexity of the models and brings in the question of overfitting. Further, there is insufficient clinical evidence to support that one model has better predictive accuracy over the other in the commonly used range of radiation dose per fraction [25, 29]. Therefore, The LQ model is still endorsed by many experts around the world [20].

4.10.2 Specific Radiobiological Considerations in the Ultrahypofractionated Approach

A growing body of evidence demonstrates that at a dose of 10 Gy or higher in each fraction, radiation can portend vascular damage leading to indirect cell death [34]. Ultrahypofractionated radiation, or SBRT, drives a complex cascade of biological processes. For example, SBRT promotes antitumor immunity by enhancing the diversity of the T-Cell Receptor (TCR) repertoire and lymphocyte trafficking to the TME via the upregulation of adhesion molecules on tumor-associated vasculature and increased chemokine signaling. SBRT can also recruit $T_{regs.}$ and myeloid-derived suppressor cells (MDSCs) into the TME and can contribute to adaptive immune resistance via interferon-ɣ (IFN-ɣ) mediated upregulation of PD-L1 expression on tumor cells and MDSCs. These have been clearly described in their seminal paper by Marciscano et al. [35]. Therefore, it is paramount to define an optimal time-dose-fraction regimen, which can maximize the immune response while minimizing the treatment-induced side effects.

The effect of tumor hypoxia is pivotal to the concept of modern hypofractionated radiation regimens. Studies have shown that tumor hypoxia leads to a large difference in the calculated level of cell killing, even for highly fractionated regimens. However, with decreasing number of fractions, a compromise in tumor control is noted for the same BED or the same normal tissue damage. The worst situation is with a single dose, which gives a survival of approximately 10^{-2} compared to 10^{-5} with 30 daily fractions [20]. A logical consequence of the predicted increased importance of tumor hypoxia in SBRT alludes to the fact that hypoxic cell radiosensitizers, which largely failed with conventionally fractionated irradiation, could be a realistic option to improve the clinical outcome in with ultrahypofractionated radiation regimens.

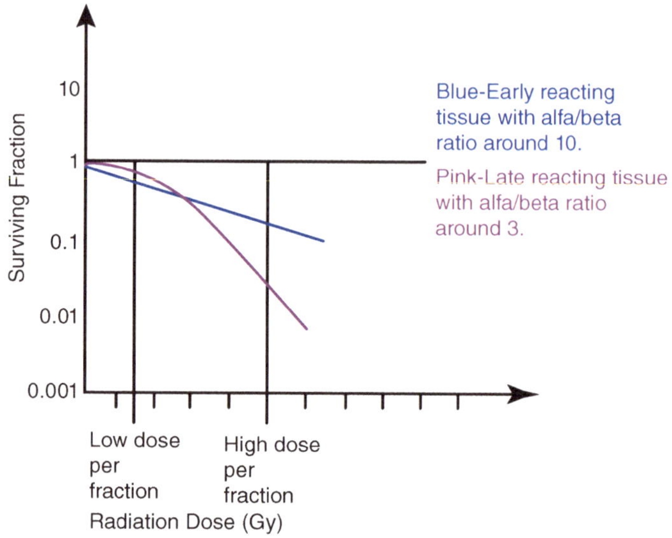

Fig. 4.12 Benefits of giving high hypofractionation for tumors with low α/β ratio. Higher dose per fraction gives better tumor kill compared to low dose per fraction

4.10.3 Hypofractionation and Toxicity

The emergence of acute toxicity during radiation therapy is a complex interplay of overall treatment time, dose per fraction, the volume of irradiated normal tissue, and the architecture of the tissue. Hypofractionated radiation-induced late toxicity depends on the α/β ratio, although this might not be the sole driving factor. Hypofractionation diminishes the role of redistribution, with it having no relevance to single-fraction treatments. Hypofractionated radiation has two potentially conflicting impacts on toxicity. It increases the risk of acute toxicity through acceleration while decreasing through lower EQD2 (equivalent dose in 2 Gy per fraction). These two contrasting effects were clinically evident in the FAST-FORWARD study in breast cancer. The study investigated a moderately hypofractionated regimen of 40 Gy in 15 fractions to two ultrahypofractionated regimens (27 Gy or 26 Gy in 5 fractions over 1 week). The EQD2 (considering an α/β ratio for acutely responding tissue of 10) were 42.2 Gy, versus 34.7 and 32.9 Gy, respectively. The observed grade 3 or higher acute Radiation Therapy Oncology Group (RTOG) toxicity rates were 13.6% versus 9.8% versus 5.8%, respectively [36]. These findings demonstrate that, in this case, the acute toxicity was reduced more by the EQD2 reduction with ultrahypofractionation relative to any countervailing increase from reduced OTT. The effect of irradiated volume of normal tissue and radiation toxicity depends not only on the dose heterogeneity but also on the position of the normal tissue along the serial-parallel axis. This defines the volume parameter (*n*). For organs with low n, higher dose region matters more, while for organs with high *n*, low-dose spillage to a larger portion of an organ matters more [22].

4.11 Radiobiology of Brachytherapy

Unlike external beam radiotherapy, brachytherapy is characterized by smaller overall treatment volumes, significant radiation dose heterogeneity (rapid dose fall-off), and generally shorter overall treatment times. Generally speaking, the radiation related dose kill is mechanistically similar to external beam radiation for photon-based brachytherapy sources which comprise the vast majority of available brachytherapy sources (for example Iridium-192, Cobalt-60). Brachytherapy is divided into low-dose rate or LDR (0.4–2 Gy/h), medium dose rate or MDR (2–12 Gy/h), high-dose rate or HDR (>12 Gy/h), and pulsed dose rate or PDR (1–4 Gy/h) based on dose rate.

Similar to external beam radiotherapy, the 4 Rs of radiobiology, namely DNA repair, cell cycle redistribution, repopulation and reoxygenation also apply to brachytherapy. DNA repair in the normal tissue is advantageous while DNA repair in the tumor is deleterious to tumor control. The fractionation with HDR brachytherapy is usually considered to be advantageous with cell cycle redistribution. Cells in the more radioresistant phase of the cell cycle might move to the more radiosensitive phase of the cell cycle for consecutive fractions of HDR brachytherapy. With protracted fractionation schemes seen with external beam radiotherapy, repopulation of tumor cells toward the end of the treatment schedule is generally considered to be deleterious to tumor control. This is considered less of a concern with brachytherapy given the short overall treatment times and generally high doses of radiation per fraction used in HDR brachytherapy. Tumor kill in between fractions of HDR brachytherapy usually helps to debulk the tumor allowing better oxygenation and increase the DNA damage.

In addition, the dose-rate effect is also relevant for brachytherapy treatment when considering low-dose rate, HDR and pulsed dose rate brachytherapy. In general, the dose rate effect implies that lower dose rates lead to decreased tumor control and lower side effects while higher dose rates lead to increased tumor control but increased normal tissue complications.

Historically, most of the data regarding the results of radiotherapy has been obtained from the low-dose rate experiences as opposed to HDR in modern times. HDR brachytherapy is generally considered to be more convenient and equivalent to low-dose rate brachytherapy for disease control. The repair of normal tissue during low-dose rate brachytherapy is likely to be superior to HDR brachytherapy. Fractionation during HDR brachytherapy has been used to offset long term side effects due to normal tissue complications [38].

Pulsed dose rate brachytherapy incorporates the convenience of HDR brachytherapy and the radiobiologic advantages of low-dose rate brachytherapy [39]. The intervals between radiation pulses can be utilized to provide nursing care to the patient. These intervals also facilitate tissue repair thereby mimicking low-dose rate brachytherapy.

4.11.1 Clinical Implications

1. HDR brachytherapy is widely used and involves the temporary placement of radioactive sources in the vicinity of the tumor, while LDR uses permanent implants which deliver radiation over a longer period of time.
2. As the dose rate increases, there is higher SLD, and hence, the total dose needs to be reduced to maintain lower levels of late effects; the rule of thumb for conversion from LDR to HDR is to multiply the dose by 0.6.
3. Image guidance helps us to spare normal tissues as much as possible and has resulted in reduced late effects of radiation, such as radiation proctitis in cervical cancer.
4. BED is more familiar to clinicians and easier to understand, in order to decide the dose prescription for various fractionation schedules.
5. Brachytherapy can be given during EBRT to reduce the overall treatment time; care should be taken to calculate the cumulative dose while respecting the normal tissues. Both EBRT and Brachytherapy should not be done on the same day as it affects the repair of sublethal damage in the normal tissues.
6. The concepts of BED and EQD2 are useful to calculate the combined dose from EBRT and brachytherapy as in cervical/head and neck cancer.
7. They are also useful for comparing different brachytherapy schedules used at various centers. BED formulae are different for EBRT and for brachytherapy.

4.12 Hyperthermia

Hyperthermia refers to the use of heat to enhance the effect of radiation or chemotherapy. There are different modes of action on a tumor. Firstly, it causes the denaturation of proteins causing the death of the cell. Secondly, radiation acts on cells in the G2 M phase, and hyperthermia acts on those in the S phase. Thirdly, tumors retain heat for a longer time compared to normal tissues due to lesser dissipation. Fourthly, there is no difference in sensitivity to heat between aerobic and anaerobic cells. Lastly, it inhibits the repair of both sublethal and potentially lethal damage by inactivating crucial DNA repair pathways.

Thermal enhancement ratio (TER) is the ratio of doses of X-rays required to produce a given level of biological damage without and with the application of heat. The data shows a consistent pattern of increasing TER with increasing the temperature, up to a value of 2 for one-hour heat treatment at 43 °C. Adverse effects of hyperthermia would be the clinical concern when external application of heat may cause surface burns. Whole-body hyperthermia can cause swelling, blood clots, bleeding, and systemic shock.

It is difficult to heat tumor tissue volumes with uniformity and precision. There is no standardized equipment available, and techniques for measuring temperature and the actual definition and calculation of thermal dose remain significant challenges which have been addressed in literature [40].

4.12.1 Clinical Implications

1. Clinical use of hyperthermia is mostly for cervical cancer and recurrent head and neck cancers, and a moderate temperature of 39–42 °C is both safe and effective.
2. The most commonly used methods of local heating are microwave and radiofrequency heating.
3. Technical challenges are in the areas of selectively heating the tumor and measuring the exact temperature and the associated side effects.
4. With chemotherapy, it is used for peritoneal involvement in refractory ovarian and colorectal cancer with surgery, both in one setting called hyperthermic intraperitoneal chemotherapy (HIPEC).

4.13 Radiation Protection and Accidents

4.13.1 Radiation Protection

No dose of radiation is safe, and it is very important to use radiation only if indicated and also to protect the radiation workers from the deleterious effects of radiation. There are two types of effects seen with exposure to radiation. For example, radiation mucositis or enteritis. It is not seen during the initial weeks of treatment; it starts after a particular dose and then onward and increases in severity with an increase in the dose. So, this has a threshold below which it does not happen, and above that dose, the probability increases, and severity also increases. This is called the deterministic effect. The other example is radiation-induced carcinogenesis or heritable effects of radiation, which is called stochastic effect. It can happen at any dose, and on increasing the dose, the probability of occurrence increases while the severity remains similar to those due to other causes.

To reduce the stochastic effects, ALARA (as low as reasonably achievable) principle comes into the picture and forms the basis for Radiation protection. This is achieved by three basic rules; time, distance, and shielding (TDS). The radiation workers should spend minimum time and be at the optimum distance from the radiation source and use the appropriate shielding methods. The International Commission on Radiation Protection (ICRP) was established in the second International Congress Radiology meeting, which was held in Stockholm in 1928. Three basic principles of radiological protection are to be followed as per ICRP such as justification (should be used only if required), optimization (following ALARA), and respecting dose limits for occupational radiation workers, patients, and the public as depicted in Table 4.4 [41, 42].

4.13.2 Clinical Implications

1. Stochastic effects have no threshold and hence can happen at any dose above which the

chances of occurrence increases while the severity does not as opposed to deterministic effects which have a threshold above which both the chances and severity increases with increase in dose.
2. No practice involving exposure to radiation should be adopted unless it produces a sufficient benefit to the exposed individuals or to society. Justification in identifying the benefits versus risks for the patient; for example, with one pelvic CT scan the exposure is 20 mGy while it is zero if you do an MRI, which uses nonionizing radiation.
3. Radiation safety tips for occupational radiation workers:
   (a) Follow the TDS principle.
   (b) Wear the lead apron, thyroid shield, lead hand gloves.
   (c) Wear the personal monitoring device such as a TLD badge at the chest level during working hours to monitor the whole body equivalent dose. The TLD badge should be worn beneath the apron. Also, wear the wrist and head TLD badges during the interventional radiology procedures.
4. Radiation safety tips for patients and the public:
   (a) Quality assurance of the equipment as per national agency, such as Atomic Energy Regulator Board (AERB) in India.
   (b) Collimate the X-ray beam to avoid unnecessary exposure.
   (c) The X-ray room doors should be closed during the exposures.
   (d) If a patient needs assistance during the X-ray exposure, the assistant will wear the lead aprons and thyroid shield during the whole procedure.
   (e) Avoid Re-dos with and without contrast.
5. Radiation safety tips for the Pediatric age group:
   (a) Use lead aprons, thyroid shields, and gonad shields, and expose only the part to be investigated.
   (b) Use alternate modalities such as MRI, ultrasound instead of X-ray and CT scan.
   (c) Protocols of automatic exposure control (AEC) ensure pediatric exposures.
   (d) Use the X-ray equipment with higher mA and for a shorter time.

4.13.3 Radiation Accidents

Radiation accidents due to a single exposure of a very high dose to the entire or almost entire body from outside the body in a short time lead to various life threatening conditions. The effects are mainly of three types depending on the amount of dose exposed to such as hematopoietic syndrome, gastrointestinal syndrome, and cerebrovascular syndrome as shown in Table 4.5. Due to high dose, the stem cells are depleted and the vascular permeability gets altered leading to extensive edema resulting in multi organ failure [43]. LD50 (lethal dose that will kill 50% of the population) is 3.5 Gy for human beings.

The other type of radiation accidents due to exposure to smaller doses are those that happen at the place of production, during transportation, or at the clinical departments. In the present era, with the presence of advanced technology, strict adherence to the protocol, frequent quality checks, and multiple checkpoints with a robust governing body controlling the whole process, accidents are rare. Still, there is a possibility that they can happen, and a clinician must understand and know how to handle it in the event it happens.

It could be in the cobalt teletherapy room or brachytherapy room if the source gets stuck or if, by overlooking, anyone is present in the teletherapy room during the patient's treatment. In such an event, the steps to be followed are as follows:

1. The gamma zone monitor beeps, indicating excess radiation and helping in identifying the accident.
2. Follow the emergency plan given by the manufacturer. Remove the patient from the site and retract the source manually, taking the minimum possible time.
3. We will not be able to identify the actual dose received, but a rough estimate based on the

Table 4.4 Permissible dose limits for radiation workers and public as per ICRP

	The effective dose in mSv (dose equivalence, 10 mSv = 1 cGy)		The equivalent dose in mSv		
	Year	Over 5 years	The lens of the eyes	Skin	Hands and feet
Radiation worker	20	100	150	500	500
Public	1	5	15	50	NA
Pregnant radiation women	One over a period of gestation (permissible total exposure to fetus during pregnancy should be less than 5 mSv)		NA	NA	NA

Table 4.5 Acute radiation syndromes following single high-dose exposure to the entire body in a short time

	Dose	Pathology	Occurrence of the event	Treatment
Hematopoietic syndrome	2.5–5 Gy	Depletion of stem cells	Counts will be very less at around 3rd week	Antibiotics, supportive care
Gastrointestinal syndrome	10 Gy	Denudation of entire lining of GI tract	Death in 7–10 days	Supportive care
Cerebrovascular syndrome	>50 Gy	Extensive intracranial edema	Death in 24–48 h	Supportive care

duration of stay and distance from the source can be made.
4. Send the TLD monitoring or personal monitoring badge for actual exposed dose measurement.
5. Depending on the estimated dose, triage has to be done, and the persons to be kept under observation.
6. In nuclear medicine departments, decontamination need to be done [44].
7. Make a report of the accident and report to AERB as per the guidelines.
8. However small the incident, it needs to be reported so that further learning and necessary actions can be facilitated.

4.14 Heritable Effects of Radiation

The development of gonads is different in males as compared to females. In male, it is a continuous process from puberty till death, and the cells, as per hierarchy, are primary spermatocytes, secondary spermatocytes, spermatids, and spermatozoa, with spermatozoa being the most differentiated. In female, at birth, the number of oocytes is fixed, and there is no further cell division. The interstitial cells of the testis, which produce hormones, are radioresistant and hence do not lead to any changes. The ovarian follicles are in various phases of maturity and, accordingly, are sensitive to radiation. Since it is linked to changes in hormones, profound hormonal changes similar to menopause are observed following exposure to radiation.

Mutations that occur in somatic cells cause cancer, while those in the germ cell cause various heritable disorders that are seen in subsequent generations. All mutations do not lead to heritable disorders, and the risk in the general population is 0.2%/Sv, and in radiation, workers of reproductive age 0.1%/Sv.

4.14.1 Clinical Implications

1. The knowledge about the heritable effects of radiation is based on animal experiments and available data from survivors of atomic bombs and radiation accidents.
2. It takes around 10 days for maturation from primary spermatocyte to spermatozoa, and so fertility is maintained for several weeks after exposure to radiation. A testicular dose of 0.15 Gy causes oligospermia that is evident after 6 weeks; a dose higher than 0.5 Gy

results in azoospermia, and a higher than 6 Gy causes permanent sterility.
3. In women, the effect is much more evident and is therapeutically used for ovarian ablation for premenopausal breast cancer patients. The dose used ranges from 12 to 18 Gy in 4–6 fractions, with the higher dose needed for younger patients.
4. For reproductive male or female a dose > 6 Gy can cause gonadal dysfunction. In case of exposure to radiation, it is better to avoid conception for at least 6 months. This factor is only one of many other factors to be considered during counseling.
5. It is observed that therapeutic radiation does not cause new mutations, but the probability of heritable disorder may increase by 0.06/Sv for a doubling dose of 1 Gy.

4.15 Effects on Embryo and Fetus

Till today, the largest data is of 1500 children born to mothers present at Hiroshima and Nagasaki bombings and followed for over two decades. The other sources are based on exposure to medical imaging and radiation accidents. The latter is less relevant in the present era as almost everything is remote-controlled, and the effects depend on the duration of gestation and the dose received.

The gestation period of 40 weeks can be divided into preimplantation up to 1.5 weeks, organogenesis (1.5–6 weeks), and fetal period (6–40 weeks). Exposure to radiation up to 3 weeks after conception is an all-or-none phenomenon; either the embryo gets aborted or continues without any discernible effects. The most important phase is organogenesis. Severe anomalies are seen when the fetus is exposed to radiation between fourth and 11th week, causing severe anomalies such as neural tube defects, polydactyly, etc. Exposure between the 11th and 16th week causes less of anomalies but more stunted growth, mental retardation, and microcephaly. These are much milder when exposed between 16th and 20th week, and no changes are observed after the 20th week of gestation.

4.15.1 Clinical Implications

1. The growth of the fetus in the uterus mimics tumors kinetics.
2. The most sensitive period of gestation is up to 20 weeks. In the event of exposure, close follow-up to identify anomalies and if needed counseling for therapeutic abortion are to be followed. Other factors to be considered are the parent's decision, ethnic and religious background, the baby being precious, and probability of future pregnancy.
3. Women occupational workers should be made aware of the effects of radiation, and they have to report pregnancy early and follow all the precautions that are mandated by AERB.
4. While ordering imaging with X-ray, all women patients should be advised to do so during the first 2 weeks of the menstrual cycle.
5. It is mandatory to confirm that a woman is not pregnant before starting radiation therapy as a treatment.
6. It is estimated that 0.1 Gy is the threshold for mental retardation, and most diagnostic procedures result in exposure of less than 0.1 Gy. However, due to the availability of other investigations, X-ray-based imaging is to be avoided.

4.16 Radiation-Induced Cancer (RIC)

There are two different mechanisms for RIC, and it is related to the dose. The first one is related to high-dose regions, as in cervical and prostate cancer. The exposed normal tissues show atrophy which is similar to that seen in premalignant conditions, along with chronic inflammation leading to cancer. The second one is seen in the peripheral uniformly low-dose region which is characteristic of RIC, typically seen following radiation for pediatric cancer. Since the probability of RIC increases with increase in the dose, every effort should be made to restrict the dose to normal tissues as per ALARA. As seen, RIC is also related to ionizing radiation type (e.g. kV X-ray, cobalt-60, linear accelerator-based megavoltage radiation, proton etc.), the technique used (such as 2-D or image-based conformal IMRT etc.). Largely, RIC happens 10–20 years later. The present data available is based on older techniques, and we need to wait for many years to see the actual occurrence with newer techniques delivering much lower dose to normal tissues and organs. Till then, it is prudent to use a reduced dose, reduce the irradiated volume, use better techniques and tailor the treatment based on risk stratification.

Precept Box

- Advanced radiobiology research has paved the way for constant improvement in radiation therapy delivery resulting in a better therapeutic gain.
- Tumors differ from normal structures in terms of lesser repair mechanisms, lower oxygen levels, inherent radiosensitivity, and less predictable response to a standard dose.
- 4 Rs of radiation form the basis for various fractionation schedules.
- Repair of SLD and repopulation happen both in tumor and normal tissues, while reoxygenation and reassortment happen in tumor tissues.
- Hypofractionation of various types have proved their worth in cancers of the breast and prostate and in re-irradiation setting.
- RIC is more of a theoretical possibility and should not be entertained as the reason for not offering radiation. Considering the benefit versus risk and using appropriate doses and techniques is the way forward.
- Among various models, the LQ model is the most time-tested and useful guide for a clinician. Similarly, EQD2 is useful to interpret different dose-fraction regimes.
- Hypoxia is an important factor to be considered and can be taken care of with various modalities such as combining radiation with surgery/chemotherapy, fractionation, brachytherapy, etc.

Multiple Choice Questions

1. Which of the following statements is true regarding hypofractionation?
   (a) It is useful to reduce the early effects of radiation.
   (b) It reduces the overall treatment time.
   (c) It increases the late effects of radiation.
   (d) It does not depend on the α/β ratio concept.
2. Which of the following is true regarding the α/β ratio?
   (a) It helps to decide the fractionation schedules.
   (b) It helps us to predict acute tumor response.
   (c) It is the same for a particular histopathology irrespective of the site.
   (d) It helps to predict long-term outcome.
3. Which of the following statements is not true regarding hypoxia?
   (a) High dose of radiation given over seconds causes hypoxia in normal tissues.
   (b) Two types of hypoxia are diffusion-limited and perfusion-limited.
   (c) All tumors bigger than 200 μm are hypoxic.
   (d) Hypoxia is more of a concern during brachy than EBRT.
4. Radiations with an LET of more than 100 keV/μm are associated with.
   (a) Greater cell kill as energy deposition is higher.
   (b) Lesser cell kills as there is overkill.
   (c) No additional benefit as the energy gets wasted.
   (d) No additional benefit as the energy gets distributed.
5. Which of the following is not true regarding the therapeutic window?
   (a) It reflects the differential response to radiation.
   (b) The larger the window better is the therapeutic gain.
   (c) The smaller the window better is the therapeutic gain.
   (d) Advanced technology widens the therapeutic window.

6. The concept of radiation weighting factor is with reference to.
   (a) Deterministic effects of radiation
   (b) Stochastic effects of radiation
   (c) Calculations in EBRT
   (d) Calculations in brachytherapy
7. Which of the following is true regarding sublethal damage?
   (a) It is represented by the steep slope of the cell survival curve.
   (b) It happens in both tumor and normal tissue.
   (c) It is influenced by the postradiation environment.
   (d) It is not of importance in brachytherapy.
8. CNS structures are serial organs, and which of the following is true?
   (a) The dose constraint is with respect to larger volumes.
   (b) Dose constraint is with respect to smaller volumes.
   (c) The FSU is distinct from each other.
   (d) Lager volumes are to be damaged to produce side effects.
9. Which of the following is an exception for Casarett's classification regarding radiosensitivity?
   (a) RBCS
   (b) Platelets
   (c) Lymphocyte
   (d) Neutrophils
10. Which of the following does not come under accelerated fractionation?
   (a) Concomitant boost
   (b) Weekend boost
   (c) Integrated boost
   (d) CHART
11. The most sensitive phase of gestation is
   (a) Preimplantation
   (b) Organogenesis
   (c) Fatal period
   (d) Postimplantation
12. Which of the following is true regarding the heritable effects of radiation?
   (a) It is a deterministic effect.
   (b) New mutations are caused.
   (c) Probability increases with an increase in dose.
   (d) Severity increases with an increase in dose.
13. Which of the following acute radiation syndromes can be salvaged?
   (a) Cerebrovascular syndrome
   (b) Hepatobiliary syndrome
   (c) Gastrointestinal syndrome
   (d) Hematopoietic syndrome

Answers

1. (b)
2. (a)
3. (d)
4. (c)
5. (c)
6. (b)
7. (b)
8. (b)
9. (c)
10. (c)
11. (b)
12. (c)
13. (d)

Acknowledgment Illustrations (Figures) by Lithika Lavanya M, Senior Resident. Department of Radiation Oncology, Ramaiah medical college, Bengaluru, Karnataka, India.

References

1. Hall EJ, Denekemp J. Residency training in radiation oncology; radiation biology and cancer biology. Int J Radiat Oncol Biol Phys. 1992;24(5):847–9. https://doi.org/10.1016/0360-3016(92)90455-q.
2. Fowler JF. Development of radiobiology for oncology-a personal view. Phys Med Biol. 2006;51(13):R263–R86. https://doi.org/10.1088/0031-9155/51/13/R16.
3. Jones B, Dale RG. The evolution of practical radiobiological modelling. Br J Radiol. 2019;92(1093):20180097. https://doi.org/10.1259/bjr.20180097.
4. Vijayakumar S, King M, Nittala MR, Duhe RJ. Do we need a new approach to cancer biology education for radiation oncology residents? Cureus. 2021;13(12):e20662. https://doi.org/10.7759/cureus.20662.

5. Pushpa Naga CH, Janaki MG, Kirthi Koushik AS. Role of altered fractionation radiotherapy schedules in head and neck squamous cell carcinoma: rationale, evidence and clinical issues. Cancer Res Front. 2017;3(1):215–29. https://doi.org/10.17980/2017.215.
6. Bourhis J, Overgaard J, Audry H, Ang KK, Saunders M, Bernier J, et al. Meta-Analysis of Radiotherapy in Carcinomas of Head And Neck (MARCH) Collaborative Group. hyperfractionated or accelerated radiotherapy in head and neck cancer: a meta-analysis. Lancet. 2006;368(9538):843–54. https://doi.org/10.1016/S0140-6736(06)69121-6.
7. Overgaard J, Hansen HS, Specht L, Overgaard M, Grau C, Andersen E, et al. Five compared with six fractions per week of conventional radiotherapy of squamous-cell carcinoma of head and neck: DAHANCA 6 & 7 randomized controlled trial. Lancet. 2003;362(9388):933–40. https://doi.org/10.1016/S0140-6736(03)14361-9.
8. Skladowski K, Maciejewski B, Golen M, Pilecki B, Przeorek W, Tarnawski R. Randomized clinical trial on 7-days-continuous accelerated irradiation (CAIR) of head and neck cancer–report on 3-year tumor control and normal tissue toxicity. Radiother Oncol. 2000;55:101–10. https://doi.org/10.1016/S0167-8140(00)00139-0.
9. Fu KK, Pajak TF, Trotti A, Jones CU, Spencer SA, Phillips TL, et al. A Radiation Therapy Oncology Group (RTOG) phase III randomized study to compare hyperfractionation and two variants of accelerated fractionation to standard fractionation radiotherapy for head and neck squamous cell carcinomas: first report of RTOG 9003. Int J Radiat Oncol Biol Phys. 2000;48(1):7–16. https://doi.org/10.1016/S0360-3016(00)00663-5.
10. Dische S, Saunders M, Barrett A, Harvey A, Gibson D, Parmar M. A randomized multicentre trial of CHART versus conventional radiotherapy in head and neck cancer. Radiother Oncol. 1997;44(2):123–36. https://doi.org/10.1016/S0167-8140(97)00094-7.
11. Haviland JS, Owen JR, Dewar JA, et al. The UK Standardisation of Breast Radiotherapy (START) trials of radiotherapy hypofractionation for treatment of early breast cancer: 10-year follow-up results of two randomised controlled trials. Lancet Oncol. 2013;14:1086–94.
12. Hickey BE, James ML, Daly T, Soh FY, Jeffery M. Hypofractionation for clinically localized prostate cancer. Cochrane Database Syst Rev. 2019;9(9):CD011462. https://doi.org/10.1002/14651858.CD011462.pub2. PMID: 31476800; PMCID: PMC6718288.
13. Adrian MB, Joanne SH, Duncan AW, Mark AS, Abdulla A, David JB, et al. Hypofractionated breast radiotherapy for 1 week versus 3 weeks (FAST-Forward): 5-year efficacy and late normal tissue effects results from a multicentre, non-inferiority, randomised, phase 3 trial. 2020;395:1613–26.
14. Orlandi E, Palazzi M, Pignoli E, Fallai C, Giostra A, Olmi P. Radiobiological basis and clinical results of the simultaneous integrated boost (SIB) in intensity-modulated radiotherapy (IMRT) for head and neck cancer: a review. Crit Rev Oncol Hematol. 2010;73(2):111–25. https://doi.org/10.1016/j.critrevonc.2009.03.003. Epub 2009 May 5.
15. Lin B, Gao F, Yang Y, Wu D, Zhang Y, Feng G, et al. FLASH radiotherapy: history and future. Front Oncol. 2021;11:644400. https://doi.org/10.3389/fonc.2021.644400. PMID: 34113566; PMCID: PMC8185194.
16. Joiner MC, Bentzen SM. Fractionation: the linear-quadratic approach. Basic Clin Radiobiol, Fourth Edition. 2009;1:102–19.
17. Withers HR. Four R's of radiotherapy. Adv Radiat Biol. 1975;5:241–7. [cited 2022 Nov 1] [Internet] Available from: https://inis.iaea.org/search/search.aspx?orig_q=RN:7228292
18. Steel GG. Dose fractionation in radiotherapy. London: Taylor and Francis; 2007. p. 163–77.
19. Morgan SC, Hoffman K, Loblaw DA, Buyyounouski MK, Patton C, Barocas D, et al. Hypofractionated radiation therapy for localized prostate cancer: an ASTRO, ASCO, and AUA evidence-based guideline. 2018;36(34):3411–30. https://doi.org/10.1200/JCO1801097.
20. Brown JM, Carlson DJ, Brenner DJ. The tumor radiobiology of SRS and SBRT: are more than the 5 R's involved? Int J Radiat Oncol Biol Phys. 2014;88(2):254. [Internet] Feb 2 [cited 2022 Nov 1]. Available from: /pmc/articles/PMC3893711/
21. Roy S, Barragan I, Ahmed SN, Sia M, Singh J, Bahl G. Integration of radiobiological modeling and indices in comparative plan evaluation: a study comparing VMAT and 3D-CRT in patients with NSCLC. Pract Radiat Oncol. 2018;8(5):e355–63.
22. Nahum AE. The radiobiology of Hypofractionation. Clin Oncol. 2015;27(5):260–9. [Internet] May 1 [cited 2022 Nov 1] Available from: http://www.clinicaloncologyonline.net/article/S0936655515000758/fulltext
23. Uzan J, Nahum AE. Radiobiological guided optimisation of the prescription dose and fractionation scheme in radiotherapy using BioSuite. Br J Radiol. 2012;85(1017):1279–86.
24. Yarnold J, Bentzen SM, Coles C, Haviland J. Hypofractionated whole-breast radiotherapy for women with early breast cancer: myths and realities. Int J Radiat Oncol Biol Phys. 2011;79(1):1–9.
25. Proust-Lima C, Taylor JMG, Sécher S, Sandler H, Kestin L, Pickles T, et al. Confirmation of a low α/β ratio for prostate cancer treated by external beam radiation therapy alone using a post-treatment repeated-measures model for PSA dynamics. Int J Radiat Oncol Biol Phys. 2011;79(1):195–201.
26. Vogelius IR, Bentzen SM. Meta-analysis of the alpha/beta ratio for prostate cancer in the presence of an overall time factor: bad news, good news, or no news? Int J Radiat Oncol Biol Phys. 2013;85(1):89–94.

[Internet] Available from:. https://doi.org/10.1016/j.ijrobp.2012.03.004.
27. Dasu A, Toma-Dasu I. Prostate alpha/beta revisited an analysis of clinical results from 14 168 patients. Acta Oncol (Madr). 2012;51(8):963–74.
28. Sachs RK, Hahnfeld P, Brenner DJ. The link between low-LET dose-response relations and the underlying kinetics of damage production/repair/disrepair. Int J Radiat Biol. 1997;72(4):351–74. [Internet]. [cited 2022 Nov 2] Available from: https://pubmed.ncbi.nlm.nih.gov/9343102/
29. Brenner DJ. The linear-quadratic model is an appropriate methodology for determining Isoeffective doses at large doses per fraction. Semin Radiat Oncol. 2008;18(4):234–9.
30. Guerrero M, Li XA. Extending the linear–quadratic model for large fraction doses pertinent to stereotactic radiotherapy. Phys Med Biol. 2004;49(20):4825. 1 [cited 2022 Nov 2] [Internet] Available from: https://iopscience.iop.org/article/10.1088/0031-9155/49/20/012
31. Iwata H, Shibamoto Y, Murata R, Tomita N, Ayakawa S, Ogino H, et al. Estimation of errors associated with use of linear-quadratic formalism for evaluation of biologic equivalence between single and hypofractionated radiation doses: an in vitro study. Int J Radiat Oncol Biol Phys. 2009;75(2):482–8. 1 [cited 2022 Nov 2] [Internet] Available from: https://pubmed.ncbi.nlm.nih.gov/19735872/
32. Otsuka S, Shibamoto Y, Iwata H, Murata R, Sugie C, Ito M, et al. Compatibility of the linear-quadratic formalism and biologically effective dose concept to high-dose-per-fraction irradiation in a murine tumor. Int J Radiat Oncol Biol Phys. 2011;81(5):1538–43. [Internet] 1 [cited 2022 Nov 2] Available from: http://www.redjournal.org/article/S0360301611027751/fulltext
33. Mehta N, King CR, Agazaryan N, Steinberg M, Hua A, Lee P. Stereotactic body radiation therapy and 3-dimensional conformal radiotherapy for stage I non-small cell lung cancer: a pooled analysis of biological equivalent dose and local control. Pract. Radiat Oncol. 2012;2(4):288–95.
34. Shibamoto Y, Miyakawa A, Otsuka S, Iwata H. Radiobiology of hypofractionated stereotactic radiotherapy: what are the optimal fractionation schedules? 2016 [cited 2022 Nov 1]; Available from: https://academic.oup.com/jrr/article/57/S1/i76/2580492
35. Marciscano AE, Haimovitz-Friedman A, Lee P, Tran PT, Tomé WA, Guha C, et al. Immunomodulatory effects of stereotactic body radiation therapy: preclinical insights and clinical opportunities. Int J Radiat Oncol Biol Phys. 2021;110(1):35–52.
36. Brunt AM, Wheatley D, Yarnold J, Somaiah N, Kelly S, Harnett A, et al. Acute skin toxicity associated with a 1-week schedule of whole breast radiotherapy compared with a standard 3-week regimen delivered in the UK FAST-Forward Trial. Radiother Oncol. 2016;120(1):114–8.
37. Liu R, Wang X, Tian JH, Yang K, Wang J, Jiang L, Hao XY. HDR versus low dose rate intracavitary brachytherapy for locally advanced uterine cervix cancer. Cochrane Database Syst Rev. 2014;2014:10. https://doi.org/10.1002/14651858.CD007563.PUB3/MEDIA/CDSR/CD007563/IMAGE_N/NCD007563-CMP-007.01.SVG.
38. Brenner DJ, Schiff PB, Huang Y, Hall EJ. Pulsed-dose-rate brachytherapy: design of convenient (daytime-only) schedules. Int J Radiat Oncol Biol Phys. 1997;39(4):809–15. https://doi.org/10.1016/S0s360-3016(97)00097-7.
39. Huilgol NG. Hyperthermic oncology, it is time of resurrection. J Radiat Cancer Res. 2019;10:155.
40. International Commission on Radiological Protection. The 2007 recommendations of the international commission on radiological protection. New York, NY: Elsevier; 2007. ICRP publication 103.
41. Limitation of exposure to ionizing radiation. Bethesda, MD: National Council on Radiation Protection and Measurements; 1993. NCRP report no. 116.
42. Macià I, Garau M, Lucas Calduch A, López EC. Radiobiology of acute radiation syndrome. Rep Pract Oncol Radiother. 2011;16(4):123–30. https://doi.org/10.1016/j.rpor.2011.06.001. PMID: 24376969; PMCID: PMC3863296.
43. Domínguez GL, Cerezo L. Decontamination of radioisotopes. Rep Pract Oncol Radiother. 2011;16(4):147–52.

Further Reading

Eric JH, Amato JG. Radiobiology for the radiologist. 8th ed. Lippincott: Williams & Wilkins; 2018.
International Atomic Energy Agency. Radiation biology: a handbook for teachers and students, training course series no. 42. Vienna: IAEA; 2010.
Joiner MC, van der Kogel A. Basic clinical radiobiology. 4th ed. Abingdon, Oxon: CRC Press; 2009. https://www.routledge.com/Basic-Clinical-Radiobiology/Joiner-Kogel/p/book/9781444179637
Podgorsak EB. Radiation oncology physics: a handbook for teachers and students. Austria: IAEA; 2005.

Online Resources

https://www.youtube.com/watch?v=lkaNFUwf_bM
https://www.youtube.com/watch?v=LyaG0AGVwh0
https://www.aapm.org/education/vl/vl.asp?id=127
https://www.youtube.com/watch?v=Qkl1A1WXI10s

Clinical Radiation Physics

5

Tharmarnadar Ganesh, Manoj K. Semwal, Prabhakar Ramachandran, Biplab Sarkar, Raghavendra Holla, and Arjunan Manikandan

5.1 Atomic Structure

An atom is the smallest unit of an element that can exist. It is composed of a nucleus surrounded by electrons and is the building block of all matter. The nucleus of an atom is made up of protons and neutrons, and the electrons orbit around the nucleus. The number of protons and neutrons in an atom's nucleus determines the atom's atomic mass. Atoms are incredibly small and are not visible to the naked eye. In fact, it is estimated that a single atom is around one-tenth of a nanometer in size. Even though atoms are so small, they are the foundation for everything in the universe. All of the matter we can see and touch, from the human body to rocks, stars, and galaxies, is made up of atoms. The size of a nucleus is about 4–15 fm where 1 fm is equal to 10^{-15} m. The rest mass of an electron is 9.1×10^{-31} kg with a negative charge of 1.6×10^{-19} C. On the contrary, a proton has a rest mass of 1.673×10^{-27} kg with a positive charge of 1.6×10^{-19} C. The presence of protons inside the nucleus makes the nucleus positively charged and responsible for the electrons orbiting around the nucleus due to the coulomb force of attraction between the electrons and protons. An atom is usually represented by its atomic number and atomic mass. The atomic number (Z) represents the number of protons in an atom, and the mass number (A) is the summation of protons and neutrons. The difference between the two represents the total number of neutrons inside an atom.

$$\text{Total number of neutrons} = \text{Mass Number} - \text{Atomic Number} \tag{5.1}$$

An element can exist in different forms with the same atomic number, and such forms are known as isotopes. A typical example is carbon, with atomic number 6 has three isotopes, ^{12}C, ^{13}C, and ^{14}C. All three isotopes have the same atomic number, i.e., the same proton number, but the mass number is different due to different neutrons. Radium is known to have 33 isotopes, with the most common one, ^{226}Ra, extensively used between the 1900s and 1980s having a half-life of 1600 years.

T. Ganesh (✉)
Manipal Hospital, New Delhi, Delhi, India

M. K. Semwal
Radiation Oncology Centre, Army Hospital (Research & Referral), Delhi Cantonment, New Delhi, Delhi, India

P. Ramachandran
Medical Physics, Radiation Oncology, Cancer Services, Princess Alexandra Hospital, Woolloongabba, QLD, Australia

B. Sarkar
Department of Radiation Oncology, Apollo Multispeciality Hospitals, Kolkata, West Bengal, India

R. Holla
Ruby Hall Clinic, Pune, Maharashtra, India

A. Manikandan
Apollo Proton Cancer Centre, Chennai, Tamil Nadu, India

A. Munshi et al. (eds.), *Radiation Oncology – Principles, Precepts and Practice*,
https://doi.org/10.1007/978-981-97-8389-2_5

Table 5.1 Atomic species with unique properties

Descriptor	Properties	Example
Isotopes	Atoms with the same number of protons	$^{12}_{6}C$, $^{13}_{6}C$, $^{14}_{6}C$
Isotones	Atoms with the same number of neutrons	$^{39}_{19}K$, $^{36}_{16}S$ (39–19 = 36–16)
Isobars	Atoms with the same mass number	$^{24}_{11}Na$, $^{24}_{12}S$
Isomers	Compounds with same molecular formulas but differ in the arrangement of atoms	Butane, methylpropane
Isosters	Molecules with same number of atoms and *Z*	HCl and F_2 NO_2^- and O_3

Atoms that have the same mass number but a different atomic number are referred to as isobars. Typical examples of isobars are ^{40}K, ^{40}Ar, ^{40}Ca, ^{40}Cl, and ^{40}S. Those atoms that have the same neutrons are referred to as isotones, and typical examples are $^{36}{}_{16}S$, $^{37}{}_{17}Cl$, $^{38}{}_{18}Ar$, $^{39}{}_{19}K$, and $^{40}{}_{20}C$. The above elements have the same neutron number, i.e., 20, which is the difference between the mass number and the atomic number. Molecules that have the same number of atoms with the same number of electrons are known as isosters. A typical example is HCl and F_2. There are two atoms in each species, and the total number of electrons in HCl is 18 (1 + 17) and F_2 is 18 (9 + 9). Nuclear isomers, discovered by Otto Hahn are atoms that have the same atomic number and mass number but pose different radioactive properties. These atoms have the same number of neutrons, protons, and electrons. Table 5.1 lists examples of isotopes, isotones, and isobars. Avogadro's number, an important concept in atomic physics defines the number of particles in one mole of a substance. In one mole, there are $6.02214076 \times 10^{23}$ particles. Mol is an SI unit that defines the amount of substance. Atomic mass is the mass of an atom, usually expressed in atomic mass units (amu) and usually expressed at g/mol. The number of atoms/particles per gram in any given substance can be determined by taking the ratio of Avogadro's number and the atomic mass. For example, Carbon has an atomic mass of 12.0110, which means that 12.0110 g of carbon contain $6.02214076 \times 10^{23}$ atoms. In other words, 1 g of carbon contains 5.014×10^{22} atoms. Once we know the number of atoms in 1 g, the number of electrons can be determined by multiplying the resultant by the atomic number. Here the atomic number also refers to the number of electrons in an atom. In the above example, the number of electrons will be 25.07×10^{22} electrons (5.014×10^{22} atoms × 6 electrons/atom). The electrons and the nucleus are governed by the coulomb force of attraction, and in order to eject an electron from an atom, the incoming particle must have energy greater than the binding energy of the electron. The electrons orbit around the nucleus in different shells. The number of electrons in a shell is given by $2n^2$, where n is the orbit number. K, being the first orbit, has 2 electrons ($2 \times (1)^2$), L has 8 electrons ($2 \times (2)^2$) and M has 18 electrons (($2 \times (3)^2$). The binding energy of these shells is usually referred in negative sign as the incoming particles need to deposit the energy to remove from the orbit. The electrons orbiting in the shells closer to the nucleus have higher binding energy compared to the shells further away from the nucleus.

5.2 Electromagnetic Radiation

Electromagnetic radiation is a type of energy that is transmitted in the form of electromagnetic waves. These waves are a combination of electric and magnetic fields, and they can travel through empty space at the speed of light. The electromagnetic spectrum consists of a large range of different types of waves, including radio waves, microwaves, infrared radiation, ultraviolet rays and gamma rays. Each type has its own unique properties. Radio waves, for example, have longer wavelengths and lower frequencies, while gamma rays have shorter wavelengths and higher frequencies. The concept of the dual nature of wave-particles proposed by Louis de Broglie stems from quantum mechanics and it has remained an intriguing topic for physicists and scientists alike since its discovery in 1900. Wave-particle duality is the phenomenon that a particle, such as an electron, can exhibit both wave-like and particle-like behavior at different points in time. Quantum theory states that the energy (E) of a photon can be related to the frequency with the following relation.

$$E = h\nu = \frac{hc}{\lambda} \tag{5.2}$$

where ν is the frequency expressed in Hz and h is the Planck's constant (6.626×10^{-34} m^2 kg s^{-1}), and c is the velocity of light (3×10^8 m/s). From Eq. (5.2), a relation can be established between the wave nature of a particle and the mass of the particles using Einstein's mass-energy relation ($E = mc^2$).

$$E = h\nu = \frac{hc}{\lambda} = mc^2$$

$$\frac{h}{\lambda} = mc \rightarrow \lambda = \frac{\mathbf{h}}{mc} \text{ or } \lambda = \frac{\mathbf{h}}{p}$$

where $p = mc$ (momentum of a photon). If we consider a particle of mass "m" moving at a velocity "v" instead of the photon, then the above equation can be rewritten as

$$\lambda = \frac{\mathbf{h}}{mv} \tag{5.3}$$

The discovery of X-rays by Wilhelm Conrad Röntgen on November 8, 1895, revolutionized the field of medicine and aided doctors to diagnose and treat many diseases that were not within human reach. It is considered to be one of the most remarkable medical breakthroughs in human history. Röntgen was experimenting with cathode ray tubes in his laboratory when he noticed that a nearby chemically coated screen would glow whenever the tube was turned on. Further experimentation revealed that the rays emanating from the tube were able to pass through many materials, including human tissue, and he called these rays "X-rays" because they were unknown at the time. Röntgen's discovery had a profound impact on the field of medicine and is still used in hospitals and clinics today. X-rays are waves of electromagnetic radiation with a very high frequency, created by the accelerating electrons from a cathode to an anode target through a potential difference. The electrons that strike the anode at high energies slow down through the release of X-rays as they pass closer to the nucleus of the atom. In other words, a brake is applied to the electrons that pass near the nucleus and the difference between the incoming and deflected electron energy is emitted as braking radiation. These X-ray photons are also known as Bremsstrahlung or white radiation. In this process, a majority of energy is dissipated as heat, and only a fraction of the energy is converted into X-rays. The incoming electrons can also strike the orbital electrons and may eject them out. If the ejection happens in the inner orbits/ shells, the vacancy created in the shell is then filled up by the electrons from the outer shells/orbits leading to the release of characteristic radiation. The energy of the characteristic radiation emitted through this process is the difference in energy between the electron orbits participating in this process. The X-ray spectrum can be varied by adjusting the X-ray tube current and the potential difference between the cathode and anode. Fig. 5.1 shows the continuous spectrum or the Bremsstrahlung spectrum, and the two characteristic peaks K_α and K_β are generated through the transitions from the L shell to K shell and M shell to K shell, respectively.

The minimum wavelength of the continuous spectrum can be determined using the equation

$$\lambda_{\text{min}} = \frac{hc}{eV}$$

The energy of the electrons is given by $E = eV$. Upon substituting the values of h, c, and e (1.6×10^{-19} C), the minimum wavelength of the emitted X-rays for an accelerating potential of 60 kV will be 0.0207 nm.

Electromagnetic radiation has a wide range of applications, from telecommunications and medical imaging to cancer treatment and food preservation. Radio waves, for example, are used in radio and television broadcasting, as well as in cell phones and other wireless communication devices. Microwaves are used in microwave ovens to heat food, and in radar systems to detect

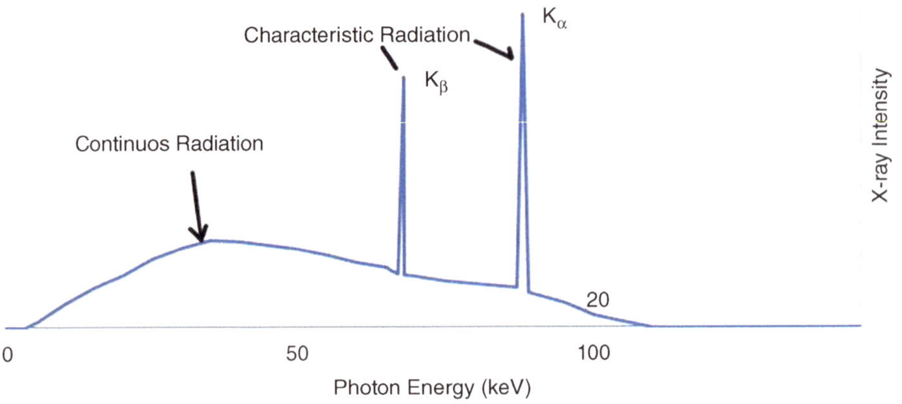

Fig. 5.1 Bremsstrahlung and characteristic X-rays

objects and measure their distance and speed. In medicine, electromagnetic radiation is used in a variety of imaging techniques, such as X-rays, computed tomography (CT) scans, and magnetic resonance imaging (MRI) scans, to produce detailed images of the inside of the body. In cancer treatment, high-energy electromagnetic radiation, such as X-rays and gamma rays, can be used to kill cancer cells and shrink tumors. While electromagnetic radiation has many useful applications, it can also be harmful if not used properly. Exposure to high levels of ionizing radiation, such as X-rays and gamma rays, can damage DNA and cause mutations, leading to cancer and other genetic diseases. It is important to limit the exposure to electromagnetic radiation to minimize the risks of developing these conditions. Overall, electromagnetic radiation is a fundamental and ubiquitous form of energy that plays a vital role in many aspects of our modern world. While its many applications have greatly benefited society, it is important to carefully manage the potential risks associated with this powerful form of energy.

5.3 Radioactivity

Radioactivity was first discovered in the late nineteenth century by the French physicist Henri Becquerel, who found that certain materials emitted strange rays that were able to penetrate opaque objects and fog photographic plates. It is a process by which certain elements, known as radioisotopes, undergo spontaneous decay to produce energetic particles and other elements. This decay process is characterized by the emission of alpha particles, beta particles, and gamma rays, which are different forms of ionizing radiation. The process of radioactivity occurs when the nucleus of an atom becomes unstable and decays, releasing energy in the form of ionizing radiation. This instability can be caused by an excess of protons or neutrons in the nucleus or by a fundamental instability in the nucleus itself. The decay process typically follows a predictable pattern known as a radioactive decay series, in which a parent radioisotope decays into a daughter isotope until a stable isotope is reached. A typical example is a $^{226}{}_{88}Ra$ series which undergoes a series of decays ending up with a stable element ^{206}Pb.

$$^{226}_{88}Ra \rightarrow ^{222}_{86}Rn \rightarrow ^{218}_{84}Po \rightarrow ^{214}_{82}Pb \rightarrow ^{214}_{83}Bi \rightarrow ^{214}_{84}Po \rightarrow ^{210}_{82}Pb \rightarrow ^{210}_{83}Bi \rightarrow ^{210}_{84}Po \rightarrow ^{206}_{82}Pb$$

In the initial years of the discovery of radium, it was quantified in terms of milligrams of radium. This was eventually replaced with "curie" which was defined as the quantity of ^{222}Rn in equilibrium with a gram of ^{226}Ra. One curie of radioactive substance emits $3.7 \times 10^{10}\ s^{-1}$. The term activity was later redefined in SI unit as "1 disintegration per second (s^{-1})" or 1 Becquerel (Bq).

$$1\,Ci = 3.7\times10^{10}\ s^{-1}\ \text{or}\ 1\,Ci = 3.7\times10^{10}\ Bq \quad (5.4)$$

$$1\,Bq = \frac{1}{3.7\times10^{10}} = 2.7\times10^{-9}\ Ci \quad (5.5)$$

The activity at any point in time is defined by

$$A = A_0\,e^{-\lambda t} \quad (5.6)$$

Similarly, the number of atoms remaining after an elapsed time "t" is defined by

$$N = N_0\,e^{-\lambda t} \quad (5.7)$$

In the above two equations, λ is the decay constant and can be determined from half-life ($T_{1/2}$).

The amount of activity left after one half-life can be determined by substituting, when $A = A_0/2$ in Eq. 5.5, $1/2 = e^{-\lambda t1/2}$.

Upon taking natural logarithms on either side, $\ln(1/2) = \ln(e^{-\lambda t1/2})$

$$\ln(2) = \lambda t_{1/2} = 0.693$$

$$\text{The decay constant } \lambda = 0.693 / t_{1/2} \quad (5.8)$$

Using the above Eq. (5.8), the decay constant for any radionuclide can be determined by knowing its half-life. For example, the decay constant for Co-60 will be 4.17×10^{-9} decays per sec (0.693/(5.27 years × 365 days × 24 h × 3600) or 0.1315 decays per year (0.693/5.27 years).

An important term that is widely used in brachytherapy is the specific activity which is the activity of 1 g of the radionuclide itself. In other words, it is defined as the activity per gram of the substance and specified as Bq g^{-1} or Ci g^{-1}. The strength of the radionuclide source used for brachytherapy is limited by its specific activity. To estimate the specific activity, we need to know the half-life and the number of atoms per gram of the radionuclide. For example, the half-life of Cobalt-60 is 5.27 years which equates to 1.662 X 10^8 secs, and one mole of Co-60 weighs contains 60 grams. In other words, 1 g of Cobalt-60 contains 1.004×10^{22} atoms ($6.023 \times 10^{23}/60$). Multiplying the total number of atoms contained in 1 gram of Cobalt-60 by the decay constant gives the total disintegrations per sec or per year. For Cobalt-60, it will be 1.32×10^{21} decays per year ($1.004 \times 10^{22} \times 0.1315$ decays per year) or 4.187×10^{12} decays per second. To determine the specific activity for Cobalt-60 in Ci, the number of atoms decaying per second needs to be divided by 3.7×10^{10} disintegration per second, which will be 1132 Ci/g (4.187×10^{12} decays per second/3.7×10^{10} disintegration per second). The half-life of a radionuclide plays an important role in determining the specific activity, the higher the half-life, the lesser the specific activity, and vice versa. So, if we compare Iridium-192, which has a half-life of 73.8 days, against Cobalt-60 (specific activity: 1132 Ci/gram), which has a half-life of 5.27 years, Iridium-192's specific activity will be 9200 Ci/g. The higher the specific activity, the smaller would be the size of the source required for clinical applications.

Radioactivity has many applications, from medical imaging and cancer treatment to electricity production in nuclear power plants. Radioisotopes are commonly used in medical imaging, such as positron-emission tomography (PET) and bone scans, to diagnose and treat diseases. In cancer treatment, radioisotopes can be used to destroy cancer cells with minimal harm to surrounding healthy tissue. In nuclear power plants, radioactivity is harnessed to produce heat, which is then used to generate steam and produce electricity. However, the use of nuclear power also carries significant risks, such as the potential for radiation leaks and the possibility of nuclear accidents. The most infamous example of a nuclear accident is the 1986 Chernobyl disaster, in which a reactor malfunction caused a catastrophic explosion and released large amounts of radioactive material into the environment.

Alpha particles are a type of ionizing radiation consisting of two protons and two neutrons bound together. These particles are produced during the

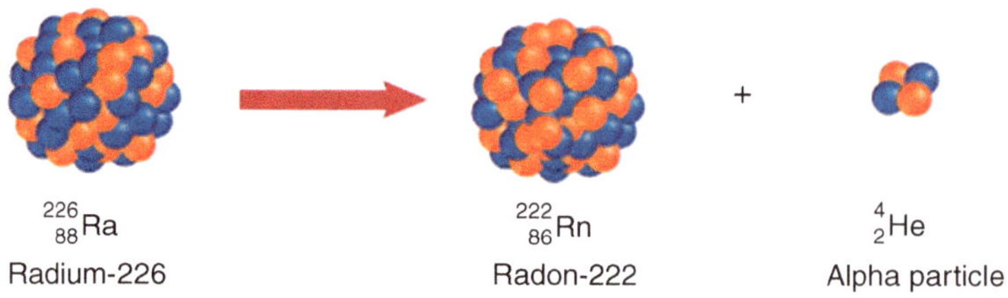

Fig. 5.2 Emission of alpha particle (helium nucleus) by Radium-226

radioactive decay of certain elements, such as uranium, actinium, thorium, and radium which are rich in neutrons. Figure 5.2 illustrates the decay of Ra-226 to Rn by the emission of an alpha particle. Here, radium has 138 neutrons and 88 protons. It decays to Rn by emitting alpha particles.

Alpha particles have a positive charge and are relatively large in size and travel just at 5–7% of the speed of light. They are easily stopped by materials such as paper or a few centimeters in air. When alpha particles are emitted from a radioactive source, they travel only a short distance before being absorbed by other atoms. This means that alpha particles are not capable of penetrating deep into the body, and they pose little threat to internal organs unless the radioactive material is ingested or inhaled. However, alpha particles can still be harmful if they are absorbed by the skin or inhaled into the lungs.

Alpha particles are commonly used in medicine for their ability to kill cancer cells. In the early 1900s, alpha-emitting radioisotopes, such as radium was used to treat cancers. The alpha particles emitted by these isotopes will destroy the cancer cells in the immediate area, while minimizing harm to healthy tissue.

In addition to their use in medicine, alpha particles are also used in other applications, such as smoke detectors and industrial radiography. In smoke detectors, a small amount of the radioactive element americium-241 is used to ionize the air in the detector. When smoke enters the detector, it absorbs the alpha particles, reducing the ionization and triggering an alarm. In industrial radiography, alpha particles are used to inspect the integrity of materials, such as metal pipes and welds, by detecting any flaws or defects in the material.

Beta rays are made up of high-energy particles, specifically electrons or positrons, which are emitted from the nucleus of an atom during radioactive decay. They are smaller and lighter and travel faster than alpha particles and can penetrate into materials causing ionization in the atoms they interact with. With beta emission, the excess neutrons in an atom are transformed into a proton and an electron (negative beta particle). The proton particle remains within the nucleus, and the electron particle is ejected as a beta particle. A classic example is the emission of beta particles due to the radioactive decay of cobalt-60 to nickel-60. Cobalt-60 emits two negative beta particles of 0.31 MeV (99.88%) and 1.48 MeV (0.12%) and eventually transforms into nickel-60 with two gamma photons (1.33 and 1.17 MeV). This ionization can cause damage to living cells, which is why exposure to beta rays is considered harmful.

Gamma rays are weightless quantum packets of energy and have very high energy. A beta or alpha particle decay is usually accompanied by the emission of gamma ray during radioactive decay. However, unlike beta and alpha particles, gamma rays are not made up of particles. Gamma rays are extremely dangerous, as they have the ability to penetrate materials and cause ionization in the atoms they interact with. This ionization can cause damage to DNA and other cellular structures, which can lead to cancer and other health effects. For this reason, it is important to avoid unnecessary exposure to gamma rays and to take precautions to protect oneself from these harmful rays.

Radioactive dating is an important application of radioactivity. By measuring the amount of a radioisotope present in a sample and comparing it to the amount of its decay products, scientists can determine the age of the sample. This technique is commonly used to date archaeological artifacts and geological samples, such as rocks and fossils. Radioactivity is also used in the field of geology to study the composition and structure of the Earth. Natural sources of ionizing radiation, such as uranium and thorium, are present in rocks and soil, and their presence can be detected using specialized instruments. By mapping the distribution of these radioactive elements, geologists can gain insight into the geologic history of an area and identify areas of geological interest, such as potential sites for oil and mineral deposits. Despite its many useful applications, the potential risks of radioactivity must also be carefully managed. Radioactive waste, such as that produced by nuclear power plants, must be properly disposed of to prevent harmful exposure to ionizing radiation. In addition, strict regulations and safety measures must be in place to prevent accidents and minimize the risks of radiation exposure to workers and the general public.

5.4 Interaction of Radiation with Matter

Why should someone interested in radiation oncology read about interaction of ionizing radiation with matter? How does it matter if an X-ray interacts with an atom in the medium by photoelectric or by Compton process? In what way would it help by understanding about these processes and the factors they depend on? The answer to all these questions is simple: *energy deposition*. If there is no interaction of radiation with matter, there is no energy deposition in the medium, and if there is no energy deposition, there is no biological effect. It is through these large numbers of interactions that a passing radiation beam deposits energy throughout the tissues, leading to atomic excitation, and ionization, finally resulting in radiobiological damage to DNA responsible for the end effect. Thus, in radiation oncology, everything revolves around energy deposited in the tissues, i.e., the dose. It is thus imperative to have an in-depth understanding of how different ionizing radiations interact with matter, what are the factors they are dependent on, and also how do they ultimately deposit the energy through secondary particles in the medium.

5.4.1 Interaction of Photons with Matter

Uncharged particles, like photons and neutrons, are called as indirectly ionizing radiations since they are not directly responsible for the energy deposition in the tissues. Instead, they are responsible for *transporting* the energy across relatively large distances in the human body, and it is only the secondary electrons that are set in motion by these photons, which are responsible for the *actual deposition of energy* in tissues.

Each photon, X-ray or γ-ray, is a packet of energy. A photon beam consisting of such photons is characterized by appropriate terminologies to describe their numbers, their energy content, how they get attenuated, the energy they transfer to the medium, and also how that energy is absorbed.

The photon fluence (Φ) is the quotient dN by da, where dN is the number of photons entering a sphere with cross-sectional area da: $\Phi = \frac{dN}{da}$ and the fluence rate, also called as flux density (ϕ) is the fluence per unit time: $\phi = \frac{d\Phi}{dt}$, dt being the time interval.

Similarly, the energy fluence (Ψ) is the quotient of dE by da, where dE is the sum of the energies of all the photons entering a sphere of cross-sectional area da, and is given as $\Psi = \frac{dE}{da}$.

If the beam is monoenergetic, then dE is easy to calculate and is given as the number of photons multiplied by the energy of each photon hν: dE = dN. $h\nu$. Energy fluence rate, also called energy flux density (ψ) is the energy fluence per unit time: $\psi = \frac{d\Psi}{dt}$.

5.4.1.1 Photon Beam Attenuation

Consider a narrow beam of monoenergetic photons incident on an absorber. A detector is placed beyond the absorber, at a fixed distance, to measure only the primary photons which do not undergo any interaction and simply pass through it. Some photons interact in the medium and are either removed from the beam (completely absorbed by the medium), or scattered from the original direction, either way resulting in the attenuation of the beam. The detector does not measure the scattered radiation. The reduction in the number of photons (dN) when the beam passes through the absorber is proportional to the incident number of photons *N* and to the thickness of the absorber dx. By including a proportionality constant, one can express it mathematically as:

$$dN = -\mu N dx \tag{5.9}$$

where μ is the proportionality constant, called the attenuation coefficient. *N* is the number of incident photons, and d*x* is the absorber thickness. The number of photons decreases as the thickness increases and hence the minus sign. The same equation is valid for the beam intensity (*I*) as well:

$$dI = -\mu I dx \tag{5.10}$$

When the thickness is expressed as a length, μ is called as the linear attenuation coefficient of the absorber material. The units of μ are 1/cm or cm^{-1}. These differential equations for attenuation can be solved to yield as follows:

$$I(x) = I_0 e^{-\mu x} \tag{5.11}$$

where $I(x)$ is the intensity of the beam transmitted through a medium of thickness x and I_0 is the intensity of the incident beam. Similar to the half-life concept, we can define the half-value layer (HVL) as the thickness of an absorber that will attenuate the initial intensity of the beam to half. From the equation, we get

$$\text{HVL} = \frac{0.693}{\mu} \tag{5.12}$$

The exponential attenuation is strictly valid for a monoenergetic beam. The HVL for such a beam remains constant irrespective of the absorber thickness. However, X-ray beams from radiation generators contain a spectrum of energies and their attenuation is not truly exponential. As the beam passes through the first HVL, more low-energy photons are preferentially removed from it and its penetrative power increases. The second HVL is greater than the first [1, 2].

Coefficients

Coefficients are useful concepts to best describe beam attenuation, energy transferred from the beam and energy absorbed by the medium.

Mass Attenuation Coefficient

The linear attenuation coefficient μ suffers from the drawback that it depends on the density of the material. The attenuation due to a given thickness *x* depends on the number of electrons present in that thickness, and hence, μ is dependent on the density of the medium. To remove the dependency on density, μ is divided by density ρ to yield $\frac{\mu}{\rho}$, called as the *mass attenuation coefficient*, a more preferred and fundamental unit [1]. Since the units of μ are cm^{-1} and ρ are g/cm^3, the mass attenuation coefficient has units of cm^2/g. The attenuation coefficient corresponds to the fraction of photons removed per unit thickness through interactions. Through these interactions photons will transfer all or part of their energy to the material. The material in turn will absorb all or only a part of this energy transferred to it.

Energy Transfer Coefficient

A photon interacting with the medium can transfer either all of its energy or only a part of it to the electrons. Subsequent to partial energy transfer to the electron, the photon travels in a different direction with reduced energy further undergoing one or more such interactions until all of its energy is converted as kinetic energy of the secondary electrons or the scattered photon escapes altogether from the material. The fraction of the energy thus transferred by the photon beam, while traversing a material, into kinetic energy of the charged particles is given by the *energy transfer coefficient* (μ_{tr}). This coefficient is given as:

$$\mu_{tr} = \frac{\overline{E}_{tr}}{h\nu}\mu \quad (5.13)$$

where $\overline{E}_{tr}$ is the average energy transferred into kinetic energy of charged particles per interaction. The mass energy transfer coefficient is obtained by dividing μ_{tr} by ρ, i.e., $\frac{\mu_{tr}}{\rho}$.

Energy Absorption Coefficient

Of the energy transferred to the material in the form of kinetic energy of the electrons, a part of it may ultimately escape. Most of the electrons will lose their kinetic energy in inelastic collisions with atomic electrons of the material, but some of them will undergo bremsstrahlung interactions in interactions with the nuclei. This energy is not deposited locally and radiated out of the volume. Hence, it is not included in the calculation of locally absorbed energy.

The *energy absorption coefficient* (μ_{en}) is thus defined as the product of energy transfer coefficient and (1-g) where g is the fraction of the energy of secondary particles lost by bremsstrahlung interactions in the material.

$$\mu_{en} = \mu_{tr}\left(1 - g\right) \quad (5.14)$$

The *mass energy absorption coefficient* is obtained by dividing μ_{en} by ρ, i.e., $\frac{\mu_{en}}{\rho}$.

In soft tissues and in low-Z material, electrons rarely undergo bremsstrahlung interactions and thus $\mu_{en} = \mu_{tr}$. However, when the kinetic energies of the secondary particles are high and the material is a high-Z one, then g becomes significant [1–3]. As pointed out earlier, it is the energy absorbed that is responsible for all the biological effects of radiation, and hence, the energy absorption coefficient assumes greater significance.

A photon beam traversing a medium suffers attenuation through five interactions: coherent scattering; photodisintegration; photoelectric effect; Compton effect; and pair production.

5.4.1.2 Photodisintegration

When a very high-energy photon interacts with the atomic nucleus, it can result in the emission of one or more nucleons. Mostly such interactions will result in the emission of neutrons by the nuclei. This type of interaction becomes important at only very high energies and is not relevant in the energy range of photons used in radiotherapy.

5.4.1.3 Coherent Scattering

Also called as classical scattering or Rayleigh scattering, the coherent scattering is a process in which no energy is transferred to the medium. The incoming photon sets an electron into oscillation when it passes close to it and the electron reradiates the energy at the same frequency. The scattered photons have the same energy as that of the incident photons and there is no energy deposition involved. Since the process is possible only in high-atomic number materials and with photons of low energy, it is ignored in radiotherapy.

5.4.1.4 Photoelectric Effect

In this interaction process, shown in Fig. 5.3, the photon is completely absorbed by the medium. If the energy of the incoming photon is higher than the atomic binding energy of an inner electron, and all of its energy is transferred to the electron which is ejected out of the atom with a kinetic energy equal to the incident photon energy minus the binding energy of the electron, $h\nu' - E_B$, where E_B is the binding energy of the electron.

The process does not end with the ejection of electron with this kinetic energy. A cascade of events take place subsequently: ejection of electron leaves the atom with a vacancy in the shell, which is immediately filled by an electron from the outer shell. In doing so, there will be further release of energy in the form of a characteristic X-ray, having an energy equal to the difference in binding energies of the two shells involved. There is also the possibility of the characteristic X-ray being completely absorbed locally within the atom, resulting in the emission of an Auger electron. When the photoelectric effect occurs in soft tissues, the characteristic photons will have very low energy, since the binding energy of the

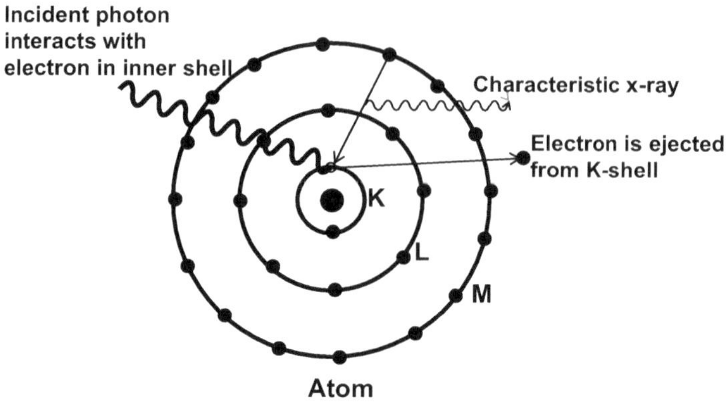

Fig. 5.3 Photoelectric effect. Incoming photon interacts with an electron in K-shell (*inner shell*). The electron is ejected out of the atom as the photoelectron. A vacancy is created. It is filled by an electron from M-shell. A characteristic X-ray is emitted

K-shell (innermost shell) is only about 0.5 keV. However, in higher-atomic number materials, the characteristic photons have higher energy, and may deposit their energies at large distances from the site of interaction.

The probability of a photon undergoing photoelectric interaction depends strongly on its energy. The mass photoelectric attenuation coefficient (τ/ρ) is inversely proportional to the cube of photon energy:

$$\tau / \rho \propto 1 / E^3 \qquad (5.15)$$

For lead, there are discontinuities at 15 and 88 keV corresponding to the binding energies of the L and K shells and they are known as absorption edges. They occur due to the sudden availability of electrons in the L-shell for the interaction process when the photon energy just equals the binding energy for that shell. Moreover, when the photon energy and binding energy are nearly equal, almost all of the interactions will be with L-shell electrons. When like-minded people meet each other, the interaction between them will be strong and easy. Likewise, when the photon energy and binding energies are equal, interaction is strong. Beyond this point, as the energy increases, the probability of interaction decreases following the inverse relationship of energy-cubed until the next discontinuity occurs. The absorption edges for water occur at less than 0.5 keV and hence are not seen in the graph.

The photoelectric absorption depends on the atomic number of the absorbing material and follows the approximate relationship:

$$\tau / \rho \propto Z^3 \qquad (5.16)$$

This strong dependency on the atomic number is the basis for many applications in diagnostic radiology. Thus, even if there is a small change in the composition of the atomic number, the photoelectric attenuation is changed significantly. Two materials with different atomic numbers will attenuate the beam to different extent, thus producing subject contrast in the image. When we combine the two equations, we get

$$\tau / \rho \propto Z^3 / E^3 \qquad (5.17)$$

The photoelectrons from photoelectric process are ejected at 90° to the direction of the incoming photon when the photon energy is low and for higher energy photons, they are emitted in a more forward direction.

5.4.1.5 Compton Effect

In this interaction process, the incoming photon interacts with electrons in the outer shells, which have negligible binding energy compared to its energy [1–6]. The electron will receive some of the energy from the photon as its kinetic energy and is ejected at an angle ϕ, and the photon, with lesser energy, is scattered at an angle θ to the direction of the incoming photon (Fig. 5.4). In the interaction, energy and momentum have to be conserved. Following this principle of conserva-

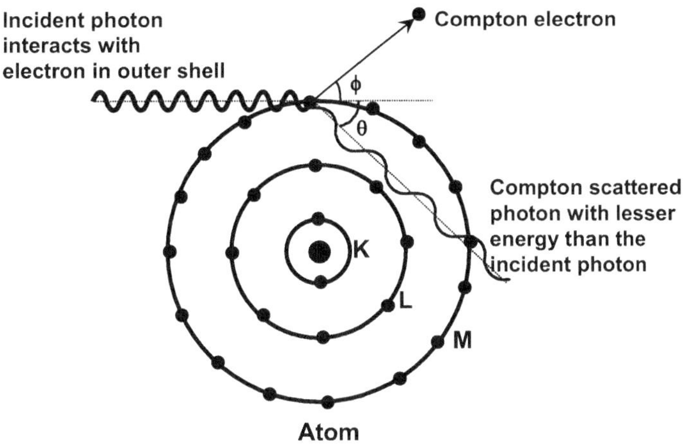

Fig. 5.4 Compton effect. Incoming photon interacts with an electron in M-shell (*outer shell*). The electron is ejected out of the atom as the Compton electron. The photon is scattered with lower energy than the original photon. Energy difference between the photons is given to the photoelectrons

tion, we can derive three important equations defining energies of scattered photon and the Compton electron:

$$E = h\nu_0 \frac{\alpha(1-\cos\theta)}{1+\alpha(1-\cos\theta)} \tag{5.18}$$

$$h\nu' = h\nu_0 \frac{1}{1+\alpha(1-\cos\theta)} \tag{5.19}$$

$$\cot(\phi) = (1+\alpha)\tan(\theta/2) \tag{5.20}$$

where $h\nu_0$ is the energy of the incident photon; $h\nu'$ is the energy of the scattered photon; E is the energy of the Compton electron ejected; $\alpha = \frac{h\nu_0}{m_0c^2}$ and m_0c^2 is the rest energy of the electron (0.511 MeV). If $h\nu_0$ is also expressed in MeV, then $\alpha = \frac{h\nu_0}{0.511}$.

Using these fundamental equations, one can arrive at some important observations that have much relevance in the practice of radiotherapy:

- When a low-energy photon undergoes Compton interaction, the scattered photon retains much of the original energy and the Compton electron receives very less energy from the incident photon.
- When a high-energy photon, having energy much higher than the rest energy of the electron, undergoes Compton effect, the Compton electron gets much of the incident photon's energy and the scattered photon carries only a fraction of the incident energy. Thus, the energy deposition is more at high photon energy since these secondary particles are the ones responsible for energy deposition.
- In the calculation of secondary barrier thickness, it is important to know the energy of the scattered photon, especially the photons scattered at 90° as they determine the thickness. It can be shown that irrespective of the incident photon's energy, the maximum energy of the photons scattered at 90° will be 0.511 MeV. Similarly, backscattered photons will have a maximum of 0.255 MeV.
- It may be noted that photons scattered at angles less than 90° will have energy greater than 0.511 MeV and as the scattered angle becomes closer to the incident photon direction, it will approach incident photon's energy.

As discussed earlier, the photoelectric effect is strongly dependent on energy, and its probability of occurring varies inversely as the cube of photon energy. As the photon energy increases, photoelectric effect decreases rapidly and the Compton effect becomes increasingly important. Though it becomes more important than the photoelectric effect, it is only on a relative scale. As the energy increases, the Compton effect also decreases. This can be explained as follows: With increasing energy, the probability of a photon undergoing an interaction continuously decreases. At very low energies, not only the interaction probability is high, but also whenever an interaction happens it is mostly by photoelectric process; rarely it will be by Compton process. At higher energies, it is vice versa. First of all, the interaction probability is reduced, and when it happens, it is mostly Compton effect.

Since the Compton interaction is between the incident photon and "free" electrons, the process is independent on the atomic number Z [1–6]. The Compton mass attenuation coefficient (σ_c/ρ) is thus independent of Z and only dependent on the number of electrons per gram. With the exception of hydrogen, all other materials have almost same electrons per gram, and thus, we can consider σ_c/ρ to be nearly the same for all materials.

5.4.1.6 Pair Production

In this interaction process, energy is converted into mass. When the incident photon, having sufficient threshold energy, passes close to the nucleus of the absorber atom, it is completely absorbed giving rise to an electron-positron pair. The threshold energy has to be 1.02 MeV since the combined rest masses of electron-positron pair comes to 1.02 MeV. Any energy in excess of 1.02 MeV is distributed between the electron and the positron in any proportion as their kinetic energies, although mostly in equal proportion. The charged particles are emitted in the forward direction (Fig. 5.5).

The positron emitted in pair production, being a charged particle, undergoes similar interactions like an electron—i.e., ionization, excitation, and bremsstrahlung. Thus, it loses all of its kinetic energy, and as it slows down finally, it is absorbed by one of the free electrons nearby giving rise to two photons, each with energy of 0.511 MeV, which is the rest mass of these particles. The photons are called annihilation photons and are emitted in opposite directions in conservation of the momentum.

The pair production attenuation coefficient (Π) is dependent on Z^2 per atom, Z per electron and Z per gram. The probability for this interaction increases as the logarithm of incident photon energy.

5.4.1.7 Relative Importance of Various Interactions

In the energy range used in radiotherapy, the total mass attenuation coefficient (μ/ρ) can be expressed as the sum of the individual mass attenuation coefficients of the three major interactions:

Total mass attenuation coefficient = Individual coefficients of (Photoelectric + Compton + Pair production)

$$(\mu / \rho) = (\tau / \rho) + (\sigma_c / \rho) + (\pi / \rho) \quad (5.21)$$

- As the energy increases, the total mass attenuation coefficient decreases. This in effect means, as the energy increases, the interaction probability decreases. In other words, with increase in energy, in absolute terms, the beam is attenuated less.
- At lower energies (<0.5 MeV), high-Z materials attenuate the beam more than the low-Z

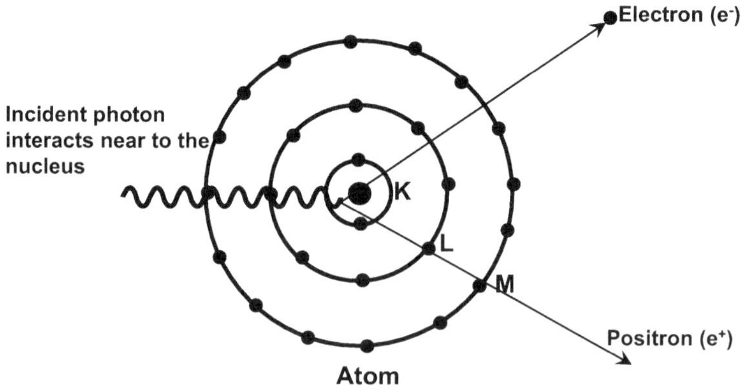

Fig. 5.5 Pair production. Incoming photon interacts near to the nucleus—it disappears completely. In its place, an electron–positron pair is created (*energy is converted into mass*)

material. The difference is significant at diagnostic energy range (<100 keV).

- In the energy range often used in therapeutic applications (1–10 MeV), there is no significant difference in the mass attenuation coefficients between the low-Z and high-Z materials.
- The difference between the low-Z and high-Z materials tend to show up only at higher energies which are not used for therapy purposes.

5.4.2 Interactions of Charged Particles

The charged particles (electrons, protons, α-particles, and nuclei) primarily interact by ionization and excitation. Although radiative interactions are possible, they are observed mostly with electrons rather than with other particles.

Charged particles interact through Coulomb force between the particle's electric field and the electric field of the orbital electrons and nuclei of the atoms of the material. While the collisions between the particles and orbital electrons result in ionization and excitation of the atoms, the collisions between particles and nuclei result in radiative interactions. Heavier particles like protons do not undergo much scattering, while electrons, being lighter undergo multiple scattering and lose their energies in many smaller steps. Heavy charged particles undergo nuclear reactions as well giving rise to radioactive nuclides.

5.4.2.1 Heavy Charged Particles

Particles like protons, mesons, α-particles, atomic nuclei are considered as "heavy" since their rest mass is much higher than the rest mass of electrons. As the charged particles travel through medium, they keep losing their energy continuously and slowing down as well. Since the rate of energy loss is dependent on the square of the particle charge and inversely proportional to the square of its velocity, the slowing down leads to a rapid increase in the rate of energy loss near the end of the particle track and drops to near-zero value beyond the range of the particle. This steep increase followed by steep drop in dose near the particle's track end is called as Bragg peak.

5.4.2.2 Electrons

Although the interactions of electrons while passing through a medium are similar to the ones experienced by heavy particles, there are some major differences between their interactions arising due to the lower mass of electrons. Being light particles, electrons undergo multiple scattering and more changes in direction of motion. This in effect smears out Bragg peak.

Electrons lose energy through excitation and ionization. In the former, energy is transferred to the atomic electron, but it is not sufficient enough to strip it off from the atom to which it is bound. The atomic electron is displaced from its stable position only to return back to its original position. In the ionizations, the atomic electron is knocked out of the atom, sometimes with sufficient energy to create an ionization track of its own. It is called as a delta ray or as secondary electron.

Radiative collisions are also possible when electrons interact with the magnetic field of the nucleus resulting in the emission of bremsstrahlung radiations. The chapter on electron beam therapy discusses in detail about electron interactions and their clinical applications.

5.4.3 Interactions of Neutrons

Neutrons, having no charge, are indirectly ionizing particles just like X-rays and γ-rays. However, their way of interaction and transfer of energy is different. Neutrons interact by (a) recoiling protons from hydrogen and recoiling heavy nuclei from other elements, and (b) nuclear disintegrations. In the former, which is much similar to a billiard-ball collision, energy is redistributed following the collision between colliding particles. When the colliding masses are similar, energy is transfer is efficient whereas with a heavier nucleus, very little energy is transferred. Thus, materials like paraffin wax and polyethylene, with more hydrogen are far more efficient neutron absorbers than lead which makes as a poor shielding material for neutrons.

In tissues, it is the recoil protons that are largely responsible for dose deposition from a high-energy neutron beam.

Readers are recommended to download and install an interactive software tool RadSim (http://radsim.weebly.com/) to see the animations of these interaction processes and also the scattering angles and energy losses.

5.5 Radiotherapy Machines: A Historical Journey and Present Perspective

The temporal distribution of radiotherapy machines can be divided into two segments. In the pre-1950 era, majority of the machines were kilovoltage X-ray machines. Post-1950, we saw the introduction of megavoltage machines: the Cobalt-60 (60-Co) teletherapy machine was introduced into clinical service in the year 1951 (Saskatoon Cancer Clinic, Saskatoon, Canada). Almost simultaneously (1953), the first 8 MV electron accelerator was introduced, while Stanford University introduced an isocentric 6 MV photon linear accelerator in 1957.

5.5.1 Deep Therapy X-ray Units

The basic structure and components of an X-ray tube are shown in Fig. 5.6. The kilovoltage units, which are hooded anode X-ray tubes, can be of three kinds depending upon their operating potential: contact units (40–50 kVp, 5–10 mA), superficial therapy units (50–150 kVp, operating at a distance of 15–20 cm from skin) and deep therapy orthovoltage units (150–500 kVp, mostly operated in the range 250–300 kVp, 10–20 mA, operating at a distance of 50 cm from skin). X-ray treatment machines that operated between 500 and 1000 kV were designated supervoltage therapy machines. Machines that operated above 1 MV were termed megavoltage machines. The Van de Graff generator was one of the first of its kind. All these are clinically obsolete now.

5.5.2 Cobalt-60 Units

Initially started with the vertical movement machines, these telecobalt units were subsequently made into isocentric (C-Arm) units with SAD of 80 to 100 cm and max field size, 36 × 36 cm^2. With the advancement of nuclear reactor technologies in the 1960s and 1970s, the size of the source was reduced. Due to higher neutron flux available in modern reactors, sources with higher strength of up to 13,000 Ci could be produced resulting in exposure rates in excess of 200 R/min at 1 m (1–2 cm dia.). Telecobalt machines, which dominated radiation therapy for nearly three decades, continue to treat many patients in the economically weaker Asian and African countries. Nonetheless, Co-60 teletherapy machines are not in clinical use in the developed countries. The relative consistency of beam output, a well-defined half-life, and the absence of small output fluctuations found in electrical (X-ray) machines are all advantages of Co-60 megavoltage machines. However, disadvantages such as source replacement after

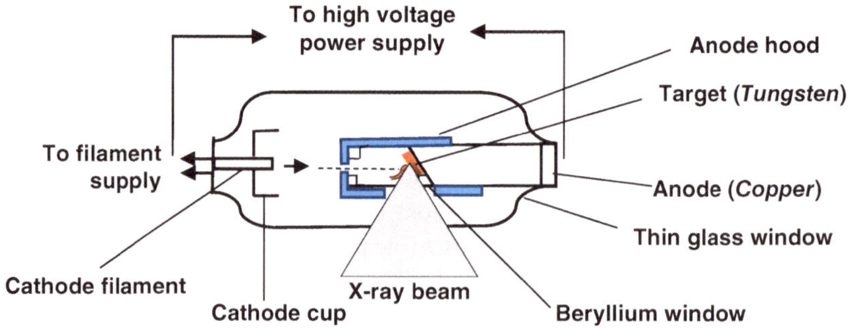

Fig. 5.6 Basic X-ray tube and its components

every 5–7 years, poor field flatness for large fields, a large penumbra due to the source's finite size, and lower depth doses when compared to megavoltage X-ray units were also encountered.

5.5.3 Gamma Knife Radiosurgery Unit and Other Advanced Cobalt Units

Aside from the C-Arm telecobalt machine, another unique machine, the gamma knife radiosurgery unit, was developed in 1968 by Swedish neurosurgeon Prof. Lars Leksell. With this specially designed unit, highly collimated beams could be delivered to single or multiple intracranial sites with submillimeter spatial accuracy. Initially limited to invasive frame radiosurgery (single fraction), now fractionated treatments are possible. A gamma knife unit is made up of three basic components: a spherical source housing, four collimator helmets, and a control couch. The 201 cobalt-60 sources with slightly higher than 30 Ci source strength were arranged in a quasi-hemispherical configuration, with all gamma rays converging at the isocenter, which is 40 cm from the sources. The target volume was positioned at the isocenter with the help of Leksell stereotactic frame. Newly loaded combined sources provided a dose rate of 300 cGy/min at the isocenter. For each helmet, 201 tungsten collimators define specified circular apertures (4, 8, 14, or 18 mm projected at the isocenter). To conform the radiation dose to the shape of the target, various combinations of aperture diameters, aperture blocking (plugging), irradiation, times, and head positions were used. A specific combination of these four parameters defines what is referred to as a "shot" in Gamma Knife terminology. The Gamma Knife unit has undergone several changes and the most recent models have less number of sources, use image guidance and cone beam compted tomography (CBCT) for patient set-up, infrared cameras to monitor intrafraction movements, as well as the ability to move the entire patient on the couch to each stereotactic x, y, and z coordinate, a large internal patient cavity for extended access to peripheral cranial anatomy, and so on.

In the 1990s, the concept of combining an MRI scanning system with a linac emerged. One such early development resulted in the ViewRay system, which combined a 0.35 T split-bore MRI that spans a radiotherapy gantry housing, three cobalt heads spaced 120° apart, and a radiotherapy gantry housing. The MRI scanner and the radiotherapy gantry shared a common isocenter, allowing for MR-based treatment localization and intrafraction treatment monitoring with MRI. The first ViewRay system also enabled practical online adaptive radiotherapy, in which patients were imaged, a new plan was generated, patient-specific QA was performed, and treatment was delivered in a time frame compatible with conventional radiotherapy treatment duration. This also marked the beginning of the first practical online adaptive radiotherapy program, with Washington University administering the first online MRI-guided adaptive radiotherapy treatment on September 8, 2014. ViewRay introduced LINAC-based machines in 2016, effectively putting an end to MRI-guided cobalt machines.

5.5.4 Linear Accelerators

The first microwave electron linear accelerator (8 MV) for medical use went into operation in 1953 at the Medical Research Council's Radiation Research Centre at Hammersmith Hospital in London. Linear accelerators have advanced in design and construction over time, and they are now the most commonly used megavoltage treatment units in clinical practice. Figure 5.7 presents a typical block diagram showing the major components of a high energy medical linac. The unit generates electrical fields by propagating microwaves (S band, 2856 MHz) through an accelerator structure, which are then used to accelerate the injected electrons along a straight line. The accelerator structure consists of a stack of cylindrical metal cavities having an axial hole through which the accelerated electrons pass. Microwaves are classified as either travelling wave or standing wave. In a travelling wave, the electrons travel with the electric field, as the field propagates through time, like a surfer riding

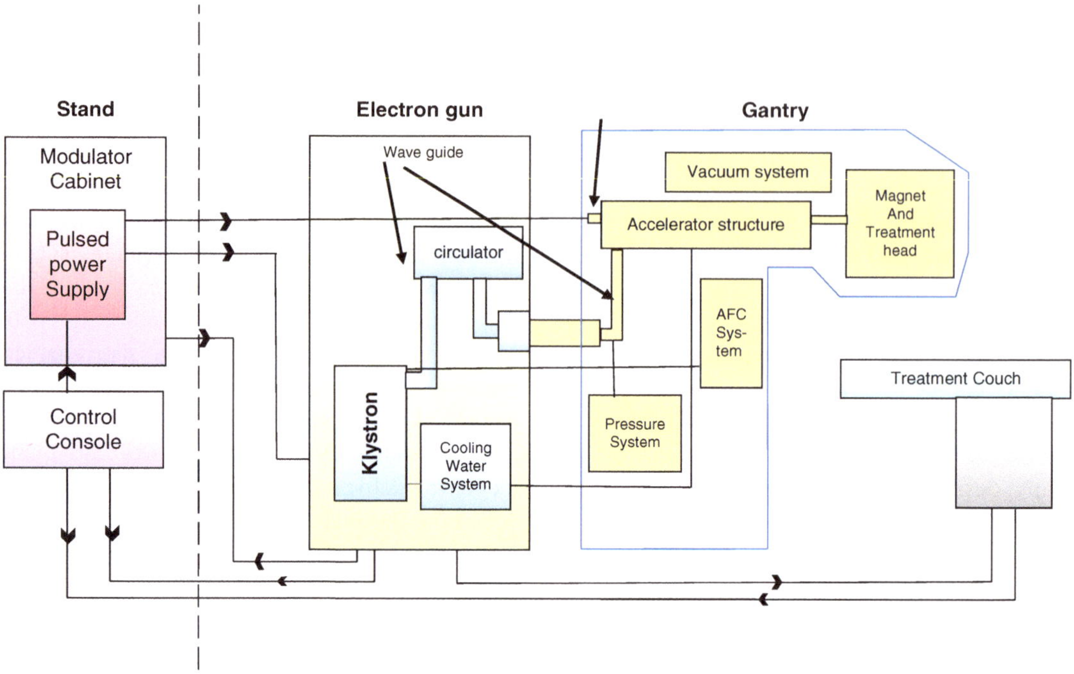

Fig. 5.7 A typical block diagram showing the major components of a high-energy medical linac

through the crest of an ocean wave. In a standing wave accelerator, reflected microwaves are used to create a standing wave. The accelerator structure in low-energy (4–6 MV) linacs is most often mounted vertically in the treatment head collinear, along with the components associated with producing, controlling, and monitoring the X-ray beam. High-energy (15–18 MV) linacs use a horizontally mounted accelerator structure with a beam-bending magnet system. Other components include the modulator used to generate pulsed DC power. Pulsed currents are used to generate microwaves and are also supplied to an electron gun to generate electrons. Before being fed into the accelerator tube, electrons are bunched, using a device known as the buncher. Thus, the electron beam consists of electron bunches in the form of a narrow pencil beam. The magnetron is a device that serves both as the source of the microwaves and as a power amplifier. The klystron is a device that amplifies the microwave power generated by a separate microwave source (RF driver). A device known as the wave guide, is used to transport the microwave generated by klystron or magnetron to the accelerator tube. A device called the circulator is used to isolate the klystron/magnetron from the reflected microwave power.

The X-ray target, fixed primary collimators, scattering foils, flattening filters, monitor ion chamber, movable secondary collimator jaws, multileaf collimator, light field localizer, and optical distance indicator are other important components in the treatment head. Figure 5.8a shows the major components inside the head of a conventional linear accelerator operating in the X-ray mode and Fig. 5.8b shows the components while the linac is operating in the electron mode. To reduce head scatter radiation, the linac head is loaded with a significant amount of shielding. The accelerator tube and target are aligned in the same line in low energy units (4–6 MeV), whereas in high energy units, the accelerator structure is placed horizontally or at an angle due to its longer length, and thus, the electrons must be bent (90° or 270°) before hitting the X-ray target. This is accomplished by the beam transport system, which includes achromatic focusing and bending magnets as well as steering.

The primary collimator is used to align X-rays toward the patient treatment plane and is placed just below the target. It also helps to reduce head

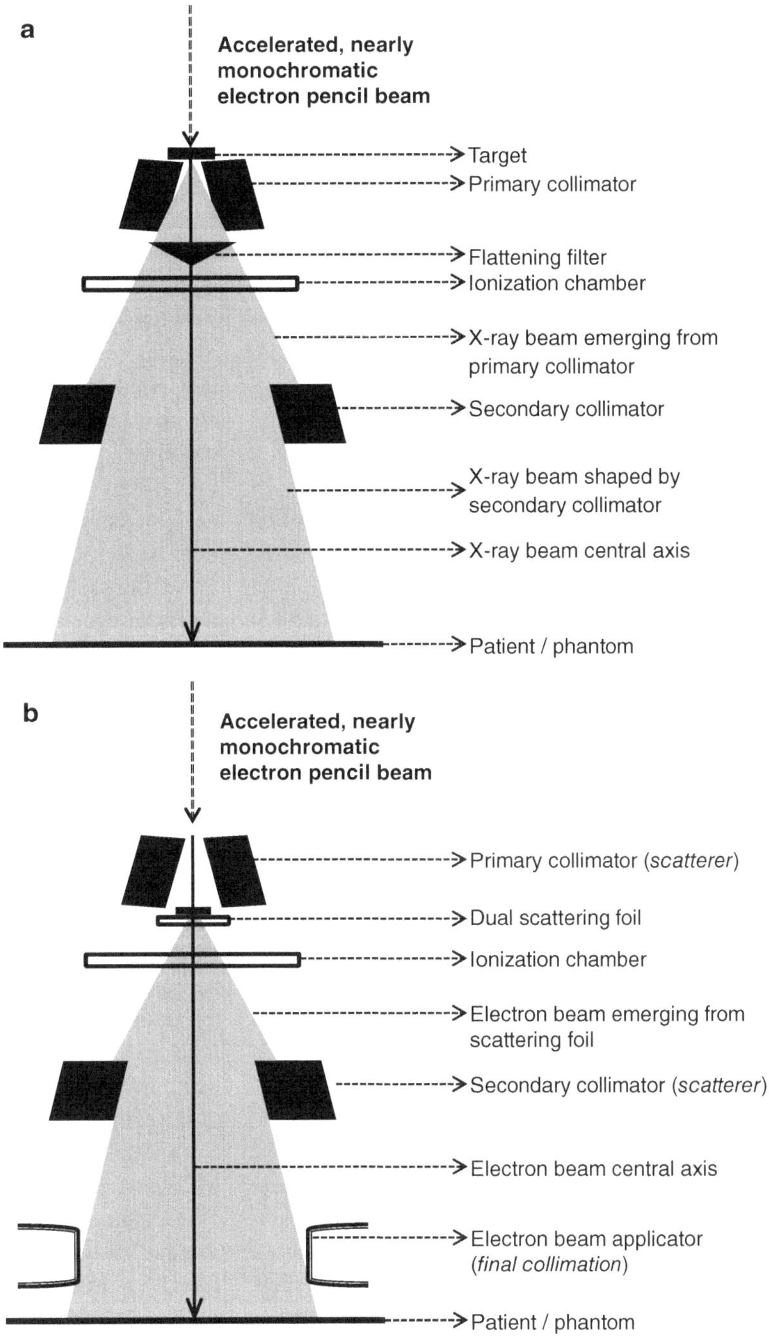

Fig. 5.8 (**a**) Linac head components in X-ray mode. (**b**) Linac head components in electron mode

leakage. Because the angular distribution of bremsstrahlung X-rays produced by megavoltage electrons incident on the target is forward peaked, a conical flattening filter is used to make the beam uniform across the field. The flattened beam is then forced into the monitor ion chamber. It is typically made up of several transmission type parallel plate chambers that cover the entire field and are used to monitor field symmetry, dose rate, and cumulative dose. Commercially available flattening filter-free (FFF) beams were introduced in 2010. After passing through the monitor chambers, the beam can be further collimated continuously and independently, with the help of secondary collimators, which consist of a pair of lead or tungsten jaws and/or multileaf collimators. Conventional linac provides a field size of 0 × 0 cm^2 to 40 × 40 cm^2 at 100 cm.

The multileaf collimator (MLC) was introduced by Takahashi in 1965. MLCs, although initially designed for beam shaping in two-dimensional (2D) and three-dimensional (3D) conformal radiotherapy, excelled in what is known as, intensity modulated radiation therapy (IMRT). In IMRT, the positions of the multileaf collimator leaves are electronically and dynamically controlled to change the shape of the field during treatment. IMRT can be delivered by moving the gantry to predefined static positions and (a) delivering multiple field segments (called segmental MLC or step-and-shoot IMRT) or (b) having the MLC leaves move across the field at varying rates while the X-ray beam is on (called dynamic or sliding window IMRT). The volumetric modulated arc therapy (VMAT) refers to treatment delivery by simultaneously moving gantry and MLC leaves. Both the gantry rotation speed and the MLC speed are varied. The technique is comparatively faster and spares the surrounding organs effectively.

The first MLC designs used were single-focused models with the MLC leaves moving in linear trajectories and thus not following beam divergence. Many of the current linacs still use this model only. Later new designs such as double-focused MLCs and multilayer MLCs were developed. In the former, the leaves followed the beam divergence thus reducing the penumbra and in the latter the multilayer offered greater flexibility in field shaping resolution, reduced leakage and greater speed. In the electron mode, the X-ray target gets retracted, and a scattering foil is aligned with the beam central line such that a pencil like electron beam is made to fall on it. The scattering foil broadens and flattens the pencil beam to produce a clinically acceptable widened beam. An electron applicator is mounted below the movable collimator jaws to provide the final field collimation.

Conventional medical linacs were outfitted with advanced imaging systems capable of CBCT in the early 2000s. These linacs are known as "image-guided radiation therapy" (IGRT) linacs. A typical unit consists of a retractable kilo-voltage X-ray source, an amorphous silicon flat-panel detector mounted perpendicular to the beam direction on a linac, a data processing software module, and an image registration tool. After comparing the daily patient set-up image with the stored reference computed tomography (CT) planning image, table adjustments can be made prior to treating the patient.

Other lesser used accelerator is microtron. The microtron is a type of electron accelerator that combines the fundamental principles of an electron linear accelerator and a cyclotron. It is possible to achieve a higher beam energy with a low-energy accelerating section by using magnets to recirculate the electron beam through a microwave accelerator cavity (or cavities) one or more times. The electron bunch must be in phase with the accelerator microwave field after each orbit in the magnet. As a result, the magnet system acts as an energy spectrometer, limiting electron energy acceptance to a narrow energy range.

Helical tomotherapy was first proposed by Mackie et al. and commercialized as a short in-line 6 MV linac mounted on a ring gantry system at a source axis distance of 85 cm. The unit delivers IMRT, and the couch is translated along the longitudinal axis through the gantry bore, just as in a helical CT. The treatment beam pitch is defined as the ratio of the distance travelled by the couch in one gantry rotation divided by the field width in the y-direction. The width of the beam in the y-direction is defined by a fixed pair of jaws during treatment and is typically 1, 2.5, or 5 cm. The beam is modulated by a 64-leaf binary MLC, whose leaves transit between open and closed states rapidly. Intensity modulation is accomplished by varying the fraction of time different leaves are opened.

The individual modulation pattern can change with angle (divided into exactly 51 projections over a full revolution). During the treatment, the gantry rotates at a constant velocity which can range between 15 and 60 s per rotation. The extent to which a treatment beam projection is modulated is characterized through what is called as the modulation factor, defined as the ratio of the maximum leaf open time to the average leaf open time for the projection. Pitch and the maximum permissible modulation factor are parameters that need to be specified by the treatment planner. Highly modulated treatments achieve greater conformality but inevitably take longer time to deliver. A helical megavoltage computed tomography (MVCT) image is taken using the onboard xenon CT system along with the 6 MV linac (detuned to 3.6 MV) with leaves fully opened. A registration software is used to verify the patient setup by comparing daily images with the prescribed CT image.

Cyberknife (CK) consists of an X-band linear accelerator mounted on an industrial robotic arm. The robotic arm provides the ability to direct a narrow beam in any orientation with respect to the target volume and was first developed for radiosurgery. The system uses two ceiling-mounted diagnostic X-ray sources and corresponding amorphous silicon image detectors mounted to the floor. The treatment is specified by the trajectory of the robot and by the number of monitor units delivered at each robotic orientation. During the patient's treatment, the CK system correlates live radiographic images with preoperative CT or MRI scans in real time to determine patient and tumor positions repeatedly over the course of treatment. The disadvantage of the CK is the prolonged treatment delivery.

Photon treatment units are evolving, and in 2017, Varian introduced ring-gantry type Halcyon machine, which was specifically created to address the challenges of radiotherapy in low resource setups. Specific features include a simpler installation process, low power consumption, easier serviceability, and simplified operations. Due to the presence of beam stoppers, the requirement for room shielding is also very limited. The treatments are also 100% IGRT, and the imaging dose (only for MV images) is included in the prescription dose. Its double-layered MLCs have demonstrated dosimetric advantages in studies. The machine does not provide electron treatment. A noteworthy point is that it has an "O"-shaped design as compared to conventional "C"-shaped units.

Other O-ring linear accelerators are MRI Linacs, available from ViewRay (MRIDIAN) and Elekta (Unity) both were introduced around 2014. MRI linacs are embedded with a MRI onboard imaging system; specialized in online adoptive treatment. The MRIDIAN system comprises a double-donut superconducting wide bore magnet with a field strength 0.345 T and a 6 MV flattening filter free (FFF) linear accelerator. Elekta Unity consists of high-field (wide bore 1.5 Tesla) MR imaging system, based on the Philips Ingenia MR scanner and 6FFF photon beam. The disadvantage is having very low patient throughput with extensive treatment time.

All O-ring linacs—Tomotherapy, Halcyon, and the MR-linacs—have a common disadvantage of having no couch rotational movement, thus restricting the use of the noncoplanar beam geometry. Additionally, majority of them do not offer a 6-dimensional couch correction as well.

5.5.5 Particle Accelerators

The first use of proton beams for radiation therapy (PBT) is credited to Wilson, who in 1946 pointed out the superior depth–dose characteristics provided by protons. PBT is seeing increasing interest worldwide because its depth–dose characteristics show advantages over those of photon beams. However, the major problem with PBT that has not been resolved over the last seven decades is the accurate calculation of the linear energy transfer value and the relative biological effectiveness (formerly considered 1.1 without much proof). In proton particle treatment facilities, either a cyclotron or a synchrotron is used to accelerate protons to sufficient intensity and energy (200–250 MeV). A cyclotron is used to accelerate particles such as protons, light ions, and deuterons to high frequencies. Alternating currents are applied to two evacuating chambers, known as Ds. Magnetic field is applied perpendicular to Ds to force the particles to move in a circular path. The particles accelerate only between the two gaps of the Ds. Once sufficient energy is achieved, it is made to exit from the edge of the container.

A synchrocyclotron changes either the magnetic field or the frequency of the electric field; a synchrotron varies both. By adjusting these parameters as the particle gains energy, it is possible to hold them on a constant path as they are accelerated. This allows the vacuum container to be a large, thin, torus. The synchrotron has the advantage of simple energy variability, whereas the cyclotron produces continuous beams with a fixed energy and higher beam intensity, making their design somewhat simpler. Beam spreading systems include passive modulation, such as the use of scattering foil, or dynamic pencil beam scanning systems to achieve the required treatment field sizes. Neutron leakage is a big concern while using scattering foils, and hence, pencil beam scanning systems are recommended.

The initial clinical proton systems used research or commercial systems for treatment, which relied on multiroom solutions. Though treatment took place, it was not on a large scale. Commercial systems helped it to deploy on a larger scale, but a significant initial cost and the size of the operation limited this type of deployment. Single-gantry proton systems brought down the initial cost and offered a scale that was easier to deploy. The first single room systems included the MEVION S250 proton therapy system and the IBA Proteus One. This system uses a superconducting synchrocyclotron that is gantry mounted, incorporates image guidance, and uses robotic patient positioning.

There are several heavy ion (carbon) facilities in Europe and Asia (Japan), and the first of these in the United States is operational in the research facilities like Los Alamos National Laboratory, but not in clinical hospitals. Heavy-ion radiation therapy requires beams at much higher energies than proton therapy. For example, a proton beam of 150 MeV can penetrate 16 cm in water. To achieve the same penetration with carbon ions, energy of 3000 MeV or 250 MeV/nucleon is needed. Synchrotrons are the only available sources for such high-energy ion beams, and they are large, complex machines requiring large facilities and similarly large capital expenditures.

Interest in heavy ions is due to two important characteristics, the first of which is their superior depth–dose characteristics. The ratio of Bragg peak dose to the entry dose is even higher than that for protons. The increased lenior energy transfer (LET) by these particles at the Bragg peak region is the other characteristic that promises huge radiobiological advantage. The combination of these two characteristics results in the potentially unique advantage of providing a high LET region that can closely conform to the target volume.

Modern neutron therapy machines use cyclotrons to accelerate protons or deuterons to energies of about 50 MeV to produce neutron beams with depth–dose characteristics equivalent to those of about 6-MV X-rays. The p, Be (protons accelerated to strike a beryllium target) reaction is used most commonly because protons are much easier to bend around the gantry of an isocentric unit, and thus, the cyclotron can be much smaller and thus less expensive.

5.6 Imaging in Radiation Oncology: Simulator

The conventional simulator mimics the functions and allowed motions of a therapy unit and uses a diagnostic X-ray tube to simulate the radiation properties of the treatment beam. It allows the beam direction and the treatment fields to be determined to encompass the projection of the target volume. Radiographic visualization of internal structures in relation to external landmarks allows special shielding devices (Cerrobend blocks) to be constructed for minimizing the dose to normal critical structures. Couch widths are similar to therapy-unit couch widths, and operating consoles feature digital displays of parameters and programmable settings for SAD, gantry angles, and field sizes. Modern simulator design has replaced the old X-ray fluoroscopy system and an image intensifier combination with the latest amorphous silicon technology flat-panel detectors.

In the 1980s and early 1990s, research led to the integration of a diagnostic CT scanner with a three-dimensional (3D) treatment-planning system, which led to the concept of virtual simulation. Such a system is now referred to as a CT simulator and consists of a CT scanner, a flat-tabletop patient position and alignment system, including an orthogonal laser system, and a digital interface to a planning system that is equipped with virtual simulation software. Modern CT simulation systems incorporate large-bore CT scanners especially designed for radiation oncology, with multislice capability, high-quality laser patient positioning and marking systems, and sophisticated virtual simulation software features.

5.7 Physical Aspects of Imaging in Radiotherapy

Imaging has become an integral part of modern day radiotherapy (RT). From diagnosis to target and normal tissue delineation for treatment planning, and on-board imaging for accurate radiation delivery, a multitude of imaging modalities has found extensive applications in modern day RT. There are some requirements from medical imaging such as electron density information and minimum geometric distortions that are uniquely important in RT. Also, fast and live imaging finds immense utility in RT. In recent times IGRT has gone beyond inter- and intrafraction motion corrections to adaptive radiotherapy [7]. Biologically guided RT (BGRT) and personalized RT are the next logical steps with ever-improving functional and molecular imaging modalities. Ultrasonography (US), CT, MRI and molecular imaging with unsealed radioisotopes have been integrated to varying degrees into the RT process. They are not cited in the same chronological order though.

Radiography and CT are based on the attenuation properties of X-rays. Therefore, they have the advantage of electron density information needed in the calculation of dose distribution. Better reconstruction algorithms, improved detector technology and continuous improvements in data processing have not only improved the anatomical image quality but have opened up a number of quantitative CT techniques for better tissue characterization with considerable dose reduction. Dual-energy CT exploits the enhanced attenuation differences among tissues owing to photoelectric absorption and thus offers improved target definition [8]. 4D-CT provides a useful estimation of tissue motion patterns.

Ultrasonagraphy is an efficient and cost-effective imaging modality. It offers real-time volumetric imaging with good soft-tissue contrast especially for targets outside the skull and lung without any radiation risk. Sound-wave reflectivity at interfaces of various tissues with different acoustic impedance properties forms the basis of this imaging. It lacks electron density information and also the resolution of other imaging modalities such as CT and MRI. On-board image guidance in RT has its origin in US. The B-Mode Acquisition and Targeting (BAT) system of the late 1990s used a 2D diagnostic US imaging system and an optical or mechanical means for tracking the position of the probe with respect to the linac. The Clarity system from Elekta leverages intramodality image matching instead of the intermodality matching one used by BAT and SonArray (Varian). By tracking the precise position and orientation of the US probe with respect to the CT scanner, the 3D US volume can be reconstructed in the reference frame of the CT. Studies comparing US with CT have shown differences within 3 mm in interfraction prostate monitoring [9, 10].

MRI is a multiparametric imaging modality unlike CT and US. The T1, T2 relaxation times and proton density differences can be optimized to provide the best possible tissue contrast. But longer imaging times and inherent geometrical distortions are some of the challenges that need to be overcome for more efficient utilization of MRI in RT. The distortions arise from inhomogeneity in the main magnet, nonlinearity in the gradient fields and the eddy currents associated with the switching of the gradient coils. On-board MR-guided radiotherapy systems have become commercially available for improved patient positioning accuracy and soft-tissue details leading to effective adaptive-RT. The on-board image guidance on these systems is implemented by 2D imaging on orthogonal views using fast continuous acquisition at a fixed 2D coordinate frame. Fast imaging sequences such as T1-weighted gradient echo technique in lung motion imaging and T2-weighted single shot fast spin echo for liver are generally utilized for the purpose. Dynamic contrast enhanced MRI and MR spectroscopy are valuable tools for providing information beyond morphological details [10].

4D-MRI has become an emerging developmental area in radiotherapy. Its successful implementation requires an integration of optimized acquisition protocols, advanced image reconstruction techniques, and sufficient hardware capabilities. Since a typical human's breathing cycle period is about 3–5 s, high temporal resolution <500 ms/frame is desired for multiphase reconstruction. In addition, high spatial resolution at the level of 1 mm isotropic voxel size might be necessary for small size stereotactic body radiotherapy targets in lung/liver treatment. To overcome the challenge of lack of electron density information in MRI various methods have been proposed to generate pseudo-CT images, also called synthetic CT (synCT) from MR images as a solution [10].

Molecular imaging using PET is increasingly being used for better tumor delineation and disease spread. The coincidence detection of annihilation photons produced by the positrons emitted from a radionuclide-tagged tracer forms the basis for tomographic imaging in the PET systems. The radionuclides such as C-11, N-13, O-15, F-18 are relatively short-lived, enabling the optimal use of imaging photons while keeping the patient radiation dose low. Furthermore, tagging of these radionuclides with biological substrates such as glucose, NH3 happens without altering their biological activity. PET is the key for moving toward BGRT. For example, hypoxic areas may be derived from PET-18F-misonidazole (F-MISO) imaging. A higher dose of radiation can then be delivered to the specific volume rather than the whole tumor in a process known as dose painting. More and more reliable imaging biomarkers are being explored to map spatial heterogeneity within the tumor and tumor response to therapy [11, 12].

3D video-based optical surface imaging has opened up new opportunities in RT. Apart from patient set-up and motion monitoring without using ionizing radiation its other applications could be in avoiding anesthesia for pediatric patients undergoing RT. Further, optical spectroscopic approaches, such as diffuse reflectance and Raman spectroscopy have demonstrated functional and molecular differences between radiation-resistant and sensitive tumors in response to radiation [13]. Such information could be useful for tailoring the dose as per the individual requirement. Electromagnetic tracking systems such as Calypso from Varian make use of electromagnetic transponders embedded within the tumor, and motion of these transponders is tracked in real-time using a detector array system.

5.8 Parameters for Point Dose Calculation in Photon Beams, Isodose Charts, Beam Profiles and Wedge Filters

When a patient is treated with ionizing radiation, it is essential to have a complete understanding of the radiation dose distribution in the patient, dose at specific points, and dose modification in the presence of beam modifiers. It is impossible to measure dose distribution directly in patients. Instead, we rely on measurements done in tissue-equivalent phantoms under valid, full-scatter conditions and using them in a dose calculation system for predicting the dose in the actual patient.

There are established international protocols to measure absorbed dose at a reference point in the most accurate manner with minimum uncertainty. From a knowledge of this dose, one can calculate the dose at any other point in the irradiated medium through a system of dosimetric functions.

Water is the often used tissue-equivalent phantom material for dose and dose distribution measurements. The choice is not only because of the fact that it closely approximates the radiation absorption and scattering of body tissues and muscles, but also it is available easily. In addition to water phantoms, solid phantoms have also been used for dosimetric measurements since they are easier to handle compared to water phan-

Table 5.2 Phantom materials and their properties

Material	Mass density (g/cm^3)	Number of electrons/g
Water [H_2O]	1.0	3.34×10^{23}
Polyethylene [CH_2]$_n$	0.92	3.44×10^{23}
Polystyrene [C_8H_8]$_n$	1.03–1.05	3.24×10^{23}
Perspex (Lucite) [$C_5O_2H_8$]$_n$	1.16–1.20	3.24×10^{23}
Paraffin [C_nH_{2n} + 2]	≈ 0.90	3.44×10^{23}
Solid water—*Commercially available*	1.0	3.34×10^{23}

toms. The ideal tissue or water equivalent phantom material should have the same effective atomic number, number of electrons per gram, and mass density. The most important requirement is the number of electrons per gram.

The properties of some of the commonly used phantom materials are given in Table 5.2.

There are many commercially available phantoms. Both homogeneous and anthropomorphic phantoms are available. The latter are useful for doing measurements to test inhomogeneity corrections.

When the photon beam is incident on a patient (or phantom), dose is absorbed all along its path and even beyond the boundaries of the radiation field. The resultant dose distribution inside the patient is dependent on various factors: beam energy, field size, depth of the point at which the dose is calculated, source to the point distance, and beam collimation among others. The dose at a reference point is established under reference conditions and the dose at other points is calculated from this reference dose point using various measured quantities that are available in tabulated formats. These quantities are the relative dose functions that express dose at various points either as ratios or as percentages of the dose at the reference point. The percentage depth dose, tissue–air ratio, tissue–phantom ratio, and tissue–maximum ratio are some of the examples of these relative dose functions.

5.8.1 Percentage Depth Dose

The dose at any depth along the central axis is described by normalizing it with respect to dose at a reference depth. The ratio of the absorbed dose at any depth d (D_d) to the absorbed dose at a reference depth d_0 (D_{d_0}) along the central axis of the beam, expressed as a percentage, is called as the percentage or percent depth dose (PDD) (Fig. 5.9).

$$\mathrm{PDD} = \frac{D_d}{D_{d_0}} \times 100\% \qquad (5.22)$$

The reference depth is at the surface for X-ray beams up to 400 kVp, and for higher energies, it is taken at the depth where the maximum dose occurs on the central axis ($d_0 = d_m$). The depth d_m increases with increasing energy. For 6 MV, it is 1.5 cm and for 15 MV X-rays, it is 2.7 cm. The dose at this point D_{d_0} is also called as the *dose maximum*, the *maximum dose*, the *given dose* and is denoted as $D_{\max}$. Beyond d_m, the PDD exponen-

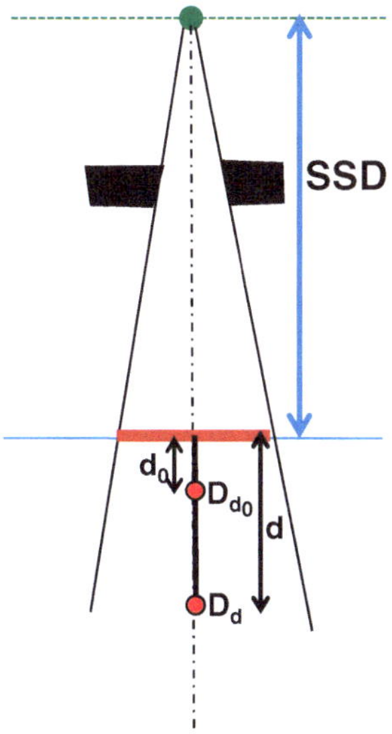

Fig. 5.9 Percentage depth dose at depth d = $\mathrm{PDD} = \frac{D_d}{D_{d_0}} \times 100\%$ D_d is the dose at depth d and D_{d_0} is the dose at the depth of dose maximum

tially falls for all energies. The PDD describes the reduction in absorbed dose due to inverse square law effect and the beam attenuation in traveling from d_m to depth d.

The PDD is dependent on the beam energy, source to surface distance (SSD), field size, and the depth d. For orthovoltage energies, the PDD is maximum (100%) at the surface and it decreases as the depth increases. However, in the megavoltage energy range, there is an initial raise in the PDD from low surface dose to dose maximum at the depth d_0. The initial low surface dose followed by the increase in dose with depth is called the buildup effect, which is responsible for the skin-sparing effect of the megavoltage beams. The PDD increases with increasing energy.

The geometric field size, defined as the projection of the beam defining collimator on a plane perpendicular to the central axis at the source to axis distance (SAD), also affects the PDD. As the field size increases, the PDD increases due to increase in scatter reaching the central axis. For a given increase in field size, the increase in PDD is larger at smaller field sizes compared to that at larger field sizes. The reason is, when a large field is further widened, only a part of the scatter arising from its periphery can reach up to the central axis where the PDD is defined. For the same reason, the PDD of a rectangular field, having the same area as that of a square field, is less compared to the PDD of the square field since the scatter arising from the periphery of the elongated side of the rectangular field cannot reach the central axis [2]. Further, the effect of field size on PDD is more pronounced for lower energy beams compared to higher energy beams since the former undergo more scattering. Percent depth dose data are measured for all clinical beams for square fields and are tabulated.

The percent depth dose increases with SSD because of the inverse square law effect. As mentioned earlier, the PDD describes the reduction in beam intensity due to inverse square law (ISL) and attenuation of the intervening tissue. When the SSD is changed, the attenuation suffered by the beam remains the same, while the ISL effect is different. Consider the PDDs at the depth of 10 cm for a 6 MV beam at 80 cm and another at 100 cm SSD. The d_m is at 1.5 cm. In moving from d_m to the depth 10 cm, the beam suffers the same attenuation by the intervening tissue for both SSDs, but the attenuations by ISL differ. For the 80 cm SSD beam, the ISL component is $[\frac{81.5}{90}]^2$, which comes to 0.82. For the 100 cm SSD beam, it is $[\frac{101.5}{110}]^2$, which comes to 0.85. Thus, for a given depth, the penetration is more and the attenuation is less for larger SSD. Accordingly, the PDD is more compared to that for smaller SSD.

For an 80 cm SAD $C^{60}o$ beam, the PDD falls by about 4.5% for every cm of tissue traversed at shallower depths and at deeper depths, the fall is only 3%. For a 100 cm SAD 6 MV beam, the PDD falls by about 4.5% for every cm of tissue traversed at shallower depths and at deeper depths, the fall is only 3%. For an 18 MV X-ray beam, the corresponding values are 3% and 2.5%, respectively.

5.8.2 Tissue–Air Ratio

Tissue–air ratio was developed for use in rotational therapy in which the tumor center is placed at the machine's isocenter and the radiation source moves in a circle about the axis of gantry rotation [3]. This setup is called as the source to axis distance (SAD) setup, and as the machine rotates, the central axis of the beam always passes through the center of the tumor. The depth of the tumor (isocenter) and along with it, the SSD will vary depending on the direction and the body contour. In the SSD setup, the machine's isocenter is placed on the surface of the patient and any rotation of the gantry would warrant tedious adjustments of the patient to align the tumor with the central axis of the beam. The SAD setup is much preferred over the SSD setup since it requires minimal human intervention and ensures maximal accuracy.

The varying SSD makes the use of PDD-based calculations in rotational therapy practically impossible. Since the PDD is dependent on SSD, corrections are to be applied to PDD every time

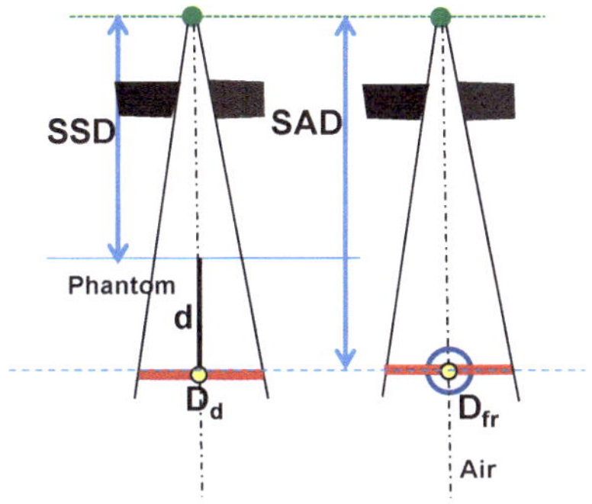

Fig. 5.10 Definition of tissue–air ratio (TAR). TAR $= \frac{D_d}{D_{fr}}$

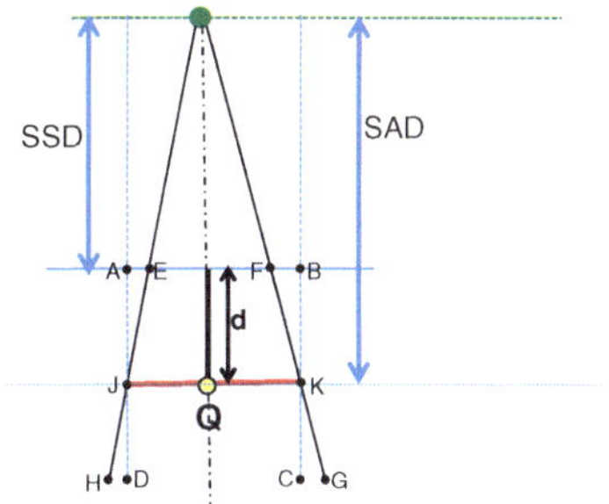

Fig. 5.11 Diagram illustrating the independence of TAR on SSD

SSD changes which is not possible. Instead, a ratio called as tissue–air ratio (TAR) was developed which is much simpler to use and is independent of SSD. It is defined as the ratio of the dose (D_d) at a given point in the phantom to the dose in free space (D_{fr}) at the same point (Fig. 5.10) and is given by

$$\text{TAR}\left(d, r_d\right) = \frac{D_d}{D_{fr}} \tag{5.23}$$

TAR is dependent on depth, field size at that depth and is independent of SSD. Since TAR is the ratio of doses at the same point in space (D_d, D_{fr}), there is no involvement of inverse square law in the photon fluence at that point. Compared to the dose in free space (D_{fr}), the dose at depth d in tissue is modified only by the attenuation by the overlying tissue and scatter. The exponential attenuation is only dependent on depth and is independent of SSD [3]. Figure 5.11 explains the independence of scatter on SSD and how TAR is independent of SSD. Consider two beams, one parallel (a beam coming from infinite distance will have no divergence and have parallel edges) and another divergent. At depth d, both beams have the same cross-sectional area, r_d. As stated above, there is no ISL effect involved in the TAR and since the two beams are attenuated by the same overlying depth d of tissue, at Q the *primary doses remain the same*. Let us see what happens to the scattered doses for the two beams: for the parallel beam, the scatter at Q comes from the area ABCD and for the divergent beam it comes from EFGH; while the divergent beam lost scatter from the triangular areas AEJ and FBK compared to the parallel beam's ABCD, the loss was compensated by the additional scatter it gained from JDH and KGC, from which the parallel beam did not get any scatter. Thus, at the point Q, the *scattered doses are also the same for both beams*. In this way, TARs (and tissue-phantom ratios [TPRs]) are independent of SSD and a single table of TARs (TPRs) can be used for all SSDs.

As the field size is increased, for a given energy and a given depth, the scatter also will increase, and hence, the TAR will also increase [1–6]. With increasing depth, the TARs continuously decrease since the ratios describe attenuation and with more overlying tissue, there is more attenuation. On the same note, with increasing energy, at a given depth and a given field size, TARs will increase since there is less attenuation with increased energy.

5.8.3 Peak Scatter Factor (Back Scatter Factor)

The peak scatter factor or back scatter factor is a special case of tissue–air ratio when the depth d is made equal to the depth of maximum dose on the central axis of the beam and is given by the equation

$$\text{PSF} = \frac{D_{\max}}{D_{fr}} = \text{TAR}\left(d_m, r_{d_m}\right) \tag{5.24}$$

where r_{d_m} is the field size at the reference depth d_m of maximum dose. Like TAR, PSF is dependent on beam quality and field size and is independent of SSD. The PSF can reach a value as high as 1.5

for orthovoltage beams, which amounts to 50% increase in dose near the surface as compared to dose in free space. However, for megavoltage beams, the values are much smaller since these radiations are forward scattering in nature. For a cobalt beam, the PSF is only about 1.036.

Both the PDD and TAR are ratios that describe the beam attenuation. It is thus obvious that they are related to each other. It can be shown that

$$\text{PDD}(d,r,f) = \text{TAR}(d,r_d).\frac{1}{\text{PSF}(r)}.\left(\frac{f+d_m}{f+d}\right)^2.100 \tag{5.25}$$

5.8.4 Scatter–Air Ratio

In the dosimetry of irregular fields, splitting the dose into its primary and scatter components is useful. Scatter–air ratio (SAR) is a ratio that helps to calculate the scattered dose in the medium and is defined as the ratio of the scattered dose at a given point in phantom to the dose in free space at the same point. The SAR has similar properties as that of TAR.

The total dose at any point in the medium is the sum of primary dose and scattered dose at that point. The scattered dose is then the total dose minus the primary dose and is given by the equation

$$\text{SAR}(d,r_d) = \text{TAR}(d,r_d) - \text{TAR}(d,0) \tag{5.26}$$

The TAR (d,0) represents the primary dose component of the total dose at that point.

Tables of SARs are available as a function of depth and field size.

5.8.5 Collimator Scatter Factor (Output Factor)

The primary dose, as the name suggests, is the dose contributed by the interaction of primary or original photons directly coming from source without undergoing any scattering. The scattered dose is due to photons scattered by the collimator system and by the phantom. However, for all practical purposes, in megavoltage photon beams, the primary dose is defined as the dose due to primary photons and photons scattered by the collimator system. Thus, the scattered dose is defined as the dose due to photons that are scattered in the phantom. The effective primary dose at a certain depth is the dose at depth minus the dose due to phantom scatter. It can also be considered as the due to a hypothetical field of zero dimensions (0×0 cm^2) since the scattering volume is reduced to zero.

If the beam output is measured in air, it will vary as a function of field size. As the field size is increased, the output increases due to increased collimator scatter. The collimator scatter factor (S_c) is defined as the ratio of the outputs measured in air for a given field size to that for a reference field size (10×10 cm^2).

5.8.6 Phantom Scatter Factor

The phantom scatter factor (S_p) is a ratio that describes the change in scatter arising in the phantom at a reference depth when the field size is changed. It is defined as the dose rate (or dose per monitor unit) at the reference depth for a given field size to the dose rate at the same point for a reference field size (usually 10×10 cm^2) with the collimator opening remaining the same. It cannot be directly measured and is indirectly determined from the total scatter factor, $S_{c,p}$ which is the product of collimator scatter and phantom scatter factors:

$$S_{c,p} = S_c .\ S_p \tag{5.27}$$

5.8.7 Tissue–Phantom and Tissue–Maximum Ratios

The TAR involves measurement of dose in free space. For energies up to Co-60, a buildup cap is used for in-air measurements. At higher energies, the size of the buildup cap required is so large that the concept of dose in free space becomes invalid. For this reason, TARs are not defined for higher energies. Instead, the concept of tissue–

phantom and tissue–maximum ratios was introduced to overcome this limitation.

Figure 5.12 illustrates the concept of tissue–phantom ratio. It is defined as the ratio of dose rate at a given depth d in phantom to the dose rate at the same point, but at reference depth d_{ref}. The reference depth d_{ref} is a point in the central axis of the beam (e.g., 10 cm depth). When d_{ref} is taken as d_{max}, the point at which the dose is maximum along the central axis, it becomes a special case of tissue–phantom ratio and is called as the tissue–maximum ratio. Like TARs, the TPRs and the TMRs make dose computations for isocentric treatments simpler. The depth of maximum dose is not constant and it reduces with increasing field size due to electron contamination. If d_{max} is taken as the reference depth, then it is obtained from the smallest field size for which the lateral electronic equilibrium is achieved.

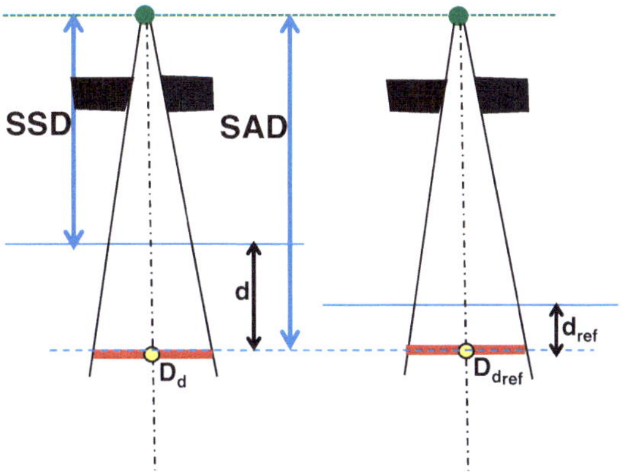

Fig. 5.12 Definitions of tissue–phantom ratio (TPR) and tissue–maximum ratio (TMR). TPR = $\frac{D_d}{D_{d_{ref}}}$, where d_{ref} is the reference depth. The TMR is a special case of TPR, where d_{ref}, the reference depth, is at the depth of dose maximum d_{max}

It is essential to understand and differentiate between the geometries of the SSD and the SAD techniques to know how to use PDD, TAR, and TPR in clinic. A point at a depth (d) of 10 cm from the surface on the central axis of a 10×10 cm^2 field is to be treated to a certain dose. The treatment setups are illustrated in Fig. 5.13.

In the SSD technique (Fig. 5.13a), the entry point is at the source-to-axis distance of f cm. The point of interest is at f + *10 cm*. The dose rate at the depth of dose maximum d_{max} and the dose rate at 10 cm on the central axis are related by PDD.

$$\text{Dose rate at the point of interest } P, D_P = \text{Dose rate at } d_{max} * \left(\frac{\text{PDD at the point of interest } P}{100}\right) \quad (5.28)$$

$$= D_{max} * \frac{\text{PDD at the point of interest } P}{100}$$

Note that due to beam divergence, though the field size is 10×10 cm^2 at 100 cm SAD, at the depth of point P, the field size will be larger. Figures 5.13b, c explain the use of TAR in calculating the dose rate at Q as:

$$\text{Dose rate at } Q, D_{Q,10} = \text{Dose rate at the same point in free space}, D_{Q,fr} * \text{TAR}\left(10\,\text{cm}, 10\times10\,\text{cm}^2\right) \quad (5.29)$$

The point Q is also at a depth of 10 cm from the surface, but it is closer to source than P. The field size 10×10 cm^2 is defined at that depth. For a high energy beam, the TAR concept is not valid and the TPR concept is used to calculate the dose rate at Q (Figs. 5.13b, d).

$$\begin{aligned}&\text{Dose rate at } Q, D_{Q,10} = \text{Dose rate at the}\\&\text{same point in space, but in phantom,}\\&\text{located at a reference depth } d_{ref},\\&D_{Q,d_{ref}} \times \text{TPR}\left(10\,\text{cm}, 10\times10\,\text{cm}^2\right)\end{aligned} \quad (5.30)$$

When the reference depth is taken as depth of dose maximum, then it is a special case of TPR and is called as TMR. The dose rate at Q is calculated from Figs. 5.13b, e.

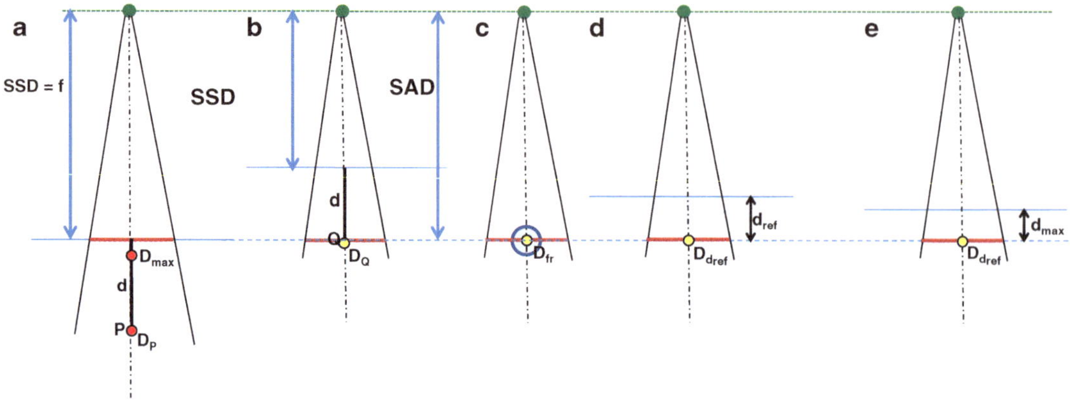

Fig. 5.13 (**a**) The use of PDD, (**b**, **c**) the use of TAR, (**b**, **d**) the use of TPR, and (**b**, **e**) the use of TMR

$$\begin{aligned}\text{Dose rate } Q, D_{Q,10} = &\text{ Dose rate at the same point in space, but in phantom,}\\ &\text{located at the depth of dose maximum } d_{\max}, D_{Q,d_{\max}}\\ &\times\ \text{TMR}\left(10\,\text{cm},\ 10\times10\,\text{cm}^2\right)\end{aligned} \tag{5.31}$$

It is clear from Fig. 5.13a–e that the PDD is influenced by both ISL and beam attenuation, whereas, the TAR and TMR are influenced only by beam attenuation. The point of interest is also closer to source in the SAD technique than in the SSD technique. Because of these reasons, for given energy, field size and treatment depth, number of monitoring units required to deliver a certain dose is always less for SAD technique as compared to SSD technique.

5.8.8 Isodose Charts

The ratios and factors used in radiotherapy, explained above, link the dose at a reference point to the dose at any arbitrary point in the irradiated medium. That is, if we have the knowledge of absorbed dose at a single reference point, with the help of these ratios and factors, we can calculate dose at any point in the irradiated medium. For example, if the dose at the reference depth is determined for a reference field size using standard protocols for any photon beam, then dose at any point in the medium can be calculated with the help of the ratios and factors explained earlier.

Isodose curves are another set of data, in graphical format, that helps to achieve the same goal. These two-dimensional curves represent the variation of dose as a function of depth along the central axis and perpendicular distance from the central axis (transverse distance). An isodose curve is a curve joining all points that receive the same dose. The dose at the reference point of dose maximum along the central axis is normalized to 100%, and a set of isodose curves, in equal decrements of percent depth dose, usually in 10% decrements, down to 30% or so, are drawn.

Such type of isodose curves, in printed form, called as isodose charts, were supplied with the treatment units in the earlier days, before computerized treatment planning became a norm. Isodose charts are of two types: SSD chart and SAD chart. In the SSD chart, the dose is normalized at the reference point of dose maximum along the central axis and it consists of isodose lines from 90% to 30% in 10% intervals. It is customary that such SSD charts are prepared only for the special case of SSD equaling SAD, although they can be prepared for any SSD. In the SAD chart, the dose is normalized to a certain depth beyond the depth of dose maximum, say like 5 cm, 10 cm or 15 cm, and the isodose lines are shown. At depths above the normalization point, dose will be higher than 100%, and at deeper depths, dose will be lesser than 100%. Isodose lines are drawn at 10% intervals starting from a value higher than 100%.

Both type of charts were used to be prepared for different field sizes. The SAD charts were prepared for different field sizes and different normalization depths.

Isodose curves can also be drawn in a transverse plane perpendicular to central axis and are useful in treatment planning.

5.8.9 Beam Profiles

A beam profile represents the dose variation across the field at a depth (Fig. 5.14). The dose is normalized to 100% at the central axis. The profile can be measured along the two major axes of the field, namely X- (cross-plane) and Y- (in-plane) axes. The profile gives information on the radiation field size, measured as the lateral distance between the 50% dose points on either side of the central axis, and the penumbra, measured as the distance between the 80% and 20% dose points on the left and right sides of the central axis. They also give a measure of field flatness and field symmetry which are critical parameters in quality assurance [1].

5.8.10 Wedge Filters and Wedge Isodose Curves

Wedge filters, or simply wedges, are wedge-shaped metallic inserts that modify the shape of the isodose curves by tilting them at different angles. They are called as beam modifiers. They are usually made up of steel or lead. In addition to such physical wedges, nonphysical wedges are also in use that mimic the same effect of tilting isodose curves through programmed movement of one of the X-ray jaws [1–6].

Figure 5.15 shows how isodose curves are tilted with the insertion of a physical wedge in the path of the beam. At the thicker end of the wedge, the curves are lifted up more compared to the thinner end. Enough caution should be exer-

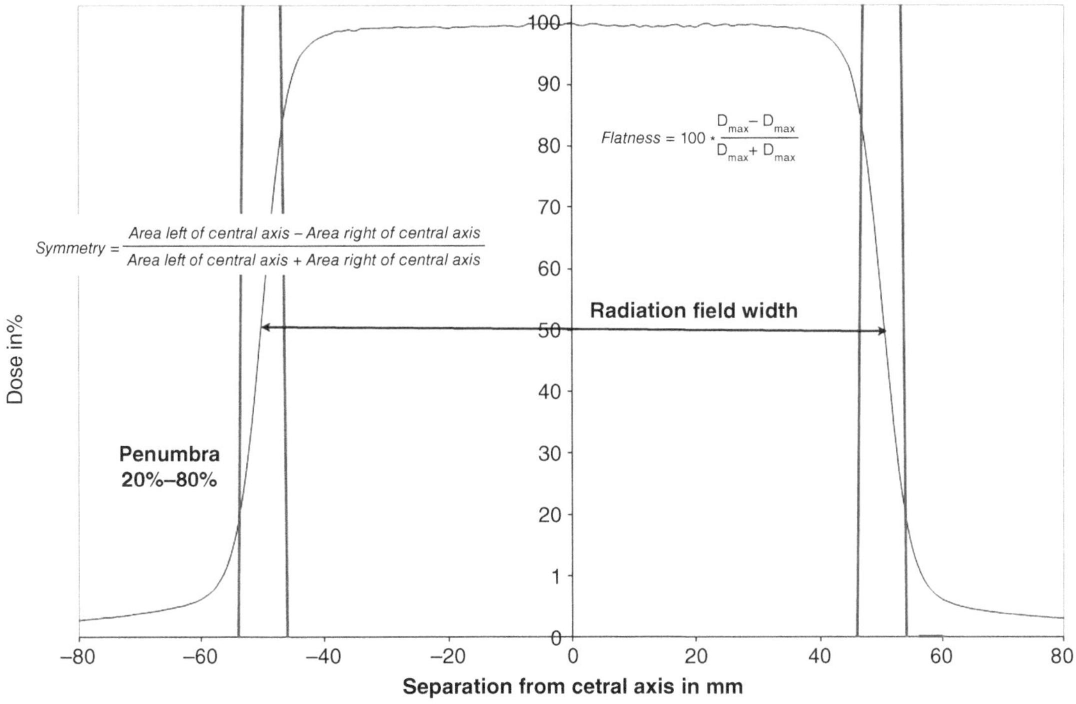

Fig. 5.14 A cross-plane dose profile showing the dose variation as a function of lateral distance on either side of the central axis. The distance between the 50% dose points on the left and right is defined as the radiation field width. Other parameters like penumbra, flatness, and symmetry that characterize the beam profile are defined

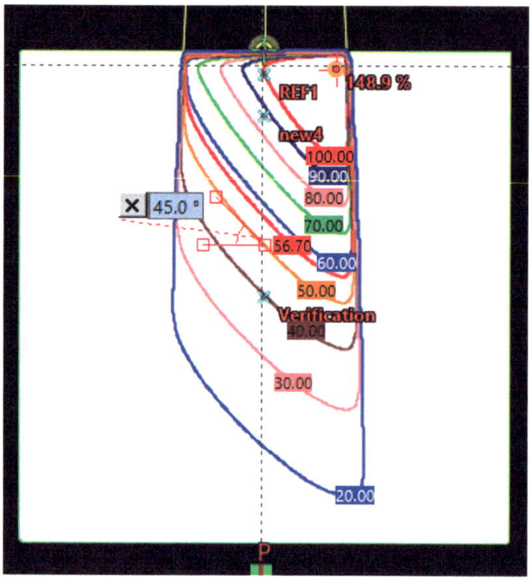

Fig. 5.15 Isodose curves for a 45° wedge filter for a ^{60}Co beam. Wedge angle is defined as the angle at which the 50% isodose curve is tilted at the central axis of the beam, or simply, as the complement of the angle that the 50% isodose line subtends with the central axis

cised in interpreting wedge isodose curves as there are multiple ways of normalizing these curves. The usual practice is to normalize at the reference point of maximum dose along the central axis, in which case, on the thinner side of the wedge, at shallower depths, isodose curves are higher than 100% since the beam is less attenuated compared to the central axis.

The physical wedges suffer from disadvantages like they are heavy to lift and insert into the collimator, they attenuate the beam, and there are possibilities for human errors and increased scatter, especially to the contralateral breast in the treatment of breast by tangential fields. The nonphysical wedges are better alternatives to physical ones. While they produce the same effect of modifying the beam like physical wedges, they do not require manual handling, they do not attenuate the beam, there is no possibility for human errors and they do not produce additional scatter. They produce the wedge effect by computer-controlled movement of one of the Y-jaws of the X-ray collimators from one end to other end while beam is ON. Either Y1, or Y2 jaw, is moved during the treatment in a preprogrammed manner to produce the desired effect. Varian's Enhanced Dynamic Wedge is one of the examples of nonphysical wedge.

5.8.10.1 Wedge Angle

The tilting of the isodose curves is characterized by the term wedge angle. In the cobalt-60 era, the wedge angle was defined as the angle through the 50% isodose curve is tilted at the central axis of the beam, or simply, as the complement of the angle that the 50% isodose line subtends with the central axis as shown in Fig. 5.15. However, at higher energies, since the 50% isodose line occurs at deeper depths, which are not clinically relevant for describing the wedge effect, the International Commission on Radiological Units and Measurements (ICRU) recommended depth of 10 cm is chosen for the definition of wedge angle [1]. Thus, wedge angle is defined as the angle between the isodose line and the normal to the central axis at 10 cm depth. When a pair of wedged fields is used, the angle between them is called as the hinge angle, ϕ. The wedge angle θ and hinge angle ϕ are related by the equation $\theta = 90 - \left(\frac{\phi}{2}\right)$ which will a give uniform dose distribution within the target volume.

5.8.10.2 Wedge Transmission Factor

The introduction of a physical wedge in the path of the beam causes the beam to be attenuated. The treatment time calculations must account for this attenuation. The wedge transmission factor, or simply the wedge factor, quantifies the beam *transmission* along the central axis and is defined as the ratio of the doses with and without wedge in place at a reference depth (e.g., 10 cm). Failure to incorporate this factor in the calculation of treatment time (or MUs) can result in serious underdosing of the target volume.

5.8.10.3 Physical Wedge Systems

Ideally, for every field size there should be an individualized physical wedge. However, it is not practically feasible. Cobalt treatment units have a set of three physical wedges for every wedge angle, covering wedged field widths up to 6, 8 and 10 cm. The 6 cm wedge is used for all fields with widths up to 6 cm, the 8 cm wedge is for widths greater than 6 cm and less than or equal to 8 cm and the 10 cm wedge is for field widths greater than 8 cm and less than or equal to 10 cm.

A universal wedge system, as the name suggests, uses a single wedge for all field widths. Although it reduces the possibility of human errors by having only one wedge for each wedge angle, the wedge geometry results in unnecessary attenuation of the beam, especially for small field sizes.

A motorized wedge, like the one in Elekta linear accelerators, is a single wedge system for all wedge angles and all field widths. It is a 60° universal wedge and can produce a wedge effect up to 30 cm field width. The wedge is inbuilt inside the linac head and is computer controlled. The wedge is driven into and away from the path of the beam by a motor, to produce a 60° wedged and an open field respectively. For intermediate wedge angles, a mixture of open and wedged fields in suitable ratios is employed. The wedge transmission factor for such a single 60° universal wedge is very small, pointing to a low efficiency.

5.9 Radiation Dosimetry

Radiation dosimetry is a science that investigates several ways for determining the amount of energy deposited in a given material by ionizing radiation, either directly or indirectly. A radiation dosimeter is a device that measures radiation quantities such as exposure, kerma, absorbed dose, or equivalent dose of ionizing radiation, either directly or indirectly. A dosimetry system is a combination of a dosimeter and a reader. To measure the absorbed dose in a medium, a radiation sensitive dosimeter must be introduced into the medium. These dosimeters measure the ionization produced in a sensitive medium.

Radiation dosimeters need to have a few desirable qualities in order to be useful. For instance, precise information of the absorbed dose to water at a specific location and its spatial distribution is important in radiotherapy, as is the potential of determining the dose to an organ of interest in the patient. Accuracy, precision, linearity, dose or dose rate independency, uniform energy response, directional independence, and high spatial resolution will all be desired dosimeter features in this application. Of course, not every dosimeter can meet every requirement. The needs of the measurement situation must be taken into consideration while selecting a radiation dosimeter and its reader.

5.9.1 Calorimetry

Calorimetry is one method of measuring absorbed dose since the heat energy absorbed by a sample is proportional to the dose delivered, and it uses exact quantification of minute temperature changes in irradiated samples to produce direct evidence of the energy absorbed during the process. For a range of megavoltage photon qualities, user ionization chambers can be calibrated against a national calorimetric standard; however, calorimetry is not always practical because it is best suited for measuring large quantities of radiation absorbed under very controlled conditions.

5.9.2 Ionization Chambers

The ionization of air, often known as exposure, was adopted as the measurable effect of photon beam radiation at the Second International Congress of Radiology in 1928. As the beam passes through a material, the ionization process produces ion pairs. The created ion pairs have some mobility when the radiation beam passes through the air, and they can be collected by passing an electric field across the air.

In radiotherapy, ionization chambers are used to determine the radiation dose. Beam calibration is another name for the process of determining the dose under standard radiation conditions. Depending on the specific needs, ionization chambers come in a variety of sizes and designs, but they all generally share the following characteristics:

(a) An ionization chamber is essentially a gas-filled cavity with a central collecting electrode and a conductive outer wall. To minimize the leakage current when a polarizing voltage is given to the chamber, the wall and the collecting electrode are separated by a high quality insulator.
(b) Typically, a guard electrode is included in the chamber to further reduce leakage. Bypassing the collecting electrode, the guard electrode intercepts the leakage current. Additionally, it ensures improved field uniformity in the chamber's active or sensitive volume, which is advantageous in charge collection.
(c) In order to account for the change in the mass of the air in the chamber volume, which varies with the ambient temperature and pressure, measurements with open air ionization chambers must be corrected for temperature and pressure.

Electrometers are instruments used to measure small currents with a magnitude of 10^{-9} A or less. In order to measure the chamber current or charge accumulated over a predetermined period of time, an electrometer used in conjunction with an ionization chamber is a high gain, negative feedback, operational amplifier with a standard resistor or a standard capacitor in the feedback path.

5.9.2.1 Free Air Ionization Chamber [1, 2]

The ionization process forms ion pairs as the beam travels through a material. When the radiation beam travels through the air, the ion pairs produced have some mobility and can be collected by applying an electric field across the air. The number of ion pairs collected is a measure of the quantity of radiation passing through the air. In this chamber, a collimated photon beam enters the chamber through the diaphragms on the one side exiting through the opposite wall. A parallel plate electrode assembly is mounted within the chamber's body with guard wires and a concentric guard ring positioned to define the air volume from which secondary charged particles are collected. The plates' separation is selected to allow for the dissipation of any residual secondary particle kinetic energy prior to the arrival of those particles at the collecting electrode, while still being sufficiently close for the applied polarizing voltage to collect all the ions generated within the defined volume. The ions produced within the collection boundary that are lost to the periphery must be balanced with externally generated ions entering this critical volume in order to meet the requirements imposed by the definition of exposure. This is the state of electronic equilibrium. Corrections are required for lack of saturation in the ionization current, the presence of any water vapor, and ionization resulting from bremsstrahlung and scattered radiation or leakage through the primary collimators.

5.9.2.2 Cylindrical (Thimble Type) Ionization Chambers

For energies up to 300 kV, the free air chamber can accurately determine the exposure. Above this energy, the electrode separation necessitates prohibitive polarizing voltages in order to collect the produced charges. The free-air chamber, which employs the concept of electronic equilibrium to provide a definitive measurement of exposure, is an impractical piece of equipment for routine clinical use due to its size. Hence, portable dosimeters that are more useful for usage in the high-energy ranges used in radiotherapy have been developed. For beam calibration in radiotherapy dosimetry, the most often used cylindrical ionization chamber is the 0.6 cm^3 chamber developed by Farmer and initially produced by Baldwin, but now readily available from several vendors. The Farmer type chamber is also referred to as a thimble chamber because of the chamber sensitive volume's thimble-like appearance. A schematic diagram of a Farmer type thimble ionization chamber is shown in Fig. 5.16.

The sensitive medium of the dosimeter is typically not the same as the medium in which it is used. Cavity theory relates the absorbed dose in the dosimeter's sensitive medium (cavity) to the absorbed dose in the surrounding medium containing the cavity. According to the ranges of secondary charged particles created by photons in the cavity medium, cavity sizes are classified as small, intermediate, or large.

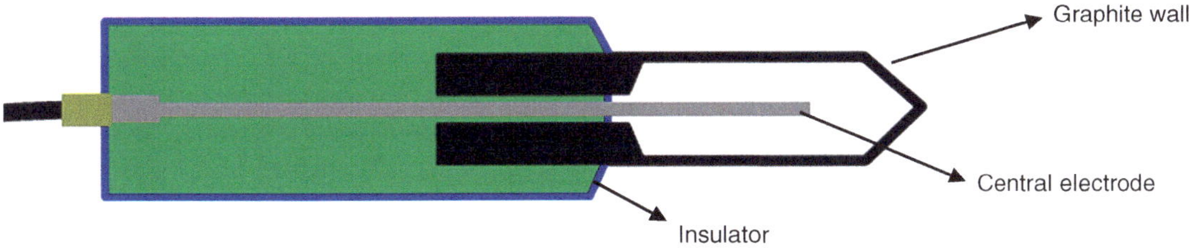

Fig. 5.16 A thimble-type cylindrical ionization chamber and its internal parts

5.9.2.3 Parallel Plate Chambers

As their name implies, parallel plate chambers are built with two parallel plates as the chamber's walls. These walls are coated with a conducting layer that forms the positive and negative electrodes. The gas layer between the walls acts as the collecting volume. Their use is recommended for absolute dose determination in electron beams of energy below 15 MeV.

The cylindrical or thimble chamber is irradiated with its central axis perpendicular to the beam direction, whereas, the plane-parallel chamber is irradiated with its axis parallel to the beam direction. In situations where the dose gradient is steep in the direction of the beam, such as in the build-up region, parallel plate chambers are preferred to cylindrical chambers. Desirable properties of these chambers are that they should be waterproof and their construction as homogeneous and water equivalent as possible.

5.9.3 Semiconductor Detectors

For relative dosimetry of photon and electron beams, semiconductor detectors are used. They are also used for the in vivo dosimetry of photon and electron beams. The electrical conductivity of a semiconductor lies in the region between that of a conductor and that of an insulator. The most commonly used semiconductor material is silicon. When the semiconductor device is irradiated, electrons and positive holes are created. The electrons are attracted toward the positively charged impurity in the n-type silicon and the positive holes diffuse toward the negatively charged impurity in the p-type silicon. This constitutes an ionization current that is proportional to the irradiated dose or dose rate. The semiconductor detector has the advantages of higher sensitivity and smaller physical size.

5.9.4 In Vivo Dosimetry

Each step in the process chain of patient receiving radiation may introduce an element of uncertainty. Therefore, it is worthwhile and even required to use in vivo dosimetry to provide a final assessment of the actual treatment for certain patient groups or for atypical treatment conditions. Thermoluminescence detectors (TLDs), semiconductor detectors (silicon diodes) and optically stimulated luminescent dosimeter (OSLD) are the types of dosimeters most commonly used for in vivo dosimetry.

5.9.5 Thermoluminescent Dosimeters

There are several crystalline materials that exhibit thermoluminescent properties. The most commonly used in radiotherapy are those based on lithium fluoride (LiF) because LiF is approximately tissue equivalent. Other TLD materials include those based on lithium borate (LiBO 4), calcium fluoride (CaF 3) and calcium sulfate (CaSO 4). TLD materials are supplied in the powder form, solid discs or rods. Impurities are introduced into the crystal and intermediate energy levels are formed within the forbidden zone. These forbidden zones are referred to as electron traps. Upon irradiation, electrons in the valence band may receive sufficient energy and raise to the conduction band. Electrons then either fall back to the valence band or get caught in an electron trap. When sufficient heat is applied to the crystals, the electron gains thermal energy and escape from the electron trap by emitting energy in the form of light. The total light output is proportional to the number of trapped electrons and which in turn is proportional to the absorbed dose. Light emitted by the TLD material is measured using a photomultiplier, which converts the light into a current signal. The advantages of TLDs are they are reusable and energy independent over a wide range of energy. They are tissue equivalent and small in sizes. But poor handling and storage can influence the precision of the dose measured.

5.9.6 Optically Stimulated Luminescent Dosimeter (OSLD)

OSLD works on a similar principle of TLDS but unlike the thermal annealing used for TL materials, an optical stimulation will provide the energy necessary to release the charges. Quantifying the amount of emitted light makes it possible to evaluate the dose. The most common phosphor for OSLDs is Al_2O_3:C (synthetic sapphire) produced by Landauer, USA. OSLDs show certain advantages including high luminescence efficiency, good precision and accuracy, stable sensitivity, speedy readout, multiple re-analyses without thermal annealing, availability of portable readers, and significantly low fading.

5.10 Quality Assurance in Radiotherapy

The International Commission on Radiological Units and Measurements recommends ±5% accuracy in dose delivery based on the dose-response curves for tumor control and normal tissue complications probability. However, the treatment delivery in radiotherapy consists of a long chain of steps starting from determining the best treatment position and selecting the most suitable patient specific immobilization device, to treatment delivery. There will be uncertainties involved in each of these steps. Thus, the requirement of ±5% overall accuracy implies requirement of accuracy much better than ±5% in each individual step. Several quality measures are taken at each step to ensure the overall accuracy is maintained in treatment delivery.

The term quality in radiotherapy refers to those product features, or in other words, clinical services which meet the stated or implied goal of effective patient care; are free from deficiencies; and conform to standards or specifications. Every radiation oncology department must set up a quality management (QM) program or quality plan that describes the specific practices, resources and activities relevant to the clinical service provided. A well-designed QM program involving experts from all related specialties shall help in ensuring that the offered clinical service consistently functions well. It shall describe the quality control tests, procedures explaining how to perform them, testing frequency, baseline values, tolerance criteria, record keeping, and the manpower required to perform those tests. A quality assurance (QA) committee consisting of a member each from radiation oncology, physics, treatment and nursing care shall be formed to oversee the QA program, to tailor the recommendations from international reports, monitor and audit the program for compliance and frame policies whenever required. The committee shall command the necessary authority from the administration to perform its responsibilities. It shall also form a policies and procedures manual.

Quality in radiotherapy service demands total commitment of the chief radiation oncologist to comprehensive quality assurance. The responsibilities of performing the actual quality assurance tests and procedures may be delegated to the staff members including radiation oncologists, physicists, radiotherapy technologists, nurses and others. Each one of the team members must be trained to perform his/her responsibilities, and know what should be done if a specific test fails or the test results exceed the tolerance from the baseline values.

Quality assurance refers to all activities, planned or systematic, necessary to instill confidence that the radiation oncology service will meet the requirements for quality care. Radiation therapy equipment are complex machines and their functional performance characteristics must be evaluated periodically to ensure that they are functioning without any deviations from those values established during commissioning of the equipment. The foundation stone for a strong QA program is laid at the time of acceptance and commissioning of the equipment for clinical use by way of establishing baseline standards. Subsequently, regular periodical QA testing is necessary for all radiotherapy equipment like linac, remote afterloading brachytherapy units, CT simulators, treatment planning systems and 3D water phantoms. The written quality manual must list out: (a) equipment parameters to be tested, (b) which instruments are to be used in the tests, (c) description of the setup geometry, (d) testing frequency (daily, monthly, annual), (e) responsible staff, (f) tolerance values, and (g) actions to be taken if tolerances are exceeded.

Various international bodies like American Association of Physicists in Medicine (AAPM) and Canadian Organization of Medical Physicists (COMP) have published guidelines and recommendations for various QA tests and their tolerances [14–16]. The AAPM came out with its Report of the Task Group (TG) No. 40 on comprehensive quality assurance for radiation oncology in 1994 [14], considered as an important reference document for QA. In the year 2009, the AAPM published its TG 142 report [15] updating some sections of the old report and adding new tests for the features which were not previously covered. Further, TG-142 prescribed tests, tolerances and their frequencies according to the types of treatments delivered with the particular machine. Besides adding recommendations for not only the new ancillary delivery technologies, the report has also added recommendations for imaging devices that have now become standard part of linear accelerators. Every department must develop a QA manual explaining the daily, monthly and annual QA tests to be carried out on their external beam therapy machines taking guidances from TG 40 and TG 142 reports and also from respective national regulatory body's stipulated requirements. Suitable forms must be developed and put in place. The forms and their contents will vary according to the machines available in the clinic and the type of clinical services offered. Each form shall have its version number. Revision history of each form like contents newly added, modified and/or deleted shall be recorded permanently. A system must be in place for periodical inspection of QA reports to check compliance to the agreed frequency of each type of tests and to take necessary steps for rectifying any unacceptable results or deteriorating trends observed in the results.

The daily QA tests for a remote afterloading high-dose rate (HDR) brachytherapy system shall include verifying source positioning accuracy, functionality of backup battery for source retraction in the event of power failure, timer accuracy, door interlocks, source indicator lights, functionality of audio-visual communication, functionality of emergency switch, availability of a radiation survey meter at the console, date and time on the console computer, etc., among other tests. Swipe tests are to be periodically done on a quarterly basis. A source exchange warrants more tests, in addition to the daily QA tests, including verification of source strength, timer linearity, timer accuracy, and timer error. Before starting the treatment for a patient, a check list has to be completed and similarly, following completion of treatment, a radiation survey of the patient has to be performed.

The treatment planning system (TPS) is another critical equipment that requires periodical QA tests. The TPS is not a simple computer system and shall be treated as a medical device. The specific QA tests shall be designed taking guidance from International Atomic Energy Agency (IAEA) Technical Report Series 430 [17]. Some of the tests include validation of entered data, validation of dose calculation accuracy in homogeneous and heterogeneous media, validity of data transfer from different imaging units, geometric accuracy of plotted data, reproducibility of test plans over period of time, software integrity, beam data integrity and integrity of plan transfer to treatment console among various other tests [18].

5.10.1 Patient-Specific QA

Patient-specific QA consists of all those procedures that ensure every patient receives the right treatment as per the physician's prescription. The initial chart checks, weekly physics chart checks, in-phantom verification of treatment plan, verification of treatment plans using portal dosimetry, in vivo dosimetry are all examples of the patient specific QA tests. There are a variety of commercially available QA devices and software solutions for verifying treatment plans of various types.

5.11 Radiation Safety in Radiotherapy

It did not take long to realize after the discovery of X-rays and natural radioactivity in the 1890s that ionizing radiation was not only useful for the diagnosis and treatment of diseases but was also harmful for the human health. The harmful effects could come through i) direct clinical effects at very high doses and ii) potential for delayed effects including genetic effects. In general, the effects of radiation can broadly be categorized as deterministic effects and stochastic effects.

Deterministic effects are called deterministic because they are certain to occur, if the absorbed dose exceeds a threshold level. The therapeutic effects of radiation are also deterministic by nature. These effects are the results of various processes, but mainly cell death and delayed cell division, caused by exposure to large radiation doses. The severity of a particular deterministic effect in an exposed individual increases with the dose above the threshold for the occurrence of the effect. Examples of deterministic effects are skin erythema (3 Gy), cataract (0.5 Gy), and lethality (whole body 3–5 Gy).

Stochastic effects on the other hand are probabilistic and random in nature such as induction of malignancies and hereditary effects without a threshold dose. They manifest after a latency period and may be epidemiologically detectable in a population. The probability of occurrence of these effects is higher for higher doses but the severity of the effect is independent of the dose. Therefore, risk factors rather than threshold doses have been estimated for stochastic effects. The examples of stochastic effects are induction of cancer and hereditary effects.

5.11.1 System of Radiation Safety in India

Laboratory studies at molecular and cellular levels, radiation effects on persons exposed to ionizing radiation for therapeutic or diagnostic purposes, radiation incidents and accidents world over have provided the necessary data for estimating radiation risks. The International Commission on Radiological Protection (ICRP) and the United Nations Scientific Committee on the Effects of Atomic Radiation (UNSCEAR) are the foremost specialized international agencies involved in assessing and reporting levels and effects of exposure to ionizing radiation. Based on the findings of such agencies International Atomic Energy Agency (IAEA) issues recommendations and guidelines on radiation safety standards [19]. Governments around the world make the findings and recommendations of these agencies as the scientific basis for establishing protective measures in their respective countries. In India, the Atomic Energy Regulatory Board (AERB) has been empowered to formulate and regulate all activities involving ionizing radiation including in the field medicine.

5.11.2 Operational Aspects of Radiation Safety in Radiotherapy

AERB has issued safety codes from time to time for ensuring radiation safety in radiotherapy practice. The recommendations contained in the safety codes are in broad alignment with the recommendations and findings of international bodies such as ICRP and IAEA. The last AERB safety code related to RT was released in the year 2011 [20]. It covers the design and manufacturing aspects of radiation equipment and sources, building design and shielding requirements, licensing process including training and assigning of roles of various radiation professionals, and regular monitoring of radiation practice in a clinical setup. Radiation oncologists, medical physicists and radiotherapy technologists/therapeutic radiographers are the main categories of radiation professionals involved in day-to-day radiotherapy practice in a clinic. Minimum qualifications for these professionals have been laid down in the safety code. All radiotherapy equipment need to be type approved by AERB before being considered for clinical applications in the country. Primarily three categories of persons, namely, the employer, the licensee(s) and the radiological safety officer (RSO) have been recognized and assigned responsibilities for ensuring safe radiation practice in an institution. The hands-on responsibility of ensuring radiological safety is that of the RSO who not only advises the employer but also remains in close liaison with the AERB for implementing the safety norms. To obtain and maintain the license to operate the RT units for clinical purposes, requirements regarding availability of radiation professionals and radiation measuring and monitoring instruments with valid calibration, records of laid down acceptance, commissioning and periodic quality assurance (QA) and safety checks of the RT units must be ensured. All this information needs to be submitted to the AERB through their web portal named eLORA (electronic licensing of radiation applications) on periodic basis.

Radiation exposures are divided into three categories namely (a) occupational exposure (radiation workers), (b) medical exposure (patients, attendants, and research volunteers), and (c) public exposure. The aim of radiation safety is to protect the public, occupational workers as well as the environment. Justification, optimization of risk and remaining within the prescribed permissible limits of dose are the three cardinal principles for protection in any radiation practice. The objective of as low as reasonably achievable (ALARA) should be followed taking economic and social factors into account. For medical exposures of patients, the optimization means management of the radiation dose to the patient commensurate with the medical purpose. For radiation workers and public, permissible dose limits have been prescribed (Table 5.3 below). In the case of medical exposures, the prescribed permissible limits do not apply. However, the exposure to patients should be controlled through delivering only the doses that are necessary to achieve the diagnostic or therapeutic objective. Potential exposures are controlled by optimizing the design of installations, equipment and operating procedures [21].

The permissible radiation levels prescribed by AERB ensure that no known deterministic effects appear on the exposed individuals and the probability of any stochastic effect is within acceptable limits. Linear–no threshold hypothesis which assumes that the risk of stochastic effects is directly proportional to the dose for all levels of dose is the working hypothesis for establishing safety standards).

For an RT installation, appropriate wall thickness of the bunker is the first basic requirement to protect the staff and the public. Depending on the

Table 5.3 Permissible dose limits prescribed by AERB based on the latest recommendations of ICRP (2007) [20, 21]

	Radiation worker	Member of public
Effective dose	20 mSv/year averaged over a period of 5 years. Max in an year is 30 mSv	1 mSv/year
Equivalent dose (lens of eye)	150 mSv/year	15 mSv/year
Equivalent dose of extremities and skin	500 mSv/year	50 mSv/year

The limits given above apply to female radiation workers also. However, once pregnancy is declared the equivalent dose limit to embryo/fetus shall be 1 mSv for the remainder of the pregnancy

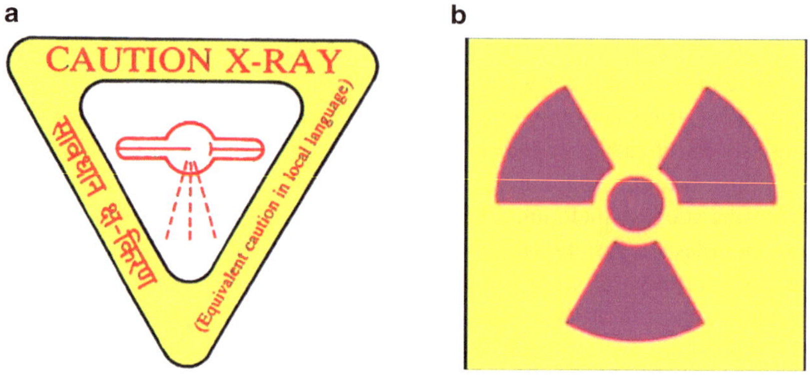

Fig. 5.17 (**a**) Radiation warning symbol for an X-ray source. (**b**) Radiation warning symbol for a radioactive source

energy of the radiation beam wall thicknesses change. Some typical values are as follows [4]: For a telecobalt unit the primary and secondary barriers of concrete should be 130 cm and 65 cm thick. For 10–25 MV photon beam, it should be 240 cm and 120 cm, respectively. Above 10 MV, there is added safety concern from neutron produced in the linear accelerator (linac) head by (X,n) and (e,n) reactions. An additional radioactivity problem to the neutron activation is the direct activation of elements in (X,n) reactions, such as oxygen-15 and nitrogen-13 with very short half-lives of a few minutes. The radioactivity in the treatment room air is removed by efficient room ventilation, which also handles the removal of ozone and noxious gases produced by photon interactions with air. Typically, there should be five to eight exchanges of air per hour in a high energy linac room. All radiation installations should have the following radiation warning symbols at their entrances to warn the general public. The first symbol (Fig. 5.17a) is for an installation housing a radiation generating equipment such as an X-ray machine or a linac and the second one (Fig. 5.17b) is for a radioactive source such as a telecobalt machine or a brachytherapy machine.

Source Security and Emergency Preparedness in Radiotherapy High activity cobalt-60 sources and Ir-192 HDR sources are extensively in use in India apart from other low activity sources. A definite security plan for physical security of the sources from known or potential threats has to be designed by all RT centers. It should include physical barriers, electronic surveillance, access control and quick detection and warning mechanism in the event of a security breach for timely action. Responsible persons for emergency preparedness need to be earmarked based on their expertise on the subject. The local police must be involved and apprised of the security plan.

5.12 Future Trends

The technological developments in the next two to three decades could prove to be a watershed in the practice of radiation oncology. The current mainstream radiotherapy delivery technology seems to have reached a stage from where only incremental improvements are expected. Improved imaging technologies, molecular diagnostic and radiation delivery aided by artificial intelligence and data analytics is likely to create a paradigm shift in the way radiation oncology is practiced today.

In the coming decade, the current trend in technological developments in MRI including 4D-MRI would help integrate it more efficiently into RT workflow as an on-board modality as well as an effective functional imaging tool [11]. CT and ultrasonography would continue evolving for better, faster and safer image generation [8–10]. Molecular and functional imaging with radionuclides would help fulfill the biologically guided and personalized radiotherapy objectives using efficient dose painting with charge particle therapy [12]. Use of optical imaging for patient set-up and motion monitoring without using ionizing radiation is likely to become more extensive. Optical spectroscopic approaches, such as diffuse reflectance and Raman spectroscopy may provide information about radiosensitivity of various tumors [13].

However, reaching submillimeter level accuracy in targeting the tumor may not be possible with the incremental trends visible in the mainstream technology due to limitation of imaging resolution, tumor/organ motion and delivery accuracy. As predicted by National Institute of Health, USA, that by 2035 around 60% of all radiation therapy procedures would be delivered using radiopharmaceuticals [12]. This new paradigm has been termed as theranostics which is a combination of the terms therapeutics and diagnostics. This term describes the combination of using one radioactive drug to identify (diagnose) and a second radioactive drug to deliver therapy to treat the main tumor and any metastatic tumors. The synergy among physicists, chemists, biologists, clinicians, data analysts and computer scientists would contribute toward reaching the new horizon. The RT professionals of the future might find it hard to believe that for well over a century radiation was sent through normal tissues to reach the tumor.

FLASH radiotherapy (FLASH RT) has received lot of attention in the recent past. In a seminal article published in 2014, Favaudon et al. claimed that 17 Gy of dose delivered by 4.5 MeV electrons at ultra-high dose rates (UHDR), i.e., dose rates in excess of 40 Gys^{-1}, dramatically spared normal tissues, and at the same time maintained the tumor control efficacy [22]. Since then, hundreds of publications have appeared and the scientific community is engaged in active research on all fronts. Yuan Gao et al., in a recent review article has outlined the biological mechanisms involved in FLASH-RT, the delivery modalities that are under development, and the in vivo experiments conducted before concluding with cautionary remarks [23].

Oxygen depletion hypothesis is the most well-explained theory for the FLASH effect. According to this theory, the differential response between FLASH-RT and conventional RT arises due to the radiochemical depletion of oxygen at UHDRs and the resultant radioresistance in the irradiated normal tissue. This can be explained as follows: Low linear energy transfer (LET) radiation cause DNA damage predominantly through radiolysis of water molecules and the subsequent generation of free radicals. These free radicals are incorporated into DNA, causing damage to them, but are largely repairable. In the presence of oxygen, free radicals get converted into peroxyl radicals, which have the potential to cause permanent damage to DNA. Peroxyl radicals are thus a more efficacious DNA damaging agent. If there is no oxygen in the immediate environment of the cell, it limits the production of peroxyl radicals and thus the extent of radiation-induced DNA damage. Treatments delivered at UHDR results in a drastic reduction in oxygen concentration in the normal cells—an extremely acute period of hypoxia—and consequently a transient radioresistance. Even though the tumor cells also experience a similar oxygen reduction, the magnitude is small. In conventional RT, delivered in longer treatment times, there is enough oxygen concentration and sufficient time for oxygen to diffuse into the irradiated region to replace the oxygen that has been lost. It has also been proposed that at UHDRs, the production of reactive oxygen species (ROS) is limited, thus promoting protection of normal tissue [24].

The conventional fractionated treatments result in the irradiation of a large proportion of circulating lymphocytes. While nearly 98.8% of the blood pool is exposed to a dose of at least 0.5 Gy or more by a 30 fractions regimen of conventional RT, in FLASH-RT, due to the extremely short time of irradiation, only a fraction of the blood pool is irradiated. The induction of chromosomal aberrations in the blood pool is directly proportional to the volume of the blood pool irradiated. Thus, in FLASH-RT, fewer lymphocytes are irradiated, resulting in reduced induction of chromosomal aberrations [24].

There are differences between the ways normal and cancer cells metabolize. The increased labile Fe in cancer cells and the presence of lower pro-oxidant burdens in normal cells have also been proposed as reasons for normal tissue sparing and isoeffective tumor killing [25].

Although many of the biological studies have showed encouraging results, there have been some negative observations too. Venkatesulu et al. observed that FLASH dose rates of 35 Gy/s do not protect mice from the detrimental side effects of irradiation [26]. Besides such contrary results, there are multiple challenges and obstacles involved in translating FLASH-RT to the stage of clinical trials. As pointed out by Gao et al., these include: there is still no such delivery system that can deliver multiple FLASH beams simultaneously; the IMRT that we commonly employ in conventional RT needs to be upgraded into subsecond scale to suit FLASH-RT; real-time adaptation to align beam and target; unclear underlying mechanism of FLASH effect; late- and overall effects of FLASH-RT is still unknown; scientific findings need to be independently verified; and accurate and reliable detector systems are to be developed [23]. The FLASH-RT certainly has got huge potentials to widen the therapeutic window in radiation treatment, but is currently riddled with several challenges before it can reach the clinical trials stage.

Multiple Choice Questions

1. The size of a nucleus is around:
   (a) 10^{-9} m
   (b) 10^{-11} m
   (c) 10^{-13} m
   (d) 10^{-15} m
2. The charge of an electron is ____________________.
   (a) 1836 times lesser than the charge of a proton
   (b) Equal to the charge of a proton
   (c) 1836 times higher than the charge of a proton
   (d) Twice the charge of a proton
3. Carbon has an atomic mass of 12.0110, and hence, 12.0110 g of carbon contain Avogadro number ($6.02214076 \times 10^{23}$) atoms. How many electrons are there in 1 g of carbon?
   (a) 5.014×10^{22}
   (b) 1×10^{23}
   (c) 3.084×10^{23}
   (d) 12.011×10^{23}
4. K_α transition refers to electron transition ____________________:
   (a) from L-shell to K-shell
   (b) from K-shell to L-shell
   (c) from M-shell to K-shell
   (d) from K-shell to M-shell
5. What will be the activity of a $C^{60}o$ source after 13.175 years from now if its current activity is 5.27 Ci? (The half-life of $C^{60}o$ is 5.27 years)
   (a) 0.9216 Ci
   (b) 0.9316 Ci
   (c) 0.9416 Ci
   (d) 0.9516 Ci
6. The decay of $C^{60}o$ to $N^{60}i$ results in the emission of one 1.33 MeV and one 1.17 MeV gamma ray photons. Which of the following statements is TRUE?
   (a) Both the photons are emitted by $\mathbf{C^{60}o}$ nucleus.
   (b) $\mathbf{C^{60}o}$ nucleus emits the 1.33 MeV photon, and the $\mathbf{N^{60}i}$ nucleus emits the 1.17 MeV photon.
   (c) $\mathbf{C^{60}o}$ nucleus emits the 1.17 MeV photon, and the $\mathbf{N^{60}i}$ nucleus emits the 1.33 MeV photon.
   (d) Both the photons are emitted by $N^{60}i$ nucleus.
7. When a high-energy photon, having energy much higher than the rest energy of the electron, undergoes Compton effect:
   (a) the scattered photon retains much of the original energy and the Compton electron receives very less energy from the incident photon
   (b) the Compton electron gets much of the incident photon's energy and the scattered photon carries only a fraction of the incident energy
   (c) The fraction of energy transferred to the Compton electron is constant irrespective of incident photon's energy
   (d) No energy is transferred to the electron and the incident photon retains all of its energy

8. Patient is to be treated by a 12 × 12 cm^2 field of 18 MV beam, at 100 cm SSD to a dose of 200 cGy at 5 cm depth. The PDD is 95.7%. If the output is 102 cGy/100 MUs at the depth of dose maximum, the number of MUs required to deliver the desired dose is:
   (a) 196
   (b) 200
   (c) 202
   (d) 205
9. On a cross-plane profile, radiation field width is defined as the:
   (a) lateral distance between the 80% dose points on either side of the central axis
   (b) lateral distance between the 50% dose points on either side of the central axis
   (c) lateral distance between the 20% dose points on either side of the central axis
   (d) lateral distance between the 10% dose points on either side of the central axis
10. In a nonphysical wedge system, the wedge effect is produced by:
   (a) the movement of both X1 and X2 jaws of the X-ray collimator system
   (b) the movement of either X1 or X2 jaw of the X-ray collimator system
   (c) the movement of both Y1 and Y2 jaws of the X-ray collimator system
   (d) the movement of either Y1 or Y2 jaw of the X-ray collimator system

Answers

1. (d)
2. (b)
3. (c)
4. (a)
5. (b)
6. (d)
7. (b)
8. (d)
9. (b)
10. (d)

Precepts

Atom is the building block of all matter. The size of the atom is roughly 1×10^{-10} m and that of nucleus is 1×10^{-15} m.

Atom is denoted as X_z^A, where X is the symbol of the element, A is the atomic mass, and Z is the atomic number.

Avogadro's number, an important concept in atomic physics defines the number of particles in one mole of a substance. In one mole, there are $6.02214076 \times 10^{23}$ particles.

The number of atoms/particles per gram in any given substance can be determined by taking the ratio of Avagadro's number and the atomic mass.

Electromagnetic radiation is a type of energy that is transmitted in the form of electromagnetic waves.

The fast moving electrons that strike the anode at high energies slow down through the release of X-rays as they pass closer to the nucleus of the atom.

The process of radioactivity occurs when the nucleus of an atom becomes unstable and decays, releasing energy in the form of ionizing radiation.

The half-life of a radionuclide plays an important role in determining the specific activity, the higher the half-life, the lesser the specific activity, and vice versa.

Alpha particles are a type of ionizing radiation consisting of two protons and two neutrons bound together. Beta rays are made up of high-energy particles, specifically electrons or positrons, which are emitted from the nucleus of an atom during radioactive decay. Gamma rays are weightless quantum packets of energy. A beta or alpha particle decay is usually accompanied by the emission of gamma ray during radioactive decay.

Photons are called as indirectly ionizing radiations. They are responsible for *transporting* the energy across relatively large distances in the human body, and it is only the secondary electrons that are set in motion by these photons, which are responsible for the *actual deposition of energy* in tissues.

Coefficients are useful concepts to best describe beam attenuation, energy transferred from the beam and energy absorbed by the medium.

Dual energy CT exploits the enhanced attenuation differences among tissues owing to photoelectric absorption and thus offers improved target definition.

The percent depth dose increases with SSD because of the inverse square law effect.

Tissue–phantom ratio is defined as the ratio of dose rate at a given depth in phantom to the dose rate at the same point, but at the reference depth (e.g., 10 cm).

Phantom scatter factor cannot be directly measured and is indirectly determined from the total scatter factor, $S_{c,p}$.

Isodose charts are of two types: SSD chart and SAD chart. In the SSD chart, the dose is normalized at the reference point of dose maximum. In the SAD chart, the dose is normalized to a certain depth beyond the depth of dose maximum.

A beam profile represents the dose variation across the field at a depth.

Nonphysical wedges produce the wedge effect by computer-controlled movement of one of the Y-jaws of the X-ray collimators from one end to other end while beam is ON.

The wedge transmission factor, or simply the wedge factor, quantifies the beam *transmission* along the central axis.

An ionization chamber is essentially a gas-filled cavity with a central collecting electrode and a conductive outer wall. They are considered as the gold standard in the measurement of ionizing radiation. The charge collected by a ionization chamber is measured with electrometers which are instruments used to measure small currents with a magnitude of 10^{-9} A or less.

Quality assurance refers to all activities, planned or systematic, necessary to instill confidence that the radiation oncology service will meet the requirements for quality care.

Deterministic effects are called deterministic because they are certain to occur, if the absorbed dose exceeds a threshold level. Stochastic effects on the other hand are probabilistic and random in nature such as induction of malignancies and hereditary effects without a threshold dose.

The recommendations contained in the safety codes published by AERB are in broad alignment with the recommendations and findings of international bodies such as ICRP and IAEA.

Justification, optimization of risk and remaining within the prescribed permissible limits of dose are the three cardinal principles for protection in any radiation practice.

Potential exposures are controlled by optimizing the design of installations, equipment and operating procedures.

For a telecobalt unit, the primary and secondary barriers of concrete should be 130 and 65 cm thick. For 10–25 MV photon beam, it should be 240 and 120 cm, respectively.

The differential response between FLASH-RT and conventional RT arises due to the radiochemical depletion of oxygen at UHDRs and the resultant radioresistance in the irradiated normal tissue.

While nearly 98.8% of the blood pool is exposed to a dose of at least 0.5 Gy or more by a 30 fractions regimen of conventional RT, in FLASH-RT, due to the extremely short time of irradiation, only a fraction of the blood pool is irradiated.

References

1. Khan FM, Gibbons JP. Khan's the physics of radiotherapy. 5th ed. Philadelphia: Lippincott Williams & Wilkins; 2014.
2. Meredith WJ, Massey JB. Fundamental physics of radiology. 3rd ed. Bristol: John Wright & Sons Ltd.; 1977.
3. Johns HE, Cunningham JR. The physics of radiology. 4th ed. Springfield: Charles C Thomas, Publisher; 1983.
4. Podgorsak EB. (Tech Ed.). Radiation oncology physics: a handbook for teachers and students. Vienna: International Atomic Energy Agency; 2005.
5. Mayles WPM, Nahum AE, Rosenwald JC, editors. Handbook of radiotherapy physics (volume 1). 2nd ed. New York: CRC Press; 2022.
6. McDermott PN, Orton CG. The physics and technology of radiation therapy. 2nd ed. Madison: Medical Physics Publishing; 2018.

7. Noid G, Zhu J, Tai A, Mistry N, Schott D, Prah D, Paulson E, Schultz C, Li XA. Improving structure delineation for radiation therapy planning using dual-energy CT. Front Oncol. 2020;10:1694. https://doi.org/10.3389/fonc.2020.01694.
8. Beaton L, Bandula S, Gaze MN, Sharma RA. How rapid advances in imaging are defining the future of precision radiation oncology. Br J Cancer. 2019;120(8):779–90. https://doi.org/10.1038/s41416-019-0412-y.
9. Camps SM, Fontanarosa D, Verhaegen F, Vanneste BG. The use of ultrasound imaging in the external beam radiotherapy workflow of prostate cancer patients. Biomed Res Int. 2018;2018 https://doi.org/10.1155/2018/7569590.
10. Western C, Hristov D, Schlosser J. Ultrasound imaging in radiation therapy: from interfractional to intrafractional guidance. Cureus. 2015;7(6):e280.
11. Chandarana H, Wang H, Tijssen RH, Das IJ. Emerging role of MRI in radiation therapy. J Magn Reson Imaging. 2018;48(6):1468–78. https://doi.org/10.1002/jmri.26271.
12. Bailey DL, Philips W, Baldock C. The future of radiotherapy is molecular. Phys Eng Sci Med. 2020;43(3):755–9. https://doi.org/10.1007/s13246-020-00917-9.
13. Dadgar S, Rajaram N. Optical imaging approaches to investigating radiation resistance. Front Oncol. 2019;9:1152. https://doi.org/10.3389/fonc.2019.01152.
14. Kutcher GJ, Coia L, Gillin M, Hanson WF, Leibel S, Morton RJ, Palta JR, Purdy JA, Reinstein LE, Svensson GK, Weller M. Comprehensive QA for radiation oncology: report of AAPM radiation therapy committee task group 40. Med Phys. 1994;21:581.
15. Klein EE, Hanley J, Bayouth J, Yin FF, Simon W, Dresser S, Serago C, Aguirre F, Ma L, Arjomandy B, Liu C. Task Group 142 report: quality assurance of medical accelerators. Med Phys. 2009;36(9, Part1):4197–212. https://doi.org/10.1118/1.3190392.
16. Canadian Partnership for Quality Radiotherapy Technical Quality Control Guidelines for Medical Linear Accelerators and Multileaf Collimators. 2016. https://www.cpqr.ca/wp-content/uploads/2017/02/MLA-2016-07-03.pdf. Accessed 23 Dec 2022.
17. Commissioning and quality assurance of computerized planning systems for radiation treatment of cancer: Technical Reports Series no. 430. International Atomic Energy Agency, Vienna, 2004.
18. Smilowitz JB, Das IJ, Feygelman V, Fraass BA, Kry SF, Marshall IR, Mihailidis DN, Ouhib Z, Ritter T, Snyder MG, Fairobent L. AAPM medical physics practice guideline 5. a.: commissioning and QA of treatment planning dose calculations—megavoltage photon and electron beams. J Appl Clin Med Phys. 2015;16(5):14–34.
19. Radiation protection and safety of radiation sources: international basic safety standards. International Atomic Energy Agency, Vienna, 2014. IAEA BSS 2014.
20. Atomic Energy Regulatory Board, Mumbai. India. AERB Safety Code No. AERB/RF-MED/SC-1 (Rev. 1): Radiation Therapy Sources, Equipment and Installations. March 2011.
21. Valentin J (ed.). The 2007 Recommendations of the international commission on radiological protection. Annals of the ICRP. Publication 103. Elsevier, 2007.
22. Favaudon V, Caplier L, Monceau V, Pouzoulet F, Sayarath M, Fouillade C, Poupon MF, Brito I, Hupé P, Bourhis J, Hall J. Ultrahigh dose-rate FLASH irradiation increases the differential response between normal and tumor tissue in mice. Sci Transl Med. 2014;6(245):245ra93.
23. Gao Y, Liu R, Chang CW, Charyyev S, Zhou J, Bradley JD, Liu T, Yang X. A potential revolution in cancer treatment: a topical review of FLASH radiotherapy. J Appl Clin Med Phys. 2022;23(10):e13790. https://doi.org/10.1002/acm2.13790.
24. Wilson JD, Hammond EM, Higgins GS, Petersson K. Ultra-high dose rate (FLASH) radiotherapy: silver bullet or fool's gold? Front Oncol. 2020;9:1563.
25. Esplen N, Mendonca MS, Bazalova-Carter M. Physics and biology of ultrahigh dose-rate (FLASH) radiotherapy: a topical review. Phys Med Biol. 2020;65(23):23TR03.
26. Venkatesulu BP, Sharma A, Pollard-Larkin JM, Sadagopan R, Symons J, Neri S, Singh PK, Tailor R, Lin SH, Krishnan S. Ultra high dose rate (35 Gy/sec) radiation does not spare the normal tissue in cardiac and splenic models of lymphopenia and gastrointestinal syndrome. Sci Rep. 2019;9(1):1–9.

6 Pathology in Oncology Practice

Tanuja Shet, Trupti Pai, Paromita Roy, and Roopa Paulose

6.1 Introduction

"As is your pathology so is your practice", nothing exemplifies this more than role of oncopathology in management of patients with cancer. Excellent quality pathology service provides maximum information to the clinician on patient management particularly when high-end sophisticated molecular techniques are unavailable. The oncopathologist has a dual role: diagnostic and theranostic. The diagnostic role includes not only labeling the disease, but also grading and staging whenever necessary. The theranostic role has evolved over the last two decades and includes the predictive and prognostic biomarker evaluation. The chapter tries to simplify this integral and intricate role of oncopathology through the journey of tissues. The authors accept that covering every site-specific cancer is beyond the scope of this chapter. The chapter tries to follow the journey of tissues in pathology and is divided into four sections, viz.:

(a) The grossing of surgical excisions: This includes steps from receiving the tissue (biopsies, excisions) to transforming them into a hematoxylin and eosin (H&E)-stained section on the glass slide for evaluation.
(b) Minimum data sets in cancer: This includes consistent format of reporting common cancers and ensures a meaningful and informative pathology report which aids clinical decisions.
(c) Frozen section: This rapid diagnostic method for real-time histopathology reporting intraoperatively and few salient points will be discussed for a complete picture.
(d) Immunohistochemistry (IHC): Although most common cancers are a morphological diagnosis, immunohistochemistry is now an innate component of diagnosis in surgical pathology. Some tumors require immunohistochemistry for diagnosis due to overlapping morphologies, unusual morphologies or unusual clinical presentation. In this era of targeted therapy, Immunohistochemistry also plays a vital role for evaluation of predictive and prognostic biomarkers.
(e) Molecular Diagnostics: Science of tumor diagnosis and prognosis in pathology has

T. Shet (✉) · T. Pai
Department of Pathology, Tata Memorial Hospital, Mumbai, Maharashtra, India

P. Roy
Department of Pathology, Tata Medical Center, Kolkata, West Bengal, India
e-mail: paromita.roy@tmckolkata.com

R. Paulose
Department of Pathology, Amrita Institute of Medical Sciences, Kochi, India
e-mail: roopapaulose@aims.amrita.edu

A. Munshi et al. (eds.), *Radiation Oncology – Principles, Precepts and Practice*,
https://doi.org/10.1007/978-981-97-8389-2_6

expanded and we now look for targets within tumor at DNA or RNA level and a snapshot of existing molecular testing methods will be covered.

6.2 Grossing and Processing of Surgical Specimens

Any specimen referred to pathology laboratory should be accompanied by a request form with patient details and clinical information in case a system for hospital electronic medical record is not available. Tissues sent for histopathological evaluation are to be sent in adequate formalin fixative to prevent autolysis due to bacterial overgrowth after surgical removal. A fixative like formalin preserves biological tissues by terminating any ongoing biochemical reactions. Though alcohol has been shown to be more effective than formaldehyde for preserving antigens such as intermediate filament proteins, it causes tissues to shrink and failure of some enzymatic markers. Many guidelines have now endorsed that 10% neutral buffered formalin is a best all-round fixative that maintains antigenicity of tissues [1, 2]. Improper fixation of tissue is the most common factor responsible for poor demonstration of biomarkers by IHC and in molecular tests in low-resource countries where the site of specimen generation and the testing laboratory are distantly located [3]. The "cold ischemia time" i.e., time from removal of the specimen till fixation, as per the best practice protocol should be 1 h or maximum 4 h. Many studies on biomarkers in breast cancer have captured the extent of damage done by a delayed fixation, e.g., a strong positive estrogen receptor (ER) may test negative if there is a >48 h delay in fixation [4]. After adequate fixation, tissues are then put through tissue processing machines to yield a dehydrated tissue which is then impregnated with paraffin which then enables cutting thin sections onto slides for staining with hematoxylin and eosin. Hence, processing of tissues takes time (12 to 36 h) depending on their size and the type of tissue processor used.

6.2.1 Basic Steps in Grossing that Are Common to Most Specimens in Oncology

The diagnosis is usually known by the time of final surgery and the key role of grossing is to assess other parameters, viz., tumor subtype if any, grading, staging and adequacy of margins of excision. Grossing is the best chance to capture all essentials in histopathology especially the margins and tumor size.

(a) Specimen orientation: A margin is the plane through which the surgeon's knife has cut a tissue, viz., the skin, mucosa or soft tissue. These cut ends or surfaces thus form margins of excision. Margins are an important aspect of any conservative cancer surgery to ensure a R0 resection. Satisfactory assessment of margins requires inking the cut ends or surfaces. In some specimens, vascular margins are also documented, e.g., nephrectomy, amputation, lung tumor specimens like lobectomy or pneumonectomy.

    Example: A three-suture orientation is usually given in breast-conserving surgery (BCS), viz., **Short** suture = **superior** aspect of specimen, **Long** suture = **lateral** aspect of specimen and **Anterior** = A black suture. It is essential to understand that specimen distortion does happen during transport and margins can be distorted [5]. The classic example of this is a BCS specimen that flattens during transport narrowing the margins in the supero-inferior and posterior plane [6].

(b) Inking the specimen surface and margin sampling: The surface of a wide excision or the outermost mucosa represents the plane through which the surgeon's knife has cut and forms the margin of excision which is inked for identification under microscope [5]. Margins are an important aspect of any conservative cancer surgery.

    *Examples

    - In wide excisions (like a BCS, or soft tissue excisions) to adequately sample the entire external surface would mean taking thousands of sections, which is impractical. Hence, we adopt a six surface sampling method whereby the anterior, posterior, superior inferior, medial and lateral surfaces are arbitrarily defined based on the orientation sutures and sampled. Inking is done with water proof inks like the Davidson's marking inks, Dhobi ink used in laundry, Fevicryl fabric paints, etc. (Fig. 6.1a). A radial or perpendicular margin with tumor and the inked surface or mucosal end is then taken so that actual distance can be measured is preferred (Fig. 6.1b).

- At flat mucosal sites like in head and neck resections a four surface margin evaluation is done usually anterior, posterior, lateral and medial (Fig. 6.1c). Mucocutaneous margins (for lip) and tracheal margin (for larynx) are however examined as shave/en face margins, so that the entire circumference can be studied. When separate margins are sent by the surgeon for frozen assessment, they are not taken again from the main specimen.
- Large amputations are sliced using a bone cutter. Bone sections are thinly sliced and fixed overnight in formalin and then put in decalcifying solution, which removes calcium and makes them soft and amenable to sectioning. Bony margins are usually assessed as shave margins from cut ends.

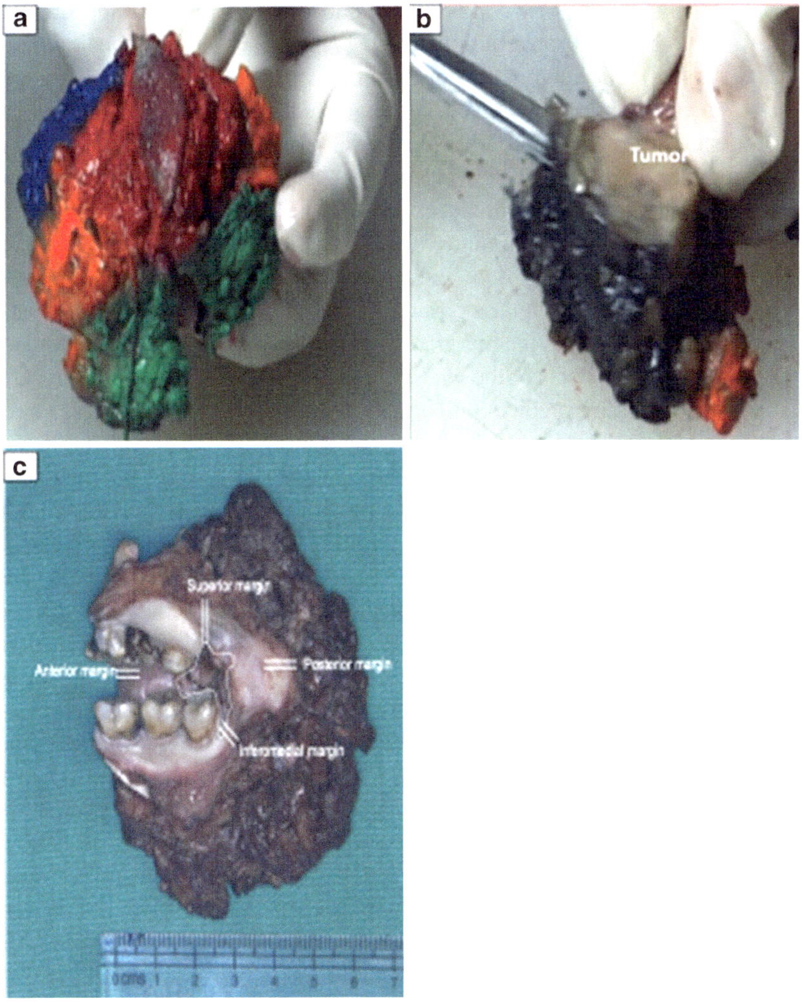

Fig. 6.1 (**a**) Breast-conserving surgery inked using six different colors for six surface for margin assessment; (**b**) Radial margin in a lumpectomy to show tumor at one end and ink surface close to it; (**c**) Mandibulectomy for buccal cancer with tumor area highlighted in center with four surface mucosa to be sampled

- Gastrointestinal resection [7–9]: In the gastrointestinal tract doughnut margins are encouraged as they cover entire circumference of the mucosa and wall. In specimens like esophagus and rectum the status of circumferential resection margin in the non-peritonealized area is essential to record. Circumferential resection margin (CRM) is the closest distance between the radial resection margin and the tumor, or areas of neural or vascular invasion or metastatic lymph node. As an example in rectal resections if the tumor is located below the anterior peritoneal reflection, the circumferential radial margin (CRM) is completely non-peritonealized (Fig. 6.2 Upper diagram) and its involvement poses a high risk for local recurrence, metastases and shorter survival but these patients will benefit from radiotherapy [7, 8]. Upper rectal tumors have a posterior and lateral non-peritonealized surface which should be evaluated (Fig. 6.2 lower diagram). For low anterior resection (LAR) and abdominoperineal resection (APR), it is important to document CRM is in the mesorectal plane, the intra-mesorectal plane, or the muscularis propria plane as it not only assesses the margin involvement, but it also reflects of the quality of surgery which in turn predicts outcome. For APR specimens, best outcome has been shown for mesorectal excision/extralevator plane and worst for muscularis propria/subsphincteric plane of excision.

(c) Tumor sampling: The pathological tumor size (pT) is critical for staging. Tumor size is noted as three dimensions, the largest dimension being most important. In specimens of oral cavity, depth of invasion (DOI) also affects the pT stage and it is most accurately assessed on microscopy (it has been described in detail later in this section). We follow a protocol of four sections to be submitted from the tumor in common cancer resection specimens. The extent of tumor sampling however varies as per organ under evaluation and also the biopsy diagnosis. Tumor sampling should try and capture the tumor size microscopically. This is important as some tumors like gastric signet ring carcinomas and lobular carcinomas are difficult to palpate and tumor size underestimation can occur during gross inspection. Blocks must be taken to assess extent of tumor invasion for staging purposes, venous invasion and to demonstrate discontinuous tumor deposits. In organs like breast evaluating whole slice of tumor with sufficient adjacent breast is best as this allows for recording factors like extensive intraduct carcinoma and lymphovascular emboli. Similarly, in colorectal tumors, full thickness sections of tumor with bowel wall are essential to for staging purposes. Sampling unique areas besides tumors is also important. For example, in head and neck, specimen's masticator space muscles are especially important to identify and take sections from to exclude stage 4b tumor. For post-chemotherapy resections the grossing should involving mapping the entire prechemotherapy tumor bed or tumor size to understand the pattern and extent of tumor regression. Recommendations for how many tumor blocks post chemotherapy vary as per site. For example, in rectum as per the recommendations of the Royal College of Pathologists, when there is no grossly visible tumor, at least five blocks should be taken from the site of the original tumor by correlating with imaging. In the event that no residual tumor is microscopically identified, then the whole area of original tumor/scarred area is sampled and examined at multiple (at least 3) deeper levels, before a complete pathological response is recorded (ypT0N0) [10].

(d) Nodal staging and Reporting lymph node metastasis**:** Macrometastases in nodes are tumor deposits >2 mm and micrometastasis is a metastasis >0.2 mm with none >2.0 mm. Isolated tumor cell clusters are small clusters of cells not larger than 0.2 mm or contain <200 cells in a single histologic cross section. Node grossing in oncopathology should

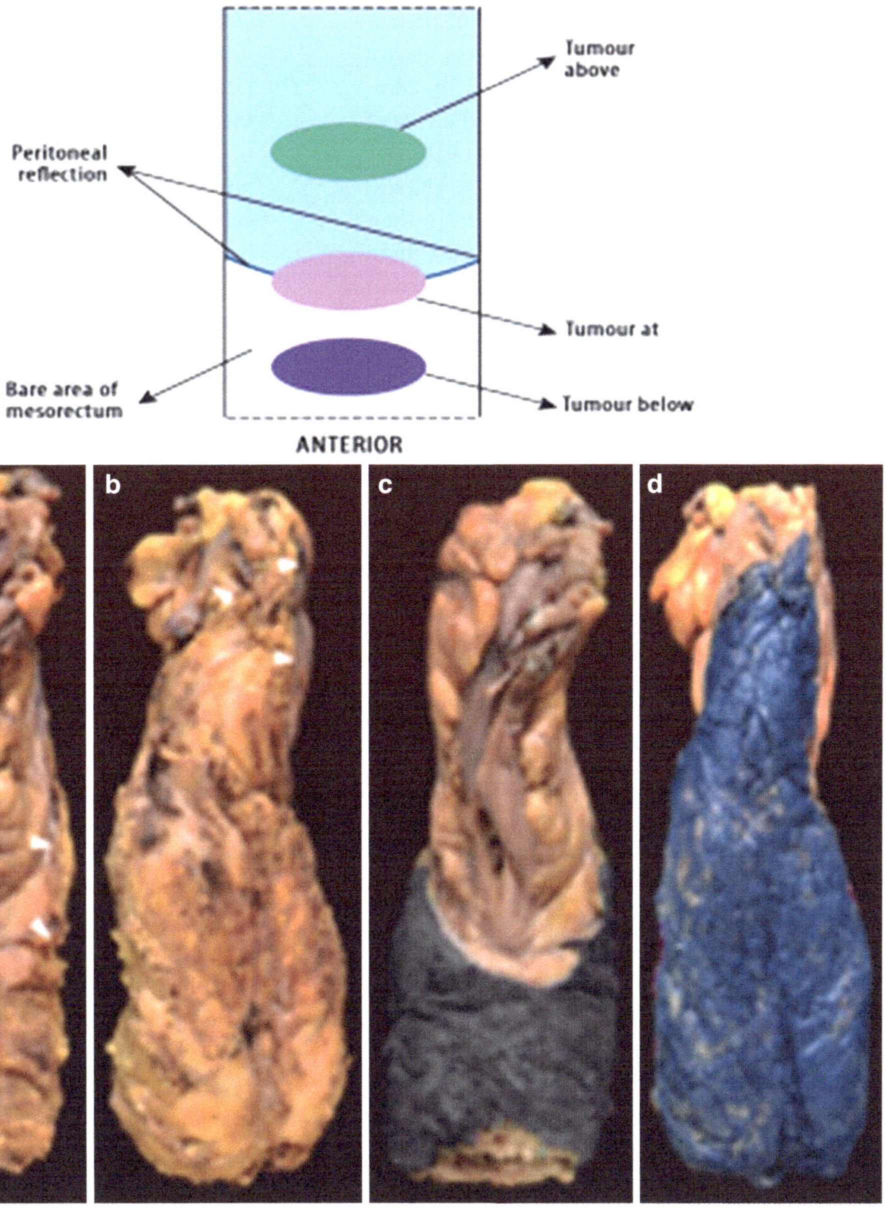

Fig. 6.2 Top: Schematic diagram depicting site of rectal of tumor in relation to the anterior peritoneal reflection. (Modified from [10]). Bottom: Low anterior resection of rectum with anterior peritoneal reflection depicted with white arrows (**a**), posterior non-peritonealized surface with peritoneal reflection depicted with white arrows (**b**), anterior non-peritonealized surface/circumferential radial margin inked black (**c**). Posterior non-peritonealized surface/radial margin inked blue (**d**)

be such that the macrometastasis are sufficiently captured. Extent of lymph node slicing depends on whether it is sentinel node or whether the node is part of a radical dissection.

Sentinel lymph node biopsy (SLNB) grossing: Most of the data on histological evaluation of SLNB has come from breast and melanoma specimens. Slicing SLNB at 2.0 mm intervals then embedding it completely ensures a high probability of detecting all metastases ≥2.0 mm [11]. It is known that step sections will pick up 9 to 15% additional metastasis in SLNB but significance of such 'occult metastasis' is unclear [12].

Nodes in radical dissections: Nodes <5 mm in size should be submitted in their entirety. Lymph nodes >5 mm in maximum size should be sliced at intervals of approximately 2 to 3 mm or less perpendicular to the long axis and submitted. Submitting perinodal adipose tissue captures the extranodal extension extent adequately.

6.3 Minimum Data Reporting Sets

The aim of minimum data sets is to provide a checklist through synoptic reporting format for a meaningful and comprehensive pathology report. It consists of three elements, viz., core (mandatory or required), conditional (recommended), optional. There are many reporting dataset guidelines like College of American Pathologists (CAP) (https://www.cap.org/protocols-and-guidelines/cancer-reporting-tools/cancer-protocol-templates), the Royal College of Pathologists (RCPath) (https://www.rcpath.org) and the International Collaboration on Cancer Reporting (ICCR) (http://www.iccr-cancer.org/datasets/) besides the AJCC staging available as a guide to the pathologist on how to go about it. General points covered in a histopathology report of cancer include the following:

1. Grade of tumor: Traditionally the biological aggressiveness or differentiation is captured in tumor grade. Differentiation is the degree to which a tumor resembles in cell of origin. In general, a three-tier grading system is used.
   Examples
   (a) The Nottingham modification of Richardson bloom scoring system for breast carcinoma is based on extent of tubule formation, nuclear grade and mitosis. Grade I tumors have >75% tubules (Fig. 6.3a), while Grade III tumors lack tubules (Fig. 6.3b).
   (b) Gleason's scoring in prostate cancer: In this, the tumor cells are graded on a scale of 1 to 5 whereby Grade 1 cells resemble normal prostate tissue (Fig. 6.3c) while cells closer to 5 are "high-grade" and have mutated so much that they barely resemble normal cells.
   (c) Squamous carcinomas and most adenocarcinomas are graded by a three tier grading system based on the degree of keratinization and gland formation respectively; well differentiated, moderately differentiated and poorly differentiated (Fig. 6.4a–d).
2. Depth of invasion (DOI): This is uniquely captured for oral cavity cancers and colonic tumors. In oral cancers DOI is calculated by dropping a perpendicular plumbline from the highest point of the arcuate line of horizon of the tumor surface (at the level of the adjacent uninvolved basement membrane), to the deepest point of invasion (Fig. 6.5). It recorded as millimeters and cut-offs of 5 and 10 mm are used for staging. In some tumors, the horizon is difficult to assess and tumor thickness should be recorded in these cases. It is important to understand that DOI may differ from the tumor thickness depending on the growth pattern of the tumor. A predominantly ulcerated tumor will have DOI more than the tumor thickness, while the reverse will be true for a proliferative tumor. In the eighth AJCC update, DOI has been incorporated into the T staging of the primary oral cancers. For colorectal cancers, the depth of tumor invasion beyond the muscularis propria, distance from CRM, etc. is recorded in mm (Fig. 6.6). The myometrial depth of invasion has to be recorded as it determines stage in endometrial adenocarcinomas.
3. Tumor growth pattern: Tumor growth sometimes recapitulates the phenotypic and genotypic traits of tumor.
   Examples

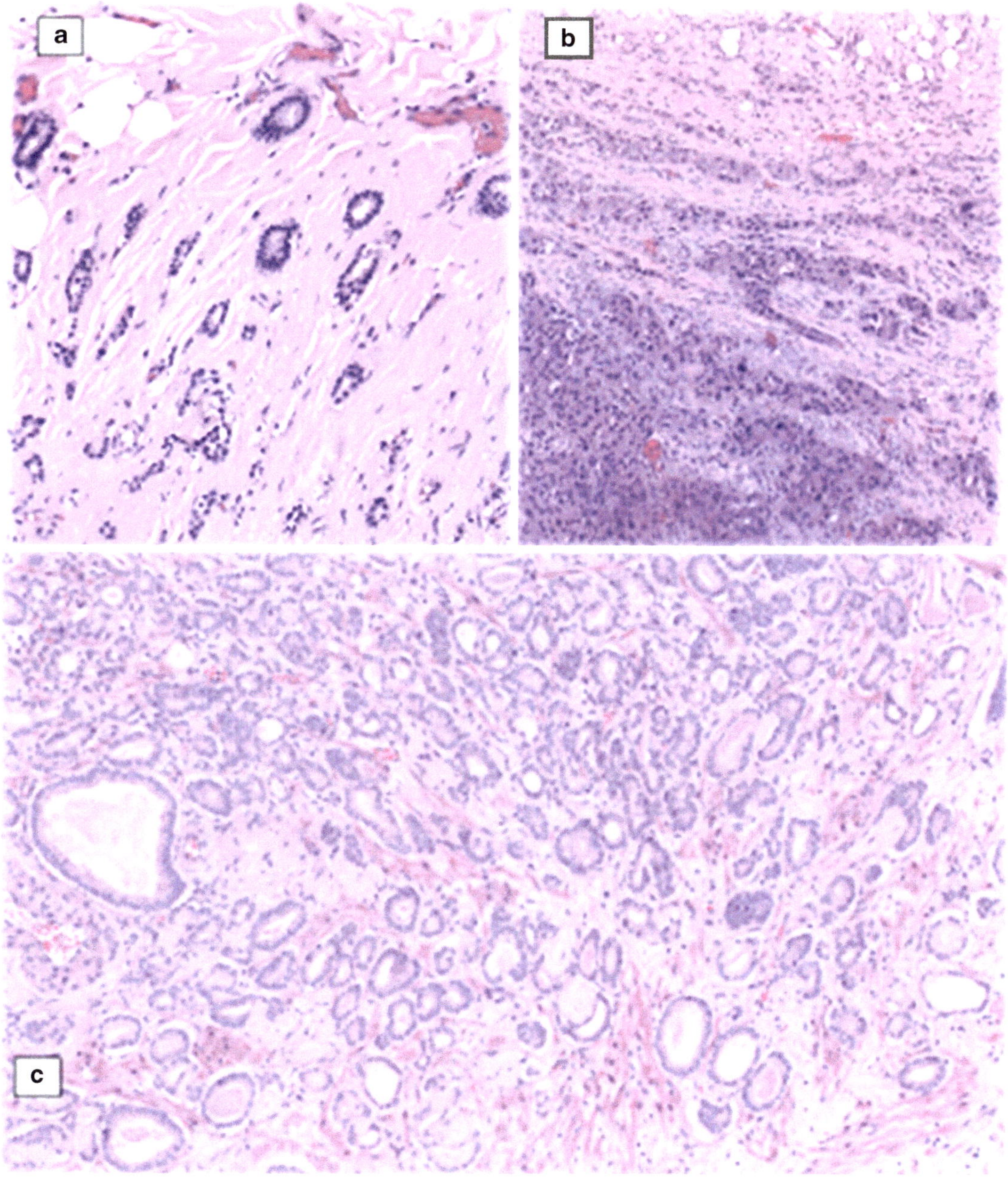

Fig. 6.3 (**a**) Grade I invasive breast carcinoma with infiltrating tubules. (**b**) Grade III invasive breast carcinoma with no tubules and just nests of cells. (**c**) Prostate adenocarcinoma low grade with infiltrating mature acini mimicking normal glands

(a) In breast histological types are based on tumor growth pattern and tumor cell arrangement as in invasive lobular carcinoma (pattern is produced due to loss of E cadherin), invasive papillary carcinoma, etc.

(b) Worst pattern of invasion (WPOI) is unique for oral and tongue tumors. It has been proved that the pattern of tumor growth at the advancing edge impacts clinical outcome, especially in early-

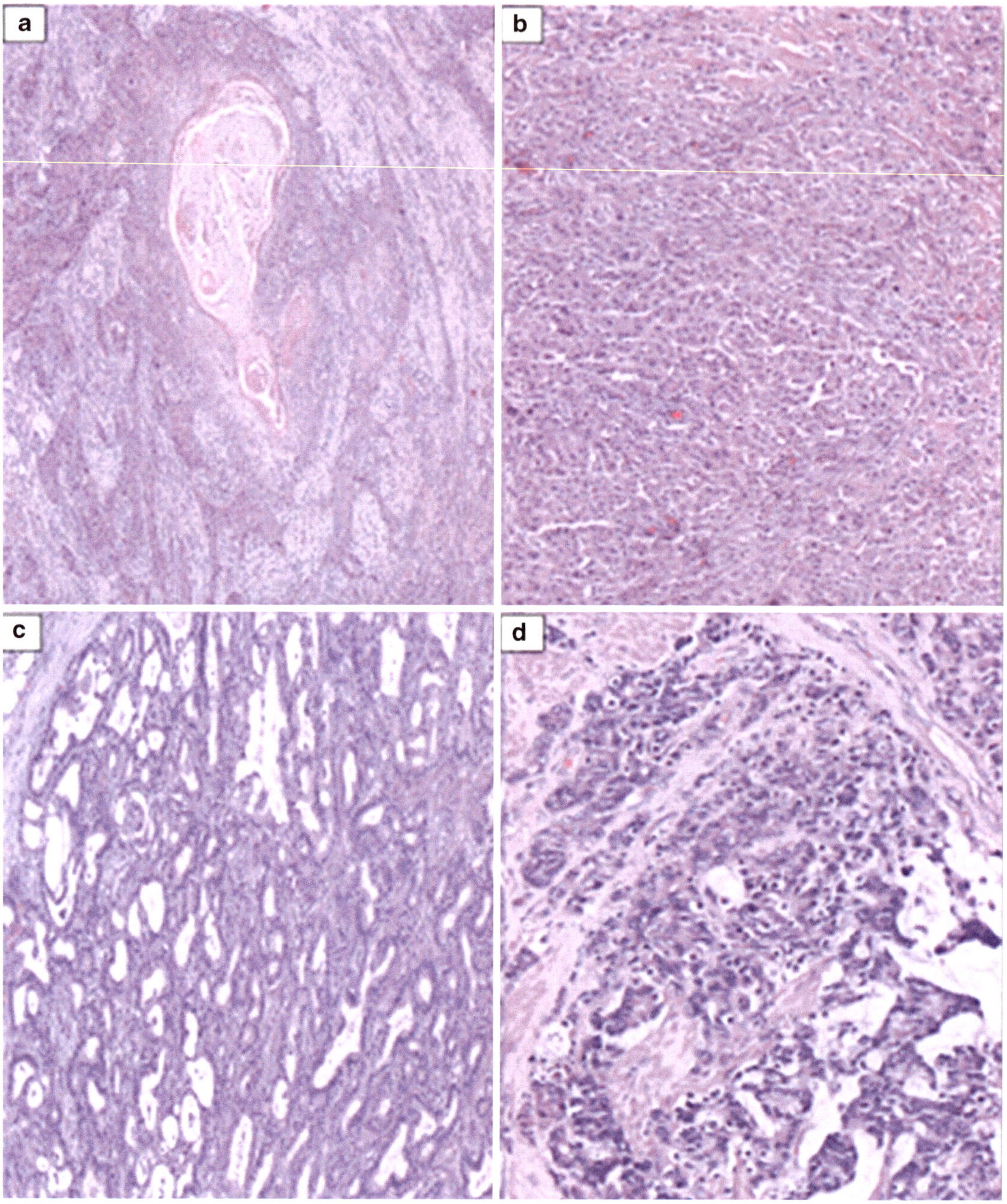

Fig. 6.4 (**a**) Well-differentiated squamous carcinoma with extensive keratinization. (**b**) Poorly differentiated squamous carcinoma which lacks obvious keratinization. (**c**) Well-differentiated adenocarcinoma with well-formed glands all throughout. (**d**) Poorly differentiated adenocarcinoma with solid nests and no well-defined glands, but extracellular mucin indicates the glandular nature

stage disease (T1 and T2) [13]. The pattern is assessed at the tumor–stroma interphase. In general, a cohesive front has better outcome compared to a non-cohesive pattern. The WPOI pattern 5 is defined as tumor dispersion of ≥1 mm between tumor satellites. The different patterns of invasion (POI) are described below (Fig. 6.7).

- POI 1—Cohesive broad pushing margin.
- POI 2—Cohesive broad pushing margins with ‘fingers’ or separate large tumor islands.
- POI 3—Noncohesive edge with invasive tumor islands of >15 cells/island.
- POI 4—Invasive tumor islands of <15 cells/island, single cell invasion and strands of tumor cells in single-file pattern, regardless of island size.

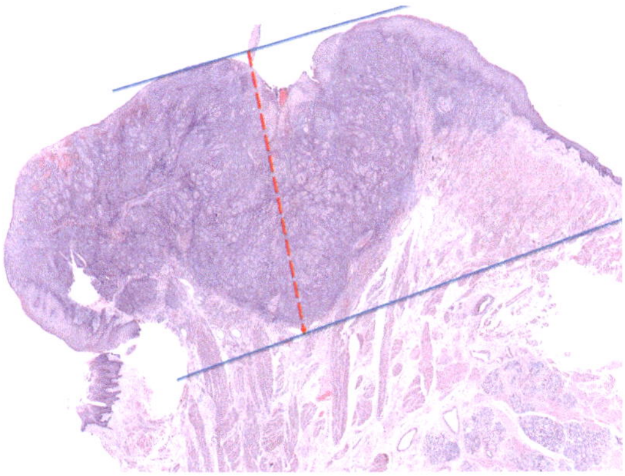

Fig. 6.5 Depth of invasion measurement in tongue carcinoma is measured as depicted by the red line

- POI 5—Tumor dispersion of ≥1 mm between tumor satellites.

4. Definition of positive margins: Margins define completeness of surgery and radiotherapy planning. A positive or a close margin indicates likelihood of residual tumor and a risk for local tumor recurrence. Each organ site has its own definition of a positive margin.
   Examples:
   (a) In BCS, the criteria defined by Society of Surgical Oncology (SSO) and the American Society for Radiation Oncology (ASTRO) guideline which is endorsed by American Society of Clinical Oncology (ASCO) and National Comprehensive Cancer Network (NCCN) says that “no tumor on ink” is adequate margin for invasive breast carcinoma and a 2 mm distance is adequate margin for pure DCIS [14].
   (b) Cut offs used for deciding close margins requiring radiotherapy in head and neck vary but in general >5 mm is considered free margin, 1–5 mm as close and <1 mm as positive [15].
5. Lymphovascular invasion (LVI) is defined as the presence of tumor cells within a definite endothelial-lined space (lymphatics or blood vessels). In breast cancer LVI beyond >1 mm tumor edge are prognostic and should be docu-

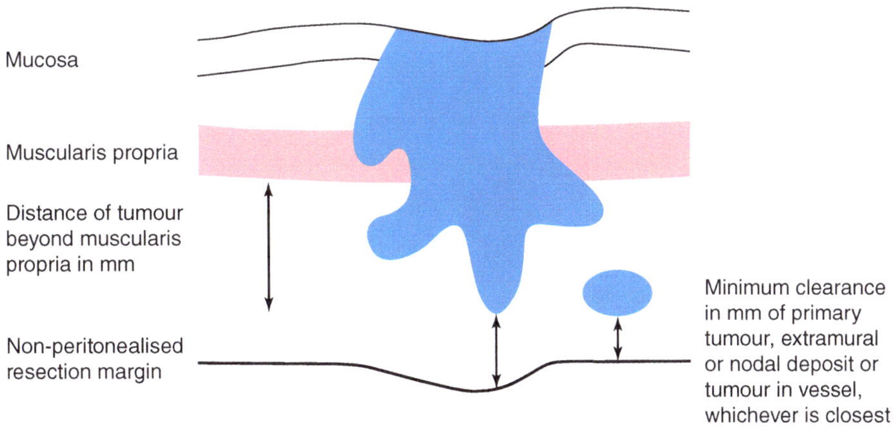

Fig. 6.6 Histological measurement of tumor beyond the muscularis propria and assessment of clearance from CRM. (Modified from [10])

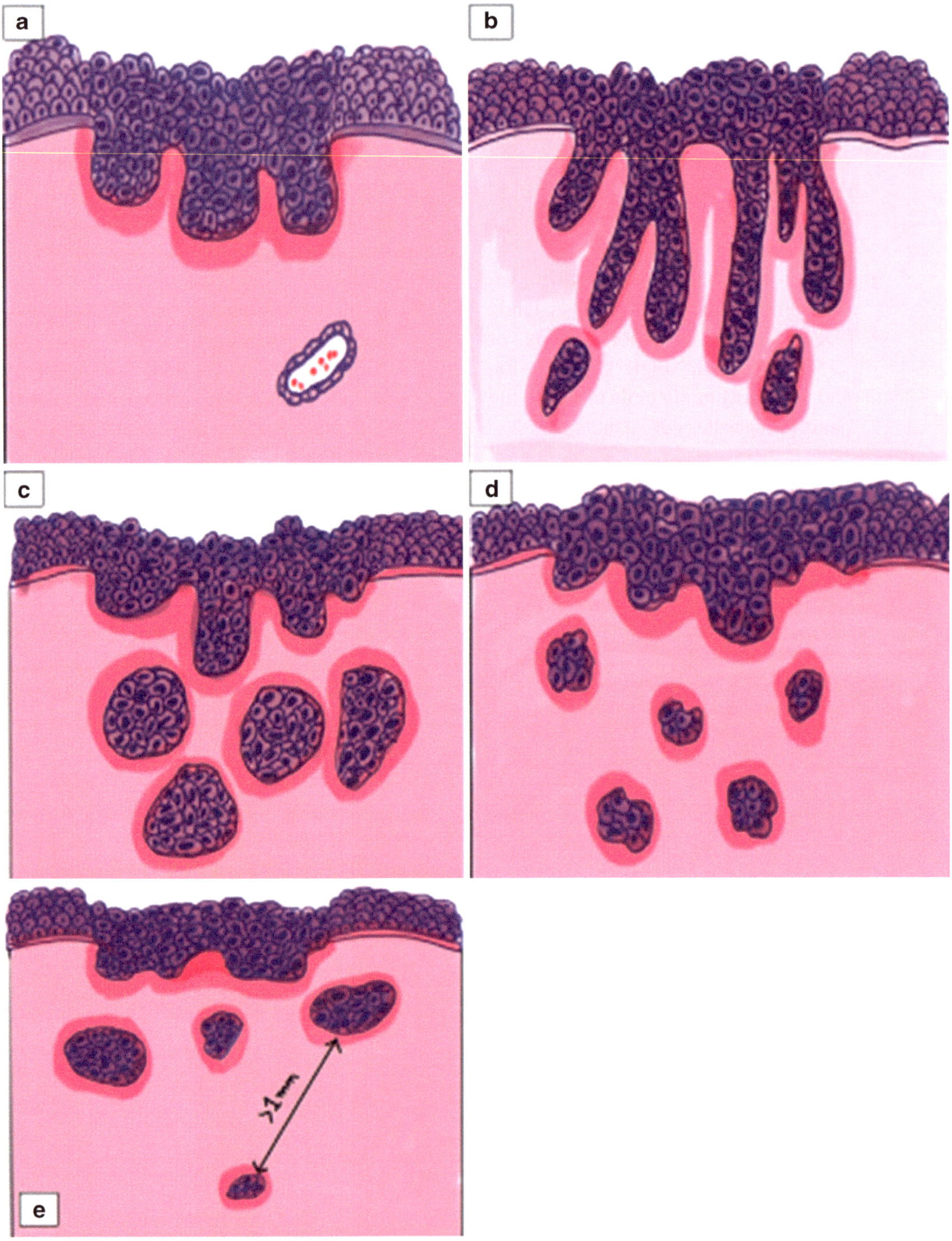

Fig. 6.7 Patterns of growth of an oral squamous carcinoma. (**a**) Cohesive growth with pushing growing edge. (**b**) Broad pushing edge but few separate tumor islands. (**c**) Invading islands composed of larger number of cells. (**d**) Smaller invasive nests and even single files. (**e**) Multiple tumor satellites separated by a distance of >1 mm. (Diagrams Courtesy Dr. Ankan Ghosh)

mented while such recommendations are not made for other organs. LVI may be confused with retraction spaces produced around tumors due to fibrosis. With use of IHC markers for endothelial cells lining lymphatics and arteries like CD31 and D249, more reproducible reporting of LVI is seen [16], but these are not routinely recommended or done to identify LVI.

6. Bone invasion: This is relevant for head and neck excisions and bone tumors. In head and neck excisions it is important to differentiate cortical erosion (which does not change stage) from full thickness invasion into medulla, which is required to stage a tumor as 4a.
7. Tumor deposits (satellites): As per TNM staging (eighth edition) [17], tumor deposit is defined as discreet discontinuous deposit of tumor seen grossly or microscopically in the pericolorectal adipose tissue within the lymphatic drainage of the primary tumor and is not histologically identified as a lymph node or vascular or neural involvement by tumor. In colorectal cancer, the presence of tumor deposits confers a worse outcome irrespective of lymph node metastases. Presence of tumor deposits in the absence of lymph node involvement by tumor is staged as N1c in the current TNM stage [17]. Cancerous nodules in the axillary fat in breast cancer without evidence of residual lymph node tissue are classified as positive axillary lymph nodes [18].

6.4 Frozen Section in Oncopathology

Frozen section is a technique in which rapid diagnosis is provided by freezing the tissue in a machine called cryostat (−15 to −25 °C). The rapid freezing results in hardening of the tissue which can thus be cut into thin sections for staining. The sections are then stained with toluidine blue and rapid HE technique for examination by the on-site pathologist. Frozen sections are the cornerstone of surgical oncology for confirming margin, node status, and primary tumor diagnosis. The sensitivity for primary diagnosis of cancer is very good, but it should be noted that rapid freezing distorts morphology, and diagnosis of some tumors, particularly sarcomas and lymphomas, leading to errors. These tumors are best diagnosed by routine processing and immunohistochemistry. Sentinel node FS evaluation in breast cancer avoids a two-step surgery, and though it has low sensitivity in detecting micrometastasis (19%), it picks up macrometastasis (75%) and reduces the percentage of patients requiring repeat surgery from 12% to 4% [19]. Frozen section nearly halves the rates of positive margin and close margins in head and neck resections which certainly translates into clinical benefits [20]. In gastric cancer frozen section of margins are mandatory for tumors with signet ring cell morphology as the gross assessment of margin can fail [21].

Frozen sections are useful for histological confirmation metastatic disease in in the liver and peritoneum for gastrointestinal and hepatobiliary cancers. Margin assessment frozen section is hence less popular for the fatty organs like breast conservative surgeries as FS sections from breast are often incomplete due to difficulty in freezing fat. However, proponents advocate that re-excision rate was reduced by a third, hence FS margin evaluation should be encouraged in BCS [22]. When used judiciously used the accuracy of FS for margins in BCS is 100% [23]. We think that gross evaluation of margins in frozen section in presence of surgeon is the best way to ensure whether FS is required in a given case. There is debate on whether cavity bed margins should be separately sent to frozen or the excision specimen sent for margin evaluation; however, most surgeons are favoring the former as there is specimen distortion during transport to frozen section room [24]. Although the diagnostic accuracy of FS pathology is not perfect, pathologist's knowledge and experience accompanied by an expert histotechnologist decide the final benefit.

6.5 Immunohistochemistry (IHC)

IHC is a method in which we detect antigen on surface of a cell with the help of an antibody grown against that antigen in non-human species. The indirect method whereby we use an additional step of secondary antibody against the primary antibody is preferred to increase the specificity of the reaction. The secondary antibody is labeled with a chromogen (usually diaminobenzidine DAB based) that produces color (usually brown) at the site of the antigen–antibody reaction within tissues.

6.5.1 Why and When Is IHC Required in Tumors?

With IHC a whole range of cell or tissue antigens ranging from amino acids and proteins to infectious agents and specific cellular populations can be analyzed [25]. IHC is indispensable in oncology, and most common applications are (1) histiogenetic identification of undifferentiated tumor; (2) subtyping of cancers like lymphomas; (3) evaluating primary site of origin in metastasis from an unknown primary; (4) predictive biomarker marker testing; (5) differentiating benign vs. malignant tumors; and (6) several research applications.

6.5.2 Approach to IHC in Tumors

The starting point is always a good hematoxylin section, and broadly there are four types of tumors. First group is of epithelial tumors or carcinomas which have their own site-specific IHC markers that are done when there is metastasis from unknown site. Second is a group of tumors derived primitive cells called round cell or malignant round cell tumor which all look so much like one another that IHC is must for classification. The third group is of lymphomas which have markers related to their cell of origin. The last group is of spindle cell tumor which chiefly includes mesenchymal tumors. There is an overlap within these groups as well. For example, some carcinomas have spindle-shaped appearance (sarcomatoid or metaplastic carcinoma), and IHC demonstration of epithelial markers will help diagnose the same (see cases illustrated from Figs. 6.8a–f and 6.9a–e). Likewise, some sarcomas can look very epithelioid or epithelial like and we need IHC to make that distinction.

Basic markers that are used in diagnosis depend on cell type like epithelial tumors are evaluated by range of cytokeratins, neuronal and glial differentiation is identified by synaptophysin/GFAP, neuroendocrine tumors are recognized by synaptophysin and chromogranin, most hematolymphoid neoplasms are lymphocyte common antigen (LCA) positive, while round cell tumors have their own repertoire of markers and all mesenchymal tumors express vimentin. However, lineage infidelity is known in tumors. For example, occasionally, a lymphoma-like anaplastic large cell and plasmablastic lymphoma can show cytokeratin (epithelial marker), and some carcinomas like renal cell or sarcomatoid carcinomas are negative for cytokeratin.

A panel-wise approach is preferred rather than reliability of one marker. To go through each marker in IHC is an impossible task for this chapter since this realm is ever expanding. Examples for common panels and their uses are given in Table 6.1. When all markers are negative vimentin is done to ensure that tissue is truly negative or has lost antigenicity as this marker always stains something. Different clones of same protein have different diagnostic implications, e.g., ALK1 used for diagnosis of ALK-positive anaplastic large cell lymphoma is different from ALK D5F3 used for detecting of ALK rearrangements in lung adenocarcinoma. As molecular diagnostic methods and next-generation sequencing methods expand, IHC profile of tumors is getting longer and Table 6.1 is bound to expand.

IHC is often used to differentiate benign vs. malignant in glandular organs that have biphasic lining of luminal and myoepithelial cells (MEC) like in breast and prostate. A combination of a nuclear and cytoplasmic stains is used to demonstrate the nuclei of MEC and the basal lamina that is deposited by MEC respectively (Fig. 6.10a–c). The nuclear marker used is p63

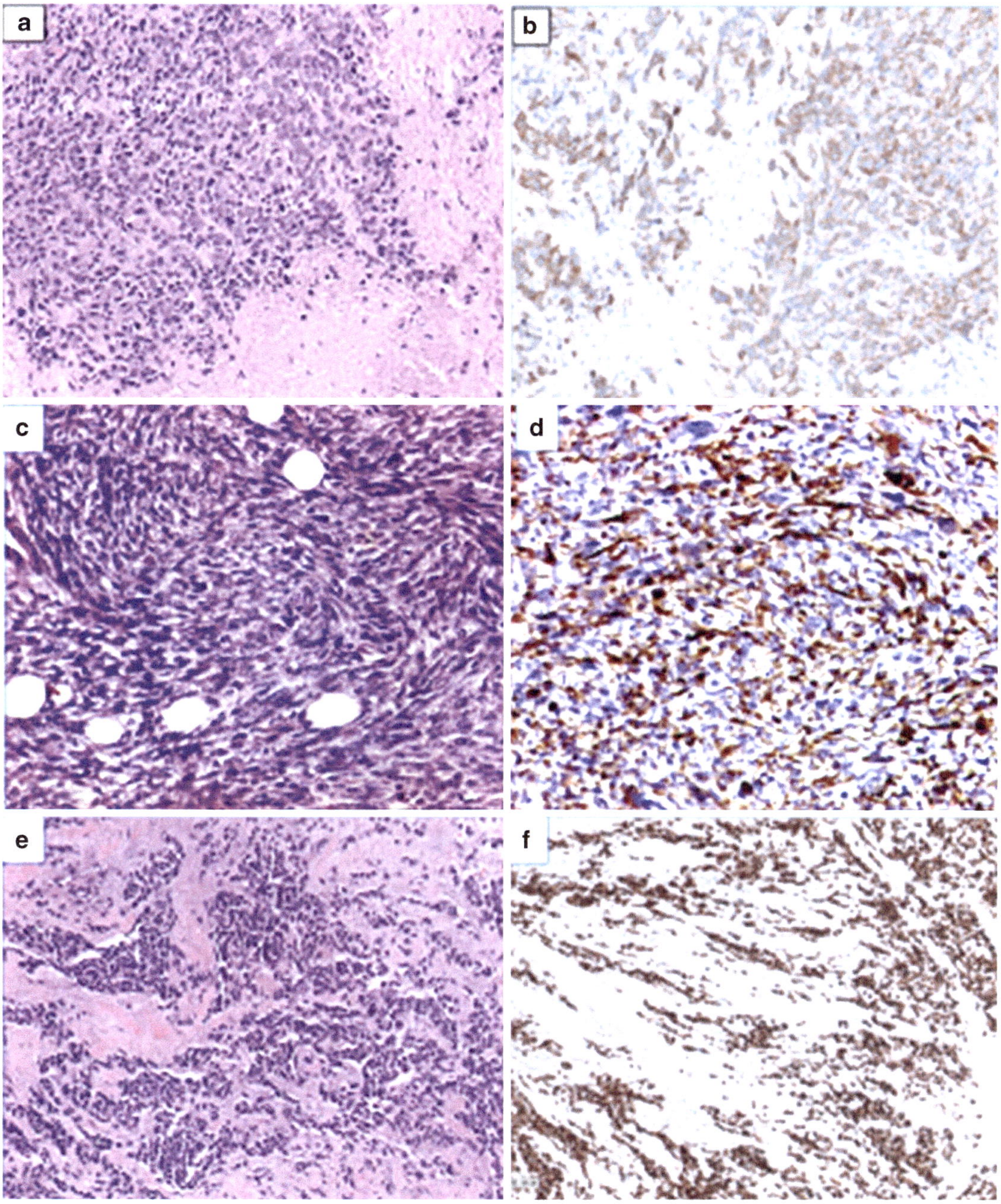

Fig. 6.8 (**a**, **b**) Undifferentiated breast tumor (left side H&E) stains with cytokeratin indicating high-grade carcinoma. (**c**, **d**) Spindle cell sarcoma stains with desmin indicating rhabdomyosarcoma. (**e**, **f**) Malignant round cell tumor with vague nests (left side, however string nuclear staining with myogenin confirms it as alveolar rhabdomyosarcoma)

and cytoplasmic markers include smooth muscle heavy chain complex (SMHCC), high-molecular-weight cytokeratins/ 34 βE12, S100, etc. The principle being that in complex well-differentiated infiltrative proliferations that mimic an infiltrating cancer, demonstration of myoepithelial cells

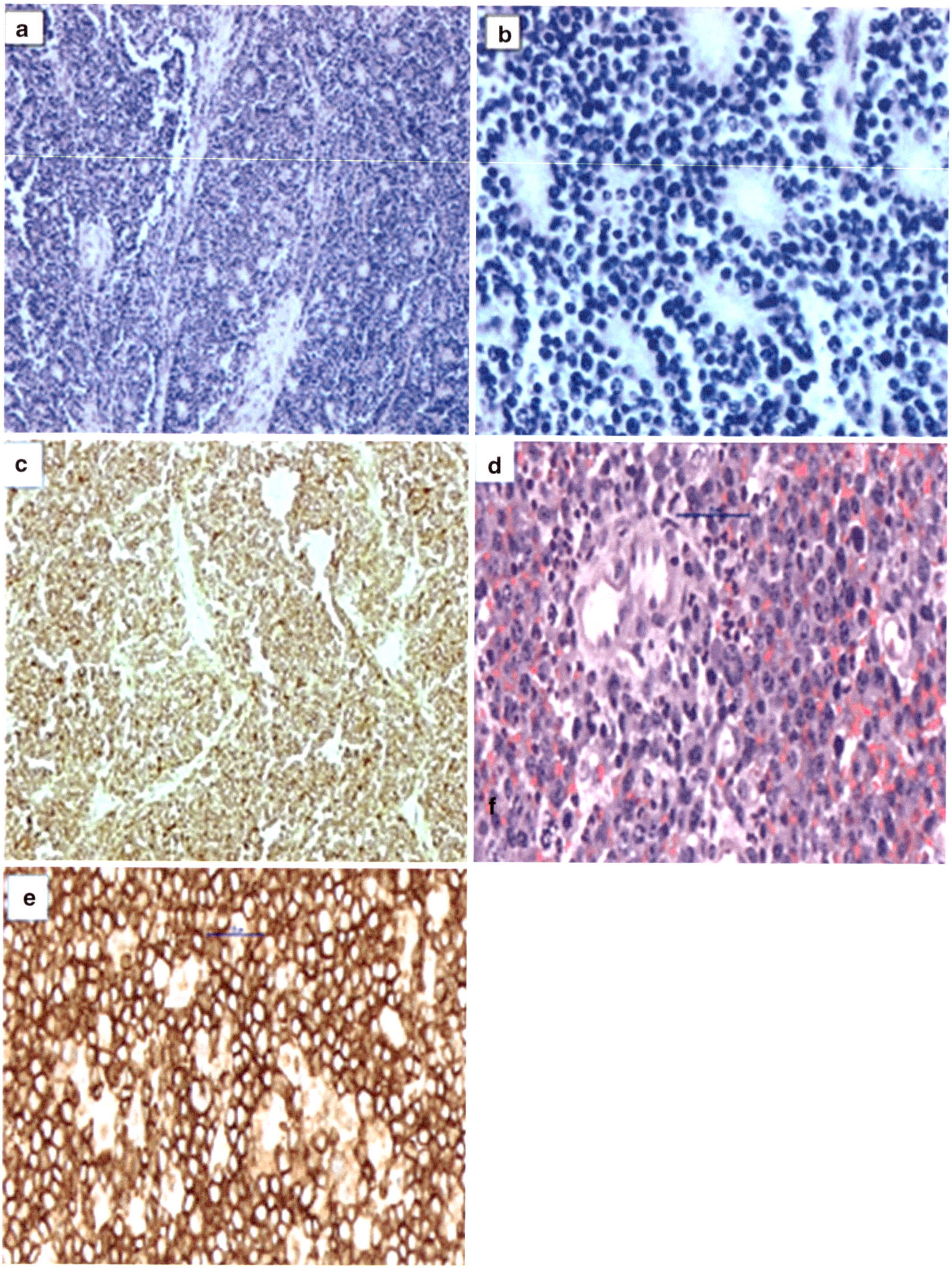

Fig. 6.9 (**a**, **b**) Rosette forming malignant round cell tumor of bone. (**c**) Staining with MIC2 membranous confirms a diagnosis of Ewing's sarcoma. (**d**) Anaplastic appearing cells in a node biopsy. (**e**) Stain with CD30 strongly and diffusely indicating an anaplastic large cell lymphoma

Table 6.1 General information on panels in immunohistochemistry

Broad category	Marker	Positive in
Undifferentiated high-grade tumor	CK (cytokeratin)	Carcinoma
	SALL4/OCT3/4	Germ cell tumors
	Vimentin	Mesenchymal tumor and sarcoma
	Melan A, S–100, HMB45, SOX10	Melanoma
Metastasis from unknown primary tumor [26]	General panel of CK7/CK20	CK7−/CK20+: Colon (75 to 95%)
		CK7+/CK20−: Lung (90%), ovary (90%)
		CK7−/Ck20−: HCC (70 to 89%), lung squamous carcinoma (50 to 90%), prostate (60 to 100%)
		CK7+/CK20+: Urothelial carcinoma (25 to 90%)
Organ-specific panel	PAX8	Renal, gynecologic cancers, thymic tumors
	TTF1	Thyroid, lung, some clones may stain pancreatic adenocarcinoma
	SATB2/CDX2	Intestinal type or colonic adenocarcinoma
	GATA3	Breast, urothelial carcinomas
	NKX3.1/ PSA/ AMACR	Prostate
	p40	Squamous carcinomas and urothelial carcinomas
	HepPAR1, arginase	Hepatocellular carcinoma
Round cell tumors	Desmin, myogenin, MyoD1	Rhabdomyosarcoma
	MIC2 (CD99), NKX2.2	Ewing's sarcoma
	Synaptophysin	Neuroblastoma
	SATB2	Small cell osteosarcoma
Lymphoma	CD20, CD19, CD79a, PAX5	B-cell lymphomas
	CD3, CD4, CD8, CD7	T-cell lymphomas
	CD30	Anaplastic large cell lymphomas, Hodgkin's lymphoma, some diffuse large B-cell lymphomas
	CD15	Hodgkin's lymphoma, myeloid sarcoma, some carcinomas
	CD10, bcl6, GCET	Germinal center markers
	MUM1	Post-germinal center marker
	CD23	Chronic lymphocytic leukemia, dendritic cells
	CD68, CD163	Histiocytic, monocytic differentiation
Spindle cell tumors [27–31]	CD34	Dermatofibroma or dermatofibrosarcoma, angiosarcoma, epithelioid sarcoma, nerve sheath tumors, solitary fibrous tumors
	S100	Neurofibroma, schwannoma
	Smooth muscle actin, H Caldesmon	Smooth muscle tumors, fibromatosis, myofibroblastic tumors
	INI1(SMARCB1)	Malignant rhabdoid tumor, epithelioid sarcoma, epithelioid MPNST, few myoepithelial carcinomas
	CD31, ERG1	Vascular tumors
	CAMTA1/ TFE3	Epithelioid hemangioendothelioma
	H3K27me3	Epithelioid MPNST

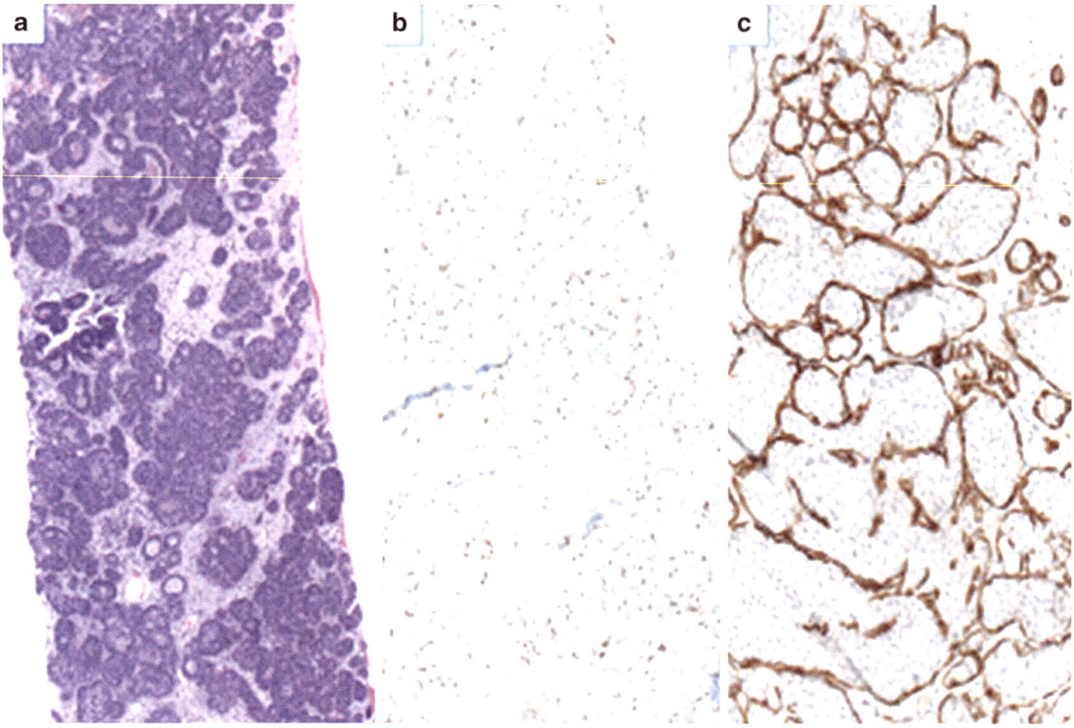

Fig. 6.10 (**a**) Core biopsy from breast lesion shows an infiltrative lesion with atypia mimicking carcinoma. However, (**b**) p63 showed retained myoepithelial cells (MEC) nuclei. (**c**) On smooth muscle heavy chain complex (SMHCC) immunostains, a complete basal lamina (brown) is highlighted around all glands which indicates an intact MEC layer, and hence diagnosis of sclerosing adenosis with an in situ carcinoma was given

(nuclei and basal lamina) generally indicates a non-invasive process.

IHC is an excellent screening test for several molecular events and has been used extensively for predictive biomarkers. All of these require good quality paraffin blocks which in turn comes from rapid fixation. Some of the biomarkers have FDA-cleared clones (Ventana 4B5 for Her2neu) which are more reproducible than the home brew assays. Ki67 a proliferation marker used ubiquitously in oncology has come back in limelight as it is used in breast as a surrogate for OncotypeDx in risk stratification for chemotherapy in early breast cancer. But Ki67 is impacted by several variables especially fixation and interpreting clinicians should be aware of it [32].

6.5.3 Is all brown positive?

There is no uniform definition for positive in an IHC result. IHC is a semi quantitative method of evaluation of antigen expression. In some diagnostic markers any extent of staining may be considered positive for some markers, whereas in some it is determined by pattern and percentage cut-offs of tumor cells expressing the antigen. For example, p53 in ovarian serous carcinoma, a percentage of above 70% or complete negative expression, is considered as indicator of mutant type p53. Semi-quantitative estimation is also used for predictive markers like ER, PR, Her2neu and Ki67 and they are at times fraught with errors as these are eyeballing estimates. For example, 2% ER in core biopsy may become <1% when put in perspective of entire tumor. The distribution of the immunohistochemical expression is also important in diagnosis of polymorphous lymphomas due to presence of microenvironment. For example, CD30 is expressed only by the neoplastic Reed-Sternberg cells in classical Hodgkin lymphoma but is negative in background cells.

Two tests that have seen explosive use in last few years is testing for PDL1 in tumor or immune cells for immunotherapy and Microsatellite testing in several cancers for immunotherapy, Lynch syndrome and prognosis as well. Microsatellite (MS) are short tandem repeats located at ends of chromosomes and generated due to an error in DNA repair. The normal tissue DNA repair system, called mismatch repair (MMR), corrects MS generation. Lack of MMR genes in tumor cells generates MS instability (MSI) and increases tumor mutation burden with better response to certain immunotherapies and impact on prognosis of tumors [33]. IHC for MMR proteins is employed to ascertain the MMR status of the tumor, i.e., MMR deficient (MSI) or proficient (MSS), using a panel of four antibodies to the two pairs of MMR proteins, namely MLH1/PMS2 and MSH2/MSH6, and quite often these are lost in pairs. MMR deficiency is seen in approximately 15% of sporadic colorectal cancer and studies have shown that they confer better prognosis and lower risk for metastases contrary to MMR proficient tumors [34]. Loss of nuclear expression of any one of the proteins on immunohistochemistry suggests MMR deficiency (MSI) which can be as a consequence of sporadic or germline mutations or due to epigenetic inactivation (Fig. 6.11a–h). When there is no loss of nuclear expression of MMR proteins, it is reported as "*No evidence of deficient mismatch repair (low probability of MSI-high)*"; and when there is loss of nuclear expression of one or more MMR proteins, it is reported as "*MSI-high or deficient mismatch repair*". Testing for MSI in CRC are for (1) prognostication as it confers a better stage adjusted prognosis, (2) to detect patients with Lynch syndrome who have germline mutations of any of the MMR genes, accounting for 5% of CRC, and (3) as a predictive tool in stage 2 colorectal cancer as studies have shown that tumors with MSI do not respond to 5 fluorouracil chemotherapy (5-FU) [33, 35, 36]. MSI testing also has a role in metastatic setting to identify those patients who will benefit with immune checkpoint inhibitors like pembrolizumab. MSI immunohistochemistry using non-validated antibody has some variability issue so validation using known MSI deficient cases is recommended. Before interpreting the loss of IHC staining, always make sure if the internal control (i.e. intervening endothelial cells, adjacent normal epithelium or admixed lymphoid cells) has stained to distinguish a false negative from true negative result.

PDL1 is a companion diagnostic IHC test wherein an IHC test used as a companion to a therapeutic drug to determine its applicability to a specific patient. PDL1-22C3 is DAKO antibody test is performed on a DAKO Autostainer that is specifically used if the drug to be given is Keytruda or pembrolizumab while PDLSP142 on Ventana Ultra platform is specifically done when Atezolizumab or Tecentriq is planned for patient. As more and more people can afford these drugs availability of multiple automated platforms for testing will be the bane for every oncology center in India.

6.5.4 Issues in IHC

The accurate demonstration of markers on IHC requires knowledge of each antibody and what it stains, rigorous validation, selecting the dilution that offers a true positive result and unmasking of antigens blocked by formalin fixation [34]. Effective antigen retrieval has been a game changer in demonstration of nuclear markers like estrogen and progesterone receptors. The chief issue is that different laboratories utilize different antibody clones and staining method, and for economic reasons, test variations occur. There are several variations to the same protein and a best fit antibody must be used to avoid false negative test [37]. The classical example is of cytokeratin and there are two types of this protein. AE1/AE3 is a Keratin cocktail that detects cytokeratins from CK1 to 8, CK10, CK14 to 16 and CK19, but does not detect CK17 or CK18. The other cytokeratin MNF116 stains CK 5, 6, 8, 17 and probably CK19. So pancreatic adenocarcinomas will be MNF11 positive more than AE1/AE3. In general, AE1/AE3 is sensitive but may stain some sarcomas and MNF116 is more specific [38].

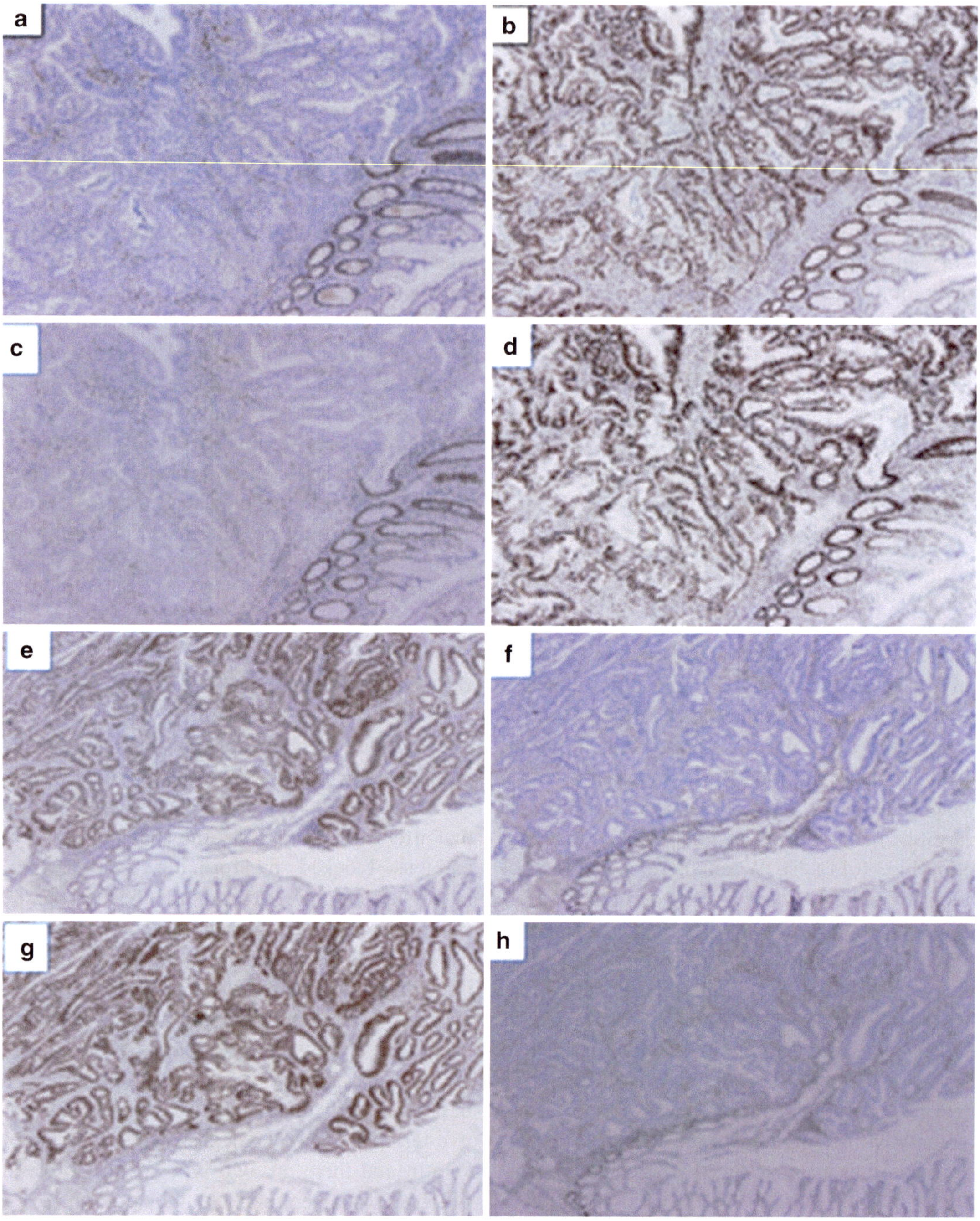

Fig. 6.11 (**a**) MLH1 absent in tumor (note normal colon serves as internal positive control). (**b**) MSH2 staining of tumor cells. (**c**) PMS2 absent in tumor cells. (**d**) MSH6 retained in tumor cells, all indicating MLH1 loss. (**e**) Another case shows MLH1 retained, (**f**) MSH2 lost, (**g**) PMS2 retained and (**h**) MSH6 lost in tumor cells indicating MSH2 loss

As a quality check an external control is run on the same slide which assesses the test specificity or internal controls that are present in tissue are evaluated to attest to the tissue antigenicity.

In concluding, IHC is no longer an ancillary test and is now integral part of oncopathology and although it is ever expanding and ready for clinic it is not a fail proof test and treating clinicians should be aware of same.

6.6 Molecular Diagnostics

The pathologist's role in today's era goes beyond routine histomorphology to include new responsibilities of molecular testing for tumors. Molecular profiling is becoming standard practice for most patients with advanced disease, replacing the historical treatment paradigm of prescribing standard chemotherapy based upon the tumor's organ of origin, histology and stage [39]. Molecular subtyping has been incorporated in several of the latest WHO tumor classifications (lung, CNS tumors, etc.).

Molecular methods could either look for abnormalities in DNA, RNA or proteins. Molecular tests can be done on any source of tumor-fresh tissue, paraffin block, cytology, FNAC material and blood. As most molecular test are performed on paraffin block, the chief issue is that if the block is poorly processed the test fails. The process of nucleic acid extraction from paraffin blocks mainly involves deparaffinization, enzymatic digestion followed by purification and alcohol precipitation [40]. The Personalized Healthcare Committee (PHC) of the CAP has established the Preanalytics for Precision Medicine Project Team (PPMPT) for evidence-based recommendations for preanalytics to ensure integrity of biomarkers during molecular testing [41, 42]. Fixation time in formalin preferably should not exceed 24 h to avoid formation of DNA-cross-linkages which compromises extraction of DNA from formalin fixed paraffin embedded tissue. Some key recommendations include cold ischemia time <1 h, fixation in 10% neutral buffered formalin with pH 7.0, no acid decalcification step, thickness of specimen not to exceed 4–5 mm while grossing, fixative volume to tissue mass ratio of 10:1, standardized tissue processing, use of low melt paraffin (60 °C) and storage of blocks in dry, pest-free environment [41, 42].

Genetic abnormalities detected in tumors that yield prognostic or etiopathogenetic information can be divided as: (A) Gene rearrangements or translocations which juxtapose otherwise distant segments of our genomic DNA leading to a chimeric gene impacts oncogenesis; (B) Gene amplification is the differential increase in a specific portion of the genome, which in protooncogenes can lead to tumorigenesis and/or tumor progression; (C) Loss of genes or deletions usually of tumor suppressor genes; and (D) Mutations in genes which is the change in a DNA sequence. A whole repertoire of methods is available to evaluate different genetic abnormalities in cancer and the method chosen depends on nature of abnormality and cost effectiveness. Traditional Sanger sequencing is restricted to the discovery of substitutions and small insertions and deletions. Mutations where the exons have considerable size and/or are widely distributed along the exome are better tested with NGS than Sanger sequencing, e.g., BRCA (BRreast Cancer gene; BRCA 1 AND 2) testing [43]. A summary of common abnormalities in cancers and methods used to test them is given in Table 6.2 [43–50]. A brief overview of molecular methods is discussed below create a better understanding on which test to do and why.

6.6.1 In Situ Hybridization (ISH)

One major limitation of traditional metaphase spreads for cytogenetics is that they require fresh tissue and proliferating cells. In situ hybridization on the other hand can be done on paraffin blocks and determines a specific DNA or RNA sequence in a tissue section (in situ) using a labeled (chromogenic, fluorescent or silver labels) complementary DNA/RNA probe. If the tags on the probes are fluorescent labeled, it is called fluorescent in situ hybridization (FISH), and if biotin or digoxigenin labeled, then chro-

Table 6.2 Genetic alterations in solid tumor [43–50]

	When	What	How	Why
Breast cancer				
HER2	In all equivocal cases by IHC	Amplification	ISH (FISH/CISH/DDISH)	Poor prognostic marker and predicts response to anti-HER2-targeted therapy
BRCA1/BRCA2	Germline testing for hereditary breast cancer TNBC in metastatic setting	Mutations	NGS	Predicts response to PARP inhibitors
Onoctype dx	Hormone-positive HER2-negative for risk stratification for need of chemotherapy	Gene expression profiling	RT-PCR	Prognostic for recurrence Predictive of chemotherapy benefit
Non–small cell lung cancer				
ALK	Advanced/metastatic NSCLC	Gene rearrangement ALK protein overexpression	FISH, NGS, RT-PCR IHC	Predicts response to ALK TKI (alectinib, crizotinib, lorlatinib)
EGFR exon 21 (L858R, L861), exon 20 (S7681), exon 18 (G719X, G719)	Advanced/metastatic NSCLC	Mutation	NGS, multiple mutation Testing	Predicts sensitivity to EGFR TKI
EGFR T790M	Advanced/metastatic NSCLC	Mutation	NGS, multiple mutation Testing	Predicts resistance to EGFR TKI and sensitivity to osimertinib
EGFR exon 19	Advanced/metastatic NSCLC	Deletion	NGS or PCR	Predicts sensitivity to EGFR TKI
EGFR exon 20	Advanced/metastatic NSCLC	Insertion mutation	NGS or PCR	Predicts resistance to EGFR TKI Mobocertinib is the first and only approved oral therapy drug specifically designed for EGFR exon 20 insertion mutations
ROS1	Advanced/metastatic NSCLC	Fusion rearrangement	FISH, NGS, RT-PCR	Predicts sensitivity to ROS TKI (Crizotinib, Entrectinib)
KRAS	Advanced/metastatic NSCLC	Mutation	Gene sequencing, PCR	Predicts resistance to EGFR TKI, poor prognosis (vs. KRAS non-mutated). Sotorasib is approved for KRAS G12C mutation
BRAF V600E	Advanced/metastatic NSCLC	Mutation	Gene sequencing	Responsive to combined BRAF and MEK inhibition

(continued)

Table 6.2 (continued)

	When	What	How	Why
MET	Advanced/metastatic NSCLC	Amplification Exon 14 skip mutation	NGS, FISH	FDA has approved 2 MET inhibitors, capmatinib and tepotinib, for the treatment of patients with advanced NSCLC harboring MET exon 14 skipping mutations
RET	Advanced/metastatic NSCLC	Fusion rearrangement	FISH, NGS, RT-PCR	Selpercatinib (FDA-approved) RET inhibitor
CNS tumors				
1p19q codeletion	Preadjuvant therapy in adult glial tumors	Deletion	FISH	Good prognosis, diagnostically helps confirm oligodendroglia tumors, predicts response to therapy
IDH1/IDH2	Preadjuvant therapy in adult glial tumors	Mutation	Sequencing, IHC	Favorable prognosis
MGMT	Preadjuvant therapy in adult glial tumors	Promoter methylation	Methylation-specific PCR	Predicts benefit from temozolomide, prognostically associated with IDH status
ATRX	IDH mutant glial tumors	Loss of expression	IHC	Seen in ~45% of anaplastic astrocytoma and ~27% of anaplastic oligoastrocytomas
TP53	Glial tumors	Mutations	NGS, IHC	Poor prognosis
H3 K27M	Diffuse midline glioma children and adults, posterior fossa ependymomas	Mutation	NGS, IHC	Aggressive behavior
WNT/SHH/ TP53-based classification	Medulloblastoma	Expression pattern	Gene expression transcriptome	Prognostic groups
Colon carcinoma				
MSI	Adenocarcinomas	Microsatellite stability	IHC, NGS	
KRAS	Metastatic adenocarcinomas	Mutation	RTPCR, NGS	Resistance to anti-EGFR therapies
BRAF V600E	Metastatic adenocarcinomas	Mutation	IHC, PCR, sequencing	Poor prognosis BRAF mutation is associated with the promoter methylation of an MMR gene, MLH1, less benefit from Pembrolizumab
PIK3CA exon 20	Metastatic adenocarcinomas	Mutation	RTPCR, NGS	Resistance to anti-EGFR therapies

(continued)

Table 6.2 (continued)

	When	What	How	Why
Gynecological carcinoma				
MSI testing	Upon diagnosis of endometrial carcinoma	Microsatellite stability	IHC, PCR, NGS	Lynch syndrome; identification of susceptibility to checkpoint inhibitors
POLE exons 9–14	Poorly differentiated endometrial carcinoma	Mutation	Sanger, NGS,	Excellent outcome
TP53	Ovarian tumors/ endometrial cancer	Mutation	NGS, IHC	Poor prognosis and need for adjuvant chemotherapy
BRCA1/BRCA2	Germline testing, recurrent ovarian cancers	Mutations	NGS	Predicts response to PARP or DDR enzyme inhibitors

Abbreviations as used in text

mogenic in situ hybridization (CISH). While FISH requires fluorescence microscope, CISH is evaluated under a light microscope since the probes produce color. Single most advantage of ISH is that it allows for morphological assessment of the labeled target gene and ensures counting only in tumor cells. FISH probes cover larger part of genome and are useful for detecting translocation in genes with wide range of break points or those with multiple translocation partners like CMYC and ALK1 [51]. The usual probe sizes that work for formalin-fixed paraffin-embedded (FFPE) tissues ranges from 20 to 500 base pairs [52]. Double hapten-labeled ISH (DDISH) uses dual chromogenic probes wherein DIG-labeled probe label the control gene, which is then detected by Naphthol Fast Red ISH detection kit and a DNP probe targets the target gene and is visualized using silver reagents e.g in Her2neu DDISH chromosome 17 appears as red signals and HER2 gene is labelled with silver which shows black signals [53]. In RNA-ISH, a single-stranded DNA probe complementary to the tested mRNA molecule is use to test for mRNA. mRNA ISH is used for detecting viruses in tissues and is faster, cheaper, easier in preparation and less toxic; however, it is affected by the poor stability of ribonucleic acid in paraffin blocks [54].

A wide range of gene deletions, amplifications, translocations and fusions can be evaluated by FISH. In general, the number of probe signals matches the gene copy number and FISH results are obtained by counting these signals in each cell. Every laboratory defines their own counting procedure including the number of analyzed cells, control slides and cut-off for an abnormal result. A significant issue in the FISH technique is the choice of the appropriate probe. Two types of probes are used in diagnostic laboratory break-apart and fusion probe. Dual-fusion probes consist of pairs two different chromosome are labeled in contrasting color and when a fused signal is produced that indicates translocation. In a break-apart probe, the two colors flank the translocation breakpoint, and hence a fused signal is normal, and when the signal breaks or splits that indicates translocation (Fig. 6.12b). Break-apart probes are popular because they can be used for a wide range of translocations where there is no need to include the second gene partner for testing. Dual-fusion probes however span longer translocation breakpoint regions and are more sensitive and specific [55]. Probes to detect copy number changes (e.g. aneusomies, deletions, or amplifications) are usually labeled in a single color and span the genomic region of interest. Such probes are either locus-specific probes (e.g. REL *and* RB1*)* or detect repetitive sequences, like chromosome enumeration probes (CEPs), that bind to centromeric or pericentromeric satellites (Fig. 6.12d). The major limitation of detecting copy number changes or deletion in tissue sections is that part of the cell can be lost during the

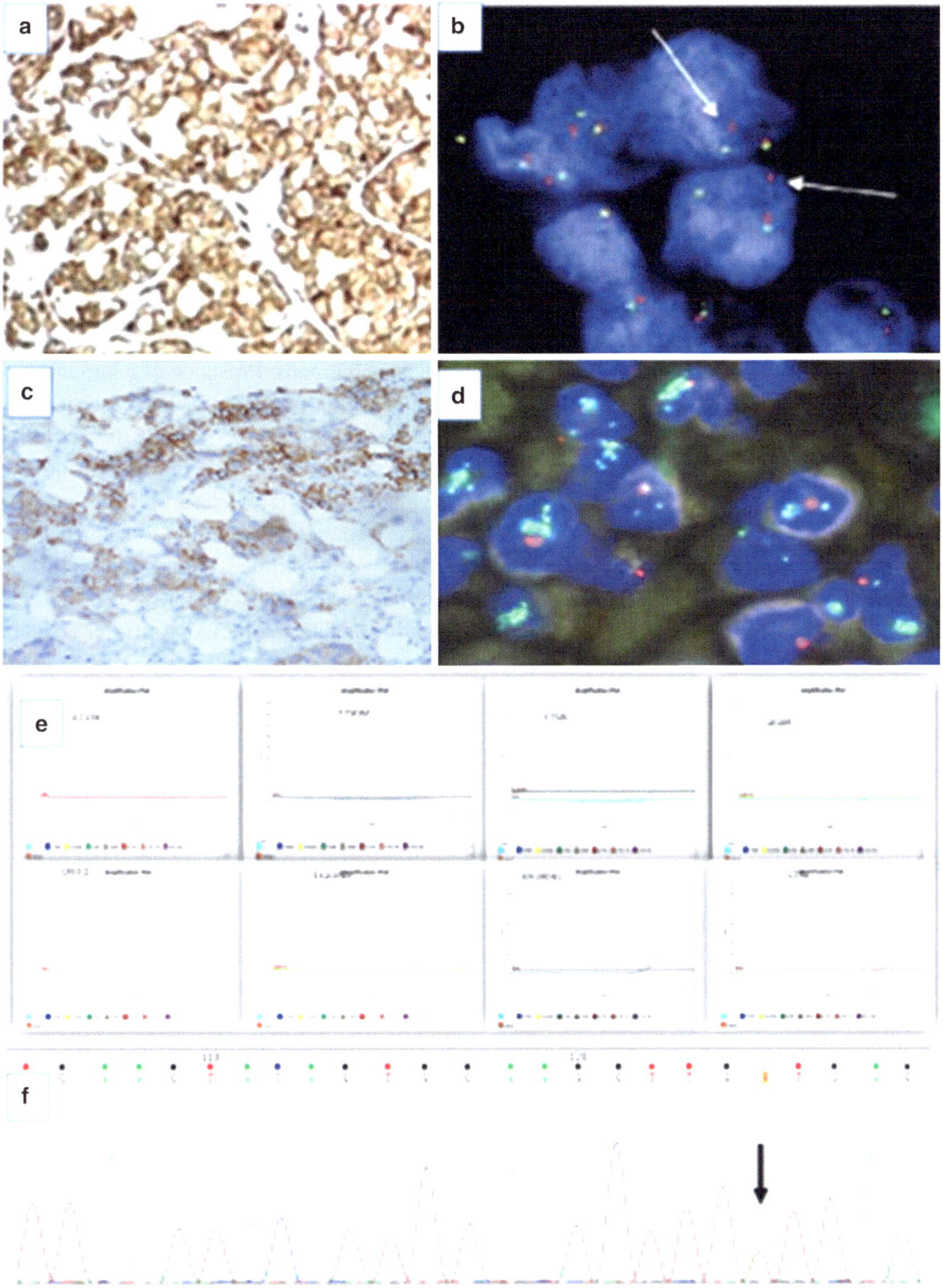

Fig. 6.12 (**a**, **b**) Case of 34-year-old male patient with lung carcinoma with signet ring cell morphology which shows diffuse, strong ALK cytoplasmic staining. (**b**) The same case showing split signal (White arrow—separate green and red signals) by Zytovision dual-color ALK break-apart probe suggestive of ALK gene rearrangement by FISH. Case of locally advanced invasive breast carcinoma showing (**c**) equivocal (score 2+) staining on HER IHC, which was reflex tested by (**d**) FISH test shows multiple green signals (>6) targeting HER2 gene and 1–2 red signals targeting the centromeric region of chromosome 17 (CEP17) with a ratio of HER2/CEP17>2, confirming the HER2 gene amplification. (**e**) Real-time plots of EGFR mutations (Exons 18–21) showing exon 19 deletion, while rest of the exons are wild-type. Exon 19 deletion is most common activating mutation in advanced non–small cell lung cancer (NSCLC) and associates with the sensitivity of EGFR TKI's. (**f**) Sanger sequencing chromatogram of a case of intestinal gastrointestinal stromal tumor (GIST) with c-kit heterozygous point mutation (arrow showing peak within the peak) in exon 11

histology sectioning process, leading to artificially induced deletions [51]. A wide range of artifacts and/or technical problems can lead to misinterpretation of ISH results [52]. Causes for False negative ISH results include low abundance of targets in tissues, overfixation leading to excessive cross-linking and suboptimal antigen preservation and low quality DNA or degraded RNA. False positive signals may also result from signals generated by diffusion of the final reporter molecule or by unquenched endogenous biotin [52, 56].

6.6.2 Polymerase Chain Reaction (PCR)

Polymerase chain reaction (PCR) is a method that amplifies and detects DNA and RNA sequences in tumor and is the cheapest and most frequently available test in a molecular pathology laboratory. In a PCR a target DNA is amplified using a pair of priming complementary sequences (oligonucleotide primers) flanking a location of interest, together with unique heat-resistant polymerases. At the end of the PCR millions of identical copies of the original specific DNA sequence are generated and migrate simultaneously, forming a single band, when applied to an electrophoretic gel. The location and size of band (compared to controls) indicates presence of target gene or translocation. If the oligonucleotide primers are labeled with fluorescent dye, the PCR product is analyzed in a computerized capillary electrophoresis instrument, which generates a graph depicting a peak of fluorescence at the migration location of the PCR product. A reverse transcriptase PCR (RT-PCR) is used to amplify a RNA (instead of DNA) target sequence and used for studying various translocations in solid tumor, viz., EWS-FLI1 is Ewing's sarcoma, SYT-SSX in synovial sarcoma, etc. [56]. The RNA sequence however has to be first converted to a double-stranded nucleic acid sequence (cDNA) by using a reverse transcriptase enzyme which is then subsequently amplified. Since RNA is not as stable as DNA for RTPCR fresh samples or good quality blocks are required. The real-time PCR (Q-PCR) allows for the real-time quantitation of a fluorescence labeled PCR product whereby a computerized Q-PCR instrument measures the amount of fluorescence emitted after each cycle which is proportional to the number of copies of the amplified target producing an amplification curve. When a certain critical copy number is reached, the amount of fluorescence increases exponentially. Fluorescence intensity is plotted against cycle number, showing the number of cycles at which the exponential increase in fluorescence happens. Presence of a larger number of copies yields exponential increase at an earlier cycle and lower "CT" (cycle threshold) values (Fig. 6.12e). Q-PCR offers rapid quantitative advantage, testing of 96 samples at one time and is less prone to contamination since the entire process of amplification and quantitation of the original target DNA is done in a single sealed tube [56]. The specificity of primer sequences is a key component of allele-specific PCR assays, besides this several preanalytical factors impact the yield of DNA/ RNA for a PCR. At least 20–200 ng of DNA should be present for successful amplification. Prolonged fixation, acid decalcification, high temperatures, degradation of tissue in blocks with storage can lead to failed PCR [56].

6.6.3 Sanger Sequencing

Sanger sequencing was the first-generation sequencing method that involved the addition of chain-terminating and fluorescently labeled dideoxynucleotides to perform sequencing of a DNA strand complementary to the interrogated template strand (Fig 6.12f). The fragments are run on a gel or capillary electrophoresis to separate them and determine the sequence many genes. However, only one sequencing reaction could be analyzed at a time, and only a limited length could be sequenced [57]. Fragment analysis is a variation of sequencing that detects changes in DNA (e.g. the length of a specific DNA sequence) or RNA to indicate the presence or absence of an inserted or deleted genomic sequence and is typically used for tests like clonality evaluation in lymphomas. As an orthogonal method, Sanger sequencing provides a means to confirm variants identified by next-generation sequencing (NGS) as is the easiest method to resolve these uncertainties and is therefore an invaluable tool in any clinical genomics laboratory. Sanger sequencing however has low sensitivity (limit of detection is ~15–20%), and variants with a lower allele frequency may be indistinguishable from noise or sequencing errors [58].

6.6.4 Next-Generation Sequencing (NGS)

With Sanger sequencing, we see a limited DNA snapshot at a time, while NGS is a massively parallel sequencing which enables us to assess thousands of reads per sample with a higher level of sensitivity and lower limit of detection. NGS sequences millions of small fragments of DNA in parallel from same sample and many samples together. Bioinformatics analyses are then used to piece together these fragments by mapping the individual reads to the human reference genome to obtain insight into unexpected DNA variation. NGS can be used to sequence entire genomes or constrained to specific areas of interest, including all 22,000 coding genes (a whole exome) or small numbers of individual genes. NGS is able to provide simultaneous screening of a variety of genomic aberrations such as single-nucleotide variants (SNVs), multiple-nucleotide variants (MNVs), small and large insertions and deletions, and copy number variation (CNVs) of the genes [59]. Several platforms are available for NGS, and Table 6.3 gives a summary of these.

6.6.4.1 Types of NGS Assays

- Whole-genome sequencing (WGS): This involves massively parallel sequencing of the entire genome which yields tremendous data output but currently is only in the research setting and not feasible for routine diagnostics.
- Whole-exome sequencing (WES): This is a technique used for sequencing all of the protein-coding regions of genes in a genome (known as exome). WES is an efficient method to identify disease-causing mutations, but it will miss DNA variations outside of the exome [61].
- Whole transcriptome sequencing (RNAseq): This technique examines the sequences of RNA and analyses the transcriptome of gene expression patterns encoded within the RNA.
- Targeted disease-specific panels (TS): Instead of obtaining lot of data that may not be clinically relevant, targeted sequencing is a popular approach to detect known and novel variants in selected sets of genes or mutational hotspots. Such panel-based testing is used for routine oncology diagnosis and management. Targeted panels may be broad, including genes for both solid and hematologic malignancies, or may be more focused for a particular type of malignancy (such as myeloid neoplasms) [62–64].

6.6.4.2 NGS Technique

Basically, NGS steps include tumor enrichment followed by target enrichment and library preparation, sequencing and data analysis. Nucleic acid isolation is a crucial first step in the NGS workflow, and it is essential to select appropriate isolation methods or kits for the same to obtain proper lysis of the cells. Yield, purity and quality of isolated nucleic acids should be assessed before proceeding to NGS library preparation. Library preparation procedure includes nucleic acid fragmentation and adaptor ligation. It includes first obtaining templates corresponding to molecules of interest for sequencing and subsequently preparing the fragments to make them compatible with the sequencing platform used by attaching adaptors. The term library refers to these fragments of DNA with flanking adaptors that are ready for sequencing. Adaptors also include short, unique sequences, known as a barcode or index, that are essentially sample identity tags. When RNA is used, cDNA synthesis is part of the library preparation workflow. Prepared libraries represent a pool or collection of DNA fragments with adapters attached to their ends which must be quantified (and normalized as needed) so that an optimal concentration of mol-

Table 6.3 Different types of sequencing platforms and the method of sequencing and their pros and cons of each platform (*GB—Giga Base pairs) [60]

Platform	Method of sequencing	Pros	Cons
Illumina	Sequencing by synthesis—Cyclic reversible termination technology Paired end sequencing	High-throughput, low cost per Gb data. High accuracy	Short read length, high instrument cost
Ion torrent	Ion semiconductor sequencing technology	Low instrumental and operational cost. Short execution time. Simple machine	Error rate not very good. Moderate cost per Gb data. More hands-on time
Pacific bioscience	Single-molecule real-time long read sequencing	Longest reading length available. Short instrument execution time	High error rate. High cost per Gb data. Many methods are still under development
Oxford Nanopore	Single-molecule real-time long read sequencing	Small, portable and low-cost instrument	High error rate. Biased errors. High cost per reading
Complete genomics BGI	Sequencing by ligation DNA nanoball sequencing mate pair library	Low cost per run, short execution time, large amount of data, error rates low	Short read length
Gene reader (Qiagen)	Sequencing by synthesis Single-nucleotide addition	High-throughput, reduced manual intervention, complete solution from extraction to data analysis	Closed system, high cost

ecules to be sequenced is loaded onto the sequencers.

Note that additional steps may be performed in different library preparation workflows. Target enrichment may be performed when sequencing is needed only for a defined set of genes (and not for WGS or RNAseq). Amplicon (PCR) or hybridization methods are used for target enrichment. Amplicon-based methods are convenient for smaller gene panels but are prone to imbalanced sequence coverage across targets and artefacts such as random sequence mismatches introduced by polymerase errors. Hybridization capture-based preparation of templates involves complementary biotinylated probes that hybridize with regions of interest, which are then isolated using streptavidin-coated magnetic beads. Hybridization capture is relatively more suitable for degenerated degraded FFPE DNA, because it can enrich even if partial overlap of capture probes with region of interest occurs, whereas amplicon PCR is unable to amplify DNA fragments if the primer binding sites are degraded. Hybridization capture prevents issues of allele dropout, generally observed using amplicon based libraries. Then comes the step of sequencing. The first step for both the Illumina and Ion Torrent platforms is to immobilize each DNA fragment and clonally amplify it. Clonal amplification is needed to generate a large enough signal for detection. The Ion Torrent uses a bead emulsion for immobilization and clonal amplification, whereas the Illumina sequencers use a flow cell.

Data analysis is the whole crux of NGS [65, 66]. The use of bar codes to tag a specimen's DNA fragments allows pooling of multiple sample, thus decreasing the sequencing cost. However, this process requires a demultiplexing step in which all reads are sorted by bar code/sample before further analysis. The demultiplexed file with raw reads is referred to as a FASTQ file. The raw data reads from the machine undergoes a series of bioinformatics processes called the bioinformatics pipeline to deliver a variant call file (VCF). If multiple reads show the same difference in allele, a variant is called. The threshold for the number or percentage of reads required before deciding on variant is determined by the laboratory and should be validated. It is critical to validate the entire NGS process from end to end (DNA extraction through the bioinformatics pipeline [59].

Based on the available evidence as well as significance on the clinical decision making, the variants have been classified into different categories as per the joint consensus and recommendations of AMP, ASCO and CAP.

- *Level A, Biomarker*: Predict response or resistance to US FDA-approved therapies for a specific type of tumor or have been included in professional guidelines as therapeutic, diagnostic and/or prognostic biomarkers for specific types of tumors.
- *Level B, Biomarker*: Predict response or resistance to a therapy based on well-powered studies with consensus from experts in the field, or have diagnostic and/or prognostic significance of certain diseases based on well-powered studies with expert consensus.
- *Level C, Biomarker*: Predict response or resistance to therapies approved by FDA or professional societies for a different tumor type (i.e. off-label use of a drug), serve as inclusion criteria for clinical trials or have diagnostic and/or prognostic significance based on the results of multiple small studies.
- *Level D, Biomarker*: Plausible therapeutic significance based on preclinical studies, or may assist disease diagnosis and/or prognosis themselves or along with other biomarkers based on small studies or multiple case reports with no consensus [58].

Reporting results in NGS recommendations: Tier-based classification of *variants is* based on the recommendations of AMP, ACMG and CAP [67]. Variant should be classified into the 4-tiered system; Tier I (variants of strong clinical significance), Tier II (variants of potential clinical significance) and Tier III (Variants of unknown clinical significance) and Tier IV (benign variants). Tiers I-III must be reported in descending order of clinical importance. It is not recommended to include Tier IV or "benign/likely benign" variants/alterations in the report; however, it should be made available. Pertinent "negatives" should be reported, in a disease-specific manner. All detected genetic alterations should be annotated and reported as designated by the HUGO Gene Nomenclature (http://www.genenames.org). Clinical significance of the variant should be clearly described in the interpretation. References such as publications and database should be appropriately cited for further information. As new technologies emerge, existing problems are exacerbated or new problems arise. NGS platforms provide vast data, but are prone to error rates (~0.1–15%) and the read lengths in some is shorter than those of traditional Sanger sequencing platforms, requiring careful examination of the importance of variants for clinical use [68]. Several other methods like liquid biopsy, tumor mutational burden, etc. are expanding the field of molecular diagnostics and tissues are becoming important sources of patient care.

6.7 Conclusion

In the era of precision medicine, oncopathology is akin to Arjuna's arrow for understanding the hidden tumor targets and the genome. Even the core biopsy, although a Lilliput of tissue, provides a Gulliver of information. The precious tissue undergoes transformation through a journey involving grossing and processing; these are sometimes frozen and undergo immunohistochemistry and molecular testing, all with the aim of giving a clearer picture on tumor diagnosis, prognosis and treatment. In this long journey, the tissue is carried on the able shoulders of the histotechnologist and the pathologist to reach fruition.

6.8 Summary

1. Oncopathologist plays an important role in both initial diagnosis of tumor, detailed examination of post-operative specimen and staging of tumor.
2. The information gained on gross and microscopic examination of tumor guides treatment

and also established prognostic and predictive factors.
3. The journey of oncopathologist is divided into four sections: grossing of surgical excision, minimum data sets in cancer, frozen section, immunohistochemistry, and molecular diagnostics.
4. The grossing of surgical excisions includes steps from receiving the tissue (biopsies, excisions) to transforming them into a hematoxylin and eosin (H&E)-stained section on the glass slide for evaluation.
5. Minimum data sets in cancer includes consistent format of reporting common cancers and ensures a meaningful and informative pathology report which aids clinical decisions.
6. Frozen section is a rapid diagnostic method for real-time histopathology reporting intraoperatively and few salient points will be discussed for a complete picture.
7. Immunohistochemistry is now an innate component of diagnosis in surgical pathology. Some tumors require immunohistochemistry for diagnosis due to overlapping morphologies, unusual morphologies or unusual clinical presentation. In this era of targeted therapy, Immunohistochemistry also plays a vital role for evaluation of predictive and prognostic biomarkers.
8. Molecular diagnostics is an important and rapidly developing area of pathology. Science of tumor diagnosis and prognosis in pathology has expanded, and we now look for targets within tumor at DNA or RNA level, and a snapshot of existing molecular testing methods will be covered.

Multiple Choice Questions

1. What statement exemplifies this more than role of oncopathology in management of patients with cancer?
   (a) As is your pathology so is your practice
   (b) As is your practice so is your pathologist
   (c) As is your surgeon so is your pathologist
   (d) As is your pathology laboratory so is your practice
2. The journey of histopathology is divided into what sections?
   (a) Grossing of surgical excision
   (b) Minimum data sets in cancer
   (c) Frozen section
   (d) All the above
3. What all steps are involved in grossing?
   (a) Specimen orientation
   (b) Specimen orientation
   (c) Tumor sampling
   (d) All the above
4. Micrometastasis in node is defined as deposit
   (a) <0.1 mm
   (b) >0.2 mm and <2 mm
   (c) >2 mm and <3 mm
   (d) >3 mm and <4 mm
5. Minimum data reporting includes all except:
   (a) Grade of tumor
   (b) Depth of invasion
   (c) Tumor growth pattern
   (d) Surgical details
6. Frozen section is done in which machine?
   (a) Thermostat
   (b) Cryostat
   (c) Hemostat
   (d) Geostat
7. Why is IHC required in tumors?
   (a) Histiogenetic identification of undifferentiated tumor
   (b) Subtyping of cancers like lymphomas
   (c) Evaluating primary site of origin in metastasis from an unknown primary
   (d) All the above
8. What is specific IHC marker for adenocarcinoma lung?
   (a) Liver
   (b) Lung
   (c) Breast
   (d) Cervix

Answers

1. (a)
2. (d)

3. (d)
4. (b)
5. (d)
6. (b)
7. (d)
8. (b)

Acknowledgments The authors thank Dr. Uma Sakhdeo, Associate Professor in Pathology, Tata Memorial Hospital, for proof reading the chapter, Dr. Ankan Ghosh, Resident in Pathology, Tata Memorial Hospital, for diagrams, and Dr. Ashish Shah for help with images.

References

1. Hammond ME, Hayes DF, Dowsett M, Allred DC, Hagerty KL, Badve S, et al. American Society of Clinical Oncology/College of American Pathologists guideline recommendations for immunohistochemical testing of estrogen and progesterone receptors in breast cancer. J Clin Oncol. 2010;28:2784–95.
2. Lee H, Douglas-Jones AG, Morgan JM, Jasani B. The effect of fixation and processing on the sensitivity of oestrogen receptor assay by immunohistochemistry in breast carcinoma. J Clin Pathol. 2002;55:236–8.
3. Shet T. Improving accuracy of breast cancer biomarker testing in India. Indian J Med Res. 2017;146:449–58.
4. Yildiz-Aktas IZ, Dabbs DJ, Bhargava R. The effect of cold ischemic time on the immunohistochemical evaluation of estrogen receptor, progesterone receptor, and HER2 expression in invasive breast carcinoma. Mod Pathol. 2012;25:1098–105.
5. Shet T, Patil A, Desai S. Breast. In: Desai S, Bal M, Rekhi B, Jambhekar NA, editors. Grossing of surgical oncology specimens—a practical guide towards complete pathology training. 1st ed; 2012. p. 56–69.
6. Graham RA, Homer MJ, Katz J, Safali H, Supran S. The pancake phenomenon contributes to the inaccuracy of margin assessment in patients with breast cancer. Am J Surg. 2002;184:89–93.
7. Quirke P, Steele R, Monson J, Grieve R, Khanna S, Couture J, et al. Effect of the plane of surgery achieved on local recurrence in patients with operable rectal cancer: a prospective study using data from the MRC CR07 and NCIC-CTG CO16 randomised clinical trial. Lancet. 2009;373:821–8.
8. Nagtegaal ID, van de Velde CJ, van der Worp E, Kapiteijn E, Quirke P, van Krieken JH, et al. Macroscopic evaluation of rectal cancer resection specimen: clinical significance of the pathologist in quality control. J Clin Oncol. 2002;20:1729–34.
9. Katti SV, Paulose RR, Malipatil B, Verma NS. Grossing and reporting of colorectal cancer resection specimens: an evidence-based approach. Indian J Cancer. 2020;57:239–52.
10. Loughrey MB, Quirke P, Shepherd NA. Standards and datasets for reporting cancers. In: Dataset for histopathological reporting of colorectal cancer; 2017. p. 3–62.
11. Weaver D. Pathology evaluation of sentinel lymph nodes in breast cancer: protocol recommendations and rationale. Mod Pathol. 2010;23:S26–32.
12. Maguire A, Brogi E. Sentinel lymph nodes for breast carcinoma: an update on current practice. Histopathology. 2016;68:152–67.
13. Xu B, Salama AM, Valero C, Yuan A, Khimraj A, Saliba M, et al. The prognostic role of histologic grade, worst pattern of invasion, and tumor budding in early oral tongue squamous cell carcinoma: a comparative study. Virchows Arch. 2021;479:597–606.
14. Moran MS, Schnitt SJ, Giuliano AE, Harris JR, Khan S, Horton J, et al. Society of Surgical Oncology–American Society for Radiation Oncology Consensus Guideline on Margins for Breast-Conserving Surgery with Whole-Breast Irradiation in Stages I and II Invasive Breast Cancer. Ann Surg Oncol. 2014;21:704–16.
15. Tasche KK, Buchakjian MR, Pagedar NA, Sperry SM. Definition of "close margin" in oral cancer surgery and association of margin distance with local recurrence rate. JAMA Otolaryngol Head Neck Surg. 2017;143:1166–72.
16. Van den Eynden GG, Van der Auwera I, Van Laere SJ, Colpaert CG, van Dam P, Dirix LY, et al. Distinguishing blood and lymph vessel invasion in breast cancer: a prospective immunohistochemical study. Br J Cancer. 2006;94:1643–9.
17. Amin MB, Edge S, Greene F, Byrd DR, Brookland RK, Washington MK, et al. AJCC cancer staging manual. 8th. ed. New York, NY: Springer; 2017. p. 252–4.
18. Connolly JL. Changes and problematic areas in interpretation of the AJCC cancer staging manual, 6th edition, for breast cancer. Arch Pathol Lab Med. 2006;130:287–91.
19. Jorns JM, Visscher D, Sabel M, Breslin T, Healy P, Daignaut S, et al. Intraoperative frozen section analysis of margins in breast conserving surgery significantly decreases reoperative rates: one-year experience at an ambulatory surgical center. Am J Clin Pathol. 2012;138:657–69.
20. Chaturvedi P, Singh B, Nair S, Nair D, Kane SV, Dcruz A, et al. Utility of frozen section in assessment of margins and neck node metastases in patients undergoing surgery for carcinoma of tongue. J Cancer Res Ther. 2012;8(S2):100–5.
21. Berlth F, Kim WH, Choi JH, Park SH, Kong SH, Lee HJ, Yang HK. Prognostic impact of frozen section investigation and extent of proximal safety margin in gastric cancer resection. Ann Surg. 2020;272(5):871–8.
22. Lombardi A, Nigri G, Maggi S, Stanzani G, Vitale V, Vecchione A, Nania A, Amanti C. Role of frozen sec-

tion in sentinel lymph node biopsy for breast cancer in the era of the ACOSOG Z0011 and IBCSG 23-10 trials. Surgeon. 2018;16:232–6.
23. Karve PV, Jambhekar NA, Desai S, Chinoy RF. Role of frozen section evaluation in patients with breast lumps: a study of 251 cases. Indian J Surg. 2005;67:248–52.
24. Horwich P, MacKay C, Bullock M, Taylor SM, Hart R, Trites J, et al. Specimen oriented intraoperative margin assessment in oral cavity and oropharyngeal squamous cell carcinoma. J Otolaryngol Head Neck Surg. 2021;50:37–49.
25. Kim SW, Roh J, Park CS. Immunohistochemistry for pathologists: protocols, pitfalls, and tips. J Pathol Transl Med. 2016;50:411–8.
26. Kandalaft PL, Gown AM. Practical applications in immunohistochemistry: carcinomas of unknown primary site. Arch Pathol Lab Med. 2016;140:508–23.
27. Rekhi B, Mridha A, Kattoor J. Small round cell lesions of bone-diagnostic approach differential diagnosis and impact on treatment. Indian J Pathol Microbiol. 2019;62:199–205.
28. Fisher C. Immunohistochemistry in diagnosis of soft tissue tumors. Histopathology. 2010;58:1001–39.
29. Shibuya R, Matsuyama A, Shiba E, Harada H, Yabuki K, Hisaoka M. CAMTA1 is a useful immunohistochemical marker for diagnosing epithelioid haemangioendothelioma. Histopathology. 2015;67(6):827–35.
30. Hornick JL, Dal Cin P, Fletcher CD. Loss of INI1 expression is characteristic of both conventional and proximal-type epithelioid sarcoma. Am J Surg Pathol. 2009;33:542–50.
31. Asano N, Yoshida A, Ichikawa H, Mori T, Nakamura M, Kawai A, Hiraoka N. Immunohistochemistry for trimethylated H3K27 in the diagnosis of malignant peripheral nerve sheath tumors. Histopathology. 2017;70:385–93.
32. Shet T. Ki67 in breast cancer: simulacra and simulation. Indian J Cancer. 2020;57:231–3.
33. Ribic CM, Sargent DJ, Moore MJ, Thibodeau SN, French AJ, Goldberg RM, et al. Tumor microsatellite-instability status as a predictor of benefit from fluorouracil-based adjuvant chemotherapy for colon cancer. N Engl J Med. 2003;349:247–57.
34. Miller RT. Avoiding pitfalls in diagnostic immunohistochemistry-important technical aspects that every pathologist should know. Semin Diagn Pathol. 2019;36:312–33.
35. Sargent DJ, Marsoni S, Monges G, Thibodeau SN, Labianca R, Hamilton SR, et al. Defective mismatch repair as a predictive marker for lack of efficacy of fluorouracil-based adjuvant therapy in colon cancer. J Clin Oncol. 2010;28:3219–26.
36. Hutchins G, Southward K, Handley K, Magill L, Beaumont C, Stahlschmidt J, et al. Value of mismatch repair, KRAS, and BRAF mutations in predicting recurrence and benefits from chemotherapy in colorectal cancer. J Clin Oncol. 2011;29:1261–70.
37. Matos LL, Trufelli DC, de Matos MG, da Silva Pinhal MA. Immunohistochemistry as an important tool in biomarkers detection and clinical practice. Biomark Insights. 2010;5:9–20.
38. Goddard MJ, Wilson B, Grant JW. Comparison of commercially available cytokeratin antibodies in normal and neoplastic adult epithelial and non-epithelial tissues. J Clin Pathol. 1991;44:660–3.
39. El-Deiry WS, Goldberg RM, Lenz HJ, Shields AF, Gibney GT, Tan AR, et al. The current state of molecular testing in the treatment of patients with solid tumors, 2019. CA Cancer J Clin. 2019;69:305–43.
40. McDonough SJ, Bhagwate A, Sun Z, Wang C, Zschunke M, Gorman JA, et al. Use of FFPE-derived DNA in next-generation sequencing: DNA extraction methods. PLoS One. 2019;14:e0211400–15.
41. Compton CC, Robb JA, Anderson MW, et al. Preanalytics and precision pathology: pathology practices to ensure molecular integrity of cancer patient biospecimens for precision medicine. Arch Pathol Lab Med. 2019;143:1346–63.
42. Buesa RJ, Peshkov MV. How much formalin is enough to fix tissues? Ann Diagn Pathol. 2012;16:202–9.
43. Dietel M, Jöhrens K, Laffert MV, Hummel M, Bläker H, Pfitzner BM, Lehmann A, Denkert C, Darb-Esfahani S, Lenze D, Heppner FL, Koch A, Sers C, Klauschen F, Anagnostopoulos I. A 2015 update on predictive molecular pathology and its role in targeted cancer therapy: a review focussing on clinical relevance. Cancer Gene Ther. 2015;22:417–30.
44. Mao C, Yang ZY, Hu XF, Chen Q, Tang JL. PIK3CA exon 20 mutations as a potential biomarker for resistance to anti-EGFR monoclonal antibodies in KRAS wild-type metastatic colorectal cancer: a systematic review and meta-analysis. Ann Oncol. 2012;23:1518–25.
45. Planchard D, Smit EF, Groen HJM, Mazieres J, Besse B, Helland Å, et al. Dabrafenib plus trametinib in patients with previously untreated $BRAF^{V600E}$-mutant metastatic non-small-cell lung cancer: an open-label, phase 2 trial. Lancet Oncol. 2017;18:1307–16.
46. Li BT, Shen R, Buonocore D, Ni A, Ginsberg MS, Ulaner GA, et al. Adotrastuzumab emtansine in patients with HER2-mutant lung cancers: resultsfrom a phase II basket trial. J Clin Oncol. 2018;36:2532–7.
47. Kristensen BW, Priesterbach-Ackley LP, Petersen JK, Wesseling P. Molecular pathology of tumors of the central nervous system. Ann Oncol. 2019;30:1265–78.
48. Haase S, Garcia-Fabiani MB, Carney S, Altshuler D, Núñez FJ, Méndez FM, et al. Mutant ATRX: uncovering a new therapeutic target for glioma. Expert Opin Ther Targets. 2018;22:599–613.
49. Northcott PA, Korshunov A, Witt H, Hielscher T, Eberhart CG, Mack S, Bouffet E, Clifford SC, Hawkins CE, French P, Rutka JT, Pfister S, Taylor

MD. Medulloblastoma comprises four distinct molecular variants. J Clin Oncol. 2011;29(11):1408–14.
50. Yemelyanova A, Vang R, Kshirsagar M, Lu D, Marks MA, Shih IM, et al. Immunohistochemical staining patterns of p53 can serve as a surrogate marker for TP53 mutations in ovarian carcinoma: an immunohistochemical and nucleotide sequencing analysis. Mod Pathol. 2011;24:1248–53.
51. Ventura RA, Martin-Subero JI, Jones M, McParland J, Gesk S, Mason DY, et al. FISH analysis for the detection of lymphoma-associated chromosomal abnormalities in routine paraffin-embedded tissue. J Mol Diagn. 2006;8:141–51.
52. Chu Y-H, Hardin H, Zhang R, Guo Z, Lloyd RV. In situ hybridization: introduction to techniques, applications and pitfalls in the performance and interpretation of assays. Semin Diagn Pathol. 2019;36:336–41.
53. Gajaria PK, Tambe S, Pai T, Patil A, Desai S, Shet T. Dual color dual hapten in situ hybridization (D-DISH)—comparision with flourescnce in situ hybridization (FISH) for Her2neu testing in breast cancer. Indian J Pathol Microbiol. 2020;63:194–9.
54. Chrzanowska NM, Kowalewski J, Lewandowska MA. Use of fluorescence in situ hybridization (FISH) in diagnosis and tailored therapies in solid tumors. Molecules. 2020;25:1864.
55. Summersgil B, Clarc J, Shipley J. Fluorescence and chromogenic in situ hybridization to detect genetic aberrations in formalin-fixed paraffin embedded material, including tissue microarrays. Nat Protoc. 2008;3:220–34.
56. Netto GJ, Saad RD, Dysert PA 2nd. Diagnostic molecular pathology: current techniques and clinical applications, part I. Proc (Bayl Univ Med Cent). 2003;16:379–83.
57. Taishan H, Chitnis N, Monos D, Dinh A. Next-generation sequencing technologies: an overview. Hum Immunol. 2021;82:801–11.
58. Hagemann IS. Overview of technical aspects and chemistries of next-generation sequencing. In: Kulkarni S, Pfeifer J, editors. Clinical genomics. San Diego, CA: Academic Press; 2015. p. 3–19.
59. Yohe S, Thyagarajan B. Review of clinical next-generation sequencing. Arch Pathol Lab Med. 2017;141:1544–57.
60. Shetty O, Gurav M, Pai T, Kumar R, Yadav S, Rane S, et al. Next generation sequencing in the clinic. In: Evidence based management of cancers in India vol XVIII (part F) beyond microscopy—pathology informatics; 2021. p. 13–25.
61. Rehm HL, Bale SJ, Bayrak-Toydemir P, Berg JS, Brown KK, Deignan JL, Friez MJ, Funke BH, Hegde MR, Lyon E, Working Group of the American College of medical genetics and genomics laboratory quality assurance Commitee. ACMG clinical laboratory standards for next-generation sequencing. Genet Med. 2013;15:733–47.
62. Richards S, Aziz N, Bale S, Bick D, Das S, Gastier-Foster J, Grody WW, Hegde M, Lyon E, Spector E, Voelkerding K, Rehm HL. ACMG laboratory quality assurance committee. Standards and guidelines for the interpretation of sequence variants: a joint consensus recommendation of the American College of Medical Genetics and Genomics and the Association for Molecular Pathology. Genet Med. 2015;17:405–24.
63. Jennings LJ, Arcila ME, Corless C, Kamel-Reid S, Lubin IM, Pfeifer J, Temple-Smolkin RL, Voelkerding KV, Nikiforova MN. Guidelines for validation of next-generation sequencing-based oncology panels: A joint consensus recommendation of the Association for Molecular Pathology and College of American pathologists. J Mol Diagn. 2017;19:341–65.
64. Wang K, Yuen ST, Xu J, Lee SP, Yan HH, Shi ST, et al. Whole-genome sequencing and comprehensive molecular profiling identify new driver mutations in gastric cancer. Nat Genet. 2014;46:573–82.
65. Li H, Durbin R. Fast and accurate long-read alignment with burrows-wheeler transform. Bioinformatics. 2010;26:589–95.
66. Roy S, Coldren C, Karunamurthy A, Kip NS, Klee EW, Lincoln SE, Leon A, Pullambhatla M, Temple-Smolkin RL, Voelkerding KV, Wang C, Carter AB. Standards and guidelines for validating next-generation sequencing bioinformatics pipelines: A joint recommendation of the Association for Molecular Pathology and the College of American Pathologists. J Mol Diagn. 2018;20:4–27.
67. Li MM, Datto M, Duncavage EJ, Kulkarni S, Lindeman NI, Roy S, et al. Standards and guidelines for the interpretation and reporting of sequence variants in cancer: A joint consensus recommendation of the Association for Molecular Pathology, American Society of Clinical Oncology, and College of American Pathologists. J Mol Diagn. 2017;19:4–23.
68. Goodwin S, McPherson J, McCombie W. Coming of age: ten years of next-generation sequencing technologies. Nat Rev Genet. 2016;17:333–51.

Further Reading

https://www.youtube.com/watch?v=VFxWpNDrgYE (Tissue Processing in Histopathology Lab)

https://www.youtube.com/watch?v=CetNjrRg7fc&t=3s (Molecular testing basics)

https://www.youtube.com/watch?v=Ehf8l3aQw00&t=2903s (Breast Pathology)

https://www.youtube.com/watch?v=jFCD8Q6qSTM (Next Generation Sequencing)

https://www.youtube.com/watch?v=tA7S75MFdNk (Cryostat tutorial)

7 Imaging in Oncology

Suyash Kulkarni, Nitin Shetty, Amit Janu, Palak Popat, Amitkumar Chaudhary, Vikas K. Pandey, and Anusheel Munshi

Medical imaging modalities have been exponentially developing in the past few decades. The results of imaging modalities heavily impact the primary diagnosis and guide treatment during the entire course of the disease, including follow-up assessments. Imaging plays multifaceted essential roles in all aspects of high-quality cancer care.

Imaging has impacted the screening, diagnosis, management, and follow-up of nearly all cancers [1, 2]. The present-day practice of radiation oncology is based on sound knowledge of radiological and structural anatomy. Significant strides in radiation oncology have been made possible by improved delineation of tumor masses by imaging techniques and more precise radiation delivery systems. The introduction of 3D conformal therapy, intensity-modulated therapy, and stereotactic radiotherapy has defined a new era in the evolution of precision radiotherapy planning. These techniques are highly dependent on onboard sound pretreatment and intratreatment imaging. In surgical oncology, imaging dramatically influences the type and extent of surgery. Further modern imaging in cancer also reduces the chances of futile surgeries. In the domain of systemic therapies, imaging is crucial for initial assessment as well as response assessment to treatment (Fig. 7.1).

This chapter presents an overview of essential imaging modalities, their indications, benefits, and pitfalls.

Most clinical imaging systems are based on the interaction of electromagnetic radiation with body tissues and fluids, with some exceptions. As such, all the imaging tools available can broadly be categorized into ionizing and nonionizing (Table 7.1).

Each diagnostic modality has its own advantages and pitfalls. Table 7.2 presents a snapshot of modern-day imaging tools.

7.1 Mammography (MG)

MG is one of the oldest investigations to detect breast pathologies, where an image is acquired by radiation exposure to the breast. It shows masses as a high-density structure against the lesser dense breast parenchyma and calcifications as small discrete foci of a far greater density. MG has evolved its acquisition from analog to digital, and with lesser radiation exposure. MG is performed under

S. Kulkarni (✉)
Department of Radiology, Tata Memorial Hospital, Mumbai, Maharashtra, India

N. Shetty · A. Janu · P. Popat · A. Chaudhary
Tata Memorial Hospital, Mumbai, Maharashtra, India

V. K. Pandey
Department of Radiation Oncology, Manipal Hospital, Delhi, India

A. Munshi
Department of Radiation Oncology, IOCI, Fortis Noida, Uttar Pradesh, India

A. Munshi et al. (eds.), *Radiation Oncology – Principles, Precepts and Practice*,
https://doi.org/10.1007/978-981-97-8389-2_7

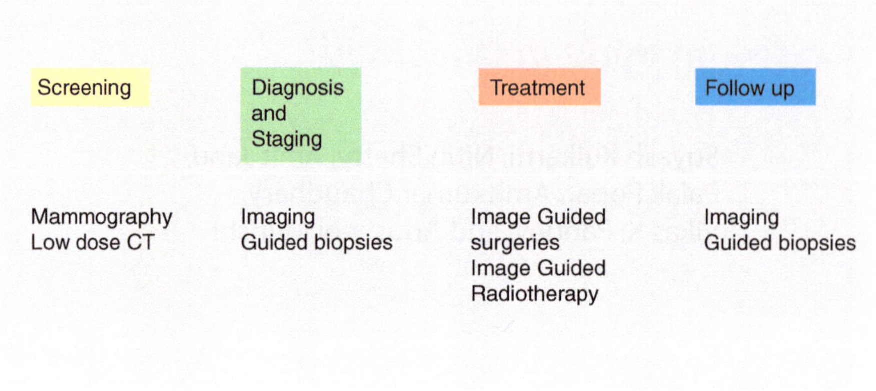

Fig. 7.1 Role of imaging in oncology

Table 7.1 Imaging tools using ionizing rays and nonionizing techniques

Imaging using ionizing rays	Imaging without using ionizing rays
Mammography	Ultrasonography
X-ray	Elastography
CT scan	Doppler
PET/CT scan	Magnetic resonance imaging
SPECT scan	

varying circumstances. Some countries follow an intensive, government-organized screening program, with women attending a 1 to 3-yearly regular MG from 40 to 50 years of age and continuing to about 74 years or beyond [3]. Most other countries follow an opportunistic mammogram for addressing symptoms, prevention or high-risk screening, and emphasizing self and clinical breast examination.

Irregular high-density masses, lobulated and spiculated margins, linear branching and pleomorphic microcalcifications, or calcifications in linear and segmental type distribution, have higher positive predictive values of representing malignancy (Fig. 7.2) [4].

Young women and those during pregnancy and lactation have a dense parenchyma, following a gradual involuting pattern over decades. Dense breast parenchyma is known to obscure masses, lowering the sensitivity of MG [5]. The latter can be addressed by additional views, spot compressions, and digital breast tomosynthesis (DBT). DBT, also called as 3D MG acquires exposures of the parenchyma over an arc with multiple shoots. These project the summation shadows from different angles. This makes it easier to distinguish overlapping parenchyma resultant asymmetries. Standardized lexicon and reporting are preferred for all radiology reporting, including mammography. The American College of Radiology-Breast Imaging Reporting and Data Systems (ACR-BIRADS) is one of the most common reporting systems [6]. A recent advance in MG imaging is the feasibility of contrast-enhanced mammography (CEM). It represents functional imaging, where perfusion and vascularity in the breast parenchyma can show enhancing lesions against a digitally subtracted parenchyma [4].

7.2 Ultrasonography

Ultrasound imaging uses sound waves and echoes to produce a picture of internal organs or masses. Ultrasound is based on acoustic waves' reflection, scattering, and frequency shift. Further, US can image tissue elasticity. Cancer tissues are generally less elastic than normal tissue. Ultrasound is one of the most common initial diagnostic imaging tools for assessing suspected cancer patients. Diagnosing tumors related to thyroid, breast, prostate, liver, pancreatic, ovarian, uterine, and kidney often involves an ultrasound scan as an initial test.

Table 7.2 A comparison of modern imaging modalities

Imaging	Advantages	Disadvantages
X-Ray	• Inexpensive	• Low sensitivity • Finding revealed only when disease more gross or advanced
Mammography	• Inexpensive	• Issue of false positivity • Radiation exposure
US	• Inexpensive • Useful in children and sick/non ambulatory patients	• Low sensitivity • Subjective in interpretation
Bone scintigraphy (bone scan)	• High sensitivity for osteoblastic lesions	• Limited to bony metastases • Challenges in metaphyseal bone metastases
CT	• Useful for bony architecture assessment (necrotic bone, bone destruction) • Can guide biopsy • Quick, commonly available • Can detect calcifications	• Ionizing radiation exposure • Less sensitive than MRI for soft-tissue lesions
MRI	• Greater sensitivity than CT for soft-tissue lesions • Lack of ionizing radiation • Especially useful for leptomeningeal or epidural extension, extent of cortical bone destruction, invasion of the viscera	• Higher cost than CT • Longer acquisition time than CT
MRS	• Measures tissue physiology	• Not very specific • Not commonly available
FDG-PET CT	• Whole body scan • Usually integrated with CT for lesion localization • Versions integrated with MRI available for even greater soft tissue localization	• Cost usually higher than CT or MRI • Radiation exposure

Ultrasonography imaging of the breast is gaining an important role in the evaluation of breast cancer [7]. It can complement MG, especially in younger patients where mammogram has less accuracy. USG has the ability to characterize hepatobiliary area and its lesions (Fig. 7.3).

The appearances of HCC in the US vary depending on the size and degree of differentiation [8]. On B-mode, HCC is hypo-echoic in more than 50% of cases, although it can be hyperechoic or of mixed echogenicity in approximately 25% of cases. In the urogenital region, US helps to detect cancer in the prostate, uterus, ovaries, bladder, and kidney. Prostate US, also called trans-rectal US, provides images of a man's prostate gland and surrounding tissue [9]. The examination typically requires the insertion of an US probe into the patient's rectum. US has gained traction as imaging for cervical cancer. Trans-vaginal and trans-rectal US give detailed images for local staging of cervical cancer as the probe can be placed close to the tumor.

Apart from these, US helps to characterize lymph nodes by their architectural changes and shape (loss of fatty hilum on USG can indicate nodal involvement) and helps diagnose and stage thyroid and neck cancers. To some extent, elastography and Doppler blood flow velocity helps to assess the tissue nature and are used in monitoring therapy response. Measurements of tissue mechanical properties with ultrasound-elastography and blood velocity using Doppler can further characterize the tissue and helps to measure therapeutic response. These properties, viz., cellularity and vascularity, are indicators of enhanced angiogenesis in tumors [10]. In models associated with breast cancer, pancreatic adenocarcinoma and colon carcinoma, the visualization of cancer using VEGFR2-targeted micro-bubble are feasible using US [11].

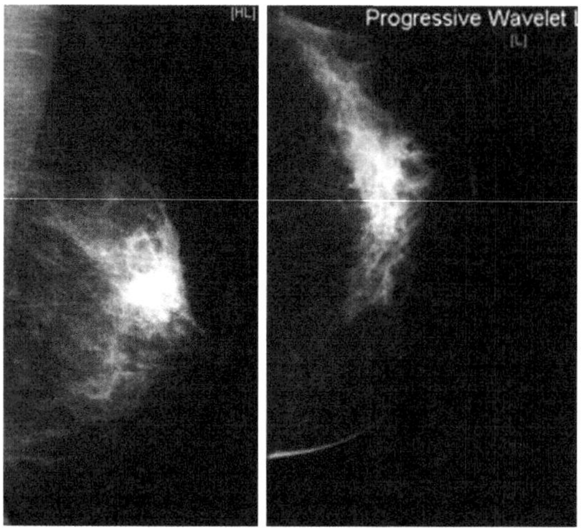

Fig. 7.2 Diffuse microcalcifications in a case of CA breast (mediolateral oblique (MLO) and craniocaudally (CC) views)

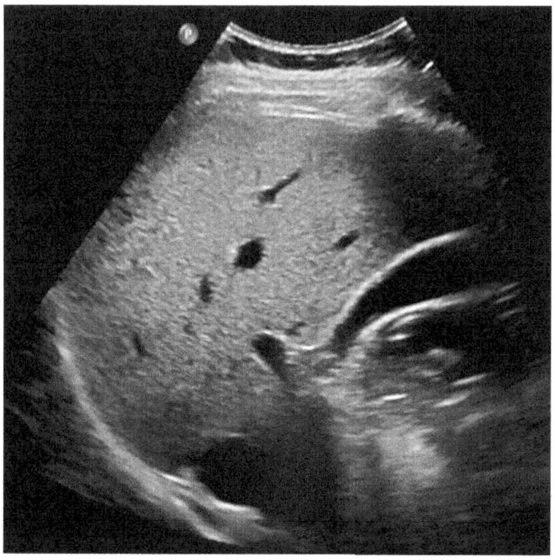

Fig. 7.3 Ultrasound of the hepatobiliary area. (Courtesy: Dr. Simran Singh, Senior Consultant, Department of Radiology, Manipal Hospital, Dwarka, New Delhi)

Ultrasound is also frequently used to guide biopsies of various regions of the body including breast nodes at various sites and soft tissues. TRUS enables the radiologist or the urologist to take TRUS-guided biopsies, which are critical for diagnosing and staging prostate cancer. When the pathological diagnosis is required, US is the imaging technique widely used worldwide to guide liver biopsy [12].

Endobronchial ultrasound (EBUS) plays a pivotal role in the minimally invasive staging of non–small cell lung cancer. The anatomical lymph node stations that are accessible via EBUS are 2R, 2L, 3P, 4R, 4L, and 7. A unique advantage of EBUS is the accessibility of stations 10R, 10L, 11R, and 11L, which are inaccessible by other invasive techniques. Endoscopic ultrasound (EUS) also provides access to 7, but in addition lymph node stations 8 and 9 can be accessed. Recently, EBUS has assisted in the diagnosis of many other disease entities, including malignancies and various infections such as tuberculosis and sarcoidosis [13].

7.3 Computed Tomography (CT)

CT with intravenous contrast is the investigation of choice for characterization as well as staging for most cancers. The widespread availability of CT scanners has aided in rapid and accurate diagnosis and staging of disease. CT Windowing, also known as gray-level mapping, is the process in which the CT image grayscale component of an image is manipulated via the CT numbers (Table 7.3). Doing this changes the appearance of the image to highlight particular structures.

Table 7.3 CT width and center of different tissues

CT examination	Width	Center (level)
Brain	190	50
Skull	3500	500
Orbits	1200	50
Abdomen	400	35
Liver	175	45
Mediastinum	325	50
Lung	2000	−500
Spinal cord	400	50
Spine	2200	400

Two important numbers are displayed on screen while visualizing a CT scan. The first is the window width (W), which describes the range of Hounsfield units displayed. The maximum window width possible is usually about 2000. The second important number is the window level (L). This is the Hounsfield number in the center of the window width [14].

In the context of lung lesions, benign nodules typically have a round shape and smooth, sharply defined borders. Malignant nodules often have an oval shape, and lobulated, irregular borders with speculations. Most adenocarcinomas are seen in the lung periphery and more commonly in the upper lobes. Adenocarcinoma is seen as a solid nodule or a part-solid nodule with some ground-glass opacity [15]. A majority of squamous cell carcinomas arise in the main, lobar, or segmental bronchi and are frequently central in location. Small cell carcinoma is the most common neuroendocrine tumor of the lung, has the propensity to arise from the main or lobar bronchi, and presents as a large hilar mass. Marked mediastinal lymph node enlargement, out of proportion to the size of the lung primary, is a classic finding.

The CT acquisition for imaging of pancreatic adenocarcinoma involves dual-phase imaging—the pancreatic parenchymal phase, 45 s after injection of IV contrast and portal venous phase at 65–75 s postcontrast administration. The contrast between the hypoattenuating adenocarcinoma and the pancreatic parenchyma is best appreciated on the pancreatic parenchymal phase. In addition, this phase is highly useful in demonstrating the relationship of the tumor with arteries and arterial variations, which have management implications. Pancreatic adenocarcinoma is usually hypodense on CT, relative to the pancreatic parenchyma, on the pancreatic parenchymal phase. A tumor contact of <180° implies abutment and >180° as encasement [16, 17]. Deformity or stenosis of a vessel is considered a vascular invasion.

Two additional concepts related to CT are unique to radiation oncology. One is the use of Cone beam CT (CBCT) as a on board imaging tool (Fig. 7.4).

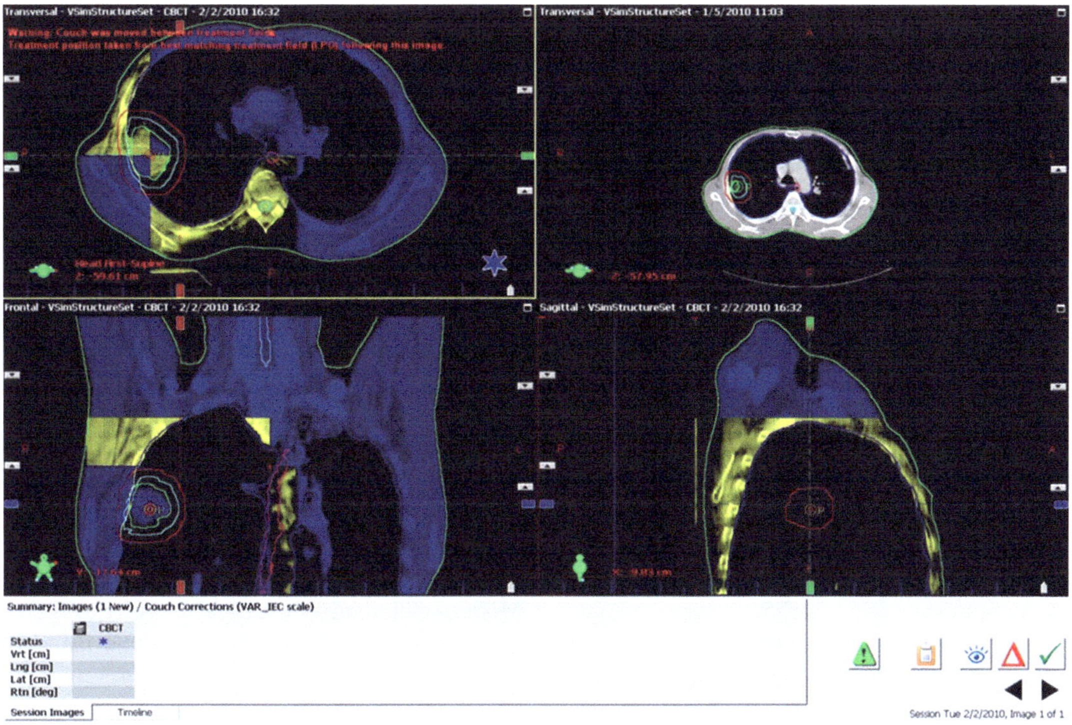

Fig. 7.4 Cone beam CT–simulation CT matching in the Linac console area before SBRT lung treatment delivery

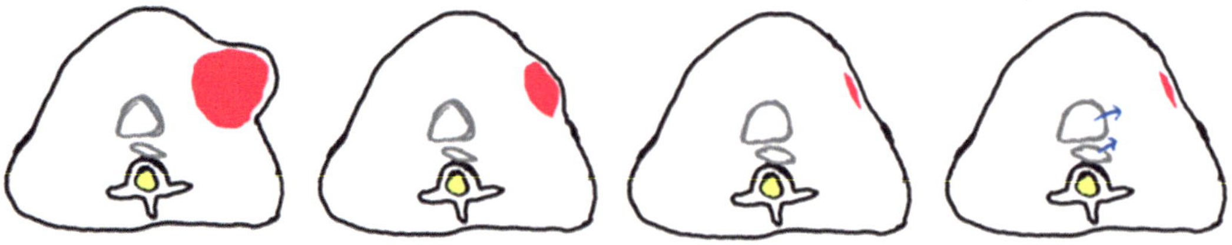

Fig. 7.5 Need for adaptive CT planning in radiotherapy. Tumors can shrink during radiotherapy with consequent changes in skin contour and shift of internal structures (blue arrows). This may need a repeat planning CT scan midway through radiotherapy

Fig. 7.6 Radiation pneumonitis in a case of CA right breast, 3 months post radiotherapy

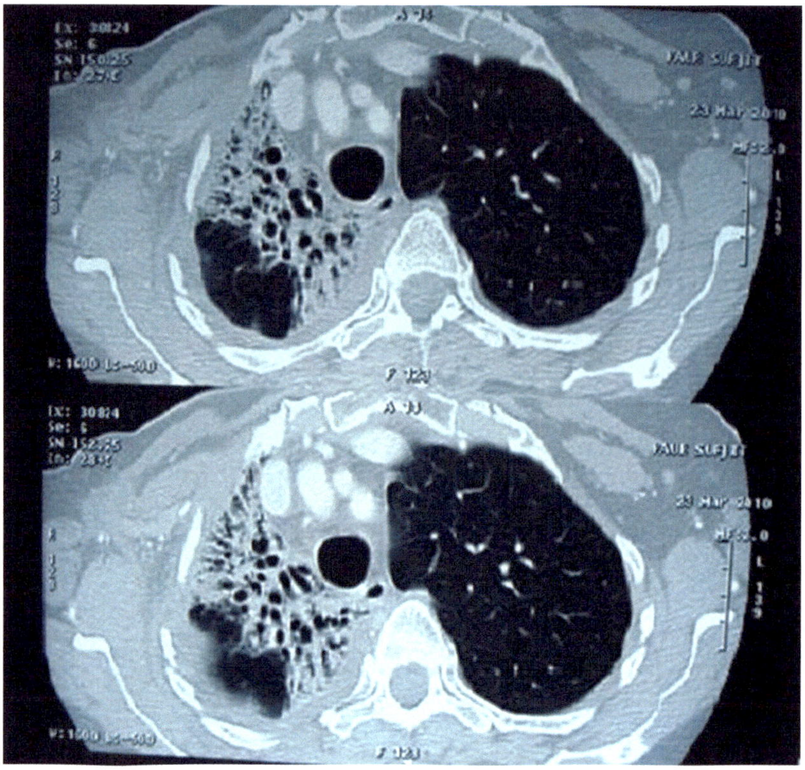

The second concept is of doing an adaptive CT. This is aimed at taking care of the changes that occur in tumors and tissues during a course of radiotherapy. If these changes appear significant, they may necessitate repeat imaging and planning as shown in Fig. 7.5.

CT is also often used in the follow-up of cancer patients to detect recurrences and posttreatment sequelae as shown in Fig. 7.6. Modern uses of CT are enlisted in Table 7.4.

CT scan is also used for the evaluation of metastatic disease, although increasingly it is being replaced by PET/CT for this purpose (Fig. 7.7). Modern radiotherapy uses 4DCT for treatment of thoracic tumors to account for motion manage-

Table 7.4 Modern uses of CT

Four-dimensional CT (4D CT), i.e., 3D plus movement synchronization	Four-dimensional CT acquisition is used for image-modulated radiotherapy (IMRT) applications in the thorax so that the radiation beams do not miss the tumor during respiratory motion
CT virtual colonography	Shown to yield similar results to optical colonoscopy for clinically important polyps/ lesions larger than 10 mm in size. Especially suitable for elderly patients
CT perfusion imaging	Based on the linear relation between the CT attenuation values (expressed by Hounsfield units) and the concentration of the contrast agent. CT perfusion imaging is used to determine therapy response. A CT perfusion study showing changes in hepatic tumor perfusion after antiangiogenic therapy
Cone beam CT (CBCT)	Frequently used in present-day linear accelerators. In one rotation of the arm, it gives a volumetric scan of a craniocaudal length of nearly 15–20 cm. This can then be matched with the images taken at the time of simulation and suitable adjustments done to couch position for matching (Fig)
CT-guided interventions	• Radiofrequency ablation of bone metastases, hepatic metastases, and HCC renal tumors • Guided brachytherapy • Alcohol injection in metastases • Nerve block for pain palliation • Guided biopsies

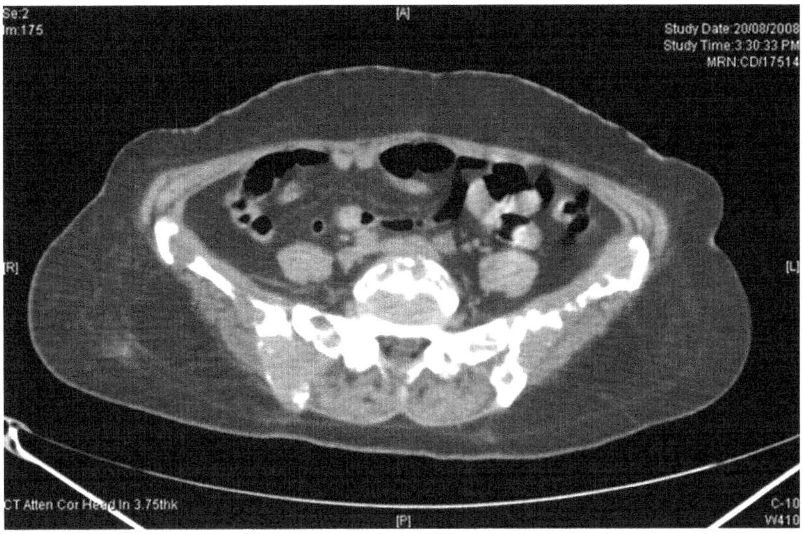

Fig. 7.7 Bony pelvic bone metastasis in a case of CA breast

ment. Table 7.5 describes terminologies used in the context of 4DCT scan imaging.

7.4 Synthetic CT

Radiation oncologists are highly dependent on CT images for radiation planning scans. One of the reasons for this is that CT provides electron density information, which a MRI generally does not provide. This also means that patients for radiation therapy planning need to often undergo two or more imaging sequences (e.g., CT and MRI). A synthetic CT or pseudo CT image is a CT image generated from an MR image instead of using a CT scanner [19]. While the idea appears simple, it requires solving several challenging problems. One of the important issues is that air and bone look almost the same in an MR image. So in essence, synthetic CT is a complex nonlinear mapping of MR values to corresponding HU values. Researchers have developed different ways to generate a synthetic CT [20]. These can be subdivided into two categories.

Table 7.5 Terminologies used in the context of 4D CT scan imaging [18]

Maximum intensity projection (MIP) [18]	A data visualization method that projects in the visualization plane the voxels with maximum intensity. Generally used for hyperdense lesions. Often used for generating summated gross tumor volume (iGTV) in 4D treatment in small lung tumors
Minimum intensity projection (MinIP) [18]	A data visualization method that enables the detection of low-density structures in a given volume. In this way, only the most hypodense structures of the volume are represented So it is the optimal tool for the detection, localization, and quantification of ground-glass opacity, mosaic attenuation, traction bronchiectasis, cystic lung disease, and linear attenuation patterns on chest CT. MinIP algorithm is particularly useful for analyzing the bile tree and pancreatic duct, which are hypodense compared to surrounding tissue
Average intensity projection (AvIP)	Generated by averaging pixel densities for all phase images of 4DCT

4DCT four-dimensional CT

First, the *conventional algorithms* which include approaches like (a) voxel-based weighted summation, (b) patch-based pseudo-CT generation, (c) multiatlas-based registration, and (d) label propagation. The second is *artificial intelligence (AI)* or *deep learning (DL)* algorithms such as UNet and CycleGANs. With regards to synthetic CT generation, DL-based algorithms have become the industry standard as they've outclassed the conventional algorithms' accuracy.

Potentially, synthetic CT scan revolutionizes the present day workflow in radiation oncology. These scans, however, are still not recommended to be used for diagnostic purposes. As is evident, the process of synthetic CT saves patients extra imaging sessions and it also completely eliminates CT–MR fusion/registration errors thereby streamlining the RT treatment workflow. Although synthetic CT is a new technology, pilot projects of adopting the MR-only RT workflow have already started. Full adoption will take some time since it is contingent on the availability of all the anatomies in synthetic CT, and not only availability but also thorough testing and commissioning.

7.5 PET/CT

Positron emission tomography (PET) provides a unique opportunity and means of imaging biological processes in vivo. PET imaging can either be used solo or integrated with CT. The final result is a three-dimensional (3D) image of the area of interest integrated with the distribution of the biological process [21].

PET uses a molecular imaging probe, a radiopharmaceutical, which targets a specific feature of malignant physiology like glycolysis, apoptosis, cell proliferation, or the expression of a specific receptor-like somatostatin receptor or prostate-specific membrane antigen (PSMA). The radiopharmaceutical typically consists of a targeting component which could be an antibody, a small molecule, or a peptide, and a signaling component which is a radioisotope capable of positron emissions such as C-11, N-13, O-15, or F-18. The radiopharmaceutical is injected into the patient, following which the tracer attaches to the specific target and the patient is then scanned by the tomographic system capable of detecting the positrons emitted by the radiopharmaceutical [22].

One of the most common radiopharmaceuticals used in oncology is F-18, labeled fluorodeoxy-glucose (FDG). PET imaging using FDG takes advantage of one of the hallmarks of cancer, namely the Warburg effect: increased glucose metabolism and conversion of glucose carbon to lactate of cancer cells as compared to normal cells [23]. This increased glucose metabolism in malignancies is mediated via an increased expression and activity of glucose transporters (GluT) in the cell membrane as well as via changes in the expression and activity of the enzyme hexokinase. This alteration in glucose metabolism represents one of the early events in carcinogenesis. Increased FDG uptake has been associated with increased mitotic activity, with the degree of uptake being a surrogate marker of biological aggressiveness. PET-based cancer imaging utilizes positron-emitting radioisotopes, created in a cyclotron, like 18F, 11C, 64Cu, 124I, 86Y, 15O, and 13N or in a generator like 68Ga. F-18 FDG is one of the most ubiquitous radiotracers in oncology today and is often used as a biological marker in radiation treatment planning.

The development of a hybrid imaging system such as PET/CT [24] enables the combination of the metabolic sensitivity of PET and the temporal and spatial resolution of CT. As a result, there has been an increased use of imaging of biomarkers to demonstrate metabolism, cell proliferation, cell migration, receptor expression, gene expression, signal transduction, hypoxia, apoptosis, angiogenesis, and vascular function. Measurements of these parameters can be used to plan therapy, give early indications of treatment response, and detect drug resistance and disease recurrence. It shows the principle of biomarker imaging with different imaging technologies.

FDG PET/CT scan can be performed using intravenous administration of iodinated contrast, termed PET-CECT, which may serve as a feasible option for staging to achieve dual information with a single investigation. When compared with CT alone, FDG PET/CT can better delineate the metabolically active tumor from the adjacent postobstructive atelectasis and consolidation (Fig. 7.8).

Thus, PET/CT is more accurate in delineating the size and extent of the tumor for accurate T staging. This has important implications for radiation therapy contouring and planning. Various advantages and disadvantages of PET/CT in different anatomical sites are enlisted in Table 7.6.

PET/CT also has the potential to reduce futile surgeries. In patients with NSCLC who are being considered for treatment with curative intent, FDG PET/CT is recommended to identify unsuspected metastases, which helps reduce the frequency of futile thoracotomies. Initial PET/CT has been found to change the stage in up to 25% of patients, and initial PET/CT has a substantial impact on the management plan in up to 40% of patients [25].

PET/CT also helps provide guidance for biopsy, particularly in patients with prior negative/inconclusive biopsies. However, false-negative results can be seen in small nodules (<8–10 mm in diameter), mucinous adenocarcinomas, and low-grade malignancies such as carcinoma in situ and minimally invasive adenocarcinoma.

7.6 PET-Based Contouring

An upcoming development in contouring is a fusion of PET/CT with the CT acquired at the time of treatment planning. The fused images can be used for accurate delineation of the gross tumor.

However, PET-based contouring can be challenging since the outline of the "hot area" is not as well defined as is the case with CT or MRI. For example, in NSCLC, some authors have taken thresholds of 15% to 50% uptake value compared to the tumor maximum whereas others have taken 40% value [26]. Another way of doing this is to take all of the volume above a designated standardized uptake value (SUV) (e.g., 2.5 in lung tumors) but again this threshold value needs to be determined for individual tumors [27, 28]. FDG-PET therefore might play an important role in GTV definition and sparing of normal tissue [28].

Figures 7.9, 7.10, 7.11, 7.12, and 7.13 depict PET/CT-based contouring of various organ sites.

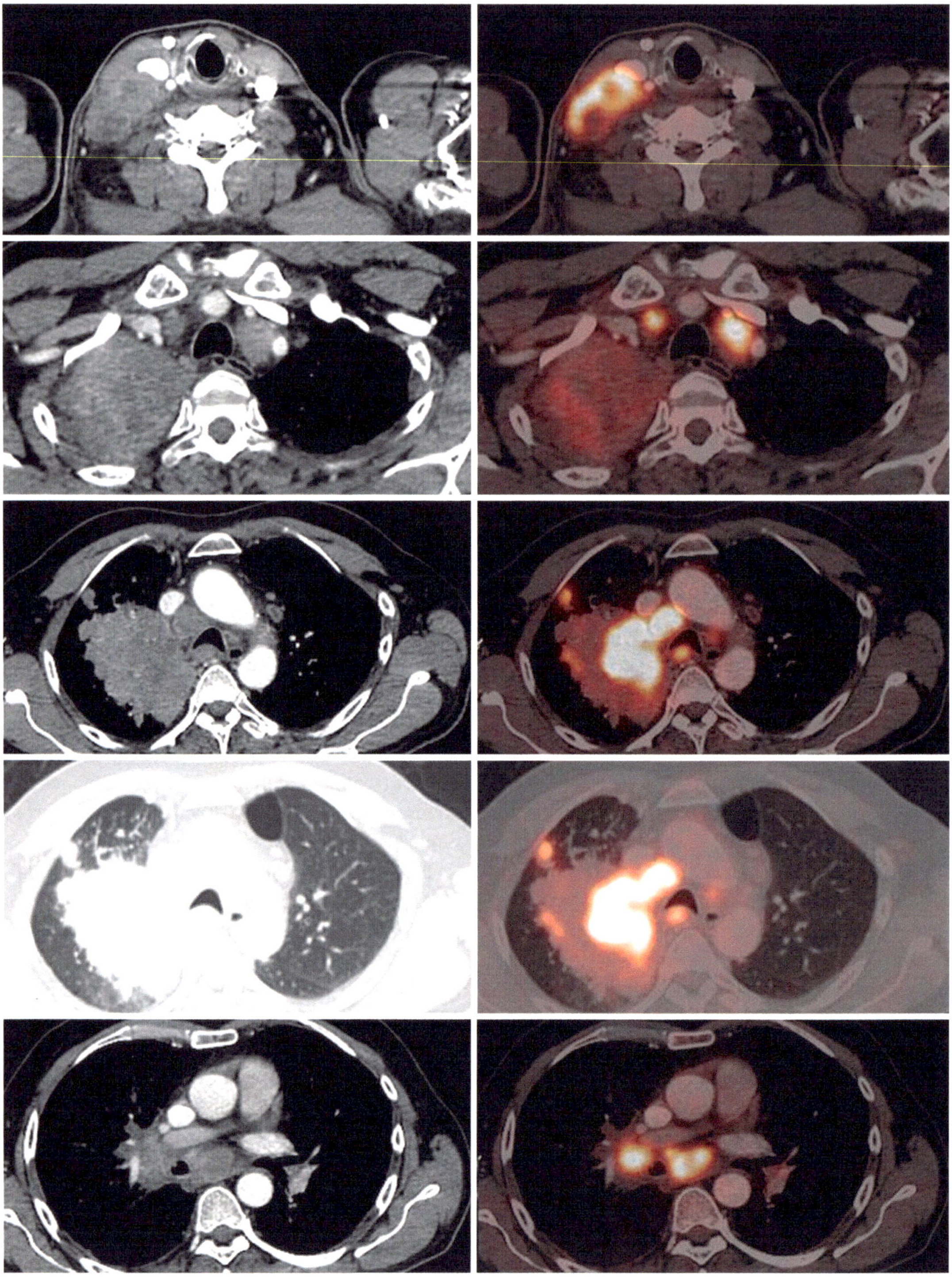

Fig. 7.8 The left panel has only CT images, making differentiation between tumor and adjacent lung collapse difficult. The right panel shows PET/CT with a clear metabolically active lung lesion and the adjacent area of collapse (non-FDG avid)

Table 7.6 Various advantages and disadvantages of PET in different anatomical sites

Site	Advantage of PET	Disadvantage of PET
Lymphomas	1. Potential for either downstaging or upstaging the disease extent so as to alter management 2. The grade and the proliferative activity of lymphoma is well correlated with the rate of tracer uptake 3. Noninvasive detection of extranodal sites, e.g., spleen, liver, bone marrow and bone 4. Assessing the chemosensitivity before high dose CT followed by autologous stem cell transplantation 5. Prognostic evaluation during and after treatment 6. The FDG-PET coregistered with CT has been used for RT conformal planning treatment	Inability to distinguish between minimal residual disease (MRD) and inflammatory lesions which also accumulate radiotracer
Brain	1. CT/MRI fusion and PET or amino acid SPECT allows a correct delineation of GTV and PTV 2. Newer radiotracers like methionine(MET) PET, IMT, and FET helps to define the GTV and differentiate tumor from normal tissue	1. Brain tumors with increased FDG uptake may be indistinguishable from normal brain tissue due to glucose uptake in normal brain 2. MET has a very short half-life of 20 min versus that of FET of 110 min, restricting the use of the former to centers with on-site cyclotrons
Head and neck	1. Marginal superiority to CT and MRI for detection of locoregional and metastatic disease. With a sensitivity of 50–96% and specificity of 88–100% 2. PET/CT is more accurate than PET or CECT alone for delineating cervical nodes in SCC of head and neck 3. Helpful in follow-up evaluation so as to differentiate viable tumor from posttreatment changes and fibrosis, thus recommending a follow-up scan at 8–12 weeks after completion of treatment 4. Useful in detecting primary in case of metastatic and unidentified primary malignancy 5. PET/CT is useful in contouring primary tumors whose borders are difficult to delineate on anatomic imaging alone such as base tongue tumors	1. May not be able to distinguish background inflammation from tangible disease due to accumulation of radiotracer at both sites 2. Limited in its ability to show hypermetabolism in cases of perineural infiltration and marrow disease

(continued)

Table 7.6 (continued)

Site	Advantage of PET	Disadvantage of PET
Esophagus	1. Early detection of disease before it is evident on CT 2. Highly specific for staging nodes outside the mediastinum, e.g., supraclavicular nodes 3. More sensitive in determining the true extent of metastatic disease to the liver, lung or bone (*Weber WA, 2004)*	1. Inability to determine exact disease extent due to poor resolution 2. Physiological uptake in the esophagus may be a limitation in detecting small, well-differentiated tumors 3. Inability to distinguish fungal infection, esophagitis secondary to CT, benign conditions like, hiatal hernia, Barrett's esophagus, and granulomatous disease from true positives
Lung	Is able to differentiate tumor from atelectasis and thus reduce the tumor volume	1. Difficulty in tumor edge definition, limited spatial resolution and tumor motion during image acquisition currently limits the accuracy of PET in target volume delineation in NSCLC without adjacent lung consolidation 2. Contoured target volumes are subject to interobserver variations. Strong collaboration is required between radiation oncologists and nuclear medicine specialists/radiologists 3. Reliable tools for tumor thresholding need to be developed 4 RT planning software systems need to accept DICOM images
	Upstaging mediastinal disease by PET can impact RT planning in order to achieve larger target volumes	
Gynecologic malignancies	1. Highly specific for demarcating para-aortic lymph nodes than CT/ MRI 2. In carcinoma ovary, it helps in accurately delineating the extent of disease outside pelvis like diaphragm, liver surface, omentum and LN 3. The PPV is nearly 90% for peritoneal carcinomatosis if a nodular diffuse pattern is seen in peritoneum	1. Marked difficulty in differentiating tumor and nodal metastases from FDG activity in nearby bowel, bladder and ureters 2. Presence of leiomyomas in the uterus may result in high false positive rates
Prostate	PSMA PET/CT is more sensitive and specific in detecting metastases in individuals with high-risk features than traditional imaging modalities like bone scintigraphy and CT scans	1. FDG is not the preferred tracer as uptake has been found to be variable and predictable 2. Urinary excretion in bladder and FDG accumulation in the bladder can obscure minimal or moderate uptake 3. FDG uptake is nonspecific and cannot differentiate between malignancy, prostatic hyperplasia, or postoperative scarring

7.6.1 Disadvantage of PET

PET has a limited resolution at present, and this contributes significantly to image blur. Other pitfalls of PET/CT include poor accuracy in the brain in view of surrounding normal FDG updates in brain tissue. Also, it is to be noted that PET/CT cannot replace invasive mediastinal staging, which should be performed for PET-negative mediastinal nodes, as well as for pathological confirmation of PET-positive nodes before any definitive management. The high false-positivity rate in mediastinal nodes on PET/CT particularly holds true for countries endemic to granulomatous diseases like tuberculosis.

To overcome the various disadvantages, novel PET/CT fusion software is available which can be used for interpretation. The majority of the lat-

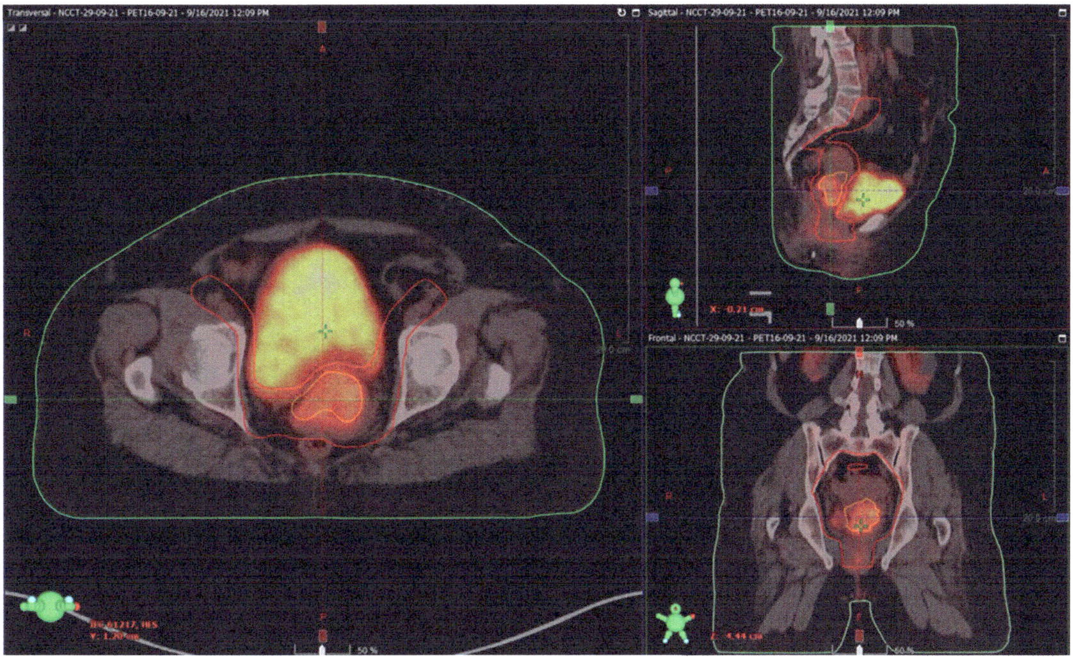

Fig. 7.9 PET/CT-based planning in a case of cervix cancer

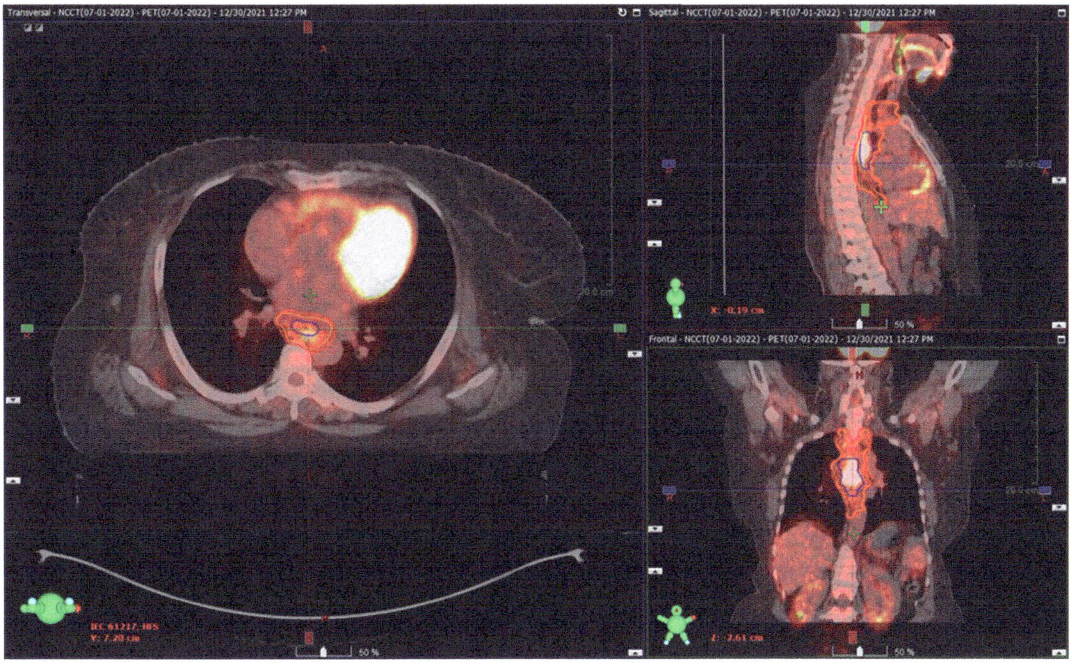

Fig. 7.10 PET/CT-based planning in a case of esophageal cancer

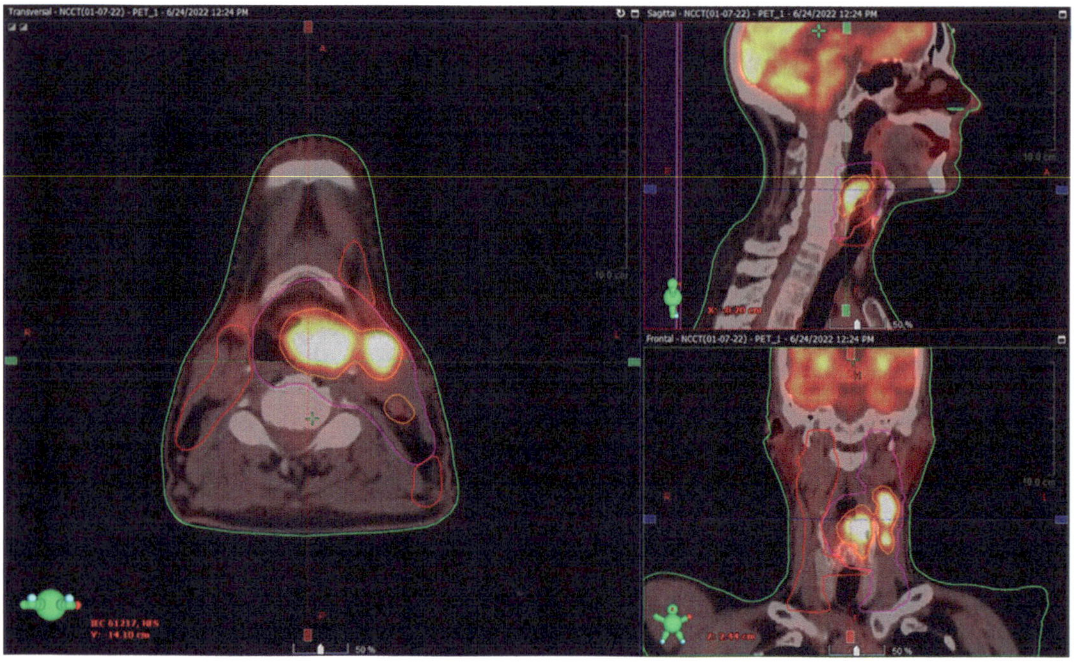

Fig. 7.11 PET/CT-based planning in a case of hypopharyngeal tumor

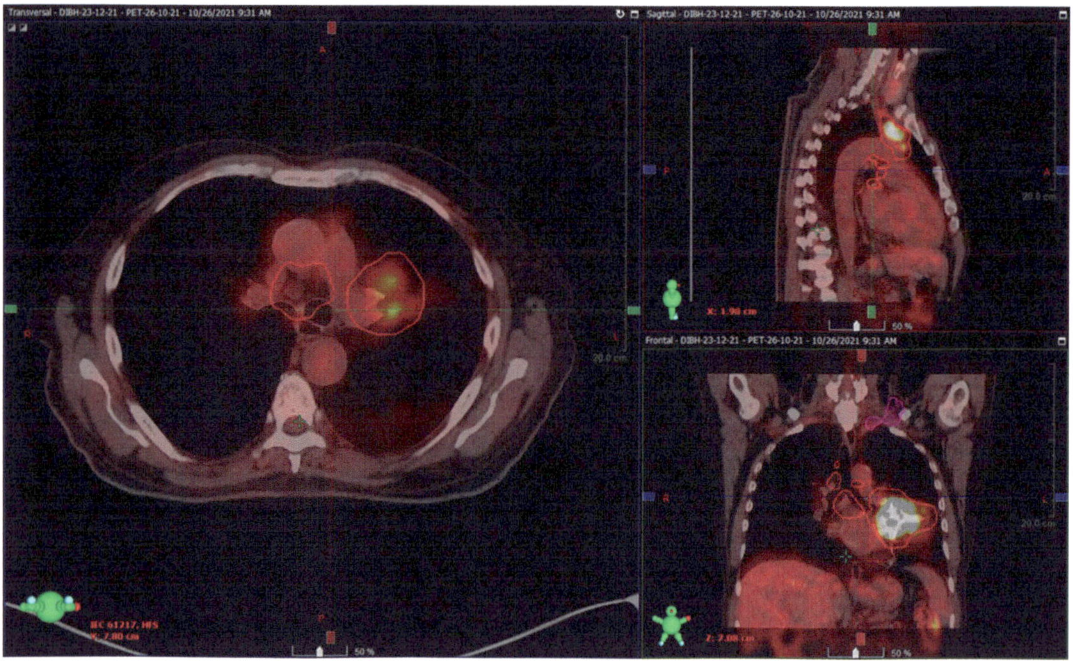

Fig. 7.12 PET/CT-based planning in a case of lung cancer

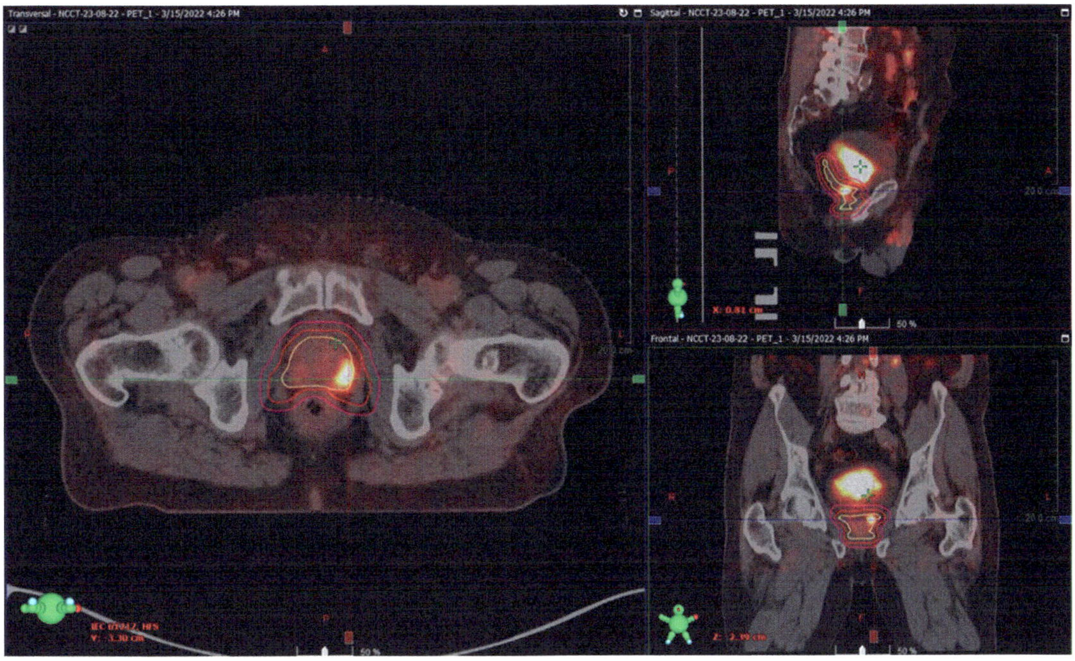

Fig. 7.13 PET/CT (PSMA)-based planning in a case of prostate cancer

est PET/CT units use thin-slice imaging along with intravenous contrast. Surti et al. [29] have tried developing ultra-thin crystal detectors which are arranged in small groups along with the development of electronics that incorporate time-of-flight calculation procedures. In the near future, the 64-slice CT units will consolidate PET as a critical element in diagnosing, staging, and management of cancer of various sites.

7.7 Magnetic Resonance Imaging (MRI)

MRI is based on the resonance of protons under the influence of the generated magnetic field. Magnetic resonance imaging (MRI) can help to classify as benign or malignant. Compared to CT, it has a higher level of spatial and temporal resolution. It is used alone or in combination with other modalities to know the extent of locoregional disease. This modality is especially important in some sites [2].

Magnetic resonance is used in cancer detection, staging, therapy response monitoring, biopsy guidance, and minimally invasive therapy guidance. Magnetic resonance imaging (MRI) offers imaging without exposure to ionizing radiation. MRI in breast imaging is of assistance with the presence of dynamic postcontrast assessment [30]. Among the noncontrast sequences, a combination of T2-weighted sequence, Fat saturated sequence, and diffusion-weighted imaging help in the characterization of the tissue, whereas T1-weighted sequence, with its fat, saturated counterpart, helps in the assessment of hemorrhagic and proteinaceous content. Diffusion with the apparent diffusion coefficient helps in assessing true cellularity and necrosis. The early enhancement of a lesion and early washout, intermediate to mildly high signal intensity on T2W sequence, and restricted diffusion together have a higher prediction of malignancy.

Gadolinium-enhanced MRI of the brain is a gold standard for lesions in the brain (Fig. 7.14) [31]. It is also the modality recommended to rule out asymptomatic brain metastases in all patients with stage II, III, and IV disease if definitive multimodality therapy is being considered. Patients with stage IB disease are less likely to have brain metastases, and MRI is optional in this setting. PET is not recommended for ruling out brain

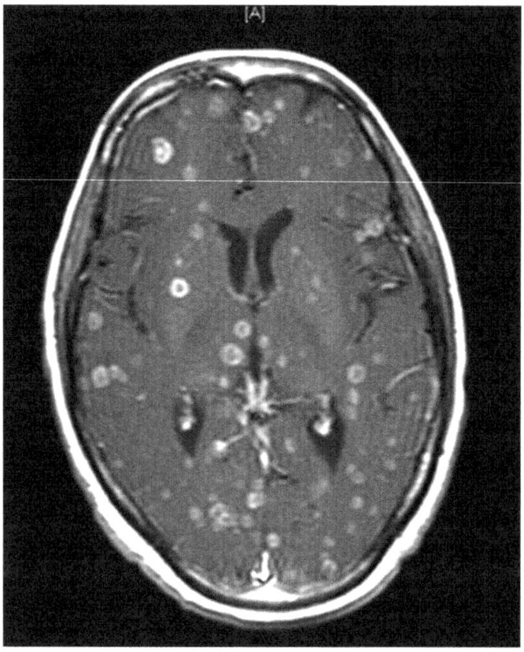

Fig. 7.14 T1 (contrast) MRI showing multiple brain metastasis in a case of CA breast

metastases, considering its poor sensitivity for small-sized lesions (<1 cm size).

MRI is less commonly performed due to lower availability of the imaging modality and higher costs. However, MRI along with MRCP can be used, especially in young patients to minimize radiation exposure. MRI along with MRCP is on par with CT scan for the detection of lesions and associated features [4]. MRI can be a problem-solving modality for cases where mass is not detected on dual phase pancreatic CT. also in patients unable to undergo contrast-enhanced CT study, MRI can provide information of pancreatic mass in such a scenario by virtue of its higher soft tissue contrast resolution [4, 5]. MRI can also be a problem-solving tool for CT-indeterminate liver lesions.

Table 7.7 MRI protocols/sequences and their characteristics

MRI protocol/sequence	
T1	Identifies blood, fat, and protein in lesions Ideal for detecting calvarial metastasis
T2	Location Extent
DWI	Lesion detection, characterization, infarct along the operative bed
SWI	Tumor bleed, calcification
Fat-suppressed Postcontrast T1	Disruption of blood–brain barrier Detection Characterization Leptomeningeal disease Surveillance
Postcontrast FLAIR	Subarachnoid hemorrhage Leptomeningeal disease Surveillance
DSC perfusion	Tumor detection Tumor characterization Optimal biopsy site Surveillance
DTI and BOLD fMRI	Surgical planning Counseling
MRS	Characterization Grading Surveillance

7.7.1 Structural Imaging

Structural imaging sequences can help detect the location of the tumor, tumor characterization, and relation with adjacent structures. MRI protocols/sequences and their characteristics are enlisted in Table 7.7. The presence of features like CSF and the vascular cleft between the tumor and underlying brain parenchyma, broad base toward the dura, gray matter interfaced tumor, and white matter enable the distinction of an extra-axial tumor from the intra-axial tumor.

7.7.2 Functional MRI (fMRI)

Blood oxygen level-dependent (BOLD) imaging, also widely known as fMRI, uses the signal from the ratio of deoxyhemoglobin to oxyhemoglobin, resulting from neuronal activity during a task-based paradigm. The technique is used for pre-surgical planning and intraoperative navigation. The commonly mapped areas are the somatosensory cortex and language areas [32]. The fMRI maps generated by the technique are overlaid on a structural image to demonstrate the relationship of eloquent areas to the tumor. MRI is the modality of choice for tumor detection, tumor characterization, estimating tumor extent and grade, optimal biopsy site / surgical planning, and surveillance.

7.7.3 Diffusion-Weighted Imaging (DWI)

DWI probes the diffusion of water molecules in the interstitium and allows assessment of tumor cellularity, peritumoral edema, regions of tumor hypoxia, integrity of white matter tracts, and postoperative injury. It can also be used for characterization of certain tumors, e.g., lymphoma, a highly cellular shows diffusion restriction; a posterior fossa tumor in a child showing restricted diffusion is highly suggestive of medulloblastoma/embryonal tumor [33].

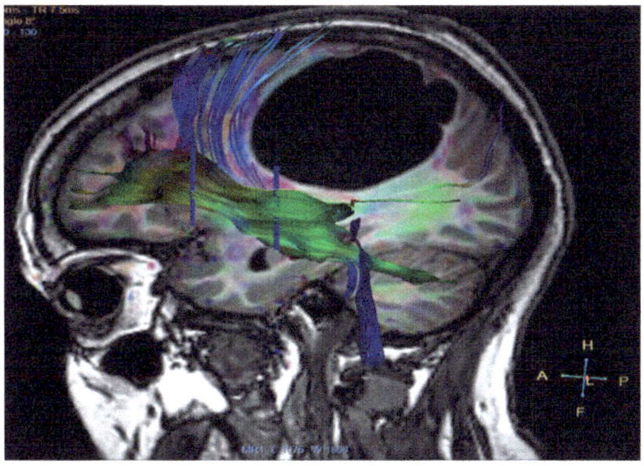

Fig. 7.15 DTI imaging showing the important fiber tracts of the brain in a patient with brain lesion. (Courtesy: Dr. Simran Singh, Senior Consultant, Department of Radiology, Manipal Hospital, Dwarka, New Delhi)

7.7.4 Diffusion Tensor Imaging (DTI)

DTI is an extension of DWI, where diffusion acquisition is made in at least 6 directions which in turn enables generation of tractograms, a trajectory of various white matter tract projections. This technique uses anisotropic diffusion to estimate the axonal (white matter) organization of the brain (Fig. 7.15).

The identification of the tracts allows preservation of the major pathways, e.g., the language tracts and the corticospinal tracts, during tumor surgery and reduce postoperative neurological deficit [34]. This is of critical important in neurosurgery and is likely to become important in radiation planning/delivery as well.

7.7.5 Perfusion Imaging (PWI)

Dynamic Susceptibility Contrast (DSC) perfusion is widely utilized perfusion sequence for tumor characterization, assessment of tumor grade and treatment response (Fig. 7.16). The percentage signal recovery following bolus of contrast can help in tumor characterization. Glial neoplasms frequently demonstrate near complete baseline recovery of the signal intensity [35]. Lymphoma shows a baseline overshoot on DSC perfusion study. For grading of gliomas, an rCBV value of 1.75 serves as a reasonable cutoff, for differentiating high grade from lower grade glio-

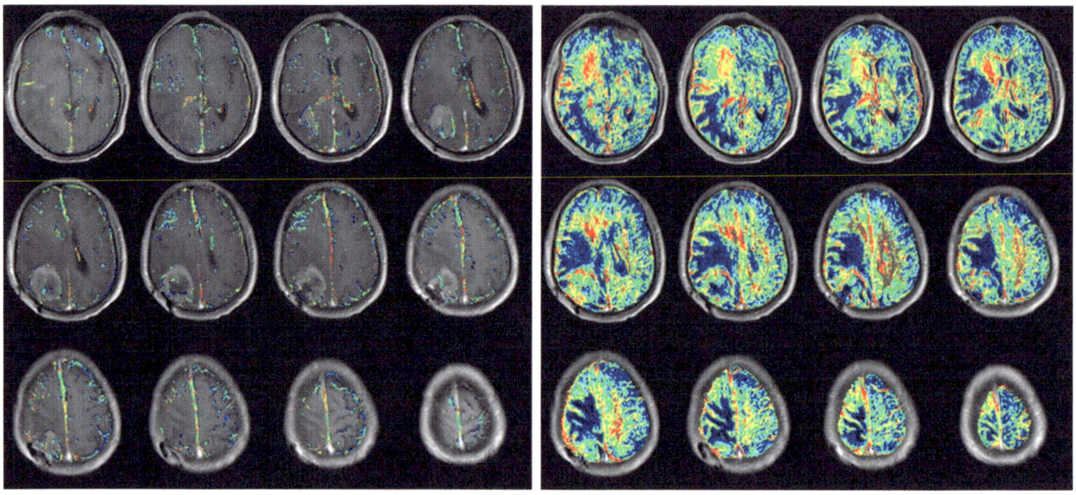

Fig. 7.16 Perfusion scan in a patient with high-grade glioma. Note the increased perfusion in the right parietal region (region of tumor) in the perfusion images

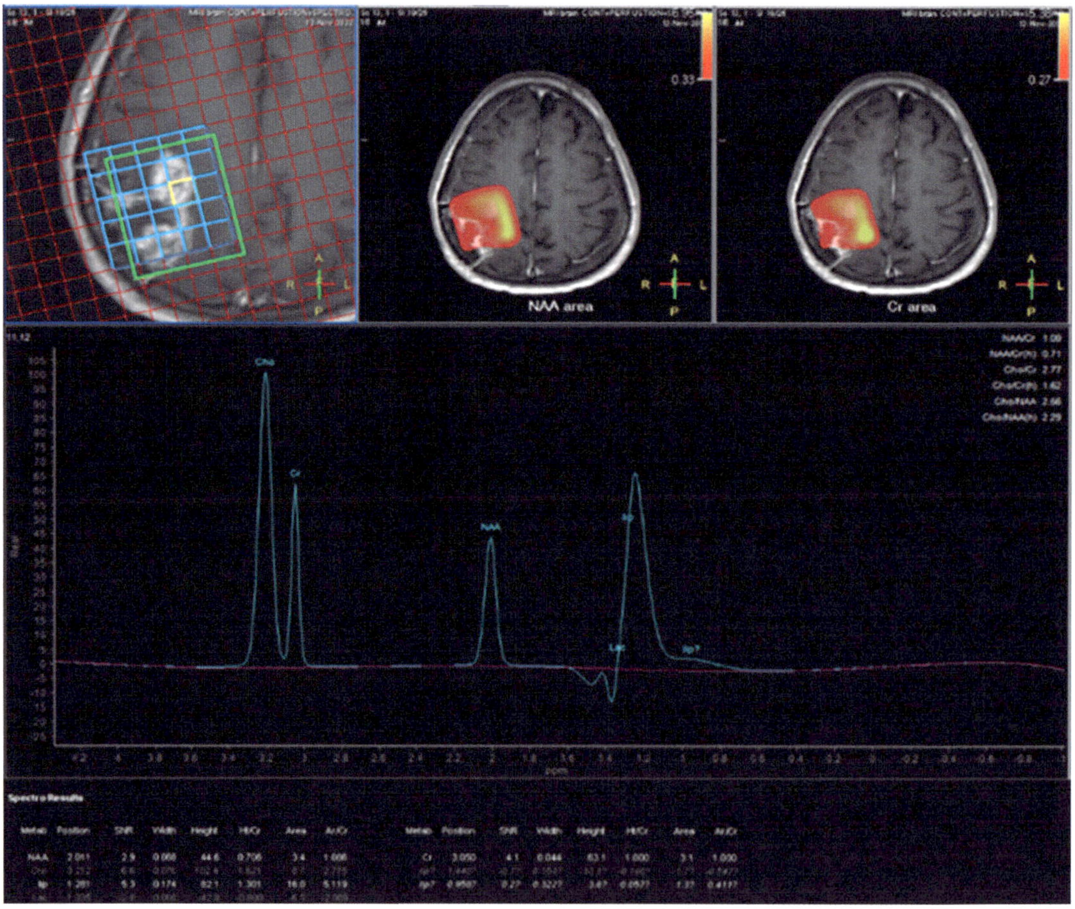

Fig. 7.17 MR spectroscopy of a glioma

mas. A reduction in rCBV of a tumor/treatment bed at follow-up scan is suggestive of response to therapy.

7.7.6 Magnetic Resonance Spectroscopy (MRS)

MRS is a noninvasive technique for measuring brain metabolites (Fig. 7.17). It is read from right to left with 0 parts per million (ppm) represented on the far right on the *x*-axis. The important metabolites used in analysis of brain tumors include N-acetyl aspartate (NAA), a marker of neuronal integrity, [36] detected at 2 ppm; Choline (Cho), marker of cell membrane turnover, at 3.2 ppm; Creatine (Cr), a "constant" cellular energy metabolite; lipid and lactate, indicator of hypoxia and necrosis, at 1.2 ppm; and myoinositol (mI), a marker of gliosis, at 3.56 ppm.

A normal spectrum demonstrates upward sloping of peaks from mI to NAA, forming the so-called Hunter's angle of approximately 45°. In brain tumors, there is elevation of choline due to increased cell membrane turnover and reduction of NAA owing to destruction of neutrons, resulting in downsloping or reversal of the Hunter's angle. Prominent lactate peak can be encountered in necrotic areas of high-grade tumors and in radionecrosis. Usefulness of MRI in various sites is enlisted in Table 7.8.

The past decade has seen the aspiration of integrated MRI-guided radiotherapy (MRIgRT) being turned into a reality. New linacs which have integrated onboard MRI linacs have ushered a new era for image guidance [37]. These MRI linacs have the ability to also provide cine intra-treatment MRI images. Advantages and disad-

Table 7.8 Usefulness of MRI in various sites

Spine	For determining the extent of extradural soft tissue in metastatic disease, evaluation of leptomeningeal disease
Brain	Both in primary as well as metastatic disease for diagnosis as well as accurately defining the number and extent of the lesion(s)
Hepatobiliary	Triple-phase MRI is especially useful in primary and metastatic disease as well as pancreatic and biliary tumors
Pelvis	Especially useful in rectal, bladder, and uterine malignancy for diagnosis and treatment planning
Breast	Useful in younger patients for evaluation and characterization of lesions

Table 7.9 MRI-guided radiotherapy (MRIgRT)

Advantages	Disadvantages
Ability to have superior-quality MRI images	Cost of equipment, special RF cage placement during installation
Ability to have functional images (can possibly guide in biological imaging, biologically guided radiotherapy)	It cannot be the only machine in a center. It needs a more conventional LINAC as a workhorse machine. Increased treatment time per patient, more manpower needed at the console area
The ability for cine online intrafraction imaging, ability to do 4-dimensional treatments	The paucity of long-term data of superiority over more conventional treatments
	Concern regarding the interaction of magnetic fields with electrons and its implication on dose deliver
	Spatial distortion associated with MRI, especially toward the periphery of the field (object pixel may move 2–4 mm from the actual position)

Table 7.10 Imaging characteristics of some primary/metastatic sites

Lung cancer SCC (squamous cell carcinoma)	More often centrally located within the lung and may grow much larger than 4 cm in diameter. Cavitation is seen in up to 82%. Commonly causes segmental or lobar lung collapse due to their central location
Lung cancer SCLC (small cell lung carcinoma)	SCLC often present with bulky hila and mediastinal lymph node masses. A mass in or adjacent to the hilum is characteristic of SCLC and the tumor may well show mediastinal invasion
Lung cancer (adenocarcinoma)	Adenocarcinomas are typically peripherally located. They can present as a solid mass or ground glass opacity
Gallbladder carcinoma	CT: Mass lesions replacing the GB, focal or diffuse thickening of the GB wall, and intraluminal mass in the GB. Mass shows heterogeneous enhancement on postcontrast study MRI mass hypointense on T1-weighted images with heterogeneously high signal on T2-weighted images. The lesion often tends to infiltrate adjacent liver parenchyma and/or peritoneum
Cholangiocarcinoma	CT: Intrahepatic cholangiocarcinoma is invariably associated with dilatation of the intrahepatic biliary radicals. Other possible features include capsular retractions due to their fibrous nature. The lesion has arterial phase hyperenhancement followed by progressive enhancement on subsequent phases
Hepatocellular carcinoma (HCC)	CT: Best evaluated on multiphase CT or DCE MRI. Typical HCC is hyperenhancing compared to the liver parenchyma on arterial phase and washes out on subsequent phases. A late-enhancing peripheral capsule is also characteristic MRI: On hepatobiliary phase MRI, a hypointense appearing lesion would imply an advanced HCC
Liver metastasis	USG: Most metastases have a targetoid appearance on sonography CT: Metastasis usually replicates the morphology of the primary tumor. For example, neuroendocrine tumor metastases are hyperenhancing in the arterial phase. Some of the metastases, e.g., GI primaries, can demonstrate peripheral washout on triple-phase CT scan
Pancreatic cancer	USG: A hypoechoic mass, dilatation of the pancreatic duct, and dilatation of the bile duct are typical features of pancreatic head tumor CT: Hypodense mass on CT that is poorly marginated, which may encase vessels and the common bile duct
High-grade glioma/ glioblastoma	MRI: Necrosis and areas of blooming on T2/SWI consistent with hemorrhage, associated with extensive perilesional edema, suggest a high-grade neoplasm, like glioblastoma. Other features include irregular thick rind of enhancement on postcontrast T1 imaging in a necrotic neoplasm is also a feature of high-grade tumor
Low-grade glioma/ oligodendroglioma	MRI: Often nonenhancing on contrast imaging. Characteristic homogeneous hyperintense signal on T2-weighted images, with suppression of the high signal on FLAIR sequence (T2-FLAIR mismatch)
CNS metastasis	MRI: Enhance on contrast imaging, often located at gray/white junction. Often tend to be multiple and bilateral and are associated with florid perilesional edema

vantages of MRI-guided radiotherapy are enlisted in Table 7.9.

The two main clinical drivers for the adoption of MRIgRT are the ability to image anatomical changes that occur before and during treatment in order to adapt the treatment approach, and to image and target the biological features of each tumor [38]. Imaging characteristics of some primary/metastatic sites are enlisted in Table 7.10.

7.8 Fluorescence-Guided Surgery

Fluorescence-guided surgery has been currently used for multiple surgical situations, including sentinel lymph node (SLN) mapping, identification of solid tumors, lymphography, angiography, and anatomical imaging during surgery.

This technique has emerged over the past decade as a promising and effective method of intraoperative cancer detection. An impressive number of fluorescently labeled antibodies, peptides, particles, and other molecules related to cancer hallmarks have been developed to illuminate target lesions. New approaches are being implemented to translate these imaging agents into the clinic, although only a few have made it past early-phase clinical trials [39].

5-ALA is a well-studied intraoperative aid for the resection of high-grade gliomas, which helps differentiate tumors from surrounding brain parenchyma. Tumor fluorescence derived from 5-aminolevulinic acid enables more complete resections of contrast-enhancing tumor, leading to improved progression-free survival in patients with malignant glioma [40].

The sentinel node is the first lymph node met by lymphatic vessels draining a tumor [41]. It is extrapolated that the absence of tumor cells in the SLN could indicate the absence of metastatic disease in other local nodes. The basic gain from SLN assessment is that extensive node dissection surgery can be avoided if the sentinel node is identified and found to be free of tumor cells. Intraoperative sentinel lymph node imaging can be performed using a hand-held X-ray detector. Radiolymphoscintigraphy confirms the location of the SLN, which is determined initially with a preoperative lymphoscintigram and intraoperative vital blue dye injection. The combination of the isotope and blue dye has a complementary effect in sentinel node localization.

7.9 Bone Scan/Bone Scintigraphy

The bone scan continues to have the most common use in oncology because of its good sensitivity and relatively low cost [42]. This modality is of special use in patients with primaries that have the propensity to spread to the bone (breast, prostate). Technetium-based radiopharmaceuticals such as 99mTc-MDP, 99mTc-MIBI, and 99mTc(V)-DMSA is used to detect metastases. With increasing access of patients to PET/CT, this investigation is less frequently used as compared to some years ago.

7.10 Radiomics

Radiomics refers to the extraction of mineable data from medical imaging. Radiomics is to radiological images as genomics is to DNA, transcriptomics to RNA, proteomics to proteins and metabolomics to metabolites. This concept has been applied within oncology (1) to improve diagnosis, (2) for prognostication, (3) for prediction of behavior of disease and (4) for alteration of decision during treatment course. Multiplicity of assessable features is an important feature in radiomics. In a study, 35 CT-based radiomic features were identified to be predictive of distant metastasis of lung cancer [43].

Some authors have shown that certain radiomic features are associated with biological gene sets, such as cell cycle phase, DNA recombination and regulation of immune system process [44]. For example, radiomics can also be indicative of various mutations of glioblastoma (GBM), such as 1p/19q deletion, MGMT methylation, TP53, EGFR, and NF1. The advantage of radiomics is its noninvasive nature and can therefore be repeated prospectively for a given patient more easily than invasive tumor biopsies. It has been suggested that radiomics could be the means to monitor dynamic tumor changes along the course of radiotherapy and to define subvolumes at risk for which dose escalation could be beneficial.

Current methods in radiomics are limited to using single images for the extraction of these textural features and may limit the applicable scope of radiomics in different clinical settings [45]. Thus, in their current form, they are not capable of capturing the true underlying tissue characteristics in high-dimensional multiparametric imaging space. Multiparametric radiological imaging is an evolving tool that is vital for the detection, characterization, and diagnosis of many different diseases and is an active area of research.

Issues with radiomics include systematic bias, nonstandardization in design of trials, inconsis-

Table 7.11 Tumor recurrence versus radiation necrosis (in brain)

Imaging modality	Tumor recurrence	Radiation necrosis
DWI/DTI	Diffusion restriction Lower ADC values High FA	No diffusion restriction Higher ADC values Low FA
DCE-MRI	Elevated CBV (usually >2.0 mL/100 g)	Decreased CBV (usually <2.0 mL/100 g)
MRS	Higher Cho/Cr ratio	Increased lactate/Cr ratio Decreased Cho/Cr ratio
PET/SPECT	Increased radiotracer uptake	No or minimal radiotracer uptake

FA fractional anisotropy, *ADC* apparent diffusion coefficient, *CBV* cerebral blood volume, *NAA* N-acetylaspartate (NAA), *Cho* choline, *Cr* creatine

tency in data generated, and paucity of collaborative research.

7.11 Post-RT/Posttreatment Imaging

Certain changes are expected on imaging in the post-RT status, and these must not be flagged as pathology. It is essential for radiologists and oncologists to be wary of these findings [46]. One of the critical issues is distinguishing tumor progression from posttreatment changes. Despite our state-of-the-art imaging techniques, serial imaging is often the most helpful and reliable noninvasive method to assess disease activity.

In the context of SBRT lung, it may sometimes become exceedingly difficulty in follow up imaging to differentiate recurrent disease from post-SBRT treatment changes [47]. To complicate the issue further, there is continued evolution of radiation fibrosis for more than 2 years. The high-risk CT features predictive for recurrence that have been described are (a) an enlarging opacity at the site of the treated primary neoplasm, (b) a sequentially enlarging opacity, (c) loss of linear margin, (d) a convex bulging margin of the radiation fibrosis, (e) disappearance of previously depicted air bronchograms, and (f) an enlarging opacity without air bronchograms (particularly after 12 months [48].

In MRI brain imaging, postradiotherapy changes may be seen as increased perilesional T2 and Flair high signal intensity and must not be mistaken for disease progression. Similarly, post-RT necrosis is seen as characteristic "swiss-cheese" appearance on postcontrast enhancement along the margins of the lesion.

On T2W imaging, there is an internal mottled low signal intensity area on representing necrosis. MR perfusion sometimes is helpful in differentiating necrosis from recurrent disease. Table 7.11 enlists imaging to differentiate tumor recurrence versus radiation necrosis (in brain). Increasing perfusion change with raised cerebral blood volume and flow in the area is more likely indicative of progressive disease [49].

An increase of radiological abnormalities that are indistinguishable from but do not constitute true tumor progression is called pseudoprogression (Fig. 7.18).

This phenomenon is an inflammatory response to chemoradiotherapy and is marked by a transient increase in enhancement and edema. Pseudoprogression is especially more frequent in patients with tumors harboring O6-methylguanine DNA methyltransferase (MGMT) promoter methylation. Features suggesting true progression are (1) enhancement outside the radiation field (2) extension across the corpus callous in tumor receiving focal radiotherapy, and (3) Subependymal enhancement. Perfusion techniques usually reveal a reduction in blood volume in areas with pseudoprogression, compared to progressive disease [50, 51].

Pseudoresponse represents a transient decrease in contrast enhancement on MRI related to diminished leakiness of the BBB following treatment. This phenomenon typically happens with antiangiogenic agents, most commonly bevacizumab. The classical feature of this finding is its short duration. Hence, if the reduction in

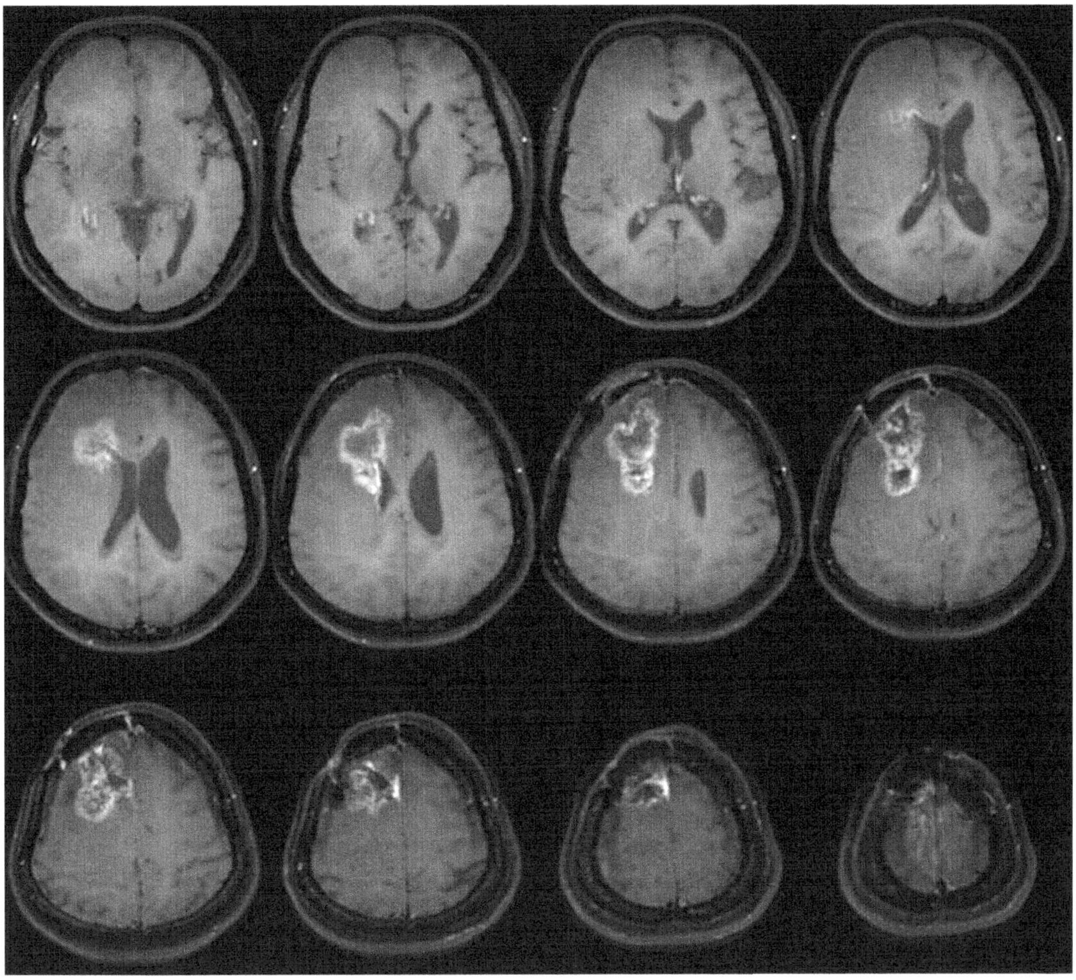

Fig. 7.18 Pseudoprogression in a case of glioma, post-RT, and temozolomide

enhancement and T2/FLAIR signal persists for >4 weeks, it should be considered as a true response.

7.12 Issues of Low-Middle-Income Countries (LMIC)

While imaging technology has evolved manifold, there are multiple challenges to the delivery of optimal cancer imaging in the developing world [52]. The most common issues are shortages of both quality imaging equipment as well as qualified general radiologists. Further issues include poor infrastructure, cultural barriers, and other obstacles. Training opportunities too are lacking.

One possible solution to the problem is remote radiology, also known as teleradiology [53]. Fortunately, the growth of the Internet has made teleradiology possible in many areas otherwise lacking quality radiologists. Also, teleradiology has enabled quick second opinions and follow-up

Summary (Precepts Box)

1. Imaging techniques have become an element in screening, determining stage, guiding cancer treatment, assessment of treatment response, and surveillance.
2. Variety of functional and anatomical imaging is available in clinical practice.
3. Mammography is oldest screening tool for breast cancer and also aids in diagnosis.
4. Image guidance tools such as CBCT have become crucial in radiation therapy. Advances include MR-guided RT and surface-guided radiotherapy.
5. Tools such as radiomics and artificial intelligence-based algorithms for imaging are likely to become increasingly important in cancer management.

assessments for the scans of cancer patients. Through teleradiology (video conferencing), radiologists can also participate in multidisciplinary tumor boards.

To summarize, imaging plays a crucial role in screening, diagnosis, staging, monitoring, and follow-up of cancer patients. Newer imaging modalities have added a new dimension to oncology imaging. Tools such as radiomics are likely to be way forward in tumor assessment and outcomes.

Multiple Choice Questions

1. What imaging modality is used for screening breast cancer?
   (a) Low-dose CT
   (b) Mammography
   (c) MRI breast
   (d) PET/CT
2. Which imaging modality does not use ionizing radiation?
   (a) Mammography
   (b) X-ray
   (c) Ultrasound
   (d) CT scan
3. What is advantage of X-rays?
   (a) Inexpensive
   (b) Negligible radiation exposure
   (c) High sensitivity
   (d) Measures atissue physiology
4. What are advantages of whole body PET/CT?
   (a) Whole body scan
   (b) Usually integrated with CT for lesion localization
   (c) Versions integrated with MRI available for even greater soft tissue localization
   (d) All the above
5. BIRADS stands for
   (a) Breast Imaging Recording and Data Systems
   (b) Breast Imaging Reporting and Description Systems
   (c) Breast Imaging Reporting and Data Systems
   (d) Bowel Imaging Recording and Data Systems
6. Which type of ultrasound is used for staging of minimally invasive NSCLC?
   (a) TRUS
   (b) EBUS
   (c) TVS
   (d) USG abdomen
7. What are modern uses of CT?
   (a) CT-guided intervention
   (b) CBCT
   (c) Four-dimensional CT
   (d) All the above
8. What is synthetic CT?
   (a) A synthetic CT or pseudo CT image is a CT image generated from an MR image instead of using a CT scanner to scan one
   (b) CT images generated on PET/CT machine
   (c) CT images generated through CT scanner
   (d) None of the above
9. Concept of radiomics is used
   (a) to improve diagnosis
   (b) in prognostication
   (c) for prediction of behavior of disease
   (d) All the above

Answers

1. (b)
2. (c)
3. (a)
4. (d)
5. (c)
6. (b)
7. (d)
8. (a)
9. (d)

References

1. Morone M, Bali MA, Tunariu N, Messiou C, Blackledge M, Grazioli L, et al. Whole-body MRI: current applications in oncology. AJR Am J Roentgenol. 2017;209:W336–49.
2. Sotoudeh H, Sharma A, Fowler KJ, McConathy J, Dehdashti F. Clinical application of PET/MRI in oncology. J Magn Reson Imaging. 2016;44:265–76.
3. Pollack EB, Mollura DR, Chong A, Harvey SC, Scheel JR. A global perspective on screening. J Breast Imaging. 2020;2:296–303.
4. Houben IPL, Vanwetswinkel S, Kalia V, Thywissen T, Nelemans PJ, Heuts EM, et al. Contrast-enhanced spectral mammography in the evaluation of breast suspicious calcifications: diagnostic accuracy and impact on surgical management. Acta Radiol. 2019;60:1110–7.
5. Kulkarni S, Singh N, Crystal P. Preoperative breast magnetic resonance imaging: applications in clinical practice. Can Assoc Radiol J. 2012;63(3):207–14.
6. Spak DA, Plaxco JS, Santiago L, Dryden MJ, Dogan BE. BI-RADS® fifth edition: a summary of changes. Diagn Interv Imaging. 2017;98:179–90.
7. Madjar H. Role of breast ultrasound for the detection and differentiation of breast lesions. Breast Care. 2010;5:109–14.
8. Alenezi AO, Krishna S, Mendiratta-Lconala M, Kielar AZ. Imaging and management of liver cancer. Semin Ultrasound CT MR. 2020;41:122–38.
9. Hari AY. An ode to ultrasonography: the window to the pelvis. Obstet Gynecol. 2018;131:732–3.
10. Folkman J. Role of angiogenesis in tumor growth and metastasis. Semin Oncol. 2002;29(29):15–8.
11. Anderson CR, Rychak JJ, Backer M, Backer J, Ley K, Klibanov AL. scVEGF microbubble ultrasound contrast agents: a novel probe for ultrasound molecular imaging of tumor angiogenesis. Invest Radiol. 2010;45:579–85.
12. Wu W, Chen MH, Yin SS, Yan K, Fan ZH, Yang W, et al. The role of contrast-enhanced sonography of focal liver lesions before the percutaneous biopsy. AJR Am J Roentgenol. 2006;187:752–761.1.
13. Fielding D, Bashirzadeh F, Nguyen P, Hodgson A, James D. Review of the role of EBUS-TBNA for the pulmonologist, including lung cancer staging. Thorac Cancer. 2010;1:44–52.
14. https://kevalnagda.github.io/ct-windowing. Accessed 15 Dec 2022.
15. Purandare NC, Rangarajan V. Imaging of lung cancer: implications on staging and management. Indian J Radiol Imaging. 2015;25:109–20.
16. Burk KS, Lo GC, Gee MS, Sahani D. Imaging and screening of pancreatic cancer. Radiol Clin. 2017;55:1223–34.
17. Al-Hawary MM, Francis IR, Chari ST, Fishman EK, Hough DM, Lu DS, et al. Pancreatic ductal adenocarcinoma radiology reporting template: consensus statement of the Society of Abdominal Radiology and the American Pancreatic Association. Radiology. 2014;270:248–60.
18. Ghonge NP, Chowdhury V. Minimum-intensity projection images in high-resolution computed tomography lung: technology update. Lung India. 2018;35:439–40.
19. Kim J, Glide-Hurst C, Doemer A, Wen N, Movsas B, Chetty IJ. Implementation of a novel algorithm for generating synthetic CT images from magnetic resonance imaging data sets for prostate cancer radiation therapy. Int J Radiat Oncol Biol Phys. 2015;91:39–47.
20. Andreasen D, Van Leemput K, Hansen RH, Andersen JA, Edmund JM. Patch-based generation of a pseudo-CT from conventional MRI sequences for MRI-only radiotherapy of the brain. Med Phys. 2015;42:1596–605.
21. Fonti R, Conson M, Del Vecchio S. PET/CT in radiation oncology. Semin Oncol. 2019;46:202–9.
22. An H, Gu Y, Yu H, Hu P, Zhang Y, Mao W, Shi H. Total-body PET/CT: current applications and future perspectives. AJR Am J Roentgenol. 2020;215:325–37.
23. Warburg O, Wind F, Negelein E. The metabolism of tumors in the body. J Gen Physiol. 1927;8:519–30.
24. Beyer T, Townsend DW, Blodgett TM. Dual-modality PET/CT tomography for clinical oncology. Q J Nucl Med. 2002;46:24–34.
25. Smoragiewicz M, Laskin J, Wilson D, Ramsden K, Yee J, Lam S, et al. Using PET-CT to reduce futile thoracotomy rates in non-small-cell lung cancer: a population-based review. Curr Oncol. 2014;21:e768–74.
26. Biehl KJ, Kong FM, Dehdashti F. 18F-FDG PET definition of gross tumor volume for radiotherapy of non–small cell lung cancer: is a single standardized uptake value threshold approach appropriate? J Nucl Med. 2006;47:1808–12.
27. Paulino AC, Johnstone PAS. FDG PET in radiotherapy planning: Pandora's box. Int J Radiat Oncol Biol Phys. 2004;59:4–5.
28. Grosu AL, Piert M, Weber WA, Jeremic B, Picchio M, Schratzenstaller U, Zimmermann FB, Schwaiger

M, Molls M. Positron emission tomography for radiation treatment planning. Strahlenther Onkol. 2005;181:483–99.
29. Surti S, Karp JS, Popescu LM, et al. Investigation of time-of-flight benefit for fully 3-D PET. IEEE Trans Med Imaging. 2006;25:529–38.
30. Mann RM, Cho N, Moy L. Breast MRI: state of the art. Radiology. 2019;292:520–36.
31. Smits M. Update on neuroimaging in brain tumours. Curr Opin Neurol. 2021;34:497–504.
32. Winfield JM, Payne GS, deSouza NM. Functional MRI and CT biomarkers in oncology. Eur J Nucl Med Mol Imaging. 2015;42:562–78.
33. Baehring JM, Fulbright RK. Diffusion-weighted MRI in neuro-oncology. CNS Oncol. 2012;1:155–67.
34. Lope-Piedrafita S. Diffusion tensor imaging (DTI). Methods Mol Biol. 2018;1718:103–16.
35. Günther M. Perfusion imaging. J Magn Reson Imaging. 2014;40:269–79.
36. Rhodes CJ. Magnetic resonance spectroscopy. Sci Prog. 2017;100:241–92.
37. Lagendijk JJ, Raaymakers BW, Raaijmakers AJ, Overweg J, Brown KJ, Kerkhof EM, et al. MRI/linac integration. Radiother Oncol. 2008;86:25–9.
38. Keall PJ, Brighi C, Glide-Hurst C, Liney G, Liu PZY, Lydiard S, Paganelli C, et al. Integrated MRI-guided radiotherapy—opportunities and challenges. Nat Rev Clin Oncol. 2022;19:458–70.
39. Mieog JSD, Achterberg FB, Zlitni A, Hutteman M, Burggraaf J, Swijnenburg RJ, . et al. Fundamentals and developments in fluorescence-guided cancer surgery. Nat Rev Clin Oncol 2022;19, 9–22.
40. Stummer W, Pichlmeier U, Meinel T, Wiestler OD, Zanella F, Reulen HJ, ALA-Glioma Study Group. Fluorescence-guided surgery with 5-aminolevulinic acid for resection of malignant glioma: a randomised controlled multicentre phase III trial. Lancet Oncol. 2006;7:392–401.
41. Mariani G, Moresco L, Viale G, Villa G, Bagnasco M, Canavese G, et al. Radioguided sentinel lymph node biopsy in breast cancer surgery. J Nucl Med. 2001;42:1198–215.
42. Nakajima K, Edenbrandt L, Mizokami A. Bone scan index: a new biomarker of bone metastasis in patients with prostate cancer. Int J Urol. 2017;24:668–73.
43. Coroller TP, Grossmann P, Hou Y, Rios Velazquez E, Leijenaar RT, Hermann G, et al. CT-based radiomic signature predicts distant metastasis in lung adenocarcinoma. Radiother Oncol. 2015;114:345–50.
44. Aerts HJ, Velazquez ER, Leijenaar RT, Parmar C, Grossmann P, Carvalho S, et al. Decoding tumour phenotype by noninvasive imaging using a quantitative radiomics approach. Nat Commun. 2014;5:4006.
45. Lambin P, Rios-Velazquez E, Leijenaar R, Carvalho S, van Stiphout RG, Granton P, et al. Radiomics: extracting more information from medical images using advanced feature analysis. Eur J Cancer. 2012;48:441–6.
46. Mattonen SA, Ward AD, Palma DA. Pulmonary imaging after stereotactic radiotherapy-does RECIST still apply? Br J Radiol. 2016;89:20160113.
47. Al-Umairi R, Tarique U, Moineddin R, Jimenez-Juan L, Kha LC, Cheung P, Oikonomou A. CT patterns and serial CT changes in lung cancer patients post stereotactic body radiotherapy (SBRT). Cancer Imaging. 2022;16(22):51.
48. Kato S, Nambu A, Onishi H, Saito A, Kuriyama K, Komiyama T, Marino K, et al. Computed tomography appearances of local recurrence after stereotactic body radiation therapy for stage I non-small-cell lung carcinoma. Jpn J Radiol. 2010;28:259–65.
49. Nichelli L, Casagranda S. Current emerging MRI tools for radionecrosis and pseudoprogression diagnosis. Curr Opin Oncol. 2021;33:597–607.
50. Ellingson BM, Chung C, Pope WB, Boxerman JL, Kaufmann TJ. Pseudoprogression, radionecrosis, inflammation or true tumor progression? Challenges associated with glioblastoma response assessment in an evolving therapeutic landscape. J Neurooncol. 2017;134:495–504.
51. Strauss SB, Meng A, Ebani EJ, Chiang GC. Imaging glioblastoma posttreatment: progression, pseudoprogression, pseudoresponse, radiation necrosis. Neuroimaging Clin N Am. 2021;31:103–20.
52. Pramesh CS, Chaturvedi H, Reddy VA, Saikia T, Ghoshal S, Pandit M, Babu KG, Ganpathy KV, Savant D, Mitera G, Sullivan R, Booth CM, National Cancer Grid. Choosing wisely India: ten low-value or harmful practices that should be avoided in cancer care. Lancet Oncol. 2019;20:e218–23.
53. Rackimuthu S, Hasan MM, Ray I, Sahito AM, Chawla P, Ghosh D. Teleradiology in India during the COVID-19 pandemic: merits, pitfalls and future perspectives. Health Policy Plan. 2022;37:1203–6.

Further Reading

https://www.coursera.org/lecture/cancer/types-of-imaging-CHcJP (Types of imaging).
https://www.youtube.com/watch?v=zQTw_PVzdOQ (mammogram technique).
https://www.youtube.com/watch?v=mOt2FeGHjaY (MRI brain sequences).
https://www.youtube.com/watch?v=gT0zV0XYKek (introduction to radiology).
https://www.youtube.com/watch?v=KZld-5W99cI (window, level, density).
https://www.youtube.com/watch?v=GHLBcCv4rqk (working of PET CT).

8 Principles of Surgical Oncology

Nita Nair, Abhay Kattepur, Shraddha Patkar, Prasanth Penumadu, Caleb Harris, and S. V. S. Deo

8.1 Introduction

Surgical oncology is a specialty that deals with the surgical resection of tumors. Oncological surgery is now recognized as a specialized field that needs dedicated training as significant changes have occurred in the past few decades. Surgery, one of the oldest modalities used to treat solid tumors, is now supplemented with adjuvant and neoadjuvant therapies with better outcomes. Oncological surgery includes the guiding principles in surgical management, improved perioperative care, expedited recovery, minimally invasive techniques, reconstruction and rehabilitation. There has also been a paradigm shift toward de-escalation in surgery by encouraging conservative/organ preservative and minimally invasive surgery to decrease morbidity without compromising disease outcomes.

Oncologic surgery is a vast topic, and its complete description is possibly beyond the purview of a single chapter. This chapter only focuses on outlining the fundamental principles of surgical oncology in the management of solid tumors.

8.2 Fundamental Surgical Principles

8.2.1 Diagnostic Biopsy in Solid Cancers

Despite advanced imaging techniques, a pathological biopsy remains the gold standard for diagnosing solid cancers. In most cases, diagnosis is made by lesser invasive procedures like a fine-needle aspiration cytology (FNAC), which can be clinician performed or image-guided. Biopsies are now increasingly recommended to diagnose solid tumors and further molecular

The original version of the chapter has been revised. A correction to this chapter can be found at https://doi.org/10.1007/978-981-97-8389-2_28

N. Nair (✉)
Consultant Breast Surgeon, Apollo Hospital, Navi Mumbai, India

A. Kattepur
Department of Surgical Oncology, Sri Devaraj Urs Academy of Higher Education and Research, Kolar, Karnataka, India

S. Patkar
Department of Surgical Oncology, Tata Memorial Centre, Homi Bhabha National Institute, Mumbai, Maharashtra, India

P. Penumadu
Medical Director and Head of the Department, Surgical Oncology, Sri Venkateswara Institute of Cancer Care and Advanced Research, Tirupati, India

C. Harris
Surgical Oncology, NEIGRIHMS, Shillong, Meghalaya, India

S. V. S. Deo
Department of Surgical Oncology, Indraprastha Apollo Hospital, New Delhi, India

A. Munshi et al. (eds.), *Radiation Oncology – Principles, Precepts and Practice*, https://doi.org/10.1007/978-981-97-8389-2_8

markers that impact treatment algorithms. The main goal of a biopsy should be help diagnose with minimal morbidity to the patient.

8.2.1.1 Types of Biopsies

1. Core needle biopsy: A small core of tissue is removed from the mass, using special needles (like J needle or Jamshidi needle for bone biopsies) to help characterize the tumors better. For example, core biopsy of breast cancers to perform immunohistochemistry for biomarkers (ER/PR/HER2 neu) for therapy. These biopsies maybe performed with image guidance in cases of impalpable or inaccessible tumors.
2. Endoscopic biopsies: These biopsies are performed for tumors within luminal organs. For example, colonoscopic biopsy, gastric biopsy.
3. Open excision/incision biopsy: (Table 8.1) Needle biopsies always may not help and occasionally require either an excision or incision biopsy to achieve a diagnosis. For example, complete excision of lymph node for diagnosis of lymphomas, or excision biopsy of oral leukoplakia and when the diagnosis is inconclusive on FNAC.

8.2.1.2 Principles of Performing Biopsies in Solid Tumors

The technique and principles of performing a biopsy vary according to the primary disease site. However, the following principles should be followed.

- Chose the least invasive option whenever feasible.
- If FNAC is inconclusive or when better characterization of tumors is needed, a biopsy is performed.
- Discrete lymph nodes in accessible locations are removed as a complete node to get adequate tissue for diagnosis.
- Core biopsy has increasingly become the method of choice.
- Use image guidance when available, especially for necrotic tumours.
- Plan incisions of biopsies in a way that these incisions can be excised within the scar of the proposed surgery.

8.2.2 Definitive Surgery

8.2.2.1 Preventive/Prophylactic Surgery

Preventive or prophylactic or risk-reducing surgery is defined as the surgical removal of an organ prior to malignant transformation or when the cancer is still in situ. It generally involves the removal of the organ at risk so that future cancer can be prevented. The most common indication is in the presence of a genetic mutation with high penetrance (Table 8.2).

Guidelines for Advocating Prophylactic Surgery

1. The genetic mutation involved in the disease process must have high penetrance.
2. There should be a reliable test available to identify patients with these mutations.
3. Removal of the organ must be associated with low morbidity and there should be replacement options available.

Table 8.1 Types of tissue biopsy

	Fine-needle aspiration biopsy	Core biopsy	Open biopsy
Invasiveness	Noninvasive	Noninvasive or invasive	Invasive
Needle size, caliber	24–25 G, hollow needle attached to a syringe	16–18 G needle	NA
Sample obtained	Small amount of sample	Core of tissue	Chunk of tissue bigger than core
Setting	OPD	OPD	OPD /OT
Anesthesia	None	LA	LA/regional/ GA
Bleeding risk	Minimal	Minimal	Sometimes appreciable
Tissue analysis	Cytological	Histological	Histological
Results	Same or next day	2–3 days	2–3 days

Table 8.2 Prophylactic cancer surgeries (reproduced from Ref. [1])

Condition	Cancer	Prophylactic surgery
Cryptorchidism	Testicular	Orchiopexy
Adenomatous polyposis coli	Colon	Total colectomy
Familial colon cancer	Colon	Colectomy
Ulcerative colitis	Colon	Colectomy
MEN syndrome 2	Thyroid (medullary)	Total thyroidectomy
Familial breast cancer	Breast	Mastectomy
Familial ovarian cancer	Ovary	Salpingo-oophorectomy
Hereditary diffuse gastric cancer	Diffuse gastric cancer	Total gastrectomy
Lynch syndrome	Endometrium, colon	Hysterectomy and colectomy

4. There should be a reliable method available that determines over time that the patient has been cured.

Examples of Prophylactic Surgeries

(a) Multiple endocrine neoplasia (MEN) syndrome [2, 3] is characterized by two or more endocrine tumors with an autosomal dominant inheritance pattern. Patients with MEN2 syndrome have an increased risk of medullary thyroid cancers requiring a prophylactic total thyroidectomy at an early age.
(b) BRCA that include BRCA 1 and BRCA 2 mutations is associated with the Hereditary Breast and Ovarian Cancer syndrome (HBOC) characterized by an increased risk of breast and ovarian cancers in affected individuals [4] warranting prophylactic mastectomy and/or salpingo-oophorectomy.
(c) Familial adenomatous polyposis (FAP) syndrome is characterized by numerous polyps in the entire colon and often the rectum. The gene responsible is the APC gene on the chromosome 5q21. A total colectomy at an early age reduces the risk of developing colon cancer. These patients also have an increased risk of periampullary cancers, desmoids and CNS tumors [5].

8.2.2.2 Curative Surgery

Surgical removal of the tumor and the draining nodal basins is possibly the most effective method to cure patients with solid cancers confined to that organ. The resection involves the removal of the primary tumor with pathologically negative margins along with removal of the draining lymph nodes.

Principles of Resection of the Primary Cancer

Factors predicting recurrence and survival in oncologic surgery and guidelines for the extent of resection of primary tumors are variable for every subsite. Definition for adequate margins, need for nodal dissections, the extent of nodal dissection and reconstructive procedures vary significantly across every tumor subtype. Knowledge and understanding of all of the above factors are required before oncologic resection of the tumor. Describing principles of resection for every subsite of cancer is beyond the scope of this chapter; hence, we have highlighted some pertinent basic surgical principles.

- To account for the microscopic disease, the primary tumor should be excised in to, including adjacent normal tissues as margins. Resection of the tumor with negative margins is the most important principle of any oncological surgery, and all efforts should be made to achieve a pathologically adequate, cancer-free margin.
- Margins that are considered adequate have been defined for most solid tumors and the surgeon should have a prior knowledge of what constitutes a negative margin before embarking on surgery for that organ/tumor subtype.
- Excessive handling, rupture or spillage should be avoided.
- All effort must be made to spare important anatomic structures (to thus reduce morbidity) without compromising margins of resection.
- De-escalation of surgery like sentinel lymph node biopsy has been developed and validated

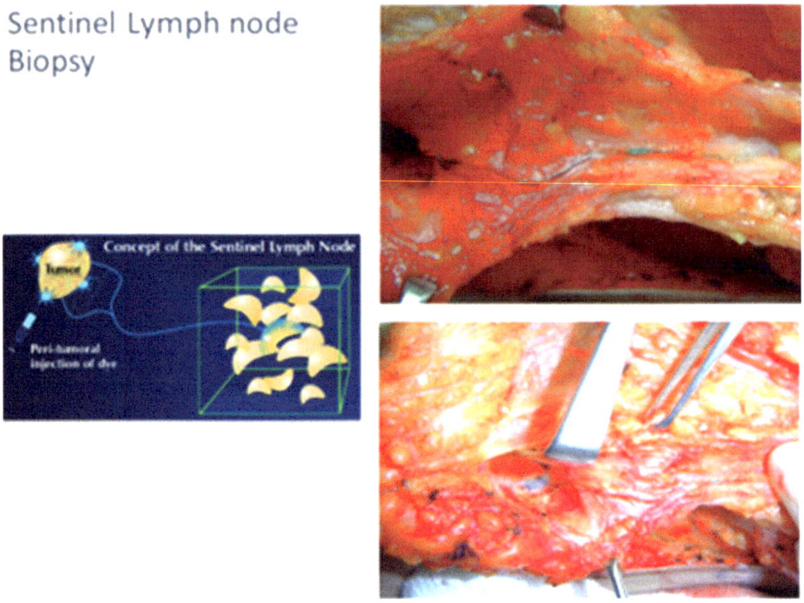

Fig. 8.1 De-escalation surgery, Sentinel node biopsy

in many tumors to reduce morbidity associated with dissection of the draining nodal basin. Sentinel lymph node is the first node which receives lymphatics from the organ it drains (sentinel means to guard) (Fig. 8.1). Cancerous growths within the organ will first involve the sentinel node before going to the remaining nodes. Therefore, if the sentinel node is found to be free of tumor, then it can be assumed that there is no disease spread to the remaining nodal basin. This reduces the need for complete nodal clearance when the sentinel node shows no disease, thus reducing morbidity associated with complete nodal clearance. The common sites for sentinel node mapping include breast, melanoma, penile, head and neck, vulva and endometrial cancers. The important prerequisite to do a sentinel node biopsy is that the node must be clinicoradiologically free of disease, i.e., cN0 status. It is important that the technique of sentinel node mapping must be validated before embarking on to assure accurate identification and low false negative rates. False negatives (when the sentinel node is negative but other nodes in the basin are positive) can be higher post neoadjuvant therapy or prior surgery in the affected organ, both of which tend to alter the normal lymphatic pathways. (The cut-off for acceptable false negative rate is ≤10%.)

- Intraoperative Radiotherapy (IORT): (Figs. 8.2, 8.3 and 8.4) Local relapse following surgical resection of solid tumor continues to affect the survival of cancer patients. During the last few decades the surgical pendulum has swung from ultra-radical resections to organ preserving conservative resections combined with radiotherapy, in the pursuit of cure for cancer. For most tumors, the likelihood of obtaining local control improves when increasing doses of radiation can be delivered. In many clinical situations the dose that can be safely delivered to tumor volume is limited by dose limiting normal tissues in close proximity to the tumor volume. Intraoperative radiotherapy (IORT) is a method where a single dose of radiation is delivered to a surgically exposed tumor bed while protecting the normal tissues. Initially, IORT was often used as a sole modality of therapy in advanced tumors. A significant amount of morbidity was also encountered due to the large (25–40 Gy) doses of IORT used during those periods. These factors led to a decline in enthusiasm for IORT. However, with the better understanding of the tumor radiobiology, improvements in radiation delivery systems,

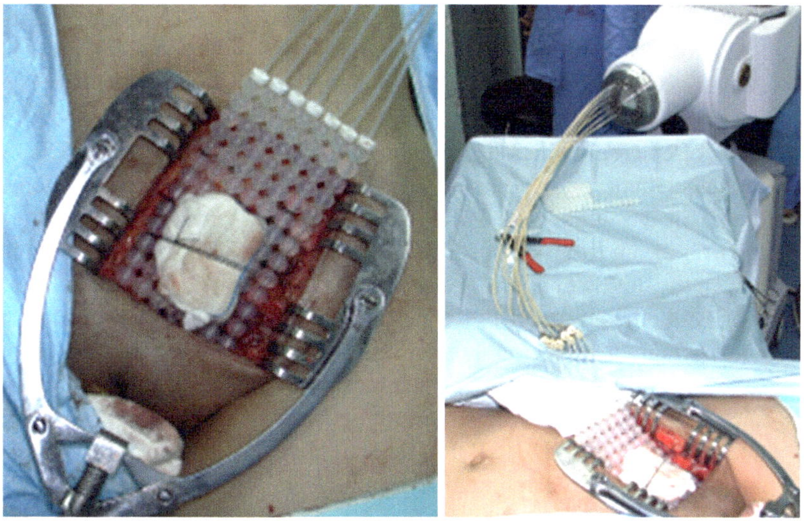

Fig. 8.2 Showing delivery of IORT (IOHDR) for a abdominal wall sarcoma

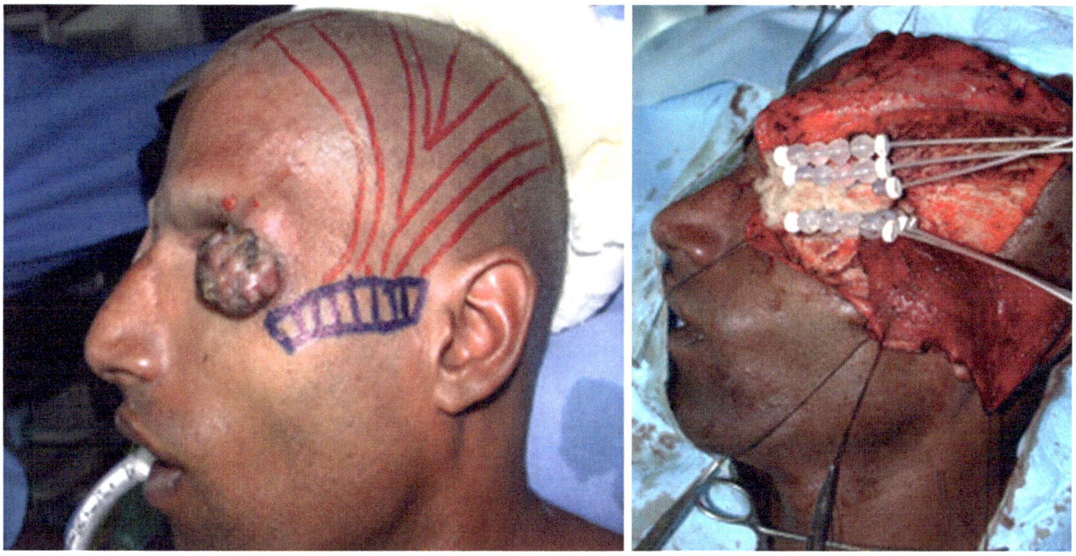

Fig. 8.3 Showing IORT for a recurrent orbital sarcoma following surgery and external beam radiotherapy

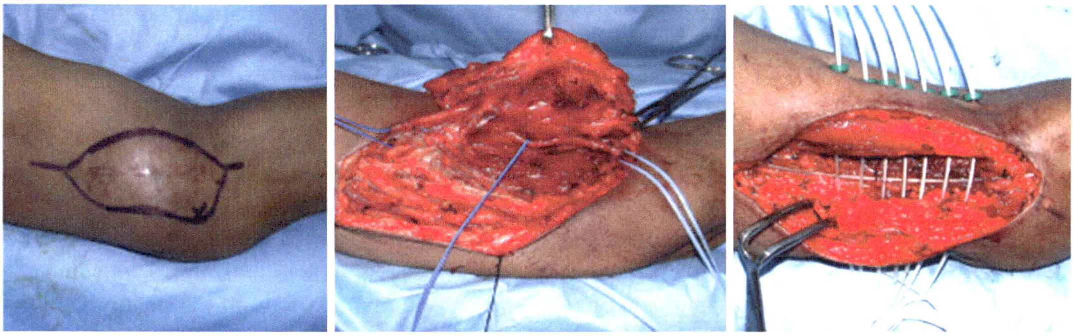

Fig. 8.4 Complex Limb salvage surgery along with brachytherapy for a recurrent extremity sarcoma

advent of high-dose rate radiotherapy, miniaturization and portability of radiotherapy delivery systems has led to a resurgence in the field of IORT. IORT can be delivered by a beam of external radiation from a mobile linear accelerator or by placing a radioactive source in the tumor bed. IORT delivered by electron beam from a linear accelerator is known as IOEBRT (intraoperative external beam radiotherapy) and IORT performed by placing high-dose rate gamma emitting radio isotope such as Ir-192 in the tumor bed is known as IOHDR (intraoperative high-dose rate radiotherapy) or HDR-IORT. Encouraging outcomes have been reported with IORT in advanced and recurrent head and neck cancers, genitourinary cancers, GI cancers, thoracic malignancies and sarcomas.

- HIPEC: Locoregional therapy for peritoneal metastases consists of CRS (cytoreductive surgery) followed by HIPEC (heated Intraperitoneal chemotherapy). In HIPEC, heated chemotherapy solution is circulated in the peritoneal cavity at a fixed flow rate for 30–120 min while maintaining an intra-abdominal temperature of 42–43 °C [1]. It is believed that the diffusion of chemotherapeutic agents has cytotoxic effect on the residual tumor deposits and that heat potentiates their cytotoxicity [2, 3]. HIPEC is not effective above a critical residual tumor size (ideally less than 2.5 mm) [4]. CRS and HIPEC is now the standard of care for treating pseudomyxoma peritonei arising from epithelial appendiceal tumors [5]. Durable control of ascites and survival can also be achieved with CRS and HIPEC in peritoneal mesothelioma, a rare entity. Epithelial subtype, lymph node status, completeness of cytoreduction and use of HIPEC were found to be prognostic factors in multivariate analyses [6]. In colorectal cancer with peritoneal metastasis, the role of CRS is established while that of HIPEC is debatable [7]. HIPEC has been used in prevention of gastric peritoneal metastasis in high-risk patients as well as in its treatment and palliation along with CRS [8]. The role of HIPEC in primary ovarian cancer is still debatable with few studies showing good results with HIPEC and others showing no benefit [9]. A benefit of CRS and HIPEC has been shown for recurrent ovarian cancer in retrospective and case control studies [10]. CRS and HIPEC can also be offered for peritoneal metastasis of mucinous ovarian tumors and neuroendocrine tumor based on the multi-institutional study by PSOGI and BIG-RENAPE groups [11]. Oxaliplatin and Mitomycin C form the backbone for the regimes for appendiceal and colorectal primaries [12], whereas cisplatin-based regimen is used for mesotheliomas and gastric and ovarian primaries [13]. With this treatment, selected patients experience significant prolongation in survival and some achieve cure [4]. The main drawbacks of CRS and HIPEC are the high morbidity associated with the procedure and the lack of level 1 evidence [14].

Minimally Invasive Surgery in Oncology (Fig. 8.5)

- Minimally invasive surgery (MIS) is slowly and steadily making its foray into oncology. Many oncological procedures are increasingly performed using minimally invasive surgery. It can be accomplished by two methods: laparoscopic or robotic, depending on the availability of expertise and equipment. These techniques benefit from lesser postoperative pain and early discharge without compromising the oncological safety. But they come at the expense of increased operative cost and time and the need for specialized training and expertise to perform the procedure. Most pelvic and thoracic surgeries can be performed via minimal invasive approach. The role of MIS in cervical cancer stands controversial and is definitely not indicated for tumor more than 4 cm as studies have shown detrimental effects using MIS in these patients [15].
- Robotic surgery: A surgical robot is a self-powered, computer-controlled device that can be programmed to aid in the positioning and manipulating surgical instruments, enabling the surgeon to carry out more complex tasks.

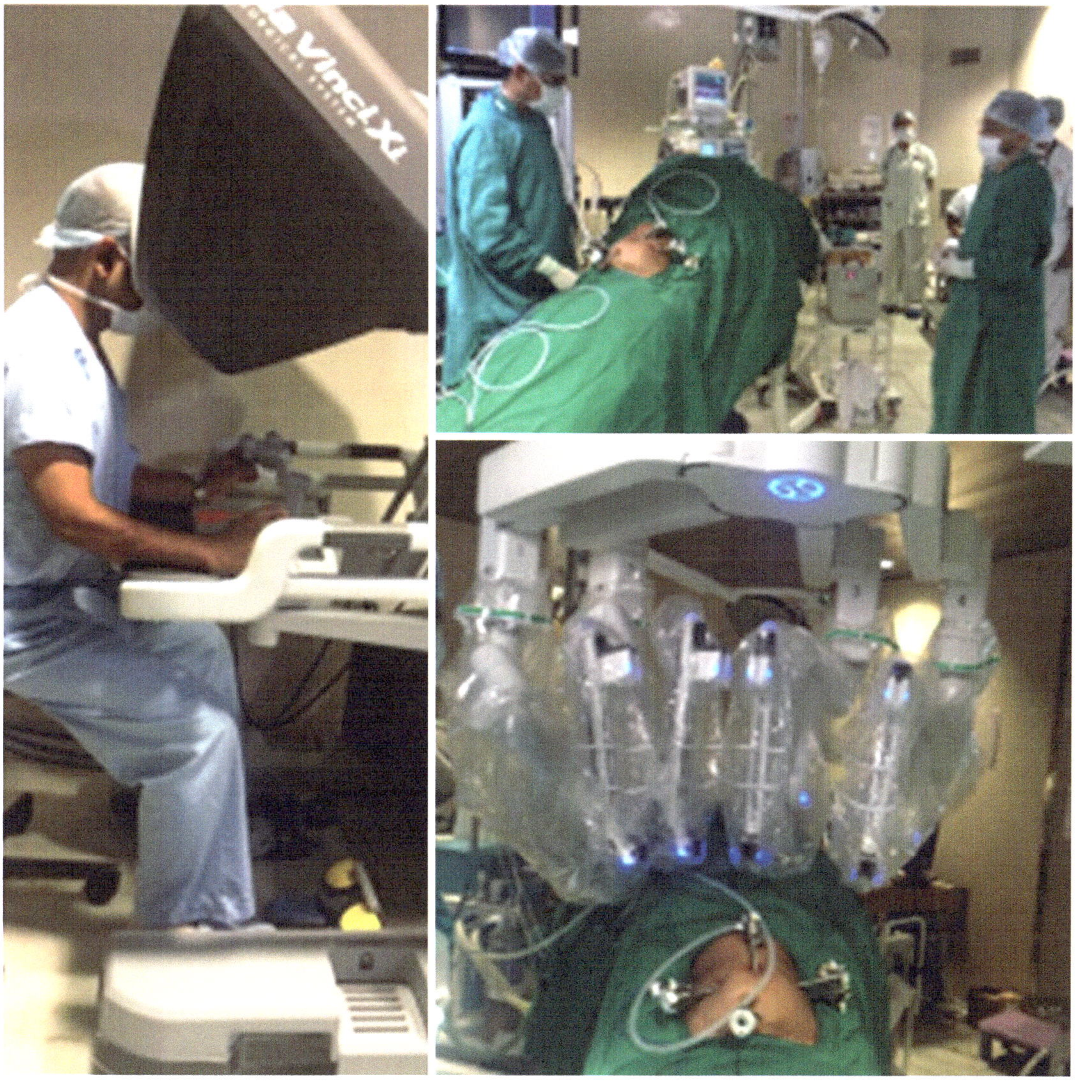

Fig. 8.5 Robotic surgery OT setup

The use of robots during MIS was first described in 1985 and the technology has evolved over the years. They are not intended to act independently from human surgeons or replace them. They are also called master-slave manipulators as their work is governed by surgeons controlling them. The two important FDA-approved robotic systems are the *da Vinci* surgical system (intuitive Surgical, Mountain View, California) and the ZEUS system (Computer Motion, Gelota, California).

Applications

- Urology: Radical prostatectomy for prostate cancer, radical cystectomy for bladder cancer and partial nephrectomy for renal masses.
- Gynecology: Radical hysterectomy for endometrial, cervical and ovarian cancers.
- Head and neck surgery: Trans-oral robotic surgery (TORS) for oropharyngeal lesions and thyroidectomy for thyroid cancers.
- General surgery: colorectal cancers, pancreatectomy for pancreatic cancers, adrenalec-

tomy for adrenal lesions, hepatectomy for liver lesions and gastrectomy for stomach cancers. Also, antireflux surgeries and cholecystectomies are being performed.
- Thoracic surgery: for lobectomy for lung lesions, esophagectomy for esophageal cancers and thymectomy.

Robotic Surgery and Surgical Education
Robotic surgery is a new medium to acquire surgical skills through simulated operations which can be done via the robot. Surgeons can use these robots to practice operations on 3-dimensional, virtual visual simulations and soft tissue models whose textures closely match the human tissues through haptic feedback. Image-guided simulations thereby allow surgeons to practice procedures on 3-dimensional reconstructions of the anatomy of actual patients who are being planned for surgery. Guidance can be offered via tele-monitoring. This can improve the learning curve, acquire surgical skill and reduce patient related errors (Table 8.3).

The Extent of Surgical Resection

The importance of classifying the extent of surgery stems from the fact that in most solid cancers, completeness of resection is the most important prognostic factor that predicts recurrence risk. These can be categorized as

- R0: complete removal of the tumor; no gross or microscopic residual disease.
- R1: complete removal of tumor with presence of microscopic residual disease.
- R2: incomplete tumor removal (gross macroscopic residual disease).

Patients undergoing R1 or R2 resections need additional adjuvant therapy to sterilize the margins to avoid local recurrences. However, those undergoing R2 resections have poor oncologic outcomes. For example, pancreatic cancer patients who undergo R2 resections show poor survival rates compared to those undergoing R0 resections [6–8]. However, in certain tumors like ovarian cancer, tumor debulking and leaving back minimal residual disease followed by adjuvant chemotherapy has shown acceptable results [9].

Margin status is an important prognostic factor. It implies presence of tumor cells at the resected margin. The risk of local recurrence increases when margin positivity occurs. Margin positivity can be either focal or diffuse. Positive margins often need revision surgery or adjuvant radiotherapy (with or without chemotherapy)

Table 8.3 Surgical techniques

	Open surgery	Laparoscopic surgery	Robotic surgery
Position of surgeon	Next to patient	Next to patient	At console
Cost	Variable, depends on the nature of surgery	Expensive	Expensive
Vision	3 dimensional	Traditionally 2 dimensional, although newer systems are 3 dimensional	3 dimensional
Ergonomics for surgeon	Good	Poor	Very good
Surgeon fatigue	Yes	Yes	Minimal/no
Operator Tremors	May be seen	Yes	No
Degrees of freedom	7	4	7
Operative time	least	Moderate	longest
Hospital stay	Longer	Shorter	Shorter
Postoperative pain	Substantial	Lesser	Lesser
Return to normal activity	Late	Early	Early
Recurring costs for hospital	Minimal	High	Very high
Haptic feedback	Yes	No	No
Learning curve	Variable	Higher than open	Higher than open

depending on the tumor subsite and other adverse prognostic factors. The amount of adjacent normal tissue needed to be incorporated so as to obtain a negative margin depends on the tumor site, histology and/or prior therapy. The general rule of the thumb is to obtain at least 1 cm of tumor free margin all around while operating. Lesser margins (more than 1 mm) are acceptable in breast and pancreatic cancers. Larger margins (>5–10 cm) are the standard in esophageal and gastric cancers where submucosal spread is common. Qualitative and molecular margins are some of the different ways to assess margin status. Qualitative margins include the presence of fascia, aponeurosis, vascular adventitia, joint capsule, periosteum, etc. at the resected end which equate to a 2–3 cm of margin depending on the thickness of these tissue. This concept is useful in soft tissue sarcoma surgeries. Molecular margins depict genetic changes at the adjacent "normal mucosa" such as p53 or EGFR mutations, suggesting that although histomorphologically the margins may be normal, genetic alterations present in them can serve as precursors for the development of new cancers in the adjacent epithelia. This is applicable in head and neck cancers. Radio-opaque clips are often placed at the tumor bed to mark the margins.

Curative Surgery Post Neoadjuvant Therapy (CT and or RT)

For patients receiving neoadjuvant therapy, either chemotherapy or chemoradiotherapy for locally advanced cancers, surgery would ultimately be needed for tumor extirpation. One must be careful to optimize these patients preoperatively (physiologically and nutritionally). It is also essential to time the surgery appropriately considering the type of neoadjuvant therapy given.

The important cancers for which neoadjuvant chemotherapy is administered include breast, ovary, lung, esophageal and gastric cancer subsites. Likewise, chemoradiotherapy in the preoperative setting is indicated for cancers of the lung, esophagus and rectum. The principles of surgery in these patients parallel those undergoing primary surgery with few caveats.

- The initial tumor stage determines the extent of resection at presentation. Special care must be taken to tackle so-called "adhesions" between the tumor and surrounding structures that could contain microscopic tumor deposits.
- Residual disease after surgery may warrant further treatment in the form of chemotherapy.

Surgery after neoadjuvant treatment is often associated with slightly higher morbidity when compared to surgery in the upfront setting. This is partly related to the treatment induced relative tissue anoxia, increase in fibrosis surrounding the tumor, lesser peritumoral vascularity, microvascular changes in the adjacent blood vessels and persistent effects on wound healing and nutrition.

Moreover, adhesions are more common after treatment especially radiotherapy. At times the tissues may be densely adhered to each other (e.g., frozen pelvis) which may prelude dissection of the affected organ from surrounding structures. This may be partly overcome by aligning to appropriate time frame after neoadjuvant therapy (commonly 3–4 weeks after chemotherapy and 6–8 weeks after radiotherapy) so that the tissues are conducive for surgery. Higher incidence of lymphorrhoea is also common after neoadjuvant therapy especially in the axillary and inguinal nodal basins.

8.2.2.3 Resection of Metastatic Disease

The most common sites of metastases from solid tumors include the bone, lungs, liver and brain. The presence of distant metastases suggests a poor prognosis for the patients. The mainstay of treatment in this setting is systemic chemotherapy. Surgery plays a role in a subset of patients and is commonly indicated for improving oncological outcomes or for palliation.

Surgical Resection of the Primary Tumor

The decision to resect a primary tumor which has already metastasized should be arrived at by a multidisciplinary team. It is important to understand the natural history of the disease, and

Table 8.4 Primary cancers which may be considered for resection even in the metastatic setting

Primary cancer	Characteristics of metastasis	Treatment for primary	Treatment for metastasis	Reference
Breast cancer	Oligo metastasis	Mastectomy/breast conservative surgery	Surgery/radiotherapy	[19]
	Disseminated metastasis	Role of surgery only palliative	Palliative RT/ systemic	[20]
Colorectal cancer	Unilobar liver metastasis	Colonic resection; anterior resection/ abdominoperineal resection (± Neoadjuvant CT/RT)	Surgery/ablative therapy	[21]
Non-small cell lung cancer	Oligo metastasis	Lobectomy	Cerebral metastasis excision+RT; adrenalectomy	[22]
Bone sarcoma	Limited pulmonary metastasis	Limb salvage surgery/amputation	Pulmonary metastatectomy	[23]
Renal cell carcinoma	Single major organ metastasis	Cytoreductive nephrectomy	Resection/systemic therapy only	[24]

only those cancers with acceptable survival following resection of the primary should be considered for surgery. It is important to note that in general, the metastases should be treatable with curative intent too (surgery/chemoradiation) (Table 8.4).

General principles of resection of metastatic disease

- Resection has better results for oligometastases rather than multiple metastases.
- Complete resection of the metastases should be deemed feasible prior to surgery.
- The primary cancer should be controlled or possible to resect at the time of the metastatectomy.
- Disease-free interval (DFI), proliferation index, etc. play an important role in determining outcomes after metastatectomy. Longer the DFI, better are the outcomes.
- Metachronous tumors have better outcomes than synchronous presentation.
- Patient and tumor-related factors should be taken into consideration prior to deciding on metastatectomy.

The resection of primary tumors which are symptomatic has been dealt with under Palliative Surgery (Sect. 8.2.2.5).

8.2.2.4 Reconstructive Surgery in Oncology

After removal of the tumor following principles of oncology, restoring cosmesis and function is important in surgeries involving the head and neck region especially. Functional rehabilitation after ablative surgery for the oral cavity and laryngopharynx often poses a big challenge. However, with improvements in microvascular surgery and prosthetic implants, reconstruction of complex defects is now feasible. The following principles are apt (Table 8.5):

- Preoperative planning along with an estimation of the functional and cosmetic deficit is mandatory.
- As far as possible, "like tissues should be replaced with like." For example, reconstruction after resection of the mandibular should be done using a bone flap rather than just a muscle-based flap in the ideal scenario.

8.2.2.5 Palliative Surgery

In such cases, surgery is used primarily for obtaining pain relief or restoration of function. The benefit of surgery is that it may improve the quality of life of some patients with advanced cancers. Examples include (1) palliative bypass for intestinal obstruction, (2) palliative mastec-

Table 8.5 Reconstructive options

	Skin grafts	Local flaps	Free flaps
Subtypes	Spilt thickness grafts Full thickness grafts	Axial flaps Random flaps	Myocutaneous Fasciocutaneous Osteomyocutaneous
Composition	Epidermis and variable thickness of dermis	Skin with underlying fascia +/− muscle	Skin + fascia +/− muscle +/− bone
Blood supply	Absent; depends on recipient site	Present	Present
Cosmesis	Poor especially partial thickness grafts	Better	Best texture and color match
Pressure dressing	Needed	Not needed	Not needed
Bridging defects	No	Yes	Yes
Expertise	No	No	Yes (microvascular anastomosis)
Time for harvest	Quick	Slower than grafts	Time consuming
Radiotherapy use on tissue	No	Yes	Yes
Recipient bed	Healthy, well-vascularized tissue	Not necessary	Not necessary

tomy for fungating/bleeding breast cancers, (3) removal of tumorous masses causing intractable pain, (4) resection of metastases causing impending fractures in weight bearing bones like the femur, tibia, etc., (5) decompression and fixation of spinal metastases causing neurological impairment, (6) resection of life threatening brain metastases, etc.

The potential risk of higher morbidity and mortality (in view of advanced cancer and poor performance status) following surgery in these patients must be carefully balanced with the benefit likely to be gained.

8.3 Assessment of Operative Risk

Operative risk comprises of mortality and major morbidity postsurgery. *Operative mortality is usually defined as mortality occurring within 30 days of any surgical procedure*. Morbidity and common complications of surgery are well documented for most oncologic procedures (Table 8.6). Various classification systems are available to grade complications postsurgery and validated across subsites. Examples include the ISGLS (International Study Group of Liver Surgery) for grading of posthepatectomy liver failure. However, the most commonly used system to grade complications is the Clavien and Dindo classification of surgical morbidity [12] (Table 8.7). This classification has been validated across many oncologic solid tumor surgeries [13, 14].

A precise knowledge of expected complications and the risk entailed in a surgical procedure is crucial to enable the patient and caregiver to make an informed choice balancing the risk and benefits of the surgery. Understanding quality of life issues postsurgery also influences decision making, especially with extensive morbid surgeries, e.g., surgery for oral cavity cancer.

8.3.1 Prediction of Operative Mortality

Operative risk assessment includes various components such as patient-related factors, performance status, surgical factors, anesthetic techniques etc. The American Society of Anesthesiologists (ASA) has developed a classification/grading of the physical fitness of patients undergoing surgery that correlates with risk of postoperative outcomes [16]. This is the commonest used system for risk prediction of postoperative mortality. Patients are graded from ASA I to V. The risk of mortality increases as the ASA grade increases (Table 8.8).

Other risk assessment systems have also been developed. For example, the POSSUM

Table 8.6 Definition and grading of complications

Group	Definition	Grades	
ISGLS	Deterioration (increase in INR and bilirubin levels) in the synthesis, excretion and detoxification functions of the liver after liver surgery (day 5 or later)	A	Impaired liver function
		B	Deviation from the expected postoperative course, no need for invasive support
		C	Multiple organ failure requiring invasive support

Table 8.7 Clavien–Dindo grading of postoperative complications

Grade	Definition
Grade I	Any deviation from normal postoperative course with no need for pharmacological, radiological, surgical or endoscopic interventions Drugs which can be used: antiemetics, antipyretics, analgesics, diuretics, electrolytes Also includes wounds which have been opened bedside for drainage
Grade II	Use of drugs (other than that listed in grade I), parenteral nutrition and blood and blood products
Grade IIIa	Require surgical/endoscopic/radiological intervention not under GA
Grade IIIb	Require surgical/endoscopic/radiological intervention under GA
Grade IVa	Life threatening complication needing ICU care with single organ dysfunction (includes dialysis)
Grade IVb	Life threatening complication needing ICU care with multiple organ dysfunction
Grade V	Death of the patient
Suffix "d"	Added to indicate "disability" when patient suffers any of the above complications at the time of discharge

GA general anesthesia

Table 8.8 ASA grading of risk

Class	Description
I	Patient is a completely healthy fit patient
II	Patient has mild systemic disease
III	Patient has severe systemic disease that is not incapacitating
IV	Patient has incapacitating disease that is a constant threat to life
V	A moribund patient who is not expected to live 24 h with or without surgery
E	Emergency surgery, E is placed after the Roman numeral

Table 8.9 Common surgical morbidity/complications

Site	Complications
Breast	Seroma, paresthesia over inner aspect of arm, lymphedema, flap or wound infection
Head and neck	Bleeding, chyle leak, carotid blowout, seroma, wound infection, nerve paresis/palsies
Gastrointestinal	Bleeding, infection, abscess formation, anastomotic leaks, atelectasis, burst abdomen
Genitourinary	Bleeding, infection, urosepsis, ureteric/bladder injuries, urinary leaks
Gynecological	Bleeding, infection, seroma formation, urinary/bladder injuries, urinary fistulae
Thoracic	Pulmonary: atelectasis/effusion, infection, bleeding, cardiac arrhythmias, chyle leak, blowouts

(Physiology and Operative Severity Score for enUmeration of Mortality and morbidity) risk stratification system was described for patients undergoing surgical procedures. This score has been validated in many solid tumor surgeries with a good predictive value for postoperative morbidity and mortality [17, 18]. Most scoring systems use a combination of criteria (Table 8.9).

1. Age of the patient: Increasing age is directly proportional to the risk of postoperative complications. The risk of postoperative mortality is 30–35 times higher in patients more than 70 years of age compared to their younger counterparts.
2. Comorbidities The risk of complications increases with increasing number of co morbidities. For example; diabetic patients are predisposed to risk of infections, electrolyte imbalances.

3. Prior chemotherapy/radiotherapy: Increasing number of patients with cancer receive chemotherapy or radiotherapy in a neo-adjuvant setting. Chemotherapy due to its immunosuppressive effects increases the risk of infections whereas preoperative radiotherapy is fraught with chances of increased wound related complications.
4. Increased BMI: Obese patients with a high body mass index (BMI) are prone to respiratory complications, thromboembolic events and impaired wound healing with a 1–1.7 times increased mortality risk.
5. Poor nutritional status: Serum albumin levels correlate directly with postoperative outcomes especially for major gastrointestinal surgeries. Low albumin levels affects wound healing with increased risk of anastomotic dehiscence.
6. Cancer cachexia: Extreme malnutrition characterized by weight loss and sarcopenia is seen in terminal cancers leading to poor outcomes even after palliative surgeries.
7. Addictions: Smoking increases the risk of postoperative mortality by almost double. It hampers wound healing due to impaired ciliary epithelium functioning resulting in infections and pulmonary infections.

8.4 Recent Advances/Ongoing Research

Oncology as a whole is an ever evolving field. Recent advances in various surgical options range from de-escalation to technical expertise.

Some of the examples are cited below

1. Changing the algorithm sequence for specific tumor types to use neo adjuvant options and obtain pathological complete response in an attempt to improve survival.
2. De-escalate surgery Post NACT: Studies are evaluating role of MRI and tumor bed biopsy to predict pCR and avoid surgery in rectal, esophageal cancer after NACTRT and breast cancer after NACT.
3. Sentinel Node Biopsy (SNB): While the role of SNB has been established in breast cancer, it is being studied in cancers of the Endometrium, Head and Neck Cancers, etc. The role of SNB in breast cancer is expanding to post NACT N0 breast cancer also.
4. HIPEC: Hyperthermic intraperitoneal chemotherapy is being increasingly used in various malignancies of the peritoneum. While this has enabled curative treatment for peritoneal surface malignancies such as low-grade mucinous neoplasms which were considered unresectable, its role in ovarian, colorectal and gastric cancer is evolving.
5. Quality of Life (QoL): While traditionally all surgical trials looked at survival (overall survival/disease-free survival) as end points, there is increasing focus on QoL as the survival outcomes have improved for most cancers.
6. Fertility Preservation: With improving outcomes for cancers, there has been a trend toward offering fertility preservation to those young and those desirous of having children. However, the disease biology should guide this decision, and further studies would identify subsets wherein this can be safely offered.

8.5 Summary

1. Surgical oncology is a specialized branch which deals with guiding principles of surgical management of tumors, improved perioperative care, expedited recovery, minimally invasive techniques, reconstruction and rehabilitation.
2. Biopsy is first step in establishing diagnosis and further molecular testing.
3. Surgery can be performed with preventive/prophylactic (where risk of cancer is exceptionally high), curative or palliative intent.
4. Surgical removal of whole tumor with draining nodal regions is the most effective way to cure patients with solid tumors.
5. Surgical margins and lymph node status are important prognostic indicators.

6. Surgical techniques have evolved from open surgeries to laparoscopic approaches and the recent robotic surgeries.
7. Surgery can also be performed to ease symptoms with palliative intent.
8. Cosmetic reconstruction after extensive surgery is important. Reconstruction option includes skin graft, local flap and free flap.
9. Proper integration of surgery with radiation and systemic therapy is the key to effective oncology treatment.

Multiple Choice Questions

1. All the following are true for metastasectomy EXCEPT?
   (a) Metastatic sites must by completely resectable
   (b) Metachronous tumors have poorer outcomes than synchronous tumors
   (c) Longer disease-free intervals have better outcomes
   (d) Performed in patients with good life expectancy
2. Which of the following is NOT true regarding principles for biopsy for cancers?
   (a) Lesser invasive options are performed initially
   (b) Image guidance improves yield
   (c) Biopsy is usually performed prior to any imaging
   (d) Biopsy tract incorporated during final surgery
3. All of the following are associated with increased risk of colorectal cancer EXCEPT?
   (a) FAP
   (b) Lynch syndrome
   (c) Inflammatory bowel disease
   (d) MEN-2
4. Sentinel lymph node biopsy for nodal staging is standard of care in the following cancers EXCEPT?
   (a) Melanoma
   (b) Pancreas
   (c) Breast
   (d) Endometrium
5. Which of the following statements is FALSE regarding intraoperative radiotherapy (IORT)?
   (a) Has been used in advanced and recurrent head and neck cancers, genitourinary cancers, GI cancers, sarcomas, gastrointestinal cancers
   (b) Accomplished using electrons or gamma rays
   (c) Improves local control by sterilizing at risk margins
   (d) Easy to perform with minimal costs
6. The concept of "qualitative" margins in sarcomas was propounded by
   (a) Halstead
   (b) Billroth
   (c) Kawaguchi
   (d) Blumgart
7. According to Clavien–Dindo classification for morbidity, category IVA is
   (a) surgical/endoscopic/radiological intervention performed under GA
   (b) life-threatening complication needing ICU care with multiple organ dysfunction
   (c) life-threatening complication needing ICU care with single organ dysfunction
   (d) perioperative mortality
8. A patient with severe systemic disease that is not incapacitating falls into which ASA category?
   (a) ASA II
   (b) ASA III
   (c) ASA IV
   (d) None
9. Operative mortality is death within
   (a) 15 days of surgery
   (b) 30 days of surgery
   (c) 60 days of surgery
   (d) 90 days of surgery

Answers

1. (b)
2. (c)
3. (d)
4. (b)
5. (d)
6. (c)
7. (c)
8. (b)
9. (b)

References

1. Govindan R. DeVita, Hellman and Rosenberg's cancer: principle and practice of oncology. 9th ed. Philadelphia: Lippincott Williams and Wilkins.
2. Schimke RN. Multiple endocrine neoplasia. N Engl J Med. 1986;314:1315–6.
3. Thakker RV. Multiple endocrine neoplasia type 1 (MEN1) and type 4 (MEN4). Mol Cell Endocrinol. 2014;386(1-2):2–15.
4. Hall MJ, Reid JE, Burbidge LA, Pruss D, et al. *BRCA1* and *BRCA2* mutations in women of different ethnicities undergoing testing for hereditary breast-ovarian cancer. Cancer. 2009;115(10):2222–33.
5. Nieuwenhuis MH, Vasen HFA. Correlations between mutation site in APC and phenotype of familial adenomatous polyposis (FAP): a review of the literature. Crit Rev Oncol Hematol. 2007;61:153–61.
6. Raut CP, Tseng JF, Sun CC, Wang H, et al. Impact of resection status on pattern of failure and survival after pancreaticoduodenectomy for pancreatic adenocarcinoma. Ann Surg. 2007;246(1):52–60.
7. Köninger J, Wente MN, Müller-Stich BP, di Mola FF, et al. R2 resection in pancreatic cancer-does it make sense? Langenbecks Arch Surg. 2008;393(6):929–34.
8. Bachmann J, Michalski CW, Martignoni ME, Büchler MW, Friess H. Pancreatic resection for pancreatic cancer. HPB (Oxford). 2006;8(5):346–51.
9. Bristow RE, Tomacruz RS, Armstrong DK, Trimble EL, Montz FJ. Survival effect of maximal cytoreductive surgery for advanced ovarian carcinoma during the platinum era: a meta-analysis. J Clin Oncol. 2002;20(5):1248–59.
10. Belal A, Salah E, Hajjar W, El-Foudeh M, et al. Pulmonary metastatectomy for soft tissue sarcomas: is it valuable? J Cardiovasc Surg (Torino). 2001;42(6):835–40.
11. Pawlik TM, Schulick RD, Choti MA. Expanding criteria for resectability of colorectal liver metastases. Oncologist. 2008;13:51–64.
12. Dindo D, Demartines N, Clavien PA. Classification of surgical complications: a new proposal with evaluation in a cohort of 6336 patients and results of a survey. Ann Surg. 2004;240(2):205–13.
13. Wang W, Babu SR, Wang L, Chen Y, Tian B, He H. Use of Clavien-Dindo classification in evaluating complications following pancreaticoduodenectomy in 1,056 cases: a retrospective analysis from one single institution. Oncol Lett. 2018;16(2):2023–9.
14. Bansal A, Sankhwar S, Goel A, Kumar M, Purkait B, Aeron R. Grading of complications of transurethral resection of bladder tumor using Clavien–Dindo classification system. Indian J Urol. 2016;32:232–7.
15. Ramirez PT, Frumovitz M, Pareja R, Lopez A, Vieira M, Ribeiro R, Buda A, Yan X, Shuzhong Y, Chetty N, Isla D, Tamura M, Zhu T, Robledo KP, Gebski V, Asher R, Behan V, Nicklin JL, Coleman RL, Obermair A. Minimally invasive versus abdominal radical hysterectomy for cervical cancer. N Engl J Med. 2018;379(20):1895–904. Epub 2018 Oct 31. https://doi.org/10.1056/NEJMoa1806395.
16. Dripps RD. New classification of physical status. Anesthesiology. 1963;24:111.
17. Lubke T, Monig SP, Holscher AH. Evaluation of POSSUM scoring system in patients with gastric cancer undergoing D2-gastrectomy. BMC Surg. 2005;5:8.
18. Makino Y, Ishida K, Kishi K, Kodama H, Miyawaki T. The association between surgical complications and the POSSUM score in head and neck reconstruction: a retrospective single center study. J Plast Surg Hand Surg. 2018;52(3):153–7.
19. Soran A, Ozmen V, Ozbas S, Karanlik H, Muslumanoglu M, Igci A, et al. Randomized trial comparing resection of primary tumor with no surgery in stage IV breast cancer at presentation: protocol MF07-01. Ann Surg Oncol. 2018;25:3141–9.
20. Jacobson G, Kaidar-Person O, Haisraely O, Galper S, Rabin T, Shahadi D, et al. Palliative radiation therapy for symptomatic advance breast cancer. Sci Rep. 2021;5(11):5282.
21. Boudjema K, Locher C, Sabbagh C, Ortega-Deballon P, Heyd B, Bachellier P, Métairie S, Paye F, Bourlier P, Adam R, Merdrignac A, Tual C, Le Pabic E, Sulpice L, Meunier B, Regimbeau JM, Bellissant E, METASYNC Study Group. Simultaneous versus delayed resection for initially resectable synchronous colorectal cancer liver metastases: a prospective, open-label, randomized, controlled trial. Ann Surg. 2021;273(1):49–56. https://doi.org/10.1097/SLA.0000000000003848.
22. Opitz I, Patella M, Payrard L, Perentes JY, Inderbitzi R, Gelpke H, Schulte S, Diezi M, Gonzalez M, Krueger T, Weder W. Prognostic factors of oligometastatic non-small-cell lung cancer following radical therapy: a multicentre analysis. Eur J Cardiothorac Surg. 2020;57(6):1166–72. https://doi.org/10.1093/ejcts/ezz384.
23. Cariboni U, De Sanctis R, Giaretta M, Voulaz E, Morenghi E, Colombo P, Novellis P, Bottoni E, Errico V, Crepaldi A, Marrari A, Navarria P, Ascolese AM, Testori A, Scorsetti M, Quagliuolo V, Santoro A, Alloisio M. Survival outcome and prognostic factors after pulmonary metastasectomy in sarcoma patients: a 18-year experience at a single high-volume referral center. Am J Clin Oncol. 2019;42(1):6–11. https://doi.org/10.1097/COC.0000000000000476.
24. Zhang Y, Hu J, Yang J, Xie Y, Chen Z, Shangguan W, Han J, He W, Yang J, Zheng Z, Zhong Q, Zhu D, Xie W. Selection of optimal candidates for cytoreductive nephrectomy in patients with metastatic clear cell renal cell carcinoma: a predictive model based on SEER database. Front Oncol. 2022;12:814512. PMID: 35127544; PMCID: PMC8814440. https://doi.org/10.3389/fonc.2022.814512.

Principles of Chemotherapy and Chemoradiation

9

Arun Chandrasekharan, Dilip Harindran Vallathol, Abdul Malik Manzar, Vanita Noronha, and Kumar Prabash

9.1 Introduction

Concurrent chemoradiotherapy (cCRT) has improved outcomes and cancer cure rates in multiple malignancies and is used as the standard of care in neoadjuvant, primary (definitive), or adjuvant settings. Chemotherapy and targeted agents mostly increase the efficacy of radiation (RT). It has become the mainstay of treating locally advanced solid tumors aiding organ preservation, avoiding surgeries, and as an adjuvant treatment protocol. Chemotherapy agents with radiosensitizing properties (i.e., cisplatin, carboplatin, capecitabine, 5-fluorouracil, and taxanes) and several novel agents (e.g., epidermal growth factor receptor inhibitors (EGFR) and several other targeted therapies) have shown promising interactions with radiation. cCRT can be used with the intent of tackling distant metastasis, improving local control, and even survival by making cells more sensitive to radiation.

There are additive and supra-additive mechanisms in radiosensitizing. Multiple mechanisms underlie radiosensitizing properties of chemotherapeutic agents and include increased radiation damage, inhibition of DNA repair, cell-cycle synchronization, increased cytotoxicity against hypoxic cells, inhibition of prosurvival pathways, and abrogation of rapid tumor cell repopulation. Radioresistance occurs through multiple mechanisms, such as a large tumor burden, hypoxia, rapid tumor cell repopulation, as well as the constitutive or acquired activation of radioresistance signaling pathways.

9.2 Theoretical Framework for Therapy: The "Steel Paradigm"

Steel Peckham hypothesized, in 1979, a framework describing the interaction of chemotherapy and RT [1]. Spatial cooperation is the term used to describe the mechanism whereby acts locoregionally and chemotherapy acts against distant metastasis without interaction between agents.

The interaction between radiation and chemotherapy can have various effects (Fig. 9.1). Radiosensitization can be additive or supra-

A. Chandrasekharan (✉)
Medical Oncology, Aster MIMS, Kozhikode, Kerala, India

D. H. Vallathol
Medical Oncology, MVR Cancer Centre, Calicut, Kerala, India

A. M. Manzar
Radiation Oncology, Aster MIMS, Calicut, Kerala, India

V. Noronha
Department of Radiation Oncology, Tata Memorial Hospital, Mumbai, Maharashtra, India

K. Prabash
Department of Medical Oncology, Tata Memorial Hospital, Mumbai, Maharashtra, India

A. Munshi et al. (eds.), *Radiation Oncology – Principles, Precepts and Practice*,
https://doi.org/10.1007/978-981-97-8389-2_9

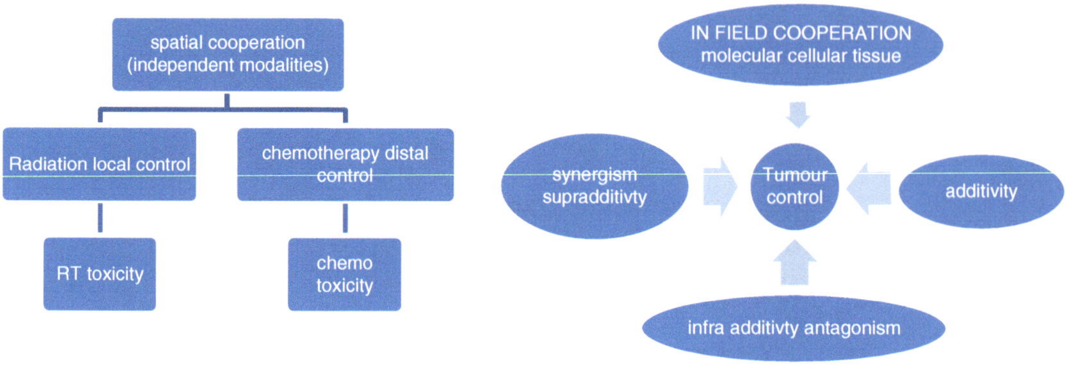

Fig. 9.1 Rationale for adding chemo to radiation. Both mechanisms contribute synergistically to tumor control. Infra-additivity protects tumor

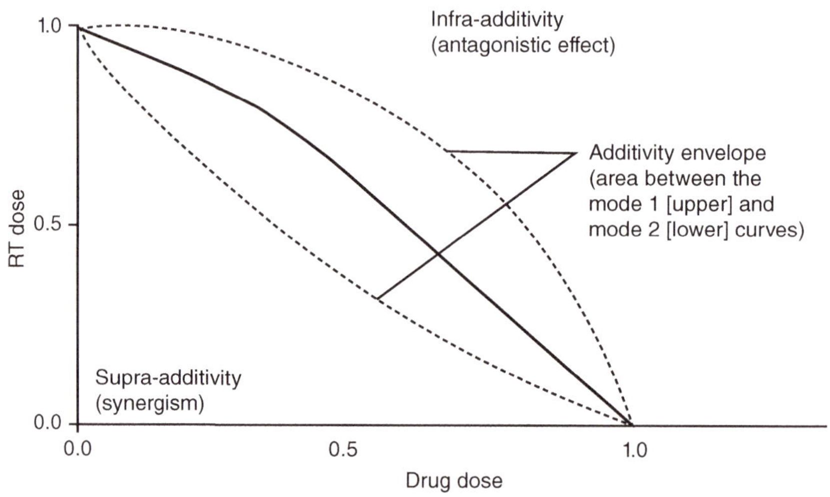

Fig. 9.2 Schematic isobologram showing the combination of radiation and a systemic agent. The scales represent an isoeffective level. The solid line is the line of additivity. When the isobole lies below the line of additivity, the combinatorial effect of radiation and drug is synergistic. When the isobole lies above the line of additivity, the combined effect of radiation and drug is protective or antagonistic. The envelope represents the combined standard errors of the combinations

additive; the interaction within the radiation field leads to increased killing of cells (cytotoxic activity) either to the same degree as (additive) or greater (supra-additive) using both modalities sequentially (Figs. 9.1 and 9.2). Radiation sensitizers must lack inherent cytotoxic effects.

Infra-additive drugs have the property of radioprotection. This causes the cytotoxic effect of radiation to decrease, both on the tumor and surrounding normal tissue. One such agent, amifostine, is an organic thiophosphate. This decreases cisplatin-induced or radiation-induced toxicity acting as a scavenger of free radicals. It does this by binding to and neutralizing organoplatinums and alkylating agents, thus inhibiting DNA adduct formation.

9.2.1 Concurrent Chemoradiotherapy: Determining the Additive Effects

Steel and Peckham have described an isobologram analysis calculated on an isoeffect concept for cCRT interaction (Fig. 9.2).

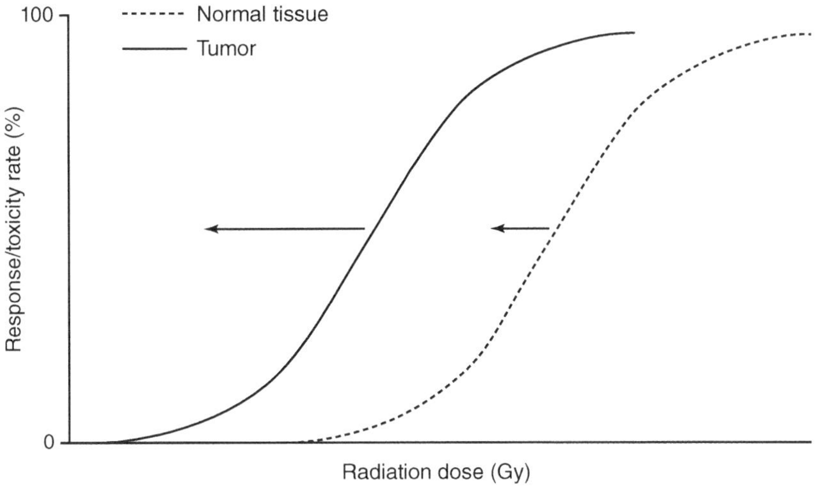

Fig. 9.3 Schematic dose–response curves for tumor and normal tissue damage with radiation. The offset between the two curves indicates the therapeutic range. Chemoradiotherapy leads to a shift of both curves to the left, ideally with a stronger shift of the tumor curve (as indicated by the longer arrow), increasing the efficacy of treatment (radiation enhancement)

This is created when the dose of each chemotherapy agent and RT is plotted with each other. This produces a cytotoxic effect (isoeffect) on the axes of increasing the dose of each agent. Two curves are created called modes 1 and 2 (Fig. 9.2). The mode 1 curve is derived when we assume chemotherapy and RT act independent of each other and is generated by plotting the dose of radiation against the dose of chemotherapy that produces an equal effect to the difference between the given cytotoxic action and the action of the given dose of radiation.

The mode two curve is generated when we assume that RT and chemotherapy have near-similar mechanisms of action. The curves are created by plotting the radiation dose against the chemotherapy doses that is needed to increase the radiation effect to the selected cytotoxic effect. There are multiple points on both of these lines which are calculated when varying the radiation dose and deriving the doses of chemotherapy. The empirical survival curve is generated for a particular combination of chemotherapy and RT.

Mostly, this concept of additivity has little clinical implication and use, since the prediction of additivity always does not emerge into clinical relevance.

The use of this method is mostly for making a hypothesis that has to be confirmed empirically.

The therapeutic ratio is used to express the radiosensitizing capability of chemotherapy. This ratio is created from sigmoid-shaped dose–response curves, calculated by making plots of the response of tissues on the ordinate axis versus the chemotherapy/RTdoses on the abscissa (Fig. 9.3).

9.2.1.1 Quantification of the Chemotherapy and Radiation Interaction

This dose–effect relationship is sigmoidal, enabling estimation of the therapeutic index (ratio). This is known as a ratio between doses (RT, chemotherapy) that produce the same level (probability) of antitumor efficacy and normal tissue damage. For the therapeutic ratio to be beneficial, it has to be positive (>1); individual drugs or their combination must be more effective against malignancy than normal tissue. There has to be a balance between a given level of antitumor efficacy and tolerable normal tissue toxicity, which gives a measure of the therapeutic ratio.

Table 9.1 Mechanisms of radioresistance. *EGFR: Epidermal growth factor receptor*

Affected process	Mechanisms	Comment
Increased burden from cancer cell	Tumor size is inversely correlated with tumor response A radiation-induced cell kill is a random event; more the cell more chance they escape cell kill	Surgery is done initially to cause zero kill, then other modalities follow
Cancer microenvironment/hypoxia	Radical oxygen species generated from O_2 due to RT is cytotoxic. Hypoxia due to 1. Increased interstitial pressure causing hypoperfusion, hypoxia acidosis 2. Cancer-related anemia-local hypoxia	Hypoxic cells are 2.5–3 times less radiosensitive than normoxic cell Chemo can increase the radiation effect 1. Through reoxygenation second to tumor shrinkage 2. By killing hypoxic cells selectively 3. Radiosensitization of hypoxic celsl to radiation
Inherent or acquired tumor resistance	Mutated p53 DNA repair gene amplification, increased levels of reactive oxygen species scavengers, activation of prosurvival oncogenes EGFR, Cmet"	Breaks in radiotherapy lead to the development of radioresistance, as they repopulate
Repopulation	Regrowth of tumor cells between doses of radiotherapy or chemotherapy. Accelerated repopulation leads to treatment failure and the emergence of radioresistance	Accelerated radiation schemes can control repopulation Antimetabolites with activity in the S phase inhibit repopulation EGFR inhibitors can block cell proliferation between fractions

9.2.1.2 Mechanisms of Radiation Resistance or Failure

Tumors have evolved numerous survival mechanisms to overcome radiation damage. Table 9.1 summarizes some widely accepted mechanisms.

Simplistically, it's predicted that larger tumors shall have a stochastic chance that few cells are resistant and will survive radiation damage. Moreover, it's understood that hypoxic tumor cells have increased resistance to radiation; many studies have shown that hypoxic cells are 2.5–3.0 times more resistant to radiation toxicity compared to normoxic cells [2–5]. As shown in Fig. 9.4, the different phases of the cell-cycle influence the radiosensitivity of cells.

One of the most commonly cited underlying processes in treatment failure [6, 7] is tumor cell repopulation, which causes rapid proliferation between RT fractions. Even though it is not fully understood, the combination of excess growth factors and the survival of resistant clones causes the rapid emergence of treatment resistance [8–10].

Some cancer cells are intrinsically radioresistant, while others evolve mechanisms of radioresistance during cCRT. Intrinsic radioresistance is noted in radiobiology when there is an increased surviving fraction of cancer cells after exposure to 2 Gy of RT; little increments in radioresistance lead to significant logarithmic decreases in the final cell kill after RT. In some cancers, evaluation of radioresistance in pretreatment biopsies forecasts for radiosensitivity and clinical results, but these results have not been reproducible widely.

9.2.2 General Mechanisms of Chemotherapy and Radiotherapy Interaction

Several major interactions between chemotherapy and radiation will be explained here and are also summarized in Table 9.2. Many chemotherapeutic agents have multiple modes of action taking place at the same time.

Firstly, DNA damage can be induced by both chemotherapy and RT. Ionizing radiation induces DNA base damage, SSB (single-strand breaks), DSB (double-strand breaks), and alkali

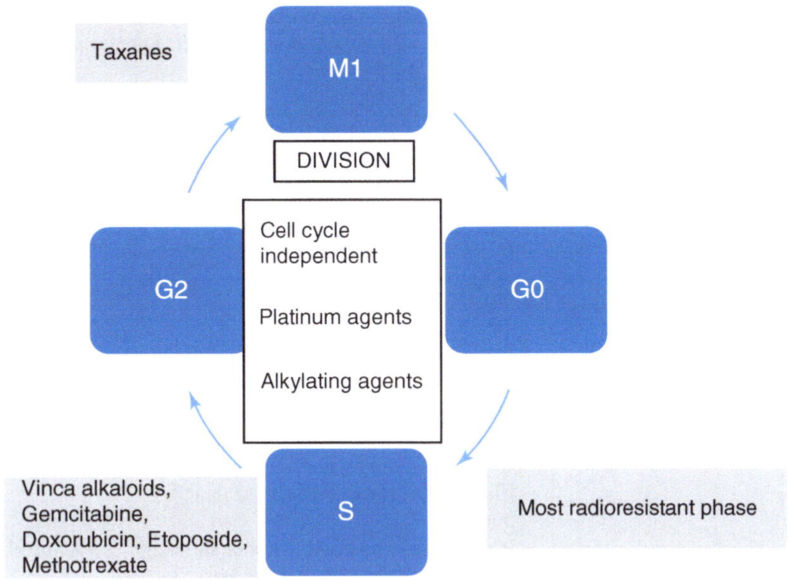

Fig. 9.4 Schematic of the cell cycle with respect to chemotherapy

labile sites. All of these errors can be rapidly repaired except for DSBs, which, if not repaired, are considered lethal [11, 12]. The cisplatin gets combined with DNA and RNA in near proximity to a radiation-induced single-strand break that acts synergistically making the defect significantly very difficult to repair (Fig. 9.5) [13, 14].

Secondly, chemotherapy can inhibit postradiation damage repair. DNA synthesis and repair share common pathways; it provides the rationale for using DNA synthesis inhibitors and radiation as a means of inducing cytotoxic damage to tumor cells. Agents affecting nucleoside and nucleotide metabolism could inhibit the repair of radiation-induced DNA damage in patients and are among the most potent radiation sensitizers. Examples of such agents include fluoropyrimidines, thymidine analogs, gemcitabine, and hydroxyurea.

Thirdly, administered concurrently, radiotherapy and chemotherapy often target different phases of the cell cycle and may cooperate to produce an additive effect (i.e., cytokinetic cooperation/ synchronization). The intrinsic radiosensitivity of tumor cells lies with the phase of cell cycle; cells in the S phase are deemed to be most radioresistant, and cells in the G2–M phase of the cell cycle are considered most radiosensitive (Fig. 9.4) [15, 16]. There are some drugs (e.g., taxanes) that are able to synchronize with the cell cycle of tumor cells, allowing increased efficacy of subsequent radiotherapy (called synchronization or cell-cycle pooling). This process was successfully shown in vitro, although the concept—especially its applicability in vivo—remains controversial.

Fourthly, the increased resistance of hypoxic cells to radiation (Table 9.1) means that hypoxic cell sensitizers may be beneficial [17]. Hypoxia is common in many cancers and was shown to be a marker of aggressive clinical behavior and poor prognosis [18]. Chemotherapy can help eliminate these resistant cells and increase efficacy of radiotherapy via multiple mechanisms (Table 9.2).

Fifthly, repopulation of proliferating tumors is mostly mediated by the increased expression of growth factors and its receptors, increasing activity of downstream signaling pathways, and the presence of activating mutations in genes involved at all levels of the signaling pathways. There are also agents that target the S phase of the cell cycle for, e.g., 5-FU, irinotecan, and hydroxyurea, and also those that inhibit proliferation and/or growth factor pathways, such as EGFR inhibitors, which may be effective in preventing tumor cell repopulation, thus radiosensitizing tumor cells.

Table 9.2 The mechanisms of interaction between chemotherapy and radiotherapy. *DNA: Deoxyribonucleic acid, RNA: Ribonucleic acid, 5FU: 5-Fluorouracil, RT: radiation, EGFR: Epidermal growth factor receptor*

Process affected	Mechanism	Drug examples
Increased radiation damage	Incorporation of chemo drug into DNA/ RNA	5FU incorporation into DN, increases susceptibility to RT damage Cisplatin cross-links with DNA or RNA intrastrand/ interstrand linkages work for both hypoxic and oxygenated cells
Inhibition of DNA repair and process	Interference with the DNA repair process after irradiation	Halogenated pyrimidine Nucleoside analogs Cisplatin, methotrexate, etoposide, hydroxyurea, doxorubicin
Cell-cycle interference (cytokinetic cooperation and synchronization)	Cytotoxic chemo and radiation are cell cycle specific and proliferating cells are always susceptible Reorientation of cells into the G2–M phase will make it radiosensitive Elimination of radioresistant cells in S phase	Taxanes cause cell-cycle arrest via tubulin stabilization. Nucleoside analogs (Gemcitabine Fludarabine), Etoposide, Methrotrexate
Enhanced activity against the hypoxic cell	Reoxygenation second to tumor shrinkag., Hypoxic cells are 2.5–3 times less sensitive than the normoxic cell. Chemotherapy can also eliminate hypoxic cells	Paclitaxel, Mitomycin
Radiotherapy enhancement by preventing repopulation	Systemic options can slow rapid proliferation, which could be the basis for repopulation	Most chemotherapeutic agents Antimetabolites with activity in the S phase EGFR inhibitors impede cell proliferation between RT fractions
Inhibition of prosurvival and poor prognosis markers	Targeted therapies block signaling pathways that might be responsible for radioresistance and poor prognosis	EGFR inhibitors
Hyper radiation sensitivity	Head and neck squamous cancer cells resistant to standard fraction cCRT can be resensitized to cCRT by using smaller fraction sizes	Effect demonstrated for taxane-based cCRT including paclitaxel as well as docetaxel. Low-dose fraction radiation

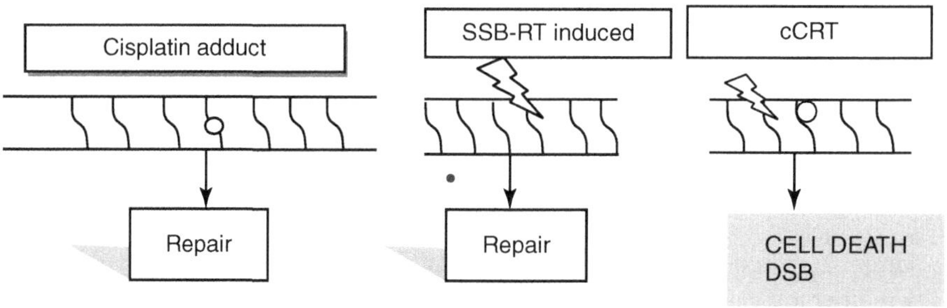

Fig. 9.5 Increased DNA damage—cisplatin, radiation alone, cCTRT. *SSB : Single-strand break, RT: Radiotherapy, cCRT: Concurrent chemoradiotherapy*

In addition to antiproliferative effects that prevent tumor cell repopulation, EGFR inhibitors and antiangiogenic agents can block signaling pathways that are responsible for aggressive tumor biology, poor prognosis, and radioresistance.

Although the exact mechanisms of various targeted therapies vary and are generally poorly understood, preclinical and clinical data support the rationale for radiosensitizing properties as a consequence of inhibiting the detrimental effects of the agents' targets.

Finally, some tumors resistant to "standard" chemoradiotherapy respond to alterations in radiation fractionation. This phenomenon is called as hyper radiation sensitivity and is observed at radiation doses >1 Gy. In preclinical trials, both paclitaxel and docetaxel have overcome radioresistance along with low-dose fractionated RT regimens.

Table 9.3 lists the patterns of practice of chemoradiation in various body sites.

Some drugs are commonly used in concurrent chemoradiation. Some others are avoided in view of their toxicities when used along with radiation (Table 9.4).

9.2.3 Some Site-Specific Examples of Chemoradiation

Chemoradiation is practiced in several sites in neoadjuvant, definitive and adjuvant forms. Some site specific examples have been discussed. Further details would be discussed in site specific chapters of the book.

9.2.3.1 Chemoradiation in Anal Cancer

The earliest effective use of cCRT was in the management of anal cancers. In the seminal paper by Nigro et al. in 1973, three patients were treated with neoadjuvant 5-FU and mitomycin-C. Two of them had complete histological response after abdominoperineal resection, while one patient refused surgery, but had no signs of recurrence even after 14 months [19]. Thus, the "Nigro" regimen was used widely, consisting of 5-FU (1000 mg/m^2 per day by continuous infusion days 1 through 4 and 29 through 32), mitomycin (10 to 15 mg/m^2 on day one only), and intermediate-dose RT (30 Gy). Because of the complete pathological response seen in the three trial patients, efforts were directed to strategies that could preserve the anal sphincter. The same group then treated patients with anal cancer with cCRT and then proceeded with abdominal perineal surgery in those who had a positive biopsy post radiation. A large majority of patients were cured fully with cCRT without surgery, with 5-year overall survival of 67% and 5-year colostomy-free survival of 59%. Many other trials confirmed these findings and have resulted in 5-year overall survival rates of 72–89% and local failure rates of 14–37% [20–26]. The "Nigro regimen" is the current gold standard for the management of squamous cell carcinoma of the anal canal.

Chemoradiation versus radiation alone was compared in 2 randomized trials, the UKCCR (ACT) and the EORTC trial, both of which showed reductions in local failure and higher pathological response with chemoradiation [27, 28]. Owing to the difficulty in administration of continuous infusion of 5-FU, capecitabine has been used as an alternative, and the results have been found to be acceptable [29, 30]. The role of mitomycin has been questioned as it is not a radiosensitizer and is not very effective in squamous cancers, and has significant bone marrow, renal and pulmonary toxicity. A joint RTOG and EORTC trial looked at responses to cCRT with and without mitomycin and concluded that mitomycin is required for the disease-free survival benefit in spite of added toxicity [31].

Table 9.3 Chemoradiation in site-specific cancers

Definitive chemoradiation	Neoadjuvant chemoradiation	Adjuvant (postop) chemoradiation
Head and neck cancers	Rectal cancer	Head and neck cancers
Esophageal cancers (squamous cell carcinoma)	Esophageal cancers (squamous cell carcinoma and adenocarcinoma)	Glioblastoma multiforme
Anal cancers		
Lung cancer (small cell lung cancer and nonsmall cell lung cancer)		

Table 9.4 Chemotherapy drugs for concurrent chemoradiation. *5FU: 5 Flurouracil*

Commonly used drugs for concurrent chemoradiation	Drugs avoided for concurrent chemoradiation
Cisplatin	Adriamycin
Carboplatin	
Capecitabine	
Paclitaxel	
Temozolomide	
5FU	

Mitomycin was also compared to cisplatin, as the latter is considered more active in squamous cancers. The RTOG 98–11 trial substituted cisplatin for mitomycin, but final results favored the mitomycin combination with regards to disease-free survival, overall survival, and colostomy-free survival [32]. However, the ACT-II randomized trial showed similar disease outcomes for both cisplatin and mitomycin and lesser hematological toxicity with cisplatin [33]. Considering both these trials, mitomycin still remains the standard of care, but cisplatin (60 mg/m^2 on days 1 and 29) could be used as a valid substitute in case of the nonavailability of mitomycin.

9.2.3.2 Chemoradiation for Lung Cancer

Stage III non-small cell lung cancer (NSCLC) represents a very heterogeneous disease in terms of patient and tumor characteristics. The presence of nodal disease in NSCLC is considered a high risk for both local and distant recurrence, and surgical resection as the sole treatment modality is considered inadequate. The most common approach to managing patients with confirmed N2 disease is Concurrent Chemoradiation (cCRT) which is defined as a combination of platinum doublet-based chemotherapy and a radical dose of radiotherapy [34]. Within this group of N2 patients, there is a highly selected subset of patients who may be considered for surgery following preoperative chemotherapy and/or cCRT [35]. For patients with N3 or T4 surgically unresectable tumors, the standard approach to management is cCRT [36].

The optimal treatment plan for a patient with NSCLC requires an accurate assessment of their overall fitness, medical comorbidities, cardiorespiratory reserve, genomic background, tumor stage, and mutation status [34]. Certain aspects of cCRT in NSCLC remain controversial, and a multidisciplinary team (MDT) approach with experts is required for decision making.

Patients with unresectable stage III NSCLC live longer if they receive chemotherapy before thoracic irradiation [sequential (sCRT)] or cCRT. Early clinical studies of cCRT in NSCLC examined whether a sequential approach to treatment delivery was useful to maximize both locoregional and micrometastatic disease control, at the same time minimizing the risks of cumulative toxicity. Despite modest improvements with sCRT, two large phase 3 randomized studies demonstrated superiority of cCRT over sCRT.

The Radiation Therapy Oncology Group trial (RTOG 9410) randomized 610 patients with unresectable stage III NSCLC to either concurrent or sequential use of chemotherapy with radiotherapy. Patients who received cCRT with once-daily radiation demonstrated improved overall survival (OS) compared to the other groups (mOS, 17.0 vs. 14.6 months; HR 0.81; 95% CI 0.66–0.996) [37]. A second phase 3 randomized study from Japan also revealed similar results in favor of cCRT. Two subsequent meta-analyses were conducted in this regard [38]. The meta-analysis published in 2010 analyzed six trials (1024 patients) comparing cCRT versus sCRT, which revealed a significant benefit in OS (HR 0.74; 95% CI 0.62–0.89) in favor of cCRT. This survival improvement equated to a 10% absolute survival benefit at 2 years for cCRT [39]. A subsequent meta-analysis (1205 patients) by Auperin et al. analyzed randomized trials directly comparing cCRT versus sCRT and reiterated the superiority of cCRT in terms of improvement of OS (HR 0.84; 95% CI 0.74–0.95; $p = 0.004$) [40].

In all of these studies, the cCRT led to improvements in efficacy, at the cost of increased acute toxicity, compared with sCRT but the late toxicities were similar [40, 41]. The most common acute toxicities were esophagitis, radiation pneumonitis, and hematological toxicities. The late toxicities include cardiotoxicity and pulmonary fibrosis. Several strategies to minimize toxicities have been tried like using low-dose chemotherapy schedules and altering the dosage and delivery mode of radiotherapy [42, 43].

The optimal chemotherapy regimen and radiation fractionation is still a matter of debate. Platinum based chemotherapy regimens are most widely used. It may be combined with etoposide, vinorelbine, or other vinca alkaloids, based on most comparative clinical trials of cCRT and sCRT when treatment has curative intent [39, 40]. In case comorbidities are present, cisplatin may be replaced by carboplatin [35]. In terms of radiotherapy, approaches have included hyperfractionation, accelerated fractionation, hypofractionation, or dose escalation. The standard dose of radiotherapy is 60 Gy in 30 daily fractions which was established by RTOG 7301 study [44]. The RTOG 0617 study was first Phase 3 trial to permit intensity modulated radiotherapy (IMRT) in NSCLC and demonstrated that IMRT had similar survival and locoregional control rates to 3D-conformal radiotherapy, but lower rates of grade ≥3 pneumonitis and lower radiation doses to the heart, thereby improving quality of life [45].

In yet another mode of delivery, cCRT has been tried as a neoadjuvant approach. Meta-analysis of trials with neoadjuvant cCRT in stage III NSCLC led to improvements in local control, tumor downstaging and pathological complete response (pCR). However, the combination of radiotherapy and chemotherapy was not superior to neoadjuvant chemotherapy alone in terms of progression-free survival and 5-year survival [46]. But several of these trials were small and clearly underpowered to detect meaningful differences in outcomes. There is no role for adjuvant cCRT at this point in time in NSCLC.

The role of cCRT in small cell lung cancer (SCLC) is more prominent as compared to NSCLC. SCLC is divided into limited stage (LS-SCLC) and extensive stage (ES-SCLC) as per the classification devised by Veterans Administration Lung Study Group [47]. The standard curative therapy for LS-SCLC is cCRT followed by Prophylactic Cranial Irradiation (PCI). An individual patient data meta-analysis from 13 randomized trials that included 2103 evaluable patients with LS-SCLC showed that the relative risk for death observed among patients receiving cCRT, in comparison with chemotherapy alone, was 0.86 (95% confidence interval [CI], 0.78–0.94; $p = 0.001$). The absolute difference in survival at 3 years was 5.4% ± 1.4% [48].

The optimal timing of concurrent radiation therapy is generally agreed to be during the first or second cycle of chemotherapy as supported by data from three meta-analyses [49, 50]. Radiation schedule and total dose for SCLC have been topics of continuous debate. As a general consensus, the usual dose and fractionation should be delivered as either 1.5 Gy twice daily to a total of 45 Gy or 1.8–2.0 Gy daily to 60–70 Gy [51]. Even though the twice-daily schedule has shown improvements in survival, it has not been widely accepted in clinical practice due to practical challenges, especially the higher rates of acute toxicity [52].

In LS-SCLC, platinum based chemotherapy with etoposide is the preferred combination. Several phase II trials and a retrospective review have investigated the role of replacing cisplatin with carboplatin [53, 54]. These have shown similar outcomes but definite phase 3 data are lacking. Given the curative intent of chemoradiotherapy, cisplatin is considered the standard therapy unless the patient cannot tolerate or has a contraindication to cisplatin-based therapy. But based on the demographics, higher prevalence of multiple comorbidities among the eligible population precludes the use of cisplatin based chemotherapy in clinical practice [55].

9.2.3.3 Chemoradiation in Cervical Cancers

Cervical cancer continues to be one of the most common cancers among females especially in the low- and middle-income countries [56]. The International Federation of Gynecology and Obstetrics (FIGO) staging was based mainly on clinical examination with the addition of certain procedures that were allowed by FIGO to change the staging. In 2018, this was revised by the FIGO to allow imaging and pathological findings, where available, to assign the stage [57].

Concurrent chemoradiotherapy (cCRT) is regarded as the standard treatment for locally advanced uterine cervical cancer (LACC), including stage Ib2-IVa disease. The Manchester system divides cervix into reference points [58]:

1. Point A—2 cm superior to the external cervical os and 2 cm lateral to the cervical canal which is point of limiting dose tolerance.
2. Point B—5 cm from midline at the level of point A which represents the dose delivered to the obturator lymph node.

The primary goal of External Beam Radiotherapy (EBRT) is to sterilize local disease and to shrink the tumor to facilitate subsequent intracavitary radiotherapy (ICRT). The ability of radiotherapy to cure locally advanced cervical cancer is limited by the size of the tumor, because the doses required to treat large tumors exceed the limit of toxicity in normal tissue [59].

ICRT is also an extremely important component of curative treatment of cervical cancer since it delivers a high central dose to the primary tumor and reduced doses to adjacent normal organs owing to sharp dose fall-off. Three dose rate systems, namely low-dose-rate (LDR), high-dose-rate (HDR), or pulsed-dose-rate (PDR) yield comparable survival rates [60]. The dose is usually prescribed to Point A or to high-risk clinical target volume if image-based planning is used. If ICRT is not feasible either due to distorted anatomy or inadequate dosimetry, then interstitial brachytherapy should be considered [61]. Currently, the recommendation is to complete the entire protocol of EBRT and brachytherapy within 8 weeks [62].

The combination of EBRT and ICRT maximizes the likelihood of locoregional control while minimizing the risk of treatment complications. The standard practice of radiotherapy is administering approximately 50 Gy (total dose in 20 to 25 fractions) radiation to the entire pelvis followed by intracavitary brachytherapy [63].

Historically, several agents have been used concurrently with radiotherapy. First of these was hydroxyurea and misonidazole [64]. But with the advent of platinum agents, they have become the standard of care as they improved outcomes compared to other agents. The chemotherapy regimen used with radiotherapy is weekly cisplatin at a dose of 40 mg/m^2 with radiotherapy [65].

In the year 1999 and later, several randomized studies demonstrated a therapeutic advantage of cCRT over radiotherapy alone. In these studies, cisplatin was used either alone or with 5-fluorouracil [66–68]. A recent meta-analysis of 13 trials confirmed an absolute benefit of 6% in overall survival (OS) at 5 years with cCRT (HR = 0.81, $p < 0.001$). A larger survival benefit was seen for the two trials in which chemotherapy was administered after cCRT with an absolute improvement of 19% at 5 years. Survival benefit was seen for both platinum-based (HR = 0.83, $p = 0.017$) and nonplatinum-based (HR = 0.77, $p = 0.009$) regimens, but no difference was seen in the size of the benefit by RT or chemotherapy dose or scheduling. There was also reduction in local and distant recurrence and progression. The benefit was higher for patients with early-stage disease. With cCRT, a 5-year OS rate of 66% and disease-free survival of 58% was achieved in these patients with locally advanced disease. Acute hematological and gastrointestinal toxicities were more frequent in the cCRT group [69]. The findings of this meta-analysis were further supported by a recent Cochrane review which confirmed the benefit of cCRT in terms of survival, whether or not platinum was used with absolute benefits of 10% and 13%, respectively. The effect was greater in trials including a high proportion of early stage patients [70, 71].

A recently published randomized controlled trial from a tertiary care center in India showed the superiority of cCRT over neoadjuvant chemotherapy followed by surgery in locally advanced cervical cancers (5 year disease-free survival 69.3% vs. 76.7%) [72]. Consequently, this has been reflected in the national cancer grid guidelines that outlines the treatment of cervical cancer. It reiterates the use of cCRT as standard of care for stage 1B2 to 3B disease [73].

9.2.3.4 Chemoradiation in CNS Tumors

Glioma denotes primary brain tumors, and is classified as per the cell of origin into astrocytic tumors (astrocytoma, anaplastic astrocytoma and glioblastoma), oligodendrogliomas, ependymomas, and mixed gliomas. They are the commonest malignant primary tumors of the brain [1, 74]. Glioblastoma multiforme is the most malignant and frequently occurring type of primary astrocytoma and accounts for approximately 60% of adult brain tumors [2, 75].

The World Health Organization classifies gliomas according to grade. Grade I gliomas have low proliferative potential whereas grade II to IV gliomas are highly malignant and invasive [3, 76].

The prime modality of treatment for gliomas is surgery and attempting an R0 resection. Adjuvant treatment after surgery involves radiotherapy and chemotherapy.

The EORTC 26981–22981/NCIC CE3 by Stupp et al. conducted a trial on newly diagnosed, histologically confirmed glioblastoma comparing radiotherapy alone or radiotherapy plus continuous daily temozolomide. At a median follow-up of 28 months, the median survival was 14.6 months with radiotherapy plus temozolomide and 12.1 months with radiotherapy alone [77]. On long term follow up, the overall survival at 5 years was 9.8% (6.4–14.0) with temozolomide versus 1.9% (0.6–4.4) with radiotherapy alone (hazard ratio (HR) 0.6, 95% CI 0.5–0.7; $p < 0.0001$) [78].

The RTOG 9402 trial randomized patients with anaplastic oligodendrogliomas and astrocytomas to procarbazine, lomustine, and vincristine (PCV) plus radiotherapy (RT) versus RT alone. Even though the there was no difference in the median survival between two cohorts in the overall population, patients with 1p/19q codeletion in the tumors had a better median survival (PCV plus RT: 14.7 vs. 2.6 years, HR = 0.36, 95% CI, 0.23 to 0.57, $p < 0.001$; RT: 7.3 vs. 2.7 years, HR = 0.40, 95% CI, 0.27 to 0.60, $p < 0.001$) and had significantly more benefit with PCV + RT (14.7 vs. 7.3 years; HR = 0.59; 95% CI, 0.37 to 0.95; $p = 0.03$) [79].

Currently, the standard protocol followed is conventional radiotherapy of 60 Gy in 2-Gy fractions over 6 weeks in combination with temozolomide as per the Stupp protocol in view of ease of delivery of treatment.

Adult medulloblastoma is uncommon. Molecular characterization forms an important part in risk stratification. The treatment involves a multimodality approach consisting of radiotherapy, surgery and chemotherapy. Concurrent weekly vincristine during craniospinal RT is widely used in pediatric protocols and often adopted in the treatment of adult medulloblastoma patients. Hematologic toxicity is rare, and therefore, it is feasible in conjunction with craniospinal RT [80]. Several other agents like hydroxyurea, platinum agents, etoposide and temozolomide have been tried as radiosensitizers.

9.2.3.5 Chemoradiation in Head and Neck Cancer

The role of chemoradiation in head and neck cancer is mostly in the locally advanced setting, both in the definitive and adjuvant setting, and helps improve functional outcomes and cure rates. Chemotherapy can be administered as induction chemotherapy prior to RT or chemoradiation or as concurrent chemotherapy at the same time as RT (cCRT). The Veterans Affairs Laryngeal Cancer Study Group [81, 82] trial and multiple other randomized trials established the role of incorporating both chemotherapy and RT, in patients with squamous cell head and neck cancer.

The Meta-Analysis of Chemotherapy on Head and Neck Cancer (MACH-NC) was updated in 2021. Published or unpublished randomized trials including patients with nonmetastatic carcinoma randomized between 1965 and 2016 were eligible [83]. These trials compared curative locoregional treatment (LRT) to LRT + CT or adding another timing of CT to LRT + CT (main question), or compared induction CT + radiotherapy to radiotherapy + concomitant (or alternating) CT (secondary question). Individual patient data (IPD) were collected and OS was the main endpoint. For the main question, 101 trials (18,951 patients, median follow-up of 6.5 years) were analyzed. Around 90% of the patients had stage III or IV disease. Interaction between treatment effect on OS and the timing of CT was significant ($p < 0.0001$), the benefit being limited to concomitant CT (HR: 0.83, 95% CI [0.79; 0.86]; 5-year absolute benefit of 6.5% (3.6%). Efficacy decreased as patients age increased ($p\_trend = 0.03$). OS was not increased by the addition of induction (HR = 0.96 [0.90; 1.01]) or adjuvant CT (1.02 [0.92; 1.13]). Efficacy of induction CT decreased with poorer performance status ($p\_trend = 0.03$). Eight trials (1214 patients) confirmed the superiority of concomitant CT on OS (HR = 0.84 [0.74; 0.95], $p = 0.005$). The update of MACH-NC confirmed the benefit and superiority of the addition of concomitant CT for nonmetastatic head and neck cancer.

Induction chemotherapy followed by cCRT versus cCRT alone was compared in four major trials each with different conclusions. The Spanish trial showed no survival benefit for induction chemotherapy followed by cCRT compared to cCRT alone as probably most patients could not complete high doses of total cisplatin (525 mg/m^2 or 600 mg/m^2). However, in a subgroup analysis, those who completed the treatment as per protocol had a significant benefit in the PFS [81]. In the Italian phase II/III trial there was an improvement in the PFS and OS in the induction chemotherapy plus cCRT arm compared with cCRT alone [84]. The DeCIDE trial which used a chemoradiation protocol consisting of docetaxel, fluorouracil, and hydroxyurea failed to show any overall survival benefit of adding TPF induction chemotherapy to concurrent therapy [82]. Lastly, the French GORTEC 2007–02 trial compared a cCRT regimen of carboplatin and fluorouracil versus an induction chemo protocol of TPF followed by cetuximab and RT. Induction was complicated by high rates of febrile neutropenia (17%), treatment related mortality (6.6%) and inability to receive RT as planned [85]. The trial showed significantly lower rates of distant metastasis for induction chemotherapy, but similar rates for locoregional control, PFS and OS.

The final role of induction chemotherapy remains controversial as per the conflicting results of these trials, and different regimens used. The use of this should largely be restricted to a certain subgroup of patients who have bulky nodal disease and T3, T4 lesions as per the sub group analysis of Italian and DeCIDE trials [82, 84].

9.2.4 Definitive Chemoradiation (Head and Neck)

For concurrent definitive chemoradiation, cisplatin remains the gold standard. Cisplatin is preferred for patients with good performance status. Efficacy of cisplatin is dose dependent and doses of >200 mg/m^2 are more effective than lower doses. Two main schedules of cisplatin are used in practice; the weekly 40 mg/m^2 dose and 3 weekly 100 mg/m^2 bolus dose. Randomized trials have compared these 2 regimens. The JCOG 1008 showed better tolerance and noninferior OS for the weekly cisplatin regimen, although the trial population had mostly postoperative patients [86]. The phase 3 trial conducted in Tata Memorial Hospital in India compared 30 mg/m^2 weekly dose to 100 mg/m^2 every 3 weeks [87]. The results showed that weekly cisplatin had worse locoregional control than the 3 weekly regimen, but similar OS and better tolerability. The primary end point of noninferiority was not met, but questions remained, as the dose of cisplatin used was lower (30 mg/m^2) than the standard dose used in a majority of centers. Also, the majority of patients had oral cavity tumors (87%) and received adjuvant chemoradiation (93%), thus decreasing applicability to other head and neck subsites and in the definitive setting. Other retrospective studies have looked at the administration of weekly cisplatin in various different populations and have shown favorable results [88–90]. However, as of today, the gold standard remains the 3 weekly 100 mg/m^2 bolus dose.

Carboplatin may be used as an alternative agent in those who have contraindications for the use of cisplatin, albeit less efficacious [91, 92]. A weekly dose of AUC 1.5 to 2 has been used either alone or in combination with fluorouracil [93].

Cetuximab, which is a monoclonal antibody that inhibits EGFR is a less preferred radiosensitizer for chemoradiation in those who have good performance status but ineligible for cisplatin. Though the Bonner trial showed improved OS with the addition of cetuximab to RT, patients ineligible for cisplatin were excluded from this study [94]. Current available data do not support use of cetuximab with RT in older patients with significant comorbidities and such patients usually receive RT alone. Cetuximab plus RT showed inferior survival when compared to cisplatin plus RT in fit patients, regardless of the schedule of administration of cisplatin [95, 96].

Another monoclonal antibody approved for use in India along with RT is nimotuzumab, which acts against the extracellular domain of EGFR. The phase 3 study conducted in India, in patients with locally advanced head and neck cancers fit for cCRT, randomized patients to receive cCRT with cisplatin (30 mg/m^2) or cCRT with cisplatin (30 mg/m^2) and nimotuzumab. The addition of nimotuzumab improved PFS (HR, 0.69; $p = 0.004$), LRC (HR, 0.67; $p = 0.006$), and DFS (HR, 0.71; $p = 0.008$) and had a trend toward improved OS (HR, 0.84; $p = 0.163$) [97].

9.2.5 Postoperative Chemoradiation (Head and Neck)

In the adjuvant setting also, cisplatin remains the chemotherapy sensitizer of choice both and can be used in both schedules; 100 mg/m^2 bolus 3 weekly or 40 mg/m^2 weekly (as discussed above) [86, 87]. For those ineligible for cisplatin, carboplatin may be used, though some may prefer RT alone. Cetuximab has not been used in the postoperative adjuvant setting and is thus not routinely recommended. Patient selection is paramount when choosing cisplatin along with radiation. The absolute and relative contraindications to the use of cisplatin in the adjuvant and definitive treatment of head and neck cancer are detailed in Table 9.5 [98].

9.2.6 Preoperative Chemoradiation in Esophageal Cancer

The primary curative modality for lower esophagus GEJ tumors is surgery, although long-term outcomes are often unsatisfactory, even for microscopically complete (R0) resection [99]. This has led to exploring multimodality treatment approaches like neoadjuvant and adjuvant approaches. This article will cover the influence of neo-adjuvant treatment in locally advanced esophageal adenocarcinoma.

To improve systemic control and local control rates (higher pathological complete response), NACT-RT has been studied in several randomized trials, including gastroesophageal junction (GEJ) adenocarcinoma. Of the several trials, three trials (DUTCH CROSS trial [100] Irish trial [101] and CALBG9781 [102]) have shown statically significant overall survival (OS) benefit.

The landmark and largest trial on NACT-RT was the Dutch CROSS trial [2]. In this trial, 366 patients with resectable esophagus tumors and GEJ (75% adenocarcinoma and 23% squamous cell carcinoma [SCC]) were randomly assigned to NACT-RT, followed by surgery versus surgery alone. Neoadjuvant CT-RT consisted of radiotherapy (RT) of 41.4 Gray in 23 fractions daily over 5 weeks along with concurrent weekly paclitaxel (50 mg/m^2) and carboplatin (area under the curve [AUC] of concentration X 2 times). NACT-RT was well tolerated. The R0 resection was higher in NACT-RT (92% versus 69%). Pathological complete response was observed in 29% of patients in the NACT-RT arm (23% adenocarcinoma and 49% SCC); Median OS was better in the NACT-RT arm (49.4 months vs. 24 months). The OS rates at 2 years and 5 years in the NACT-RT arm were 67% and 47%, respectively, which was better than the surgery alone arm (2 years—50% and 5 years—34%). There was no significant difference in postop complications in both arms. Almost 94% of patients underwent surgery in the NACT-RT arm. Grade 3 or

Table 9.5 Absolute and relative contraindications to cisplatin in definitive or postoperative treatment of locally advanced head and neck cancer, modified from Ahn et al. [98]

Clinical condition	Relative contraindications	Absolute contraindications
Performance status	ECOG score = 2	ECOG score ≥ 3
Biological age	According to geriatric assessment and screening tools	ND
Renal dysfunction	Creatinine clearance 50–60 mL/min	Creatinine clearance <50 mL/min
Hearing impairment	Hearing loss or tinnitus grade = 1 or 2[a]	Hearing loss or tinnitus grade = ≥3[a]
Neuropathy	Grade = 1[a]	Grade = ≥2[a]
Marrow, hepatic, respiratory, and cardiovascular, dysfunctions	Grade 2[a] or Child–Pugh score = B[c]	Grade ≥3[a] or Child–Pugh score = C[c]
Other comorbidities	Insulin-dependent diabetes mellitus, recurrent (pulmonary) infections, severe psychiatric disorders interfering with treatment compliance	Life-threatening conditions such as uncontrolled systemic infection or autoimmune disease
HIV/AIDS	CD4 count 200–350/μL[d]	CD4 count <200/μL[d]
Nutritional status	Involuntary weight loss ≥20%	ND
Pregnancy and lactation	ND	First trimester of pregnancy[e], lactation not recommended
Hypersensitivity to platinum agents	ND	Allergy to agents that contain platinum[f] or mannitol
Previous platinum therapy	>200 mg/m^2 or > 3 cycles of TPF induction	ND
Drug interactions	Concomitant use of nephrotoxic drugs	ND
Socioeconomic status	Impaired social and economic support	ND

HIV/AIDS human immunodeficiency virus infection/acquired immune deficiency syndrome, *ECOG* Eastern Oncology Cooperative Group, *TPF* docetaxel, cisplatin, 5-fluorouracil, *ND* not defined
[a]Based on the National Cancer Institute Common Toxicity Criteria version 4.0
[b]Repeated audiometry exams may be indicated during the treatment
[c]For hepatic impairment
[d]World Health Organization definition
[e]Fetal exposure to radiation, irrespective of the duration of pregnancy, increases the risk on developing malignancies in childhood and in addition is associated with abortion and intrauterine death. Therefore, radiotherapy is preferably postponed until after delivery
[f]If a skin test does not rule out cross-reactions among platinum agents

more hematological toxicity was observed in 7% of patients in the NACT-RT arm. The results of the Dutch CROSS trial changed the approach to treating esophageal and GEJ tumors and became the standard of care. In spite of impressive results, there were questions if neo-adjuvant chemotherapy (NACT) would be superior to NACT-RT, especially in esophageal adenocarcinomas.

The CROSS trial was challenged by phase II/III German FLOT-4 trial [102], published in 2019. In this trial, 716 patients, histologically proven resectable gastric (44%) and GEJ (56%) adenocarcinoma (classified according to Siewert), were randomized to four cycles of preoperative and four cycles of postoperative FLOT regimen (docetaxel 50 mg/m^2 plus oxaliplatin 85 mg/m^2 and leucovorin 200 mg/m^2 with short term infusional 5FU 2600 mg/m^2 as a 24-h infusion, all on day 1, administered every 2 weeks) versus epirubicin-based triplet therapy (ECF/ECX). At a median follow-up of 43 months, the median OS was 50 months in FLOT versus 35 months favoring the FLOT regimen. A pronounced effect was seen with signet ring cells tumors. Perioperative complications were similar. Significantly more Grade 3 and 4 nausea, diarrhea, vomiting, neutropenia, and infections were seen with FLOT regimen and only 46% patients completed all cycles of chemotherapy.

Recently, the 10 years' data of CROSS trial [103] was published in 2021, which showed that after a median follow-up of 157 months, the absolute 10-year overall survival (OS) benefit was 13% (38% vs. 25%). NACT-RT reduced the risk of mortality from esophageal cancer (HR, 0.60; 95% CI, 0.46 to 0.80).

After the publication of the FLOT trial, many oncologists have been treating their patients with this strategy of chemotherapy and surgery in esophageal adenocarcinoma. However, CROSS is a very effective strategy for squamous cell carcinoma as well as adenocarcinoma.

Ongoing trials to explore future strategies include the NEO-AEGIS trial (Fig. 9.6) [104] directly compared the CROSS NACT-RT versus MAGIC/FLOT-based NACT. In the preliminary report present in ASCO 2021, 377 patients with cT_{2-3} N_{0-3} M_0 AEG were randomly assigned to CROSS or perioperative CT with a median follow-up of 24.5 (1–92) months. There were a total of 143 deaths (70 CROSS, 73 MAGIC/FLOT arm). While the CROSS arm had more favorable pCR rates, R0 resections and toxicity profile, the 3-year estimated survival probability was 56% (95% CI 47–64) CROSS and 57% (95% CI 48–65) MAGIC/FLOT (Table 9.6).

Other ongoing trials include the OPEC trial [105], which compares FLOT with the CROSS trial for adenocarcinoma, the TopGear trial [106] which compares prechemotherapy versus NACT-RT, and the RACE trial [107] which is a German-based prospective phase III trial comparing FLOT plus CTRT plus surgery versus FLOT plus surgery.

To conclude, concurrent chemoradiotherapy is a highly effective locoregional treatment modality for solid tumors. These concepts can be applied as a radical approach or combined with surgery as adjuvant/induction or neoadjuvant modality. With our increasing knowledge of the molecular basis of chemotherapy and radiation interactions, we have obtained new insights into how to best use and potentially combine agents with radiation, with benefits already proven in patients.

Therapeutic ratios have to be improved by exploring newer molecules to tackle hypoxia and for use with concurrent radiation. Concurrent chemoradiation has been extensively used and perfected in almost all sites for treating locally advanced solid malignancies. Further clinical trials and newer molecules have to be tested in order to increase the current cure rates in resistant cancers and malignancies with poorer prognoses. With a better understanding of chemo, radiobiology at cellular levels, and use of advanced simulations using AI, newer treatment modalities can be tailored and implemented faster, cutting short

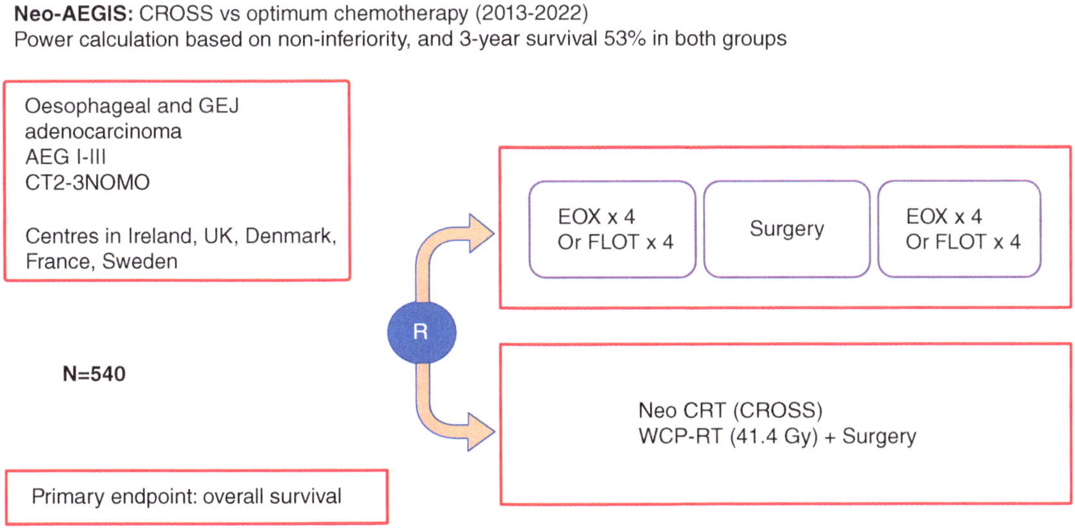

Fig. 9.6 Protocol of Neo-AEGIS trial

Table 9.6 Comparison of toxicities of MAGIC/FLOT versus CROSS arms of Neo-AEGIS trial

	Arm A (MAGIC/FLOT)	Arm B CROSS
R0 (negative margins)	82%	95%
ypN0	44.5%	60.1%
Tumor regression grades 1 and 2	12.1%	41.7%
Pathologic complete response	5%	16%
Neutropenia (Gr3/4)	14.1%	2.8%
Neutropenic sepsis	2.7%	0.6%
Postoperative in-hospital deaths	3%	3%
Postoperative pneumonia/ARDS	20%/0.6%	16%/4.3%
Anastomotic leak	12%	11.7%

time required for preclinical trials to approved users.

Key Points

1. Concomitant chemoradiation can be used with the intent of tackling distant metastasis, improving local control and even survival by making cells more sensitive to radiation.
2. In localized anal cancer, 5-FU/Capecitabine combined with mitomycin along with radiation is the standard of care.
3. Standard of care for treatment of unresectable stage 3 nonsmall cell lung cancers is concurrent chemoradiation.
4. The commonly used chemotherapy regimen in nonsmall cell lung cancer is weekly paclitaxel (50 mg/m^2) and carboplatin (AUC2).
5. The mainstay of treatment of small cell lung cancer-limited stage is chemoradiation.
6. Chemoradiation is the treatment of choice of carcinoma cervix for Stage 1B2 to Stage 3 disease using a combination of EBRT and ICRT.
7. Cisplatin remains the ideal chemotherapy agent used with radiation in head and neck cancer.
8. Patients unfit for cisplatin in head and neck cancer can be considered for less intense regimens like carboplatin or EGFR inhibitors (cetuximab or nimotuzumab).

Multiple Choice Questions

1. If a patient is suspected of having N0/N1 disease but intraoperatively is found to be having N2 disease, the so-called incidental N2 disease, then which of the following will be the best course of treatment?
   (a) Abandon the procedure and treat with chemoradiation
   (b) Complete the procedure and treat with adjuvant chemotherapy
   (c) Using intraoperative radiotherapy (IORT) and completing the resection
   (d) Doing a thorough mediastinal node clearance and treating with adjuvant chemoradiation
2. Which one of the following statements is true of small cell lung cancer (SCLC)?
   (a) Surgery is the most important treatment modality
   (b) Limited stage disease describes disease less than T2 N1 M0
   (c) Approximately 80% of patients respond to chemotherapy
   (d) Prophylactic cranial irradiation has been shown to prolong survival
3. A 65-year-old man with a 40 pack-year history of smoking presents to the emergency room with shortness of breath. Chest radiograph demonstrates a left lower lobe mass and left pleural effusion. CT reveals a 3-cm left lower lobe mass, left hilar fullness, and a moderate left pleural effusion. Biopsy of the mass and thoracentesis are both positive for adenocarcinoma. Staging studies do not reveal any distant metastases. Which of the

following is the next best step in his management?
   (a) Referral to a thoracic surgeon
   (b) Radiation to the chest
   (c) Concurrent chemotherapy and radiation
   (d) Platinum-based doublet therapy
4. Papilloma viruses are characterized by what genetic feature?
   (a) A genome of ds circular DNA encoding around 8 ORF
   (b) A virus with RT able to integrate and cause cancer
   (c) A variable ds RNA virus
   (d) A large DNA containing virus
5. Most cervical cancers arise:
   (a) From endocervix
   (b) From ectocervix
   (c) From os of cervix
   (d) At the junction of ecto and endocervix
6. Which of the following best reflects the evidence on chemoradiotherapy (chemotherapyand radiotherapy given simultaneously for a short duration of time) for treating bulky early-stage cervical cancer?
   (a) There is no evidence of any survival benefit from adding chemotherapy to radiotherapy alone
   (b) Several RCTs have found that chemoradiotherapy improves overall survival and progression-free survival compared with radiotherapy, when used either with or after hysterectomy
   (c) Several RCTs have found that chemoradiotherapy improves progression-free survival but not overall survival compared with radiotherapy, when used either with or after hysterectomy
7. Which cumulative dose of cisplatin is the most effective in head and neck cancer?
   (a) >100 mg/m^2
   (b) >150 mg/m^2
   (c) >200 mg/m^2
   (d) >400 mg/m^2
8. Which of these is an absolute contraindication for the use of concurrent cisplatin with radiation?
   (a) Creatinine clearance <50 mL/min
   (b) Mild hearing loss
   (c) Grade 1 neuropathy
   (d) ECOG status 2
9. Cetuximab, used as a radiosensitizing agent in head and neck cancer, works by targeting which receptor?
   (a) VEGF
   (b) EGFR
   (c) mTOR
   (d) PD-1
10. When phase in the cell cycle is the most resistant to radiation?
   (a) G1
   (b) S
   (c) G2
   (d) M
11. All the factors below are responsible for radioresistance, except
   (a) Reoxygenation
   (b) Repopulation
   (c) Redistribution
   (d) Repair

Answers

1. (a)
2. (b)
3. (d)
4. (a)
5. (d)
6. (b)
7. (c)
8. (a)
9. (b)
10. (b)
11. (a)

References

1. Steel GG, Peckham MJ. Exploitable mechanisms in combined radiotherapy-chemotherapy: the concept of additivity. Int J Radiat Oncol Biol Phys. 1979;5:85–91.
2. Coleman CN. Hypoxia in tumours: a paradigm for the approach to biochemical and physiologic heterogeneity. J Natl Cancer Inst. 1988;80:310–7.

3. Gray LH, Conger AD, Ebert M, Hornsey S, Scott OCA. The concentration of oxygen dissolved in tissues at the time of irradiation as a factor in radiotherapy [internet]. Br J Radiol. 1953;26:638–48.
4. Kumar P. Tumour hypoxia and anemia: impact on efficacycacy of radiation therapy. Semin Hematol. 2000;37(4 Suppl 6):4–8.
5. Overgaard J, Hansen HS, Andersen AP, et al. Misonidazole combined with split-course radiotherapy in the treatment of invasive carcinoma of larynx and pharynx: report from the DAHANCA 2 study. Int J Radiat Oncol Biol Phys. 1989;16:1065–8.
6. Fowler JF, Lindstrom MJ. Loss of local control with prolongation in radiotherapy. Int J Radiat Oncol Biol Phys. 1992;23:457–67.
7. Trott KR, Kummermehr J. What is known about tumour proliferation rates to choose between accelerated fractionation or hyperfractionation? Radiother Oncol. 1985;3:1–9.
8. Krause M, Ostermann G, Petersen C, et al. Decreased repopulation as well as increased reoxygenation contribute to the improvement in local control after targeting of the EGFR by C225 during fractionated irradiation. Radiother Oncol. 2005;76:162–7.
9. Grillo-Ruggieri F, Mantello G. Is one Gy equal to one Gy after treatment interruptions? Rays. 2004;29:275–8.
10. Schmidt-Ullrich RK, Contessa JN, Dent P, et al. Molecular mechanisms of radiation-induced accelerated repopulation. Radiat Oncol Investig. 1999;7:321–30.
11. Bennett CB, Lewis AL, Baldwin KK, Resnick MA. Lethality induced by a single site-specific double-strand break in a dispensable yeast plasmid. Proc Natl Acad Sci U S A. 1993;90:5613–7.
12. Nickoloff JA, Hoekstra MF. DNA damage and repair: DNA repair in higher eukaryotes, vol. 2. Berlin: Springer Science & Business Media; 1998.
13. Begg AC. Cisplatin and radiation: interaction probabilities and therapeutic possibilities. Int J Radiat Oncol Biol Phys. 1990;19:1183–9.
14. Yang LX, Douple EB, O'Hara JA, Wang HJ. Production of DNA double-strand breaks by interactions between carboplatin and radiation: a potential mechanism for radiopotentiation. Radiat Res. 1995;143:309–15.
15. Sinclair WK, Morton RA. X-ray sensitivity during the cell generation cycle of cultured Chinese hamster cells. Radiat Res. 1966;29:450–74.
16. Terasima T, Tolmach LJ. Changes in x-ray sensitivity of HeLa cells during the division cycle. Nature. 1961;190:1210–1.
17. Vaupel P. Tumour microenvironmental physiology and its implications for radiation oncology. Semin Radiat Oncol. 2004;14:198–206.
18. Semenza GL. Targeting HIF-1 for cancer therapy. Nat Rev Cancer. 2003;3:721–32.
19. Nigro ND, Vaitkevicius VK, Considine B Jr. Combined therapy for cancer of the anal canal: a preliminary report. Dis Colon Rectum. 1974;17(3):354–6.
20. Leichman L, Nigro N, Vaitkevicius VK, et al. Cancer of the anal canal. Model for preoperative adjuvant combined modality therapy. Am J Med. 1985;78:211–5.
21. Doci R, Zucali R, La Monica G, et al. Primary chemoradiation therapy with fluorouracil and cisplatin for cancer of the anus: results in 35 consecutive patients. J Clin Oncol. 1996;14:3121–5.
22. Peiffert D, Seitz JF, Rougier P, et al. Preliminary results of a phase II study of high-dose radiation therapy and neoadjuvant plus concomitant 5-fluorouracil with CDDP chemotherapy for patients with anal canal cancer: a French cooperative study. Ann Oncol. 1997;8:575–81.
23. Gerard JP, Ayzac L, Hun D, et al. Treatment of anal canal carcinoma with high dose radiation therapy and concomitant fluorouracil-cisplatinum. Long-term results in 95 patients. Radiother Oncol. 1998;46:249–56.
24. Sischy B, Doggett RL, Krall JM, et al. Definitive irradiation and chemotherapy for radiosensitization in management of anal carcinoma: interim report on Radiation Therapy Oncology Group Study No. 8314. J Natl Cancer Inst. 1989;81:850–6.
25. Allal A, Kurtz JM, Pipard G, et al. Chemoradiotherapy versus radiotherapy alone for anal cancer: a retrospective comparison. Int J Radiat Oncol Biol Phys. 1993;27:59–66.
26. Martenson JA, Lipsitz SR, Lefkopoulou M, et al. Results of combined modality therapy for patients with anal cancer (E7283). An Eastern Cooperative Oncology Group Study. Cancer. 1995;76:1731–6.
27. Northover J, Glynne-Jones R, Sebag-Montefiore D, et al. Chemoradiation for the treatment of epidermoid anal cancer: 13-year follow-up of the first randomised UKCCCR anal cancer trial (ACT I). Br J Cancer. 2010;102:1123–8.
28. Bartelink H, Roelofsen F, Eschwege F, et al. Concomitant radiotherapy and chemotherapy is superior to radiotherapy alone in the treatment of locally advanced anal cancer: results of a phase III randomized trial of the European Organization for Research and Treatment of Cancer Radiotherapy and Gastrointestinal Cooperative Groups. J Clin Oncol. 1997;15:2040–9.
29. Glynne-Jones R, Meadows H, Wan S, et al. EXTRA—a multicenter phase II study of chemoradiation using a 5 day per week oral regimen of capecitabine and intravenous mitomycin C in anal cancer. Int J Radiat Oncol Biol Phys. 2008;72:119–26.
30. Bhargava PG, Engineer R, Ramaswamy A. Et efficacycacy and tolerability of capecitabine and mitomycin-C based concurrent radiotherapy in patients with anal canal cancer. Ann Oncol. 2020;31:S1283.
31. Flam M, John M, Pajak TF, et al. Role of mitomycin in combination with fluorouracil and radiotherapy, and of salvage chemoradiation in the definitive non-

surgical treatment of epidermoid carcinoma of the anal canal: results of a phase III randomized intergroup study. J Clin Oncol. 1996;14:2527–39.
32. Gunderson LL, Winter KA, Ajani JA, et al. Long-term update of US GI intergroup RTOG 98-11 phase III trial for anal carcinoma: survival, relapse, and colostomy failure with concurrent chemoradiation involving fluorouracil/mitomycin versus fluorouracil/cisplatin. J Clin Oncol. 2012;30:4344–51.
33. James RD, Glynne-Jones R, Meadows HM, et al. Mitomycin or cisplatin chemoradiation with or without maintenance chemotherapy for treatment of squamous-cell carcinoma of the anus (ACT II): a randomised, phase 3, open-label, 2×2 factorial trial [internet]. Lancet Oncol. 2013;14:516–24.
34. Conibear J, AstraZeneca UK Limited. Rationale for concurrent chemoradiotherapy for patients with stage III non-small-cell lung cancer. Br J Cancer. 2020;123(Suppl 1):10–7.
35. Eberhardt WEE, De Ruysscher D, Weder W, et al. 2nd ESMO consensus conference in lung cancer: locally advanced stage III non-small-cell lung cancer. Ann Oncol. 2015;26:1573–88.
36. Postmus PE, Kerr KM, Oudkerk M, et al. Early and locally advanced non-small-cell lung cancer (NSCLC): ESMO clinical practice guidelines for diagnosis, treatment and follow-up. Ann Oncol. 2017;28(suppl_4):iv1–21.
37. Curran WJ Jr, Paulus R, Langer CJ, et al. Sequential vs. concurrent chemoradiation for stage III non-small cell lung cancer: randomized phase III trial RTOG 9410. J Natl Cancer Inst. 2011;103:1452–60.
38. Furuse K, Fukuoka M, Kawahara M, et al. Phase III study of concurrent versus sequential thoracic radiotherapy in combination with mitomycin, vindesine, and cisplatin in unresectable stage III non-small-cell lung cancer. J Clin Oncol. 1999;17:2692–9.
39. O'Rourke N, Roqué I, Figuls M, Farré Bernadó N, Macbeth F. Concurrent chemoradiotherapy in non-small cell lung cancer. Cochrane Database Syst Rev. 2010;(6):CD002140.
40. Aupérin A, Le Péchoux C, Rolland E, et al. Meta-analysis of concomitant versus sequential radiochemotherapy in locally advanced non-small-cell lung cancer. J Clin Oncol. 2010;28:2181–90.
41. O'Rourke N, Macbeth F. Is concurrent chemoradiation the standard of care for locally advanced non-small cell lung cancer? A review of guidelines and evidence. Clin Oncol. 2010;22:347–55.
42. Koning CC, Wouterse SJ, Daams JG, Uitterhoeve LL, van den Heuvel MM, Belderbos JS. Toxicity of concurrent radiochemotherapy for locally advanced non-small-cell lung cancer: a systematic review of the literature. Clin Lung Cancer. 2013;14:481–7.
43. Maguire J, Khan I, McMenemin R, et al. SOCCAR: a randomised phase II trial comparing sequential versus concurrent chemotherapy and radical hypofractionated radiotherapy in patients with inoperable stage III non-small cell lung cancer and good performance status. Eur J Cancer. 2014;50:2939–49.
44. Perez CA, Stanley K, Rubin P, et al. A prospective randomized study of various irradiation doses and fractionation schedules in the treatment of inoperable non-oat-cell carcinoma of the lung. Preliminary report by the radiation therapy oncology group. Cancer. 1980;45:2744–53.
45. Bradley JD, Paulus R, Komaki R, et al. Standard-dose versus high-dose conformal radiotherapy with concurrent and consolidation carboplatin plus paclitaxel with or without cetuximab for patients with stage IIIA or IIIB non-small-cell lung cancer (RTOG 0617): a randomised, two-by-two factorial phase 3 study. Lancet Oncol. 2015;16:187–99.
46. Guo SX, Jian Y, Chen YL, Cai Y, Zhang QY, Tou FF. Neoadjuvant Chemoradiotherapy vesus chemotherapy alone followed by surgery for Resectable stage III non-small-cell lung cancer: a meta-analysis. Sci Rep. 2016;6:34388.
47. Shepherd FA, Ginsberg RJ, Haddad R, et al. Importance of clinical staging in limited small-cell lung cancer: a valuable system to separate prognostic subgroups. The University of Toronto Lung Oncology Group. J Clin Oncol. 1993;11:1592–7.
48. Pignon JP, Arriagada R, Ihde DC, et al. A meta-analysis of thoracic radiotherapy for small-cell lung cancer. N Engl J Med. 1992;327:1618–24.
49. De Ruysscher D, Pijls-Johannesma M, Vansteenkiste J, Kester A, Rutten I, Lambin P. Systematic review and meta-analysis of randomised, controlled trials of the timing of chest radiotherapy in patients with limited-stage, small-cell lung cancer. Ann Oncol. 2006;17:543–52.
50. Pijls-Johannesma MCG, De Ruysscher D, Lambin P, Rutten I, Vansteenkiste JF. Early versus late chest radiotherapy for limited stage small cell lung cancer. Cochrane Database Syst Rev. 2005;2004:CD004700.
51. Turrisi AT 3rd, Kim K, Blum R, et al. Twice-daily compared with once-daily thoracic radiotherapy in limited small-cell lung cancer treated concurrently with cisplatin and etoposide. N Engl J Med. 1999;340:265–71.
52. Movsas B, Moughan J, Komaki R, et al. Radiotherapy patterns of care study in lung carcinoma. J Clin Oncol. 2003;21:4553–9.
53. Spigel DR, Hainsworth JD, Burris HA, et al. Long-term follow-up of limited stage small cell lung cancer patients treated with carboplatin-based chemotherapy and radiotherapy by the Minnie Pearl Cancer Research Network (MPCRN). J Clin Orthod. 2004;22(14_suppl):7222.
54. Baas P, Belderbos JSA, Senan S, et al. Concurrent chemotherapy (carboplatin, paclitaxel, etoposide) and involved-field radiotherapy in limited stage small cell lung cancer: a Dutch multicenter phase II study. Br J Cancer. 2006;94:625–30.
55. Stinchcombe TE, Gore EM. Limited-stage small cell lung cancer: current chemoradiotherapy treatment paradigms. Oncologist. 2010;15:187–95.
56. Sung H, Ferlay J, Siegel RL, et al. Global cancer statistics 2020: GLOBOCAN estimates of incidence

and mortality worldwide for 36 cancers in 185 countries. CA Cancer J Clin. 2021;71:209–49.
57. Bhatla N, Berek JS, Cuello Fredes M, et al. Revised FIGO staging for carcinoma of the cervix uteri. Int J Gynaecol Obstet. 2019;145:129–35.
58. Tod MC, Meredith WJ. A dosage system for use in the treatment of cancer of the uterine cervix. Br J Radiol. 1938;11:809–24.
59. Bhatla N, Aoki D, Sharma DN, Sankaranarayanan R. Cancer of the cervix uteri. Int J Gynaecol Obstet. 2018;143(Suppl 2):22–36.
60. Sharma DN, Rath GK, Gandhi AK, Kumar P. Low-dose-rate versus high-dose-rate versus pulsed-dose-rate Intracavitary brachytherapy in cervical carcinoma: a Monoinstitutional comparative study. Int J Radiat Oncol. 2015; https://www.redjournal.org/article/S0360-3016(15)01978-1/abstract.
61. Sharma DN, Rath GK, Thulkar S, Kumar S, Subramani V, Julka PK. Use of transrectal ultrasound for high dose rate interstitial brachytherapy for patients of carcinoma of uterine cervix. J Gynecol Oncol. 2010;21(1):12–7.
62. Perez CA, Grigsby PW, Castro-Vita H, Lockett MA. Carcinoma of the uterine cervix. I. Impact of prolongation of overall treatment time and timing of brachytherapy on outcome of radiation therapy. Int J Radiat Oncol Biol Phys. 1995;32:1275–88.
63. Monk BJ, Tewari KS, Koh W-J. Multimodality therapy for locally advanced cervical carcinoma: state of the art and future directions. J Clin Oncol. 2007;25:2952–65.
64. Stehman FB, Bundy BN, Thomas G, et al. Hydroxyurea versus misonidazole with radiation in cervical carcinoma: long-term follow-up of a gynecologic oncology group trial. J Clin Oncol. 1993;11:1523–8.
65. Petrelli F, De Stefani A, Raspagliesi F, Lorusso D, Barni S. Radiotherapy with concurrent cisplatin-based doublet or weekly cisplatin for cervical cancer: a systematic review and meta-analysis. Gynecol Oncol. 2014;134:166–71.
66. Keys HM, Bundy BN, Stehman FB, et al. Cisplatin, radiation, and adjuvant hysterectomy compared with radiation and adjuvant hysterectomy for bulky stage IB cervical carcinoma. N Engl J Med. 1999;340:1154–61.
67. Peters WA 3rd, Liu PY, Barrett RJ 2nd, et al. Concurrent chemotherapy and pelvic radiation therapy compared with pelvic radiation therapy alone as adjuvant therapy after radical surgery in high-risk early-stage cancer of the cervix. J Clin Oncol. 2000;18:1606–13.
68. Whitney CW, Sause W, Bundy BN, et al. Randomized comparison of fluorouracil plus cisplatin versus hydroxyurea as an adjunct to radiation therapy in stage IIB-IVA carcinoma of the cervix with negative para-aortic lymph nodes: a gynecologic oncology group and southwest oncology group study. J Clin Oncol. 1999;17:1339–48.
69. Chemoradiotherapy for Cervical Cancer Meta-Analysis Collaboration. Reducing uncertainties about the effects of chemoradiotherapy for cervical cancer: a systematic review and meta-analysis of individual patient data from 18 randomized trials. J Clin Oncol. 2008;26:5802–12.
70. Green J, Kirwan J, Tierney J, et al. Concomitant chemotherapy and radiation therapy for cancer of the uterine cervix. Cochrane Database Syst Rev. 2005;(3):CD002225.
71. Nandakumar A, Kishor Rath G, Chandra Kataki A, et al. Concurrent chemoradiation for cancer of the cervix: results of a multi-institutional study from the setting of a developing country (India). J Glob Oncol. 2015;1:11–22.
72. Gupta S, Maheshwari A, Parab P, et al. Neoadjuvant chemotherapy followed by radical surgery versus concomitant chemotherapy and radiotherapy in patients with stage IB2, IIA, or IIB squamous cervical cancer: a randomized controlled trial. J Clin Oncol. 2018;36:1548–55.
73. Chopra SJ, Mathew A, Maheshwari A, et al. National Cancer Grid of India consensus guidelines on the management of cervical cancer. J Glob Oncol. 2018;4:1–15.
74. Alcantara Llaguno SR, Parada LF. Cell of origin of glioma: biological and clinical implications. Br J Cancer. 2016;115:1445–50.
75. Hanif F, Muzaffar K, Perveen K, Malhi SM, Simjee SU. Glioblastoma Multiforme: a review of its epidemiology and pathogenesis through clinical presentation and treatment. Asian Pac J Cancer Prev. 2017;18:3–9.
76. Louis DN, von Deimling A. Grading of diffuse astrocytic gliomas: Broders, Kernohan, Zülch, the WHO… and Shakespeare. Acta Neuropathol. 2017;134:517–20.
77. Stupp R, Mason WP, van den Bent MJ, et al. Radiotherapy plus concomitant and adjuvant temozolomide for glioblastoma. N Engl J Med. 2005;352:987–96.
78. Stupp R, Hegi ME, Mason WP, et al. Effects of radiotherapy with concomitant and adjuvant temozolomide versus radiotherapy alone on survival in glioblastoma in a randomised phase III study: 5-year analysis of the EORTC-NCIC trial. Lancet Oncol. 2009;10:459–66.
79. Cairncross G, Wang M, Shaw E, et al. Phase III trial of chemoradiotherapy for anaplastic oligodendroglioma: long-term results of RTOG 9402. J Clin Oncol. 2013;31:337–43.
80. Brandes AA, Franceschi E, Tosoni A, Blatt V, Ermani M. Long-term results of a prospective study on the treatment of medulloblastoma in adults. Cancer. 2007;110:2035–41.
81. Department of Veterans Affairs Laryngeal Cancer Study Group, Wolf GT, Fisher SG, et al. Induction chemotherapy plus radiation compared with surgery plus radiation in patients with advanced laryngeal cancer. N Engl J Med. 1991;324:1685–90.

82. Spaulding MB, Fischer SG, Wolf GT. Tumour response, toxicity, and survival after neoadjuvant organ-preserving chemotherapy for advanced laryngeal carcinoma. The Department of Veterans Affairs Cooperative Laryngeal Cancer Study Group. J Clin Oncol. 1994;12:1592–9.
83. Lacas B, Carmel A, Landais C, Wong SJ, Licitra L, Tobias JS, et al. Meta-analysis of chemotherapy in head and neck cancer (MACH-NC): an update on 107 randomized trials and 19,805 patients, on behalf of MACH-NC Group. Radiother Oncol. 2021;156:281–93.
84. Forastiere AA, Zhang Q, Weber RS, et al. Long-term results of RTOG 91-11: a comparison of three nonsurgical treatment strategies to preserve the larynx in patients with locally advanced larynx cancer. J Clin Oncol. 2013;31:845–52.
85. Haddad R, O'Neill A, Rabinowits G, et al. Induction chemotherapy followed by concurrent chemoradiotherapy (sequential chemoradiotherapy) versus concurrent chemoradiotherapy alone in locally advanced head and neck cancer (PARADIGM): a randomised phase 3 trial. Lancet Oncol. 2013;14:257–64.
86. Kiyota N, Tahara M, Fujii H, et al. Phase II/III trial of post-operative chemoradiotherapy comparing 3-weekly cisplatin with weekly cisplatin in high-risk patients with squamous cell carcinoma of head and neck (JCOG1008). J Clin Orthod. 2020;38(15_suppl):6502.
87. Noronha V, Joshi A, Patil VM, et al. Once-a-week versus once-every-3-weeks cisplatin chemoradiation for locally advanced head and neck cancer: a phase III randomized noninferiority trial. J Clin Oncol. 2018;36(11):1064–72.
88. Blanchard P, Bourhis J, Lacas B, et al. Taxane-cisplatin-fluorouracil as induction chemotherapy in locally advanced head and neck cancers: an individual patient data meta-analysis of the meta-analysis of chemotherapy in head and neck cancer group. J Clin Oncol. 2013;31:2854–60.
89. Oliva M, Huang SH, Xu W, et al. Impact of cisplatin dose and smoking pack-years in human papillomavirus–positive oropharyngeal squamous cell carcinoma treated with chemoradiotherapy. Eur J Cancer. 2019;118:112–20.
90. Forastiere AA, Maor M, Weber RS, et al. Long-term results of intergroup RTOG 91–11: a phase III trial to preserve the larynx—induction cisplatin/5-FU and radiation therapy versus concurrent cisplatin and radiation therapy versus radiation therapy. J Clin Oncol. 2006;24(18_suppl):5517.
91. Fountzilas G, Ciuleanu E, Dafni U, et al. Concomitant radiochemotherapy vs radiotherapy alone in patients with head and neck cancer: a Hellenic Cooperative Oncology Group Phase III Study. Med Oncol. 2004;21:95–107.
92. Lokich J, Anderson N. Carboplatin versus cisplatin in solid tumours: an analysis of the literature. Ann Oncol. 1998;9:13–21.
93. Douple EB, Richmond RC, O'Hara JA, Coughlin CT. Carboplatin as a potentiator of radiation therapy. Cancer Treat Rev. 1985;12 Suppl A:111–24.
94. Bonner JA, Harari PM, Giralt J, et al. Radiotherapy plus cetuximab for squamous-cell carcinoma of the head and neck. N Engl J Med. 2006;354:567–78.
95. Bernier J, Russi EG, Homey B, et al. Management of radiation dermatitis in patients receiving cetuximab and radiotherapy for locally advanced squamous cell carcinoma of the head and neck: proposals for a revised grading system and consensus management guidelines. Ann Oncol. 2011;22:2191–200.
96. Bonner JA, Harari PM, Giralt J, et al. Radiotherapy plus cetuximab for locoregionally advanced head and neck cancer: 5-year survival data from a phase 3 randomised trial, and relation between cetuximab-induced rash and survival. Lancet Oncol. 2010;11:21–8.
97. Patil VM, Noronha V, Joshi A, et al. A randomized phase 3 trial comparing nimotuzumab plus cisplatin chemoradiotherapy versus cisplatin chemoradiotherapy alone in locally advanced head and neck cancer. Cancer. 2019;125:3184. https://doi.org/10.1002/cncr.32179.
98. Ahn M-J, D'Cruz A, Vermorken JB, et al. Clinical recommendations for defining platinum unsuitable head and neck cancer patient populations on chemoradiotherapy: a literature review. Oral Oncol. 2016;53:10–6.
99. Sung H, et al. Global cancer statistics 2020: GLOBOCAN estimates of incidence and mortality worldwide for 36 cancers in 185 countries. CA Cancer J Clin. 2021;71:209–49.
100. van Hagen P, Hulshof MC, van Lanschot JJ, Steyerberg EW, van Berge Henegouwen MI, Wijnhoven BP, et al. Preoperative chemoradiotherapy for esophageal or junctional cancer. N Engl J Med. 2012;366:2074–84.
101. Tepper J, Krasna MJ, Niedzwiecki D, Hollis D, Reed CE, Goldberg R, et al. Phase III trial of trimodality therapy with cisplatin, fluorouracil, radiotherapy, and surgery compared with surgery alone for esophageal cancer: CALGB 9781. J Clin Oncol. 2008;26:1086–92.
102. Al-Batran SE, Homann N, Pauligk C, Goetze TO, Meiler J, Kasper S, Kopp HG, Mayer F, et al. Perioperative chemotherapy with fluorouracil plus leucovorin, oxaliplatin, and docetaxel versus fluorouracil or capecitabine plus cisplatin and epirubicin for locally advanced, resectable gastric or gastro-oesophageal junction adenocarcinoma (FLOT4): a randomised, phase 2/3 trial. Lancet. 2019;393:1948–57.
103. Eyck BM, van Lanschot JJB, Hulshof MCCM, van der Wilk BJ, Shapiro J, van Hagen P, et al. Ten-year outcome of neoadjuvant chemoradiotherapy plus surgery for esophageal cancer: the randomized controlled CROSS trial. J Clin Oncol. 2021;39:1995–2004.

104. Reynolds JV, Preston SR, O'Neill B, et al. Neo-AEGIS (Neoadjuvant trial in Adenocarcinoma of the Esophagus and Esophago-Gastric Junction International Study): Preliminary results of phase III RCT of CROSS versus perioperative chemotherapy (Modified MAGIC or FLOT protocol). J Clin Oncol. 2021;39(suppl 15):Abstr 4004. https://meetinglibrary.asco.org/record/195929/abstract. Accessed on Nov 11, 2022.
105. Hoeppner J, Lordick F, Brunner T, Glatz T, Bronsert P, Rothling N, et al. ESOPEC: prospective randomized controlled multicenter phase III trial comparing perioperative chemotherapy (FLOT protocol) to neoadjuvant chemoradiation (CROSS protocol) in patients with adenocarcinoma of the esophagus (NCT02509286). BMC Cancer. 2016;16:503.
106. Leong T, Smithers BM, Haustermans K, Michael M, Gebski V, Miller D, Zalcberg J, Boussioutas A, et al. TopGear: a randomized, phase III trial of perioperative ECF chemotherapy with or without preoperative Chemoradiation for Resectable gastric cancer: interim results from an international, intergroup trial of the AGITG, TROG, EORTC and CCTG. Ann Surg Oncol. 2017;24:2252–8.
107. Lorenzen S, Biederstadt A, Ronellenfitsch U, Reissfelder C, Monig S, Wenz F, et al. RACE-trial: neoadjuvant radiochemotherapy versus chemotherapy for patients with locally advanced, potentially resectable adenocarcinoma of the gastroesophageal junction—a randomized phase III joint study of the AIO, ARO and DGAV. BMC Cancer. 2020;20:886.
108. Walsh TN, Noonan N, Hollywood D, Kelly A, Keeling N, Hennessy TP. A comparison of multimodal therapy and surgery for esophageal adenocarcinoma. N Engl J Med. 1996;335:462–7.

Further Reading

https://www.youtube.com/watch?v=iluT8GzcN_8 (Chemoradiation in cervical cancer).

https://www.youtube.com/watch?v=gr3frQLgOoo (Chemoradiation in glioblastoma).

https://www.onclive.com/view/role-of-chemoradiation-in-resectable-pancreatic-cancer (Chemoradiation in pancreatic cancer).

https://www.youtube.com/watch?v=dLSyv0Ttmrs (Minimising side effects of chemoradiation).

10 Integrating Immunotherapy and Targeted Therapy with Radiation Therapy

Srinivas Chilukuri, Utpal Gaikwad, Sujith Kumar Mullapally, Santanu Samanta, and Pranshu Mohindra

10.1 Introduction

Immunotherapy has recently emerged as an effective systemic therapy for several malignancies and is rapidly being incorporated in treatment protocols for most malignancies [1]. There is an increased research regarding combining conventional therapies with immunotherapy to further enhance its effectiveness. Radiation therapy is one such treatment which has shown a great deal of promise as a combination agent. Radiation therapy is an important modality in cancer care in various forms of definitive, neoadjuvant, adjuvant, or palliative treatment. However, the precise mechanism of how radiation and immunotherapy benefit one another remains hazy.

S. Chilukuri (✉)
Department of Radiation Oncology, Apollo Proton Cancer Centre, Chennai, Tamil Nadu, India

U. Gaikwad
Krupamayi Hospitals, Aurangabad, Maharashtra, India

S. K. Mullapally
Department of Medical Oncology, Apollo Proton Cancer Centre, Chennai, Tamil Nadu, India

S. Samanta
Department of Radiation Oncology, University of Arkansas for Medical Sciences, Little Rock, AR, USA

P. Mohindra
Department of Radiation Oncology, University Hospital Cleaveland Medical Centre, Cleaveland, OH, USA

It is a well-established fact that radiation contributes to the local control, but radiation clearly seems to affect the systemic tumor control through several complex immune mechanisms. The local effect in the form of inducing cell death resulting from DNA damage triggers activation of $CD8^+$ T cells. Also by altering the tumor microenvironment, it induces systemic immune effects [2]. However, radiated tumor cells upregulate negative feedback pathways such as those regulated by immune checkpoint proteins that can reduce the immune response. Immune checkpoint inhibitors given along with radiation can overcome this and lead to enhanced systemic immune response. This biological synergism enables "spatial co-operation" through induced systemic immune response against out of field tumors (abscopal effect) as well as "cytotoxic enhancement" via immune-mediated increased cell kill among irradiated tumor sites. Therefore, these effects of radiation can be enhanced with immunotherapy agents. More importantly, several preclinical and clinical evidence in various tumor sites seem to indicate that effects of immunotherapy are enhanced due to the addition of radiation therapy. Typically, "immune cold" tumors that may not respond to immunotherapy agents, may transform to "immune hot" tumors after radiation therapy [3]. Immunotherapy is also often associated with transient pseudoprogressions leading to worsening of symptoms. Addition of radiation by reducing the tumor

A. Munshi et al. (eds.), *Radiation Oncology – Principles, Precepts and Practice*,
https://doi.org/10.1007/978-981-97-8389-2_10

burden allows one to buy time till immune therapy-related responses can be realized, therefore affecting "temporal modulation." Immunotherapy also can potentially promote normal tissue protection through TGF-beta inhibition-mediated suppression of radiation-induced fibrosis.

Use of checkpoint inhibitors combined with radiation therapy has been explored on a large scale, which has led to FDA approval of multiple immunotherapy agents for use with radiation therapy. The most robust evidence emerged with the PACIFIC trial which was a randomized trial evaluating addition of checkpoint inhibitor durvalumab to chemoradiation [4]. The combination demonstrated improvement in progression free survival and overall survival with an acceptable toxicity profile making this a new standard of care in locally advanced non-small cell lung cancers. Similarly, several ongoing trials are testing various combinations in several disease sites. These trials include a variety of immunotherapy agents and are testing combinations with varying dose, fractionations and volumes of radiation therapy. In the following paragraphs, the unique radiation therapy-immunotherapy agent interactions are summarized.

10.2 Immunogenic Effects of Radiation Therapy

Cancer cells in the body evolve from "immune elimination" where tumor cells are recognized and destroyed by the host immune system, to immune equilibrium and then "immune escape" where tumor tries to evade the host immune surveillance by several mechanisms, either by escaping detection (by downregulating MHC class I expression) or by avoiding destruction (creating immunosuppressive tumor microenvironment [5].

10.2.1 Increased Antigenicity

Radiation increases tumor cell antigenicity by releasing neoantigens generated from radiation-induced DNA damage for immune recognition. Tumor cells accumulate mutant nucleic acids and protein because of radiation-induced DNA damage, which has increased propensity for triggering immunogenic responses. The presence of the calcium-binding protein calretinin is a key signal to phagocytosis and radiation therapy helps to promote translocation of calreticulin from the endoplasmic reticulum to plasma membrane [6]. The other distinct cellular and molecular events that are triggered by radiation includes extracellular release of ATP, extracellular release of high mobility group box 1 (HMGB1) and production of cytokines such as type 1 interferon (IFN-1). These molecular signals collectively known as "damage-associated molecular patterns (DAMPs)" bind to the receptor of the bystander cells in the tumor microenvironment and help to recruit the immune cells which cause immunogenic cell death. Local radiation may also promote macrophage differentiation that acts as pro immunogenic component of a strong T-cell-driven immune response [7]. These complex interactions and immunomodulatory effects of radiation along with various factors affecting them are shown in Figs. 10.1 and 10.2.

10.2.2 Upregulation of MHC

In addition to modifying antigen release, radiation also upregulate the expression of MHC molecules on cancer cell to facilitate presentation of antigen to cytotoxic T cell. Radiation therapy induces MHC class I expression that enhances antigen processing in tumor cells and increases presentation of antigen-derived peptides on MHC class I molecules [2] Irradiated tumor cells often express mutations that generate "neoantigens" and neoepitopes that potentially act as targets for more robust immune response. Theoretically, an attempt to generate immune response to self-antigens is often weak, since body negatively selects the response as self-tolerance. Hence, generating neoantigens or neoepitopes may bypass this limitation to produce a robust immune response [8].

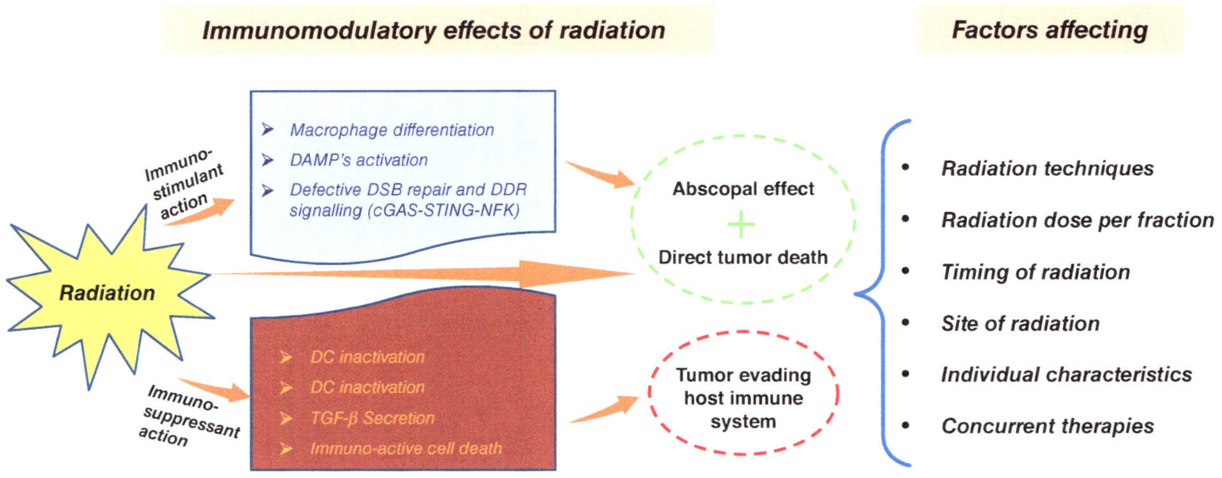

Fig. 10.1 Immunomodulatory effects radiation and factors affecting them. *DSB* double-strand break, *DDR* DNA damage repair, *DC* dendritic cell, *NK* natural killer cell, *TGFβ* tissue growth factor beta

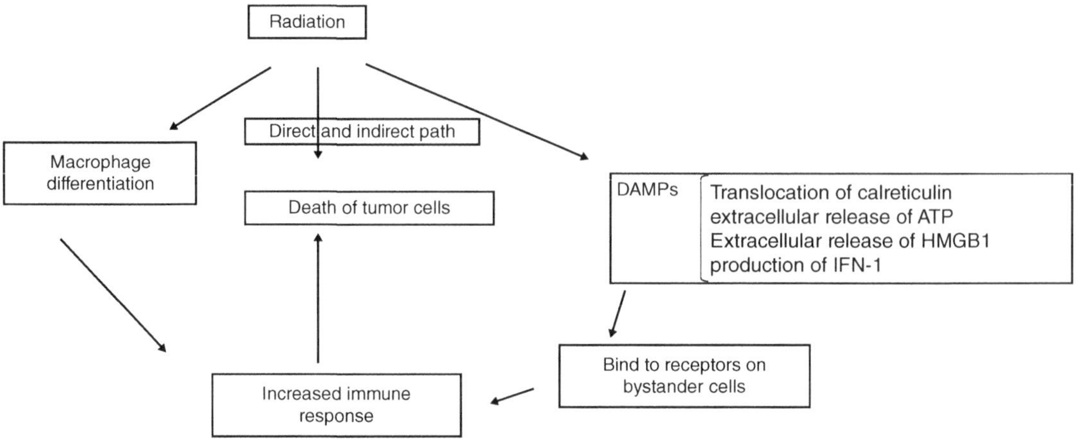

Fig. 10.2 Radiation-induced upregulation of MHC and its effects. *IHC* immunohistochemistry complex, *DAMPs* damage-associated molecular patterns, *ATP* adenosine triphosphate, *HMGB-1* high mobility group box 1, *INF-1* Interferon 1

The ability to enhance tumor antigen presentation also makes radiation a potential modality to be combined with Chimeric Antigen Receptor (CAR) T-cell therapies especially for solid tumors. For solid tumors generating CAR could be difficult due to challenges in target selection [9]. Theoretically, since radiation enhances the antigen presentation and induces the expression of MHC class I molecules, solid tumors can also be potentially targeted with CAR T cells.

10.2.3 Activation of Dendritic Cells

Dendritic cells are also considered to be potent antigen-presenting cells for their specialized dendritic morphology that serves to link innate and adaptive immunity. Irradiated tumor cells activate dendritic cells through DAMPs as mentioned above, utilizing co stimulatory molecules CD70 and CD86 on dendritic cells. Additionally, radiation also act to strengthen tumor cross presentation of dendritic cells that may promote to proliferate t cell response.

Activation of NK cells: Another important role is tumor cell kill is played by NK cells which are cytotoxic lymphocytes that take part in innate immunity. Activation of NK cells often after radiation help to promote increase secretion of cytokines that enhance cytotoxicity via Genasyme B, Perforin, FAS, and ADCC. Several NK cell-based immunotherapies have been developed based on this mechanism. These radiation effects on MHC, DC affecting tumor cells are summarized in Fig. 10.2.

10.3 Modulation of Tumor Microenvironment

Radiation can also help to reprogram the tumor microenvironment from immune suppressive to immune stimulatory state. Tumor microenvironment is often a balance maintained by multiple agents and the effect of radiation can be like a double edged sword. On the one hand, radiation stimulates secretion of cytokines which improves immunogenicity, promotes T-cell transformation that indirectly causes immune suppression of the tumor and promotes antitumor immunity of dendritic cells [10]; but on the other hand, it leads to secretion of cytokines such as TGF-beta that induces immunosuppression. TGF-beta triggers immunosuppressive microenvironment through suppressing the cytotoxic $CD8^+$ T cells, $CD4^+$ T cells, promoting Treg transformation and inhibiting NK cells [11]. TGF-beta also triggers activation of VEGF transcription which encourages tumor growth and metastases [12]. In a preclinical study, combining radiation with TGF-beta inhibitors led to increased T-cell priming and decreased tumor growth and metastases. Addition of anti-programmed cell death protein-1 (PD-1) therapy led to enhancement of these effects [13].

10.3.1 Activation of cGAS-STING Innate Immune Response

Radiation also activates antitumor responses by triggering the stimulator of interferon genes (STING)-mediated DNA sensing pathway. This pathway is a part of the innate immune response that protects host from pathogens and is essential for host immunity. Radiation damage to tumors releases mutant DNA within cytoplasm that upregulate type I IFN gene via STING pathway and leads to dendritic cells sensing irradiated cancer cells to mount adoptive immune response [14].

Recognizing the importance of STING pathway in antitumor immunity has led to several studies investigating therapeutic potential of STING agonists. Currently there is ongoing clinical trial evaluating the safety and efficacy of STING agonist MIW815 for patients with advanced lymphoma [15].

10.3.2 Lymphopenia with Radiation Therapy

Lymphopenia is a significant toxicity from most systemic therapy agents, but can also be induced by radiation, especially when large volumes of bone marrow or circulatory blood volume is in the irradiated volume. The lethal dose required to reduce the survival fraction of circulating lymphocytes by 90% is only 3 Gy primarily mediated through apoptosis with exposure to as low as 1 Gy of radiation can destroy mature circulating lymphocytes [16, 17]. Radiation-induced lymphopenia (RIL) has been linked with overall oncological outcomes in many different cancers. In a secondary analysis of 2073 newly diagnosed glioblastoma patients enrolled into five clinical trials, treated with temozolomide chemoradiotherapy alone or in combination with experimental agents, inferior OS was noted with high-grade (grade 3, 4) lymphopenia [18]. In a meta-analysis of esophageal cancer patients treated with chemoradiotherapy, RIL was associated with lower pathologic complete response, OS and PFS [19]. Similar negative outcomes have been described in head and neck cancers, lung cancers and hepatocellular cancers [20–22].

While the predominant lymphopenia effect is from systemic therapy and from direct irradiation of functioning marrow, additional radiation dosimetric parameters have also been linked with lymphopenia supporting the hypothesis that radiation is a critical contributor. For instance, in patients with glioblastoma, volume of brain exposed to intermediate radiation doses of 15–30 Gy (Brain$V_{15\text{–}30\ Gy}$) has been linked with grade 3 or higher lymphopenia [23]. A small phase-2 randomized trial of 50 patients from PGIMER, India, also noted improved survival with the smaller volumes [24]. An analysis of breast cancer patients demonstrated that volumetric RT technique spreading the low and intermediate radiation doses to a larger volume can increase the risk of RIL [25]. Extending this further, secondary analysis of glioblastoma patients treated with protons or photons from MDACC, demonstrated that grade 3 or higher lymphopenia was significantly reduced with proton therapy [26]. Analysis of esophageal cancer patients treated with neoadjuvant chemoradiotherapy has also demonstrated lower grade 4 lymphopenia with proton therapy when compared with intensity modulated radiation therapy (IMRT) [27]. Effective dose to the circulating immune cells (EDIC) was identified as another such predictor of radiation-induced lymphopenia in thoracic, breast and brain radiotherapy.

In the modern era of increased immunotherapy use as part of multimodality therapy, radiation-induced lymphopenia could possibly blunt immunotherapy response. Indeed, lower absolute lymphocyte count (ALC) has been linked to shorter time to progression in patients receiving nivolumab or pembrolizumab, while risk of immune related adverse events is higher in patients with higher baseline ALC.

10.4 Clinical Evidence for Combining Immunotherapy and Radiation

10.4.1 Immune–Radiotherapy Combination; Key Retrospective Studies

Most of the studies have evaluated immune checkpoint inhibitors, either PD-1/PD-L1 inhibitors or CTLA-4 inhibitors. The most important initial concern has been the safety of this combination in various settings. In patients with metastatic melanomas, immune checkpoint inhibitors were found to be safe in patients who were irradiated for extracranial metastases [28, 29]. In another study in newly diagnosed brain metastases from various tumors, stereotactic brain irradiation to varying volumes was found to be safe. However, there was an increased incidence of symptomatic brain necrosis which was especially more pronounced in patients with melanomas. Therefore, it is considered safe to combine immunotherapy for most sites except for intracranial lesions especially in the setting of radiosurgery.

10.4.2 Immune–Radiotherapy Combination; Key Prospective Studies

Initial proof-of-principle trial by Golden et al. demonstrated that RT combined with granulocyte-macrophage colony-stimulating factor (GM-CSF) does produce a high objective abscopal effect in patients with metastatic solid cancers and this trial provides the proof of principle for the several clinical trials using this combination [30]. Shaverdian et al. investigated patients included in KEYNOTE-001 trial and studied specifically the impact of radiation [31]. They showed that patients who received radiation prior to pembrolizumab had better PFS and OS compared to those who did not receive radiation therapy with acceptable toxicities [31]. Similarly, Formenti et al. demonstrated that some of the patients of NSCLC who did not respond initially to CTLA-4 inhibitor therapy had a discernible response with some achieving prolonged survival [15]. Similarly, other checkpoint inhibitors such as atezolizumab have been tested for safety and efficacy in single-arm phase 2 studies and have shown reasonable safety profile. However, the most robust evidence was noted in PACIFIC trial which was described earlier. Addition of Durvalumab showed persisting PFS and OS benefit on longer follow-up [4]. Although, a pooled analysis of two randomized trials in which patients with NSCLC were treated with pembrolizumab and radiation showed that there was doubling of overall response rates, the trial did not meet the prespecified endpoint criteria for a meaningful clinical benefit. Pembrolizumab given as a lead-in dose, concomitantly with chemoradiotherapy and then as adjuvant treatment, did not yield statistically significant improvement in event-free survival versus placebo in patients with locally advanced head and neck cancers, according to an unpublished study [32]. However, when evaluating patients based on PD-L1 expression, event-free survival favored those with a combined positive score ≥ 1 vs. patients with a combined positive score < 1 (HR, 1.09; 95% CI, 0.56 to 2.11) [32]. Other ongoing phase 3 clinical trials on radiation therapy with immunotherapy have been summarized in Table 10.1.

Table 10.1 Clinical trials using radiation and immunotherapy

Trials	Design	Phase	Immunotherapy	Endpoints
NCT05111197	Local SBRT for residual hypermetabolic lesion with locally advanced lung cancer responder to immunotherapy	3	Anti-PD-1 and PD-L1	Overall survival (OS)
NCT05223647	Chemoimmunotherapy plus thoracic radiotherapy in extensive small cell lung cancer	3	Durvalumab	OS
NCT03867175	Immunotherapy with or without SBRT in stage IV non-small cell lung cancer	3	Pembrolizumab	PFS
NCT02617589	Nivolumab compared to temozolomide with radiation for GBM	3	Nivolumab	OS
NCT04929041	Addition of radiation therapy to usual immunotherapy for stage IV non-small cell lung cancer who are PD-L1 negative	3	Multiple agents	PFS and OS
NCT03721341	SBRT for comprehensive treatment of 4–10 oligometastatic tumors	3	Multiple agents	OS
NCT03774732	PD-1 inhibitor with chemotherapy with concurrent irradiation at varied tumor sites in advanced non-small cell lung cancer	3	Pembrolizumab	OS
NCT03952585	Deintensified radiation therapy with immunotherapy in early stage HPV-positive oropharyngeal cancer	3	Nivolumab	PFS and quality of life
NCT0381102	Addition of atezolizumab to usual chemoradiation treatment for limited stage small cell lung cancer	3	Atezolizumab	OS
NCT01810913	Docetaxel cetuximab and atezolizumab to usual chemoradiotherapy for high-risk cancer	3	Atezolizumab	DFS
NCT03391869	Nivolumab and ipilimumab with or without consolidation therapy in stage IV lung cancer	3	Nivolumab and ipilimumab	OS
NCT00861614	Immunotherapy in advanced prostate cancer	3	Ipilimumab	OS
NCT04402788	Radiation therapy to atezolizumab for extensive small cell cancer (RAPTOR trial)	2, 3	Atezolizumab	PFS in phase 2 and OS in phase 3
NCT03775265	Chemoradiotherapy with or without atezolizumab in locally advanced muscle invasive bladder cancer	3	Atezolizumab	Bladder intact event-free survival
NCT03793179	Testing the timing of pembrolizumab alone or with chemotherapy for nonsmall cell lung cancer	3	Pembrolizumab	OS
NCT04944914	Camrelizumab plus SBRT for oligometastatic nasopharyngeal cancer	3	Camrelizumab	Median progression-free survival
NCT04396860	Ipilimumab and nivolumab plus radiation compared to standard treatment for MGMT unmethylated glioblastoma	3	Ipilimumab and nivolumab	PFS and OS
NCT05111197	Immunotherapy + SRT/SBRT vs. immunotherapy in metastatic NSCLC	3	Any anti-PD-1 or anti-PD-L1 immunotherapy	1-year OS
NCT03867175	SBRT + immunotherapy vs. immunotherapy in metastatic NSCLC	3	Pembrolizumab	5-year PFS
NCT04929041	Immunotherapy +/-chemotherapy vs. immunotherapy +/− chemotherapy + SBRT in PD-L1-negative metastatic NSCLC	3	Ipilimumab/ nivolumab/ pembrolizumab	5-year PFS, OS

(continued)

Table 10.1 (continued)

Trials	Design	Phase	Immunotherapy	Endpoints
NCT03774732	Immunotherapy + chemotherapy + SBRT vs. immunotherapy + chemotherapy in metastatic NSCLC	3	Pembrolizumab	2-year OS
NCT03391869	Immunotherapy vs. immunotherapy + local consolidation therapy (surgery +/− RT) in metastatic NSCLC	3	Nivolumab and ipilimumab	1-year OS
NCT03827577	Chemotherapy +/− targeted therapy +/− immunotherapy based on PD-L1 expression vs. local ablative therapy + chemotherapy +/− targeted therapy +/− immunotherapy based on PD-L1 expression in metastatic NSCLC	3	Anti-PD-L1 immunotherapy	5-year OS

SBRT stereotactic body radiotherapy, *SRT* stereotactic radiotherapy, *PD-1* program death receptor 1, *PD-L1* Program death receptor ligand 1, *RT* radiotherapy

10.5 Optimizing the Combination

Although the theoretical rationale of the synergism of the combination of immunotherapy and radiation therapy is well established, the optimal combination is not yet known. Several parameters such as dose/fractionation, number of tumor site/s, sequence, timing of radiation, volume of irradiation and the type of immunotherapy agents apart from individual immune characteristics could influence this synergism and more clinical data is needed to establish the right combination [33].

10.5.1 Optimizing the Combination: Timing, Dose per Fraction, Sites of Radiation

10.5.1.1 Dose Fractionation

Improved therapeutic ratio with optimizing dose fractionation has been the cornerstone in radiation oncology [34]. Physical and radiobiological effects of varying fractionation schedules on tumors and normal tissues have been extensively studied. In addition, effects of fractionation on immune function have also been studied. Though it is well known that low radiation dose is immunostimulant and chemoattractant, the higher dose suppresses immune system. There are multiple interrelated cellular responses occurring in between these extremes that affects immunomodulatory effects of radiation [2, 35, 36]. The ideal dose fractionation is the one which is safe and that maximizes the antitumor immune response.

Various dose fractionation schemes have been tested for their immunogenic potential when given in combination with immune checkpoint inhibitors (ICI). In a preclinical study by Dewan et al., effect of fraction size on the immunomodulatory effects of ICI was demonstrated [37]. In this mice model, investigators demonstrated that enhanced effects of radiation at primary site and abscopal (immunomodulatory) effects at distant site were more common with fractionated regimen than single fraction or no radiation cohort [37]. After this initial study, fractionated radiation with moderate hypofractionation was commonly used in multiple studies evaluating the same. Another preclinical study showed that 7.5 Gy/fraction was better than 5 Gy/fraction in maintaining low Treg numbers in mice bearing B16-ovalbumin murine melanoma [38]. However, Lugade et al. showed that 15 Gy single dose could elicit greater immunogenicity than 3 Gy in 5 fractions [39]. Later, it was also shown that higher doses (>15 Gy per fraction) may increase the proportion of splenic Tregs suppressing the immune response, which counteracts the immune-stimulant effect of radiation. Similarly, other preclinical studies have also shown that moderate hypofractionation, have better overall immunogenic efficacy than conventional or extreme hypofractionation regimes, while conventional fractionation is more immunogenic than extreme hypofractionation [9].

Over the period with increasing use of combination therapy especially in palliative setting, moderately hypofractionated radiation has been the most preferred regimen. In a randomized phase I/II trial for lung and liver metastases from NSCLC, there were better out of field ORRs (38% vs. 10%) and longer progression free survival with 50 Gy in 4 fractions compared to 45 Gy in 15 fractions in combination with pembrolizumab [40]. In various preclinical and clinical studies, some have corroborated the same finding that within moderate hypofractionation, increase in dose per fraction was better, others have found contradictory results [41–44]. These divergent results could be due to the variation in complex interrelated interactions of various radiation characteristics (type, dose, volume, fractionation, dose rate), tumor characteristics deciding immune response, and thus it is important to consider treatment (immunotherapy, radiation) and tumor-related factors prior to deciding optimal combination therapy [45].

In addition to immunogenicity, it is important to note that with high dose per fraction there is theoretically a higher risk of radiation related toxicities. In a multicentric study of metastatic patients receiving palliative radiation with immunotherapy, Bang et al. noted higher equivalent dose in 2 Gy fractions was associated with a higher adverse events (AEs) ($p = 0.01$) [46]. Though this study involved retrospective audit of heterogeneous group of metastatic cancers such as lung cancer, melanoma, renal cell carcinoma, one must consider toxicity profile of a combination therapy with hypofractionated radiation with respect to site and volume of irradiation. In a nutshell, moderately hypofractionated or clinically used SBRT regimen seems to induce higher immunogenicity; however, the optimal dose per fraction should be based on site, volume and anticipated toxicities associated with radiation therapy [47]. These potential benefits and limitations of combination therapy (radiation and immunotherapy) are summarized in Fig. 10.3a. In the future, several ongoing prospective trials exploring this will inform the optimal regimen.

Size of radiation target/volume of irradiated area also affects radiation immunogenicity. Larger volumes of irradiation are associated with higher immune suppression and higher toxicities. Similarly, higher linear energy transfer (LET) and higher radiobiological effectiveness (RBE) has been associated with higher immunogenic response. In this context, proton therapy and carbon ion therapy in view of higher LET, higher RBE, sharper dose fall off and significantly lower integral dose may be theoretically superior to modern photon based techniques such as IMRT and VMAT [48]. Particle therapy has been shown to induce higher immunogenicity and leads to lesser immune suppression compared to photon-based techniques. The particle therapy could also limit the normal tissue exposure and therefore lower toxicities compared to photon-based approaches in combination with immunotherapy agents [49].

10.5.1.2 Site/S of Irradiation

Mathematic modeling that estimates the focal therapy activated T cells showed that not all metastatic sites participate in the systemic immune activation equally, and therefore the success in triggering immune activation depends on the selection of metastatic site for radiation therapy.

For example, radiation to bony (osseous) lesion showed higher responses than radiation to other visceral lesions or primary lesion [50, 51]. Similarly, in an open labeled phase I trial, evaluating ipilimumab and SBRT in patients with multiple liver and/or lung metastases found that there was higher T-cell activation with liver irradiation compared to irradiation of lung [52]. It is hypothesized that low dose irradiation to overcome immunosuppressive tumor microenvironment for access to $CD8^+$ T cells and radiation to all the metastatic sites might activate more shared tumor activated antigens with irradiation of different tumor lesions. Results from multiple retrospective and a few prospective studies strengthen the hypothesis [53]. Therefore, we need to carefully consider the site/sites of irradiation appropriately, or choose to irradiate all metastatic sites when appropriate.

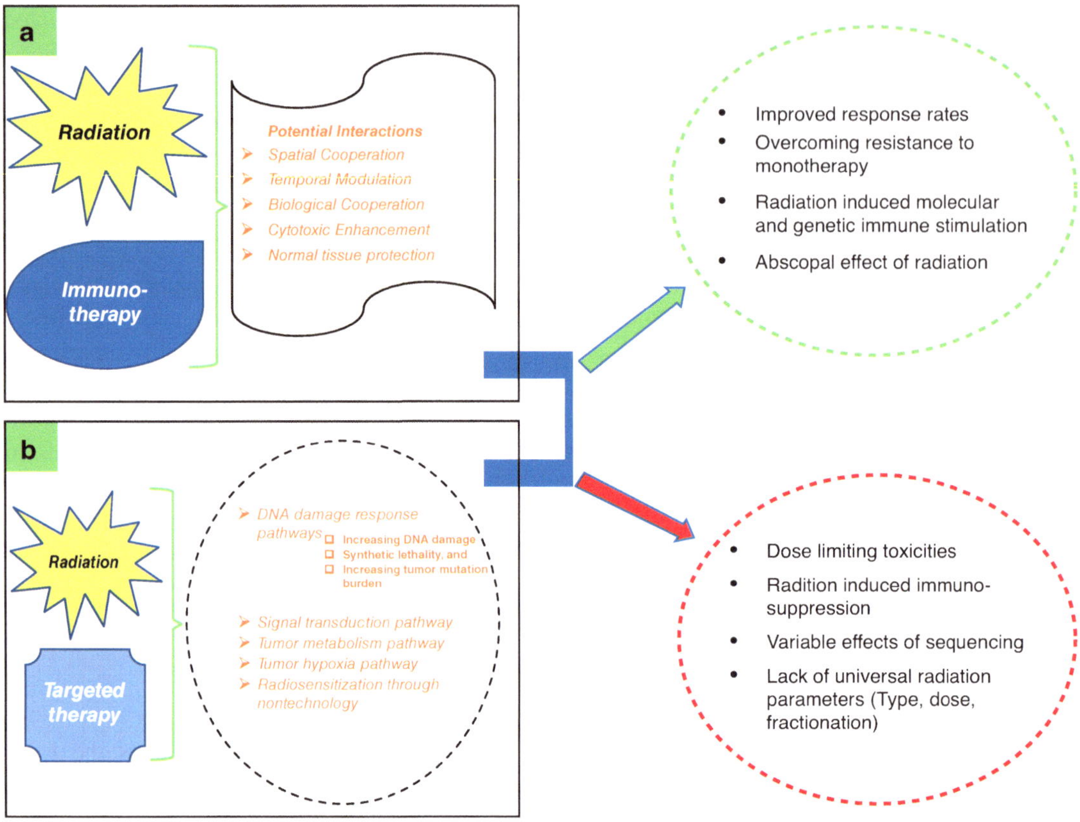

Fig. 10.3 Combination therapy (immunotherapy and radiation): benefits and limitations. (**a**) Radiation and immunotherapy combination—plausible potential interactions. (**b**) Radiation and targeted therapy combination—plausible mechanism of action/pathways

10.5.1.3 Timing of Radiation

For achieving good synergism in combination therapy, immunostimulatory effect of radiation should boost immunotherapy action and tumoricidal action of immunotherapy agent should enhance the radiation effect. In addition to these desired effects, it is also important to know radiation and immunotherapy agent's individual toxicity profile, and sequence the combination treatment accordingly as any treatment gap due to overlapping toxicities will impact overall treatment time and therefore the outcomes.

Mechanism of action of anti-CTLA-4 agents warrants their use before radiation, while anti-PD-1/PD-L1 agents are better used in concurrent or adjuvant setting in combination therapy [37, 54]. Anti-PD-L1 followed by radiation can limit antitumor response by radiation-induced T-cell apoptosis, while radiation followed by anti-PD-L1 shows better antitumor response.

Most of the literature available point toward the approach of either concurrent or delayed administration of ICI agents after irradiation. The enhanced recruitment of T cells induced by radiation can destroy tumor cells at the primary site as well as metastatic sites. PACIFIC trial in a phase 3 randomized trial setting, has shown that this approach of using ICI after chemoradiation improves long-term outcomes with acceptable toxicities in patients with inoperable lung cancers [4]. Similar approach was used in other studies that confirmed the safety and efficacy of this approach in patients of lung cancer [55, 56]. On deeper post hoc analysis of PACIFIC trial patients, it was found that outcomes were better if Durvalumab was given concurrently or within 14 days of completion of radiation [57]. However, a retrospective study showed that ICI given after 21 days of SBRT had a longer overall survival compared to patients receiving ICIs within first 21 days of SBRT. This could be due to several known and unknown confounders, and there is a need for more robust data. Trials such as PACIFIC-2, KEYLINK-012 and Checkmate-73 L are expected to deduce the best strategy to sequence the combination therapies [58–60].

10.5.2 Selection of Immunotherapy Agent

Radiation therapy has been more often combined with PD-1 inhibitors rather than with CTLA-4 inhibitors. In a retrospective analysis comparing these drugs in the setting of metastatic NSCLC in combination with SBRT, PD-1 inhibitors resulted in a higher PFS compared to CTLA-4 inhibitors (80% vs. 31% at 12 months, $p = 0.02$) without any significant difference in OS [28]. The question of safety and efficacy of single agent versus dual immunotherapy agents is controversial. However, combination regimen (Nivolumab and Ipilimumab) along with radiation (in the setting of multisite RT in metastatic NSCLC) has shown significantly higher grade 3 toxicities compared to patients receiving single agent (Nivolumab) [29].

10.5.3 Combination of Immunotherapy with Brachytherapy

Almost all preclinical and clinical studies have utilized EBRT in combination with immunotherapy and the role of brachytherapy as a radiation modality has only been recently explored. While EBRT necessitates treatment with large volume of normal tissue getting treated, brachytherapy has the potential to reduce radiation doses to normal tissues and pertinent lymphoid organs from low doses of radiation.

Encouraged by the preclinical studies, [61] ongoing clinical trials are looking at the role of Y90 for liver metastasis in combination with checkpoint blockade [62]. Another brachytherapy–immunotherapy combination study, which is actively enrolling patients, is testing the safety and efficacy of adding anti-PD-1 checkpoint blockade to standard chemoradiotherapy in patients with advanced cervical cancer [63].

10.5.4 Toxicities of the Combination Therapy

When considering sequential and or concurrent immune–radiotherapy combination, it is important to assess toxicity profile of each so that combination therapy is well tolerated. Immunotherapy agents in general have systemic toxicities such as pneumonitis, colitis, hepatitis, endocrinopathies, cardiovascular abnormalities, dermatological toxicities, etc., while radiation therapy being local therapy causes side effects pertinent to treatment region; hence, immunotherapy agents with overlapping toxicities are better avoided. Also, it has been proposed that immune checkpoint inhibitors may lead to immune-mediated inflammatory response in regions receiving radiation therapy in the past akin to "radiation recall reactions" [64]. Therefore, one must have a lower threshold for suspecting these toxicities such as pneumonitis or hepatitis when higher volumes of respective organs have been irradiated in the past. So far in multiple preclinical and clinical studies combination of immunotherapy and radiotherapy is well tolerated with rare grade 3 or 4 events which necessitate multiple interventions. The rate of toxicities varies widely based on several patient, tumor and treatment-related factors. Long-term data from several ongoing trials especially adjuvant trials may increase our knowledge in the future.

10.6 Radiation and Targeted Therapy: Synergism and Biological Principles

10.6.1 Introduction

Combining radiosensitizers like chemotherapy with radiation therapy is an established strategy and this approach has been the standard of care for several cancer sites. However, despite targeted therapy making a significant impact on management of ever increasing list of cancers such as lung adenocarcinoma, renal cell carcinoma, breast cancer, melanomas, etc., the optimal combination of radiation therapy and targeted therapies has not been established. This combination needs to be leveraged as with increasing adoption of molecular profiling both at presentation and at metastatic progression, it is likely that more and more patients will be treated with targeted therapies. "Radiation precision medicine" aims to utilize the genomic data to identify patients who could harbor genetic signatures that would benefit from the use of optimal radiation and targeted therapy combinations. Although this is an attractive hypothesis, very little research has explored this possibility.

Combining targeted therapy with radiation therapy should be scientifically based on molecular architecture, radiobiology, pharmacological principles and immune responses to the drug and radiation therapy. The expected clinical endpoints are that of disease control and balance between tumor control and normal tissue toxicity [65].

10.6.2 Preclinical Rationale for Combination of Targeted Therapy with Radiotherapy

Radiation therapy and targeted therapy as a combination, like we discussed earlier, is a fascinating combination therapy and has not be explored to its fullest potential till date. Various mechanisms explaining rationale basis of this combination are discussed below and summarized in Fig. 10.3b.

10.6.2.1 Targeting DNA Damage Response Pathways

Chemotherapeutic agents such as Cisplatin and 5 Fluorouracil are considered potent radiosensitizers due to their effect on enhancing DNA damage. Therefore, it would be synergistic to add radiation with targeted agents that play a critical role in DNA damage response (DDR). A number of genes such as PARP1, WEE1, ATM, CHK1, ATR and DNA-PK are among the key mediators of DDR. Novel DDR inhibitors if combined with radiation could promote cell-specific radiosensitization by following ways:

1. Increasing DNA damage to such levels that could trigger apoptosis or death,
2. By synthetic lethality, and.
3. Increasing tumor mutation burden (triggered by DNA damage).

This radiosensitizer action thereby enhances tumor antigenicity and T-cell-mediated cell kill. Examples of such inhibitors are PARP inhibitors and WEE1 inhibitors such as Adavosertib (that can overcome the G2M checkpoint which is important in p53 mutant cancer cells which lack the G1 checkpoint). Induction of replication stress is another mechanism through which DDR inhibitors exert their effect especially in the context of cMyc and KRAS mutations.

10.6.2.2 Targeting Signal Transduction Pathway

Signal transduction pathways which are commonly altered in cancers can be targeted for a selective radiosensitization effect. For example, EGFR, ERBB2, PI3K/AKT, etc. inhibitors have been used concurrently with radiation therapy and have shown varying benefits. Growth factors such as these are key to the proliferation of cancer cell and inhibition of apoptosis, and contribute to radioresistance. Similarly, androgen deprivation therapy used in intermediate- and high-risk prostate cancers contributes to radiosensitization through targeting signal transduction pathways. ADT contributes to this effect through several mechanisms including relieving hypoxia, inactivation of androgen receptor signaling-mediated DNA damage and response-mediated radioresistance.

10.6.2.3 Targeting Tumor Metabolism

Mitochondrial metabolism is essential for survival of cancer cells and therefore is a target to suppress the proliferation and inducing apoptosis. Preclinical evidence supports use of Glutamine metabolism/glutaminase inhibitors such as Telaglenstat in combination with chemotherapy or other targeted therapies in lung and colorectal cancers. Similarly, this has been tested along with radiation in head neck squamous cell carcinoma and lung cancer cell lines and have shown to result in radiosensitization and increase in efficacy of radiation by 30%.

10.6.2.4 Targeting Tumor Hypoxia

Tumor hypoxia is a known mediator of radioresistance, and a lot of research has been dedicated to targeting tumor hypoxia. Historically, oxygen mimetics such as nitroimidazole compounds, hypoxia, hypoxia-activated cytotoxic chemotherapeutic/targeted drugs and hyperbaric oxygen have been used and tested in clinical trials. Recently, antiangiogenic-targeted therapies such as VEGF inhibitors and PI3K inhibitors that have shown to normalize blood flow to the tumors have been tested along with radiation. VEGF inhibitors have been tested in brain tumors and sarcomas as radiosensitizers, whereas PI3K inhibitors have been tested in hypoxic non-small cell lung cancers. Another approach for reducing tumor hypoxia is through altering tumor metabolism via inhibition of oxidative phosphorylation. Atovaquone, antimalarial by inhibiting mitochondrial complex III, has been shown to have a significant radiosensitization effect in preclinical models.

10.6.2.5 Radiosensitization Through Nanotechnology

Nanoparticle delivery enhances radiation effect by improving the local deposition of radiation and increasing the local production of oxygen free radicals leading to DNA damage and cell cycle arrest. Silver nanoparticles with or without surface modified with polyethylene glycol have been shown to result in radiosensitization irrespective of normoxic or hypoxic status of cancer cells.

10.6.3 Clinical Evidence for Combining Targeted Therapy with Radiation

Various studies addressing combination of targeted therapies with radiation are summarized in Table 10.2. Of these, studies showing positive statistically significant benefit with addition of targeted therapy are added in Table 10.2a, and few of them are discussed in detail below. Few studies could not achieve planned statistical benefit and are compiled in Table 10.2b. In a meta-analysis by Tchelebi et al., 13 prospective randomized trials with a total of 5678 patients who received radiotherapy with or without targeted therapy were included [66]. The trials included cancers of the head and neck (6 trials, 3295 patients), esophagus (3 trials, 762 patients), lung (2 trials, 550 patients) and brain (2 trials, 1542 patients) [66]. Three studies evaluated a small molecule and radiotherapy in 949 patients, and 10 studies evaluated antibodies and radiotherapy in 4729 patients. Overall survival was not improved by the addition of receptor tyrosine kinase inhibitors to radiotherapy-based treatment (hazard ratio 1.02, 95% confidence interval 0.90 to 1.15, $p = 0.76$) but was found to have increased grade 3 toxicity (relative risk 1.18, 95% confidence interval 1.06 to 1.33, $p = 0.009$) [66]. The conclusion was that addition of tyrosine kinase inhibitors to radiotherapy does not improve survival and worsens toxicity. The limitations of this trial included inclusion of only conventionally fractionated radiotherapy and not stereotactic radiotherapy, there was no biomarker analysis involved, only concurrent TKI use was included and bias inherent in the involved studies [66].

Site-wise clinical evidence for the use of combination in several cancers have been summarized as follows.

10.6.3.1 Head and Neck Cancers

Head and neck squamous cell carcinoma (HNSCC) are known to have high expression of epidermal growth receptor factors (EGFR). Anti-EGFR antibodies like cetuximab and nimotuzumab have been used clinically in both curative and metastatic HNSCC. Cetuximab with radiotherapy was found to improve overall survival and locoregional control versus radiation alone in cisplatin ineligible patients [67]. However, when compared to cisplatin concurrent with radiation, cetuximab was proven to be inferior even in human papilloma virus-positive HNSCC which

Table 10.2 Trials evaluating benefit of various targeted therapy agents with radiation or chemoradiation

Author	Year	Site	Treatment	Targeted agent used
(a) Trials with positive results				
Bonner et al.	2006	HNSCC	RT	Cetuximab
Sperduto et al.	2013	NSCLC	RT	Erlotinib
Ang et al.	2014	HNSCC	CRT	Cetuximab
Gilbert et al.	2014	Glioblastoma	CRT	Bevacizumab
Chinot et al.	2014	Glioblastoma	CRT	Bevacizumab
Bradley et al.	2015	NSCLC	RT	Cetuximab
Harrington et al.	2015	HNSCC	CRT	Lapatinib
Giralt et al.	2015	HNSCC	CRT	Panitumumab
Crosby et al.	2017	Esophagus	CRT	Cetuximab
Suntharalingam et al.	2017	Esophagus	CRT	Cetuximab
Mehanna et al.	2018	HNSCC	RT	Cetuximab
Geoffrois et al.	2018	HNSCC	CRT	Cetuximab
Wu et al.	2018	Esophagus	CRT	Erlotinib
(b) Trials with no significant benefit of combination therapy				
Kelly et al.	2008	NSCLC	CRT	Gefitinib; no benefit
Tao et al.	2007	HNSCC	RT	Cetuximab
Salma et al.	2011	HNSCC	CRT	Bevacizumab
Govindan et al.	2011	NSCLC	CRT	Cetuximab
Lefebvre et al.	2013	HNSCC	RT	Cetuximab
Sweigert et al.	2016	HNSCC	CRT	Cetuximab
Hammell et al.	2016	Pancreas	CRT	Gemcitabine
Edelman et al.	2017	NSCLC	CRT	Panitumumab
Gillison et al.	2018	HNSCC	RT	Cetuximab

RT radiotherapy, *CRT* chemoradiotherapy, *NSCLC* non-small cell lung cancer, *HNSCC* Head and neck squamous cell carcinoma/cancer

are considered to be a distinct subset with better prognosis [68]. This was also proven in De ESCALATE HPV trial, where in low-risk oropharyngeal HPV-positive cancers, there was detrimental effect on tumor control with no benefits in toxicity [69]. Thus, cisplatin remains the standard of care for concurrent radiotherapy in HNSCC. Cetuximab is useful in cisplatin ineligible patients and also in palliative care settings in combination with chemotherapy. Higher skin toxicities and electrolyte imbalances with cetuximab are noted as major side effects.

Nimotuzumab, a fully humanized anti-EGFR antibody, also has been tried concurrent with definitive radiation. Nimotuzumab with cisplatin was found to have better locoregional control when compared with cisplatin alone with radiation with no additional side effects over cisplatin alone [70].

10.6.3.2 Lung Cancer

In the current era of targeted therapy for non-small cell lung cancer (NSCLC), there has been attempts to use concurrent targeted therapy with definitive radiation. High rates of Grade 2 or more radiation pneumonitis are reported with concurrent use of first generation TKIs like erlotinib (37.5–44%) [71] and third-generation TKIs like osimertinib (64%). The hypothesis is that anti-EGFR antibodies are present in alveolar cells and also promote G2M phase arrest in irradiated cells preventing normal tissue repair during concurrent chemoradiation [72]. Hence, currently, concurrent TKIs with thoracic radiation are not recommended and should be done in context of clinical trials only.

However, in Stage IV lung cancer with molecular alterations like EGFR mutation or ALK rearrangement, there is evidence of better local control of the lung primary and metastatic sites [73]. In SINDAS trial, a phase III randomized study evaluating the role of first line TKI with or without radiation therapy for synchronous oligometastatic EGFR-mutated NSCLC, addition of TKI with RT significantly improved both PFS and OS when compared to TKI alone. The median OS was increased to 25.5 months versus 17.7 months with TKI alone ($p < 0.001$). The dose of radiation was 25–40 Gy in 5 fractions to all metastatic sites and the primary lung mass, depending on tumor size and location. Skin reactions was most frequent adverse event (grade 3–4 was 15% in both arms) but dose reduction and discontinuation were less (<5%). TKI with RT arm had 7.5% grade 3–4 pneumonitis [74]. International consensus on metastatic NSCLC also recommends the addition of RT in molecular alteration driven oligo-focal NSCLC which include oligometastatic, oligo-residual and oncoprogressive cancer to improve survival [75].

10.6.3.3 Breast Cancer

Trastuzumab, anti-HER2 antibody, is integral part of neoadjuvant and adjuvant treatment protocols in HER2-positive breast cancer. In adjuvant settings, after breast conservation or mastectomy, the use of trastuzumab with adjuvant radiation has been studied and there is no increased incidence of pneumonitis, cough, dyspnea or acute skin reactions [76]. In HERA trial, post mastectomy radiation (PMRT) for 1 to 3 positive lymph nodes had better locoregional relapse free survival (97% vs. 90%) when compared to no PMRT ($p = 0.04$) and a nonsignificant improved overall survival (HR 0.6, $p = 0.06$) [77].

Recently, retrospective analysis of patients in 3321 patients who received Trastuzumab in HERA trial showed similar rates of cardiotoxicity among patients irrespective of left or right sided PMRT when compared to no PMRT group [78]. Similarly, low rates of skin toxicity and low cardiac toxicity was seen with use of trastuzumab with adjuvant hypofractionated radiotherapy [79].

10.6.3.4 Brain Tumors

Primary Brain Tumors (Gliomas)

Bevacizumab, an anti-vascular endothelial growth factor (VEGF) antibody, was tested concurrently with radiation and temozolomide in glioblastomas. Bevacizumab was started from fourth week and continued for 12 cycles as maintenance cycles. There was progression free survival was higher than temozolomide alone (10.7 vs. 7.3 months) but no difference in overall survival was seen. Worse quality of life and increased symptom burden was noted with increase in rates of thromboembolic events, intestinal perforation, neutropenia and hypertension in bevacizumab group [80]. However, in re-irradiation setting, concurrent use of bevacizumab was found to be feasible [81] and improves overall survival with mild increase in toxicities in multiple retrospective series [82].

Brain Metastases

Brain is considered as a sanctuary site for dissemination of cancers in view of the blood brain barrier that often limit the penetration of systemic agents especially chemotherapy. Newer molecules like targeted therapy especially the third-generation drugs like Osimertinib, Alectinib, etc., cross the blood brain barrier. Radiation delivery advances like stereotactic radiotherapy combined with targeted therapy has been used to provide better local control with lesser toxicities.

Targeted therapy like TKIs combined with radiation is often beneficial with improvement in median overall survival with better local control and PFS irrespective of radiation modality and without additional clinical safety concerns especially if stereotactic radiation is used [83].

Studies have used TKIs irrespective of the biology of primary or specific to molecular alteration in the primary. As expected, results are better with a personalized treatment approach. In NSCLC, concurrent use of EGFR TKI with whole brain radiotherapy improves overall survival and local control with minimal increase in adverse effects [84]. Addition of SRS to systemic therapy has improved the median progression free survival and local control without a decline in cognitive function noted with whole brain RT [85].

Brain metastases from melanoma were conventionally treated with whole brain radiotherapy with poor local control and higher cognitive adverse effects. Currently, the use of targeted therapy with SRS is preferred in view of the relative radioresistance and better long term QOL. The median dose in most studies was 20 or 21 Gy. Targeted therapy was given before, concurrently, or after radiotherapy. BRAF and MEK inhibitors are most common drugs used. Fatigue, dermatitis, cutaneous rash, cognitive changes bleeding, radiation necrosis, headache, diarrhea, nausea, etc., are reported [86].

10.6.4 PARP Inhibitors with Radiotherapy

Synthetic lethality due to PARP inhibitors when given concurrently with radiotherapy has shown to induce tumor kill in preclinical studies [87]. There are multiple studies that have tested whether addition of PARP inhibitors will be feasible and beneficial. In a review of 12 clinical studies, Barcellini et al. [88] concluded that despite of limited studies and heterogeneity of evaluated tumor and patient populations, combination approach is feasible but clinical efficacy is unproven yet. Different studies have pointed out the same and current use of PARP inhibitors is limited to clinical trials [89].

There are other targeted therapies which have shown enhancement of radiation effects in preclinical studies and are being investigated in clinical trials like losartan, MEK inhibitors, etc. Discussion regarding the same is beyond the scope of this chapter.

10.6.5 Toxicities of Combination Targeted Therapy and Radiotherapy

Targeted therapy is known to have some unique toxicity profile different from the conventional chemotherapy. This is because many of them especially the oral tyrosine kinase inhibitors inhibit multiple kinases like VEGF, EGF, FGF, etc. The most common toxicities include cutaneous rash, diarrhea, fatigue, anemia, skin hypopigmentation. The benefits of advanced radiation techniques like image guided radiotherapy, stereotactic radiotherapy, etc. is in ability to deliver high doses to relatively smaller areas with decrease in normal tissue toxicity. This has helped in reducing the toxicities due to combination strategies.

10.7 Conclusion

Combining Immunotherapy and or targeted therapy with radiation therapy is a promising approach, which can deliver better results in term of response rates, overall survival and better quality of life. However, in spite of having strong biological rationale, the results from various trials have not been positive possibly due to crossplay of different pathways in the cells. Future studies will be focusing on optimizing the combination therapy (immunotherapy and or targeted therapy with radiation therapy) to improve clinical outcomes without adding clinically significant toxicities. Benefits of modern precise radiation techniques such as image-guided intensity modulated radiation therapy (IG-IMRT), modern particle therapy (protons, carbon ions) as a part of combination is another fascinating unexplored aspect, that needs to addressed and robust clinical evidence generation can have significant impact on the future oncology practice.

10.8 Summary

1. Immunotherapy combined with radiotherapy enhances local and systemic control by enabling spatial co-operation and cytotoxic enhancement.
2. Immunogenic effects of radiation are mediated by increased antigenicity, upregulation of major histocompatibility complex and activation of dendritic cells.
3. Most robust data for immune-radiotherapy at present is through PACIFIC trial, which showed OS and PFS benefit with addition of Duralumab.
4. Studies are needed to optimize combination of radiotherapy with immunotherapy with

respect to timing, dose fractionation and immunotherapy agents to be used.
5. Systemic and local toxicities due to combination should be monitored in combination therapy.
6. Future of radiotherapy includes "Radiation precision medicine," that utilizes genomic data to identify patients who would benefit from optimal radiotherapy and targeted therapy combination.
7. Preclinical rationale for combining targeted therapy and radiotherapy includes targeting DNA damage response pathways, targeting signal transduction pathways, tumor metabolism, tumor hypoxia and radiosensitization.
8. In spite of strong biological rationale, further studies are needed to establish the use of radiotherapy with immunotherapy, while minimizing toxicities.

Multiple Choice Questions

1. Term abscopal effects was coined by
   (a) R. H. Mole
   (b) H. McArthur
   (c) S. Formenti
   (d) X. S. Wang
   (e) R. Mohan
2. Radiation induces
   (a) Abscopal effect
   (b) Bystander effects
   (c) Immunosuppression
   (d) Immunostimulation
   (e) All of the above
3. Effects of radiation are
   (a) Increased antigenicity
   (b) Upregulation of MHC
   (c) Activation of dendritic cells
   (d) Activation of cGAS STING response
   (e) All of the above
4. Which of the following is true regarding immunotherapy and radiation combination?
   (a) Anti-CTLA-4 agents are preferably used before radiation in combination therapy
   (b) Anti-PD-1/PD-L1 agents are better used in concurrent or adjuvant setting in combination therapy
   (c) Immunotherapy agent's sequencing does not affect outcomes in combination therapy
   (d) Both a and b
   (e) None of the above
5. Please choose correct ascending immunogenic radiation dose fractionation sequence (least to highest).
   (a) Standard fractionation, moderate hypofractionation, extreme hypofractionation
   (b) Extreme hypofractionation, standard fractionation, moderate hypofractionation
   (c) Moderate hypofractionation, extreme hypofractionation, standard fractionation
   (d) Extreme hypofractionation, moderate fractionation, standard fractionation
   (e) Standard fractionation, extreme hypofractionation, moderate hypofractionation
6. Novel DNA damage response (DDR) inhibitors when combined with radiation promote cell-specific radiosensitization by which of the following ways?
   (a) Increasing DNA damage triggering apoptosis or cell death
   (b) By synthetic lethality
   (c) Increasing tumor mutation burden (triggered by DNA damage)
   (d) All of the above (a, b, c)
   (e) None of the above
7. Which of the following mechanisms supports combination of targeted therapy with radiotherapy?
   (a) Targeting DNA damage response pathways
   (b) Targeting signal transduction pathway
   (c) Targeting tumor metabolism
   (d) Targeting tumor hypoxia
   (e) All of the above
8. Which of the following is correct with respect to modern particle therapy (proton therapy, carbon ion therapy) compared to standard radiation therapy (IMRT)?
   (a) Higher immunostimulatory effect
   (b) Less immunosuppression
   (c) Anti-metastatic potential
   (d) a and b only

(e) All of the above (a, b, and c)

9. All of the following are correct except
   (a) Type of radiation technique, dose per fraction, and timing of radiation affects its immunogenicity
   (b) Particle therapy shows overall higher immunogenicity compared to standard radiation theoretically
   (c) Site of radiation treatment has no impact on its immunogenic response
   (d) Sequencing of immunotherapy agents depends on their mechanism of action
   (e) Combination of immunotherapy and radiation carries risk of overlapping toxicities

Answers

1. (a)
2. (e)
3. (e)
4. (d)
5. (b)
6. (d)
7. (e)
8. (e)
9. (c)

References

1. Tang J, Pearce L, O'Donnell-Tormey J, Hubbard-Lucey VM. Trends in the global immuno-oncology landscape. Nat Rev Drug Discov. 2018;17(11):783–4.
2. Formenti SC, Demaria S. Systemic effects of local radiotherapy. Lancet Oncol. 2009;10(7):718–26.
3. Kumari S, Mukherjee S, Sinha D, Abdisalaam S, Krishnan S, Asaithamby A. Immunomodulatory effects of radiotherapy. Int J Mol Sci. 2020;21(21):8151.
4. Antonia SJ, Villegas A, Daniel D, Vicente D, Murakami S, Hui R, et al. Durvalumab after Chemoradiotherapy in stage III non-small-cell lung cancer. N Engl J Med. 2017;377(20):1919–29.
5. Kalbasi A, June CH, Haas N, Vapiwala N. Radiation and immunotherapy: a synergistic combination. J Clin Invest. 2013;123(7):2756–63.
6. Golden EB, Pellicciotta I, Demaria S, Barcellos-Hoff MH, Formenti SC. The convergence of radiation and immunogenic cell death signaling pathways. Front Oncol. 2012;2:88.
7. Arina A, Beckett M, Fernandez C, Zheng W, Pitroda S, Chmura SJ, et al. Tumor-reprogrammed resident T cells resist radiation to control tumors. Nat Commun. 2019;10(1):3959.
8. Mapara MY, Sykes M. Tolerance and cancer: mechanisms of tumor evasion and strategies for breaking tolerance. J Clin Oncol. 2004;22(6):1136–51.
9. Newick K, O'Brien S, Moon E, Albelda SM. CAR T cell therapy for solid tumors. Annu Rev Med. 2017;68:139–52.
10. Reits EA, Hodge JW, Herberts CA, Groothuis TA, Chakraborty M, Wansley EK, et al. Radiation modulates the peptide repertoire, enhances MHC class I expression, and induces successful antitumor immunotherapy. J Exp Med. 2006;203(5):1259–71.
11. Yang L, Pang Y, Moses HL. TGF-β and immune cells: an important regulatory axis in the tumor microenvironment and progression. Trends Immunol. 2010;31(6):220–7.
12. Kubiczkova L, Sedlarikova L, Hajek R, Sevcikova S. TGF-β—an excellent servant but a bad master. J Transl Med. 2012;10(1):183.
13. Vanpouille-Box C, Alard A, Aryankalayil MJ, Sarfraz Y, Diamond JM, Schneider RJ, et al. DNA exonuclease Trex1 regulates radiotherapy-induced tumour immunogenicity. Nat Commun. 2017;8(1):15618.
14. Constanzo J, Faget J, Ursino C, Badie C, Pouget JP. Radiation-induced immunity and toxicities: the versatility of the cGAS-STING pathway. Front Immunol. 2021;12:680503. https://doi.org/10.3389/fimmu.2021.680503.
15. Formenti SC, Rudqvist NP, Golden E, Cooper B, Wennerberg E, Lhuillier C, et al. Radiotherapy induces responses of lung cancer to CTLA-4 blockade. Nat Med. 2018;24(12):1845–51.
16. Sellins KS, Cohen JJ. Gene induction by gamma-irradiation leads to DNA fragmentation in lymphocytes. J Immunol. 1987;139(10):3199–206.
17. Cozzarini C. Radiation induced lymphocyte apoptosis: an effective way of "tailoring" radiotherapy to the right patients only? EBioMedicine. 2015;2(12):1852–3.
18. Le Rhun E, Oppong FB, Vanlancker M, Stupp R, Nabors B, Chinot O, et al. Prognostic significance of therapy-induced myelosuppression in newly diagnosed glioblastoma. Neuro Oncol. 2022;24(9):1533–45.
19. Dai D, Tian Q, Yu G, Shui Y, Jiang H, Wei Q. Severe radiation-induced lymphopenia affects the outcomes of esophageal cancer: a comprehensive systematic review and meta-analysis. Cancer. 2022;14(12):3024.
20. Campian J, Sarai G, Ye X, Marur S, Grossman SA. The association between severe treatment-related lymphopenia and progression free survival in patients with newly diagnosed squamous cell head and neck cancer. Head Neck. 2014;36(12):1747–53.
21. De B, Ng SP, Liu AY, Avila S, Tao R, Holliday EB, et al. Radiation-associated lymphopenia and outcomes of patients with Unresectable hepatocellular carcinoma treated with radiotherapy. J Hepatocell Carcinoma. 2021;8:57–69.

22. Upadhyay R, Venkatesulu BP, Giridhar P, Kim BK, Sharma A, Elghazawy H, et al. Risk and impact of radiation related lymphopenia in lung cancer: a systematic review and meta-analysis. Radiother Oncol J Eur Soc Ther Radiol Oncol. 2021;157:225–33.
23. Rudra S, Hui C, Rao YJ, Samson P, Lin AJ, Chang X, et al. Effect of radiation treatment volume reduction on lymphopenia in patients receiving chemoradiotherapy for glioblastoma. Int J Radiat Oncol Biol Phys. 2018;101(1):217–25.
24. Kumar N, Kumar R, Sharma SC, Mukherjee A, Khandelwal N, Tripathi M, et al. Impact of volume of irradiation on survival and quality of life in glioblastoma: a prospective, phase 2, randomized comparison of RTOG and MDACC protocols. Neurooncol Pract. 2020;7(1):86–93.
25. Ellsworth SG. Field size effects on the risk and severity of treatment-induced lymphopenia in patients undergoing radiation therapy for solid tumors. Adv Radiat Oncol. 2018;3(4):512–9.
26. Mohan R, Liu AY, Brown PD, Mahajan A, Dinh J, Chung C, et al. Proton therapy reduces the likelihood of high-grade radiation-induced lymphopenia in glioblastoma patients: phase II randomized study of protons vs photons. Neuro Oncol. 2020;23(2):284–94.
27. Routman DM, Garant A, Lester SC, Day CN, Harmsen WS, Sanheuza CT, et al. A comparison of grade 4 lymphopenia with proton versus photon radiation therapy for esophageal cancer. Adv Radiat Oncol. 2019;4(1):63–9.
28. Chen D, Menon H, Verma V, Guo C, Ramapriyan R, Barsoumian H, et al. Response and outcomes after anti-CTLA4 versus anti-PD1 combined with stereotactic body radiation therapy for metastatic non-small cell lung cancer: retrospective analysis of two single-institution prospective trials. J Immunother Cancer. 2020;8(1):e000492.
29. Cascone T, William WN, Weissferdt A, Lin HY, Leung CH, Carter BW, et al. Neoadjuvant nivolumab (N) or nivolumab plus ipilimumab (NI) for resectable non-small cell lung cancer (NSCLC): clinical and correlative results from the NEOSTAR study. J Clin Oncol. 2019;37(15_suppl):8504.
30. Golden EB, Chhabra A, Chachoua A, Adams S, Donach M, Fenton-Kerimian M, et al. Local radiotherapy and granulocyte-macrophage colony-stimulating factor to generate abscopal responses in patients with metastatic solid tumours: a proof-of-principle trial. Lancet Oncol. 2015;16(7):795–803.
31. Shaverdian N, Lisberg AE, Bornazyan K, Veruttipong D, Goldman JW, Formenti SC, et al. Previous radiotherapy and the clinical activity and toxicity of pembrolizumab in the treatment of non-small-cell lung cancer: a secondary analysis of the KEYNOTE-001 phase 1 trial. Lancet Oncol. 2017;18(7):895–903.
32. Machiels JP, Tao Y, Burtness B, Tahara M, Licitra L, Rischin D, et al. Pembrolizumab given concomitantly with chemoradiation and as maintenance therapy for locally advanced head and neck squamous cell carcinoma: KEYNOTE-412. Future Oncol. 2020;16(18):1235. https://moh-it.pure.elsevier.com/en/publications/pembrolizumab-given-concomitantly-with-chemoradiation-and-as-main. Accessed 16 Nov 24.
33. Vatner RE, Cooper BT, Vanpouille-Box C, Demaria S, Formenti SC. Combinations of immunotherapy and radiation in cancer therapy. Front Oncol. 2014;4:325. https://doi.org/10.3389/fonc.2014.00325.
34. Thames HD. Early fractionation methods and the origins of the NSD concept. Acta Oncol. 1988;27(2):89–103.
35. Jeong H, Bok S, Hong BJ, Choi HS, Ahn GO. Radiation-induced immune responses: mechanisms and therapeutic perspectives. Blood Res. 2016;51(3):157–63.
36. de Andrade CH, Villar RC. Radiotherapy and immune response: the systemic effects of a local treatment. Clinics. 2018;73:e557s.
37. Dewan MZ, Galloway AE, Kawashima N, Dewyngaert JK, Babb JS, Formenti SC, et al. Fractionated but not single-dose radiotherapy induces an immune-mediated abscopal effect when combined with anti-CTLA-4 antibody. Clin Cancer Res. 2009;15(17):5379–88.
38. Schaue D, Ratikan JA, Iwamoto KS, McBride WH. Maximizing tumor immunity with fractionated radiation. Int J Radiat Oncol Biol Phys. 2012;83(4):1306–10.
39. Lugade AA, Moran JP, Gerber SA, Rose RC, Frelinger JG, Lord EM. Local radiation therapy of B16 melanoma tumors increases the generation of tumor antigen-specific effector cells that traffic to the tumor. J Immunol. 2005;174(12):7516–23.
40. Welsh J, Menon H, Chen D, Tang C, Altan M, Hess K, et al. Pembrolizumab with or without radiation therapy for metastatic non-small cell lung cancer: a randomized phase I/II trial. J Immunother Cancer. 2020;8:e001001.
41. Tsai MH, Cook JA, Chandramouli GVR, DeGraff W, Yan H, Zhao S, et al. Gene expression profiling of breast, prostate, and glioma cells following single versus fractionated doses of radiation. Cancer Res. 2007;67(8):3845–52.
42. Rubner Y, Muth C, Strnad A, Derer A, Sieber R, Buslei R, et al. Fractionated radiotherapy is the main stimulus for the induction of cell death and of Hsp70 release of p53 mutated glioblastoma cell lines. Radiat Oncol. 2014;9(1):89.
43. Kulzer L, Rubner Y, Deloch L, Allgäuer A, Frey B, Fietkau R, et al. Norm- and hypo-fractionated radiotherapy is capable of activating human dendritic cells. J Immunotoxicol. 2014;11(4):328–36.
44. Deng L, Liang H, Xu M, Yang X, Burnette B, Arina A, et al. STING-dependent cytosolic DNA sensing promotes radiation-induced type I interferon-dependent antitumor immunity in immunogenic tumors. Immunity. 2014;41(5):843–52.
45. Demaria S, Guha C, Schoenfeld J, Morris Z, Monjazeb A, Sikora A, et al. Radiation dose and fraction in immunotherapy: one-size regimen does not fit

all settings, so how does one choose? J Immunother Cancer. 2021;9(4):e002038.
46. Bang A, Wilhite TJ, Pike LRG, Cagney DN, Aizer AA, Taylor A, et al. Multicenter evaluation of the tolerability of combined treatment with PD-1 and CTLA-4 immune checkpoint inhibitors and palliative radiation therapy. Int J Radiat Oncol Biol Phys. 2017;98(2):344–51.
47. Deloch L, Derer A, Hartmann J, Frey B, Fietkau R, Gaipl US. Modern radiotherapy concepts and the impact of radiation on immune activation. Front Oncol. 2016;6:141. https://doi.org/10.3389/fonc.2016.00141.
48. Mohan R, Grosshans D. Proton therapy—present and future. Adv Drug Deliv Rev. 2017;109:26–44.
49. Lee HJ, Zeng J, Rengan R. Proton beam therapy and immunotherapy: an emerging partnership for immune activation in non-small cell lung cancer. Transl Lung Cancer Res. 2018;7(2):180–8.
50. Zs M, Ei G, Lr W, Pm C, Cm H, Js K, et al. Tumor-specific inhibition of in situ vaccination by distant untreated tumor sites. Cancer Immunol Res. 2018;6(7):825. https://pubmed.ncbi.nlm.nih.gov/29748391/. Accessed 31 Oct 2022.
51. Pike LRG, Bang A, Mahal BA, Taylor A, Krishnan M, Spektor A, et al. The impact of radiation therapy on lymphocyte count and survival in metastatic cancer patients receiving PD-1 immune checkpoint inhibitors. Int J Radiat Oncol Biol Phys. 2019;103(1):142–51.
52. Tang C, Welsh JW, de Groot P, Massarelli E, Chang JY, Hess KR, et al. Ipilimumab with stereotactic ablative radiation therapy: phase I results and immunologic correlates from peripheral T cells. Clin Cancer Res. 2017;23(6):1388–96.
53. Kang J, Demaria S, Formenti S. Current clinical trials testing the combination of immunotherapy with radiotherapy. J Immunother Cancer. 2016;4:51.
54. Young KH, Baird JR, Savage T, Cottam B, Friedman D, Bambina S, et al. Optimizing timing of immunotherapy improves control of tumors by Hypofractionated radiation therapy. PLoS One. 2016;11(6):e0157164.
55. von Itzstein MS, Gonugunta AS, Mayo HG, Minna JD, Gerber DE. Immunotherapy use in patients with lung cancer and comorbidities. Cancer J. 2020;26(6):525–36.
56. Geng Y, Zhang Q, Feng S, Li C, Wang L, Zhao X, et al. Safety and efficacy of PD-1/PD-L1 inhibitors combined with radiotherapy in patients with non-small-cell lung cancer: a systematic review and meta-analysis. Cancer Med. 2021;10(4):1222–39.
57. Faivre-Finn C, Vicente D, Kurata T, Planchard D, Paz-Ares L, Vansteenkiste JF, et al. Four-year survival with Durvalumab after Chemoradiotherapy in stage III NSCLC—an update from the PACIFIC trial. J Thorac Oncol. 2021;16(5):860–7.
58. Bradley JD, Nishio M, Okamoto I, Newton MD, Trani L, Shire NJ, et al. PACIFIC-2: phase 3 study of concurrent durvalumab and platinum-based chemoradiotherapy in patients with unresectable, stage III NSCLC. J Clin Oncol. 2019;37(15_suppl):TPS8573.
59. Merck Sharp & Dohme LLC. A phase 3 study of Pembrolizumab (MK-3475) in combination with concurrent chemoradiation therapy followed by Pembrolizumab with or without Olaparib vs concurrent chemoradiation therapy followed by Durvalumab in participants with unresectable, locally advanced, stage III non-small cell lung cancer (NSCLC). clinicaltrials.gov; 2022. Report No.: NCT04380636. https://clinicaltrials.gov/ct2/show/NCT04380636. Accessed 30 Oct 2022.
60. De Ruysscher D, Ramalingam S, Urbanic J, Gerber DE, Tan DSW, Cai J, et al. CheckMate 73L: a phase 3 study comparing Nivolumab plus concurrent chemoradiotherapy followed by Nivolumab with or without Ipilimumab versus concurrent chemoradiotherapy followed by Durvalumab for previously untreated, locally advanced stage III non-small-cell lung cancer. Clin Lung Cancer. 2022;23(3):e264–8.
61. Wang Q, Li S, Qiao S, Zheng Z, Duan X, Zhu X. Changes in T lymphocyte subsets in different tumors before and after radiotherapy: a meta-analysis. Front Immunol. 2021;12:648652. https://doi.org/10.3389/fimmu.2021.648652.
62. Wang C, Park J, Ouyang C, Longmate JA, Tajon M, Chao J, et al. A pilot feasibility study of Yttrium-90 liver radioembolization followed by Durvalumab and Tremelimumab in patients with microsatellite stable colorectal cancer liver metastases. Oncologist. 2020;25(5):382–e776.
63. Royal Marsden NHS Foundation Trust. Phase I Study of the anti-PD1 immune checkpoint inhibitor Pembrolizumab and Platinum in combination with radical radiotherApY in Cervix cAncer. [Internet]. clinicaltrials.gov; 2019. Report No.: NCT03144466. https://clinicaltrials.gov/ct2/show/NCT03144466. Accessed 22 Nov 2022.
64. Teng F, Li M, Yu J. Radiation recall pneumonitis induced by PD-1/PD-L1 blockades: mechanisms and therapeutic implications. BMC Med. 2020;18(1):275.
65. Selzer E, Kornek G. Targeted drugs in combination with radiotherapy for the treatment of solid tumors: current state and future developments. Expert Rev Clin Pharmacol. 2013;6(6):663–76.
66. Tchelebi LT, Batchelder E, Wang M, Lehrer EJ, Drabick JJ, Sharma N, et al. Radiotherapy and receptor tyrosine kinase inhibition for solid cancers (ROCKIT): a meta-analysis of 13 studies. JNCI Cancer Spectr. 2021;5(4):pkab050.
67. Bonner JA, et al. Radiotherapy plus Cetuximab for squamous-cell carcinoma of the head and neck. N Engl J Med. 2006;354:567. https://doi.org/10.1056/nejmoa053422.
68. Radiotherapy plus cetuximab or cisplatin in human papillomavirus-positive oropharyngeal cancer (NRG Oncology RTOG 1016): a randomised, multicentre, non-inferiority trial. Lancet. https://www.thelancet.com/journals/lancet/article/piis0140-6736(18)32779-x/fulltext. Accessed 11 Oct 2022.
69. Mehanna H, Robinson M, Hartley A, Kong A, Foran B, Fulton-Lieuw T, et al. Radiotherapy plus cispla-

tin or cetuximab in low-risk human papillomavirus-positive oropharyngeal cancer (De-ESCALaTE HPV): an open-label randomised controlled phase 3 trial. Lancet. 2019;393(10166):51–60.
70. Patil VM, Noronha V, Joshi A, Agarwal J, Ghosh-Laskar S, Budrukkar A, et al. A randomized phase 3 trial comparing nimotuzumab plus cisplatin chemoradiotherapy versus cisplatin chemoradiotherapy alone in locally advanced head and neck cancer. Cancer. 2019;125(18):3184–97.
71. Zhuang H, Yuan Z, Chang JY, Wang J, Pang Q, Zhao L, et al. Radiation pneumonitis in patients with non–small-cell lung cancer treated with Erlotinib concurrent with thoracic radiotherapy. J Thorac Oncol. 2014;9(6):882–5.
72. Jia W, Guo H, Jing W, Jing X, Li J, Wang M, et al. An especially high rate of radiation pneumonitis observed in patients treated with thoracic radiotherapy and simultaneous osimertinib. Radiother Oncol. 2020;152:96–100.
73. Zheng L, Wang Y, Xu Z, Yang Q, Zhu G, Liao X, et al. Concurrent EGFR-TKI and thoracic radiotherapy as first-line treatment for stage IV non-small cell lung cancer harboring EGFR active mutations. Oncologist. 2019;24(8):1031–e612.
74. Wang XS, Bai YF, Verma V, Yu RL, Tian W, Ao R, et al. Randomized trial of first-line tyrosine kinase inhibitor with or without radiotherapy for synchronous oligometastatic EGFR-mutated non-small cell lung cancer. J Natl Cancer Inst. 2023;115:742.
75. Zhu Z, Ni J, Cai X, Su S, Zhuang H, Yang Z, et al. International consensus on radiotherapy in metastatic non-small cell lung cancer. Transl Lung Cancer Res. 2022;11(9):1763. https://tlcr.amegroups.com/article/view/68265. Accessed 22 Sep 2022.
76. Halyard MY, Pisansky TM, Dueck AC, Suman V, Pierce L, Solin L, et al. Radiotherapy and adjuvant Trastuzumab in operable breast cancer: tolerability and adverse event data from the NCCTG phase III trial N9831. J Clin Oncol. 2009;27(16):2638–44.
77. Abi Jaoude J, de Azambuja E, Makki M, Tamim H, Tfayli A, Geara F, et al. Post-mastectomy radiation therapy in human epidermal growth factor receptor 2 positive breast cancer patients: analysis of the HERA trial. Int J Radiat Oncol Biol Phys. 2020;106(3):503–10.
78. Bachir B, Anouti S, Jaoude JA, Kayali M, Tfayli A, Azambuja ED, et al. Evaluation of cardiotoxicity in HER-2 positive breast cancer patients treated with radiation therapy and Trastuzumab. Int J Radiat Oncol Biol Phys. 2021;111(3):S7.
79. De Santis MC, Bonfantini F, Di Salvo F, Fiorentino A, Riboldi VM, Di Cosimo S, et al. Trastuzumab and Hypofractionated whole breast radiotherapy: a victorious combination? Clin Breast Cancer. 2018;18(3):e363–71.
80. Gilbert MR, Dignam JJ, Armstrong TS, Wefel JS, Blumenthal DT, Vogelbaum MA, et al. A randomized trial of bevacizumab for newly diagnosed glioblastoma. N Engl J Med. 2014;370(8):699–708.
81. Chan J, Jayamanne D, Wheeler H, Khasraw M, Wong M, Kastelan M, et al. The role of large volume re-irradiation with bevacizumab in chemorefractory high grade glioma. Clin Transl Radiat Oncol. 2020;22:33–9.
82. Helis CA, Prim SN, Cramer CK, Strowd R, Lesser GJ, White JJ, et al. Clinical outcomes of dose-escalated re-irradiation in patients with recurrent high-grade glioma. Neurooncol Pract. 2022;9(5):390–401.
83. Tallet AV, Dhermain F, Rhun EL, Noël G, Kirova YM. Combined irradiation and targeted therapy or immune checkpoint blockade in brain metastases: toxicities and efficacy. Ann Oncol. 2017;28(12):2962–76.
84. Cai Y, Wang JY, Liu H. Clinical observation of whole brain radiotherapy concomitant with targeted therapy for brain metastasis in non-small cell lung cancer patients with chemotherapy failure. Asian Pac J Cancer Prev. 2013;14(10):5699–703.
85. Singh R, Lehrer EJ, Ko S, Peterson J, Lou Y, Porter AB, et al. Brain metastases from non-small cell lung cancer with EGFR or ALK mutations: a systematic review and meta-analysis of multidisciplinary approaches. Radiother Oncol J Eur Soc Ther Radiol Oncol. 2020;144:165–79.
86. Ge Y, Che X, Gao X, Zhao S, Su J. Combination of radiotherapy and targeted therapy for melanoma brain metastases: a systematic review. Melanoma Res. 2021;31(5):413–20.
87. Hernandez AL, Young CD, Bian L, Weigel K, Nolan K, Frederick B, et al. PARP inhibition enhances radiotherapy of SMAD4-deficient human head and neck squamous cell carcinomas in experimental models. Clin Cancer Res. 2020;26(12):3058–70.
88. Barcellini A, Loap P, Murata K, Villa R, Kirova Y, Okonogi N, et al. PARP inhibitors in combination with radiotherapy: to do or not to do? Cancer. 2021;13(21)):5380.
89. Jannetti SA, Zeglis BM, Zalutsky MR, Reiner T. Poly(ADP-Ribose)Polymerase (PARP) inhibitors and radiation therapy. Front Pharmacol. 2020;11:170. https://doi.org/10.3389/fphar.2020.00170.

Further Reading

https://pubmed.ncbi.nlm.nih.gov/19573801/
https://pubmed.ncbi.nlm.nih.gov/27660705/
https://pubmed.ncbi.nlm.nih.gov/3291904/
https://pubmed.ncbi.nlm.nih.gov/33827904/
https://pubmed.ncbi.nlm.nih.gov/35906199/
https://pubmed.ncbi.nlm.nih.gov/32335187/
https://www.ncbi.nlm.nih.gov/pmc/articles/PMC5844965/

11 Multidisciplinary Care in Oncology

Sushmita Ghoshal, Nagarjun Ballari, Swarupa Mitra, Bidhu K. Mohanti, and Anusheel Munshi

Preamble

Cancer is not a single disease but a spectrum of malignant disorders of various sites, histologies, and stages. Progress in scientific knowledge and technology has revolutionized the traditional concepts of diagnosis, prognosis, and treatment of cancer. The traditional armamentarium of cancer has been surgery, radiotherapy, and systemic chemotherapy. Historically, cancer was treated with a single modality, a classic example being Halsted mastectomy. It was soon realized that single modality treatment is unlikely to give efficient results in this disease. In recent years, the complexity of cancer management has necessitated the needs for optimal coordination between healthcare specialists and other skilled staff of an oncology team or cancer center, such that each diagnosed cancer patient would receive suitable approaches for cancer therapy and supportive care. This chapter gives an overview of present-day multidisciplinary care in oncology practice [1].

11.1 The Need and the Concept of Multidisciplinary Team (MDT)

Cancer treatment techniques have been modified to minimize side effects, and multimodality treatments are increasingly being used for better outcomes. A combination of radiation therapy and chemotherapy increases the scope and efficacy of organ preservation strategies. The newer agents for systemic therapy, including immunotherapy, have catapulted the specialty of Medical Oncology to the forefront of cancer care. The plethora of current scientific literature may be overwhelming for the practitioner and perplexing for the patients. Therefore, it is only logical that all the specialties working in the field of cancer treatment use a common platform to create and consolidate the best possible treatment plan for a given patient. This is the essence of multidisciplinary cancer care [2]. It is further recognized that multimodality cancer management has increased the opportunities for extending the various practices of cancer surgery, radiotherapy and systemic therapy. Yet,

S. Ghoshal (✉)
Department of Radiotherapy and Oncology, PGIMER, Chandigarh, India

N. Ballari
Radiotherapy, HBCHRC, Mullanpur, Punjab, India

S. Mitra
Radiation Oncology, Fortis Memorial Research Institute, Gurugram, Haryana, India

B. K. Mohanti
Academic and Head of Radiation Oncology, Bagchi Sri Shankara Cancer Centre & Research Institute, Bhubaneswar, Odisha, India

A. Munshi
Department of Radiation Oncology, IOCI, Fortis Noida, Uttar Pradesh, India

A. Munshi et al. (eds.), *Radiation Oncology – Principles, Precepts and Practice*,
https://doi.org/10.1007/978-981-97-8389-2_11

the pursuit of the anticancer therapy can prove to be futile, deleterious and costly for many cancer patients; more so when palliative care can improve the pain and symptom relief and the quality of life of those patients for the remaining life with positive benefits. The importance of the multidisciplinary team in providing an adequate service to the patient and his/her family in the Middle-East countries has shown that oncologists and nurses can improve team building, and obtain training for unmet needs of patients like delivery of timely palliative care [3].

11.2 Members of the Multidisciplinary Team (MDT)

The multidisciplinary team (MDT) comprises a core group of specialists from varied disciplines. These include surgical oncology, radiation oncology, medical oncology, plastic surgery, radiology, hematology, pathology, nuclear medicine, and nursing [4].

Regular MDT meetings provide a platform for this core group to discuss each patient. The key discussion items remain: (1) radiographic features, (2) pathological findings, (3) diagnostic and/or therapeutic options, (4) best approach for each patient, (5) appropriate sequencing of the treatment modalities, (6) integration of evidence-based guidelines and treatment options and (7) discussion of clinical trial findings relevant to that case [1, 5].

There can be additional members who provide advice to the MDT but may not attend meetings regularly. These professionals include interventional radiology, palliative care, anesthesia/intensive care, teenage/young adult cancer services, oncology pharmacy, geriatric oncology, rehabilitation, clinical psychology/psychiatry, immunology, occupational therapy, dietetics, physiotherapy, and transplant services. Occasionally, primary care doctors, community nurses, and social workers may also be involved [6].

11.3 Role of Diagnostic Specialists

Besides surgical, radiation, and medical oncologists, specialists in diagnostics are a critical component of any multidisciplinary team (Fig. 11.1).

Pathology of cancer is not limited to morphological description only. A typical oncopathologist's report will provide information regarding morphological diagnosis and immunohistochemistry of varying extents based on the existing scenario. In some patients, further molecular/genetic testing may be required. This information is crucial for the MDT.

Another important team member is the radiologist. A radiologist is essential for the interpretation of images and the extent of disease and for establishing the staging of disease. The radiologist is also useful in the clinical course to the radiation oncologist for delineation of the tumor extent for conformal and stereotactic radiation. Similarly, intervention radiology can be essential in a tissue diagnosis of distant suspicious areas, potentially altering the aim of management between curative and palliative.

Nuclear Medicine specialist with his armamentarium of PET and SPECT is indispensable for any oncology MDT. PET CT has been proven to establish the staging of disease and reduce futile radical procedures. Further, it is a crucial modality for response assessment to disease and in-patient follow-up. However, interpretation of images, including posttreatment scans, can often be challenging, and it is here that the inputs of the specialist are much required. In LMIC, diseases such as tuberculosis are rampant and may play confounders.

11.3.1 Role of Surgical Oncologist

While the surgical oncologist forms the core of the surgical team in MDT, other specialty surgeons' role is also important. These include ENT, neurosurgery, and vascular surgery, among oth-

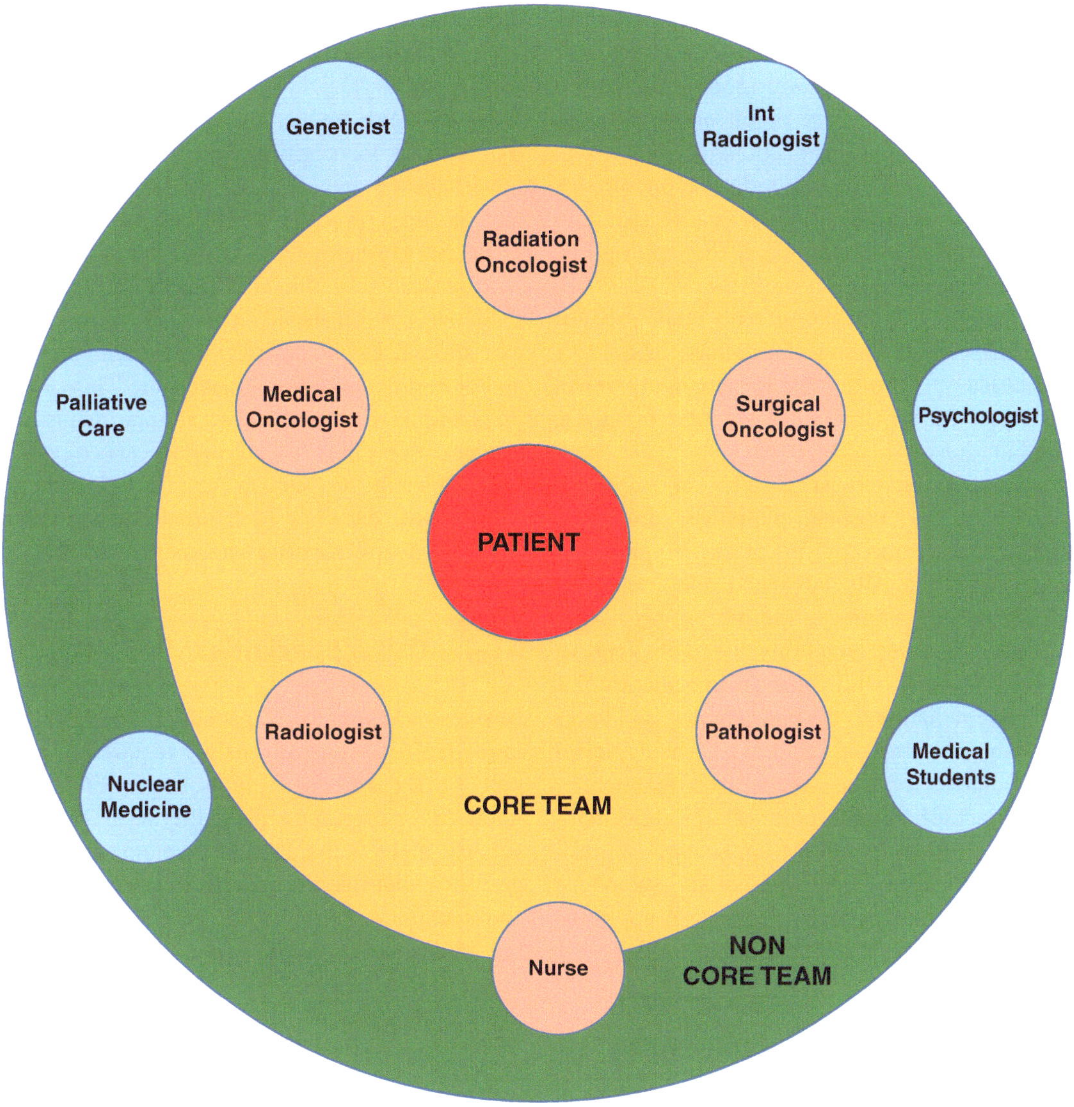

Fig. 11.1 Core and noncore teams in MDT

ers. Breast surgery may often need the help of plastic surgery for (1) reconstruction of the breast and (2) lymphedema sparing procedures such as lymphovenous anastomosis. Surgeries near the base of the skull region may need the help of an ENT surgeon as well as a neurosurgery team besides the surgical oncologist. Surgeries near important vessels such as larger arteries or IVC often require the presence and support of a vascular surgeon [7].

11.3.2 Role of Radiation Oncologist

It has been estimated that more than 60% of all cancer patients receive radiotherapy during the course of their illness. The proportion may be higher where patients present with advanced disease or other facilities are not readily available. Radiotherapy can be radical (curative), adjuvant, or used for symptom relief. All modern radiotherapy techniques are heavily image dependent and often require the inputs of a radiologist. This can be especially true for image based contouring and newer modalities such as MR Linacs. Sites where organ preservation is crucial such as anal canal, bladder, and larynx required a critical role of a radiation oncologist. Techniques and doses of postoperative radiation are constantly evolving, and the appropriate protocol can be discussed in the MDT. When radiotherapy is used along with other modalities, proper sequencing of treatment should be scheduled. It is important to accurately select patients who are likely to benefit from these combinations.

Radiation oncologist is also an important member of the hospital palliative care team. Palliative radiotherapy is often required for patients with locally advanced and metastatic disease. A short course of radiotherapy is proven to relieve pain and symptoms due to obstruction. Even in patients with disseminated metastases, palliative radiation to the brain, bones, and lungs can enhance symptom relief if indicated. Newer developments to make personalized treatment of cancer include the incorporation of radiogenomics, bioinformatics, and the application of artificial intelligence. The radiation oncologist involved in translational research can apply the acquired knowledge in future clinics.

11.3.3 Role of a Medical Oncologist

The domain of medical oncology has rapidly evolved over the past few years. From simple chemotherapy agents, the focus has shifted to target-based therapy and immunotherapy, molecular testing and next-generation sequencing (NGS). Chemotherapy is often given in a neoadjuvant fashion in locally advanced lesions to downstage them and enables further treatment by surgeon or a radiation oncology [8]. Integration and sequencing of chemotherapy/targeted therapy with radiation therapy and surgery is continuously evolving, requiring critical inputs from the medical oncologist.

11.3.4 Role of Palliative Care Specialist

Cancer patients frequently have relapses either in the form of locoregional recurrences or distant metastasis. In spite of all the advances in present day oncology practice, the disease may become refractory to treatment. Palliative therapy therefore has a crucial role in overall cancer management. Palliative care is usually provided by palliative care specialists who have received special training and/or certification in palliative care. They focus on physical, emotional, social, and spiritual issues cancer patients may face during the cancer experience. In some sites such as lung, early institution of palliative care has been proved to favorable influence overall survival outcomes as well [9].

11.4 Other Specialties

Besides the above core specialists, several other individual and group teams are required to be a part of MDT at frequent points. Examples include the need for speech therapist after head and neck surgery, a physiotherapy after a brain tumor surgery or a dietitian during chemotherapy or head and neck radiotherapy. Other specialists such as the Nutritionist, Psycho-Oncologist, Physiotherapist and Clinical Trial Expert are also important members of the MDT.

11.4.1 A Typical MDT Tumor Board Discussion

A 56 years male, chronic smoker, tobacco chewer, and alcoholic presented to the surgical oncology OPD of the hospital with complaints of difficulty in swallowing for solid diet, change in voice, and running nose. An outside noncontrast CT scan of neck had suggested lesion in left pyriform fossa with left-sided cervical lymphadenopathy. A biopsy from left pyriform fossa suggested squamous cell carcinoma grade II. The tissue block was sent for review in the hospital. Slides and

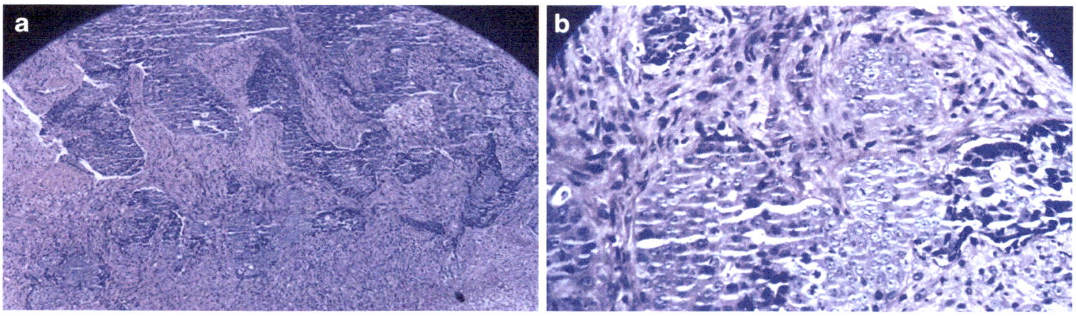

Fig. 11.2 (**a**, **b**) Sections show an infiltrating tumor comprising of nests of undifferentiated cells with moderate nuclear pleomorphism and hyperchromasia in the background of fibroblastic stroma (H&E, 10× and 40×)

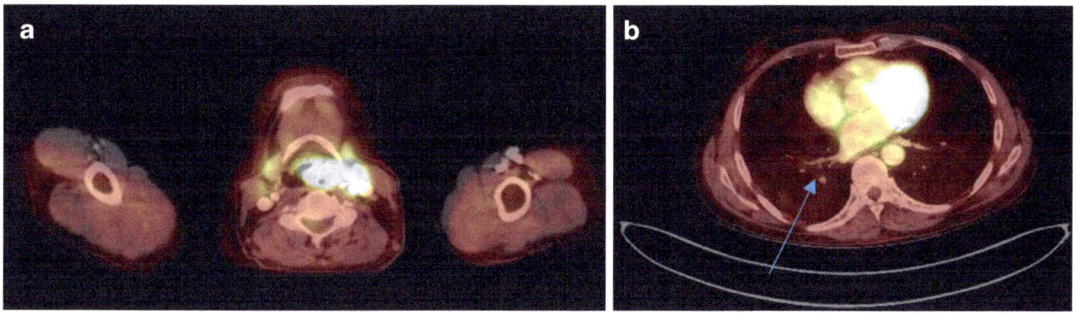

Fig. 11.3 (**a**, **b**) PET CT images showing pyriform sinus lesion and a suspicious lesion in lung

block review revealed poorly differentiated carcinoma and also there were some spindle cells like features seen (Fig. 11.2a,b). Immunohistochemistry was positive for Pan-CK, EMA, synaptophysin and negative for vimentin, p40, p63, CD56, CK7, and CK20. Final histopathology was suggestive of poorly differentiated carcinoma with neuroendocrine differentiation.

The whole body PET-CT suggested a lesion in the left pyriform fossa associated with left level II and III cervical lymph nodes—primary disease, with a subcentimeter fibronodular lesion in right lung, suspicious for metastases (Fig. 11.3a, b).

The case was brought to the MDT tumor board by the Surgical Oncologist.

Surgical Oncologist: Histopathology review and IHC has revealed an unusual neoplasm.

Pathologist: IHC strongly suggests that this is not a squamous cell carcinoma, the usual lesion in this location. This is likely to be a neuroendocrine carcinoma based on the morphology finding of and IHC.

Nuclear Medicine: The primary lesion is seen pyriform fossa with left level II and III cervical lymphadenopathy. The lung lesion is suspicious but too small to characterize.

Interventional radiologist: It would be difficult to take biopsy or FNAC from lesion, the chance of false negativity is high in view of the small sub cm size of the lesion.

Radiation Oncologist: Can we then give the benefit of doubt and proceed with radical intent, keeping the lung lesion on follow-up. However, we need to note that laryngeal small cell neuroendocrine carcinomas are particularly aggressive tumors, with a 5-year survival of no more than 5–10%.

Medical Oncologist: We can proceed with radical radiation with concurrent chemotherapy.

Surgical Oncologist: The histology is unusual, and few case reports and small series are there in literature. Can be taken up for surgery in case the response to chemoradiation is suboptimal and there is persistent residual disease after chemoradiation in follow up scan.

Final Tumor Board decision (documented): Concurrent chemoradiotherapy and keep close follow-up for suspicious lung lesion.

Table 11.1 How can MDT improve outcomes?

1	By reviewing their processes and the quality of their decisions regularly and seeking to continuously improve their practice
2	The team members can monitor compliance with evidence-based guidelines and streamline the resources for improved management strategies and enhanced cost-effectiveness
3	Shortening the waiting times for patients
4	Recruiting patients into clinical trials
5	Updating their required inputs from molecular pathology at periodic intervals
6	Having a policy for patient engagement in the individual care and policy development for the team
7	Periodic audit of their patient data, enabling further improvement in workflows
8	Patient management is evidence-based, and there is continuity of care
9	Regular meetings facilitate communication flow between all those involved in the treatment of the patient
10	Opportunity for education and learning to its members and trainee doctor
11	Also, improve the Well-being and work satisfaction of individual team members

11.5 Goals of Multidisciplinary Team (MDT)

The goal of multidisciplinary cancer care is to increase survival and improve the quality of survivorship. Each member of the MDT plays a critical role right from the time the patient presents in the out-patient department. The initials goals of MDT include establishment of the diagnosis of the patient, aided by pathology and radiology teams. Subsequently a care plan of the patient needs to be put in place. Treatment of cancer is often associated with acute and long-term toxicities that influence compliance and quality of life of patients. Appropriate interventions before, during and after treatment may reduce the frequency and severity of these toxicities. The availability of oncology nurses in the team improves the quality of wound dressing, stoma care, and lymphedema management. The trained nurses can be a bridge between the clinicians and patients, sharing information, and giving practical suggestions. They can also play the role of patient advocates Experts in the fields of physiotherapy, occupational therapy, speech and language pathology, and preparation of prosthesis help in the rehabilitation of patients who undergo mutilating surgeries. For head and neck cancer patients who are candidates for radiotherapy, specialized dental care before, during and after treatment helps to prevent dental caries and radio-necrosis of the mandible [10].

The impact of nutrition management throughout cancer treatment cannot be overemphasized. The disease itself and its treatment result in nutritional imbalance, causing weight loss, debility, fatigue, and poor tolerance of therapy. Dietary consultations, particularly the use of ingredients easily available in their kitchens, help patients and their caregiver's tide over the crisis [11]. The psychosocial aspects of cancer care must also be addressed by the MDT. It has been documented that newly diagnosed cancer patients, those on treatment, and survivors experience a wide range of emotional turbulence that require the active intervention of experts. Patients and their caregivers also worry about their catastrophic health expenditure and distress financing. For holistic care, the inclusion of psychologists/counselors in the team could be of immense help [12]. Access to palliative care facilities and early integration of palliative care with cancer-directed treatment should be the way forward. In order to improve outcomes, the collaboration of clinical and translational research personnel in the MDT may reduce the gap between existing practice and basic research (MDT).

MDTs improve decision-making, improve coordination, and reduce waiting times. Its construction and sustenance are challenging as it needs care coordination, interdepartmental communication, and stakeholders' active participation, including patients and all professionals. MDTs also facilitate further learning among the residents and junior faculty [13].

Unless the involved professionals make the MDTs a critical part of their working agenda and prioritize the meticulous organization of MDTs, the benefits wouldn't be realized. In a nutshell,

Table 11.2 Evidence for the utility of a MDT

Outcomes assessed	Study	Level of evidence	Total cases (*N*)	Site of cancer	The difference in MDT meeting arm and control arm (with respect to given outcome)
Overall survival	M. F. Back et al. [14]	3b	67	Glioma	Not significant (18.7 vs. 11.9 months, $P = 0.11$)
Overall survival	A. E. Martin-Ucar et al. [15]	4	240	Lung	Not significant
Overall survival	Peng et al. [16]	3a	30,814	Colorectal	In favor of MDT group (HR, 0.81; 95% CI, 0.69–0.94; $P = 0.005$)
Overall survival	M. R. Stephens et al. [17]	4	144	Esophagus	5 years (52% vs. 10%, $P < 0.001$)
Overall survival	Zhao et al. [18]	4	212	Esophagus	In favor of MDT group (HR, 0.59; 95% CI, 0.38–0.92; $P = 0.019$)
Overall survival	L. M. Forrest et al. [19]	4	243	Lung	Median (6.6 months vs. 3.2 months)
Overall survival	Zhu et al. [20]	4	422	Metastatic castration resistant prostate cancer	Median overall survival with and without MDT: 39.7 vs. 27.0 months; HR, 0.549; $P = 0.001$)
Overall survival	E. J. Junor et al. [21]	4	533	Ovarian	In favor of MDT group
Overall survival	R. O. Dillman and S. D. Chico et al. [22]	4	16,035	All cancers	5 years (71% vs. 63%, $P < 0.001$)
Quality of life	P. V. Murray et al. [23]	2b	88	Lung	Not significant
Patient experience	M. Gabel et al. [24]	4	269	Breast	Improved in MDT group, $P < 0.001$
Rate of intervention	L. M. Forrest et al. [19]	4	243	Lung	Patients receiving chemo (23% vs. 7%)
Rate of intervention	A. G. Davison et al. [25]	4	112	Lung	30% increase in resection in favor of MDT
Rate of intervention	A. E. Martin-Ucar et al. [15]	4	240	Lung	Increase in resection (23.4% vs. 12.2%)
Rate of intervention	A.-M. Bouvier et al. [26]	3b	2935	Colorectal	Increase in trial recruitment (10.3% vs. 5.1%)
Time to intervention	M. Gabel et al. [24]	4	269	Breast	Time to treatment (29.6 vs. 42.2 days)
Time to intervention	A. G. Davison et al. [25]	4	112	Lung	Not significant
Time to intervention	M. F. Back et al. [14]	3b	67	Glioma	Not significant
Accuracy of staging	A. R. Davies et al. [27]	3b	118	Upper GI	MDT improved staging accuracy
Expenses per patient	D. J. Fader et al. [28]	4	208	Melanoma	MDT saved USD 1600 per patient
Decision quality	F. Kee et al. [29]	4	50	Lung	NSD, team discussion did not improve the quality of decision making overall
Psychological morbidity of team members	R. Haward et al. [30]	5	72	Breast	Lower prevalence of psychiatric morbidity (15.7% vs. 26.6%; $P < 0.005$

Levels of evidence as defined by Oxford Centre for Evidence-Based Medicine (1a: systematic review of RCTs, 1b: individual RCT, 1c: all or none, 2a: systematic review of cohort studies, 2b: individual cohort study (including a low quality RCT), 2c: "Outcomes" Research, 3a: systematic review of case-control studies, 3b: individual Case-Control Study, 4: case-series (and poor quality cohort and case–control studies), 5: expert opinion without explicit critical appraisal) there is an urgent need to embrace this culture change to enable all the involved to understand individual roles and move forward together.

The various ways in which MDT can improve outcomes are given in Table 11.1.

11.6 Evidence for the Utility of MDT

Many studies have been done to test the utility and possible benefit of having a MDT (Table 11.2). Overall, the number of published studies reporting positive impact of cancer MDTs is more than the studies that failed to show benefits. It needs to be acknowledged that the design of nearly all of the studies has been poor. The studies often used before-and-after designs (outcomes before and after establishment of MDT), considered as weak evidence for establishing causal associations. Such designs often have multiple potential confounders. For example, along with establishment of MDT meetings, there could have been other organizational changes like streamlining of care pathways and improvement of infrastructure of hospital, making the interpretation of causal links between the MDT meetings and observed outcomes extremely challenging.

11.7 Legal Aspects

The question of culpability and responsibility over decisions made by the MDT is a complicated and undecided issue. As of now, it seems logical that all members present in the tumor board meeting are responsible for the MDT's decision. However, it is not clear if the final responsibility of decision-making is of the "lead clinician" or the MDT [31].

For example, what happens when the MDT jointly approves a decision which is clearly against the current evidence? Such uncertainties mean that there is a risk of legal action against the MDT as a group. Till the legal issues get clarified, clinicians are advised to improve their medicolegal awareness and proceed with due caution.

11.8 MDT Tumor Boards

Tumor boards are a natural extension of the working of MDT, where all the stakeholders convene at regular intervals to discuss cases. At the same time, it is desirable to hold the meeting physically; the COVID-19 pandemic created unprecedented challenges for healthcare systems [32]. In the present context, it also forced a rapid shift from in-person to virtual MDT meetings [33].

Fortunately, with the existing in-house IT support and the existing infrastructure, this transition took little time. Through various online platforms, it was also possible to "share screen," enabling visualization of radiology images and pathology images [34].

Further, the virtual MDT meetings provide increased opportunities for participation and interaction and enable continued communication, high-quality discussions, and effective coordination of patient care. Many clinicians believe that virtual MDTs are likely to be the future of cancer care. Strategies to optimize remote meetings

Table 11.3 Key performance indicators of MDT tumor boards

1	Regularity and punctuality of the tumor boards
2	The implementation rate of the tumor board decisions
3	Concurrence rate between tumor board decisions and existing evidence

Table 11.4 Predominant challenges in MDT

1	Fear of losing patients (and revenue) for individual specialty members, especially in private set ups
2	Inhibitions regarding the possible discussion on a prior treatment that was inadequate or inappropriate
3	Ensuring that MDT recommendations are followed in the practice
4	Ensuring active patient participation in the MDT process
5	Striking the right balance of educational and care delivery objectives
6	Ensuring limited exposure of the MDT members to medicolegal liability
7	Lack of time availability with consultants

include information technology (IT) support and management of distractions.

There are suggested parameters to evaluate the robustness of the systems of MDT tumor boards. Some of these are given in Table 11.3.

There remain some persistent hurdles and barriers to implementation of the concept of MDT and corresponding tumor boards (Table 11.4). These barriers can be of varied nature [35].

To summarize the nature of MDT is complex. Evaluation of the benefit of MDT meetings in an objective manner is complex. There is a paucity of high-quality evidence for their effectiveness, which is not surprising. The potential of MDT is immense but the challenge is to realize it to the maximum possible. Further research needs to be done to overcome the attitudinal and technological barriers to MDT. New technologies such as artificial intelligence-based platforms and driven engines are likely to enhance the scope and robustness of MDT and MDT meetings with enhanced reliability, efficiency, transparency, and accountability.

Precepts Box

1. The multidisciplinary team (MDT) comprises a core group of specialists from varied disciplines.
2. Critical team members of oncology MDT include surgical oncologist, medical oncologist, radiation oncologist, radiologist, and pathologist, among others.
3. MDT has been shown to improve clinical outcomes.
4. Multiple challenges to the concept of MDT and tumor boards remain.
5. Online engines and AI have the potential to revolutionize MDT tumor board concept.

Multiple Choice Questions

1. What is the essence of multidisciplinary cancer care?
   (a) A common platform to create and consolidate the best possible treatment plan for a given patient
   (b) Patients discuss among themselves for their treatment
   (c) Doctors and hospital management have a meeting for patient management
   (d) Doctor and lawyers discuss about patient treatment
2. Multidisciplinary team (MDT) comprises of
   (a) Radiation oncologist
   (b) Surgical oncologist
   (c) Medical oncologist
   (d) All of above
3. What percentage of patient need radiotherapy during their treatment?
   (a) 20%
   (b) 30%
   (c) 60%
   (d) 70%
4. Newer developments to make personalized treatment of cancer include
   (a) Radio genomics
   (b) Bioinformatics
   (c) Application of artificial intelligence
   (d) All of above
5. NGS stands for
   (a) New generation sequencing
   (b) Next-generation sequencing
   (c) New genomic sequencing
   (d) Next genomic sequencing
6. What is goal of multidisciplinary cancer care?
   (a) To increase survival and improve the quality of survivorship
   (b) To have a social gathering to relax busy schedule of doctors
   (c) To ask patient to take second opinion
   (d) Improve the local recurrence
7. How does MDT improve outcomes?
   (a) Prolonging the waiting times wwfor patients
   (b) By reviewing their processes and the quality of their decisions regularly and seeking to continuously improve their practice
   (c) Removing patients from clinical trials
   (d) Regular meetings facilitate good relations between members
8. What study assessed overall survival benefit based on MDT?
   (a) P. V. Murray et al.
   (b) G. Davison et al.

(c) E. J. Junor et al.
(d) M. Gabel et al.

9. What are key performance indicators of tumor board meeting?
   (a) Regularity and punctuality of the tumor boards
   (b) The implementation rate of the tumor board decisions
   (c) Concurrence rate between tumor board decisions and existing evidence
   (d) All of the above
10. What are predominant challenges of MDT?
   (a) Ensuring that MDT recommendations are followed in the practice
   (b) Increasing patients for individual specialty members may, especially in private set ups
   (c) Arrange for platform to gather consultants
   (d) Encouragement regarding the possible discussion on a prior treatment that was inadequate or inappropriate

Answers

1. (a)
2. (d)
3. (c)
4. (d)
5. (b)
6. (a)
7. (b)
8. (c)
9. (d)
10. (a)

References

1. Hanna N. Leveraging the virtual landscape for effective multidisciplinary care. In: Presentation of the ASCO Annual Meeting, 4 June 2021.
2. Munshi A, Sengar M. Multidisciplinary care in oncology: are we united? Acta Oncol. 2011;50:314–6.
3. Silbermann M, Pitsillides B, Al-Alfi N, Omran S, et al. Multidisciplinary care team for cancer patients and its implementation in several middle eastern countries. Ann Oncol. 2013;24 Suppl 7(Suppl 7):vii41–7.
4. Dubois CA, Borgès Da Silva R, Lavoie-Tremblay M, Lespérance B, Bentein K, Marchand A, et al. Developing and maintaining the resilience of interdisciplinary cancer care teams: an interventional study. BMC Health Serv Res. 2020;20:1039.
5. Berghmans T, Lievens Y, Aapro M, Baird AM, Beishon M, Calabrese F, et al. European cancer organisation essential requirements for quality cancer care (ERQCC): lung cancer. Lung Cancer. 2020;150:221–39.
6. Oxford University Hospitals NHS Foundation Trust. Multidisciplinary teams (MDT). 2021. https://www.ouh.nhs.uk/haematology/team/mdt.aspx. Accessed 18 Nov 2022.
7. Baláž P, Gürlich R, Havlůj L, Klézl P, Turyna R, et al. The role of the vascular surgery in oncosurgery. Cas Lek Desk. 2020;159:203–9.
8. Sandals M, Demski K, Ni Y, Konig V, Leach B, Mazzei E, et al. The impact of clinical decision making in a molecular tumour board at a tertiary care centre. Abstract 3128. J Clin Oncol. 2021;39(15 Suppl):3128.
9. Temel JS, Greer JA, Muzikansky A, Gallagher ER, Admane S, Jackson VA, Dahlin CM, Blinderman CD, Jacobsen J, Pirl WF, Billings JA, Lynch TJ. Early palliative care for patients with metastatic non-small-cell lung cancer. N Engl J Med. 2010;363:733–42.
10. Devi S, Singh N. Dental care during and after radiotherapy in head and neck cancer. Natl J Maxillofac Surg. 2014;5:117–25.
11. Munshi A, Pandey MB, Durga T, Pandey KC, Bahadur S, Mohanti BK. Weight loss during radiotherapy for head and neck malignancies: what factors impact it? Nutr Cancer. 2003;47:136–40.
12. Mathew B, Mohanti BK, Tewari S, Kabra V, Gulia P, Bajpai P, Munshi A. Collusion: the facade and its implications on total pain management in palliative care. Indian J Palliat Care. 2021;27:176–9.
13. Wihl J, Rosell L, Carlsson T, Kinhult S, Lindell G, Nilbert M, et al. Medical and nonmedical information during multidisciplinary team meetings in cancer care. Curr Oncol. 2021;28:1008–16.
14. Back MF, Ang EL, Ng WH, See SJ, Lim CC, Tay L, et al. Improvements in quality of care resulting from a formal multidisciplinary tumour clinic in the management of high-grade glioma. Ann Acad Med Singapore. 2007;36(5):347–51.
15. Martin-Ucar AE, Waller DA, Atkins JL, Swinson D, O'Byrne KJ, Peake MD. The beneficial effects of specialist thoracic surgery on the resection rate for non-small-cell lung cancer. Lung Cancer. 2004;46(2):227–32.
16. Peng D, Cheng YX, Cheng Y, et al. Improved overall survival of colorectal cancer under multidisciplinary team: a meta-analysis. Biomed Res Int. 2021;2021:5541613.
17. Stephens MR, Lewis WG, Brewster AE, Lord I, Blackshaw GR, Hodzovic I, et al. Multidisciplinary team management is associated with improved out-

comes after surgery for esophageal cancer. Dis Esophagus. 2006;19(3):164–71.
18. Zhao S, Qi W, Chen J, et al. Role of a multidisciplinary team in administering radiotherapy for oesophageal cancer. BMC Cancer. 2020;20(1):974.
19. Forrest LM, McMillan DC, McArdle CS, Angerson WJ, Dunlop DJ. An evaluation of the impact of a multidisciplinary team, in a single centre, on treatment and survival in patients with inoperable non-small-cell lung cancer. Br J Cancer. 2005;93(9):977–8.
20. Zhu S, Chen J, Ni Y, Zhang H, Liu Z, Shen P, et al. Dynamic multidisciplinary team discussion can improve the prognosis of metastatic castration-resistant prostate cancer patients. Prostate. 2021;81(11):721–7.
21. Junor EJ, Hole DJ, Gillis CR. Management of ovarian cancer: referral to a multidisciplinary team matters. Br J Cancer. 1994;70(2):363–70.
22. Dillman RO, Chico SD. Cancer patient survival improvement is correlated with the opening of a community cancer center: comparisons with intramural and extramural benchmarks. J Oncol Pract. 2005;1(3):84–92.
23. Murray PV, O'Brien ME, Sayer R, Cooke N, Knowles G, Miller AC, et al. The pathway study: results of a pilot feasibility study in patients suspected of having lung carcinoma investigated in a conventional chest clinic setting compared to a centralised two-stop pathway. Lung Cancer. 2003;42(3):283–90.
24. Gabel M, Hilton NE, Nathanson SD. Multidisciplinary breast cancer clinics: do they work? Cancer. 1997;79(12):2380–4.
25. Davison AG, Eraut CD, Haque AS, Doffman S, Tanqueray A, Trask CW, et al. Telemedicine for multidisciplinary lung cancer meetings. J Telemed Telecare. 2004;10(3):140–3.
26. Bouvier AM, Remontet L, Jougla E, Launoy G, Grosclaude P, Buémi A, et al. Place of multidisciplinary consulting meetings and clinical trials in the management of colorectal cancer in France in 2000. Gastroenterol Clin Biol. 2007;31(3):286–91.
27. Davies AR, Deans DAC, Penman I, Plevris JN, Fletcher J, Wall L, et al. The multidisciplinary team meeting improves staging accuracy and treatment selection for gastro-esophageal cancer. Dis Esophagus. 2006;19(6):496–503.
28. Fader DJ, Wise CG, Normolle DP, Johnson TM. The multidisciplinary melanoma clinic: a cost outcomes analysis of specialty care. J Am Acad Dermatol. 1998;38(5 I):742–51.
29. Kee F, Owen T, Leathem R. Decision making in a multidisciplinary cancer team: does team discussion result in better quality decisions? Med Decis Making. 2004;24(6):602–13.
30. Haward R, Amir Z, Borrill C, Dawson J, Scully J, West M, et al. Breast cancer teams: the impact of constitution, new cancer workload, and methods of operation on their effectiveness. Br J Cancer. 2003;89(1):15–22.
31. Ross T, Pawa N. The multidisciplinary team—who is liable when things go wrong? Eur J Surg Oncol. 2020;46:95–7.
32. Rajasekaran RB, Whitwell D, Cosker TDA, Gibbons CLMH, Carr A, et al. Will virtual multidisciplinary team meetings become the norm for musculoskeletal oncology care following the COVID-19 pandemic?—experience from a tertiary sarcoma center. BMC Musculoskelet Disord. 2021;22:18.
33. Kerawala C, Riva F, Paleri V, et al. Videoconferencing for multidisciplinary team meetings in the coronavirus disease era—human factors awareness and recognition. J Laryngol Otol. 2020;134:1118–9.
34. Simpson L, Mathew A, Wojciechowski R, et al. Characteristics of the multidisciplinary cancer conference (tumor boards) in the COVID-19 era. Abstract e13561. J Clin Oncol. 2021;39(15 Suppl):e13561. https://doi.org/10.1200/JCO.2021.39.15_suppl.e13561.
35. Maharaj AD, Evans SM, Zalcberg JR, Ioannou LJ, Graco M, Croagh D, et al. Barriers and enablers to the implementation of multidisciplinary team meetings: a qualitative study using the theoretical domains framework. BMJ Qual Saf. 2021;30:792–803.

Further Reading

https://www.youtube.com/watch?v=vjuCS7-CH3A (what is a multidisciplinary tumour board).

https://www.youtube.com/watch?v=qjVY1u_cOp8 (Multidisciplinary Tumour Board on "Ovarian Cancer"-An initiative of MGH, USA & Oncology Club, Bangladesh).

https://www.youtube.com/watch?v=xfxqd0utSUA (45th ICON Multidisciplinary Tumour Board 17 on CNS).

Teletherapy Technology and Principles

12

Biplab Sarkar and Prabhakar Ramachandran

12.1 A Brief History of Radiation (Imaging and Therapy)

The history of using radiation for imaging and therapeutic purposes began with the discovery of the Röntgen ray, more commonly termed as the X-ray.

Three events in the nineteenth century revolutionized diagnosis and treatment. These are the discovery of the X-ray (Wilhelm Conrad Röntgen: 1895), the discovery of radioactivity (Becquerel: 1896), and the isolation of Radium from pitchblende (Curies: 1898). On November 8, 1895, Wilhelm Conrad Röntgen, a professor of theoretical physics at the University of Würzburg, observed the effects of an unusual phenomenon while conducting laboratory experiments on cathode rays from a vacuum tube, building on the work of Hertz and Lenard. He observed a peculiar black line (luminescent effect) across the barium platinocyanide paper lying near a well-shielded Crookes tube which was used in the high-voltage experiment. This luminescent effect in barium could only have been caused by a type of ray, but since no light passed through the well-shielded tube; he was certain it was a different kind of ray generated from the active Crookes tube. In the following few days, he tested several properties of the newly developed ray, which he termed the X-ray, and found it had great penetration power capable of passing through cloth and wood and found its existence even at a distance of 2 m. Röntgent presented his results to the Wurzburg Physio-Medical Society on December 28, 1895, and announced his discovery at a meeting of the same society on January 23, 1896, where he publicly demonstrated X-ray imaging on the hand of Dr. Albert von Kölliker [1].

Röentgen realized immediately that these newly discovered rays could be helpful in medical applications and informed physicians he knew across Europe (January 1, 1896). The first formal diagnostic application took place in Birmingham, England, where Dr. John Hall-Edwards used X-rays to identify a needle stuck in a worker's hand and surgically removed it.

A parallel experiment by Henri Becquerel led to the discovery of Becquerel rays, and subsequent investigation by Pierre Curie Marya Sklodowska Curie uncovered natural radioactivity in Pitch Blende and Polonium by the early 1989 [2, 3].

By December 1989, Curies discovered Radium, which has greater penetration power than Po-210.

B. Sarkar (✉)
Department of Radiation Oncology, Apollo Multispeciality Hospitals, Kolkata, West Bengal, India

P. Ramachandran
Medical Physics, Radiation Oncology, Cancer Services, Princess Alexandra Hospital, Woolloongabba, QLD, Australia

A. Munshi et al. (eds.), *Radiation Oncology – Principles, Precepts and Practice*,
https://doi.org/10.1007/978-981-97-8389-2_12

Soon after the discovery of X-ray and natural radioactivity, scientists around the world engaged in experiments with these two discoveries. It was unknown who was the first to use X-rays for cancer treatment or biological experiments. In 1896, I. R. Tarkhanov in Russia substantiated the ability of ionizing radiation to cause functional and structural changes in cells.

In the same year, J. Gillman (USA) and V. Despeignes (France) attempted to treat malignant neoplasms with X-rays.

Some authors claimed the first use of these new X-rays in treating patients was accredited to Emil H. Grubbe in Chicago; he applied X-rays to advanced ulcerating breast cancer [4].

In 1903–1904, radium-226 began to be used in the treatment of patients with skin cancer and uterine cancer [5].

Although radium was an excellent source of uninterrupted radiation, the extraction of radium salts from the mineral pitchblende was a difficult process. On the contrary, X-ray tubes in those days were unstable and often had failures with limited penetration power. However, the use of both X-rays and Radium gained popularity in cancer therapy. In 1902, Radium was successfully used in Vienna for the treatment of pharyngeal cancer. In 1905, Robert Abbe from St. Luke Memorial Hospital (New York) was the first to implant radium directly into the tumor.

Technological advancements in the early part of this century led to the development of more stable X-ray apparatus capable of delivering higher energies, and considerable advances were made with the advent of high-performing transformers and Coolidge tubes in 1917. By 1920, X-ray machines were operating at energies exceeding 200 kV. By 1922, it was established that cancers like laryngeal cancers could be effectively treated with minimal sequelae and morbidity.

The development of radiation medicine is broadly categorized into four major eras: first, the era of discovery, lasting from the discovery of the X-ray to 1920; the orthovoltage era, spanning from the late 1920s through World War II; the megavoltage era, which began with higher-energy linacs for therapy in the 1950s and continues with advancements such as intensity modulated radiation therapy (IMRT) and volumetric modulated arc therapy (VMAT) [6]. Brachytherapy, which developed in parallel with external beam radiation therapy, established its own dosimetry and practice until the 1950s. Over the years, there have been several advancement in terms of after-loaders and computer-controlled delivery systems. However, the widespread application of brachytherapy is limited only to a few sites due to surgical complexity. By mid-1980s, the first hospital-based particle (proton) facility became feasible.

12.2 Technological Development in Radiotherapy

It was soon realized that radiation has adverse effects as well, as evidenced by Marie Curie's death in 1934 due to radiation sickness. The necessity of radiation protection was also realized, and measures were taken to incorporate them into clinical practice. In the late 1930s, there were growing interest in studying the effects of radiation on normal and malignant tissues. The radiobiological basis of radiation was also established through clinical observations in humans, laboratory experiments, and later, insights gained from atomic bomb survivors. The introduction of different teletherapy machines using isotopes such as Ra-226, Cs-137, Co-60 revolutionized the field of radiotherapy. Early machines, employing Radium or Cesium, had unidirectional movements in the anterior-posterior direction. Subsequently, the use of multiple fields from different directions to reduce the skin reactions and organs at risk (OAR) toxicities led to the emergence of rotational machines with Co-60 isotopes. It is estimated that nearly 50 million patients have benefited from Co-60 treatment, and it remains a viable option, although the number of such isotope units is gradually being replaced by modern linear accelerators.

This chapter focuses on a few major advancements in the teletherapy machines, particularly in linear accelerator technology, which in many aspects is highly efficient and incorporates enhanced safety features.

The most signifcant change in linear accelerators since their introduction into clinical practice in the 1950s, aside from the development of C-ARM (allowing 360° rotation centered on a single point) systems is the introduction of a multileaf collimator by Takahashi in 1965 [7]. Other noteworthy advancements are the generation of the first mega voltage beam using a Van de Graaff generator in 1937 and incorporation of CT scanners for radiotherapy treatment planning in 1973, which led to the inception of actual three-dimensional conformal radiotherapy. Additional developments include the introduction of IMRT conceptually in 1982 and its incorporation in clinical routine by many centers in 1990, helical tomotherapy in 1994, flattening filter free beams in 2006 and VMAT in 2008 [8–11]. Other developments which helped radiotherapy to establish as a specialty include the dedicated intra-cranial stereotactic radiosurgery systems, commonly known as the Gamma Knife proposed by Leksell in 1950, development of linac-based radiosurgery in 1980s, the robotic arm cranial and extra-cranial stereotactic cyberknife system in 1990, first MV and then kV Cone Beam CT in early 2000s, six dimensional robotic couch in 2010, MR guided radiotherapy in 2017, use of PET images in clinical routine practice (2010–2015) and volumetric modulated radiotherapy delivery capable ring gantry linear accelerator Halcyon in 2017 [11–14]. These developments in therapeutic techniques facilitated faster and more accurate delivery of radiation in combination with advancements in surgery, and improvements in systemic therapy. They led to the increase in the therapeutic ratio, a reduction in normal tissue complication probability, and improved tumor control, ultimately enhancing cancer survival rates[15].

12.3 Imaging, Simulation Technique, and Tumor Volume Definition

The radiation dose is delivered in single or, very rarely, multiple daily fractions over several weeks for radiobiological reasons. During radiation therapy it is critical to ensure that the patient is treated in a reproducible manner on a daily basis and that the same volume is irradiated during each fraction, particularly with modern conformal techniques. Simulation, in conjunction with on-board imaging (images captured during therapy delivery), ensures precise target volume delineation and consistent geometrical reproducibility of anatomical structures. Table 12.1 lists different imaging modalities used in radiation in radiation oncology for simulation demonstrating varying degrees of resolution and their common application in the radiation oncology space. Advancements in imaging technology have enabled modern CT scanners to achieve resolution as close to 0.25 mm, magnetic resonance imaging (MRI) systems up to 0.5 mm and positron emission tomography (PET) systems to 3 mm.

Major 2D imaging techniques in oncology include X-ray, ultrasound, and 2D angiography, whereas 3D imaging technologies include CT, MRI, PET, and magnetic resonance spectroscopy (MRS) [16]. Modern radiation treatment planning guidelines recommend the use of multiple imaging modalities for tumor definition, depending on the treatment site. While CT images are commonly used for treatment planning, MRI pro-

Table 12.1 Radiotherapy imaging modalities and their resolution

Modality	Visibility	Resolution
2D X-ray simulation	Bone image	≈0.1–0.2 mm
CT	Bone-soft tissue interface, breast, head neck, and tissue–air (lung)	~0.5–1 mm diameter
MRI	Soft tissue, brain, pelvis, head neck, liver, extremities	1–2 mm diameter
PET	Head and neck, lung, liver	4–5 mm diameter

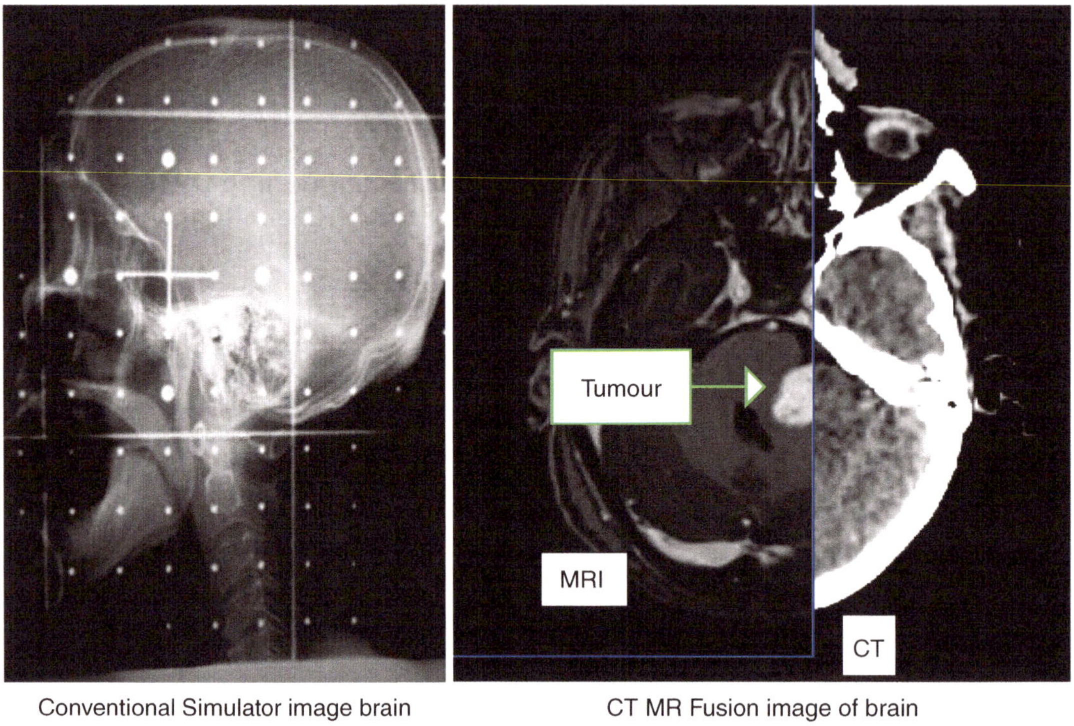

Fig. 12.1 Left panel: Conventional simulation (2D); right panel: CT simulation images with image fusion for better target volume delineation

vides superior soft-tissue definition, enhancing the overall accuracy of the process (Fig. 12.1).

PET images can help identify disease at the metabolic level before it is visible on CT or MR scans. Image registration is a useful tool for transferring disease information from MR and PET scans to CT scans for treatment planning. The detection of cancer using imaging is dependent on the concentration of tumor cells and the resolution of the imaging modality used. The contrast agent has a significant impact on CT and MRI image resolution. If the tumor cell density is around $10^6/mm^3$, it becomes easier to achieve high-resolution imaging and accurately differentiate between tumor and healthy tissue, provided that the contrast agent is effectively enhancing the signal. Although number of photon flux provided facilitates to separate small differences in tissue attenuation with 1 mm resolution in CT, however, due to physical and biological limitations, physical and biological limitations typically constrain the minimum measurable lesion size to about 3 mm [17].

Similar limitations can be found in modern MR imaging systems. In terms of detection limits, PET is the closest modality in nuclear medicine's arsenal to CT or MR imaging. The resolution limit of modern clinical PET scanners is 4 mm, which corresponds to the detection of tumors with a volume of 0.2 mL (7 mm in diameter) [17]. Different imaging modalities investigate different properties of the human body; for example, a CT scan detects the soft tissue-bone interface well because it works in the variation of electron concentration. CT scans provide a clear picture of a lung tumor. To provide good resolution in soft tissues, MRI uses variations in proton density and the alignment of these protons in the

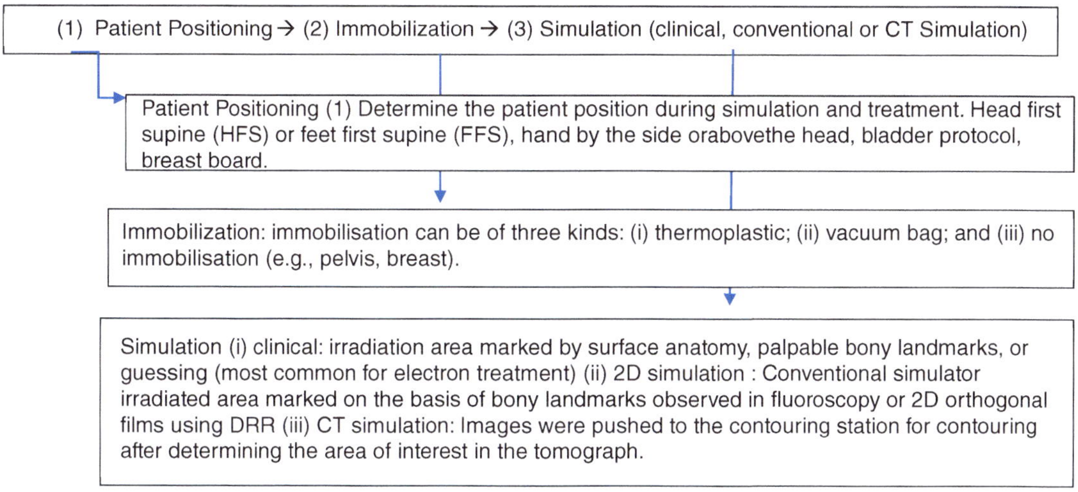

Fig. 12.2 Radiotherapy positioning, immobilization, and simulation flow chart

presence of a strong magnetic field. Figure 12.1 illustrates the difference in visibility between the MRI and CT scans for the fused image of the brain. The tumor is clearly visible in the MRI image but not in the CT image.

Flow chart (Fig. 12.2) showing some of the procedures required before external beam radiotherapy. The positioning, immobilization, and external contour measurements of the patient are carried out during the simulation or CT simulation.

Patient positioning is arguably the most important aspect of radiotherapy treatment because it ensures patient reproducibility on a daily basis. Comfort is the most critical factor in patient positioning. The most suitable treatment position is often head-first supine (HFS), where the patient is most at ease. Due to the reduction in bowel dose for obese patients, some institutions have adopted the prone position for treating pelvic patients. Other scenarios where prone positioning is preferred include the accelerated partial breast irradiation (as it reduces dose to the ipsilateral lung and heart), craniospinal irradiation, and other skull-base tumors. However, with the availability of CT simulation in modern practice, supine position for craniospinal treatment is preferred over prone position.

12.3.1 Immobilization

There are two main goals for immobilizing patients for radiotherapy simulation and delivery. The primary goal of any immobilization is to maintain the patient's position throughout treatment. The secondary goal is to provide a relaxed and comfortable position during the simulation/treatment. The human body is divided into two parts, where immobilizing the surface also immobilizes the inner tissues. This part includes the brain, head, neck, and extremities. The other segment is where immobilizing the surface does not always mean immobilizing the tissue inside, such as the thorax, abdomen, and pelvis. Several articles have been published comparing various immobilization devices; however, the focus is on the site-specific general recommendation for fractionated radiotherapy, excluding SRS/SRT and SBRT. A very rigid immobilization causes discomfort and may result in a catastrophic feeling for the patient. From the dosimetry point of view, it is essential to use materials and accessories that are comfortable for patients and are commissioned and validated by physicists. Figure 12.3 shows the use of immobilization for different sites, and Table 12.2 presents the recommended immobilization for different sites.

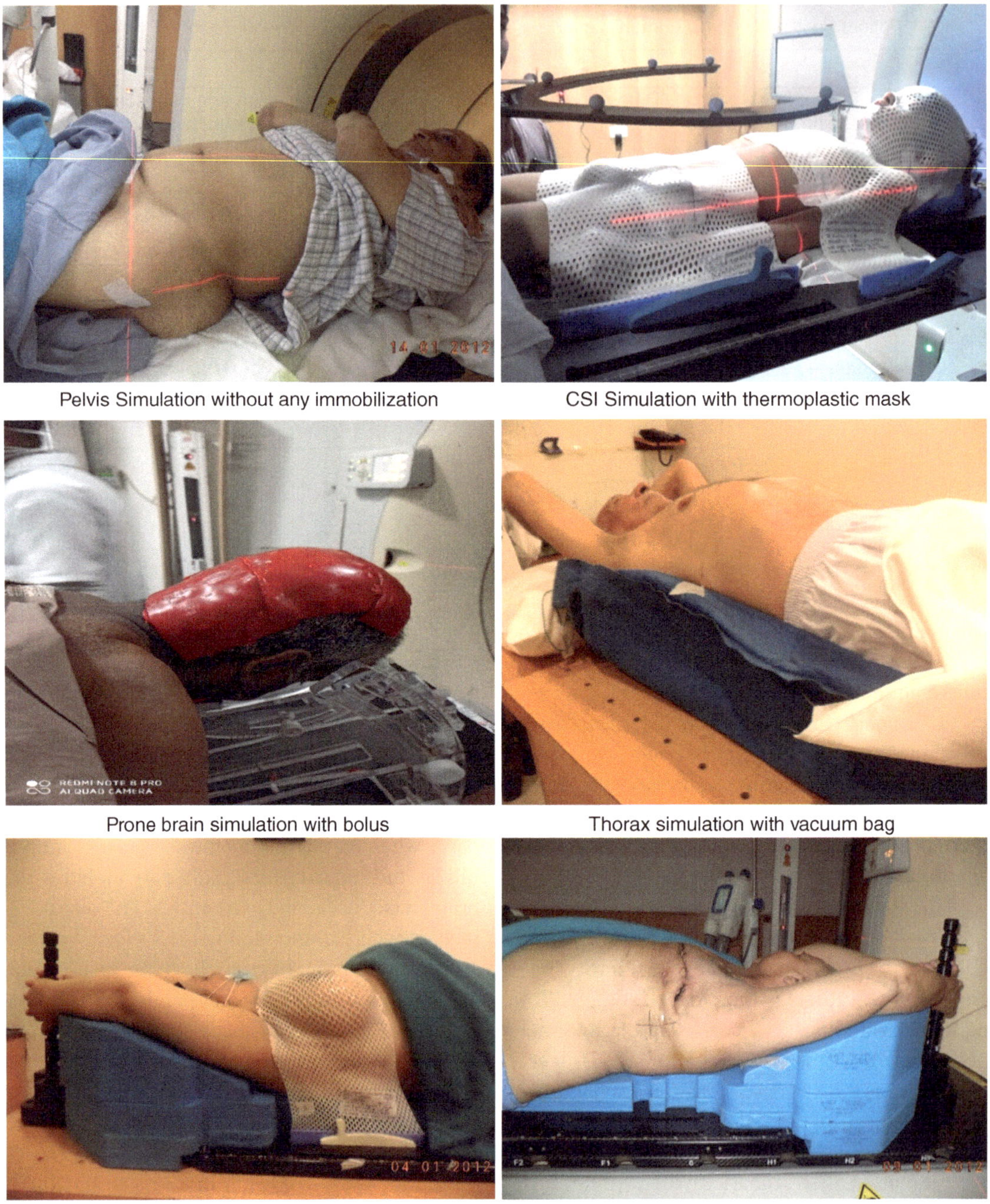

Pelvis Simulation without any immobilization

CSI Simulation with thermoplastic mask

Prone brain simulation with bolus

Thorax simulation with vacuum bag

Breast simulation : Breast Board and thermoplastic mask

Breast simulation : Breast Board

Fig. 12.3 Illustrates different immobilization for different sites: Pelvis simulation without any immobilization; CSI simulation with thermoplastic mask; Prone brain simulation with bolus; Thorax simulation with vacuum bag; Breast simulation: Breast Board and thermoplastic mask; Breast simulation: Breast Board; Lung SBRT simulation; Breast simulation with ABC

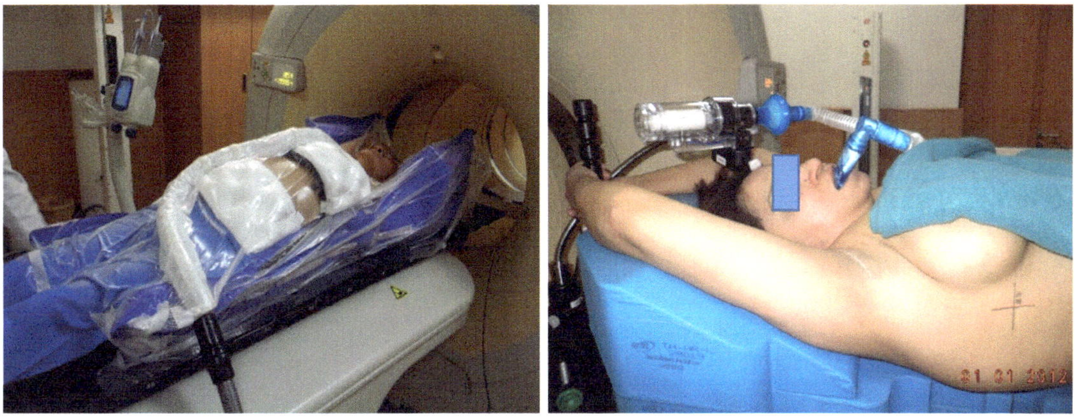

Lung SBRT Simulation   Active Breathing Coordinator (ABC)

Fig. 12.3 Continued

Table 12.2 Simulation and immobilization recommendation as a function of sites

Site	Recommended/may also be considered	Other instructions
Brain[a]	3 Clamp Thermoplastic for conventional fractionation SRS/SRT: Frameless - Double layered thermoplastic frames Frame-based: Leksell stereotactic Frame or Leksell Vantage Frame	Hands by side Pituitary: Mark the hairline and eye brow with a wire during simulation. An anterior superior oblique beam needs to be placed in between these two marks to avoid alopecia
H&N[a]	Head–neck board (all in one board) 3 or 5 Clamp thermoplastic	Hands by side. Neck neutral. For larynx cases neck may be extended
Breast and chest wall[a]	(i) Supine breast board + (ii) 4 Clamp thermoplastic for (1) sagging breast—to reduce the skin fold and avoid missing the field (iii) 2 clamp thin chest wall (width ≤10 mm at the thinnest part) required dose build-up (iv) Active breathing coordinator controlled or deep inspiration breath hold simulation (for left side only) (v) Prone position—Prone breast board no thermoplastic	Whichever is more comfortable: Ipsilateral hand abducted, contralateral hand preferably abducted or by side Head tilted in the opposite direction (to reduce dose from SCF field to esophagus) Some people take neck in neutral position with no specific clinical reason. Neck neutral condition may lead to serious esophagitis during radiotherapy
Thorax (lung, esophagus)	Conventional fractionation: Two-clamp or four clamp thermoplastics, Vacloc and abdominal compressor (SBRT) SBRT LUNG: Free breathing with full body or hemi body vacuum bag: Use time-averaged 4D scans or slow scans or produce MIP (maximum intensity projection for generating ITV) CT which will be fused with free-breathing 3D planning CT. For dose calculation, consider using Average Intensity Project (AIP). To reduce lung motion, use a thermoplastic or Vac-loc with an abdominal compressor[a]. Consider using Free-Breathing CT for pre-treatment verification and register the scans with the cone beam CT (CBCT). DIBH: Appropriate coaching (2–3 days) is required to ensure the patient is comfortable with the device	For esophagus: Hands by side For upper lobe lung, hands by side, for middle and lower lobe and large tumors hands over the head For thorax, hands should not be placed over chest

(continued)

Table 12.2 (continued)

Site	Recommended/may also be considered	Other instructions
Abdomen (liver, stomach)	Conventional fractionation: Two clamp of four clamp thermoplastics, Vac-lok, prone belly board SBRT: Abdomen compressor[b]. MIP	Hands over chest. To avoid hands by side
Pelvis (cervix, prostate, vagina)	1. No immobilization required, knee rest is sufficient 2. Thermoplastic 3. Vac-lok: Prostate SBRT or hypofractionation	Hands over chest
CSI	1. For stable, cooperative and adult patient—5 clamp head–neck mask covering brain to chest level 2. For unstable patients without anesthesia: Brain thorax+ abdomen-pelvis thermoplastic	Hands by side
TBI/TMI	1. VMAT/IMRT: Head thermoplastic +full body Vac-lok 2. Conventional standing and rice bag technique: No additional immobilization required	

[a]Vacuum Bag strictly not recommended
[b]Recommended but not mandatory

12.3.2 Simulation Technique

12.3.2.1 2D Simulation

1. Despite enormous technological advances in radiotherapy techniques, a large proportion of cancer patients in underdeveloped and developing countries continue to receive radiotherapy with surface marking based on anatomical landmarks. The field border is marked on the patient's body during this process, and the beam arrangement, number of fields, field size, gantry angle, collimator angle, and wedge angle, if necessary, are derived empirically from prior experiences. Figure 12.4 depicts a typical head and neck simulation undergoing treatment in the Co-60 machine. Currently, the most common clinical simulation is applied for the electron treatment of visible or palpable tumors.
2. Prior to the introduction of CT-based planning, and in many places still today, radiotherapy treatment planning is done using a dedicated C-ARM conventional simulator. Except for the beam energy, a traditional X-ray-based simulator closely resembles the therapy machine (Co-60 or LINAC). The simulator, which has a 125 KV X-ray and an image intensifier, can provide both fluoroscopic and static images. The treatment fields are marked on orthogonal radiographs to determine the irradiation area. Virtual multileaf collimator (MLC) blocking can be provided by modern simulators and transferred to therapy machines (Co-60/LINAC).
3. 3D Simulation: CT Simulation and Radiotherapy Planning: With the advent of the CT scan, radiotherapy planning was transformed from two to three dimensions, allowing advanced 3D TPS to calculate 3D dose distribution. The introduction of 3D imaging has accelerated radiotherapy advancement for accurate target delineation, dose calculation, and patient position verification prior to therapy delivery. It also contributes to the realistic and robust development of patient immobilization in radiotherapy practice. The identification of different treatment volumes such as gross tumor volume (GTV), clinical target volume (CTV), and planning target volume is

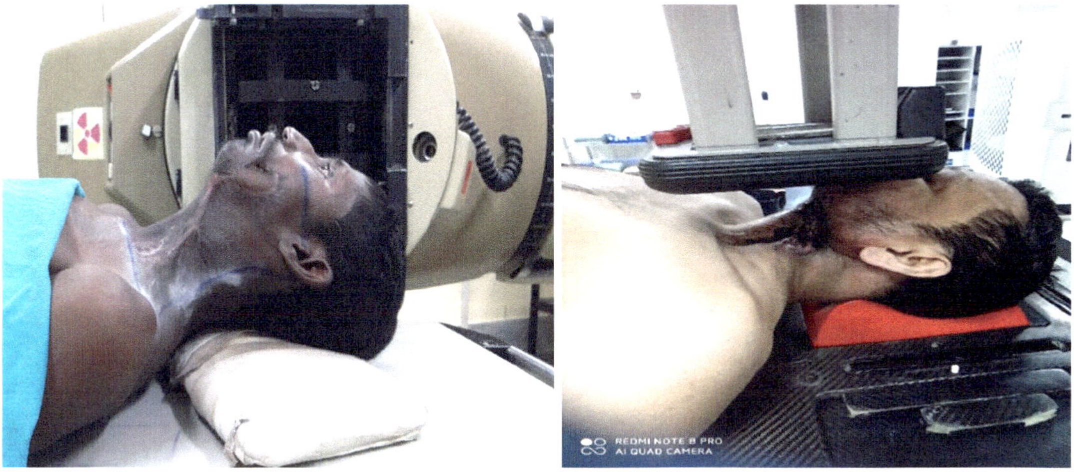

Fig. 12.4 Clinical simulation in a Co-60 machine (left panel) and for electron setup in linear accelerator (right panel)

also possible with 3D visualization of the tumor volume (PTV).

3D Simulation Technique: The patient is immobilized and placed on the CT simulator couch. If the gross position\ of the target is known, a CT external laser system (sagittal, coronal, and axial: three wall or ceiling mounted lasers) is used to identify a specific position, preferably intersecting in the center of the tumor volume, known as the isocenter. This position is marked with radio opaque beads (BB). This marking on the patient's surface or mold is the only connection between the CT simulator and the radiotherapy treatment machine. The topogram, or scout image, is used to determine the area of interest; in most cases, the preferred CT slice thickness is 3 mm. Extracranial stereotaxy (lung, liver, spine): 2 mm, cranial stereotaxy: 1 mm (for small lesions); 1.5 mm or 2 mm (maximum). The recommended slice thickness for palliative, long-term tumors and TBI/TMI is 5 mm.

12.3.3 Image Fusion and Contouring

Image fusion and contouring are required for the visualization of the target and the organ at risk. If any organ is not delineated in any form of intensity modulated plan, it will receive a large dose with no dose or fluence modulation; thus, it is recommended to draw all functional organs and targets. The definition of target volumes is given in ICRU 50/62/83 reports; however for the compleateness brifly mentined below. Visual (gross) or palpable tumor volume (GTV); depending upon the site, there will be a subclinical disease that is identified as a clinical target volume (CTV). Subclinical disease cannot be detected using imaging (CT or MRI) and must be calculated pathologically for each site or disease, which is often done via autopsy. Figure 12.5 (left panel) depicts the various target volumes. CTV is a geometrical margin that may encompass all subclinical disease and contain no tumor cells beyond it. In addition, to incorporate internal motion Internal target volume (for the lung and liver, where internal or breathing motion moves the target volume), then a setup margin to produce the planning target volume. In some cases, CT cannot completely or partially visualize the target, as shown in the right panel of Fig. 12.5, so other imaging modalities such as MR and PET are used to draw the GTV [18]. In MR imaging, various sequences are available. The most common sequences required for brain tumor contouring are T1-contrast (gadolinium), T2-flair, 3D sequence, perfusion images, and the CISS sequence for drawing the cochlea.

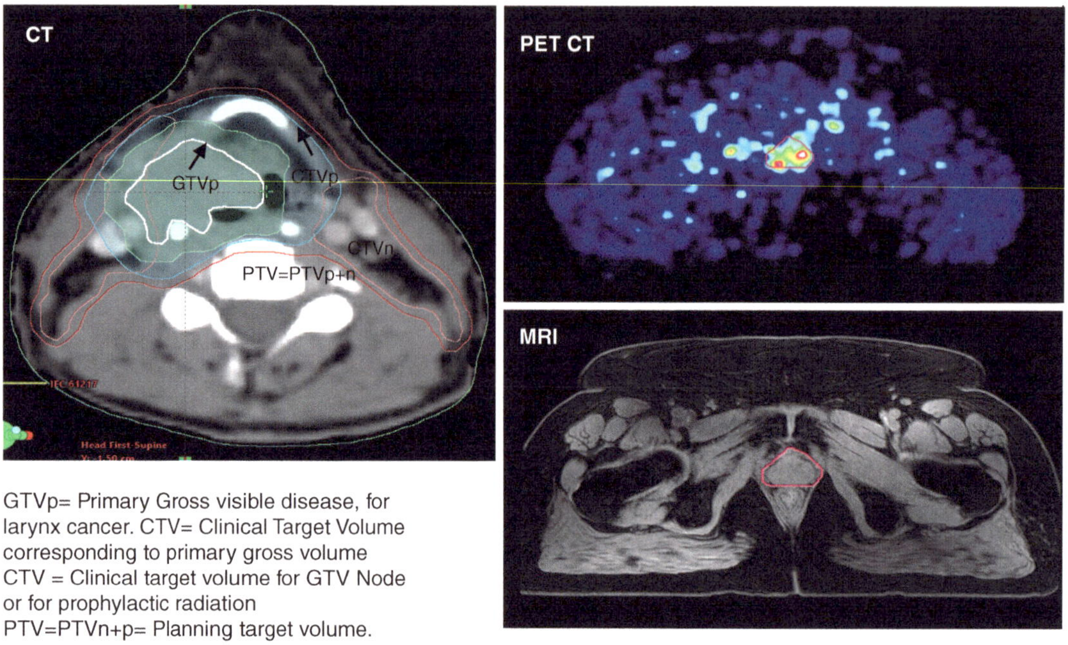

Fig. 12.5 Target definition in CT, PET-CT, and MRI. GTVp = Primary gross visible disease, for larynx cancer. CTV = Clinical target volume corresponding to primary gross volume. CTV = Clinical target volume for GTV Node or for prophylactic radiation. PTV = PTVn + p = Planning target volume

12.4 Planning and Dose Calculation

12.4.1 Manual Dose Calculation Algorithm

The absorbed dose in a phantom/patient at any given depth is determined by beam energy, depth, field size, distance from the radiation source, collimation, and the phantom material or incident medium composition. A small geometric field size assumes that the depth dose is the result of primary photons and that any other contributions are close to zero. Secondary scattered photons from the collimator increase with larger field sizes, contributing to an increase in dose in the traversing medium. The equivalent square field is used to determine the field size in most manual calculations. It is defined as a square with the same area-to-perimeter ratio as the given shape.

$$\text{Rectangles}: \frac{A}{P} = \frac{a \times b}{2(a+b)},$$

$$\text{Squares}: \frac{A}{P} = \frac{axa}{4xa} = \frac{a}{4}$$

where A and P are the area and perimeter of the field, respectively, and a and b are the sides of the rectangular field. Using the above equations, the equivalent square field size for a 10 cm × 15 cm rectangular field size is found to be 12 cm. The output, PDD, and TMR measured under specific conditions for a 10 cm × 15 cm field will be approximately equivalent to those measured for a 12 cm × 12 cm field. Since it is practically difficult to measure these parameters for all rectangular field size combinations, the use of the equivalent square field becomes a convenient approach for manual calculations.

The photon fluence is affected by the inverse square law and attenuation. In general, the output at isocenter is calibrated to 1 cGy/MU for isocen-

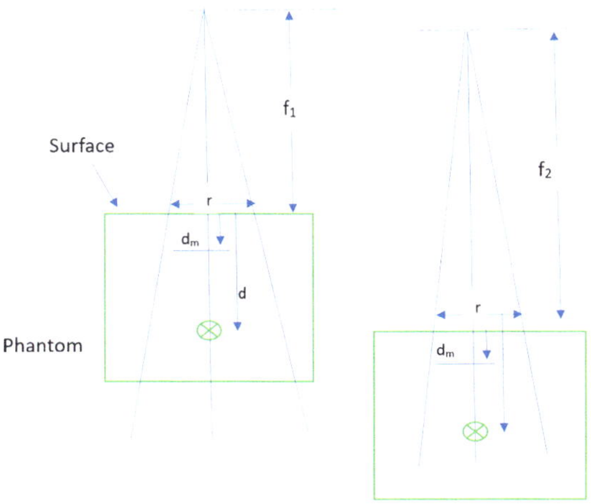

Fig. 12.6 Calculation of Mayneord Factor

tric reference conditions whereas for source to surface distance (SSD) reference conditions, the output is defined at Dmax. The depth of dose maximum (Dmax) is a function of energy and field size. The typical max values at standard field size (10 × 10 cm²) for Co-60, 4 MV, 6 MV, 10 MV, 15 MV, 18 MV are 0.5 cm, 1 cm, 1.5 cm, 2 cm, 3 cm and 3.3 cm, respectively. In clinical therapy, the desired SSD is a comprise between needing a large enough distance to mitigate the effects of the source dimensions, but short enough so that the dose rate is not diminished. The change in the PDD from differing SSD sizes is approximated by the Mayneord Factor (Fig. 12.6):

PDD is a function of SSD, which makes it challenging from a practical standpoint to use varied SSDs for treatment plans involving more than one treatment field. As the SSD may vary depending on the surface geometry and the rotation of the gantry, the tissue-air ratio (TAR) can be used, removing the dependence on the SSD correction. This is defined as the ratio of the dose at a point in the phantom to the dose in free space at the same distance, for a given field size r_d:

$$TAR\left(d,r_{\mathrm{d}}\right)=\frac{D_{\mathrm{d}}}{D_{fs}}$$

One of the drawbacks of TAR is the issue encountered due to the size of the build-up cap while measuring the dose in free space for small field sizes at higher energies. For example, the build-up cap will be 6 cm in diameter (water equivalent) for 15 MV, which leads to loss of lateral charged particle equilibrium for field size less than or equal to 6 cm × 6 cm². The issue with in-air measurement can be resolved with the use of tissue phantom ratio (TPR). TPR is defined as the ratio of the dose at a point in the phantom to the dose at the same distance from the source at a reference depth. If the reference depth is at the point of maximum dose, then it is termed the tissue maximum ratio TMR.

$$\mathrm{TPR}\left(d,r_{\mathrm{d}}\right)=\frac{D_{\mathrm{d}}}{D_{t0}},\mathrm{TPR}=\mathrm{TMR}$$

$$\text{if } t0 \text{ is the depth of maximum dose}$$

As stated previously, increasing the field size results in more secondary scatter from the collimators. This could be measured separately with the help of the collimator scatter factor S_c which quantifies the scatter from varied field sizes produced by the collimator. Additionally, the scattered radiation may originate from the phantom/patient itself as the field size is changed; thus, a phantom scatter factor S_p is used.

For fixed SSD treatments, percentage depth doses can be used when determining the monitor units MU required:

$$\mathrm{MU}=\frac{\mathrm{TD}\times 100}{K\times \mathrm{PDD}_{\mathrm{d}}\times S_{\mathrm{c}}\left(r_{\mathrm{c}}\right)\times S_{\mathrm{p}}\left(r\right)\times\left(\frac{\mathrm{SCD}}{\mathrm{SSD}+t_0}\right)^2}$$

where TD is the dose in cGy, K is 1 cGy/MU, collimator scatter factor for field sizes at the SAD, and SCD is the source to calibration point.

For isocentric SAD treatments, TMR/TPR is used instead:

$$\mathrm{MU}=\frac{\mathrm{ID}\times 100}{K\times \mathrm{TMR}\left(d,r_{\mathrm{d}}\right)\times S_{\mathrm{c}}\left(r_{\mathrm{c}}\right)\times S_{\mathrm{p}}\left(r_{\mathrm{d}}\right)\times\left(\frac{\mathrm{SCD}}{\mathrm{SAD}}\right)^2}$$

Clinical radiotherapy planning—Dose calculation is done on a point at a certain depth using the output (cGy/MU) and depth dose data for fixed source to surface distance (SSD) treatment. Any organ (viz. Eye) that needs to be shielded from radiation is performed by placing lead blocks mounted on a shadow tray. Alternatively, in modern radiotherapy this can be accomplished

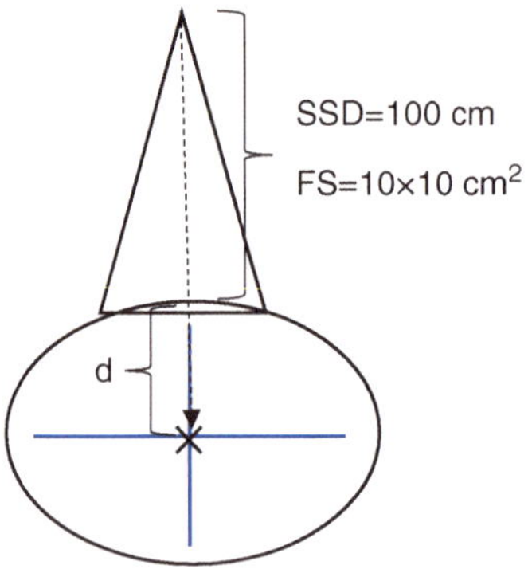

Fig. 12.7 Single field: dose calculation at point X

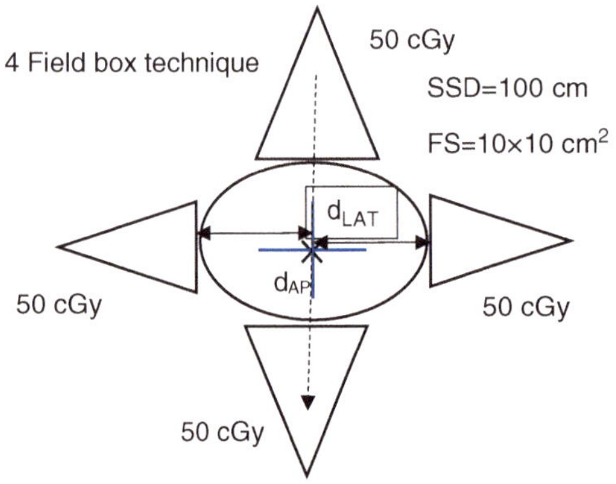

Fig. 12.8 4 Field box technique

with the help of multi-leaf collimator for most cases.

1D or Point Dose Calculation Summary

Typical example of point dose calculation for 6 MV photon beam is shown in Fig. 12.7.

A field size of 10 × 10 cm^2 at depth *d* cm (e.g. = 10 cm)

Prescription dose =200 cGy, Beam energy 6 MV

Daily absorbed dose at a 10 cm depth is calculated as

PDD at 10 cm for 10 × 10 cm^2 at 10 cm depth = 67.1%

Typical Output for 10 × 10 cm^2 with SSD = 100 cm = 1 cGy/MU

$$\text{Treatment time} = \frac{\text{prescription dose}}{\text{Output} \times \text{PDD}(10\text{cm})}$$

$$= \frac{200\text{cGy}}{\frac{1\text{cGy}}{\text{MU}} \times 67.1\%} = 298.1\ \text{MU}$$

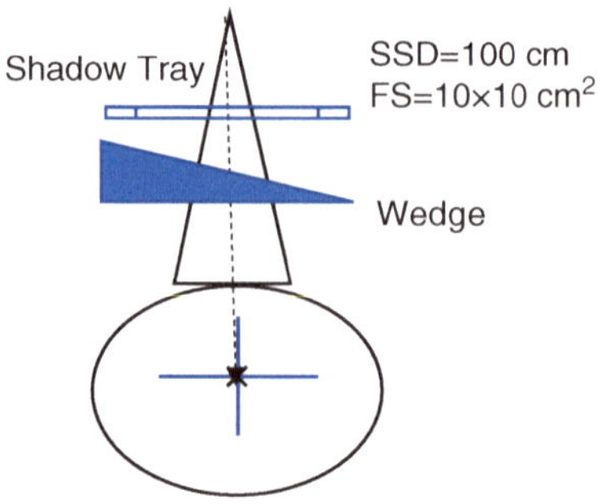

Fig. 12.9 Shadow tray and wedge

For Co-60, the output unit will be expressed in Gy/min or cGy/min depending on the source strength. The PDD for 10 × 10 cm^2 at 80 cm SSD is 56.7% for a 150 RMM (Röntgen/min at 1 m), the output of the machine is around 98.4 cGy/min. Upon applying the above values into the treatment time equation, the treatment time equates to 3.6 min.

For a prescription dose of 200 cGy, 4 fields have to be calculated for 50 cGy each with their respective PDD at the given depth, as shown in Fig. 12.8. For dose calculation PDD for individual fields need to be used.

For any other accessories (like shadow tray or wedge) in beam path, tray factor (≈0.98 for Cobalt-60, may vary depending on the energy and the thickness of the tray for other treatment conditions) and wedge factor (≈0.269 for 60° motorized wedge) will be multiplied in denominator. Insertion of any beam modifying device (accessories) will increase the treatment time (Fig. 12.9).

The dose calculation is performed at a single point, usually at the midplane, using the output, depth dose or tissue maximum ratio value for an equivalent square field size. It is possible to do an isocentric treatment, with a fixed source to axis distance (SAD) with such a simulation.

12.4.2 Computational Algorithms

The main drawbacks of manual dose calculations include the potential for human calculation errors and the challenges of handling complex treatment plans, which can be tedious to calculate consistently by hand. Computer-based algo-

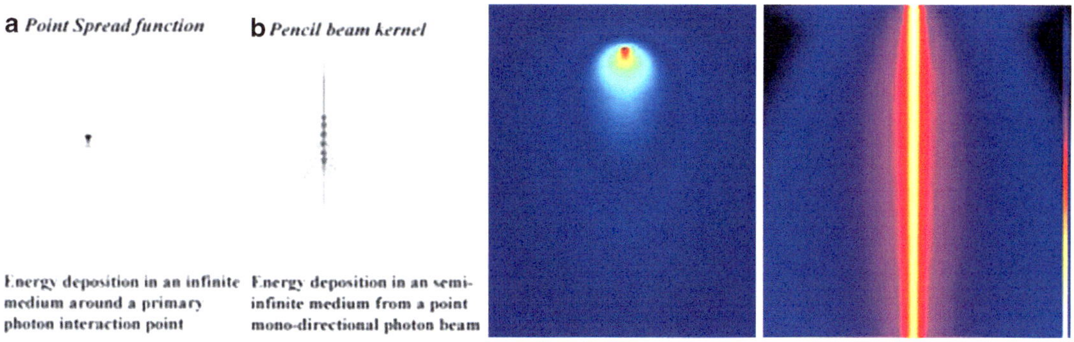

Fig. 12.10 Pictorial representation dose spread kernel. (**a**) and (**b**) illustrate a spherical spread pattern centered around the interaction point and a narrow, elongated deposition along the beam path, resembling a pencil-shaped profile, respectively.

rithms are now used to improve the accuracy of dose calculation using treatment planning systems. A key aspect of the modern radiotherapy journey is the assurance of quality, accuracy and consistency, which can be achieved via simulation and computerized dosimetry planning.

Dose calculation algorithms are constructed with the intent to predict the dose delivered to tissue as accurately as possible. However, radiation interactions with matter can become complex when accounting for heterogeneous densities and in particular at tissue-air or tissue-bone or lung-tissue or lung-bone interfaces. As a result, approximations are often employed in physics models to address these complexities. The main method of dose calculation involves separating the ionizing radiation into two components: primary radiation (photons and electrons originating from the radiation source) and scatter, known as scatter kernels or dose spread kernels (DSK).

12.4.3 Dose Spread Kernels

Dose spread kernels (DSK) are defined as the pattern of spread from the point or site of interaction. Monte Carlo simulations are commonly used to calculate secondary particle spread and energy deposition. At a point of interest, kernels contain information about the contributions from upstream scattering points. To obtain the dose, the kernels are convolved with the correlating total energy released per unit mass (TERMA) (Fig. 12.10).

From the above images, the first image depicts the paths of a point-spread function; the following pencil beam kernel is an integration of multiple points along the trajectory. The colored images demonstrate this in a colored heat map.

12.4.3.1 TERMA

TERMA is the second component of dose algorithms to input data relating to the primary radiation fluence throughout the material. TERMA is defined as the total energy release per unit mass (the loss of energy from uncharged primaries), and is the product of the monoenergetic beam fluence Φ at a localised point r and the local mass attenuation coefficient $\frac{\mu}{\rho}$:

$$T_E(r) = \frac{\mu}{\rho}(r,E)\Phi_E(r)$$

For the entire beam energy spectrum at a local point, TERMA is determined through integration:

$$T(r) = \int_E \frac{\mu}{\rho}(r,E)\Phi_E(r)\,dE$$

12.4.3.2 Monte Carlo

Monte Carlo is a sophisticated dose calculation method that involves repeated random sampling of generated tracked particles and their interactions in matter.

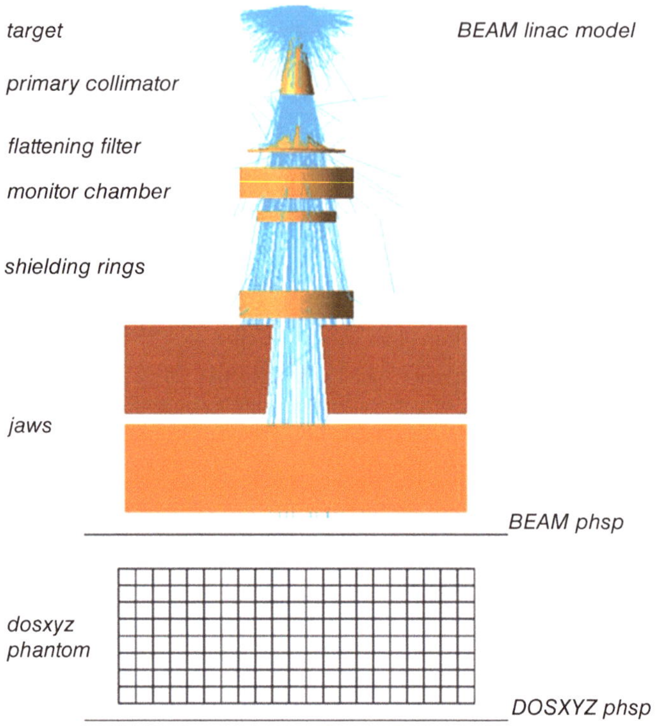

Fig. 12.11 Monte Carlo simulation using EGSnrc Monte Carlo code and DOSxyz dose calculation in phantom

A Monte Carlo simulation requires input parameters: the physical parameters of the LINAC and the phantom, the mathematical equations that dictate the interactions with matter, and the cut-off energies of secondary and/or tertiary particles generated. Physical parameters can include the object dimensions, the material type and density, and the electron density.

For a single particle history, a random number generator (RNG) initiates the location, direction, and energy from the modeled LINAC source. The particle is transported to the first site of interaction, where the type of interaction and the production of secondary particles are determined. The particle history continues until a cut-off energy is reached and the energy is deposited in the local region. This process can be repeated as many times as desired. The image below illustrates this process using the EGSnrc Monte Carlo code (Fig. 12.11). The particle location, direction, and energy from the LINAC model is saved in a phase-space file using the BEAMnrc program. This file is then imported into DOSxyz for beam analysis and/or dose calculation using a phantom model.

Because Monte Carlo simulates real physical processes, the dose calculations are exceptionally accurate. The accuracy is limited by the geometric detail of the LINAC and phantom models, assuming sufficient histories are used. The biggest drawback for Monte Carlo algorithms is the computation time; multiple billions of particle histories are require deven for low doses, which can take an hour or more. To balance speed and accuracy in dose calculations, Monte Carlo is often used to generate dose spread kernels.

Fortunately, recent studies focus on improving Monte Carlo times, for example via GPU acceleration. This can reduce a VMAT plan computation time down to minutes.

12.4.4 Pencil Beam Convolution Algorithm

Pencil beam convolution is a method of dose computation that involves use of pencil beam kernel functions to compute the dose along a narrow beam through the medium. The point scatter kernel is integrated along the line of propagation

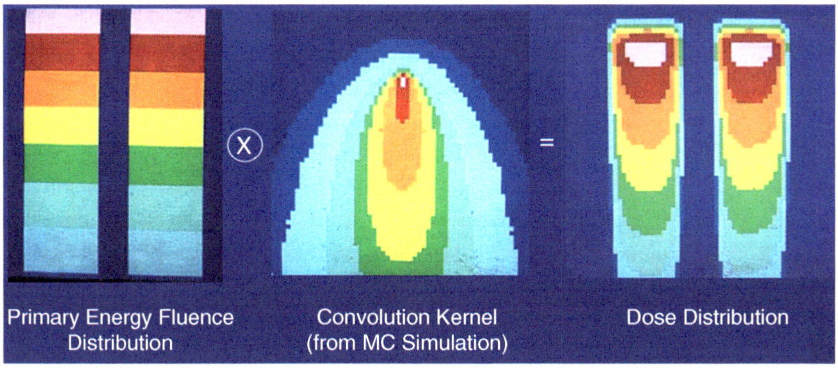

Fig. 12.12 Pencil beam convolution algorithm: primary energy fluence, convolution kernel and dose distribution

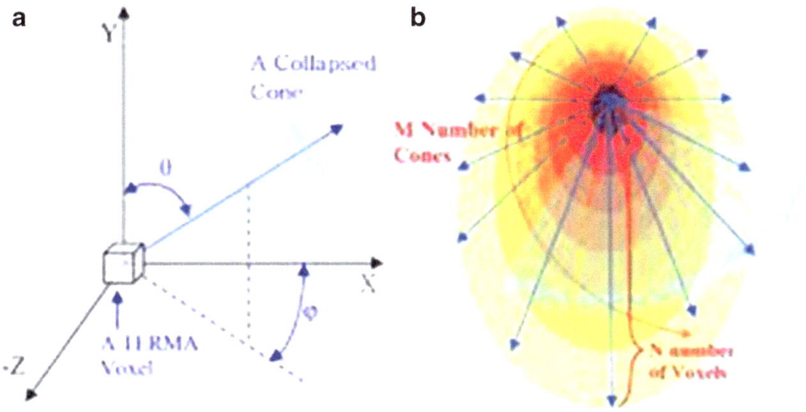

Fig. 12.13 Collapsed cone convolution. (**a**) Energy transport via a collapsed cone; (**b**) Dose distribution using multiple cones

of the narrow beam, creating the pencil beam kernel (Fig. 12.12). This is then superimposed with the TERMA to determine the dose. For the dose at a single point within the medium using a single pencil beam, it can be mathematically represented with:

$$D(x,y,z) = \iint T(x',y')\,K(x',y';x,y,z)\,dx'dy'$$

where x, y, and z are the coordinates of the point, x' and y' are the 2D entry coordinates of the pencil beam. This is simplified with convolution integrals to create the dose map:

$$D(x,y,z) = \iint T(x',y',z)\,K(x-x',y-y')\,dxdy$$

$$D = T \otimes K$$

A main benefit of pencil beam convolution is its speed; it computes significantly faster than either Monte Carlo, which requires heavy computer processing power, or manual calculations, which is tedious. This is crucial in treatment planning systems, which require processing times to be as low as possible. For a homogeneous phantom, this method is ideal. However, pencil beam convolution suffers in heterogeneous tissues as the algorithm fails to consider changes in electron density laterally from the central axis. As such, regions going from higher densities to lower densities (such as the lungs) tend to overestimate the dose. Kernel scaling of the path length can be determined by the ratio of electron densities between tissues and water, but this is done only along the direction of propagation from the central axis.

12.4.5 Collapsed Cone Convolution

Collapsed cone convolution (CCC) is an alternative method to the pencil beam that aims to improve the dose estimation for inhomogeneous tissues. For this method, the site of interaction with the primary radiation is the apex of cones that radially spread outward. Additionally, the primary beam consists of a finite number of polar coordinate angles (Fig. 12.13). It depicts how cones spread from the initial site of interaction; the energy within each cone is collapsed into the central line. Each line acts as the central axis for its corresponding cone, where the total energy within the cone is collapsed into the axis line. Figure 12.13b demonstrates the energy deposition pattern formed the combined effect of M collapsed cones from a voxel, spreading dose to N voxels. The overlapping cones create a spherical energy distribution which is a spread of scattered radiation.

A benefit of using a CCC algorithm is that it improves the accuracy of the dose contribution from lateral scatter. Furthermore, the computational time can be reduced; the time is proportional to the number of cones used per interaction. Unfortunately, the effectiveness of CCC is reduced at long distances from the cone apex.

12.4.6 Anisotropic Analytical Algorithm

Anisotropic analytical algorithm (AAA) is another method that improves upon the pencil beam convolution method. AAA is a 3D pencil beam convolution algorithm that uses separate modeling for primary photons, scattered photons, scattered electrons, and contaminating electrons. Separate convolutions are used for the models and are superimposed to determine the dose distribution. The clinical broad beam is also separated into multiple smaller beamlets, which are superimposed for the final dose distribution.

In this algorithm, heterogeneous tissues are accounted for anisotropically, i.e., radiological scaling in the direction of the beamlet is accounted for in the energy deposition density function below, and electron-density-based scaling of lateral photon scatter. For radiological scaling of the beamlet:

$$z' = \int_{z}^{0} \frac{\rho(t)}{\rho_{\text{water}}} dt$$

where ρ is the electron density. The following image is an example of anisotropic scaling of an incident beamlet β. The lateral scattered photons are scaled via electron density values.

As with the CCC method, AAA considers the heterogeneous tissue types in both lateral and longitudinal directions. Compared to CCC, AAA has a relatively shorter calculation time and better accuracy. However, AAA still lacks the accuracy of Monte Carlo methods.

12.4.7 Acuros XB

Acuros XB (AXB) is an alternative method to Monte-Carlo-kernel-based dose computations that involves solving linear Boltzmann transport equations (LBTE). The LBTE is a form of the Boltzmann equation that assumes that radiation has no self-interaction—only interacting with matter.

Rather than correcting the density of voxels based on kernels in water, ABX explicitly models interactions with matter using density and material type.

It has the benefit of lacking statistical noise in the calculated dose compared to Monte Carlo but providing similar results and significantly faster computation times. AXB doses were found to strongly agree with MC while providing far better dose calculations in heterogeneous tissues compared to PBC, CCC, and AAA.

12.5 2D Simulation and 2D Dose Calculation

It is possible to manually generate an isodose distribution in the irradiated area using the isodose charts sometimes available from the vendor. Nevertheless, it's a labor-intensive and time-consuming process with very limited accuracy. Computer controlled dose calculation in 2D was available, which was the primitive version of the modern treatment planning system (TPS).

12.6 3D Simulation and 3D Dose Calculation

The treatment planning system's dose calculation accuracy has also evolved over time, from the pencil beam algorithm (1952) to the most advanced Monte Carlo (2005) dose calculation engines. Today, 3D planning is an essential component of radiotherapy. A three-dimensional conformal dose distribution, in general, denotes a dose distribution that follows the target volumes. Several methods, including 3D CRT, forward and backward IMRT, volumetric modulated arc therapy, and tomotherapy, can achieve a well-confirmed distribution.

In this section, we will discuss various delivery techniques for different sites and which technique would be the best choice in terms of dosimetric attribute and cost effectiveness. Historically, 3DCRT began with the customized shielding block and was later replaced by the simple and convenient multileaf collimator (MLC) [7]. Despite the fact that MLC was introduced in 1965, there was considerable debate until 1990 as to whether choice blocks or MLC were superior, and only a few linear accelerators were built with MLC built in. However, block storage and fabrication were the most difficult issues, resulting in a decrease in the use of the Cerrobend block. Following that, the MLCs were computer controlled, and Varian introduced its dynamic MLC, which was built into all machines with an inbuilt MLC, eliminating the need for Cerrobend blocks. Blocks are currently only used in Cobalt teletherapy machines, which do not have a place in MLC.

MLC-shaped fields are simple, where a target volume is irradiated by fewer fields shaped by the MLC. The best-known example is the four field box technique for pelvis irradiation or bi-lateral and lower anterior neck treatment for head and neck cancers.

Despite the fact that multiple field 3DCRT for head, neck, and pelvis cases was common with MLC or by using Cerrobend blacks from 1995 to 2005, the rise of IMRT reduced the use of multiple field 3DCRT or compensator-based IMRT delivery because the latter two techniques were very labor-intensive and time consuming. Therefore, we will restrict our discussion to the clinically available techniques of step and shoot, dynamic IMRT, and volumetric modulated arc therapy.

12.6.1 Radiotherapy Planning

Radiotherapy planning depends upon the judicious choice of the beam/arc angle. There is something called saturation dose distribution, use of more beams or large arcs does not increase the dosimetric quality of the radiotherapy plans, especially target conformity and OAR dose. The use of excess beams or arcs raises the MU, resulting in low dose spillage. Classically, 3DCRT beam angles were developed over years, and the beam/arc angles should be restricted to classical 3DCRT beam angles for VMAT with some modifications [19–21]. Simple planning techniques will be demonstrated in this section, dose constraints and prescription can be found in site specific chapters. The most dependable dose constraints for various sites are EMBRACE (cervix), START-Trail for hypo-fractionated breast, CHHip-Trail for hypo-fractionated prostate, and QUANTECH for the rest [22–25].

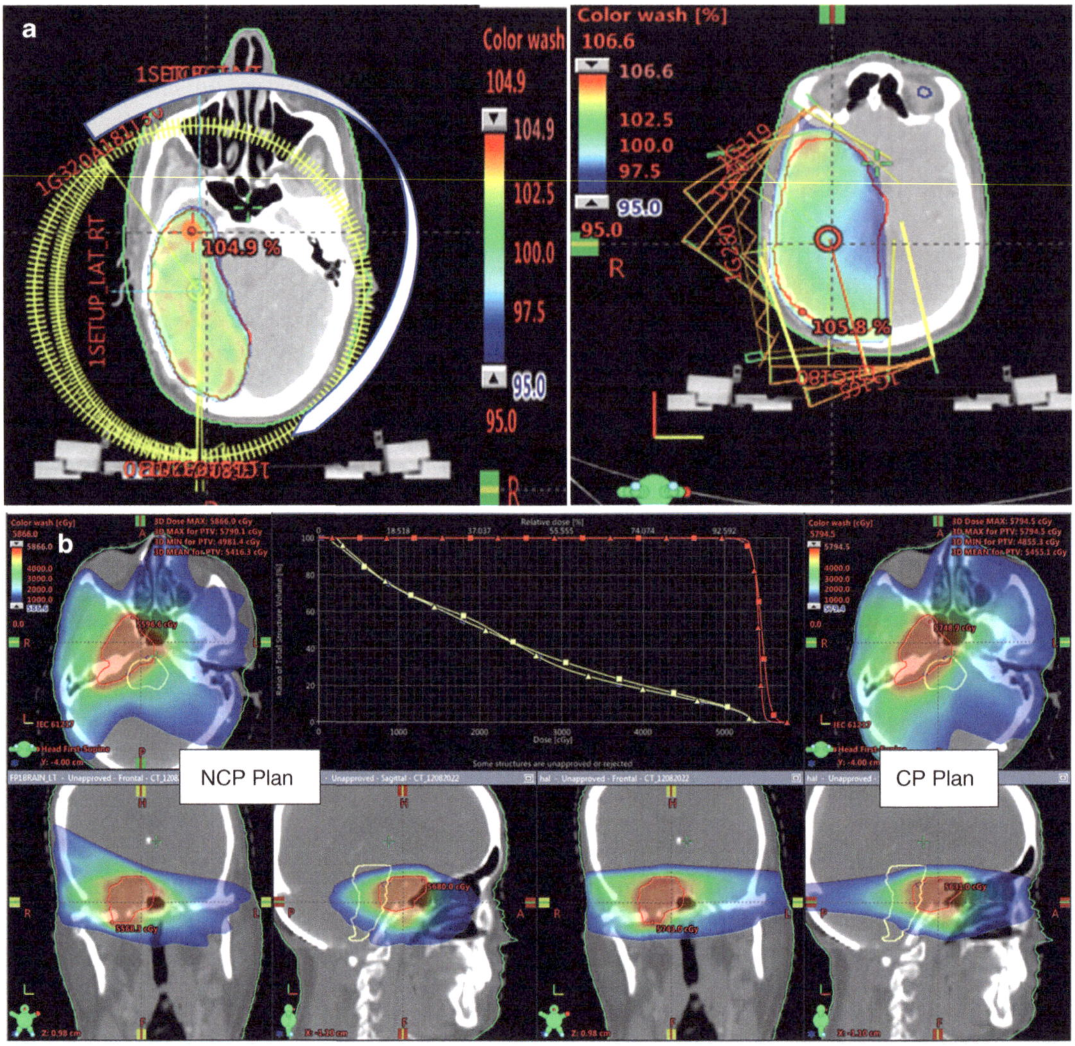

Fig. 12.14 (**a**) Arc angle: Coplanar Arc 150°–320° clockwise (CW) 150°–320° counter clockwise (CCW)+. Non-coplanar Arc at Table 30°: 181°–320° CW + CCW. Left panel shows a full rotation of the coplanar beam; however, 320°–150° CW was blocked during optimization. This feature available in Eclipse/Varian Planning system. For Elekta/Monaco beam have to be divided as 150°–179° and 181°–320°. (**b**) Increase in the low dose spillage due to use of non-coplanar beam with no improvement in OAR dose or target conformity / homogeneity

Brain Beam energy used includes 6 MV, 6 MV FFF. Because different brain tumors have different histopathologies, the dose and fractionation regimen vary greatly, ranging from 18 Gy of prophylactic cranial irradiation for pediatric acute lymphoblastic leukemia to 60 Gy in 30F for glioblastoma. For conventional 2 Gy/day dose fractionation regimens, the brain OARs (optic nerve, optic charisma, and brainstem) are serial structure and have a dose limitation of 54 Gy to 0.2 cc volume. Hippocampus is a serial-parallel organ with a dose limitation of ≤30 Gy. The general principle of brain irradiation is to keep the beams and arcs unidirectional for the ipsilateral tumor, touching until the midline, and to spare as much of the contralateral brain as possible. Figure 12.14a shows the VMAT (left panel) and

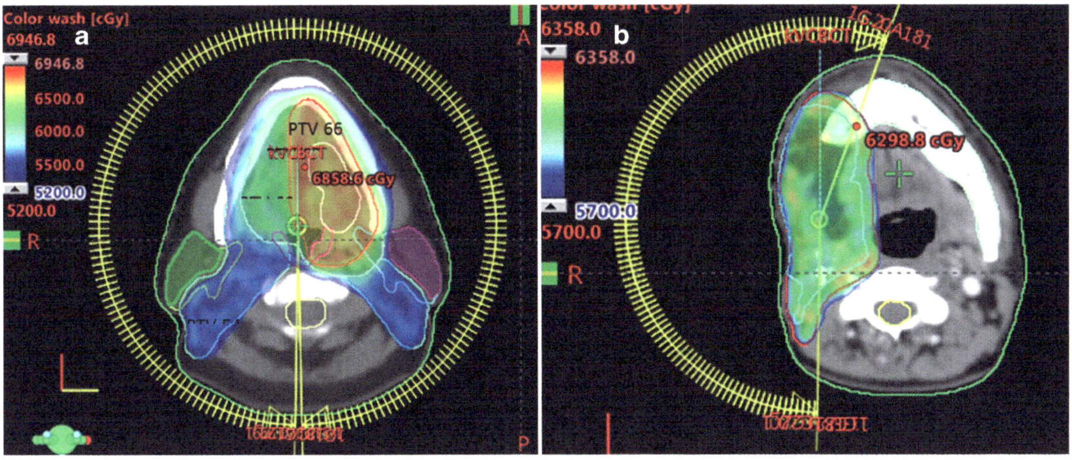

Fig. 12.15 (**a**) Full (×2) arc VMAT for SIB treatment of Ca tongue 66 Gy, 60 Gy, and 54 Gy in 30 F. (**b**) Partial arc (200°)×2 for ipsilateral neck irradiation (60 Gy/30 F) in post-OP buccal mucosa

3DCRT dose distribution (right panel) for a target touching the midline of the brain.

For PTVs crossing the midline or central target (pituitary), the beam/arc angle will be from both sides of the brain to improve dose distribution conformity; however, every effort should be made to restrict the beam/arc in the side of the target, so the other part receives less radiation. An abrupt increase in beam/arc angle is strongly not recommended. A non-coplanar (NCP) beam can be used for stereotactic radiosurgery, and reirradiation cases. In general, the NCP beams contribute only slightly to increasing conformity or homogeneity for a large target volume that is well separated from the OAR NCP beams. Targeting OARs that are embedded in or invaded by OAR, NCP beams might be helpful to restrict the dose to OARs; however, use of NCP beams must be judicious as it increases the low-dose spillage (supplementary material online: Fig. 12.14b).

Head Neck

In general, beam energies of 6 MV and/or 6 MV FFF are suitable for Head and Neck cases. A large number of sites are included in this section. For nasopharynx or diseases with cranial extension, brainstem, optic nerve, optic charisma, and lens will be included in the serial OARs. Parallel organs include the parotid (Parotid_NOS), tongue, hypopharynx, swallowing structure, constrictor muscle, mandible, and lips [25]. If any parallel organ is found to be inside the GTV/PTV volume, the dose should not be compromised on the dose. However, all efforts should be made to spare the serial organs. For ipsilateral neck irradiation, the beam should be confined to the same direction; in other cases, a full arc should be preferable. A minimum of 2 arcs (CW and CCW) with complementary collimator angles is required for good target-dose confirmation. IMRT offers a near similar dose distribution for HN cases. The beam angles are $G = 180°$, 232°, 284°, 336°, 28°, 80°, 132°, this arrangement reduces the dose to the oral cavity. For oral cavity and lip cancer beam arrangement starts from 0°, 52°, 104°, 156°, 208°, 260°, 312°. Delivery time is a bit higher in dynamic IMRT, and step-and-shoot (SnS) IMRT takes a substantially higher time than D-IMRT/VMAT. VAMT is the most commonly used treatment for all types of HN cancer. The NOS portion of serial organs such as the parotid, larynx, hypopharynx, and so on should be optimized (Fig. 12.15a, b).

3D-CRT have exactly same beam angles ($G = 180°$, 232°, 284°, 336°, 28°, 80°, 132°, or 0°, 52°, 104°, 156°, 208°, 260°, 312°) further supported by the supplementary fields. For MLC shaped fields with FiF or wedged lateral ports

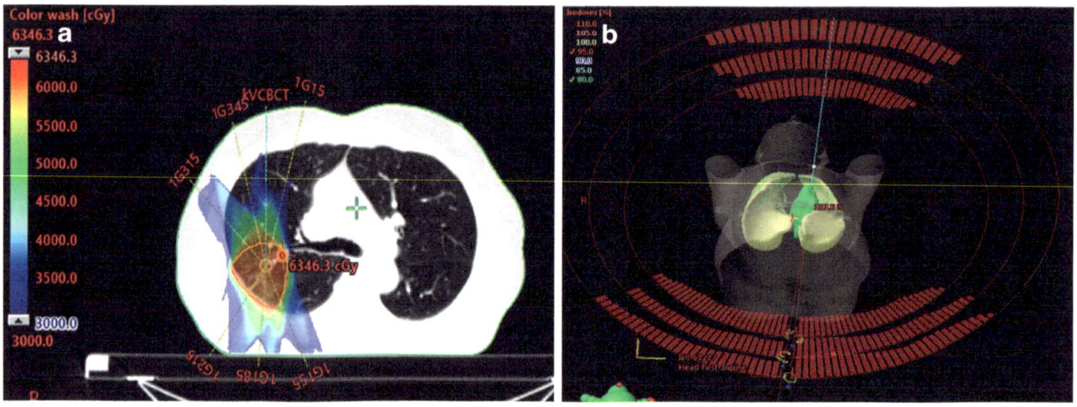

Fig. 12.16 (**a**) Lung IMRT. (**b**) Esophagus VMAT

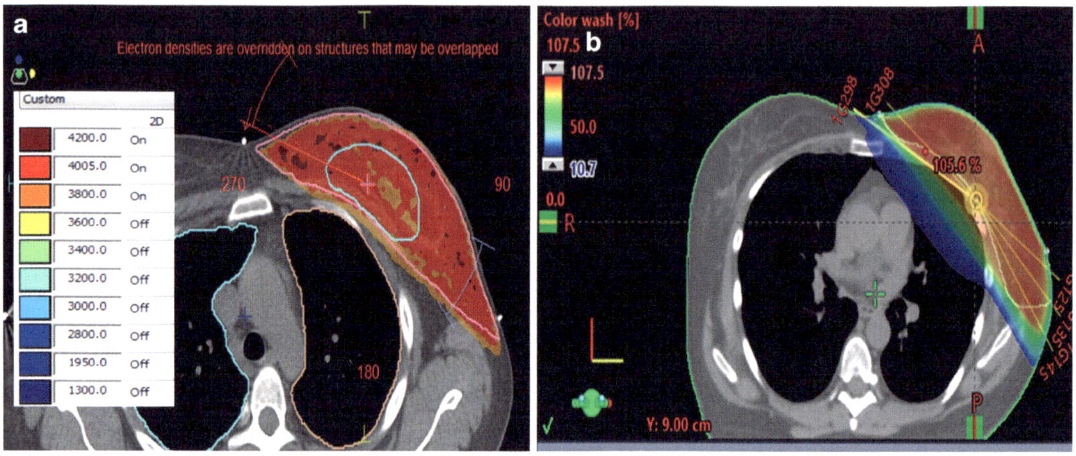

Fig. 12.17 (**a**) t-VMAT dose distribution for left breast. (**b**) t-IMRT dose distribution for left breast

and lower anterior neck fields, up to 46 Gy in 23 F, followed by off cord photon and on-cord electron up to 60 Gy and 6 Gy.

Thorax For thorax cases, beam energies of 6 MV or 6 MV FFF can be utilized. Thorax cases include the esophagus and lung; for the esophagus and a centralized lung lesion, the beam angle should be anterior-posterior biased; lateral beams can be used for peripheral lung lesions. The basic rationale of the dose plan for thorax cases is to reduce the dose to the lungs, especially on the contralateral side. For thorax, in Varian Eclipse, preferred technique is IMRT as short arcs VMAT do not produce good dose distribution; however, for Elekta Monaco, short arcs (320°–0°–40° × 2 + (140°–180°–220°) are recommended. 3DCRT also have the similar arc angle (Fig. 12.16a: IMRT and 12.16b: VAMT).

Breast For breast cases, beam energies of 6 MV or 6 MV FFF are typically used. Left Breast/chest wall (CW): OAR: heart, left lung, contralateral breast, and Esophagus for SCF beam, right breast/CW OARs are right lung, contralateral breast. Breast/CW (±SCF ± Axilla±IMC chain) can be planned using 3D-CRT with the wedge, field in field IMRT (FiF), tangential IMRT, Tangential(t) VMAT, and half arc VMAT. However, half arc VMAT using 180° (or near to that) is not advisable to use. However, some planning systems like Eclipse do not give a good dose distribution for small arc (≈30°) VMAT; hence, people use a half arc VMAT for

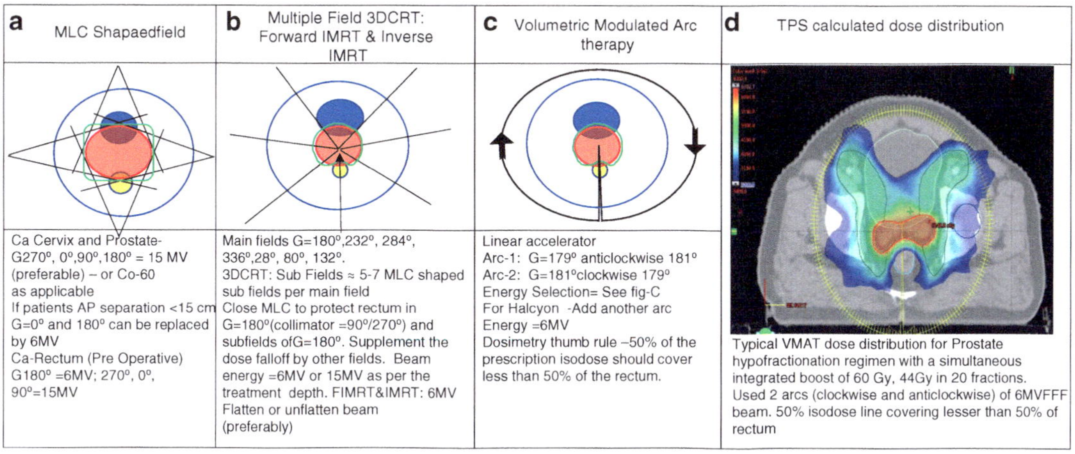

Fig. 12.18 Dose regime: Ca cervix/endometrium—46 Gy/23# or 50 Gy/25# followed by brachytherapy 3 × 7 Gy for cervix and 2 × 6 Gy for endometrium. CA prostate—74 Gy in 37 fractions, hypofractionation regimen with a simultaneous integrated boost of 60 Gy, 44 Gy in 20 fractions. CA rectum post preoperative or nonoperable cases 45–50.4 Gy in 1.8–2.0 Gy per day. Ca rectum pre-perative cases 25Gy in 5 fractions followed by surgery next week

these kinds of planning systems. The use of a half arc is a violation of classical 3DCRT beam geometry and the contribution of the low dose is also controversial. In such a scenario, tangential IMRT is preferred over the full ac VMAT. Figure 12.17 shows a typical Breast plan with tVMAT and tIMRT beam arrangement and dose distribution. Both medial and lateral tVMAT arc angles are around 30° (CW + CCW). tIMRT uses two medial tangential beams with a spread of 15°(CW)–30° (Breast), and three lateral tangential beams with maximum separation between two extreme beams less than 30°. Figure 12.17a, b show the t-VMAT and t-IMRT beam arrangement and dose distribution respectively.

Doing a VMAT for SCF/Axillary field is a complex process; hence, it is done manually by clinical marking. For IMRT, SCF is done by adding a field to the same isocenter after IMRT optimization and dose calculation. To avoid geometrical miss due to breathing motion it's essential to give a 2 cm flash margin to all kinds of plans (3DCRT, t-IMRT, t-VMAT, HA-VMAT). The other detail of the breast/CW planning can be found in recent dosimetric studies [26–28].

Pelvis There are several techniques for dose delivery with varying degree of dose conformity to the target. The techniques are MLC Shaped field more commonly known as the 4 field box technique, 3DCRT, IMRT, and VMAT 4 field box still used as the most common technique worldwide, including in developed countries. Beam angles, beam energies, and dose distribution are summarized in Fig. 12.18. The scope includes radiotherapy treatment for the Ca Cervix, endometrium, rectum, prostate, and other solid tumors of the pelvic area. For cases like cervix and endometrium external beam radiation are followed by brachytherapy treatment. While prostate has a simultaneous integrated boost or sequential boost until 74 Gy.

12.6.2 Craniospinal Irradiation

Craniospinal irradiation is a specialized treatment for medulloblastoma and PNET patients that involves irradiating the entire neuroaxis as well as the entire brain. There are several methods for simulation and irradiation of the CSI, starting with the conventional simulator-based technique described by Munshi et al. [29]. We will only discuss CT simulation, 3DCRT, and VAMT-based techniques. Figure 12.19a, b shows the CSI simulation technique, in which CT slices are abstained from the brain to mid-thigh level. The contouring of the entire neuroaxis, as well as

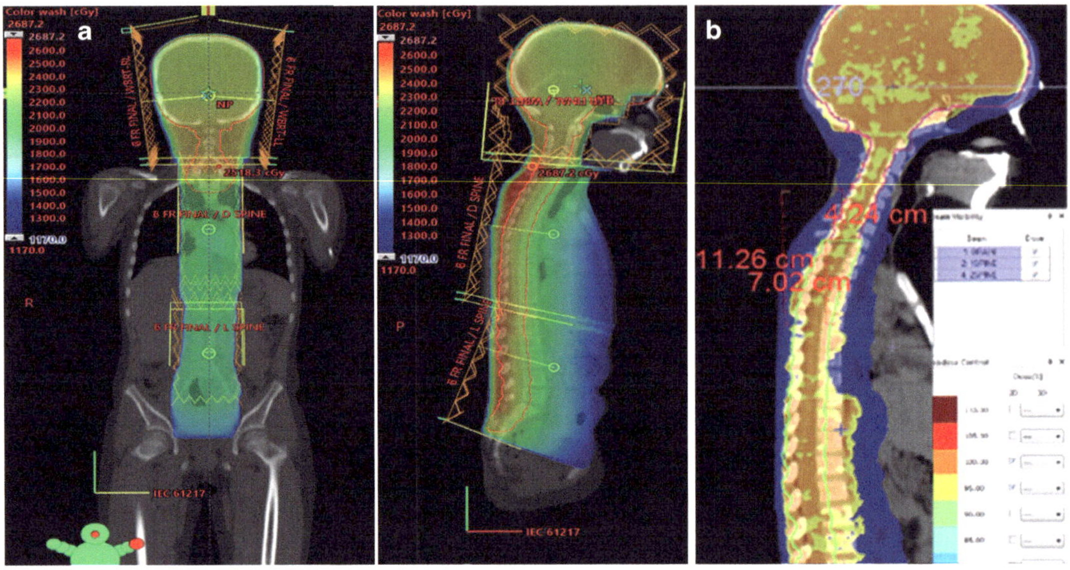

Fig. 12.19 (**a**) 3DCRT plan with junction matching. (**b**) VMAT plan

the entire brain and boost area identified in the MRI scan (often the posterior fossa), is defined. High and intermediate risk patients receive 34.2 Gy–36 Gy, while low-risk cases are irradiated with 23.4 Gy at a fraction size of 1.8 Gy. Junction matching is done for the 3DCRT plan to avoid any underdoses or overdoses. A junction dose feature is also used to reduce the possibility of overdosing or underdosing due to geometrical error in the patient's positioning, in addition to jaw and MLC calibration. This process is time consuming, labor-intensive, and complex for the delivery and image matching. 3DCRT (Fig. 12.19a) contains two lateral beams of 6 MV at gantry ≈86° and ≈274° to match the beam convergence at the anterior side to reduce eye dose. To match the inferior divergence of the brain ports, a small couch angle [30] is provided. The caudal border of the brain port matched with the cranial border of the superior spine field placed at table 90°/270° at G = 180° ± ≈ 5°. Divergence of the superior spine port and the brain port, adjusted by the first's gantry angle and the second's collimator angle. Similarly, if required, an additional spine field (posterior spine field) is added to cover the target volume at table 90°/270° at G = 180° ± gantry angle required to match the divergence between adjacent spine fields. Two or three junction shifts of 5 mm to 1 cm done during the planning. For VMAT (Fig. 12.19b) two cranial full arc (eye blocked), 2 upper and 2 lower spine arcs (arc length 140° symmetric about 180°). The collimator angle is complementary between the arcs at each isocenter. Optimization and dose calculation were carried out, with no junction shift required. VMAT based CSI plan is easy to do and deliver. It is not recommended to use the full arc in spinal isocenters [30–34].

12.7 Quality Assurance and Quality Check in Radiotherapy

This section talks about quality assurance and quality checks during the radiotherapy process. Quality assurance can be divided into two categories: patient specific QA and machine specific QA. Different medical physics and regulatory bodies have published extensive machine specific and patient specific QA protocols. For machine specific QA, the most common machine specific QA protocol is TG-142 [35]. A series of comprehensive reports are available from the Atomic Energy Regulatory Board (India), which includes Chap. 5 (RADIATION THERAPY

Table 12.3 Tests

	Tolerance
Daily	
X-ray and electron output constancy	2%
Door interlock and audio visual monitor	Functional
Localization lasers	1.5 mm
Optical distance indicator	2 mm
Monthly	
X-ray, central axis dosimetry parameters (PDD,TMR)	1% from baseline
Emergency off switches, interlocks, electron cone interlocks	Functional
Light/radiation fields coincidence	2 mm or 1% on side
Gantry/collimator angle indicators	1°
Cross-hair centering	1 mm
Jaw symmetry	1 mm
X-ray beam flatness constancy, electron beam flatness constancy	1% from baseline
X-ray and electron symmetry	
Electron central parameters (R50)	1 mm from baseline
Latching of wedges, blocking trays	Functional
X-ray and electron symmetry	1% from baseline
Wedge position, tray position, applicator position	2 mm
Back-up monitor constancy	2%
Field light intensity	Functional
Annual	
X-ray and electron output constancy	2%
Table-top sag and vertical travel of table.	2 mm
Transmission factor constancy for all treatment accessories	2%
Collimator rotation isocenter, gantry rotation isocenter, couch rotation isocenter	1 mm from baseline
Coincidence of collimator, gantry, couch axis with isocenter,	2 mm
Coincidence of radiation and mechanical isocenter	2 mm
X-ray output constancy vs. gantry angle, electron output constancy vs. gantry angle	1% from baseline
Off-axis factor constancy vs. gantry angle, arc mode, safety interlocks	1% from baseline

SOURCES, EQUIPMENT AND INSTALLATIONS CODE NO. AERB/RF-MED/SC-1) and the test tolerance limits are available in "ACCEPTANCE/QUALITY ASSURANCE TESTS FOR MEDICAL LINEAR ACCELERATOR": Document number: RPAD/ACC/QA/04 (Online Document 2) [36]. Table 12.3 summarizes key points and their tolerance limits for linear accelerators capable of delivering IMRT.

Patient specific QA includes plan verification in the record and verification (R&V) system, plan checking and planned dose/fluence verification. Planned dose verification methods exhibit multiple layers of complexity in their methodology, starting with (1) dose-to MU verification with an ionization chamber in a water or solid water phantom [37, 38]. This is one of the most dependable QA techniques and must be performed on IMRT or VMAT plans. This technique ensures the correspondence of the planned MU with the prescribed dose. Further, fluence verification can be carried out using film, EPID, an ionization chamber array (iMatrixx, Sever29), diode array (Delta-4, Arc Check), etc. These are relative measurements that, when combined with point dose verification, complete the QA process [39, 40].

12.8 Conclusion

Radiation oncology is a fast-growing specialty. It has seen tremendous technological advancement in the last two decades, and the overall process has become complicated and heavily reliant on technology. These improvements reflect the outcome, in terms of increased survival and lesser morbidity in terms of early and late chemoradiotherapy toxicity. This chapter discusses the history and chronological development of radiotherapy, as it is difficult to find a rational correlation between modern techniques and the stages of development without these two facts. This chapter also covers the basic simulation and imaging, dose calculation in different dimensions of space, computer-based 3D dose calculation, site-wise radiotherapy planning, and QA/QC. The references can be used to extend the reasoning further.

12.9 Summary

1. Landmark developments in radiation oncology that facilatated automation and modernization include the introduction of MLC (1967), IMRT (1982), helical Tomotherapy (1994), and VMAT (2008).
2. 2D X-ray resolution is 1 mm, CT and MRI resolution is 3 mm, and PET resolution is 4 mm. MRI is suitable for soft tissue, whereas CT is a suitable imaging modality for bones or high-density materials. PET is preferred to determine metabolic activity.
3. The radiotherapy steps are the following: (1) determination of disease site and area of interest, (2) patient positioning, (3) immobilization, (4) simulation (clinical, 2D conventional, 3D-CT simulation, and time-gated 3D simulation), (5) contouring, (6) RT planning, (7) patient-specific QA, (8) patient on-couch positional verification, and (9) treatment delivery.
4. Dose calculation can be 1D (single point) using the output factor PDD/TMR, 2D (multiple points), or computer-based 3D dose calculation.
5. A good radiotherapy plan can help improve the dose escalation and reduce the doses to the organs at risk. Side effects, the risk of secondary cancer, and issues related to disease control can all be minimized with careful planning.
6. For IMRT and VMAT, the total solid angle at the isocenter should be somewhat equivalent to 3DCRT. The number of IMRT/VMAT beams/arc angles and their entry should not be significantly greater than that of 3DCRT.
7. For an ipsilateral target, beams should be placed ipsilaterally, especially in cases involving the lung or breast.
8. Electron treatment, preferably utilizing SSD technique (field size defined at the surface), can be administered based on either maximum dose or a surface dose criteria.
9. An increased number of beams or arcs may not improve the dose homogeneity or conformity, but it may increase the low dose spillage. The selection of beams and arcs must be prudent and consistent.
10. Quality assurance procedures can be categorized based on different time intervals: daily, monthly, quaterly, six-monthly and annual QA. Output measurements are required to be conducted daily prior to the commencement of treatment.

Multiple Choice Questions

1. Which imaging system gives a better visibility of the soft tissues?
   (a) CT
   (b) MRI
   (c) PET
2. Discovery of X-ray was done in which year?
   (a) 1895
   (b) 1896
   (c) 1901
3. Radioactivity was discovered by
   (a) The Curie Couple

(b) Becquerel
(c) Albert Einstein

4. In a conventional simulator, what kind of imaging we get?
   (a) One dimensional
   (b) Two dimensional
   (c) Three dimensional
5. What is the accuracy of the cross-wire centering in an IMRT capable linear accelerator?
   (a) 1 mm diameter
   (b) 1.5 mm diameter
   (c) 2 mm diameter
6. What is the annual X-ray and electron output constancy?
   (a) 1% from base line
   (b) 2% from base line
   (c) 3% from base line
7. In a 3DCRT, craniospinal irradiation junction shift is done to circumvent
   (a) Field overlap
   (b) Field gap
   (c) Both
   (d) For better conformity of the dose distribution to target
8. Difference between 4 field box technique and IMRT is
   (a) Better dose conformity
   (b) Better dose homogeneity
   (c) Both
9. If a serial organ partially is inside the target volume and the prescription dose is higher than the organ tolerance, then which of the following should be done:
   (a) Compromise the target dose and make the serial OAR dose with in tolerance dose
   (b) Deliver the target dose without compromising serial organ dose
   (c) Tradeoff between target and OAR dose to maximize and minimize respectively
10. For ipsilateral neck irradiation, if one parotid falls partially within the gross target volume / planning target volume
    (a) We should not restrict any dose to parotid and give full dose to the target volume
    (b) Draw the remaining parotid outside the planning target volume and reduce its dose to 26 Gy
    (c) Save the parotid first followed by target dose coverage

Answers

1. (b)
2. (a)
3. (b)
4. (b)
5. (a)
6. (b)
7. (c)
8. (a)
9. (a)
10. (a)

References

1. Rontgen W. On a new kind of rays. In: Proceedings of the Würzburg Physico-Medical Society; 1895.
2. Becquerel H. Sur les radiations invisibles emises par les selsd'uranium. CR Acad Sci (Paris). 1896;122:689–94.
3. Curie P, Curie M. Sur une substance nouvelle radioactive, contenue dans la pechblende. C R Acad Sci. 1898;127:175.
4. Deeley TJ, Hale BT. The past seventy-five years in radiotherapy. Br J Radiol. 1973;46(550):906–10.
5. Wei WI, Sham JS. Nasopharyngeal carcinoma. Lancet. 2005;365(9476):2041–54.
6. Brady LW, Wazir DE, Perez CA. Perez & Brady's principles and practice of radiation oncology. Philadelphia: Lippincott Williams & Wilkins; 2013.
7. Takahashi S. Conformation radiotherapy-rotation techniques as applied to radiography and radiotherapy of cancer. Acta Radiol. 1965;242:1–42.
8. Brahme A, Roos J-E, Lax I. Solution of an integral equation encountered in radiation therapy Phys. Med Biol. 1982;27:1221–9.
9. Bortfeld T, Burkelbach J, Boesecke R, Schlegel W. Methods of image reconstruction from projections applied to conformation radiotherapy. Phys Med Biol. 1990;35:1423.
10. Vassiliev ON, Titt U, Pönisch F, Kry SF, Mohan R, Gillin MT. Dosimetric properties of photon beams from a flattening filter free clinical accelerator. Phys Med Biol. 2006;51(7):1907.
11. Otto K. Volumetric modulated arc therapy: IMRT in a single gantry arc. Med Phys. 2008;35(1):310–7.
12. Leksell L. An historical vignette. In: Radiosurgery: baseline and trends. New York: Raven Press; 1992. p. 257–61. Adler JR. Stereotactic surgery and radio-

surgery madison. Medical Physics Publication; 1993. Frameless radiosurgery; pp. 237–48.
13. Mukherji SK, Miracle AC. Conebeam CT of the head and neck, part 2: clinical applications. Am J Neuroradiol. 2009;30(7):1285–92.
14. la Fougere C, Suchorska B, Bartenstein P, Kreth FW, Tonn JC. Molecular imaging of gliomas with PET: opportunities and limitations. Neurooncology. 2011;13(8):806–19.
15. Arnold M, Rutherford MJ, Bardot A, Ferlay J, Andersson TM, Myklebust TÅ, Tervonen H, Thursfield V, Ransom D, Shack L, Woods RR. Progress in cancer survival, mortality, and incidence in seven high-income countries 1995–2014 (ICBP SURVMARK-2): a population-based study. Lancet Oncol. 2019;20(11):1493–505.
16. Pysz MA, Gambhir SS, Willmann JK. Molecular imaging: current status and emerging strategies. Clin Radiol. 2010;65(7):500–16.
17. Erdi YE. Limits of tumor detectability in nuclear medicine and PET. Mol Imaging Radionucl Ther. 2012;21(1):23.
18. Munshi A, Ganesh T, Gupta RK, Vaishya S, Patir R, Sarkar B, Khataniar N, Bansal K, Rastogi K, Mohanti BK. Perfusion magnetic resonance imaging in contouring of glioblastoma patients: preliminary experience from a single institution. J Cancer Res Ther. 2020;16(6):1488.
19. Ray A, Sarkar B. Small bowel toxicity in pelvic radiotherapy for postoperative gynecological cancer: comparison between conformal radiotherapy and intensity modulated radiotherapy. Asia Pac J Clin Oncol. 2013;9(3):280–4.
20. Goswami J, Patra NB, Sarkar B, Basu A, Pal S. Dosimetric comparison between conventional and conformal radiotherapy for carcinoma cervix: are we treating the right volumes? South Asian J Cancer. 2013;2(3):128.
21. Sarkar B, Munshi A, Manikandan A, Roy S, Ganesh T, Mohanti BK, Pradhan A. A low gradient junction technique of craniospinal irradiation using volumetric-modulated arc therapy and its advantages over the conventional therapy. Cancer/Radiothérapie. 2018;22(1):62–72.
22. Pötter R, Tanderup K, Kirisits C, de Leeuw A, Kirchheiner K, Nout R, Tan LT, Haie-Meder C, Mahantshetty U, Segedin B, Hoskin P. The EMBRACE II study: the outcome and prospect of two decades of evolution within the GEC-ESTRO GYN working group and the EMBRACE studies. Clin Transl Radiat Oncol. 2018;9:48–60.
23. Haviland JS, Owen JR, Dewar JA, Agrawal RK, Barrett J, Barrett-Lee PJ, Dobbs HJ, Hopwood P, Lawton PA, Magee BJ, Mills J. The UK standardisation of breast radiotherapy (START) trials of radiotherapy hypofractionation for treatment of early breast cancer: 10-year follow-up results of two randomised controlled trials. Lancet Oncol. 2013;14(11):1086–94.
24. Dearnaley D, Syndikus I, Mossop H, Khoo V, Birtle A, Bloomfield D, Graham J, Kirkbride P, Logue J, Malik Z, Money-Kyrle J. Conventional versus hypofractionated high-dose intensity-modulated radiotherapy for prostate cancer: 5-year outcomes of the randomised, non-inferiority, phase 3 CHHiP trial. Lancet Oncol. 2016;17(8):1047–60.
25. Bentzen SM, Constine LS, Deasy JO, Eisbruch A, Jackson A, Marks LB, Ten Haken RK, Yorke ED. Quantitative analyses of normal tissue effects in the clinic (QUANTEC): an introduction to the scientific issues. Int J Radiat Oncol Biol Phys. 2010;76(3):S3–9.
26. Balaji K, Subramanian B, Yadav P, Radha CA, Ramasubramanian V. Radiation therapy for breast cancer: literature review. Med Dosim. 2016;41(3):253–7.
27. Munshi A, Sarkar B, Anbazhagan S, Giri UK, Kaur H, Jassal K, Ganesh T, Kalyan MB. Short tangential arcs in VMAT based breast and chest wall radiotherapy lead to conformity of the breast dose with lesser cardiac and lung doses: a prospective study of breast conservation and mastectomy patients. Australas Phys Eng Sci Med. 2017;40(3):729–36.
28. Jagsi R, Griffith KA, Moran JM, Matuszak MM, Marsh R, Grubb M, Abu-Isa E, Dilworth JT, Dominello MM, Heimburger D, Lack D. Comparative effectiveness analysis of 3D-conformal radiation therapy versus intensity modulated radiation therapy (IMRT) in a prospective multicenter cohort of patients with breast cancer. Int J Radiat Oncol Biol Phys. 2022;112(3):643–53.
29. Munshi A, Jalali R. A simple technique of supine craniospinal irradiation. Med Dosim. 2008;33:1–5. https://doi.org/10.1016/j.meddos.2007.03.004.
30. Fogliata A, Bergström S, Cafaro I, et al. Cranio-spinal irradiation with volu- metric modulated arc therapy: a multi-institutional treatment experience. Radiother Oncol. 2011;99:79–85.
31. Lee YK, Brooks CJ, Bedford JL, Warrington AP, Saran FH. Development and evaluation of multiple isocentric volumetric modulated arc therapy technique for craniospinal axis radiotherapy planning. Int J Radiat Oncol Biol Phys. 2012;82:1006–12.
32. Biswal SS, Sarkar B, Goyel M. Comparative dosimetric, setup margin and treatment time analysis between ring gantry and C-arm linear accelerators for VMAT based craniospinal irradiation plans. J Cancer Res Ther. 2024;20:224–31.
33. Studenski MT, Shen X, Yu Y, et al. Intensity-modulated radiation therapy and volumetric-modulated arc therapy for adult craniospinal irradiation—a comparison with traditional techniques. Med Dosim. 2013;38:48–54.
34. Sarkar B, Munshi A, Ganesh T, Manikandan A, Mohanti BK. Dosimetric comparison of short and full arc in spinal PTV in volumetric-modulated arc therapy-based craniospinal irradiation. Med Dosim. 2020;45(1):1–6.
35. Klein EE, Hanley J, Bayouth J, Yin FF, Simon W, Dresser S, Serago C, Aguirre F, Ma L, Arjomandy B, Liu C. Task group 142 report: qual-

ity assurance of medical accelerators a. Med Phys. 2009;36(9Part1):4197–212.
36. 5 Radiation therapy sources, equipment and installations code no. AERB/RF-MED/SC-1 (Rev. 1). https://www.aerb.gov.in/images/PDF/CodesGuides/RadiationFacility/MedicalApplicationsInvolvingRadiation/4.pdf.
37. Sarkar B, Ghosh B, Mahendramohan S, Basu A, Goswami J, Ray A. Optimized point dose measurement for monitor unit verification in intensity modulated radiation therapy using 6 MV photons by three different methodologies with different detector-phantom combinations: a comparative study. J Med Phys. 2010;35(3):144.
38. Dong L, Antolak J, Salehpour M, Forster K, O'Neill L, Kendall R, Rosen I. Patient-specific point dose measurement for IMRT monitor unit verification. Int J Radiat Oncol Biol Phys. 2003;56(3):867–77.
39. Manikandan A, Sarkar B, Nandy M, Sureka CS, Gossman MS, Sujatha N, Rajendran VT. Detector system dose verification comparisons for arc therapy: couch vs. gantry mount. J Appl Clin Med Phys. 2014;15(3):41–53.
40. Bedford JL, Hanson IM, Hansen VN. Comparison of forward- and back-projection in vivo EPID dosimetry for VMAT treatment of the prostate. Phys Med Biol. 2018;63:025008.

Further Reading

AERB safety code: radiation therapy sources, equipment and installations. https://www.aerb.gov.in/images/PDF/CodesGuides/RadiationFacility/MedicalApplicationsInvolvingRadiation/4.pdf.

Jimenez RB, Wright JL, Bellon JR. Beyond mean heart dose: cardiac metrics for the modern era. Int J Radiat Oncol Biol Phys. 2022;114(1):140–2.

Manikandan A, Sarkar B, Holla R, Vivek TR, Sujatha N. Quality assurance of dynamic parameters in volumetric modulated arc therapy. Br J Radiol. 2012;85(1015):1002–10.

Miften M, Olch A, Mihailidis D, Moran J, Pawlicki T, Molineu A, Li H, Wijesooriya K, Shi J, Xia P, Papanikolaou N. Tolerance limits and methodologies for IMRT measurement-based verification QA: recommendations of AAPM Task Group No. 218. Med Phys. 2018;45(4):e53–83.

Radiotherapy shielding calculation: NCRP report no. 11. https://ncrponline.org/publications/reports/ncrp-reports-151/.

Brachytherapy Technology and Principles

13

Ankita Mehta, Raghavendra Hajare, Rajendra Bhalavat, Shyam Shrivastava, and Umesh Mahantshetty

13.1 Introduction

With a history spanning almost a century, brachytherapy was the first conformal radiation technique ever developed [1]. It is commonly utilized in gynecological, prostate, head and neck, breast malignancies and sarcomas. Nevertheless, it has a wide range of applications, including lung, hepatobiliary, gastrointestinal and central nervous system malignancies [2, 3]. It has practically been utilized in radical (curative) settings for primary lesion for all early-stage tumors, and also in palliative settings for recurrent/metastatic disease for all accessible tumors either as a single modality or in combination with other treatment modalities.

It has the ubiquitous potential to deliver high-dose radiation to the target region without traversing normal tissue, which makes it one of the most conformal radiation techniques. With advancements in the field of external beam radiotherapy, there was a subtle drift away from brachytherapy. Recently, interest in brachytherapy has been renewed by the advent of sophisticated planning techniques and rapid technological advances. The key advancements are the incorporation of imaging guidance, stepping source technology with the potential for intensity modulation and the developments in quality assurance. This has led to higher precision in treatment delivery and improvement in the outcomes while limiting the toxicities [4]. This chapter focuses on the underlying principles and physical and technological aspects of brachytherapy.

13.2 Physical and Radiobiological Effects of Brachytherapy

It has a characteristic steep dose gradient with an exponential rate of dose fall off in accordance with the inverse square law [5]. This is attributable to the relatively much lower energy of radiation emitted from brachytherapy sources, within the keV range in comparison to the MeV range of energy used in teletherapy. The dose distribution is highly

A. Mehta
Radiation Oncology, Kasturba Medical College, Manipal, Manipal Academy of Higher Education, Manipal, Karnataka, India

R. Hajare · U. Mahantshetty (✉)
Radiation Oncology, Homi Bhabha Cancer Hospital and Research Centre, Visakhapatnam, Andhra Pradesh, India

R. Bhalavat
Radiation Oncology, Jupiter Hospital, Thane, Maharashtra, India

S. Shrivastava
Radiation Oncology, HCG-ICH Khubchandani Cancer Centre, Mumbai, Maharashtra, India

A. Munshi et al. (eds.), *Radiation Oncology – Principles, Precepts and Practice*,
https://doi.org/10.1007/978-981-97-8389-2_13

heterogeneous, with the target region receiving a median dose of 150–200% relative to the prescribed dose and a dose gradient at the periphery of 5–20% per millimeter. The high central dose along with hypofractionation significantly enhances the radiobiological effectiveness in the central region harboring the tumor, delivering a very high dose to the potentially active (subclinical disease) in the target area and sparing the normal tissue at the periphery to a higher extent in comparison to teletherapy [6]. These properties are attributable to high local tumor control rates with brachytherapy and minimal normal tissue toxicities for small volume tumors. However, this also necessitates stringent accuracy in treatment planning and delivery as geometric deviations of even few millimeters can have considerable impact on the dosimetry and possibly tumor control.

13.3 Physical Principles of Brachytherapy

13.3.1 Properties of Brachytherapy Source

The radiation utilized for brachytherapy is emitted from the disintegration of a radioactive source. The phenomenon by which atoms with unstable nuclei transform into a stable state by releasing energy in the form of radiation is known as radioactivity. The specifications of source strength are defined by the quantification of radioactive material and the output of radiation in terms of:

1. Half-life: It is the time taken for the decay of radioactive atoms to half of their original number.
2. Specific activity: it is the number of atoms disintegrating per unit time per unit mass of radioactive material. The SI unit is the Becquerel or Bq (1 Bq = 1 disintegration/s), but curie is the unit in common use.
3. Exposure rate: It is defined as the charge rate produced in air by the emission of indirectly ionizing radiation per unit mass of air. The SI unit is C/kg-h, but a commonly used unit is roentgen per hour (R/h).
4. Air Kerma Strength (KERMA—Kinetic Energy Released in Materials): It is the measure of energy that is transferred to charged particles per unit mass in a volume of air at a reference distance of 1 m. The SI unit is Gy/h.

13.3.2 Properties of Radionuclide

There are several underlying physical properties that are desirable for the selection of a radioactive material as a brachytherapy source:

1. Energy: It should be in an optimal range, being low enough to provide a steep dose gradient and minimize radiation protection requirements but high enough to deliver dose at a certain depth, minimize scatter and avoid energy deposition in bone by photoelectric effect. The majority of sources emit energy in the form of gamma rays. A few exceptions are beta ray emitting sources used for superficial treatments (Strontium-90, yttrium-90), neutron emitting and electronic x-ray emitting sources.
2. Specific activity: It should be high for that the source dimensions to be smaller and versatile for implants and to allow faster treatment delivery.
3. Half-life: It should be in an optimal range and suited according to the facility requirements. A large half-life supports the working life over a longer duration and minimizes the decay correction, while a small half-life carries the advantage of a shorter treatment duration.
4. Physical form: It should be available in insoluble and nontoxic forms with no gaseous disintegration products. It should be possible to manufacture in various shapes and sizes, including rigid tubes, needles, spheres, ribbons and flexible wires. Sources are constructed with varying thicknesses and types of encapsulating material. This prevents leakage of radioactive material and absorbs nonpenetrating radiation (beta rays, alpha rays and low-energy photons) that would increase the surface doses without any contribution to the therapeutic effect.

Table 13.1 Physical characteristics of various brachytherapy sources

Source	Energy (MeV)	Half-life	Decay	Source form	Application
Cesium 137	0.662	30 years	β, γ	Tubes and needles	LDR intracavitary and interstitial
Cesium 131	0.030	9.69 days	EC[a] (γ)	Seeds	LDR permanent implants
Iridium 192	0.397	73.8 days	β, γ	Seeds, wires, hairpin, ribbon, slugs	LDR temporary interstitial, intravascular, HDR intracavitary and interstitial
Cobalt 60	1.25	5.26 years	β, γ	Encapsulated spheres, pellets, needles, slug	HDR Intracavitary and interstitial
Iodine 125	0.028	59.6 days	γ	Seeds	Permanent interstitial
Palladium 103	0.020	17 days	EC[a] (γ)	Seeds	Permanent interstitial
Gold 198	0.412	2.7 days	β, γ	Seeds	Permanent interstitial
Strontium/yttrium 90	2.24	28.9 years	β	Plaque	Superficial ocular lesions
Californium 252	0–13 (mean—2.3)	2.65 years	α	Seeds	Intracavitary and interstitial brachytherapy

[a] *EC* electron capture

13.3.3 Other Considerations

Apart from the physical properties, the selection of a source is determined by facility requirements, availability, cost effectiveness and indications. Radium-226, a naturally occurring radionuclide, was the first source to be utilized. Later, with the introduction of other sources, it became obsolete owing to its long half-life (1626 years) and complex decay with the emission of radioactive salts and gases. However, several widely used quantities for source strength specification and treatment prescription are defined in terms of radium equivalent. The development of miniature and highly flexible sources that can be used in after-loading devices has been one of the important technological developments in brachytherapy. The physical properties and applications of presently available brachytherapy radionuclides are described in Table 13.1.

13.4 Classifications

Based on the application method, dosage rate and duration of the treatment, the various brachytherapy approaches are broadly categorized as:

13.4.1 Application Technique

1. *Intracavitary brachytherapy*—placing applicators in a body cavity in close proximity to the target tissue such as vaginal or uterine cavity.
2. *Interstitial brachytherapy*—implanting interstitial needles into the target tissues such as breast and buccal mucosa.
3. *Surface mold or plaque brachytherapy*—placing applicators on skin or mucosal surface adjacent to the target tissue such as skin lesions over nose, scalp
4. *Intraluminal brachytherapy*—placing single source line applicators in lumen to treat the surface and adjacent underlying tissue such as esophagus, rectum
5. *Intravascular brachytherapy*—placing applicators inside blood vessels, especially for preventing restenosis in vessels.

13.4.2 Dose Rate

According to the dose delivery rate, three definitions were originally proposed in ICRU Report 38 [7].

1. *Low-dose rate (LDR)*: 0.4–2 Gy/h
2. *Medium-dose rate (MDR)*: 2–12 Gy/h

3. *High-dose rate (HDR)*: 12 Gy/h

In the 1960s, HDR brachytherapy was widely adopted due to the much shorter treatment duration compared to several days with LDR brachytherapy that reduced hospitalization and allowed for displacement of normal tissues away from the source to a greater extent. It also offered the potential of dose optimization by altering dwell position and dwell times. The LDR brachytherapy had been in practice for nearly more than a century and is radio biologically sought to be more effective for tumor killing and normal tissue repair. However, high-quality evidence has proven comparable outcomes and toxicity of HDR and LDR brachytherapy [8].

The other two forms are ultra-low-dose rate (ULDR) and pulsed dose rate (PDR). The ULDR ranges from 0.01 to 0.3 Gy/h that is delivered with permanent seed implants. In PDR, short pulses of high-dose radiation are delivered intermittently, typically in 1–4 h intervals to simulate radiobiological effect of LDR treatment.

13.4.3 Treatment Duration

1. *Temporary brachytherapy*—applicators are left in patient for specified time till they deliver the prescribed dose.
2. *Permanent brachytherapy*—implants are permanently placed that deliver treatment at a very low-dose rate over several weeks to months. The initial source strength is selected so that the prescribed dose is delivered when the radioactivity has decayed to a negligible level.

13.5 Brachytherapy Processes (Fig. 13.1)

13.5.1 Prebrachytherapy Evaluation

Proper case selection and comprehensive evaluation are the foremost prerequisites for having good outcomes with brachytherapy. This includes a clinical examination and imaging as indicated to depict the tumor topography and its relations. This preliminary assessment determines the application technique, complexity and challenges of performing the procedure, and defines the target region and potential organs at risk (OARs).

13.5.1.1 Clinical Examination

This should include a thorough assessment with documentation and three-dimensional diagrammatic depiction of the findings. The placement of radio opaque markers at the confines of tumor aids in defining the target region and is particularly important in patients who are planned for brachytherapy prior to other modality of treatment as initial tumor volume should be accounted regardless of the subsequent response. Examination under anesthesia offers a more comprehensive assessment owing to relaxation of surrounding musculature and is preferred for assessment of gynecological and head and neck cancers.

13.5.1.2 Imaging

The indication and choice of imaging modality depends on the site, extent of disease as well as the resource availability. A thorough understanding of the imaging findings and anatomy is essentially important. In patients who undergo other modality of treatment prior to brachytherapy, comparison of the baseline and prebrachytherapy imaging can be useful for adaptive planning with high doses are delivered to the residual disease at the time of brachytherapy. This approach has been pioneered during the last decade for the treatment of cervical cancer. It can also be potentially utilized for anorectal and head and neck cancers.

13.5.2 Application

The placement of applicators defines the positioning of sources in relation to target volume and dose-limiting tissues, and also the intersource spacing. It is the key determinant of quality of plan and hence the treatment outcome. It is the one of the most challenging issues faced in practicing brachytherapy. Dose optimization with modulation of dwell positions and dwell times to

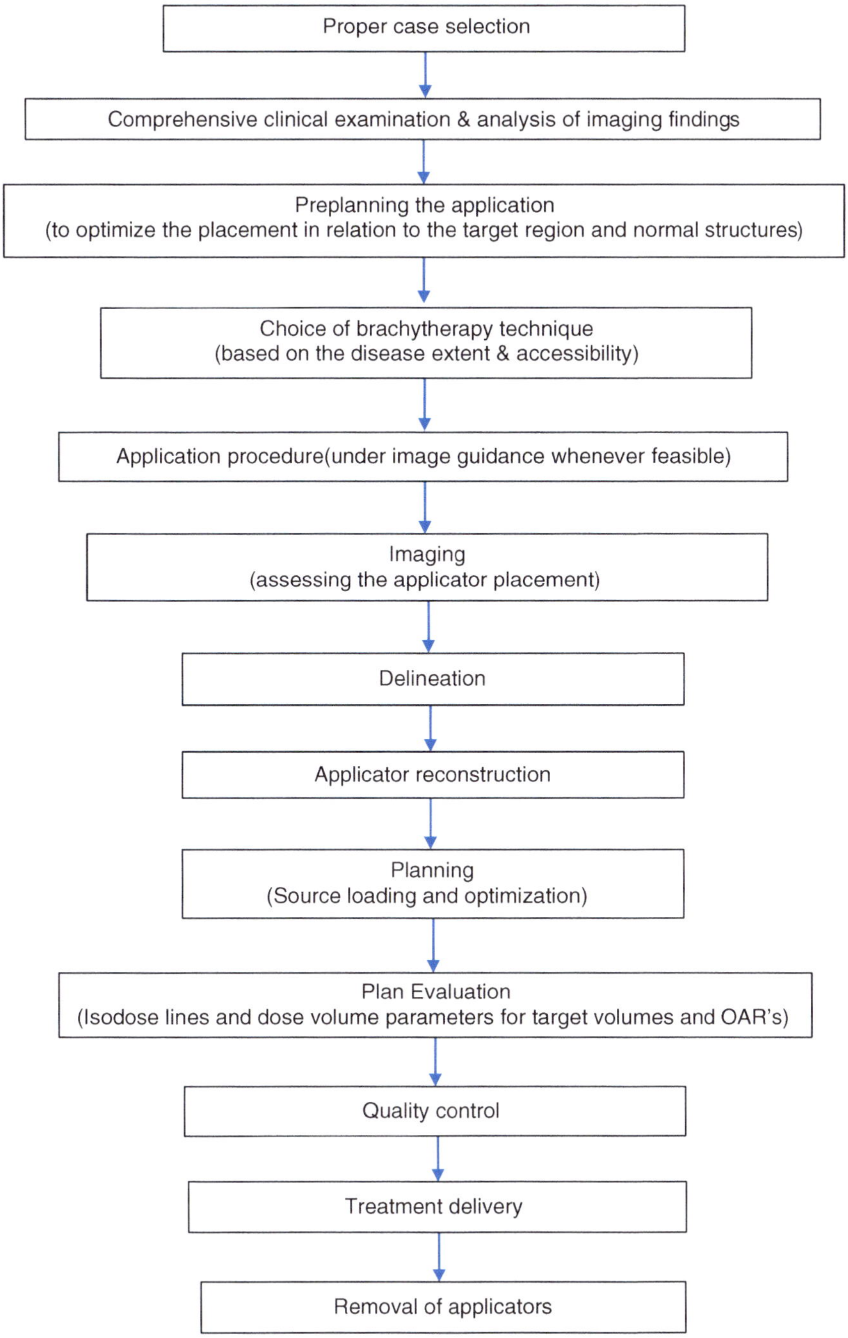

Fig. 13.1 Flowchart depicting the workflow for brachytherapy treatment

improve dose coverage cannot compensate for improper application geometry beyond a certain extent. Also, higher degree of optimization can increase the high-dose regions in the vicinity of applicators and the organs at risk doses. Over the years, customized applicators and templates have been developed for various sites that help in attaining a predefined geometry.

13.6 Principles of Application

13.6.1 Intracavitary Brachytherapy

This method involves placing applicators in a cavity near the target tissue. It forms the cornerstone of treatment for cervical cancer. It is also a suitable modality for selected cases of head and neck cancers (nasal cavity, nasopharynx, trachea, oral cavity, external auditory canal, sinuses, and orbits), breast cancers, and central nervous system malignancies.

13.6.1.1 Cervical Cancer

Intracavitary brachytherapy in cervical cancer accounts for the vast majority of all brachytherapy procedures performed worldwide [9]. The most widely used intracavitary applicators are the classic tandem and the ovoid or ring applicator. Application should preferably be performed under anesthesia as it dilates the cervical canal, along with adequate relaxation of the pelvic floor and vaginal muscles for vaginal packing and enhances the reproducibility for subsequent applications [10]. The procedure commences with some part preparation and Foley catheterization. The endocervical canal is appropriately dilated and uterine cavity sounded (preferably under image guidance) and a flanged tandem of appropriate length is inserted. Then optimal-sized vaginal applicators are placed (ovoid/ring/cylinder) that are determined by the vaginal disease extent and capacity.

In recent years, the combination of intracavitary and interstitial brachytherapy has increasingly been utilized given the potential for better dose optimization and improvement in outcomes [11]. In combined application, interstitial needles are inserted just lateral to the vaginal applicator with 4–5 cm of length advanced in the parametrium (Fig. 13.2). Lastly, vaginal packing is done and a rectal tube is inserted to keep its lumen collapsed, followed by securing of the applicator position with a bandage or perineal strap. In the past two decades, numerous novel customized applicators have been introduced with modifications of the indigenous tandem and ovoid applicators to act as templates for interstitial needles (Vienna, Utrecht, Geneva, and Venezia) (Fig. 13.3). The factors determining the choice of application technique are described in Table 13.2.

13.6.2 Balloon Brachytherapy

It is used in postoperative settings in breast and central nervous system malignancies. In breast cancer, a mammosite applicator is placed in the

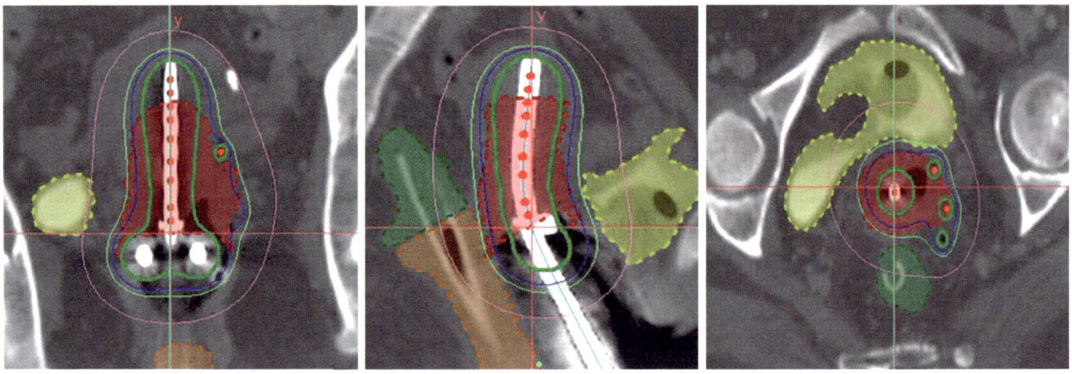

Fig. 13.2 Combination of intracavitary and interstitial brachytherapy (free hand interstitial needles on the left side). Representative coronal, sagittal, and axial CT images with isodose distributions

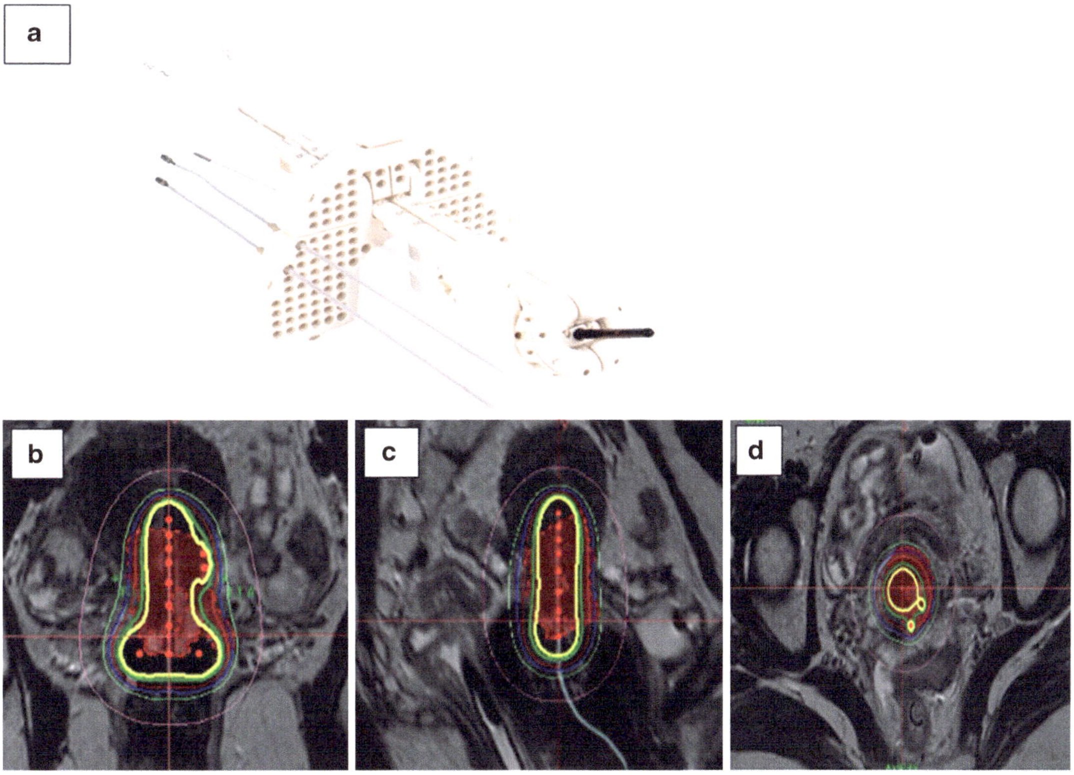

Fig. 13.3 Venezia applicator (**a**)—hybrid applicator for intracavitary and interstitial brachytherapy. It comprises of a tandem and two lunar ovoids having holes as a template for interstitial needles and also has additional vaginal caps and perineal template. Representative coronal (**b**), sagittal (**c**) and axial MRI (**d**) images with isodose distributions

Table 13.2 Selection of brachytherapy application technique for cervical cancer

Technique	Indication
Intracavitary	No residual disease, or residual disease confined to the cervix, medial parametrium and/or upper vagina
Intracavitary and interstitial	Large volume tumors and/or extension beyond the medial third of parametrium Optimizing OARs doses for small volume tumors
Interstitial	Tumor extension of >3.5 cm at the level of the vaginal applicator or >2.5 cm at the level of point A Vault cancers involving parametria Reirradiation for central recurrence

lumpectomy cavity either intraoperatively or in postoperative settings. It is ideally suited for patients who have adequate parenchyma surrounding the lumpectomy cavity for inflation of the balloon to a diameter of between 4 and 6 cm [12]. The applicator consists of an expandable balloon with single or multilumen source catheters. The multilumen or strut-based applicators offer more flexibility in source positioning and hence allow better dose optimization [13, 14]. The balloon should have a minimum skin distance of 0.3 cm with good conformance of the surface to the walls of the lumpectomy cavity and a symmetrical expansion.

The GliaSite system is used in malignant brain tumors and brain metastases in primary and recurrent settings. It comprises of an inflatable balloon catheter that is placed in the surgical cavity and the latter instilled with liquid I125 or embedded Cs-131 seeds are placed intraoperatively [15, 16].

13.6.3 Intraluminal Brachytherapy

Intraluminal brachytherapy (ILBT) is a type of intracavitary brachytherapy that is used in hollow cylindrical organs such as the bronchus, esophagus, or bile duct, as well as the blood vessel. It involves the placement of a single source line in the lumen. It is commonly used in palliative settings for the treatment of esophageal or bronchial cancers or occasionally as a boost after external beam radiation therapy. The tumor is visualized under the bronchoscope/endoscope, and the cranial and caudal extents are marked with radio opaque clips. After the insertion of the guide wire, the scope is removed and brachytherapy catheter is advanced till its distal end is beyond the lowermost extent of disease.

13.6.4 Interstitial Brachytherapy

It has a wide range of applications and is commonly utilized in head and neck cancers, breast, prostate, and gynecological cancers and soft tissue sarcomas. It requires a certain level of training, skill and expertise, with the complexity of the procedure dependent on the tumor location, tissue fibrosis and the proximity to critical structures. Imaging done prior to the procedure can be used for preplanning of implants where delineation and depiction of the desired dose objectives serve as a surrogate to define the ideal position of needles. In patients who have received another modality prior to brachytherapy (external beam radiation, surgery, or chemotherapy), the implant volume should be large enough to cover the original extent of disease.

13.6.4.1 Procedure

The implant geometry and boundaries are determined by the tumor topography. A single plane implant is suitable for tumors with a thickness of up to 1 cm, and two parallel planes are required for tumors having a thickness of up to 2.5 cm. Volume implants are required for tumors having a thickness of more than 2.5 cm. Depending on the tumor's confines, volume implants can take the shape of a sphere, cube, or cylinder.

The implant may be done in a free-hand manner or using templates, or a combination of both, depending on the extent and site of disease. The use of a template guides the needle along a predefined path that helps in attaining a good implant geometry. It is commonly used in gynecological, prostate and breast cancers.

Before the implant procedure, the relationships and course of nearby normal structures and large blood vessels should be thoroughly analyzed to avoid injury. After piercing through the skin and underlying fascia, the interstitial needles are gradually advanced with the direction being guided by the palpation of tips in underlying tissues till the exit. Ideally, the needles should be uniformly spaced to obtain a uniform and homogenous coverage of the target region. After the insertion of the catheters, the needles are removed and finally, the position of the plastic catheters is secured at the exit end with beads and buttons. The figures below demonstrate the workflow of interstitial brachytherapy (Figs. 13.4, 13.5 and 13.6).

13.6.5 Surface Mold Brachytherapy

It is utilized for the treatment of superficial cancers that are typically less than 5 mm in thickness.

It is commonly used for cancers of the scalp, face, pinna, external auditory canal, lip, buccal mucosa, hard palate and orbital cavity after exenteration. It can be delivered either by predesigned or customized applicators. The predesigned applicators from Leipzig and Valencia are ideally suited for small, regular and well-circumscribed skin cancers. They have varying sized surface cups for radioactive sources with a body made up of tungsten alloy to shield nearby structures.

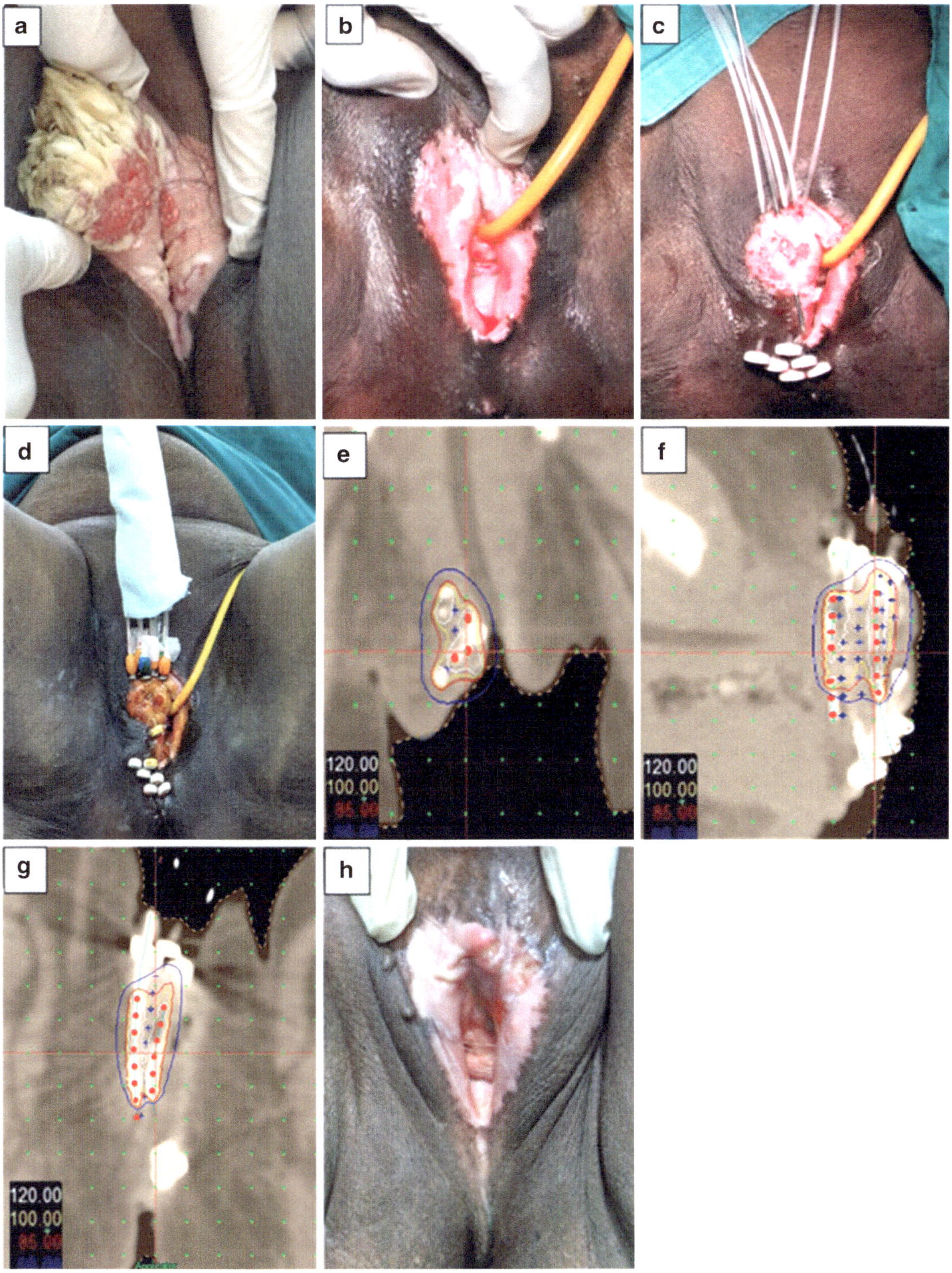

Fig. 13.4 Vulvar cancer treated by EBRT followed by brachytherapy boost. (**a**) At presentation, (**b**) Post EBRT, (**c** and **d**) Catheter placements, (**e**–**g**) Representative coronal, sagittal, and axial CT images with isodose distributions, (**h**) Post treatment 3 months

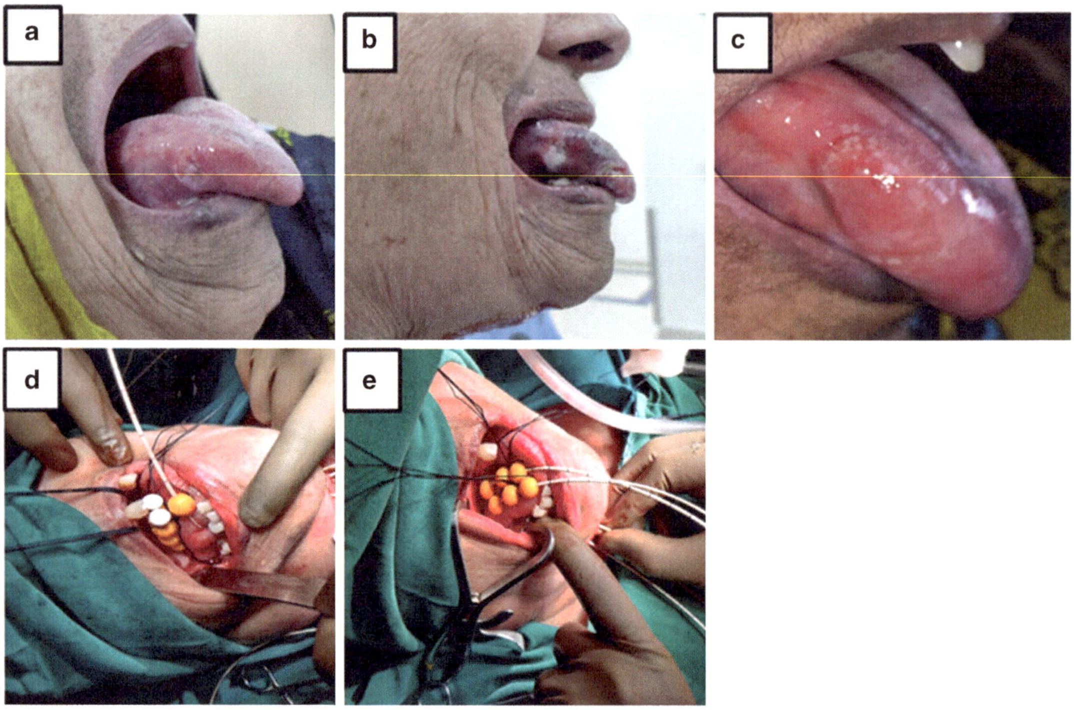

Fig. 13.5 Early ulceroproliferative T1N0 lesion of right lateral border of tongue (treated by brachytherapy alone). (**a**) Clinical photograph at diagnosis; (**b**) Immediate postremoval-confluent mucositis; (**c**) Clinical photograph—3 months after completion of treatment depicting complete response; (**d**, **e**) Double planes interstitial implant with plastic beads on dorsum of tongue to improve the mucosal coverage (Bhalavat's technique). Additional single tube passed through lateral surface of tongue (to correct lateral surface underdose)

Customized molds are utilized for complex shaped irregularities and large volume tumors. They are fabricated from a variety of materials such as thermoplastic sheet, wax, Perspex and plaster of Paris. The mold can be planar, linear or cylindrical shapes that are depending upon the shape of the target volume. The external contour of the tumor is replicated over a tissue equivalent material that acts as a frame. The distance between the mold and skin or mucosa should be at least 5 mm to avoid surface overdosage at the surface but not beyond 1.5 cm. The frame is then converted into the actual mold by either layering of wax or using it as a negative impression to create the actual mold. The area to be treated along with its orientations is marked on the mold, and then catheters are secured. The mold should snugly fit to the surface in a reproducible manner, with catheters encompassing the whole target within a certain margin.

A variation of the classic variety, sandwich type molds are particularly useful for lip, pinna and floor of mouth cancers. The target tissue is encompassed by mold on both of its surfaces. This improves the dose homogeneity with a reduction in the high-dose volumes [17].

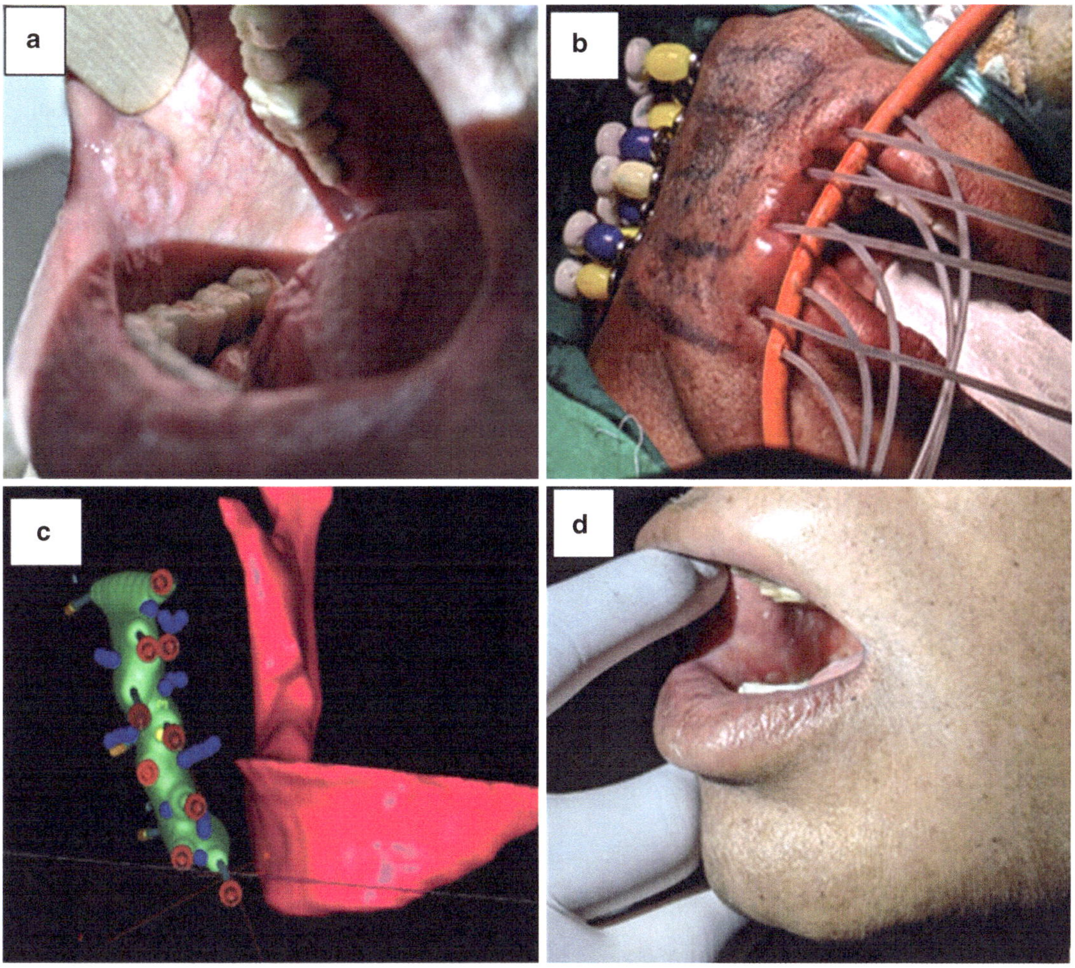

Fig. 13.6 Early superficial T2N0 lesion of buccal mucosa (treated by EBRT to the primary and draining neck nodes followed by brachytherapy boost to the primary site). (**a**) Clinical photograph at diagnosis; (**b**) Two planes interstitial implant; (**c**) 100% dose (green color), reconstructed catheters with ends (blue & red); mandible contour (pink); (**d**) Clinical photograph—3 months after completion of treatment depicting complete response

13.7 Brachytherapy Treatment Planning

This comprises of defining the target region and organs at risk, reconstruction of the applicators, dose calculation, optimization and plan evaluation. The brachytherapy planning has evolved a long way from traditional manual planning to point based dosimetry to the present standard of care, image-guided adaptive brachytherapy.

13.7.1 Manual Brachytherapy Planning

Historically, brachytherapy planning commenced with specifying dose prescription in terms of exposure and exposure rate with the use of radium equivalent dosimetry for LDR intracavitary gynecologic brachytherapy (Paris, Stockholm and Manchester systems). The planning of interstitial brachytherapy was based on

Table 13.3 Details of various intracavitary and interstitial brachytherapy systems

Intracavitary brachytherapy systems			
Dosimetric system	**Stockholm system (1910)**	**Paris system (1912)**	**Manchester system (1930)**
Applicator	Uterine rubber tube and vaginal boxes	Uterine tube and vaginal colpostats connected by metallic spring	Uterine tube and vaginal ovoids
Treatment duration	2–3 fractions of 20–30 h each, delivered 1 week apart	Single fraction lasting over 5–8 days	2 fractions of 72 h each, delivered 1 week apart
Loading pattern of uterine & vaginal sources	Unequal	Equal	Unequal
Prescription system	Milligram/hour	Milligram/hour	Point based prescription system
Total dose	6500–7100 mg-h 4500 by vaginal boxes	7000–8000 mg-h Over 5 days	7500 R at point A
Interstitial brachytherapy system			
Dosimetric system	**Manchester system (1930)**	**Quimby system (1930)**	**Paris system (1960)**
Type of implant	Geometrical volume implants	Volume implants	Single and double plane implants
Loading pattern	Nonuniform distribution of sources based on the target volume dimensions with more strength in the periphery	Uniform distribution of sources of equal strength	Uniform strength sources are kept parallel and equidistant with centers in same plane
Dose distribution	Uniform dose distribution	Non uniform dose distribution (25–30% higher in center than periphery)	Basal dose rate calculated by the mean of the minimum dose between adjacent sources along central plane Dose is prescribed at 85% isodose of the basal dose

generalized mathematical rule systems (Paterson-Parker, Quimby system, Paris system). These rules, based on tables and nomograms, determine the required number, location, and treatment time of sources to provide adequate dose coverage to the estimated target volume. Contrast radiographic films were used to localize the applicator. The normal tissue point doses were analyzed based on their approximate location in relation to the applicator.

These systems were based on precise calculations and derivations and still form the underlying basis for application in settings of contemporary image-guided planning. However, a major limitation was that the description of the anatomy and dose distribution was approximative. The broad principles of these systems are discussed in Table 13.3.

13.7.2 X-Ray-Based Planning

It was the first imaging modality to be incorporated for brachytherapy planning. The equipment used was an x-ray simulator or jigs with known geometry and reference points. It serves the purpose of assessment of applicator placement and localizing the target region with the aid of radio-opaque markers. It is useful for analyzing the geometry of intracavitary brachytherapy, localizing the seeds and needle configuration for interstitial implants and designing catheters for intraluminal brachytherapy. The configuration of the sources is determined with respect to applicator position, bony anatomy, surgical clips and contrast-filled catheters. The planning is based on standard rules of source loading with a point-based system for dose prescription and organs at risk doses. The major limitations of X-ray-based planning were the lack of soft tissue details and anatomic correlations with point-based planning.

13.7.3 Image-Guided Brachytherapy

The incorporation of advanced imaging modalities in brachytherapy practice has significantly improved the treatment accuracy and clinical outcomes with a reduction in toxicity [18]. Imaging has a key role in various aspects that involve preplanning the application, real-time guidance during the application, applicator reconstruction, delineation and treatment planning.

13.7.3.1 Ultrasound-Based Planning

It is a widely available, portable, time and cost-effective modality that has increasingly been utilized in recent years for image-guided brachytherapy. However, it is highly operator dependent and has a learning curve for acquisition and interpretation. The advanced modes have largely improved its utility with the avoidance of vascular injury by Doppler study and treatment planning with three-dimensional and contrast-enhanced studies. It is particularly used for prostate, gynecological, and breast malignancies and has a potential role for a wide range of malignancies, including head and neck, biliary tract and anal canal malignancies [17].

Real-time guidance with transabdominal and transrectal ultrasound reduces the chances of uterine perforation by facilitating selection of an appropriate curvature and length of tandem and the injury to normal tissue and vasculature by interstitial needle. It has proven to be quite useful for delineation and treatment planning, with findings comparable to those of the MR study [18–20].

In prostate cancer implants, transrectal ultrasound-based preplanning determines the ideal seed and needle positions according to the anatomy and relations of the prostate gland that predetermines the estimated regions of interest and desired dose objectives. It also provides excellent intraoperative real-time visualization to guide the placement in accordance with the preplan [21].

In breast cancers, ultrasonography is useful for interstitial implant preplanning with visualization of the lumpectomy cavity or seroma in open cavity surgeries and areas of scarring in closed cavity surgeries. During the procedure, it provides real-time guidance for needle insertion in relation to the lumpectomy bed, skin, and chest wall and ascertains the implant symmetry [22].

13.7.3.2 CT-Based Planning

It was the first modality to be used for image-guided brachytherapy and has been widely adopted across various sites due to its wide accessibility, time and cost effectiveness [23]. Compared to the two-dimensional planning that had been done for decades, it introduced the concept of volume-based planning with significantly improved outcomes and a lower toxicity profile of brachytherapy. It is the preferred modality in interstitial brachytherapy of breast cancers for all processes, including preplanning, delineation and treatment planning [22]. It is the recommended modality for postimplant dosimetry in prostate cancer [24].

However, compared to MRI-based planning, the soft tissue details are limited, which can possibly lead to an over or underestimation of the target volumes and organs at risk. Therefore, collaboration of clinical findings and placement of radio-opaque markers becomes necessary. Another limitation is the image artifact from high-density materials such as applicators or seeds, which leads to erroneous data that can potentially affect the delineation.

13.7.3.3 MRI-Based Planning

It is the ideal modality for image-guided adaptive brachytherapy because of the excellent soft tissue details. The prospects of planning and the outcomes of brachytherapy have demonstrated enormous improvement with MRI-based planning in the past decade, particularly in cervical and prostate cancers [25].

In gynecological malignancies, MRI precisely defines the organs at risk and target volumes with differentiation into subvolumes and offers the potential for adaptive planning. This has translated into better outcomes and a lower toxicity profile. It can also be used as the sole modality for treatment planning [26]. In prostate cancers, MRI offers excellent visualization of the gland and normal structures that limit the OAR doses, along with the potential for dose differentiation in target volume. However, it is not useful for postimplant planning in prostate cancer because of its limited ability for applicator reconstruction [27].

It is an efficient modality for application preplanning, but ultrasound is better suited for this purpose because of its portability and real-time guidance. In spite of the better outcomes, MR-based brachytherapy has not been widely adopted because of limited availability, relatively longer acquisition times, susceptibility to motion artifacts and higher costs involved.

The incorporation of advanced imaging modalities in brachytherapy practice has significantly improved the treatment accuracy, clinical outcomes with reduction in the toxicity [18]. Imaging has a key role in various aspects that involves preplanning the application, real-time guidance during the application, applicator reconstruction and delineation.

13.7.3.4 PET-Based Planning

Currently, there is no clearly established role of PET in brachytherapy planning. It is proposed that it might have a potential role in depicting the tumor shrinkage in response adaptive brachytherapy and optimizing the doses to target volume and organs at risk [28].

13.7.4 Applicator Reconstruction

This step involves assembling the geometry of the applicator relative to the patient's anatomy, determining the source positions and pathway. The optimal imaging modality for reconstruction is determined based on the applicator. Single orthogonal pairs of radiographic films are suitable for reconstruction of single catheter endobronchial and intraluminal applicators and vaginal brachytherapy cylinders. The three-dimensional applicators should be ideally reconstructed by three-dimensional imaging studies.

It is one of the key determinants of accurate treatment planning as errors in this step can introduce major deviations in dosimetry [29]. The reconstruction quality is enhanced with a lower slice thickness of cross-sectional imaging and with the axis of imaging being perpendicular to the orientation of the applicators. Appropriate applicator commissioning is essential to minimize the involved uncertainties and errors. The reconstruction of the applicator can be performed by direct reconstruction, automated reconstruction using library plans, or a combination of these two methods.

13.7.5 Delineation

The definition of volumes for organs at risk and target volumes in brachytherapy is analogous to that of external beam radiotherapy. However, delineation in brachytherapy requires utmost precision and accuracy because of the sharp dose gradients that can considerably impact the dosimetry of target volume and OARs. In addition to the imaging findings, clinical findings and anatomical information should always be taken into account for the delineation of the target volume. The definition of various target volumes is as follows:

1. Gross tumor volume (GTV)—It includes the gross macroscopic disease at the time of brachytherapy.
2. Clinical target volume (CTV)—It includes the volume that can potentially harbor microscopic disease. In patients who receive EBRT and/or chemotherapy prior to brachytherapy, it can be differentiated into subvolumes based on the differential response pattern that indicates the potential burden of microscopic disease.

   High-risk CTV: It includes the gross disease at the time of brachytherapy with whole of its subsite or with a certain margin based on the pattern of disease spread.

   Intermediate *risk CTV: It* includes the volume harboring gross disease at initial diagnosis that has regressed by the time of brachytherapy along with a certain margin around the high risk CTV. The initially involved anatomical regions are superimposed on the topography at the time of brachytherapy.

Low-risk CTV—It includes the entire region that can potentially harbor microscopic disease.

3. Planning target volume (*PTV)—Generally*, it is not taken into account for brachytherapy planning because of the minimal motion and setup uncertainties involved. It needs to be taken into consideration when there are uncertainties regarding consistency of source position (moving sources, fractionated techniques), applicator positioning (intracavitary applications, permanent implants) and target motion (intraluminal brachytherapy).

13.7.6 Dose Calculation

The dose distribution in planning systems is computed by the summation of the dose-rate contribution, per second, from each dwell position. With the information on source air Kerma strength and the dwell times, the total dose received at any point in the contoured structures can be calculated by adding the dose contributions from all dwell positions, which is the dwell time times the dose-rate contribution scaled with the strength of the radiation source.

The total dose at a point is determined by the summation of the dose contributions from all dwell positions. That is calculated by dwell time multiplied with the dose-rate contribution scaled with the strength of the source. Finally, the dose is scaled with respect to the source strength decay. The most widely used calculations of the dose-rate contributions are generally based on the AAPM TG43 formalism [5, 30].

13.7.7 Optimization

The objective of dose optimization is to conform the minimum prescription isodose line to the target volume with avoidance of organs at risk. It comprises of two steps, selective activation of dwell positions and the determination of relative dwell times. It can be done by various methods.

1. *Manual method*—The dwell positions and relative dwell times are manually iterated till the desired objectives are achieved.
2. *Geometric optimization*—It is based on the principle of inverse square law. Based on the implant geometry, the dwell time at a particular dwell position is modulated based on its distance from other dwell positions that determines their dose contribution.
3. *Graphical optimization*—It is based on the graphical modulation of isodose lines where a selected isodose line can be shifted into the desired position by "drag and drop" function and the dwell positions and times automatically modified.
4. *Inverse optimization*—It is based on optimization of dwell times and position using an inverse planning algorithm to achieve the prespecified objectives. It offers the advantage of all the objectives being simultaneously taken into account in the planning process.

After optimization, the dose distributions should be carefully analyzed for any steep gradients that can potentially lead to underdosage of the target volume or high-dose volumes in the OARs and undefined region.

13.7.8 Plan Evaluation

It involves analyzing the dose volume parameters with respect to the prespecified objectives in terms of target volume dose coverage, homogeneity and OAR's constraints. The tools used are isodose lines, dose volume parameters and the histograms. The complex dose heterogeneity of brachytherapy plan is depicted by the cumulative and differential dose volume histograms. The acceptable dose limits for target volumes and OARs are defined by dose volume parameters. A certain set of dose and volume specifications have been defined by ICRU reports and various recommendations to ensure uniform standards of practice and reporting.

Apart from the dose volume parameters, it is extremely crucial to analyze the isodose lines in cross-sectional imaging for any marginal under coverage of target volume, or any high-dose regions in undefined regions that may be clinically relevant. The recommendations for dose volume reporting of image-guided brachytherapy in cervical cancer are defined in ICRU Report 89 and for interstitial, surface mold and intraluminal brachytherapy in ICRU Report 58 [6, 31].

13.7.9 Reporting

13.7.9.1 Intracavitary Brachytherapy

The integrated doses from EBRT and brachytherapy are taken into account for evaluation of the target volume and organs at risk. The ICRU 89 has defined three levels of reporting of dose volume parameters. The level 1 is the basic minimum standard, level 2 is the advanced standard while level 3 is for research purpose only [6].

Level 1

- Dose reporting

TRAK
Point A dose
Recto-vaginal reference point dose
D 0.1 cm^3, D2 cm^3 for the bladder, rectum

Level 2

- Dose reporting for defined volumes

D98%, D90%, D50% for High-risk CTV
D98%, D90% for Intermediate-risk CTV (if used for prescription)
D98% for GTV residual
D98% for pathological lymph nodes

OARs

- Bladder reference-point dose

D0.1 cm^3, D2 cm^3 for the sigmoid
D 2 cm^3 for the bowel
Intermediate- and low-dose parameters for the bladder, rectum, sigmoid and bowel
Vaginal point doses at level of sources (lateral at 5 mm)
Lower and mid-vagina doses in relation to postero-inferior border of the pubic symphysis (PIBS, PIBS ±2 cm)

13.7.9.2 Interstitial Brachytherapy

1. Description of volumes

   GTV, CTV, PTV and the treated volume.

   Treated Volume

   The volume encompassed by an isodose surface that has been selected or specified as being appropriate to achieve the purpose of treatment. This should ideally encompass the entire CTV.

2. Description of Dose Distribution

   Prescribed Dose

   The dose which the physician intends to enter in the patient's treatment chart.

   Minimum Target Dose

   The minimum dose at the periphery of the CTV. Ideally, it should be equal to the minimum dose adequate for treatment and entirely encompass the CTV.

   Mean Central Dose

   The arithmetic mean of the local minimum doses between sources, in the central plane, or in the central planes if there are more than one.

   High-Dose Volumes

   The volume encompassed by the isodose corresponding to 150% of the mean central dose around the sources in any plane parallel to the central plane where a high-dose region is suspected. The maximum dimension of the largest region in all planes calculated should be reported.

   Low-Dose Volumes

   The volume within the clinical target volume, encompassed by an isodose corresponding to 90% of the prescribed dose. The maximum dimension of the low-dose volume in any plane calculated should be reported.

   Dose Uniformity Parameters

   Dose homogeneity index—the ratio of minimum target dose to the mean central dose.

Other definitions that are commonly in use, although not initially proposed in ICRU 58 are

1. Dose nonuniformity ratio—the ratio of the volume receiving more than 1.5 times the prescribed dose to the volume receiving the prescribed dose.
2. Conformal index (COIN)—$PTV_{PD}/V_{PTV} * PTV_{PD}/V_{PD}$.

   This takes into account the coverage of the PTV by the prescribed dose and also the unwanted irradiation of normal tissues outside the PTV. PTV_{PD} is the volume inside the PTV irradiated by the prescribed dose (PD), V_{PTV} is the volume of PTV and the V_{PD} is the volume receiving at least the prescribed dose.
3. OAR's doses—These are reported according to the site-specific recommendations.

13.7.9.3 Intraluminal Brachytherapy

1. Dose prescription and the point of prescription in terms of distance from source line and position along the source line. The dose is generally prescribed at 0.5 cm depth from the mucosal surface along the central plane or at a distance of 1–1.5 cm from source depending on the applicator circumference.
2. Minimum target dose, if different from the prescribed dose.
3. Dose at 1 cm from axis of the source line.
4. Dose at surface of applicator in contact with the tissue.

The doses are calculated and reported with respect to the central plane of the applicator which is defined as the plane perpendicular to the axis of the organ lumen, midway between the proximal and distal ends of the PTV.

13.7.9.4 Surface Mold Brachytherapy

1. Prescription dose and point of dose prescription
2. Dose at 5 mm in tissue at the center of the applicator
3. Minimum target dose
4. Description of high dose at tissue surface in contact with applicator, usually near the center of the applicator

13.7.10 Removal of Applicators

The removal of applicator may be done in the treatment room or in operating room under sedation or anesthesia. Patient should be in comfortable position and in a well-lit area. Applicators are removed gently generally in the reverse order as inserted (the tube/needle inserted first is removed at the last and the last inserted tube/needle/gauze packing is removed first). Patient should be thoroughly and carefully examined especially in the implanted area for any injury or tissue damage (e.g., vaginal tear), bleeding, infection, unusual swelling, etc. Removed applicator and its components should be carefully counted and sent for cleaning and sterilization. Single use components should be disposed as hospital policy. Patient may be shifted to appropriate resting area as per hospital SOP and then discharged from hospital accordingly.

13.8 Potential and Future Scope of Brachytherapy

The past two decades have witnessed enormous advancements in the field of brachytherapy, including the genesis of customized applicators; the incorporation of image guidance for delineation and planning; and refinements in treatment delivery and quality assurance. This has led to remarkable improvements in clinical outcomes, particularly in cervical and prostate cancers. The active areas of ongoing research include image-guided robotic-based systems with automatic needle insertions; knowledge-based planning; electromagnetic tracking for quality assurance; self-shielded applicators; intensity modulated brachytherapy; and biological optimization. Sophisticated 3D printing of applicators and mannequin will augment training and skill development in brachytherapy and possibly development of individualized novel applicators according to need and suitable anatomical site [32].

It is a promising modality for head and neck, breast, gastrointestinal malignancies and soft tissue sarcomas. However, it is not being practiced to its full potential for other sites apart from cervical cancer. The declining trend in its utilization is particularly attributable to the complexities involved and the skill and expertise required. Widespread adoption of brachytherapy across various sites with promotion of research and training may enlighten its potential role and future possibilities [33].

Precept Box (Take Home Points)

- Brachytherapy has the ubiquitous potential to deliver high dose of radiation to the target region without traversing normal tissue.
- The energy emitted from brachytherapy sources is quite low, within the keV range that yields a characteristic steep dose gradient based on the principle of inverse square law.
- The heterogeneous dose distribution with a high central dose and relatively much lesser peripheral dose enhances the therapeutic ratio for small volume tumors.
- The treatment planning and delivery needs to be highly stringent as slight inaccuracies may have considerable impact on the dosimetry.
- The suitability of a radionuclide as a brachytherapy source is determined by its underlying physical properties (energy, half-life, specific activity and physical form).
- The various brachytherapy approaches are classified based on the application technique (intracavitary, interstitial, surface mold, intraluminal and intravascular), dose rate (low, medium and high) and treatment duration (temporary, permanent).
- A comprehensive clinical examination with documentation and three-dimensional diagrammatic depiction of the findings and a thorough understanding of the imaging features is essential to determine the application technique, define the treatment volumes and understand the complexities and potential risks involved.
- The quality of application in terms of intersource spacing and placement in relation to the target volume and organs at risk is the key determinant of the quality of brachytherapy plan.
- The most widely performed brachytherapy procedure worldwide is intracavitary brachytherapy for cervical cancer. Recent evidences have proved the combination of intracavitary and interstitial brachytherapy to be particularly advantageous for tumors with large volumes, those extending beyond the medial third of parametrium, and also for small volume tumors for optimizing the OAR's doses.
- The inception of image guide adaptive brachytherapy in cervical cancer had been a major progress that has largely improved the treatment outcomes. The underlying concept is differentiating the doses to various subvolumes based on the response patterns to EBRT and potential disease burden.
- The incorporation of advanced imaging modalities (ultrasonography, MRI) has largely refined the practice of brachytherapy. They are useful at various steps including preplanning the application, real-time guidance for application, precisely defining the volumes in delineation and adaptive planning.
- The scope for further development include knowledge based planning, robotic application, intensity modulated brachytherapy, electromagnetic tracking for quality assurance and biological optimization.

Multiple Choice Questions

1. The vast majority of brachytherapy sources emit energy in the form of:
   (a) X rays
   (b) Beta rays
   (c) Gamma rays
   (d) Alpha rays
2. The dose nonuniformity ratio is defined as the ratio of the volume receiving more than __ times the prescribed dose to the volume receiving prescribed dose.
   (a) 1.5
   (b) 2
   (c) 2.5
   (d) 3
3. Mold brachytherapy is ideally suited for tumors with thickness up to
   (a) 3 mm
   (b) 5 mm
   (c) 8 mm
   (d) 1 cm
4. The ICRU 58 has defined the dose volume recommendations for all of the following techniques except
   (a) Intracavitary
   (b) Interstitial
   (c) Intraluminal
   (d) Surface mold
5. The intermediate-risk CTV in cervical cancer does not include:
   (a) Residual disease at the time of brachytherapy
   (b) Whole cervix
   (c) Disease at initial diagnosis
   (d) Whole parametrium
6. The rules for implantation and dosimetry of planar interstitial brachytherapy implants were traditionally defined in:
   (a) Paris system
   (b) Quimby system
   (c) Manchester system
   (d) Stockholm system
7. According to ICRU 58, the low-dose volume is defined as the volume of CTV encompassed by an isodose corresponding to __% of the prescribed dose:
   (a) 90
   (b) 85
   (c) 95
   (d) 80
8. The ICRU report 89 has recommended the reporting of total reference air Kerma in which level of reporting?
   (a) Level 1
   (b) Level 2
   (c) Level 3
   (d) None of the above
9. The low-dose rate brachytherapy ranges from
   (a) 0.2–4 Gy/h
   (b) 0.4–2 Gy/h
   (c) 2–4 Gy/h
   (d) 0.2–2 Gy/h
10. In interstitial brachytherapy, the mean of the local minimum doses between sources in central planes is termed as:
   (a) Mean central dose
   (b) Minimum target dose
   (c) Prescribed dose
   (d) None of the above

Answers

1. (c)
2. (a)
3. (b)
4. (a)
5. (d)
6. (a)
7. (a)
8. (a)
9. (b)
10. (a)

References

1. Dale RG, Jones B. The clinical radiobiology of brachytherapy. Br J Radiol. 1998;71(845):465–83.
2. Nag S. High dose rate brachytherapy: its clinical applications and treatment guidelines. Technol Cancer Res Treat. 2004;3(3):269–87.

3. Ferreira C, Sterling D, Reynolds M, Dusenbery K, Chen C, Alaei P. First clinical implementation of GammaTile permanent brain implants after FDA clearance. Brachytherapy. 2021;20(3):673–85.
4. Lee CD. Recent developments and best practice in brachytherapy treatment planning. Br J Radiol. 2014;87(1041):20140146.
5. Nath R, Anderson LL, Luxton G, Weaver KA, Williamson JF, Meigooni AS. Dosimetry of interstitial brachytherapy sources: recommendations of the AAPM Radiation Therapy Committee Task Group No. 43. American Association of Physicists in Medicine. Med Phys. 1995;22(2):209–34.
6. International Commission on Radiation Units and Measurements. Prescribing, recording, and reporting brachytherapy for cancer of the cervix (ICRU Report 89), Bethesda; 2016.
7. International Commission on Radiation Units and Measurements. Dose and volume specification for reporting intracavitary therapy in gynecology: (ICRU Report 38), Bethesda; 1985.
8. Viani GA, Manta GB, Stefano EJ, de Fendi LI. Brachytherapy for cervix cancer: low-dose rate or high-dose rate brachytherapy—a meta-analysis of clinical trials. J Exp Clin Cancer Res. 2009;28(1):47.
9. Guedea F, López Torrecilla J, Londres B, Ventura M, Bilbao P, Borràs J. Patterns of care for brachytherapy in Europe: updated results for Spain. Clin Transl Oncol. 2012;14(1):36–42.
10. Smith M, Todd J, Symonds R. Analgesia for pelvic brachytherapy. Br J Anaesth. 2002;88:270–6.
11. Rogowski P, Rottler M, Walter F, Saicic S, et al. Clinical outcome of combined intracavitary/interstitial brachytherapy using a hybrid applicator in locally advanced cervical cancer. Gynecol Oncol. 2022;166(3):576–81.
12. Edmundson GK, Vicini FA, Chen PY, Mitchell C, Martinez AA. Dosimetric characteristics of the MammoSite RTS, a new breast brachytherapy applicator. Int J Radiat Oncol Biol Phys. 2002;52:1132–9.
13. Yashar CM, Scanderbeg D, Kuske R, et al. Initial clinical experience with the Strut-Adjusted Volume Implant (SAVI) breast brachytherapy device for accelerated partial-breast irradiation (APBI): first 100 patients with more than 1 year of follow-up. Int J Radiat Oncol Biol Phys. 2011;80(3):765–70.
14. Smith GL, Xu Y, Buchholz TA, Smith BD, et al. Brachytherapy for accelerated partial-breast irradiation: a rapidly emerging technology in breast cancer care. J Clin Oncol. 2011;29(2):157–65.
15. Nakaji P, Smith K, Youssef E, Thomas T, et al. Resection and surgically targeted radiation therapy for the treatment of larger recurrent or newly diagnosed brain metastasis: results from a prospective trial. Cureus. 2020;12(11):e11570.
16. Welsh J, Sanan A, Gabayan AJ, Green SB, et al. GliaSite brachytherapy boost as part of initial treatment of glioblastoma multiforme: a retrospective multi-institutional pilot study. Int J Radiat Oncol Biol Phys. 2007;68(1):159–65.
17. Banerjee S, Kataria T, Gupta D, Goyal S, et al. Use of ultrasound in image-guided high-dose-rate brachytherapy: enumerations and arguments. J Contemp Brachytherapy. 2017;9(2):146–50.
18. Van Dyk S, Schneider M, Kondalsamy-Chennakesavan S, Bernshaw D, Narayan K. Ultrasound use in gynecologic brachytherapy: time to focus the beam. Brachytherapy. 2015;14(3):390–400.
19. Sharma DN, Rath GK, Thulkar S, et al. Use of transrectal ultrasound for high dose rate interstitial brachytherapy for patients of carcinoma of uterine cervix. J Gynecol Oncol. 2010;21:12–7.
20. Van Dyk S, Kondalsamy-Chennakesavan S, Schneider M, Bernshaw D, Narayan K. Comparison of measurements of the uterus and cervix obtained by magnetic resonance and transabdominal ultrasound imaging to identify the brachytherapy target in patients with cervix cancer. Int J Radiat Oncol Biol Phys. 2014;88(4):860–5.
21. Blasko JC, Mate T, Sylvester JE, Grimm PD, Cavanagh W. Brachytherapy for carcinoma of the prostate: techniques, patient selection, and clinical outcomes. Semin Radiat Oncol. 2002;12(1):81–94.
22. Strnad V, Major T, Polgar C, Lotter M, et al. ESTRO-ACROP guideline: interstitial multi-catheter breast brachytherapy as Accelerated Partial Breast Irradiation alone or as boost—GEC-ESTRO Breast Cancer Working Group practical recommendations. Radiother Oncol. 2018;128(3):411–20.
23. Viswanathan AN, Creutzberg CL, Craighead P, McCormack M, et al. International brachytherapy practice patterns: a survey of the Gynecologic Cancer Intergroup (GCIG). Int J Radiat Oncol Biol Phys. 2012;82(1):250–5.
24. Nag S, Beyer D, Friedland J, Grimm P, Nath R. American Brachytherapy Society (ABS) recommendations for transperineal permanent brachytherapy of prostate cancer. Int J Radiat Oncol Biol Phys. 1999;44(4):789–99.
25. Tanderup K, Viswanathan AN, Kirisits C, Frank SJ. Magnetic resonance image guided brachytherapy. Semin Radiat Oncol. 2014;24(3):181–91.
26. Viswanathan AN, Szymonifka J, Tempany-Afdhal CM, O'Farrell DA, Cormack RA. A prospective trial of real-time magnetic resonance-guided catheter placement in interstitial gynecologic brachytherapy. Brachytherapy. 2013;12(3):240–7.
27. Blanchard P, Ménard C, Frank SJ. Clinical use of magnetic resonance imaging across the prostate brachytherapy workflow. Brachytherapy. 2017;16(4):734–42.
28. Salem A, Salem AF, Al-Ibraheem A, Lataifeh I, Almousa A, Jaradat I. Evidence for the use PET for radiation therapy planning in patients with cervical cancer: a systematic review. Hematol Oncol Stem Cell Ther. 2011;4(4):173–81.
29. Schindel J, Zhang W, Bhatia SK, Sun W, Kim Y. Dosimetric impacts of applicator displacements and applicator reconstruction-uncertainties on 3D image-guided brachytherapy for cervical cancer. J Contemp Brachytherapy. 2013;5(4):250–7.

30. Perez-Calatayud J, Ballester F, Das RK, Dewerd LA, et al. Dose calculation for photon-emitting brachytherapy sources with average energy higher than 50 keV: report of the AAPM and ESTRO. Med Phys. 2012;39(5):2904–29.
31. International Commission on Radiation Units and Measurements. Dose and volume specification for reporting interstitial therapy (ICRU 58), Bethesda; 1997.
32. Mourya A, Aggarwal LM, Choudhary S. Evolution of brachytherapy applicators for the treatment of cervical cancer. J Med Phys. 2021;46(4):231–43.
33. Cunha JAM, Flynn R, Bélanger C, Callaghan C, et al. Brachytherapy future directions. Semin Radiat Oncol. 2020;30(1):94–106.

Further Reading

https://acsjournals.onlinelibrary.wiley.com/doi/full/10.3322/caac.21578
https://link.springer.com/chapter/10.1007/978-3-319-44827-5_2
https://pubmed.ncbi.nlm.nih.gov/33299441/
https://pubmed.ncbi.nlm.nih.gov/696400/
https://www.thegreenjournal.com/article/S0167-8140(21)06248-4/fulltext

Electron Beams

14

Tharmarnadar Ganesh
and Prabhakar Ramachandran

14.1 Introduction

Clinical electron beams have a sharp dose fall-off beyond a specific depth and definite range that is unique to each beam energy, making them a preferred modality for treating superficial lesions with sensitive structures beneath. In modern radiation oncology, about 3–10% of the patients would require electron beam therapy.

Although their depth–dose characteristics make them an attractive choice in treating superficial lesions, there are several factors that add to the complexity in the clinical use of electron beams. Everyone has to understand their limitations and be aware of potential situations where their characteristics may change drastically (Table 14.1).

14.2 Production of Electron Beams

In addition to photon beams, conventional C-arm type medical linear accelerators (linac) can generate multiple energy electron beams ranging from 4 to 20 MeV. Although the buyer has the option to choose up to eight electron beams and their energies, limiting it to five beams with energies of 6, 9, 12, 15, and 18 MeV is considered most optimal, which can cater to all clinical needs.

Electrons are produced in the electron gun by thermionic emission and are injected into the accelerating structure of the linac. All the electrons are accelerated uniformly as they move from one end to the other end of the structure to the desired energy level (say, 6 or 18 MeV) by microwaves. The electron beam that exits the accelerating structure window is a monoenergetic pencil beam of about 3 mm in diameter, which should not be confused with the clinical electron beam that enters the patient. This pencil beam then passes through a bending magnet, which bends it by nearly 270°. In the photon mode, this beam is made to strike a target, while in the electron mode, it is made to strike a suitably designed dual-scattering foil that scatters the 3 mm diameter beam into a broad beam up to 25 cm × 25 cm field size [1]. It is pertinent here to highlight the important difference between the beam currents in the linac's photon mode and electron mode operations. Typically, in the electron mode, the beam current is of the order of 0.1 μA, while in the photon (x-ray) mode it is about 1000-times higher at around 100 μA [1]. The reasons are the extremely poor conversion efficiency associated with the production of

T. Ganesh (✉)
Manipal Hospital, New Delhi, Delhi, India

P. Ramachandran
Medical Physics, Radiation Oncology, Cancer Services, Princess Alexandra Hospital, Woolloongabba, QLD, Australia

A. Munshi et al. (eds.), *Radiation Oncology – Principles, Precepts and Practice*,
https://doi.org/10.1007/978-981-97-8389-2_14

Table 14.1 Advantages and limitations of clinical electron beams

Advantages	Disadvantages
Very useful for skin and superficial cancers. High skin doses associated with electron beams can be advantageous in some situations	Lack of sharpness at beam edges as compared to photons
Multiple energies are available in regular linacs to meet different clinical requirements. There is zero dose beyond the practical range of electrons	Dosimetry issues can be challenging
In the cases where treatment by electron beam is indicated, it is often irreplaceable with any other noninvasive technique	In a clinic, typically <10% of the cases are only suitable for treatment by electron beams
	Treating with very small fields is a challenge
	Dose calculations by treatment planning system may not be accurate, especially in the presence of inhomogeneities

x-rays at the target as well as the loss of x-ray beam intensity in the flattening filter. Additionally, photon beams are poor energy depositors in tissue since nearly 30–60% of the photon beam passes through a patient that is 30 cm thick, and whatever energy that is deposited by the photon beam is spread throughout the patient. The electron beams do not have to suffer either conversion inefficiency or attenuation in the flattening filter. Besides these, electrons are efficient energy depositors, depositing all of their energy within the first few centimeters (cm) of entering the patient's skin surface [2]. Because of these reasons, the linac has to operate at a higher electron beam current in the photon mode as compared to that in the electron mode.

In the early linacs of the 1980s, only a single scattering foil [3, 4], made up of thick, high atomic number material, was employed to broaden the pencil beam. The design called for solid-walled electron applicators to generate additional scatter in order to improve field homogeneity. The use of thicker, higher-atomic number material as the scatterer resulted in increased bremsstrahlung contamination of the beam.

At around the same time, scanning electron beam systems were also in use. They used magnetic fields to sweep the small, pencil-shaped electron beam across the desired field area (just like painting a wall uniformly with brush strokes). Since the electron beam did not have to traverse any scattering medium, the resultant clinical beam had minimal bremsstrahlung contamination [4].

While scanning systems soon disappeared due to a few mishaps that occurred in clinics, single-scattering foil systems gave way to dual-scattering foil systems [1, 4, 5]. Such a system consisted of two foils separated by a few centimeters. The first foil, made of high-atomic number material, widens the beam through multiple scattering with minimal energy loss, while the second foil, thicker in its axis, flattens the beam by preferentially scattering the electrons to the periphery. Its central, thicker portion is in the form of a high-atomic number "button" on a low-atomic number foil. Since the scattering is more effective, the electron applicators are open-walled.

The foil thicknesses are chosen as a compromise between being thick enough to produce acceptable levels of electron scattering while being thin enough to ensure that the energy loss and the additional bremsstrahlung contamination of the beam are not too high. As this balance of electron interactions is energy-dependent, each electron energy requires its own foil, although, in practice, this is not always the case.

14.3 Interaction of Electrons with Matter

Electrons are charged particles with Coulomb electric force around them. When they move through the medium, they interact by either elastic interactions in which no energy is lost or inelastic interactions in which some energy is lost. Elastic collisions result in simple scattering of the electrons (a change in the direction of the incident electron) with no loss of energy. In inelastic collisions, electrons interact with either orbital electrons, resulting in collision loss, or with atomic nuclei, leading to radiative loss of energy (bremsstrahlung). The loss of energy through collision interactions results in ionizations and excitations of the atoms in the medium. Each electron particle will have to undergo a large number of interactions following a highly zigzag path, before losing all of its kinetic energy [6].

The rate of energy loss by the electrons in the absorbing medium is described as the stopping power of the medium: collision (ionization) stopping power for collision interactions and radiative stopping power for radiative interactions. The stopping powers of electrons are dependent on various factors: the velocity and energy of the electrons, besides the electron density and the atomic number of the medium through which they pass.

Since collision loss is through electron-electron interaction, and since low-atomic number materials have more electrons per gram, collision interactions will predominate in low-atomic number materials. Even in high-atomic number materials, in the clinically relevant electron energies, they will still predominate over radiative interactions, although their magnitude will be less compared to that in low-atomic number materials.

14.4 Characteristics of Electron Beams

The most important feature that makes the use of an electron beam clinically attractive is its sharp fall-off in dose as it passes through a medium (Fig. 14.1). This is primarily due to the loss of content energy per unit length as it passes through the medium. The electron depth–dose curve is characterized by several parameters: the surface dose (D_s) is the dose at 0.5 mm depth, D_m is the dose maximum along the central axis, D_x is the dose due to x-ray contamination (bremsstrahlung tail), R_{100} is the depth of dose maximum, R_t is the therapeutic range (usually taken as 90%, 85%, or 80%), R_{50} is the depth of 50% of dose maximum, and R_p is the practical range which is given as the nominal energy in MeV divided by two in centimeters; thus, the practical range R_p of a 10 MeV electron beam would be 5 cm. G_o is the normalized dose gradient that characterizes the dose falloff.

Unlike photon beams, the surface dose for electron beams increases with increasing energy.

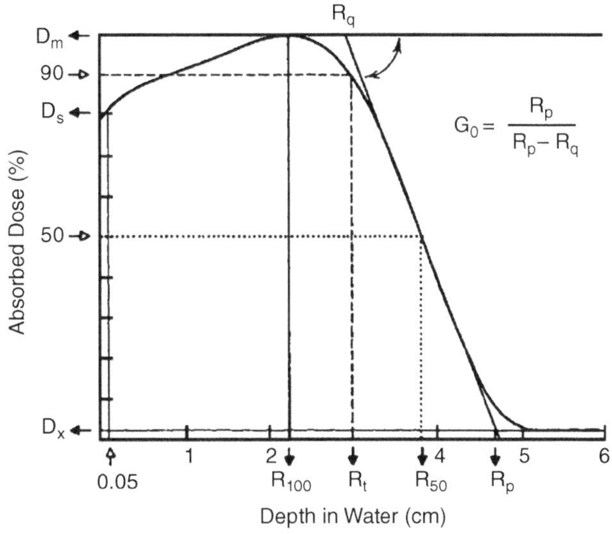

Fig. 14.1 Typical central axis depth–dose curve for an electron beam showing the different parameters that characterize the curve. (Ref [7])

The dose absorbed at the surface of the water medium is somewhere around 78% for a 6 MeV beam and can go up to 95% for high-energy electron beams (Table 14.2).

This is explained as follows: (i) when electrons enter the surface, they travel in the direction of the beam only for the first few mm; (ii) as they interact with the medium, they undergo multiple scattering and their paths become oblique to the beam direction; and (iii) because they are travelling obliquely, their actual path length increases compared to the path length measured along the beam direction (similar to a person taking a detour when moving from one point to another and ends up covering more distance instead of walking straight), (iv) low-energy electron beams experience more scattering than high-energy beams; (v) this results in more pronounced dose deposition for low-energy beams at D_m, i.e., more electrons deposit dose at D_m, resulting in a relatively lower normalized surface dose; (vi) the effect is less for high-energy electron beams because they experience comparatively less scattering and only fewer electrons deposit dose at D_m, and the normalized surface dose is higher as a result.

The shape of the central axis depth-dose (CADD) curves and the associated parameters change as the electron beam energy changes (Fig. 14.2). The D_m increases initially with increasing energy, and after 15 MeV or so, it decreases with a further increase in energy. Higher energies display a broad dose maximum, and it becomes difficult to identify a single point as D_m. The bremsstrahlung tail ranges from less than 1% for low energies to about 5% for high-energy beams. The depths of D_{90}, D_{50}, and R_p continually increase with increasing energy. Thumb rules are often used in electron beam therapy. The depths (in cm) of 80% and 90% of the dose maximum are approximately equal to 1/2.8 and 1/3.2 of the electron energy expressed in MeV [8]. The most useful depth for dose prescription, defined as the therapeutic range (R_t), is typically defined at 90% isodose line depth. For a given energy, these depths and the dose at the surface vary between machines, even if they are from the same vendor and are of the same model. The energy of electron beams at any given depth z (cm) can be estimated [9] using the formula:

$$E_z\left(\mathrm{MeV}\right) = 1 - \frac{z}{R_p}$$

The electrons lose energy at the rate of 2 MeV/cm in tissue, and the dose beyond the practical range (R_p) is contributed only by the x-ray contamination. The relationship $E_0 = 2.33 \times R_{50}$ can be used to approximate the mean energy of electron beams at the surface.

Isodose curves are lines connecting all points that receive the same dose. With the dose at the depth of dose maximum normalized to 100%, isodose curves are drawn for various percentage levels, usually in steps of 10% downward from 100% (Fig. 14.3). Electron isodose curves are markedly different from those of photon beams in many ways. Electrons undergo complex scattering, and the beam expands below the surface. The shape of the curves is dependent on many factors, such as energy, field size, collimation, and the isodose level. In general, for low-energy beams, the curves bulge out, while for high-energy beams, the curves either constrict or bulge out depending on the isodose level.

Table 14.2 Central axis depth–dose characteristics of electron beams for various energies and field sizes (*Data from a Varian TrueBeam machine*)

Energy	Surface dose (%)	Depth of dose maximum (d_m) (mm)			Depth of 90% dose (mm)			Depth of 80% dose (mm)			Depth of 50% dose (mm)			Depth of 10% dose (mm)			Bremsstrahlung tail (%)		
Applicator (*cm × cm*)		6 × 6	10 × 10	15 × 15	6 × 6	10 × 10	15 × 15	6 × 6	10 × 10	15 × 15	6 × 6	10 × 10	15 × 15	6 × 6	10 × 10	15 × 15	6 × 6	10 × 10	15 × 15
6 MeV	78.4	12.5	12.0	12.5	17.3	17.2	17.3	19.3	19.2	19.3	23.3	23.3	23.5	29.5	29.5	29.6	0.8	0.8	0.8
9 MeV	81.5	19.5	20.5	20.0	27.2	27.1	27.2	29.8	29.8	29.9	33.5	35.4	35.5	43.4	43.4	43.4	1.2	1.3	1.2
12 MeV	89.3	27.0	26.0	27.0	38.0	37.8	38.5	41.9	41.5	42.2	49.4	49.2	49.5	59.9	59.7	60.0	2.0	2.0	2.1
15 MeV	90.7	29.5	31.5	32.0	46.4	48.1	48.1	51.8	53.0	53.0	62.0	62.4	62.6	75.3	75.5	75.6	3.1	3.2	3.3
18 MeV	93.3	26.5	25.0	27.5	51.9	55.6	56.0	59.5	62.6	63.0	73.3	74.7	74.8	91.0	91.2	91.4	4.5	4.6	4.7

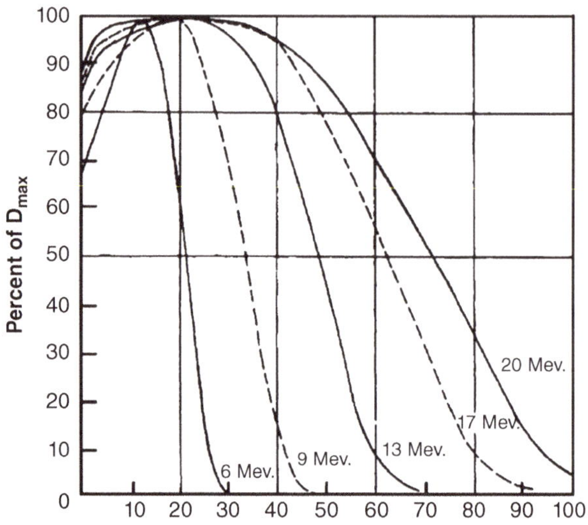

Fig. 14.2 Central axis depth–dose curves for electron beams of different energies. (Ref [7])

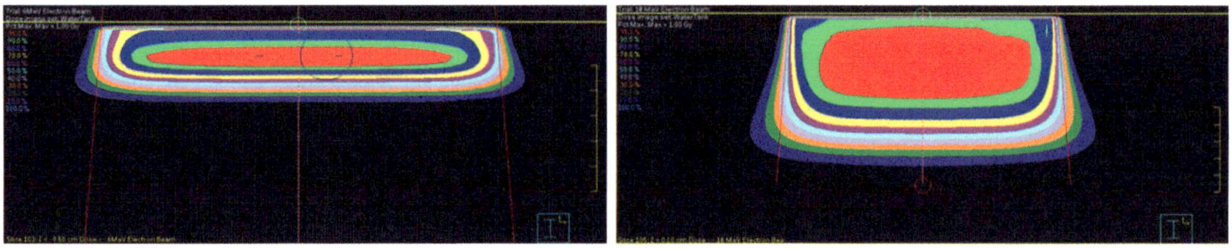

Fig. 14.3 Isodose curves for a 6 MeV beam (left) and an 18 MeV beam (right), both of field size 14 cm × 14 cm

14.4.1 Effects of Field Size

As the Table 14.2 shows, increasing the field size has practically no or minimal effect on the depth–dose characteristics of electron beams of all energies. However, decreasing the field size can affect the depth dose characteristics significantly, depending on the energy, as shown in the Fig. 14.4. When the field size is smaller than the practical range of the electron beam's energy, the following changes can occur: (i) the depth at which the dose maximum occurs will shift toward the surface; (ii) the depths of D_{90}, D_{80}, D_{50} will also shift toward the surface; (iii) the output at the dose maximum will change. The changes are more pronounced for higher energy electron beams.

When using a combination of a small field size and a high energy electron beam, the depth–dose characteristics of the beam should be measured to ensure that the intended dose prescriptions would be achieved. The practical range remains unaffected due to changes in field size [7, 10].

14.4.2 Effects of SSD

Unlike photon beams, changes in SSD do not affect the depth–dose curves of electron beams [11]. When the SSD is varied between 100 and 115 cm, the changes observed in depth are less than a few mm. However, the shapes of the isodose curves change dramatically at extended SSDs. There is a constriction of the higher isodose curves (like 90%) and widening of the lower isodose curves beyond the field edges, resulting in significantly increased penumbral width (Fig. 14.5). When treating patients at SSDs higher than the nominal SSD, these facts must be considered in the design of portals to ensure proper coverage of target volumes and avoid dose to nearby structures.

Besides the change in the shape of the isodose curves, an increase in SSD would decrease the output in cGy/MU. With photons, this change in output with SSD is estimated accurately using the inverse square law (ISL). But, since the electron beams do not appear to originate from a

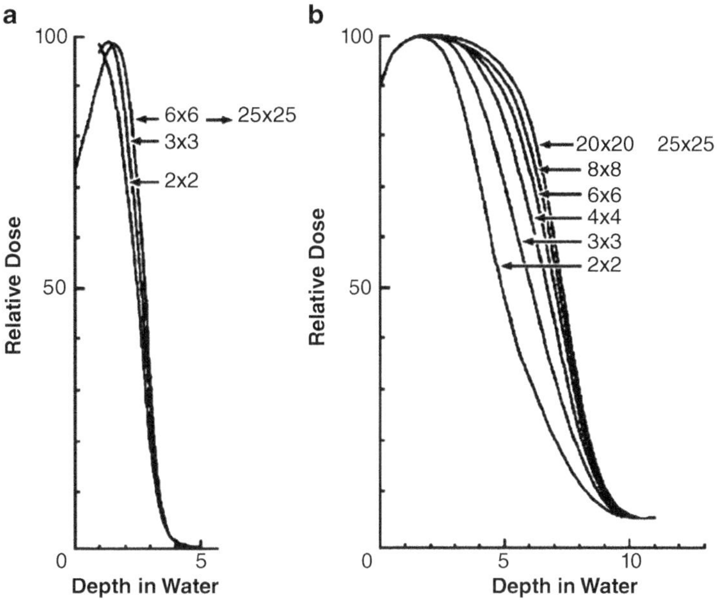

Fig. 14.4 Changes in depth–dose curves for different field sizes for (**a**) 7 MeV and (**b**) 18 MeV electron beams, 100 cm SSD

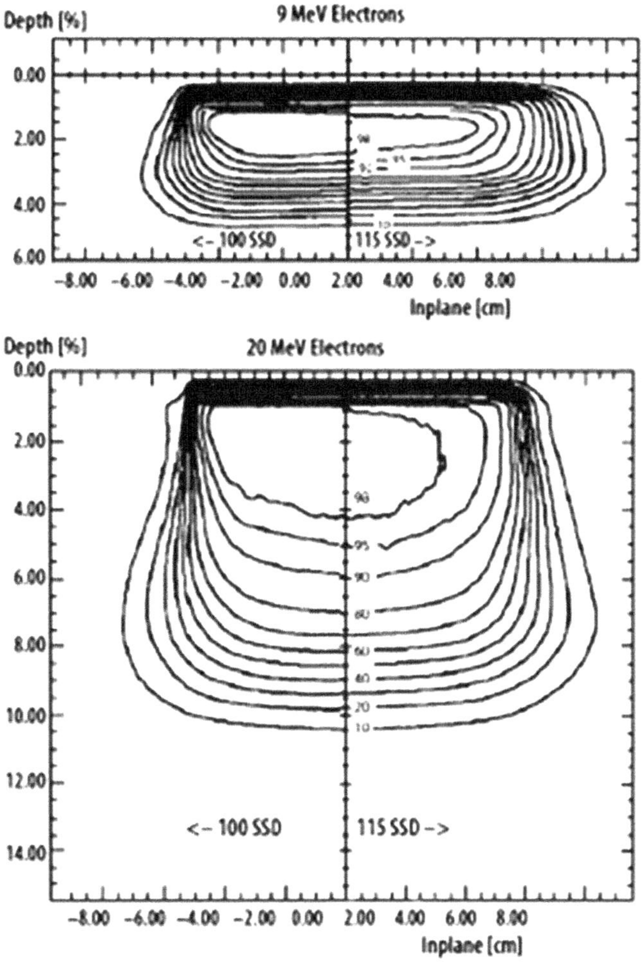

Fig. 14.5 Changes in shapes of the isodose curves as the SSD is varied from 100 cm (on the left part of the figure) to 115 cm (on the right part of the figure) for 9 and 20 MeV electron beams. (Ref [7])

single "source," simple ISL cannot be employed to find the output. Methods like virtual SSD and effective SSD [6, 9] have been successfully used to account for increased SSD. These are tedious methods since they are to be determined for every energy-applicator combination involving many measurements. Instead, whenever there is a need to treat a patient under nonstandard conditions, the best alternate way is to go to the machine and make relative measurements of the output at the new SSD with respect to the output at the nominal SSD for the reference open field size of 10 cm × 10 cm in a solid water phantom and do appropriate calculations.

14.5 Clinical Usage

Electron beams are employed at several clinical sites for treating superficial lesions (Table 14.3). Every time they are used, it is invariably as a single direct field, except in total skin electron therapy. When the incident surface is flat and the treated volume is soft tissue without any inhomogeneities, then it is fairly easy to estimate the dose at the desired depth through the interpretation of the CADD and isodose curves. However, in practical scenarios, this is rarely the case, and often beams are incident in an oblique direction, on an irregular surface, and with inhomogeneities present in the path of the beam. Each one of these has the potential to alter the dose significantly if not accounted for.

14.5.1 Selection of Energy, Field Size, and Prescription of Dose

The beam direction is selected in such a way that the beam is incident normally on a flat surface. It is always desirable to take the patient on the treatment table and determine the beam angle with the applicator in place. Subsequently, a CT-simulation is done with the patient lying in the treatment position. The CT images are transferred to a treatment planning system where a 3-D model of the patient is built using these images. The treatment depth, including the set-up margin and safety margin, is decided along the chosen beam direction on the patient's 3-D model. The choice of energy is based on this treatment depth and the departmental policy. One can select the energy in such a way that the 90% isodose line lies at the treatment depth and the dose is prescribed at the same 90% isodose level. In this way, the entire target volume would receive the full prescription dose, but at the same time, some part of the target volume would receive 111% of the prescription dose. The alternative is to prescribe the dose at 100% isodose level while keeping the treatment depth at the 90% isodose level. In this way, no part of the target volume would receive any dose higher than 100%, but the target would be encompassed by only 90% of the prescription dose. The choice of prescribing at either 100% or 90% isodose level, and deciding which isodose level should cover the target volume are based on the departmental policy. Readily available beam data in the form of a table (Table 14.2) would help in choosing the appropriate energy.

Table 14.3 Clinical sites for electron treatment

Skin cancers (SCC, BCC)[12]	Radical treatment or postoperative (in indicated cases)
Keloids[13]	Postoperative
Breast[14]	As boost after whole-breast radiotherapy
TSET[15]	Cutaneous lymphomas
Intraoperative electrons [16]	Used for sites such as breast (as APBI or boost), pancreas, retroperitoneal soft tissue sarcoma or sites such as pelvis (for treating small residual disease)
Intraoperative electrons (FLASH) [17]	Upcoming for breast as APBI/boost

SCC squamous cell carcinoma, *BCC* basal cell carcinoma, *TSET* total skin electron therapy, *APBI* accelerated partial breast irradiation

If any critical structures lie below the treatment depth, then all possible precautions should be taken to keep the dose less than the tolerance dose through careful calculations. Similarly, if there are inhomogeneities lying above the target volume in the path of the beam, then their effects must be carefully considered before choosing the energy

Field size is chosen after carefully considering the constriction in the widths of 90–80% isodose lines and also the bulging of the lower isodose lines outside the field edge. As a thumb rule, the field edge shall be at least 1 cm away from the edge of the target volume, inclusive of set-up and safety margins [7].

14.5.2 Beam Obliquity

Electron beam depth–dose curves and isodose curves are valid only when the beam is incident normally on a flat surface. When the beam is incident obliquely, as is common in many clinical situations, these fundamental properties of the beam are altered, often significantly. When the beam is incident at an oblique angle to the surface, the depth–dose curve is shifted toward the surface, thus reducing the depth coverage (Fig. 14.6).

In other words, for an obliquely incident beam, the 90% dose occurs at shallower depths and thus compromises the dose coverage of the target volume. As the angle of obliquity increases, especially beyond 30°, the shift toward the surface is greater. Additionally, the D_m can increase by as much as 20% compared to the dose at the d_m for a normally incident beam. Electron beams with lower energies are more affected. In real situations, the problem is even more complex due to the large air gaps caused by the sloping surfaces and the resultant reduction in output due to the inverse square law. It is for these reasons the beam angle should be optimally selected to

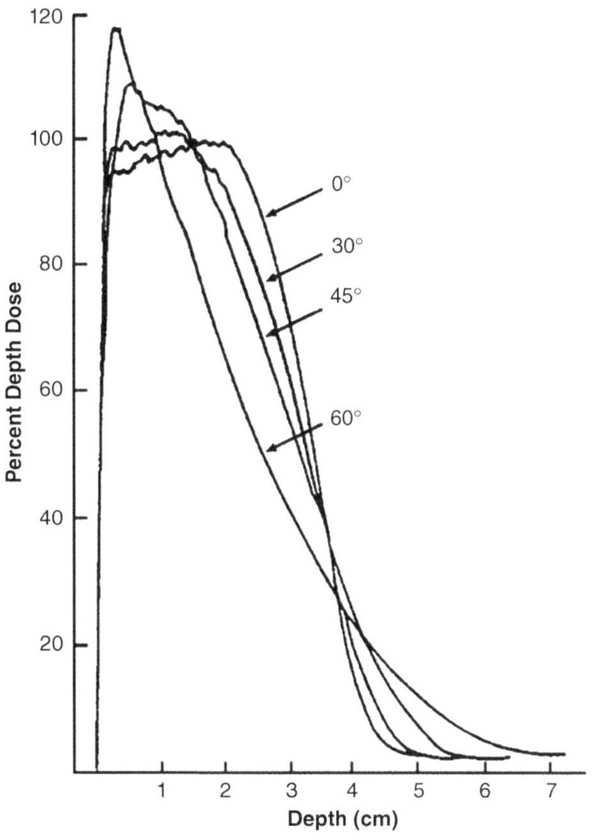

Fig. 14.6 Changes in central axis depth–dose curve for a 9 MeV electron beam incident on water phantom at different angles. (Ref [7])

reduce obliquity and avoid associated dose uncertainties.

14.5.3 Irregular Surfaces

In many clinical situations, like in the treatment of lesions in the nose, eye, ear, and areas of surgical excision, the electron beam has to fall on highly irregular surfaces. Such irregular surfaces alter the shape of the electron beam's isodose curves in a significant way. When the irregularity is in the shape of a sharp outwardly projection, the electrons are scattered away from it resulting in dose hot spots just adjacent to it on either side. Surface irregularity in the form of a sharp depression would cause the electrons to scatter inward, thus causing hot spots just below the depression. The magnitude of these hot spots can be as high as 30%. The use of tissue equivalent bolus material to smoothen such irregularities can reduce the size and magnitude of these hot spots.

14.5.4 Tissue Inhomogeneities

Tissue homogeneities, like with the photon beams, pose a major hurdle with the electron beams as well in accurately estimating the dose within the inhomogeneities, at the interface boundaries, and beyond that. Inhomogeneities in the form of air cavities, lungs, and bones cause perturbations in the electron beam dose distributions. While it may be difficult to determine the dose distributions within and in the vicinity of small inhomogeneities, for large inhomogeneities, the doses beyond the inhomogeneities may be estimated by simple correction methods.

The simplest method to correct for the presence of inhomogeneity, which can be manually performed, is the coefficient of equivalent thickness (CET) method [6, 8, 18, 19] which scales the thickness of the inhomogeneity by its density relative to water and thus determines its CET. The CET of any material is simply its electron density relative to the electron density of water. In effect, it is the mass density of the inhomogeneity. The density of the lung is 0.25 g/cm^3, and when it is expressed relative to water, which has a density of 1 g/cm^3, the CET is 0.25. Thus, if a beam passes through 10 cm of lung tissue, it is equivalent to a beam passing through 2.5 cm of tissue. For solid bone, the CET is about 1.5–1.8 whereas for spongy bone, the value is around 1.1 [19]. The effective water equivalent depth concept is used to calculate the dose at any point lying beyond an inhomogeneity. If the point is at a depth of d cm from the surface and the thickness of the inhomogeneity is z cm, then the point may be considered as lying at an effective water equivalent depth of $d_{\text{eff}} = d - z(1 - \text{CET})$, where d_{eff} is the effective water equivalent depth and CET is the coefficient of equivalent thickness for that inhomogeneity. Once d_{eff} is known, dose can be inferred from the dose distribution data measured in water.

Problem A point lying at a depth of 3 cm from the surface has to be treated by electrons. Of the total depth 3 cm, there are two inhomogeneities in the form of 0.5 cm of compact bone and 0.5 cm of spongy bone. What should be the right choice of energy to adequately cover this depth?

Depth of the point = 3 cm
Of this, 0.5 cm is compact bone (CET = 1.65) and 0.5 cm spongy bone (CET = 1.1)
Water equivalent depth = d_{eff} = 3 cm − 0.5 cm * (1–1.65) − 0.5 cm * (1–1.1)= 3 cm + 0.33 cm + 0.05 cm = 3.38 cm

From Table 14.2, the choice of energy should be 12 MeV with the *dose prescribed at d_m (2.7 cm)*. With 90% depth dose occurring at 3.8 cm, the point would be covered by 95% dose. It is cautioned here that the CET method is an oversimplification of the effect of inhomogeneity, and the bones may not have the same density uniformly throughout their dimensions. The dose distribution in the real clinical scenario may depend on many other factors, like the size and thickness of the inhomogeneity and the dose perturbation due to altered scatter conditions. In the soft tissues lying directly under a sufficiently wide bone, the dose would be reduced due to the increased attenuation within the bone, thus shifting the isodose curves toward the surface. The doses at the edges are also altered—outside of the edges of the bone, the dose would be slightly increased by about 5%, and inside, at the edges, the dose would be decreased by the same amount due to increased scatter outward of the heterogeneity. If the inhomogeneity is lung tissue or air cavities, the dose distribution beyond the region of inhomogeneity will be drastically changed, increasing the dose to underlying structures up to several centimeters. Failure to take into account such changes in dose would result in high doses to these structures. Air cavities adjacent to a block of soft tissue being treated can also alter the dose distribution. Electrons tend to scatter more into low-density regions like the lungs and air cavities. But electrons thus scattered from tissues into adjacent air cavities are not replaced by electrons scattered from the air back into the tissues being treated. This results in lower doses to target volume near such edges.

In light of all these facts, it is always advisable to assess the desired dose distribution information through treatment planning with the patient's CT data in a treatment planning system having an accurate electron dose calculation algorithm.

14.5.5 Bolus and Its Use in Electron Beams

In treatment with electron beams, the bolus is used for (1) making an irregular surface flat, (2) lifting the isodose lines up toward the surface for protecting the structures underlying the target volume from receiving high doses, and (3) to increase the surface dose when required [6, 7].

When the electron beam is incident on an irregular surface, isodose lines below the surface will more or less follow the external contour, thus making the dose within the target volume nonuniform. The irregular surface can be made flat by the use of a suitably designed bolus, improving the dose uniformity within the target volume.

In certain clinical situations, a need may arise to protect the sensitive structures lying directly beneath the target volume from receiving significant doses. Treatment of intramammary lymph nodes along with chest wall irradiation combined with a need to protect the underlying lung tissues is one such situation. A single field wide-enough to cover both the intramammary nodes and the chest wall is preferred, but the energy in that case would be 12 MeV to adequately cover the nodal region. This would unacceptably irradiate underlying lung tissues in the chest wall region to significant doses. A suitably designed bolus placed over only the chest wall region would lift the higher isodose curves toward the surface, protecting the lung tissues while adequately covering the nodal region.

The surface dose in low-energy electron beams is typically around 70–80% and would thus underdose this region if the lesion involves the skin. A few millimeters thick tissue-equivalent bolus would help to increase the surface dose to the desired value.

The ideal bolus material should have similar stopping power and scattering power as that of the soft tissue [6]. Any bolus material before its clinical use should be checked for these properties by comparing the dose distribution in a bolus with that in a water medium. Readily available thin wax sheets, honey wax, paraffin wax, and gauze pieces soaked in water are some of the economically cheaper examples. There are several commercially available bolus materials like Lucite, Superflab, Superflex, and many more [20]. There are a few thumb rules to be followed while using the bolus: the minimum width of the bolus shall be 2 cm, and the bolus shall not abruptly end within the treatment field, which will cause dose perturbation due to scattering. For boluses placed entirely within the field, sharp edges should be avoided. Instead, whenever possible, the bolus should be extended beyond the field edge(s). Care should be taken to place the bolus in close contact with the skin surface. If there are large air gaps, the dose will be perturbed due to electron scattering outside. Modern treatment planning systems allow the "custom digital design" of the boluses of any desired thickness, shape, and density. While this may be desirable in many situations, one should keep in mind that such digital boluses are ideal boluses, and it will be impossible to fabricate such perfect boluses in the real world. Instead, the patient should be CT-simulated with a fabricated bolus in place, and treatment planning should be done on these CT images. While using bolus, the nominal SSD should be set to the surface of the bolus and not at the skin level below the bolus. In this way inadvertent errors in MU calculation can be avoided.

14.5.6 Adjacent Fields

In the clinical use of electron beams, the use of adjacent fields must be avoided. If unavoidable situations arise, then adequate care must be taken to ensure there are no unacceptable high doses to any structure as a result of field overlapping, no critical organs are in the vicinity of the field junction, and beams are parallel to each other along the edges. Dose distributions for various field separations shall be assessed across the entire target volume before a decision is taken in this regard [6, 8, 19].

The usual practice is to abut the fields at the surface since electron fields are used to treat only superficial depths and hot spots can largely be accepted at these depths. If possible, junction feathering can help reduce the magnitude of the hot spots.

If an electron field is to be abutted with a photon field on the surface, the hot spot will occur on the photon field side at a depth where the bulging of the electron field occurs [21]. The situation becomes worse when the SSD for the electron field is higher than the nominal SSD. At the extended SSD, there will be an increased scattering of electrons due to the large air gap between the applicator and the patient surface, leading to larger hot and cold spots.

14.5.7 Combination of Electron and Photon Beams

High-energy photon beams spare the skin, while electron beams have a sharp dose fall-off. In the treatment of some selected sites like internal mammary nodes, a combination of photon and electron beams offers the benefits of both worlds. The energies and the weights of the respective beams are selected upon careful consideration of the isodose curves for the combined fields. It is customary to combine the beams in the ratio of 3:1, favoring the electron beam.

14.5.8 External Electron Shielding Using Lead and Cerrobend

Electron beams are shaped by open-ended applicators, often called electron applicators or electron cones. Most applicators are square, but rectangular and circular applicators are also used. The field size ranges from 4 cm × 4 cm to 25 cm × 25 cm. However, the area to be irradiated is rarely regular. These applicator-defined regular fields can be shaped by either lead or Cerrobend metal to the desired size and shape. The use of lead for shielding has been largely discontinued due to lead poisoning. Instead, Cerrobend, a mixture of bismuth (50%), lead (26.7%), tin (13.3%), and cadmium (10%), is used. The advantage of Cerrobend metal is its low melting temperature of 70 °C as against the lead's 327.5 °C. The current trend is to replace Cerrobend blocks with 3D-printed tungsten carbide for electron shielding [22].

Two different methods are followed to shape the fields: (1) use of simple lead sheets, cut into various shapes, by placing them either directly on the skin to shield the desired area, or by placing them at the applicator's end and using the shadow cast by these sheets on the light field projection on the patient's skin; (2) use of custom-fabricated shields and inserting them as an applicator insert at its end.

The lead thickness required, in millimeters (mm), to adequately shield the electron beam is calculated as the nominal energy of the electron beam in MeV divided by two. If a low-melting point alloy like Cerrobend is used, then the practice is to multiply the thickness calculated for lead by 1.2 [9] and add another 1 mm for additional safety to calculate the total thickness. It must be ensured that the fabricated shielding has a minimum thickness calculated as above since less than adequate thickness can possibly result in a higher dose than the dose in open field. For example, while using lead shield for a 12 MeV electron beam, its thickness shall be at least 6 mm and if the shield is made of Cerrobend metal, it shall be at least 7.2 mm (20% more than the lead thickness) plus 1 mm which is 8.2 mm.

Whenever the electron field-shaping shielding blocks are placed away from the skin, it results in a widened penumbra and compromised coverage of the target volume. Hence, the best practice is to place them directly on the skin to keep the penumbra smaller. However, this may not be possible in all the scenarios due to various reasons like the patient's position, the patient's comfort, the weight of the shields, etc. Linac manufacturers provide special kits for fabricating custom-made shields using Cerrobend low-melting point alloy that can be attached to the electron applicators as "applicator inserts."

The desire to shield normal structures should not lead to situations where the depth–dose characteristics of the electron beam and the point of dose specification are significantly altered. The output and depth–dose characteristics defined for any given electron beam are only applicable for field sizes that satisfy the lateral scatter equilibrium condition. When the field size is smaller than this, electron beams show a marked reduction in dose at depth (see Fig. 14.4). Such changes can be very rapid for low-energy beams with small beam diameters and can result in severe underdosing or overdosing depending on the situation. This lateral scatter equilibrium is established for field sizes greater than the equivalent radius defined by the relation the $R_{eq} = 0.88\ (E)^{1/2}$ [6]. The output needs to be corrected if the treatment field size is smaller than R_{eq}. In such situations, it is desirable to measure the depth dose, the output factor, and the isodose distribution to avoid underdosing of the target volume. Every possible care must be taken when smaller field sizes and shields (cutouts) are employed in clinics.

14.5.9 Internal Shielding

In many clinical situations, it may become necessary to use an internal shield to protect underlying deep structures while treating the target volume above. These situations pose special challenges, as the direct use of shielding without proper precautions can lead to possible overdosing of tissues lying upstream near the shield. In the treatment of eyelid lesions, internal shielding is used to protect the eye lens and eye. However, at the tissue–metal interface, the backscatter from the metal can increase the dose to the tissue significantly [23].

This dose increase is due to electron backscatter at the tissue–metal interface upstream and is dependent on the beam energy at the interface; it can be as high as 30–70% in the electron energy range of 1–20 MeV [19]. The electron backscatter factor (EBF) is given by an empirical formula derived from experimental data,

$$\text{EBF} = 1 + 0.735e^{-0.052\bar{E}_d}$$

where $\bar{E}_d$ is the average energy of the electrons incident on the interface.

The EBF is the ratio of the dose at the interface in the presence of lead to the dose at the same point in a homogeneous tissue-equivalent medium.

There are a few important points to be noted regarding electron backscatter:

1. In the electron energy range of 1–25 MeV, the distance (range) to which backscattered electrons can travel is about 1–2 g/cm^2 of polystyrene.
2. As one moves away from the interface upstream, i.e., toward the source, the dose falls off exponentially, which means that the first few mm of the target volume in the direction of the backscatter that is in contact with the tissue–metal interface is the region at high risk.

The problem of enhanced dose due to backscattered electrons can be effectively addressed by adding a few millimeters of low-atomic-number material like wax or aluminum between the metal and the tissue upstream. Commercially available eye shields, which can be used as internal shields, are made of tungsten and coated with dental acrylic to reduce backscatter. These eye shields are designed for energies up to 9 MeV only and shall not be used with high-energy beams [24].

14.5.10 Electron Arc Therapy

Electron arc therapy is a highly specialized technique for treating superficial structures with rotational electron arc fields. The technique is well-suited for treating lesions in the skull or in any other region of the body where it is cylindrical in shape, like in the case of the postmastectomy chest wall, abdomen, etc.

Figure 14.7 shows a schematic setup for an electron arc field [25]. The X-ray collimators are usually set for a field size of 10 cm × 40 cm at the isocenter. A special "applicator" in flush with the collimator face is used instead of the usual applicator extending up to the isocenter. A Cerrobend aperture defining the arc field is inserted into this applicator. The isocenter is placed well beyond the lesion depth in order to cover the entire width of the target volume in a single arc span. But the depth of the isocenter will change in the superior and inferior planes, since the external body contour changes shape in the superior-inferior direction.

For each electron energy, all dosimetric data, such as dose per monitor unit, complete depth–dose characteristics, beam profiles in the major planes for the range of aperture widths and lengths, bremsstrahlung dose, and bremsstrahlung profile, are accurately measured for static fields [26]. Even a small error in the measured data for static fields will result in large errors in the calculated dose for arced fields, as the data for arced fields are often obtained through the integration of stationary beam profiles. Data can also be obtained through direct measurements on a cylindrical phantom.

With the patient lying in the supine position, a custom-fabricated, tertiary field-defining Cerrobend cast, similar to an armor plate, is placed over the patient's body and supported by the treatment table. A patient-specific bolus is placed on the patient's chest wall to limit the penetration of the beam. The cast has a central field-defining open area and normal tissue protecting Cerrobend shielding all around.

Arc field width is usually 4–8 cm at the isocenter to ensure normal incidence of the beam throughout the arc span [25, 27]. The width does not remain constant and varies along the long

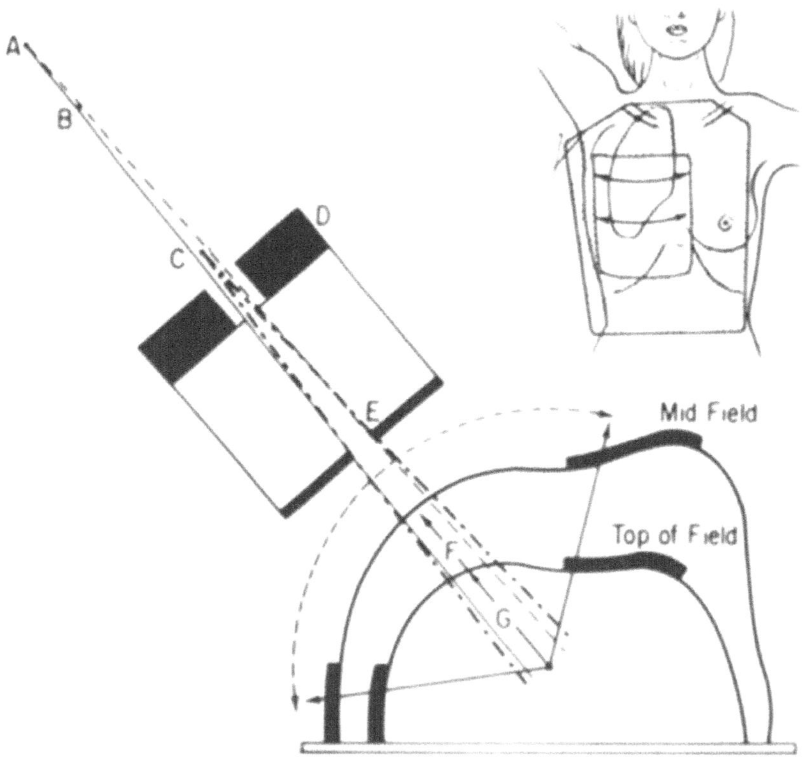

Fig. 14.7 Electron arc therapy setup for chest wall irradiation. (**a**) is the x-ray target position (not used in the electron modes); (**b**) is the electron scattering foil; (**c**) is the effective source position while using Cerrobend collimator (**e**); (**d**) is the primary photon collimator; (**f**) is the depth of the isocenter in the central plane (mid-field level) from the body contour; (**g**) is the depth of the isocenter at the top of the field; the four thick black strips are the tertiary collimators. (Ref [25])

axis of the field. The isocenter is chosen so that it is at the same distance from the body contour for all angles, and it lies beyond the maximum range of the electrons to avoid any accumulation of electron dose. In the axial plane, the arc beam is swept beyond the treatment field edge until the entire beam is integrated, and in the cranio-caudal plane, the arc field edge extends several cm outside the tertiary collimation to achieve a sharper penumbra (Fig. 14.8).

Dose distributions for electron arced fields differ from that for the static field in terms of: (1) reduced surface dose, (2) D_m occurring at deeper depths, (3) steeper dose gradients beyond D_m, and (4) increased bremsstrahlung dose in the region around the isocenter, which is usually five times the value for the static field. There is no difference in the penumbra for arced fields. The clear advantage of the electron arc field technique is that it can treat difficult lesions lying under sloping surfaces without having to resort to abutting fields and with near-ideal isodose distributions. However, there are numerous drawbacks that make this technique rarely practiced in clinics: (i) electron arc therapy remains a complex treatment technique requiring several accessories and measurements; (ii) there are too many variables including the dependence of electron arc on energy, arc length, width, and beam modifiers; (iii) in spite of significant advances in photon beam treatment planning and the availability of linacs with electron arc therapy capabilities, currently no commercially available treatment planning system can do dose calculation for this technique; (iv) the efforts that go into the commissioning of this technique are equivalent to the commissioning of all electron energies in a linac.

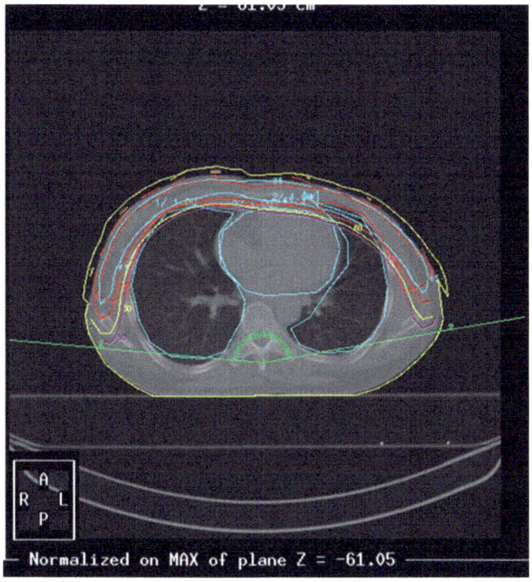

Fig. 14.8 Electron arc therapy in a patient with bilateral breast cancer (postbilateral mastectomy)

14.5.11 Total Skin Electron Irradiation

In the treatment of superficial lesions like mycosis fungoides and cutaneous lymphomas, electron beams with 2–9 MeV energy have been successfully employed [6, 7]. Treatments are delivered by a specialized technique called total skin electron irradiation, either by a translational technique or by a large field technique. In the former, the patient lying horizontally is moved head-to-toe, across a wide enough beam to cover the transverse dimensions of the patient, while in the latter, a standing patient is treated with multiple broad beams at large SSDs. The translational technique is rarely followed in the current period, and the other large field technique is described below.

Large, wide electron fields from the linac are further widened by scattering in the intervening air at large SSDs like 4 m [28]. The fields attain a Gaussian intensity distribution with a profile width (50–50%) of 1 m. Gerbi et al. [19] showed that a suitable combination of multiple fields, angled between 10° and 20° above and below horizontal, can give a uniform field of height 1 m. More such fields can be combined to get an even larger uniform field, sufficient to treat a standing patient from skull to foot. A careful assessment of the x-ray contamination of the beam is necessary to estimate the whole body photon dose to the patient.

This particular technique has been adopted successfully by various institutions with their own unique modifications to suit their machine, bunker dimensions, and preferences. Of these, the modified Stanford technique has become widely popular and is practiced by many institutions. Six fields (one anterior, one posterior, and four oblique fields) are positioned equidistantly around the circumference of the patient [29]. Each field has two beams: one pointing above and one pointing below the horizontal direction. In front of the patient, a tall stand having an acrylic plate of thickness 1 cm is used to provide additional scatter.

When the beam is incident obliquely, electrons will have to traverse longer path lengths in reaching a point. As a consequence, for an oblique beam, the entire depth–dose curve and the d_m will shift toward the surface. Multiple, large fields directed from different angles would show a shift in dose distribution toward the surface. It is possible to achieve ±10% over a large part of the body surface using this technique. However, some self-shielded areas like the vertex, inner thighs, axillae, and scrotum would require an additional boost. The bremsstrahlung dose would be twice that from the single field.

With the availability of high-dose-rate mode in all modern linacs, this special technique has become considerably easier to deliver. A dedicated electron energy and wide-aperture insert, in combination with wider jaw settings enabled through an interlock, helps select high-dose rate mode, with dose rates reaching up to 2400 monitor units per minute.

In vivo dosimetry through thermoluminescent dosimeters, film or other suitable in vivo detectors can be used to examine accuracy in dose delivery besides assessing low- and high-dose regions.

14.5.12 FLASH Radiotherapy by Electrons

FLASH radiotherapy refers to radiotherapy treatments delivered at ultra-high dose rates that are several thousand times higher than the dose rates used in conventional radiotherapy. Such FLASH radiotherapy treatments have proven to dramatically reduce normal tissue toxicities while retaining antitumor efficacy, thus widening the therapeutic window [30–34]. FLASH effects have been observed with electrons (4–6 MeV), protons, and photons in biological samples. However, there are several technical challenges involved when it comes to delivering dose to a target volume in less than 200 ms, an important requirement for FLASH. Since it takes several seconds for energy changes in proton beam therapy, treating a 3D volume of tumor tissue by spread-out Bragg peak (SOBP) in less than 200 ms is not a easy task to achieve. Photons exhibit FLASH effects, but only at keV energies and in extremely small field sizes making clinical use impossible at this time. Only electron beams have shown some promise. A 6-MeV electron beam from a dedicated machine, the Oriatron eRT6, was used in the treatment of T-cell cutaneous lymphoma in a 75-year-old patient [35]. Intensive research is going on to use very high-energy electron (VHEE) beams for treating deep-seated tumors [31].

Key Points (Precepts Box)

1. Electrons have steep dose falloff and limited penetration which make them attractive for treating superficial lesions. Lower-energy beams have sharper falloff.
2. Energy range clinically used: 4–20 MeV.
3. Electrons interact with medium through ionization and excitation (collision interaction), thus transferring energy to the medium and through bremsstrahlung (radiative interaction).
4. Electrons lose energy at the rate of 2 MeV per cm of soft tissue traversed.
5. The skin sparing effect with electron beams is only marginal. Surface dose increases with increase in electron energy. Lesions involving skin may require use of bolus.
6. All modern-day linacs use dual scattering foils.

7. Oblique incidence of electron beams has to be avoided which will otherwise result in increased dose to d_m and shift in depth–dose curve toward surface.
8. For large slabs of inhomogeneities, CET method can be used to calculate effective depth.
9. Abutting electron fields produce hot spots on either side due to bulging of the isodose curves.
10. Abutting electron–photon fields will result in high doses on the photon side of the field.
11. Field shaping can be done with lead or Cerrobend, a low-melting point alloy.
12. Minimum thickness of lead required in mm = Energy of the electron beam incident on the shielding (in MeV)/2.
13. Internal shielding is possible, but care must be taken to take care of the back-scattered electrons.
14. No treatment planning system can calculate dose distributions for electron arc therapy.

Multiple Choice Questions

1. In modern linacs, for generating clinically acceptable electron beams, which of the following technologies is employed?
   (a) Single scattering foil
   (b) Dual scattering foil
   (c) Scanning beam
   (d) Multiple scattering foil
2. While using internal shielding with electron beams, the enhanced dose due to backscatter occurs at the:
   (a) Skin surface
   (b) Tissue-shielding proximal interface
   (c) Shielding-tissue distal interface
   (d) Exit surface
3. What is the dose absorbed at the surface of the water medium for a 6 MeV electron and 6 MV photon beam, respectively?
   (a) 78% and 19%
   (b) 95% and 78%
   (c) 19% and 95%
   (d) 19% and 78%
4. The range of 12 MeV electrons in water is:
   (a) 12 cm
   (b) 6 cm
   (c) 3 cm
   (d) 25 cm
5. A dose of 20 Gy needs to be delivered at a point located at 4 cm from the surface. Of the 4 cm from the surface, 1 cm is of bone (CET = 1.6) and 1 cm of low density (CET = 0.25). Determine the optimal energy of the electron beam that adequately covers this depth at 90% isoline?
   (a) 9 MeV
   (b) 6 MeV
   (c) 12 MeV
   (d) 18 MeV
6. How much energy do electrons lose per cm of soft tissue traversed?
   (a) 1 cm
   (b) 2 cm
   (c) 3 cm
   (d) 4 cm
7. Abutting electron fields produce hot spots on either side due to:
   (a) Shallow isoline
   (b) Bulging of the isodose curves
   (c) Shielding
   (d) The presence of bone
8. In the electron mode, the x-ray jaws are:
   (a) Fully closed
   (b) Fully retracted
   (c) Set to a predefined position that is unique for every energy-applicator combination
   (d) Set to 25 cm × 25 cm for all energy-applicator combinations

9. The percent depth dose at 6 cm for a 6 MeV electron beam is approximately
   (a) 90%
   (b) 60%
   (c) 30%
   (d) <1%
10. While treating a patient with 12 MeV electron beam, it is desired to shield a part of the field with a cutout made of low melting point alloy like Cerrobend. The thickness of the shielding required is:
   (a) 6 mm
   (b) 7.2 mm
   (c) 8.2 mm
   (d) 12.2 mm

Answers

1. (b)
2. (b)
3. (a)
4. (b)
5. (c)
6. (b)
7. (b)
8. (c)
9. (d)
10. (c)

References

1. Karzmark CJ, Nunan CS, Tanabe E. Medical electron accelerators. McGraw-Hill; 1993. p. 144, 163.
2. Rogers DW. Notes on the structure of radiotherapy depth-dose distributions. Carleton Laboratory for Radiotherapy Physics Report CLRP. 2009;7(01).
3. Bjärngard BE, Piontek RW, Svensson GK. Electron scattering and collimation system for a 12-MeV linear accelerator. Med Phys. 1976;3(3):153–8.
4. Bova FJ. Clinical electron beam physics. In: Smith AR, editor. Radiation therapy physics. Berlin: Springer; 1995. p. 123–38.
5. Bieda MR, Antolak JA, Hogstrom KR. The effect of scattering foil parameters on electron-beam Monte Carlo calculations. Med Phys. 2001;28(12):2527–34.
6. Khan FM, Gibbons JP. Khan's the physics of radiation therapy. 5th ed. Philadelphia: Lippincott Williams & Wilkins; 2014. p. 256–308.
7. Gerbi BJ, Kirova YM, Orecchia R. Clinical applications of high-energy electron beams. In: Levitt SH, Purdy JA, Perez CA, Poortmans P, editors. Technical basis of radiation therapy practical clinical applications. 5th ed. Berlin: Springer; 2012. p. 157–96.
8. Chang DS, Lasley FD, Das IJ, Mendonca MS, Dynlacht JR. Basic radiotherapy physics and biology. New York: Springer; 2014. p. 111–23.
9. Faiz MK, Karen PD, Kenneth RH, Gerald JK, et al. Clinical electron-beam dosimetry: report of AAPM Radiation Therapy Committee Task Group No. 25. Med Phys. 1991;18(1):73–109.
10. Strydom WY, Parker WI, Olivares MA. Electron beams: physical and clinical aspects. In: Podgorsak EB, editor. Radiation oncology physics: a handbook for teachers and students. Vienna: International Atomic Energy Agency (IAEA); 2005. p. 273–99.
11. Saw CB, Pawlicki T, Korb LJ, Wu A. Effects of extended SSD on electron-beam depth-dose curves. Med Dosim. 1994;19(2):77–81.
12. Veness M, Richards S. Role of modern radiotherapy in treating skin cancer. Australas J Dermatol. 2003;44(3):159–68.
13. Maarouf M, Schleicher U, Schmachtenberg A, Ammon J. Radiotherapy in the management of keloids clinical experience with electron beam irradiation and comparison with X-ray therapy. Strahlenther Onkol. 2002;178(6):330–5.
14. Kirova YM, Campana F, Fournier-Bidoz N, Stilhart A, Dendale R, Bollet MA, Fourquet A. Postmastectomy electron beam chest wall irradiation in women with breast cancer: a clinical step toward conformal electron therapy. Int J Radiat Oncol Biol Phys. 2007;69(4):1139–44.
15. Hoppe RT. Mycosis fungoides: radiation therapy. Dermatol Ther. 2003;16(4):347–54.
16. Orecchia R, Veronesi U. Intraoperative electrons. Semin Radiat Oncol. 2005;15(2):76–83. WB Saunders
17. Calvo Manuel FÁ, Serrano J, Solé C, Cambeiro M, Palma J, Aristu J, Garcia-Sabrido JL, Cuesta MA, Del Valle E, Lapuente F, Miñana B. Clinical feasibility of combining intraoperative electron radiation therapy with minimally invasive surgery: a potential for electron-FLASH clinical development. Clin Transl Oncol. 2023;25(2):429–39.
18. Gahbauer R, Landberg T, Chavaudra J, Dobbs J, Gupta N, Hanks G, Horiot JC, Johansson KA, Möller T, Naudy S, Purdy J. ICRU report 71: prescribing, recording, and reporting electron beam therapy. J ICRU. 2004;4(1):5–9.
19. Gerbi BJ, Antolak JA, Deibel FC, Followill DS, Herman MG, Higgins PD, Huq MS, Mihailidis DN, Yorke ED, Hogstrom KR, Khan FM. Recommendations for clinical electron beam dosimetry: supplement to the recommendations of Task Group 25. Med Phys. 2009;36(7):3239–79.
20. Jabbari N, Khalkhali HR. Developing equations to predict surface dose and therapeutic interval in bolused electron fields: a Monte Carlo Study. Radiat Phys Chem. 2017;136:64–70.

21. Johnson JM, Khan FM. Dosimetric effects of abutting extended source to surface distance electron fields with photon fields in the treatment of head and neck cancers. Int J Radiat Oncol Biol Phys. 1994;28(3):741–7.
22. Leat CJ, Ramachandran P, Langton CM. Tungsten carbide and LMPA electron cutouts: comparison and validation using Monte Carlo modelling and measurement of dose. Biomed Phys Eng Express. 2020;7(1):015007.
23. Lambert GD, Sandland MR, Whitton AC, Doughty D. Combining backscattered electrons and low energy photons to improve the dose distribution to an eyelid. Int J Radiat Oncol Biol Phys. 1985;11(3):617–20.
24. Shiu AS, Tung SS, Gastorf RJ, Hogstrom KR, Morrison WH, Peters LJ. Dosimetric evaluation of lead and tungsten eye shields in electron beam treatment. Int J Radiat Oncol Biol Phys. 1996;35(3):599–604.
25. Leavitt DD. Physics of electron arc therapy. In: Smith AR, editor. Radiation therapy physics. Berlin: Springer; 1995. p. 140–53.
26. Alfonso R, La Fuente LD, Larrinaga E, Ascencion Y, Alonso JL, Bianchini S. Dosimetric commissioning of electron arc therapy. In: Proceedings of XIV Workshop on Nuclear Physics. VIII International Symposium on Nuclear and Related Techniques. WONP-NURT 2013. CEADEN, Cuba. https://inis.iaea.org/records/47t8v-6t934
27. Leavitt DD, Stewart JR, Moeller JH, Lee WL, Takach GA Jr. Electron arc therapy: design, implementation and evaluation of a dynamic multi-vane collimator system. Int J Radiat Oncol Biol Phys. 1989;17(5):1089–94.
28. Li HH, Altman MB, Wooten HO. Special procedures and equipment. Quality and safety in radiation oncology: implementing tools and best practices for patients, providers, and payers. 2016.
29. Anacak Y, Arican Z, Bar-Deroma R, Tamir A, Kuten A. Total skin electron irradiation: evaluation of dose uniformity throughout the skin surface. Med Dosim. 2003;28(1):31–4.
30. Favaudon V, Caplier L, Monceau V, Pouzoulet F, Sayarath M, Fouillade C, et al. Ultrahigh dose-rate FLASH irradiation increases the differential response between normal and tumor tissue in mice. Sci Transl Med. 2014;6:245ra93. https://doi.org/10.1126/scitranslmed.3008973.
31. Gao Y, Liu R, Chang CW, Charyyev S, Zhou J, Bradley JD, Liu T, Yang X. A potential revolution in cancer treatment: a topical review of FLASH radiotherapy. J Appl Clin Med Phys. 2022;23:e13790.
32. Bourhis J, Montay-Gruel P, Jorge PG, Bailat C, Petit B, Ollivier J, Jeanneret-Sozzi W, Ozsahin M, Bochud F, Moeckli R, Germond JF. Clinical translation of FLASH radiotherapy: why and how? Radiother Oncol. 2019;139:11–7.
33. Hughes JR, Parsons JL. FLASH radiotherapy: current knowledge and future insights using proton-beam therapy. Int J Mol Sci. 2020;21(18):6492.
34. Matuszak N, Suchorska WM, Milecki P, Kruszyna-Mochalska M, Misiarz A, Pracz J, Malicki J. FLASH radiotherapy: an emerging approach in radiation therapy. Rep Pract Oncol Radiother. 2022;27(2):343–51.
35. Bourhis J, Sozzi WJ, Jorge PG, Gaide O, Bailat C, Duclos F, Patin D, Ozsahin M, Bochud F, Germond JF, Moeckli R. Treatment of a first patient with FLASH-radiotherapy. Radiother Oncol. 2019;139:18–22.

Further Reading

https://vimeo.com/78553521
https://www.youtube.com/watch?v=8Icetn2W2O8
https://www.youtube.com/watch?v=HjYw674kd-4&t=10s
https://www.youtube.com/watch?v=kzvE5_zwnNU
https://www.youtube.com/watch?v=SfKML8wWtCU&t=2937s

Reirradiation in Radiotherapy

15

Sanjoy Chatterjee, Tapesh Bhattacharya, Shaurav Maulik, Rimpa Achari, Shikha Goyal, Shagun Mishra, Selvamani Backianathan, and Sarbani Ghosh-Laskar

15.1 Introduction

Targeting a localized tumor with radiation therapy remains a logical option following relapse recurrence or following a second primary diagnosis. However, factors that essentially determine the use of treatment do not depend on feasibility. Rather, the clinical benefits versus risks, supported by scientific data, should provide further information to the clinician for making the optimal decision for treatment. Reirradiation therefore must weigh the risks associated with normal tissue damage against the associated probability of tumor control. The use of radiation therapy and normal tissue effects of radiation to such therapy has not changed, but the volume of normal tissue being irradiated during initial therapy has been reduced with the use of more conformal techniques. A decision to offer reirradiation should always be made on a personalized basis after detailed discussion with the patient on expected benefits versus risk of increased late morbidities. A multidisciplinary approach is highly recommended as the perceived risk–benefit ratio is complex and requires detailed case by case analysis. Many factors can influence the decision for reirradiation (Table 15.1).

Every case which is considered for reirradiation is unique. As such, each patient needs consideration on individual merits per contemporary treatment options available. Newer imaging techniques help early detection of small recurrences as well as accurate delineation. Conformal treatment techniques delivering high doses while minimizing morbidity have enabled more patients to be considered for reirradiation.

S. Chatterjee (✉)
FRCR, Tata Medical Center, Kolkata, West Bengal, India

T. Bhattacharya · S. Maulik · R. Achari
Radiation Oncology, Tata Medical Center, Kolkata, West Bengal, India

S. Goyal
Radiation Oncology, PGI, Chandigarh, India

S. Mishra
Radiation Oncology, SGPGI, Lucknow, Uttar Pradesh, India

S. Backianathan
Radiation Oncology, CMC, Vellore, Tamil Nadu, India

S. Ghosh-Laskar
Radiation Oncology, TMH, Mumbai, Maharashtra, India

A. Munshi et al. (eds.), *Radiation Oncology – Principles, Precepts and Practice*,
https://doi.org/10.1007/978-981-97-8389-2_15

Table 15.1 Factors influencing decision for reirradiation

Patient factors	Tumor factors	Treatment factors
Performance status	Anatomical location of the recurrence	Availability of IMRT and IGRT
Quality of life	Volume of disease to be treated	Availability of dose volume parameters for past radiotherapy
Associated comorbidities	Disease-free interval	Ability of planning systems to do fusion of images
Support available for physical and mental issues	Intrinsic radiosensitivity	
Presence of inherent radiation repair defects		
Time gap between primary RT and reirradiation		

15.2 Radiobiology/Physics of Reirradiation

Using the linear quadratic model, one needs to calculate the dose received by a particular organ following fractionated photon therapy. Irradiated tissues with an acute endpoint (e.g., tumor control for squamous carcinoma) could be assigned an α/β of 10 and those for late events like fibrosis, myelopathy, etc., could be assigned a α/β of 3. A period of time is necessary for repair after any course of radiotherapy. For most sites this is usually allowed to be up to 12 months with a minimum of 6 months as mandatory period after weighing the risks versus benefits. Before approving a reirradiation treatment plan it is necessary that the cumulative dose to a particular organ at risk is within the acceptable limits. For this it may be necessary to calculate the equivalent dose in 2 Gy per fraction (EQD2) for the initial treatment and for the reirradiation one of the options for the clinician is to prescribe reirradiation treatment based on EQD2 estimates.

15.3 Reirradiation in Brain Metastasis

Brain metastases (BM) are the most common intracranial tumors in adults [1]. It develops in approximately one-third of cancer patients during their course of illness [2]. The most common primary sites are lung cancers, breast cancers, colon cancers, and malignant melanomas [3, 4].

Table 15.2 Reirradiation options for patient with brain metastasis

WBRT
SRS/SRT
WBRT followed by SRS/SRT
Surgery
Surgery followed by SRS/SRT

Recent advancements in systemic anticancer treatments including targeted agents and immune checkpoint inhibitors, have increased overall survival in a substantial number of patients with brain metastasis. Consequently, intracranial local failures after brain radiotherapy are observed more frequently [5, 6]. The aim of treating recurrent brain metastasis, especially those that have undergone brain irradiation, is to improve the patient's quality of life by improving local control while minimizing the risk of significant neurotoxicity. Without any further treatment, the prognosis of recurrent brain metastasis is extremely poor with a median overall survival of a few weeks to months, neurological sequelae being the most common cause of death. Salvage options for treating progressive recurrent brain metastasis include surgery, radiotherapy, and systemic treatments. In the majority of cases, surgery is often difficult because of the deep location of the metastatic deposits and its close proximity to functional areas; reirradiation being an efficient alternative option [7].

There are different permutations and combinations of reirradiation schedules (Table 15.2). These depend on the initial modality of radiation,

Table 15.3 Results of WBRT followed by WBRT

Authors	Sample size	Initial WBRT dose	Second WBRT dose	Median duration gap between 2 sessions	Response rates	Median Survival after re-WBRT
Sadikov et al. 2007 [14]	72	20 Gy/5 Fr	25 Gy/10 Fr	9.6 months	40%	4.1 months
Son et al. 2012 [15]	17	Median 35 Gy @2–3 Gy/ fraction	21.6 Gy@1.5–2 Gy/ fraction	Mean 17.6	80%	5.2 months
Ozgen et al. 2013 [16]	28	Median 30 Gy	Median 25 Gy	9.5 months	39%	3 months
Wong et al. 1996 [17]	86	30 Gy/10 fractions	20 Gy/10 fractions	7.6 months	70%	4 months

site of recurrence, the volume of metastasis, overlap with the previous irradiation volumes, the time elapsed between two radiation sessions, patient compliance, and availability of optimum resources. Extracranial control of the disease, performance status of the patients, and expected lifespan are the key factors that determine the justification of reirradiation and its different permutations and combinations [8].

The cumulative biologically effective dose ($BED_{cumulative}$) summating $BED_{initial}$ plus $BED_{reirradiation}$ is a useful tool in determining the reirradiation dose regimes. A BED cumulative of 200 Gy_2 is the maximum recommended dose to limit radiation-induced brain necrosis [9–11].

15.4 Reirradiation in Different Permutations and Combinations

15.4.1 WBRT Followed by WBRT

After an initial WBRT dose of 30 Gy/10 fractions/2 weeks, a dose of 20–25 Gy/10–15 fractions is an accepted regimen of re-WBRT used in different studies [12, 13]. The median duration of the gap between two sessions of radiation ranges from more than 7 to 15 months in those studies. Despite different response rates (complete response, partial response, and stable disease) ranging from 27% to 80%, the median overall survival time in these studies was not longer than 5 months. A majority of the patients suffer from grade 1 or 2 acute toxicities. Late effects are difficult to determine because only a very few patients remain alive at a longer follow-up [14–17]. The results of WBRT followed by WBRT are displayed in Table 15.3.

15.4.2 SRS Followed by SRS/SRT

The pivotal role of ablative radiotherapy in the form of stereotactic radiosurgery (SRS) or stereotactic radiotherapy (SRT) for brain metastasis is well recognized especially in a limited number of brain metastasis with a total volume of the metastasis <20 cc [18]. SRS or SRT is now preferred over WBRT in order to maximize local control and minimize the probability of neurocognitive impairment [19, 20]. It provides an excellent 1-year local control of around 70–90% across all the histologies [7, 21–23]. Reirradiation with WBRT can lead to frequent and rapid neurocognitive decline with poor quality of life [19, 20]. SRS or SRT can be performed efficiently for the management of new recurrent brain lesions to defer whole-brain radiotherapy, especially for the limited number of metastasis with the extracranial disease being adequately controlled. The two major concerns are the risk of major adverse events in the form of symptomatic brain necrosis and the effectiveness of ablative reirradiation in treating radioresistant clones [24, 25].

Table 15.4 SRS followed by SRS

Authors	Sample size	Initial SRS dose (median)	Second SRS/ SRT dose (median)	Median duration gap between 2 sessions	Local control	Median overall survival after re-SRS	Severe toxicities
Terakedis et al. 2014 [27]	37	18 Gy	18 Gy	9 months	Not reported	8.3 months	Radionecrosis 16%
Rana et al. 2017 [28]	28	24 Gy	26.5 Gy	9.7 months	1 year 88.3%	22 months	Radionecrosis 18%
Balermpas et al. 2018 [29]	31	18 Gy	19 Gy	16.5 months	1 year 79.5%	1 year 61.7%	Radionecrosis 16.1%
Dincoglan et al. 2019 [30]	30	20 Gy	21 Gy	13.5 months	1 year 83.3%	23 months 1 year 76%	Radionecrosis 13.3%
Iorio Morin et al. 2019 [31]	56	20 Gy	18 Gy	13 months	1 year 68%	14 months after second SRS	Radionecrosis 5%

Table 15.5 SRS followed by WBRT

Author	Sample size	WBRT dose	SRS dose	Median gap between two sessions	Median survival after SRS	Severe toxicities
Caballero et al. 2012 [32]	310	Varies from 19.8 Gy to 60 Gy Most common 30 Gy/10 Fr	17–18.5 Gy	8.1 months	8.4 months	Not reported
Maranzano et al. 2012 [33]	69	30 Gy/10 Fr—49% 20 Gy/5 Fr—29%	20 Gy	11 months	10 months	6% Radionecrosis
Huang et al. 2017 [34]	56	Median dose 40 Gy/20 Fr (range 30–50 Gy)	17 Gy	10.5 months	10.8 months	Not reported
Yomo et al. 2013 [35]	77	Not reported	20 Gy	10.6 months	8.2 months	Not reported

A 12 months local control ranges between 60% and 90% provided the extracranial disease is adequately controlled. The rates of symptomatic brain necrosis lie between 9% and 24% as reported in different series [26–31]. The results of SRS followed by SRS/SRT are displayed in Table 15.4.

15.4.3 SRS After Previous Whole-Brain Radiotherapy

Several studies have explored the role of SRS/SRT for salvage treatment of patients with recurrent brain metastasis who were previously treated with WBRT. In most studies, around half of the patients had stable disease after SRS/SRT with a median overall survival of more than 6 months in the favorable group [32–35]. The results of SRS after WBRT are displayed in Table 15.5.

15.5 Particle Therapy in Reirradiation of Brain Metastasis

Though there is no randomized controlled trial comparing proton beam therapy and photon-based SRS for the first-line treatment of brain metastasis. The study by Atkins et al. reported a similar lower rate of local recurrence and acute toxicities with distant intracranial relapse being the most common mode of failure [36]. Although theoretically the role of particle therapy in this regard looks promising and proton therapy has been explored in treating brain metastasis in the first-line setting, clinical outcome data in the reirradiation settings is yet to emerge. Carbon ion radiotherapy because of its excellent conformality and higher LET and RBE, was used as a novel approach in radioresistant primary intracranial tumors; however, in the setting of first-line or second-line treatment of brain metastasis, it is yet to be explored.

Key Points

- With recent advancements in systemic treatment, patients with brain metastasis are surviving longer.
- After initial brain radiation, a reasonable proportion of patients are surviving long enough to suffer intracranial recurrence either infield, marginal or distant.
- Patients who received initial WBRT or SRS can be reirradiated with SRS or WBRT for recurrence in different permutations and combinations.
- The key determinant factors to decide reirradiation include KPS, life expectancy, primary cancer, and its histology, the volume of the target, number of metastasis, the gap between two sessions of radiation and extracranial control of the disease.
- SRS/SRT is now preferred over WBRT for reirradiation to prevent neurocognitive deterioration.
- A BED cumulative of 200 Gy_2 is the maximum recommended dose to limit radiation-induced brain necrosis.
- Radiation-induced symptomatic brain necrosis varies between 5% and 25% as reported in the literature.
- At least a minimum gap of 6 months is required between the two sessions of radiation for its safe delivery and favorable outcome provided the extracranial disease is well-controlled.
- Though treating recurrent radioresistant clones is a challenge, the role of particle therapy in the reirradiation of brain metastasis is yet to emerge.

15.6 Reirradiation for Recurrent Primary Brain Tumors

Histology, grade, and location of a primary malignant brain tumor predominantly steer the natural history of a primary brain neoplasm. According to the recent CBTRUS data, the most commonly occurring primary malignant brain tumor is glioblastoma (14.3% of all tumors and 49.1% of malignant tumors), and the most common non-malignant tumor is meningioma (39.0% of all tumors and 54.5% of non-malignant tumors) [37]. Despite postoperative temozolomide-based chemoradiation, almost all glioblastomas progress [38, 39]. Anaplastic astrocytomas and other lower grade diffuse gliomas will also progress, albeit after a longer interval, as a function of their molecular types [40, 41]. Treatment choices for recurrent glioma are limited. Redo surgery, salvage chemotherapy, and reirradiation decisions are principally guided by the progression-free interval, neurological performance status and the molecular signature of the glioma [42]. Reirradiation for recurrent gliomas has limited prospective data and consequently, its adoption in clinical treatment algorithms is fraught with the fear of radionecrosis that may occur after months to years post treatment. Contemporary practice of radiation oncology is characterized by multimodality, multiparametric image guidance integral to both oncology decision making and the radiation therapy delivery chain. Incorporation of functional imaging into treatment planning, and advancement of software and hardware systems ensure high precision radiation treatments with the potential to improve the therapeutic ratio of reirradiation [43–45]. The purpose of the present section is to give an overview of available clinical and radiobiology knowledge in glioma reirradiation at primary recurrence.

Almost all clinical information in glioma reirradiation comes from multiple, heterogeneous, non-comparative studies involving nearly 2000 patients [46]. Various linear accelerator, cyberknife and gamma knife based external beam reirradiation techniques, namely, three dimensional conformal, intensity modulated and stereotactic radiotherapy has been variably used. Almost all studies used MRI coregistration for delineation of reirradiation target volumes; some studies have incorporated metabolic or functional imaging. The median PTV margin from gross tumor volume (GTV) identified on T1 postcontrast MRI and at times from high signal intensity fluid-attenuated inversion recovery (FLAIR) was 3 mm (range, 0–3 mm) in most studies. Reported reirradiation doses have been 24–36 Gy with a daily fraction size of 1.8–6.0 Gy delivered after a median interval of 11.6 months (range, 3–19 months) from the first course of radiation therapy [47–51]. The reported overall survival rate at 6 and 12 months, across trials being 73% (95% confidence interval (CI) 69–77%) and 36% (95% CI 32–40%) respectively; the progression-free survival rate at 6 and 12 months being 43% (95% CI 35–50%) and 17% (95% CI 13–20%) respectively [46, 52]. There is some data on feasible and safe particle therapy-based brain reirradiation [53, 54].

The normal brain is known to have a low repair capacity, reflected in an α/β ratio of 2.0 Gy [26–28]. The cumulative biologic effective tolerance dose (BED), which is the summation of the initial and reirradiation BED converted to normalized total dose (NTD) as the total dose delivered in 2 Gy fractions, also expressed as EQD2 is best considered to understand reirradiation protocols. Clinical studies report a median EQD2 of 81.6–101.9 Gy in conventionally fractionated, and 90.0–137.2 Gy in hypofractionated stereotactic techniques [9, 46–50, 55]. At reirradiation, there is no dose–response [above or below 36 Gy in 2 Gy equivalent doses (EQD2)] in the external beam reirradiation studies but a short fractionation regimen ($\leq$5 fractions) seems to provide a superior 6 months PFS and brachytherapy studies report a longer OS [52]. Radionecrosis, reported as Grade 3+ adverse event, is seen in 7% (95% CI 4–10%), found to occur at a cumulative NTD of >100 Gy and there was no correlation between the time interval between primary and reirradiation and the chance of developing radionecrosis [46, 51, 56]. Various chemotherapy agents like lomustine, bevacizumab, or temozolomide are often variably used during reirradiation. Temozolomide is known to be a radiosensitizer and a cumulative NTD of >100 Gy should not be exceeded [57, 58]. In addition to high cumulative NTD as described, the chance of reirradiation-induced brain radionecrosis is reported to be 28% in a series that combined hyperbaric oxygen at retreatment indicating a radiosensitizing effect in well-oxygenated normal brain tissue, and it is best avoided [59].

In conclusion, reirradiation is a feasible palliative option for well selected patients with recurrent malignant glioma. Glioma reirradiation decisions are mostly individualized and the best reported outcomes are usually seen in younger (<50 years) patients with a good performance status, more than a 1 year interval between the courses of radiation, receiving 30–36 Gy of fractionated radiation to smaller target volumes (20–30 mL) and a cumulative NTD of >100 Gy should not be exceeded [60–65]. Multimodal image co-registration is an integral step to ensure accurate target delineation and modern radiotherapy techniques are assuming technological standards to prevent radiation induced injury to previously treated normal brain volumes. The incidence and post treatment progression of non-glial primary malignant brain tumors needing re irradiation are less common and the biological principles discussed above are applicable.

15.7 Reirradiation in Head and Neck Cancers

Recurrence of tumor within an irradiated field often results from radio resistant clones, requiring high dose reirradiation for optimal tumor control. Spinal cord and brain stem are often the organs put at highest priority during head and neck cancer reirradiation. Ang et al. [66] have suggested a score-based risk stratification using individual treatment biological effective dose (BED), alongside cumulative BED and gap between irradiation to predict the risk of myelopathy associated with reirradiation in monkeys. Such experiments provide guidance; however, the final decision needs to be taken by the clinician after detailed discussions with the patient.

Patients with second primaries or locoregional recurrences in treatment fields are often poor surgical candidates due to medical fitness or disease extent. Though technically resectable, surgical intervention in such scenarios may lead to significant morbidity or even death.

In the setting of unresectable disease, treatment outcomes with palliative systemic therapy with an intent to palliate have been used. Such therapy using immunomodulation, systemic chemotherapy with or without cetuximab should be considered and side effects of each along with perceived benefits clearly explained to the patient. Reirradiation with a view to get long term tumor control and in some cases cure, has high risks of toxicities [67–70].

When surgical salvage is not possible, the use of concurrent chemoradiation have been supported by two randomized studies. The RTOG 9610 and 9911 studies [71, 72] provide the strongest evidence for reirradiation in unresectable cancers. Addition of paclitaxel and carboplatin-based radiation with hyperfractionated radiotherapy in the RTOG 9911 study provided a 25.9% survival at 2 years at the expense of 8% fatality. Similar reports from single centers including that from Gustave Roussy [73] support the strategy. It may be noted that treatment fields in most studies included the area at highest risk with a 1 cm margin to avoid excess toxicities.

In summary, reirradiation is a risky but commonly used strategy to enhance local control and disease-free survival rates in recurrent head and neck cancers. Surgical salvage followed by adjuvant reirradiation may provide the best chance of cure for eligible patients. In those that cannot be treated with surgery concurrent chemoradiation with a platinum containing regimen could provide the best chance of survival. Treatment decision making is often complex and requires a multidisciplinary approach.

15.8 Reirradiation of Pelvic Malignancies

Treating a pelvic malignancy is challenging in following prior pelvic radiation therapy. This decision to reirradiate is usually taken when alternative options for locoregionally confined disease such as surgical resection alone have been ruled out due to high complication risk or disease extent, or systemic therapy options are either non-curative or unable to provide symptomatic relief. Literature regarding optimum technique-dose-fractionation-target volume or normal tissue toxicity is sparse and mostly derived from small, single institution retrospective data; hence, most recommendations can only give a broad outline and case-based individualization is necessary.

The most common pelvic malignancies considered for reirradiation for relapse include gynecologic (cervical, endometrial) cancers, prostate cancer and rectal cancers. Outcome data on pelvic reirradiation is heterogeneous (patient selection, technique, dose, intent of treatment, evaluation parameters and toxicity reporting) with scant prospective studies. Reirradiation has traditionally been used as a palliative treatment in most instances for locally recurrent non-metastatic as well as metastatic tumors and may add to quality of life in carefully selected cases. If surgical resection or exenteration is part of treatment plan for recurrence then intraoperative radiotherapy (brachytherapy, intraoperative electrons, low energy photons) is often a good option, offering the possibility of direct visualization of tumor bed, physical separation or shielding of pelvic viscera from the reirradiation region. Adjuvant irradiation (brachytherapy or EBRT) in the setting of R1/R2 resection potentially improves local control as well as survival. Sites such as vault or parametrial recurrences are amenable to interstitial brachytherapy. Nodal recurrences may require additional systemic therapy or nodal dissection. Limited volume nodal recurrences (1–3 nodes) may be addressed with SABR or with conventional fractionation.

Pelvic reirradiation carries a high risk of acute as well as serious late morbidity in the form of vaginal or ureteral stenosis, urethral or ureteral strictures causing hydronephrosis, fistula formation (rectovaginal, vesicovaginal, cutaneous), adhesions of bowel loops causing recurrent obstruction and perforation of hollow viscera such as duodenum or small bowel. There are no reliable data on tolerance to reirradiation or dose constraints for normal tissues. Data on acute reacting tissues such as gastrointestinal tract may give some pointers on tolerance to a second course. It is evident that recovery is mainly limited by vascular changes leading to consequential late effects. Late reacting tissues like urinary bladder show a stronger component of consequential effect and lesser recovery.

Various techniques to reduce toxicity in reirradiation include (a) physical separation from organs at risk such as using intra-abdominal or pelvic slings to separate bowel from site of recurrence if surgery is being performed prior to reirradiation [74], (b) use of hyaluronic spacer gels injected anterior to anterior rectal wall in prostate reirradiation [75], (c) irradiation of partial volumes guided by biologic imaging (PET-CECT) or high soft tissue contrast imaging such as MRI (such as partial prostate or dominant intraprostatic nodule irradiation, (d) using brachytherapy or highly conformal techniques such as SABR, and (e) altered patient positioning such as prone position during rectal or posterior abdominal reirradiation.

15.8.1 Rectal Cancer

Pelvic recurrence is observed most commonly in presacral region, at anastomotic sites and less often in nodes only. Most recurrences are locally advanced and upfront surgery is often not feasible. Single modality treatments such as reirradiation and chemotherapy are essentially palliative with symptom control rates of around 75% while combined modality incorporating downsizing with chemoradiation and eventual surgery may give a reasonable possibility of improved survival (3 year overall survival of 38.3% in operated vs. 23.8% in non-operated patients) [76]. Various techniques including IMRT, VMAT, particle therapy (proton, carbon ion), etc., with conventional fractionation may be appropriate with concurrent radiosensitizing chemotherapy agents such as 5FU and capecitabine and hyperthermia. Hyperfractionated accelerated EBRT and endorectal brachytherapy have been used in neoadjuvant setting with encouraging local control and survival. Surgery with R1/R2 resection may benefit further from incorporation of intraoperative brachytherapy. Patients who achieve pathologic complete response have better survival than those who do not. Late toxicity rates of 5–15% have been reported (higher in combined modality therapies) although morbidity reporting is not consistent. Patients who have nodal relapses only and limited to small regions can be treated with SABR without irradiating large bowel volumes, and thus lower morbidity [77, 78]. The French group GRECCAR has initiated the first phase 3 trial to investigate rectal cancer reirradiation, comparing neoadjuvant chemotherapy with and without sequential reirradiation (30.6 Gy with concurrent capecitabine), the primary outcome being R0 resection rates [79].

15.8.2 Prostate Cancer

Local recurrence after prior prostatic irradiation may affect nearly 30–50% patients on follow up. A majority of these patients are considered for androgen deprivation therapy (ADT) (continuous or intermittent) or additional systemic therapies based on disease extent and hormone sensitive status. With more sensitive and wider availability of modern diagnostic techniques such as multiparametric MRI, PSMA PET-CECT, and ultrasensitive PSA assays, more and more recurrences are detected at local stage. Local salvage options for these recurrences, if disease is confined to prostate alone, are limited to local ablative therapies (cryoablation, high frequency ultrasound), radiotherapy (brachytherapy, SABR, conventionally to moderately hypofractionated EBRT) or surgery. Even limited volume recurrences after radical radiotherapy doses to prostate (>70 Gy) are difficult to salvage surgically. Issues such as volume to be irradiated, dose-volume parameters, association with ADT are still heterogeneous and not clearly agreed upon [80]. Consideration for reirradiation in prostate cancer may have to be made for biochemical, local or nodal relapse. Australian and New Zealand Radiation Oncology Genito-Urinary group made recommendations for patients with recurrent prostate cancer, emphasizing that treatment decisions need to be individualized to each scenario and preferably clinical trial inclusion [81]. Local salvage therapy (including reirradiation) can be offered to good performance status patients with only local recurrence (biopsy proven), life expectancy longer than 10 years and willingness to accept reirradiation-related adverse effects. Patients with favorable prognostic factors such as initial PSA <10 ng/mL, initial stage T1/T2 and Gleason score of 7 or less, benefit the most from salvage local therapy. Brachytherapy is an attractive option for reirradiation of localized prostate recurrences since it makes it possible to deliver high radiation doses confined to small volumes, with minimum collateral damage from excess dose to adjacent organs at risk [82]. However, it is an invasive procedure and needs both fitness for anesthesia as well as skill for optimal results. Patients who are not fit for the same either technically or medically may still be considered for reirradiation by SABR which to a large extent may mimic the dose fall-off with brachytherapy, or with image guidance using conventional or moderate hypofractionation [83].

15.8.3 Gynecologic Cancer

Recurrences in gynecological cancers (cervical or endometrial) needing reirradiation may be central, lateral or nodal. Upfront surgical management often involves exenteration with high probability of R2 resection as well as unacceptable morbidity. Interstitial brachytherapy (preferably with image guidance) is effective for central recurrences while combined modality (surgery and intraoperative brachytherapy may be appropriate in lateral pelvic wall recurrences [84]. Nodal recurrences may be salvaged with SABR. In the American Brachytherapy Society review of various reirradiation techniques for reirradiation in gynecologic cancers, brachytherapy was most common [85]. Overall, small recurrences (<2–4 cm), long reirradiation interval and doses >40–50 Gy had better outcomes. Local control at 1–5 years was 44–88%, and overall survival at 2–5 years was 39.5–82%. Variable grade 3–4 late toxicities ranging from 0% to 43% were reported, with most publication reporting >15% serious toxicity.

Table 15.6 lists recent reirradiation series with outcomes and toxicities in pelvic malignancies.

15.8.4 Breast Cancer

Locoregional recurrence rates after radiotherapy for breast conservation surgery (BCS) or mastectomy is approximately 5–15%. Chest wall/ipsilateral breast accounts for 60–95% of all recurrences [107]. It is necessary to identify whether the recurrence is from the original tumor or it is a new primary. This is done based on clinical grounds, location, mammographic appearance, histological type, hormone and Her2 receptor expression and DNA flow cytometry [108].

The treatment of LRR is challenging and needs to be multidisciplinary including surgical resection, systemic therapy (chemotherapy, hormonal therapy, targeted therapy, immunotherapy) based on receptor status and reirradiation. Local RT is an excellent option after a good complete surgical excision of the local recurrence and it has shown overall survival benefit. However, a survey among Italian Breast group found that only 0.6% offer re-RT for breast cancer recurrences [109].

Adverse prognostic features include time to recurrence >48 months, lymphovascular invasion, ER negative status, high grade histology in the recurrent tumor, and close or positive margins after resection of the recurrence. Patients with 2 or more of these high-risk features have poor survival, warranting more aggressive therapy at the time of disease recurrence [110]. A novel, simple and feasible size classification (rClasses 0–IV) for Locally recurrent breast cancer has been suggested by Notter et al. rClass 0-I has the best local control after complete response and rClass IV the worst overall survival (OS) [111].

Reirradiation modalities could be either with photons, electrons or charged particle therapy such as protons. The treatment delivery could be either with Conformal field in field (FiF)

Table 15.6 Reirradiation series with outcomes and toxicities in pelvic malignancies

Author Year Study type Duration	Number of patients	RT technique, dose, volume	Other therapy	Previous RT technique, dose, volume re-RT interval	Outcome	Toxicity
Rectum						
[77] 2021 Retrospective 2015–2020	69 (81 lesions) 86% nodes 10% R1 resection margin postop 97% had 1–2 sites	SABR (robotic, VMAT, IMRT) Dose 30 Gy/5 Fr Median GTV 13.4 cc	10% had surgery prior to SABR (R1)	Median previous dose 44.2 Gy_{10}, 43.2 Gy_3 Median cumulative dose 84.2 Gy_{10}, 97.2 Gy_3	Median FU 28 months Median PFS 12.1 months Median OS 38.7 months 2 year OS 77% 58.3% deaths due to local recurrence 42.6% had local relapse at death or last follow up	NR
[86] Retrospective 2010–2018	35 80% were cT3-4 N0, 20% were N+	3D-CRT Dose 30–30.6 Gy 14 patients received IORT	Concurrent capecitabine Post-RT surgery after a median time of 11 weeks (34.3% Exenteration, 31.4% APR, 6% others) 42.9% had R0 resection, 5.7% unresectable	Median re-RT interval 25 months Median RT dose 50 Gy (with chemo)	Median FU 53 months 60% had partial response after RT 3-year OS 64.9% 5-year OS 21.3% 3-year & 5-year DFS 19.0%	5.7% G3 toxicity after chemo/RT and 25.7% after surgery
[87] 2014–2018	14	Carbon ion radiotherapy Median dose 60 Gy RBE (range 35–76.8) in median 16 fractions (15–20) GTV: T1C enhancing tumor CTV: 5–10 mm margin PTV: 3–10 mm margin GTV median volume 154.6 cc		Median re-RT interval 65 months (14–139 months) Prev RT: 45–76 Gy (median 45) 1 had received brachy boost 20 Gy One had recorded SBRT 30/6 at first recurrence	Median FU 18 months 1-year LC 78%, 2-LC 52% Pelvic relapse in 6 (close to bowel) 1-year OS 100%, 2-year OS 76.2% 1-year MFS 64.3% 2-year MFS 43%	No G $\geq$ 3 acute/late reaction nor pelvic infections

[88] 2020 2006–2016	67	3D-CRT/ IMRT/VMAT Dose 30.0 Gy–45.0 Gy (HFRT 1.2–1.5 Gy BD or 2 Gy OD) GTV based on CT/MRI/PET and clinical examination CTV: GTV + 0.8–1.0 cm PTV: CTV+ 0.7–1 cm or ITV + 0.5–1.0 cm Median GTV 107 cm^3	67% received concurrent chemo 56% underwent resection (R0 in 75%)	Median re-RT interval 29 months	Median OS 15.6 months (39.2 months for operated) 3-year OS 32%. Resectable patients had higher OS (62% vs. 16%, median 39.2 months vs. 8.9 months) Smaller GTV <107 cm^3 in those without metastases had better OS (44% vs. 21%)	Late G3–4 GI 7%, GU 14%
[89] Retrospective 2001–2012	102	HF accelerated RT (39 Gy, 1.5 Gy BD) 86% 3D-CRT PTV = GTV + 1.5–2.5 cm margin 22 patients underwent IORT during surgery—Median dose 10 Gy	91% concurrent chemo 66% underwent surgery after RT (67% R0)	Median re-RT interval 30 months Median prior RT dose 50.4 Gy	Median FU 28 months 3-year LC 40% 3-year OS 39% Resectability was prognostic for both LC and OS	3-year G3/4 toxicity 34%
[90] Retrospective 2007–2011	10 Recurrence or second primary in rectum	Endorectal brachy—HDR (preoperative) Dose 26 Gy in 4 fractions (once daily)	7 patients underwent surgery (LAR 1, APR 5, transanal excision 1) 85.7% R0	Median re-RT interval 11 years Prior RT: EBRT in 6, EBRT+brachy in 2 and Brachy in 2	Path CR in 3/10 Median FU 9 months 6/7 patients undergoing surgery were alive at last follow up	No acute ≥G3 toxicity
[91] 2008 Prospective phase 2 2000–2005	24 (7 had metastases) 80% unresectable 8% R2 12% R1	Median RT dose 39.6 Gy (1.8 Gy/Fr)	Concurrent 5FU infusion Pelvic regional hyperthermia twice a week during RT	Median re-RT interval 34 months	3-year LC 15% 3-year OS 30% Pain response 70%	G3 acute GI 1.2% Late GI 4.2%

(continued)

Table 15.6 (continued)

Author Year Study type Duration	Number of patients	RT technique, dose, volume	Other therapy	Previous RT technique, dose, volume re-RT interval	Outcome	Toxicity
[92] Prospective phase 2 1997–2001	59 Biopsy proven pelvic-only recurrence	Dose 30 Gy (1.2 Gy BD) to PTV2 and 10.8 Gy boost to PTV1 GTV PTV1: GTV + 2 cm PTV2: GTV + 4 cm	Concurrent 5FU infusion daily. Adjuvant chemotherapy raltitrexed 5 cycles (50.8%) 50.8% underwent surgery (35.6% R0, 5.1% R1)	Median re-RT interval 27 months Median prior RT dose 50.4 Gy Prior surgery—Anterior resection 76.3%, APR 23.7% 74.6% received prior chemo	Median FU 36 months Median OS 42 months 5-year OS 39.3% 5-year LC 38.8% 5-year DFS 29.2%	G3 GI acute toxicity 5.1% 11.8% late toxicity (2 skin fibrosis, 2 impotence, 2 urinary complications requiring nephrostomy, 1 small bowel fistula requiring surgical diversion)
Prostate						
[93] 2012–2019	44 PET-MRI proven recurrence	SABR (VMAT) Dose 35 Gy/7 Fr/1.5 weeks (BED 151.7 $Gy_{1.5}$) GTV: Defined on MRI CTV: GTV + 3 mm PTV: CTV + 3 mm Median PTV 44 cc	Short-term ADT if PSADT <12 months (27% received ADT concurrently)	Median interval for re-RT 60 months Prior RT—64% 3D-CRT, 36% IMRT Median BED (ab 1.5) of prev RT 177.3 Gy	Median FU 25.4 months 2-year LC 90.1% 2-year bRFS 58.3% 2-year CRFS 67.9% BCF lower for re-RT interval > 4 years; LC correlated with BED at first EBRT >177 Gy	4% late ≥G3 GU tox
[94] Prospective trial (interim results) 2016–2019	25 Biopsy-proven PET and MRI positive localized recurrence	SABR (IMRT/VMAT–FFF) Dose 36–38 Gy/6 fractions Median GTV volume 2.7 cc	None	Median re-RT interval 8.3 years	Median follow-up 25 months 1-year PET response (CMR)—92% 2-year FFBF 80% (75% failures were prostate only)	1 acute grade 3 GI toxicity (4%) due to a rectal ulcer overlying the hydrogel.

[95] Retrospective 2009–2018	82 Prostate-only recurrence with prostate size<60 cc	USG-guided interstitial pulsed-dose rate brachytherapy 30 Gy (single pulse dose 0.6 Gy/h, 24 h per day) × 2 sessions, 4 weeks apart—Total reference dose 60 Gy (EQD2 = 71.5 Gy-eq)	43.9% received ADT	Median re-RT interval 48 months (3–245) Previous therapy EBRT 52%, Sx + EBRT 29%, brachytherapy (LDR) 19%	Median FU 49 months 5-year overall cumulative local recurrence rate 17.7% No diff according to risk group PSA-free survival 67.3%, 70.4%, and 63.8% for low-, intermediate-, and high-risk group, respectively	No grade 3 GI late effects Urinary incontinence G3 4.9% Urinary frequency/ urgency G3 2.4% G3 urinary retention 6.1%
[96] Retrospective 2009–2018	50 Biopsy proven	SABR 34 Gy/5 Fr CTV: Prostate and contiguous extension No PTV margin Brachy-like dose distribution (PTV V150 was 15–25%)	14% received ADT	Median re-RT interval 8.1 years Median prior dose 75.6 Gy 92% prior 3D-CRT/ IMRT/proton, rest brachy/SABR/ salvage prostate bed RT	Median FU 44 months 5f b DFS 60% 5-year LC 94%	5 years ≥G3 GU toxicity 8% 82% patients lost potency (median time 12 months)
[97] Retrospective 2012–2017	23 PET-proven local recurrence	SABR (VMAT, FFF) Dose 25 Gy/5 Fr GTV = CTV: Recurrent nodule + whole prostate/tumor bed PTV: CTV + 5 mm margin (3 mm posteriorly) Median CTV volume 50.3 cc	8 patients received concurrent ADT	Median re-RT interval 90 months Median prior RT: 74 Gy Previous RT: 65.2% radical, 34.8% postop/salvage	Median FU 33 months Median bRFS 19 months 2-year bRFS 41.7% Median LC 30 months 2-year LC 58.1%	1 G3 hematuria 1 G3 late toxicity (urethral obstruction)

(continued)

Table 15.6 (continued)

Author Year Study type Duration	Number of patients	RT technique, dose, volume	Other therapy	Previous RT technique, dose, volume re-RT interval	Outcome	Toxicity
[98] 2019 Retrospective 2009–2016	64 Local recurrence of prostate	SABR Median dose 30 Gy/5 Fr (BED 150 Gy) GTV: MR/PET-based recurrent volume CTV: Variable (whole prostate/ gross lesion with margins) PTV: CTV + 5 mm margin (3 mm posteriorly)	25% received concurrent ADT	Median re-RT interval 99.7 months Prior EBRT dose 70.2 Gy median ~86% treated with 3D-CRT (others—Brachy, IMRT, combination)	Median FU 26.1 months 2-year LC 75% 2-year cRFS 40% 2-year bRFS 53%	1 acute G3 GU, 1 late G3 GU toxicity
[99] Retrospective 2011–2015	28 Prostate-only recurrence	HDR-BT (10) or SABR (18) Dose 35 Gy/5 Fr	4/28 received concurrent ADT	HDR-BT: 70% had received EBRT SABR: 83.3% had received LDR BT	HDR-BT: Median FU 22.5 months 90% had biochemical response, 50% sustained bNED bRFS in responders 55.6% SABR: Median FU 14.5 months 55.6% sustained bNED bRFS in responders 66.7%	HDR-BT: 1 G3 GU toxicity (transient) SABR: 1 G4 GU toxicity (transient)
[100] Retrospective 2011–2014	21 In-field prostate recurrence (10/21 after adjuvant EBRT) 18F-PET and MRI defined recurrence	Robotic SABR 36.25 Gy/5 Fr GTV: MRI-defined disease CTV: 1–2 mm margin PTV: 5 mm margin (3 mm posteriorly)	2/21 patients received ADT	Median re-RT interval 111 months Median prior EBRT dose 71.1 Gy	Median FU 11.7 months 1-year bRFS 83.3% 1 in-field progression 2 had biochemical progression from distant metastases	No ≥G3/4 acute or late toxicity
Gynecologic						
[101] Retrospective (multi-institutional) 2008–2017	26 Primary sites endometrium (20) Cervix (4) Vulva (1) vagina (1)	HDR-ISBT 2–6 Fr HR-CTV 34.6 cc (median) Median HR-CTV D90 29.1 Gy	None	Median re-RT interval 20.3 months All had prior EBRT and 16/26 had prior brachy Previous RT EQD2 63 Gy	Median FU 24 months CR in 64%, PR in 19% 69% recurred (local in 54%) 2-year LC 50% 2-year PFS 38% 2-year OS 78%	7.4% G3 Toxicities (vaginal stenosis)

[102] Retrospective	20 (21 sites) Locoregional recurrence (Gyn malignancies) 20 LNs (pelvic, para-aortic, abdominal) 1 LN + acetabular metastases Primary sites: Cervix (30.0%) Uterus (55.0%) Vulva (5.0%) Vagina (5.0%) Ovary (5.0%)	SABR MC dose 40–45 Gy in 5 Fr (3 received SABR as boost after EBRT Cumulative BED_{10} including EBRT dose 82.7 Gy (median) Median GTV 9.7 cc	42.8% received chemo before or after re-RT	Median re-RT interval 30.2 months Prior EBRT technique IMRT in 57.1%, 3D-CRT in 23.8% Median prior EBRT dose 45 Gy	Median FU 31.2 months 3-year LC 61.4% 3-year DMFS 44% 3-year OS 51.9% Median OS 64.5 months $BED_{10} \geq 82.7$ Gy predictive of LC (3-year in-field LC 83.3% vs. 36.0%)	3 late ≥G3 toxicities (small bowel obstruction in 2, ureteral stenosis in 1) 3-year risk of ≥G2 toxicity 38.1% and ≥G3 14.3%
[103] Retrospective 2008–2016	16 Unresectable and isolated recurrent LN (gynecological cancers)	Carbon ion RT Dose 48–57.6 Gy (RBE) in 12–16 Fr Median tumor size 27 mm Bowel dose constrained at 20–30% of prescribed dose	None	Median re-RT interval 27.5 months Median prior dose 50 Gy/25 Fr	Median FU 37 months 3-year LC 94% 3-year DFS 55% 3-year OS 74%	No acute toxicity No ≥G3 late toxicity
[104] Retrospective 1995–2012	52 Central pelvic recurrence of carcinoma cervix	HDR-ISBT Dose 42 Gy/7 Fr (BD) GTV: Based on T2 MRI CTV: GTV + upper third of vagina Median tumor diameter 22.5 mm	42% received chemo before ISRT	Median re-RT interval 13 months	Median FU 55.6 months Median OS 32 months 5-year OS 52.6% LC 76.9% Tumor size, FIGO stage and disease-free interval prognostic for survival. Patients in ISRT group survived better than palliative (no ISRT) patients if <2 risk factors	25% had ≥G3 toxicity (VVF, RVF, hematuria, bowel obstruction, genital skin ulceration)

(continued)

Table 15.6 (continued)

Author Year Study type Duration	Number of patients	RT technique, dose, volume	Other therapy	Previous RT technique, dose, volume re-RT interval	Outcome	Toxicity
[105] Retrospective 1998–2014	45 Pelvic recurrence of carcinoma cervix	HDR (ISBT) 4–6 Gy/Fr × 4–6 Fr	4 patients received EBRT before HDR-ISRT 13 received chemo (6 concurrent)	Median re-RT interval 36 months	67% had CR 5-year OS 52% 5-year DFS in responders 45% Survival was higher in responder (5 years 73% vs. 0%)	33% had G3–4 toxicity With 14 fistulas— Higher rates with concurrent chemotherapy
[106] Retrospective 2008–2018	22 Vaginal recurrence of endometrial cancer	Definitive Brachy (ICRT/ISBT) ± EBRT (IGRT) 50% had additional EBRT Brachy: HDR, Ir-192 (interstitial for thickness >5 mm) Median HR-CTV 23.2 cc MC Brachy regime 22.5–27.5 Gy in 5 Fr (BD) EBRT dose 30.6 Gy to primary, 45–50.4 Gy with SIB to gross nodes CTV: Entire vagina, paravaginal tissues, pelvic nodes PTV: CTV + 0.7–1 cm margin	None	Median reirradiation interval 26.6 months Prior RT 55% had prior brachy 23% had prior EBRT Rest had combined treatment Median prior brachy regime 21 Gy/3 Fr Median prior EBRT dose 45 Gy	Median FU 27.6 months 3-year LC 65.8% 3-year DFS 40.8% 3-year OS 68.1%	No ≥ G3 acute or late GI/GU 1 G3 left ureteral stricture needing DJ stent

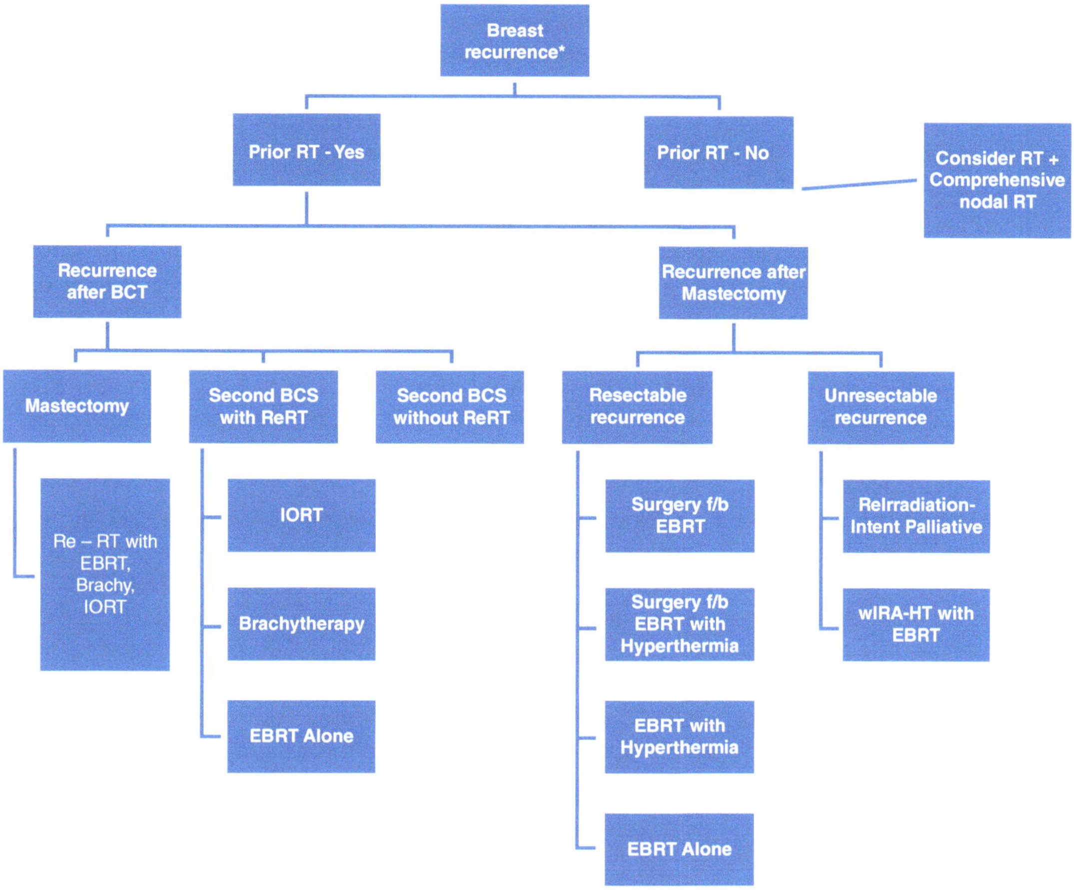

Fig. 15.1 Approaching reirradiation in breast malignancies (*IORT* intraoperative brachytherapy, *RT* Radiotherapy, *EBRT* external beam radiotherapy, *HT* hyperthermia, *wIRA* water-filtered infrared A)

techniques or VMAT with or without respiratory gating, rrachytherapy, intraoperative RT, hyperthermia with or without EBRT [111–116]. The re-RT doses range from 29.4 to 60 Gy, cumulative EQD2 doses <110 Gy resulted in local control rates ranging from 43% to 96%. There are no consensus guidelines with regards to fractionation. Most of the published literature favors conventional fractionation and the majority of evidence for hypofractionation in the Salvage BCS followed by partial breast irradiation scenario [113] and NRG-RTOG 1104 [114]. The evidence for hypofractionation is sparse in this setting [107]. The treatment options for breast cancer recurrences are summarized in Fig. 15.1.

In summary, re-RT is an excellent option for patients who relapse locally after initial radiation. It offers good local control rates with acceptable toxicities. The outcomes with reirradiation with modern systemic therapy and advanced treatment delivery modalities need more data to mature. Proton therapy has a dosimetric advantage of effective sparing of lungs and heart and whether it translates to clinical benefit needs to be evaluated in future.

15.9 Reirradiation for Gastrointestinal Malignancies

Radiotherapy plays an important role in the management of gastrointestinal (GI) cancers, and is used in the both definitive and adjuvant setting for patients with primary tumors in esophageal, gastric, pancreatic, and hepatobiliary regions [117–121]. Even with advanced radiotherapy techniques, systemic therapy, and surgical options available, many patients experience locoregional recurrent (LRR) disease, or develop a new primary cancer within a previously irradiated field. Retreatment by any modality poses a challenge both in terms of effectiveness and safety. In this setting, management options frequently include one or a combination of re-RT, systemic therapy, occasional salvage surgery, or supportive care. There is a paucity of data regarding the role of re-RT with external-beam radiation in the management of recurrent GI tumors, and even less describing outcomes for de novo GI cancers developing in fields of previous external-beam radiation, leaving this population of patients with markedly unmet oncologic needs. Dose schemes depend on a number of factors: total dose in the first RT series, indication of the reirradiation (symptom control versus. Locally ablative treatment), location and distance to organs at risk as well as technical approaches of the treating institution.

15.10 Reirradiation in Carcinoma Esophagus

With recent improvements in treatments, 10-year overall survival rates with the trimodality treatment approach have reached 38% [122]. Despite optimal management with trimodality therapy, nearly 49% of patients with esophageal cancer develop disease recurrence either (local or distant) within 5 years [122]. Importantly, up to 22% of these failures have been described as locoregional failures (LRFs) only. These include intraluminal or anastomotic with/out regional nodal disease. In the definitive setting local recurrence rate as a first site has been reported to be approximately in 44–61% patients [123, 124]. Regional lymph nodal failure is common in the surgery group while failure at primary site is common in the definitive setting [125]. Subsequent treatment-related decision-making becomes complex, particularly in the setting of a previous surgery like esophagectomy, while in the definitive setting, local recurrence is frequently masqueraded by strictures. For these locoregional recurrences, systemic therapy alone is not a curative treatment, and surgical resection carries high-risk even in the hands of experts [126]. There are no consistent recommendations or guidelines on how to salvage LRFs, though surgical salvage and high dose brachytherapy, external beam radiotherapy with photons and protons have been investigated. Most of these patients will die in 1 year without treatment. The rate of salvage esophagectomy in patients treated with primary definitive radiotherapy with curative intent has been reported to be 4–29%. However, a high hospital mortality rate (about 8–15%) has been reported [127]. The combination of cisplatin (CDDP) and 5-fluorouracil (5-FU) is frequently used with palliative intent with reported response rates between 6 and 8 months [128].

HDR intraluminal esophageal brachytherapy (HDR-ILBT) offers a highly conformal treatment with the possibility of delivering a relatively high radiation dose to the tumor volume while also delivering a relatively low dose to the surrounding normal tissues, particularly the lung, the spinal cord and the heart. According to the American Brachytherapy Society guidelines, good candidates for endoluminal HDR brachytherapy are patients with a primary tumor up to 10 cm in length, confined to the esophageal wall, within the thoracic esophagus, no tracheoesophageal fistula or tracheal involvement and without regional lymph node or systemic metastases [129]. Various different dose regimes have been used to apply esophageal endoluminal brachytherapy; and single doses between 4 and 6 Gy have been most commonly reported [130]. Several risk factors for the development of severe brachytherapy-associated toxicities have been reported, and it has been suggested that the risk increases with the use of smaller applicators (<8 mm), total doses delivered to the mucosal surface exceeding 12 Gy and the application of concomitant chemotherapy [131, 132]. Table 15.7 shows reirradiation using brachytherapy for advanced/recurrent esophageal carcinoma.

Table 15.7 Reirradiation using brachytherapy for advanced/recurrent esophageal carcinoma

Author, year	No.	Reirradiation interval, months	Reirradiation dose	Results	Toxicity
Wong Hee Kam et al. 2015 [133]	30	Not reported	5–7 Gy HDR × 4–6 fractions Prescription depth: 5 mm from the applicator surface	Median LPRS = 9.8 months. OS: 31.5% OS: 17.5% @ 2 years. Median OS, 9.8 months	Severe toxicity = 7 patients (23%), severe stenosis = 2 persistent fistula = 4 infections = 1 Death-5 patients
Folkert et al. 2013 [131]	09	Not reported	Median dose of 12 Gy HDR (range, 10–15 Gy) × 3 fractions Prescription depth: 7 mm (range, 4–10 mm)	Freedom from failure at 18 months = 11.1% OS at 18 months = 50%	Severe toxicity = 7.1% Stricture = 1 Fistula = 1 Death = 0
Nonoshita Takeshi 2007 [134]	06	Median = 3.5 months	Median total dose was 20 Gy HDR (range 20–32 Gy) × 3–4 fractions Prescription depth = 10 mm	Median OS = 30.0 months Median local control = 12.5 months	Severe toxicity = 0%
Nicolay et al. 2016 [130]	20	Not reported	4–6 Gy HDR in 2–4 fractions Prescription depth = 10 mm	2-year recurrence-free survival 6% Median OS = 14 months 3-year OS 6%	Fistula = 3%
Sharma et al. 2002 [135]	21	Median = 12.5 months	6 Gy × 2 fractions HDR Prescription depth = 10 mm	Median OS = 5.9 months Median dysphagia-free survival = 10 months	Overall complications = 8 (38%) Esophageal stricture = 5 Ulceration = 3

External beam re-RT plays a prominent role in the treatment of locally recurrent esophageal cancer and has been reported to have beneficial effects on palliation of symptoms and occasional cure [136]. A salvage esophagectomy is a high-risk surgical procedure because of fibrous changes in the mediastinum after radiation and difficulties of anastomosis as a result of the irradiated gastric tube [137]. A study by Chen et al. [138] analyzed the treatment modalities for recurrent esophageal cancer following definitive radiotherapy and compared the survival of patients who received salvage surgery, chemoradiation and best supportive care. The median survival period was 11.2 months in patients who received salvage CRT which was comparable to those patients who underwent salvage surgery, with 5-year survival rate of 4.6%. In contrast, all of the patients who received best supportive care died within 1 year, the median survival was only 6.2 months. When delivering re-RT to any area of the aerodigestive tract, late formation of strictures and fistulas become of paramount concern and studies report increase incidence of tracheo-esophageal fistula after reirradiation. A recent systematic review and meta-analysis revealed that the median overall survival and local failure-free survival for patients after reirradiation were 13.9 months (95% CI 4.18–46.51), and 11.01 months (95% CI 5.99–20.22) respectively. Esophageal perforation (pooled incidence: 21%; 95% CI 11–31), tracheoesophageal fistula (pooled incidence: 30%, 95% CI 2–58), and radiation pneumonitis (pooled incidence: 4%; 95% CI 2–6) have been reported [139]. The factors that influence the efficacy of re-RT are also of interest. The optimal radiation dose for recurrent esophageal cancer has not been determined.

Studies have reported a radiation dose ≥45 Gy and conformal RT to be associated with a better prognosis [140, 141], while concurrent use of chemotherapy increases toxicity [142] with mixed results reported in studies [125, 140, 141]. Regional nodal recurrence treated with re-RT has a better survival compared to primary failure (LN vs. PF, mean 28 vs. 6 months, $p = 0.004$) and time to recurrence>12 months has better outcomes (14 vs. 11 months, $p = 0.037$) than <12 months [143, 144]. Patients with re-RT interval ≤ 12 months, esophageal stenosis and disappearance of fat space between tumor and adjacent **tissue** have a higher risk of perforation. Patients with esophageal stenosis might be treated with stenting/dilatation, which is a risk factor for esophageal perforation with re-RT [145]. Taking into account the heterogeneity of patient population and treatment intent, individualized treatment options should be offered to each patient. Studies on the effectiveness of reradiotherapy in patients with recurrent esophageal cancer are listed in Table 15.8.

The role of proton RT in the management of esophageal cancer is under active investigation. A prospective study phase IIB [150] trail demonstrated that reirradiation using protons is feasible and has encouraging symptom control rates, modest radiation-related toxicity. The median time to symptom recurrence was 10 months while median survival of 14 months has been reported with low toxicity.

15.11 Reirradiation in Upper Abdominal Malignancies: Pancreas and Liver

The results of reirradiation to the abdominal area are sparse and this may be due to skepticism of physicians to reirradiate the abdominal area, as many organs, including the duodenum, stomach, small bowel, and liver, are known to be radiosensitive [151–153]. Radical reresection represents the most effective treatment and could be considered as the first therapeutic choice. Unfortunately, less than half of the patients with local recurrence show cases of resectable disease and surgery may not always be possible due to patient comorbidities and clinical conditions [154]. Reirradiation to the abdomen could potentially play a role in palliation of symptoms or local control but historically has been associated with poor outcomes. The availability of modern radiation techniques including stereotactic RT (SBRT), intensity-modulated radiation therapy (IMRT), and/or proton therapy (PT), allow for highly conformal dose distribution and sparing of the surrounding organs at risks (OARs). Safety has been the primary concern in reirradiation because calculating cumulative dose for organs at risk (OARs) is difficult and an excessive dose application to such organs may result in serious complications. Experts in upper abdominal SBRT would generally agree that sparing gastrointestinal organs must be prioritized ahead of planning target volume (PTV) coverage given the potential for mucosal injury and upper GI bleeding [155]. However, the degree of forgiveness of previous dose and techniques to minimize toxicity vary widely. There are limited data on the extent of repair in the bowel and the impact of reirradiation. The cumulative dose constraints for reirradiation for small bowel have been reported to be as high as 90–98 Gy_3 with no grade 3–5 toxicities, but caution must be exercised given the lack of long-term follow-up and small numbers [156]. A recent review [154] reported a pooled result of severe toxicity (≥G3) of 12% (95% CI 7.6–19%) in reirradiation of upper abdominal malignancies.

15.11.1 Reirradiation in Pancreatic Cancer

The most common pattern of failure following treatment of localized disease is distant, with up to 75% of patients developing metastatic failure [157] while locoregional failure is seen in 30–40% patients. The locoregional failure rate are reported to be as high as 42% even after a prior course of radiation [158, 159]. The presence of locoregional failure can lead to a wide range of complications, like biliary and bowel obstruction, cholangitis, gastrointestinal bleeding, and pain due to celiac plexopathy, potentially increasing morbidity and mortality. Current therapeutic

Table 15.8 Studies on the effectiveness of reradiotherapy in patients with recurrent esophageal cancer

Author, year	No.	Pattern of recurrence	Reirradiation interval, months	Initial dose	Reirradiation dose	Results	Toxicity
Liang Hong et al. 2018 [136]	39	Local/nodes	16 months	Median 62 Gy (40–76 Gy)	Median = 50 Gy (21–70)	Local failure = 21.3% OS = 21 months 3-year OS = 13%	Fistula = 9% Pneumonitis = 24% Pleural effusion = 15.1%
Fakhrian et al. 2012 [140]	21	Local/nodes	19 months	Definitive (8 patients) = 54 Gy (40–59) Neoadjuvant (13 patients) = 40 Gy (30–45 Gy)	Median 45 (30–68)	Median OS = 12 months 3-year OS = 19% Median recurrence-free interval = 8 months Recurrence-free survival 3 years = 15%	Fistula = 1 patient Stenosis = 1 patient Pneumonitis = 1 patient
Chen Y 2014 [138]	36	Local/nodes	14.6 months	54–63 Gy	Median dose = 50.4 Gy	Median OS = 11.2 months 3-year OS = 12.2%	Pneumonitis = 2.8% Fistula = 5 Perforation = 2 Death = 1
Xu et al. 2016 [146]	47	Local/nodes	26 months	>50 Gy	Median 58 Gy (26–64)	Dysphagia relief = 57.1% Median OS = 17 months 3-year OS 17%	Fistula = 4 Stenosis = 1 Bleeding/death = 1
Zhao Kaikai et al. 2020 [125]	252	Local/nodes	20 months	72 Gy (42.5–84)	Median = 60 Gy (12–86.3)	Median OS = 13 months 2-year OS = 21.8%	Perforation = 21.4% Pneumonitis = 12.8%
Kim Young Suk 2012 [141]	10	Local	15.6 months	50–50.4 Gy (50.4–63)	46.5 (44–50.4)	Not reported	Perforation/fistula = 3
Katano 2017 [147]	6	Local	17.4 months	Median dose = 50.4 Gy	30–50.4	Median OS = 13.6 months	Grade 3 dysphagia = 1
Zhi-guo Zhou 2015 [148]	55	Local	12 months	Median = 61.2 Gy	54 (18–66 Gy)	Median = 5 months 1-year OS = 16.4%	Perforation/ fistula = 20% Pneumonitis = 5.45%
Yamaguchi Shinsaku 2011 [149]	31	Local	10 months	Median = 60 Gy	36–40 Gy	Local control median = 8.8 months, median OS = 8 months	Perforation = 4 Stenosis = 1 Hemorrhage = 1

approaches for patients who develop isolated local recurrence following conventionally-fractionated radiotherapy include palliative chemotherapy and best supportive care. Select patients are eligible for reresection, yet locally recurrent disease is often unresectable as a consequence of vascular involvement, post-radiation fibrosis, or poor performance status [160]. Median survival following reresection is 11.4 months, while in-hospital morbidity and mortality are 20% and 7%, respectively [161, 162]. Palliative chemotherapy offers modest local control but is the main driver for survival [163]. The impact of re-RT in pancreas on survival is less well documented due to increased propensity of pancreatic cancer to disseminate early [164]. However, improved local control can translate into decreased morbidity and can also potentially lead to increased time off of systemic therapy and associated toxicities [165, 166]. Some centers advocate BID re-RT regimen (39 Gy at 1.5 Gy twice daily) with progression-free survival at 1 year of 38% [167]. One potential option is reirradiation with conformal techniques such as stereotactic body radiation therapy (SBRT). Although SBRT has been used in the reirradiation setting across multiple disease sites, there are few small retrospective studies investigating the use of SBRT for reirradiation in locally recurrent pancreatic cancer [168–171]. Patient selection is imperative, as toxicity with SBRT reirradiation can be significant due to the proximity of radiosensitive stomach and bowel. For example, the duodenum is known to be vulnerable to radiation, and excessive radiation to the duodenum may result in upper gastrointestinal (GI) bleeding, which could be fatal [151]. Smaller local recurrences may be ideal candidates for SBRT reirradiation and that peri-SBRT chemotherapy should be included in the treatment plan when feasible [172]. D2cc of the duodenum and a preexisting duodenal ulcer have been found to be significantly related to late severe toxicities; therefore, extreme caution is required to reduce the duodenal dose as much as possible for patients who have gastrointestinal morbidity [173]. Patients with a progression-free interval of greater than 9 months prior to isolated local recurrence or progression may be most suitable for reirradiation with SBRT51, as they appear to have a better prognosis with survival that is long enough for local control to be of potential benefit. Studies on reirradiation for pancreatic cancer are shown in Table 15.9.

15.12 Reirradiation in Liver Metastasis and Hepatocellular Carcinoma

Patients with a diagnosis of primary liver tumors (namely, hepatocellular carcinoma [HCC] and intrahepatic cholangiocarcinoma [ICC]) treated with recent radiation techniques have reported excellent local control with 2 year local control of more than 80% for HCC [175, 176] and 1-year local control of more than 84% for ICC [174, 177, 178]. These cancers commonly progress elsewhere in the liver and subsequent reirradiation poses a challenge. The control rates of liver metastases, depending upon the primary site of disease have been variable with 1-year local control rates between 20% and 95% [179–181]. For considering reirradiation of the liver, a careful assessment of the patient's baseline liver function is required in addition to the prior doses of radiation to the bile duct, and gastrointestinal luminal structures. Most data for reirradiation in liver malignancies is available from proton series [182, 183]. The volume of liver irradiated is especially crucial in HCC where patients usually have baseline cirrhosis and there is a much higher risk of hepatic decompensation than patients with liver metastases who usually have normal liver function. The ability of liver regeneration as a parallel organ has been studied primarily in surgical settings following partial hepatectomy; however, studies have also shown regeneration of unirradiated liver regions regenerating after protons and SBRT [184, 185]. As with other GI malignancies, generating a composite plan of the combined dose from the initial and retreatment courses to assess mucosal dosing is suggested. With careful selection the quoted incidence of radiation-induced liver disease (RILD) has been

Table 15.9 Studies on reirradiation for pancreatic cancer

Author, year	No.	Technique	Reirradiation interval, months	Initial dose	Reirradiation dose	Results	Toxicity
Haque et al. 2009 [174]	13	IMRT/3D-CRT	26 (5–83)	45 Gy (30–50.4 Gy)	24–48 Gy @1.5 Gy twice daily	Median freedom from local progression = 14 months Median OS = 14 months	Bleeding: 8% Duodenal stricture = 8%
Lominska Chris E 2012 [168]	28	SBRT	At least 2 months	50.4 Gy (41.4–70.2)	22.5 (20–30) @7 Gy (4–8)	Median OS = 5.9 months Freedom from local progression at 1 years = 70%	Obstruction = 1 patient Perforation = 1 patient
Wild AT 2013 [169]	18	SBRT	13.1 (7.8–17.5)	50.4 Gy (45 Gy–50.4 Gy)	25 (20–27)	Median OS = 8.8 months Freedom from local progression at 1 years = 70%	Bowel obstruction = 6%
Dagoglu Nergiz 2016 [170]	30	SBRT	18 (4–316)	SBRT 24 Gy/3 fractions (24–30 Gy) EBRT 50.4 Gy (46.8–54)	25 (24–36)	Median OS = 14 months 2-year local control = 78%	Bowel obstruction = 7%
Koong AJ 2017 [171]	23	SBRT	13 (2–32)	50.4 Gy (30 Gy–60 Gy)	25 (20–33) 1 patient = 25 Gy/1 fraction	Median OS = 8.5 months	GI toxicity = 8.7% Bleeding = 1 patient Fistula = 1 patient
Lee Jason Joon Bock 2019 [173]	42	3D-CRT/IMRT	15.2 (3.8–69.3)	50.0 Gy (range, 30.0–60.0 Gy)	45.0 Gy (range, 15.0–75.0 Gy).	Median OS = 12.5 months Failure-free survival = 9.2 months	GI toxicity = 4 3 bleeding 1 perforation
Reddy AV 2022 [172]	27	SBRT	19 (4.2–43.1)	SBRT 33 Gy (32.5–36 Gy) EBRT 50.4 Gy (18–55)	24–30 Gy@5–6 Gy/fraction	Median OS = 18.3 months Median local progression-free survival = 16.2 months	Grade 3 = 4%

reported to be 4.1% in studies of reirradiation. These studies recommend limiting the volume of unirradiated liver to at least 800 cc below 15 Gy [186]. Strategies to mitigate liver motion (breath-hold techniques, abdominal compression) and stomach filling are strongly recommended [155]. Finally, reirradiation should be considered for patients with HCC with good liver function (e.g., Child–Pugh score of A5) and for metastatic patients with a favorable systemic disease burden and good performance status. Studies on reirradiation in liver metastasis and hepatocellular carcinoma are shown in Table 15.10.

15.12.1 Reirradiation in Carcinoma Lung

Locally advanced non-small lung cancer (NSCLC) is often treated with a combination of surgery, systemic therapy, and radiation therapy. The pattern of failure for locally advanced NSCLC following chemoradiotherapy is frequently distant. The rates of local failures in the thorax, as seen in RTOG 0617 [188], are reported to be as high as 30–50% at 2 years [189, 190]. The isolated rates of locoregional recurrences are reported to be 25% [191]. Patents with recurrence are largely treated with systemic therapy alone and have a poor prognosis [192]. Systemic chemotherapy is given with palliative intent, with a median overall survival of less than 10 months and less than 30% of patients achieve an objective response which is similar to de-novo stage IV disease and thereby rendering re-RT, a more attractive treatment option [193]. With the advent of targeted and immune therapy the outcomes are improving, and some of these patients live long enough to develop with limited recurrence who can be offered re-RT, the incidence is likely to increase with better survivals. The tolerances of organs at risk (OAR) including heart, lung, great vessels, trachea/bronchi, and esophageal must be respected, as these are predictors of toxicity risk. Radiation pneumonitis is a major risk after reirradiation of lung tumors and severe fibrosis, fistula, and bleeding can also be noted as a result of the very high cumulative doses delivered to critical organs [193]. There is also no consensus on the optimal modality to deliver reirradiation. Proton therapy, SBRT, and IMRT are now common techniques for reirradiation with each modality providing advantages dependent on the specific scenario. Although stereotactic body radiation therapy can play an important role for recurrences after prior radiation therapy, particularly for isolated primary tumor recurrences that are more peripherally located and away from critical mediastinal structures [194]. Treatment-related toxicities can be severe while considering re-RT with SBRT especially in centrally-located recurrences [195–197]. The correlation between the central position of the tumor and reirradiation toxicity has also been documented by few studies [198] and a recent systematic review [199]. Hence, for the majority of centrally-located recurrent cases, conventional fractionation or a milder (≥8–10 fractions) hypofractionation regimen should be considered. Strong evidence on the minimal duration between definitive chemoradiation and potential reirradiation is also currently lacking with most series reporting a median of 12–18 months [200].

A small retrospective series found that delivering a higher dose of reirradiation, a dose which would often not be achievable with 3D-CRT due to OAR dose, was associated with improved survival [201]. Particle therapy may further reduce toxicities and/or enable safer re-RT dose escalation compared with 3D-CRT and IMRT [202, 203]. Higher BED has been associated with improved survival both for conventional fractionation and SBRT in the re-RT setting [204, 205]. A radical reirradiation cumulative dose should ideally be $BED_{10} > 100$ Gy for NSCLC. High-dose reirradiation results in median overall survival of about 17 months in selected patients, with median time to progression of about 10 months [193]. The pooled 2-year locoregional progression-free survival rate has been reported to be 47.2% [199].

Radiation-induced toxicities after re-RT remain a concern. In a pooled analysis of 14 studies, De Ruysscher et al. reported Grade 3–4 lung toxicity was seen in 0–21% with an average of 7% of patients exhibiting grade 3–4 lung toxic effects and 2% grade 3 esophagitis [193]. Grade

Table 15.10 Reirradiation in liver metastasis and hepatocellular carcinoma

Author, year	No.	Technique	Reirradiation interval, months	Initial dose	Reirradiation dose	Results	Toxicity
Susan G.R. McDuff 2017 [186]	47 (Mets+primary)	SBRT/ IMRT/ protons	9 months	Mean: 64.3 Gy [EQD2], $\alpha/\beta = 10$) 5–6 fractions or 15 fractions	Mean: 65.0 Gy 5–6 fractions or 15 fractions	Median OS = 14 months 1-year LF: 46.4% 1-year estimates of LF for liver metastases and HCC were 61.0% and 32.5%, respectively	4.1% nausea & vomiting 4.1% RILD
Seung Won Seol 2015 [187]	45 (HCC)	3D-CRT	13.8 months	32.5 Gy_{10} to 93.8 Gy_{10}	30 to 55 Gy (median, 45 Gy) in daily 2–15 Gy (median, 3 Gy) fractions	OS rates at 1- and 2-year 57% and 38% Median OS = 11.2 months	1 patient = duodenal perforation 1 patient = grade 3 pneumonitis

5 (lethal) bleeding was reported in 12 patients (3%; range 0–12%). Evans et al. reported a 25% risk of grade 5 aortic toxicity occurring with composite doses to the aorta greater than 120 Gy to 1 cm^3 [206]. For esophagus, prior dose escalation studies and limited reirradiation data suggest that the threshold for fistula starts around 85–90 Gy EDQ2 [207]. A recent systematic review reported data on acute toxicity were analyzed in all studies: ≥G3 acute toxicity was observed in 135 patients (10.5%). There was no difference in ≥G3 acute toxicity rate with respect to different radiation techniques, cumulative dose and reirradiation total dose and fractionation 84. A study attempted to estimate the tolerance of thoracic organs-at-risk (OAR) to reirradiation by calculating accumulated dose distributions using deformable image registration and dose conversion to EQD_2 As a conclusion, a total dose of 100 Gy_3 was reported to be safe in the setting of reirradiation [208]. The American Radium Society (ARS) and American College of Radiology (ACR) [209] have given re-RT composite dose constraints in 2 Gy equivalent doses of: esophagus V60 < 40% and DMax <100–110 Gy, lung V20 < 40%, heart V40 < 50%, aorta/great vessels DMax <120 Gy, trachea and proximal bronchial tree DMax <110 Gy, spinal cord DMax <57 Gy, and brachial plexus DMax <85 Gy.

With all thoracic reirradiation, the common good prognostic signs are small volume recurrences, good performance status, good lung function tests (FEV_1 > 1) and a longer interval between initial treatment and relapse (>12 months) [210]. The median survivals reported in SABR re-RT studies are higher and the difference is partially due to SABR being used in patients with low volume disease and conventional fractionation being used for patients with larger relapses or nodal involvement. Studies using carbon ions for reirradiation have recently been reported with encouraging results [229, 230]. Studies reporting reirradiation with conventional fractionation are shown in Table 15.11. Studies detailing stereotac-

Precepts Box

1. Treatment approaches: Recurrence in previously irradiated area is a therapeutic challenge.
   (a) Salvage surgery, if feasible, is preferred.
   (b) Reirradiation, with or without chemotherapy or supportive care, may be offered.
2. Patient selection is critical. Factors to be considered include performance status, age, co-morbidities, dose of prior radiotherapy and radiation fields, time since prior treatment, and extent and location of tumor.
3. Resected disease without high-risk features—reirradiation—can be avoided.
4. Unresectable, locally recurrent disease without distant metastases treatment options includes reirradiation with or without concurrent chemotherapy or SBRT.
5. Palliative reirradiation can be offered for symptomatic relief or those who may experience symptoms from local progression.
6. Toxicity from reirradiation can be considerable.
7. It is critical to select patients judiciously and provide robust supportive care.

Table 15.11 Studies reporting reirradiation with conventional fractionation

Author, year	No.	Technique	Reirradiation interval, months	Initial dose	Reirradiation dose	Results	Toxicity
Wu et al. 2003 [211]	23	3D-CRT	13 (6–42)	30–78 (66)	46–60 (51)	Median OS = 14	G3 late pulmonary fibrosis ($n = 2$) No G4–5 toxicity
Kruser et al. 2014 [212]	13	IMRT	(19.1)	NSCLC: 30–80.5 (57), SCLC: 12–54 (45)	NSCLC: Median for radical = 56 ($n = 9$), SCLC: Range 40–45 for radical ($n = 4$)	Median OS = 11.7	Grade 2 pneumonitis (25%) Bronchostenosis (25%)
Tada et al. 2005 [200]	19	Unknown	5–60 (16)	50–69.6	50 Gy/25 fractions	Median OS = 7.1	Grade 3 pneumonitis (5.2%) grade 2 esophagitis (16%)
Ohguri T et al. 2012 [213]	33	Unknown	7.9 months (1.1–28.2)	Median RT dose: 70 Gy (30–85 Gy)	RT + HT: Median dose 50 Gy (29–70 Gy) + median applications 5 (2–12)	Median LC: 12.1 months Median OS: 18.1 months	Pleuritis G3 (3%)
Tetar 2015 [214]	30	IMRT/4DCT	51 months (5–189)	Median RT dose: 59.8 Gy (24–70)	Median 60 Gy (39–66 Gy)	LPFS: 6.7 months Median OS: 13.5 months	Grade 3 esophagitis (3.3%) Grade 5: 3 patients (bleeding)
Sumita K 2016 [215]	21	Not reported	26 months (7–87.8)	Median 60 Gy (51.8–87.5) (NSCLC) 43.1 Gy (SCLC)	Median 60 Gy (54–87.5) (NSCLC) Median 50 Gy (50–87.5)(SCLC)	Median LPFS: 12.9 months Median OS: 31.4 months	Pneumonitis G3: 4.7%
Hong JH 2019 [216]	31	IMRT	15.1 months (4.4–56.3)	Median 60 Gy (45–66 Gy)	Median 64.5 Gy for NSCLC Median 45 Gy for SCLC	LC 2 years: 43.7% Median OS: 20.4 months OS 2 years: 39.4%	Late pericarditis G3: 3.2%
Schlampp I 2019 [217]	62	IMRT	12 months (1–62)	Median 60 Gy (36–70 Gy)	Median dose 38.5 Gy (20–60)	Median LPFS 6.5 months Median OS 9.3 months	Pneumonitis G3: 3.2% Pneumonitis G5: 1.6% Esophageal stenosis G3: 1.6% Trache-oesophageal fistula G4: 1.6% Pneumothorax G4: 1.6%

Table 15.12 Studies detailing stereotactic ablative radiotherapy (SABR)/protons reirradiation

Author, year	No.	Technique	Reirradiation interval, months	Initial dose	Reirradiation dose	Results	Toxicity
Reyngold M 2013 [204]	39	SBRT	37 (1–180)	61 (30–80)	70.4 Gy (BED10)	Median OS = 22 months Median LRFS = 13.8 months	Grade 2/3 pneumonitis (23%) Grade 2/3 chest wall pain (18%) Grade 4 skin toxicity (2.5%)
Kelly 2010 [218]	36	SBRT	22 (0–92)	61.5 (30–79.2)	50 Gy in 4 fractions or 40 Gy in 4 fractions	Local control 92% at 2 Year OS 2 years: 59%	Grade 3 pneumonitis (19.4%) Grade 3 esophagitis (8.3%) Grade 3 chest wall ulcers (8.3%)
Trakul 2012 [219]	15	SBRT	16 (5–80)	BED 87.5 (60–112.5)	80 (60–112.5) Gy	1 year LC: 65.5% 1-year OS: 58.2%	Chest wall toxicity = 6.7%
Trovo M 2014 [220]	17	SBRT	18 months (1–60)	50–60 Gy/20–30 Fr	30 Gy	LC 1 year: 86% OS 2 years: 29% Median OS: 19 months	Pneumonitis G3: 23% Pneumonitis G5: 5.8% Fatal hemoptysis: 5.8%
Hearn 2014 [221]	10	SBRT	14.8 months (9.9–26.3 months)	50 Gy/5 fractions (8 patients) 30 Gy/1 fractions (1 patient) 34 Gy/1 fractions (1 patient)	(50–60)	Not reported	No Grade 3/4
Kilburn 2014 [222]	23	SBRT	18 (6–61)	66 Gy/33 Fractions (median	50 in 10 fractions	LC = 80% at 1 year Median OS = 21 months	Grade 2/3 pneumonitis (9%) Grade 5 aortoesophageal fistula (3%)
Patel 2015 [223]	26	SBRT	8 (3–26)	61.2	Median BED 48 Gy	Median OS = 14 months, 1-year LC = 78.6%	No Grade 3/4

Repka MC 2017 [224]	20	SBRT	30.8 months (2.6–93.6)	Median 63 Gy	SBRT: Median 35 Gy (25–45 Gy)/5 Fr	LC 1 years: 30% OS 1 years: 45% Median OS: 12 months	Hemoptysis G5: 5%
Ceylon 2017 [225]	28	SBRT	14 (4–56)	54 (40–57)	45 Gy in 3 fractions or 30 Gy in 5 fractions	LPFS 2 years: 37% Median LC (BED <48 Gy): 13 months Median LC (BED >48 Gy): 48 months Median OS: 21 months OS 2 years: 42%	No > G3 toxicity
Ren C 2018 [226]	67	SBRT	16 (2–96)	56 (30–120)	3D (40–60) SBRT (40–50)	Not reported	18 G3 pneumonitis 1 G4 pneumonitis
Sumodhee S 2019 [198]	46	SBRT	22.6 (6.2–101.5)	66 Gy (44–70)	60 (40–75)	4-year LC and LRC: 44.2 and 14.7%, 4-year OS: 30.8%, Median local control = 36.3 Median OS = 21.8 months	Grade 5 = 4.4% 2 = hemoptysis/ alveolitis
McAvoy SA et al. 2013 [227]	33	Protons	36 months (1–376)	Median RT dose: 63 Gy (40–74 Gy)	Median 66 Gy RBE (16.4–75 Gy) (RBE)	Median LC: 18 months Median OS: 11.1 Mo OS 2 years: 33%	Pulmonary ≥ G3 (7) Tracheal necrosis G4 (2) Esophagitis G4 (1) Pericarditis G3 (1)
McAvoy SA et al. 2014 [227]	102	Protons	17 months (1–376)	Median RT dose: 70 EQD2 Gy (33–276 EQD2 Gy)	Median 70.7 EQD2 Gy	Median LPFS: 11.4 Months OS 2 years: 32.6% Median OS 14.7 Months	Pneumonitis G3: 6.9% Pneumonitis G4: 2.9% Esophagitis G3: 4.9% Esophagitis G4: 2%
Chao 2017 [228]	57	Protons	19 (4–151) months	Not reported	66.6 (30–74)	In field failure: 16% Regional failure: 9% Median OS: 14.9 months OS 2 years: 43%	Toxicity ≥ G3: 39%
Karube 2017 [229]	29	Carbon-ion	20 (8–99)	66 Gy	54–72 Gy RBE/12 Fr	LC 2 years: 66.9% OS 2 years: 69%	Pneumonitis G3: 3.4%
Hayashi K 2018 [230]	95	Carbon-ion	17 months (22–119)	Median 52.8 Gy RBE	Median 66 Gy RBE/12–16 Fr	LC 2 years: 54% Median LC: 13 months OS 2 years: 61.9% Median OS: 18 months	Bronchopleural fistula G5: 1% Pneumonitis G4: 1% Pneumonitis G3: 1% Chest wall pain G3: 1%

tic ablative radiotherapy (SABR)/protons reirradiation are shown in Table 15.12.

Multiple Choice Questions

1. Which of the following is true in the treatment of recurrent head and neck cancer?
   (a) Recurrences within the treatment field cannot be salvaged due to proximity to critical structures.
   (b) Surgical salvage in nonlaryngeal sites are not practiced.
   (c) Surgical salvage has a chance of cure in excess of 65%.
   (d) Surgical salvage can lead to recurrences in up to 50% cases.
   (e) Adjuvant chemotherapy reduces the chances of distant metastases.
2. When calculating dose for reirradiation, which of the following is true?
   (a) BED from initial treatment may not help if irradiating after 3 years.
   (b) An a/b ratio of 10 is to be considered for the spinal cord.
   (c) Initial treatment EQD2 of tumor is of utmost importance and determines the prescribed dose to the second cancer.
   (d) Cumulative EQD2 of spinal cord determines the maximal permissible dose to the tumor.
   (e) Reirradiation at 5.5 months postinitial therapy is safe and justified as shown in animal experiments.
3. Which of the following is true regarding reirradiation for head and neck cancers?
   (a) Adjuvant reirradiation post surgery is unnecessary.
   (b) Benefit of adjuvant reirradiation is justified by strong level II data.
   (c) Adjuvant reirradiation improves overall survival.
   (d) Salvage reirradiation without surgery leads to a 25% 2-year overall survival.
   (e) Salvage reirradiation with hyperfractionated radiotherapy is not any better than conventional fractionation based on cross trial data comparison.
4. The alpha beta ratio (α/β ratio) based on preclinical and clinical data for normal central nervous system tissue is estimated to be:
   (a) 2 Gy
   (b) 3 Gy
   (c) 5 Gy
   (d) 10 Gy
5. Which of the following statements regarding reirradiation for recurrent malignant glioma is correct?
   (a) Computerized tomography-based radiation planning is the standard for target delineation.
   (b) T1 weighted postcontrast abnormalities in the magnetic resonance imaging (MRI) of brain must be targeted at reirradiation.
   (c) There is no dose–effect benefit seen beyond radiation doses of 30–36 Gy fractionated at 1.8–6.0 Gy per fraction.
   (d) The median incidence of radionecrosis reported across reirradiation studies is 30%.
6. Which statement is correct about radionecrosis after brain reirradiation?
   (a) The median incidence of radionecrosis reported across reirradiation studies is 30%.
   (b) Temozolomide enhances radiosensitization to cause radionecrosis always and should never be used with reirradiation.
   (c) Symptomatic radionecrosis is always treated by radical surgery.
   (d) Brain reirradiation concurrent with hyperbaric oxygen therapy increases the chance of radionecrosis.
7. The median 12-month progression-free survival rate after reirradiation for recurrent high-grade glioma is:
   (a) 11% (95% CI 07–13%)
   (b) 17% (95% CI 13–20%)
   (c) 19% (95% CI 15–22%)
   (d) 21% (95% CI 17–24%)
8. Which of the following chemotherapy agents has been combined with brain reirradiation for high-grade gliomas?
   (a) Bevacizumab

(b) Temozolomide
(c) Paclitaxel
(d) All of the above

Answers

1. (d)
2. (d)
3. (e)
4. (a)
5. (b,c)
6. (d)
7. (b)
8. (d)

References

1. Lagerwaard FJ, Levendag PC, Nowak PJ, Eijkenboom WM, Hanssens PE, Schmitz PI. Identification of prognostic factors in patients with brain metastases: a review of 1292 patients. Int J Radiat Oncol Biol Phys. 1999;43(4):795–803.
2. Gavrilovic IT, Posner JB. Brain metastases: epidemiology and pathophysiology. J Neuro-Oncol. 2005;75(1):5–14.
3. Achrol AS, Rennert RC, Anders C, Soffietti R, Ahluwalia MS, Nayak L, et al. Brain metastases. Nat Rev Dis Prim. 2019;5:5. https://doi.org/10.1038/s41572-018-0055-y.
4. Gu L, Qing S, Zhu X, Ju X, Cao Y, Jia Z, Shen Y, Cao F, Fang F, Zhang H. Stereotactic radiation therapy (SRT) for brain metastases of multiple primary tumors: a single institution retrospective analysis. Front Oncol. 2019;9:135.
5. Larkin J, Chiarion-Sileni V, Gonzalez R, Grob JJ, Cowey CL, Lao CD, Schadendorf D, et al. Combined nivolumab and ipilimumab or monotherapy in untreated melanoma. N Engl J Med. 2015;373(1):23–34.
6. Reck M, Rodríguez-Abreu D, Robinson AG, Hui R, Csőszi T, Fülöp A, Gottfried M, et al. Pembrolizumab versus chemotherapy for PD-L1–positive non–small-cell lung cancer. N Engl J Med. 2016;375(19):1823–33.
7. Kocher M, Soffietti R, Abacioglu U, Villà S, Fauchon F, Baumert BG, Fariselli L, et al. Adjuvant whole-brain radiotherapy versus observation after radiosurgery or surgical resection of one to three cerebral metastases: results of the EORTC 22952-26001 study. J Clin Oncol. 2011;29(2):134–41.
8. Chidambaram S, Pannullo SC, Schwartz TH, Wernicke AG. Reirradiation of recurrent brain metastases: where do we stand? World Neurosurg. 2019;125:156–63.
9. Mayer R, Sminia P. Reirradiation tolerance of the human brain. Int J Radiat Oncol Biol Phys. 2008;70(5):1350–60.
10. Liu L, Bassano DA, Prasad SC, Hahn SS, Chung CT. The linear-quadratic model and fractionated stereotactic radiotherapy. Int J Radiat Oncol Biol Phys. 2003;57(3):827–32.
11. Shigematsu N, Kunieda E, Kawaguchi O, Takeda A, Ihara N, Yamashita S, Kubo A, Ito H. Indications of stereotactic irradiation for brain lesions. Acta Oncol. 2000;39(5):597–603.
12. Borgelt B, Gelber R, Kramer S, Brady LW, Chang CH, Davis LW, Perez CA, Hendrickson FR. The palliation of brain metastases: final results of the first two studies by the Radiation Therapy Oncology Group. Int J Radiat Oncol Biol Phys. 1980;6(1):1–9.
13. Hoskin PJ, Crow J, Ford HT. The influence of extent and local management on the outcome of radiotherapy for brain metastases. Int J Radiat Oncol Biol Phys. 1990;19(1):111–5.
14. Sadikov E, Bezjak A, Yi Q-L, Wells W, Dawson L, Millar B-A, Laperriere N. Value of whole brain re-irradiation for brain metastases—single centre experience. Clin Oncol. 2007;19(7):532–8. https://doi.org/10.1016/j.clon.2007.06.001.
15. Son CH, Jimenez R, Niemierko A, Loeffler JS, Oh KS, Shih HA. Outcomes after whole brain reirradiation in patients with brain metastases. Int J Radiat Oncol Biol Phys. 2012;82:e167.
16. Ozgen Z, Atasoy BM, Kefeli AU, Seker A, Dane F, Abacioglu U. The benefit of whole brain reirradiation in patients with multiple brain metastases. Radiat Oncol. 2013;8:186.
17. Wong WW, Schild SE, Sawyer TE, Shaw EG. Analysis of outcome in patients reirradiated for brain metastases. Int J Radiat Oncol Biol Phys. 1996;34(3):585–90. https://doi.org/10.1016/0360-3016(95)02156-6.
18. Kocher M, Wittig A, Piroth MD, Treuer H, Seegenschmiedt H, Ruge M, Grosu A-L, Guckenberger M. Stereotactic radiosurgery for treatment of brain metastases. Strahlenther Onkol. 2014;190:521. https://doi.org/10.1007/s00066-014-0648-7.
19. Soffietti R, Kocher M, Abacioglu UM, Villa S, Fauchon F, Baumert BG, Fariselli L, et al. A European Organisation for Research and Treatment of Cancer phase III trial of adjuvant whole-brain radiotherapy versus observation in patients with one to three brain metastases from solid tumors after surgical resection or radiosurgery: quality-of-life results. J Clin Oncol. 2013;31(1):65–72.
20. Brown PD, Jaeckle K, Ballman KV, Farace E, Cerhan JH, Keith Anderson S, Carrero XW, et al. Effect of radiosurgery alone vs radiosurgery with whole brain radiation therapy on cognitive function in patients with 1 to 3 brain metastases. JAMA. 2016;316:401. https://doi.org/10.1001/jama.2016.9839.
21. Sahgal A, Aoyama H, Kocher M, Neupane B, Collette S, Tago M, Shaw P, Beyene J, Chang EL. Phase 3 tri-

als of stereotactic radiosurgery with or without whole-brain radiation therapy for 1 to 4 brain metastases: individual patient data meta-analysis. Int J Radiat Oncol Biol Phys. 2015;91(4):710–7.
22. Aoyama H, Shirato H, Tago M, Nakagawa K, Toyoda T, Hatano K, Kenjyo M, et al. Stereotactic radiosurgery plus whole-brain radiation therapy vs stereotactic radiosurgery alone for treatment of brain metastases. JAMA. 2006;295(21):2483–91. https://doi.org/10.1001/jama.295.21.2483.
23. Combs SE, Schulz-Ertner D, Thilmann C, Edler L, Debus J. Treatment of cerebral metastases from breast cancer with stereotactic radiosurgery. Strahlenther Onkol. 2004;180(9):590–6.
24. Sneed PK, Mendez J, Johanna GM, den Hoek V-v, Seymour ZA, Ma L, Molinaro AM, Fogh SE, Nakamura JL, McDermott MW. Adverse radiation effect after stereotactic radiosurgery for brain metastases: incidence, time course, and risk factors. J Neurosurg. 2015;123:373. https://doi.org/10.3171/2014.10.jns141610.
25. Kuwahara Y, Tomita K, Urushihara Y, Sato T, Kurimasa A, Fukumoto M. Association between radiation-induced cell death and clinically relevant Radioresistance. Histochem Cell Biol. 2018;150(6):649–59. https://doi.org/10.1007/s00418-018-1728-z.
26. Lucia F, Touati R, Crainic N, Dissaux G, Pradier O, Bourbonne V, Schick U. Efficacy and safety of a second course of stereotactic radiation therapy for locally recurrent brain metastases: a systematic review. Cancers. 2021;13(19):4929. https://doi.org/10.3390/cancers13194929.
27. Terakedis BE, Jensen RL, Boucher K, Shrieve DC. Tumor control and incidence of radiation necrosis after reirradiation with stereotactic radiosurgery for brain metastases. J Radiosurg SBRT. 2014;3:21–8.
28. Rana N, Pendyala P, Cleary RK, Luo G, Zhao Z, Chambless LB, Cmelak AJ, Attia A, Stavas MJ. Long-term outcomes after salvage stereotactic radiosurgery (SRS) following in-field failure of initial SRS for brain metastases. Front Oncol. 2017;7:279.
29. Balermpas P, Stera S, von der Grün JM, Loutfi-Krauss B, Forster M-T, Wagner M, Keller C, et al. Repeated in-field radiosurgery for locally recurrent brain metastases: feasibility, results and survival in a heavily treated patient cohort. PLoS One. 2018;13(6):e0198692. https://doi.org/10.1371/journal.pone.0198692.
30. Dincoglan F, Sager O, Demiral S, Gamsiz H, Uysal B, Onal E, Ekmen A, Dirican B, Beyzadeoglu M. Fractionated stereotactic radiosurgery for locally recurrent brain metastases after failed stereotactic radiosurgery. Indian J Cancer. 2019;56:151–6. https://doi.org/10.4103/ijc.ijc_786_18.
31. Iorio-Morin C, Mercure-Cyr R, Figueiredo G, Touchette CJ, Masson-Côté L, Mathieu D. Repeat stereotactic radiosurgery for the management of locally recurrent brain metastases. J Neurooncol. 2019;145:551. https://doi.org/10.1007/s11060-019-03323-8.
32. Caballero JA, Sneed PK, Lamborn KR, Ma L, Denduluri S, Nakamura JL, Barani IJ, McDermott MW. Prognostic factors for survival in patients treated with stereotactic radiosurgery for recurrent brain metastases after prior whole brain radiotherapy. Int J Radiat Oncol Biol Phys. 2012;83(1):303–9.
33. Maranzano E, Trippa F, Casale M, Costantini S, Anselmo P, Carletti S, Principi M, Caserta C, Loreti F, Giorgi C. Reirradiation of brain metastases with radiosurgery. Radiother Oncol. 2012;102:192. https://doi.org/10.1016/j.radonc.2011.07.018.
34. Huang Z, Sun B, Shen G, Cha L, Meng X, Wang J, Zhou Z, Wu S. Brain metastasis reirradiation in patients with advanced breast cancer. J Radiat Res. 2017;58(1):142–8.
35. Yomo S, Hayashi M. The efficacy and limitations of stereotactic radiosurgery as a salvage treatment after failed whole brain radiotherapy for brain metastases. J Neurooncol. 2013;113:459. https://doi.org/10.1007/s11060-013-1138-y.
36. Atkins KM, Pashtan IM, Bussière MR, Kang KH, Niemierko A, Daly JE, Botticello TM, et al. Proton stereotactic radiosurgery for brain metastases: a single-institution analysis of 370 patients. Int J Radiat Oncol Biol Phys. 2018;101:820. https://doi.org/10.1016/j.ijrobp.2018.03.056.

Primary Brain Tumour References

37. Ostrom QT, Cioffi G, Waite K, Kruchko C, Barnholtz-Sloan JS. CBTRUS statistical report: primary brain and other central nervous system tumors diagnosed in the United States in 2014-2018. Neuro Oncol. 2021;23:iii1–105.
38. Stupp R, Mason WP, van den Bent MJ, Weller M, Fisher B, Taphoorn MJB, et al. Radiotherapy plus concomitant and adjuvant temozolomide for glioblastoma. N Engl J Med. 2005;352:987–96.
39. Stupp R, Hegi ME, Mason WP, van den Bent MJ, Taphoorn MJB, Janzer RC, et al. Effects of radiotherapy with concomitant and adjuvant temozolomide versus radiotherapy alone on survival in glioblastoma in a randomised phase III study: 5-year analysis of the EORTC-NCIC trial. Lancet Oncol. 2009;10:459–66.
40. Ceccarelli M, Barthel FP, Malta TM, Sabedot TS, Salama SR, Murray BA, et al. Molecular profiling reveals biologically discrete subsets and pathways of progression in diffuse glioma. Cell. 2016;164:550–63.
41. Louis DN, Perry A, Wesseling P, Brat DJ, Cree IA, Figarella-Branger D, et al. The 2021 WHO classification of tumors of the central nervous system: a summary. Neuro Oncol. 2021;23:1231. https://doi.org/10.1093/neuonc/noab106.

42. McBain C, Lawrie TA, Rogozińska E, Kernohan A, Robinson T, Jefferies S. Treatment options for progression or recurrence of glioblastoma: a network meta-analysis. Cochrane Database Syst Rev. 2021;5:CD013579.
43. Combs SE, Widmer V, Thilmann C, Hof H, Debus J, Schulz-Ertner D. Stereotactic radiosurgery (SRS): treatment option for recurrent glioblastoma multiforme (GBM). Cancer. 2005;104:2168–73.
44. Rieken S, Habermehl D, Giesel FL, Hoffmann C, Burger U, Rief H, et al. Analysis of FET-PET imaging for target volume definition in patients with gliomas treated with conformal radiotherapy. Radiother Oncol. 2013;109:487–92.
45. De Maria L, Terzi di Bergamo L, Conti A, Hayashi K, Pinzi V, Murai T, et al. CyberKnife for recurrent malignant gliomas: a systematic review and meta-analysis. Front Oncol. 2021;11:652646.
46. Kazmi F, Soon YY, Leong YH, Koh WY, Vellayappan B. Re-irradiation for recurrent glioblastoma (GBM): a systematic review and meta-analysis. J Neurooncol. 2019;142:79–90.
47. Kim HK, Thornton AF, Greenberg HS, Page MA, Junck L, Sandler HM. Results of re-irradiation of primary intracranial neoplasms with three-dimensional conformal therapy. Am J Clin Oncol. 1997;20:358–63.
48. Kim B, Soisson E, Duma C, Chen P, Hafer R, Cox C, et al. Treatment of recurrent high grade gliomas with hypofractionated stereotactic image-guided helical tomotherapy. Clin Neurol Neurosurg. 2011;113:509–12.
49. Shi W, Blomain ES, Siglin J, Palmer JD, Dan T, Wang Y, et al. Salvage fractionated stereotactic re-irradiation (FSRT) for patients with recurrent high grade gliomas progressed after bevacizumab treatment. J Neurooncol. 2018;137:171–7.
50. Sahebjam S, Forsyth PA, Tran ND, Arrington JA, Macaulay R, Etame AB, et al. Hypofractionated stereotactic re-irradiation with pembrolizumab and bevacizumab in patients with recurrent high-grade gliomas: results from a phase I study. Neuro Oncol. 2021;23:677–86.
51. Kim IH. Appraisal of re-irradiation for the recurrent glioblastoma in the era of MGMT promotor methylation. Radiat Oncol J. 2019;37:1–12.
52. Choi M, Zabramski JM. Re-irradiation using brachytherapy for recurrent intracranial tumors: a systematic review and meta-analysis of the literature. Cureus. 2020;12:e9666.
53. Rieken S, Habermehl D, Haberer T, Jaekel O, Debus J, Combs SE. Proton and carbon ion radiotherapy for primary brain tumors delivered with active raster scanning at the Heidelberg Ion Therapy Center (HIT): early treatment results and study concepts. Radiat Oncol. 2012;7:41.
54. Adeberg S, Harrabi SB, Verma V, Bernhardt D, Grau N, Debus J, et al. Treatment of meningioma and glioma with protons and carbon ions. Radiat Oncol. 2017;12:193.
55. Kondziolka D, Shin SM, Brunswick A, Kim I, Silverman JS. The biology of radiosurgery and its clinical applications for brain tumors. Neuro Oncol. 2015;17:29–44.
56. Maranzano E, Anselmo P, Casale M, Trippa F, Carletti S, Principi M, et al. Treatment of recurrent glioblastoma with stereotactic radiotherapy: long-term results of a mono-institutional trial. Tumori. 2011;97:56–61.
57. Voynov G, Kaufman S, Hong T, Pinkerton A, Simon R, Dowsett R. Treatment of recurrent malignant gliomas with stereotactic intensity modulated radiation therapy. Am J Clin Oncol. 2002;25:606–11.
58. Chakravarti A, Erkkinen MG, Nestler U, Stupp R, Mehta M, Aldape K, et al. Temozolomide-mediated radiation enhancement in glioblastoma: a report on underlying mechanisms. Clin Cancer Res. 2006;12:4738–46.
59. Kohshi K, Yamamoto H, Nakahara A, Katoh T, Takagi M. Fractionated stereotactic radiotherapy using gamma unit after hyperbaric oxygenation on recurrent high-grade gliomas. J Neurooncol. 2007;82:297–303.
60. Hudes RS, Corn BW, Werner-Wasik M, Andrews D, Rosenstock J, Thoron L, et al. A phase I dose escalation study of hypofractionated stereotactic radiotherapy as salvage therapy for persistent or recurrent malignant glioma. Int J Radiat Oncol Biol Phys. 1999;43:293–8.
61. Vordermark D, Kölbl O, Ruprecht K, Vince GH, Bratengeier K, Flentje M. Hypofractionated stereotactic re-irradiation: treatment option in recurrent malignant glioma. BMC Cancer. 2005;5:55.
62. Fokas E, Wacker U, Gross MW, Henzel M, Encheva E, Engenhart-Cabillic R. Hypofractionated stereotactic reirradiation of recurrent glioblastomas: a beneficial treatment option after high-dose radiotherapy? Strahlenther Onkol. 2009;185:235–40.
63. Combs SE, Niyazi M, Adeberg S, Bougatf N, Kaul D, Fleischmann DF, et al. Re-irradiation of recurrent gliomas: pooled analysis and validation of an established prognostic score-report of the Radiation Oncology Group (ROG) of the German Cancer Consortium (DKTK). Cancer Med. 2018;7:1742–9.
64. Yazici G, Cengiz M, Ozyigit G, Eren G, Yildiz F, Akyol F, et al. Hypofractionated stereotactic reirradiation for recurrent glioblastoma. J Neurooncol. 2014;120:117–23.
65. Dincoglan F, Beyzadeoglu M, Sager O, Demiral S, Gamsiz H, Uysal B, et al. Management of patients with recurrent glioblastoma using hypofractionated stereotactic radiotherapy. Tumori. 2015;101:179–84.

Head and Neck Cancer References

66. Ang KK, Jiang GL, Feng Y, Stephens LC, Tucker SL, Price RE. Extent and kinetics of recovery of

occult spinal cord injury. Int J Radiat Oncol Biol Phys. 2001;50:1013–20.
67. Tortochaux J, Tao Y, Tournay E, Lapeyre M, Lesaunier F, Bardet E, et al. Randomized phase III trial (GORTEC 98-03) comparing re-irradiation plus chemotherapy versus methotrexate in patients with recurrent or a second primary head and neck squamous cell carcinoma, treated with a palliative intent. Radiother Oncol. 2011;100:70–5.
68. Bachar GY, Goh C, Goldstein DP, O'Sullivan B, Irish JC. Long-term outcome analysis after surgical salvage for recurrent tonsil carcinoma following radical radiotherapy. Eur Arch Otorrinolaringol. 2010;267:295–301.
69. Goodwin WJ Jr. Salvage surgery for patients with recurrent squamous cell carcinoma of the upper aerodigestive tract: when do the ends justify the means? Laryngoscope. 2000;110:1–18.
70. Janot F, de Raucourt D, Benhamou E, Ferron C, Dolivet G, Bensadoun R-J, et al. Randomized trial of postoperative reirradiation combined with chemotherapy after salvage surgery compared with salvage surgery alone in head and neck carcinoma. J Clin Oncol. 2008;26:5518–23.
71. Spencer SA, Harris J, Wheeler RH, Machtay M, Schultz C, Spanos W, et al. Final report of RTOG 9610, a multi-institutional trial of reirradiation and chemotherapy for unresectable recurrent squamous cell carcinoma of the head and neck. Head Neck. 2008;30:281–8.
72. Langer CJ, Harris J, Horwitz EM, Nicolaou N, Kies M, Curran W, et al. Phase II study of low-dose paclitaxel and cisplatin in combination with split-course concomitant twice-daily reirradiation in recurrent squamous cell carcinoma of the head and neck: results of Radiation Therapy Oncology Group Protocol 9911. J Clin Oncol. 2007;25:4800–5.
73. De Crevoisier R, Bourhis J, Domenge C, Wibault P, Koscielny S, Lusinchi A, et al. Full-dose reirradiation for unresectable head and neck carcinoma: experience at the Gustave-Roussy Institute in a series of 169 patients. J Clin Oncol. 1998;16:3556–62.
74. Höckel M, Knapstein PG, Kutzner J. A novel combined operative and radiotherapeutic treatment approach for recurrent gynecologic malignant lesions infiltrating the pelvic wall. Surg Gynecol Obstet. 1991;173:297–302.
75. Fiorentino A, Giaj Levra N, Mazzola R, Fersino S, Ricchetti F, Ballario R, et al. Stereotactic body radiation therapy with hydrogel spacer: a salvage reirradiation strategy for prostate cancer recurrence. Minerva Urol Nefrol. 2016;68:464–6.
76. Lee J, Kim CY, Koom WS, Rim CH. Practical effectiveness of re-irradiation with or without surgery for locoregional recurrence of rectal cancer: a meta-analysis and systematic review. Radiother Oncol. 2019;140:10–9.
77. Johnstone P, Okonta L, Aitken K, Holmes J, Harrison M, Harji D, et al. A multicentre retrospective review of SABR reirradiation in rectal cancer recurrence. Radiother Oncol. 2021;162:1–6.
78. Murray LJ, Lilley J, Hawkins MA, Henry AM, Dickinson P, Sebag-Montefiore D. Pelvic re-irradiation using stereotactic ablative radiotherapy (SABR): a systematic review. Radiother Oncol. 2017;125:213–22.
79. Denost Q, Frison E, Salut C, Sitta R, Rullier A, Harji D, et al. A phase III randomized trial evaluating chemotherapy followed by pelvic reirradiation versus chemotherapy alone as preoperative treatment for locally recurrent rectal cancer—GRECCAR 15 trial protocol. Color Dis. 2021;23:1909–18.
80. Zerini D, Jereczek-Fossa BA, Ciabattoni A, Mirri A, Bertoni F, Fersino S, et al. PROLAPSE: survey about local prostate cancer relapse salvage treatment with external beam re-irradiation: results of the Italian Association of Radiotherapy and Clinical Oncology (AIRO). J Cancer Res Clin Oncol. 2020;146:2311–7.
81. Lieng H, Hayden AJ, Christie DRH, Davis BJ, Eade TN, Emmett L, et al. Radiotherapy for recurrent prostate cancer: 2018 recommendations of the Australian and New Zealand Radiation Oncology Genito-Urinary group. Radiother Oncol. 2018;129:377–86.
82. Chatzikonstantinou G, Zamboglou N, Rödel C, Zoga E, Strouthos I, Butt SA, et al. High-dose-rate brachytherapy as salvage modality for locally recurrent prostate cancer after definitive radiotherapy: a systematic review. Strahlenther Onkol. 2017;193:683–91.
83. Munoz F, Fiorica F, Caravatta L, Rosa C, Ferella L, Boldrini L, et al. Outcomes and toxicities of re-irradiation for prostate cancer: a systematic review on behalf of the Re-Irradiation Working Group of the Italian Association of Radiotherapy and Clinical Oncology (AIRO). Cancer Treat Rev. 2021;95:102176.
84. Llewelyn M, Taylor A. Re-irradiation of cervical and endometrial cancer. Curr Opin Oncol. 2017;29:343–50.
85. Sturdza A, Viswanathan AN, Erickson B, Yashar C, Bruggeman A, Feddock J, et al. American Brachytherapy Society working group report on the patterns of care and a literature review of reirradiation for gynecologic cancers. Brachytherapy. 2020;19:127–38.
86. Dijkstra EA, Mul VEM, Hemmer PHJ, Havenga K, Hospers GAP, Muijs CT, et al. Re-irradiation in patients with recurrent rectal cancer is safe and feasible. Ann Surg Oncol. 2021;28:5194–204.
87. Barcellini A, Vitolo V, Cobianchi L, Peloso A, Vanoli A, Mirandola A, et al. Re-irradiation with carbon ion radiotherapy for pelvic rectal cancer recurrences in patients previously irradiated to the pelvis. In Vivo. 2020;34:1547–53.
88. Al-Haidari G, Skovlund E, Undseth C, Rekstad BL, Larsen SG, Åsli LM, et al. Re-irradiation for recurrent rectal cancer—a single-center experience. Acta Oncol. 2020;59:534–40.

89. Tao R, Tsai CJ, Jensen G, Eng C, Kopetz S, Overman MJ, et al. Hyperfractionated accelerated reirradiation for rectal cancer: an analysis of outcomes and toxicity. Radiother Oncol. 2017;122:146–51.
90. Chuong MD, Fernandez DC, Shridhar R, Hoffe SE, Saini A, Hunt D, et al. High-dose-rate endorectal brachytherapy for locally advanced rectal cancer in previously irradiated patients. Brachytherapy. 2013;12:457–62.
91. Milani V, Pazos M, Issels RD, Buecklein V, Rahman S, Tschoep K, et al. Radiochemotherapy in combination with regional hyperthermia in preirradiated patients with recurrent rectal cancer. Strahlenther Onkol. 2008;184:163–8.
92. Valentini V, Morganti AG, Gambacorta MA, Mohiuddin M, Doglietto GB, Coco C, et al. Preoperative hyperfractionated chemoradiation for locally recurrent rectal cancer in patients previously irradiated to the pelvis: a multicentric phase II study. Int J Radiat Oncol Biol Phys. 2006;64:1129–39.
93. Matrone F, Revelant A, Fanetti G, Polesel J, Chiovati P, Avanzo M, et al. Partial prostate re-irradiation for the treatment of isolated local recurrence of prostate cancer in patients previously treated with primary external beam radiotherapy: short-term results of a monocentric study. Neoplasma. 2021;68:216–26.
94. Bergamin S, Eade T, Kneebone A, Booth J, Hsiao E, Schembri GP, et al. Interim results of a prospective prostate-specific membrane antigen-directed focal stereotactic reirradiation trial for locally recurrent prostate cancer. Int J Radiat Oncol Biol Phys. 2020;108:1172–8.
95. Schönle N, Strnad V, Lotter M, Kreppner S, Fietkau R. Long-term results of a protocol-based ultrasound-guided salvage brachytherapy as re-irradiation for local recurrent prostate cancer. Radiother Oncol. 2020;150:201–5.
96. Fuller D, Wurzer J, Shirazi R, Bridge S, Law J, Crabtree T, et al. Retreatment for local recurrence of prostatic carcinoma after prior therapeutic irradiation: efficacy and toxicity of HDR-like SBRT. Int J Radiat Oncol Biol Phys. 2020;106:291–9.
97. D'Agostino GR, Di Brina L, Mancosu P, Franzese C, Iftode C, Franceschini D, et al. Reirradiation of locally recurrent prostate cancer with volumetric modulated arc therapy. Int J Radiat Oncol Biol Phys. 2019;104:614–21.
98. Jereczek-Fossa BA, Rojas DP, Zerini D, Fodor C, Viola A, Fanetti G, et al. Reirradiation for isolated local recurrence of prostate cancer: mono-institutional series of 64 patients treated with salvage stereotactic body radiotherapy (SBRT). Br J Radiol. 2019;92:20180494.
99. Mbeutcha A, Chauveinc L, Bondiau P-Y, Chand M-E, Durand M, Chevallier D, et al. Salvage prostate re-irradiation using high-dose-rate brachytherapy or focal stereotactic body radiotherapy for local recurrence after definitive radiation therapy. Radiat Oncol. 2017;12:49.
100. Janoray G, Reynaud-Bougnoux A, Ruffier-Loubière A, Bernadou G, Pointreau Y, Calais G. Stereotactic body re-irradiation therapy for locally recurrent prostate cancer after external-beam radiation therapy: initial report. Cancer Radiother. 2016;20:275–81.
101. Raziee H, D'Souza D, Velker V, Barnes E, Taggar A, Mendez L, et al. Salvage re-irradiation with single-modality interstitial brachytherapy for the treatment of recurrent gynaecological tumours in the pelvis: a multi-institutional study. Clin Oncol. 2020;32:43–51.
102. Ling DC, Vargo JA, Burton SA, Heron DE, Beriwal S. Salvage curative-intent reirradiation stereotactic body radiation therapy for isolated pelvic and/or paraortic recurrences of gynecologic malignancies. Pract Radiat Oncol. 2019;9:418–25.
103. Shiba S, Okonogi N, Kato S, Wakatsuki M, Kobayashi D, Kiyohara H, et al. Clinical impact of re-irradiation with carbon-ion radiotherapy for lymph node recurrence of gynecological cancers. Anticancer Res. 2017;37:5577–83.
104. Mabuchi S, Takahashi R, Isohashi F, Yokoi T, Okazawa M, Sasano T, et al. Reirradiation using high-dose-rate interstitial brachytherapy for locally recurrent cervical cancer: a single institutional experience. Int J Gynecol Cancer. 2014;24:141–8.
105. da Silva VTM, Fortuna Diniz AP, Martins J, Cursino K, Esteves SCB, Teixeira JC. Use of interstitial brachytherapy in pelvic recurrence of cervical carcinoma: clinical response, survival, and toxicity. Brachytherapy. 2019;18:146–53.
106. Ling DC, Vargo JA, Glaser SM, Kim H, Beriwal S. Outcomes after definitive re-irradiation with 3D brachytherapy with or without external beam radiation therapy for vaginal recurrence of endometrial cancer. Gynecol Oncol. 2019;152(3):581–6. https://doi.org/10.1016/j.ygyno.2018.12.022. Epub 2018 Dec 29. PMID: 30600093; PMCID: PMC7422921.
107. Harms W, Budach W, Dunst J, Feyer P, Fietkau R, Haase W, et al. DEGRO practical guidelines for radiotherapy of breast cancer VI: therapy of locoregional breast cancer recurrences. Strahlenther Onkol. 2016;192:199–208.
108. Vicini FA, Antonucci JV, Wallace M, Gilbert S, Goldstein NS, Kestin L, et al. Long-term efficacy and patterns of failure after accelerated partial breast irradiation: a molecular assay-based clonality evaluation. Int J Radiat Oncol Biol Phys. 2007;68:341–6.
109. Gregucci F, Fozza A, Falivene S, Smaniotto D, Morra A, Daidone A, et al. Present clinical practice of breast cancer radiotherapy in Italy: a nationwide survey by the Italian Society of Radiotherapy and Clinical Oncology (AIRO) Breast Group. Radiol Med. 2020;125:674–82.
110. Panet-Raymond V, Truong PT, Alexander C, Lesperance M, McDonald RE, Watson PH. Clinicopathologic factors of the recurrent tumor predict outcome in patients with ipsilateral breast tumor recurrence. Cancer. 2011;117:2035–43.

111. Notter M, Thomsen AR, Nitsche M, Hermann RM, Wolff HA, Habl G, et al. Combined wIRA-hyperthermia and hypofractionated re-irradiation in the treatment of locally recurrent breast cancer: evaluation of therapeutic outcome based on a novel size classification. Cancers. 2020;12:606. https://doi.org/10.3390/cancers12030606.
112. Fattahi S, Ahmed SK, Park SS, Petersen IA, Shumway DA, Stish BJ, et al. Reirradiation for locoregional recurrent breast cancer. Adv Radiat Oncol. 2021;6:100640.
113. Hannoun-Levi J-M, Ihrai T, Courdi A. Local treatment options for ipsilateral breast tumour recurrence. Cancer Treat Rev. 2013;39:737–41.
114. Arthur DW, Winter KA, Kuerer HM, Haffty B, Cuttino L, Todor DA, et al. Effectiveness of breast-conserving surgery and 3-dimensional conformal partial breast reirradiation for recurrence of breast cancer in the ipsilateral breast: the NRG Oncology/RTOG 1014 phase 2 clinical trial. JAMA Oncol. 2020;6:75–82.
115. Datta NR, Puric E, Klingbiel D, Gomez S, Bodis S. Hyperthermia and radiation therapy in locoregional recurrent breast cancers: a systematic review and meta-analysis. Int J Radiat Oncol Biol Phys. 2016;94:1073–87.
116. Chen I, Botty Van den Bruele AM, Gillespie EF, Mueller BA, Xu AJ, Cuaron J, et al. Salvage of locally recurrent breast cancer with repeat breast conservation using 45 Gy hyperfractionated partial breast re-irradiation. Breast Cancer Res Treat. 2021;188:409–14.
117. Eyck BM, van Lanschot JJB, Hulshof MCCM, van der Wilk BJ, Shapiro J, van Hagen P, et al. Ten-year outcome of neoadjuvant chemoradiotherapy plus surgery for esophageal cancer: the randomized controlled CROSS trial. J Clin Oncol. 2021;39:1995–2004.
118. Macdonald JS, Smalley SR, Benedetti J, Hundahl SA, Estes NC, Stemmermann GN, et al. Chemoradiotherapy after surgery compared with surgery alone for adenocarcinoma of the stomach or gastroesophageal junction. N Engl J Med. 2001;345:725–30.
119. Gastrointestnal Tumor Study Group. Further evidence of effective adjuvant combined radiation and chemotherapy following curative resection of pancreatic cancer. Cancer. 1987;59:2006–10.
120. Ben-Josef E, Guthrie KA, El-Khoueiry AB, Corless CL, Zalupski MM, Lowy AM, et al. SWOG S0809: a phase II intergroup trial of adjuvant capecitabine and gemcitabine followed by radiotherapy and concurrent capecitabine in extrahepatic cholangiocarcinoma and gallbladder carcinoma. J Clin Oncol. 2015;33:2617–22.
121. Hong TS, Wo JY, Yeap BY, Ben-Josef E, McDonnell EI, Blaszkowsky LS, et al. Multi-institutional phase II study of high-dose hypofractionated proton beam therapy in patients with localized, unresectable hepatocellular carcinoma and intrahepatic cholangiocarcinoma. J Clin Oncol. 2016;34:460–8.
122. Shapiro J, van Lanschot JJB, Hulshof MCCM, van Hagen P, van Berge Henegouwen MI, Wijnhoven BPL, et al. Neoadjuvant chemoradiotherapy plus surgery versus surgery alone for oesophageal or junctional cancer (CROSS): long-term results of a randomised controlled trial. Lancet Oncol. 2015;16:1090–8.
123. Herskovic A, Martz K, al-Sarraf M, Leichman L, Brindle J, Vaitkevicius V, Cooper J, Byhardt R, Davis L, Emami B. Combined chemotherapy and radiotherapy compared with radiotherapy alone in patients with cancer of the esophagus. N Engl J Med. 1992;326(24):1593–8. https://doi.org/10.1056/NEJM199206113262403. PMID: 1584260.
124. Araújo CM, Souhami L, Gil RA, Carvalho R, Garcia JA, Froimtchuk MJ, Pinto LH, Canary PC. A randomized trial comparing radiation therapy versus concomitant radiation therapy and chemotherapy in carcinoma of the thoracic esophagus. Cancer. 1991;67(9):2258–61. https://doi.org/10.1002/1097-0142(19910501)67:9<2258::aid-cncr2820670908>3.0.co;2-g. PMID: 1707338.
125. Zhao K, Si Y, Sun L, Meng X, Yu J. Efficacy and toxicity of re-irradiation for esophageal cancer patients with locoregional recurrence: a retrospective analysis. Radiat Oncol. 2020;15(1):243. https://doi.org/10.1186/s13014-020-01685-2. PMID: 33087143; PMCID: PMC7576837.
126. Markar SR, Karthikesalingam A, Penna M, Low DE. Assessment of short-term clinical outcomes following salvage esophagectomy for the treatment of esophageal malignancy: systematic review and pooled analysis. Ann Surg Oncol. 2014;21(3):922–31.https://doi.org/10.1245/s10434--013-3364-0. Epub 2013 Nov 9. PMID: 24212722.
127. Jamel S, Markar SR. Salvage esophagectomy: safe therapeutic strategy? J Thorac Dis. 2017;9(Suppl 8):S799–808. https://doi.org/10.21037/jtd.2017.05.09.
128. Bleiberg H, Conroy T, Paillot B, Lacave AJ, Blijham G, Jacob JH, Bedenne L, Namer M, De Besi P, Gay F, Collette L, Sahmoud T. Randomised phase II study of cisplatin and 5-fluorouracil (5-FU) versus cisplatin alone in advanced squamous cell oesophageal cancer. Eur J Cancer. 1997;33(8):1216–20. https://doi.org/10.1016/s0959-8049(97)00088-9. PMID: 9301445.
129. Gaspar LE, Nag S, Herskovic A, Mantravadi R, Speiser B. American Brachytherapy Society (ABS) consensus guidelines for brachytherapy of esophageal cancer. Clinical Research Committee, American Brachytherapy Society, Philadelphia, PA. Int J Radiat Oncol Biol Phys. 1997;38(1):127–32. https://doi.org/10.1016/s0360-3016(97)00231-9. PMID: 9212013.
130. Nicolay NH, Rademacher J, Oelmann-Avendano J, Debus J, Huber PE, Lindel K. High dose-rate endoluminal brachytherapy for primary and recurrent esoph-

ageal cancer: experience from a large single-center cohort. Strahlenther Onkol. 2016;192(7):458–66. https://doi.org/10.1007/s00066-016-0979-7. Epub 2016 May 31. PMID: 27245818.

131. Folkert MR, Cohen GN, Wu AJ, Gerdes H, Schattner MA, Markowitz AJ, Ludwig E, Ilson DH, Bains MS, Zelefsky MJ, Goodman KA. Endoluminal high-dose-rate brachytherapy for early stage and recurrent esophageal cancer in medically inoperable patients. Brachytherapy. 2013;12(5):463–70. https://doi.org/10.1016/j.brachy.2012.12.001. Epub 2013 Feb 20. PMID: 23434221.
132. Gaspar LE, Qian C, Kocha WI, Coia LR, Herskovic A, Graham M. A phase I/II study of external beam radiation, brachytherapy and concurrent chemotherapy in localized cancer of the esophagus (RTOG 92-07): preliminary toxicity report. Int J Radiat Oncol Biol Phys. 1997;37(3):593–9. https://doi.org/10.1016/s0360-3016(96)00591-3. PMID: 9112458.
133. Wong Hee Kam S, Rivera S, Hennequin C, Lourenço N, Chirica M, Munoz-Bongrand N, Gornet JM, Quéro L. Salvage high-dose-rate brachytherapy for esophageal cancer in previously irradiated patients: a retrospective analysis. Brachytherapy. 2015;14(4):531–6. https://doi.org/10.1016/j.brachy.2015.02.392. Epub 2015 Apr 20. PMID: 25906950.
134. Nonoshita T, Sasaki T, Hirata H, Toh Y, Shioyama Y, Nakamura K, Honda H. High-dose-rate brachytherapy for previously irradiated patients with recurrent esophageal cancer. Radiat Med. 2007;25(8):373–7. https://doi.org/10.1007/s11604-007-0152-4. Epub 2007 Oct 26. PMID: 17952540.
135. Sharma V, Mahantshetty U, Dinshaw KA, Deshpande R, Sharma S. Palliation of advanced/recurrent esophageal carcinoma with high-dose-rate brachytherapy. Int J Radiat Oncol Biol Phys. 2002;52(2):310–5. https://doi.org/10.1016/s0360--3016(01)01822-3. PMID: 11872275.
136. Hong L, Huang YX, Zhuang QY, Zhang XQ, Tang LR, Du KX, Lin XY, Zheng BH, Cai SL, Wu JX, Li JL. Survival benefit of re-irradiation in esophageal cancer patients with locoregional recurrence: a propensity score-matched analysis. Radiat Oncol. 2018;13(1):171. https://doi.org/10.1186/s13014-018-1122-y. PMID: 30201005; PMCID: PMC6131819.
137. Tachimori Y. Role of salvage esophagectomy after definitive chemoradiotherapy. Gen Thorac Cardiovasc Surg. 2009;57(2):71–8. https://doi.org/10.1007/s11748-008-0337-5. Epub 2009 Feb 12. PMID: 19214447.
138. Chen Y, Lu Y, Wang Y, Yang H, Xia Y, Chen M, Song H, Li T, Li D, Wang J, Li S, Wang J. Comparison of salvage chemoradiation versus salvage surgery for recurrent esophageal squamous cell carcinoma after definitive radiochemotherapy or radiotherapy alone. Dis Esophagus. 2014;27(2):134–40. https://doi.org/10.1111/j.1442-2050.2012.01440.x. Epub 2012 Oct 22. PMID: 23088212.
139. Lan K, Chen J. Efficacy and safety of re-irradiation for locoregional esophageal squamous cell carcinoma recurrence after radiotherapy: a systematic review and meta-analysis. Radiat Oncol. 2022;17(1):61. https://doi.org/10.1186/s13014-022-02019-0. PMID: 35346285; PMCID: PMC8962014.
140. Fakhrian K, Gamisch N, Schuster T, Thamm R, Molls M, Geinitz H. Salvage radiotherapy in patients with recurrent esophageal carcinoma. Strahlenther Onkol. 2012;188(2):136–42. https://doi.org/10.1007/s00066-011-0023-x. Epub 2012 Jan 6. PMID: 22218502.
141. Kim YS, Lee CG, Kim KH, Kim T, Lee J, Cho Y, Koom WS. Re-irradiation of recurrent esophageal cancer after primary definitive radiotherapy. Radiat Oncol J. 2012;30(4):182–8. https://doi.org/10.3857/roj.2012.30.4.182. Epub 2012 Dec 31. PMID: 23346537; PMCID: PMC3546286.
142. Werner-Wasik M, Pequignot E, Leeper D, Hauck W, Curran W. Predictors of severe esophagitis include use of concurrent chemotherapy, but not the length of irradiated esophagus: a multivariate analysis of patients with lung cancer treated with nonoperative therapy. Int J Radiat Oncol Biol Phys. 2000;48(3):689–96. https://doi.org/10.1016/s0360--3016(00)00699-4. PMID: 11020565.
143. Kim DW, Raoof S, Lamba N, Lee G, Bitterman DS, Mahal AR, Sanford NN, Lam MB, Mamon HJ. Definitive re-irradiation of locally recurrent esophageal cancer after trimodality therapy in patients with a poor performance status. Mol Clin Oncol. 2020;13(1):27–32. https://doi.org/10.3892/mco.2020.2044. Epub 2020 May 11. PMID: 32454972; PMCID: PMC7241232.
144. Yamashita H, Jingu K, Niibe Y, Katsui K, Matsumoto T, Nishina T, Terahara A. Definitive salvage radiation therapy and chemoradiation therapy for lymph node oligo-recurrence of esophageal cancer: a Japanese multi-institutional study of 237 patients. Radiat Oncol. 2017;12(1):38. https://doi.org/10.1186/s13014-017-0780-5. PMID: 28219406; PMCID: PMC5319190.
145. Bick BL, Song LM, Buttar NS, Baron TH, Nichols FC, Maldonado F, Katzka DA, Enders FT, Topazian MD. Stent-associated esophagorespiratory fistulas: incidence and risk factors. Gastrointest Endosc. 2013;77(2):181–9. https://doi.org/10.1016/j.gie.2012.10.004. Epub 2012 Dec 11. PMID: 23245798.
146. Xu X, Wang Z, Jiang S, Shang Y, Wu Y. Evaluating the optimal re-irradiation dose for locally recurrent esophageal squamous cell carcinoma after definitive radiotherapy. Radiat Oncol. 2019;14(1):191. https://doi.org/10.1186/s13014-019-1402-1. PMID: 31684983; PMCID: PMC6827256.
147. Katano A, Yamashita H, Nakagawa K. Re-irradiation of locoregional esophageal cancer recurrence following definitive chemoradiotherapy: a report of 6 cases. Mol Clin Oncol. 2017;7(4):681–6. https://doi.

org/10.3892/mco.2017.1384. Epub 2017 Aug 18. PMID: 29046800; PMCID: PMC5639416.
148. Zhou ZG, Zhen CJ, Bai WW, Zhang P, Qiao XY, Liang JL, Gao XS, Wang SS. Salvage radiotherapy in patients with local recurrent esophageal cancer after radical radiochemotherapy. Radiat Oncol. 2015;10:54. https://doi.org/10.1186/s13014-015-0358-z. PMID: 25888966; PMCID: PMC4351944.
149. Yamaguchi S, Ohguri T, Imada H, Yahara K, Moon SD, Higure A, Yamaguchi K, Yoshikawa I, Harada M, Korogi Y. Multimodal approaches including three-dimensional conformal re-irradiation for recurrent or persistent esophageal cancer: preliminary results. J Radiat Res. 2011;52(6):812–20. https://doi.org/10.1269/jrr.11066. Epub 2011 Oct 21. PMID: 22020080.
150. Lin SH, Hobbs BP, Verma V, Tidwell RS, Smith GL, Lei X, Corsini EM, Mok I, Wei X, Yao L, Wang X, Komaki RU, Chang JY, Chun SG, Jeter MD, Swisher SG, Ajani JA, Blum-Murphy M, Vaporciyan AA, Mehran RJ, Koong AC, Gandhi SJ, Hofstetter WL, Hong TS, Delaney TF, Liao Z, Mohan R. Randomized phase IIB trial of proton beam therapy versus intensity-modulated radiation therapy for locally advanced esophageal cancer. J Clin Oncol. 2020;38(14):1569–79. https://doi.org/10.1200/JCO.19.02503. Epub 2020 Mar 11. PMID: 32160096; PMCID: PMC7213588.
151. Kelly P, Das P, Pinnix CC, Beddar S, Briere T, Pham M, Krishnan S, Delclos ME, Crane CH. Duodenal toxicity after fractionated chemoradiation for unresectable pancreatic cancer. Int J Radiat Oncol Biol Phys. 2013;85(3):e143–9. https://doi.org/10.1016/j.ijrobp.2012.09.035. Epub 2012 Nov 29. PMID: 23200173.
152. Kavanagh BD, Pan CC, Dawson LA, Das SK, Li XA, Ten Haken RK, Miften M. Radiation dose-volume effects in the stomach and small bowel. Int J Radiat Oncol Biol Phys. 2010;76(3 Suppl):S101–7. https://doi.org/10.1016/j.ijrobp.2009.05.071. PMID: 20171503.
153. Cheng JC, Wu JK, Huang CM, Huang DY, Cheng SH, Lin YM, Jian JJ, Yang PS, Chuang VP, Huang AT. Radiation-induced liver disease after radiotherapy for hepatocellular carcinoma: clinical manifestation and dosimetric description. Radiother Oncol. 2002;63(1):41–5. https://doi.org/10.1016/s0167--8140(02)00061-0. PMID: 12065102.
154. Caravatta L, Fiorica F, Rosa C, Boldrini L, Alitto AR, Nardangeli A, Dionisi F, Bianco L, Munoz F, Lupattelli M, Mantello G, Genovesi D, Massaccesi M, Re-Irradiation Working Group of the Italian Association of Radiotherapy and Clinical Oncology (AIRO). Role of upper abdominal reirradiation for gastrointestinal malignancies: a systematic review of cumulative dose, toxicity, and outcomes on behalf of the Re-Irradiation Working Group of the Italian Association of Radiotherapy and Clinical Oncology (AIRO). Strahlenther Onkol. 2020;196(1):1–14. https://doi.org/10.1007/s00066-019-01519-5. Epub 2019 Oct 4. PMID: 31586232.
155. Owen D, Lukovic J, Hosni A, Crane CH, Hong TS, Dawson LA, Velec M, Lawrence TS. Challenges in reirradiation of intrahepatic tumors. Semin Radiat Oncol. 2020;30(3):242–52. https://doi.org/10.1016/j.semradonc.2020.02.004. PMID: 32503790.
156. Abusaris H, Hoogeman M, Nuyttens JJ. Re-irradiation: outcome, cumulative dose and toxicity in patients retreated with stereotactic radiotherapy in the abdominal or pelvic region. Technol Cancer Res Treat. 2012;11(6):591–7. https://doi.org/10.7785/tcrt.2012.500261. Epub 2012 May 7. PMID: 22568625; PMCID: PMC4527454.
157. Winter JM, Tang LH, Klimstra DS, Liu W, Linkov I, Brennan MF, D'Angelica MI, DeMatteo RP, Fong Y, Jarnagin WR, O'reilly EM, Allen PJ. Failure patterns in resected pancreas adenocarcinoma: lack of predicted benefit to SMAD4 expression. Ann Surg. 2013;258(2):331–5. https://doi.org/10.1097/SLA.0b013e31827fe9ce. PMID: 23360922; PMCID: PMC4431700.
158. Baine MJ, Sleightholm R, Lin C. Incidence and patterns of locoregional failure after stereotactic body radiation therapy for pancreatic adenocarcinoma. Pract Radiat Oncol. 2019;9(1):e29–37. https://doi.org/10.1016/j.prro.2018.06.006. Epub 2018 Jul 3. PMID: 30612720.
159. Zhu X, Ju X, Cao Y, Shen Y, Cao F, Qing S, Fang F, Jia Z, Zhang H. Patterns of local failure after stereotactic body radiation therapy and sequential chemotherapy as initial treatment for pancreatic cancer: implications of target volume design. Int J Radiat Oncol Biol Phys. 2019;104(1):101–10. https://doi.org/10.1016/j.ijrobp.2019.01.075. Epub 2019 Jan 23. PMID: 30684663.
160. Strobel O, Hartwig W, Hackert T, Hinz U, Berens V, Grenacher L, Bergmann F, Debus J, Jäger D, Büchler M, Werner J. Re-resection for isolated local recurrence of pancreatic cancer is feasible, safe, and associated with encouraging survival. Ann Surg Oncol. 2013;20(3):964–72.https://doi.org/10.1245/s10434--012-2762-z. Epub 2012 Dec 12. PMID: 23233235.
161. Kleeff J, Reiser C, Hinz U, Bachmann J, Debus J, Jaeger D, Friess H, Büchler MW. Surgery for recurrent pancreatic ductal adenocarcinoma. Ann Surg. 2007;245(4):566–72. https://doi.org/10.1097/01.sla.0000245845.06772.7d. PMID: 17414605; PMCID: PMC1877037.
162. Zacharias T, Oussoultzoglou E, Jaeck D, Pessaux P, Bachellier P. Surgery for recurrence of periampullary malignancies. J Gastrointest Surg. 2009;13(4):760–7. https://doi.org/10.1007/s11605--008-0769-3. Epub 2008 Dec 3. PMID: 19050979.
163. Suzuki S, Furukawa T, Oshima N, Izumo W, Shimizu K, Yamamoto M. Original scientific reports: clinicopathological findings of remnant pancreatic cancers in survivors following curative resections of pan-

creatic cancers. World J Surg. 2016;40(4):974–81. https://doi.org/10.1007/s00268-015-3353-5. PMID: 26589594; PMCID: PMC4767846.
164. Haeno H, Gonen M, Davis MB, Herman JM, Iacobuzio-Donahue CA, Michor F. Computational modeling of pancreatic cancer reveals kinetics of metastasis suggesting optimum treatment strategies. Cell. 2012;148(1–2):362–75. https://doi.org/10.1016/j.cell.2011.11.060. PMID: 22265421; PMCID: PMC3289413.
165. Hashimoto K, Ueno H, Ikeda M, Kojima Y, Hagihara A, Kondo S, Morizane C, Okusaka T. Do recurrent and metastatic pancreatic cancer patients have the same outcomes with gemcitabine treatment? Oncology. 2009;77(3–4):217–23. https://doi.org/10.1159/000236022. Epub 2009 Sep 4. PMID: 19729980.
166. Xue P, Kanai M, Mori Y, Nishimura T, Uza N, Kodama Y, Kawaguchi Y, Takaori K, Matsumoto S, Uemoto S, Chiba T. Comparative outcomes between initially unresectable and recurrent cases of advanced pancreatic cancer following palliative chemotherapy. Pancreas. 2014;43(3):411–6. https://doi.org/10.1097/MPA.0000000000000050. PMID: 24622071.
167. Hunt A, Das P, Minsky BD, Koay EJ, Krishnan S, Herman JM, Taniguchi C, Koong A, Smith GL, Holliday EB. Hyperfractionated abdominal reirradiation for gastrointestinal malignancies. Radiat Oncol. 2018;13(1):143. https://doi.org/10.1186/s13014-018-1084-0. PMID: 30086784; PMCID: PMC6081944.
168. Lominska CE, Unger K, Nasr NM, Haddad N, Gagnon G. Stereotactic body radiation therapy for reirradiation of localized adenocarcinoma of the pancreas. Radiat Oncol. 2012;7:74. https://doi.org/10.1186/1748-717X-7-74. PMID: 22607687; PMCID: PMC3441204.
169. Wild AT, Hiniker SM, Chang DT, Tran PT, Khashab MA, Limaye MR, Laheru DA, Le DT, Kumar R, Pai JS, Hargens B, Sharabi AB, Shin EJ, Zheng L, Pawlik TM, Wolfgang CL, Koong AC, Herman JM. Re-irradiation with stereotactic body radiation therapy as a novel treatment option for isolated local recurrence of pancreatic cancer after multimodality therapy: experience from two institutions. J Gastrointest Oncol. 2013;4(4):343–51. https://doi.org/10.3978/j.issn.2078-6891.2013.044. PMID: 24294505; PMCID: PMC3819776.
170. Dagoglu N, Callery M, Moser J, Tseng J, Kent T, Bullock A, Miksad R, Mancias JD, Mahadevan A. Stereotactic body radiotherapy (SBRT) reirradiation for recurrent pancreas cancer. J Cancer. 2016;7(3):283–8. https://doi.org/10.7150/jca.13295. PMID: 26918041; PMCID: PMC4747882.
171. Koong AJ, Toesca DAS, von Eyben R, Pollom EL, Chang DT. Reirradiation with stereotactic body radiation therapy after prior conventional fractionation radiation for locally recurrent pancreatic adenocarcinoma. Adv Radiat Oncol. 2017;2(1):27–36. https://doi.org/10.1016/j.adro.2017.01.003. PMID: 28740913; PMCID: PMC5514250.
172. Reddy AV, Hill CS, Sehgal S, He J, Zheng L, Herman JM, Meyer J, Narang AK. Efficacy and safety of reirradiation with stereotactic body radiation therapy for locally recurrent pancreatic adenocarcinoma. Clin Oncol (R Coll Radiol). 2022;34(6):386–94. https://doi.org/10.1016/j.clon.2021.12.014. Epub 2021 Dec 31. PMID: 34974972.
173. Lee JJB, Choi SH, Im JH, Seong J. Clinical safety and efficacy of salvage reirradiation for upper abdominal malignancies. Strahlenther Onkol. 2019;195(6):526–33. https://doi.org/10.1007/s00066-018-01420-7. Epub 2019 Jan 28. PMID: 30689025.
174. Haque W, Crane CH, Krishnan S, et al. Reirradiation to the abdomen for gastrointestinal malignancies. Radiat Oncol. 2009;4:55. Published 2009 Nov 18. https://doi.org/10.1186/1748-717X-4-55.
175. Hong TS, Wo JY, Yeap BY, Ben-Josef E, McDonnell EI, Blaszkowsky LS, Kwak EL, Allen JN, Clark JW, Goyal L, Murphy JE, Javle MM, Wolfgang JA, Drapek LC, Arellano RS, Mamon HJ, Mullen JT, Yoon SS, Tanabe KK, Ferrone CR, Ryan DP, DeLaney TF, Crane CH, Zhu AX. Multi-institutional phase II study of high-dose hypofractionated proton beam therapy in patients with localized, unresectable hepatocellular carcinoma and intrahepatic cholangiocarcinoma. J Clin Oncol. 2016;34(5):460–8. https://doi.org/10.1200/JCO.2015.64.2710. Epub 2015 Dec 14. PMID: 26668346; PMCID: PMC4872014.
176. Bush DA, Kayali Z, Grove R, Slater JD. The safety and efficacy of high-dose proton beam radiotherapy for hepatocellular carcinoma: a phase 2 prospective trial. Cancer. 2011;117(13):3053–9. https://doi.org/10.1002/cncr.25809. Epub 2011 Jan 24. PMID: 21264826.
177. Jung DH, Kim MS, Cho CK, Yoo HJ, Jang WI, Seo YS, Paik EK, Kim KB, Han CJ, Kim SB. Outcomes of stereotactic body radiotherapy for unresectable primary or recurrent cholangiocarcinoma. Radiat Oncol J. 2014;32(3):163–9. https://doi.org/10.3857/roj.2014.32.3.163. Epub 2014 Sep 30. PMID: 25324988; PMCID: PMC4194299.
178. Kopek N, Holt MI, Hansen AT, Høyer M. Stereotactic body radiotherapy for unresectable cholangiocarcinoma. Radiother Oncol. 2010;94(1):47–52. https://doi.org/10.1016/j.radonc.2009.11.004. Epub 2009 Dec 4. PMID: 19963295.
179. Rusthoven KE, Kavanagh BD, Cardenes H, Stieber VW, Burri SH, Feigenberg SJ, Chidel MA, Pugh TJ, Franklin W, Kane M, Gaspar LE, Schefter TE. Multi-institutional phase I/II trial of stereotactic body radiation therapy for liver metastases. J Clin Oncol. 2009;27(10):1572–8. https://doi.org/10.1200/JCO.2008.19.6329. Epub 2009 Mar 2. PMID: 19255321.
180. Hong TS, Wo JY, Borger DR, Yeap BY, McDonnell EI, Willers H, Blaszkowsky LS, Kwak EL, Allen JN, Clark JW, Tanguturi S, Goyal L, Murphy JE,

Wolfgang JA, Drapek LC, Arellano RS, Mamon HJ, Mullen JT, Tanabe KK, Ferrone CR, Ryan DP, Iafrate AJ, DeLaney TF, Zhu AX. Phase II study of proton-based stereotactic body radiation therapy for liver metastases: importance of tumor genotype. J Natl Cancer Inst. 2017;109(9) https://doi.org/10.1093/jnci/djx031. PMID: 28954285.

181. Lee MT, Kim JJ, Dinniwell R, Brierley J, Lockwood G, Wong R, Cummings B, Ringash J, Tse RV, Knox JJ, Dawson LA. Phase I study of individualized stereotactic body radiotherapy of liver metastases. J Clin Oncol. 2009;27(10):1585–91. https://doi.org/10.1200/JCO.2008.20.0600. Epub 2009 Mar 2. PMID: 19255313.
182. Oshiro Y, Mizumoto M, Okumura T, Fukuda K, Fukumitsu N, Abei M, Ishikawa H, Takizawa D, Sakurai H. Analysis of repeated proton beam therapy for patients with hepatocellular carcinoma. Radiother Oncol. 2017;123(2):240–5. https://doi.org/10.1016/j.radonc.2017.03.004. Epub 2017 Mar 30. PMID: 28366501.
183. Hashimoto T, Tokuuye K, Fukumitsu N, Igaki H, Hata M, Kagei K, Sugahara S, Ohara K, Matsuzaki Y, Akine Y. Repeated proton beam therapy for hepatocellular carcinoma. Int J Radiat Oncol Biol Phys. 2006;65(1):196–202. https://doi.org/10.1016/j.ijrobp.2005.11.043. Epub 2006 Mar 24. PMID: 16563656.
184. Farach A, Quesada J, Teh BS. Liver regeneration following repeat SBRT. J Gastrointest Oncol. 2015;6(2):E2–6. https://doi.org/10.3978/j.issn.2078-6891.2014.082. PMID: 25830045; PMCID: PMC4311083.
185. Ohara K, Okumura T, Tsuji H, Chiba T, Min M, Tatsuzaki H, Tsujii H, Akine Y, Itai Y. Radiation tolerance of cirrhotic livers in relation to the preserved functional capacity: analysis of patients with hepatocellular carcinoma treated by focused proton beam radiotherapy. Int J Radiat Oncol Biol Phys. 1997;38(2):367–72. https://doi.org/10.1016/s0360--3016(96)00587-1. PMID: 9226325.
186. McDuff SGR, Remillard KA, Zheng H, Raldow AC, Wo JY, Eyler CE, Drapek LC, Goyal L, Blaszkowsky LS, Clark JW, Allen JN, Parikh AR, Ryan DP, Ferrone CR, Tanabe KK, Wolfgang JA, Zhu AX, Hong TS. Liver reirradiation for patients with hepatocellular carcinoma and liver metastasis. Pract Radiat Oncol. 2018;8(6):414–21. https://doi.org/10.1016/j.prro.2018.04.012. Epub 2018 Apr 26. PMID: 29937235.
187. Seol SW, Yu JI, Park HC, Lim DH, Oh D, Noh JM, Cho WK, Paik SW. Treatment outcome of hepatic re-irradiation in patients with hepatocellular carcinoma. Radiat Oncol J. 2015;33(4):276–83. https://doi.org/10.3857/roj.2015.33.4.276. Epub 2015 Dec 30. PMID: 26756027; PMCID: PMC4707210.
188. Bradley JD, Hu C, Komaki RR, Masters GA, Blumenschein GR, Schild SE, Bogart JA, Forster KM, Magliocco AM, Kavadi VS, Narayan S, Iyengar P, Robinson CG, Wynn RB, Koprowski CD, Olson MR, Meng J, Paulus R, Curran WJ Jr, Choy H. Long-term results of NRG Oncology RTOG 0617: standard- versus high-dose chemoradiotherapy with or without cetuximab for unresectable stage III non-small-cell lung cancer. J Clin Oncol. 2020;38(7):706–14. https://doi.org/10.1200/JCO.19.01162. Epub 2019 Dec 16. PMID: 31841363; PMCID: PMC7048161.
189. Albain KS, Swann RS, Rusch VW, Turrisi AT 3rd, Shepherd FA, Smith C, Chen Y, Livingston RB, Feins RH, Gandara DR, Fry WA, Darling G, Johnson DH, Green MR, Miller RC, Ley J, Sause WT, Cox JD. Radiotherapy plus chemotherapy with or without surgical resection for stage III non-small-cell lung cancer: a phase III randomised controlled trial. Lancet. 2009;374(9687):379–86. https://doi.org/10.1016/S0140-6736(09)60737-6. Epub 2009 Jul 24. PMID: 19632716; PMCID: PMC4407808.
190. Aupérin A, Le Péchoux C, Rolland E, Curran WJ, Furuse K, Fournel P, Belderbos J, Clamon G, Ulutin HC, Paulus R, Yamanaka T, Bozonnat MC, Uitterhoeve A, Wang X, Stewart L, Arriagada R, Burdett S, Pignon JP. Meta-analysis of concomitant versus sequential radiochemotherapy in locally advanced non-small-cell lung cancer. J Clin Oncol. 2010;28(13):2181–90. https://doi.org/10.1200/JCO.2009.26.2543. Epub 2010 Mar 29. PMID: 20351327.
191. Curran WJ Jr, Paulus R, Langer CJ, Komaki R, Lee JS, Hauser S, Movsas B, Wasserman T, Rosenthal SA, Gore E, Machtay M, Sause W, Cox JD. Sequential vs. concurrent chemoradiation for stage III non-small cell lung cancer: randomized phase III trial RTOG 9410. J Natl Cancer Inst. 2011;103(19):1452–60. https://doi.org/10.1093/jnci/djr325. Epub 2011 Sep 8. Erratum in: J Natl Cancer Inst. 2012 Jan 4;104(1):79. PMID: 21903745; PMCID: PMC3186782.
192. Noble J, Ellis PM, Mackay JA, Evans WK, Lung Cancer Disease Site Group of Cancer Care Ontario's Program in Evidence-based Care. Second-line or subsequent systemic therapy for recurrent or progressive non-small cell lung cancer: a systematic review and practice guideline. J Thorac Oncol. 2006;1(9):1042–58. PMID: 17409993.
193. De Ruysscher D, Faivre-Finn C, Le Pechoux C, Peeters S, Belderbos J. High-dose re-irradiation following radical radiotherapy for non-small-cell lung cancer. Lancet Oncol. 2014;15(13):e620–4. https://doi.org/10.1016/S1470-2045(14)70345-6. Epub 2014 Nov 24. PMID: 25456380.
194. Kumar SS, McGarry RC. Management of local recurrences and regional failure in early stage non-small cell lung cancer after stereotactic body radiation therapy. Transl Lung Cancer Res. 2019;8(Suppl 2):S213–21. https://doi.org/10.21037/tlcr.2019.09.06. PMID: 31673526; PMCID: PMC6795579.
195. Peulen H, Karlsson K, Lindberg K, Tullgren O, Baumann P, Lax I, Lewensohn R, Wersäll P. Toxicity

after reirradiation of pulmonary tumours with stereotactic body radiotherapy. Radiother Oncol. 2011;101(2):260–6. https://doi.org/10.1016/j.radonc.2011.09.012. Epub 2011 Nov 5. PMID: 22056534.
196. Bezjak A, Paulus R, Gaspar LE, Timmerman RD, Straube WL, Ryan WF, Garces YI, Pu AT, Singh AK, Videtic GM, McGarry RC, Iyengar P, Pantarotto JR, Urbanic JJ, Sun AY, Daly ME, Grills IS, Sperduto P, Normolle DP, Bradley JD, Choy H. Safety and efficacy of a five-fraction stereotactic body radiotherapy schedule for centrally located non-small-cell lung cancer: NRG Oncology/RTOG 0813 trial. J Clin Oncol. 2019;37(15):1316–25. https://doi.org/10.1200/JCO.18.00622. Epub 2019 Apr 3. PMID: 30943123; PMCID: PMC6524984.
197. Kramer GW, Gans S, Ullmann E, van Meerbeeck JP, Legrand CC, Leer JW. Hypofractionated external beam radiotherapy as retreatment for symptomatic non-small-cell lung carcinoma: an effective treatment? Int J Radiat Oncol Biol Phys. 2004;58(5):1388–93. https://doi.org/10.1016/j.ijrobp.2003.09.087. PMID: 15050314.
198. Sumodhee S, Bondiau PY, Poudenx M, Cohen C, Naghavi AO, Padovani B, Maneval D, Gal J, Leysalle A, Ghalloussi H, Otto J, Doyen J. Long term efficacy and toxicity after stereotactic ablative reirradiation in locally relapsed stage III non-small cell lung cancer. BMC Cancer. 2019;19(1):305. https://doi.org/10.1186/s12885-019-5542-3. PMID: 30943943; PMCID: PMC6448259.
199. Maddalo M, D'Angelo E, Fiorica F, Argenone A, Scricciolo M, Cozzi S, Nardangeli A, Dionisi F, Costantino G, Vagge S, Pontoriero A, Donato V, Massaccesi M, Re-irradiation Study Group of the Italian Association of Radiation Oncology. Thoracic re-irradiation with 3D-conformal or more advanced techniques: a systematic review of treatment safety by the Re-irradiation Study Group of the Italian Association of Radiation and Oncology AIRO. Crit Rev Oncol Hematol. 2021;167:103500. https://doi.org/10.1016/j.critrevonc.2021.103500. Epub 2021 Oct 21. PMID: 34688894.
200. Tada T, Fukuda H, Matsui K, Hirashima T, Hosono M, Takada Y, Inoue Y. Non-small-cell lung cancer: reirradiation for loco-regional relapse previously treated with radiation therapy. Int J Clin Oncol. 2005;10(4):247–50.https://doi.org/10.1007/s10147--005-0501-1. PMID: 16136369.
201. Käsmann L, Janssen S, Baschnagel AM, et al. Prognostic factors and outcome of reirradiation for locally recurrent small cell lung cancer-a multicenter study. Transl Lung Cancer Res. 2020;9(2):232–8. https://doi.org/10.21037/tlcr.2020.01.19.
202. Troost EGC, Wink KCJ, Roelofs E, Simone CB 2nd, Makocki S, Löck S, van Kollenburg P, Dechambre D, Minken AWH, van der Stoep J, Avery S, Jansen N, Solberg T, Bussink J, de Ruysscher D. Photons or protons for reirradiation in (non-)small cell lung cancer: results of the multicentric ROCOCO *in silico* study. Br J Radiol. 2020;93(1107):20190879. https://doi.org/10.1259/bjr.20190879. Epub 2019 Dec 20. PMID: 31804145; PMCID: PMC7066965.
203. Verma V, Rwigema JM, Malyapa RS, Regine WF, Simone CB 2nd. Systematic assessment of clinical outcomes and toxicities of proton radiotherapy for reirradiation. Radiother Oncol. 2017;125(1):21–30. https://doi.org/10.1016/j.radonc.2017.08.005. Epub 2017 Sep 20. PMID: 28941560.
204. Reyngold M, Wu AJ, McLane A, Zhang Z, Hsu M, Stein NF, Zhou Y, Ho AY, Rosenzweig KE, Yorke ED, Rimner A. Toxicity and outcomes of thoracic re-irradiation using stereotactic body radiation therapy (SBRT). Radiat Oncol. 2013;8:99. https://doi.org/10.1186/1748-717X-8-99. PMID: 23617949; PMCID: PMC3651392.
205. McAvoy S, Ciura K, Wei C, Rineer J, Liao Z, Chang JY, Palmer MB, Cox JD, Komaki R, Gomez DR. Definitive reirradiation for locoregionally recurrent non-small cell lung cancer with proton beam therapy or intensity modulated radiation therapy: predictors of high-grade toxicity and survival outcomes. Int J Radiat Oncol Biol Phys. 2014;90(4):819–27. https://doi.org/10.1016/j.ijrobp.2014.07.030. Epub 2014 Sep 11. PMID: 25220718.
206. Evans JD, Gomez DR, Amini A, Rebueno N, Allen PK, Martel MK, Rineer JM, Ang KK, McAvoy S, Cox JD, Komaki R, Welsh JW. Aortic dose constraints when reirradiating thoracic tumors. Radiother Oncol. 2013;106(3):327–32. https://doi.org/10.1016/j.radonc.2013.02.002. Epub 2013 Feb 28. PMID: 23453540; PMCID: PMC3921976.
207. Meijneke TR, Petit SF, Wentzler D, Hoogeman M, Nuyttens JJ. Reirradiation and stereotactic radiotherapy for tumors in the lung: dose summation and toxicity. Radiother Oncol. 2013;107(3):423–7. https://doi.org/10.1016/j.radonc.2013.03.015. Epub 2013 May 3. PMID: 23647748.
208. Schröder C, Stiefel I, Tanadini-Lang S, Pytko I, Vu E, Guckenberger M, Andratschke N. Re-irradiation in the thorax—an analysis of efficacy and safety based on accumulated EQD2 doses. Radiother Oncol. 2020;152:56–62. https://doi.org/10.1016/j.radonc.2020.07.033. Epub 2020 Jul 24. PMID: 32717358.
209. Simone C, Amini A, Chetty I, Choi J, Chun S, Donington J, Edelman M, Higgins K, Kestin L, Mohindra P, Movsas B, Rodrigues G, Rosenzweig K, Rybkin I, Shepherd A, Slotman B, Wolf A, Chang J. American Radium Society (ARS) and American College of Radiology (ACR) approriate use criteria systematic review and guidelines on reirradiation for non-small cell lung cancer (NSCLC). Int J Radiat Oncol Biol Phys. 2020;108(2):E48–9. https://doi.org/10.1016/j.ijrobp.2020.02.583.
210. Fischer-Valuck BW, Robinson CG, Simone CB 2nd, Gomez DR, Bradley JD. Challenges in re-irradiation in the thorax: managing patients with locally recurrent non-small cell lung cancer. Semin Radiat Oncol.

2020;30(3):223–31. https://doi.org/10.1016/j.semradonc.2020.02.001. PMID: 32503787
211. Wu KL, Jiang GL, Qian H, Wang LJ, Yang HJ, Fu XL, Zhao S. Three-dimensional conformal radiotherapy for locoregionally recurrent lung carcinoma after external beam irradiation: a prospective phase I-II clinical trial. Int J Radiat Oncol Biol Phys. 2003;57(5):1345–50.https://doi.org/10.1016/s0360--3016(03)00768-5. PMID: 14630272.
212. Kruser TJ, McCabe BP, Mehta MP, Khuntia D, Campbell TC, Geye HM, Cannon GM. Reirradiation for locoregionally recurrent lung cancer: outcomes in small cell and non-small cell lung carcinoma. Am J Clin Oncol. 2014;37(1):70–6. https://doi.org/10.1097/COC.0b013e31826b9950. PMID: 23357968; PMCID: PMC4284072.
213. Ohguri T, Imada H, Yahara K, Moon SD, Yamaguchi S, Yatera K, Mukae H, Hanagiri T, Tanaka F, Korogi Y. Re-irradiation plus regional hyperthermia for recurrent non-small cell lung cancer: a potential modality for inducing long-term survival in selected patients. Lung Cancer. 2012;77(1):140–5. https://doi.org/10.1016/j.lungcan.2012.02.018. Epub 2012 Mar 23. PMID: 22445656.
214. Tetar S, Dahele M, Griffioen G, Slotman B, Senan S. High-dose conventional thoracic re-irradiation for lung cancer: updated results. Lung Cancer. 2015;88(2):235–6. https://doi.org/10.1016/j.lungcan.2015.02.008. Epub 2015 Feb 18. PMID: 25736570.
215. Sumita K, Harada H, Asakura H, Ogawa H, Onoe T, Murayama S, Nakamura S, Tanigawa N, Takahashi T, Nishimura T. Re-irradiation for locoregionally recurrent tumors of the thorax: a single-institution, retrospective study. Radiat Oncol. 2016;11:104. https://doi.org/10.1186/s13014-016-0673-z. PMID: 27485533; PMCID: PMC4971641.
216. Hong JH, Kim YS, Lee SW, Lee SJ, Kang JH, Hong SH, Hong JY, Cheon G. High-dose thoracic re-irradiation of lung cancer using highly conformal radiotherapy is effective with acceptable toxicity. Cancer Res Treat. 2019;51(3):1156–66. https://doi.org/10.4143/crt.2018.472. Epub 2018 Nov 29. PMID: 30514067; PMCID: PMC6639224.
217. Schlampp I, Rieber J, Adeberg S, Bozorgmehr F, Heußel CP, Steins M, Kappes J, Hoffmann H, Welzel T, Debus J, Rieken S. Re-irradiation in locally recurrent lung cancer patients. Strahlenther Onkol. 2019;195(8):725–33. https://doi.org/10.1007/s00066-019-01457-2. Epub 2019 Apr 1. PMID: 30937509.
218. Kelly P, Balter PA, Rebueno N, Sharp HJ, Liao Z, Komaki R, Chang JY. Stereotactic body radiation therapy for patients with lung cancer previously treated with thoracic radiation. Int J Radiat Oncol Biol Phys. 2010;78(5):1387–93. https://doi.org/10.1016/j.ijrobp.2009.09.070. Epub 2010 Apr 8. PMID: 20381271; PMCID: PMC3401019.
219. Trakul N, Harris JP, Le QT, Hara WY, Maxim PG, Loo BW Jr, Diehn M. Stereotactic ablative radiotherapy for reirradiation of locally recurrent lung tumors. J Thorac Oncol. 2012;7(9):1462–5. https://doi.org/10.1097/JTO.0b013e31825f22ce. PMID: 22895143.
220. Trovo M, Minatel E, Durofil E, Polesel J, Avanzo M, Baresic T, Bearz A, Del Conte A, Franchin G, Gobitti C, Rumeileh IA, Trovo MG. Stereotactic body radiation therapy for re-irradiation of persistent or recurrent non-small cell lung cancer. Int J Radiat Oncol Biol Phys. 2014;88(5):1114–9. https://doi.org/10.1016/j.ijrobp.2014.01.012. PMID: 24661664.
221. Hearn JW, Videtic GM, Djemil T, Stephans KL. Salvage stereotactic body radiation therapy (SBRT) for local failure after primary lung SBRT. Int J Radiat Oncol Biol Phys. 2014;90(2):402–6. https://doi.org/10.1016/j.ijrobp.2014.05.048. Epub 2014 Jul 10. PMID: 25017480.
222. Kilburn JM, Kuremsky JG, Blackstock AW, Munley MT, Kearns WT, Hinson WH, Lovato JF, Miller AA, Petty WJ, Urbanic JJ. Thoracic re-irradiation using stereotactic body radiotherapy (SBRT) techniques as first or second course of treatment. Radiother Oncol. 2014;110(3):505–10. https://doi.org/10.1016/j.radonc.2013.11.017. Epub 2014 Jan 17. PMID: 24444530; PMCID: PMC4006197.
223. Patel NR, Lanciano R, Sura K, Yang J, Lamond J, Feng J, Good M, Gracely EJ, Komarnicky L, Brady L. Stereotactic body radiotherapy for re-irradiation of lung cancer recurrence with lower biological effective doses. J Radiat Oncol. 2015;4(1):65–70. https://doi.org/10.1007/s13566-014-0175-2. Epub 2014 Dec 10. PMID: 25774244; PMCID: PMC4348501.
224. Repka MC, Aghdam N, Kataria SK, Campbell L, Suy S, Collins SP, Anderson E, Lischalk JW, Collins BT. Five-fraction SBRT for ultra-central NSCLC in-field recurrences following high-dose conventional radiation. Radiat Oncol. 2017;12(1):162. https://doi.org/10.1186/s13014-017-0897-6. PMID: 29052514; PMCID: PMC5649069.
225. Ceylan C, Hamacı A, Ayata H, Berberoglu K, Kılıç A, Güden M, Engin K. Re-irradiation of locoregional NSCLC recurrence using robotic stereotactic body radiotherapy. Oncol Res Treat. 2017;40(4):207–14. https://doi.org/10.1159/000457129. Epub 2017 Mar 9. PMID: 28324876.
226. Ren C, Ji T, Liu T, Dang J, Li G. The risk and predictors for severe radiation pneumonitis in lung cancer patients treated with thoracic reirradiation. Radiat Oncol. 2018;13(1):69. https://doi.org/10.1186/s13014-018-1016-z. PMID: 29661254; PMCID: PMC5902864.
227. McAvoy SA, Ciura KT, Rineer JM, Allen PK, Liao Z, Chang JY, Palmer MB, Cox JD, Komaki R, Gomez DR. Feasibility of proton beam therapy for reirradiation of locoregionally recurrent non-small cell lung cancer. Radiother Oncol. 2013;109(1):38–44. https://doi.org/10.1016/j.radonc.2013.08.014. Epub 2013 Sep 7. PMID: 24016675.

228. Chao HH, Berman AT, Simone CB 2nd, Ciunci C, Gabriel P, Lin H, Both S, Langer C, Lelionis K, Rengan R, Hahn SM, Prabhu K, Fagundes M, Hartsell W, Mick R, Plastaras JP. Multi-institutional prospective study of reirradiation with proton beam radiotherapy for locoregionally recurrent non-small cell lung cancer. J Thorac Oncol. 2017;12(2):281–92. https://doi.org/10.1016/j.jtho.2016.10.018. Epub 2016 Nov 5. PMID: 27826034.
229. Karube M, Yamamoto N, Tsuji H, Kanematsu N, Nakajima M, Yamashita H, Nakagawa K, Kamada T. Carbon-ion re-irradiation for recurrences after initial treatment of stage I non-small cell lung cancer with carbon-ion radiotherapy. Radiother Oncol. 2017;125(1):31–5. https://doi.org/10.1016/j.radonc.2017.07.022. Epub 2017 Aug 4. PMID: 28784437.
230. Hayashi K, Yamamoto N, Karube M, Nakajima M, Tsuji H, Ogawa K, Kamada T. Feasibility of carbon-ion radiotherapy for re-irradiation of locoregionally recurrent, metastatic, or secondary lung tumors. Cancer Sci. 2018;109(5):1562–9. https://doi.org/10.1111/cas.13555. Epub 2018 Apr 6. PMID: 29498145; PMCID: PMC5980300.

Further Reading

https://www.youtube.com/watch?v=5387lbP7hOc (Reirradiation – General aspects)

https://www.youtube.com/watch?v=8BfSteom8Hw (Practical Aspects of Reirradiation Spine SBRT)

https://www.youtube.com/watch?v=tDdau-4kvCg (Reirradiation in recurrent Head Neck Cancers)

https://www.youtube.com/watch?v=w03S02P6vx4 (Reirradiation in Glioblastoma)

Early and Late Toxicities of Radiotherapy

16

Biswa Mohan Biswal, Lau Fen Nee,
and Gokula Kumar Appalanaido

16.1 Introduction

In clinical radiotherapy (RT), there is always simultaneous exposure of normal and cancer-bearing cells. Radiation oncologist tries to minimize normal tissue exposure and at the same time maximize dose to the tumor-bearing tissue. Despite advances in RT equipment and treatment techniques, inevitably some normal tissues will be irradiated leading to toxicity. Although most of the toxicities are manageable, in some clinical situation, the side-effects could be life-long and irreparable causing prolonged suffering, cosmetic deficiencies, and tremendous psychological trauma [1]. The risk of late effects on normal tissue (LENT) and radiation-induced second malignancy (RISM) ranges from 5% to 10% and 2% to 10% respectively in adults treated with curative RT [2]. Addition of radiosensitizers and chemotherapy drugs during radiation improve survival however increases acute toxicities significantly [3]. Modern cancer immunotherapy incorporated in RT to increase immune-mediated cell kill could also increase end organ toxicities [4]. Modern radiation techniques, namely, altered fractionation and flash RT, too increase toxicities further [5]. Heavy ion and proton beam therapy promise to limit complications in CNS tumors in children, but acute complications are similar like photon beam therapy.

Understanding molecular biology and gene expression study could explain details of pathogenesis of radiation complications, change in tumor microenvironment, interaction of vascularity and tumor immunity. Elucidation of molecular events could provide investigators strategies to alleviate RT related complications in future. Radiobiologists strive to investigate further means to limit RT related toxicities. Radiation exposure to dividing normal tissue could lead to early reaction during or within few weeks after RT, known as acute effects. However, organ complication may also emerge years after RT, known as late-effects. Second cancers (RISM), though rare, could develop in exposed tissue many decades later especially among people who are subsequently cured and live long enough to develop cancer, e.g., younger patients. Another distinct group of radiation-induced effects called "acute radiation syndrome" brought about by sudden high dose exposure to the whole body such as from nuclear accidents or war, is beyond the scope of this chapter. Briefly, the three categories in acute radiation syndrome are (1)

B. M. Biswal (✉)
Department of Radiotherapy and Oncology, KPJ Ipoh Specialist Hospital, Ipoh, Perak, Malaysia

L. F. Nee
Department of Radiotherapy and Oncology, National Cancer Institute, Putrajaya, Malaysia

G. K. Appalanaido
Department of Radiotherapy and Oncology, Advanced Medical and Dental Institute (IPPT), University Science Malaysia, Penang, Malaysia

A. Munshi et al. (eds.), *Radiation Oncology – Principles, Precepts and Practice*,
https://doi.org/10.1007/978-981-97-8389-2_16

cerebrovascular syndrome (dose levels >100 Gy, death within 1–2 days), (2) gastrointestinal syndrome (>5–100 Gy, death within 3–10 days), and (3) hematopoietic syndrome (>1–5 Gy, potential death within weeks to 2 months) [6].

16.2 Mechanism of Acute and Late Radiation Toxicities

RT is primarily a local treatment and the radiation-induced toxicities seen will be mainly manifested by the local and nearby tissues which received most of the radiation dose. However, the entry and exit radiation beams also give rise to irradiation of tissues some way off the intended area and one should be aware that these tissues may also sustain damage and show toxicities if the dose received is significant. Different types of tissues may not respond the same way to the radiation received, leading to differences seen in the speed of onset of the tissue's toxicity manifestation, its severity as well as its natural progression and outcome. The response of a tissue/organ to radiation depends on the tissue/organ factors, treatment factors, and host factors. The tissue/organ factors include (1) the inherent radiation sensitivity of the tissue's individual cells, (2) the tissue kinetics or proliferative organization (H-type/early responding tissue or F-type/late-responding tissue), (3) the structural organization of the tissue [parallel or serial organ; (Fig. 16.1)], and (4) the tissue microenvironment, e.g., vascularity, oxygen tension, and pH milieu. The treatment factors include (1) total radiation dose, (2) dose per fraction, (3) overall treatment time, (4) time interval between fractions, (5) volume of tissue irradiated, (6) concurrent agents, e.g., chemotherapy or radioprotector, (7) previous irradiation's residual damage, and (8) previous surgery. The host factors include (1) comorbidities especially those affecting the small vessels, e.g., diabetes, hypertension, connective tissue disease, (2) genetic susceptibility to radiation, and (3) additional insults, e.g., smoking, additional trauma to irradiated site (sharp dental prosthesis, fish bones, spicy food, etc.). The tolerance dose is the maximum dose that is deemed associated with an acceptable tissue complication probability (Fig. 16.2). The tolerance therefore would be influenced by the factors mentioned. The dose constraints for RT planning optimization are guided by the tolerance for the various tissue endpoints. Frequently used guides are the QUANTEC or Emami tolerances and constraints [7, 8].

Radiation toxicities are broadly categorized into acute and late effects. However, occasionally, an intermediate wave of subacute effects may be seen in some organs. Acute effects occur during or shortly after RT while late effects manifest after a longer latent period [9]. The conventional but arbitrary cut-off point is taken as 90 days after the start of RT. Acute effects quickly worsen to a peak but eventually recovers completely while late effects worsen progressively over time and usually do not recover, leaving lifelong residual damage. A comparison between acute and late effects is given in Table 16.1. Most late effects typically take years to manifest, even as long as 10–20 years; hence, long-term follow-up of a patient is important to pick them up. In some organs, there may be intermediate subacute effect phase which manifests about 3–6 months after RT, progresses somewhat but eventually usually recovers, e.g., subacute lung pneumonitis from lung irradiation, somnolence syndrome from brain irradiation, Lhermitte's syndrome from spinal cord irradiation. As acute and late effects have different pathogenesis, the acute effects are not related to the late effects. Hence, the severity of acute effects in a tissue/organ does not lead on nor predict for the risk and severity of late effects in the same tissue/organ, e.g., a mild acute skin effect does not necessarily mean mild late skin effect; in fact, the late skin effect may be severe. The exception is the consequential late effect (CLE) in which an acute effect is so severe that it does not recover but persists to become a late effect, e.g., an area of skin necrosis developing during RT which does not ever recover or mouth dryness during head and neck RT which never recovers. Another peculiarity with acute effects is the "radiation recall phenomenon." The radiation recall phenomenon is an acute inflammatory reaction confined to previously irradiated

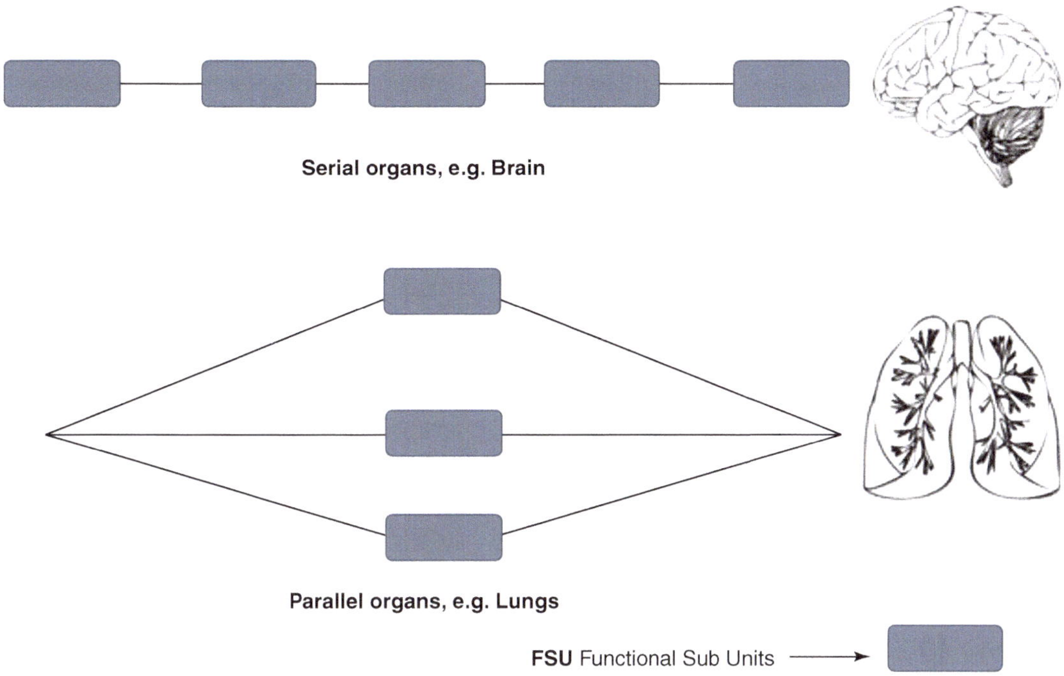

Fig. 16.1 Illustration showing arrangement of functional subunits (FSU) in serial (brain) and parallel (lungs) organs

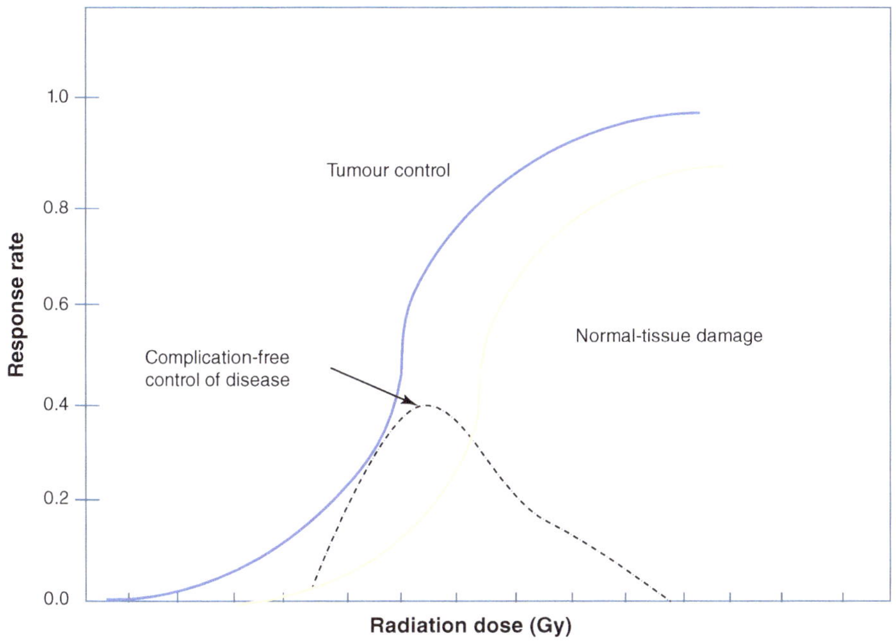

Fig. 16.2 Modified Holthusen's curve showing probability of tumor control (blue), normal-tissue damage (yellow), and complication-free control of disease in dotted line

Table 16.1 Comparison of acute and late radiation effects

Characteristics	Acute effects	Late effects
Time of onset	During or shortly after RT. Timing is dependent on the lifespan of the target cells, not dependent on dose	Long latent period of many months to years. Latency is dependent on dose; higher dose brings on the effect faster
Natural progression	Reversible, usually complete recovery	Irreversible, worsen with time
Severity	Dependent on dose	Dependent on dose
Types of tissue	Rapidly renewing tissues, e.g., epithelium, bone marrow	Slowly renewing tissues, e.g., brain, liver, lung
Pathogenesis	Parenchymal cell loss, inflammation	Complex vascular damage, stromal fibrosis, cell loss
α/β ratio	High	Low
Fractionation sensitivity (spared by lower dose per fraction)	Low	High
Overall treatment time impact	High	None

area that can be triggered when a systemic agent (e.g., chemotherapy, targeted drug) is administered after RT [10]. The reaction is reversible with the withdrawal of the triggering agent. All radiation side effects should be managed optimally so that the patient can finish RT without unnecessary interruptions. Preventive measures should also be taken especially since late effects are not reversible. Even though irreversible, late effects should still be picked up and the patient managed and rehabilitated to reduce impairment on quality of life.

16.2.1 Pathogenesis of Acute and Late Effects

The general pathogenesis of acute and late effects is described here. The specific effects in each organ will be described in more detail under the relevant RT site. The organ specific effects depend very much on the target parenchymal cells, the types of functional subunits (FSUs) of the tissue and the FSUs structure in the organ. RT induces DNA damage through the generation of intermediary reactive oxygen species ("indirect action") or through direct ionization of DNA ("direct action"). The radiation damage activates various cellular signaling pathways which lead to activation of cell death signal cascade, pro-inflammatory and pro-fibrotic cytokines, vascular injury and activation of the coagulation cascade. Unrepaired radiation damage which leads to cell death do so predominantly due to mitotic catastrophe (reproductive death), though other mechanisms of cell death, e.g., apoptosis, autophagy, senescence, and necrosis, may play a role. Understanding the pathogenesis of these effects helps in selecting the appropriate measures to prevent, mitigate, treat, and rehabilitate them.

Acute effects induced by radiation are caused by both acute inflammatory response (edema/swelling, redness, tenderness) and parenchymal cell loss. Some of the earliest acute effects to manifest may be seen within a few hours of radiation and these are usually not due to cell killing but to inflammatory reaction, vascular leakage and edema mediated by radiation-induced inflammatory cytokines. The irradiated endothelial cells, macrophages, parenchymal cells and even fibroblasts release substances such as interleukin-1 (IL-1), IL-6, tumor necrosis factor-α (TNF-α), and cyclooxygenase-2 (COX-2). Subsequently, sphingomyelinases, ceramide, and the caspase cascade are activated. Common examples of these symptoms are nausea and fatigue occurring within a day of brain or abdomen irradiation, or acute skin erythema and edema from breast irradiation. However, the main cause of acute effects is due to the inability of the injured stem cells to renew and replace the depleting functional cells (from ongoing differentiation and cell loss). This is most apparent in tissues with high turnover, e.g., epithelium of skin or intestine, bone marrow. The overall turnover time (lifespan) of the tissue determines the onset time of the early response. Progressive hypoplasia of the tissue occurs and may be complicated by secondary problems such as infection on top of skin desquamation or oral mucositis. Eventually, the tissue heals due to cellular repopulation from the surviving stem cells in the irradiated volume or stem cells migrating in from outside the irradiated volume.

Late effects are contributed by depletion of parenchymal cells, vascular changes and fibrosis. They tend to occur in tissues with slow turnover of cells, e.g., liver, lung, heart, kidney, brain, muscle, subcutaneous tissue, muscular wall of intestine. The pathological lesions include atrophy, necrosis, fibrosis as well as vascular damage causing hypoxia, thrombosis and ischemia. The symptoms will be organ-specific, with organ failure being the worst possible outcome. The molecular reaction is induced during the irradiation but the processes evolve over years and eventually manifest clinically particularly when equilibrium in the tissue is disturbed. An intercurrent event which calls on the tissue to proliferate may cause the cells to die in mitosis in large numbers and poor healing brings about the late effects, e.g., tooth extraction in a previously irradiated mouth causing osteoradionecrosis. Chronic radiation effects develop through complex interacting processes. There is a complex interaction of the parenchymal cells, stromal cells (fibroblasts), vascular endothelial cells as well as macrophages, mediated by cascades of cytokines and growth factors. Consequently, parenchymal cell loss can lead to atrophy of its supplying vasculature; fibroblast differentiation leads to fibrosis; vascular damage leading to loss of capillaries, telangiectasia and fibrin leakage which in turn promotes collagen deposition; leukocyte adhesion and thrombi formation lead to blockage of lumen, hypoxia, and ischemia, hence contributing to the downstream parenchymal cell loss [11]. The common late effect of fibrosis which can occur in many organs is modulated by the transforming growth factor-β (TGF-β) which stimulates the differentiation of fibroblasts to fibrocytes with subsequent excessive collagen synthesis and deposition.

Understanding the pathogenesis helps to devise appropriate intervention strategies to prevent, treat or alleviate these radiation effects. For example, inflammatory effects can be managed with steroids or anti-inflammatory drugs. Reactive oxygen species may be scavenged by radioprotector drugs like amifostine and superoxide dismutases. Combination of pentoxifylline and vitamin E helps to improve blood flow and is used to manage problem of radiation fibrosis and necrosis. Ischemic effects may be mitigated by hyperbaric oxygen (HBO) (Table 16.2).

16.2.2 Documentation of Acute and Late Effects

The importance of assessing the patient thoroughly and documenting the acute and late effects as they happen cannot be overemphasized. The use of a standardized grading and scoring scale for these toxicities will make comparisons between results of different centers easier. Examples of widely-used grading systems are the CTCAE (Common Terminology Criteria for Adverse Events) system and RTOG/EORTC (Radiation Therapy and Oncology Group/ European Organization for Research and Treatment of Cancer) toxicity classification [12, 13]. The basic progressive changes in severity accounting for grade 0 through grade 5 is given in Table 16.3.

Table 16.2 Management of common toxicities related to radiation therapy

No	Organ	Toxicities	Prevention and treatment	Tolerance for whole volume (Gy) and endpoints[a]
1	Brain	– Acute: Edema, somnolence – Late: Necrosis, cognitive decline	– Dexamethasone – Hippocampal sparing treatment, dexamethasone, bevacizumab, hyperbaric oxygen (HBO), surgery, memantine, donepezil	24: Somnolence syndrome 55: Necrosis
2	Optic nerve	– Acute: Optic neuritis, pain – Late: Neuropathy, blindness	– Analgesics, dexamethasone	54: Optic neuropathy
3	Brain stem	– Acute: Raised ICP – Late: Necrosis, CN palsy	– Steroids – Steroids, bevacizumab	54: Necrosis, cranial neuropathy
4	Spinal cord	– Acute: Lhermitte's syndrome – Late: Myelopathy	– Steroids – Steroids, rehabilitation, anticoagulant, pentoxifylline, HBO	35: Lhermitte's syndrome 50: Myelopathy
5	Lens	– Late: Cataract	– Surgery (intraocular lens replacement)	5: Cataract
6	Retina	– Acute: Retinitis – Late: Retinopathy, blindness	– Analgesics, steroids – Laser photocoagulation, intravitreal bevacizumab, pentoxifylline, HBO	50: Blindness
7	Ear	– Acute: Otitis media, effusion – Late: Deafness	– Decongestant, grommet insertion, antibiotic	30: Serous otitis media
8	Skin	– Acute: Erythema, dry and moist desquamation, ulcer – Fibrosis, pigmentation, telangiectasia, ulcer	– Skin care, strataXRT, avoid sun and irritants, aqueous cream, steroid creams (hydrocortisone 1%, mometasone 1%), triethanolamine cream, dressing, analgesics – Pentoxifylline and vitamin E, HBO, skin graft	60: Moist desquamation
9	Salivary gland	– Acute: Sticky saliva, dryness – Late: Permanent xerostomia	– Oral care, amifostine – Oral care, pilocarpine, artificial saliva, sugar-free gum, transplant	10–20: Transient xerostomia 25: Permanent xerostomia
10	Oral mucosa	– Acute: Erythema, mucositis – Late: Ulcer, fibrosis	– Oral care, dental assessment prior to RT, mouthwash, honey, nystatin, nutrition support, analgesics (aspirin, benzydamine, NSAIDs, lignocaine viscous, opiates) – Oral care, palifermin, laser therapy	20: Acute ulcerative mucositis 60–70: Atrophy/ fibrosis
11	Mandible/ teeth	– Late: ORNJ, dental caries	– Dental assessment before RT, oral care, antibiotic, pentoxifylline and vitamin E +/– clodronate, surgery (sequestrectomy, vascularized composite tissue transfer), HBO	50: ORNJ

(continued)

Table 16.2 (continued)

No	Organ	Toxicities	Prevention and treatment	Tolerance for whole volume (Gy) and endpoints[a]
12	Pharyngeal muscle	– Acute: Odynophagia – Late: Stricture, ulcer, dysphagia, regurgitation	– Benzydamine throat spray, lignocaine viscous, nutritional support, swallowing exercises – Speech/swallowing therapy	Dmean 60: Aspiration
13	Breast	– Acute: Mastitis, erythema, edema (swelling), desquamation – Late: Skin/fat necrosis, telangiectasia, breast distortion, atrophy, fibrosis	– Loose clothing, skin care, aqueous cream, arm physiotherapy – Pentoxifylline + vitamin E	60: Fibrosis/atrophy
14	Lymphatics	– Late: Lymphedema, heavy limbs	– Compression device/dressing, physiotherapy, decongestive therapy, skin care, rehabilitation	70: Node atrophy 90: Lymph vessel sclerosis
15	Esophagus	– Acute: Esophagitis, dysphagia – Late: Fibrosis, stricture, perforation	– Amifostine, prophylactic glutamine, soft diet, nutritional support, mucaine gel, lignocaine viscous, sucralfate, antacids, antifungals, analgesics – Swallowing therapy, PEG-tube, endoscopic dilatation, stents	40–45: Dysphagia 55: Ulcer/fistula
16	Lungs	– Acute: Pneumonitis – Late: Fibrosis, respiratory failure	– Special RT techniques (respiratory gating and tumor tracking, DIBH), dexamethasone, antibiotics – Prednisolone	12–14: Whole lung RT, pneumonitis
17	Heart	– Acute: Pericarditis, myocarditis, palpitations, arrhythmia – Late: Pericardial fibrosis, arrhythmia, cardiomyopathy, coronary syndromes	– Special RT techniques (DIBH), NSAIDs (e.g. ibuprofen) – Specialized cardiac intervention and care	20: Transient arrhythmia, ECG changes 40: Cardiomyopathy, myocardial infarction
18	Stomach	– Acute: Gastritis, nausea, vomiting – Late: Ulcer, fibrosis, bleeding telangiectasia	– Antiemetics (5-HT3 antagonists, prochlorperazine), antacids, avoid irritant food, nutritional support – Prednisolone, coagulation, formalin, surgery	20: Atony 50: Ulcer
19	Small bowel	– Acute: Enteritis, malabsorption – Late: Ulcer, stricture, obstruction, adhesion	– Full bladder protocol during pelvic RT, antiemetics, dexamethasone, hydration, nutritional support – Anti-inflammatory drugs, surgery	30: Malabsorption 40: Ulcer/ obstruction
20	Large bowel	– Acute: Colitis, diarrhea, pain, flatulence, ileus – Late: Chronic colitis, ulcer, stricture, fistula, perforation, bleeding telangiectasia	– Low fiber/low fat diet, rehydration, nutritional support, anti-diarrheal drug (diphenoxylate/atropine, loperamide), octreotide – Anti-inflammatory drugs, aspirin, coagulation, YAG or argon laser, HBO, surgery	10–20: Diarrhea, pain 45: Ulcer/ obstruction

(continued)

Table 16.2 (continued)

No	Organ	Toxicities	Prevention and treatment	Tolerance for whole volume (Gy) and endpoints[a]
21	Rectum	– Acute: Proctitis, bloody discharge, tenesmus – Late: Chronic proctitis, bleeding telangiectasia, ulcer, fistula, stricture, incontinence	– Rectal spacer, low fiber/low fat diet, rehydration, anti-diarrheal drug, analgesics – Steroid enema, sucralfate enema, mesalazine suppositories, anti-inflammatory drugs, aspirin, antioxidants (vitamins A, E, C), antibiotic (metronidazole), argon plasma coagulation, formalin therapy, HBO, YAG or argon laser, surgery	50: Acute proctitis 60: Chronic proctitis, ulcer
22	Bladder	– Acute: Cystitis, dysuria, frequency, urgency – Late: Chronic cystitis, fibrosis, contracture, reduced capacity, obstruction, bleeding telangiectasia, fistula	– NSAIDs, treat infection, urinary alkalinizers (ural, mist potassium citrate, cranberry juice) – Systemic therapy (glucosamine, pentosan polysulfate, estrogen), intravesical therapy (alum, botulinum toxin, hyaluronic acid, formalin), formalin bladder washout, argon plasma coagulation, YAG laser, HBO, surgery	20–35: Acute cystitis 65: Atrophy/ulcer
23	Kidney	– Acute: Nephritis – Late: Nephropathy, chronic nephritis, renal failure	– Analgesics – Kidney shielding, ACE inhibitors or angiotensin or (AII) receptor blocker mitigation, dialysis, transplant	20: Late nephropathy
24	Gonads	– Acute and late: Azoospermia, sterility, menopause	– Testicular shielding, ovariopexy (to protect or displace from radiation beams)	1.5: Testis permanent sterility 2.5: Ovary permanent sterility
25	Bone marrow	– Acute: Low cell count, hypoplasia – Late: Fibrosis, aplastic anemia	– Growth factors, transfusion, bone marrow transplant	2: Transient hypoplasia 4: Lethal aplasia
26	Liver	– Acute: Hepatitis, ascites, veno-occlusive disease (VOD) – Late: Hepatocellular loss, fibrosis, vein thrombosis, liver failure	– Anticoagulants, thrombolytics, steroids – Liver transplant	30: VOD
27	Thyroid	– Acute: Thyroiditis – Late: Hypothyroidism	– Steroids, analgesics – Thyroxine replacement	45: Thyroiditis
28	Lachrymal system	– Acute: Dry eye, conjunctivitis – Late: Stricture, epiphora, dry eye, corneal ulcer	– Steroids, analgesics, eye drops – Artificial tear	40: Dry eye, ulceration
29	Hair	– Acute: Hair thinning – Late: Alopecia	– Minoxidil, botulinum toxin, hair transplant	40: Transient alopecia

[a] Adapted from *Basic Clinical Radiobiology* 4th Edition (2009) and Emami (1991, 2013); Disclaimer: The tolerance of whole volume dose mentioned in the table is a representative dose; Readers are encouraged to follow specific QUANTEC dose guidelines for individual organs

Table 16.3 Basis for most toxicity scoring systems

Grade	Severity of effects
0	No change from baseline
1	Mild. No impact on activities of daily life (ADL)
2	Moderate. Mild ADL impact, outpatient management
3	Severe. Severe ADL impact, hospitalization
4	Life threatening
5	Lethal

16.3 Specific Radiation Toxicities

16.3.1 Radiation Toxicities of Central Nervous System (CNS)

The CNS may be apparently less sensitive to radiation compared to other late responding organs, but the sequelae can be severe and devastating, e.g., paralysis from myelopathy. The magnitude of toxicity is influenced by dose per fraction, total radiation dose, addition of radiosensitizing chemotherapy, volume of tissue irradiated, re-irradiation, proximity of the critical structure to the tumor and extreme young age (baby's developing brain). Irradiation of entire brain leads to neuropsychological and cognitive deficiencies. Partial volume irradiation of the brain and spinal cord to a high dose could also lead to toxicity. The earliest acute effects in the brain are usually edema giving rise to raised intracranial pressure, therefore the symptoms of headache, nausea, vomiting, and fatigue. A subacute effect of somnolence syndrome from transient demyelination is quite commonly seen 3–6 months later. However, the most important injuries to brain are the late effects, mainly of radiation necrosis which usually appear 2–3 years later, manifesting as seizures, functional defect or cognitive problems. Changes within the first year are usually in the white matter, while vasculopathies and gray matter changes predominate in the later years. Three main cell types in the brain are involved in the pathogenesis: neurons, glial cells, and vascular endothelial cells. Except for a few neuronal tissues with potential for neurogenesis, most CNS tissues have limited mitotic capability. Therefore, damage could be contributed by radiation-induced oxidation of membrane lipid bilayer, microvascular permeability, cell–cell junctional re-arrangement, mitochondrial alteration, and oxidative stress. Radiation mainly results in coagulation necrosis of white matter, cerebral blood vessels, axonal demyelination, and damage to the endothelial cells of vessels. Leukoencephalopathy, a progressive late toxicity, occurs due to overproduction of myelin in oligodendrocytes. Hippocampal tissue of the brain has high potentiality for neurogenesis; therefore, protection of hippocampus during RT could preserve cognitive function and memory following cranial RT. Hypofractionated RT and very high dose radiosurgery in small lesions have specific threshold for radiation damage. When diameter of SRS-treated lesion increases from 2 to 3 cm, the tolerance dose could decrease from 24 to 15 Gy [14]. Spinal cord shows similar response to radiation with the 3 waves of acute inflammation, subacute transient demyelination (Lhermitte's sign) and late vascular and gray matter necrosis (risk of permanent myelopathy).

Radiation-induced optic neuropathy (RION) is a concern for treatment of intracranial tumors near sella turcica. The optic nerves and chiasm are very thin, measuring <5 mm in diameter. Though the tolerance is usually limited to 45 Gy in conventional fractionated radiotherapy, the risk of RION appears to rise only after 60 Gy [15]. Using 50 Gy in conventional fractionation, the possible risk of visual deficit is 5% according to Emami et al. [16]. However, in QUANTEC review, the chance of optic neuritis was <2% for Dmax of <55 Gy in same conventional fractionation [17]. In proton beam radiation, the tolerance limit ranges from 55 to 60 CGE. Additionally, the risk of RION is influenced by older age, hypertension, diabetes mellitus and chemotherapy.

The other neurological structures such as brainstem, spinal cord and internal ear get radiation exposure during treatment of head and neck cancers. According to QUANTEC, conventional fractionation dose up to 54 Gy to entire brainstem is safe with very minimal toxicity [18]. In spinal cord (full cord cross section) conventional fractionation total dose of 50 Gy, 60 Gy, and 69 Gy are associated with 0.2%, 6%, and 50% rate of myelopathy. Single fraction maximum point dose to thecal sac of 10 Gy [19] is safe whereas RTOG0631 recommends D_{10} and $D_{0.35}$ mL of 10 Gy and Dmax of 14 Gy for the treated spinal cord. For auditory apparatus, conventional fractionation dose should be limited to <35 Gy–<45 Gy to avoid sensorineural hearing loss [20].

16.3.2 Radiation Toxicities of Head and Neck

Surgery, RT, chemotherapy, and immunotherapy are used in the multidisciplinary management of head and neck cancers. Out of all modalities, radiation therapy plays a significant role in the management of head and neck cancer. About 50% of all head and neck cases need RT in the course of illness [21]. RT is used as a sole modality, adjuvant therapy or as part of palliative care. Nasopharyngeal cancer, the most common cancer found in China and Southeast Asia are predominantly treated with RT with more than 90% 5-year cure rates in early stages with good quality of life [22]. Cancer of the lips, tongue, palatine tonsil, and larynx in organ-confined stages are treated with radiation therapy for organ and function preservation. Radiation treatment allows for near normal cosmetic outcome compared to surgery.

Radiation toxicities in this area result in significant morbidity and at times mortality during and after treatment. Mucosal lining from nasal cavity, oral cavity, oropharynx, hypopharynx, and larynx are exposed during RT treatment of above subsites. In addition, other innocent nearby organs are also frequently exposed, e.g., pituitary gland, temporal lobes, brainstem, spinal cord, visual pathways, eyes, auditory apparatus, salivary glands, dental complex, taste buds, temporomandibular joint, masticatory muscles, pharyngeal muscles, and thyroid gland. Despite careful planning and optimization, these normal structures get exposed to radiation leading to acute and late complications. The magnitude of complications (toxicities) depends on location, tumor volume, stage, body mass index, comorbidities (diabetes, connective vascular diseases), RT techniques, fractionations and addition of cytotoxic chemotherapy agents [23].

Acute toxicities though self-limiting could pose tremendous pain, dysphagia, malnutrition, dehydration, stridor and low quality of life. They may be severe enough to result in treatment breaks and discontinuation of effective RT especially in developing countries [24]. Late effects resulting in complications like osteoradionecrosis of jaw (ORNJ), pharyngeal muscle fibrosis, brain necrosis, optic neuropathy and hyposalivation could be untreatable despite advances in management of such complications. Therefore, radiation oncologist should be aware of and understand these debilitating and potentially fatal toxicities, so that appropriate preventive measure or early supportive care treatment can be used to avoid such outcomes. The most common toxicities are radiation dermatitis, mucositis, loss of taste, dysphagia, laryngitis, xerostomia, deafness, dental resorption, and ORNJ. The common complications are described below.

16.3.2.1 Radiation-Induced Skin Reaction (RISR)

Skin with its appendages is invariably exposed to radiation during external beam RT. Despite technological advances epidermis, dermis, underlying fat and skin appendages receive dose of radiation which causes acute and late toxicities.

Radiation-induced skin reaction (RISR) is one of the most common toxicities during practice of RT especially in head and neck, breast, axilla and inguinal area (Fig. 16.3). The incidence of RISR is up to 95% including all grades of toxicity. The evolution of the acute toxicities is reflected by the NCI CTC scores of 4 grades: (1) dry desquamation with erythema, (2) brisk erythema with patchy moist desquamation mainly at skin folds, (3) moist desquamation at non-skin fold areas, and (4) ulceration, bleeding, and necrosis. Late changes are chronic ulceration, fibrosis, telangiectasia, keratosis, and skin cancer. The progression of skin toxicity epitomizes acute and late reaction to radiation. Initial erythema and increased pigmentation caused by acute

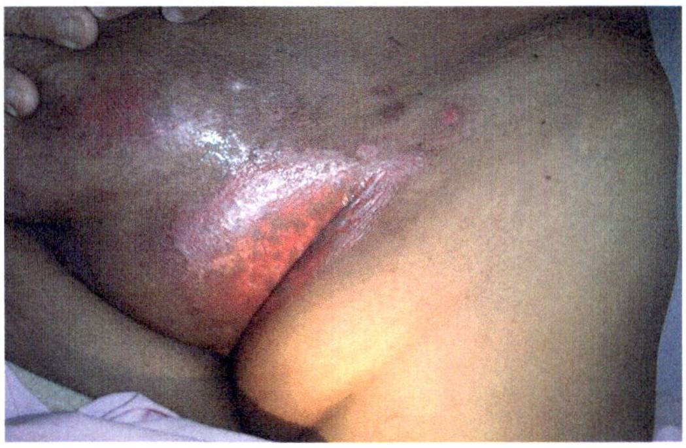

Fig. 16.3 Moist desquamation over inguinal region during external radiotherapy. Moist and skin fold regions of body are vulnerable to skin toxicity

inflammation is followed by dry and moist desquamation caused by damage to stem cells in the epidermal basal layer preventing replacement of the cells shed from the outer layer; the onset time of desquamation of 3–4 weeks correlates with the lifespan of the mature cells. When radiation ends, the residual surviving stem cells repopulate and heal the wound. In chronic phases of progressive oxidative stress, the late effect is from the deeper dermal and subcutaneous layer with fibroblasts giving rise to fibrosis and skin contraction/induration, as well as late vascular injury causing telangiectasia. Hair epilation occurs after 3 weeks and regrowth takes place after 1–3 months.

Skin toxicity is influenced by RT techniques. Megavoltage RT is known to be skin sparing but advances in techniques from 3D-Conformal to IMRT increases skin dose and acute effects. Helical tomotherapy may also increase skin dose compared to Field-in-Field IMRT. Orthovoltage and electron beam also increase skin toxicity.

The risk factors for dermatological toxicities are extrinsic and intrinsic. The extrinsic factors include total dose, fractionation scheme, use of bolus, volume of tissue exposed. The intrinsic factors are inherent radiosensitivity of tissue, connective vascular diseases and genetic influence (ataxia telangiectasia syndrome, Gorlin syndrome, etc.).

16.3.2.2 Radiation-Induced Oral Mucositis (RIOM)

Radiation-induced oral mucositis is the most common toxicity of RT to the head and neck region. It is essential to differentiate between *oral stomatitis* and *radiation mucositis* [25]. Oral stomatitis ensues due to chemotherapy-induced loss of oral epithelium, whereas RIOM is due to radiation-induced inflammation of the oral mucosa. About 50–70% of cancers in this region develop RIOM during radical RT; however, the prevalence of RIOM could be almost 100% when combined with concurrent cytotoxic chemotherapy [26]. RIOM not only cause sore mouth; they are also associated with loss of taste (dysgeusia), pain, dryness, and difficulty in swallowing. These symptoms lead to significant decrease in food and fluid intake leading to dehydration, malnutrition and loss of body weight. Head and neck RT could lead to >5% loss of weight until eighth week or more than 7.5% of body weight at the end of 12 weeks after RT [27]. The magnitude of RIOM is influenced by fractionation factors, area of radiation exposure, radiation techniques, poor dental hygiene, other trauma to the mucosa (e.g., sharp dentures) and addition of chemotherapy or targeted therapy. Mucosal pain is very debilitating leading to constant psychological distress and poor quality of life.

The pathobiology of radiation-induced oral mucositis has been described by Sonis et al. as five distinct phases of (1) initiation, (2) damage response, (3) amplification, (4) ulceration, and (5) healing (Fig. 16.4) [28, 29]. It typically evolves as erythema, patchy mucositis, confluent mucositis and ultimately ulceration (the basis for its severity grading) (Fig. 16.5). Understanding these phases will assist oncologists to manage RIOM effectively. Loss of protective barrier of mucosa also promotes bacterial and fungal invasion and further organism-induced inflammation. However, eventually new cells grow from the basal layers of epithelium in the healing phase. On the other hand, proliferation of fibroblasts and vascular injury cause late toxicities like fibrosis and telangiectasia.

Modified fractionation schedules of hyperfractionation (smaller than 2 Gy dose per fraction, more than 1 fraction a day at interfraction interval of at least 6 h) or accelerated fractionation (more than 10 Gy per week) are expected to improve local control rate and reduce late radiation

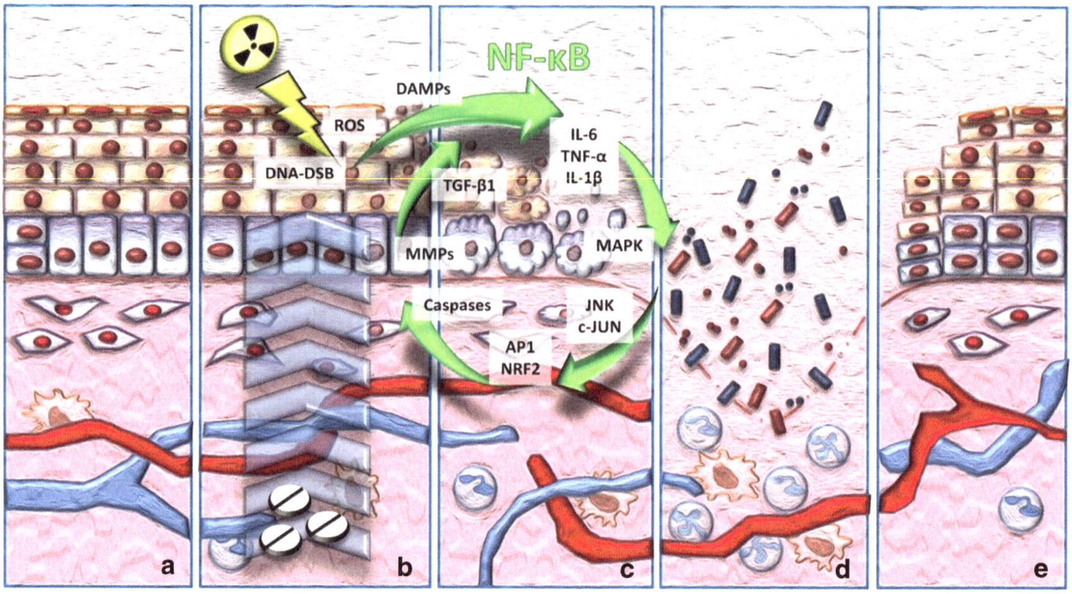

Fig. 16.4 Illustration demonstrating pathobiology of radiation mucositis: (**a**) normal tissue, (**b**) initiation phase and primary injury response, (**c**) amplification of the injury signal, (**d**) ulceration, (**e**) tissue re-epithelialization and healing. (Reproduced from Pulito C, Cristaudo A, La Porta C, Zappen S, Blandino G, Morrone A, Strano S. Oral mucositis: the hidden side of cancer therapy. J Experimental Clin Cancer Res 2020; 39; 210)

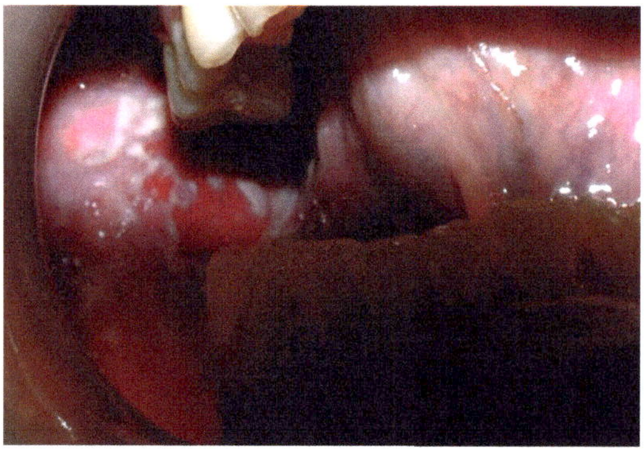

Fig. 16.5 Grade-IV radiation mucositis on fifth week of radiotherapy involving right buccal mucosa opposite to molar teeth

complications but at the cost of more acute radiation mucositis which may be dose-limiting. In a phase-III clinical trial on advanced head and neck cancer patients randomized to either conventional fractionation (70 Gy/35F/7 weeks) or accelerated fractionation schedule (64 Gy/34F/23 days), higher grade-3 radiation mucositis rate was reported in the experimental arm (75%) compared to conventional arm (23%) [30]. In a recent meta-analysis of 6 studies on head and neck cancers comparing accelerated versus conventional fractionation, significant increase in oral mucositis rate was found in the accelerated fractionated group [31].

Addition of systemic agents increases the rate of acute mucositis. The current management trend in locally advanced head and neck cancer is concurrent chemoradiotherapy to exploit radiosensitivity properties of specific drugs. The prevalence of RIOM is about 97.1% in concurrent chemoradiotherapy patients [26]. In a study, anti-EGFR monoclonal antibody nimotuzumab was observed as significant factor in the development of symptomatic grade-3 mucositis [32]. In another study using anti-EGFR antibody cetuximab, the rate of grade-3 RIOM was in 78.6% of patients [33]. In severe complicated cases of RIOM, the basal cell layer of epithelium could not be replenished leading to mucosal fibrosis and functional deformity of oropharyngeal anatomy.

16.3.2.3 Radiation-Induced Xerostomia

Salivary glands are invariably exposed to radiation during the treatment of head and neck cancers. Parotid and submaxillary glands are the major source of saliva contributing 90% of the volume of about 1.5 L/day. Parotid gland predominantly supplies 65% of stimulated saliva whereas submaxillary glands contribute 25% unstimulated saliva. Sublingual gland provides 2–5%, whereas minor salivary glands in oral cavity, hard palate, and soft palate are responsible for remaining 5% of saliva. Reduction of salivary flow or thick saliva starts from second week to post RT period and continue to decline over several months (Fig. 16.6). The serous acinar cells undergo interphase death by apoptosis, accounting for the fast onset of response. In certain conditions salivary flow could recover over 12–18-month period after RT. At the cellular level the acute component of radiation xerostomia is due to damage to the plasma membrane of the acinar cells; the above effect downregulates saliva secretion via receptor-mediated signaling. The late and persistent xerostomia is secondary to the damage to progenitor stem cells of salivary gland. Animal study on xerostomia shows four distinct phases of (1) reduction in water content of saliva, (2) degradation of acinar cells leading to decreased production of amylase, (3) actual loss of acinar cells due to loss of stem cells, and (4) recovery of damage but continuing deterioration of function due to damage to ducts, vessels and nerves [34]. Xerostomia through changing the oral milieu promotes the secondary effects of oral infection and dental caries.

16.3.2.4 Osteoradionecrosis of Jaw (ORNJ)

Osteoradionecrosis of jaw is an avascular necrosis of the bone especially at the angle of mandible. ORNJ is defined as "an area of clinically exposed necrotic bone that was previously irradiated and non-healing over a period of 3 months without evidence of recurrence." High dose radiation in least vascular area of bone leads to loss of vascularity and progression to necrosis and cutaneous sinus against the lesion. RT induces an endarteritis that results in tissue hypoxia, hypocellularity, and hypovascularity (3Hs), which in turn causes tissue breakdown and chronic non-healing wounds [35]. The main factor causing this ischemic necrosis is radiation-induced obliteration of the inferior alveolar artery. Surgical trauma,

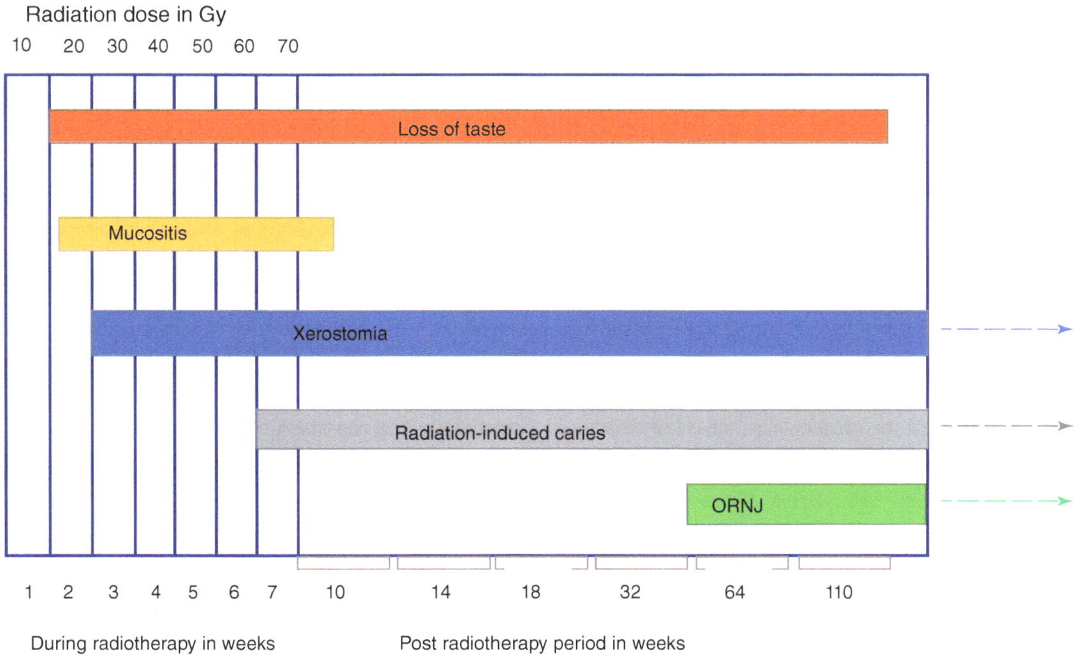

Fig. 16.6 Duration of acute and late toxicities of radiotherapy in head and neck region

dental extraction, implant placement could precipitate ORNJ after exposure to radiation [36]. The symptoms include pain, discharge and progressive deformity of the jaw bone. The incidence of ORN is 7.5% in head and neck cancer patients receiving RT at a median time of occurrence of 27 months. The ORNJ are staged according to Karagozoglu et al. to 3 well-defined stages for inter-observer communication [37]. Dose volume histogram (DVH) parameters such as V30 and V70 values are significantly higher in ORN. In a study of 616 patients receiving RT to the mandible developed ORN in 56 patients [38]. In their analysis, primary tumor site, pre-RT mandible surgery, dental extraction and V60 > 14% were proposed as risk factors for ORNJ.

16.3.2.5 Other Toxicities

Besides the common toxicities described above, other radiation side effects are also experienced including pharyngeal muscle dysfunction, hypothyroidism, trismus, temporomandibular joint fibrosis, carotid artery stenosis, middle ear effusion, and hypopituitarism.

16.3.3 Radiation Toxicities of Breast and Soft Tissue

16.3.3.1 Soft-Tissue Fibrosis and Breast Cosmesis

Hypofractionated whole breast RT after breast conservation surgery (BCT) in early breast cancer is now a standard practice. Acute effects of breast swelling from edema may impact on the accuracy of radiation treatment set-up and homogeneity of dose received. Acute skin effects are as described earlier; inframammary skin folds and axilla are particularly susceptible (Fig. 16.7). The most common late effects are breast shrinkage, edema, retraction and telangiectasia. Subcutaneous fibrosis due to RT leads to breast induration which continues for many years. Fat necrosis and edema contribute to post-radiation breast induration. Cosmesis is an endpoint of interest in breast treatment. A study on BCT breast treatment using 40 Gy in 15 daily fractions and in suitable patients boost dose of 9 Gy in 3 fractions using 3D conformal technique reported 68% grade 0–1 and 15% grade 2 skin toxicity with late grade-1 and 2 skin toxicity observed among 18% and 1% patients; however, G0/1 subcutaneous toxicity was observed in 98% and grade 2 toxicity among 2% of their patient population [39]. Studies suggest breast size, obesity, breast volume, and dosimetric inhomogeneity could contribute to breast radiation toxicity.

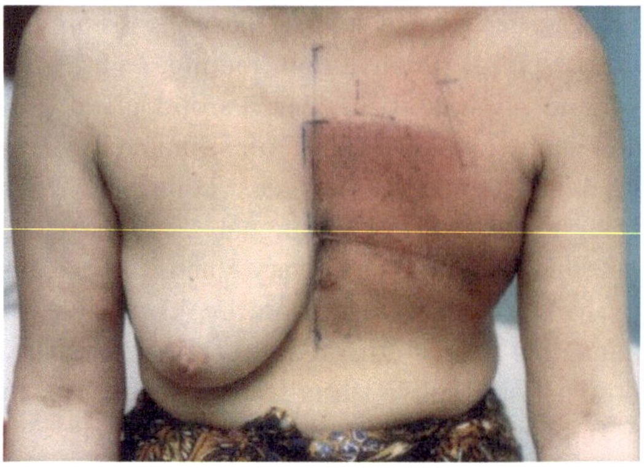

Fig. 16.7 Extensive erythema following chest wall irradiation in a case of breast cancer undergoing adjuvant radiotherapy

Higher skin and subcutaneous tissue dose of radiation could lead to fibrosis, subcutaneous tissue scarring and tissue necrosis. The above toxicity is similar to wound healing with fibrosis. The pathophysiology of radiation-induced breast fibrosis is inflammation. Acute tissue injury and hardening of exposed skin and inflammatory process induce inflammation, fibroblast proliferation, extracellular matrix deposit, and vasculogenesis.

Almost 80% of early breast cancers on BCT showed good to excellent cosmesis on long-term follow-up. Radiotherapy on a postmastectomy breast with reconstruction very often carries risk of fibrosis, hence poorer cosmesis. In the case of implant-based reconstructed breast, fibrosis leads to capsular contracture and 15–30% of implant failures. Apart from the type of reconstruction, timing of RT (immediate or delayed) also influences toxicity.

16.3.3.2 Lymphedema

Lymphedema is a complication of surgery in solid tumors requiring extensive lymph node dissection. Extensive lymph node removal leads to poor lymphatic drainage. Most common etiology for arm lymphedema is breast cancer with axillary lymph node dissection; inguinal block dissection in cancer of vulva and cutaneous melanoma may cause lower limb edema. Further post-operative RT fibrosis in above regions could contribute further risk for arm or leg lymphedema respectively. The development of lymphedema could be noticeable toward the end of RT or years after RT.

Risk of lymphedema after breast treatment ranges from 5% after sentinel lymph node sampling with or without RT to about 20–30% after axillary lymph node dissection followed by regional nodal RT. Symptoms includes swelling, pain, restriction in movement, heaviness of limbs and skin weeping (lymphorrhea).

16.3.4 Radiation Toxicities of Thorax

16.3.4.1 Radiation Toxicities of Esophagus

Radiation esophagitis is one of the most common dose-limiting acute radiation toxicities during thoracic RT. Treatment of lung cancer, breast cancer, esophageal cancer, mediastinal tumors etc. invariably include large extent of esophagus leading to significant toxicity. Use of accelerated RT, and concurrent systemic chemotherapy further increases incidence of esophagitis, leading to significant morbidity and mortality. People of Caucasian origin, elderly, poor general condition, low body mass index, esophageal erosion and specific germline polymorphism increase the incidence of radiation esophagitis. In the past, use of 2D RT usually involved significant proportion of esophagus causing severe acute and late radiation toxicities resulting in painful dysphagia, malnutrition, dehydration and even death. Radiation-induced epithelial inflammation, thinning of basal cell layer, mucosal denudation leads to the acute effects while muscle wall fibrosis, vascular damage and epithelial cell loss lead to ulceration and stricture as late toxicity. Secondary infection with candidiasis may compound the acute effects and thus require prompt treatment. The incidence rate of significant acute esophagitis was 1.3% but increased to 14–49% in concurrent chemo-radiotherapy (CCRT) [40]. Modern RT techniques like IMRT and IGRT reduce normal tissue toxicities. In 2D era, concurrent chemo-radiotherapy leads to 75% grade-2, 22% grade-3 and 2% grade-4 toxicity [41]. However, using IMRT technique, esophageal toxicities in CCRT reduced to 32%, 17%, and 0.9% grade 2, grade 3, and grade 4 respectively [42].

16.3.4.2 Radiation-Induced Lung Injury (RILI)

Injury to lung must be given attention in treatment of lung cancer, esophageal cancer, mediastinal tumors, breast cancer, other thoracic malignancies, total body irradiation for leukemia and whole lung irradiation for Ewing sarcoma. Lung tissues are very radiosensitive; however, as an organ, it has a high volume effect due to the parallel pattern of its FSUs (i.e., the pulmonary alveoli). Tolerance is low when large volume is irradiated but small volume may withstand very high dose of radiation. So lung injuries are compensated, thus clinically not manifested until critical level of injury is reached. RILI is divided into acute radiation pneumonitis (weeks to 6 months), subacute radiation pneumonitis (6–12 months) and late radiation fibrosis (>year). They are controlled by complex signaling cascades.

In acute radiation pneumonitis, there is Type II pneumocytes death, endothelial damage and congestion of alveoli leading to exudation that prevent optimal diffusion of oxygen resulting in hypoxia. In subacute radiation pneumonitis, the increased release of cytokines like TGF-β and IL-6 later encourages proliferation of fibroblasts, vascular leakage and accumulation of fluid in alveolar capillaries. The above pathology leads to progressive fibrosis and fibrothorax. The radiation pneumonitis is reversible with treatment (high dose steroid); however, radiation fibrosis is permanent and progressive. Clinical manifestation include x-ray changes, cough, dyspnea, chest pain, and in severe cases, respiratory failure.

The risk factors for development of RILI are smoking, COPD, interstitial lung disease (ILD), autoimmunity, old age, concomitant systemic chemotherapy, tamoxifen, and supraclavicular irradiation in postoperative adjuvant RT of breast cancer. Addition of systemic therapy, which is radiosensitizing or has inherent lung toxicity, e.g., bleomycin, cyclophosphamide, anti-EGFR, immunotherapy, is expected to increase RILI [43]. However, the risk is modifiable by treatment factors like keeping to RT planning constraints and techniques like IGRT, respiratory gating and deep inspiration breath hold (DIBH).

When combined with immunotherapy, radiation regulates the T-cell receptor, enhances its diversity, and has a potent synergistic effect [44]. RT may also, in rare cases, induce an *abscopal effect*.

Pulmonary toxicity in the KEYNOTE-001 phase 1 trial was noted to be increased in patients given pembrolizumab who also had prior RT (13% vs. 1%) but this increase was limited to mild grade toxicity [45]. Rate of grade 3–4 pneumonitis was similarly not increased in another study (PACIFIC) testing durvalumab in adjuvant setting post-RT [46]. From these trials of immunotherapy after fractionated RT, the toxicity specifically of interstitial lung disease (ILD) seems to be mild when present. However, the risk of ILD in the concomitant use of SBRT and immunotherapy is significantly increased especially when the planning target volumes are large and two or more lung lobes are targeted [47].

Immunotherapy alone causes significant incidence of ILD. Hence, in combination with RT, a high index of suspicion is needed to differentiate between RT-induced pneumonitis/fibrosis against ILD purely caused by the drug. Clinical judgment based on the dose of radiation, dose volume parameters and timing of symptom onset, give some clue. Immunotherapy-induced ILD is usually expected within 90 days of discontinuation of the drug. However, delayed immune-related events (DIRE) in immunotherapy is not uncommon [48].

16.3.4.3 Radiation-Induced Heart Disease (RIHD)

During RT treatment of breast cancer, lung cancer, mediastinal lymphoma or thymoma, part of the heart is invariably included in the high dose volume. Any extent of heart or its components exposed to radiation has consequences ranging from electrocardiographic changes to early cardiac death after RT, encompassing pericardial disease, rhythm disorder, cardiac dysfunction, cardiomyopathy, congestive failure, valvular disease and coronary syndromes like ischemic disease and myocardial infarct [49]. Early effect is limited to acute pericarditis. Cardiac toxicity is mainly from late effects, histopathologically characterized by loss of cardiomyocytes, dense fibrosis (of myocardium, pericardium, and valves), coronary atherosclerosis, and microvascular damage. Dose escalation study in RTOG-0617 demonstrated that higher escalated radiation dose (74 Gy) was inferior to standard dose (60 Gy) in the treatment of lung cancer in terms of overall survival and toxicity [50, 51]. The cause for lower survival was due to death related to RT-induced cardiac events.

Cardiac risk is increased in patients with underlying heart disease and with the addition of cardiotoxic drugs like anthracyclines, taxanes, and trastuzumab which are widely used in breast cancer. Breast cancer patients also usually need adjuvant RT to the primary site and lymphatics, with cardiac risk a particular concern for left-sided breast cancer. Techniques such as deep inspiration breath holding (DIBH) [52] (Fig. 16.8) and prone position technique could bring heart away from anterior chest wall [53].

16.3.5 Radiation Toxicities of Abdomen and Pelvis

Abdomen and pelvic viscera are exposed to radiation during treatment of malignancies in liver, pancreas, para-aortic nodes, retroperitoneal sarcomas, lumbosacral spine, rectum, prostate, endometrium, and cervix. The radiation exposure of individual viscera and their response depends upon location of the tumor, distortion due to surgery, extent of disease, RT technique, field size, total dose, and fractionation schedule.

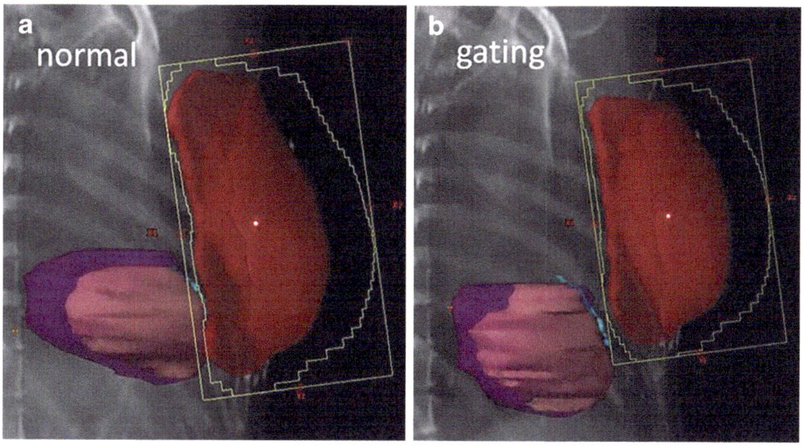

Fig. 16.8 Digital reconstruction from two treatment plans; (**a**) planning with normal breathing without deep-inspiration breath-hold (DIBH); (**b**) the same patient planned using gating with DIBH.PTV in red, heart in purple, left ventricle in bright purple, and left anterior descending artery contoured in light blue color. (Reproduced from Piroth MD, Baumann R, Budach W, Dunst J et al. Heart toxicity from breast cancer radiotherapy. Current findings, assessment and prevention. Strahlentherapie Onkologie 2019;195:1–12)

16.3.5.1 Radiation Enteropathy

Small and large bowels are exposed during treatment of rectal cancer, para-aortic lymphatics, genitourinary cancer, and retroperitoneal sarcoma. Fixed loop of bowels from postoperative adhesion are particularly susceptible to receiving higher dose compared to mobile loops. Treating with a full bladder protocol in pelvic malignancies is a common practice to displace the bowels out of the RT fields. Release of factors from irradiated stomach such as serotonin, neurokinin-1 act on central receptors causing nausea and vomiting, dyspepsia. Increased initial bowel motility (borborygmi) is followed by an atonic phase, then the phase of epithelial denudation causing water and electrolyte loss (dehydration and diarrhea), malabsorption, and loss of barrier for infection. Chronic radiation enteritis is a vexing condition affecting about 5–55% patients following radiation therapy [54]. It results from muscular wall fibrosis and obliterative endarteritis in the epithelium, manifesting as chronic ulcer, intestinal perforation, obstruction, fistula formation and malabsorption.

16.3.5.2 Radiation Proctopathy

Patients normally present with combination of symptoms such as urgency of stool, rectal bleeding, mucoid stool, diarrhea, fecal incontinence, bloating, intestinal borborygmi and flatulence. In later stages new fragile blood vessels (telangiectasia) develop in mucosal scar with loss of wall elasticity leading to hemorrhagic proctitis, stricture, fistula, intestinal obstruction, and fecal incontinence from sphincter dysfunction.

Risk factors for radiation proctopathy include tumor factors like location, size of tumor, extent of subclinical disease, and treatment factors such as prior pelvic surgeries, extent of radiation coverage, RT techniques and concurrent treatment with chemotherapy, e.g., 5-fluorouracil or capecitabine, which are commonly used in rectal cancer and are inherently diarrhea-inducing. Over the last few years, there are tremendous advances in technology like IMRT, IMPT, IGRT, and hypofractionated SBRT. IMRT consistently reduces incidence of acute effects in rectum and late effects up to 1 year post RT. However, there is no evidence of reduction in late effects (radiation proctopathy) beyond 1 year in the literature. A recent study from TMH Mumbai showed significant reduction in >grade-2 rectal toxicity from 42.4% to 21.1% (HR 0.46; $p < 0.001$) when using image-guided IMRT versus conventional 3DCRT in postoperative cervical cancer. The cumulative late toxicity (>grade-2) was also reduced from 48.9% to 28.1% at a median follow up of 46 months [55]. The risk of late toxicities is influenced greatly by patient factors such as comorbidities like diabetes, connective vascular disorder, inflammatory bowel diseases, HIV, low body mass index (<18.5 kg/m^2) and chronic tobacco use [56].

There should be high index of suspicion of radiation late effects rather than rectal tumor recurrence as the cause of chronic symptoms manifesting many years later, from endoscopic features without a need for biopsy. Biopsy may precipitate worse late complication and needs to be done with caution [57].

16.3.5.3 Radiation Toxicities of Bladder

Injury to the bladder, ureter, and urethra is very common during RT treatment of pelvic tumors. The urothelium has a long turnover time of many months. Acute effects of bladder are due to mucosal hyperemia and edema mediated by change in prostaglandin metabolism as well as urothelial barrier function impairment allowing secondary infection. The early effects show a strong consequential component leading on to chronic effects of mucosal breakdown, detrusor muscle fibrosis, sphincter damage, vascular injury and telangiectasia. Radiation cystitis presents as early irritative symptoms of frequency, urgency, dysuria and late symptoms from bladder contracture, urethral stricture, chronic hemorrhagic cystitis.

Curative RT for cancer of prostate, rectum, cervix, and endometrium commonly expose urinary bladder to radiation. The rate of toxicities seen are dependent on the proximity of bladder to the tumor, volume irradiated, total dose and dose fractionation and RT technique.

In prostate cancer RT, total dose in standard EBRT techniques vary between 74 and 84 Gy. From literature of EBRT and interstitial RT, incidence of significant grade 3 urinary toxicities ranged from 5% to 13% with very low (1%) rate of grade-4 urinary toxicities [58, 59].

In bladder cancer RT, from the Ontario Cancer Registry, total dose of conventional RT is usually 60 Gy, with 45% of patients experiencing change in bladder habit and 14% moderate to severe impairment of bladder function. Radiation cystitis is very common during and after RT. The toxicities vary from urinary blood clots (18%), incontinence (20%) and urinary frequency (50%) at 3 years [60]. In rectal cancer RT, bladder complications are surprisingly very low, about 2–4% grade 1/2 toxicity [61].

In cervical cancer, the standard radical treatment includes both EBRT (40–50 Gy) and intracavitary RT (20–40 Gy) with concurrent chemotherapy which would increase both GI and GU toxicities. Severe radiation cystitis is noted among 7.7% of patients at 3 years and 9.3% at 5 years. The bladder toxicity increases at a rate of 0.34% per year of follow up (11% at 10 years; 13% at 15 years and 14% at 20 years) [62–64]. Urethral stricture and radiation cystitis are two most commonly encountered toxicities. In contrast lower doses in endometrial cancer keep bladder toxicity rates low [65].

16.4 Radiation Toxicities Specific to Brachytherapy

Brachytherapy delivers a precise dose of radiotherapy in and around radioactive source with steep dose fall-off that is superior to any other radiation modalities. As most centers around the globe have shifted from low dose rate (LDR) brachytherapy toward high-dose-rate brachytherapy (HDR-IBT) using the afterloader system due to convenience and radiation protection issues, the word brachytherapy in this section will mean HDR-IBT and vice-versa. The reported grade 3–4 delayed vaginal toxicity such as bleeding and stenosis in patients who underwent adjuvant HDR brachytherapy to a dose of 21 Gy in 3 fractions to upper 3–5 cm of vaginal vault in PORTEC-2 trial is less than 2% [66]. In a 20-year comprehensive literature review by Delishaj et al., there was 27.7% grade 1–2 late toxicity, specifically bleeding, fibrosis, telangiectasia, stenosis, narrow vagina, and dryness [67]. Vaginal stenosis and dyspareunia are the major concerns especially in the sexually active woman. While the use of vaginal dilators for patients who have had prior pelvic radiotherapy is encouraged by some of the treating clinicians, this practice is not backed by randomized controlled data and review of evidence showed some concern about its safety [68]. While rectal toxicity is commonly reported in gynecological brachytherapy, ureteric radiation injury is a bigger concern that causes significant pain and quality of life issues in woman. The rate of grade 3 urinary toxicity is around 15.1% and dose of 23.1 Gray in 2 gray EqD2 to 0.1 cc of urethra predicted for the 10% probably of it developing [69].

The most common toxicity of prostate brachytherapy is the urethral injury especially the intraprostatic urethra and to a lesser extent the anterior rectal wall. The experience of the brachytherapist and also with recent advancement in the use of hydrogel spacer technology that pushes rectum away from prostate gland, rectal toxicity is further minimized. Ureteric edema requiring catheterization following low-dose-rate brachytherapy (permanent seed implants) may occur in 7–19.7% of patients [70]. The peak incidence of bladder irritation and increased frequency is around 2–10 weeks after brachytherapy, which last up to 1 year and eventually around 75% of the patients will recover [71]. The estimated chronic effects of brachytherapy (more than 12 months after brachytherapy) is in the range of 5–6% for incontinence which increased to 13% if patient have had a trans-ureteric resection of prostate (TURP), 1–2% risk of stricture, 1–2% risk of hematuria, and 4–14% risk of impotence [72].

Endorectal HDR brachytherapy can be used as a boost after external beam irradiation in medically inoperable patients. CL Chiang et.al treated 18 patients with 3 different dose level (10 Gray in single fraction, 20 Gray in 2 fractions and 30 Gray in 3 fractions after completing a course of external beam radiotherapy. The median EqD2 (2 Gy equivalent dose) for α/β (alpha/beta) ratio of 10 Gy is 66 Gy. Approximately 83.3% of patients developed acute proctitis; however, the treatment was well tolerated with only 2 of them having grade 3 proctitis [73].

Liver brachytherapy has some distinct advantages as compared to other local liver directed minimally invasive procedures such as radiofrequency ablation or microwave ablation, namely the absence of heat sink effect, less issues with the tumor size and also the ability to treat subdiaphragmatic tumors [74, 75]. Dose constraints for HDR-IBT liver can be adopted from the work done by Mohnike et al. In general, dose to one-third of normal liver is kept below 5 Gy and in case of cirrhotic liver two-third of normal liver or 700 cc is essential to be below 5 Gy in single fraction [76]. The most feared toxicity of liver HDR-IBT is the radiation induced liver disease (RILD) which is infrequently fatal and the only way to avoid it is by exercising high vigilance to the dose volume parameters while delivering brachytherapy treatment [77]. It is estimated that a single hepatic subunit can tolerate up to 10 Gy of radiation in single fraction [78]. While there is evidence that biliary tract and central hilar region can tolerate single fraction of 20 Gy without serious toxicity, acute edema may present obstructive jaundice [79].

Cautious use of esophageal intraluminal brachytherapy delivering 12 Gy in single fraction is shown to result in good palliation and quality of life [80]. Combination of EBRT and fractionated intraluminal brachytherapy boost (4 Gy × 5 fractions) could result in 85% grade-2 radiation esophagitis in esophageal cancer [81]. Therefore, dose per fraction is one of the important risk factors for esophageal toxicity. Sharma et al. from Tata Memorial Hospital, Mumbai, reported 12% life-threatening complication such as tracheo-esophageal fistula after brachytherapy [82]. Esophageal stricture is a late effect of esophageal radiotherapy which may require repeated esophageal dilatation.

Interstitial brachytherapy is an important component in the management of head and neck cancers especially oral tongue. Early toxicities of interstitial brachytherapy are mucositis, xerostomia, dysphagia, odynophagia and oral pain which are self-limiting and resolve over 2–6 months post treatment. However, late occurrence of complications such as oral non-healing ulceration (26%) and bone exposure (15%) is not unusual even after a decade [83]. Appropriate oral care and custom-made mandible protector acting as a spacer may reduce the risk of osteoradionecrosis of jaw (ORNJ) [84].

16.5 Radiation-Induced Second Malignancy (RISM)

Radiation-induced second malignancy is a stochastic effect (in contrast to the earlier described deterministic effects) in which there is no threshold radiation dose for its development, its severity is independent of dose but the probability of occurrence increases with dose (Table 16.4). RISM usually occurs in organs/tissues in the high dose volume but some may also occur in the low dose (2 Gy) volume. Cells which incorrectly repair their radiation-induced DNA damage with stable mutations and viable have the potential for transforming into cancerous lesion in future. Chronic radiation damage, especially chronic inflammation, in tissues which received high doses may also predispose their cells to carcinogenesis. Due to the long latency of RISM, younger age at irradiation will also increase its risk.

Having survived their primary malignancy, about 17–19% of patients develop second cancer, mainly due to 3 factors, i.e., continuing cancer-predisposing lifestyle/environment (e.g., smoking), genetic susceptibility (e.g., Rb or BRCA mutation) and treatment with RT or chemotherapy [85]. RISM is a rare late toxicity accounting for only about 5% of all treatment-related second cancers. Following curative RT in solid cancers, the rate of RISM after control of primary malignancy is 0.2–1.0% per year. It has bimodal occurrence. The first peak occurs 1–3 years after RT, mostly hematological malignancy such as leukemia. The second peak occurs 10 years after RT as solid malignancy, e.g., sarcoma, breast cancer, and thyroid cancer.

Modern radiation techniques like IMRT or VMAT contribute more low-dose volume, which are potentially carcinogenic radiation to normal tissues than older RT techniques. Therefore, long-term follow-up with the new techniques will provide further data on risk of RISM. Proton beam RT compared to photon beam RT results in lower risk of RISM than photon RT due to its unique radiation distribution and lower integral dose. Therefore, it's imperative to formulate special follow up policy for young cancer survivors to undergo regular follow-up to detect second malignancy early for optimal management.

16.6 Summary

Acceptable levels of radiation toxicities are anticipated outcome of effective RT. Trade-off between tumor lethal dose and tissue tolerance creates individualized treatment plan for a specific anatomical site. Over the last decade, technological progress leads to very conformal radiation delivery to the tumor volume with limited dose exposure to the surrounding organs at risk. Modern radiobiological understanding provides insights for newer tissue dose constraints for individual end-organs, allowing radiation oncologists to deliver very high dose radiation to the tumor in the vicinity of sensitive normal tissue. Use of modern techniques, applying optimization, respecting normal tissue constraints as well as prompt treatment should result in minimum acute or late toxicities with increased tumor control.

Table 16.4 Comparison of deterministic and stochastic effects

Effects	Deterministic	Stochastic
Threshold dose	Yes	No
Severity	Increases with dose	Not related to dose
Probability of occurrence	Increases with dose	Increases with dose
Type of effect	Acute and late effects	Late effect with very long latency
	Acute radiation dermatitis, cardiomyopathy	Radiation carcinogenesis

Multiple Choice Questions

1. Which of the following late complication is not a consequence of fibrosis in the organ/tissue?
   (a) Neck stiffness
   (b) Myelopathy
   (c) Rectal stricture
   (d) Bladder contracture
2. The cytokine responsible for late organ fibrosis is
   (a) Interleukin-1 (IL-1)
   (b) Tumor necrosis factor-α (TNF-α)
   (c) Transforming growth factor-β (TGF-β)
   (d) Cyclooxygenase-2 (COX-2)
3. Which of the following is NOT an example of acute effects?
   (a) Diarrhea from pelvic radiotherapy
   (b) Thrombocytopenia from craniospinal radiotherapy
   (c) Pneumonitis occurring a month after completion of breast radiotherapy
   (d) Congestive heart failure from thoracic radiotherapy
4. Compared to toxicity of esophageal fistula, acute esophagitis has
   (a) Lower α/β ratio
   (b) More sparing with reducing dose per fraction
   (c) Less severity with shortened overall treatment time
   (d) A shorter latency
5. These mitigating interventions are used to target the pathogenesis of late effects EXCEPT
   (a) Amifostine
   (b) Pentoxifylline + vitamin E
   (c) Hyperbaric oxygen
   (d) Steroids
6. The incorrect statement regarding radiation-induced oral mucositis is
   (a) Radiation-induced mucositis is influenced by age
   (b) Oral stomatitis and radiation mucositis are synonyms
   (c) Accelerated fractionation results in higher radiation induced oral mucositis
   (d) Mucosal fibrosis is caused by damage to basal cell layer of mucosal epithelium
7. Regarding osteoradionecrosis, which statement is true?
   (a) Osteoradionecrosis usually occurs at the angle of mandible
   (b) Hypoxia, hypovascularity, and hypocellularity are induced by radiation-endarteritis
   (c) Radiation-induced obliteration of inferior alveolar artery is an important factor in pathogenesis
   (d) Incidence of osteoradionecrosis in head and neck cancer is less than 1%
8. Regarding radiation-induced skin reaction, all statements are correct except
   (a) Moist desquamation is caused by damage to stem cells in basal cell layer of epidermis
   (b) Onset of desquamation at 3–4 weeks correlate with lifespan of epithelial cells
   (c) Skin fibrosis, induration and telangiectasia due to damage of stem cells
   (d) Onset of hair epilation starts at the second-week of fractionated radiotherapy
9. With regard to radiation pneumonitis in breast or chest wall irradiation in breast carcinomas, available evidence in the literature suggests:
   (a) Smokers are likely to have lower risk of radiation pneumonitis
   (b) Concomitant use of Tamoxifen with radiotherapy increases the risk of radiation pneumonitis
   (c) Use of higher energy photon beam (>10 MV) is associated with bronchiolitis obliterans organizing pneumonia (BOOP) syndrome in which the lung injury occurs outside the radiation field
   (d) Use of high tangents defined as within 2 cm from humeral head for breast carcinoma irradiation in conjunction with regional nodal irradiation is known to increase the risk of radiation pneumonitis

10. A 56-year-old man known to be chronic smoker of 40 pack years was diagnosed with AJCC stage III, TNM8 T3 (8 cm), N1, M0 non-small cell lung cancer. He was planned for a course of definitive concurrent chemoradiotherapy followed by adjuvant immune-modulating agent with Durvalumab. With regard to his planned treatment, current evidence suggests:
    (a) Severe pneumonitis (grade 3 and 4) is significantly increased with the use of adjuvant Durvalumab after chemoradiotherapy
    (b) Esophagus is the most likely dose limiting organ in delivering the radical dose of radiotherapy
    (c) Use of hyperfractionation (delivering 2 fractions of 1.5 Gy, cumulative 3 Gy per day to a total dose of 54 Gy) will reduce the risk of esophageal toxicity
    (d) Immune-mediated mechanism plays and important role in progressive lung fibrosis after radiation therapy

Answers

1. (b)
2. (c)
3. (d)
4. (d)
5. (a)
6. (b)
7. (d)
8. (d)
9. (a)
10. (d)

References

1. Bansal M, Mohanti BK, Shah N, Chaudhry R, et al. Radiation related morbidities and their impact on quality of life in head and neck cancer patients receiving radical radiotherapy. Qual Life Res. 2004;13(2):481–8.
2. Mohanti BK, Bansal M. Late sequelae of radiotherapy in adults. Support Care Cancer. 2005;13(10):775–80.
3. Niyazi M, Maihoefer C, Krause M, Rödel C, Budach W, Belka C. Radiotherapy and "new" drugs-new side effects? Radiat Oncol. 2011;6:177.
4. Modi C, Berim L, Isserow L, Malhotra J, et al. Combining radiation therapy and immunotherapy for lung cancers: a narrative review. Shanghai Chest. 2021;5:10.
5. Lin B, Gao F, Yang Y, Wu D, et al. FLASH radiotherapy: history and future. Front Oncol. 2021;11:644400.
6. Hall EJ, Giaccia AJ. Acute radiation syndrome. In: Radiobiology for the radiologist. 7th ed. Philadelphia: Lippincott Williams & Wilkins; 2012. p. 114.
7. Bentzen SM, Constine LS, Deasy JO, Eisbruch A, et al. Quantitative Analyses of Normal Tissue Effects in the Clinic (QUANTEC): an introduction to the scientific issues. Int J Radiat Oncol Biol Phys. 2010;76(3 Suppl):S3–9.
8. Emami B. Tolerance of normal tissue to therapeutic radiation. Rep Radiother Oncol. 2013;1:35–48.
9. Dorr W. Pathogenesis of normal-tissue side-effects. In: Joiner MC, van der Kogel AJ, editors. Basic clinical radiobiology. London: Hodder Arnold; 2009. p. 169–89.
10. Burris HA 3rd, Hurtig J. Radiation recall with anticancer agents. Oncologist. 2010;15(11):1227–37.
11. Stone HB, Coleman CN, Anscher MS, McBride WH. Effects of radiation on normal tissue: consequences and mechanisms. Lancet Oncol. 2003;4(9):529–36.
12. Trotti A, Colevas AD, Setser A, Rusch V, et al. CTCAE v3.0: development of a comprehensive grading system for the adverse effects of cancer treatment. Semin Radiat Oncol. 2003;13(3):176–81.
13. Cox JD, Stetz J, Pajak TF. Toxicity criteria of the Radiation Therapy Oncology Group (RTOG) and the European Organization for Research and Treatment of Cancer (EORTC). Int J Radiat Oncol Biol Phys. 1995;31(5):1341–6.
14. Shaw E, Scott C, Souhami L, Dinapoli R, et al. Single dose radiosurgical treatment of recurrent previously irradiated primary brain tumors and brain metastases: final report of RTOG protocol 90-05. Int J Radiat Oncol Biol Phys. 2000;47(2):291–8.
15. Parsons JT, Bova FJ, Fitzgerald CR, Mendenhall WM, Million RR. Radiation optic neuropathy after megavoltage external-beam irradiation: analysis of time-dose factors. Int J Radiat Oncol Biol Phys. 1994;30(4):755–63.
16. Emami B, Lyman J, Brown A, Coia L, et al. Tolerance of normal tissue to therapeutic irradiation. Int J Radiat Oncol Biol Phys. 1991;21(1):109–22.
17. Mayo C, Martel MK, Marks LB, Flickinger J, Nam J, Kirkpatrick J. Radiation dose-volume effects of optic nerves and chiasm. Int J Radiat Oncol Biol Phys. 2010;76(3 Suppl):S28–35.
18. Mayo C, Yorke E, Merchant TE. Radiation associated brainstem injury. Int J Radiat Oncol Biol Phys. 2010;76(3 Suppl):S36–41.
19. Sahgal A, Ma L, Gibbs I, Gerszten PC, et al. Spinal cord tolerance for stereotactic body radiotherapy. Int J Radiat Oncol Biol Phys. 2010;77(2):548–53.

20. Bhandare N, Jackson A, Eisbruch A, Pan CC, et al. Radiation therapy and hearing loss. Int J Radiat Oncol Biol Phys. 2010;76(3 Suppl):S50–7.
21. Yeh S-A. Radiotherapy for head and neck cancer. Semin Plast Surg. 2010;24(2):127–36.
22. Guo Q, Lu T, Lin S, Zong J, et al. Long-term survival of nasopharyngeal carcinoma patients with Stage II in intensity-modulated radiation therapy era. Jpn J Clin Oncol. 2016;46(3):241–7.
23. Pereira I, Vasconcelos B. Management of head and neck radiation therapy complications. Clin Res Trials. 2020;6.
24. Muzumder S, Srikantia N, Udayashankar AH, Kainthaje PB, John Sebastian MG. Burden of acute toxicities in head-and-neck radiation therapy: a single-institutional experience. South Asian J Cancer. 2019;8(2):120–3.
25. Biswal BM. Current trends in the management of oral mucositis related to cancer treatment. Malays J Med Sci. 2008;15(3):4–13.
26. Givens DJ, Karnell LH, Gupta AK, Clamon GH, et al. Adverse events associated with concurrent chemoradiation therapy in patients with head and neck cancer. Arch Otolaryngol Head Neck Surg. 2009;135(12):1209–17.
27. Cacicedo J, Dal Pra A, Alongi F, Navarro A. Impact of weight loss in patients with head and neck carcinoma undergoing radiotherapy: is it an underestimated phenomenon? A radiation oncologist's perspective. Eur J Clin Nutr. 2015;69(7):757–60.
28. Sonis ST. A biological approach to mucositis. J Support Oncol. 2004;2(1):21–32; discussion 5–6.
29. Pulito C, Cristaudo A, Porta CL, Zapperi S, et al. Oral mucositis: the hidden side of cancer therapy. J Exp Clin Cancer Res. 2020;39(1):210.
30. Bourhis J, Lapeyre M, Tortochaux J, Rives M, et al. Phase III randomized trial of very accelerated radiation therapy compared with conventional radiation therapy in squamous cell head and neck cancer: a GORTEC trial. J Clin Oncol. 2006;24(18):2873–8.
31. Matuschek C, Haussmann J, Bölke E, Gripp S, et al. Accelerated vs. conventionally fractionated adjuvant radiotherapy in high-risk head and neck cancer: a meta-analysis. Radiat Oncol. 2018;13(1):195.
32. Rawat S, Tandan H, Patel S, Chaudhari S. Safety and efficacy of nimotuzumab with concurrent chemoradiotherapy in unresectable locally advanced squamous cell carcinoma of head and neck: an Indian rural hospital experience. South Asian J Cancer. 2019;8(1):52–6.
33. Yokota T, Onoe T, Ogawa H, Hamauchi S, et al. Distinctive mucositis and feeding-tube dependency in cetuximab plus radiotherapy for head and neck cancer. Jpn J Clin Oncol. 2015;45(2):183–8.
34. Zeilstra LJ, Vissink A, Konings AW, Coppes RP. Radiation induced cell loss in rat submandibular gland and its relation to gland function. Int J Radiat Biol. 2000;76(3):419–29.
35. Marx RE. Osteoradionecrosis: a new concept of its pathophysiology. J Oral Maxillofac Surg. 1983;41(5):283–8.
36. Shaw R, Butterworth C, Tesfaye B, Bickerstaff M, et al. HOPON (Hyperbaric Oxygen for the Prevention of Osteoradionecrosis): a randomised controlled trial of hyperbaric oxygen to prevent osteoradionecrosis of the irradiated mandible: study protocol for a randomised controlled trial. Trials. 2018;19(1):22. https://doi.org/10.1186/s13063-017-2376-7.
37. Karagozoglu KH, Dekker HA, Rietveld D, de Bree R, et al. Proposal for a new staging system for osteoradionecrosis of the mandible. Med Oral Patol Oral Cir Bucal. 2014;19(5):e433–7.
38. Kubota H, Miyawaki D, Mukumoto N, Ishihara T, et al. Risk factors for osteoradionecrosis of the jaw in patients with head and neck squamous cell carcinoma. Radiat Oncol. 2021;16(1):1.
39. Ciammella P, Podgornii A, Galeandro M, Micera R, et al. Toxicity and cosmetic outcome of hypofractionated whole-breast radiotherapy: predictive clinical and dosimetric factors. Radiat Oncol. 2014;9:97.
40. Byhardt RW, Scott C, Sause WT, Emami B, et al. Response, toxicity, failure patterns, and survival in five Radiation Therapy Oncology Group (RTOG) trials of sequential and/or concurrent chemotherapy and radiotherapy for locally advanced non-small-cell carcinoma of the lung. Int J Radiat Oncol Biol Phys. 1998;42(3):469–78.
41. Werner-Wasik M, Paulus R, Curran WJ Jr, Byhardt R. Acute esophagitis and late lung toxicity in concurrent chemoradiotherapy trials in patients with locally advanced non-small-cell lung cancer: analysis of the radiation therapy oncology group (RTOG) database. Clin Lung Cancer. 2011;12(4):245–51.
42. Palma DA, Senan S, Oberije C, Belderbos J, et al. Predicting esophagitis after chemoradiation therapy for non-small cell lung cancer: an individual patient data meta-analysis. Int J Radiat Oncol Biol Phys. 2013;87(4):690–6.
43. Arroyo-Hernández M, Maldonado F, Lozano-Ruiz F, Muñoz-Montaño W, Nuñez-Baez M, Arrieta O. Radiation-induced lung injury: current evidence. BMC Pulm Med. 2021;21(1):9.
44. Twyman-Saint Victor C, Rech AJ, Maity A, Rengan R, et al. Radiation and dual checkpoint blockade activate non-redundant immune mechanisms in cancer. Nature. 2015;520(7547):373–7.
45. Shaverdian N, Lisberg AE, Bornazyan K, Veruttipong D, et al. Previous radiotherapy and the clinical activity and toxicity of pembrolizumab in the treatment of non-small-cell lung cancer: a secondary analysis of the KEYNOTE-001 phase 1 trial. Lancet Oncol. 2017;18(7):895–903.
46. Antonia SJ, Villegas A, Daniel D, Vicente D, et al. Durvalumab after chemoradiotherapy in stage III non–small-cell lung cancer. N Engl J Med. 2017;377(20):1919–29.
47. Tian S, Switchenko JM, Buchwald ZS, Patel PR, et al. Lung stereotactic body radiation therapy and

concurrent immunotherapy: a multicenter safety and toxicity analysis. Int J Radiat Oncol Biol Phys. 2020;108(1):304–13.
48. Couey MA, Bell RB, Patel AA, Romba MC, et al. Delayed immune-related events (DIRE) after discontinuation of immunotherapy: diagnostic hazard of autoimmunity at a distance. J Immunother Cancer. 2019;7(1):165.
49. Dess RT, Sun Y, Matuszak MM, Sun G, et al. Cardiac events after radiation therapy: combined analysis of prospective multicenter trials for locally advanced non-small-cell lung cancer. J Clin Oncol. 2017;35(13):1395–402.
50. Bradley JD, Paulus R, Komaki R, Masters G, et al. Standard-dose versus high-dose conformal radiotherapy with concurrent and consolidation carboplatin plus paclitaxel with or without cetuximab for patients with stage IIIA or IIIB non-small-cell lung cancer (RTOG 0617): a randomised, two-by-two factorial phase 3 study. Lancet Oncol. 2015;16(2):187–99.
51. Roach MC, Bradley JD, Robinson CG. Optimizing radiation dose and fractionation for the definitive treatment of locally advanced non-small cell lung cancer. J Thorac Dis. 2018;10(Suppl 21):S2465–73.
52. Piroth MD, Baumann R, Budach W, Dunst J, et al. Heart toxicity from breast cancer radiotherapy: current findings, assessment, and prevention. Strahlenther Onkol. 2019;195(1):1–12.
53. Taylor CW, Nisbet A, McGale P, Darby SC. Cardiac exposures in breast cancer radiotherapy: 1950s–1990s. Int J Radiat Oncol Biol Phys. 2007;69(5):1484–95.
54. Lefevre JH, Amiot A, Joly F, Bretagnol F, Panis Y. Risk of recurrence after surgery for chronic radiation enteritis. Br J Surg. 2011;98(12):1792–7.
55. Chopra S, Gupta S, Kannan S, Dora T, et al. Late toxicity after adjuvant conventional radiation versus image-guided intensity-modulated radiotherapy for cervical cancer (PARCER): a randomized controlled trial. J Clin Oncol. 2021;39(33):3682–92.
56. Doi H, Ishimaru F, Tanooka M, Inoue H, et al. Body mass index can affect gastrointestinal and genitourinary toxicity in patients with prostate cancer treated with external beam radiation therapy. Oncol Lett. 2014;7(1):209–14.
57. Vanneste BG, Van De Voorde L, de Ridder RJ, Van Limbergen EJ, Lambin P, van Lin EN. Chronic radiation proctitis: tricks to prevent and treat. Int J Colorectal Dis. 2015;30(10):1293–303.
58. Lawton CA, Won M, Pilepich MV, Asbell SO, et al. Long-term treatment sequelae following external beam irradiation for adenocarcinoma of the prostate: analysis of RTOG studies 7506 and 7706. Int J Radiat Oncol Biol Phys. 1991;21(4):935–9.
59. Peeters ST, Heemsbergen WD, van Putten WL, Slot A, et al. Acute and late complications after radiotherapy for prostate cancer: results of a multicenter randomized trial comparing 68 Gy to 78 Gy. Int J Radiat Oncol Biol Phys. 2005;61(4):1019–34.
60. Hoskin PJ, Rojas AM, Phillips H, Saunders MI. Acute and late morbidity in the treatment of advanced bladder carcinoma with accelerated radiotherapy, carbogen, and nicotinamide. Cancer. 2005;103(11):2287–97.
61. Sauer R, Becker H, Hohenberger W, Rödel C, et al. Preoperative versus postoperative chemoradiotherapy for rectal cancer. N Engl J Med. 2004;351(17):1731–40.
62. McIntyre JF, Eifel PJ, Levenback C, Oswald MJ. Ureteral stricture as a late complication of radiotherapy for stage IB carcinoma of the uterine cervix. Cancer. 1995;75(3):836–43.
63. Takeshi K, Katsuyuki K, Yoshiaki T, Teppei S, et al. Definitive radiotherapy combined with high-dose-rate brachytherapy for Stage III carcinoma of the uterine cervix: retrospective analysis of prognostic factors concerning patient characteristics and treatment parameters. Int J Radiat Oncol Biol Phys. 1998;41(2):319–27.
64. Eifel PJ, Levenback C, Wharton JT, Oswald MJ. Time course and incidence of late complications in patients treated with radiation therapy for FIGO stage IB carcinoma of the uterine cervix. Int J Radiat Oncol Biol Phys. 1995;32(5):1289–300.
65. Elliott SP, Malaeb BS. Long-term urinary adverse effects of pelvic radiotherapy. World J Urol. 2011;29(1):35–41.
66. Nout RA, Smit VT, Putter H, Jürgenliemk-Schulz IM, et al. Vaginal brachytherapy versus pelvic external beam radiotherapy for patients with endometrial cancer of high-intermediate risk (PORTEC-2): an open-label, non-inferiority, randomised trial. Lancet. 2010;375(9717):816–23.
67. Delishaj D, Barcellini A, D'Amico R, Ursino S, et al. Vaginal toxicity after high-dose-rate endovaginal brachytherapy: 20 years of results. J Contemp Brachytherapy. 2018;10(6):559–66.
68. Miles T, Johnson N. Vaginal dilator therapy for women receiving pelvic radiotherapy. Cochrane Database Syst Rev. 2010(9):CD007291-CD.
69. Amsbaugh MJ, Bhatt N, Hunter T, Gaskins J, et al. Computed tomography planned interstitial brachytherapy for locally advanced gynecologic cancer: outcomes and dosimetric predictors of urinary toxicity. Brachytherapy. 2016;15(1):49–56.
70. Crook J, Patil N, Wallace K, Borg J, et al. A phase III randomized trial of the timing of meloxicam with iodine-125 prostate brachytherapy. Int J Radiat Oncol Biol Phys. 2010;77(2):496–501.
71. Budäus L, Bolla M, Bossi A, Cozzarini C, et al. Functional outcomes and complications following radiation therapy for prostate cancer: a critical analysis of the literature. Eur Urol. 2012;61(1):112–27.
72. Crook J, Lukka H, Klotz L, Bestic N, Johnston M. Genitourinary cancer disease site group of the cancer care ontario practice guidelines I. Systematic overview of the evidence for brachytherapy in clinically localized prostate cancer. CMAJ. 2001;164(7):975–81.
73. Chiang C, Lee S, Lee VW, Yeung CS, et al. Toxicity outcome of endorectal brachytherapy boost in medically inoperable patients. Strahlenther Onkol. 2020;196(11):993–7.

74. John RG, Ho F, Appalanaido GK, Chen D, et al. Can radiotherapy finally "go live" in the management of liver metastases? Hepatoma Res. 2020;6:56.
75. Appalanaido GK, Bahajjaj SIBZ-A, Shukor SA, Ahmad MZ, Francis HCH. Case report—staged brachytherapy achieving complete metabolic response in unresectable oligometastatic colorectal cancer to the liver. Oxford Med Case Rep. 2021;2021(4)
76. Mohnike K, Wieners G, Schwartz F, Seidensticker M, et al. Computed tomography-guided high-dose-rate brachytherapy in hepatocellular carcinoma: safety, efficacy, and effect on survival. Int J Radiat Oncol Biol Phys. 2010;78(1):172–9.
77. Liang SX, Huang XB, Zhu XD, Zhang WD, et al. Dosimetric predictor identification for radiation-induced liver disease after hypofractionated conformal radiotherapy for primary liver carcinoma patients with Child-Pugh Grade A cirrhosis. Radiother Oncol. 2011;98(2):265–9.
78. Rühl R, Lüdemann L, Czarnecka A, Streitparth F, et al. Radiobiological restrictions and tolerance doses of repeated single-fraction hdr-irradiation of intersecting small liver volumes for recurrent hepatic metastases. Radiat Oncol. 2010;5:44.
79. Collettini F, Gebauer B. CT-guided high dose rate brachytherapy ablation of liver metastases. Imaging Med. 2013;5:383–8.
80. Homs MY, Steyerberg EW, Eijkenboom WM, Tilanus HW, et al. Single-dose brachytherapy versus metal stent placement for the palliation of dysphagia from oesophageal cancer: multicentre randomised trial. Lancet. 2004;364(9444):1497–504.
81. Vuong T, Szego P, David M, Evans M, et al. The safety and usefulness of high-dose-rate endoluminal brachytherapy as a boost in the treatment of patients with esophageal cancer with external beam radiation with or without chemotherapy. Int J Radiat Oncol Biol Phys. 2005;63(3):758–64.
82. Sharma V, Agarwal J, Dinshaw K, Nehru RM, et al. Late esophageal toxicity using a combination of external beam radiation, intraluminal brachytherapy and 5-fluorouracil infusion in carcinoma of the esophagus. Dis Esophagus. 2000;13(3):219–25.
83. Abdalmear MM, Tomita M, Shibuya H. Long-term results of brachytherapy in early mobile tongue cancer with 10-year minimal follow-up. Anticancer Res. 2013;33(1):317–24.
84. Miura M, Takeda M, Sasaki T, Inoue T, et al. Factors affecting mandibular complications in low dose rate brachytherapy for oral tongue carcinoma with special reference to spacer. Int J Radiat Oncol Biol Phys. 1998;41(4):763–70.
85. Morton LM, Onel K, Curtis RE, Hungate EA, Armstrong GT. The rising incidence of second cancers: patterns of occurrence and identification of risk factors for children and adults. Am Soc Clin Oncol Educ Book. 2014:e57–67.

Proton and Heavy Ion Therapy

17

Siddhartha Laskar, Jifmi Jose Manjali,
Srinivas Chilukuri, Tapesh Bhattacharya,
Dayanand Sharma, and Rakesh Jalali

17.1 Introduction

Radiotherapy (RT) has continuously evolved through last century as one of the pillars of cancer treatment [1]. Immediately after the discovery of X-rays by Wilhelm Röentgen in 1895, radiation was experimentally used in the treatment of various tumors (Fig. 17.1) [2]. Since then, the indications of RT have broadened, and dose-fractionation regimes in tumors were established. Approximately at the same period, Ernest Rutherford discovered protons in 1917 (Fig. 17.2) [3]. In 1930, Ernest O. Lawrence and his associates invented the cyclotron to accelerate proton to the energy high enough for cancer treatment applications (Fig. 17.3) [4]. He won the 1939 Nobel Prize for physics for this work. In 1931, he founded the Radiation Laboratory, later called the Lawrence Berkeley National Accelerator Laboratory (LBNL), where the first synchrocyclotron was developed a decade later and was used to accelerate heavy ions as well (Fig. 17.4) [5].

S. Laskar (✉) · J. J. Manjali
Tata Memorial Centre, Homi Bhabha National Institute (HBNI), Mumbai, Maharashtra, India

S. Chilukuri · D. Sharma · R. Jalali
Apollo Proton Centre, Chennai, Tamil Nadu, India

T. Bhattacharya
Tata Medical Centre, Kolkata, West Bengal, India

Early in the twentieth century, Rutherford hypothesized that the alpha particles, and therefore atoms, contained a positively charged central structure of large size that was surrounded by negatively charged particles [6]. Subsequently, Sir James Chadwick did extensive experiments to study the interaction of the mysterious radiation on different atoms and range of the protons produced. He concluded that the new radiation was not gamma rays, but that of uncharged particles with the same mass as proton and named them neutrons. He was awarded the Nobel Prize in Physics in 1935 for this discovery [7].

Though Bragg peak was proposed in 1903 by William Bragg to describe the pronounced peak on the curve which plots the energy loss of ionizing radiation during its travel through matter, it was not until 1946 when Robert R. Wilson recommended the use of this characteristic of protons for cancer treatment. He described the preferential deposition of its energy at the end of its path and then completely stops. Combining multiple Bragg peaks of different energies in order to spread the proton stopping region over the tumor volume is called Spread Out Bragg Peak (SOBP). This technique is used to direct the beam onto the tumor and spare the surrounding normal tissues. Known as ***"the father of proton therapy,"*** Robert Wilson was later responsible for the development of Fermi Laboratory and became its founding director (Fig. 17.5).

A. Munshi et al. (eds.), *Radiation Oncology – Principles, Precepts and Practice*,
https://doi.org/10.1007/978-981-97-8389-2_17

Fig. 17.1 Wilhelm Conrad Röntgen (27 March 1845–10 February 1923), Nobel Prize 1901 (Physics), German physicist who discovered X-rays

Fig. 17.2 Ernest Rutherford (30 August 1871–19 October 1937), Nobel Prize 1908 (Chemistry)

Fig. 17.3 Ernest O. Lawrence (8 August 1901–27 August 1958), Nobel Prize 1939 (Physics)

The Massachusetts General Hospital (MGH)-Harvard Cyclotron Laboratory was opened in 1961, and then the Institute for Theoretical and Experimental Physics (ITEP) in Moscow in 1967 [5]. Over the following decades, many centers developed around the world. By the 1970s and 1980s, several technological advancements resulted in continued data accumulation on selected tumor sites. These included ocular, base skull, brain and pediatric tumors. The simultaneous development of technologies in imaging modalities like computed tomography (CT) and magnetic resonance (MR), accurate planning and dose calculation algorithms and advanced treatment planning techniques, helped in the delivery of precise and safe particle RT. Initially, passive scattering was the technique used for proton therapy delivery, with a uniform beamlet being scattered by some material. Despite its various

Fig. 17.4 The 184" cyclotron (year 1942)

Fig. 17.5 Robert R. Wilson (4 March 1914–16 January 2000)

advantages of being less expensive and easy to implement, it was associated with several disadvantages like inability to spare the normal tissues proximal to the target and high neutron contamination. Hence, pencil beam scanning (PBS) was developed by Paul Scherrer Institute in 1984 as a unique technique of delivering proton therapy. This further led to the adoption of intensity modulated proton therapy (IMPT) which is the most advanced form of PBS and akin to intensity modulated RT (IMRT) in photons.

Similar to protons, heavy ions have a finite range in matter. Heavy ions were of much interest due to their sharp Bragg peak (though a significant fragmentation tail is present for ions heavier than oxygen), very less lateral scatter and the most important—its higher radiobiological effectiveness (RBE) compared to proton and X-rays. Due to this higher biological effectiveness, they are more effective against hypoxic tumors which are radioresistant to X-rays and protons [8, 9]. Between 1977 and 1992, multiple early phase clinical trials were conducted with these ions where around 2400 and 400 patients with skull base tumors were treated with helium and neon ions, respectively. In 1993, the LBNL was decommissioned due to financial difficulties and ageing equipment, making it the last time the USA had a therapeutic heavy ion beam for humans [10]. Meanwhile, Japan initiated the project to construct the heavy ion medical accelerator (HIMAC) in the National Institute of Radiological Sciences (NIRS, Chiba), now called as QST hospital, as part of the national strategy and initiated the world's first attempt to use heavy

Table 17.1 Important milestones in the evolution of particle therapy

Year	Milestone
1903	William Bragg describes the pronounced ***peak*** (named after him) on the curve which plots the energy loss of ionizing radiation during its travel through matter
1917	Ernest Rutherford discovered ***proton*** (named it "protos" in Greek)
1930	Ernest Lawrence and colleagues invented the ***cyclotron*** and used it to accelerate protons
1932	Walther Wilhelm Georg Franz Bothe and Herbert Becker created neutron beams and the subsequent work by Joliot-Curies and Sir James Chadwick formally identified the ***neutron***
1946	Robert R. Wilson ***proposed the use of proton beam*** for cancer treatment
1954	Cornelius Tobias and John Lawrence treated ***the first patient*** with a pituitary tumor with protons at the University of Berkeley (USA)
1973 and 1977	***Heavy ion research*** at Lawrence Berkeley National Accelerator Laboratory (LBLN, USA)—Treatment with neon and carbon ions, respectively
1984	Paul Scherrer Institute (Switzerland) introduces ***pencil beam scanning*** in proton therapy
1989 and 1990	***First hospital based proton therapy centers*** started in Clatterbridge (UK), followed by Loma Linda (USA). The ***first rotating gantry for proton*** was installed in Loma Linda
1994	***CIRT*** used for treatment of intractable cancers in HIMAC, Chiba
1997	GSI Helmholtz Centre for Heavy Ion research (Germany) used ***raster scanning*** for CIRT
2009	HIT (Germany) introduced the first ***rotating gantry*** for CIRT

USA United States of America, *UK* United Kingdom, *CIRT* carbon ion radiotherapy, *HIMAC* heavy ion medical accelerator, *GSI Gesellschaft für Schwerionenforschung* (German for Society for Heavy Ion Research), *HIT* Heidelberg Ion Therapy Center

ion in medical use in 1984. After 10 years of construction, HIMAC started treatment of intractable cancers using Carbon Ion RT (CIRT) in 1994 [10, 11]. Raster scanning which uses strong magnets to scan a narrow ion beam across the tumor while modulating the beam energy, was developed and implemented by GSI in 1997 [11, 12]. Subsequently, Heidelberg Ion Therapy Centre (HIT) in Germany introduced the first rotating gantry for CIRT.

Boron Neutron Capture Therapy (BNCT) is yet another widely researched aspect. Shortly after the discovery of neutron, Goldhaber and Taylor described BNCT, where boron is bombarded with a neutron and produces an alpha particle and lithium ion [13, 14]. In the 1990s, there was a renewed interest worldwide in BNCT with the invention of epithermal neutron sources. Unfortunately, BNCT centers were closed due to possible economic and political reasons.

The milestones in the development of particle therapy have been summarized in Table 17.1.

Presently, according to Particle Therapy Co-Operative Group (PTCOG), there are over 100 proton therapy centers which are functional and around 30 centers under construction [15]. Also, there are around 14 centers with a functional carbon ion facility worldwide, with a couple of them under construction. India has currently two proton therapy centers which are functional – one at Apollo Proton center, Chennai, and another at the Tata Memorial Center, Mumbai.

17.2 Classification of Particles

The study of the fundamental (or elementary) particles and forces that constitute matter and radiation is called Particle Physics. Till the former half of the twentieth century, proton and electron were considered elementary particles. However, elementary particles of the universe were described and confirmed by scientists worldwide in various stages throughout the latter half of the twentieth century, thus resulting in the Standard Model (Table 17.2).

The Standard Model of particle physics is the theory describing the fundamental forces (electromagnetic, strong and weak interactions) of the universe and classifies the elementary particles (Fig. 17.6). The current model has been finalized in the mid-1970s through the experimental confirmation of

Table 17.2 Timeline for the discovery of sub-atomic and elementary particles

Year	Event
1895	X-rays discovered by Wilhelm Röentgen
1897	Electron discovered by J. J. Thomson
1899	Alpha particle discovered by Ernest Rutherford
1900	Gamma ray discovered by Paul Villard in uranium decay
1911	Atomic nucleus identified by Ernest Rutherford
1917	Proton discovered by Ernest Rutherford
1932	Neutron discovered by James Chadwick
1932	Antielectron (or positron), the first antiparticle discovered by Carl D. Anderson
1937	Muon (or mu lepton) discovered by Seth Neddermeyer, Carl D. Anderson, J. C. Street, and E. C. Stevenson
1947	Pion (or pi meson) discovered by C. F. Powell's group
1947	Laon (or K meson) discovered by George Dixon Rochester and Clifford Charles Butler
1947	$\Lambda 0$ discovered during a study of cosmic-ray interactions
1955	Antiproton discovered by Owen Chamberlin, Emilio Segrè, Clyde Wiegand, Thomas Tpsilantis
1956	Electron antineutrino detected by Federick Reines and Clyde Cowan
1962	Muon neutrino (or mu neutrino) shown to be a distinct entity by Leon Lederman
1964	Xi baryon discovery at Brookhaven National Laboratory
1969	Partons (internal constituents of hadrons) observed in deep inelastic scattering experiments between protons and electrons at Stanford Linear Accelerator Center; this was eventually associated with quark model and thus constitutes the discovery of up, down and strange quarks
1974	J/Ψ meson discovered by group led by Burton Richter and Samuel Ting, demonstrating the existence of the charm quark
1975	Tau discovered by a group headed by Martin Perl
1977	Upsilon meson discovered at Fermilab, demonstrating the existence of the bottom quark
1979	Gluon observed indirectly in three-jet events by Deutsches Elektronen-Synchrotron (DESY)
1983	W and Z bosons discovered by Carlo Rubbia, Simon van der Meer, and the European Organization for Nuclear Research (CERN) UA1 collaboration
1995	Top quark discovered at Fermilab
2000	Tau neutrino observed at Fermilab
2012	A particle with predicted characteristics of Higgs boson discovered in experiments conducted at CERN

This Table is adapted from Physics encyclopedia

quarks [3]. Elementary particles are broadly divided into fermions (matter) and bosons (force carriers) on the basis of various characteristics like color charge, spin, etc. Fermions are elementary particles of matter and have a half integer spin. There are 12 fermions, each of which has a corresponding *antiparticle*. For example, the antiparticle of electron is called positron. They are further classified into quarks and leptons on the basis of their interactions. Quarks participate in strong interactions, while leptons interactions are weak. Proton and neutron are examples of baryons. Bosons are force carriers and are classified into gauge and scalar bosons.

17.3 Physical Properties of Particles

By virtue of their properties like Bragg peak, increased linear energy transfer (LET) and increased radiobiological effectiveness (RBE), particles (like proton, carbon and neutron) impart safer and more effective RT in certain scenarios. For photons, photoelectric effect and pair production occur at low and high energies. Compton effect occurs in the energy range between photoelectric and pair production processes, which forms most of the therapeutic energy range. Unlike uncharged particles like photons, electrons and other charged particles interact in a fun-

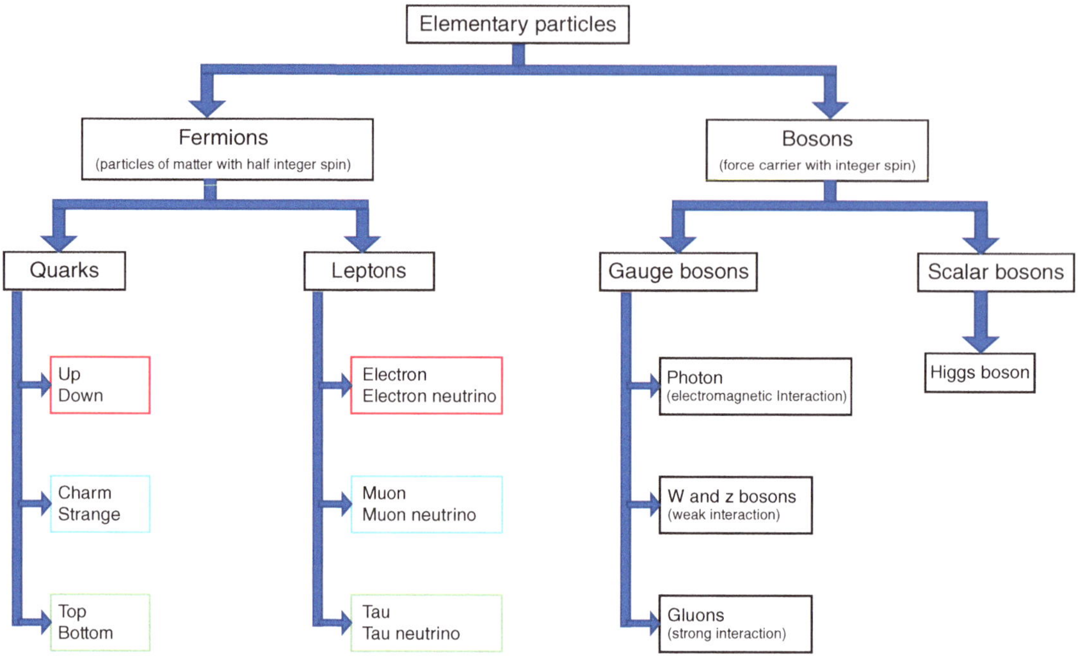

Fig. 17.6 Elementary particles

damentally different manner due to the associated Coulomb field around it.

The differences in their respective masses can significantly alter the dose distributions. Electrons interact with inelastic collisions with atomic electrons (resulting in excitation or ionization of these electrons) and atomic nuclei (resulting in a scattering of the electron and emission of gamma ray). Due to its low rest mass compared to the matter it interacts with, electrons undergo multiple deflections following a very tortuous path and do not have a distinct Bragg peak. Also, electrons deposit their maximum dose at very shallow depths. Unlike electrons, heavier particles have a straighter course (less scattering), a distinct Bragg peak and a greater depth of penetration. Table 17.3 summarizes the mode of interactions of various particles.

17.3.1 Charged Particles

Charged particles have an electric field around them which interacts with the electric field of the surrounding matter. This results in loss of energy, scattering of the primary particle, gamma rays and fragmentation of particles and atomic nuclei, which in turn lead to dose deposition. Due to their larger mass compared to electrons, there are certain salient features for particles:

- Loses only small fraction of energy in each interaction and deposits dose in multiple steps (as opposed to electron where it loses energy in few large interactions)
- Mostly continues to travel in the original path and with small scattering angles when deflection occurs (as opposed to electron where significant deflections occur)
- Negligible loss through Bremsstrahlung radiation (as opposed to electron where at times Bremsstrahlung ray is the predominant mode)

These characteristics confer minimal entrance dose and greater depth of penetration for particles.

17.3.1.1 Bragg Peak

Charged particles have a pronounced peak on the curve that plots energy of the ionizing radiation

Table 17.3 Mode of interaction of particles with matter

Incident particle	Characteristic	Interacting particle in matter	Mode of interaction	Ejectiles	Influence on incident particle
Photon	Uncharged	Bound electrons	Photoelectric effect	Photoelectron	
		Free electrons	Compton effect	Compton electron	
		Nuclei	Pair production		
Electron	Charged	Electrons	Inelastic Coulombic scattering	Secondary electrons	Original trajectory
		Nuclei	Inelastic Coulombic scattering	Gamma ray	Deflected trajectory
Protons	Charged	Electrons	Inelastic Coulombic scattering[a]	Secondary electrons	Original trajectory
		Nuclei	Elastic Coulombic scattering[b]	Recoil nucleus	Deflected trajectory
		Nuclei	Non-elastic nuclear reactions[c]	Secondary protons, heavier ions, neutrons, gamma rays	Removal of incident proton
		Nuclei	Inelastic Coulombic scattering	Bremsstrahlung rays	Deflected trajectory
Carbon ion	Charged	Electrons	Inelastic Coulombic scattering[a]	Secondary electrons	Original trajectory
		Nuclei	Non-elastic nuclear reactions (including fusion)[c]	Protons, alpha particle, lithium, boron, beryllium; fluorine & sodium radioactive isotopes	Removal of incident carbon ion
		Nuclei	Elastic Coulombic scattering[b]	Recoil nucleus	Deflected trajectory
		Nuclei	Inelastic coulombic scattering	Bremsstrahlung or gamma rays	Deflected trajectory
Neutron	Uncharged	Nuclei	Elastic scattering	Recoil nuclei	Deflected trajectory
		Nuclei	Inelastic scattering	Gamma ray	Deflected trajectory
		Nuclei	Capture reaction	Gamma rays, alpha particle and lithium ions	Removal of incident neutron

[a] Phenomenon that decides the range of the particle
[b] Phenomenon that decides the lateral penumbra (less common for carbon ions)
[c] Phenomenon that causes fragmentation tail for heavy ions (less common in protons)

while they travel through matter. This is called a Bragg peak. The depth dose profile exhibits a plateau region (build-up) with a sharp increase of dose in a well-defined, energy-dependent depth (Bragg peak) and steep dose fall-off beyond the maximum. Figure 17.7 shows the depth-dose graphs for the various particles. Bragg peak is one of the unique characteristics that provides the advantages of better normal tissue sparing surrounding (proximal and distal) the target, scope for dose-escalation and an overall decreased integral dose. In turn, this reduces normal tissue toxicities and second malignancies. Due to increased deposition of energy at the Bragg peak, the RBE in the region is higher compared to photons (1.1 for protons and 2–3.5 for carbon ions). Nonetheless, these advantages come with certain downsides. The sharp fall-off of the dose after the Bragg peak increases uncertainties and inability to estimate the dose at a particular region during treatment planning. Also, the beam is very sensitive to anatomical changes. While small variations in the position do not affect photon dose distribution to a great extent, this may lead to gross underdosage of the target and overdosage to normal tissues. These *drawbacks* can be addressed by ensuring high quality simulation scan with minimal or no artifacts, motion management to estimate organ motions, identifying the composition of all materials in the field and regular monitoring for anatomical changes during the course of treatment.

17.3.1.2 Modes of Interaction of Charged Particles

As alluded to previously, charged particles interact with matter using Coulombic interactions with electrons (inelastic scattering) and atomic nuclei (inelastic and elastic scattering; non-elastic nuclear reactions), due to the electric field around them (Table 17.3).

Inelastic Coulombic scattering due to interaction with orbital electrons is the predominant mode of energy loss of charged particles. The amount of energy lost by a particle per unit length while passing through matter is called ***stopping power (SP)***. *This determines the range of a particle in a medium*. SP depends on various factors of incident particle and the absorbing material (Table 17.4).

- Mass of incident particle—As the mass of the incident particle increases, lesser is the amount of energy lost in each event and the amount of energy lost due to radiative losses (Bremsstrahlung). As the mass of protons and other charged particles are much greater than electrons, it is considered that the SP is independent of the mass.
- Velocity of incident particle—The relationship between SP and particle is variable according to three regions. In low kinetic and relativistic energy ranges, SP increases with increase in velocity. However, the SP is inversely proportional to the square of the velocity in the intermediate energy range (partly responsible for the Bragg peak).
- Charge of the incident particle—SP is proportional to the square of the charge. Hence, a particle with higher charge will have a smaller range in a medium than a particle with similar kinetic energy but a lesser charge.
- Electron density of the absorbing material—SP is directly proportional to electron density. The decreasing order of electron densities for tissues is bone> muscle>liver>adipose tissue>lung [16].
- Mean ionization energy of the absorbing material is the mean energy that can be transferred to the atom. The SP decreases, as the mean ionization energy increases. The decreasing order of mean ionization energies is fat>bone>muscle = lung>water [17].
- Polarization/mass density—SP decreases in the compressed state of the absorber than its gaseous form. This dependency is negligible

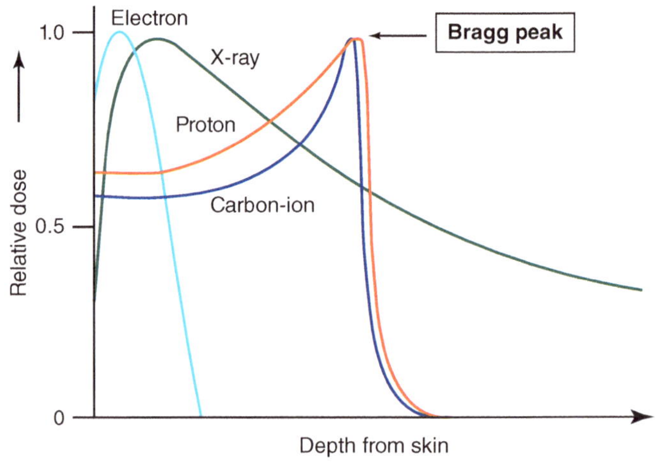

Fig. 17.7 Depth-dose curves for various particles and photons

Table 17.4 Dependencies of stopping power

Factor	Characteristic	Relationship
Atomic electron	Mass	Inversely proportional (independent for protons and other ions)
	Velocity	Inversely proportional to the square of velocity
	Charge	Directly proportional to the square of the charge
Absorbing material	Electron density	Directly proportional
	Mean ionization energy	Inversely proportional
	Polarization/mass density	More in gaseous forms than solid forms of the same material (only relevant for particles in relativistic range energies)

for particles that are heavier than electron in the low to intermediate energy range.

Knowing all the dependencies, SP can be estimated using the Bethe-Bloch equation.

$$\mathrm{SP} = 0.3071 z^2 \frac{Z}{A\beta^2}\left[\ln\frac{2m_e c^2}{I} + \ln\frac{\beta^2}{1-\beta^2} - \beta^2\right]$$

where z, Z, A, β, m_e, c and I represent the incident particle charge, atomic number of the absorber, atomic mass of absorber, the ratio of the incident particle's velocity-to-the speed of light, mass of an electron, speed of light and the absorber's mean ionization/excitation energy, respectively. The SP for different tissues is summarized in ICRU 49 [18]. For treatment planning, the SP of tissues is estimated from the Hounsfield unit values from the planning CT using a calibration curve. This calibration curve is based on a stoichiometric technique developed by Schneider et al. and is integrated into the Monte Carlo equation in the treatment planning system (TPS). Despite its usefulness in treatment planning and dose calculation, one of the main drawbacks is the degeneracy of the curve, i.e., same SP ratio given to different HU values. This is leads to the range uncertainty in particle radiotherapy and needs to be dealt with caution (discussed in section on planning).

The dose delivered can be calculated from stopping power, density of the tissue and the fluence of the radiation (number of incident particles traversing an area) using the below formula.

$$\text{Dose} = \text{fluence} \times \frac{\text{stopping power}}{\text{density}}$$

Elastic Coulomb scattering occurs when the incident particle interacts with the electric field around the atomic nuclei of the matter. This can occur multiple times, hence also called Multiple Coulomb scattering. *This determines the lateral spread of the beam (lateral penumbra).* The lateral spread of the incident particle depends on the water equivalent depth of penetration, the incident particle's charge, momentum and velocity [19, 20]. The inferences from these dependencies are that:

- Greater the momentum of the particle, lesser is the lateral spread,
- Lateral spread increases with increasing depth and charge.

Therefore, heavier particles like carbon ions have a sharper penumbra as compared to protons.

Radiobiological Effectiveness (RBE)—Radiation cell kill is mediated via single and double-strand breaks (SSB and DSB), with the latter being more lethal. The probability of DSB depends on density of energy deposition sites (directly proportional). Linear energy transfer (LET) is defined as the density of energy deposition along the path of a charged particle and the unit used is keV/μm. The LET increases along the path of the particle and is maximum at the Bragg peak, after which it decreases rapidly.

Ionized particles are known to produce dense ionization, and therefore, more DSBs. The RBE can be calculated along the beam path using the dose per fraction, LET and α/β of a tissue. Like LET, RBE also varies along the beam path.

17.3.1.3 Proton Therapy Physics

Protons follow principles of charged particles with certain peculiarities that are highlighted below

- Range straggling—The Bragg curve of a proton beam consists of an initial build-up region, a Bragg peak and rapid fall of dose. As the energy increases and the protons penetrate deeper, there is a longitudinal spread (distal penumbra) of the Bragg peak. This is called range straggling (Fig. 17.8). This is caused by the accumulation of various interactions with differing energy losses by the incident beam as they travel through tissues. The clinical implication of this in pencil beam scanning are that shallow targets will require more discrete energy layers than for an equally sized target at a deeper location. This in turn will lead to longer delivery times and more susceptibility to organ motion for shallow targets. Nevertheless, distal target margins produced in treatment planning is not made to conform to them owing to proton range uncertainties.
- Lateral penumbra of protons increases with increasing depth of penetration. This spread is much sharper than that seen in photons. The clinical implication of this feature is that shallow targets get much conformal dose coverage than targets at a depth.
- Modes of interaction—Protons mainly interact with matter by inelastic Coulombic scattering due to atomic electrons and elastic Coulombic scattering due to atomic nuclei (Table 17.3 and Fig. 17.9).
- RBE—The variability of RBE along the path of a proton beam is considered to be minimal, at least compared to our abilities to measure the same. Though studies have challenged the adoption of a single value specially at the end of the Spread Out Bragg Peak (SOBP), the current recommendation is to use a generic RBE of 1.1 throughout the proton path [21].

17.3.1.4 Carbon Ion Therapy Physics

Carbon and other heavy ions differ from proton due to their significantly greater charge and mass. These differences cause the fundamental differences in the way these ions interact with matter.

- Lesser entry dose and minimal range straggling—Predominantly, carbon ions interact with atomic electrons and occur as inelastic Coulombic scattering (Table 17.3 and Fig. 17.10). Due to the large difference in the mass between carbon ion and an atomic electron, the energy loss per single interaction is very minimal (minimal entry dose). However,

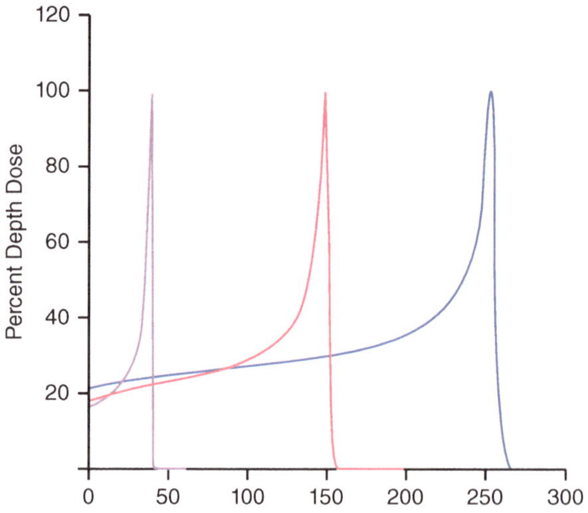

Fig. 17.8 Range straggling (note the wider Bragg peak for higher energy protons)

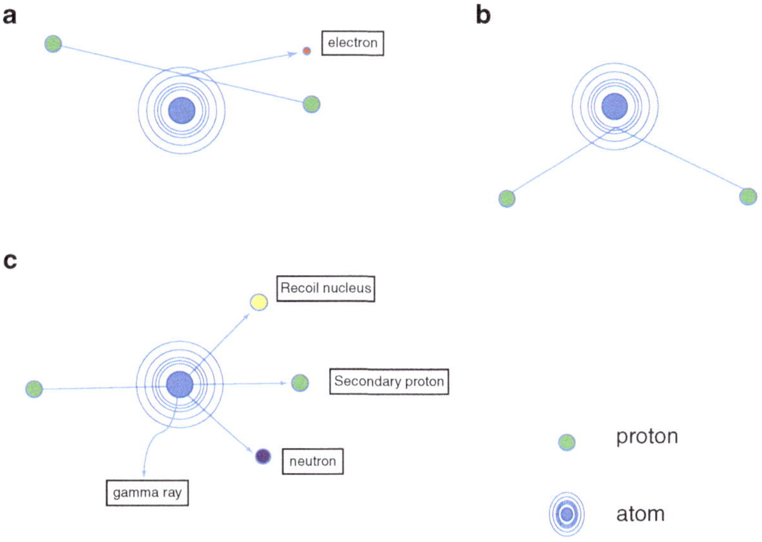

Fig. 17.9 (a) Inelastic Coulombic scattering, (b) Rutherford scattering or Coulomb scattering, and (c) non-elastic nuclear reactions

numerous such interactions gradually decrease the velocity of the carbon ion as it travels in matter. The large mass of the carbon ion also results in minimizing the number and extent of deflection, thus conferring a much straighter path. Range straggling is also minimal due limited energy loss variation along the path of the incident beam between different particles. Thus, carbon ions have a well-defined range.

- Fragmentation tail—Carbon ions interact with atomic nuclei via non-elastic nuclear reaction including fusions. These nuclear interactions result in the incident particle being shattered into multiple small fragments which deposit dose locally or more frequently into few high energy fragments that travel significant distance and have additional interactions. The trajectory of the particles is dependent on their mass, with heavier fragments moving forward peaked (in line with incident beam) and lighter ones having an angular trajectory.
- Sharp lateral penumbra—As elastic Coulombic scattering due to atomic nuclei is lesser than happens in proton, carbon ions have a sharper lateral penumbra. This advantage is more pronounced at a depth as the lateral penumbra of a carbon ion beam is relatively constant, unlike protons where there is an increase with depth. Also, radiation losses due to Bremsstrahlung rays are negligible.
- Stopping power is explained by the Bethe—Bloch equation. Since the SP is dependent on the square of the charge, a carbon ion would penetrate a shorter distance when compared with a proton with the same kinetic energy.
- LET and RBE—Unlike protons, a generic RBE value cannot be considered for the whole path of a carbon ions as the variations are substantial. While the LET is relatively less at the entry region due to higher velocity, there is a sudden increase in LET as carbon ions approach their Bragg peak.
- Charged particles heavier than carbon are not routinely used in clinics due to their shallower depth of penetration, more prominent fragmentation tails and higher does in the build region (not sparing the normal tissue proximal to the Bragg peak). Helium is potentially very useful in radiotherapy as:
  - Its physical properties are superior to those of protons.
  - It has a reduced level of lateral scattering, but also a low RBE
  - The peak: plateau ratio in helium is similar to carbon ions, hence they are good in both radio-sensitive and radio-resistant tumors
  - Reduced dose halo (peripheral dose around the pencil beam caused by scattering of the

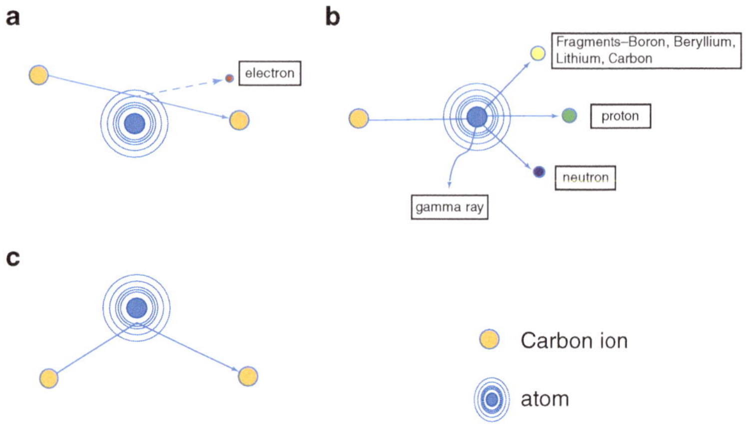

Fig. 17.10 (a) Inelastic Coulombic scattering, (b) non-elastic nuclear reactions, and (c) Coulomb scattering

primary particles)—making it attractive in pediatric patients as there are lesser concerns about late effects

- Oxygen ions are
  - Useful in radiotherapy specially in radioresistant tumors as they have a good LET in the SOBP to overcome hypoxia and hence reduce the oxygen enhancement ratio (OER)
  - Considered to have a high LET (high peak: plateau ratio) and therefore, increase risk of toxicity in non-malignant tissue.

17.3.2 Uncharged Particle (Neutron Therapy Physics)

Neutrons are uncharged particles and carry a mass similar to protons. They are classified on the basis of their energy: thermal, epithermal, intermediate and fast neutrons (in the order of increasing energy). While primary neutrons refer to the beams produced from neutron sources, secondary neutrons are those produced due to interaction of high energy photons or heavy particles with matter. Neutrons interact with atomic nuclei as opposed to orbital electrons in the case of photons. They have three modes of interaction with atomic nuclei—elastic scattering, inelastic scattering and neutron capture (Fig. 17.11). Neutron energy is the major determinant of the particular mode of interaction in matter. While intermediate and fast neutrons (energy above 10 keV) cause elastic and inelastic scattering, neutron capture occurs when thermal and epithermal neutrons (energy less than 10 keV) interact with matter.

17.4 Biological Properties of Particles

DNA damage caused by ionizing radiation (IR) can be simple, isolated and repairable or complex, clustered and irreparable. The two types of DNA damage are single-strand breaks (SSB) and double-strand breaks (DSBs). While SSBs are simple and efficiently repaired using base excision repair, complex DSBs are irreparable in general [22]. The mechanisms involved in repairing DSBs are non-homologous end-joining (NHEJ) and homologous recombination (HR) [23, 24]. Clustered DNS damages (SSBs/DSBs) are a challenge for DNA repair mechanisms, and are the cause of cell death due to radiotherapy. The factor that determines the type of DNA damage is the mean distance between two ionization events and the physical parameter that correlates with this feature is linear energy transfer (LET). When the mean distance between two ionization events is larger than the diameter of a DNA helix, the probability for a DSB is less (low LET radiation) and when this distance is smaller, the probability for a DSB is higher (high LET radiation). Hence, high LET radiations cause clustered, complex DSBs that are irreparable [25–27]. While X-rays and electrons are low LET radiations, the LET of charged particles

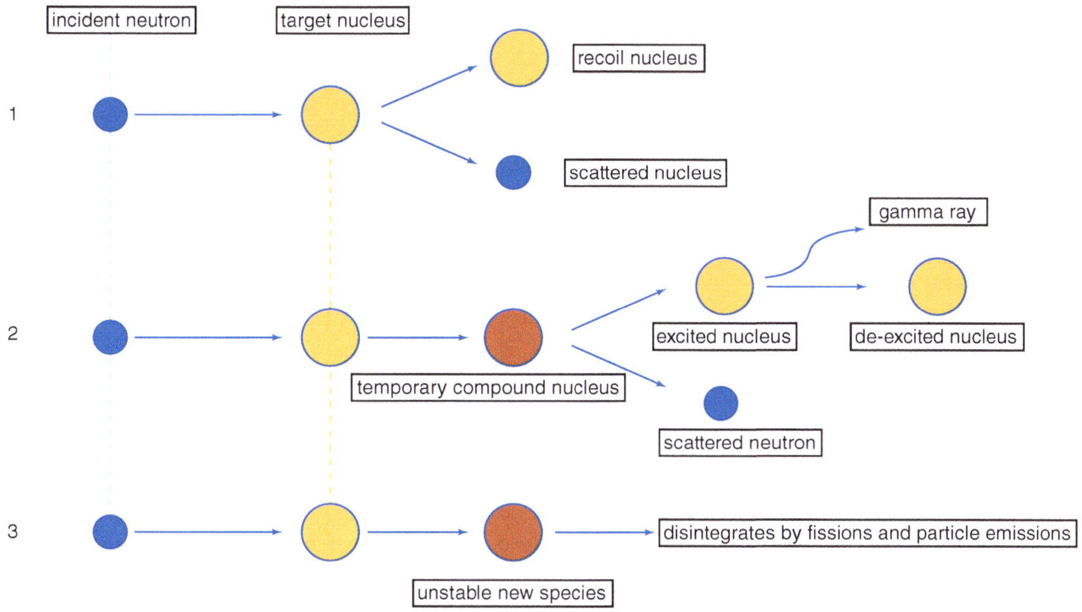

Fig. 17.11 Interaction of the uncharged particles (neutron-proton)

depends on the energy. As an approximation, *protons are considered low LET* ions as the LET is only high enough toward the end of the Bragg peak to produce clustered DNA damage, where the dose is extremely low. Heavy charged particles (like carbon) have sparse ionization in the initial entry region (low LET) and dense ionization as they approach their Bragg peak (high LET) (Fig. 17.12). Thus, high LET radiation is highly effective and has an RBE that is 3–5 times (at Bragg peak) that of low LET radiation. It is noteworthy that 70% of the DSBs produced by high LET particles are complex as opposed to 20% for low LET particles [28–32].

17.4.1 Oxygen Effect

Low LET particles depend on free radical mediated cell kill where the oxygen stabilizes these free radicals. Unlike low LET radiations, high LET radiation causes DNA damage with direct interaction and oxygen has a minor role to play in its effects. Oxygen Enhancement Ratio (OER) of a particular radiation is the ratio of the dose required to cause cell kill in the absence of oxygen to the dose required in its presence. Multiple studies show that the OER for heavy ions decreases when the LET reaches around 100 keV/μm, reaching unity at a value close to 500 keV/μm [33–35]. This property is highly useful in treating hypoxic tumors which are radioresistant to conventional low LET RT with an OER of 2–3.

17.4.2 Mechanisms of Cell Death

Evolving literature on the mechanisms of cell death caused by high LET particles suggest that the existence of alternate pathways of cellular signaling. Accumulation of cell cycle regulators like TP53 and CDKN1A levels was both dose and LET dependent [36]. High LET radiation leads to a different gene activation and signaling compared to low LET, and ultimately resulting in different mechanisms of cell death [37]. Furthermore, the radiation induced apoptosis caused by high LET radiation was p-53 independent [38, 39]. Additional in vitro studies suggest an upregulation of genes involved in TGF-β signaling pathway [40, 41].

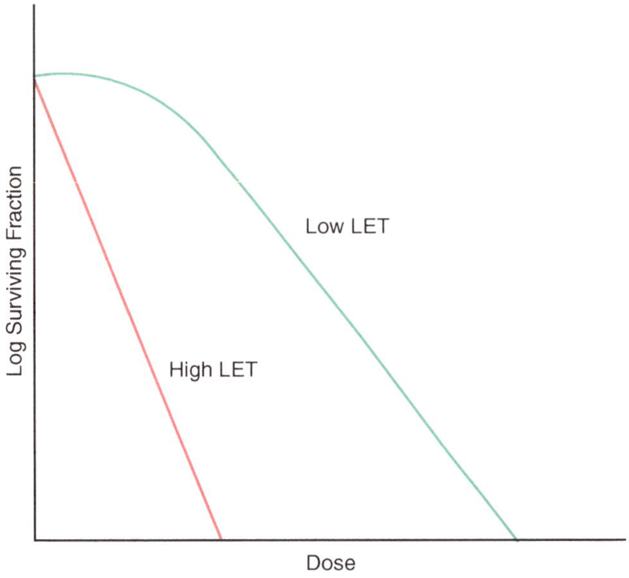

Fig. 17.12 Cell survival curves for high-linear energy transfer (LET) particles and low-LET X-rays. High-LET radiation results in steeper survival curves with minimal or no shoulder, indicating reduced sublethal damage repair capacity

17.4.3 Radiobiological Effectiveness (RBE)

RBE is the ratio of the dose of a reference radiation to that of a test radiation for the same cell kill. The factors that influence RBE is dose rate, cell line, end point (10% survival or 1% survival) and dose (if the dose-response curve of either radiation is non-linear). Unlike low-LET particles, quantitative experiments conducted on different cell lines to estimate the RBE of high LET radiation have suggested that the use of a constant number is not possible and varies from 2 to 3.5 [42–46].

ICRU 93 recommends the use of RBE-weighted dose for carbon ions, while the choice of RBE model is left to the treating institution [47].

Furthermore, the biological properties of charged particles are hot topic of research [48–51].

17.5 Particle Therapy Systems

Particle therapy systems have three main components—particle accelerators, beam transport systems and treatment delivery unit.

17.5.1 Production and Particle Accelerators

From the first cyclotron developed Dr. Ernest Lawrence in 1930, particle accelerator technology has undergone numerous developments over the subsequent decades. The present-day accelerators are compact in size, so as to be easily accommodated around existing hospitals and medical facilities situated in urban neighborhoods. The particle accelerators used in current practice are isochronous cyclotrons, synchrocyclotron and synchrotrons. The choice of a particular accelerator depends on the required dose rate, intended throughout of the center and the cost involved.

17.5.1.1 Cyclotron and Isochronous Cyclotrons

Cyclotrons were invented for the purpose of accelerating charged particles as linear accelerators cannot accelerate them to high enough energies using a practical length of the accelerating chamber. Cyclotrons work on two principles—charged particles get accelerated by an electric field (similar to linear accelerators) and charged particles follow a circular path when under a magnetic field. A cyclotron accelerates the particles in a spiral pattern under a constant magnetic field, with electric field being applied during each revolution to increase the speed. The centripetal force provided by the magnetic field applied perpendicular to plane of particle movement is called the Lorentz force. The rotational frequency can be calculated, which is a constant for a given magnetic field.

Figure 17.13 illustrates a schematic diagram of a cyclotron. At the center, there are two semicircular disks (called dees) which are placed perpendicular to the magnetic field and with a gap in between them which is connected to a sinusoidal RF generator. The frequency of the RF generator is matched with the angular frequency of the particles. The particles get accelerated whenever they pass the gap. As RF generator alternated the electric field, they get accelerated twice per revolution. As the particles are accelerated (increase in speed and energy), the radius of their trajectories increases and they tend to move in an enlarging spiral path. They then reach their highest energy in their outermost trajectory possible, and thereby exit the cyclotron.

The classic cyclotrons accelerate particles to a range of 10–20 MeV. As the speed of the particle approaches that of light (relativistic range), the resting mass of the particle also increases. This in turn decreases the rotational frequency of the particle, which is no longer in resonance with the RF generator frequency. Hence, will not be accelerated by the dee voltage. To overcome this limitation, an ***isochronous cyclotron*** will increase its magnetic field strength by a calculated factor to achieve the radius to maintain the isochronism condition. With increase in the magnetic strength, the particle's trajectory deviates from the mid-plane of the cyclotron and eventually gets lost by colliding with the interior vacuum walls of the cyclotron. A focusing magnetic field in the opposite direction is needed to keep the particle in mid-plane. Since increasing and decreasing the magnetic field at the same was impractical, an improvisation was suggested in 1938 by Llewellyn H. Thomas [52]. He designed an *azimuthally varying field (AVF) cyclotron* where the magnetic field is modulated in the azimuthal direction as well which applies a net vertical force on the particles and keeps them captured in a net spiral accelerating path. Later, the symmetrical straight sectors were replaced with spirally shaped sectors (Fig. 17.14) [53]. Most of the present-day proton therapy centers use this version of the isochronous cyclotron. The average magnetic field strength of commercial isochronous cyclotrons is around 2.5 T with a diameter of around 4–5 m and height of 2.5 m. *Superconducting magnets* are used routinely to shrink the size of cyclotrons. Figure 17.15 shows a cyclotron after installation.

17.5.1.2 Synchrocyclotron

The concept of synchrocyclotron was independently reported from USSR Academy of Sciences in 1944 and UC Berkeley in 1945 [54, 55]. In a synchrocyclotron, as the particle traveling speed reaches relativistic range, the RF generator frequency changes to match the sagging particle revolution frequency (cf. isochronous cyclotron). Thus, the particle experiences a continuous acceleration during each revolution. This results in a much lower beam production duty cycle from a synchrocyclotron, when compared to a continuous current produced by an isochronous cyclotron.

17.5.1.3 Synchrotron

Conceptualized around the same time of a synchrocyclotron, the first proton synchrotron was built in 1952 [54, 55]. The concept of a synchrotron was formulated to overcome the drawback of limited particle energy for a given magnet size in a cyclotron and synchrocyclotron. The two conditions followed in a synchrotron are that the magnetic field synchronizes with the particle energy and the electric field (RF) synchronizes with the particle orbital frequency (which in-turn depends on the magnetic field and the particle energy). A pulsed beam current is used to maintain the synchronism of the electric and magnetic fields. As a synchrotron is not capable of accelerating particles from zero initial velocity, pre-accelerated particles from a linear accelerator are injected into the system. This injection of parti-

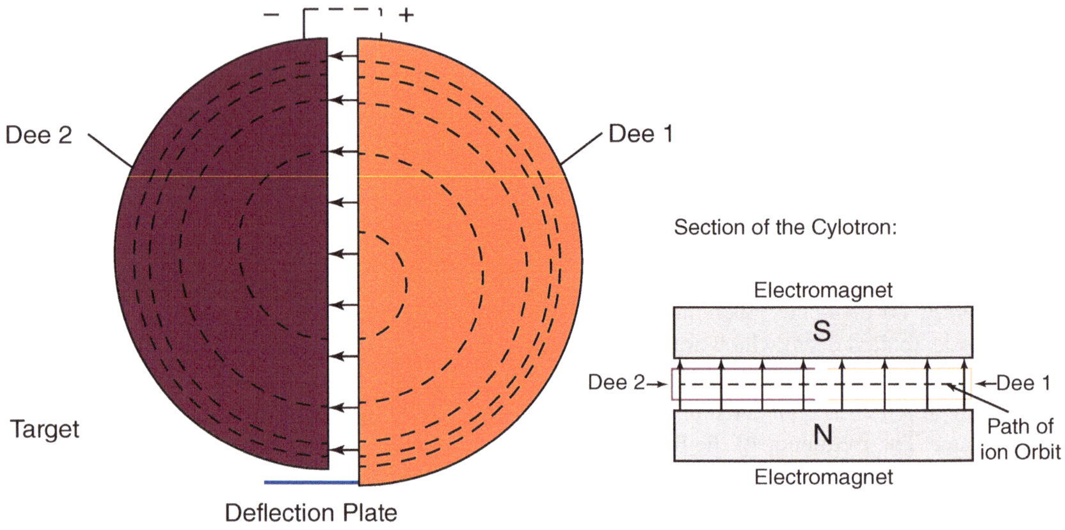

Fig. 17.13 Schematic diagram of a cyclotron

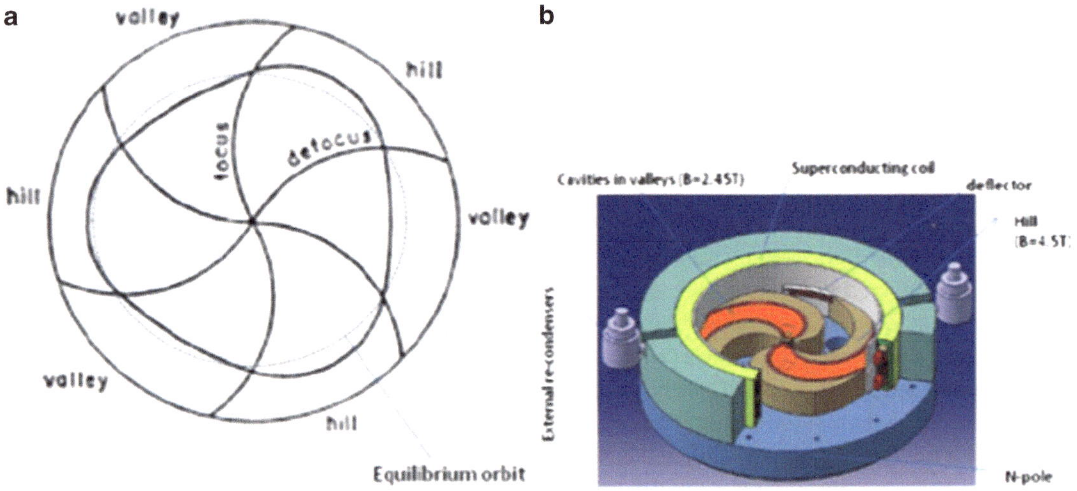

Fig. 17.14 (**a**) The trajectory of a particle in an isochronous cyclotron is influenced by alternating regions of strong and weak azimuthal magnetic fields, known as "sectors" or "hills and valleys." These regions provide axial focusing, counteracting the defocusing effect caused by the radially increasing magnetic field. (**b**) Schematic view of the median plane in the Carbon/Proton superconducting cyclotron

cles is to be synchronized with the RF frequency. Once the protons are accelerated to the required energies, they are extracted from the ring using a RF kicker signal. Bending and focusing magnets of various types (dipole, quadruple and sextuple magnets) arranged in the focusing lattice are used for focusing the deviating protons.

17.5.2 Energy Selection System (ESS)

An energy selection process is needed for the accelerators to a treat a 3-dimensional tumor volume conformally. Cyclotrons and synchrocyclotrons deliver *monoenergetic beams* (230 MeV), and therefore require external devices to select the appropriate energy to be delivered at a par-

Fig. 17.15 IBA 230 MeV resistive isochronous cyclotron. (Picture courtesy IBA)

ticular spot. The treatment energies range from 70 to 230 MeV for protons. Energy selection system uses degraders to reduce the energy to the appropriate range and a series of dipole magnets to filter out the unwanted energies so as to provide a single energy for treatment. *Unlike other circular accelerators, synchrotron does not require an energy selection system as the particles are extracted depending on the number of revolutions in the ring.* The typical treatment energies range from 70 to 230 MeV for protons and 120 to 430 MeV/amu for carbons.

Though linear accelerators were initially considered to be impractical for particle acceleration, CERN has recently developed a LIGHT proton linear accelerator [56]. There are several unique features to this accelerator. The energy selection is performed on a pulse-to-pulse basis which confers an extremely small energy switching time and makes it ideal for pencil beam scanning delivery. Unlike the circular accelerators, a LIGHT accelerator does not need to use a de-escalation delivery scheme (deepest layer treated followed by the next deeper layer) for treating layers of the tumor as the energy switching time is extremely small. Furthermore, as the system does not need a beam degrader, there is no beam loss. This in turn reduces thickness of shielding needed while constructing the facility. Due to all these characteristics, LIGHT proton linear accelerator is promising for future accelerators.

17.5.3 Beam Transport System (BTS) and Treatment Delivery System

The particles extracted from the accelerator pass through the ESS and then are transported using a beam line to the respective gantry. Figure 17.16 is a schematic diagram of a proton therapy facility with three major components. The BTS consists of a series of bending and focusing magnets that are used to transport the beam, at the end of which an energized switching magnet diverts the beam to the treatment nozzle in the particular gantry room. Automatic steering and tuning are used to maintain high-precision beam characteristics. A single accelerator can be connected to around 1–5 gantry rooms, which helps using the accelerator efficiently. *Therefore, only a single gantry can deliver the beam at a particular time.* The gantry can be fixed or rotating (partial or complete arc) which houses elements to shape and modulate the beam according to the technique used.

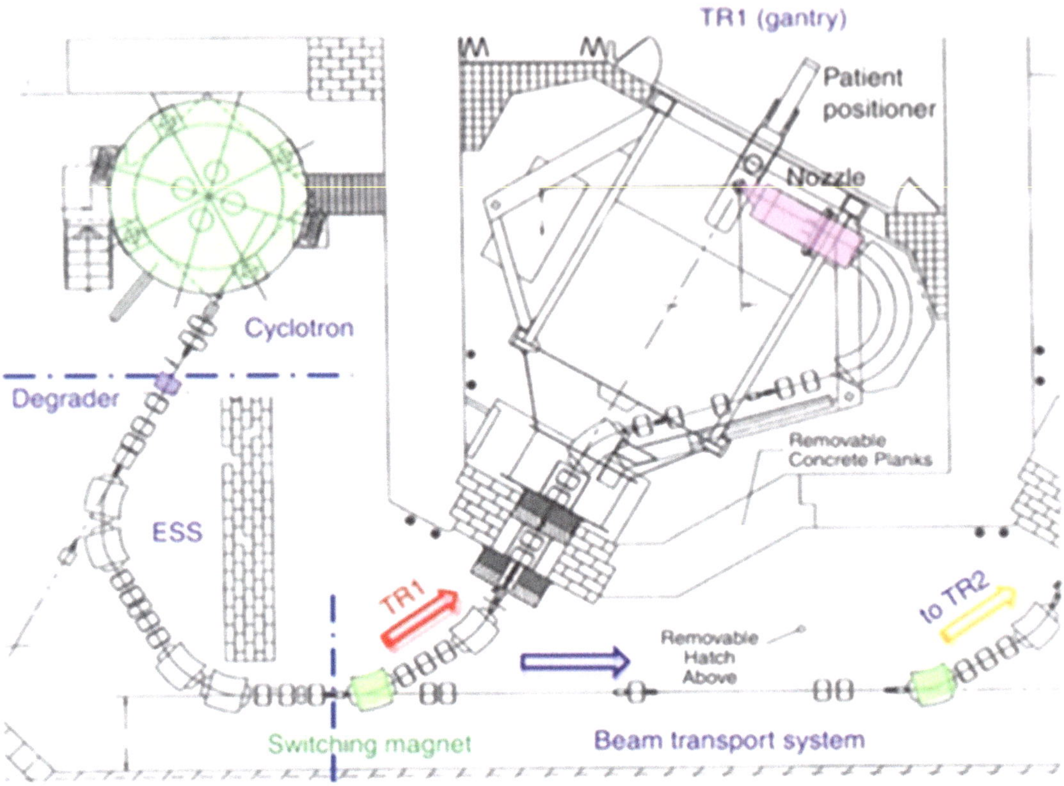

Fig. 17.16 Schematic diagram of a proton therapy facility

17.6 Treatment Delivery Techniques

The unique depth dose deposition characteristics sets particle therapy apart from conventional radiotherapy. As alluded to before, particles have a pronounced Bragg peak for dose deposition when they travel through matter, spanning through a few millimeters. However, since tumors have variable dimensions, spread of the Bragg peak of a single energy will not be able to treat them homogenously. ***Spread Out Bragg Peak (SOBP)*** technology helps in overcoming this. It is the uniform dose region created at a depth formed by the optimal stacking of multiple depth dose curves of Bragg peaks of different energies (Fig. 17.17). The energies are modulated according to the tumor thickness in the direction of the beam. The lateral and depth profiles of an SOBP beam (red color) are compared against those of a photon beam (cyan color) in Fig. 17.18. While protons have a sharp penumbra in the lateral dose profile, photons show increased low doses far from the target. In the depth dose profile, SOBP has a higher entrance dose than photons which is due to the summation of the plateau doses of different energy Bragg curves. Larger the target thickness in the direction of the beam, higher will be the entrance dose. Hence, protons have a reduced skin sparing effect than photons. Furthermore, one important consideration for SOBP in the clinical scenarios is the phenomenon of distal end degradation [57]. As the beam arrangements in routine planning are such that the beams enter obliquely or pass through inhomogenous mediums (air, bones), the Bragg peaks at the proximal and distal edges of the target would be of varying depths of penetration. This leads to range uncertainties in proton planning (discussed later).

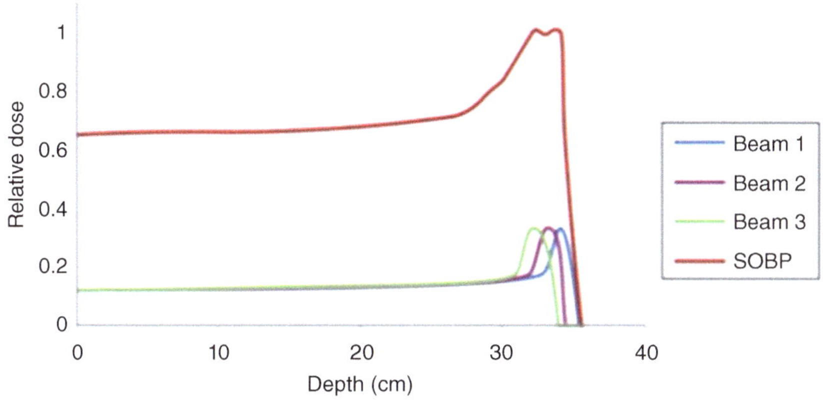

Fig. 17.17 Stacking of multiple depth–dose curves to generate spread-out Bragg peak (SOBP)

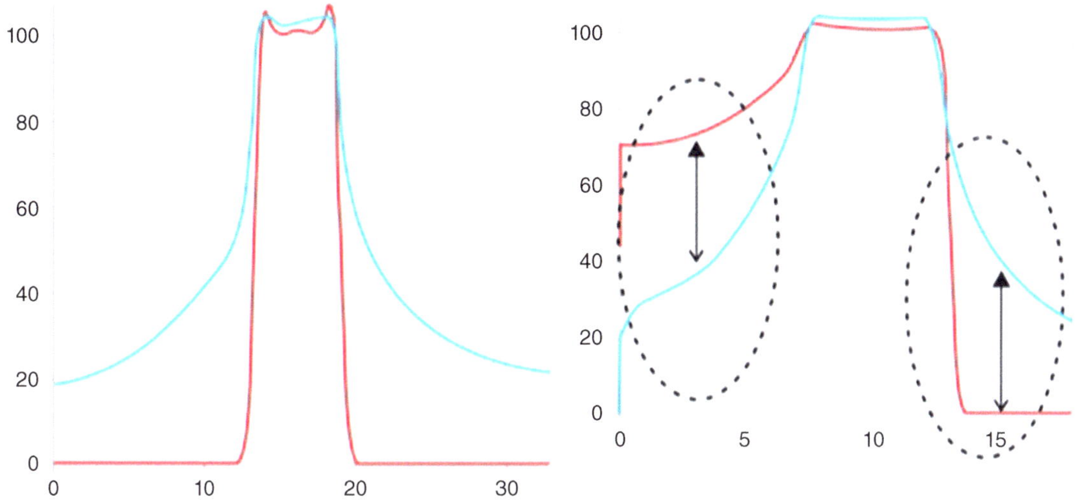

Fig. 17.18 The lateral and depth profiles of a beam have Spread-out Bragg Peak (red color) are compared against those of a photon beam (cyan color)

17.6.1 Proton Delivery Techniques

There are three delivery techniques of proton delivery, namely passive scattering (PS-PT), uniform scanning (US-PT) and pencil beam scanning (PBS-PT). Broadly speaking, PS-PT scatters the proton beam across the cross-section of the tumor, while the latter two techniques scan the beam across the tumor using magnets (principle of Lorentz forces).

17.6.1.1 Passive Scattering (PS-PT)

PS-PT is a proton delivery technique where a monoenergetic beam is modulated and modified by various external accessories to treat a tumor. For decades, PS-PT was the most commonly employed technique before scanning techniques were invented, due to its relatively simple treatment geometry. The beam passes through accessories in the following order (Fig. 17.19):

- Energy selector system (ESS)—A carbon double wedge ESS adjusts the monoenergetic beam that exits from the accelerator. It reduces this high proton energy to give a beam with a

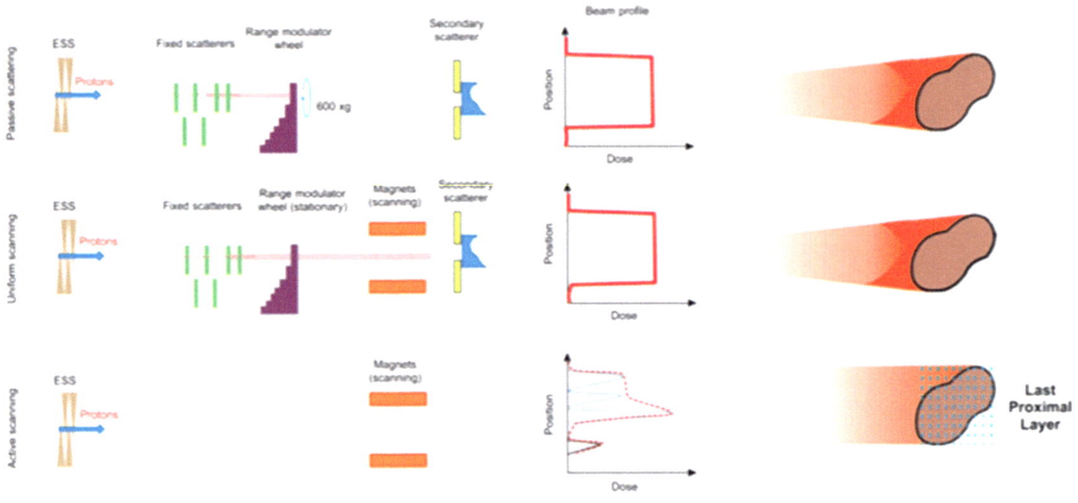

Fig. 17.19 Passive scattering proton delivery technique

maximum energy such that the protons reach the distal end of the tumor to be treated.

- Degrader or range modulator wheel spreads the beam with a single maximum energy into that with different energies, so as to dispose the dose through the tumor *in the direction of the beam*. The degrader creates the SOBP needed to treat different layers of the tumor. The different energies are not delivered simultaneously, but all energies are delivered in one cycle of the modulator wheel (typical frequency of 10 Hz).
- Scatterer—The beam that is extracted from the accelerator is narrow and has a spot size of 4 mm. Physical scatterers, as the name suggests, scatters the beam uniformly in a plane that is *orthogonal to the direction of the beam*. For small tumors, a single scatterer spreads the beam sufficiently. However, double scatterers are required for broadening the beam to cover larger tumors (maximum diameter of 25 cm). The techniques are then called **single scattering and double scattering**.
- Collimators or apertures are used to shape the edges of the beam laterally, as none of the tumors are perfectly spherical in shape. They are usually made of brass.
- Compensator helps the beam to conform the distal edge of the tumor. Unfortunately, PS-PT does not allow proximal tumor conformity.

In treatment planning for PS-PT, the appropriate beam angles are selected for the gantry and couch along with the specific range and modulation of the SOBP to ensure target coverage. The apertures and compensators are automatically generated in the TPS. However, after plan completion, the hardware needs to be fabricated with the use of an in-house machinist, materials (brass, Lucite, wax) and the time to fabricate the devices. This is one of the main disadvantages of PS-PT which made it cumbersome to use; the other disadvantages being inability for proximal tumor conformity and increased chances of nuclear fragmentation due to many accessories in the beam line.

17.6.1.2 Uniform Scanning (US-PT)

US-PT is a hybrid of PS-PT and pencil beam scanning. As described above, the beam in US-PT is also modified and modulated by an ESS, range modulator, scatterer, collimator, and compensator. However, here the range modulator remains stationary while protons of a single energy is delivered at a time. It then adjusts to treat the next energy levels (Fig. 17.19). The spacing between each energy layer depends on the width of the Bragg peaks. A broad beam of around 5 cm spot size is magnetically steered across the entire treatment area to form a flat field. Unlike pencil beam scanning, the beam is not turned off while changing focus in the tumor or switching the energy. It continuously scans with a frequency of around 30 Hz.

17.6.1.3 Pencil Beam Scanning (PB-PT)

PBS-PT or active scanning is proton dose-delivery system where a focused beam of protons of variable intensity and energy is magnetically scanned across the field. Multiple pencil beams with spot sizes varying between 4 and 10 mm diameter are super-positioned and are carefully weighted and optimized to deliver a target dose to a tumor. The dose from these individual pencil beams accumulate to give an intended homogenous broad beam distribution [57, 58]. PBS can be executed in three different ways, namely discrete spot scanning or continuous line scanning and raster scanning (further reading suggested in [59, 60]).

Unlike the previous two proton delivery techniques, additional hardware is not required in PBS-PT (Fig. 17.19). The series of wedges in the ESS is used to degrade the energy in a stepwise manner, negating the need of a range modulator. As the magnets can focus the pencil beams through the entire cross section and precisely control them at the edges of the tumor, scatterer and collimator is not required. Furthermore, the ability to control the energy (and therefor the depth of penetration of the protons) results in not needing range compensators and the resulting beams can conform to the proximal and distal shapes of the target. The treatment planning in PBS is very complex and requires inverse planning.

The advantages of PBS are that it

- Increases dose conformality with better sparing of normal tissue proximal to target
- Reduces the need for expensive beam shaping accessories
- Has higher beam utilization efficiency (~100%), resulting in minimal neutron scatter. This in turn requires less shielding and lower cost of infrastructure compared to PS-PT.

Despite the advantages of PBS, there are a few disadvantages. PBS is very sensitive where set-up uncertainties, patient motion and interplay effect lead to gross dose perturbation. Though repainting strategies (segments of an energy layer is delivered multiple times) and robust optimization are employed to account for the same, it still remains a major concern. PBS is the most time dependent of the three modalities where each energy layer is scanned through and then layer-switching pauses (on the order of 1–5 s) occur between each energy layer.

17.6.2 Carbon Ion Delivery Techniques

Similar to proton therapy, the carbon ion delivery systems available are passive scattering (PS) and pencil beam scanning (PBS). To reduce the excessive dose to proximal normal tissue in PS, a method called layer-stacking was proposed [61, 62]. In this, the tumor is divided into thin slices of 2.5 mm. The variable SOBP needed for the target volume produced by the mini-ridge filters and range shifters is deposited longitudinally through the tumor in a step-wise manner. These thin SOBP slices are shaped laterally by multi-leaf collimators (MLCs). Similar to passive scattering, HIMAC developed in-house broad beam techniques like single and spiral beam wobbling methods [63]. A wobbling magnet along with scatterers was used to treat cross section of the tumor. However, these methods also suffered from the disadvantages of PS.

PBS with all its advantages is the preferred technique in carbon ion delivery also. GSI (Germany) developed the first pencil beam scanning method for carbon ions. However, due to its sensitivity to organ motion, Furakawa et al. proposed and developed respiratory phase-controlled rescanning (PCR) method so that PBS could be applied for respiratory moving targets [64, 65].

17.7 Treatment Planning

Similar to the process for photons, treatment planning in protons constitutes immobilization, CT imaging, co-registration with other imaging modalities, target and organ-at-risk (OAR) contouring, radiation portal selection, optimization, dose calculation, analyzing the resultant dose distribution and treatment verification [66]. Every step in this process is important so as to ensure the intended dose delivery.

Immobilization—Site-specific immobilization techniques are employed. As in photon therapy, these devices maintain the patient in a position reproducibly, minimizing the risk of motion and maximizing comfort. Comfortable position is all the more important as proton treatment delivery time is much longer. As proton range is very sensitive to any kind of matter that comes in its path, *minimizing the thickness gradient* of immobilization devices is crucial. Hence, conventional immobilization devices are to be modified and the material is substituted with carbon composites.

CT imaging—CT imaging is needed for target/OAR delineation as well as dose calculation. One important consideration in proton therapy planning is the use of contrast agents for planning CT. Contrast agents have a higher HU value than the normal tissue. This will overestimate the proton dose required, as the patients will not be treated with contrast injection in every fraction. Hence, non-contrast planning CTs are used. If contrast material is thought to be indispensable in a particular case, then it is advisable to take two scans—one with contrast (for delineation) and one without (for dose calculation). Axial mode CT images are preferred over spiral mode, due to better spatial resolution. A slice thickness of 2.5 mm is advisable for better longitudinal accuracy.

Delineation of structures and co-registration with other imaging modalities—The recommendations for delineation of targets in proton therapy are absolutely similar to that in photon RT. However, extreme caution must be observed as marginal misses or overdosing of OARs is theoretically possible due to the rapid fall-off in proton therapy compared to photon therapy. Co-registration with MR and PET-CT can be performed where required.

Radiation portal selection—Patient positioning couch and parts of the delivery system like gantry nozzle dictate many of the possible field orientations and constraints. Portal selection is decided by visualizing the target through the path of protons of particular beam orientations (beam's eye view, BEV) where target coverage is optimized and critical structures are avoided. An important consideration when selecting portal orientation is the resulting airgap between the compensator and the patient which may degrade the penumbra because of scatter. Also, beam angles that may result in sharp gradients must be avoided, such as those tangential to bone, air cavities or metal implants.

Optimization—PBS techniques need optimization for spot position and beam weight.

- Spot position optimization is performed by the computational resources on the basis of the target volumes delineated. Depending on the depths of the target scanned, the spot sizes of the pencil beams may vary between 3 mm and 1.2 cm sigma.
- Beam weight optimization—There are two ways to optimize the dose—single-field optimization (SFO) and multi-field optimization (MFO). The features have been tabulated in Table 17.5. While single-field uniform dose refers to the optimization method using a sin-

Table 17.5 Types of planning using pencil beam scanning

Single field optimization (SFO)	Multifield optimization (MFO) or intensity modulated proton therapy (IMPT)
Optimize the weights on a field-to-field basis, i.e., each field is optimized individually to deliver a fraction of the prescribed dose to the entire target volume	Spot weights are optimized together to create an integrated uniform target dose
Less sparing of critical structures	Better sparing of critical structures
Less sensitive to set-up errors	More sensitive to set-up errors

gle field (SFO), the term intensity modulated proton therapy (IMPT) is associated with MFO.

Dose calculation is a vital step to ensure quality of care to patients undergoing proton therapy. Currently, the two main dose calculation algorithms available for proton therapy planning is pencil beam (PB) and Monte Carlo (MC) algorithms [67]. PB algorithms are analytical approximations where each spot in a proton PBS beam is decomposed into a subset of narrow beamlets commonly referred to as pencil beams. The dose is calculated according to tissue inhomogeneities encountered by the spot in the central axis. This makes PB insensitive to lateral inhomogeneities and in addition, it is unable to account for the low dose envelope (called nuclear halo) of primary protons due to multiple scattering and nuclear fragments. Despite the disadvantages, PB is regarded as the standard practice and the there are two variants that are commercially available [68, 69].

MC algorithms use random sampling of interaction cross-sections that represent the underlying physics to simulate individual particles trajectories [70]. The particles are tracked individually until their energy fall below a certain threshold when all the energy is assumed to be deposited locally. A sufficiently large number of source particle histories have to be run for good statistical precision in the dose distribution. MC methods are potentially very accurate and are in literature often referred to as the gold standard for dose calculation [67, 70, 71]. MC algorithms are very slow as they are performed on several existing research-based codes for routine dose calculation. RaySearch Laboratories AB (Sweden) and Varian Medical Systems (USA) have modified and developed commercial MC dose calculation algorithms for proton spot scanning which are much faster.

Analysis of dose distribution and evaluation—Once the plan is generated, the dose volume histograms are evaluated. Also plan comparisons are made, to select the best plan.

17.7.1 Range Uncertainties

As alluded to before, range uncertainty is a major challenge in proton therapy. The end-of-range is the region where there is a sharp dose fall-off. Any undershoot of the proton range can lead to an underdosing of the distal edge of the tumor. Furthermore, it can cause dose-inhomogeneities in the target as well. The various factors that affect the range uncertainty are described in Table 17.6 [72, 73]. As PBS does not treat the entire target volume simultaneously, any change in the density along the path of beam can lead to dose perturbations, thereby potentially overtreating some areas of adjacent normal tissues or undertreating areas of the tumor volume. This phenomenon has been termed the "interplay effect" [74]. To overcome these uncertainties, certain methods have been proposed.

- **Robustness Optimization and Evaluation—** There are two ways of accounting for the robustness of plans.
  - Robustness Optimization—Each spot is scored for its robustness for maintaining overall sensitivity to potential plan degradation and optimizes them so as to optimally satisfy the specified criteria simultaneously under all scenarios [75, 76]. The spots with poor robustness (high sensitivity to plan degradation) will be penalized by iteratively decreasing and eliminating their intensity. The metric maintained for modeling setup errors is in the order of 3 mm along with range uncertainties compounded to 3% [77].
  - Robustness Evaluation—The goal is to assess the resilience of a dose distribution to uncertainties as a part of the assessment of a treatment plan. This is done my recalculating the dose with set-up errors, internal motion and HU conversion errors. In analogy with the photon domain, where the DVH of the PTV is the "worst-case" representation of the DVH of the CTV, for proton treatment planning one may select the

Table 17.6 Factors causing range uncertainty

Treatment planning	HU-stopping power calculation	Improving algorithms
	Binning and reconstruction errors	Improving algorithms
	Metal artifacts	• Use metal reduction tools • Avoid beams passing through the metals • Account for their atomic number accurately • Select the shortest path through these metal implants
Treatment implementation	Anatomical changes	• Robust optimization and evaluation • Repeat imaging and planning if changes are noticed on volumetric on-board imaging or interim check CT
	Set-up errors	• Robust immobilization devices • Robust optimization & evaluation
	Organ motion (Interplay effect)	• Motion management • 4D robust optimization and evaluation • Repainting or rescanning

CTV DVH that corresponds to the overall worst-case scenario for plan evaluation.

- Repainting or rescanning is a way where each proton energy layer is treated more than once to give a cumulative intended dose, so as to achieve a statistical averaging of the motion effects. While volumetric repainting is delivered by repetitive scanning through the whole target volume, each energy layer is rescanned more than once before switching to the next energy layer in layered repainting [78, 79].
- Metal artifacts—As these are of high atomic number, they may cause dose perturbations. There are multiple ways to overcome/ the impact of such metal implants (Table 17.6).
- RBE based planning—Though the RBE of protons is simplistically assumed as a constant 1.1, it may vary substantially along the path of proton beam leading to unforeseen local failures or toxicities. This is dependent on tumor and patient factors. Empirical and biophysical models have been developed to predict variable proton RBE in tissues for better clinical decisions and understand unexpected toxicities [80].

17.7.2 Treatment Planning in Carbon Ion Therapy

Typically, treatment planning is the same as that in proton. The only major difference being the RBE variability. The treatment planning system must take into account all factors that account for RBE variability like particle type, incident energy, penetration depth, fraction dose, clinical endpoint and so forth. For an RBE modeling in passive beam delivery system, the RBE-weighted depth dose curves are stored as a beam library for the SOBP; these depth dose curves are scaled by the prescribed dose during dose calculation. However, pencil beam scanning delivery system requires a more general modeling. As described earlier (section on RBE), local effect model (LEM) and microdosimetric kinetic model (MKM) are the two well described models developed by British Standards Institution (BSI) and NIRS.

17.8 Motion Management

Moving targets are challenge in RT, especially in particle therapy due to the drastic dose perturbations it can cause as discussed in the previous section. Motion management techniques used in photon therapy are employed in particle therapy as well.

17.9 Treatment Verification

Image guided RT (IGRT) is necessary for treatment verification. 2D orthogonal X-rays is still the commonest mode of on-board imaging.

Volumetric imaging is available as CT on rails and on-board cone beam CTs.

17.10 Clinical Use of Particles in Various Cancers (with Available Evidence)

Particle therapy has gained impetus over the past decades because of its ability to spare critical structures in close proximity to the target. This substantially reduces the dose to normal tissues, improving outcomes and lowering acute and long-term toxicities. The role of particle therapy in various sites of cancers is discussed below.

17.10.1 Pediatric Cancers

Apart from normal tissue sparing, particle therapy has been employed extensively in children for the advantages of reduced growth perturbation and risk of malignancy.

Central nervous system (CNS) Tumors—High and standard risk medulloblastoma (MB) mandate craniospinal irradiation (CSI) that entails RT to the whole neural axis followed by boost to the tumor bed/posterior fossa. Proton therapy has been shown to reduce long-term cardiac, pulmonary and gastrointestinal toxicities for medulloblastoma patients by 25–50% relative to photon treatment [81]. In children, ependymomas are mostly intracranial, with fourth ventricle being the most common subsite. Proton therapy helps in reducing the doses to the surrounding critical structure such as brainstem, especially with dose-escalation in the current studies to a dose of 59.4 Gy to tumor bed [82]. In germ cell tumors, RT has an indispensable role in local therapy. While whole ventricular RT (WVRT) is indicated in localized pure germinoma, CSI is required in disseminated pure germinoma or non-seminomatous germinoma along with chemotherapy. Since germinomas tend to have a very good survival, long-term toxicity reduction needs to be high priority along with disease cure [83]. Craniopharyngioma is the third most common brain tumor in children which is benign and has an excellent prognosis after treatment. However, children often suffer from long term complications of tumor directed therapy due to the proximity of the tumor to the optic apparatus, pituitary and hypothalamus. Proton therapy helps in reducing the toxicities of treatment [84]. Selected studies in the above tumors have been summarized in Table 17.7 [81–83, 85].

Non-CNS Tumors—Role of proton therapy has been studied mainly in the following cancers—Ewing's sarcoma (EWS), rhabdomyosarcoma (RMS), and neuroblastoma (NB). The above tumors can occur in various sites throughout the body. Hence, proton therapy helps in reducing side effects to growing bones, neural, cardiovascular, and reproductive organs [86].

Table 17.7 Selected studies for particle therapy in pediatric CNS tumors

Author	Year	Tumor	*N*	Particles	Dose (GyE)	Outcome
Yock et al.	2016	Medulloblastoma	59	Proton	23.4 (CSI) + 30.6 (boost)	Cumulative incidence of ototoxicity gr 3–4: 12% and neuroendocrine deficit: 55%, 5-year PFS: 80% & OS: 83%
McDonald et al.	2008	Ependymoma	17	Proton	54	At median follow-up of 26 months, LC: 86% & OS: 89%
McDonald et al.	2011	Germ cell tumor	22	Proton	18–36 (WVRT/CSI) +18–36 (boost)	At median follow-up of 28 months, LC: 100% & OS: 100%
Jimenez et al.	2021	Craniopharyngioma	77	Proton	52.2	Acute gr 3 toxicities: 4%, severe long-term toxicities were still prevalent

N number of patients, *Gy E* gray equivalent, *CSI* craniospinal irradiation, *PFS* progression-free survival, *OS* overall survival, *LC* local control, *WVRT* whole ventricular radiotherapy

Table 17.8 Selected studies for particle therapy in pediatric non-CNS cancers

Author	Year	Tumor	*N*	Particles	Dose (GyE)	Outcome
Rombi et al.	2012	EWS of pelvis, trunk, head, base skull	30	Proton	54	3-year LC 86%
Kharod et al.	2020	EWS of base skull and cranium	25	Proton	54–55.8	4-year LC: 96%, 4-year OS: 92%
Uezono et al.	2020	EWS pelvis	35	Proton	54–64.8	3-year LC: 92%, 3-year OS: 83%
Ladra et al.	2014	RMS (all sites)	57	Proton	36–50.4	5-year LC: 78%, 5-year OS: 81%
Mizumoto et al.	2018	RMS of HN, para-meningeal, prostate, other sites	55	Proton	36–60	2-year OS: 84.8%
Indelicato et al.	2020	RMS of the pelvis	31	Proton	50.4	5-year LC: 83%, 5-year OS: 84%
Buszek et al.	2021	RMS of all sites	94	Proton	36–50.8	4-year LC: 85%, 4-year OS: 71%
Hill-Kayser et al.	2019	Neuroblastoma	45	Proton	21.6	3-year LC: 97%, 3-year OS: 89%
Hishiki et al. et al.	2018	Neuroblastoma	50	Proton	19.8 (30.6 for residual disease)	3-year OS 69.5%

N number of patients, *GyE* gray equivalent, *EWS* Ewing's sarcoma, *LC* local control, *OS* overall survival, *RMS* rhabdomyosarcoma, *HN* head and neck

Presently, proton therapy is frequently used in unresectable EWS of the pelvis and head neck region where dose escalation is considered [87]. Multiple studies have reported outcomes with acceptable toxicities for various sites of treatment in RMS. Similarly, proton therapy has been studied in multiple retrospective series for treatment of the primary site (post-maximum safe resection) as well as for metastatic Metaiodobenzylguanidine (MIBG)-avid sites in neuroblastoma.

Selected studies in the above tumors have been summarized in Table 17.8 [86–94].

17.10.2 Chordoma and Chondrosarcoma

Skull base and spinal chordoma/chondrosarcomas have been of interest in particle therapy due to their proximity to critical structures and radioresistant nature. Studies have reported a local control (LC) of 70–80% with the use of particle therapy in skull base chordomas [95, 96]. A meta-analysis performed for the use of carbon ions reported acceptable toxicities compared to proton therapy with the most common toxicities being mucositis and skin reactions [97]. However, another study from NIRS reported higher grade 3 toxicities with carbon ion therapy, for which a higher dose per fraction and passive scattering delivery system were cited as the possible reasons [98]. Innalfi et al. reported that the local recurrence for such tumors occurred at the interface of tumor and critical structures, where a lower dose was accepted so as to respect the latter [99]. Treating post-operative cases with metallic implants with particle therapy is a challenge for various reasons (explained in previous section) and the clinical results are worse [100]. Similarly, studies in proton and carbon ion therapy have shown encouraging results in spinal chordomas with a 5-year LC of 75–80% [101, 102]. Chondrosarcoma (CS) is a rare tumor and the treatment of the same face similar challenges as that in chordomas (location and radio-resistance). A meta-analysis of studies for particle therapy in CS has reported a 5-year LC of 90% (Lu et al.).

Salient studies for the above tumors have been summarized in Table 17.9 [95, 96, 99, 101–106].

Table 17.9 Selected studies for particle therapy in chordomas and chondrosarcomas (skull base and spine)

Author	Year	Tumor	*N*	Particles	Dose (GyE)	Outcome
Fung et al.	2017	Skull base chordomas	106	Photon + proton, proton alone	68.4–73.8	5-year LC: 75.1%
Deraniyagala et al.	2014	Skull base chordoma	33	Proton	77.4–79.4	2-year LC: 86%, OS: 92%
Schulz-Ertner et al.	2007	Skull base chordoma	96	Carbon ion	60–70	5-year LC: 70% & OS: 88.5%
Uhl et al.	2014	Skull base chordoma	155	Carbon ion	60	5-year LC: 72% % OS: 85%
Iannalfi et al.	2020	Skull base chordoma	135	Proton, carbon ions	70.4–74	5-year LC was 84% (proton) & 71% (carbon ion)
Chen et al.	2013	Spinal chordoma	24	Proton	77.4	5-year LC: 79.8% & OS: 78.1%
Imai et al.	2016	Spinal chordoma	188	Carbon ion	64–73.6	5-year LC: 77.2% & OS: 81.1%
Weber e al.	2016	Chondrosarcoma	77	Proton	64–76	8-year LC: 89.7% & OS: 93.5%
Uhl et al.	2014	Chondrosarcoma	79	Carbon	60	5-year LC: 88% & OS: 97.5%

N number of patients, *GyE* gray equivalent, *LC* local control, *OS* overall survival

17.10.3 Ocular Malignancies

Ocular particle beam radiotherapy, most commonly proton therapy, for uveal lelanoma (UM) is considered as the gold standard of care given the excellent local tumor control and eye preservation compared to other modalities [107–113]. Patients who planned for proton therapy for any tumor undergo tantalum marker placements around the tumor which are MR compatible and useful for X-ray localization. The help of an ophthalmologist, measurements between the markers and also markers and anatomical structures of the eye are documented. The tumor basal dimensions are also noted. Simulation is done using a thermoplastic mold and bite block or chin rest in the seated position. A specialized software is used for planning called Eyeplan (commercial solutions available from RayStation and Varian). It includes eye and tumor modeling, gaze angle optimization, aperture design, dose–volume histogram (DVH) analysis, marker alignment and daily image-guided positioning assessment. The TPS uses a geometrical model scaled to the individual anatomy. Verification is done using orthogonal X-rays where the 3D positioning is confirmed with operative notes/drawings. Most centers have a dedicated passive scattering horizontal beam line. UM proton therapy is delivered as a hypo-fractionated course to a dose of 60–70 Gy in 4–5 fractions given daily [107]. Multiple studies with large number of patients have reported a LC of 96% and eye preservation rate of 90% at 5 years [111–113]. Useful vision (>20/200 Snellen) with proton irradiation is reported to be in the range of 40–65% at 3 years [112, 113].

Vascular tumors such as choroidal hemangiomas have been treated with proton therapy to a dose of 12–20 Gy in 4–8 fractions [114, 115]. Depending on histology, choroidal metastases can be treated to a dose of 20–45 GyE in 2–4 fractions with a good local control for oligometastatic disease [116].

17.10.4 Sarcoma and Soft Tissue Malignancies

Particle therapy in sarcoma has been focused in cases of unresectable disease. With a need of dose-escalation in these cases, the superior dose

Table 17.10 Selected studies for particle therapy in unresectable sarcoma

Author	Year	Tumor	*N*	Particles	Dose (GyE)	Outcome
Schmitt et al.	1989	Unresectable sarcoma	94	Neutron	16.8	5-year LC: 56%
Kamada et al.	2002	Unresectable sarcoma	57	Carbon ion	52.8–73.6	3-year LC: 73%, 3-year OS: 46%
Imai et al.	2018	Unresectable sarcoma	128	Carbon ion	64–73.6	5-year LC: 65%, 5-year OS: 39%
Demizu et al.	2017	Unresectable sarcoma	91	Carbon ion, proton	70.4	3-year LC: 92%, 3-year OS: 83%
Kepka et al.	2005	Unresectable sarcoma	112	Proton	25–87.5	5-year LC: 45%, 5-year OS: 35%
Weber et al.	2007	Unresectable sarcoma	13	Proton	50.4–76	4-year LC: 74.1%

N number of patients, *GyE* gray equivalent, *LC* local control, *OS* overall survival

distribution and increased radiobiologic effectiveness of particles were pitched as the possible advantages over conventional therapy. Considering its rare incidence, most studies were a missed baggage of all histologies in sarcoma. The initial studies were using heavy ions like neutron and Helium ions, which later shifted toward carbon ions and protons (Table 17.10) [117–122]. One of the earliest studies was that using neutron therapy in patients' post-surgery (R0/R1 resection) or with gross disease. Though the overall 5-year LC for those with gross disease was 56%, it varied with the tumor size (76% in <10 cm vs. 42% in >10 cm) [117]. Carbon ion and proton therapy studies shown favorable LC rates in unresectable sarcoma (5-year LC of 60–70%), which is numerically better than that achieved with photon therapy. Gr 3–4 complications in skin and soft tissue were observed in patients, though majority were acute and recovered completely [118, 119]. Another study identified the surface area that received 60 GyE > 20 cm^2 as the strongest independent prognostic factor ($p = 0.041$) [123].

17.10.5 Head Neck Cancers (HNC)

Particle therapy in HNC offers the advantage of protecting OARs and reducing significant reduction in toxicities—mucositis, xerostomia, osteoradionecrosis (ORN) of mandible, and neural injuries. Some of the salient studies have been tabulated in Table 17.11 [124–131]. This in turn makes dose-escalation and re-irradiation feasible while maintaining the quality of life (QOL). Particle therapy in oral cavity cancers is mostly an extrapolation from the treatment of oropharyngeal cancers (OPC). A Japanese study reported very low incidence of ORN when treated with protons [132]. Apart from acceptable LC and reduced toxicities with using carbon ion therapy in oral cavity cancers, Ikawa et al. reported no effect of respectability on LC. This could be probably due to the higher RBE of carbons [124]. Particle therapy has good potential in sparing critical structures like brainstem, optic apparatus and pituitary gland in nasopharyngeal cancers [125, 133]. Chan et al. have reported a decrease in radiographic temporal lobe changes, endocrine function and ORN [134]. Similarly, particle therapy in sino-nasal sinus cancers has reduced incidence of toxicities [135].

Re-irradiation in recurrent HNC is a highly challenging scenario. Particle therapy is very useful in this scenario to deliver tumoricidal doses with minimal toxicities. Clinical studies have shown that proton and carbon ion therapy significantly reduced toxicities such as skin reactions, esophagitis (Table 17.11).

17.10.6 Central Nervous System (CNS) Cancers

Conventional RT is known to trigger an adaptive response where plasticity and motility is promoted in CNS tumor cells. High LET radiation is shown to have an ability to mitigate this adaptive response, though further studies are needed to prove this hypotheses [136]. Also, dose escalated RT in high grade tumors like glioblastoma help to

Table 17.11 Selected studies for particle therapy in head neck cancers

Author	Year	Tumor	*N*	Particles	Dose (GyE)	Outcome
Ikawa et al.	2019	Oral cavity cancers	76	Carbon ion	57.6–70.4	3-year LRC: 86.8% & OS: 78.4%
Blanchard et al.	2016	Oropharyngeal cancers	150	Proton: Photon (2:1 case matched)	66–70	3-year LRC (NS): 97.8% (proton) & 93.5% (photon) Significant difference in gr 3 weight loss and G-tube dependence
Chan et al.	2012	Nasopharyngeal cancers	23	Proton	70	2-year LRC: 100% & OS: 100%
Holliday et al.	2015	Nasopharyngeal cancers	30	Proton: Photon (1:2 case matched)	70	G-tube placement rate (SS): 20% (proton) vs. 65% (photon) and increased mean dose to oral cavity was correlated with higher rate of GT placement (SS)
Akbaba et al.	2019	Nasopharyngeal cancers	26	Carbon + photon	74	2-year LC: 95% & OS: 100%
Yu et al.	2019	Sino-nasal cancers	69	Proton	60	3-year LC: 92.9% & OS: 100%
Romesser et al.	2016	Recurrent tumors	92	Proton	60	1-year LRF: 25.1% & OS: 65.2%
Held et al.	2019	Recurrent tumors	229	Carbon ion	51	Median PFS: 24.2 months & OS: 26.1 months

N number of patients, *GyE* gray equivalent, *LC* local control, *NS* non-significant, *SS* statistically significant, *LRC* locoregional control, *OS* overall survival, *LRF* locoregional failure, *PFS* progression-free survival

overcome hypoxic environments, where particle therapy would be a good modality to choose as it causes reduced normal tissue toxicities [137, 138]. Particle therapy is dosimetrically superior in reducing the dose to normal tissues in case of longitudinal targets such as the case in CSI [139]. Clinical studies in the most common brain tumors such as meningioma, glioblastoma and low-grade glioma (LGG) have been enlisted in Table 17.12 [140–147]. Since meningiomas have a good long-term survival, proton therapy is used in order to preserve cognition and Quality of Life (QOL). Higher amounts of toxicities were seen with meningiomas with cavernous sinus involvement; however, it is not certain whether it is due to treatment or the disease itself (124). Clinical results with a higher dose (proton therapy or photon-proton combination) in grade II/III meningioma were better than those with lower doses [148]. Also, the toxicities associated with a dose >60 Gy was minimal [149]. Gliomas (gr I–III) often need adjuvant treatment and particle therapy has been studied in these situations as well [150, 151]. Also, particle therapy has aided in conducting dose escalation studies in glioblastoma with minimum toxicities [147].

17.10.7 Thoracic Cancers

Similar to other sites, particle therapy has emerged as an exciting treatment strategy in thoracic tumors for their dosimetric and clinical advantages. By virtue of their location, these tumors originate and invade in close proximity to critical structures such as the heart, thoracic cord, esophagus and uninvolved lung. Despite rapid advances in planning and execution of particle therapy on moving targets like 4D dynamic dose calculation, 4D optimization and rescanning, lung radiotherapy is still challenging since such conformal target planning comes at the expense of the baseline plan robustness. Multiple studies in lung cancer have shown a dosimetric reduction in lung parameters, especially the reduction in low dose bath in the setting of concurrent chemotherapy [152, 153]. Multiple studies have been conducted in lung cancer using proton and car-

Table 17.12 Selected studies on particle therapy for brain tumors

Author	Year	Tumor	*N*	Particles	Dose (GyE)	Outcome
Halasz et al.	2011	Meningioma gr I–II	50	Proton	13 (SRS)	3-year LC: 94%
Combs et al.	2013	Meningioma gr I–III	107	Proton + carbon ion boost	54	2-year LC gr I: 100%, gr II-III: 33%
Murray et al.	2018	Meningioma gr I–III	96	Proton	57	5-year LC: 86.4%, OS: 88.2%
El Shafie et al.	2018	Meningioma gr I–III	110	Protons, carbon ions	47	5-year PFS: 96.6%, 5-year OS—96.2%
Shih et al.	2015	LGG gr II	20	Proton	54	3-year PFS: 85%, 5-year PFS: 40%
Hasegawa et al.	2012	LGG gr II	14	Carbon ion	50.4–55.2	5-year OS: 43%, 10-year OS: 36%
Mizumoto et al.	2010	Glioblastoma	20	Proton	50.4	2-year LC: 15.5%, 2-year OS: 45.3%
Adelberg et al.	2017	HGG	66	Photon/proton	50.4 + 10 (boost)	2-year OS: 40%

N number of patients, *GyE* gray equivalent, *SRS* stereotactic surgery, *LC* local control, *OS* overall survival, *LGG* low-grade glioma, *HGG* high-grade glioma

bon ions have concluded that particle therapy is well tolerated and effective (Table 17.13) [154–162]. However, the only *phase II randomized controlled trial (RCT)* conducted in locally advanced lung cancer and comparing photon vs. proton therapy was negative [157].

Particle therapy has been explored in esophageal cancers in an effort to minimize collateral dose to OARs and decreasing toxicities such as cardiopulmonary and other adverse effects. A RCT compared photon versus proton where there was a significant difference in the risk and severity in adverse effects [161].

17.10.8 Gastrointestinal Cancers

The management in borderline and locally advanced (unresectable) pancreatic cancer consists of local RT with concurrent and sequential chemotherapy, resulting in poor survival outcomes. As local recurrence or progression is a significant pattern of failure, use of IMRT and stereotactic body radiotherapy (SBRT) to deliver escalated doses to the tumor bed/tumor have been proposed. Conventional RT is limited due to the surrounding critical structures. Hence, particle therapy has been pitched in for treatment of pancreatic cancer for better sparing of critical structures along with dose escalation. Salient studies have been summarized in Table 17.14 [163–166].

17.10.9 Breast Cancers

Particle therapy in particle breast irradiation (PBI) has been investigated as a means to minimize the ipsilateral breast dose as well as heart, lung and contralateral breast exposure compared to external-beam photon therapy, with the benefit of being a non-invasive modality compared to brachytherapy options. Furthermore, particle therapy has been used for whole-breast irradiation when regional lymphatics are targeted along with it. Studies in these settings have been summarized in Table 17.15 [167–170].

17.10.10 Genitourinary and Gynecological Cancers

Hypofractionation has been shown to improve the therapeutic ratio in prostate cancer. Particle therapy has been used in prostate cancer to decrease toxicities while delivering moderate and extreme hypofractionation.

Proton therapy provides the dosimetric advantage in reducing bladder/bowel toxicities and

Table 17.13 Selected studies for particle therapy in thoracic cancers

Author	Year	Tumor	*N*	Particles	Dose (GyE)	Outcome
Ohnishi et al.	2020	Early lung cancer	669	Proton	BED 74–131	3-year LC: 90%, 3-year OS: 80%
Chang et al.	2017	Early lung cancer	35	Proton	87.5	5-year LC: 85%, 5-year OS: 28%
Chang et al.	2017	Locally advanced lung cancer	64	Proton	74	5-year LRF: 28%, 5-year OS: 29%
Liao et al.	2018	Locally advanced lung cancer	57	Proton, photon	66 or 74	1-year LF: 11%
Karube et al.	2016	Early lung cancer	70	Carbon ion	28–50 in single fraction	5-year LC: 86%, 5-year OS: 40%
Takahashi et al.	2015	Locally advanced lung cancer	62	Carbon ion	68–72	2-year LC: 93%, 2-year OS: 52%
Xi et al.	2017	Esophageal cancer	132	Proton	50.4	5-year LC: 60%, 5-year OS: 42%
Lin et al.	2020	Esophageal cancer	46	Proton, photon	50.4	3-year OS: 51%
Akutsu et al.	2012	Esophageal cancer (neoadjuvant)	31	Carbon ion	28.8–36.8	pCR; 39%, 5-year PFS: 62%

N number of patients, *GyE* gray equivalent, *BED* biologically equivalent dose, *LC* local control, *OS* overall survival, *LRF* locoregional failure, *pCR* pathological complete response, *PFS* progression-free survival

Table 17.14 Selected studies for particle therapy in pancreatic cancers

Author	Year	Tumor	*N*	Particles	Dose (GyE)	Outcome
Hong et al.	2014	Early pancreatic cancer (pre-operative)	48	Proton	25	2-year LC: 84% (resected patients), 2-year OS: 42%
Murphy et al.	2019	Borderline resectable pancreatic cancer	48	Proton, photon	25–30	2-year PFS: 43% (resected: 55%), 2-year OS: 56% (resected: 72%) R0 resection rate: 65%
Murphy et al.	2019	Locally advanced pancreatic cancer	49	Proton	25–30 or 50.4	Median PFS: 17.5 months, median OS: 31.4 months, R0 resection rate: 69%
Kawashiro et al.	2018	Locally advanced pancreatic cancer	72	Proton	52.8–55.2	2-year LC: 76%, 2-year OS: 46%

N number of patients, *GyE* gray equivalent, *LC* local control, *OS* overall survival, *PFS* progression-free survival

Table 17.15 Selected studies for particle therapy in breast cancer

Author	Year	Tumor	*N*	Particles	Dose (GyE)	Outcome
Bush et al.	2014	Breast cancer (PBI)	100	Proton	40	5-year BTRFS: 97%, 5-year OS: 95%
Galland -Girodet et al.	2014	Breast cancer (PBI)	98	Proton	32	7-year LR: 7%
Verma et al.	2017	Breast cancer (WBI)	91	Proton	50.4	Isolated LR in 2 patients, Acute dermatitis gr 2: 77% and gr 3: 5%
Thorpe et al.	2020	Breast cancer (WBI)	82	Proton	50.4	Isolated LR in 2 patients, Gr3 adverse effects (dermatitis and breast pain): 7.3%

N number of patients, *GyE* gray equivalent, *PBI* partial breast irradiation, *BTRFS* breast tumor recurrence-free survival, *OS* overall survival, *LR* local recurrence, *WBI* whole breast irradiation

Table 17.16 Selected studies in genitourinary and gynecological cancers

Author	Year	Tumor	*N*	Particles	Dose (GyE)	Outcome
Slater et al.	2019	Prostate cancer (LR)	100	Proton	60	3-year BC: 99.3%, Acute toxicities ≥2: 1.7% (GI), 16% (GU)
Henderson et al.	2017	Prostate cancer (LR & IR)	215		70–72.5	5-year BC: 98.3 (LR) & 92.7(IR), No acute toxicities (GI & GU)
Kagei et al.	2003	Uterine cervical cancer	25	Proton to primary tumor (avoiding brachytherapy)	37–101	5-year LC: 75%, Late toxicities ≥gr 4: 4%
Wakatsuki et al.	2014	Uterine cervical cancer	22	Carbon ion	39 (WP) + 15 (residual disease)	5-year LC: 68.2% & OS: 50%, No acute (≥gr 2) or late (≥gr 3) toxicities

N number of patients, *GyE* Gray equivalent, *LR* low risk, *BC* biochemical control, *IR* intermediate risk, *GI* gastrointestinal, *GU* genito-urinary, *WP* whole pelvis, *re-RT* re-irradiation

bone marrow suppression in the context of chemoradiation in uterine cervical cancers. Though there is potential for the use of particle therapy in re-irradiation for gynecological malignancies, there is paucity of data at this time. Selected studies conducted in prostate and uterine cervical cancers have been summarized in Table 17.16 [171–174].

17.11 Economic Considerations

Among the particles used in clinical scenarios, proton has been utilized the most. There has been an exponential increase in the number of proton therapy centers worldwide, especially in the US, which was spurred on the superior dose distribution leading to expected benefits such as reduced toxicities and increased prospects of dose escalation. However, despite the touted advantages of proton therapy, the straggling clinical evidence to support the superiority of proton therapy over conventional RT has questioned the quick adoption of this technology [175, 176]. Indeed, this technology has been often scrutinized by national payers on the same grounds. Unable to remain financially viable and facing bankruptcy, many centers have had to undergo major restructuring.

Cost of construction remains one of the major contributors toward the financial burden of starting a proton therapy center. It includes the expense for large geographic area, adequate shielding, accelerators, multi-story gantries and transportation beam lines. Apart from the capital investment, the other facets of financial demands are that of operating a successful proton therapy facility. They include increased energy requirements, staff (radiation oncologists, physicists, radiotherapy technicians, maintenance staff) and on-site engineering support. This requires the center to have full treatment schedules to remain financially solvent. Furthermore, this reflects in the high reimbursements to offset high capital investment, quality assurance and operational costs [177].

All the more, there is no denying of the potential use of this technology and numerous studies have been conducted to establish the cost-effectiveness (CE) in these cases. In CE studies, the cost of intervention is weighed against the estimated medical costs for the management of their probable adverse effects. Major outcomes of these studies include the quality and quantity of further survival, termed quality-adjusted life-years (QALYs), as well as the costs associated with each QALY gained from each RT modality, termed the incremental cost-effectiveness ratio (ICER). Sensitivity analyses are also performed to know the variability with difference in the input parameters. A review for CE studies for proton versus photon therapy recommended that patients who are most likely to be associated with greater CE are those who experience a high toxicity due to close proximity of critical OARs and who are likely to have a long life expectancy to be able to manifest these side effects [178]. The dosimetric potential can be optimized when such patients are selected. The treatment volume (e.g., CSI in brain tumors) must be considered as an important factor in estimating side effects. Patients with poor prognosis tumors are likely to have a limited CE with proton therapy as they may not live long enough to enjoy the improvement in QALYs. Though proton therapy can be useful for reirradiation cases, this group can possibly be a fourth category depending on anatomical location and life expectancy. Another systematic review concluded that proton therapy is cost-effective in pediatric cancers (owing to decrease in toxicities with a relatively long life) and head and neck cancers [179]. Furthermore, the subset of patients with a left sided breast cancers and large locally advanced lung cancers were cost effective in reducing the toxicities. However, prostate cancer was not cost-effective. Despite these conclusions, the limitations of CE studies must be acknowledged—high dependence on the quality of existing data (toxicities), inability to encompass all potential costs, variation between countries and current studies are based on passive scattering technology (while scanning is used currently).

In conclusion, particle therapy has a potential use in the treatment of cancer. However, the success of such capital-intensive therapies depends on the judicious use of the same on the basis of clear evidence from clinical and CE studies, and collaboration of patients, physicians, national payers and policy makers. National stakeholders must approach this technology and optimize the use of the same so that maximum patients derive its benefits.

Summary (Precepts Box)

1. Heavy particle therapy is a special form of EBRT, including protons, neutrons and carbon ions. The present era of particle therapy is focused mainly on protons.
2. Particles are classified as fermions (matter and bosons (force carriers)).
3. By direct ionization effects, Bragg peak, increased LET and RBE particles impart safer and more effective RT. The main interaction is Coulombic interaction with orbital electrons and atomic nuclei.
4. Protons are considered low LET ions as the LET is only high enough toward the end of the Bragg peak to produce clustered DNA damage, hence making them an ideal candidate for focused RT where the normal tissue lying proximal to the target is treated with low LET (reduced normal tissue toxicities), while the target is treated with high LET (increased cell kill).
5. Particle therapy systems have three main components—particle accelerators (cyclotron and isochronous cyclotrons, synchrocyclotron and synchrotron), energy selection system and beam transport and delivery systems.
6. Absorbed dose deposited by the proton beam increases at the end of the proton range, referred to as Brag Peak. Bragg peak is too narrow to irradiate a target volume, spread out Bragg peak (SOBP) technique is used to overcome this.
7. Three delivery techniques of protons include passive scattering, uniform scanning and pencil beam scanning. Carbon ion delivery techniques include passive scattering and pencil beam scanning.
8. Treatment planning is similar to photons.
9. Motion management techniques are critical and similar to that used in photon therapy.
10. At present time, the clinical superiority of protons compared with photons is established in some pediatric populations and in rare situations when normal structures are in close proximity to treatment volumes. Trials are underway to determine role of protons.

Multiple Choice Questions

1. X-ray was discovered by Wilhelm Röentgen in
   (a) 1895
   (b) 1896
   (c) 1897
   (d) 1795
2. Who discovered protons?
   (a) Ernest O Lawrence
   (b) Ernest Rutherford
   (c) Hans Geiger
   (d) Ernest Marsden
3. Sir James Chadwick discovered
   (a) Photons
   (b) Protons
   (c) Nucleus
   (d) Neutrons
4. Who is the father of proton therapy?
   (a) Robert Wilson
   (b) John Lawrence
   (c) Cornelius Tobias
   (d) William Bragg
5. RBE of protons is
   (a) 0.1
   (b) 1.1
   (c) 2.1
   (d) 3.1
6. Stopping power depends on
   (a) Mass of incident particle
   (b) Velocity of incident particle
   (c) Both the above
   (d) None of the above
7. What are components of particle therapy systems?
   (a) Particle accelerators
   (b) Beam transport systems
   (c) Treatment delivery unit
   (d) All the above
8. First cyclotron was developed in
   (a) 1930
   (b) 1940
   (c) 1950
   (d) 1980
9. A monoenergetic beam is modulated and modified by various external accessories to a treat a tumor proton treatment delivery technique called
   (a) Passive scattering
   (b) Pencil beam scanning
   (c) Uniform scanning
   (d) Variable scanning

Answers

1. (a)
2. (b)
3. (d)
4. (a)
5. (b)
6. (c)
7. (d)
8. (a)
9. (a)

References

1. Atun R, Jaffray DA, Barton MB, Bray F, Baumann M, Vikram B, et al. Expanding global access to radiotherapy. Lancet Oncol. 2015;16(10):1153–86.
2. Halperin EC. Perez & Brady's principles and practice of radiation oncology, 7e. 2019 [cited 2022 Aug 22]. https://oncology.lwwhealthlibrary.com/book.aspx?bookid=2524.
3. Oerter R. The theory of almost everything: the standard model, the unsung triumph of modern physics. Penguin Group; 2006.
4. Lawrence EO, Livingston MS. The production of high speed protons without the use of high voltages. Phys Rev. 1931;38(4):834.
5. Giap H, Giap B. Historical perspective and evolution of charged particle beam therapy. Transl Cancer Res. 2012 [cited 2022 Aug 4];1(3). https://tcr.amegroups.com/article/view/595.
6. Geiger H, Rutherford E. The scattering of α-particles by matter. Proc R Soc Lond Ser Contain Pap Math Phys Character. 1910;83(565):492–504.
7. Chadwick J. The existence of a neutron. Proc R Soc Lond Ser Contain Pap Math Phys Character. 1932;136(830):692–708.
8. Linstadt DE, Castro JR, Phillips TL. Neon ion radiotherapy: results of the phase I/II clinical trial. Int J Radiat Oncol Biol Phys. 1991;20(4):761–9.
9. Castro JR. Results of heavy ion radiotherapy. Radiat Environ Biophys. 1995;34(1):45–8.
10. Pompos A, Foote RL, Koong AC, Le QT, Mohan R, Paganetti H, et al. National effort to re-establish heavy ion cancer therapy in the United

States. Front Oncol. 2022 [cited 2022 Aug 10];12. https://www.frontiersin.org/articles/10.3389/fonc.2022.880712.
11. Jäkel O, Kraft G, Karger CP. The history of ion beam therapy in Germany. Z Für Med Phys. 2022;32(1):6–22.
12. Haberer T, Becher W, Schardt D, Kraft G. Magnetic scanning system for heavy ion therapy. Nucl Instrum Methods Phys Res A. 1993;(330):296–305.
13. Taylor HJ, Goldhaber M. Detection of nuclear disintegration in a photographic emulsion. Nature. 1935;135:341.
14. Kruger PG. Some biological effects of nuclear disintegration products on neoplastic tissue. Proc Natl Acad Sci U S A. 1940;26(3):181–92.
15. PTCOG—Home. [cited 2022 Aug 10]. https://www.ptcog.ch/index.php.
16. ICRU Report 46, Photon, electron, proton and neutron interaction data for body tissues—ICRU. [cited 2022 Aug 22]. https://www.icru.org/report/photon-electron-proton-and-neutron-interaction-data-for-body-tissues-report-46/.
17. ICRU—Report 37—stopping powers for electrons and positrons | Engineering360. [cited 2022 Aug 22]. https://standards.globalspec.com/std/367449/REPORT%2037.
18. Deasy J. ICRU Report 49, stopping powers and ranges for protons and alpha particles. Med Phys. 1994;21(5):709–10.
19. Highland VL. Some practical remarks on multiple scattering. Nucl Instrum Methods. 1975;129(2):497–9.
20. Eyges L. Multiple scattering with energy loss. Phys Rev. 1948;74(10):1534–5.
21. Deluca P, Wambersie A, Whitmore G. Prescribing, recording, and reporting proton-beam therapy. J ICRU. 2007;7:1–210.
22. Mullins EA, Rodriguez AA, Bradley NP, Eichman BF. Emerging roles of DNA glycosylases and the base excision repair pathway. Trends Biochem Sci. 2019;44(9):765–81.
23. Scully R, Panday A, Elango R, Willis NA. DNA double-strand break repair-pathway choice in somatic mammalian cells. Nat Rev Mol Cell Biol. 2019;20(11):698–714.
24. Shrivastav M, De Haro LP, Nickoloff JA. Regulation of DNA double-strand break repair pathway choice. Cell Res. 2008;18(1):134–47.
25. Timm S, Lorat Y, Jakob B, Taucher-Scholz G, Rübe CE. Clustered DNA damage concentrated in particle trajectories causes persistent large-scale rearrangements in chromatin architecture. Radiother Oncol. 2018;129(3):600–10.
26. Zhang X, Ye C, Sun F, Wei W, Hu B, Wang J. Both complexity and location of DNA damage contribute to cellular senescence induced by ionizing radiation. PLoS One. 2016;11(5):e0155725.
27. Averbeck NB, Ringel O, Herrlitz M, Jakob B, Durante M, Taucher-Scholz G. DNA end resection is needed for the repair of complex lesions in G1-phase human cells. Cell Cycle Georget Tex. 2014;13(16):2509–16.
28. Lomax ME, Folkes LK, O'Neill P. Biological consequences of radiation-induced DNA damage: relevance to radiotherapy. Clin Oncol R Coll Radiol G B. 2013;25(10):578–85.
29. Ward JF. Complexity of damage produced by ionizing radiation. Cold Spring Harb Symp Quant Biol. 2000;65:377–82.
30. Malouff TD, Trifiletti DM. Principles and practice of particle therapy | Wiley. Wiley.com. [cited 2022 Aug 4]. https://www.wiley.com/en-us/Principles+and+Practice+of+Particle+Therapy-p-9781119707523.
31. Ando K, Goodhead DT. Dependence and independence of survival parameters on linear energy transfer in cells and tissues. J Radiat Res (Tokyo). 2016;57(6):596–606.
32. Suzuki M, Kase Y, Kanai T, Ando K. Change in radiosensitivity with fractionated-dose irradiation of carbon-ion beams in five different human cell lines. Int J Radiat Oncol Biol Phys. 2000;48(1):251–8.
33. Furusawa Y, Fukutsu K, Aoki M, Itsukaichi H, Eguchi-Kasai K, Ohara H, et al. Inactivation of aerobic and hypoxic cells from three different cell lines by accelerated (3)He-, (12)C- and (20)Ne-ion beams. Radiat Res. 2000 [cited 2022 Aug 23];154(5). https://pubmed.ncbi.nlm.nih.gov/11025645/.
34. Tobias CA, Blakely EA, Alpen EL, Castro JR, Ainsworth EJ, Curtis SB, et al. Molecular and cellular radiobiology of heavy ions. Int J Radiat Oncol Biol Phys. 1982;8(12):2109–20.
35. Barendsen GW, Koot CJ, Van Kersen GR, Bewley DK, Field SB, Parnell CJ. The effect of oxygen on impairment of the proliferative capacity of human cells in culture by ionizing radiations of different LET. Int J Radiat Biol Relat Stud Phys Chem Med. 1966;10(4):317–27.
36. Fournier C, Wiese C, Taucher-Scholz G. Accumulation of the cell cycle regulators TP53 and CDKN1A (p21) in human fibroblasts after exposure to low- and high-LET radiation. Radiat Res. 2004;161(6):675–84.
37. Chang PY, Bjornstad KA, Rosen CJ, McNamara MP, Mancini R, Goldstein LE, et al. Effects of iron ions, protons and X rays on human lens cell differentiation. Radiat Res. 2005;164(4 Pt 2):531–9.
38. Iwadate Y, Mizoe J, Osaka Y, Yamaura A, Tsujii H. High linear energy transfer carbon radiation effectively kills cultured glioma cells with either mutant or wild-type p53. Int J Radiat Oncol Biol Phys. 2001;50(3):803–8.
39. Tsuboi K, Moritake T, Tsuchida Y, Tokuuye K, Matsumura A, Ando K. Cell cycle checkpoint and apoptosis induction in glioblastoma cells and fibroblasts irradiated with carbon beam. J Radiat Res (Tokyo). 2007;48(4):317–25.
40. Zhou J, Wang G, Chen Y, Wang H, Hua Y, Cai Z. Immunogenic cell death in cancer therapy: present and emerging inducers. J Cell Mol Med. 2019;23(8):4854–65.

41. Higo M, Uzawa K, Kawata T, Kato Y, Kouzu Y, Yamamoto N, et al. Enhancement of SPHK1 in vitro by carbon ion irradiation in oral squamous cell carcinoma. Int J Radiat Oncol Biol Phys. 2006;65(3):867–75.
42. Kraft G, Krämer M, Scholz M. LET, track structure and models. A review. Radiat Environ Biophys. 1992;31(3):161–80.
43. Friedrich T, Scholz U, Elsässer T, Durante M, Scholz M. Calculation of the biological effects of ion beams based on the microscopic spatial damage distribution pattern. Int J Radiat Biol. 2012;88(1–2):103–7.
44. Kanai T, Endo M, Minohara S, Miyahara N, Koyama-ito H, Tomura H, et al. Biophysical characteristics of HIMAC clinical irradiation system for heavy-ion radiation therapy. Int J Radiat Oncol Biol Phys. 1999;44(1):201–10.
45. Kanai T, Matsufuji N, Miyamoto T, Mizoe J, Kamada T, Tsuji H, et al. Examination of GyE system for HIMAC carbon therapy. Int J Radiat Oncol Biol Phys. 2006;64(2):650–6.
46. Hawkins RB. A microdosimetric-kinetic theory of the dependence of the RBE for cell death on LET. Med Phys. 1998;25(7 Pt 1):1157–70.
47. ICRU Report 93: prescribing, recording, and reporting light ion beam therapy—ICRU. [cited 2022 Aug 23]. https://www.icru.org/report/icru-report-93-prescribing-recording-and-reporting-light-ion--beam-therapy/.
48. Bassler N, Jäkel O, Søndergaard CS, Petersen JB. Dose- and LET-painting with particle therapy. Acta Oncol Stockh Swed. 2010;49(7):1170–6.
49. Scifoni E, Tinganelli W, Weyrather WK, Durante M, Maier A, Krämer M. Including oxygen enhancement ratio in ion beam treatment planning: model implementation and experimental verification. Phys Med Biol. 2013;58(11):3871–95.
50. Formenti SC, Demaria S. Radiation therapy to convert the tumor into an in situ vaccine. Int J Radiat Oncol Biol Phys. 2012;84(4):879–80.
51. Huang Y, Dong Y, Zhao J, Zhang L, Kong L, Lu JJ. Comparison of the effects of photon, proton and carbon-ion radiation on the ecto-calreticulin exposure in various tumor cell lines. Ann Transl Med. 2019;7(20):542.
52. Thomas LH. The paths of ions in the cyclotron I. Orbits in the magnetic field. Phys Rev. 1938;54(8):580–8.
53. Kerst DW. Constant frequency cyclotrons with spiral ridged poles. University of Illinois and Midwestern Universities Research Association: University of Illinois; 1955. Report No.: 64.
54. McMillan EM. The synchrotron—a proposed high energy particle accelerator. Phys Rev. 1945;68(5–6):143–4.
55. Veksler VI. A new method of acceleration of relativistic particles. J Phys. 1945;9:153.
56. Degiovanni ASP, Ungaro D. A linear accelerator for proton therapy. Chicago, IL, USA.
57. Urie M, Goitein M, Holley WR, Chen GT. Degradation of the Bragg peak due to inhomogeneities. Phys Med Biol. 1986;31(1):1–15.
58. Smith A, Gillin M, Bues M, Zhu XR, Suzuki K, Mohan R, et al. The M. D. Anderson proton therapy system. Med Phys. 2009;36(9):4068–83.
59. Kooy HM, Clasie BM, Lu HM, Madden TM, Bentefour H, Depauw N, et al. A case study in proton pencil-beam scanning delivery. Int J Radiat Oncol Biol Phys. 2010;76(2):624–30.
60. Klimpki G, Zhang Y, Fattori G, Psoroulas S, Weber DC, Lomax A, et al. The impact of pencil beam scanning techniques on the effectiveness and efficiency of rescanning moving targets. Phys Med Biol. 2018;63(14):145006.
61. Sun XS, Li YJ, Liu JY, Liao WT, Wu C, Pu YH. Shortening the delivery time of proton therapy by real-time compensation method with raster scanning. Nucl Sci Tech. 2022;33(6):73.
62. Kanai T, Kanematsu N, Minohara S, Komori M, Torikoshi M, Asakura H, et al. Commissioning of a conformal irradiation system for heavy-ion radiotherapy using a layer-stacking method. Med Phys. 2006;33(8):2989–97.
63. Futami Y, Kanai T, Fujita M, Tomura H, Higashi A, Matsufuji N, et al. Broad-beam three-dimensional irradiation system for heavy-ion radiotherapy at HIMAC. Nucl Instrum Methods Phys Res Sect Accel Spectrometers Detect Assoc Equip. 1999;430(1):143–53.
64. Noda K. Beam delivery method for carbon-ion radiotherapy with the heavy-ion medical accelerator in Chiba. Int J Part Ther. 2016;2(4):481–9.
65. Hara Y, Furukawa T, Mizushima K, Inaniwa T, Saotome N, Tansho R, et al. Commissioning of full energy scanning irradiation with carbon-ion beams ranging from 55.6 to 430MeV/u at the NIRS-HIMAC. Nucl Instrum Methods Phys Res Sect B Beam Interact Mater At. 2017;406:343–6.
66. Furukawa T, Hara Y, Mizushima K, Saotome N, Tansho R, Saraya Y, et al. Development of NIRS pencil beam scanning system for carbon ion radiotherapy. Nucl Instrum Methods Phys Res Sect B Beam Interact Mater At. 2017;406:361–7.
67. Bussière MR, Adams JA. Treatment planning for conformal proton radiation therapy. Technol Cancer Res Treat. 2003;2(5):389–99.
68. Saini J, Traneus E, Maes D, Regmi R, Bowen SR, Bloch C, et al. Advanced proton beam dosimetry part I: review and performance evaluation of dose calculation algorithms. Transl Lung Cancer Res. 2018;7(2):171–9.
69. Hong L, Goitein M, Bucciolini M, Comiskey R, Gottschalk B, Rosenthal S, et al. A pencil beam algorithm for proton dose calculations. Phys Med Biol. 1996;41(8):1305–30.
70. Schaffner B, Pedroni E, Lomax A. Dose calculation models for proton treatment planning using a dynamic beam delivery system: an attempt to include density heterogeneity effects in the analytical dose calculation. Phys Med Biol. 1999;44(1):27–41.
71. Paganetti H, Jiang H, Parodi K, Slopsema R, Engelsman M. Clinical implementation of full Monte Carlo dose calculation in proton beam therapy. Phys Med Biol. 2008;53(17):4825–53.

72. Paganetti H. Monte Carlo simulations will change the way we treat patients with proton beams today. Br J Radiol. 2014;87(1040):20140293.
73. Albertini F, Hug EB, Lomax AJ. Is it necessary to plan with safety margins for actively scanned proton therapy? Phys Med Biol. 2011;56(14):4399–413.
74. Paganetti H. Range uncertainties in proton therapy and the role of Monte Carlo simulations. Phys Med Biol. 2012;57(11):R99–117.
75. Vyfhuis MAL, Onyeuku N, Diwanji T, Mossahebi S, Amin NP, Badiyan SN, et al. Advances in proton therapy in lung cancer. Ther Adv Respir Dis. 2018;12:1753466618783878.
76. Liu W, Zhang X, Li Y, Mohan R. Robust optimization of intensity modulated proton therapy. Med Phys. 2012;39(2):1079–91.
77. Zhang X. A review of the robust optimization process and advances with Monte Carlo in the proton therapy management of head and neck tumors. Int J Part Ther. 2021;8(1):14–24.
78. Liu C, Patel SH, Shan J, Schild SE, Vargas CE, Wong WW, et al. Robust optimization for intensity modulated proton therapy to redistribute high linear energy transfer from nearby critical organs to tumors in head and neck cancer. Int J Radiat Oncol Biol Phys. 2020;107(1):181–93.
79. Engwall E, Glimelius L, Hynning E. Effectiveness of different rescanning techniques for scanned proton radiotherapy in lung cancer patients. Phys Med Biol. 2018;63(9):095006.
80. Seco J, Robertson D, Trofimov A, Paganetti H. Breathing interplay effects during proton beam scanning: simulation and statistical analysis. Phys Med Biol. 2009;54(14):N283–94.
81. Mein S, Kopp B, Vela A, Dutheil P, Lesueur P, Stefan D, et al. How can we consider variable RBE and LETd prediction during clinical practice? A pediatric case report at the Normandy Proton Therapy Centre using an independent dose engine. Radiat Oncol. 2022;17(1):23.
82. Yock TI, Yeap BY, Ebb DH, Weyman E, Eaton BR, Sherry NA, et al. Long-term toxic effects of proton radiotherapy for paediatric medulloblastoma: a phase 2 single-arm study. Lancet Oncol. 2016;17(3):287–98.
83. MacDonald SM, Safai S, Trofimov A, Wolfgang J, Fullerton B, Yeap BY, et al. Proton radiotherapy for childhood ependymoma: initial clinical outcomes and dose comparisons. Int J Radiat Oncol Biol Phys. 2008;71(4):979–86.
84. MacDonald SM, Trofimov A, Safai S, Adams J, Fullerton B, Ebb D, et al. Proton radiotherapy for pediatric central nervous system germ cell tumors: early clinical outcomes. Int J Radiat Oncol Biol Phys. 2011;79(1):121–9.
85. Winkfield KM, Linsenmeier C, Yock TI, Grant PE, Yeap BY, Butler WE, et al. Surveillance of craniopharyngioma cyst growth in children treated with proton radiotherapy. Int J Radiat Oncol Biol Phys. 2009;73(3):716–21.
86. Jimenez RB, Ahmed S, Johnson A, Thomas H, Depauw N, Horick N, et al. Proton radiation therapy for pediatric craniopharyngioma. Int J Radiat Oncol Biol Phys. 2021;110(5):1480–7.
87. Rombi B, DeLaney TF, MacDonald SM, Huang MS, Ebb DH, Liebsch NJ, et al. Proton radiotherapy for pediatric Ewing's sarcoma: initial clinical outcomes. Int J Radiat Oncol Biol Phys. 2012;82(3):1142–8.
88. Uezono H, Indelicato DJ, Rotondo RL, Mailhot Vega RB, Bradfield SM, Morris CG, et al. Treatment outcomes after proton therapy for Ewing sarcoma of the pelvis. Int J Radiat Oncol Biol Phys. 2020;107(5):974–81.
89. Kharod SM, Indelicato DJ, Rotondo RL, Mailhot Vega RB, Uezono H, Morris CG, et al. Outcomes following proton therapy for Ewing sarcoma of the cranium and skull base. Pediatr Blood Cancer. 2020;67(2):e28080.
90. Ladra MM, Szymonifka JD, Mahajan A, Friedmann AM, Yong Yeap B, Goebel CP, et al. Preliminary results of a phase II trial of proton radiotherapy for pediatric rhabdomyosarcoma. J Clin Oncol. 2014;32(33):3762–70.
91. Mizumoto M, Murayama S, Akimoto T, Demizu Y, Fukushima T, Ishida Y, et al. Preliminary results of proton radiotherapy for pediatric rhabdomyosarcoma: a multi-institutional study in Japan. Cancer Med. 2018;7(5):1870–4.
92. Indelicato DJ, Rotondo RL, Krasin MJ, Vega RBM, Uezono H, Bradfield S, et al. Outcomes following proton therapy for Group III pelvic rhabdomyosarcoma. Int J Radiat Oncol Biol Phys. 2020;106(5):968–76.
93. Buszek SM, Ludmir EB, Grosshans DR, McAleer MF, McGovern SL, Harrison DJ, et al. Disease control and patterns of failure after proton beam therapy for rhabdomyosarcoma. Int J Radiat Oncol Biol Phys. 2021;109(3):718–25.
94. Hill-Kayser CE, Tochner Z, Li Y, Kurtz G, Lustig RA, James P, et al. Outcomes after proton therapy for treatment of pediatric high-risk neuroblastoma. Int J Radiat Oncol Biol Phys. 2019;104(2):401–8.
95. Hishiki T, Matsumoto K, Ohira M, Kamijo T, Shichino H, Kuroda T, et al. Results of a phase II trial for high-risk neuroblastoma treatment protocol JN-H-07: a report from the Japan Childhood Cancer Group Neuroblastoma Committee (JNBSG). Int J Clin Oncol. 2018;23(5):965–73.
96. Fung V, Calugaru V, Bolle S, Mammar H, Alapetite C, Maingon P, et al. Proton beam therapy for skull base chordomas in 106 patients: a dose adaptive radiation protocol. Radiother Oncol. 2018;128(2):198–202.
97. Deraniyagala RL, Yeung D, Mendenhall WM, Li Z, Morris CG, Mendenhall NP, et al. Proton therapy for skull base chordomas: an outcome study from the University of Florida Proton Therapy Institute. J Neurol Surg Part B Skull Base. 2014;75(1):53–7.
98. Lu VM, O'Connor KP, Mahajan A, Carlson ML, Van Gompel JJ. Carbon ion radiotherapy for skull

base chordomas and chondrosarcomas: a systematic review and meta-analysis of local control, survival, and toxicity outcomes. J Neuro-Oncol. 2020;147(3):503–13.
99. Koto M, Hasegawa A, Takagi R, Fujikawa A, Morikawa T, Kishimoto R, et al. Risk factors for brain injury after carbon ion radiotherapy for skull base tumors. Radiother Oncol. 2014;111(1):25–9.
100. Iannalfi A, D'Ippolito E, Riva G, Molinelli S, Gandini S, Viselner G, et al. Proton and carbon ion radiotherapy in skull base chordomas: a prospective study based on a dual particle and a patient-customized treatment strategy. Neuro Oncol. 2020;22(9):1348–58.
101. DeLaney TF, Liebsch NJ, Pedlow FX, Adams J, Dean S, Yeap BY, et al. Phase II study of high-dose photon/proton radiotherapy in the management of spine sarcomas. Int J Radiat Oncol Biol Phys. 2009;74(3):732–9.
102. Chen YL, Liebsch N, Kobayashi W, Goldberg S, Kirsch D, Calkins G, et al. Definitive high-dose photon/proton radiotherapy for unresected mobile spine and sacral chordomas. Spine. 2013;38(15):E930–6.
103. Imai R, Kamada T, Araki N, Working Group for Bone and Soft Tissue Sarcomas. Carbon ion radiation therapy for unresectable sacral chordoma: an analysis of 188 cases. Int J Radiat Oncol Biol Phys. 2016;95(1):322–7.
104. Schulz-Ertner D, Karger CP, Feuerhake A, Nikoghosyan A, Combs SE, Jäkel O, et al. Effectiveness of carbon ion radiotherapy in the treatment of skull-base chordomas. Int J Radiat Oncol Biol Phys. 2007;68(2):449–57.
105. Uhl M, Mattke M, Welzel T, Roeder F, Oelmann J, Habl G, et al. Highly effective treatment of skull base chordoma with carbon ion irradiation using a raster scan technique in 155 patients: first long-term results. Cancer. 2014;120(21):3410–7.
106. Weber DC, Badiyan S, Malyapa R, Albertini F, Bolsi A, Lomax AJ, et al. Long-term outcomes and prognostic factors of skull-base chondrosarcoma patients treated with pencil-beam scanning proton therapy at the Paul Scherrer Institute. Neuro Oncol. 2016;18(2):236–43.
107. Uhl M, Mattke M, Welzel T, Oelmann J, Habl G, Jensen AD, et al. High control rate in patients with chondrosarcoma of the skull base after carbon ion therapy: first report of long-term results. Cancer. 2014;120(10):1579–85.
108. Hrbacek J, Mishra KK, Kacperek A, Dendale R, Nauraye C, Auger M, et al. Practice patterns analysis of ocular proton therapy centers: the International OPTIC Survey. Int J Radiat Oncol Biol Phys. 2016;95(1):336–43.
109. Wang Z, Nabhan M, Schild SE, Stafford SL, Petersen IA, Foote RL, et al. Charged particle radiation therapy for uveal melanoma: a systematic review and meta-analysis. Int J Radiat Oncol Biol Phys. 2013;86(1):18–26.
110. Allen AM, Pawlicki T, Dong L, Fourkal E, Buyyounouski M, Cengel K, et al. An evidence based review of proton beam therapy: the report of ASTRO's emerging technology committee. Radiother Oncol. 2012;103(1):8–11.
111. Toyama S, Tsuji H, Mizoguchi N, Nomiya T, Kamada T, Tokumaru S, et al. Long-term results of carbon ion radiation therapy for locally advanced or unfavorably located choroidal melanoma: usefulness of CT-based 2-port orthogonal therapy for reducing the incidence of neovascular glaucoma. Int J Radiat Oncol Biol Phys. 2013;86(2):270–6.
112. Munzenrider JE, Verhey LJ, Gragoudas ES, Seddon JM, Urie M, Gentry R, et al. Conservative treatment of uveal melanoma: local recurrence after proton beam therapy. Int J Radiat Oncol Biol Phys. 1989;17(3):493–8.
113. Dendale R, Lumbroso-Le Rouic L, Noel G, Feuvret L, Levy C, Delacroix S, et al. Proton beam radiotherapy for uveal melanoma: results of Curie Institut-Orsay proton therapy center (ICPO). Int J Radiat Oncol Biol Phys. 2006;65(3):780–7.
114. Fuss M, Loredo LN, Blacharski PA, Grove RI, Slater JD. Proton radiation therapy for medium and large choroidal melanoma: preservation of the eye and its functionality. Int J Radiat Oncol Biol Phys. 2001;49(4):1053–9.
115. Chan RVP, Yonekawa Y, Lane AM, Skondra D, Munzenrider JE, Collier JM, et al. Proton beam irradiation using a light-field technique for the treatment of choroidal hemangiomas. Ophthalmologica. 2010;224(4):209–16.
116. Levy-Gabriel C, Rouic LLL, Plancher C, Dendale R, Delacroix S, Asselain B, et al. Long-term results of low-dose proton beam therapy for circumscribed choroidal hemangiomas. Retina. 2009;29(2):170–5.
117. Tsina EK, Lane AM, Zacks DN, Munzenrider JE, Collier JM, Gragoudas ES. Treatment of metastatic tumors of the choroid with proton beam irradiation. Ophthalmology. 2005;112(2):337–43.
118. Schmitt G, Mills EE, Levin V, Pape H, Smit BJ, Zamboglou N. The role of neutrons in the treatment of soft tissue sarcomas. Cancer. 1989;64(10):2064–8.
119. Kamada T, Tsujii H, Tsuji H, Yanagi T, Mizoe J, Miyamoto T, et al. Efficacy and safety of carbon ion radiotherapy in bone and soft tissue sarcomas. J Clin Oncol. 2002;20(22):4466–71.
120. Imai R, Kamada T, Araki N. Carbon ion radiotherapy for unresectable localized axial soft tissue sarcoma. Cancer Med. 2018;7(9):4308–14.
121. Demizu Y, Jin D, Sulaiman NS, Nagano F, Terashima K, Tokumaru S, et al. Particle therapy using protons or carbon ions for unresectable or incompletely resected bone and soft tissue sarcomas of the pelvis. Int J Radiat Oncol Biol Phys. 2017;98(2):367–74.
122. Kepka L, DeLaney TF, Suit HD, Goldberg SI. Results of radiation therapy for unresected soft-tissue sarcomas. Int J Radiat Oncol Biol Phys. 2005;63(3):852–9.

123. Weber DC, Rutz HP, Bolsi A, Pedroni E, Coray A, Jermann M, et al. Spot scanning proton therapy in the curative treatment of adult patients with sarcoma: the Paul Scherrer institute experience. Int J Radiat Oncol Biol Phys. 2007;69(3):865–71.
124. Yanagi T, Kamada T, Tsuji H, Imai R, Serizawa I, Tsujii H. Dose-volume histogram and dose-surface histogram analysis for skin reactions to carbon ion radiotherapy for bone and soft tissue sarcoma. Radiother Oncol. 2010;95(1):60–5.
125. Ikawa H, Koto M, Demizu Y, Saitoh JI, Suefuji H, Okimoto T, et al. Multicenter study of carbon-ion radiation therapy for nonsquamous cell carcinomas of the oral cavity. Cancer Med. 2019;8(12):5482–91.
126. Blanchard P, Garden AS, Gunn GB, Rosenthal DI, Morrison WH, Hernandez M, et al. Intensity-modulated proton beam therapy (IMPT) versus intensity-modulated photon therapy (IMRT) for patients with oropharynx cancer—a case matched analysis. Radiother Oncol. 2016;120(1):48–55.
127. Chan A, Adams JA, Weyman E, Parambi R, Goldsmith T, Holman A, et al. A phase II trial of proton radiation therapy with chemotherapy for nasopharyngeal carcinoma. Int J Radiat Oncol Biol Phys. 2012;84(3):S151–2.
128. Holliday EB, Garden AS, Rosenthal DI, Fuller CD, Morrison WH, Gunn GB, et al. Proton therapy reduces treatment-related toxicities for patients with nasopharyngeal cancer: a case-match control study of intensity-modulated proton therapy and intensity-modulated photon therapy. Int J Part Ther. 2015;2(1):19–28.
129. Akbaba S, Held T, Lang K, Forster T, Federspil P, Herfarth K, et al. Bimodal radiotherapy with active raster-scanning carbon ion radiotherapy and intensity-modulated radiotherapy in high-risk nasopharyngeal carcinoma results in excellent local control. Cancers. 2019;11(3):E379.
130. Yu NY, Gamez ME, Hartsell WF, Tsai HK, Laramore GE, Larson GL, et al. A multi-institutional experience of proton beam therapy for sinonasal tumors. Adv Radiat Oncol. 2019;4(4):689–98.
131. Romesser PB, Cahlon O, Scher ED, Hug EB, Sine K, DeSelm C, et al. Proton beam re-irradiation for recurrent head and neck cancer: multi-institutional report on feasibility and early outcomes. Int J Radiat Oncol Biol Phys. 2016;95(1):386–95.
132. Held T, Windisch P, Akbaba S, Lang K, El Shafie R, Bernhardt D, et al. Carbon ion reirradiation for recurrent head and neck cancer: a single-institutional experience. Int J Radiat Oncol Biol Phys. 2019;105(4):803–11.
133. Takayama K, Nakamura T, Takada A, Makita C, Suzuki M, Azami Y, et al. Treatment results of alternating chemoradiotherapy followed by proton beam therapy boost combined with intra-arterial infusion chemotherapy for stage III-IVB tongue cancer. J Cancer Res Clin Oncol. 2016;142(3):659–67.
134. Sio TT, Lin HK, Shi Q, Gunn GB, Cleeland CS, Lee JJ, et al. Intensity modulated proton therapy versus intensity modulated photon radiation therapy for oropharyngeal cancer: first comparative results of patient-reported outcomes. Int J Radiat Oncol Biol Phys. 2016;95(4):1107–14.
135. Chan AW, Liebsch LJ, Deschler DG, Adams JA, Vrishali LV, McIntyre JF, et al. Proton radiotherapy for T4 nasopharyngeal carcinoma. J Clin Oncol. 2004;22(14_suppl):5574.
136. Patel SH, Wang Z, Wong WW, Murad MH, Buckey CR, Mohammed K, et al. Charged particle therapy versus photon therapy for paranasal sinus and nasal cavity malignant diseases: a systematic review and meta-analysis. Lancet Oncol. 2014;15(9):1027–38.
137. Moncharmont C, Levy A, Guy JB, Falk AT, Guilbert M, Trone JC, et al. Radiation-enhanced cell migration/invasion process: a review. Crit Rev Oncol Hematol. 2014;92(2):133–42.
138. Boehling NS, Grosshans DR, Bluett JB, Palmer MT, Song X, Amos RA, et al. Dosimetric comparison of three-dimensional conformal proton radiotherapy, intensity-modulated proton therapy, and intensity-modulated radiotherapy for treatment of pediatric craniopharyngiomas. Int J Radiat Oncol Biol Phys. 2012;82(2):643–52.
139. Adeberg S, Harrabi SB, Bougatf N, Verma V, Windisch P, Bernhardt D, et al. Dosimetric comparison of proton radiation therapy, volumetric modulated arc therapy, and three-dimensional conformal radiotherapy based on intracranial tumor location. Cancers. 2018;10(11):E401.
140. Seravalli E, Bosman M, Lassen-Ramshad Y, Vestergaard A, Oldenburger F, Visser J, et al. Dosimetric comparison of five different techniques for craniospinal irradiation across 15 European centers: analysis on behalf of the SIOP-E-BTG (radiotherapy working group). Acta Oncol Stockh Swed. 2018;57(9):1240–9.
141. Halasz LM, Bussière MR, Dennis ER, Niemierko A, Chapman PH, Loeffler JS, et al. Proton stereotactic radiosurgery for the treatment of benign meningiomas. Int J Radiat Oncol Biol Phys. 2011;81(5):1428–35.
142. Combs SE, Welzel T, Habermehl D, Rieken S, Dittmar JO, Kessel K, et al. Prospective evaluation of early treatment outcome in patients with meningiomas treated with particle therapy based on target volume definition with MRI and 68Ga-DOTATOC-PET. Acta Oncol. 2013;52(3):514–20.
143. Murray FR, Snider JW, Bolsi A, Lomax AJ, Walser M, Kliebsch U, et al. Long-term clinical outcomes of pencil beam scanning proton therapy for benign and non-benign intracranial meningiomas. Int J Radiat Oncol Biol Phys. 2017;99(5):1190–8.
144. El Shafie RA, Czech M, Kessel KA, Habermehl D, Weber D, Rieken S, et al. Clinical outcome after particle therapy for meningiomas of the skull base: toxicity and local control in patients treated with active rasterscanning. Radiat Oncol. 2018;13(1):54.
145. Shih HA, Sherman JC, Nachtigall LB, Colvin MK, Fullerton BC, Daartz J, et al. Proton therapy for

low-grade gliomas: Results from a prospective trial. Cancer. 2015;121(10):1712–9.
146. Hasegawa A, Mizoe J, Tsujii H, Kamada T, Jingu K, Iwadate Y, et al. Experience with carbon ion radiotherapy for WHO Grade 2 diffuse astrocytomas. Int J Radiat Oncol Biol Phys. 2011;83:100–6.
147. Mizumoto M, Tsuboi K, Igaki H, Yamamoto T, Takano S, Oshiro Y, et al. Phase I/II trial of hyperfractionated concomitant boost proton radiotherapy for supratentorial glioblastoma multiforme. Int J Radiat Oncol Biol Phys. 2010;77(1):98–105.
148. Adeberg S, Bernhardt D, Harrabi SB, Uhl M, Paul A, Bougatf N, et al. Sequential proton boost after standard chemoradiation for high-grade glioma. Radiother Oncol. 2017;125(2):266–72.
149. Hug EB, Devries A, Thornton AF, Munzenride JE, Pardo FS, Hedley-Whyte ET, et al. Management of atypical and malignant meningiomas: role of high-dose, 3D-conformal radiation therapy. J Neurooncol. 2000;48(2):151–60.
150. Chan AW, Bernstein KD, Adams JA, Parambi RJ, Loeffler JS. Dose escalation with proton radiation therapy for high-grade meningiomas. Technol Cancer Res Treat. 2012;11(6):607–14.
151. Harrabi SB, Bougatf N, Mohr A, Haberer T, Herfarth K, Combs SE, et al. Dosimetric advantages of proton therapy over conventional radiotherapy with photons in young patients and adults with low-grade glioma. Strahlenther Onkol. 2016;192(11):759–69.
152. Dennis ER, Bussiere MR, Niemierko A, Lu MW, Fullerton BC, Loeffler JS, et al. A comparison of critical structure dose and toxicity risks in patients with low grade gliomas treated with IMRT versus proton radiation therapy. Technol Cancer Res Treat. 2013;12(1):1–9.
153. Berman AT, Teo BKK, Dolney D, Swisher-McClure S, Shahnazi K, Both S, et al. An in-silico comparison of proton beam and IMRT for postoperative radiotherapy in completely resected stage IIIA non-small cell lung cancer. Radiat Oncol. 2013;8:144.
154. Hoppe BS, Huh S, Flampouri S, Nichols RC, Oliver KR, Morris CG, et al. Double-scattered proton-based stereotactic body radiotherapy for stage I lung cancer: a dosimetric comparison with photon-based stereotactic body radiotherapy. Radiother Oncol. 2010;97(3):425–30.
155. Ohnishi K, Nakamura N, Harada H, Tokumaru S, Wada H, Arimura T, et al. Proton beam therapy for histologically or clinically diagnosed stage I non-small cell lung cancer (NSCLC): the first nationwide retrospective study in Japan. Int J Radiat Oncol Biol Phys. 2020;106(1):82–9.
156. Chang JY, Zhang W, Komaki R, Choi NC, Chan S, Gomez D, et al. Long-term outcome of phase I/II prospective study of dose-escalated proton therapy for early-stage non-small cell lung cancer. Radiother Oncol. 2017;122(2):274–80.
157. Chang JY, Verma V, Li M, Zhang W, Komaki R, Lu C, et al. Proton beam radiotherapy and concurrent chemotherapy for unresectable stage III non-small cell lung cancer: final results of a phase 2 study. JAMA Oncol. 2017;3(8):e172032.
158. Liao Z, Lee JJ, Komaki R, Gomez DR, O'Reilly MS, Fossella FV, et al. Bayesian adaptive randomization trial of passive scattering proton therapy and intensity-modulated photon radiotherapy for locally advanced non-small-cell lung cancer. J Clin Oncol. 2018;36(18):1813–22.
159. Karube M, Yamamoto N, Nakajima M, Yamashita H, Nakagawa K, Miyamoto T, et al. Single-fraction carbon-ion radiation therapy for patients 80 years of age and older with stage I non-small cell lung cancer. Int J Radiat Oncol Biol Phys. 2016;95(1):542–8.
160. Takahashi W, Nakajima M, Yamamoto N, Yamashita H, Nakagawa K, Miyamoto T, et al. A prospective nonrandomized phase I/II study of carbon ion radiotherapy in a favorable subset of locally advanced non-small cell lung cancer (NSCLC). Cancer. 2015;121(8):1321–7.
161. Xi M, Xu C, Liao Z, Chang JY, Gomez DR, Jeter M, et al. Comparative outcomes after definitive chemoradiotherapy using proton beam therapy versus intensity modulated radiation therapy for esophageal cancer: a retrospective, single-institutional analysis. Int J Radiat Oncol Biol Phys. 2017;99(3):667–76.
162. Lin SH, Hobbs BP, Verma V, Tidwell RS, Smith GL, Lei X, et al. Randomized phase IIB trial of proton beam therapy versus intensity-modulated radiation therapy for locally advanced esophageal cancer. J Clin Oncol. 2020;38(14):1569–79.
163. Akutsu Y, Yasuda S, Nagata M, Izumi Y, Okazumi S, Shimada H, et al. A phase I/II clinical trial of preoperative short-course carbon-ion radiotherapy for patients with squamous cell carcinoma of the esophagus. J Surg Oncol. 2012;105(8):750–5.
164. Hong TS, Ryan DP, Borger DR, Blaszkowsky LS, Yeap BY, Ancukiewicz M, et al. A phase 1/2 and biomarker study of preoperative short course chemoradiation with proton beam therapy and capecitabine followed by early surgery for resectable pancreatic ductal adenocarcinoma. Int J Radiat Oncol Biol Phys. 2014;89(4):830–8.
165. Murphy JE, Wo JY, Ryan DP, Jiang W, Yeap BY, Drapek LC, et al. Total neoadjuvant therapy with FOLFIRINOX followed by individualized chemoradiotherapy for borderline resectable pancreatic adenocarcinoma. JAMA Oncol. 2018;4(7):963–9.
166. Murphy JE, Wo JY, Ryan DP, Clark JW, Jiang W, Yeap BY, et al. Total neoadjuvant therapy with FOLFIRINOX in combination with losartan followed by chemoradiotherapy for locally advanced pancreatic cancer: a phase 2 clinical trial. JAMA Oncol. 2019;5(7):1020–7.
167. Kawashiro S, Yamada S, Okamoto M, Ohno T, Nakano T, Shinoto M, et al. Multi-institutional study of carbon-ion radiotherapy for locally advanced pancreatic cancer: Japan Carbon-ion Radiation Oncology Study Group (J-CROS) study 1403 pancreas. Int J Radiat Oncol Biol Phys. 2018;101(5):1212–21.

168. Bush DA, Do S, Lum S, Garberoglio C, Mirshahidi H, Patyal B, et al. Partial breast radiation therapy with proton beam: 5-year results with cosmetic outcomes. Int J Radiat Oncol Biol Phys. 2014;90(3):501–5.
169. Galland-Girodet S, Pashtan I, MacDonald SM, Ancukiewicz M, Hirsch AE, Kachnic LA, et al. Long-term cosmetic outcomes and toxicities of proton beam therapy compared with photon-based 3-dimensional conformal accelerated partial-breast irradiation: a phase 1 trial. Int J Radiat Oncol Biol Phys. 2014;90(3):493–500.
170. Verma V, Iftekaruddin Z, Badar N, Hartsell W, Han-Chih Chang J, Gondi V, et al. Proton beam radiotherapy as part of comprehensive regional nodal irradiation for locally advanced breast cancer. Radiother Oncol. 2017;123(2):294–8.
171. Thorpe CS, Niska JR, Anderson JD, Girardo ME, McGee LA, Hartsell WF, et al. Acute toxicities after proton beam therapy following breast-conserving surgery for breast cancer: multi-institutional prospective PCG registry analysis. Breast J. 2020;26(9):1760–4.
172. Slater JM, Slater JD, Kang JI, Namihas IC, Jabola BR, Brown K, et al. Hypofractionated proton therapy in early prostate cancer: results of a phase I/II trial at Loma Linda University. Int J Part Ther. 2019;6(1):1–9.
173. Henderson RH, Bryant C, Hoppe BS, Nichols RC, Mendenhall WM, Flampouri S, et al. Five-year outcomes from a prospective trial of image-guided accelerated hypofractionated proton therapy for prostate cancer. Acta Oncol. 2017;56(7):963–70.
174. Kagei K, Tokuuye K, Okumura T, Ohara K, Shioyama Y, Sugahara S, et al. Long-term results of proton beam therapy for carcinoma of the uterine cervix. Int J Radiat Oncol Biol Phys. 2003;55(5):1265–71.
175. Wakatsuki M, Kato S, Ohno T, Karasawa K, Ando K, Kiyohara H, et al. Dose-escalation study of carbon ion radiotherapy for locally advanced squamous cell carcinoma of the uterine cervix (9902). Gynecol Oncol. 2014;132(1):87–92.
176. Hancock J. For cancer centers, proton therapy's promise is undercut by lagging demand. The New York Times. 2018 [cited 2022 Aug 23]. https://www.nytimes.com/2018/04/27/business/proton-therapy-finances.html.
177. Zietman AL. Can proton therapy be considered a standard of care in oncology? Lessons from the United States. Br J Cancer. 2019;120(8):775–6.
178. Elnahal SM, Kerstiens J, Helsper RS, Zietman AL, Johnstone PAS. Proton beam therapy and accountable care: the challenges ahead. Int J Radiat Oncol Biol Phys. 2013;85(4):e165–72.
179. Verma V, Shah C, Rwigema JCM, Solberg T, Zhu X, Ii CBS. Cost-comparativeness of proton versus photon therapy. Chin Clin Oncol. 2016;5(4):10.
180. Verma V, Mishra MV, Mehta MP. A systematic review of the cost and cost-effectiveness studies of proton radiotherapy. Cancer. 2016;122(10):1483–501.

Further Reading

https://www.youtube.com/watch?v=6z-WVA-4VR0
https://www.youtube.com/watch?v=c-0MOXV-vf8
https://www.youtube.com/watch?v=Hkmp8bjKmG0
https://www.youtube.com/watch?v=jSgnWfbEx1A

18 Special Techniques: Total Body Irradiation, T-SET, IORT, and Hyperthermia

Tejinder Kataria, Shyam Singh Bisht, Susovan Banerjee, Kaliyaperumal Venketesan, Manoj Tayal, Soumya Sarkar, and Tamilselvan

18.1 Introduction

Recent advances in radiation techniques have brought a change in the delivery of treatment for specific disease sites as well as in mitigating radiation toxicity while increasing the control rates. Some examples include total body irradiation (TBI) for hematologic cancers as a preparation for marrow or stem cell transplantation, total skin electron therapy (T-SET), intra-operative radiotherapy (IORT), and hyperthermia. This chapter includes an introduction to each technique, dose, planning, clinical application, and a review of key studies.

T. Kataria (✉)
Division of Radiation Oncology, Medanta Cancer Institute, Medanta, Gurgaon, Haryana, India

Guru Gobind Singh Indraprastha University, New Delhi, India

S. S. Bisht · S. Banerjee
Division of Radiation Oncology, Medanta Cancer Institute, Medanta, Gurgaon, Haryana, India

K. Venketesan · Tamilselvan
Division of Radiation Oncology, Medanta, Gurgaon, Haryana, India

Scientist D, DRDO, Vishakhpatnam, Andhra Pradesh, India

S. Sarkar
Department Of Radiation Oncology, Tata Medical Centre Kolkatta, Kolkatta, West Bengal, India

M. Tayal
Department of Radiation Oncology, Max Hospital, Patparganj, New Delhi, India

18.2 Total Body Radiation (TBI)

18.2.1 Introduction

Total body irradiation (TBI) is a special technique in which a patient's whole body is treated with radiation. TBI is used primarily as part of the treatment for patients with acute myeloid leukemia (AML) or acute lymphoid leukemia (ALL) undergoing hematopoietic stem cell transplantation (HCT) [1]. The aim of TBI is to suppress the patient's immune system and prevent the rejection of donor bone marrow after a bone marrow transplant using donor marrow. Total body irradiation (TBI) is a treatment modality for hematologic [2] oncologic, and immunologic diseases, and it has been used as a conditioning regimen in conjunction with Chemotherapy, primarily for bone marrow transplantation (BMT) or peripheral blood stem cell transplantation [2, 3]. Through two major mechanisms, allogeneic stem cell transplantation cures hematologic malignancies. The pre-transplantation conditioning regimen directly kills leukemic cells, eradicating or reducing the malignancy to a minimal residual disease state. A graft-versus-tumor effect is the second mechanism. The relative importance of these two mechanisms varies by disease: graft-versus-tumor effects are most important in chronic myelogenous leukemia (CML) and less important, but still demonstrable, in acute

A. Munshi et al. (eds.), *Radiation Oncology – Principles, Precepts and Practice*,
https://doi.org/10.1007/978-981-97-8389-2_18

myelogenous leukemia (AML) and acute lymphoblastic leukemia (ALL) [4].

The goals of TBI are (i) eradicating diseased marrow, (ii) reducing tumor burden, (iii) immunosuppression, and (iv) depleting the BM to allow physical space for engraftment of healthy donor marrow [5]. TBI may be particularly important in the setting of matched-unrelated donor transplants when adequate immune suppression is essential [6].

Leukemia cells have a D0 value in the range of 0.8–1.5 Gy and are as radiosensitive as bone marrow cells [7].

18.2.2 Dose and Fractionation

The TBI can be treated either as a single fraction or multiple fractions. The dose-limiting organ is generally regarded as the lung, which can develop radiation-induced interstitial pneumonitis [3]. The total TBI dose should not exceed more than 12 Gy. Numerous dosages and fractionation have been applied to TBI. Typical examples are,

- 2 Gy in a single fraction
- 12 Gy in 6 fractions
- 8 Gy in 8 fractions

For multiple fractions, TBI will be delivered as two fractions in a day with a gap of 6–8 h. The higher doses may contribute to improved control of the disease, but the trials have provided evidence that higher doses cause a rapid increase in the rate of severe complications as well [8].

18.2.2.1 Delivery Techniques

TBI is primarily performed on standard linear accelerators. The most common way to administer TBI is in a standing or lie-down position with increased source-to-skin distance (SSD). Patients are treated from a single field if the size of the treatment room allows it. Otherwise, multiple fields are necessary. Anterior and posterior fields are used, with the dose prescribed to mid-plane. Most centers use photon energies between 6 and 10 MV.

Differences in separation along the patient length can result in dose inhomogeneities that can exceed 10–20%. In vivo dosimetry should be used to measure doses at multiple locations on the patient's body with the aim of achieving uniform doses within 5%. Most treatment plans require the protection of certain sensitive structures, especially the lungs. The lung block can be designed to cover the central part of the lung by approximately 1–2 cm between the edges of the shadow of the lung and the edges of the block [7]. TBI is a highly complex and specialized technique that presents unique challenges. Great attention should be paid to patient positioning, beam direction, dose calculation, organ shielding, shield verification, tissue compensation, and quality assurance details. The TBI method described above is based on point measurements and can result in unexpected dose inhomogeneities [9]. More centers are using computed tomography (CT)-based 3D planning (Fig. 18.1) for better target coverage and reduced dose to organs at risk (OARs).

TBI can be delivered in multiple ways, which depends on the radiotherapy equipment available in the hospital. The most common treatment technique is to deliver horizontal radiation with the patient positioned at an extended SSD (approximately 4 m). In the conventional radiotherapy machine, the head can be rotated to 45° to achieve maximum field size, and the patients may then be treated either in the standing position or lying supine on the couch (Fig. 18.2). Normally, the patients are positioned behind the Plexiglas barrier which helps in the skin-sparing effect of high-energy photons [10]. The radiotherapy equipment used to deliver TBI is a telecobalt unit and linear accelerator, and the beam energy used is Co-60, 6 MV X-rays, and 10 MV X-rays.

18.2.2.2 AP-PA Technique

In the AP-PA technique, the patients will be in a standing position, and the technique gives a better uniform dose along the longitudinal axis [3]. The patients will stand behind the Plexiglas barrier facing the beam with front and back for each field. The Plexiglas barrier with a thickness of 1 cm is used for skin-sparing and to hold the shielding blocks to spare critical organs (usually lungs and sometimes kidneys).

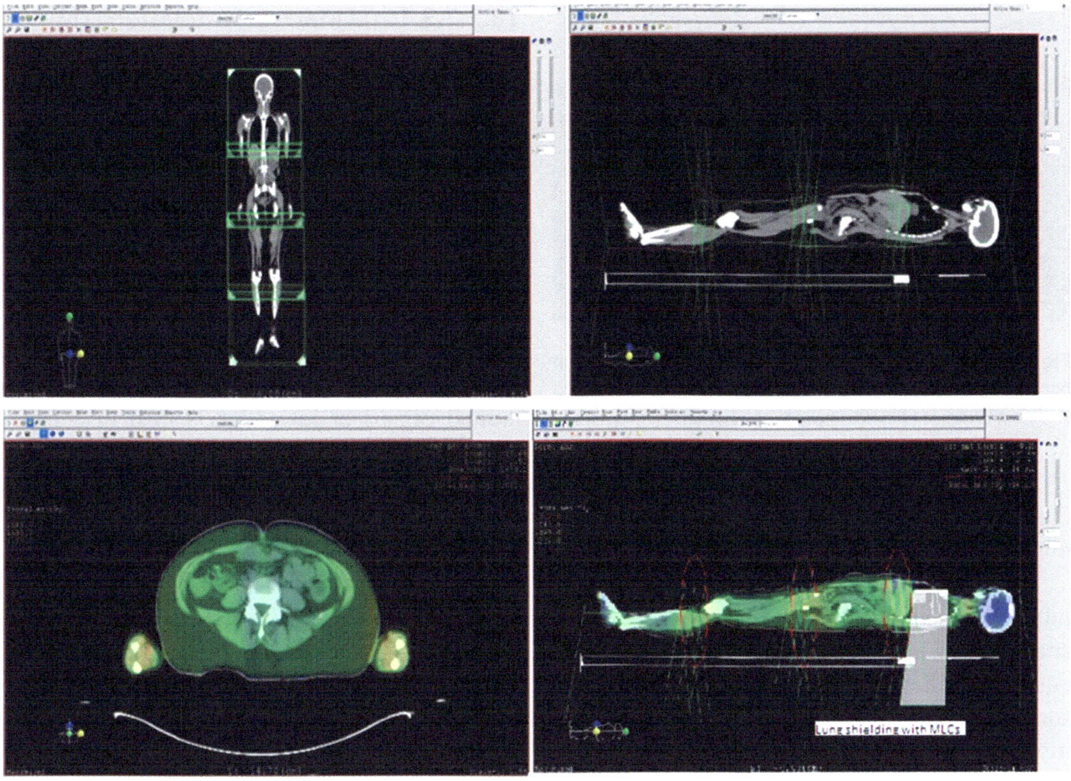

MLC: Multileaf Collimator

Courtesy: Medanta The Medicity, Gurugram. India

Fig. 18.1 Computed tomography (CT)-based 3D planning

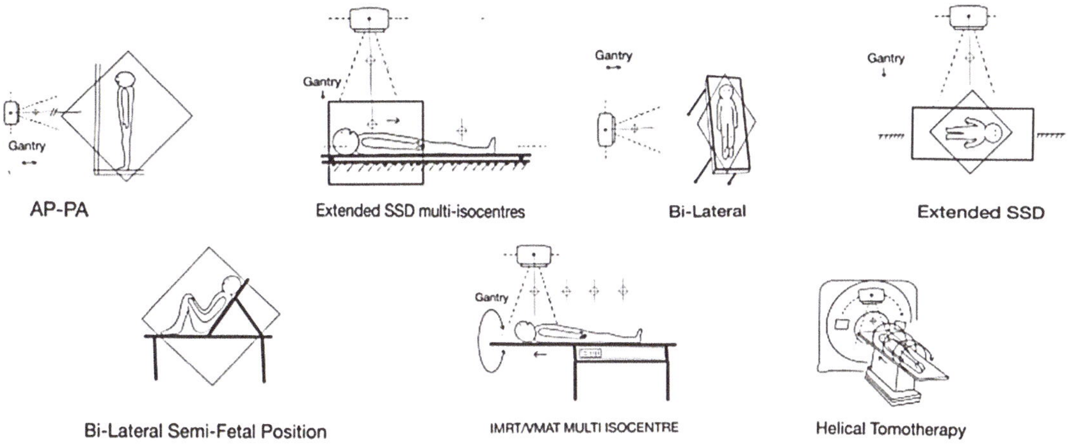

Fig. 18.2 Examples of TBI technique

18.2.2.3 Bilateral Technique

The bilateral technique involves the supine position of the patients, which is comfortable for the patients, but the variation of the dose is greater along the path of the beams. The lead or brass compensator is placed in the tray at the head of the radiotherapy equipment to improve dose uniformity. The patient is positioned on

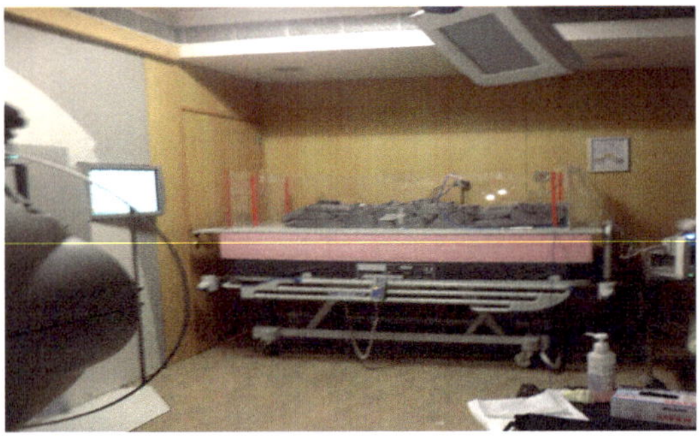

Fig. 18.3 TBI using box at extended SSD. The gap between the box lateral wall and patient is tightly filled with the fillers (in present case rice bags)

the TBI couch, either lying or in a semi-fetal position [10].

In one of the bilateral TBI treatment techniques, the patient is placed in a top open perspex box, which is filled with nearly tissue equivalent fillers (e.g., rice bags) to ensure uniform dose distribution across the patient's craniocaudal length. The gantry was set to 270° or 90° and the collimator was set to 45° to irradiate the patient from the side. Figure 18.3 depicts a typical example of such treatment: a patient is placed 333.5 cm away from an SSD. After completion of the treatment on one side, the patient is rotated from a head-first to a feet-first condition to treat the other side, keeping the gantry and collimation in the same position. Treatment SSD may vary depending upon the room size. Online supplementary material 1 and 2 provides the TBI check list and optically stimulated luminescent dosimeter (OSLD) measurement results.

18.2.2.4 VMAT and Tomotherapy

In recent years, TBI has frequently been treated in beam modulation arc techniques like VMAT and tomotherapy [11, 12], which incorporates the advantages of CT planning. Whole-body CT scans, with the patient in the standard treatment position, were used for treatment planning and several multileaf collimator fields were used to optimize dose uniformity to the whole body while sparing critical organs like lungs and kidneys [13].

18.2.2.5 Total Marrow Irradiation (TMI) and Lymphoid Irradiation (TMLI)

TMI and TMLI emerged as a result of helical tomotherapy's ability to perform targeted radiotherapy with improved dose conformation to targeted structures while sparing organs at risk (OARs) [14]. Targeted total marrow irradiation using tomographic intensity-modulated radiation therapy by Wong et al. [15] predicted the potential to escalate the dose to bone (and containing marrow) up to 20 Gy while maintaining doses to normal organs at lower levels than in conventional TBI to 12 Gy.

Recently Shahid et al. demonstrated the VMAT-based TMLI in ring gantry linear accelerator [16]. Patients need to be changed from head first to feet first position to treat the complete target volume. Although the dose distribution and organ at risk doses they have reported are higher than tomotherapy results, it is comparable with other VMAT-based techniques [17]. Figure 18.4 shows the VMAT-based TMLI dose distribution.

18.2.2.6 Treatment Planning

TBI treatment planning has to guarantee a sufficiently high and homogeneous dose in the target volume under optimal sparing of the lungs, the organ most at risk. In TBI, treatment planning normally involves prescribing the dose to the point at which the dose is specified as an essential condition. It has been recommended to specify the dose at the center of the abdomen at the level of the umbilicus, and the conventional treatment time calculation method is normally used to deliver the prescribed dose to the point [1, 10].

In the AP-PA technique, lung blocks can be designed from a radiograph obtained while the patient is in the standing position. The blocks cover the central portion of the lung, with approximately 1 or 2 cm between the edge of the lung shadow and the edge of the lung of the blocks on the film (fig). The lung blocks that transmit 50% of the primary beam are used [7].

In the bilateral TBI technique, there will be a difference in separation along the patient's length, resulting in dose heterogeneity that can exceed

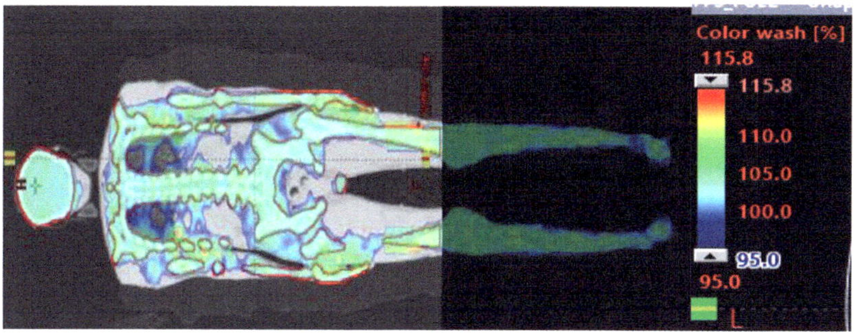

Fig. 18.4 TMLI dose distribution: HFS plan + FFS plan+ HFS junction plan. (Image adapted from Shahid et al. with permission)

10–20%. Measurements of thickness are made at the following patient anatomic points: top of the head, forehead, chin, suprasternal notch, xiphoid, umbilicus, central axis, pelvis, thigh, knees, calves, ankles, and toes. The measurements are used to make a customized patient-specific compensator (lead or brass) placed in the beam path at the head of the radiotherapy equipment to achieve dose homogeneity [1, 6]. The arms of the patients are kept laterally close to the body at the level of mid AP thickness and it is believed that the hands are masking the lung image [3].

In the beam modulated arc technique like VMAT and helical tomotherapy, the TBI treatment planning involves customized immobilization for patients using either full body Vac-lock or thermoplastic molds, followed by whole-body planning CT scan (both head first supine and feet first supine), contouring, treatment planning in the treatment planning system, patient-specific QA and treatment delivery. Due to limitations in the treatment length with the modulation arc technique, two plans were generated, one for treatment head to mid-thigh region and the other for treatment from foot to remainder lower extremities. The plans are generated with the uniform prescription dose to the PTV and to achieve the lung mean dose of 70–80% of the prescription dose [7].

18.2.2.7 Dosimetry

All dosimetric [11] measurements have to be performed under physical and technical conditions used in TBI, especially the calibration of the TBI dose monitor and the calibration of dosimeters for absolute and relative TBI dosimetry, taking into account all influences of scatter radiation from floor or walls and the reduced scatter dose from the distant treatment head, so as influences by dose modifiers or technical modifications [18].

The in vivo measurements can generally be used to measure entrance and exit doses on the different regions of the patient's skin. The point of measurement is chosen such that they allow for verification of the dose delivered at the point of dose specification (usually mid-abdomen) and estimation of the homogeneity of the dose distribution of the patient. Either diodes or thermoluminescent dosimeters (TLD) are used for in vivo measurement [1].

In the modulation arc technique, the QA for the treatment plan must be performed prior to treatment delivery to ensure the planning dose can be delivered with high accuracy [11]. In general, the QA plans are generated in the phantom, small-volume ion chambers are used for point dose verification, and either radio-sensitive films or radiation array detectors are used for planar dose verification.

18.2.2.8 Complications of TBI

Leukemia cells have a D0 value in the range of 0.8–1.5 Gy and are as radiosensitive as bone marrow cells. Acute toxicity includes nausea, vomiting, diarrhea, stomatitis, temporary loss of taste, parotitis, and rash. Late toxicity includes interstitial pneumonia (IP), hepatic vein obstructive disease (VOD), cataracts, infertility, hormone-related diseases, ototoxicity (osteoporosis), growth retardation, and secondary malignant tumors [19].

18.2.2.9 Challenges of Newer Techniques

TBI delivered using large-field IMRT approaches at standard SSDs introduces new challenges and unknowns as multiple abutting fields can result in dose uncertainties at the junction sites [19]. TMI and TMLI require long contouring time (up to 8 h for HT-TMI and 12–16 h for HT-TMLI). According to studies, the average in-room time is 2 h, but this can be reduced to 90 min with practice, and the average beam on time is 50 min [14].

18.3 Total Skin Electron Therapy

18.3.1 Introduction

Cutaneous lymphomas are a group of disorders characterized by the localization of malignant lymphocytes to the skin and the most prevalent entity in the group of primary cutaneous lymphomas is mycosis fungoides (MF), epidermotropic CTCL (cutaneous T-cell lymphoma). Approximately two-thirds of these lymphomas are of T-cell origin, and the pathogenesis and management of both T- and B-cell cutaneous lymphomas are characterized by the proliferation of small and medium T cells with cerebriform nuclei [20–26]. Megavoltage X-rays beam treatment of such superficial tumors is complicated by photon transport properties like build-up/ exit and integral dose. These beams can treat shallow tumors, however, exposing normal tissues to radiation. Low-energy X-rays are also used for skin irradiation. They presently have limited usage as the results were less than encouraging because it was difficult to treat the entire skin area adequately. There were maximum field size and field-junction limitations, and it was not possible to treat to an adequate depth without a large X-ray integral dose. Megavoltage electron beams represent an important treatment modality in modern radiotherapy, providing a unique option for the treatment of superficial tumors. The electron beam central axis depth dose curve exhibits a high surface dose, with dose build-up at a certain depth referred to as the dose maximum and a sharp dose fall-off. This characteristic of the electron beam to treat a spectrum of skin diseases while treating the entire body skin with the electron is named Total skin electron therapy (TSET). TSET is used when other forms of therapy, such as UVA light, psoralen (PUVA), UVB narrow band, topical steroids, or other systemic agents, are ineffective [8].

European Organization for Research and Treatment of Cancer (EORTC) Cutaneous Lymphoma Task Force recommends the following objectives: (1) align the distribution of dose to the target volume, (2) be practical, comfortable, and efficient for the patient, (3) provide sufficient doses within the target volume, (4) reliably attain cutaneous remission, (5) minimize toxicity, (6) produce beneficial long-term clinical results, and (7) accommodate repeated administration as required. Three principal TSEI methods, which fulfill the EORTC recommendations, include (1) large electron field techniques, (2) rotational techniques, and (3) techniques involving patient shift during irradiation. The EORTC recommendations for the dosimetry and the mode of treatment implementation should be based on the following technological requirements like geometric conditions of the alignment of the patient, the number of treatment fields, and dose fractionation [27–34].

18.3.1.1 Six-Dual-Field Irradiation Technique

This method is used in most of the recent TSET treatments (Figs. 18.5 and 18.6). The six patient orientations with six angular orientations about a vertical axis, three each on an alternating day.

The angulation is spaced at 60° intervals, resulting in a variety of dose that has a 60° periodicity. The following consideration should be given to accurate patient positioning. (a) Correct orientation toward the beam. (b) Joints in the limbs at appropriately large angles. (c) Slightly bending the torso. (d) Hips to expose folds in the lower trunk. (e) unfolding of skin with the use of tape, bras, extreme positioning, netting, and Styrofoam wedges, can minimize patch treatments [32].

18.3.1.2 Technical Limitations and Tolerance

The optimal source-to-patient distance is between 3 and 8 m. If the treatment distance is less than 3 m, it is difficult to obtain dose homo-

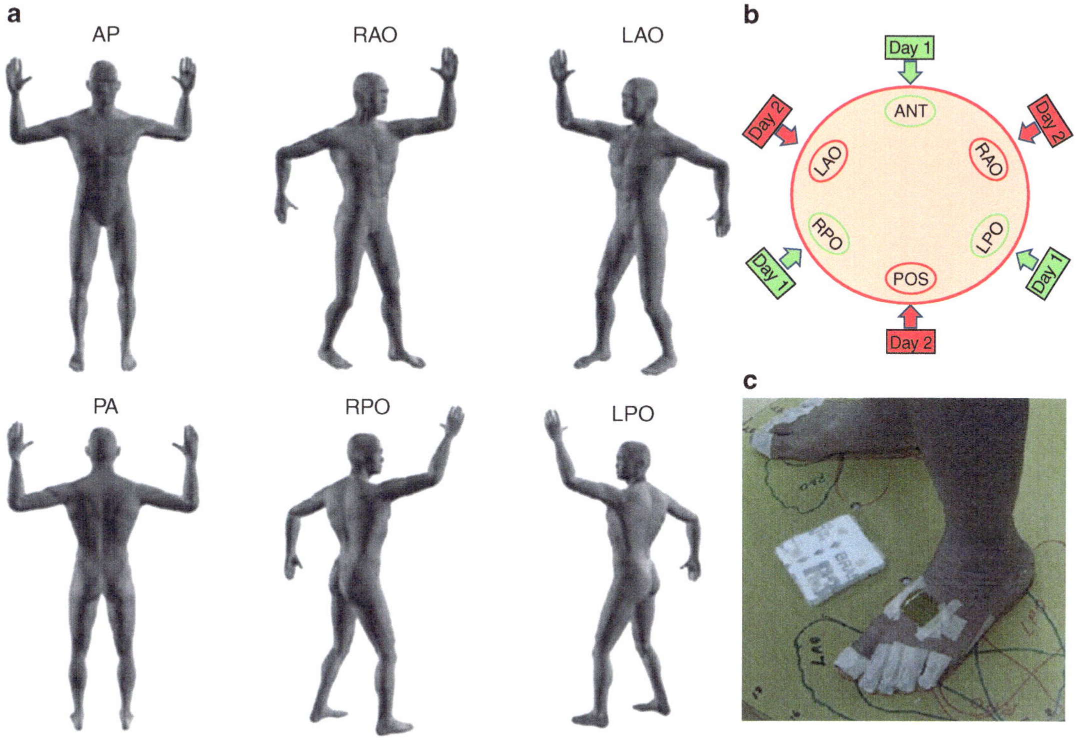

Fig. 18.5 (**a**) Six-patient position during TSET. (**b**) Day 1 and Day 2 treatment positions were shown in the leg frame. (**c**) Leg positions during the six-dual field technique

geneity and problem limit photon contamination. The dose rate is reduced, and the time taken for radiation is increased in the case of a treatment distance greater than 8 m. The dose in the air should be homogeneous within 10% for the entire vertical and lateral dimensions of the beam and the junction between fields that constitute a single beam. The short treatment distance treatment is possible with the use of beam-scattering devices, but these produce greater bremsstrahlung or X-ray photon contamination. Photon contamination should be less than 0.7 Gy for a full course of TSET to avoid hematologic sequelae. Linac operating parameters should be the dose rate of more than 1 Gy/min or higher and the long diagonal of the field, which improves the vertical dose uniformity. The beam scatterer should be used for the provision of large, uniform, low-energy electron fields for TSET and to reduce the beam energy. It can be a thin material used primarily to spread out the beam. Interposed material has the ability of scatters the incident electron beam and reduces contaminant megavoltage X-ray background [33].

$$Ep,\mathrm{o} = 1.95R_{\mathrm{p}} + 0.48 \qquad (18.1)$$

The treatment field size at the patient treatment plane should be 200 cm in height by 80 cm in width to encompass the largest patient. The electron beam uniformity is ±8%, and a horizontal uniformity is of ±4% over the central 160 cm × 60 cm area. The penetration depth is varied with the stage and type of disease. It may vary over the body (5–15 mm or more at the 50% isodose surface for most lesions), and multiple electron beam energies are used to cover the different ranges of depths. The dose uniformity should be maintained in height and width of the patient for the distance of 2–5 m (according to the treatment distance). Frequent exchange of the air in the treatment room is essential for avoiding ozone production for shielding in the megavoltage X-ray treatment rooms adequate for TSET therapy which involves bringing a large fluence of energetic electrons.

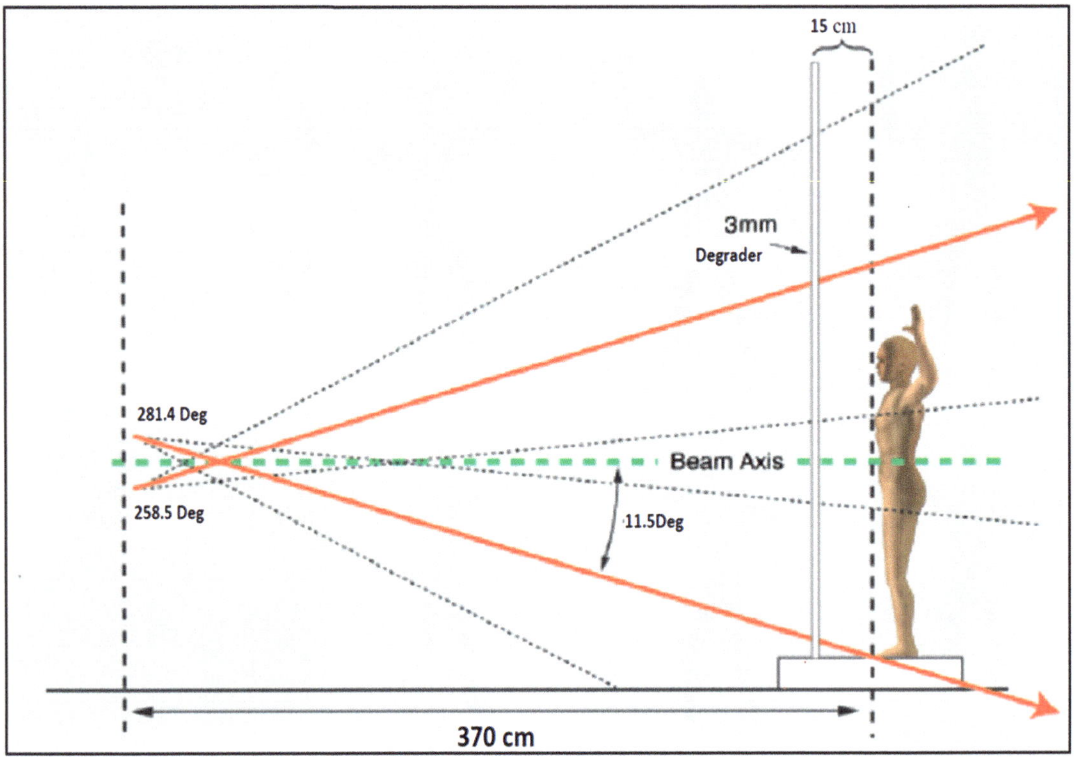

Fig. 18.6 Six dual-field techniques show the SSD, gantry angles, and degrader position (setup positioning parameters from Medanta The Medicity)

18.3.1.3 Dosimetry and Instrumentation

The dosimetry should be performed at shallow depths over a large area in the patient treatment plane (Fig. 18.7). The short ranges of the electrons necessitate special attention if the dosimetry and calibration. Suitable detectors and instrumentation is used to attain accurate absorbed dose results. By observing appropriate precautions and exercising care, valid dosimetry data for use in patient treatment can be obtained. For performing the dosimetry, ionization chambers, film, thermoluminescent materials, fricke dosimeters, and electron collectors (Faraday cups) can be used [34].

In these, the ionization chambers (parallel plate) are generally used to measure the absolute dose and depth dose curves. Solid phantoms like polystyrene can be used because their electron density is very close to that of tissue. Table 18.1 shows the steps involved in the TSET dosimetry. Beam specifications for 4, 6, and 8 MeV electron beams at extended SSD in the central axis were tabulated in Table 18.2.

Factor A and Factor B are calculated using the chamber and cylindrical water-equivalent phantoms. Factor A is defined as the ratio between the summation of anterior readings at the two oblique gantry angles (e.g., 252.5° and 287.5°) (chamber kept at central axis) and central position reading (gantry angle 270°) (Eq. 18.2).

$$\textbf{\textit{Factor A}} = \frac{\begin{array}{c}\text{AP1}\left(\text{GA} - 252.5^{\circ}\right)\\ +\text{AP2}\left(\text{GA} - 287.5^{\circ}\right)\end{array}}{\begin{array}{c}\text{Central position meter}\\ \text{reading (direct output}\\ \left(\text{GA} - 270^{\circ}\right)\end{array}} \tag{18.2}$$

$$\textbf{\textit{Factor B}} = \frac{\begin{array}{c}\text{Total charge collected}\\ \text{by all 12 fields}\end{array}}{\begin{array}{c}\text{Charge collected}\\ \text{by dual anterior}\\ \text{fields}\left(\text{AP1} + \text{AP2}\right)\end{array}} \tag{18.3}$$

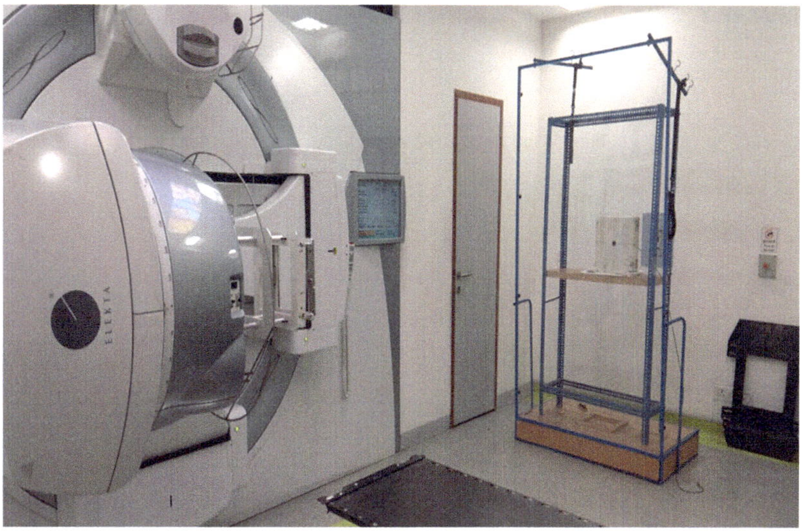

Fig. 18.7 Output measurement at the SSD of 3.8 m using the parallel plate chamber and RW3 plastic water phantoms. (Image from Department of Radiation Oncology, Medanta, The Medicity, Gurgaon, India)

Table 18.1 Steps involved in TSET dosimetry; standard source to skin distance (SSD) denotes the 100 cm from virtual electron source; extended SSD denotes the treatment distance (3–5 m)

Step	SSD	Measurement	Correction	Detector/Phantom
Step 1	Standard	Absolute dose	Adjust to 1 cGy per MU at Zref	Ionization chamber/water
Step 2	Standard	Profiles (in-plane and cross plane) in water	Correct the flatness/ symmetry	Ionization chamber/water
Step 3	Extended	Profiles (in-plane and cross plane) in the air over the treatment region	Verify the flatness and symmetry within ±10% from CAX	Ionization chamber/plastic water
Step 4	Extended	Finding a factor A	Verify with other electron energy	Ionization chamber/ cylindrical phantom
Step 5	Extended	Finding a factor B	Verify with other electron energy	Ionization chamber/ cylindrical phantom
Step 6	Extended	The dose rate at the dose maximum for the corresponding energy	Second verification with another available detector	Ionization chamber/plastic water
Step 7	Calculate the monitor unit for the corresponding energy by using the factors A, B, and dose rate at extended SSD for the prescribed dose			

Table 18.2 Beam specification of 4, 6, and 8 MeV at the treatment distance of 3.8 m at central axis for 20 × 20 cm^2 applicator (Data obtained from Medanta-The Medicity, Elekta Infinity®)

PMMA beam spoiler	3 mm thickness degrader			6 mm thickness degrader		
Energy	4 MeV	6 MeV	8 MeV	4 MeV	6 MeV	8 MeV
Max (mm)	6	10	15	3	7	12
Surface dose (in %)	87.9	85.4	86.8	96.0	89.4	88.9
Output (cGy/MU)	0.02908	0.03934	0.04698	0.02754	0.0385	0.04594

$$B = \frac{\begin{array}{c}\mathrm{AP1} + \mathrm{AP2} + \mathrm{LAO1} + \mathrm{LAO2} + \mathrm{RAO1} \\ +\mathrm{RAO2} + \mathrm{PA1} + \mathrm{PA2} + \mathrm{LPO1} \\ +\mathrm{LPO2} + \mathrm{RPO1} + \mathrm{RPO2}\end{array}}{\left(\mathrm{AP1} + \mathrm{AP2}\right)}$$

Factor B is defined as the ratio between the total charge collected by all 12 fields and charge collected by dual anterior fields (Eq. 18.3).

Table 18.3 In vivo measurement location during TSET delivery

In vivo dosimetry points					
1	Forehead	11	Rt. arm	21	Top of the feet Lt.
2	Nape (post) neck	12	Lt. palm (v)	22	Lt. knee lateral aspect
3	Vertex (top of head)	13	Rt. palm (v)	23	Rt. thigh medial lesion
4	Face	14	Lt. palm back	24	Belly
5	Rt. shoulder and	15	Rt. palm back	25	Lt. breast under surface
6	Lt. shoulder ant.	16	Umbilicus	26	Sternum
7	Lt. shoulder post.	17	Back	27	Ant. neck
8	Lt. axilla	18	Lt. flank	28	Thigh Lt. medial high
9	Rt. axilla	19	Lt. thigh lateral		
10	Lt. arm	20	Lt. thigh medial		

18.3.1.4 In Vivo Dosimetry in TSET

In vivo dosimetry is used to evaluate and verify the delivered dose to the planned dose. American Association of Physicists in Medicine (AAPM), TG-62, Diode In-Vivo Dosimetry for Patients Receiving External Beam Radiation Therapy. In vivo dosimetry is supplementary to a routine clinical QA program, the recent developments in understanding issues in treatment delivery and treatment adaptation have led to an increased importance of the safety and quality efficacy as an in vivo dosimetry program. For in vivo dosimetry, the following detectors could be used. Surface dosimeters like thermoluminescent dosimeters (TLD), optically stimulated luminescent dosimeters (OSLD), diode, and metal-oxide-semiconductor field-effect transistor (MOSFET) are placed on the patient's surface to measure the entrance dose. They can also be used more generally to verify skin dose [35, 36].

Calibration of in vivo dosimeters is mandatory to calculate the accurate, delivered dose. In this calibration process, each type of detector requires a special procedure to account for dose linearity, fading, energy response, and dose rate-related correction factors. Some detectors are needed to correct the beam-dependent correction factors like SSD, field size, compensators, and angle of incidence.

In TSET, in vivo dosimetry is mandatory to verify the delivered dose and provide the area where a further dose is needed. Table 18.3 shows the crucial region for the in vivo measurement points in the patient. In vivo dosimetry measurements are important for TSET for two reasons: (1) determination of the dose to the patient's skin, and (2) verification that the prescribed dose to the patient's skin is accurate. The actual uniformity of the dose delivered to the patient's skin, however, may vary significantly from that measured in air, so measurement of the actual skin-dose distribution is required.

18.3.1.5 Patient Shielding Requirements

Eye: Globe of the eye <15% of the prescribed skin surface dose. A dose of 26 Gy is administered to the eyelids while shielding; Protective goggles for the eyes are used for 5 weeks of therapy, and then special contact lenses are used so the eyelids can be treated for the final 4 weeks [28].

Skin: high degree of curvature should not receive a dose of 120% (e.g., nose, wrists, forearms, penis, and scrotum); Thin structures: Four or more multiple fields contribute the dose to the hands and ears, but these regions will not receive more than 120%; Pacemakers and Ostomy sites: place small shields over, but this may compromise efficacy [33].

18.3.1.6 Dose and Dose Prescription

The recommended dose prescription is 36 Gy over 9 weeks, 4 Gy per week, and 4 days per week [19]. The effectiveness of TSET has been confirmed by multiple large single and multi-institutional series. Overall response rates range from 94.7% to 100% range using 30–36 Gy dose. There is a well-established linear dose-dependent relationship, and Jones et al. reported significantly higher CR in stage IA–III patients treated with 35 vs. 30 Gy (85% vs. 64%) [37].

18.3.1.7 Regional Patch/Boost Treatments

The patient is typically standing during the TSET treatment, and the soles are unexposed to the radiation. So, the soles must be separately treated in standard SSD with an adequate dose. The shielded regions of the skin may also receive skin surface doses well below 80%. The regions which are difficult to access, like inframammary, pannicular, perianal, buttocks, thighs, inner thighs, and perineal skin regions during extended SSD treatment, also needed the additional dose, which can be quantified using in vivo measurements. The unfolding of the skin can minimize patch treatments of other regions of the skin. It increases the dose by 10–40% within folds and at shadowed regions.

Rotational treatment is somewhat better in achieving a homogeneous skin dose but requires careful coordination of rotational speed and treatment delivery time [35].

A variety of beam energies were used when the technique was first developed; in current use, 4–6 MeV electrons have become the standard. Some accelerators offer a special high-dose rate electron mode designed for TSET treatments at extended distances [38, 39].

18.3.1.8 Toxicity Evaluation and Management

Weekly assessment includes evaluating the skin for texture and color change, edema, bullae, ulceration, or infection. Evidence of bullae or infection might require a treatment interruption of 1–2 weeks. With 36 Gy, acute side effects are erythema and dry desquamation (76%), blisters (52%), hyperpigmentation (50%), skin pain (48%), and skin infections requiring antibiotics (32%) [40]. Subacute toxicity is the loss of sweat glands leading to irregular temperature control and recovers within 9–12 months. Patients may have brittle nails and alopecia, which recovers in 3–6 months. The application of emollients and sunscreen is helpful.

18.4 Intra-operative Radiotherapy

18.4.1 Introduction

The use of adjuvant radiotherapy treatment has centered around identifying the microscopic post-operative residual disease and sterilization of the tumor bed. Margin-positive tumors (R2), R1 resection, and extranodal infiltration have been absolute indications of Radiotherapy in most cancer sites. The mandatory delay required for wound healing before post op adjuvant RT may lead to the repopulation of the tumor cells. Post-op radiation planning mandates delineation of the tumor bed with a larger margin, depending on operative notes, fiducials left during surgery, and peri-operative imaging. Any residual gross disease makes such treatments ineffective unless a high curative dose is delivered. These factors, in the post-op setting, are indicative of increased chances of acute toxicity, treatment breaks, poorer outcomes, and an increased risk of late side effects. A technique like intra-operative Radiotherapy (IORT) enables the delivery of the adequate dose precisely to the tumor site during surgery under direct visualization and, at the same time, gives an opportunity to retract normal organs from the treatment field.

18.4.1.1 IORT Technique, Principles, and Methods

Attempts to irradiate tumors intraoperatively date back to 1909 when Carl Beck brought gastric and colon cancers to the surgical incision and exposed them to Radiotherapy [41]. IORT was formally introduced in 1960 by Abe et al. from Japan, went out of favor, and is seeing a resurgence currently due to the availability of shielded mobile linear accelerators, dedicated site-specific low KV IORT devices, newer applicators, availability of HDR brachytherapy, a better understanding of single fraction dose radiobiology and publication of more literature prompting proper case selection. IORT can be delivered by Electrons, low-kV X-rays, and HDR brachytherapy depending on indications of treatment, available logistics, and the expertise of the physician. A comprehensive table has been made to understand the three treatment techniques (Table 18.4) [42, 43]. After the treatment field is identified and the best possible applicator with known dosimetry is chosen, a dose of 7–30 Gy is delivered while the patient is under anesthesia after the tumor has been delivered. The applicator is approximated on the treatment area and normal organs are retracted from the field. Treatment time depends on field size, dose, radiation generating unit, and intended depth of treatment. There is no need for imaging, contouring, and individualized planning as in other radiotherapy treatment processes.

If a dedicated setup is not available, IORT can still be practiced (although challenging) with the transfer of the anesthetized patient midway between the surgery through a sterile corridor to a sterile linear accelerator unit, then after treatment of IORT, the surgery the completed.

18.4.1.2 Clinical Results

The increasing discussion of IORT may be associated with more usage and benefit of neoadjuvant radiotherapy (NART) in clinical settings like rectum, sarcoma, and pancreatic cancers. Adverse pathological margins or residual disease post-NART can be treated more conveniently by IORT techniques under direct visualization. Clinical benefits of such treatments have been proved especially with RP sarcomas and advanced colorectal cancers where delivering an additional dose otherwise was though untenable [44, 45]. In our days of organ-preserving cancer surgeries, limited/conformal irradiation fields, recognizing a recurrence (while it is still treatable) will probably see more reirradiation episodes than ever before. A table highlighting the clinical summary is provided (Table 18.5) [42, 44–46, 49–52]. Sarcomas and colorectal cancers are sites where IORT has proven its role in advanced cancers. In early breast cancer may be the sole modality of Radiotherapy in highly selected cases [47, 53] or can be used as a boost in adjunct to whole breast Radiotherapy. In Pancreatic carcinoma, most

Table 18.4 Techniques of Intra-operative Radio-Therapy

Technique	Principle/applicators	Advantage	Disadvantage	Clinical sites
IOERT [40, 41] 2–40 Gy/min Requires shielding	Miniaturized, mobile LAs up to 12 MeV energy Circular and rectangular rigid Applicators of variable shapes and size	Greater depth of penetration and dose homogeneity Less treatment time	Rigid applicators, challenging to use in difficult sites Larger volumes require field abutment Dedicated setup	Abdominal pelvic malignancies. Head neck cancers, sarcoma
High-dose rate [40, 41] radioactive source Requires shielding	Radioactive source Flexible applicators Harrison–Anderson–Mick (HAM) applicator Superlab	Can treat uneven, Larger, narrow sites Widely available	Limited depth of penetration (0.5–1 cm depth) Prolonged treatment time	Sarcoma, abdominopelvic malignancies. Head neck cancers
Low VK X-rays at a high-dose rate [40, 42] 2 Gy/min at 1 cm No shielding	Modern low-kV (20–50 kV) mobile IORT devices Novel applicators matching target shape	Suited for spherical targets Flat, surface, and needle applicators are recently available	Limited depth of penetration of 0.5–1 cm Limited experience (except for breast)	Breast, spine metastasis Brain

Electronic brachytherapy is not discussed here as an IORT procedure

Table 18.5 Sites Treated with Intra-operative Radio-Therapy

Site	Indication	Dose/ technique	results	Toxicity	Comments
Head neck [40]	Mostly reirradiation	10–20 Gy, HDR/ IOERT	50–60% local control in 2 or 3 years FU	6–30% grade 3 Fistula, ORN, neuropathy	Attractive tool in reirradiation setting
Breast [44, 46, 47]	Radical T1–2, N0 tumors	8–21 Gy X-rays/ IOERT	3–5% local recurrence	Less skin toxicity than standard EBRT	Higher chances of local recurrence Cannot replace WBI Boost
Colorectal [40, 43]	T3/T4 Margin positive Residual disease	10–25 Gy HDR ROBERT	R0: 90–100 R1: 60 is LC % at 5 years	Wound complications. Fistula nerve toxicity	IORT provided significant benefits in local control and survival in patients with R+ resection
Sarcoma [40, 48] Limbs	Boost under direct visualization	HDR ROBERT	5-year local control—82–97% Limb perseverance (81–100%)	Wound complication (3–17%) ORN Nerve toxicity	1. Pre-op EBRT SX and IORT can improve outcomes and decrease toxicity
RP sarcoma [40, 48]	Boost under direct visualization	HDR ROBERT	5-year LC (51–89%)	Bowel and neurotoxicity	2. Maybe a very valuable modality to deliver adequate curate doses in RP sarcomas
Gynecological Cancers [40, 49]	Reirradiation settings	6–30 Gy ROBERT HDR	1-year LRC (58%)	Neurotoxicity Ureteric and bowel fistulas	Helpful in the treatment of lateral pelvic recurrence in a reirradiation setting
Brain metastasis High-grade glioma [50, 51]	Boost to EBRT	10–30 Gy	1-year LC (50–90%) for brain metastases	Radionecrosis beyond 20 Gy reported	It should be explored only in a trial setting
Pancreatic cancers [45]	Inoperable cancer pancreas	10–30 ROBERT	Median survival 8–25 months	Duodenal stricture, bleeding Dose to the spinal cord	Pre-OP CTRT/ NACT can give better results
Pediatric Tumors [40]	Goal to decrease integral dose	HDR/IOT	Heterogeneous series across age groups and sites and different histopathologies		Attractive option as EBRT has retarded / abnormal growth risk of second malignancies

available treatments have failed to provide impressive results; recent IORT guidelines document future direction [52]. IORT for head, neck, gynecological, and genitourinary cancers are generally practiced in re-irradiation settings [42, 52]. Sites like the stomach and brain are explored under trial settings as available treatments have not given satisfactory results to date.

18.4.1.3 Early and Late Toxicity

The complications of IORT are mainly based on volume irradiated, dose delivered, and site of treatment. Wound complications are most common ranging up to 30% in some series [42], especially for re-irradiation patients. In the brain, even a sharp fall of dose may not be sufficient to spare the critical structures [51]. Head-neck IORT procedures record 6–30% Grade 3 toxicities that include. Fistula, ORN, and neuropathy are comparable to reirradiation treatments by other modalities [42]. Breast IORTs are generally safe procedures and report less skin toxicity than EBRT procedures apart from boost to supplement EBRT doses can be also used as a radical technique too [42, 50]. Abdominal and pelvis toxicities like wound dehiscence, bowel fistula, bowel obstruction, urinary stricture, and late neuropathy are associated with gastrointestinal, hepatobiliary, and genitourinary procedures. Limb sarcomas IORT procedures are generally safe; however, wound complications do happen in the range of 3–15%, ORN and neuropathy are also known occasionally [42, 45].

18.4.1.4 Conclusion and Future Directions

In view of available logistics and clinical results, IORT needs to be explored as a very conformal radiotherapy treatment that can sterilize the tumor bed and improve local control and survival. This is particularly important for advanced cases undergoing surgery where an R1 or R2 resection is contemplated. In the future, IORT may see new rationale and techniques, making it more effective [46]. IORT being an ionizing therapy treatment of high dose doses may modulate the immune system toward eliminating the cancer cells. Researches are underway to use to improve the effectiveness of IORT by complementing it with nanoparticle therapy. Another direction may be to have IORT devices that produce protons or even heavy ions for better penetration, sharp dose fall, and improved biological effectiveness.

Although IORT is not considered the mainstay of radiotherapy treatment in most of the institutions of the Indian subcontinent, some dedicated machines have been commissioned for use recently and we may see a change in the near future [54, 55].

Highlights

- The principles of IORT are based on conformal treatment under direct visualization.
- IORT can be an effective way to boost EBRT doses.
- IORT can be the sole radiotherapy treatment in highly selected cases.
- IORT has a role in R1 and R2 resected diseases and recurrent and reirradiation cancers.
- IORT should be practiced in high-volume/specialized centers.
- IORT doses and volumes should be meticulously planned to prevent crippling toxicities.

18.5 Hyperthermia

18.5.1 Introduction

The ancient texts describe the induction of high body temperatures to cure various diseases, including cancer, since Egyptian civilization. The heat of the sun was considered sacred and associated with healing properties. Hot water, sand, mud baths from natural springs, hot air, and steam from volcanic caves were deployed for the treatment of different maladies [56]. Imhotep (2655–2600 BC) first described the use of heat treatments. The Edwin Smith papyrus (1700 BC) reported the use of a "fire drill" for the treatment of breast cancer [57]. Hippocrates (460–370 BC) was convinced of the curative powers of heat, and his words were, "what medicines do not heal the lance will"; what the lance does not heal, "fire will" [58].

Hyperthermia is a term that describes exogenously raising the temperature of either a body part or the whole body to induce a therapeutic effect. Modern Hyperthermia was reported by Carl D. W. Busch (1826–1881) for an advanced sarcoma of the face, treated with induction of high temperature [59]. William B. Coley developed a bacterial pyrogen in 1891 that caused erysipelas and its associated fever. He injected the toxins into the tumors and reported a 5-year survival rate increase from 28% to 64% for inoperable cancers [60]. Local Hyperthermia (42–44 BC) was used by Frans J. E. Westermark in 1898 to treat inoperable cancer cervix in Sweden [61].

George W. Crile Jr. is credited with the discovery of selective destruction of tumors when the tumor temperature is raised to 42–50 °C without damaging healthy tissues [59]. Around the same time, an improvement in tumor destruction on heating for 40 min combined with X-Rays treatment was observed [56]. The development of microwave technology after World War II introduced a revolution in the induction of Hyperthermia for deep-seated tissues (loco-regional therapy). Adaptive microwave therapy by multiple microwave antennas (a phased array) surrounding the human body can adjust the heat of focus microwave energy at the correct tumor location [61–63].

18.5.2 Preclinical Evidence

18.5.2.1 Perfusion and Oxygenation

The available evidence suggests that hyperthermia induces an increase in blood perfusion during the early phase of tumor heating, and this can lead to an increase in oxygenation of the tumor core through the microcirculation, improving the hypoxic cell oxygenation thereby assisting X-rays to be more effective, in cell-kill causing hyperthermia-induced radio-sensitization. However, a few studies increase oxygen consumption (can't Hoff's law) and may counteract the availability of oxygen through hyperperfusion [64, 65].

18.5.2.2 Cell Death

Cellular death is produced through the apoptotic pathway and unfolding of heat-labile non-histone proteins in the nucleus, impairing transcription and replication, and creating genomic instability [66].

18.5.2.3 DNA Repair Inhibition

Hyperthermia impairs post-treatment DNA excision repair & prevents non-homologous end joining, leading to BRCA-2 degradation. The letter results in reduced BRCA-2 accumulation at the double-strand break site, thereby enhancing the DNA-damaging effect of radiation and/or chemotherapy [67].

18.5.2.4 Immune Response

Hyperthermia induces immune response through various mechanisms [67–74], including

1. Increased expression of major histocompatibility complex (MHC)-1 surface receptors.
2. Enhancing efficacy of natural killer (NK) cells and CD8+ cells.
3. Secretion of heat shock proteins (HSPs, mainly HSP-70) that stimulate the antigen-presenting dendritic cells (d-APC) and NK cells.
4. NK cell lysis, macrophage activation, dendritic cell maturation, and increased production of interferon gamma lead to a direct increase in the leucocyte immunogenicity.

18.5.3 Techniques of Hyperthermia

Hyperthermia can be delivered through capacitive, radiative, infrared, microwave, or ultrasound systems. Hyperthermia can be local, loco-regional, or whole-body treatment, and the technique differs according to the nature, size, and location of the tumor. An excellent review of different delivery systems of Hyperthermia has been published by Sethi et al. [72]. Local hyperthermia applied after infusion of ferromagnetic and nanoparticles is being propagated as a successful model with biodegradable substances [75]. Mild hyperthermia (39.5–43 °C) is clinically recommended for daily treatment combined prior to radiation therapy. Thermal enhancement ratio (TER) describes the gain in tumor control with the addition of hyperthermia to radiotherapy. A gap of 1–4 h before or after radiation is recommended for best TER achievement [69].

18.5.4 Clinical Evidence

Thermoradiotherapy has been tested in multiple randomized and clinical trials for the treatment of different cancer sites [71–82], which include breast cancer [75], head and neck cancers [70], urinary bladder cancer [80], prostate cancer [78], cervix cancer [74], glioblastoma multiform [81], non-small cell lung cancer, melanomas, sarcomas, and esophageal cancers. Peeken et al. [71] and Jha et al. [74] have written an overview of various studies with their results on the use of

Hyperthermia for de novo as well as recurrent intracranial and extracranial cancers.

Datta et al. [69] published a systematic review and meta-analysis of HTRT in H&N cancers in favor of HTRT vs. RT alone. The overall complete response was 39.5% (range 31.3–46.9%) with Radiotherapy vs. 62.5% (range 33.9–83.3%) with hyperthermia and radiotherapy.

Hyperthermia has been combined with chemotherapy for locally advanced carcinoma pancreas, re-radiation for unresectable carcinoma rectum, as hyperthermia with intraperitoneal chemotherapy (HIPEC) for disseminated peritoneal carcinomatosis or ovarian cancer recurrence [70].

The concept of thermal isodose curves and cumulative equivalent minutes at 43 °C of exposure has been promulgated to have a dose unit for Hyperthermia induced cellular injury, based upon Arrhenius plots [73]. However, the final verdict is still awaited on the global use of a thermal dose unit.

18.6 Summary

1. *Total body irradiation is used as a priming technique for stem cell or bone marrow transplant, and the modern linear accelerators are fabricated to provide conformal avoidance of critical organs with the available multi-leaf collimators, thereby reducing the incidence of mucositis and pneumonitis.*
2. Total skin electron therapy is a labor-intensive and exquisite delivery of radiation for mycosis fungoides, stages 1–3. Care should be taken to take proper measurements before treatment prescription and have appropriate shielding for eyes, head, and nails. Palms, soles, groin, and scalp may need boost electron therapy in case there are plaques.
3. *The principles of IORT are based on conformal treatment under direct visualization.*

   IORT can be an effective way to boost EBRT dose; as the sole radiotherapy treatment in highly selected cases, it has a role in R1, and R2 resected diseases and recurrent and reirradiation cancers. IORT should be practiced in high-volume/specialized centers.

   IORT doses and volumes should be meticulously planned to prevent crippling toxicities.
4. *Hyperthermia acts as a radiation sensitizer; however, the dosimetry and delivery of the isotherms were a challenge. The recent developments and delivery techniques make it a promising aid for improving therapeutic tolerance.*

18.7 Conclusions and Future Directions

The special techniques chapter focuses on the established techniques of total body radiation, total skin electron therapy, and evolving techniques of intra-operative radiotherapy and hyperthermia. The available clinical and preclinical data have defined the rationale, biological principles, techniques, and outcomes for TBI and TSET. However, randomized trials with key outcomes to make IORT and hyperthermia either a part of mainstream cancer care or for residual/recurrent disease for various sites needs more work, and the reader is recommended to either participate or learn these techniques before applying them to the patients.

Multiple Choice Questions

1. Goals of TBI are all the following except
   (a) Eradicating disease marrow
   (b) Reducing tumor burden
   (c) Activation of the immune system
   (d) Prepare the bone marrow for engraftment of healthy donor marrow
2. Total dose of TBI in myeloablative regimen is
   (a) 2–8 Gy
   (b) 10–16 Gy
   (c) 20–32 Gy
   (d) 28–40 Gy
3. Different techniques of TBI include all except
   (a) Stereotactic body radiotherapy
   (b) Volumetric modulated arc therapy IMRT
   (c) Helical tomographic IMRT
   (d) Image-guided radiation therapy

4. Targeted total marrow irradiation using tomographic intensity-modulated radiation therapy is possible due to
   (a) Development of 3D conformal radiotherapy
   (b) Escalating bone and marrow dose while maintaining a low dose to normal organs
   (c) Targeted radiotherapy with increased dose to bones and normal organs
   (d) None of these
5. Commonly used photon energy for TBI is
   (a) 6 MV
   (b) 10 MV
   (c) Both of these
   (d) None of these
6. Which of the following is optimally suited for whole-body superficial skin treatment?
   (a) Low-energy electrons
   (b) Ultraviolet A-rays
   (c) Orthovoltage X-rays
   (d) Megavoltage electrons
7. Which report provides a detailed description of TSET techniques, commissioning, and in vivo dosimetry?
   (a) AAPM TG 23
   (b) ASTRO report 23
   (c) AAPM report no 23
   (d) ASTRO TG 30
8. Which of the following is not the goal of TSET?
   (a) Superficially irradiate the whole skin of the body
   (b) Irradiate skin up to a limited depth (5–15 mm)
   (c) Dose homogeneity should be within <5% at the skin surface
   (d) There are limits on the achievable dose homogeneity of about 10–15%
9. Stanford technique was initially developed by:
   (a) Karzmark at Stanford University
   (b) Specht L et al. at MSKCC
   (c) Willemze R et al. at MSKCC
   (d) Sezary et al. at Stanford University
10. Which of the following statement about TSET is wrong?
    (a) Dose prescription is 30–36 Gy.
    (b) In Modified Stanford, six dual fields are delivered over 6 days.
    (c) In modified Stanford, one fraction to the whole body is delivered in 2 days.
    (d) X-ray dose can be reduced by angling the beam axes in vertical directions.
11. IORT can be delivered by
    (a) Electrons
    (b) Low-energy X-rays
    (c) HDR brachytherapy
    (d) All of the above
12. IORT has a potential role in
    (a) Margin-positive surgery
    (b) Reirradiation radiotherapy
    (c) Boost the tumor bed after NART
    (d) All of the above
13. IORT is still not popular in research constrained set ups because
    (a) The need for dedicated logistics is not met.
    (b) IORT is not effective for Indian cancers.
    (c) Not approved by our regulatory authorities.
    (d) All of the above.
14. In breast cancers, IORT is
    (a) Sole radiotherapy treatment in few selected cases.
    (b) Associated with higher chances of Local recurrence in RCTs.
    (c) Used as a boost in adjuvant with EBRT for almost all patients.
    (d) All of the above.
15. IORT involves the following steps except
    (a) Preplanning before surgery.
    (b) Placing the appropriate applicator in the tumor bed.
    (c) Planning imaging, contouring, and treatment planning and treatment.
    (d) Completing surgery after IORT treatment is over.
16. What is the therapeutic range for hyperthermia?
    (a) 45–50 °C
    (b) 39–42 °C
    (c) 25–35 °C
    (d) None of the above
17. Mechanisms of hyperthermia-induced cellular injury include
    (a) Apoptosis
    (b) Oxygen enhancement
    (c) Induction of immune response

   (d) All of the above
18. Hyperthermia has synergistic action with radiotherapy if given
   (a) 1–4 h before or after radiation
   (b) 24 h before radiation
   (c) 24 h after radiation
   (d) Independently of radiation
19. Hyperthermia can be delivered by
   (a) Microwaves
   (b) Ultrasound
   (c) Radiofrequency waves
   (d) All of the above
20. Hyperthermia has been tested for
   (a) Cancers of head and neck
   (b) Ovarian cancer
   (c) Brain tumors
   (d) Cancer cervix
   (e) All of the above

Answers

1. (c)
2. (b)
3. (a)
4. (b)
5. (a)
6. (d)
7. (c)
8. (c)
9. (a)
10. (b)
11. (d)
12. (d)
13. (d)
14. (a)
15. (d)
16. (a)
17. (b)
18. (d)
19. (d)
20. (e)

References

1. Mayles P, Nahum A, Rosewald J-C. Handbook of radiotherapy physics, theory and practice. CRC Press; 2021. ISBN 9780367192075
2. Appelbaum FR. The influence of total dose, fractionation, dose rate, and distribution of total body irradiation on bone marrow transplantation. Semin Oncol. 1993;20:3–10; quiz 11.
3. Gilson D, Taylor RE. Total body irradiation. Report on a meeting organized by the BIR Oncology Committee, held at The Royal Institute of British Architects, London, 28 November 1996. Br J Radiol. 1997;70:1201–3.
4. Passweg JR, Tiberghien P, Cahn J-Y, et al. Graft versus leukemia effects in T-lineage and B-lineage acute lymphoblastic leukemia. Bone Marrow Transplant. 1998;21:153–8.
5. Cosset JM, Soci G, Girinsky T, et al. Radiobiological and clinical bases for total body irradiation in the leukemias and lymphomas. Semin Radiat Oncol. 1995;5:301–15.
6. Nakasone H, Fuji S, Yakushijin K, et al. Complication Working Group of Japanese Society for Hematopoietic Cell Transplantation. Impact of total body irradiation on successful neutrophil engraftment in unrelated bone marrow or cord blood transplantation. Am J Hematol. 2017;92(2):171–8. https://doi.org/10.1002/ajh.24613. PMID: 27880980.
7. Paix A, Antoni D, Waissi W, Ledoux MP, Bilger K, Fornecker L, Noel G. Total body irradiation in allogeneic bone marrow transplantation conditioning regimens: a review. Crit Rev Oncol Hematol. 2018;123:138–48. https://doi.org/10.1016/j.critrevonc.2018.01.011. Epub 2018 Feb 8. PMID: 29482775.
8. Hong S, Barker C, Klein JP, et al. Trends in utilization of total body irradiation (TBI) prior to hematopoietic cell transplantation (HCT) worldwide. Biol Blood Marrow Transplant. 2012;18:S336–7.
9. Hui SK, Das RK, Thomadsen B, et al. CT-based analysis of dose homogeneity in total body irradiation using lateral beam. J Appl Clin Med Phys. 2004;5:71–9.
10. Quast U. Whole body radiotherapy: a TBI-guideline. J Med Phys. 2006;31(1):5–12. https://doi.org/10.4103/0971-6203.25664. PMID: 21206634; PMCID: PMC3003894.
11. Springer A, Hammer J, Winkler E, et al. Total body irradiation with volumetric modulated arc therapy: dosimetric data and first clinical experience. Radiat Oncol. 2016;11:1–9.
12. Peñagarícano JA, Chao M, Van Rhee F, Moros EG, Corry PM, Ratanatharathorn V. Clinical feasibility of TBI with helical tomotherapy. Bone Marrow Transplant. 2011;46(7):929–35. https://doi.org/10.1038/bmt.2010.237. Epub 2010 Oct 11. PMID: 20935684.
13. Mackie TR, Holmes T, Swerdloff S, Reckwerdt P, Deasy JO, Yang J, Paliwal B, Kinsella T. Tomotherapy: a new concept for the delivery of dynamic conformal radiotherapy. Med Phys. 1993;20:1709–19.
14. Schultheiss TE, Wong J, Liu A, Olivera G, Somlo G. Image-guided total marrow and total lymphatic irradiation using helical tomotherapy. Int J Radiat Oncol Biol Phys. 2007;67:1259–67.

15. Wong JY, et al. Targeted total marrow irradiation using three dimensional image-guided tomographic intensity-modulated radiation therapy: an alternative to standard total body irradiation. Biol Blood Marrow Transplant. 2006;12:306–15.
16. Shahid T, Mandal S, Biswal SS, et al. Preclinical validation and treatment of volumetric modulated arc therapy based total bone marrow irradiation in halcyon™ ring gantry linear accelerator. Radiat Oncol. 2022;17(1):1–1.
17. Mancosu P, Cozzi L, Muren LP. Total marrow irradiation for hematopoietic malignancies using volumetric modulated arc therapy: a review of treatment planning studies. Phys Imaging Radiat Oncol. 2019;1(11):47–53.
18. Quast U. Dosimetry of total body irradiation. Vienna: International Atomic Energy Agency (IAEA); 1988.
19. Wong JYC, Filippi AR, et al. Total body irradiation: guidelines from the International Lymphoma Radiation Oncology Group (ILROG). Int J Radiat Oncol Biol Phys. 2018;101(3):521–9. https://doi.org/10.1016/j.ijrobp.2018.04.071.
20. Willemze R, Kerl H, Sterry W, et al. EORTC classification for primary cutaneous lymphomas. A proposal from the Cutaneous Lymphoma Study Group of the European Organization for Research and Treatment of Cancer (EORTC). Blood. 1997;90:354–71.
21. Willemze R, Meijer C. EORTC classification for primary cutaneous lymphomas: the best guide to good clinical management. Am J Dermatopathol. 1999;21:265–73.
22. Girardi M, Heald PW, Wilson LD. The pathogenesis of mycosis fungoides. N Engl J Med. 2004;350:1978–88.
23. Pandolfino TL, Siegel RS, Kuzel TM, et al. Primary cutaneous B-cell lymphoma: review and current concepts. J Clin Oncol. 2000;18:2152–68.
24. Siegel RS, Pandolfino T, Guitart J, et al. Primary cutaneous Tcell lymphoma: review and current concepts. J Clin Oncol. 2000;18:2908–25.
25. Kim EJ, Hess S, Richardson SK, et al. Immunopathogenesis and therapy of cutaneous T cell lymphoma. J Clin Invest. 2005;115:798–812.
26. Willemze R, Jaffe ES, Burg G, et al. WHO-EORTC classification for cutaneous lymphomas. Blood. 2005;105:3768–85.
27. Hoppe RT. Mycosis fungoides: radiation therapy. Dermatol Ther. 2003;16:347–54.
28. Karzmark CJ, Anderson J, Buffa A, et al. Total skin electron therapy: technique and dosimetry. AAPM Task Group 30. New York: American Institute of Physics; 1987.
29. Karzmark CJ, Loevinger R, Steele RE, et al. A technique for large-field, superficial electron therapy. Radiology. 1960;74:633–44.
30. Specht L, Dabaja B, Illidge T, et al. Modern radiation therapy for primary cutaneous lymphomas: field and dose guidelines from the International Lymphoma Radiation Oncology Group. Int J Radiat Oncol Biol Phys. 2015;92:32–9.
31. Jones GW, Kacinski BM, Wilson LD, et al. Total skin electron radiation in the management of mycosis fungoides: consensus of the European Organization for Research and Treatment of Cancer (EORTC) Cutaneous Lymphoma Project Group. J Am Acad Dermatol. 2002;47(3):364–70.
32. Smith BD, Wilson LD. Cutaneous lymphomas. Semin Radiat Oncol. 2007;17(3):158–68.
33. AAPM report no. 23. Total skin electron therapy: technique and dosimetry. New York: American Institute of Physics; 1987.
34. Weaver RD, Gerbi BJ, Dusenbery KE. Evaluation of dose variation during total skin electron irradiation using thermoluminescent dosimeters. Int J Radiat Oncol Biol Phys. 1995;33:475–8.
35. Piotrowski T, Malicki J. The rotary dual technique for total skin irradiation in the treatment of mycosis fungoides—a description of the applied method. Rep Pract Oncol Radiother. 2006;11(1):29–37.
36. Dieterich S, Ford E, Pavord D, Zeng J. Chapter 3—In-vivo dosimetry. In: Dieterich S, Ford E, Pavord D, Zeng J, editors. Practical radiation oncology physics. Elsevier; 2016. p. 30–9.
37. Jones GW, Tadros A, Hodson D, et al. Prognosis with newly diagnosed mycosis fungoides after total skin electron radiation of 30 or 35 Gy. Int J Radiat Oncol Biol Phys. 1994;28:839–45.
38. Podgorsak EB, Pla C, Pla M, et al. Physical aspects of a rotational total skin electron irradiation. Med Phys. 1983;10:159–68.
39. Reynard EP, Evans MD, Devic S, et al. Rotational total skin electron irradiation with a linear accelerator. J Appl Clin Med Phys. 2008;9:2793.
40. Lloyd S, Chen Z, Foss FM, et al. Acute toxicity and risk of infection during total skin electron beam therapy for mycosis fungoides. J Am Acad Dermatol. 2013;69:537–43.
41. Beck C. On external Roentgen treatment of internal structures (eventration treatment). NY Med J. 1909;89:621–2.
42. Pilar A, Gupta M, Laskar SG, Laskar S. Intraoperative radiotherapy: review of techniques and results. Ecancermedicalscience. 2017;11:750. https://doi.org/10.3332/ecancer.2017.750.
43. Severgnini M. Special technique overview. IORT. School of medical physics for radiation therapy. Trieste: International Centre for Theoretical Science. 31 Mar 2017. http://indico.ictp.it/event/7955/session/4/contribution/32/material/slides/1.pdf.
44. Calvo FA, Sole CV, Rutten HJ, Poortmans P, Asencio JM, Serrano J, Aristu J, Roeder F, Dries WJ. ESTRO/ACROP IORT recommendations for intraoperative radiation therapy in primary locally advanced rectal cancer. Clin Transl Radiat Oncol. 2020;25:29–36. https://doi.org/10.1016/j.ctro.2020.09.001. PMID: 33005755; PMCID: PMC7519207.
45. Roeder F, Morillo V, Saleh-Ebrahimi L, Calvo FA, Poortmans P, Ferrer Albiach C. Intraoperative radiation therapy (IORT) for soft tissue sarcoma—

ESTRO IORT Task Force/ACROP recommendations. Radiother Oncol. 2020;150:293–302. https://doi.org/10.1016/j.radonc.2020.07.019. Epub 2020 Jul 15. PMID: 32679306.
46. Paunesku T, Woloschak GE. Future directions of intraoperative radiation therapy: a brief review. Front Oncol. 2017;7:300. Published 2017 Dec 12. https://doi.org/10.3389/fonc.2017.00300.
47. Vaidya JS, Wenz F, Bulsara M, et al. Risk-adapted targeted intraoperative radiotherapy versus whole-breast radiotherapy for breast cancer: 5-year results for local control and overall survival from the TARGIT-A randomised trial. Lancet. 2014;383(9917):603–13.
48. Sethi A, Emami B, Small W Jr, Thomas TO. Intraoperative radiotherapy with INTRABEAM: technical and dosimetric considerations. Front Oncol. 2018;8:74. https://doi.org/10.3389/fonc.2018.00074.
49. Calvo FA, Krengli M, Asencio JM, Serrano J, Poortmans P, Roeder F, Krempien R, Hensley FW. ESTRO IORT Task Force/ACROP recommendations for intraoperative radiation therapy in unresected pancreatic cancer. Radiother Oncol. 2020;148:57.
50. Fastener G, Gaisberger C, Kaiser J, Scherer P, Ciabattoni A, Petoukhova A, Sperk E, Poortmans P, Calvo FA, Sedlmayer F, Cristina Leonardi M. ESTRO IORT Task Force/ACROP recommendations for intraoperative radiation therapy with electrons (IOERT) in breast cancer. Radiother Oncol. 2020;149:150. https://doi.org/10.1016/j.radonc.2020.04.059.
51. Cifarelli CP, Jacobson GM. Intraoperative radiotherapy in brain malignancies: indications and outcomes in primary and metastatic brain tumors. Front Oncol. 2021;11:768168. https://doi.org/10.3389/fonc.2021.768168.
52. Banerjee S, Goyal S, Kataria T, Gupta D, Patro KC, Malik A, Bisht SS, Abhishek A. Re-irradiation in gynecological cancers, present experiences and future hopes. J Radiat Oncol. 2018;7(3):205–11.
53. Veronesi U, Orecchia R, Maisonneuve P, et al. Intraoperative radiotherapy versus external radiotherapy for early breast cancer (ELIOT): a randomised controlled equivalence trial. Lancet Oncol. 2013;14(13):1269–77.
54. Vaidya JS, Vaidya UJ, Baum M, Bulsara MK, Joseph D, Tobias JS. Global adoption of single-shot targeted intraoperative radiotherapy (TARGIT-IORT) for breast cancer—better for patients, better for healthcare systems. Front Oncol. 2022;12:3667.
55. Seegenschmiedt MH, Fessenden P, Vernon CC, editors. Thermoradiotherapy and thermochemotherapy, vol. 1. Biology, physiology and physics. Berlin: Springer; 1995.
56. Szasz A, Szasz N, Szasz O. Oncothermia: principles and practices. Dordrecht: Springer; 2010.
57. Webster JG, editor. Encyclopedia of medical devices and instrumentation. Hoboken, NJ: Wiley; 2006.
58. Vander Vorst A, Rosen A, Kotsuka Y. RF/microwave interaction with biological tissues. Hoboken, NJ: Wiley; 2006.
59. Bickels J, Kollender Y, Merinsky O, Meller I. Coley's toxin: historical perspective. Isr Med Assoc J. 2002;4(6):471–2.
60. Dahl O, Overgaard J. A century with hyperthermic oncology in Scandinavia. Acta Oncol. 1995;34(8):1075–83.
61. Grey AW. History of biological effects and medical applications of microwave energy. IEEE Trans Microwave Ther Tech. 1984;MTT-32:882–1200.
62. Fenn AJ. Adaptive phased array thermotherapy for cancer. Artech House; 2009.
63. Thews O, Li YP, Kelleher DK, Chance B, et al. Microcirculatory function, tissue oxygenation, micro-regional redox status, and ATP distribution in tumors upon localized infrared-A-hyperthermia at 42°C. Adv Exp Med Biol. 2003;530:237–47. https://doi.org/10.1007/978-1-4615-0075-9-23.
64. Overgaard J, Grau C, Lindegaard JC, Horsman MR. The potential of using hyperthermia to eliminate radioresistant hypoxic cells. Radiother Oncol. 1991;20(Suppl 1):113–6. https://doi.org/10.1016/0167-8140(91)90197-0.
65. Harmon BV, Corder AM, Collins RJ, et al. Cell death induced in a murine mastocytoma by 42-47 C heating in vitro: evidence that the form of death changes from apoptosis to necrosis above a critical heat load. Int J Radiat Biol. 1990;58(5):845–58.
66. Oei AL, Vriend LEM, Crezee J, et al. Effects of hyperthermia on DNA repair pathways: one treatment to inhibit them all. Radiat Oncol. 2015;10:165. https://doi.org/10.1186/s13014-015-0462-0.
67. Toraya-Brown S, Fierlng S. Local tumor hyperthermia as immunotherapy for metastatic cancer. Int J Hyperth. 2014;30(8):531–9.
68. Overgaard J. Simultaneous and sequential hyperthermia and radiation treatment of an experimental tumour and its surrounding tissue in vivo. Int J Radiat Biol. 1980;6:1507–17.
69. Datta NR, Rogers S, Ordonez SG, et al. Hyperthermia and radiotherapy in the management of head and neck cancers; a systematic review and meta-analysis. Int J Hyperth. 2015;32(1):31–40.
70. Fotopoulou C, Cho CH, Kraetschell R, et al. Regional abdominal hyperthermia combined with systemic chemotherapy for treatment of patients with ovarian cancer relapse: results of a pilot study. Int J Hyperth. 2010;26(2):118–26.
71. Peeken JC, Vaupel P, Combs SE. Integrating hyperthermia into modern radiation oncology:what evidence is necessary? Front Oncol. 2017;7:132. https://doi.org/10.3389/fonc.2017.00132. PMID: 28713771.
72. Sethi M, Chakravarti SK. Hyperthermia techniques for cancer treatment: a review. Int J Pharmtech Res. 2015;8(6):292–9.
73. Burchardt E, Roszak A. Hyperthermia in cervical cancer-current status. Rep Pract Oncol Radiother. 2018;23(6):595–603. https://doi.org/10.1016/j.rpor.2018.05.006. Epub 2018 Jun 15. PMID: 30534024 PMCID: PMC6277277.

74. Jha S, Sharma PK, Malviya R, et al. Hyperthermia role and risk factor for cancer treatment. Achiev Life Sci. 2016;10(2):161–7.
75. Beating breast cancer with hyperthermia Cancer Institute. https://www.youtube.com/watch?v=e1Dj5AJcoDK. Last Accessed 26 Jan 2022.
76. How hyperthermia jump starts the immune modulation effect to fight cancer. https://www.youtube.com/watch?V=6G8k1vXC-bw. Last Accessed 26 Jan 2022.
77. Putting the heat on tumors. https://www.youtube.com/watch?v=PaUc6cyaKH4.
78. Prostatic cancer treatment/hyperthermia in combination with HDR in first UMMC patient. https://www.youtube.com/watch?v=EdSqhbLIVfo. Last Accessed 26 Jan 2022.
79. Intravenous magnetic nanoparticle hyperthermia-Video abstract 43770. https://www.youtube.com/watch?v=IpaAoDW3nlc. Last Accessed 26 Jan 2022.
80. Hyperthermia for the treatment of bladder cancer. https://www.youtube.com/watch?gu3CzXEgmlA. Last Accessed 26 Jan 2022.
81. Hyperthermia treatment for inoperable brain tumor at Barnes-Jewish Hospital. https://www.barnesjewish.org/Medical-Services/Neurology-Neurosurgery/Brain-and-Spine-Tumors/Hyperthermia-Treatment-for-Brain-Tumors. Last Accessed 26 Jan 2022.
82. Systematic review of hyperthermia for recurrent breast cancer. https://www.youtube.com/watch?v=bOVisuSLREA. Last Accessed 26 Jan 2022.

Further Reading

https://www.youtube.com/watch?V=6G8k1vXC-bw
https://www.youtube.com/watch?v=bOVisuSLREA
https://www.youtube.com/watch?v=e1Dj5AJcoDK
https://www.youtube.com/watch?v=EdSqhbLIVfo
https://www.youtube.com/watch?v=RuM3M0wM_1E

Special Techniques: II SRS, SRT, and SBRT

19

Anusheel Munshi

Stereotaxy is a high-precision technique of radiation therapy used in oncology [1, 2]. Stereotaxy combines the advantages of image guidance and highly conformal and highly hypofractionated radiation dose delivery to treat tumors over a short time. The advent of the gamma knife in the 1950s was the first step toward stereotaxy, and this was aptly termed stereotactic radiosurgery (SRS). Subsequently various organizations have given their definitions, with most limiting the term to treatment deliver within 1–5 fractions. For patients treated with precision technique, but using a higher number of fractions, the term stereotactic radiotherapy (SRT) is used. Cranial stereotaxy paved the way for extracranial stereotaxy that included sites such as the spine, lung, and liver [3]. For extracranial stereotaxy, the terms stereotactic body radiotherapy (SBRT) and stereotactic ablative radiotherapy (SABR) have been used. One of the first clinical experiences was initiated at Karolinska University Hospital, Sweden, in 1991 and was closely followed by Japan in 1994. This technique not only decreased the duration of treatment but also matched the survival rates of surgically treated patients [4, 5]. Over the last few years, there has been a dramatic rise in research as well as practice of stereotaxy (Fig. 19.1).

Are there any crucial differences between stereotaxy and conventional radiation? SBRT and SRS represent a combination of several steps related to immobilization, planning and treatment delivery (Table 19.1) (Fig. 19.2).

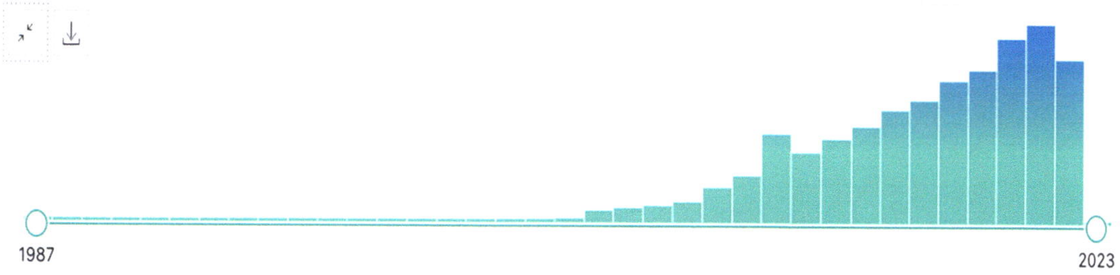

Fig. 19.1 Source: Pubmed search on November 30, 2022, with the keyword "SBRT"

A. Munshi (✉)
Department of Radiation Oncology, IOCI,
Fortis Noida, Uttar Pradesh, India

A. Munshi et al. (eds.), *Radiation Oncology – Principles, Precepts and Practice*,
https://doi.org/10.1007/978-981-97-8389-2_19

Table 19.1 Salient features of SBRT/SRS that distinguish it from conventional radiation

Integration of modern imaging, simulation, treatment planning, and delivery technologies
Much higher dose per fraction
More number of beams used
Frequent use of non-coplanar beams
Small or no beam margins for penumbra
Use of inhomogeneous dose distributions/dose-painting techniques, including IMRT
Integration of modern imaging, simulation, treatment planning, and delivery technologies
Much higher dose per fraction

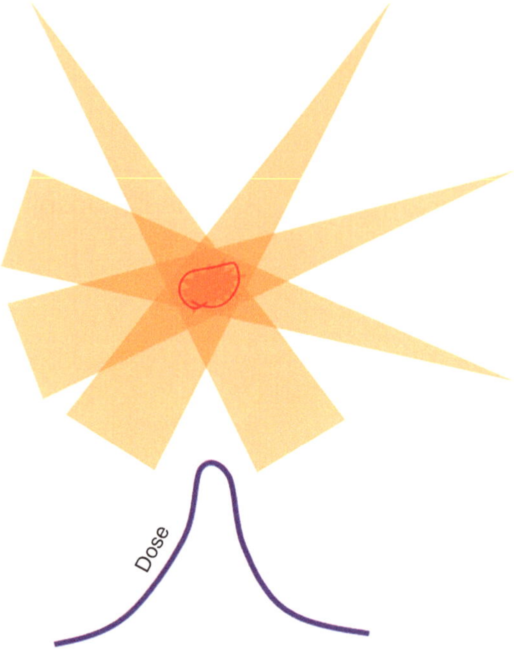

Fig. 19.2 Multiple beams fired on the target result in a central hot area with rapid fall off of dose beyond

19.1 SBRT and Hypo SRT: Radiobiology

The traditional principle of radiation therapy has been that more fractionation translated into improved radiotherapy efficacy. In SBRT and SRS, high doses per fraction are used (usually >6 Gy per fraction). These high doses per fraction in SRT and SRS lead to lethal/supralethal effects. This subsequently causes ablation and direct endothelial apoptosis. Organs at risk are prevented from serious damage by rapid fall off of dose. For different dose fractionation schedules, the EQD2 would vary depending on the alpha/beta value (Fig. 19.3).

Being local treatment modalities, SBRT, surgery, and radiofrequency ablation (RFA) can occasionally seem competitors for achieving the same goal. However, more often than not, a given clinical situation clearly indicates a natural front-runner (Table 19.2).

Fig. 19.3 EQD2 as a function of the total dose, dose per fraction, and alpha/beta

	LR		Late Effects			
Alpha/Beta (Gy)	12	10	3	2	1.5	1
Dose Schedule	EQD2	EQD2	EQD2	EQD2	EQD2	EQD2
74/37#	74	74	74	74	74	74
64/32#	64	64	64	64	64	64
60/30#	60	60	60	60	60	60
55/20#	58	58.5	63	65	66	68
54/3#	115	126	226	270	300	342

Table 19.2 Difference between surgery and SBRT

	Surgery	SBRT
Type of procedure	Diagnostic, staging, and therapeutic	Only therapeutic
	Inpatient procedure	Outpatient procedure
	Invasive	Noninvasive
	Advantage of complete removal of the primary tumor	Does not remove the tumor but damages the DNA of tumor cells and prevents them from multiplying
No. of days involved in the treatment	Extended recovery period	Can vary from 1 day to 10 days
Side effects/complications	Bleeding, infection, post-operative side effects	Related to the site being treated
Mortality	Very low	Uncommon (usually lower than surgery)
Local control	Good to excellent	Comparable to surgery
Overall survival	Good to excellent	Comparable to surgery
Cause-specific survival	Good to excellent	Comparable to surgery

19.2 Sites Where SRS/SRT/SBRT Is Commonly Used

1. Thorax—Primary lung carcinoma (early stage), lung metastases (solitary)
2. Spine—Metastases (solitary or multiple)
3. Gastrointestinal—Hepatocellular carcinoma, liver metastases, pancreatic carcinoma
4. Genitourinary—Prostate carcinoma, renal cell carcinoma
5. Cranial lesions (Benign lesions such as meningioma, pituitary adenoma, acoustic neuroma, or malignant such as brain metastasis or a recurrent glioma)
6. Nonspinal bone metastasis

Table 19.3 gives the suitable indications/criteria for SBRT/SRS of different sites.

Table 19.3 Suitable indications/criteria for SBRT/SRS of different sites

Lung SBRT	Liver SBRT	Spinal SBRT	Cranial SRS	Pancreatic SBRT	Prostate SBRT
Metastatic disease					
≤3 lung metastasis (and controlled systemic disease)	≤3 metastases; no lesion >6 cm No lesion immediately adjacent to the GI tract (distance >6 mm) Adequate pre-treatment baseline liver function. Child–Pugh A or early Child–Pugh B	Localized spine metastasis from C1 to L5 (solitary metastasis or 2 separate spine levels or 3 separate sites) and each of the separate sites must have a maximal involvement of 2 contiguous vertebral bodies. Spinal compression (only if there is a gap of >3 mm between the spinal cord and the edge of an epidural lesion). Paraspinal mass (<5 cm) contiguous with spine metastasis. Intractable pain	Brain metastasis (usually 1–3 lesions)	NA	NA
Primary disease					
T1 N0 M0/ T2 N0 M0 primary tumor Medically inoperable	≤6 cm lesion, Child–Pugh A	NA	Meningioma, pituitary adenoma, acoustic neuroma, recurrent glioma, trigeminal neuralgia	Locally advanced unresectable pancreatic cancer	Early/ intermediate-risk prostate cancer

19.3 GTV, CTV, and PTV Concepts in SRS and SBRT

Generally, no CTV margin is given in stereotaxy cases. CTV is generally considered equal to GTV. No additional margin is given for a microscopic disease. This approach is indirectly supported by data from trials/ excellent local control.

The PTV is created by adding a margin to GTV/CTV. It is based on institutional experience and shall vary according to several factors. An important factor in SBRT is the technique of 4D treatment (Breath-hold, Gating, MIP) (Fig. 19.4).

Factors to be taken into account include set-up uncertainties, internal organ motion, respiration, tumor location in the lung, and fixation to adjacent structures.

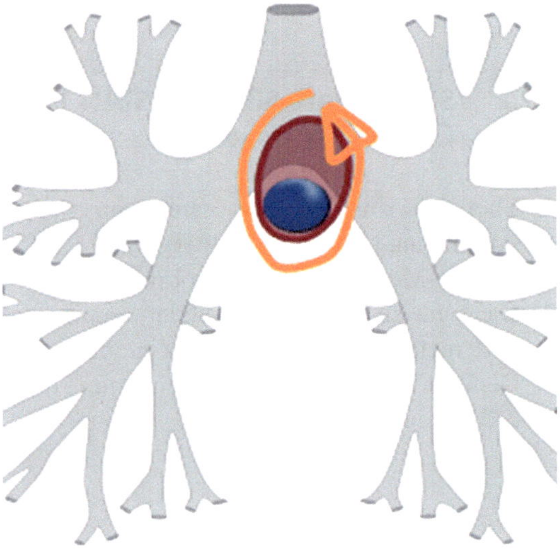

Fig. 19.4 Concept of maximum intensity projection (MIP); MIP image aims to capture all the possible positions where the high-density region has moved during respiratory cycle. (Blue= tumour, Crimson Red= envelope within which tumour moves, orange arrow shows the motion trajectory)

19.4 Dosimetry Issues

SBRT and SRS typically treat small volumes. Cranial SRS can sometimes treat very small volumes (e.g., small acoustic neuromas), and this can bring in special dosimetry issues. As per ICRU 91, the "Absorbed dose is prescribed to the isodose surface that covers an optima percentage volume of PTV while optimally restricting the dose to the PRV" Median dose to GTV is a very robust parameter and needs to be frequently used in reporting. PTV min and PTV median are other parameters that should be reported. For dose reporting, the report recommends the following metrics:

(a) D50% (PTV median absorbed dose): ICRU 91 recommends a CTV be defined for each case [6]. The D50% can also be reported for the CTV. In the specific case of peripheral lung lesions (for which the absorbed-dose distribution is strongly affected by tissue-density variations), a dose to a target that does not include uninvolved lung parenchyma, D50% (GTV / CTV) should be systematically reported.
(b) The SRT near-maximum dose, Dnear-max: For PTV larger or equal to 2 cm^3, the *near-max* volume represents 2% of the PTV, as recommended in ICRU Report 83 (ICRU, 2010) (D2%). For a PTV of less than 2 cm^3, near-max is an absolute volume of 35 mm^3, in which case $D35\ mm^3$ is reported.
(c) The SRT near-minimum dose, Dnear-min: For PTV larger or equal to 2 cm^3, the *near-min* volume represents 2% of the PTV, as recommended in ICRU Report 83 (ICRU, 2010) (D98%). For PTV of less than 2 cm^3, near-min is an absolute volume of 35 mm^3, in which case $D\ (V\text{—}35\ mm^3)$ is reported.

It is therefore recommended to always report both the dose level as well as the volume percentage. It is also important to report D98%, D50%, and D2% for the PTV as well as the CTV. Regarding the OAR, it is important to keep the dose gradient beyond PTV as sharp as possible. Occasionally, to save an OAR, it is a useful exercise to make a substructure and put a dose objective on it so as the steer away the high dose to the area where the clinician/physicist would like to have it (Fig. 19.5).

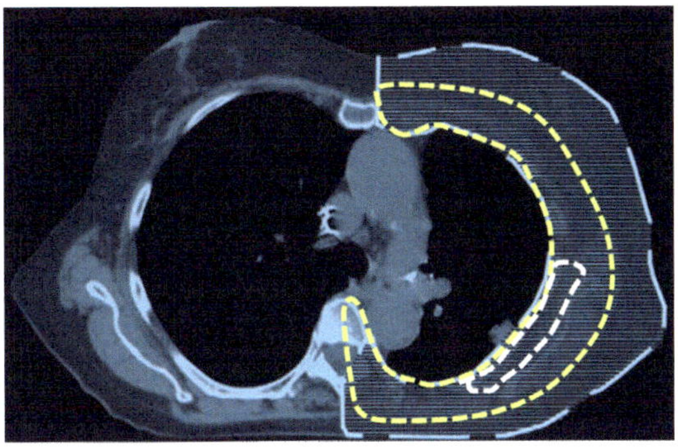

Fig. 19.5 Lung SBRT: Note the typical contour of the chest wall (yellow dotted line) and a smaller portion of the chest wall drawn near the tumor (white dotted line)

Historically, SBRT fraction regimens were delivered on alternate days. There is low-level evidence suggesting that alternate-day regimens have a better therapeutic ratio. In cases where the patient is keen, or treatment needs to be completed early, daily fractionation can be used. Cranial SRS/SRT is generally delivered on consecutive days.

19.5 Concepts of Motion and 4D

Respiratory motions, in their simplest forms, represent a sine wave (Fig. 19.6). Physiological respiratory motion of primary lung tumors may reduce the chances of obtaining an optimal local control rate after RT.

Every center needs to define a threshold of tumor motion where 4 D radiotherapy should be considered. This varies between 5 mm to 7.5 mm [7]. In general, the techniques employed for 4 D treatment include breath hold, respiratory gating and tracking. Internal gating using internal tumor motion surrogates such as implanted fiducial markers. External gating uses external respiratory surrogates such as makers placed on the patient's abdomen. Time weighted treatments figure the position where tumor spends >70–80% of the entire motion cycle and treat in that zone.

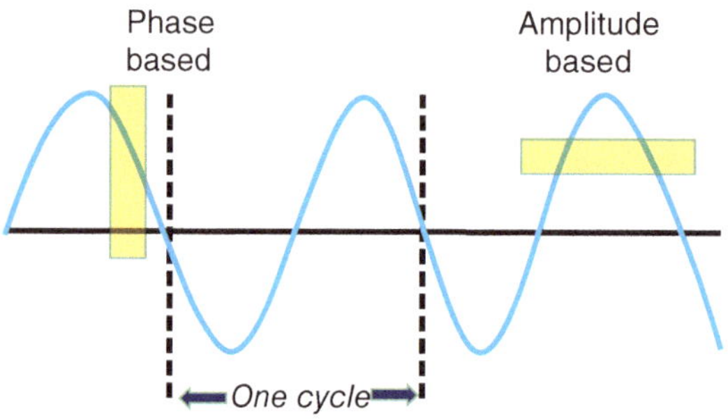

Fig. 19.6 The concept of the respiratory cycle, phase, and amplitude

Inspiratory Phase	Expiratory Phase
Less Time spent	More Time spent
Lung expanded V 20 low	Lung compressed V 20 higher
Less density of lung	More density of lung

Fig. 19.7 Differences between inspiratory and expiratory phase of lung for treatment planning and delivery

Further, it needs to be decided whether the selected phase of respiration should lie in end-inspiration or end-expiration [8].

Respiratory gating can be phase based or amplitude based (Fig. 19.6).

Average movement in normal breathing can vary depending on the location of the lesion in lung [9]. For upper lobe the lesion can move up to nearly 0.5 cm. For the lower lobe, the lesions can move between 1.5 and 4.0 cm. For the middle lobe lesion, the motion can be between 0.5 and 2.5 cm, while for the hilar lesion, the movement can be up to 1.0–1.5 cm. In a study from Tata Memorial Hospital, 100 consecutive patients with lung cancer planned for radical RT/chemoradiation were accrued and 87 patients were eligible for analysis at RT conclusion. BH was well tolerated by most lung cancer patients with minimum median BH time of at least 16 s in any of the three phases of respiration. Respiratory training improved mean BH time while consistently maintaining DIBH and DEBH times throughout the course of radiotherapy [10].

Concept of 4D simulation and treatment is relevant for SBRT treatment of thorax and abdomen, sites which have maximum positional effects related to respiration. Figure 19.7 gives the important differences between inspiratory and expiratory phase of Lung for treatment planning and delivery.

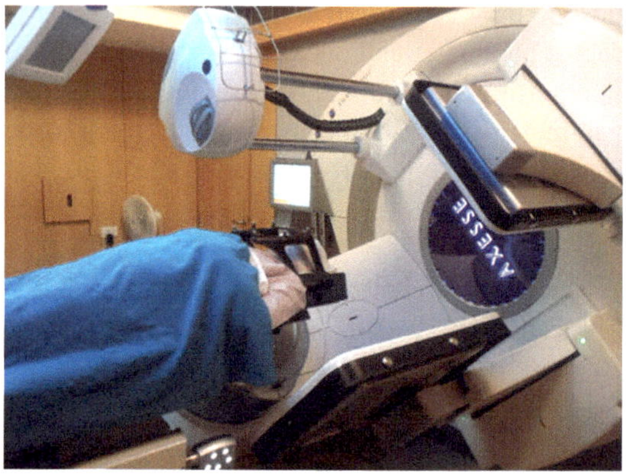

Fig. 19.8 Acquisition of cone beam CT scan

19.6 Image Guidance During SRS/ SBRT

Various tools are available image guidance during SRS/SBRT. These include traditional methods like Electronic portal imaging devices, which were based on MV imaging and had poor resolution. More modern linear accelerators are equipped with on board imaging devices such as cone beam CT (Fig. 19.8) which allow matching of images to the CT taken at the time of simulation. Other systems include surface-guided radiotherapy (SGRT) which is based on mapping of patients' skin surface by means of lasers.

19.7 MRI-Guided RT

MR-guided radiotherapy (MRgRT) has made online adaptive radiotherapy (ART) possible. This allows for both high-resolution three-dimensional (3D) magnetic resonance imaging (MRI) scans, allowing for accurate daily visualization of the disease and OARs, as well as real-time cine imaging to accurately gate a radiation treatment in the proper phase of the respiration cycle [11, 12].

Having one or more critical OARs in close vicinity to the PTV region is one of the strongest indications of using MRI guided RT (Fig. 19.9).

In these cases, only slight changes in volumes or relative positions between the PTV and the OARs can result in significant differences in both PTV coverage and OAR dose. One important downside with MRI is ensuring the spatial fidelity of the image data, causing shifts in the images of structures. This depends on the main magnetic field strength, orientation, and gradient field strength. For lung tissue, this effect can result in maximum distortions of up to 4 mm, although mean geometric distortions are substantially lower (<1 mm). Such effects are more prominent toward the periphery of the field [13, 14].

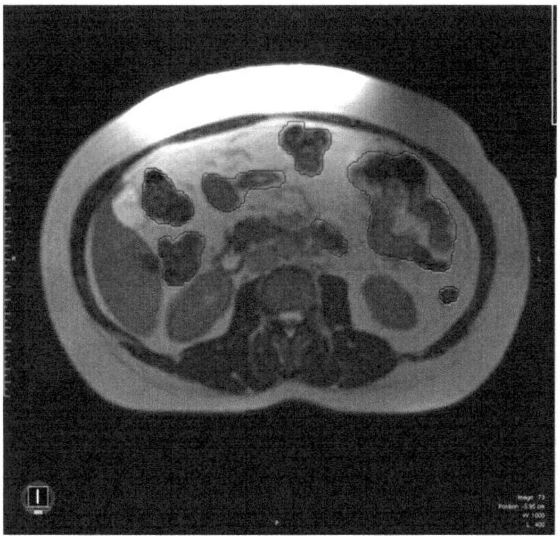

Fig. 19.9 A liver metastasis very close to bowels treated with SBRT on 0.35 Tesla MR-Linac (Courtesy Prof Banu Atalar, Maslak Acibadem Hospital, Istanbul, Turkey)

19.8 Recommendations for Any Centre Practicing SBRT/SRS

Several authors and academic bodies have made recommendations for centers practicing SBRT (Table 19.4).

A snapshot of the common fractionation used in SRS or SBRT is given in Table 19.5. The variation in dose fractionation is necessitated primarily be the vicinity of the OARs and the patient's performance status.

Table 19.4 Mandatory/highly recommended for any center practicing SBRT/SRS

Written departmental protocol covering all mandatory aspects of SBRT practice
Site specific SBRT implementation based on multidisciplinary team of Radiation Oncologists, Physicists and Radiation technologists
General radiotherapy quality assurance system involving reporting, monitoring, and correcting process deviations
Appropriate on-board imaging systems
Method for assessment of respiratory motion and factoring this in planning and delivery of treatment

Table 19.5 Common fractionation schedules used in SRS/SBRT

Lung	Liver	Spine	Brain metastasis	Pancreas	Prostate
54 Gy/3#	45 Gy/3#	16–24 Gy/1#	16–24 Gy/1#	40 Gy/5#	36.25 Gy/5#
60 Gy/5#	50 Gy/5#	27 Gy/3#	30 Gy/5#	33 Gy/5#	
50 Gy/5#	35 Gy/5#	24 Gy/3#	27 Gy/3#		
48 Gy/6#					
35 Gy/7#					

Note: Total dose and dose per fraction for each site is dependent on the local site and the proximity of organs at risk (OARs)

19.9 Lung SBRT

Carcinoma lung is the most common cause of cancer and mortality worldwide. Nearly 1.4 million new cases of lung cancer are diagnosed every year. The traditional management of early non-small cell lung cancer (NSCLC) carcinoma lung patients is surgery. It should be noted that even in early disease, 20–30% of patients are inoperable because of their comorbidities. Traditionally, these inoperable patients received conventional radical radiotherapy and had relatively poor two-year survival rates ranging from 20% to 30%. In the case of NSCLC, the advantages of high-precision and shorter overall treatment duration with this technique have translated into improved control and survival rates as well, as compared to conventional and radiotherapy.

Most patients who have received SBRT in various studies had T1 or early T2 lesions with no mediastinal or systemic metastasis. Earlier studies included only peripheral disease patients for SCRT since there were reports of severe complications for patients with central lesions. However, there is now emerging literature for treating central lesions closer to the mediastinum. However, it is to be noted that it is appropriate to avoid high-dose fractions when tumors are close to the mediastinum. Lower prescription doses (i.e., 12 Gy × 4 or 10 × 5) should be avoided for squamous cell carcinomas if normal tissue tolerances are met [15] (Table 19.6).

The usual patients who are sent to radiation oncology for SBRT include those with compromised PFT, Prior major resection as pneumonectomy, compromised FEV1 (<40%), significant IHD/ Echo abnormalities, pulmonary vascular diseases, and associated comorbidities.

19.10 SBRT Lung: A Simulation and Planning Workflow

- Supine position, and arms generally placed either above the head.
- Immobilization device is used depending on the site to be targeted. A Vac-Loc bag is useful in most instances.
- Certain positioning devices may also be used to make the patient comfortable (e.g., knee rest).
- CT scan is acquired with or without contrast with a slice thickness of less than 3 mm for radiotherapy planning.
- A 4DCT (motion) scan is acquired and decision for need of motion strategy taken.
- Delineation of target volumes.

  • GTV—gross tumor volume as seen on CT scan and other imaging (toggling between the lung and mediastinal windows).
  • PTV—depending on the immobilization method and the 4 D technique used. This can vary between 4 mm to 1 cm.

Based on the location of the lesion, it can be classified into peripheral, central, and ultracentral (Table 19.7 and Fig. 19.10).

Daily kVCBCT is used for onboard imaging (OBI) during treatment (Fig. 19.11) (Table 19.8).

There are a large number of prospective and retrospective studies assessing the role of SBRT in carcinoma lung. Most of the studies included early-stage tumors (≤5–6 cm), with very few

Table 19.6 Indications for lung SBRT

Routine indications for lung SBRT	Cautionary indications for lung SBRT
NSCLC (histopathology proven)	Lesions larger than 5 cm
Small lesions (<5 cm)	Multifocal lesions
No nodes/distant metastasis	Recurrent lesions post RT/post SBRT
Patient preference for SBRT as compared to surgery	Centrally located lesions
Compromised performance status	Severe ILD
	Consolidation SBRT boost after CTRT

NSCLC Non-small cell lung cancer, ILD Interstitial Lung Disease, CTRT Chemoradiation

Table 19.7 Comparative locations of peripheral, central and ultracentral lung tumours

Peripheral	Central	Ultracentral
Lies beyond the zone of a uniform 2-cm expansion around PBT	Lies within or touching the zone of a uniform 2-cm expansion around the PBT OR Lesion immediately adjacent to the pleura covering the mediastinum or pericardium	Lies within or is touching the zone of a uniform, 1-cm expansion around the PBT OR Lesion immediately touches/invades mediastinum, trachea, bronchus, or esophagus

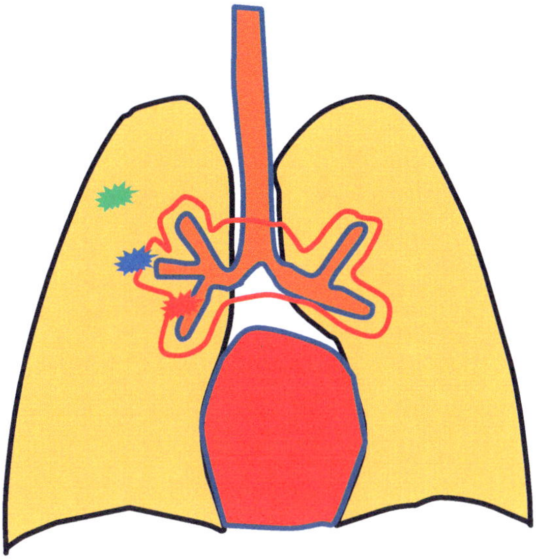

Fig. 19.10 The concept and representative location of peripheral (green), central (blue), and ultracentral (red) lesions

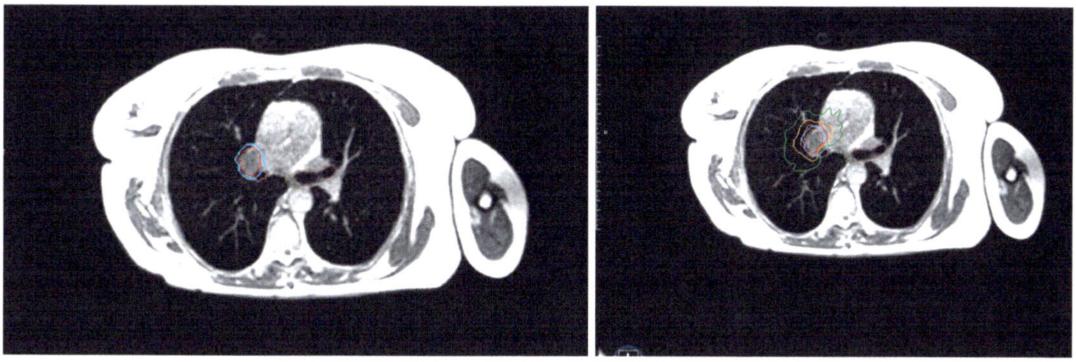

Fig. 19.11 Target volume and SBRT plan of a central lung tumor treated with 0.35 Tesla MR-Linac. (Courtesy Prof Banu Atalar, Maslak Acibadem Hospital, Istanbul, Turkey)

including T3, metastatic, and even recurrent lesions, primarily which were inoperable due to medical reasons [16–19]. The age group of subjects ranged from 50 to 78 years (median); with follow-up ranging from 11 to 90 months (median). The majority of them have been single institutional trials. The number of patients per study has varied from 30 to 250. The dose delivered and fractionation have varied from 15 Gy in a single fraction (#) to 70 Gy/10#. The

Table 19.8 Constraints for thoracic SBRT

Description	Constraint	1 fraction	3 fractions		5 fractions		8 fractions		Reference
		Mandatory	Optimal	Mandatory	Optimal	Mandatory	Optimal	Mandatory	
Heart	D_{max} *(0.5 cc)*	–	<24 Gy	<26 Gy	<27 Gy	<29 Gy	<50 Gy	<60 Gy	UK SABR Consortium [6], LungTECH [7]
Heart/pericardium	<15 cc	–	–	24 Gy	–	32 Gy	–	–	Timmerman et al. [14]
	MPD	–	–	30 Gy	–	38 Gy	–	–	
Normal lungs *(Lungs-GTV)*	$V_{20\,Gy}$	–	–	<10%	–	<10%	–	<10%	UK SABR Consortium [6], LungTECH [7]
Lungs	1500 cc	–	–	10.5 Gy	–	12.5 Gy	–	–	Timmerman et al. [14]
	1000 cc	–	–	11.4 Gy	–	13.5 Gy	–	–	
Trachea and bronchus	D_{max} *(0.5 cc)*	–	<30 Gy	<32 Gy	<32 Gy	<35 Gy	<32 Gy	<44 Gy	UK SABR Consortium [6], LungTECH [7]
	<4 cc	8.8 Gy	–	15 Gy	–	18 Gy	–	–	Timmerman et al. [14]
	MPD	22 Gy	–	30 Gy	–	38 Gy	–	–	
Brachial plexus	D_{max} *(0.5 cc)*	–	<24 Gy	<26 Gy	<27 Gy	<29 Gy	<27 Gy	<38 Gy	UK SABR Consortium [6], LungTECH [7]
	<3 cc	14.4 Gy	–	22.5 Gy	–	30 Gy	–	–	Timmerman et al. [14]
	MPD	16 Gy	–	24 Gy	–	32 Gy	–	–	
Chest wall	D_{max} *(0.5 cc)*	–	<37 Gy	–	<39 Gy	–	<39 Gy	–	UK SABR Consortium [6], LungTECH [7]
	$D_{30\,cc}$	–	<30 Gy	–	<32 Gy	–	<32 Gy	—	
Great vessels	D_{max} *(0.5 cc)*	–	–	<45 Gy	–	<53 Gy	–	–	UK SABR Consortium [6], LungTECH [7]
	<10 cc	31 Gy	–	39 Gy	–	47 Gy	–	–	Timmerman et al. [14]
	MPD	37 Gy	–	45 Gy	–	53 Gy	–	–	

D_{max} is the near-point maximum dose, defined in this case as $D_{0.5\,cc}$, which is the minimum dose to the 0.5 cc volume of the organ receiving the highest doses
MPD is the maximum point dose permissible for the concerned organ
$V_{20\,Gy}$ is the percentage volume of the organ receiving a dose of 20 Gy or higher
$D_{30\,cc}$ is the minimum dose to 30 cc of the organ that receives the highest doses

commonly used dose fractionation schedules are 54–60 Gy/3#. For lesions abutting the ribs, suggested fractionations could be 60 Gy/5#. For central lesions, the dose schedules include and 50 Gy/5#. For ultracentral lesions, dose schedules like 48 Gy/6# or 35 Gy/5# have been uses. Usually, not more than three fractions are delivered per week (alternate day treatment schedule). However recent literature does not report excess complications with the daily treatment schedule. Currently, the overall three-year survival rates with SBRT have ranged between 50% and 70%, with local control rates of >85% in most series. While the norm is to give SBRT fraction on alternate day, daily fractionation has also been used in studies [20]. Efforts are also on to make SBRT a single fraction treatment, and this would be a remarkable achievement [21].

The predominant toxicities related to SBRT lung include lung pneumonitis, rib fractures and esophagitis. These, in most cases are self-limiting. The chance of life threatening toxicities is usually 1–2% and can include hemorrhage and bronchial/esophageal necrosis.

In one of the earliest reports from India, between December 2007 and December 2010, 8 patients of early stage (T1-2N0M0) underwent Lung SBRT. Response and toxicity were assessed in 3 monthly follow-up visits. The median duration of follow-up was 18 months (range 8–44 months). The mean dose to the PTV was 99.5% (range 97.7–102.1%). Seven patients survived at 1.5 years, and one patient died due to unrelated causes. One patient had grade 2 pneumonitis. No toxicities of grade 3 or higher were identified [22].

SBRT has shown promise in the management of early NSCLC patients with poor performance status. These results have closely matched the results obtained by surgical resection. It therefore heralds a paradigm shift in the management of early-stage lung cancer, presenting the option of effective treatment by a simple, noninvasive technique.

19.11 Spine SBRT

Spinal metastases are the most common tumors of the spine. These are most commonly found as bone metastasis. Occasionally they present with symptoms of spinal canal invasion and cord compression. The following are the common criteria employed for spinal SBRT:

- Involvement of ≤3 adjacent vertebrae or multiple nonadjacent levels (separated by at least one vertebral level)
- Absence or minimal epidural disease
- Absence of spinal instability signs
- No significant compression fracture

The target volume for Spinal SBRT can be (a) Vertebral body only, (b) Vertebral body and pedicle only (c) Posterior elements "arc" only (Fig. 19.12).

The target volume is chosen by the treating radiation oncologist based on the extent of tumor involvement and the available techniques. A gap of 3 mm or more between the edge of the epidural

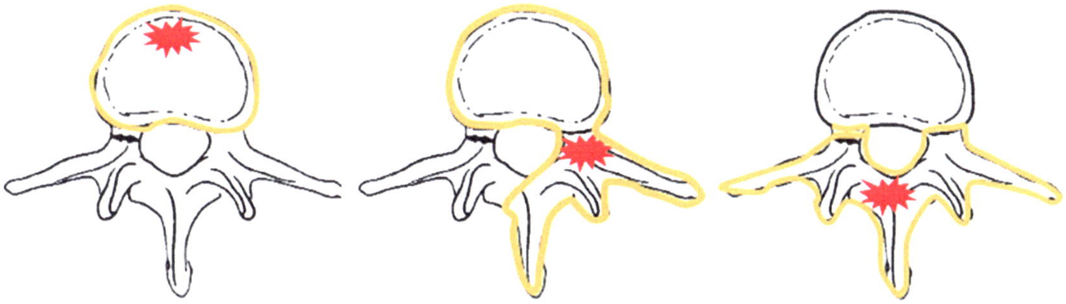

Fig. 19.12 Suggested target volumes for vertebral SBRT, based on the location of the disease

metastasis and the edge of the spinal cord is recommended in the presence of epidural extension. Concurrent SABR and chemotherapy administration is not recommended as it can increase the risk of adverse effects to adjacent organs [23]. For follow-up of treated patients, a clinically validated radiographic response criterion for spinal SABR is lacking. There is no international consensus on how to classify tumor response. The two most commonly used scales include Response Evaluation Criteria in Solid Tumors, Version 1.1, and The University of Texas MD Anderson Cancer Center's criteria.

As is the case in many other sites, caution is needed in interpreting magnetic resonance imaging changes postradiotherapy. In such situations, changes (including pseudoprogression) can be hard to distinguish from tumor progression [24].

Radiation myelopathy (RM), vertebral compression fracture (VCF), and epidural disease progression are recognized as important risks following SBRT for spine metastases. A systematic review of the literature using MEDLINE and a review of the bibliographies of reviewed articles on SBRT for spinal metastases were conducted [25]. The risk of local failure at 1 year was 21.4% (range: 12–27%). Of these, 67% occurred within the epidural space. The risk of RM following SBRT was found to be dependent on the maximum dose to the spinal cord and estimated to be ≤5%. The crude risk of VCF was 13.7%. On average, 45% of these were surgically salvaged.

In a study reported from India, 76 lesions in 52 patients with spinal metastasis received SBRT during the period July 2010 to December 2012 [26]. The median dose prescribed to involved vertebra (planning target volume) was 24 Gy (range 24–27 Gy) in a median of three fractions (range 1–3). Uninvolved elements were prescribed 21 Gy in three fractions. At a median follow-up of 8.48 months (range 3–40 months), 1 year local control and overall survival was 94 and 68%, respectively. Complete pain relief was seen in 90% of patients with malignant epidural compression (MEC) and 93.75% without MEC (P = NS) and neurological improvement was seen in 60% of patients in both groups of patients (Fig. 19.13).

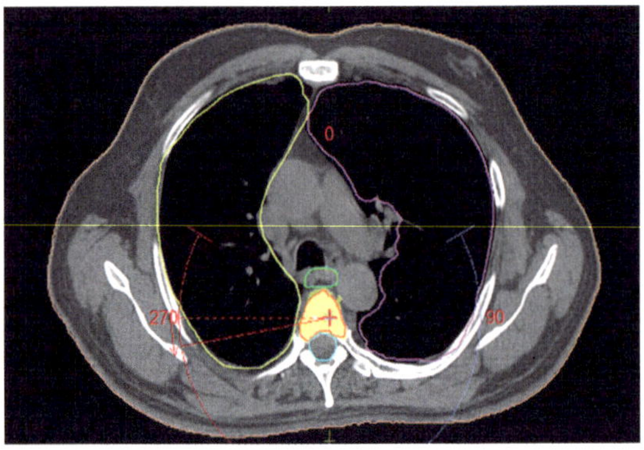

Fig. 19.13 A SBRT spine case. The case was planned rotational arc technique

19.12 Liver SBRT

Stereotactic body radiation therapy (SBRT) in the context of liver can be used for management of metastatic disease as well a primary hepatocellular carcinoma (HCC). For primary HCC, this technique results in excellent local control rates with favorable toxicity profile despite being predominantly used in heavily pretreated patients or those unsuitable for other local therapies. SBRT may be used as a sole treatment or in combination with other local therapies as well as a bridging strategy for patient awaiting liver transplants.

SBRT of liver is also a well-established treatment for metastasis to the liver. Metastatic disease of the liver is a common life-threatening complication encountered by patients with cancer. Patients with resected metastatic colorectal cancer have 5-year survival rates of 25–40%. However, surgery can often be technically challenging and is an invasive procedure. Only 10–20% of metastatic colorectal cancer patients are candidates for surgical resection. As such, SBRT has emerged as an effective and safe therapy for patients with liver oligometastatic disease (OMD). Typically, patients for SBRT liver have their metastatic disease confined to the liver. Patients with extrahepatic disease can also be taken if the hepatic metastasis is considered the most life-limiting component of their disease. Adequate liver function and a life expectancy of at least 6 months are other essentials. Also, patients with less than 1000 cm^3 of the uninvolved liver are best excluded from Liver SBRT.

The usual criteria for Liver SBRT are given in Table 19.9.

For dosimetry, all patients are required to preserve at least 700 cc of normal liver. It is recommended that no more than 30% of the normal liver receive more than 27 Gy and no more than 50% of the normal liver receive over 24 Gy. At least 700 cc of the normal liver should not receive more than 15 Gy [27] (Fig. 19.14) (Table 19.10).

For radiation planning, a biphasic contrast treatment planning computerized tomography (CT) scan is recommended. Additional imaging studies, such as magnetic resonance imaging and positron emission tomography, are sometimes fused with the planning CT scan for more accu-

Table 19.9 Selection criteria for liver SBRT

	Category of patient		
Selection criteria	Suitable	Cautionary	Unsuitable
Lesion number	<3	4	>4
Lesion diameter (cm)	1–3	>3 and ≤ 6	>6
Free liver volume (cc)	>1000	<1000 and ≥ 700	<700
Liver function	Child–Pugh A	Child–Pugh B	Child–Pugh C
Distance from OARs (in mm)	>8	5–8	<5

SBRT stereotactic body radiation therapy, *OARs* Organs at risk

Table 19.10 Constraints for gastrointestinal/liver SBRT

	Constraint	1 fraction	3 fractions		5 fractions		
Description		Mandatory	Optimal	Mandatory	Optimal	Mandatory	Reference
Duodenum	D_{max} *(0.5 cc)*	–	–	<22.2 Gy	—	<35 Gy	3 Fr: AAPM [1] 5 Fr: ABC-07 [8] SPARC Protocols [9]
	$D_{1\,cc}$	–	–	–	<33 Gy	–	
	$D_{5\,cc}$	–	–	<16.5 Gy	<25 Gy	–	
	$D_{9\,cc}$	–	–	–	<15 Gy	–	
	$D_{10\,cc}$	–	–	<11.4 Gy	—	< 25 Gy	
	<5 cc	8.8 Gy	–	15 Gy	–	18 Gy	Timmerman et al. [14]
	MPD	16 Gy	–	24 Gy	–	32 Gy	
Stomach	D_{max} *(0.5 cc)*	–	–	<22.2 Gy	<33 Gy	<35 Gy	3 Fr: AAPM [1] 5 Fr: ABC-07 [8] SPARC Protocols [9]
	$D_{5\,cc}$	–	–	–	<25 Gy	–	
	$D_{10\,cc}$	–	–	<16.5 Gy	–	<25 Gy	
	$D_{50\,cc}$	–	–	–	<12 Gy	<25 Gy	
	<10 cc	13 Gy	–	21 Gy	–	28 Gy	Timmerman et al. [14]
	MPD	16 Gy	–	24 Gy	–	32 Gy	
Small bowel	D_{max} *(0.5 cc)*	–	–	<25.2 Gy	<30 Gy	<35 Gy	3 Fr: AAPM [1] 5 Fr: ABC-07 [8] SPARC Protocols [9]
	$D_{5\,cc}$	–	–	<17.7 Gy	<25 Gy	–	
	$D_{10\,cc}$	–	–	–	–	<25 Gy	
Jejunum & ileum	<5 cc	9.8 Gy	–	16.2 Gy	–	19.5 Gy	Timmerman et al. [14]
	MPD	19 Gy	–	27 Gy	–	35 Gy	
Common bile duct	D_{max} *(0.5 cc)*	–	<50 Gy	–	<50 Gy	–	3 Fr: AAPM [1] 5 Fr: ABC-07 [8] SPARC Protocols [9]

(continued)

Table 19.10 (continued)

Description	Constraint	1 fraction	3 fractions		5 fractions		Reference
		Mandatory	Optimal	Mandatory	Optimal	Mandatory	
Esophagus	D_{max} *(0.5 cc)*	–	–	<25.2 Gy	<32 Gy	<34 Gy *(<40 Gy for 8 fr)*	3 Fr: AAPM [1] 5 Fr: ABC-07 [8] SPARC Protocols [9] LungTECH for 8 fr [7]
	<5 cc	14.5 Gy	–	21 Gy	–	27.5 Gy	Timmerman et al. [14]
	MPD	19 Gy	–	27 Gy	–	35 Gy	
Large bowel	D_{max} *(0.5 cc)*	–	–	<28.2 Gy	–	<32 Gy	3 Fr: AAPM [1] 5 Fr: ABC-07 [8] SPARC Protocols [9]
Colon	<20 cc	11 Gy	–	20.4 Gy	–	25 Gy	Timmerman et al. [14]
	MPD	22 Gy	–	30 Gy	–	38 Gy	
Rectum	D_{max} *(0.5 cc)*		–	<28.2 Gy	–	<32 Gy	AAPM [1]
	<20 cc	11 Gy	–	20.4 Gy	–	25 Gy	Timmerman et al. [14]
	MPD	22 Gy	–	30 Gy	–	38 Gy	
Normal liver *(Liver-GTV)*	$V_{10\,Gy}$	–	–	–	<70%	–	3 Fr: AAPM [1] Wulf et al. [10, 11] Rusthoven et al. [12] 5 Fr: ABC-07 [8] SPARC Protocols [9]
	Mean dose	–	–	–	<13 Gy	<15.2 Gy	
	$D_{50\%}$	–	< 15 Gy	–	–	–	
	Dose to ≥ 700 cc	–	<15 Gy	<19.2 Gy	–	–	
Liver	700 cc	–	–	17.1 Gy	–	21 Gy	Timmerman et al. [14]
Kidneys *(individual and combined)*	Mean dose	–	–	–	<10 Gy	–	3 Fr: AAPM [1] 5 Fr: ABC-07 [8] SPARC Protocols [9]
	Dose to ≥ 200 cc	–	–	<16 Gy	–	–	
If solitary kidney or if one kidney mean dose > 10 Gy	$V_{10\,Gy}$	–	–	–	<10%	<45%	5 Fr: ABC-07 [8] SPARC Protocols [9]
Renal hilum/vascular trunk	<2/3rd volume	10.6 Gy	–	18.6 Gy	–	23 Gy	Timmerman et al. [14]
Renal cortex *(both)*	200 cc	–	–	14.4 Gy	–	17.5 Gy	

D_{max} is the near-point maximum dose, defined in this case as $D_{0.5\,cc}$, which is the minimum dose to the 0.5 cc volume of the organ receiving the highest doses

$D_{1\,cc}$, $D_{5\,cc}$, $D_{9\,cc}$, $D_{10\,cc}$ and $D_{50\,cc}$ are the minimum doses to the specified volume of the organ (1 cc, 5 cc, etc.) that receive the highest doses

$V_{10\,Gy}$ is the percentage volume of the organ receiving a dose of 10 Gy or higher

Dose to ≥700 cc and ≥200 cc are the maximum doses to the specified volume of the organ (700 cc, 200 cc) that receive the lowest doses

If the total kidney volume is <200 cc, or if treating renal or adrenal lesions, then total dose to contralateral kidney should be <16 Gy and aim should be to minimize spillage into ipsilateral kidney if possible

MPD is the maximum point dose permissible for the concerned organ

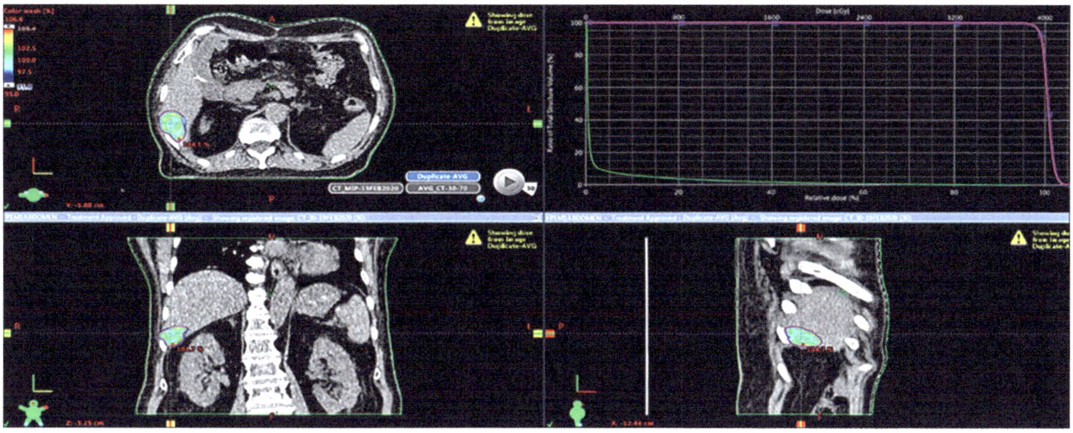

Fig. 19.14 Prescription dose color wash for SBRT liver in a case of metastatic disease

rate delineation of the gross target volume. The dose per fraction and total dose were determined using the dose volume histogram of organs at risk. Various fractionation schedules exist for Liver SBRT, including, with a preferred schedule of 50 Gy in 5 Gy fractions over 2 weeks. The normal liver is usually defined as the volume of the liver not involved by gross tumor (total liver volume). The dose to 60% of the liver volume should not exceed 30 Gy. For those patients with a history of hepatitis or cirrhosis, the dose to 70% of the volume should not exceed 27 Gy. Data from single institutions and phase 1 and 2 studies have confirmed the safety and efficacy of various SBRT dose regimens. These reveal nearly 56% to 100% 2-year actuarial LC rates and limited severe toxicity. Grade 3 or higher toxicity rates after SBRT liver are observed in <5% of all cases and nearly all of these toxicities are hepatobiliary [28]. Factors influencing SBRT dose include disease volume, number of lesions, and location of the lesions (e.g., hilar or perihilar lesions would have to be given a lesser dose and smaller fraction size). GTV volume and maximum diameter were associated with an increased risk of death for every 10 mL or for every 1 cm increase in the longest tumor diameter. Serious late SBRT-related complications, including life-threatening toxicity, could occur even months after treatment. These include ulceration, stenosis, fibrosis, and even necrosis [28, 29].

19.13 Prostate SBRT

Prostate cancer is a slow growing disease with an alpha/beta value of 1.5 to 2.5 Gy. This value may actually be lower than that of surrounding OARs (e.g., rectum alpha/beta <3 Gy). This suggests that the therapeutic ratio of external beam radiotherapy is likely to improve by hypofractionation. Usual eligibility criteria for patients include clinical stage T1-T2b, Gleason Sum <6 and PSA <10 ng/ml [30]. Some institutes follow a practice of ultrasound-guided insertion of 3 fiducial gold seeds transperineally followed by a planning CT scan. The planning and all treatments are performed in the supine position with a comfortably full bladder and empty rectum. Often the bladder protocol is followed. This involves patients being asked to empty their bowels and bladder 1 h. before simulation and treatment and drink 250–500 cc of water. A custom vacuum lock bag is used for pelvic immobilization.

The clinical target volume (CTV) is only the prostate. Generally, the seminal vesicles (SV) are not a part of CTV although some groups include proximal 1 cm of SV in intermediate-risk patients. Generally an isotropic PTV margin is given (nearly 4–5 mm, except posteriorly where it is 3–4 mm). The rectum is typically contoured as a single solid organ from the bottom of the ischium to the sigmoid flexure. The bladder and penile bulb are also contoured as single solid organs [31]. Planning objectives aim that volume of CTV receiving 35 Gy (CTV V35 Gy) should receive >99%, and PTV V33.25 Gy >99%. The maximal dose (Dmax) should be kept at <105%. The normal tissue DVH constraints are rectum V28 < 40%, rectal V32 < 33%, bladder V32 < 40%, and penile bulb V20 < 90% (Table 19.11).

Stereotactic ablative body radiotherapy (SABR) for low-risk localized prostate cancer is feasible, well-tolerated, and has an excellent long-term outcome [32]. SABR also allows greater patient convenience and can allow treatment of an increased number of patients in the radiation therapy department. Two hundred and thirty males were treated with Cyberknife SBRT to a dose of 35 Gray (Gy) or 36.25 Gy in five consecutive days [33]. All patients had a Gleason score of six and a PSA < 10 ng/ml. Median follow-up was 9 years. The 10-year actuarial biochemical disease-free survival was 93.7% and the local control was 98.4%.

In the HYPO-RT-PC trial, 1200 intermediate/high-risk prostate cancer patients were taken. Conventional # (39 × 2 Gy) was compared to ultrahypofractionation (UHF) (7 × 6.1 Gy, alternate day) [34]. Failure-free survival was similar in both arms (84% at 5 years). There was increased physician reported RTOG grade 2 or worse toxicity at end of RT and at 1-year follow-up in UHF arm. Rest of parameters of toxicity were same in both arms. In the pace B trial, low- or intermediate-risk patients were taken. SBRT was 36.25 Gy in 5 # (either given daily or alternate day). Seventy-three percent patients had fiducial placement [35]. In this study, extreme hypofractionation with SBRT did not increase either the gastrointestinal toxicity or the genitourinary toxicity. This schedule has therefore become a common schedule for SBRT prostate. There has not been any significant difference reported between alternate day or once weekly treatments in control rates or toxicity outcomes [36].

Attempts have been made to extrapolate SBRT to high-risk prostate cancer, giving 35–37.5 Gy in 5# to prostate and gross node and 25 Gy in 5# to elective pelvic nodal region on alternate days [37] with promising early outcomes. Similarly SBRT prostate has been investigated for postoperative adjuvant as well as salvage RT with encouraging preliminary data [38]. More mature data of such clinical scenarios is awaited.

Prostate is also a site commonly used for treatment in MRI linacs (Fig. 19.15) Department of Radiation Oncology.

Table 19.11 Constraints for pelvic SBRT

Description	Constraint	1 fraction Mandatory	3 fractions Mandatory	5 fractions Mandatory	Reference
Bladder	$D_{15\,cc}$	–	< 16.8 Gy	< 18.3 Gy	AAPM [1]
	D_{max} *(0.5 cc)*	–	< 28.2 Gy	< 38 Gy	
Bladder wall	< 15 cc	8.7 Gy	15 Gy	18.3 Gy	Timmerman et al. [14]
	MPD	22 Gy	30 Gy	38 Gy	
Penile bulb	$D_{3\,cc}$	–	<21.9 Gy	<30 Gy	AAPM [1]
	D_{max} *(0.5 cc)*	–	< 42 Gy	<50 Gy	
	<3 cc	14 Gy	21.9 Gy	30 Gy	Timmerman et al. [14]
	MPD	34 Gy	42 Gy	50 Gy	
Ureter	D_{max} *(0.5 cc)*	–	<40 Gy	<45 Gy	BR001 [13]

D_{max} is the near-point maximum dose, defined in this case as $D_{0.5\,cc}$, which is the minimum dose to the 0.5 cc volume of the organ receiving the highest doses

$D_{3\,cc}$ and $D_{15\,cc}$ are the minimum doses to the specified volume of the organ (3 cc, 15 cc) that receive the highest doses

MPD is the maximum point dose permissible for the concerned organ

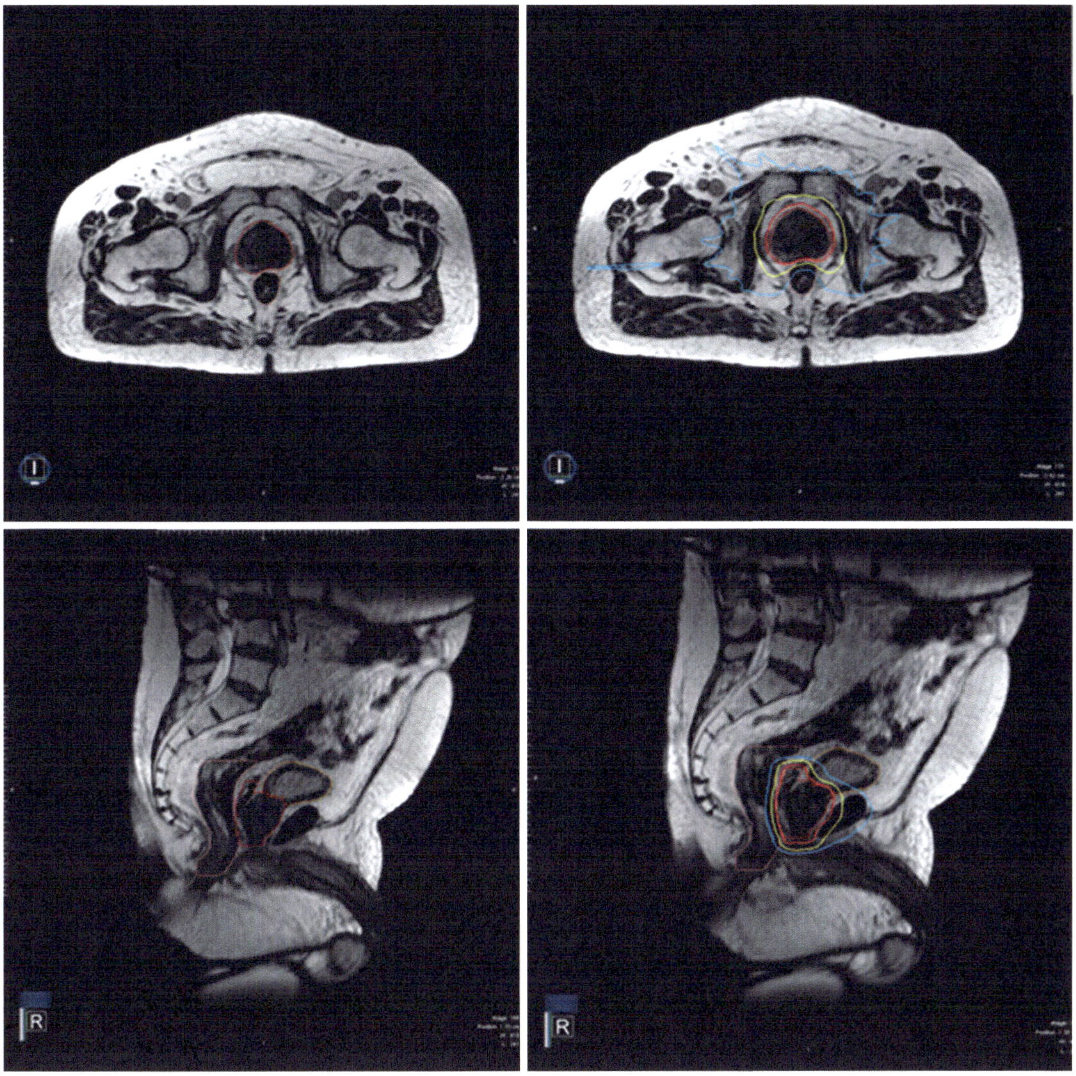

Fig. 19.15 Intermediate prostate cancer treated with SBRT on an 0.35 Tesla MR-Linac. (Courtesy Prof Banu Atalar, Maslak Acibadem Hospital, Istanbul, Turkey)

19.14 Adrenal SBRT

The adrenal gland is a common site for metastatic spread. About 50% of melanomas, 30–40% of breast and lung cancers, and 10–20% of renal and gastrointestinal tumors metastasize to the adrenal glands. In SBRT planning for adrenal metastasis, a PET scan is fused with the planning CT scan when available. The average adrenal movement of 1 cm can exist in the craniocaudal direction and <6 mm in circumferential directions and this needs to be factored. Often, based on the motion pattern, a 7–10-mm margin is added to the gross tumor volume to generate the planning target volume (PTV) [39]. Treatment is prescribed to the 100% isodose line, with the 80% line covering the PTV. The prescribed dose in literature has ranged from 50 Gy in 10 fractions to 48 Gy in 8 fractions.

In a report from Hokkaido University, 10 patients received SBRT after placement of internal fiducial markers, with most receiving 48 Gy in eight fractions and 80% isodose covering the PTV. No patient developed appreciable toxicity during follow-up (range, 5–21 months). Five tumors achieved a complete response, two had a partial response, and four were stable, according to the RECIST. The 1-year overall survival and local control rate was 78% and 100%, respectively.

In another study, of the 30 patients, 14 with five or fewer metastatic lesions (including adrenal) underwent SBRT The 1-year survival, local control, and distant control rate was 44%, 55%, and 13%, respectively [40]. In the follow-up, the evaluation of the response is done using the Response Evaluation Criteria in Solid Tumors (RECIST) on an abdominal CT or PET scan.

19.15 Renal Cell Carcinoma (RCC) SBRT

Patients with larger (T1b, >4 cm) renal cell carcinoma (RCC) not suitable for surgery have few treatment options because thermal ablation is less effective in this setting. SBRT represents an effective, safe, and nephron-sparing alternative for large RCC. Currently, dose-escalation to 48 Gy in 3–4 fractions are effective and well tolerated. Emerging role of immune therapies in RCC patients warrants further studies to explore interactions between SBRT and immune response.

After SBRT, there can be a decrease in eGFR decreased by 8–10 mL/min. Three to 5% patients may require dialysis. Grade 1–2 toxicity can be experienced by 20–30% patients. Local control rates are >95% [41]. However, larger series of SBRT to renal lesions are still awaited.

19.16 Locally Recurrent Head and Neck SCC

Stereotactic body radiation therapy is a viable treatment option for patients with unresectable, recurrent head and neck cancer. In selected cases, significant tumor control rates are achievable with minimal severe toxicity Stereotactic body radiotherapy is a promising but still evolving strategy for the treatment of locally recurrent, unresectable head and neck cancers following definitive radiotherapy. Typically doses of 35–40 Gy in 5–6 fractions are employed. A study included 414 patients with unresectable rSCCHN: 217 treated with IMRT and 197 with SBRT [42]. The unadjusted 2-year OS rate was 35.4% for IMRT and 16.3% for SBRT ($P < 0.01$). Among SBRT patients, RPA classification retained an independent association with OS. Further subset analysis demonstrated comparable OS when ≥35 Gy was delivered with SBRT to small tumor volumes.

It is now known that late carotid rupture is a rare but concerning late adverse event. Proximity to vessels and involvement of skin are important risk factors in this regard. Still, reirradiation using stereotactic body radiotherapy can offer patients a potential logistical advantage compared with other palliative treatments, because it requires just 1 to 2 weeks of treatment.

The literature on SBRT in primary nonmetastatic head neck cancer is sparse and evolving. Potential applications of SBRT in primary untreated nonmetastatic head neck cancer are being explored [43]. Similarly SBRT boost following external beam radiotherapy could emerge as an important technique of dose escalation.

19.17 Nonvertebral Bone Metastasis SBRT

Bone metastases are frequent in cancer patients. They can be distressingly painful or may cause pathological fractures. The use of CTV (Clinical Target Volume) in such scenarios has not been defined. Doses and fractions used have varied from 15 to 24 Gy/1 fraction to 24–50 Gy in 3–5 fractions, with local control being around 90% with a low rate of side effects [44]. More research in volume delineation determining whether or not to use CTV and the role of MRI in volume contouring, optimal doses, and fractionation according to histology and a reliable response assessment tool.

19.18 Pancreas SBRT

Most pancreatic lesions (75%) are located in the head of the pancreas, while the rest (25%) are located in the tail region. For locally advanced unresectable lesions, the options include chemotherapy alone, chemotherapy followed by chemoradiation, chemotherapy followed by SBRT or SBRT followed by chemotherapy. Patients with an inflammatory abscess in the pancreas or those with wide infiltration into duodenum or stomach would not be appropriate candidates for SBRT.

For simulation, it is generally not advisable to use abdominal compression, since it may actually push the duodenum close to the pancreas. The critical OARs include the stomach duodenum and the small bowel. In the Stanford experience of 167 patients, single fraction (25 Gy) had much higher toxicity than 6.6 Gy × 5# [45].

Most studies of SBRT pancreas have been Phase II studies [46, 47]. These studies had approximately 15–20 patients, and reported a median OS of 11–12 months. The reported local control post SBRT was 80–90%. For five fractions the general duodenal constraint is D1cc < 26 Gy. In some situations, when the critical OARs are away, the dose can be escalated to 8 Gy × 5 # [48]. In case of frank duodenal infiltration, SBRT could still have a palliative goal with reduced doses such as 5–6 Gy × 5#. Similarly, in case of nodal involvement, the target volume increases significantly and so the dose may be reduced to 25 Gy in 5#.

OAR constraints and target coverage need to be balanced to optimize control rates with acceptable toxicity [45–49]. There is some risk of bleeding post SBRT pancreas. The risks of gastric/duodenal bleed are around 3–5%, and the risk of grade ≥ 3 GI toxicity is around 4–5%.

In view of the dynamically changing nature of the duodenal position, it is not uncommon to find the duodenum not abutting the tumor in simulation images and then find it closely abutting the tumor in the onboard imaging. As such a good onboard imaging (CBCT or on board MRI) is critical for pancreatic SBRT. Further it is also useful to have a motion management strategy in place. The additional benefit of SBRT in the overall scheme of things is that the radiation gets over early and so systemic therapy can be initiated early on. In case chemotherapy has been given, it is advisable to wait for 1–2 weeks before starting SBRT [49]. In view of proximity of structures such as duodenum and the daily variation in these structures modalities such as MRI guided RT are also utilized for treatment of these lesions (Fig. 19.16).

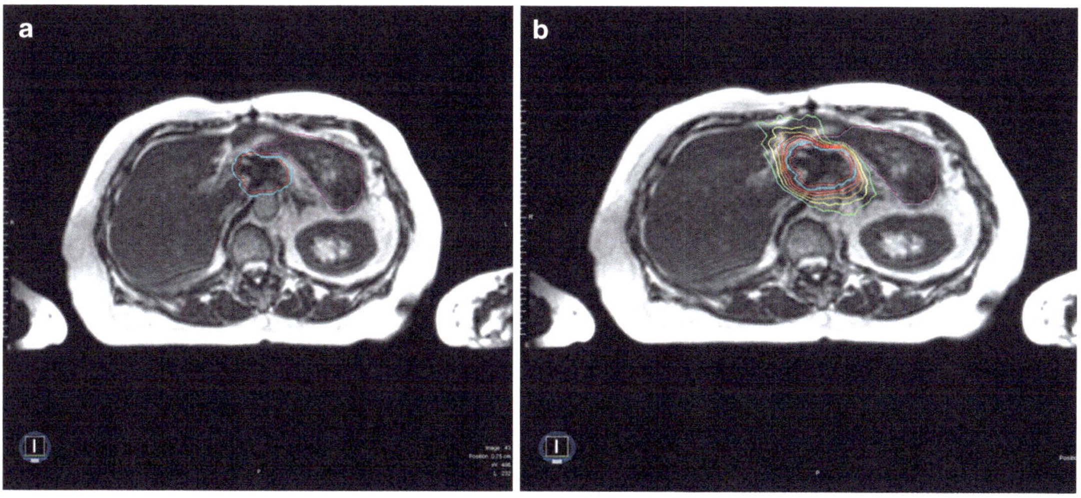

Fig. 19.16 The GTV and PTV of a pancreatic cancer (**a**) and the isodoses (**b**). Patient was treated with SBRT on a 0.35 Tesla MR-Linac. (Courtesy Prof Banu Atalar, Maslak Acibadem Hospital, Istanbul, Turkey)

In most studies, the 1-year freedom from local progression rate post SBRT has been nearly 80%. And the 1-year OS and PFS rates were are around 60% and 40% respectively. In around 70% patients, distant metastasis was the first site of failure. Local progression was the first site of failure in <15% cases [50].

19.19 Brain SRS/SRT

SRS in brain can be employed for a number of indications (Table 19.12). One of the common indications for brain SRS is metastasis to the brain. The brain is a common site of metastasis. Causes of death in patients with BM include systemic progression (one-third of patients), intracranial progression (one-third of patients), and combination (one-third of patients). Crucial for predicting the individual prognosis is Karnofsky's performance status, age, extracranial tumor activity, number of brain metastasis, primary tumor type/ molecular subtype [51]. Table 19.12 gives a comparison of approaches (SRS versus open surgery) in the context of brain metastasis.

Initially, brain SRS was done using an invasive frame as immobilization. Stereotaxy for brain is now commonly done in modern linear accelerators (Figs. 19.17 and 19.18).

This particular set has a posterior thermoplastic component (hidden behind head), an anterior component which also allows the placement of nasal bridge and the mouth bite (see arrows) as shown in Fig 19.18a and the final thermoplastic mask anteriorly. All these are intended to give robustness to immobilization.(For typical cranial stereotaxy, only the face has the anterior thermoplastic mask).

Table 19.12 SRS versus open surgery in brain metastasis

SRS	Open surgery
Minimal mass effect	Mass effect (especially in posterior fossa)
Eloquent areas/ noneloquent areas	Noneloquent areas
Typically 1–3 lesions	Typically one lesion
Tumor <30 mm	Surgically accessible tumor, esp. if >30 mm
Deep lesion	Seizure control
Radioresistant histology	Uncertain diagnosis

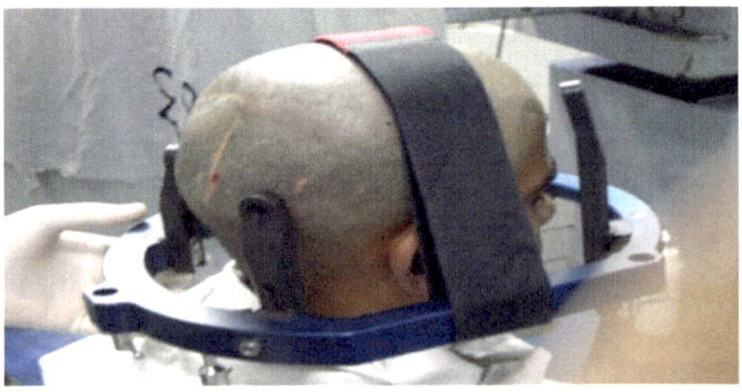

Fig. 19.17 Stereotaxy with invasive frame (for a brain lesion)

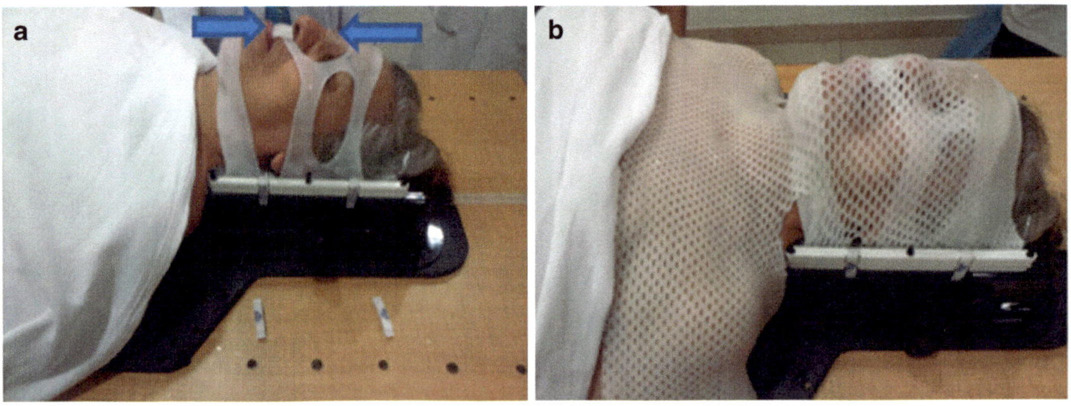

Fig. 19.18 (**a** and **b**): Frameless stereotaxy (for stereotaxy of lower cranial/upper neck region)

19.19.1 Frameless Stereotaxy

In recent years, focus has moved to stereotaxy without a rigid frame, also known as frameless stereotaxy (Fig 19.18). The advantage is patient comfort, ease of on board imaging and the flexibility of using multiple fractions.

For effective SRS, the cumulative volume of metastasis should ideally be <10 cc [52].

Studies have explored the efficacy and outcomes of radiosurgery in patients with greater number of lesions. In a multicentric Japanese study, 1–10 newly diagnosed brain metastases (largest tumor <10 mL in volume and < 3 cm in longest diameter) were analyzed. The total cumulative volume was ≤15 mL. KPS score of 70 or higher were taken in the study [53]. Overall survival did not differ between the patients with two to four tumors and those with five to ten (HR 0.97, 95% CI 0.81–1.18, $P = 0.78$).

Approach for patient with metastasis close to critical structures/larger lesions has also been studied. Potential toxicities with single fraction SRS in larger lesions include necrosis, edema and steroid dependence.

Two hundred eighty-nine consecutive patients with brain metastases >2.0 cm who received SRS were analyzed. There was better Local Control with 3# (3 × 9 Gy) [54] (Fig. 19.19) (Table 19.13).

Tumors of 21–40 mm were 7.3 to 16 times more likely to develop grade 3–5 neurotoxicity compared to tumors <20 mm [55]. Other variables significantly associated with grade 3–5 neurotoxicity were tumor dose and Karnofsky Performance Status. The actuarial incidence of radionecrosis was 5%, 8%, 9%, and 11% at 6, 12, 18, and 24 months.

Postoperative SRS also has an important role in controlling brain metastasis. A study assigned completely resection of 1–3 brain metastases (max dia 4 cm) to either SRS treatment of the resection cavity (within 30 days of surgery) or observation (OBS) [56]. The 12-month local tumor recurrence-free rates were 43% (OBS) and 72% (SRS) ($P = 0.015$).

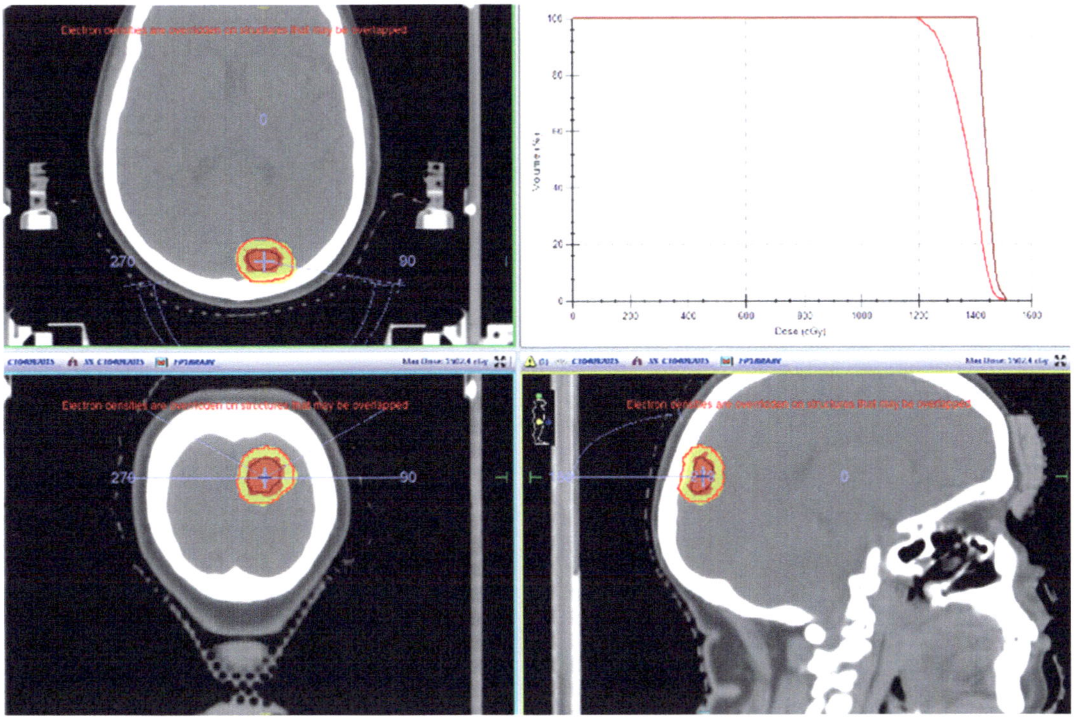

Fig. 19.19 Cranial metastasis treated by SRS

Table 19.13 Stereotactic SRS for brain metastasis (dose fractionation). (Based on RTOG 9005 study)

SRS (single fraction)—Recommendations on the 80% isodose		
24 Gy	Volume	<=20 mm or <=4.3 cm^3 18 Gy
18 Gy	Volume	21–30 mm or 4.3–14.1 cm^3
15 Gy	Volume	31–40 mm or > 14.1 cm^3

Note: SRS dose is inversely related to tumor size. The dose has to be adapted to tumor location, organs at risk, and volume of V12

A new prognostic marker for brain metastasis has been recently described by some researches. iBMV = BM number at the first stereotactic radiosurgery divided by the Time (in years) since the initial primary diagnosis [57]. Patients with a score of <2 perform better than patient with score >2. Radiosurgery has also been used for small recurrent gliomas. Typically, doses of 25–35 Gy in 5–6 fractions have been used in such situations (Table 19.14).

Table 19.14 Constraints for cranial SRS/SRT

Description	Constraint	1 fraction		3 fractions		5 fractions		8 fractions		Reference
		Optimal	Mandatory	Optimal	Mandatory	Optimal	Mandatory	Optimal	Mandatory	
Optic pathway	D_{max} *(0.1 cc)*	–	<8 Gy	–	<15 Gy	–	<22.5 Gy	–	25 Gy	AAPM [1], Hiniker et al. [2]
	<0.2 cc	–	8 Gy	–	15 Gy	–	20 Gy	–	–	Timmerman et al. [14]
	MPD	–	10 Gy	–	19.5 Gy	–	25 Gy	–	–	
Brainstem	D_{max} *(0.1 cc)*	<10 Gy	<15 Gy	<18 Gy	<23.1 Gy	<23 Gy	<31 Gy	–	–	AAPM [1], Tamura et al. [3]
	<1 cc	–	10 Gy	–	18 Gy	–	26 Gy	–	–	Timmerman et al. [14]
	MPD	–	15 Gy	–	23 Gy	–	31 Gy	–	–	
	MPD	12.5 Gy	15–20 Gy							Mayo Clinic [16]
Lens	D_{max} *(0.1 cc)*	<1.5 Gy	–	–	–	–	–	–	–	Hanna et al. [4]
Cochlea	Mean	<4 Gy	<9 Gy	–	< 17.1	–	<25 Gy	–	–	AAPM [1], Tamura et al. [3]
	MPD	–	12 Gy	–	20 Gy	–	27.5 Gy	–	–	Timmerman et al. [14]
Orbit	D_{max} *(0.1 cc)*	<8 Gy	–	–	–	–	–	–	–	Hanna et al. [4]
Normal brain *(whole brain-PTV)*	$D_{10\,cc}$	<12 Gy	–	–	–	–	–	–	–	Hanna et al. [4]
	$D_{50\%}$	<5 Gy	–	–	–	–	–	–	–	
Spinal canal *(including medulla)*	D_{max} *(0.1 cc)*	<10 Gy	<14 Gy	<18 Gy	<21.9 Gy	<23 Gy	<30 Gy	<25 Gy	<32 Gy	AAPM [1], Grimm et al. [5] UK SABR Consortium [6] LungTECH [7]
	$D_{1\,cc}$	<7 Gy	–	<12.3 Gy	–	<14.5 Gy	–	–	–	
Spinal cord	<0.25 cc	–	10 Gy	–	18 Gy	–	22.5 Gy	–	–	Timmerman et al. [14]
	<1.2 cc	–	7 Gy	–	11.1 Gy	–	13.5 Gy	–	–	
	MPD	–	14 Gy	–	22 Gy	–	30 Gy	–	–	
	D_{max} *(<0.035 cc)*	–	14 Gy	–	–	–	–	–	–	RTOG 0613 [15]
	D_{max} *(<0.35 cc)*	–	10 Gy	–	–	–	–	–	–	
	D_{max} *(<1.2 cc)* [SBRT only]	–	7 Gy	–	–	–	–	–	–	

(continued)

Table 19.14 (continued)

Description	Constraint	1 fraction		3 fractions		5 fractions		8 fractions		Reference
		Optimal	Mandatory	Optimal	Mandatory	Optimal	Mandatory	Optimal	Mandatory	
Cauda equina and sacral plexus	D_{max} *(0.1 cc)*	–	<16 Gy	–	<24 Gy	–	<32 Gy	–	–	AAPM [1]
	$D_{5\ cc}$	–	<14 Gy	–	<22 Gy	–	<30 Gy	–	–	
Cauda equina	<5 cc	–	14 Gy	–	21.9 Gy	–	30 Gy	–	–	Timmerman et al. [14]
	MPD	–	16 Gy	–	24 Gy	–	34 Gy	–	–	
	D_{max} *(<0.035 cc)*	–	16 Gy	–	–	–	–	–	–	RTOG 0613 [15]
	D_{max} *(<5 cc)*	–	14 Gy	–	–	–	–	–	–	
Sacral plexus	<3 cc	–	14.4 Gy	–	22.5 Gy	–	30 Gy	–	–	Timmerman et al. [14]
	MPD	–	16 Gy	–	24 Gy	–	32 Gy	–	–	
	D_{max} *(<0.035 cc)*	–	18 Gy	–	–	–	–	–	–	RTOG 0613 [15]
	D_{max}*(<5 cc)*	–	14.4 Gy	–	–	–	–	–	–	

D_{max} is the near-point maximum dose, defined in this case as $D_{0.1\ cc}$, which is the minimum dose to the 0.1 cc volume of the organ receiving the highest doses. If the volume is specified (like 0.035 cc, 0.35 cc, 1.2 cc, 5 cc), then it is the maximum dose to the specified volume of the organ receiving the highest dose

$D_{1\ cc}$, $D_{1.2\ cc}$, $D_{5\ cc}$ and $D_{10\ cc}$ are the minimum doses to the specified volume of the organ (1 cc, 1.2 cc, 5 cc, 10 cc) that receive the highest doses

$D_{50\%}$ is the median dose to the volume (equal to the minimum dose to the 50% of the volume receiving the highest doses)

MPD is the maximum point dose permissible for the concerned organ

19.19.2 SRS for Benign Cranial Lesions

Finally, radiosurgeries are used in more benign conditions such as meningioma, pituitary adenoma, acoustic neuroma and trigeminal neuralgias [58–60] (Table 19.15).

In a large meta-analysis related to radiosurgery in meningiomas, the estimate of disease control rate ranged from 87.0% to 100.0% at 5 years (67.0% to 100.0% at 10 years). The PFS rate ranged 78.0–98.9% at 5 years and 53.1–97.2% at 10 years, respectively. The overall symptom control was 92.3%, while the overall toxicity was 8.1% [61].

In a meta-analysis of radiosurgery for acoustic neuroma, the mean deterioration ratio for patients with serviceable hearing before treatment was 49% for SRS and 45% for fractionated stereotactic radiotherapy (FSRT), respectively. The risk for facial nerve deterioration was 3.6% for SRS and 11.2% for FSRT and for trigeminal nerve deterioration 6.0% for SRS and 8.4% for FSRT [62].

In a meta-analysis of pituitary adenoma, overall tumor control and five-year progression-free survival (PFS) estimate after one GK procedure for NFPA was 93% (95% CI 89–97%) and 95% (95% CI 91–99%), respectively. In case of secreting pituitary adenomas, overall remission (cure without need for medication) estimates were 45% (95% CI 35–54%) for GH-secreting adenomas, 64% (95% CI 0.52–0.75%) for ACTH-secreting adenomas, and 34% (95% CI: 19–48%) for PRL-secreting adenomas [63].

In a study analyzing trigeminal neuralgias, the results for 65 studies (6461 patients) were reported: 45 Gamma Knife RS (GKS) studies (5687 patients [88%]), 11 linear accelerator (LINAC) RS studies (511 patients [8%]), and 9 CyberKnife RS (CKR) studies (263 patients [4%]). With the exception of one prospective study, all studies were retrospective. The mean maximal doses were 71.1–90.1 Gy (prescribed at the 100% isodose line) for GKS, 83.3 Gy for LINAC, and 64.3–80.5 Gy for CKR (the latter two prescribed at the 80% or 90% isodose lines, respectively). Recurrence rates ranged from 0% to 52.2% (mean 24.6%, median 23%) for GKS, from 19% to 63% (mean 32.2%, median 29%) for LINAC, and from 15.8% to 33% (mean 25.8%, median 27.2%) for CKR [64] (Table 19.16).

Table 19.15 Single fraction SRS doses in benign brain lesions

Indication	SRS dose (single fraction)
Meningioma	12–16 Gy
Pituitary adenoma	12–20 Gy (nonfunctional) 18–30 Gy (functional)
Acoustic neuroma	12–16 Gy
Trigeminal neuralgia	50–70 Gy

Table 19.16 Challenges for SRS, SRT, and SBRT

- Issue of dose-volume constraints of varied hypofractionated regimens.
- Optimum motion management for each patient.
- Adequate reporting/long-term FU.
- Issue of dose-volume constraints of varied hypofractionated regimens.
- Data collaboration and joint publications.

19.20 Upcoming Advances in SBRT

Radiation has potential immunostimulatory effects. These effects amplify antitumor response when combined with immune checkpoint inhibition. Local radiotherapy can therefore lead to systemic immunity through an in-situ vaccination effect. The resultant responses can happen in distant nonirradiated sites and are known as abscopal effects. The mechanism of such effects is not completely understood. One possible hypothesis has suggested that DNA damage leads to enhanced cross-priming of naive T cells [65]. This technique of treatment, combining SBRT and immunotherapy is known as iSBRT.

Cardiac SBRT is an upcoming field related to use of SBRT as a tool in atrial and ventricular cardiac arrhythmias. The aim of treatment with SBRT is to produce a scaring of the aberrant electrical pathway that is causative in producing atrial or ventricular arrhythmias. High doses of radiotherapy used in a single fraction are capable of causing endothelial damage to the small vessels and have the capability to affect neuronal/electrical pathways of the heart. All studies used a uniform cardiac dose of 25 Gy [66]. Primary SBRT for the Breast is another emerging indication for SBRT for women wishing to avoid surgery [67]. Mature data on this is awaited.

Other advances in radiation therapy that are likely to influence and possibly improve outcomes include Proton SBRT, FLASH RT, and Artificial Intelligence (AI) based SBRT. The ultimate aim is to deliver personalized SBRT where each patient receives customized doses based on the tumor's molecular and other characteristics [68].

19.21 Published Guidelines for SBRT

Different organizations and academic bodies have issues guidelines from time to time. These include ESTRO ACROP consensus guideline for non-small cell lung cancer, EORTC recommendations for high-precision radiotherapy, lung cancer, SEOR SBRT-SG stereotactic body radiation therapy consensus guidelines for nonspine bone metastasis, UK Consensus on Normal Tissue Dose Constraints for Stereotactic Radiotherapy, and others [69–80].

Various SBRT protocols for different sites give an insight into the SBRT processes, techniques, and constraints in greater detail (Table 19.17).

Table 19.17 RTOG SBRT protocols

Site	RTOG protocol
Lung	RTOG 0236
Lung	RTOG 0813
Lung	RTOG 0915
Spine	RTOG 0631
Liver	RTOG 0438
Liver	RTOG 1112
Pancreas	RTOG 9704
Prostate	RTOG 0938

Precept Box

1. SBRT, SRT, and SRS are high-precision techniques for treatment of cranial and extracranial sites.
2. Most common sites for stereotaxy include brain metastasis, small early lung lesions, early prostate cancer, and vertebral metastasis.
3. Techniques such as surgery and radiofrequency ablation (RFA) are alternative techniques for local treatment of disease.
4. Defining dose constraints for different organs at risk for different dose fractionations remains a challenge.
5. Basic infrastructure and protocols related to SBRT need to be implemented for all centers practicing SBRT.

Multiple Choice Questions

1. One Salient feature of SBRT which distinguishes it from conventional radiation includes:
   (a) Lower dose per fraction
   (b) More number of beams
   (c) Large beam margins for penumbra
   (d) No use of non-coplanar beams
2. What Clinical Target Volume (CTV) margins are recommended for stereotaxy cases?
   (a) 1–2 mm
   (b) 3 mm
   (c) 4 mm
   (d) CTV is equal to GTV
3. What parameters are used for dose reporting in stereotaxy?
   (a) D50%
   (b) Dear min
   (c) Dear max
   (d) All the above
4. Average movement is maximum for which lobe of lung?
   (a) Upper
   (b) Middle

   (c) Lower lobe
   (d) All have same range of movement
5. What is common fractionation for brain SRS?
   (a) 33 Gy/5 fractions
   (b) 16–24 Gy/1 fraction
   (c) 40 Gy/5 fractions
   (d) 54 Gy/3 fractions
6. Which of the following is routine indication for lung SBRT?
   (a) Multifocal lesions
   (b) Centrally located lesion
   (c) Small lesion (<5 cm)
   (d) Consolidation SBRT boost after CTRT
7. Ultracentral lung lesions are defined as
   (a) Lies beyond the zone of a uniform 2-cm expansion around PBT
   (b) Lies within or touching the zone of a uniform 2-cm expansion around the PBT
   (c) Lies within or is touching the zone of a uniform, 1-cm expansion around the PBT
   (d) All the above
8. What is optimal dose constrain of heart, Dmax (0.5 cc) for 3 fraction SBRT?
   (a) <24 Gy
   (b) <26 Gy
   (c) <27 Gy
   (d) <28 Gy
9. How many contiguous vertebrae can be treated in spine SBRT?
   (a) <2
   (b) ≤3
   (c) ≤4
   (d) <6

Answers

1. b
2. d
3. d
4. c
5. b
6. c
7. c
8. a
9. b

References

1. Lartigau E. Stereotactic body radiotherapy. BMJ. 2011;343:d4286.
2. Munshi A, Krishnatry R, Banerjee S, Agarwal JP. Stereotactic conformal radiotherapy in non-small cell lung cancer—an overview. Clin Oncol (R Coll Radiol). 2012;24:556–68.
3. Guckenberger M, Andratschke N, Alheit H, et al. Definition of stereotactic body radiotherapy: principles and practice for the treatment of stage I non-small cell lung cancer. Strahlenther Onkol. 2014;190:26–33.
4. Japa Onishi H, Shirato H, Nagata Y, Hiraoka M, Fujino M, Gomi K, Karasawa K, Hayakawa K, Niibe Y, Takai Y, Kimura T, Takeda A, Ouchi A, Hareyama M, Kokubo M, Kozuka T, Arimoto T, Hara R, Itami J, Araki T. Stereotactic body radiotherapy (SBRT) for operable stage I non-small-cell lung cancer: can SBRT be comparable to surgery? Int J Radiat Oncol Biol Phys. 2011;1:81.
5. Tandberg DJ, Tong BC, Ackerson BG, Kelsey CR. Surgery versus stereotactic body radiation therapy for stage I non–small cell lung cancer: a comprehensive review. Cancer. 2018;124:667–78.
6. https://www.icru.org/report/icru-report-91-prescribing-recording-and-reporting-of-stereotactic-treatments-with-small-photon-beams-2/. Last accessed 30 Nov 2022.
7. AAPM TG 76. https://www.aapm.org/pubs/reports/rpt_91.pdf. Last accessed 30 Nov 2022.
8. Wink N, Panknin C, Solberg TD. Phase versus amplitude sorting of 4D-CT data. J Appl Clin Med Phys. 2006;7:77–85.
9. Seppenwoolde Y, Shirato H, Kitamura K, Shimizu S, van Herk M, Lebesque JV, Miyasaka K. Precise and real-time measurement of 3D tumor motion in lung due to breathing and heartbeat, measured during radiotherapy. Int J Radiat Oncol Biol Phys. 2002;53:822–34.
10. Tibdewal A, Munshi A, Pathak R, Misra S, Daptardar A, Singh V, Agarwal JP. Breath-holding times in various phases of respiration and effect of respiratory training in lung cancer patients. J Med Imaging Radiat Oncol. 2015;59:520–6.
11. Henke L, Kashani R, Yang D, et al. Simulated online adaptive magnetic resonance-guided stereotactic body radiation therapy for the treatment of oligometastatic disease of the abdomen and central thorax: characterization of potential advantages. Int J Radiat Oncol Biol Phys. 2016;96:1078–86.
12. Henke L, Kashani R, Robinson C, et al. Phase I trial of stereotactic radiation therapy (SMART) for the treatment of oligometastatic or unresectable primary malignancies of the abdomen. Radiother Oncol. 2017;126:519–26.
13. Moerland M, Beersma R, Bhagwandien R, Wijrdeman H, Bakker C. Analysis and correction of geometric distortions in 1.5 T magnetic resonance images for

use in radiotherapy treatment planning. Phys Med Biol. 1995;40:1651.

14. Stanescu T, Wachowicz K, Jaffray D. Characterization of tissue magnetic susceptibility-induced distortions for MRIgRT. Med Phys. 2012;39:7185–93.
15. Shiue K, Cerra-Franco A, Shapiro R, Estabrook N, Mannina EM, Deig CR, et al. Histology, tumor volume, and radiation dose predict outcomes in NSCLC patients after stereotactic ablative radiotherapy. J Thorac Oncol. 2018;13:1549–59.
16. Onishi H, Kuriyama K, Komiyama T, et al. Clinical outcomes of stereotactic radiotherapy for stage I non-small cell lung cancer using a novel irradiation technique: patient self-controlled breath-hold and beam switching using a combination of linear accelerator and CT scanner. Lung Cancer. 2004;45:45–55.
17. Oshiro Y, Aruga T, Tsuboi K, et al. Stereotactic body radiotherapy for lung tumors at the pulmonary hilum. Strahlenther Onkol. 2010;186(5):274–9. https://doi.org/10.1007/s00066-010-2072-y. Epub 2010 Apr 26
18. Dunlap NE, James M, Paul W, et al. Size matters: a comparison of T1 and T2 peripheral non-small-cell lung cancers treated with stereotactic body radiation therapy (SBRT). Thorac Cardiovasc Surg. 2010;140:583–9.
19. Chen Y, Guo W, Lu Y, Zou B. Dose-individualized stereotactic body radiotherapy for T1-3N0 non-small cell lung cancer: long-term results and efficacy of adjuvant chemotherapy. Radiother Oncol. 2008;88:351–8.
20. Stauder MC, Macdonald OK, Olivier KR, Call JA, Lafata K, Mayo CS, et al. Early pulmonary toxicity following lung stereotactic body radiation therapy delivered in consecutive daily fractions. Radiother Oncol. 2011;99:166–71.
21. Videtic GM, Stephans KL, Woody NM, Reddy CA, Zhuang T, Magnelli A, Djemil T. 30 Gy or 34 Gy? Comparing two single-fraction SBRT dose schedules for stage I medically inoperable non-small cell lung cancer. Int J Radiat Oncol Biol Phys. 2014;90:203–8.
22. Kundu S, Mathew A, Munshi A, Prabhash K, Pramesh CS, Agarwal JP. Stereotactic body radiotherapy in early-stage non-small cell lung cancer: first experience from an Indian Centre. Indian J Cancer. 2013;50:227–32.
23. Bahig H, Simard D, Létourneau L, Wong P, Roberge D, Filion E, et al. A study of pseudoprogression after spine stereotactic body radiation therapy. Int J Radiat Oncol Biol Phys. 2016;96:848–56.
24. Tseng CL, Eppinga W, Charest-Morin R, Soliman H, Myrehaug S, Maralani PJ, et al. Spine stereotactic body radiotherapy: indications, outcomes, and points of caution. Global Spine J. 2017;7:179–97.
25. Chang JH, Shin JH, Yamada YJ, Mesfin A, Fehlings MG, Rhines LD, Sahgal A. Stereotactic body radiotherapy for spinal metastases: what are the risks and how do we minimize them? Spine (Phila Pa 1976). 2016;41(Suppl 20):S238–45.
26. Anand AK, Venkadamanickam G, Punnakal AU, Walia BS, Kumar A, Bansal AK, Singh HM. Hypofractionated stereotactic body radiotherapy in spinal metastasis—with or without epidural extension. Clin Oncol (R Coll Radiol). 2015;27:345–52.
27. Chmura SJ, Winter K, Salama JK, et al. Phase I trial of stereotactic body radiation therapy (SBRT) to multiple metastatic sites: a NRG oncology study. Int J Radiat Oncol Biol Phys. 2018;102:S68–9.
28. Andolino DL, Forquer JA, Henderson MA, Barriger RB, Shapiro RH, Brabham JG, et al. Chest wall toxicity after stereotactic body radiotherapy for malignant lesions of the lung and liver. Int J Radiat Oncol Biol Phys. 2011;80:692–7.
29. van der Pool AE, Mendez Romero A, Wunderink W, Heijmen BJ, Levendag PC, Verhoef C, et al. Stereotactic body radiation therapy for colorectal liver metastases. Br J Surg. 2010;97:377–82.
30. Haque W, Butler EB, Teh BS. Stereotactic body radiation therapy for prostate cancer-a review. Chin Clin Oncol. 2017;6(Suppl 2):S10.
31. Cha E, Elguindi S, Onochie I, Gorovets D, Deasy JO, Zelefsky M, Gillespie EF. Clinical implementation of deep learning contour autosegmentation for prostate radiotherapy. Radiother Oncol. 2021;159:1–7.
32. Zhao X, Ye Y, Yu H, Jiang L, Cheng C, Guo X, Ju X, Zhu X, Zhang H. Five-year outcomes of stereotactic body radiation therapy (SBRT) for prostate cancer: the largest experience in China. J Cancer Res Clin Oncol. 2021;147(12):3557–64.
33. Katz A. Stereotactic body radiotherapy for low-risk prostate cancer: a ten-year analysis. Cureus. 2017;9(9):e1668.
34. Widmark A, Gunnlaugsson A, Beckman L, Thellenberg-Karlsson C, Hoyer M, Lagerlund M, et al. Ultra-hypofractionated versus conventionally fractionated radiotherapy for prostate cancer: 5-year outcomes of the HYPO-RT-PC randomised, non-inferiority, phase 3 trial. Lancet. 2019;394:385–95.
35. Brand DH, Tree AC, Ostler P, van der Voet H, Loblaw A, Chu W, et al. Intensity-modulated fractionated radiotherapy versus stereotactic body radiotherapy for prostate cancer (PACE-B): acute toxicity findings from an international, randomised, open-label, phase 3, non-inferiority trial. Lancet Oncol. 2019;20:1531–43.
36. Zilli T, Jorcano S, Bral S, Rubio C, Bruynzeel AME, Oliveira A, Abacioglu U, Minn H, Symon Z, Miralbell R. Once-a-week or every-other-day urethra-sparing prostate cancer stereotactic body radiotherapy, a randomized phase II trial: 18 months follow-up results. Cancer Med. 2020;9:3097–106.
37. Telkhade T, Murthy V, Kanala TS, Mathew JM, Phurailatpam R, Mokal S, Chourasiya D, Panigrahi G, Krishnatry R. Safety and efficacy of ultra-hypofractionation in node-positive prostate cancer. Clin Oncol (R Coll Radiol). 2021;33(3):172–80.
38. Ballas LK, Luo C, Chung E, Kishan AU, Shuryak I, Quinn DI, Dorff T, Jhimlee S, Chiu R, Abreu A, Jennelle R, Aron M, Groshen S. Phase 1 trial of SBRT

to the prostate fossa after prostatectomy. Int J Radiat Oncol Biol Phys. 2019;104(1):50–60.
39. Katoh N, Onimaru R, Sakuhara Y, et al. Real-time tumor-tracking radiotherapy for adrenal tumors. Radiother Oncol. 2008;87:418–24.
40. Chawla S, Chen Y, Katz AW, Muhs AG, Philip A, Okunieff P, Milano MT. Stereotactic body radiotherapy for treatment of adrenal metastases. Int J Radiat Oncol Biol Phys. 2009;75:71–5.
41. Siva S, Correa RJM, Warner A, Staehler M, Ellis RJ, Ponsky L, Kaplan ID, Mahadevan A, Chu W, Gandhidasan S, Swaminath A, Onishi H, Teh BS, Lo SS, Muacevic A, Louie AV. Stereotactic ablative radiotherapy for ≥T1b primary renal cell carcinoma: a report from the International Radiosurgery Oncology Consortium for Kidney (IROCK). Int J Radiat Oncol Biol Phys. 2020;108(4):941–9.
42. Vargo JA, Ward MC, Caudell JJ, Riaz N, Dunlap NE, Isrow D, et al. A multi-institutional comparison of SBRT and IMRT for definitive reirradiation of recurrent or second primary head and neck cancer. Int J Radiat Oncol Biol Phys. 2018;100:595–605.
43. Swain M, Ghosh-Laskar S. Stereotactic body radiotherapy (SBRT) for primary non-metastatic head and neck cancer: when less is enough. Oral Oncol. 2021;116:105265.
44. De la Pinta C. SBRT in non-spine bone metastases: a literature review. Med Oncol. 2020;37:119.
45. Pollom EL, Alagappan M, von Eyben R, Kunz PL, Fisher GA, Ford JA, Poultsides GA, Visser BC, Norton JA, Kamaya A, Cox VL, Columbo LA, Koong AC, Chang DT. Single- versus multifraction stereotactic body radiation therapy for pancreatic adenocarcinoma: outcomes and toxicity. Int J Radiat Oncol Biol Phys. 2014;90:918–25.
46. Koong AC, Christofferson E, Le QT, Goodman KA, Ho A, Kuo T, Ford JM, Fisher GA, Greco R, Norton J, Yang GP. Phase II study to assess the efficacy of conventionally fractionated radiotherapy followed by a stereotactic radiosurgery boost in patients with locally advanced pancreatic cancer. Int J Radiat Oncol Biol Phys. 2005;63:320–3.
47. Schellenberg D, Goodman KA, Lee F, Chang S, Kuo T, Ford JM, Fisher GA, Quon A, Desser TS, Norton J, Greco R, Yang GP, Koong AC. Gemcitabine chemotherapy and single-fraction stereotactic body radiotherapy for locally advanced pancreatic cancer. Int J Radiat Oncol Biol Phys. 2008;72:678–86.
48. Oar A, Lee M, Le H, Hruby G, Dalfsen R, Pryor D, Lee D, Chu J, Holloway L, Briggs A, Barbour A, Chander S, Ng SP, Samra J, Shakeshaft J, Goldstein D, Nguyen N, Goodman KA, Chang DT, Kneebone A. Australasian Gastrointestinal Trials Group (AGITG) and Trans-Tasman Radiation Oncology Group (TROG) guidelines for Pancreatic Stereotactic Body Radiation Therapy (SBRT). Pract Radiat Oncol. 2020;10:e136–46.
49. Herman JM, Chang DT, Goodman KA, Dholakia AS, Raman SP, Hacker-Prietz A, Iacobuzio-Donahue CA, Griffith ME, Pawlik TM, Pai JS, O'Reilly E, Fisher GA, Wild AT, Rosati LM, Zheng L, Wolfgang CL, Laheru DA, Columbo LA, Sugar EA, Koong AC. Phase 2 multi-institutional trial evaluating gemcitabine and stereotactic body radiotherapy for patients with locally advanced unresectable pancreatic adenocarcinoma. Cancer. 2015;121:1128–37.
50. Jung J, Yoon SM, Park JH, Seo DW, Lee SS, Kim MH, Lee SK, Park DH, Song TJ, Ryoo BY, Chang HM, Kim KP, Yoo C, Jeong JH, Kim SC, Hwang DW, Lee JH, Song KB, Jo YY, Park J, Kim JH. Stereotactic body radiation therapy for locally advanced pancreatic cancer. PLoS One. 2019;14(4):*e0214970.*
51. Berghoff AS, Schur S, Füreder LM, Gatterbauer B, Dieckmann K, Widhalm G, Hainfellner J, Zielinski CC, Birner P, Bartsch R, Preusser M. Descriptive statistical analysis of a real life cohort of 2419 patients with brain metastases of solid cancers. ESMO Open. 2016;1:e00002.
52. Routman DM, Bian SX, Diao K, Liu JL, Yu C, Ye J, Zada G, Chang EL. The growing importance of lesion volume as a prognostic factor in patients with multiple brain metastases treated with stereotactic radiosurgery. Cancer Med. 2018;7(3):757–64.
53. Yamamoto M, Serizawa T, Shuto T, Akabane A, Higuchi Y, Kawagishi J, et al. Stereotactic radiosurgery for patients with multiple brain metastases (JLGK0901): a multi-institutional prospective observational study. Lancet Oncol. 2014;15:387–95.
54. Minniti G, Scaringi C, Paolini S, Lanzetta G, Romano A, Cicone F, Osti M, Enrici RM, Esposito V. Single-fraction versus multifraction (3 × 9 Gy) stereotactic radiosurgery for large (>2 cm) brain metastases: a comparative analysis of local control and risk of radiation-induced brain necrosis. Int J Radiat Oncol Biol Phys. 2016;95:1142–8.
55. Shaw E, Scott C, Souhami L, Dinapoli R, Kline R, Loeffler J, Farnan N. Single dose radiosurgical treatment of recurrent previously irradiated primary brain tumors and brain metastases: final report of RTOG protocol 90-05. Int J Radiat Oncol Biol Phys. 2000;47:291–8.
56. Mahajan A, Ahmed S, McAleer MF, Weinberg JS, Li J, Brown P, Settle S, Prabhu SS, Lang FF, Levine N, McGovern S, Sulman E, McCutcheon IE, Azeem S, Cahill D, Tatsui C, Heimberger AB, Ferguson S, Ghia A, Demonte F, Raza S, Guha-Thakurta N, Yang J, Sawaya R, Hess KR, Rao G. Post-operative stereotactic radiosurgery versus observation for completely resected brain metastases: a single-centre, randomised, controlled, phase 3 trial. Lancet Oncol. 2017;18:1040–8.
57. Yamamoto M, Aiyama H, Koiso T, Watanabe S, Kawabe T, Sato Y, Higuchi Y, Kasuya H, Barfod BE. Validity of a recently proposed prognostic grading index, brain metastasis velocity, for patients with brain metastasis undergoing multiple radiosurgical procedures. Int J Radiat Oncol Biol Phys. 2019;103:631–7.
58. Minniti G, Osti MF, Niyazi M. Target delineation and optimal radiosurgical dose for pituitary tumors. Radiat Oncol. 2016;11:135. Braunstein S, Ma L. Stereotactic

radiosurgery for vestibular schwannomas. Cancer Manag Res. 2018;10:3733–40.
59. Marchetti M, Sahgal A, De Salles AAF, et al. Stereotactic radiosurgery for intracranial noncavernous sinus benign meningioma: international stereotactic radiosurgery society systematic review, meta-analysis and practice guideline. Neurosurgery. 2020;87:879–90.
60. Pokhrel D, Sood S, McClinton C, et al. Linac-based stereotactic radiosurgery (SRS) in the treatment of refractory trigeminal neuralgia: detailed description of SRS procedure and reported clinical outcomes. J Appl Clin Med Phys. 2017;18:136–43.
61. Pinzi V, Biagioli E, Roberto A, Galli F, Rizzi M, Chiappa F, et al. Radiosurgery for intracranial meningiomas: a systematic review and meta-analysis. Crit Rev Oncol Hematol. 2017;13:122–34.
62. Persson O, Bartek J Jr, Shalom NB, Wangerid T, Jakola AS, Förander P. Stereotactic radiosurgery vs. fractionated radiotherapy for tumor control in vestibular schwannoma patients: a systematic review. Acta Neurochir. 2017;159:1013–21.
63. Albano L, Losa M, Barzaghi LR, Niranjan A, Siddiqui Z, Flickinger JC, et al. Gamma knife radiosurgery for pituitary tumors: a systematic review and meta-analysis. Cancers (Basel). 2021;13:4998.
64. Tuleasca C, Régis J, Sahgal A, De Salles A, Hayashi M, Ma L, et al. Stereotactic radiosurgery for trigeminal neuralgia: a systematic review. J Neurosurg. 2018;130:733–57.
65. Deng L, Liang H, Xu M, et al. STING-dependent cytosolic DNA sensing promotes radiation-induced type I interferon-dependent antitumor immunity in immunogenic tumors. Immunity. 2014;41:843–52.
66. Munshi A. Ablative radiosurgery for cardiac arrhythmias—a systematic review. Cancer Radiother. 2021;25(4):373–9.
67. Shibamoto Y, Murai T, Suzuki K, Hashizume C, Ohta K, Yamada Y, et al. Definitive radiotherapy with SBRT or IMRT boost for breast cancer: excellent local control and cosmetic outcome. Technol Cancer Res Treat. 2018;17:1533033818799355.
68. Vanpouille-Box C, Demaria S, Formenti SC, et al. Cytosolic DNA sensing in organismal tumor control. Cancer Cell. 2018;34:361–78.
69. Dunne EM, Sahgal A, Lo SS, Bergman A, Kosztyla R, Dea N, Chang EL, Chang UK, Chao ST, Faruqi S, Ghia AJ, Redmond KJ, Soltys SG, Liu MC. International consensus recommendations for target volume delineation specific to sacral metastases and spinal stereotactic body radiation therapy (SBRT). Radiother Oncol. 2020;145:21–9.
70. Jereczek-Fossa BA, Marvaso G, Zaffaroni M, Gugliandolo SG, Zerini D, Corso F, Gandini S, Alongi F, Bossi A, Cornford P, De Bari B, Fonteyne V, Hoskin P, Pieters BR, Tree AC, Arcangeli S, Fuller DB, Franzese C, Hannoun-Levi JM, Janoray G, Kerkmeijer L, Kwok Y, Livi L, Loi M, Miralbell R, Pasquier D, Pinkawa M, Scher N, Scorsetti M, Shelan M, Toledano A, van As N, Vavassori A, Zilli T, Pepa M, Ost P, on the behalf of the European Society for Radiotherapy, Oncology Advisory Committee on Radiation Oncology Practice (ESTRO ACROP). Salvage stereotactic body radiotherapy (SBRT) for intraprostatic relapse after prostate cancer radiotherapy: an ESTRO ACROP Delphi consensus. Cancer Treat Rev. 2021;98:102206.
71. Guckenberger M, Andratschke N, Dieckmann K, Hoogeman MS, Hoyer M, Hurkmans C, Tanadini-Lang S, Lartigau E, Méndez Romero A, Senan S, Verellen D. ESTRO ACROP consensus guideline on implementation and practice of stereotactic body radiotherapy for peripherally located early stage non-small cell lung cancer. Radiother Oncol. 2017;124(1):11–7.
72. Lopez-Campos F, Cacicedo J, Couñago F, García R, Leaman-Alcibar O, Navarro-Martin A, Pérez-Montero H, Conde-Moreno A. SEOR SBRT-SG stereotactic body radiation therapy consensus guidelines for non-spine bone metastasis. Clin Transl Oncol. 2022;24(2):215–26.
73. Simone CB 2nd, Bogart JA, Cabrera AR, Daly ME, DeNunzio NJ, Detterbeck F, Faivre-Finn C, Gatschet N, Gore E, Jabbour SK, Kruser TJ, Schneider BJ, Slotman B, Turrisi A, Wu AJ, Zeng J, Rosenzweig KE. Radiation therapy for small cell lung cancer: an ASTRO clinical practice guideline. Pract Radiat Oncol. 2020;10(3):158–73.
74. Hanna GG, Murray L, Patel R, Jain S, Aitken KL, Franks KN, van As N, Tree A, Hatfield P, Harrow S, McDonald F, Ahmed M, Saran FH, Webster GJ, Khoo V, Landau D, Eaton DJ, Hawkins MA. UK consensus on normal tissue dose constraints for stereotactic radiotherapy. Clin Oncol (R Coll Radiol). 2018;30(1):5–14.
75. Oar A, Lee M, Le H, Hruby G, Dalfsen R, Pryor D, Lee D, Chu J, Holloway L, Briggs A, Barbour A, Chander S, Ng SP, Samra J, Shakeshaft J, Goldstein D, Nguyen N, Goodman KA, Chang DT, Kneebone A. Australasian Gastrointestinal Trials Group (AGITG) and Trans-Tasman Radiation Oncology Group (TROG) guidelines for Pancreatic Stereotactic Body Radiation Therapy (SBRT). Pract Radiat Oncol. 2020;10(3):e136–46.
76. Redmond KJ, Robertson S, Lo SS, Soltys SG, Ryu S, McNutt T, Chao ST, Yamada Y, Ghia A, Chang EL, Sheehan J, Sahgal A. Consensus contouring guidelines for postoperative stereotactic body radiation therapy for metastatic solid tumor malignancies to the spine. Int J Radiat Oncol Biol Phys. 2017;97(1):64–74.
77. Videtic GM, Donington J, Giuliani M, Heinzerling J, Karas TZ, Kelsey CR, Lally BE, Latzka K, Lo SS, Moghanaki D, Movsas B. Stereotactic body radiation therapy for early-stage non-small cell lung cancer: executive summary of an ASTRO evidence-based guideline. Pract Radiat Oncol. 2017;7:295–301.
78. Cox BW, Spratt DE, Lovelock M, Bilsky MH, Lis E, Ryu S, Sheehan J, Gerszten PC, Chang E, Gibbs I, Soltys S, Sahgal A, Deasy J, Flickinger J, Quader M, Mindea S, Yamada Y. International Spine Radiosurgery

Consortium consensus guidelines for target volume definition in spinal stereotactic radiosurgery. Int J Radiat Oncol Biol Phys. 2012;83(5):e597–605. https://doi.org/10.1016/j.ijrobp.2012.03.009. Epub 2012 May 19. PMID: 22608954.

79. Michael TM, Louis SC, Paul O. Normal tissue tolerance dose metrics for radiation therapy of major organs. Semin Radiat Oncol. 2007;17:131–40.
80. Munshi A, Rastogi K. Management of cancer during the COVID-19 pandemic: practical suggestions for the radiation oncology departments. Cancer Res Stat Treat. 2020;3(Suppl 1):S115–8.

Online Resources

https://econtour.org/references(contouring)

https://www.youtube.com/watch?v=6TXUYmOQhTk (SBRT in non metastatic Lung cancer)

https://www.youtube.com/watch?v=ZpK3JhVUNDk (SBRT for Prostate cancer)

https://www.youtube.com/watch?v=O0U8jUuH-Nk (Targets and Treatment planning in SRS and SBRT)

Quality of Life in Radiation Therapy

20

Aruna Turaka, Yousra Izzuddeen K. N., Lachlan McDowell, Bhanu Prasad Venkatesulu, and Bobby Nagendra Koneru

20.1 Introduction and Definition of QoL in Cancer

Cancer causes a high degree of morbidity and mortality for most of the patients. The cancer patient goes through a trajectory of physical symptoms, diagnosis, treatment and survivorship. Quality of life (QoL) assessment is a direct and patient-centered method in the cancer care delivery which is aimed to identify the outcomes, way beyond the standard scales of loco-regional control, disease-free survival, overall survival, morbidities, and death. Health-related QoL indicates the patients' perception of their health [1].

Assessment of a cancer patient's QoL can provide helpful information on disease- and treatment-related effects and also on a patient's overall experiences. QoL is a subjective multidimensional construct which aims to evaluate at a minimum: physical, social and psychological domains as well as disease- and treatment-related symptoms. In 1992, Gotay [2] defined that "QoL is a state of well-being which is a composite of 1. the ability to perform everyday activities which reflect physical, psychological and social well-being, and 2. patient satisfaction with level of functioning and the control of disease and/or treatment related symptoms." Thereon, the interest in incorporating QoL as an end point in clinical studies of cancer treatment has evolved in the last 30 years.

Subsequently, cancer clinical trials started to integrate of QoL to evaluate the different cancer therapies and to supplement standard end-points such as response and survival. The main modalities of cancer treatments, such as, surgery, radiotherapy, chemotherapy and newer immunotherapy can cure or substantially prolong life for many cancers, yet the disease and treatment-related mor-

A. Turaka (✉)
Department of Radiation Oncology, Nazareth Hospital Pennsylvania, Philadelphia, PA, USA

Department of Radiation Oncology, Penn State Health Milton S Hershey Medical Center, Philadelphia, PA, USA

Yousra Izzuddeen K. N.
Department of Radiation Oncology, All India Institute of Medical Sciences, New Delhi, India

Specialist Radiation Oncology, Burjeel Medical City, Abhu Dhabi, Abhu Dhabi

L. McDowell
Peter MacCallum Cancer Centre, University of Melbourne, Parkville, VIC, Australia

Department of Radiation Oncology, Princess Alexandria hospital, Brisbane, Australia
e-mail: Lachlan.McDowell@petermac.org

B. P. Venkatesulu
Department of Radiation Oncology, Loyola University, Chicago, IL, USA

Department of Radiation Oncology, University of Pittsburgh Medical Center, Pittsburgh, PA, USA

B. N. Koneru
The Leonard C. Ferguson Cancer Center at FHN Memorial Hospital, Freeport, IL, USA
e-mail: bobbykoneru@gopog.com

A. Munshi et al. (eds.), *Radiation Oncology – Principles, Precepts and Practice*,
https://doi.org/10.1007/978-981-97-8389-2_20

bidities become significant for most of these patients. For example, 50-year-old lady with breast cancer and underlying coronary disease would require more than 6 months of treatment duration consisting of surgery and postoperative chemo-hormone therapy and radiotherapy; a 10-year-old with childhood leukemia would undergo nearly 1 year of chemotherapy requiring multiple hospital visits and out-of-school stay. "How are you feeling today?"–a small inquiry like this by a doctor or nurse in the ward or out-patient can encourage a patient to relate about himself/herself. There is growing recognition that patient-reported outcome (PRO) measures—encompassing, for example, health-related quality of life (HRQOL)—can complement traditional biomedical outcome measures (e.g., survival, disease-free survival) in conveying important information for cancer care decision-making [3]. This is reflected in a growing body of biomedical literature on QoL.

In 2020, The Setting International Standards in Analyzing Patient-Reported Outcomes and Quality of Life (SISAQOL) Endpoints Data Consortium was formed to establish recommendations for patient reported outcomes analysis [4]. Hence the assessment of QoL in cancer patients is a subjective PRO, which enables the patient (and in some cases the caregivers) for reporting of symptoms, functions, and improvements experienced by the patient from the time of diagnosis and during the course of disease trajectory in his/her lifetime. This chapter aims to equip the physicians, researchers and healthcare providers with the essential aspects of QoL related to the radiotherapy in cancer and as well QoL interconnected to the use of combined modalities in cancer management.

20.1.1 Radiotherapy-Related QoL Issues

Radiotherapy can be used both for cure and palliation of cancers. For curative purposes, radiation may be used alone (e.g., in head and neck cancer, cervix, prostate) or in combination with surgery (head and neck cancers, endometrial tumors) or chemotherapy (laryngeal, anal, rectal, breast, lung and esophageal cancer). Radiotherapy may provide palliation in most cases involving pain, bleeding, and obstruction; for growth restraint of a tumor; and for early pain relief in bone metastasis. New technology has allowed safe delivery of high doses of radiotherapy and use of intensity-modulated radiation therapy (IMRT) and higher techniques with conformal planning has resulted in improved sparing of normal tissues. In the last 30 years, modern radiotherapy practices are able to control tumor; preserve long-term functioning such as voice in laryngeal, sphincter function in anal cancers; and avoid organ-removing surgery by combining radiotherapy with chemotherapy or, hormone therapy in head and neck, breast, rectal, prostate cancers [5]. Assessment of QOL after treatment for cancer by any of the modalities—radiotherapy, systemic therapy, surgery, or combined modalities—is extremely challenging as it can be confounded by several personal factors like age, comorbidities, gender, mental health as well as disease-related factors like disease type, stage and response to treatment often making it difficult to quantify the long-term impact on the QOL of these patients. It is recognized that clinical trials are incorporating the QOL measurement of participants, usually within a defined follow-up period. However, the patient's QoL on disease progression or till the end of life is infrequently reported [6]. In order, to understand the patient's perspectives, the QoL assessment of cancer patients in routine real life cancer care, outside of clinical trials, requires to be encouraged by clinicians and researchers.

20.2 Commonly Used QOL Tools in Cancer

Historically, HRQOL was measured using Karnofsky's performance score (KPS), introduced by David A. Karnofsky and Joseph H. Burchenal in 1949 to assess the functional performance status of cancer patients receiving chemotherapy, has remained a simple and valuable scale for more than 70 years [7]. For the last 50 years or so, the health status of the public and those who are sick, has become significant for the healthcare systems worldwide. The general health status instruments and scales such as EQ-5D (EuroQoL) [8] and

Table 20.1 General health status assessment scale

QoL questionnaire modules	Scales and items
EQ-5D (EuroQoL)	6 questions: physical functions (2—mobility, self-care), social functioning (1—usual activities), symptoms (1—pain/discomfort), mental functioning (1—anxiety/depression), health state thermometer
SF-36 (Medical outcomes study, MOS SF-36), version 2.0	36-item questionnaire: General health; 8 scales/domains (physical functioning, role limitations due to physical issues, bodily pain, general health, vitality, social functioning, role limitations due to emotional problems, mental health); 2 summary measures that aggregate scales.

Table 20.2 Assessment of performance status-KPS[a] and ECOG[b] scales [Ref. 10]

ECOG Performance Status	Karnofsky Performance Status
0—Fully active, able to carry on all pre-disease performance without restriction	100—Normal, no complaints; no evidence of disease
	90—Able to carry on normal activity; minor signs or symptoms of disease
1—Restricted in physically strenuous activity but ambulatory and able to carry out work of a light or sedentary nature, e.g., light house work, office work	80—Normal activity with effort, some signs or symptoms of disease
	70—Cares for self but unable to carry on normal activity or to do active work
2—Ambulatory and capable of all self-care but unable to carry out any work activities; up and about more than 50% of waking hours	60—Requires occasional assistance but is able to care for most of personal needs
	50—Requires considerable assistance and frequent medical care
3—Capable of only limited self-care; confined to bed or chair more than 50% of waking hours	40—Disabled; requires special care and assistance
	30—Severely disabled; hospitalization is indicated although death not imminent
4—Completely disabled; cannot carry on any self-care; totally confined to bed or chair	20—Very ill; hospitalization and active supportive care necessary
	10—Moribund
5—Dead	0—Dead

[a]Karnofsky D, Burchenal J, The clinical evaluation of chemotherapeutic agents in cancer. In: MacLeod C, ed. Evaluation of Chemotherapeutic Agents. New York, NY: Columbia University Press; 1949:191–205
[b]Zubrod C, et al. Appraisal of methods for the study of chemotherapy in man: Comparative therapeutic trial of nitrogen mustard and thiophosphoramide. *Journal of Chronic Diseases*; 1960:11:7–33

SF-36 [9] are most widely used across spectrum of diseases, in order to capture and assess the patient-reported outcomes (PROs). These two modules are shown in Table 20.1.

The use of standard criteria to assess the cancer patients across various socio-geographic situations [10]—such as out-patient clinics, hospitals, and cancer centers by treating physicians and health care teams—can be done in a quick, uniform and simple manner by Karnofsky Performance Status (KPS, introduced in 1949) scale or Eastern Cooperative Oncology Group (ECOG, introduced in 1960) scale [11]. For each cancer patient, any one of these two scales should be the starting point for QoL record before initiation any ant-cancer therapy such as surgery, chemotherapy and radiotherapy or for palliative care. The KPS and ECOG scales can be seen in a comparative format in Table 20.2.

The measurement of overall HRQOL for cancer patients is done by using generic modules which include psychometric tools. The two most validated generic modules, developed in 1993, are: (1) European Organization for Research and Treatment of Cancer Quality of Life Questionnaire Version 3.0 (EORTC QLQ-C30 v. 3.0) [12]; and (2) The Functional Assessment of Cancer Therapy-General (FACT-G v.4) [13]. The brief details of these two modules are shown in Table 20.3. The EORTC QLQ-C30 covers five functional scales (physical, role, cognitive, emotional and social), three symptom scales (fatigue, pain, nausea and vomiting), a

Table 20.3 Cancer-specific generic QoL assessment instruments

QoL questionnaire modules	Scales and items
*EORTC QLQ-C30, V.3.0 [12]	30-item questionnaires: Global health (2); Functional Scales: Physical functioning (5), Role functioning (2), Emotional functioning (4), Cognitive functioning (2), Social functioning (2); Symptom's scales/items: Fatigue (3), Nausea/vomiting (2), Pain (2); Global QOL (1 each: dyspnea, insomnia, appetite, constipation, diarrhea, financial)
FACT-G, Version 4 [13]	27-Item questionnaires: 4 domains/scales: functional wellbeing, emotional wellbeing, social/family wellbeing, physical wellbeing

global health status scale and a number of single-item symptom scores (dyspnea, loss of appetite, insomnia, constipation and diarrhea) as well as perceived financial burden, which had been validated [12]. The subjective responses from patients (or, caregivers) are collected by Likert method, and subsequently converted with the aid of Scoring Manual [14], to scores ranging from 0 to 100, with higher functional scores representing better health and higher score for symptom scales representing higher symptom burden. The other frequently used questionnaire is the Functional Assessment of Cancer Therapy Scale (FACT-G version 4) [13]. The FACT-G covers four dimensions, including physical, social, emotional and functional well-being in 27-items with a 7-day recall. In the last three decades, the cancer-specific QoL research has expanded to develop the instruments which can assess the QoL for a particular cancer site or type (e.g., brain tumor, head and neck cancer, lung cancer etc.) and cancer condition (e.g., palliative care, bone metastasis, financial toxicity, etc.). These commonly used cancer site/condition-specific instruments are shown in Table 20.4.

The basic concepts of QOL assessment and the models are depicted diagrammatically in Fig. 20.1. It is recommended that the generic modules and the instruments suitable for a cancer site or condition should be applied together to derive the patient's subjective experiences in a comprehensive manner. In order to evaluate the patient's subjective PRO, the doctors, nurses, health system specialists, researchers, and social scientists should be trained and skilled to use these scales, modules and instruments of QoL during the cancer care trajectory.

20.3 QoL-Related to Common Cancers and Conditions (Overview/Systemic Review)

Incorporation of site-specific QoL questionnaires has become a standard approach to measure the HRQOL over the past 2–3 decades to evaluate the benefit of newer oncological treatments (radiation therapy, chemotherapy/immunotherapy) over standard treatment methods, e.g., for breast cancer—EORTC-QLQ-BR23, FACT-B; prostate cancer—EORTC-QLQ-PR25. Although there are many validated cancer sites or condition-specific tools available, it is imperative to use the validated tools (Tables 20.1, 20.2, 20.3 and 20.4). Every reader of this chapter, i.e., the physician, nurse, researcher and social scientist, is further encouraged to study in-depth the various cancer-related QoL measurement tools currently in use [15, 16]. For example, in order to assess the QoL in a HNSCC (head and neck squamous cell carcinoma) patient receiving concurrent chemo-radiotherapy, EORTC QLQ C30 should be combined with EORTC QLQ H&N 43 at baseline(before treatment), at mid-course/end of treatment course, at follow-up 3, 6, 12 months.

20.3.1 CNS Tumors and Brain Metastases

The assessment of QOL in tumors of the central nervous system and brain metastasis, is especially challenging, due to the baseline neurocognitive deficits associated with the brain lesions itself, which in turn affects the QoL. Moreover, in most patients with brain metastasis, in view of their very short survival, assessing QoL becomes especially important, as this would be the most important parameter based on which further treatment would be determined. Baseline neurocognitive function has been strongly associated with QoL in patients after whole brain radiother-

Table 20.4 Cancer site and condition-specific QoL assessment instruments[a] [Ref. 15–20]

Cancer site/condition	QoL questionnaire modules	Scales and items
Brain tumor & brain metastasis	EORTC QLQ-BN20	20-Item questionnaire—for assessment of future uncertainty, visual disorder, motor dysfunction and communication deficit, including 7 single-items which record headaches, seizures, drowsiness, hair loss, itchy skin, weakness of legs and bladder control
	FACT-Br	23 items to assess brain cancer-specific concerns including concentration, memory, seizures, eyesight, hearing, speech, personality, expression of thoughts, weakness, coordination and headaches
Bone metastases & bone pain	EORTC QLQ-BM22	22 items divided into 4 scales: painful sites (5 items), painful characteristics (3 items), functional interference (8 items) and psychosocial aspects (6 items)
	FACT-BP	16-item questionnaire, 5-point Likert-type scale, for assessing cancer-related bone pain; from various metastatic cancers
Head and neck cancer	EORTC QLQ-H&N 43	43-item questionnaires: 7 domains/scales: pain (4), swallowing (4), senses (2), speech (3), social eating (4), social contact (5), sexuality (2); Single items specific to HNC: teeth, trismus, dry mouth, sticky saliva, cough, felt ill, pain killers, nutritional supplements, feeding tube, weight loss, weight gain
	FACT- H& N, version 4	39 questionnaires: Physical well-being (7), Social/family well-being (7), Emotional well-being (6), Functional well-being (7), H&N specific (12) 4 dimensions (16 items): Physical, Social/family, emotional and functional well-being
	FACT- NP, version 4 (Nasopharyngeal cancer)	43 questionnaires: Physical well-being, social/family well-being, emotional well-being, functional well-being, nasopharyngeal subscale
Esophageal cancer	ERTC QLQ-OES18	18 items: dysphagia, feeding, reflux, chest/abdominal pain
	FACT-E	17 items—4 dimension: Physical, social/family, emotional and functional well-being
Stomach cancer	EORTC-QLQ-STO22	22-item questionnaire has five scales (dysphagia, pain, diet, symptoms, emotional problems), and four single items (dry mouth, body image, and hair loss (two items))
	FACT-Ga	Gastric cancer patients 18 years and older; 46 items, 5-point Likert-type scale
Colorectal malignancies	EORTC QLQ-CR29	29 items—5 functional and 18 symptom scales. Four subscales (urinary frequency, blood and mucus in stools, body image) and 19 single items (urinary incontinence, dysuria, buttock pain, stoma, sexual interest, impotence, fecal incontinence)
	FACT- C	37 items: 5 domains and 4 dimensions from FACT-G (physical, social/family, emotional and functional well-being) and an additional CRC module
Breast cancer	EORTC QLQ-BR23	23 items: to assess body image, sexual functioning, sexual enjoyment, future perspective, systemic therapy side effects, breast symptoms, arm symptoms and upset by hair loss
	FACT-B	37-item instrument designed to measure five domains of HRQOL in breast cancer patients: Physical, social, emotional, functional well-being as well as a breast cancer subscale (BCS)

(continued)

Table 20.4 (continued)

Cancer site/ condition	QoL questionnaire modules	Scales and items
Prostate cancer	EORTC QLQ-PR25	25-items: 6 domains: Urinary symptoms, Incontinence, bowel symptoms, hormonal treatment related symptoms, sexual activity and sexual functions
	FACT-P	39 items: consists of 12 items of prostate cancer subscale (PCS) + FACT-G (27 items) for the total score
Palliative care in cancer	EORTC QLQ-C15-PAL	15-item version (adapted from EORTC QLQ C30) to assess physical and emotional function, pain, fatigue, nausea/vomiting, appetite, dyspnea, constipation, sleeping difficulties and overall QoL
	McGill QOL-Expanded (MRQOL-E) [17]	21 items: 8 domains (physical, psychological, existential, social, healthcare, cognitive functioning, (feeling like a) burden, and environment); comprehensive assessment of people with life-limiting illness
Cancer in elderly (geriatric cancer patients)	EORTC QLQ-ELD 14	Diagnosed >70 years age: Five scales (mobility, worries about others, future worries, maintaining purpose and illness burden) and two single items (joint stiffness and family support)
Financial toxicity	Comprehensive Score for financial Toxicity (COST) [18]	11 items to measure the financial burden and concerns and worry of cancer patients
Cancer survivors	Short-Form Survivor Unmet Needs Survey (SF-SUNS) [19]	30 items: to measure the well-being and needs of cancer survivors
	Patient Centered Survivorship Care Index (PC-SCI) [20]	42 items: comprises 7 multi-item factors that reflect cancer survivors' priorities for survivorship care: Information and support in survivorship, medical home, patient engagement, care coordination, help navigating insurance, care transitions, and prevention and wellness services.

Ref. [15]. https://qol.eortc.org/questionnaires/
Ref. [16]. https://www.facit.org/measures-language-availability
Ref. [17]. Cohen SR et al. BMC Palliat Care. 2019; 8:92. doi: 10.1186/s12904-019-0473-y
Ref. [18]. de Souza JA et al. Cancer. 2014;120:3245–53. doi: 10.1002/cncr.28814
Ref. [19]. Campbell HS et al. Cancer 2014;1071–9. doi: 10.1007/s00520-013-2061-7
Ref. [20]. Mead KH et al. BMC Health Serv Res. 2021;21:1353. doi: 10.1186/s12913-021-07356-6
[a]For comprehensive PRO purposes, these cancer site-specific instruments should be applied conjointly with the cancer generic module

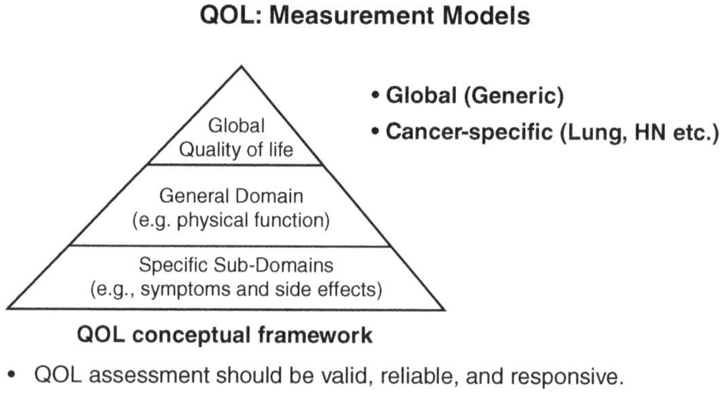

Fig. 20.1 Quality of life assessment tools

apy (WBRT) [21]. Historically, the Mini Mental Status Examination (MMSE), which reviews a range of cognitive functions including orientation, recall, attention, calculation, language manipulation, and praxis, was designed to assess patients who had stroke. This was then extrapolated later to patients with brain tumors. However, it was later shown that this scale was inadequate to assess and define neurocognitive function in patients with brain metastasis after radiotherapy [22]. The most often used QoL questionnaires are the EORTC-QLQ-BN20 and FACT-Br. The EORTC QLQ-BN20 is a 20-item questionnaire comprising assessment of future uncertainty, visual disorder, motor dysfunction, and communication deficit, as well as seven single-items which include headaches, seizures, drowsiness, hair loss, itchy skin, weakness of legs and bladder control. This is used along with the EORTC QLQ-C30 generic module. FACT-Br is a brain cancer-specific questionnaire to be used along with FACT-G generic instrument. Both tools together can assess general physical, social/family, emotional and functional well-being along with the brain cancer-specific concerns like concentration, memory, seizures, eyesight, hearing, speech, personality, expression of thoughts, weakness, coordination, and headaches. The QLQ-BN20 and the FACT-Br are both valid and reliable tools that have been used extensively in the primary brain cancer population. Choice between the two tools should consider each instrument's individual strengths and weaknesses [23]. Although there is no single superior tool, QLQ-BN20 may provide evaluation of QoL in trials including functional and physical domains while considering factors including tumor location, seizures, and functional status that may influence QoL. Together combined with QLQ-C30, for the patients with poor prognosis brain metastases who received WBRT, it could capture maintenance of nearly all QOL scores 1 month after WBRT thereby showing that treatment intervention likely played a symptom-stabilizing role and prevented QOL deterioration in the palliative setting [24] (Fig. 20.2a, b, Ref. [24], Springer Link). For children and adults with brain tumors, the published works from India have given focus to the significance of QoL measurements in order to assess the benefits of radiotherapy [25, 26]. Children with brain tumors are highly vulnerable, with their growing brain being affected by the disease- and treatment-related factors. Modern radiotherapy techniques are able to deliver radiation doses with greater precision to the tumor and its surrounding region. A randomized clinical trial (RCT) of stereotactic conformal radiotherapy compared with conventional radiotherapy, with similar radiation doses, while evaluating for clinically meaningful end points showed significantly higher global intelligence quotient (IQ) and lower neurocognitive and endocrine dysfunction on long-term follow-up [25]. In adults with primary brain tumors, QoL assessment done before starting adjuvant radiotherapy or combined chemo-radiotherapy, utilized the generic QLQ-C30 and CNS-specific QLQ-BN20 instruments. The key findings showed that elderly patients, those with high grade gliomas, patients with illiteracy and lower financial condition had poorer QOL scores [26]. These salient findings on the QOL of brain tumors are shown in Table 20.5; which reveal that certain key interventions can be guided by these findings.

20.3.2 Bone Metastases

The increasing survival with improved cancer treatments has given rise in the incidence of bone metastases (BMs) in cancer patients with solid tumors and hematological malignancies. One estimate of the prevalence showed that approximately 330,000 cancer patients in the USA seek the various treatments for BMs [27]. The bone metastasis is commonly seen in cancers of breast, lung, prostate, thyroid; and in lymphomas, leukemias and multiple myeloma (where tumor cells invade the bones). Patients with BMs are at an increased risk of experiencing morbidity due to bone pain, limitation in mobility and other complications. There have been number of tools used to assess QOL in patients with BMs in various studies in the early 2000s; which can be either generic or specific to bone metastasis (Tables 20.2, 20.3 and 20.4). Various treatment modalities are utilized to manage BMs, depending on site and extent of lesion, pathologic features and

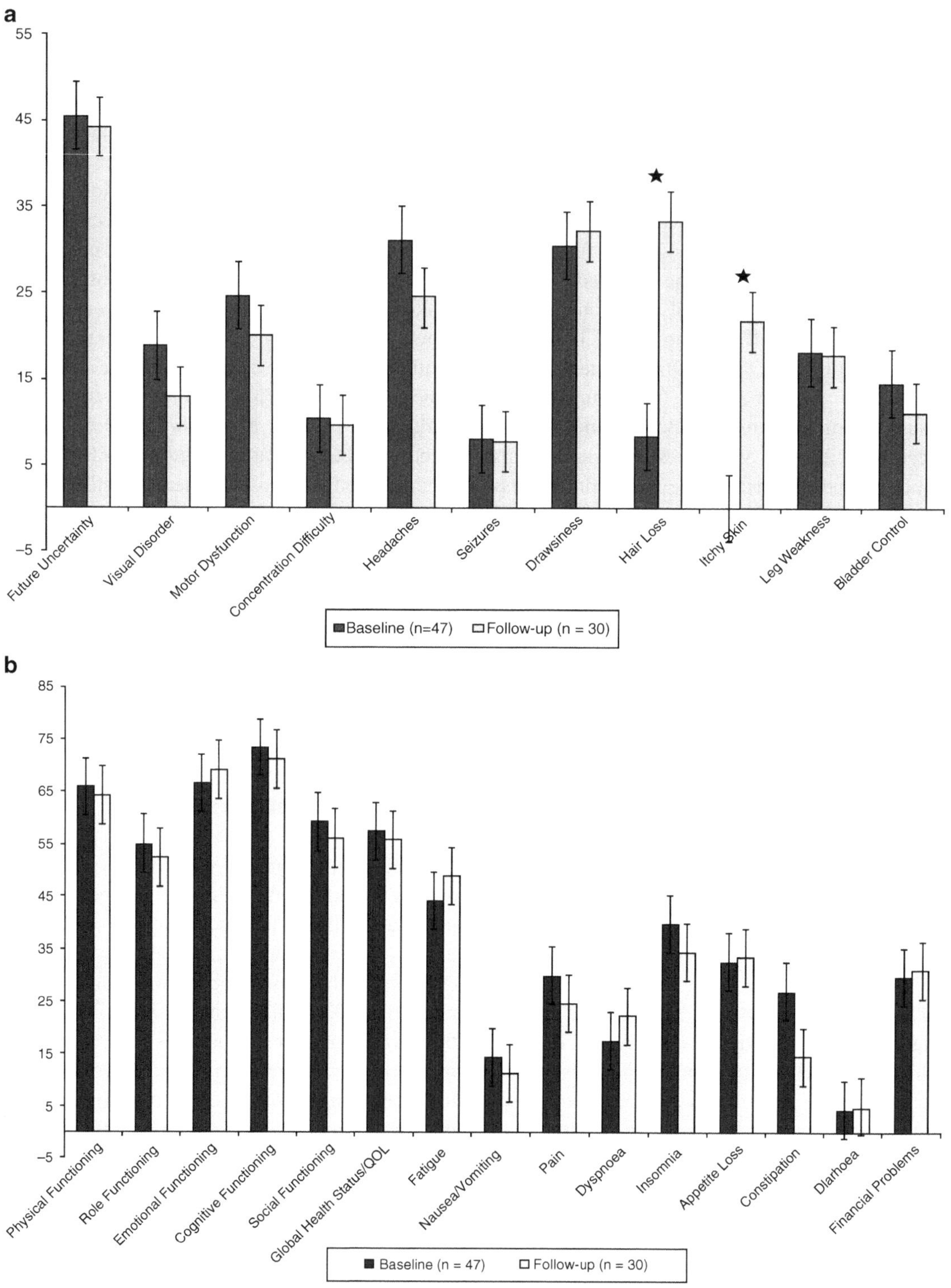

Fig. 20.2 (**a**) Brain metastasis: EORTC QLQ-BN20 mean scores before and after radiotherapy. A higher score indicates worse symptomatology for all scales; bars represent standard errors; stars indicate statistical significance. [Ref. 24, Fig. 1, by Springer's publications]. (**b**) 2 EORTC QLQ-C30 mean scores before and after radiotherapy. Higher score indicates worse symptomatology for all symptom scales. A higher score indicates better functioning/quality of life for all functioning and quality of life scales; bars indicate standard errors; no items were statistically significant. [Ref. 24, Fig. 1, by Springer's publications]

Table 20.5 QOL assessments done in children and adults from India (Ref. [25, 26])

Children (n = 200 patients) [25]	Adults (n = 243 patients) [26]
RCT: compared stereotactic conformal radiotherapy (SCRT) vs. conventional radiotherapy (ConvRT), with same radiation dose.	Prospective study
Key findings	Key findings
Global Intelligence Quotient (IQ) Score: higher after SCRT	Global QOL Health Score: lower for HGG, compared to LGG
Performance Quotient: higher after SCRT	Elderly: lower functional scale scores
Anxiety and Depression scores: lower after SCRT	Illiterate and Lower Socio-economic status: poor QOL scores
Endocrine dysfunction: lower after SCRT	Lower Performance Status (KPS 70) = Poor QOL scores
Similar overall survival in both groups	Type of surgery (biopsy/complete excision) and site of tumor did not show any impact on QOL score

Note: *LGG* low-grade glioma, *HGG* high-grade glioma, *RCT* randomized clinical trial

life expectancy. Although surgical fixation, chemotherapy/hormone therapy, radionuclide therapy and pain relief drugs are given as multidisciplinary approach in different combinations for the suitable patients, palliative radiotherapy remains the most effective modality to control bone pain, reduce bone loss, and improve the QOL [28, 29]. Forty percent of patients would show pain reduction and better QOL at day 10 after radiotherapy with further improvements in QOL at day 42 in responders. A single 8-Gy radiotherapy dose, compares favorable to 20 Gy in 5 fractions or 30 Gy in 10 fractions, should be offered to all patients, more so for those with poor survival [29, 30]. In prospective studies and clinical trials, the commonly used QOL tools to evaluate the bone metastases are Brief Pain Inventory, Edmonton Symptom Assessment System, EORTC QLQ-C15-PAL, QLQ-BM22 and FACT-BP [28]. It is noteworthy that clinical management for BMs, more so when palliative radiotherapy is delivered, is to be done by inclusion of QoL assessment with an appropriate module before and if feasible after the radiotherapy. Such an approach in clinical practice can reduce unnecessary pain relief prescription, increase non-drug measures such as physiotherapy and improve the subjective PRO.

20.3.3 Head Neck Cancer

Head and neck cancer (HNC) is primarily a disease of low- and middle-income countries (LMICs), accounting for 67% of total HNC burden and 82% of HNC-specific deaths [31]. The prevailing scenario shows high burden, late presentation, improper or suboptimal treatments, low access to care and poorer survival compared to developed countries. Efforts are ongoing to bridge such disparities and increase the collaboration between high-income countries (HIC) and LMIC, in order to improve the management and research [32].

Patients diagnosed with head and neck cancer (HNC) face some of the most significant HRQOL challenges of all cancer patients. They are often confronted with HNC-specific problems, such as oral dysfunction, swallowing and speech impediments. These disorders have a distinct impact on the QOL of patients [33]. Across the spectrum of head and neck cancers, radiation therapy forms a vital modality in many patients' treatment, including those treated in the definitive, adjuvant and palliative setting. Many HRQOL domains are highly relevant to HNC patients and their care givers and clinicians, including global QoL score and also assessments of physical, role, functioning, emotional, cognitive and social functioning, among others (shown in Tables 20.2, 20.3 and 20.4). During high-dose radiation therapy schedules for HNC patients across continents including LMIC, the HRQOL studies involving longitudinal trajectories typically show large detriments during and into the early post-treatment period when compared to baseline scores [33–36]. It is observed that 6 weeks to 3 months after the radiotherapy (RT) or chemo-radiotherapy (CRT) course, the acute radiation or chemo-

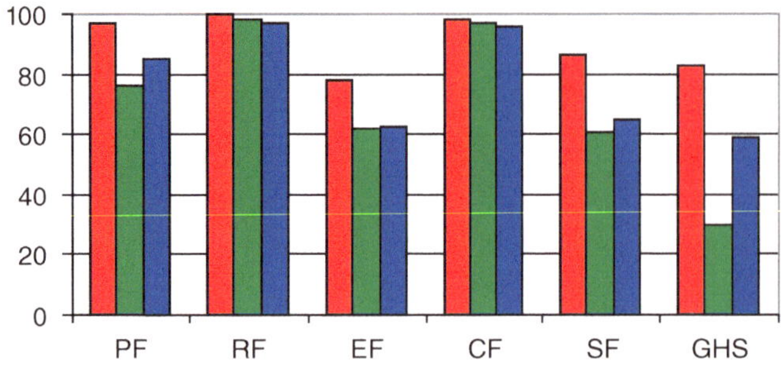

Fig. 20.3 Head and neck cancer receiving radical radiotherapy: quality of life (EORTC QLQ-C30) mean scores of five function scales and global health status for head and neck cancer patients, showing change over time [Ref. 36]. Note: QOL functional scales; *PF* physical function, *RF* role function, *EF* emotional function, *CF* cognitive function, *SF* social function, *GHS* global health status

radiation morbidities impact the QoL function scales and increase the symptom distress [34–36]. This is shown in Fig. 20.3. With time, however, usually in the range of 6–12 months, many of these outcomes will come to approximate their baseline levels, particularly for the global health status measures [33].

20.3.3.1 Advances in Radiation Therapy Technology

The last two to three decades have seen significant improvements in radiation therapy planning and delivery, offering the promise of reduced toxicity and improved HRQL for survivors. One of the major steps in this evolution was the development of three-dimensional (3-D) image-based highly conformal techniques such as intensity modulated radiotherapy (IMRT). There have now been multiple studies comparing toxicity and HRQOL profiles between these approaches, which generally favor IMRT, at least in terms of xerostomia. Huang TL et al. from Taiwan report favorable HRQOL results with IMRT compared to three-dimensional radiotherapy in a prospective, longitudinal randomized study in patients treated with definitive radiotherapy for nasopharyngeal cancer [37]. Global QoL was similar between the two groups, but differences were noted in role, emotional and social functioning. With regards to the symptom scales, patients reported improved dry mouth and mouth opening in the IMRT arm at various time points, and exhibited improved scores for sticky saliva, swallowing, senses, sexuality, felling ill, insomnia and pain.

20.3.3.2 Definitive (Chemo) radiotherapy

Definitive chemoradiotherapy or radiotherapy is most commonly used for patients with cancers of the pharyngeal axis, including the nasopharynx (NPC), oropharynx (OPC), hypopharynx (HPC) and the larynx, as the non-surgical curative approaches for organ preservation [32, 33, 37]. NPC and OPC are the most intensively studied sites with respect to HRQOL outcomes and are included to illustrate the QOL in patients treated with definite RT or CRT.

NPC is an endemic disease throughout Southeast Asia but is relatively rare in India, except in North-East region. NPC is most frequently treated with high-dose chemoradiotherapy, with or without neo-adjuvant chemotherapy (NACT). The high dose of RT for disease control, combined with concurrent and/ or NACT, has several adverse effects to impact the QOL, which can be mitigated by IMRT [38]. A comprehensive overview of HRQOL outcomes in NPC cohorts has been recently published, which correlated various socio-demographic, patient, tumor and treatment factors with HRQOL outcomes, particularly in endemic populations [38]. Newer approaches, including adaptive RT planning and risk-adapted volume for RT dose prescription can offer the potential for improving HRQOL in the future. Since NPC is a disease of the children and young adults, timely understanding of the tumor, host, and treatment factors and their adverse impact on several aspects of long-term survivors becomes a significant part of the anti-cancer therapy. Longitudinal QOL assessments can address the unmet need of the NPC survivors [38].

It is well known that OPC consists of two distinct biological entities, based on human papilloma virus (HPV) testing, i.e., HPV-related OPC and non-HPV-related OPC. Patients presenting with low–intermediate-risk HPV-positive disease have excellent disease outcomes and global research efforts are focusing on de-escalation of treatment modality, such as surgery, RT or CRT to reduce therapeutic adverse effects and better the HRQOL profile. Patients with HPV-positive and negative disease have been reported to have different HRQL trajectories [39]. Many HPV-positive patients harbor a low burden of HNC-specific symptoms at baseline with excellent HRQL scores, equivalent or better than reported by the age-matched general population, and better than those with HPV-negative disease. HPV-positive OPC patients are reported to show increased morbidities to chemoradiotherapy, with more significant deficits in their QOL functions compared to their HPV-negative counterparts during and after treatment [39, 40]. However, with time, the HPV-positive cohort will typically recover their initial HRQL advantage.

20.3.3.3 Adjuvant

Adjuvant post-operative radiation therapy (PORT) may be indicated following resection of HNC for any number of adverse features with the goal of reducing locoregional recurrence, and for those in the post-operative stage at highest risk, concurrent CRT may also be indicated. Patients receiving adjuvant treatment will face additional HRQL challenges incurred from surgery, including outcomes such as body image distress and functional loss, which will be dependent on the site and extent of resection and the success of any reconstructive procedures [41]. It is clinically significant to note that adjuvant CRT has higher distress in term of QOL domains as compared to PORT [42]. Majority of the HNC patients after surgery and adjuvant PORT or CRT are concerned about oral function, saliva secretion and swallowing ability.

20.3.3.4 Palliative

Approximately 30–50% of radiotherapy practice is delivered as palliative RT. Radiation therapy remains an important modality to reduce locoregional tumor progression and provide quick symptom relief in patients with incurable HNC [43, 44]. Many palliative schedules are available to the radiation oncologist which must be individualized to the patient, in order to achieve the pain relief and improve the voice, breathing, and swallowing functions for remaining life. Improved HRQOL scores for functional and symptom scales have been observed with short course palliative dose schedules, randomized for 30 Gy in 2 weeks versus 20 Gy in 1 week, with a median overall survival of 7 months [45]. The benefits of palliative radiotherapy were well demonstrated in this study of advanced stage IV HNC patients treated at Lucknow, India.

20.3.4 Breast

Radiation therapy is commonly used for early-stage breast cancer with breast conservation approach and in locally advanced breast cancer. The evolution of hypofractionation in breast cancer has significant benefits in terms of reduction in overall treatment time and is convenient for patients with similar tumor control outcomes. Multiple studies have shown that hypofractionation is associated with reduced skin toxicity and similar if not improved QOL compared to conventional fractionated radiation therapy [46]. There are various QOL tools to assess patients undergoing treatment with breast cancer, in women who have undergone breast reconstruction, as well as patients with metastatic breast cancer (shown in Table 20.4). EORTC QLQ-BR23 originally developed in 1996 is a 23-domain questionnaire answered on a 4-point Likert scale that assesses side effects related to different treatment modalities, body image, sexuality, and future perspective. With respect to radiation-related side effects, this scale has questions related to the cosmesis such as skin changes, sensitivity of breast, swelling and pain over the breast, pain over shoulder as well as lymphedema. Given the rapid advances in the therapeutic arsenal against breast cancer, an updated QLQ-BR45 has been in development for clinical use. This questionnaire further expands domains to 45 questions with more emphasis on the systemic side effects of cytotoxic chemotherapy and questions on altered sweating and questions on cosmesis address some of the impact of radiation and surgery. Both BR23 and BR45* aim more to address the physical aspects of side effects of cancer therapy with little to no emphasis on psychological or emotional wellbeing of the patient [47]. Functional Assessment of Chronic Illness Therapy General Questionnaire, including Breast Cancer Supplement (FACT-B) questionnaire is another commonly used tool to assess QOL in breast cancer patients which has a 37-item scale. FACT-B in addition to assessing breast cancer-related QOL issues also places emphasis on emotional and functional wellbeing (Table 20.4). EORTC QLQ-BRECON23 addresses specific questions in women who have undergone breast reconstruction. In a systematic review of 182 breast cancer trials, HRQOL was assessed as an endpoint in only 38 (20.8%) of the studies and as a primary endpoint in only 10.9% of the studies. Among 22 trials that had a positive primary endpoint, 18 had a significant benefit in QOL favoring the experimental arm [48]. Hence, for women diagnosed with breast cancer, the QOL assessment during disease and survivorship can add immense value to the clinical management.

20.3.5 Lung

Radiation forms an important component in the multi-modality treatment approaches in lung cancer. The contemporary utility of SBRT in early-stage lung cancer, definitive chemo-radiation in locally advanced lung cancer and evolution of ablative therapy in the oligometastatic stage together form the paradigm. QOL metrics assumes paramount importance during the diagnosis and treatment trajectory, as the cancer survivorship improves, and for the appropriate integration of palliative care (shown in Tables 20.2, 20.3 and 20.4). EORTC QLQ-LC13 and Functional Assessment of Cancer Treatment-Lung (FACT-L) are the most commonly used modules to assess QOL in lung cancer clinical trials. QLQ-LC13 published in 1994 comprises multi-item as well as single item measures of lung cancer associated symptoms such as cough, blood in sputum, pain, difficulty in breathing and side effects from chemotherapy and radiation such as hair loss, neuropathy, mouth sores and dysphagia. Given the ongoing major advances in lung cancer treatment, an updated version—QLQ-LC29, which retains 12 of the original questions from QLQ-LC13 with 17 new questions that address existential issues and additional side effects of surgery, systemic therapies and radiation, has been validated [49]. Functional Assessment of Cancer Treatment-Lung (FACT-L) is a 36 item questionnaire with the Lung cancer subscale having 10 items, and questions to address the side effects of cancer directed therapy and smoking habit. In a secondary analysis of RTOG 0617 trial from the USA that compared high-dose radiation to standard dose radiation in locally advanced lung cancer, FACT-L as a patient reported outcome showed that patients who received high-dose radiation had clinically meaningful decline in QOL at 3 months. The baseline QOL was an independent prognostic factor for overall survival, with higher global QOL having improved survival. In addition, at 12 months, IMRT resulted in lesser decline in QOL compared to 3D-CRT giving insights into impact of radiation technology in improving QOL in lung cancer patients [50]. In a small phase 2 study from Delhi comparing CRT with conventional dose of 60 Gy in 6 weeks versus 48 Gy in 20 fractions over 4 weeks showed the benefit of hypofractionation in improvement for emotional function, cognitive function and chest pain; when the QOL tools QLQ-C30 and QLQ-LC13 were jointly used to demonstrate positive correlations [51]. In view of complexity of the disease and its symptom burden, lung cancer patients will always show dynamic changes which can be reflected by QOL measurements.

20.3.6 Gynecological Cancers

Assessment of QOL among gynecologic cancer patients provides supplementary information for physicians for selecting anti-cancer and supportive-care therapy. Managing QOL in gynecologic cancer patients requires careful consideration of a variety of issues, many of which revolve around the treatment modalities employed and major side effects induced by the therapeutic interventions, as well as disease-associated factors that can negatively affect QOL [52]. For these women diagnosed in LMIC, the place of residence, educational status, marital status, payment type, cancer type and stage of the disease can affect their global QOL, financial distress, pain and fatigue domains, as reported in this study from Ethiopia by Ayana et al. [52].

Gynecological cancer patients undergoing radiation represent a heterogeneous cohort with each subsite have shown different experiences of the patients due to their unique spectrum of side effects. Collaborative studies have become relevant to understand the QoL issues faced by patients in different geographic and socio-cultural regions. A longitudinal prospective QoL analysis showed that patients' general QoL and functioning were impaired before treatment, due to several tumor-related symptoms, and functioning and general QoL returned to near normal post-therapy after 6 months or more [53].

EORTC has subsite's specific questionnaires with QLQ-CX24 for cervical cancer, QLQ-EN24 for endometrial cancer, QLQ-OV28 for ovarian cancer, QLQ-VU34 for vulvar cancer as well as QOL scales specific to certain side effects such as QLQ-PRT20 for radiation proctitis, and QLQ-SHQ-22 for sexual health. QLQ-CX24 was developed in a multicultural and multidisciplinary setting and is based on 3 multi-item scale structure of symptom experience, body image, and sexual/vaginal functioning. Indian women newly diagnosed with locally advanced cervical cancer treated by CRT, were assessed by combining the QLQ-C30 and QLQ-CX24 tools. The mean global health score after 6 months of treatment was significantly higher than the pre-treatment score (59.5 vs. 50.1). Fatigue, pain, insomnia, and appetite loss improved but episodes of diarrhea increased after therapy [54]. Similar to the EORTC, Functional Assessment of Cancer Therapy has site specific questionnaires for cervix (FACT-C), Vulva (FACT-V), Ovary (FACT-O), Endometrium (FACT-En). PORTEC-3 trial reported long-term QOL results comparing chemoradiation vs. radiation for high-risk endometrial cancer. Long-term morbidities and QOL assessments were done by using EORTC QLQ-C30 and CX24/OV28 subscales. The results at a median of 6+ years noted a long-lasting, clinically relevant, negative impact of chemoradiation therapy on toxicity and HRQOL, most importantly persistent peripheral sensory neuropathy and lower limb weakness. These long-term QOL data are essential for patient information and shared decision-making, even when outcomes show significantly improved overall survival and failure-free survival with chemoradiotherapy versus radiotherapy alone [55].

20.3.7 Prostate

Curative treatment options for early-stage prostate cancer (low-, intermediate-, and high-risk groups) including surgery alone, radiation therapy (RT) alone (external beam or interstitial implant with brachytherapy alone), or combination of RT with androgen deprivation therapy (ADT) give the outcomes of a long-life expectancy. The treatment side effects, both acute and late urinary (GU) and bowel (GI) morbidities, are seen after surgery or RT [56]. The ADT-related metabolic syndrome and sexual dysfunction with psychological distress can become pronounced in a sizeable proportion of patients. All these features together have an impact on HRQOL in older men. Use of ADT results in drop of testosterone levels which can impair cognitive function with associated depression and a risk for suicide because of a rise in the plasma levels of Abeta (beta amyloid), a protein involved in the pathogenesis of cognitive disorders, like Alzheimer's dementia [57]. Recently, hypofractionation regimen with short RT duration is widely adopted in clinical practice, including SBRT with equivalent clinical outcomes. The study done at Fox Chase Cancer Center, USA, comparing the standard (76 Gy in 2 Gy per fraction) with moderate hypofractionation regimen (70.2 Gy in 2.7 Gy per fraction) showed no decline in global QoL, no increase in risk for GI or GU side effects or decline in GU/GI-related QoL with hypofractionation, measured using Expanded Prostate cancer Index Composite (EPIC) questionnaire at 0, 6, 12, 24 months [56]. In a secondary analysis on the QoL outcomes (measured using the PCSS, Prostate cancer symptom scale, and EORTC QLQ-C30) measured at baseline, up to 10 years and at 15 years), the RCT(HYPO-RT-PC) from the Swedish research council, compared ultra-hypofractionation regimen of 42.7 Gy in 7 fractions (3 days per week over 2.5 weeks) with conventional fractionation of 78 Gy in 39 fractions (5 days per week for 8 weeks). The assessments showed higher acute toxicities for the ultra-hypofractionated RT course, but the long-term patient-report QoL outcomes up to 6 years after completion of treatment were similar in both RT dose regimens [58].

20.3.8 Rectal and Anal Canal Cancers

Radiation therapy (RT) to the pelvis for rectal and anal malignancies can have an impact on QoL due to fecal incontinence, urinary symptoms, and sexual dysfunction. The incremental improvements in surgical and RT techniques (pre- or postoperative RT), like intensity modulated radiation therapy (IMRT) with daily image guidance, laparoscopic surgeries with TME for LAR, and management of chemotherapy toxicities are leading to less pelvic pain, dermatological, and late bowel and urinary problems [59]. Thus, there is an increase in emphasis to include patient-related outcomes (PROs) with HRQOL tools in the ongoing clinical trials. A systematic literature review of relevant QoL articles between 2014 and 2019 done by Neibart et al. described various tools used to measure QoL for rectal and anal canal malignancies: EORTC QLQ-C30 for general questionnaire, EORTC QLQ-CR29, FACT-C, EORTC QLQ-ANL27 as disease-specific tools and Wexner continence grade scale, FIQL (Fecal incontinence quality of life) scales for assessment of fecal incontinence [60].

Definitive concurrent chemo-radiation therapy is the treatment of choice for anal canal cancers and has shown increased sphincter preservation rates, reducing the need for permanent colostomy from APR. Pelvic chemoradiation therapy (CRT) is the curative therapy for non-metastatic anal cancer, resulting in excellent survival rates. However, anal cancer survivors may experience diminished QoL (QOL) due to late toxicities from pelvic CRT. By using FACT-G (7 item version), three themes most mentioned by respondents were persistent toxicity effecting QOL (82%), insufficient upfront information about CRT (56%), and gratitude toward care received (35%) [61].

20.3.9 RT, Cancer Therapy Toxicities—Impact on QoL

The acute and late side effects of radiotherapy vary differently based on the primary site, in combination with surgery or chemotherapy and the RT technique used. Site-specific QOL questionnaires described in the previous sections help to assess QOL related to radiotherapy and/or combined anti-cancer therapy for each disease site. QOL studies on patients who received pelvic radiotherapy in prostate, rectal, anal and gynecologic cancers showed increased fatigue in women than men, while sexual functions were more affected in men. Chemotherapy was associated with more diarrhea and worse global QoL. The significant toxicities which linger post-treatment, such as cystitis, proctitis, urethral and perineal pain are collectively addressed with a term, "pelvic radiation disease" [62]. A multi-institutional Brazilian study in HNC comparing difference in QOL between different techniques of radiotherapy, measured by EORTC QLQ-C30 and EORTC H&N43, showed that patients treated with IMRT have improved HRQOL compared to those patients with 3D-CRT, who in turn may have better HRQOL than those treated with 2D-RT [63]. Most survivors of lung cancer have undergone invasive treatment (surgery, radiation therapy, and/or chemotherapy) and carry a higher comorbidity burden than survivors of other cancers. Physical and mental components of HRQOL seem to be most at risk for decline, necessitating to develop appropriate interventions after lung cancer therapy [64].

20.3.10 QOL in Geriatric Population

The QoL is a patient-centered evaluation with patient perceptions being the gold standard of assessment. The HRQOL issues involve all adults ≥18 years, however there is more concern for the elderly cancer patients. Cancer in elderly people is increasing with an aging population world-wide. Particular attention should be paid to the older patient with cancer, a group who may have increased vulnerability in HRQOL outcomes following cancer therapy. Approximately 50% of all neoplasms occur in persons aged 65 years or older and will increase with the expansion of the older population. The expectancy of an active life is 10 years and 8 years respectively for persons aged 65 and 70 years. In the last two decades, it has been recognized that radiation therapy fulfills all the requirements to be used with curative or palliative intent in nearly every case of cancer in the elderly, and modern techniques and equipment to deliver radiotherapy treatment permit a large sparing of normal tissues [65].

In 2020, the average life expectancy in India is 69.9 years, compared to 77.4 years in the USA and 84.6 years in Japan [66]. In developed countries, those over 65 years are considered as elderly, whereas in developing countries, due to lower life expectancy, it may be reasonable to consider those over 60 as elderly. In choosing the most appropriate treatment for the elderly cancer patient, one has to take into consideration performance status and life expectancy. Although in some cases an aggressive approach to older individuals may irreversibly compromise their HRQOL, more often tumor control and improvement in QoL can be achieved concomitantly. The optimal approach to cancer is multidisciplinary and this approach should be routine in elderly. A good QoL should be a primary goal in the treatment of elderly patients with cancer. Thus, assessment of health-related QOL in elderly patients with cancer remains an important area of clinical research, with specific implications, both in measuring and in evaluating the results in this setting [67]. A multidimensional evaluation tool testing the possible comorbidities and their relevance to life expectancy and QoL, the cognitive and emotional functions, and the socioeconomic and family situation of the patient should be accounted (Table 20.4).

Age itself should not be a contraindication to radiation therapy. Radiotherapy plays an important role in the control of loco-regional tumors in elderly patients in whom surgery is contraindicated, due to age-related comorbid conditions (psychiatric disorders, cardiovascular, CNS diseases like Parkinsonism, dementia, other organ dysfunction etc.). A prospective thesis study done from India to assess the QoL and radiation morbidity in patients above 60 years and below 50 years of age during and after radical dose of radiation in HNC using EORTC QLQ C-30 showed a decline in physical functioning, social and emotional functioning during RT for elderly patients compared with younger age ([68], shown in Table 20.6 and Fig. 20.4a, b). In the real world of cancer management, where those at 60 + years present with certain common malignancies of lung, HNC, breast, gynecologic and hematologic, the complexity of diagnosis, anti-cancer modalities and supportive care should be discussed and counseled adequately to the elderly cancer patients and their family care givers. In a review article, Fitch et al. emphasized the importance of QOL measurement in older adults with cancer to include comprehensive geriatric assessment (CGA) and PROs apart from routine tools such as functional health, adverse symptoms, and global QOL in the clinical practice and research [69].

Table 20.6 Comparative values of QOL in different phases, elderly versus young patients with head and neck cancer treated with definitive radiation therapy [Ref. 68]

	Phase I			Phase II			Phase III			Phase IV		
Functional scales	Group A	Group B	*P* value	Group A	Group B	*P* value	Group A	Group B	*P* value	Group A	Group B	*P* value
Physical functioning (PF)	69.39	77.50	**0.035**	40.00	50.00	**0.008**	60.00	62.50	0.434	61.36	64.00	0.357
Role functioning (RF)	61.22	63.75	0.747	34.69	38.75	0.483	53.06	57.50	0.221	54.55	58.57	0.417
Emotional functioning (EF)	68.37	69.15	0.877	48.28	49.80	0.813	63.94	64.97	0.839	65.87	66.38	0.917
Cognitive functioning (CF)	70.07	79.17	**0.001**	45.92	47.92	0.417	56.12	61.25	**0.026**	57.20	62.85	**0.027**
Social functioning (SF)	73.47	77.09	0.123	42.52	45.00	0.296	54.08	59.58	**0.009**	59.47	61.91	0.270
Global health status (GHS)	64.79	66.87	0.444	34.01	35.62	0.659	55.10	57.92	0.236	56.06	58.57	0.301

Note: 4 phases: Pre-treatment phase I, during RT-phase II, 1-month post-RT-phase III and 6 months' post-RT-phase IV. *P* value <0.05 is statistically significant

20.4 Benefits in Using QOL in Cancer Management

QoL is an important concept in the field of health and medicine, more so in cancer care process during the last three decades. Traditionally, biomedical endpoints such as survival, morbidities, recurrence etc. have been the focus to improve cancer diagnosis and treatment modalities; whereas PROs like QOL measurement and the patient's lived experience as a narrative are equally important parameters in our biomedical world [70, 71]. In a real-world setting, it is now proven that the overall QOL score has a strong prognostic value for overall survival for a number of populations of patients with cancer above and beyond that provided by clinical factors and sociodemographic variables [72]. With these recent developments, the cancer physicians, scientists, nurses and cancer social workers can improve the overall cancer care by engaging with the patients to learn from their QOL and narrative structures.

The continuous development of QOL tools and the incorporation of HRQOL assessment as mandatory into clinical trials, in countries like the USA, Europe, China, Japan, South Korea, and Australia, have shown new benefits in cancer management. The findings of physical, psychological, social, and financial function scores; and changes in symptom burden during cancer therapy and cancer survivorship have given new information. Patients with good functional scores and low symptom burden at pre-treatment stage can recover faster to anti-cancer therapy. Those who show return to pre-treatment QOL scores, or, improvement in the scores, within 6 months to 1 year after anti-cancer therapy have higher cancer survival and lesser disease-failure [73, 74]. The positive correlation of QOL scores with progression-free survival and the decreased risks of morbidity and mortality in cancer patients are becoming scientifically validated.

In most of the LMIC including India, the significance of clinical cancer research, including QOL assessment in cancer trials, are yet to be adopted routinely. Desai et al. reported that only 5% of the global cancer resources were spent in LMICs causing high mortality-to-income ratio,

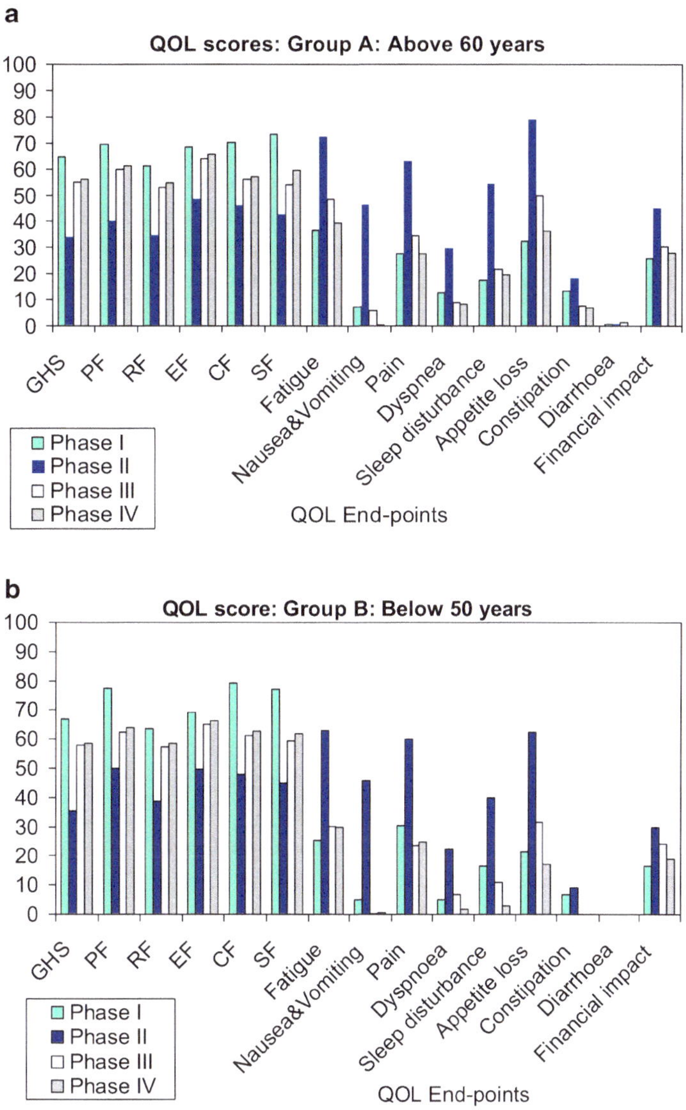

Fig. 20.4 (**a**) Head and neck cancer: The pictorial representation of the QOL scores (on scale of 0–100) is shown for group A (above 60 years). [Ref. 68]. (**b**) Head and neck cancer: The pictorial representation of the QOL scores (on scale of 0–100) is shown for group B. [Ref. 68]

with limited access to clinical trials compared with high-income countries [75].

There are challenges in adopting QOL into clinical trials across the globe, more so in LMIC, with varying administrative, technical and workflow-related issues. These can be broadly due to certain hurdles and reasons:

1. Doctors, nurses and health researchers require to understand the necessities of QOL assessment.
2. Costs and resources needed for implementation, impact on clinic workflow, staff participation/controlling or monitoring by physician to be put in place.
3. Two-thirds of most common malignancies like breast, cervix uteri, lung, head and neck and stomach present in advanced stages (stages III and IV) where diagnosis and cancer treatments take precedence over PROs.
4. Access to multi-disciplinary cancer management team remains fragmented, combined with low socio-cultural acceptance of cancer diagnosis and therapy, and the poor perception by patients and caregivers about cancer as a disease entity.

Cancer survivors in LMICs in Southeast Asia have impaired HRQOL and substantial proportions have psychological distress. Patients with advanced cancer stages at diagnosis and those in a poor socioeconomic position are most at risk of such poor outcomes. Supportive interventions for cancer patients that address wider aspects of patient wellbeing are needed, as well as policies that address financial and other barriers to timely treatment. The patient-reported outcomes including longitudinal QOL assessment can mitigate many of these prevailing issues [76].

20.5 Conclusions

With the recent improvements in various treatments including advancement in radiation therapy techniques, surgery and chemotherapy/immunotherapy, cancer is considered as a chronic disease for several malignancies, giving rise to long-term survival. An important aspect of patient-centered care and self-management has been defined as "a persons' ability to manage the symptoms and consequences of living with a chronic condition, including treatment, physical, social and lifestyle changes" [77]. HRQOL is a broader concept of general QoL and can directly influence various factors including the new diagnosis of cancer. Various instruments used to assess QoL to obtain PROs can help to collaborate among different specialty physicians and thus, address the late side effects and monitor the long-term disease outcomes.

20.6 Summary

- QOL assessment is increasingly being realized as an important end point in many recent large trials across the world and the earliest QOL tools used in Oncology are EORTC QLQ-C30 and FACT-G.
- There are two broad types of QOL tools—Generic (across all types of cancer) and cancer-specific (for particular cancer or cancer-related condition). QOL assessment of any specific site should include both the generic and specific tools of that respective questionnaire.
- The newer radiation therapy techniques like IMRT causes lesser decline in QOL compared to 3D-CRT.
- The EORTC subsite-specific scale assesses three multi-item scale structure of symptom experience, body image and sexual/vaginal functioning in patients with gynecological malignancies.
- Long-term QOL assessment is challenging due to confounding factors like age, sex, comorbidities, stage of disease as well as lack of compliance, follow up and missing data.
- Older patients are less likely to receive standard of care compared to younger patients due to the associated comorbid conditions and a comprehensive geriatric assessment (CGA) including QOL can help to address the needs of these patients prior to starting cancer treatments.
- Best palliative care can be provided for patients with locally advanced or end-stage cancers by including multidisciplinary team based on the patient symptom burden from QoL assessments and then to reduce the suffering due to physical and psychosocial issues.

Multiple Choice Questions

1. What age is considered optimal for elderly in QoL studies in Western countries and India?
   A. 65 years, 60 years
   B. 70 years, 65 years
   C. 66 years, 61 years
   D. 75 years, 70 years
   E. 60 years, 65 years
2. All the following are generic questionnaires except:
   A. IBCSG
   B. SF-36
   C. EORTC QOL-C30
   D. FACT-B

3. The following questionnaires are used for assessment of the general QOL in cancer patients except:
   A. ECOG
   B. EORTC QLQ-C30
   C. FACT-Br
   D. FACT-G
4. Which of the following statement about the EORTC QLQ-C30 is true?
   A. A high score on functional scale represents better QOL.
   B. A high score on symptom scale represents better QOL
   C. A low score on functional scale shows better QOL
   D. A high score on functional and symptom scale represents a better QOL.
5. The MMSE (Mini Mental Status Examination) which is used to assess cognitive function was first designed for which of the following clinical condition?
   A. Brain tumor
   B. Brain metastasis
   C. Stroke
   D. Psychiatric illness
6. All of the following are questionnaires used to assess QoL in patients specifically with bone metastasis except:
   A. BOMET-QOL
   B. Fact-Br
   C. EORTC-QLQ-BM 22
   D. FACT-BP
7. Which HRQL instrument has become increasingly prioritized to compare and contrast outcomes in HPV-associated oropharyngeal cancer de-escalation studies?
   A. FACT-G
   B. EQ-5D
   C. MDADI
   D. MDASI-HN
8. In many definitive chemoradiotherapy trials, what is the typical time frame for global or general measures to return to baseline values (using mean scores of different arms)?
   A. 1–2 weeks
   B. 1–3 months
   C. 3–4 months
   D. 6–12 months
   E. Global measures never recover to baseline values
9. What differences in HRQL trajectories have been observed between HPV-positive and HPV-negative patient cohorts?
   A. No differences
   B. HPV-positive patients typically have lower baseline HRQL and are more vulnerable to impacts during CRT
   C. HPV-positive patients typically have higher baseline HRQL and are more vulnerable to impacts during CRT
   D. HPV-negative patients typically have lower baseline HRQL and are more vulnerable to impacts during CRT
10. In HNC patients, technical improvements in radiotherapy technique have been shown
    A. To improve all HRQL domains
    B. To improve some HRQL domains
    C. To show no impact in HRQL domains
11. Radiotherapy can provide an HRQL benefit to patients with incurable disease
    A. Yes
    B. No
    C. Not sure
12. In the MD Anderson Cancer Center randomized study by Shaitelman et al. that assessed QOL in conventionally fractionated vs. hypofractionated whole-breast irradiation, what was the scale used to assess QOL?
    A. (FACT-B) Questionnaire
    B. Breast Cancer Treatment Outcome Scale (BCTOS)
    C. Harvard Breast Cosmesis Scale (HBCS),
    D. EORTC QLQ-BR23
13. FACT-B assesses all the following functional domains except:
    A. Physical
    B. Psychological
    C. Emotional
    D. Social

14. In the RTOG 0617 randomized study of high-dose vs. standard dose radiation, which of the following was associated with improved QOL?
    A. Use of IMRT
    B. Patients' management through multidisciplinary clinics
    C. Use of motion management techniques
    D. PET-based contours of the target
15. EORTC QLQ-LC29 is an updated version of the EORTC QLQ-LC13 scale. What specific subscale has been added?
    A. Immunotherapy complications
    B. Targeted therapy complications
    C. Surgical complications
    D. Radiation complications
16. In the TIME-C randomized study of IMRT vs. 3D-CRT in postoperative endometrial and cervical cancer patients, what scale was used to measure bowel function?
    A. QLQ-CX24
    B. QLQ-EN24
    C. EPIC score
    D. FACT-C
17. PORTEC-2 trial which compared vaginal brachytherapy vs. external beam radiation in stage I high–intermediate-risk endometrial cancer patients noted worsening QOL in EBRT for which of the following?
    A. Bowel function
    B. Sexual function
    C. Bladder function
    D. A and C
    E. All the above
18. What serum marker is used to estimate the risk of suicide, depression in neurocognitive QoL assessment tools for prostate cancer?
    A. Serum PSA
    B. Serum testosterone
    C. Serum estradiol
    D. Abeta (beta amyloid)
    E. IL-10 & IL-12
19. What is true regarding patient-reported QoL outcomes (PROs) in HYPO-RT-PC trial estimated using a validated PCSS and EORTC QLQ-C30 questionnaire, a secondary endpoint at 6 years? (A noninferior, phase-III randomized clinical trial by Swedish research council comparing ultra-hypofractionation to conventionally fractionated radiotherapy for intermediate, high-risk prostate carcinoma).
    A. Acute GI toxicities (7 of 10 bowel symptoms) were higher in ultra-hypofractionation at the end of radiotherapy.
    B. Acute urinary symptoms/problems and sexual functioning symptoms were higher in the ultra-hypofractionation arm.
    C. At 6 years, overall urinary problems, overall sexual bother and global health/QoL was higher in the ultra-hypofractionation arm.
    D. No significant difference in overall urinary problems, overall sexual bother and global health/QoL was higher in the ultra-hypofractionation arm at 6 years.
    E. A &D
20. What latest disease site-specific QoL tools are used to measure HRQOL for rectal and anal canal malignancies?
    A. EORTC QLQ-C30
    B. EORTC QLQ-CR29, FACT-C, EORTC QLQ-ANL27
    C. EORTC QLQ-C30, QLQ-CR38
    D. A & C
    E. B & C
21. The most important challenge related to assessment of QOL after cancer-related therapy is:
    A. Confounding factors like age, sex, comorbidities, etc.
    B. Compliance
    C. Missing data
    D. All of the above
22. Which of the following statements is false?
    A. The global QOL scores for patients with sphincter preserving low anterior resection were similar to patient with abdominoperineal resection with colostomy.
    B. There were better functional scores for those patients who underwent colostomy.
    C. There were better functional scores for those who underwent sphincter preservation surgery
    D. All of the above statement are false

23. In assessment of QOL postchemotherapy in early lung cancers, the QOL returned to baseline in.
    A. 2 months
    B. 3 months
    C. 6 months
    D. 1 year
24. Which of the following statements are true with regards to QOL after radiotherapy in rectal cancers?
    A. Fatigue was more common in women than in men.
    B. Sexual functions were more affected in women than in men.
    C. Bowel symptoms were worse in men than in women.
    D. There was no difference in any of the parameters between sexes.
25. What are the outcomes of QoL among elderly versus younger head and neck cancer patients in the study from India?
    A. QoL of elderly patients was lower compared to younger patients at phase II (during RT course) in terms of physical, social and emotional functioning levels.
    B. Global health status (GHS) score was higher for younger patients than elderly at all phases and declined markedly in phase II for both groups.
    C. The elderly experienced more symptoms compared to younger patients and had more financial problems.
    D. All of the above are true
    E. None

Answers

1. A
2. D
3. C
4. A
5. C
6. B
7. C
8. D
9. C
10. B
11. A
12. A
13. B
14. A
15. C
16. C
17. D
18. D
19. E
20. B
21. D
22. C
23. B
24. A
25. D

References[1]

1. Mohanti BK. Quality of life. In: Bahadur S, Thakar A, Mohanti B, editors. Carcinoma of the larynx and hypopharynx. Singapore: Springer; 2019. https://doi.org/10.1007/978-981-13-3110-7_17.
2. Gotay CC, Korn EL, McCabe MS, Moore TD, Cheson BD. Quality-of-life assessment in cancer treatment protocols: research issues in protocol development. J Natl Cancer Inst. 1992;84(8):575–9. https://doi.org/10.1093/jnci/84.8.575. PMID: 1556768.
3. Lipscomb J, Gotay CC, Snyder CF. Patient-reported outcomes in cancer: a review of recent research and policy initiatives. CA Cancer J Clin. 2007;57(5):278–300. https://doi.org/10.3322/CA.57.5.278. PMID: 17855485.
4. Coens C, Pe M, Dueck AC, Sloan J, Basch E, Calvert M, et al. Setting International Standards in Analyzing Patient-Reported Outcomes and Quality of Life Endpoints Data Consortium. International standards for the analysis of quality-of-life and patient-reported outcome endpoints in cancer randomised controlled trials: recommendations of the SISAQOL Consortium. Lancet Oncol. 2020;21(2):e83–96. https://doi.org/10.1016/S1470-2045(19)30790-9. PMID: 32007209.
5. Brady LW, Markoe AM, Fisher SA, Micaily B. Cancer cure with organ preservation using radiation therapy. Acta Oncol. 1988;27(3):195–202. https://doi.org/10.3109/02841868809093525.
6. Haslam A, Herrera-Perez D, Gill J, Prasad V. Patient experience captured by quality-of-life measurement in oncology clinical trials. JAMA Netw Open. 2020;3(3):e200363. https://doi.org/10.1001/jamanetworkopen.2020.0363.

[1]* The article on providing the QOL assessment for cancer patients.

7. Péus D, Newcomb N, Hofer S. Appraisal of the Karnofsky Performance Status and proposal of a simple algorithmic system for its evaluation. BMC Med Inform Decis Mak. 2013;13:72. https://doi.org/10.1186/1472-6947-13-72.
8. Devlin NJ, Brooks R. EQ-5D and the EuroQoL Group: past, present and future. Appl Health Econ Health Policy. 2017;15(2):127–37. https://doi.org/10.1007/s40258-017-0310-5.
9. Ware JE Jr, Sherbourne CD. The MOS 36-item short-form health survey (SF-36). I. Conceptual framework and item selection. Med Care. 1992;30(6):473–83. PMID: 1593914.
10. https://ecog-acrin.org/resources/ecog-performance-status/.
11. Zubrod C, et al. Appraisal of methods for the study of chemotherapy in man: comparative therapeutic trial of nitrogen mustard and thiophosphoramide. J Chronic Dis. 1960;11:7–33.
12. *Aaronson NK, Ahmedzai S, Bergman B, Bullinger M, et al. The European Organization for Research and Treatment of Cancer QLQ-C30: a quality-of-life instrument for use in international clinical trials in oncology. J Natl Cancer Inst. 1993;85(5):365–76. https://doi.org/10.1093/jnci/85.5.365.. PMID: 8433390.
13. *Cella DF, Tulsky DS, Gray G, Sarafian B, et al. The Functional Assessment of Cancer Therapy scale: development and validation of the general measure. J Clin Oncol. 1993;11(3):570–9. https://doi.org/10.1200/JCO.1993.11.3.570.
14. https://qol.eortc.org/manuals/.
15. https://qol.eortc.org/questionnaires/.
16. https://www.facit.org/measures-language-availability.
17. Cohen SR, Russell LB, Leis A, Shahidi J, et al. More comprehensively measuring quality of life in life-threatening illness: the McGill Quality of Life Questionnaire—expanded. BMC Palliat Care. 2019;18(1):92. https://doi.org/10.1186/s12904-019-0473-y.
18. de Souza JA, Yap BJ, Hlubocky FJ, Wroblewski K, Ratain MJ, Cella D, Daugherty CK. The development of a financial toxicity patient-reported outcome in cancer: the COST measure. Cancer. 2014;120(20):3245–53. https://doi.org/10.1002/cncr.28814.
19. Campbell HS, Hall AE, Sanson-Fisher RW, Barker D, Turner D, Taylor-Brown J. Development and validation of the Short-Form Survivor Unmet Needs Survey (SF-SUNS). Support Care Cancer. 2014;22(4):1071–9. https://doi.org/10.1007/s00520-013-2061-7.
20. Mead KH, Wang Y, Cleary S, Arem H, Pratt-Chapman ML. Defining a patient-centered approach to cancer survivorship care: development of the patient centered survivorship care index (PC-SCI). BMC Health Serv Res. 2021;21(1):1353. https://doi.org/10.1186/s12913-021-07356-6.
21. Li J, Bentzen SM, Li J, Renschler M, Mehta MP. Relationship between neurocognitive function and quality of life after whole-brain radiotherapy in patients with brain metastasis. Int J Radiat Oncol Biol Phys. 2008;71:64–70. https://doi.org/10.1016/j.ijrobp.2007.09.059.
22. Meyers CA, Wefel JS. The use of the mini-mental state examination to assess cognitive functioning in cancer trials: no ifs, ands, buts, or sensitivity. J Clin Oncol. 2003;21:3557–8. https://doi.org/10.1200/JCO.2003.07.080. PMID: 12913103.
23. Chow R, Lao N, Popovic M, Chow E, Cella D, et al. Comparison of the EORTC QLQ-BN20 and the FACT-Br quality of life questionnaires for patients with primary brain cancers: a literature review. Support Care Cancer. 2014;22:2593–8. https://doi.org/10.1007/s00520-014-2352-7. PMID: 25015058.
24. Chen E, Nguyen J, Zhang L, et al. Quality of life in patients with brain metastases using the EORTC QLQ-BN20 and QLQ-C30. J Radiat Oncol. 2012;1:179–86. https://doi.org/10.1007/s13566-012-0016-0.
25. Jalali R, Gupta T, Goda JS, Goswami S, et al. Efficacy of stereotactic conformal radiotherapy vs conventional radiotherapy on benign and low-grade Brain tumors: a randomized clinical trial. JAMA Oncol. 2017;3:1368–76. https://doi.org/10.1001/jamaoncol.2017.0997.
26. Budrukkar A, Jalali R, Dutta D, Sarin R, et al. Prospective assessment of quality of life in adult patients with primary brain tumors in routine neurooncology practice. J Neuro-Oncol. 2009;95:413–9. https://doi.org/10.1007/s11060-009-9939-8.
27. Hernandez RK, Adhia A, Wade SW, O'Connor E, et al. Prevalence of bone metastases and bone-targeting agent use among solid tumor patients in the United States. Clin Epidemiol. 2015;7:335–45. https://doi.org/10.2147/CLEP.S85496.
28. Koufopoulou C, Mosa E, Charalampakis N, Nikolaou M, et al. Evaluation of quality of life outcomes following palliative radiotherapy in bone metastases: a literature review. J BUON. 2019;24(5):1747–60.
29. McDonald R, Ding K, Brundage M, Meyer RM, et al. Effect of radiotherapy on painful bone metastases: a secondary analysis of the NCIC clinical trials group symptom control trial SC.23. JAMA Oncol. 2017;3(7):953–9. https://doi.org/10.1001/jamaoncol.2016.6770.
30. Agarawal JP, Swangsilpa T, van der Linden Y, Rades D, Jeremic B, Hoskin PJ. The role of external beam radiotherapy in the management of bone metastases. Clin Oncol (R Coll Radiol). 2006;18(10):747–60. https://doi.org/10.1016/j.clon.2006.09.007.
31. Fagan JJ, Noronha V, Graboyes EM. Making the best of limited resources: improving outcomes in head and neck cancer. Am Soc Clin Oncol Educ Book. 2021;41:1–11. https://doi.org/10.1200/EDBK_320923.
32. Wibonele BK, Smith BD, Altonji S, Kaplan S, Cho J, Lee WT. Head and neck cancer research collaborations between the United States and low- and middle-income countries: 10-year publication analysis. Head Neck. 2021;43(8):2395–404.

33. van Nieuwenhuizen AJ, Buffart LM, Langendijk JA, Vergeer MR, et al. Health-related quality of life and overall survival: a prospective study in patients with head and neck cancer treated with radiotherapy. Qual Life Res. 2021;30:1145–53. https://doi.org/10.1007/s11136-020-02716-x.
34. Chawla S, Mohanti BK, Rakshak M, Saxena S, Rath GK, Bahadur S. Temporal assessment of quality of life of head and neck cancer patients receiving radical radiotherapy. Qual Life Res. 1999;8(1–2):73–8. https://doi.org/10.1023/a:1026476928283.
35. Yucel B, Akkaş EA, Okur Y, Eren AA, Eren MF, Karapinar H, Babacan NA, Kiliçkap S. The impact of radiotherapy on quality of life for cancer patients: a longitudinal study. Support Care Cancer. 2014;22(9):2479–87. https://doi.org/10.1007/s00520-014-2235-y.
36. Bansal M, Mohanti BK, Shah N, Chaudhry R, Bahadur S, Shukla NK. Radiation related morbidities and their impact on quality of life in head and neck cancer patients receiving radical radiotherapy. Qual Life Res. 2004;13(2):481–8. https://doi.org/10.1023/B:QURE.0000018491.80646.bc.
37. Huang TL, Chien CY, Tsai WL, Liao KC, et al. Long-term late toxicities and quality of life for survivors of nasopharyngeal carcinoma treated with intensity-modulated radiotherapy versus non-intensity-modulated radiotherapy. Head Neck. 2016;38(Suppl 1):E1026–32. https://doi.org/10.1002/hed.24150.
38. McDowell L, Corry J, Ringash J, Rischin D. Quality of life, toxicity and unmet needs in nasopharyngeal cancer survivors. Front Oncol. 2020;10:930. https://doi.org/10.3389/fonc.2020.00930.
39. Ringash J, Fisher R, Peters L, Trotti A, O'Sullivan B, Corry J, Kenny L, Nuyts S, Wratten C, Rischin D. Effect of p16 status on the quality-of-life experience during chemoradiation for locally advanced oropharyngeal cancer: a substudy of randomized trial Trans-Tasman Radiation Oncology Group (TROG) 02.02 (HeadSTART). Int J Radiat Oncol Biol Phys. 2017;97(4):678–86. https://doi.org/10.1016/j.ijrobp.2016.03.017.
40. Tasoulas J, Lenze NR, Farquhar D, Schrank PT, Shen C, Shazib MA, Singer B, Patel S, Grilley Olson JE, Hayes DN, Gulley ML, Chera BS, Hackman T, Olshan AF, Weiss J, Sheth S. The addition of chemotherapy to adjuvant radiation is associated with inferior survival outcomes in intermediate-risk HPV-negative HNSCC. Cancer Med. 2021;10(10):3231–9. https://doi.org/10.1002/cam4.3883.
41. Oskam IM, Verdonck-de Leeuw IM, Aaronson NK, Kuik DJ, et al. Quality of life as predictor of survival: a prospective study on patients treated with combined surgery and radiotherapy for advanced oral and oropharyngeal cancer. Radiother Oncol. 2010;97(2):258–62. https://doi.org/10.1016/j.radonc.2010.02.005.
42. Mair MD, Nair S, Nikam S, Nair D, Agarwal JP, Chaturvedi P. Longitudinal and cross-sectional assessment of quality of life in surgically treated advanced (T4) cancer of the buccal mucosa. Oral Surg Oral Med Oral Pathol Oral Radiol. 2017;124(6):529–36. https://doi.org/10.1016/j.oooo.2017.09.006.
43. Fabian A, Domschikowski J, Letsch A, Schmalz C, Freitag-Wolf S, Dunst J, Krug D. Use and reporting of patient-reported outcomes in trials of palliative radiotherapy: a systematic review. JAMA Netw Open. 2022;5(9):e2231930. https://doi.org/10.1001/jamanetworkopen.2022.31930.
44. Mohanti BK, Umapathy H, Bahadur S, Thakar A, Pathy S. Short course palliative radiotherapy of 20 Gy in 5 fractions for advanced and incurable head and neck cancer: AIIMS study. Radiother Oncol. 2004;71(3):275–80. https://doi.org/10.1016/j.radonc.2004.03.009.
45. Prakash S, Chakrabarti D, Kumar R, Agrawal MM, et al. Palliative radiotherapy: a one-week course in advanced head and neck cancer—quality of life outcomes. BMJ Support Palliat Care. 2022:bmjspcare-2021-002908. https://doi.org/10.1136/bmjspcare-2021-002908.
46. Meattini I, Saieva C, Miccinesi G, Desideri I, et al. Accelerated partial breast irradiation using intensity modulated radiotherapy versus whole breast irradiation: health-related quality of life final analysis from the Florence phase 3 trial. Eur J Cancer. 2017;76:17–26. https://doi.org/10.1016/j.ejca.2017.01.023.
47. *Bjelic-Radisic V, Cardoso F, Cameron D, Brain E, et al. An international update of the EORTC questionnaire for assessing quality of life in breast cancer patients: EORTC QLQ-BR45. Ann Oncol. 2020;31(2):283–8.
48. Marta GN, Moraes FY, Leite ETT, Chow E, et al. A critical evaluation of quality of life in clinical trials of breast cancer patients treated with radiation therapy. Ann Palliat Med. 2017;6(Suppl 2):S223–32.
49. Koller M, Shamieh O, Hjermstad MJ, Hornslien K, et al. European Organisation for Research and Treatment of Cancer (EORTC) Quality of Life Group; EORTC Lung Cancer Group; European Society of Thoracic Surgeons. Psychometric properties of the updated EORTC module for assessing quality of life in patients with lung cancer (QLQ-LC29): an international, observational field study. Lancet Oncol. 2020;21(5):723–32. https://doi.org/10.1016/S1470-2045(20)30093-0.
50. Movsas B, Hu C, Sloan J, Bradley J, et al. Quality of life analysis of a radiation dose-escalation study of patients with non-small-cell lung cancer: a secondary analysis of the Radiation Therapy Oncology Group 0617 randomized clinical trial. JAMA Oncol. 2016;2(3):359–67. https://doi.org/10.1001/jamaoncol.2015.3969.
51. Roy S, Pathy S, Mohanti BK, Raina V, et al. Accelerated hypofractionated radiotherapy with concomitant chemotherapy in locally advanced squamous cell carcinoma of lung: evaluation of response, survival, toxicity and quality of life from a Phase II randomized study. Br J Radiol. 2016;89(1062):20150966. https://doi.org/10.1259/bjr.20150966.

52. Ayana BA, Negash S, Yusuf L, Tigeneh W, Haile D. Health related quality of life of gynaecologic cancer patients attending at Tikur Anbesa Specialized Hospital (TASH), Addis Ababa, Ethiopia. BMC Womens Health. 2018;18(1):7. https://doi.org/10.1186/s12905-017-0507-7.
53. Kirchheiner K, Pötter R, Tanderup K, Lindegaard JC, et al. EMBRACE Collaborative Group. Health-related quality of life in locally advanced cervical cancer patients after definitive chemoradiation therapy including image guided adaptive brachytherapy: an analysis from the EMBRACE Study. Int J Radiat Oncol Biol Phys. 2016;94(5):1088–98. https://doi.org/10.1016/j.ijrobp.2015.12.363.
54. Dahiya N, Acharya AS, Bachani D, Sharma D, Gupta S, Haresh K, Rath G. Quality of life of patients with advanced cervical cancer before and after chemoradiotherapy. Asian Pac J Cancer Prev. 2016;17(7):3095–9.
55. Post CCB, de Boer SM, Powell ME, Mileshkin L, et al. Long-term toxicity and health-related quality of life after adjuvant chemoradiation therapy or radiation therapy alone for high-risk endometrial cancer in the randomized PORTEC-3 trial. Int J Radiat Oncol Biol Phys. 2021;109(4):975–86. https://doi.org/10.1016/j.ijrobp.2020.10.030.
56. Turaka A, Zhu F, Buyyounouski MK, Horwitz EM, et al. Conventional versus hypofractionated IMRT: results of late GI and GU toxicity and quality of life from a phase III trial. Int J Radiat Oncol Biol Phys. 2010;78(3):S67. (Abstract # 142). https://doi.org/10.1016/j.ijrobp.2010.07.188.
57. Yuan F, Chang D, Jing M, Zhu B, You Y. Effectiveness of cognitive behavioural therapy on quality of life in patients with prostate cancer after androgen deprivation therapy: a protocol for systematic review and meta-analysis. BMJ Open. 2021;11(11):e049314. https://doi.org/10.1136/bmjopen-2021-049314.
58. Fransson P, Nilsson P, Gunnlaugsson A, Beckman L, et al. Ultra-hypofractionated versus conventionally fractionated radiotherapy for prostate cancer (HYPO-RT-PC): patient-reported quality-of-life outcomes of a randomised, controlled, non-inferiority, phase 3 trial. Lancet Oncol. 2021;22(2):235–45. https://doi.org/10.1016/S1470-2045(20)30581-7.
59. Chen TY, Wiltink LM, Nout RA, Meershoek-Klein Kranenbarg E, et al. Bowel function 14 years after preoperative short-course radiotherapy and total mesorectal excision for rectal cancer: report of a multicenter randomized trial. Clin Colorectal Cancer. 2015;14(2):106–14. https://doi.org/10.1016/j.clcc.2014.12.007. Epub 2014 Dec 31
60. Neibart SS, Manne SL, Jabbour SK. Quality of life after radiotherapy for rectal and anal cancer. Curr Colorectal Cancer Rep. 2020;16(1):1–10. https://doi.org/10.1007/s11888-019-00448-w.
61. Corrigan KL, De B, Rooney MK, Ludmir EB, et al. Patient-reported outcomes after chemoradiation in patients with anal cancer: a qualitative analysis. Adv Radiat Oncol. 2022;7(4):100986. https://doi.org/10.1016/j.adro.2022.
62. Sammour T, Kahokehr AA. Pelvic radiation disease. Clin Colon Rectal Surg. 2022;35(3):204–11. https://doi.org/10.1055/s-0041-1740041.
63. Santos M, Oliveira E Silva LF, Kohler HF, Curioni O, et al. Health-related quality of life outcomes in head and neck cancer: results from a prospective, real-world data study with Brazilian patients treated with intensity modulated radiation therapy, conformal and conventional radiation techniques. Int J Radiat Oncol Biol Phys. 2021;109(2):485–94. https://doi.org/10.1016/j.ijrobp.2020.09.044.
64. Vijayvergia N, Shah PC, Denlinger CS. Survivorship in non-small cell lung cancer: challenges faced and steps forward. J Natl Compr Cancer Netw. 2015;13(9):1151–61. https://doi.org/10.6004/jnccn.2015.0140.
65. Olmi P, Cefaro GA, Balzi M, Becciolini A, Geinitz H. Radiotherapy in the aged. Clin Geriatr Med. 1997;13(1):143–68.
66. https://data.worldbank.org/indicator/SP.DYN.LE00.IN?locations=IN. Accessed 28 Nov 2022.
67. Di Maio M, Perrone F. Quality of life in elderly patients with cancer. Health Qual Life Outcomes. 2003;1:44. https://doi.org/10.1186/1477-7525-1-44.
68. Turaka A, Mohanti BK, Chander S, Deo SV, et al. Quality of Life Assessment in patients above 60, below 50 years of age during and after curative radiotherapy in head and neck cancers. J Clin Oncol. 2006;24(18_suppl):5552. https://doi.org/10.1200/jco.2006.24.18_suppl.5552.
69. Fitch MI, Strohschein FJ, Nyrop K. Measuring quality of life in older people with cancer. Curr Opin Support Palliat Care. 2021;15(1):39–47. https://doi.org/10.1097/SPC.0000000000000535.
70. Haraldstad K, Wahl A, Andenæs R, Andersen JR, et al. A systematic review of quality of life research in medicine and health sciences. Qual Life Res. 2019;28(10):2641–50. https://doi.org/10.1007/s11136-019-02214-9.
71. Mohanti BK. Lessons from narrative medicine: cancer care will improve with narrative oncology. Indian J Cancer. 2021;58(4):615–8. https://doi.org/10.4103/ijc.IJC_809_20.
72. Husson O, de Rooij BH, Kieffer J, Oerlemans S, et al. The EORTC QLQ-C30 summary score as prognostic factor for survival of patients with cancer in the "real-world": results from the Population-Based PROFILES Registry. Oncologist. 2020;25(4):e722–32. https://doi.org/10.1634/theoncologist.2019-0348.
73. Hwang TJ, Gyawali B. Association between progression-free survival and patients' quality of life in cancer clinical trials. Int J Cancer. 2019;144(7):1746–51. https://doi.org/10.1002/ijc.31957.
74. Sitlinger A, Zafar SY. Health-related quality of life: the impact on morbidity and mortality. Surg Oncol Clin N Am. 2018;27(4):675–84. https://doi.org/10.1016/j.soc.2018.05.008.

75. Desai A, Sirohi B, Mathew A. Clinical trial access in low- and middle-income countries: a case study on India. Cancer Investig. 2021;39(9):685–9. https://doi.org/10.1080/07357907.2021.
76. ACTION Study Group. Health-related quality of life and psychological distress among cancer survivors in Southeast Asia: results from a longitudinal study in eight low- and middle-income countries. BMC Med. 2017;15(1):10. https://doi.org/10.1186/s12916-016-0768-2.
77. McCorkle R, Ercolano E, Lazenby M, Schulman-Green D, et al. Self-management: enabling and empowering patients living with cancer as a chronic illness. CA Cancer J Clin. 2011;61(1):50–62. https://doi.org/10.3322/caac.20093.

Palliative Care in Cancer

21

Sushma Bhatnagar and Saurabh Vig

21.1 Introduction

The World Health Organization (WHO) defines "Palliative Care as an approach that improves the quality of life of patients (adults and children) and their families who are facing life-threatening illnesses" [1]. This field or specialty has its roots in the concept of holistic care for a patient and its caregivers. Palliative Care(PC) by definition finds its application in all branches of medicine dealing with chronic illnesses which negatively affect the quality of life of patients and their caregivers. However, this specialty has originated and has its roots in the oncological setting providing care for chronically ill cancer patients.

It is estimated that each year nearly 60 million people are in need of palliative care. Palliative care in cancer patients should be the key component in low- and middle-income countries (LMICs), since more than 70% of all cancer deaths worldwide occur in LMIC due to advanced stage at diagnosis, limited access to management, and a very short survival time from cancer diagnosis [2].

This chapter aims to impart knowledge and give acquaintance to the reader on the origins of this specialty, its journey in cancer patient care, and its role in the holistic management of an oncology patient.

S. Bhatnagar (✉) · S. Vig
Department of Onco-Anesthesia and Palliative Medicine, Dr. B.R.A. Inst. Rotary Cancer Hospital, A.I.I.M.S., New Delhi, India

21.2 History of Palliative Medicine

Palliative care is a relatively new branch in the field of medicine and finds its humble origin in the concept of hospice care and care of the dying patient. This branch as a concept originated in the late 1960s and has progressed by leaps and bounds to become an integral part of holistic cancer patient care. The major landmarks and milestones in its growth and establishment as a specialty are summarized in Table 21.1.

From Table 21.1, we can see that palliative care grew in leaps and bounds over a short span of 30–40 years from a concept with hospice care to a separate recognized medical specialty. However, despite this impressive progress seen worldwide, the establishment of this branch in India has gone through its own hurdles. The major developments have been seen only in the recent past. Table 21.2 summarizes the history and growth of palliative medicine in India.

Thus, as we can see from Table 21.2, palliative care in India has had a slow but steady growth in both the legislative side of recognition of this branch, and a calibrated integration into the health care system along with the academic and clinical sides. Hence, Palliative Care or Palliative Medicine as a specialty is now an established branch with teaching and training in both government and private sector. The humble beginning of palliative care in India since the 1990s has been

A. Munshi et al. (eds.), *Radiation Oncology – Principles, Precepts and Practice*,
https://doi.org/10.1007/978-981-97-8389-2_21

Table 21.1 Overview of the history of Palliative Care [3, 4]

Year	Event	Importance
1967	Opening of St. Christopher's Hospice in London by Dame Cicely Saunders	• Birth of the Hospice movement out of which palliative care was ultimately born
1969	Elisabeth Kubler Ross published a book titled "On Death and Dying."	• The first description of the 5 stages of grief • This book and the author's lectures raised public awareness about the care of patients near the end of life
1974	First hospice was established in the United States by Florence Wald	• Marked the beginning of the hospice movement in America
1974	Dr. Balfour Mount a surgical oncologist coined the term "Palliative Care."	• Hospice care came under the umbrella of palliative care and the concept of medical care from the time of diagnosis of a life-threatening illness came to be known in the form of "Palliative Care."
1980	The federal government in the USA mandated hospice care to be covered under Medicare	• Acceptance of end-of-life care under conventional medical care, thus, an alternative to home and volunteer-based system of hospice care
1990	The World Health Organization (WHO) recognized palliative care as a distinct specialty	• Path established for the worldwide spread of palliative medicine and hospice care
2006	American Board of Medical Specialities (ABMS) and Accreditation Council for Graduate Medical Education (ACGME) recognized Palliative Care as a separate specialty	• Path cleared for the introduction of palliative care as a specialty subject across America

narrated, to record the eminent persons, places, and milestones [10].

21.3 Principles and Policies of Palliative Care [11, 12]

Palliative care in its basic essence is nothing but good clinical practice. A patient-centric holistic approach keeping in mind the individual needs of every patient, caregiver concerns, spiritual beliefs of the patient and family, and the socio-economic background of the family should be kept in mind and the goals of comfort care should be aligned with these.

The basic aim of palliative care is to relieve the pain of every patient in a totalitarian concept, i.e., pain in the form of physical agony, spiritual suffering, the psychological, spiritual, and practical struggles of the patient and the caregiver [13]. Also, patients' individual preferences and religious, cultural, and social values are to be kept in mind while providing comfort care in any form, be it pain relief, symptomatic support at the end of life, or social and mental support in the form of counseling. Keeping this approach in mind the basic principles in medical ethics, i.e., *Autonomy, Beneficence, Nonmaleficence, and Justice* [14] apply equally well in palliative care.

- Autonomy—this principle in its basic essence states that any patient with preserved cognition to understand his or her medical condition and make rational decisions should have the power to make decisions on their course of treatment or in other words give consent to any planned treatment after being made aware of their choices. This principle thus mandates a palliative care physician to give full disclosure of the disease stage and the available treatment options to a patient with intact cognition and decision-making abilities so that he or she can make informed choices on their course of treatment.
- Beneficence—this principle emphasizes the obligation of a physician to act for the benefit of every patient. What makes it different from the principle of nonmaleficence is that this principle calls not just to prevent harm but also to act proactively for the benefit of every patient and promote their welfare.

Table 21.2 Overview of the history of Palliative Care in India [5–9]

Year	Event	Importance
1984	Amendment in the National cancer control program—made "pain relief" a basic service to be provided at the primary health center level	• Recognition of the importance of pain relief in oncology at the legislative level
1985	Formulation of the Narcotic Substances and Psychotropic Substances (NDPS) act	• Though an initial step, but made opioid procurement and storage cumbersome. This led to a fall in morphine consumption in India in the following years
1986–87	Palliative Care Clinic at Ahmedabad and Bangalore "Shanti Avedana Sadan" hospice opened in Mumbai	• Opioid (oral morphine) for cancer pain relief • Birth of the hospice care movement in India
1993	Indian doctors trained in United Kingdom Formation of Pain and Palliative Care Society (PPCS) in Calicut	• Training at Oxford under WHO-Macmillan Cancer Support • Percolation of the concept of palliative care in the medical fraternity • IAPC body constituted at Varanasi
1994	Formation of the Indian Association of Palliative Care (IAPC)	
2003	Formulation of non-government organization "Pallium India" with the help of the WHO	• Aim to disseminate palliative care services and improve access to palliative care outside Kerala
2008	Formal integration of a Palliative care policy into the health care policy in Kerala	• Kerala becomes the first state of India to have a formal policy for palliative care in the health care system and legislation
2010	The medical council of India recognized Palliative care as a separate specialty	• Ground-breaking milestone to establish palliative care as a separate specialty
2012	– A national program in palliative care was launched and the 12th 5-year plan makes a special provision for palliative care in the care of cancer patients – The First MD course in palliative care started at TATA Memorial Hospital, Mumbai	• Palliative care makes a way in the central government [policy, thus its importance in healthcare is further reinforced • Beginning of the integration of palliative care into the medical teaching curriculum
2014	First amendment in the NDPS act	• Enabled registered medical institutes and practitioners to procure morphine under a single license, thus overcoming the major drawback of multiple licensing for morphine procurement in the initial NDPS act
2021	– The second amendment in the NDPS act – Diplomate in National Board (DNB) palliative medicine program started	• Further streamlining of opioid licensing procurement and storage • A major step in the initiation of teaching and training of palliative medicine in the private and corporate health care sector

- Nonmaleficence—the obligation of a physician to not harm a patient. This principle morally binds a physician not to carry out any task or procedure which may not be useful or may be harmful to any patient. The practical application of this principle in palliative medicine is that it helps in planning end-of-life care decisions, narrowing down, and withholding futile interventions for a patient nearing death. The obligation to relieve suffering guides a palliative care physician to continue pain relief, dyspnea management, plan nutrition, and hydration, and avoid futile interventions like ventilator support, vasopressors, etc. in a patient suffering from a terminal disease nearing the end of life.
- Justice—this principle states that there should be equitable distribution of medical resources for all patients and a fair, appropriate and justifiable treatment should be done for all patients.

Added to these four basic principles or pillars of Biomedical Ethics, there are two more aspects which form the cornerstone of medical practice:

- Dignity—the patient and the persons treating the patient have the right to dignity.
- Truthfulness and honesty—the concept of informed consent and truth telling.

All these put together, these six values comprise the operational parts of Ethics in Palliative Care [15].

Physicians and nurses encounter dilemmas and difficulties in their practice of Palliative Care. Guided by the above principles and values, individual policies to implement palliative care may be formulated with minor adjustments as per regional practices and culture. The delivery of palliative care and medical ethics are complementary, and use of the two together maximizes the protection and satisfaction available to the vulnerable patient and family members [15]. To implement and practice palliative care in an institute, hospital, or individual clinic the first and foremost requirement is to integrate the concept of palliative care into the basic patient care in any area. Once this is done the practice of palliative care can easily be started. The following section describes the basic plan to integrate palliative care into a functional medical setup.

21.4 Integration of Palliative Care with Oncology

It is a known fact that palliative care in oncology improves patients' quality of life, and brings about a positive change in the outlook toward the disease for both the patient and the caregivers. Quality palliative care even brings about an improvement in overall survival. In a path-breaking study by Temel and colleagues [16] early palliative care in patients diagnosed with advanced stage non-small cell lung carcinoma significantly improved the median survival rates as compared to the standard practice of palliative care at end of life alone. Numerous other pieces of evidence in the literature show that timely integration of palliative care into oncology brings about a positive change in the quality of life of the patient and caregiver. However, how to integrate palliative care into an oncological setup is the health system challenge in many parts of the world.

Various conceptual models are given in literature that describes the integration of palliative care into the oncological practice of cancer-directed therapy (CDT) such as radiotherapy, chemotherapy and surgery. In addition, some clinical models are described by some medical centers which have worked in their setup to integrate palliative care into oncological practice.

Theoretically or conceptually, four models are described to integrate palliative care into oncology [17]. Each model aims to answer certain basic questions, namely,

- Who should receive a palliative care referral?
- When should palliative care be introduced?
- Which setting is most appropriate for palliative care delivery?

The salient features of each of these four conceptual models are described below.

1. Time-based model of integration—this model emphasizes the timing of integration of palliative care in a cancer patient undergoing CDT. Literature has time and again emphasized the importance of early palliative care in improving patient outcomes [18, 19]. The essence of this model is that palliative care should be started as early as possible as soon as a patient is diagnosed with cancer. The extent of involvement of palliative care is expected to increase and finally focus on end-of-life care and advanced care planning if the patient progress to advanced stage malignancy with limited life expectancy.
2. Provider-based or "Palli-centric" model—this model talks about providing palliative care according to the hospital setup and the need of the patient, which may be simple or complex palliative care needs. This model explains the delivery of palliative care in three forms—primary, secondary, and tertiary palliative care.

These divisions are made according to the needs of a patient, and accordingly the hospital is set up to deal with the patients. Key points in each of these forms of palliative care delivery are:

- Primary palliative care—palliative care is provided by the oncologist or the primary physician with no specialist involvement. The primary caregiver needs to have basic knowledge of palliative care needs and the site of delivery can be an outpatient setting. No inpatient palliative care or specialized interventions to provide pain relief etc. are done. For these complex care needs patient is referred to a palliative care specialist.
- Secondary palliative care—specialist palliative care teams are involved in the care of the patient. The setting of care may be inpatient or outpatient (OPD) care. A qualified palliative care specialist with a team of physiotherapists, dieticians, counselors, etc. is involved in patient care.
- Tertiary palliative care—patients with the most complex palliative care needs (e.g., acute pain crises, palliative care emergencies like malignant spinal cord compression, malignant bowel obstruction, massive pleural effusion or ascites causing dyspnea, etc.) are managed. This is the highest level of palliative care and thus requires a palliative care specialist and has to be provided as an inpatient level care preferably in tertiary care hospitals and academic institutes. Also, education and research in palliative care are best done in this type of setup.

3. Issue-based or the "Onco-centric" model—this concept states an oncologist manages the day-to-day palliative care needs of a patient and takes referrals whenever necessary. The pattern of referral to a palliative care specialist or an allied subspecialty differs and two conceptual models are described:
   - Congress model of referral—in this model, the oncologist or the primary care physician makes multiple referrals to various specialists for numerous supportive care needs of a patient beyond his or her expertise in management. Multiple referrals like to a pain physician for analgesia, a psychiatrist for anxiety or depression, a dietician, physiotherapist, etc. are potentially confusing for a patient and result in multiple lines of management which may be conflicting with one another and expensive. Thus, this type of referral model is generally not encouraged to be put into practice.
   - Integrated care model of referral—this is the alternate and more practical referral model to the "congress model of referral." In this model, the oncologist refers the patient to a specialized palliative care team early in the course of a malignancy. Palliative care needs, supportive care, and specialized referrals if any are taken care of by this team from here on. The composition of palliative care may differ as per institutional practices and available resources but a palliative care physician is the core member of this team. Other members like a clinical psychologist, trained palliative care nurse, dietician, physiotherapist, etc. are desirable.
   - This model follows the principle of the palliative care team or unit being a single contact point for the patient for any supportive care need. Thus, this model avoids the drawback of the congress model and avoids the confusion of multiple referrals and conflicting opinions for the patient. Also, this model allows for the integration of palliative care at the earliest possible step in the journey of care of a cancer patient.
4. System-based or "patient-centric" model of integration—this model follows a patient's journey into a health care system during the treatments (CDT) for a malignancy. The basic idea in this model follows the path of an integrated system of referral from time to time and a single palliative care team looking after the patient. The patient should be encouraged to visit the palliative care team early and regularly.

These theoretical models aim to answer the "who, when, and where" of palliative care integration. On the ground level, the pattern of integration and palliative care practice is decided and delivered by the clinical service settings. The above-described models of integration can be applied to various cancer care settings, two main clinical settings described in the literature are outpatient palliative care and embedded care model. The salient features of each of these two settings are:

- Outpatient palliative care clinics—the main theme of palliative care practice here is an outpatient palliative care clinic. Major cancer centers across the world began palliative care services by starting dedicated outpatient clinics where patients sent for palliative care referral are seen by a specialist palliative care multidisciplinary team. This model has been well-accepted worldwide, and palliative care across major centers around the world is more or less based on this model. Often, "Supportive care clinic" is a better-accepted nomenclature instead of palliative care clinic among the doctors and healthcare professionals. This nomenclature eliminates the stigma attached to the name palliative care and its association with death and dying.
- Embedded care model—this model aims to embed palliative care in the main oncology clinics and not as a separate outpatient area. The added advantage here is the earliest access to palliative care for every patient in the oncology clinic and ease of communication between the oncologist and the palliative care physician. This model despite its benefit over the outpatient model has not been that well-accepted worldwide as it is resource-intensive as every center may not have multiple palliative care physicians to be present at multiple oncology clinics plus this can burden the optimal interaction time with oncologists for such a combined clinic.

Our practice—we at the Institute Rotary Cancer Hospital (IRCH), All India Institute of Medical Sciences (AIIMS), have put in practice the "Patient-Centric Integrated Supportive Care Model." We have a six-bedded inpatient palliative care ward, outpatient palliative care clinics, and emergency palliative care services [20]. For our cancer management at the National Cancer Institute, the new dedicated cancer center of AIIMS, in addition to the Patient-Centric Integrated Supportive Care Model, we are trying the embedded care model where a palliative care physician is present with the medical and radiation oncologist in the disease management clinics where every new case in OPD is worked up and a multidisciplinary care plan is formulated at the same setting.

21.5 Clinical Practice of Palliative Care [21, 22]

Practicing palliative care in an oncology setup requires in-depth knowledge of the disease trajectory and management options of various malignancies, the expected complications associated with oncological therapies like surgery, chemotherapy, and radiotherapy; their management and after-effects on patients' quality of life. Further to this, the palliative care physician needs to be well versed in the symptomatic management and supportive care needs at the end of life for the patient and caregivers. A holistic approach keeping in mind the concept of "total pain" [13] and to assess the patient's physical, mental, social, and spiritual well-being plus caregivers' concerns is the Center Stage and basic foundation for the practice of palliative care. Thus, palliative care practice in oncology is essentially a team-based approach to looking after the overall well-being of a patient and ensuring a smooth transition into end-of-life care if the disease progresses beyond cure.

Literature has described the composition [22] of a palliative care team. A typical palliative care team should ideally include a trained palliative care physician, palliative care nurse, physiotherapist, pharmacist, social worker, clinical psychologist, and bereavement counselor. This team-based approach is vital in both hospital-based and at community/home-based palliative care.

The main components in the clinical practice of palliative care may be broadly categorized as:

(a) Communication
(b) Management of common symptoms
(c) Psycho-social care

21.5.1 Communication

Effective communication with the patient and caregiver regarding disease stage, prognosis, and plan of care is the cornerstone of good palliative care practice. Basics of two-way clear communication with the patient and caregivers, establishing clear goals of care with regular review to give realistic hope, and avoidance of any misunderstanding or mistrust in the treating team are the basic minimum that every palliative care physician should strive for.

The literature describes *four major domains of communication* [22], these are, namely, *communicating bad news, negotiating goals of care, advanced care planning, and withdrawing or withholding aggressive life supporting measures at the end of life*. The basics of good communication are establishing eye contact, asking patient-centric open-ended questions, asking what the patient wants, and listening to the patient's concerns and hope from treatment, thus overall achieving a "patient centric" modus operandi for counseling, and further creating the bond and trust. Some additional points which are to be kept in mind in each of these domains are:

- *Communicating bad news and Negotiating Goals of care*—"SPIKES," a six-step protocol, has been devised and described in the literature as a useful tool to aid clinicians in communicating bad news to a patient [23]. The basic goal of this protocol is to address the four important objectives in breaking bad news namely gathering information (from the patient), disclosing medical information, providing support after giving the news, and collaborating with the patient (and family caregiver) to formulate a plan of medical care. ***S****—Setting up the meeting*. The place of meeting should be quiet, have comfortable seating, and should be spacious to adjust family members. The meeting to break a piece of distressing news should be planned and free from interruptions. ***P****—patients' Perception*—this step concentrates on finding out what the patient knows and understands about his or her disease condition. ***I****—an Invitation* from the patient for full disclosure. The patient should be ready and express a desire to know about his or her condition. ***K****—Knowledge,* and information to the patient. Palliative care physicians should make an honest effort to explain the medical pathology and underlying disease stage of each patient in terms and language best understood by the patient and the caregivers. ***E****—Empathy*. Give adequate time for understanding and coming to terms with the gravity of the situation after full disclosure of information. Be a patient listener to patient's or caregiver's responses and empathize with the family. This helps in building trust and a positive rapport with the whole family. ***S****—Strategize and Summarize*—the discussion should conclude with formulating a plan of care for the future with realistic goals and patients' active involvement in charting out a supportive care plan.
- *Advanced care planning and Withholding/ Withdrawing life supportive care* [24, 25]—. Dealing with probable or imminent death and knowing that one is suffering from advanced cancer beyond curative treatment is one big fact that has to be dealt with by a cancer patient and caregivers. A patient has a right to make informed choices at the end-of-life (EOL) stage about general aspects of care as well as other as other important aspects like a place of death, invasive treatments, life-supporting measures in the terminal stage, surroundings of death, etc. However, a patient at the end of life may not be mentally able to comprehend and opine about such important aspects of his or her care. Here comes the important role of a palliative care physician and the concept of advance care planning. The patient should be empowered

through proper counseling and focused knowledge on the disease stage, the future course of an incurable stage, and the problems expected to occur as the disease will progress. The patient should be empowered to understand the benefits of palliative care even at EOL stage, and not to be afraid or be in denial of the inevitable but rather should plan for death. Some important aspects like a place of death at home or hospital, terminal sedation before death, and appointing a surrogate decision-maker in advance keeping in mind that patient himself or herself might not be mentally capable of making these important decisions toward the end of life all come under the concept of "Advanced Care Planning."

Advanced care planning if done properly under the supervision of the palliative care physician will ultimately lead to a good death free from improper use of life-prolonging intensive care in terminal stage. The patient being near family members, in home surrounding, to have a peaceful end with or without terminal sedation is medically proven to have minimal psychological trauma on the caregivers and would make bereavement care easier for the palliative care team. World-wide steps have been taken in this direction to bring about a peaceful death in terminally sick patients. In India recently Kasturba Medical College, Manipal, has come out with the "BLUE MAPLE" document [26] to evaluate and plan limitation of life-sustaining treatment at the end of life and formulate a care plan for a terminally sick patient. All India Institute of Medical Sciences, Delhi, has also formulated an end-of-life care policy for terminally sick patients [27]. These efforts should be spread across the countries in Asia and Africa in order to minimize the hospital deaths of cancer patients and for patients with other terminal EOL stage of a disease to be aware about the pain, trauma, economic burden and futility of intensive medical care for the irreversible condition.

21.5.2 Management of Common Symptoms [21, 28, 29]

The second major aspect of the practice of palliative care in oncology is the management of common distressing symptoms in cancer patients. This is the domain where the application of medicine in relieving common distressing symptoms like pain, nausea, vomiting, etc. brings about an instant change and improvement in patients' quality of life. Major symptoms complex which need attention by a palliative care physician may be divided into:

(a) Cancer pain
(b) Other non-pain symptoms

21.5.2.1 Cancer Pain

Cancer pain relief, as the key foundation of palliative care, was brought out by the WHO for the global attention in 1986 (Fig. 21.1 [30]). In hindsight, this could galvanize the action plan of Professor Jan Stjernsward the then Chief of Cancer Unit in WHO for addressing the pain relief with WHO-ladder and to establish palliative care for the vast majority of incurable cancer patients at that time. Since then, the progress in pain research and pharmacotherapy of pain has made several modifications in our clinical approaches to cancer pain relief. The non-drug interventions in cancer pain management, such as radiotherapy for pain in bone metastasis, are considered equally important in palliative care [31, 32].

Pain is the most common and the most distressing symptom experienced by cancer patients. Cancer pain is multifactorial in origin and may be visceral, somatic, neuropathic, or bony in origin. A detailed discussion on all aspects of cancer pain management is beyond the scope of this chapter. A brief description of the basic approach to cancer pain management, use of opioids, and use of interventions will be discussed. The basic management of cancer pain is done along the lines of the World Health Organization (WHO) stepladder management of pain [30, 32, 33]. This ladder describes a step-wise incremental approach for oral analgesics of various classes

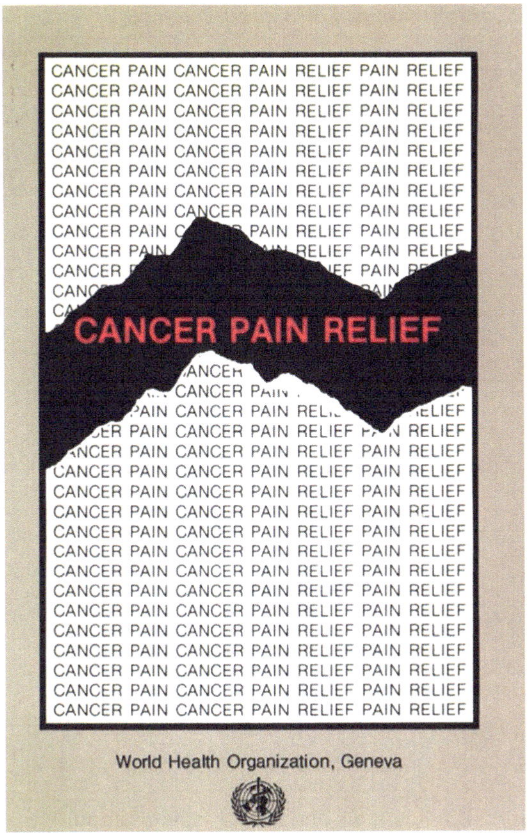

Fig. 21.1 Cancer pain relief document, WHO, 1986

with increasing severity of pain. The basic essence of management according to the WHO step ladder is to give oral analgesics by the mouth, by the clock, i.e., dosing round the clock, as per the severity of pain and the drug prescription individualized to the needs of the patients. Interventions or nerve blocks described originally as the fourth step in the WHO analgesic ladder can be used at any step in the pain ladder as per the current concept [34].

Opioids are the cornerstone in managing cancer pain. Morphine is the main drug that is successful as an oral analgesic in the majority of cancer pain patients. Some major issues restricting the free use of opioids for cancer pain management are optimization of morphine dose and safe use of opioids from procurement to dispensing [35].The morphine prescription for cancer pain relief requires dynamic assessment to observe changes in pain score, adverse side effects and breakthrough and to assess the need for any concomitant use other approaches for palliation—such as systemic chemotherapy (e.g., for superior vena caval syndrome in lung cancer),radiotherapy (e.g., for relief of tumor compression) or surgery (e.g., tracheostomy or colostomy). Well-managed opioid prescriptions are the cornerstone of a complex environment requiring multidisciplinary dynamic assessments integrated into the patient's care pathway [36]. Recent literature shows continued efforts in this direction to make the use of opioids in cancer pain easy and safe. Newer methods to calculate starting dose of morphine-like intravenous (IV) slow infusion versus IV bolus doses to achieve optimum pain control with minimum adverse effects have been described [37]. Also, in our department, we have made effort to define an "Opioid Circle of Safety" which explains a series of steps from opioid procurement storage to use aiming at maximizing the safe use of opioids with maximum benefit. This document has been released as a pictograph and can be freely used to promote the safe use of opioids [38]. The physicians, nurses, and clinical researchers should acquire the knowledge and skills to manage the cancer pain effectively.

21.5.2.2 Other Non-pain Symptoms

A battery of other distressing symptoms may commonly affect a cancer patient and hurt the quality of life. Table 21.3 lists these and gives an overview of their potential etiologies and management. For the effective palliative care practice it is imperative to learn further about the distressing symptoms and their management. The Handbook for Certificate Course in Essentials of Palliative Care by IAPC is a very useful resource material [39].

21.5.3 Psycho-Social Care

The existence of psycho-social issues is common in patients who are referred for palliative care. Patient undergoing palliative care should be given the timely symptom management for pain, physical symptoms and in equal measures for the psy-

Table 21.3 Common distressing non-pain symptoms and their management

Symptom	Etiology	Management
Cachexia/anorexia	Long-standing chronic disease, malignancy-related	Course of megestrol acetate or dexamethasone
Anxiety/depression	The multifactorial, progressive course of the disease, financial psychological stress of cancer treatment, etc.	Counseling, psychological and psychiatric follow-up if needed
Constipation	Commonly due to opioids prescribed for pain, or malignancy-related (bowel adhesions, abdominal growth, etc.)	Laxatives based on etiology of constipation (stimulant laxative for opioid-induced, bulk-forming if poor oral intake, etc.)
Nausea and vomiting	Treatment-related—chemotherapy, opioids, etc.	Antiemetic based on etiology (ondansetron for chemotherapy-related, metoclopramide for opioid-related, etc.)
Dyspnea	Malignant pleural effusion or massive ascites, lung metastasis or primary lung cancer, lower respiratory tract infection	Tapping of pleural or ascitic fluid, treat the underlying infection if any, in terminal cases—low-dose morphine, supportive measures like an extra fan to blow air on face for subjective relief
Delirium/altered sensorium	Brain metastasis, metabolic anomalies (uremia, hyponatremia), etc.	Measures to decrease intracranial pressure (steroids, osmotic diuretics, etc.)
Neurological issue—Motor/sensory loss, loss of bladder bowel sensation	Malignant spinal cord compression	Urgent surgical intervention or radiotherapy if needed, steroids, spine stabilization

cho-social distress. The relief provided for psychological issues and the attention given to the social aspects can achieve the real world improvements in the day to day living of our patients [40].

Patients with advanced incurable cancer and those undergoing prolonged anti-cancer therapy can experience several psycho-social issues, such as:

1. Anxiety, depression
2. Fear and agitation
3. Feelings of isolation and sadness
4. Social and financial hardships during cancer trajectory of diagnosis and treatments

Clinical studies have demonstrated that the medical interventions for psychological features like depression, and anxiety and the psychological counseling for emotional wellbeing of the patient (and caregiver) show positive benefits on the quality of life of those undergoing palliative care [41].

Addressing psychosocial issues would enhance treatment compliance, physician–patient relationship, treatment efficacy, and quality of life. The hospitals and health systems should recognize the importance of integrating psycho-oncology services in cancer care and attempts to define the various roles that a psycho-oncologist and social worker can play across the entire cancer trajectory [42].

21.6 Economic Burden of Cancer and the Role of Palliative Care

The biomedical research in anti-cancer therapy and the new technologies have increased the survival of most of the cancer patients. Yet, it is well recognized that the economic burden of cancer management has substantially increased in the last 20 years. Approximately one-half of individuals with cancer face personal economic burdens associated with the disease and its treatment, a problem known as financial toxicity (FT). FT more frequently affects socioeconomically vulnerable individuals and leads to subsequent adverse economic and health outcomes [43]. For many cancer patients and their families, the out-of-pocket expenditures (OOPE) due to futile anti-cancer therapy may require to be monitored and palliative care should be instituted in a timely manner. The research toward the understanding of economic aspects of cancer management is gaining ground in Asia and Africa [44]. The role of palliative care in cancer management has become essential, so that it is suitable integration with oncological practices can reduce patient's OOPE and create a healthy cancer care trajectory. Access to palliative care within healthcare systems of LMICs needs to be improved. A collaborative approach has shown that the North East Regions of India, which has high cancer burden and cancer mortality, could increase its palliative care facilities [45]. The implementation of this model for LMIC regions with similar health systems and sociocultural and economic contexts would benefit thousands of cancer patients.

21.7 Teaching Training and Research in Palliative Care

Teaching and disseminating knowledge about this new branch are an essential duty of every cancer specialist, palliative care physician and cancer nurse. In India, efforts to spread knowledge about this branch and empower various oncological setups to establish and practice palliative care have borne fruits in the recent past [46]. Also, these efforts are visible in the form of starting academic courses MD and DNB in palliative medicine across India (Table 21.2). Simultaneously, a focus toward methodological research in palliative care has been recognized [47]. The cancer specialists, nurses, and scientists can engage in palliative care research to gain insight and bring improvements for their cancer patients.

21.8 Summary Box

Palliative care is an indispensable part of oncology practice. An integrated approach with timely involvement of a palliative care specialist in the treatment of a cancer patient leads to better patient satisfaction and relief from distressing symptoms; in addition, it may also improve overall survival. This branch, although recent in origin, has progressed in leaps and bounds and is now a must in oncological practice.

Multiple Choice Questions

1. First hospice care center was established in India in which year?
   A. 1987
   B. 1986
   C. 1990
   D. 1992
   Ans B.
2. Which of the following is a strong opioid?
   A. Morphine
   B. Tramadol
   C. Codeine
   D. Tapentadol
   Ans. A
3. Latest amendment in NDPS act came in which year?
   A. 2022
   B. 2021
   C. 2020
   D. 2014
   Ans. B
4. The communication skills in attending to the palliative care patient can be done in a professional manner, when you learn which of the following approach?
   A. DATES
   B. POKER
   C. SPIKES
   D. All of Above
   Ans. C
5. Which of the following statement is TRUE about the need for palliative care (PC) in cancer patients?
   A. Only cancer pain relief is delivered in PC
   B. Quality of life is never improved by PC
   C. Integration of PC with active cancer treatments can benefit the cancer patient
   D. Patients undergoing PC can experience increased physical distress.
   Ans. C

References

1. Palliative care [Internet]. [cited 2022 May 28]. Available from: https://www.who.int/news-room/fact-sheets/detail/palliative-care.
2. List JM, O'Connor JM. How should low- and middle-income countries motivate equity in cancer prevention and control? AMA J Ethics. 2020;22(2):E147–55. https://doi.org/10.1001/amajethics.2020.
3. Loscalzo MJ. Palliative care: an historical perspective. Hematology. 2008;2008(1):465.
4. Saunders C. The evolution of palliative care. J R Soc Med. 2001;94(9):430–2.
5. Shanmugasundaram S, Chapman Y, O'Connor M. Development of palliative care in India: an overview. Int J Nurs Pract. 2006;12(4):241–6.
6. Bhatnagar S. Diplomate in National Board Palliative Medicine. Indian J Palliat Care. 2021;27(1):1–2.
7. NDPS (Amendment) Bill, 2021 [Internet]. Drishti IAS. [cited 2022 Jun 1]. Available from: https://www.drishtiias.com/daily-updates/daily-news-analysis/ndps-amendment-bill-2021.
8. Khosla D, Patel FD, Sharma SC. Palliative care in India: current progress and future needs. Indian J Palliat Care. 2012;18(3):149–54.
9. Salins N. Specialist palliative medicine training in India. Indian J Palliat Care. 2015;21(3):257.
10. https://www.palliativecare.in/beholden-to-the-humble-beginning-of-palliative-care-in-india/. Cited on 12 Dec 2022.
11. O'Neill B, Fallon M. ABC of palliative care: principles of palliative care and pain control. BMJ. 1997;315(7111):801.
12. General principles of palliative care - UpToDate [Internet]. [cited 2022 Jun 1]. Available from: https://www.uptodate.com/contents/table-of-contents/palliative-care/general-principles-of-palliative-care.
13. Ong CK, Forbes D. Embracing Cicely Saunders's concept of total pain. BMJ. 2005;331(7516):576.
14. Beauchamp TL, Childress, JF. Principles of biomedical ethics. 8th ed. New York: Oxford University Press; 2013.
15. Mohanti BK. Ethics in palliative care. Indian J Palliat Care. 2009;15(2):89–92. https://doi.org/10.4103/0973-1075.58450.
16. Temel JS, Greer JA, Muzikansky A, Gallagher ER, Admane S, Jackson VA, et al. Early palliative care for patients with metastatic non–small-cell lung cancer. N Engl J Med. 2010;363(8):733–42.
17. Hui D, Bruera E. Models of integration of oncology and palliative care. Ann Palliat Med. 2015;4(3):89–98. https://doi.org/10.3978/j.issn.2224-5820.2015.04.01.
18. Zhi WI, Smith TJ. Early integration of palliative care into oncology: evidence, challenges and barriers. Ann Palliat Med. 2015;4(3):122–31. https://doi.org/10.3978/j.issn.2224-5820.2015.07.03.
19. Rowland K, Schumann SA. PURLs. Palliative care: earlier is better. J Fam Pract. 2010;59(12):695–8.
20. Patel A, Bhatnagar S, Ratre B. Patient centric integrative supportive care model at a tertiary cancer care center of India. Indian J Med Sci. 2021;73(2):275–80. https://doi.org/10.25259/IJMS_457_2020.
21. Kittelson SM, Elie MC, Pennypacker L. Palliative care symptom management. Crit Care Nurs Clin North Am. 2015;27(3):315–39.
22. Morrison LJ, Morrison RS. Palliative care and pain management. Med Clin North Am. 2006;90(5):983–1004.
23. Baile WF, Buckman R, Lenzi R, Glober G, Beale EA, Kudelka AP. SPIKES—A six-step protocol for delivering bad news: application to the patient with cancer. Oncologist. 2000;5(4):302–11.
24. Advance care planning in palliative care—Advance Care Planning Australia [Internet]. [cited 2022 Jun 8]. Available from: https://www.advancecareplanning.org.au/understand-advance-care-planning/advance-care-planning-in-specific-health-settings/advance-care-planning-in-palliative-care.
25. Berkowitz CM, Wolf SP, Troy J, Kamal AH. Characteristics of advance care planning in patients with cancer referred to palliative care. JCO Oncol Pract. 2021;17(2):e94–100.
26. BLUE-MAPLE-End-of-Life-Care-Manipal.pdf [Internet]. [cited 2022 Jun 6]. Available from: https://palliumindia.org/wp-content/uploads/2021/10/BLUE-MAPLE-End-of-Life-Care-Manipal.pdf.
27. irch-9-3-20.pdf [Internet]. [cited 2022 Jun 8]. Available from: https://aiims.edu/images/pdf/notice/irch-9-3-20.pdf.
28. Morrison RS, Meier DE. Palliative Care. N Engl J Med. 2004;350(25):2582–90.
29. Smith CB, Phillips T, Smith TJ. Using the new ASCO clinical practice guideline for palliative care concurrent with oncology care using the TEAM approach. Am Soc Clin Oncol Educ Book. 2017;(37):714–23.
30. World Health Organization. Cancer pain relief. World Health Organization; 1986. https://apps.who.int/iris/handle/10665/43944.
31. Nersesyan H, Slavin KV. Current aproach to cancer pain management: availability and implications of different treatment options. Ther Clin Risk Manag. 2007;3(3):381–400.
32. WHO guidelines for the pharmacological and Radiotherapeutic Management of Cancer Pain in adults and adolescents. Geneva: World Health Organization; 2018. PMID: 30776210.
33. Anekar AA, Cascella M. WHO Analgesic Ladder. In: StatPearls [Internet]. Treasure Island: StatPearls Publishing; 2022 [cited 2022 Jun 8]. Available from: http://www.ncbi.nlm.nih.gov/books/NBK554435/.
34. Bhaskar AK. Interventional management of cancer pain. Curr Opin Support Palliat Care. 2012;6(1):1–9.
35. Mandalà M, Moro C, Labianca R, Cremonesi M, Barni S. Optimizing use of opiates in the management of cancer pain. Ther Clin Risk Manag. 2006;2(4):447–53.

36. George B, Minello C, Allano G, Maindet C, Burnod A, Lemaire A. Opioids in cancer-related pain: current situation and outlook. Support Care Cancer. 2019;27(8):3105–18. https://doi.org/10.1007/s00520-019-04828-8.
37. Kumar V, Sirohiya P, Garg R, Gupta N, Bharti SJ, Velpandian T, et al. Comparison of two techniques (intermittent intravenous bolus morphine vs. morphine infusion) for analgesic titration in patients who had advanced cancer with severe pain: a prospective randomised study. BMJ Support Palliat Care. 2020:bmjspcare-2020-002397.
38. Bhatnagar S, Hoda W. Balancing improved opioid supply and safe use of opioids for cancer pain by using "circle of safety" Indian J Palliat Care. 2017;23(3):221–2.
39. http://palliativecare.in/wp-content/uploads/2014/09/CCEPC_Handbook.pdf.
40. Almada AL, Casquinha P, Cotovio V, Heitor Dos Santos MJ, Caixeiro A. The potential role of psychosocial rehabilitation in palliative care. J R Coll Physicians Edinb. 2018;48(4):311–7. https://doi.org/10.4997/JRCPE.2018.405.
41. Uitterhoeve RJ, Vernooy M, Litjens M, Potting K, Bensing J, De Mulder P, van Achterberg T. Psychosocial interventions for patients with advanced cancer—a systematic review of the literature. Br J Cancer. 2004;91(6):1050–62. https://doi.org/10.1038/sj.bjc.6602103.
42. Mathew B, Mohanti BK, Tewari S, Munshi A. Integrating psycho-oncology services in cancer care in India. Indian J Cancer. 2021;58(2):290–3. https://doi.org/10.4103/ijc.IJC_217_20.
43. Smith GL, Banegas MP, Acquati C, Chang S, Chino F, Conti RM, Greenup RA, Kroll JL, Liang MI, Pisu M, Primm KM, Roth ME, Shankaran V, Yabroff KR. Navigating financial toxicity in patients with cancer: a multidisciplinary management approach. CA Cancer J Clin. 2022;72(5):437–53. https://doi.org/10.3322/caac.21730.
44. Bates MJ, Namisango E, Tomeny E, Muula A, Squire SB, Niessen L. Palliative care within universal health coverage: the Malawi Patient-and-Carer Cancer Cost Survey. BMJ Support Palliat Care. 2019:bmjspcare-2019-001945. https://doi.org/10.1136/bmjspcare-2019-001945.
45. Vallath N, Rahul RR, Mahanta T, Dakua D, Gogoi PP, Venkataramanan R, Sethuraman L, Bahar HMI, Bhagabati K, Goswami D, Purushotham A. Oncology-based palliative care development: the approach, challenges, and solutions from north-east region of India, a model for low- and middle-income countries. JCO Glob Oncol. 2021;7:223–32. https://doi.org/10.1200/GO.20.00487.
46. Rao SR, Salins N, Goh CR, Bhatnagar S. Building palliative care capacity in cancer treatment centres: a participatory action research. BMC Palliat Care. 2022;21(1):101.
47. Mohanti BK. Research focus in palliative care. Indian J Palliat Care. 2011;17(Suppl):S8-S11. https://doi.org/10.4103/0973-1075.76231.

Psycho-Oncology: Actions Beyond Targeting the Tumor

22

Soumitra Shankar Datta, Rhea Daruvala, Arnab Mukherjee, V. Surendran, Seema R. Rao, and Brindha Sitaram

22.1 Introduction

Increased awareness, detection, and advances in treatment options have improved outcomes in oncology. At the same time, there are more people diagnosed with cancer, and a large proportion of the global cancer burden is from the low-and-middle-income countries (LMICs) [1]. During the course of diagnosis, treatment, and beyond, patients from LMICs face considerable emotional and social challenges that impact their life in several ways. In this chapter, in the beginning, we will cover the common psychiatric comorbidities in cancer patients and their management. Subsequent sections are devoted to patients with brain tumors, patients with delirium, management of tobacco cessation, optimizing outcomes for longer-term cancer survivors, and management of end of life. There is a special section on children, as their needs are unique.

22.2 Psychiatric Comorbidities in Cancer

Most people find it very difficult to accept the diagnosis of cancer. Many perceive it as a catastrophe and anticipate irreparable physical, functional, social loss, leading to psychological trauma, loss of control, and isolation. National Comprehensive Cancer Network (NCCN) introduced the term "distress" in the 1990s to better qualify acute cancer-related psychosocial sufferings. Distress can range from common feelings of vulnerability to disabling psychological symptoms. Emotional reactions to a stressor or "Adjustment disorders" are the most common diagnosis in patients with cancer. A recent systematic review reported the prevalence of

S. S. Datta (✉)
Department of Palliative Care and Psycho-oncology, Tata Medical Center, New Town, Kolkata, West Bengal, India

Institute of Clinical Trials & Methodology, University College London, London, UK
e-mail: ssdatta2000@yahoo.com

R. Daruvala
Narayana Health, Bengaluru, Karnataka, India

A. Mukherjee
Department of Palliative Care and Psycho-oncology, Tata Medical Center, New Town, Kolkata, West Bengal, India

V. Surendran
Department of Psycho-oncology & RCTC, Cancer Institute (WIA), Chennai, Tamil Nadu, India

S. R. Rao
Bangalore Hospice Trust, Bangalore, Karnataka, India

School of Medicine, Cardiff University, Cardiff, UK

B. Sitaram
Center of Psycho-oncology for Education and Research (COPER), Bangalore, Karnataka, India

Health Care Global Speciality Hospitals, Bangalore, Karnataka, India

A. Munshi et al. (eds.), *Radiation Oncology – Principles, Precepts and Practice*,
https://doi.org/10.1007/978-981-97-8389-2_22

Adjustment disorder in cancer patients to be 19.4% (14.5–24.8) [2]. Distress and adjustment disorders are often self-limiting and improve over time. One-third of patients undergoing radiation therapy can have disabling anxiety and depressive symptoms [3, 4]. Anxiety is very common during the pre-RT phase and generally doubles during active treatment [3]. Pre-operative anxiety in cancer and procedural anxiety is also quite common and interferes with decision making [5, 6]. Depressive symptoms remain elevated often even during post-RT follow-ups [7, 8]. The commonest risk factors are younger age and female gender, comorbid physical symptoms (side effects or pain), history of depressive illness, stage of cancer, time dependence (prior, during, following RT), and receiving chemotherapy [3]. Untreated anxiety and depressive illness may result in frequent hospital visits, longer stays, or poor adherence to treatment. Fear of recurrence can cause significant preoccupation and anxiety in survivors even years after treatment completion [9].

Acute anxiety and panic symptoms are common during imaging and radiotherapy procedures, which can lead to disruptions in the treatment [7–10]. The acute anxiety symptoms can be in the context of a preexisting anxiety disorder or be reactive to ongoing stress related to the cancer diagnosis and treatment. Sudden de novo panic symptoms may even lead to treatment abandonment.

Several studies report the prevalence of suicidal ideations to be two to ten times more in cancer patients than in the general population [11–13]. Head and neck cancer patients treated with radiation alone have higher rates of completed suicide when compared with those patients treated with surgery alone [12]. Depression, premorbid psychiatric illness, past history of suicidal attempts, pain, and poor social support are known to increase the risk of completed suicide [14].

Delirium is frequent in cancer patients and often underdiagnosed and misdiagnosed [15, 16]. Due to altered sensorium, confusion, increased drowsiness or agitation, delirium interferes with the care process and is associated with increased postdischarge mortality, institutionalization, longer stay in the hospital, and increased rehabilitation needs. Correcting the underlying cause of delirium remains the mainstay in treatment for delirium. Evaluation by trained psychiatrists is recommended for behavioral abnormalities during the treatment of cancer-related delirium [15–17].

Substance abuse and withdrawal interfere with cancer treatment due to interaction with chemotherapeutic agents, poor motivation to adequately engage in the treatment process and impaired quality of life [18, 19]. A recent systematic review highlighted the high rates of substance-related problems in patients with cancer [20]. Careful history taking with a nonjudgmental stand is important to identify unaddressed substance-related problems for cancer patients in India. Screening and counseling interventions have demonstrated to improve the pain control among patients with substance use disorder and improved quality of life for patients and their families [21]. Cancer patients with substance-related problems benefit from a referral to psycho-oncology services.

Cancer and its treatment may be associated with lowered self-esteem, cause body image difficulties, fertility problems, fatigue, and problems in intimate relationships. Cancer of particular organs (prostate, penile, testicular, ovarian, and cervical) and treatment modalities may also directly affect sexual functioning [22–25]. The reported prevalence of sexual dysfunction among women with breast cancer is as high as 32–93%, but a vast majority of them never report it to their healthcare providers [22]. A number of cancers produce visible disfigurement of parts of the body leading to problems in physical identity, communication, and interpersonal relationship [26]. Disfigurement and body image disturbance often leads to social isolation. Body image problems make breast cancer patients perceive themselves as less desirable after treatment completion [22]. Many individuals with head and neck cancer report problems in finding a new job or even going back to their previous job and difficulties in intimate relationships [27]. Patients with greater disfigurement often have more depressive symptoms. Greater self-efficacy and social support decrease the impact.

A prospective study reports women who underwent manually defined radiotherapy, developed fibrosis in the treatment area and had poorer body image [28]. More than one-fourth of patients with body image issues continue to have difficulties even many years after treatment completion [28]. It is thus important to manage psychological distress and psychiatric morbidity in cancer patients early and appropriately.

22.3 Psychological Interventions in Cancer

Cancer research in the recent past has been focusing vigorously on the psychosocial impact of the disease and its treatment, beyond the mere survival of a patient, thus paving the way for scientific application of evidence based psychological interventions in oncology. It is evident that a huge number of patients experience emotional symptoms (health anxiety, irritable mood and demoralization) or psychopathological conditions (major depression and posttraumatic stress disorder) which mandates integrated and comprehensive care by cancer care professionals. In order to increase psychological well-being and to alleviate physical symptoms, different types of psychological interventions have been developed and tested among cancer patients as a means of facilitating adjustment during the initial phase of diagnosis and treatment continuum.

Nonpharmacological interventions in oncology mainly comprises three components [29]: (1) education which includes providing information about disease and its treatment; (2) using specific psychological strategies as allowing patients verbalize their emotional pain, providing clarification where appropriate, discuss problem solving techniques, providing emotional and social support and social support; and (3) facilitate environmental change at the patients home and elsewhere. Psychological management includes making the core oncology team responsive to the emotional needs of the patient as well as referral to mental health professionals. Psychotherapeutic interventions and psychosocial rehabilitation in oncology have repeatedly proven to be beneficial in reducing the severity of somatic symptoms chiefly pain [30] in improving quality of life, wellbeing, and ability to return to work [31].

The use of integrative therapies, especially mind–body therapies, is an effective supportive care strategy during cancer treatment. Most common and proven psychological interventions in oncology include counseling, psycho-education, facilitating coping and providing psychological support, in different forms through individual sessions, group sessions, family therapy following the any of the accepted models of psychological care as cognitive-behavioral, supportive-expressive, existential, and psychodynamic approaches [32]. Managing distress in oncology, which is one of key objectives of therapy, is commonly dealt with supportive-expressive group psychotherapy [33], cognitive-behavioral therapy [34] and cognitive-existential therapy [35], meaning-centered psychotherapy [36], and mindfulness stress reduction programs [37].

Interventions that incorporate restorative approaches, supportive-expressive techniques and cognitive-behavioral psychosocial interventions have been effective in managing physical symptoms primarily fatigue and sleep difficulties [38]. While quality of life is being considered as one of the key outcome indicators of cancer treatment, researchers extensively studied the effect of interventions tailored toward improving the quality of life of patients. Behavioral therapy, social cognitive therapy including self-efficacy and self-regulation therapies [39] were found to be efficient in enhancing the quality of life by improving mood, coping and functional living.

A multidisciplinary team (nursing, psychiatry, oncology) and a combination of psychotherapeutic management (problem-solving, psychoeducation, behavioral activation, counseling techniques and pharmacological management) have proved effective in managing distress, and lowering depression rates in cancer patients. The choice of intervention usually depends on several variables, including the type and stage of cancer diagnosis, premorbid clinical psychological condition, and the availability of psycho-oncology services with trained professionals [40]. A collaborative and integrative approach involving a multidisciplinary team is the key strategy of offering holistic cancer care.

22.4 Management of Anxiety in Adults Undergoing Radiation Therapy

Radiation therapy (RT) happens in a high-tech environment with several set procedures that may be overwhelming for some patients. With quality aspects driving cancer care, the last decade has seen a surge of interest in understanding patient experience during RT.

Patients undergoing RT (Tele and/or Brachytherapy) experience anxiety and distress. Heightened distress has been noted in 41% of patients prior to starting external RT [42–44]. Although distress seems to decrease along the course of radiation [45–47], there are reports of increasing levels of distress from 61% to 89% toward the end of RT [48]. Similarly patients getting brachytherapy experience fear, anxiety and depression. Patients with gynecological cancers, undergoing brachytherapy, may be embarrassed during the treatment. One study estimated the prevalence of acute stress disorder to be 30% for gynecological cancer patients one week post brachytherapy, and 41% of them had posttraumatic stress disorder 3 months after treatment [49].

One of the worst fears of patients on RT is the thermoplastic mask [50]. "Mask anxiety" is distress, anxiety, and claustrophobia associated with a thermoplastic mask. It is experienced by 14% and 58% of patients on RT [41, 51]. Claustrophobia and side-effects of RT (pain, thick secretions) in HNC patients have been associated with increased fear of masks during RT. About 11% disruption during CT simulation and 24% rate of disruption on Day 1 of treatment has been attributed to mask anxiety [7]. It is known to decrease or sustain (remain) and increase in some patients throughout the course of RT [52].

The acute distress caused by the medical procedure is termed Procedural anxiety. It may occur in anticipation or during or interfere with completion or avoidance of the necessary procedure [53, 54]. The fear of the procedure may be rational (fear of MRI detecting tumor) or irrational (e.g., fear of suffocation during MRI or claustrophobia). Therefore, it is important to make a distinction between procedural anxiety and phobia. If the patient's fear or anxiety is about the result or outcomes of the procedure (e.g., MRI detecting a tumor) it is procedural anxiety. Phobias are excessive fears or anxiety to a specific procedure's immediate effect (e.g., fear of suffocation during MRI or claustrophobia).

22.4.1 Sources of Procedural Anxiety

A systematic review [55] on patients' experience with RT has identified several sources of procedural anxiety as summarized in Table 22.1.

22.4.2 Clinical Implications of Procedural Anxiety

Unattended psychological distress of patients can result in various problems [7, 56] in their engagement with oncology services:

1. Refusal to undergo RT
2. Poor adherence
3. Non-compliance
4. Treatment delays
5. Compromised treatment outcomes
6. Poor satisfaction and quality of life
7. Increases hospital cost and resource use

Management of procedural anxiety involves a patient-centered approach. Routine distress screening (NCCN distress management) and need assessments to facilitate referral pathways: providing information, education, and psychosocial support. Screening-based referrals to psycho-oncology services have shown 31 out of 33 patients undergoing RT had significant psychosocial symptoms [57]. Such an approach is feasible and mitigates distress, improves patient preparedness, trusting patient-professional relationship, adherence to RT, patient satisfaction, and better quality of life [43, 57–59]. Several patient-centered approaches (Table 22.2) have been shown to improve psychosocial outcomes of patients undergoing RT [65, 66].

Table 22.1 Sources of procedural anxiety during radiotherapy

A.	Misinformation and Misconceptions about Radiation: "Nuclear weapon," "Poison," "Becoming Radioactive" Terminologies to describe RT in the Indian setting: "Shock," "Current" could instill fear
B.	Time: (i) Waiting time: Unpredictable waiting time disrupts a patient's daily routine and activities. Patients find this more distressing than treatment time. (ii) Treatment Time: Treatment time appears prolonged and stressful for patients who fear isolation and have claustrophobia
C.	Treatment Environment: (i) Freezing Room temperature (ii) Hard couches (iii) Noise of the machine (iv) Sharp lights
D.	Treatment concerns: (i) Side effects: Burns, scars, hair loss, diarrhea, etc. (ii) Positioning and immobilization: Mask anxiety, claustrophobia with mask, the hard and sharp edges of the mask, etc. (iii) The tattoo marks: Social stigma (iv) Disruption of daily activities and routine
E.	Treatment preparation protocols (e.g., bladder filling, bowel evacuation). • Shame and embarrassment • Delays in treatment time lead to distress
F.	Lack of support from RT team (i) Lack of information, communication regarding the procedure and process. Use of technical jargon. (ii) Poor response to patient needs from the team. (iii) Patient perception of treating team as incompetent and unempathetic. (iv) Frequent change in treatment team members and equipment causes insecurity and anxiety.

Table 22.2 Patient-centered approaches to improve psychosocial outcomes of patients undergoing RT

Management of procedural anxiety [43, 58–63]	Benefits
1. Planning radiotherapy treatment schedules: • Create a treatment algorithm: Minimizing waiting time • Involve patients for scheduling treatment • Avoid delays and overlaps of appointment • Provide message alerts about delays • While waiting: Provide art and music therapy [48, 64]	• Reduce patient distress and disruption of daily activities • Minimizes distress and perceived waiting time
2. Trusting professional–patient relationships, open communication and continuity of care • Open communication: Addressing patient concerns empathetically • Continuity of care with the same team [43, 63]	• Alters patient's perception. The team is perceived as competent, knowledgeable, empathetic, addressing their concerns • Continuity of care (the same team) fosters trusting professional-patient relationships. Patients feel safe, confident, and reduces their anxiety
3. Mask anxiety (thermoplastic cast) • Pre-treatment education, sensory and procedural information • Psychological techniques: Relaxation, Cognitive-Behavior therapy • Medication (anxiolytics) • Making modifications in the mask and open mask (nose, eye and mouth openings)	Effective in reducing mask anxiety

Both adults and children on RT experience tremendous distress and anxiety. Unaddressed patient distress has far-reaching clinical implications, which compromises treatment outcomes and their quality of life. Several unconventional techniques including open and empathetic communication, along with psychosocial techniques have shown promising results in fostering better patient-professional relationships, reducing patient distress, improving their treatment experience and quality of life.

22.5 Neuropsychological Testing in Patients with Brain Tumors

Neuropsychological testing involves the comprehensive assessment of neurocognitive functions across multiple domains, including attention, expressive and receptive language, visuospatial functioning, learning and memory, reasoning; fine motor speed and dexterity, and emotions and behavior [67]. Neuropsychological testing in patients with brain tumors is important as it plays an essential role in the assessment and management of central nervous system tumors. Testing as assessment prior to treatment allows clinicians to establish a baseline of the patient's functioning against which they can compare his functioning post treatment. Neuropsychological testing is useful in understanding and assessing cognitive, emotional, and behavioral aspects. Based on evaluations and assessments, patients may be provided with adequate neuropsychological rehabilitation to aid in developing functionality that may be lost due to brain tumors.

The nature of deficits is most often related to the site of the tumor. For example, left hemisphere tumors commonly affect language. However, compression and displacement effects created by a brain tumor may cause more widespread damage and global neuropsychological impairment [68]. Neuropsychological assessments can therefore help evaluate the extent and nature of neuropsychological impairment which is central to planning rehabilitation and supportive care. More importantly, neuropsychological functioning has been found to be related to prognosis and tumor recurrence and may even be more sensitive in predicting early brain tumor recurrence as compared to imagining techniques [68].

Some areas that are typically assessed using neuropsychological assessments are indicated in Fig. 22.1.

Neuropsychological examinations require specialized skills and expertise of a trained clinical psychologist or a neuropsychologist. The assessments are usually time consuming and extensive. The various stages and components in a neuropsychological assessment are indicated Fig. 22.2.

Since neuropsychological evaluations can be performed for several functions including IQ, memory, reasoning, verbal skills, etc., testing every function is almost impossible. Most often only specific functions are assessed. For example in tests of intelligence, often a battery of tests may be used. A screening assessment and clinical history helps cli-

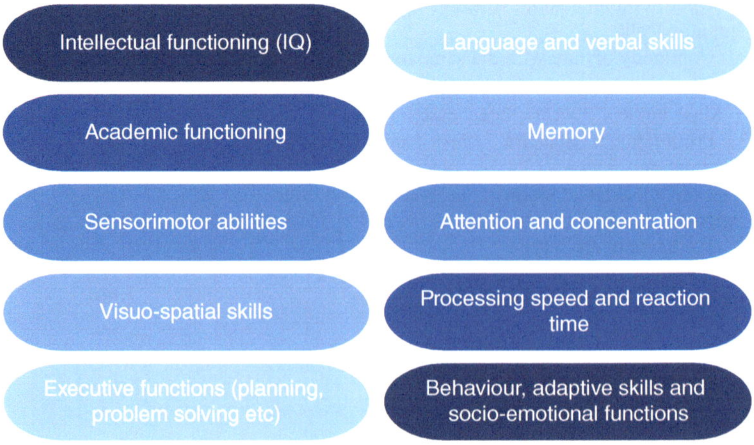

Fig. 22.1 Areas commonly assessed in neuropsychological assessments [69]

Examination of records and history	Clinical interview with patient and caregiver	Neuropsychological testing and interpretation	Feedback, rehabilitation and support
Understand the context	Explore developmental factors that may affect the current presentation (e.g. pre-existing autism)	Administer standardized neuro-psychological tests to assess the extent and nature of neuropsychological problems	Evaluate expectation of the patient and family regarding the neuro-psychological test
Clarify medical, neurological and psychiatric history	Explore personality factors, psychological/psychiatric factors that warrant clinical attention	Determine if the findings reflect specific brain-behavior relations/lesion location	Communicate test results in a language that is accessible to the patient/caregiver
Clarify family history	Clarify impact of psychiatric and cognitive symptoms on daily living	Examine the degree of cognitive strength and dysfunction	Provide feedback to referring clinician
Understand the reason for referral	Find out the neuro-behavioral signs and symptoms with their onset and progression	Integrate test findings with patient background information already collected	Discuss treatment, rehabilitation and permanent deficits with the patient

Fig. 22.2 Various components involved in a neuropsychological evaluation of a patient [70]

nicians identify probable deficits and then further administer specific tests that aid in quantifying the dysfunction or deficit. Some common neurocognitive symptoms and tests that can be administered to test them are shown in Table 22.3.

It is important to note however that there may be several tests that evaluate the same or similar functions. Clinicians should also consider the cultural suitability of specific neuropsychological tests and whether they are validated and standardized for the population to which they administer them to obtain fair results in mind the cultural aspects of the tests and whether they are validated and standardized on the population you administer them on to be able to obtain fair results. Factors such as educational status, literacy, language, etc. are important during assessment to ensure the results are free of bias.

22.6 Psychological Interventions for Children with Cancer

With the long and often intensive treatments of childhood cancer, children and families are at an inherent risk of developing and psychological issues. The psychological problems in childhood cancers have a diverse range, starting from mood disturbances, anxiety, sleep disturbances, family discord and sibling rivalry to self-harm and suicidal behaviors [71–73]. Psychological screening, assessment, and intervention are therefore key in preventing, identifying, treating and supporting children and their families with emotional and behavioral problems. We will aim to briefly address the common issues in children and their families undergoing treatment for childhood cancer and touch upon the psychological interventions used to treat the same. Psychological issues in childhood cancers can broadly be divided into three main areas: a) problems associated with various physical symptoms as pain c) psychological issues or psychiatric diagnosis during the course of treatment or beyond c) problems in intrafamilial relationships.

Medical play, therapy, education, preparation, and coping techniques help provide children with a sense of mastery over their environment and to reduce their need for sedation [74]. Additionally, given the multidisciplinary nature of treatment, nurses, radiation therapists, and other members of the clinical team can positively affect children's experience as well [75]. Another area of interest is pre-exposure therapy for acclimation, which typically includes departmental tours and audiovisual aids. More recently, efforts have been under way to use augmented reality/virtual reality (AR/VR) to provide children and their fami-

Table 22.3 Various neurocognitive functions and examples of tests used to measure these functions

Neurocognitive function measured	Tests	Specific areas measured
Cognitive screening instrument	Mini Mental State Examination (MMSE)	Registration, attention, calculation, language, recall, orientation, and ability to follow simple commands
Intellectual function (IQ)		
Alertness	Test battery of attentional performance (TAP)	Intrinsic alertness and reaction time using visual and auditory stimuli individually and together.
	d^2 test	Attention and concentration performance
	Trail Making Test, version A (TMT-A)	Visual search abilities, scanning and processing speed, intellectual flexibility, as well as executive functions
Memory	Wechsler Memory Scale revised (WMS-r)	Consists of 13 subtests
	Verbal learning and memory test (VLMT)	Declarative-episodic memory function
	Rey Osterrieth complex figure test (ROCF)	Spatial visual constructive capacity and visual memory performance
Executive functions	Version B of the Trail Making Test (TMT-B)	Ability to switch tasks
	Stroop color-word test	Sensorimotor performance test used to evaluate basic cognitive functions of reading, naming, and the ability of selective attention
	Regensburg Word Fluency Test (RWT)	Lexical fluency and semantic capacities
Visuo-spatial abilities	Block design test	
	Rey Complex Figure Test and Recognition Trial	

lies with realistic virtual experiences that provide clear expectations of their upcoming treatment. During the treatments themselves, video-based distraction therapy (VBDT) is used, and data suggest that young children who watch movies during RT are less likely to require daily sedation [76]. Based on Fig. 22.3, there may be a number of psychological issues that occur in children with cancer. A few of the commonly seen issues along with basic interventions that can be used are outlined in Table 22.4. Symptoms and their interventions have been divided based on whether they are emotional symptoms or physical symptoms. Sometimes it may even be one or the other (Table 22.5).

22.6.1 Management of Procedural Anxiety in Children

Procedural anxiety in children poses unique challenges for the RT team. Overwhelming treatment atmosphere, isolation, separation anxiety, past aversive experiences, parental worry and anxiety triggers distress in children undergoing RT [93–96]. In order to improve patient compliance, reduce treatment time and hospital stay children are subjected to sedation or anesthesia. Concerns about long term toxicities with anesthesia (neurocognitive impairments, behavioral and emotional difficulties) have prompted Children's Oncology Group Study-COGS [97] to consider alternatives that minimize distress and enhance the quality of life of children undergoing RT. Global surveys of the COGS have shown the use of several non-pharmacological age-appropriate methods (Table 22.6) by radiation centers to improve the

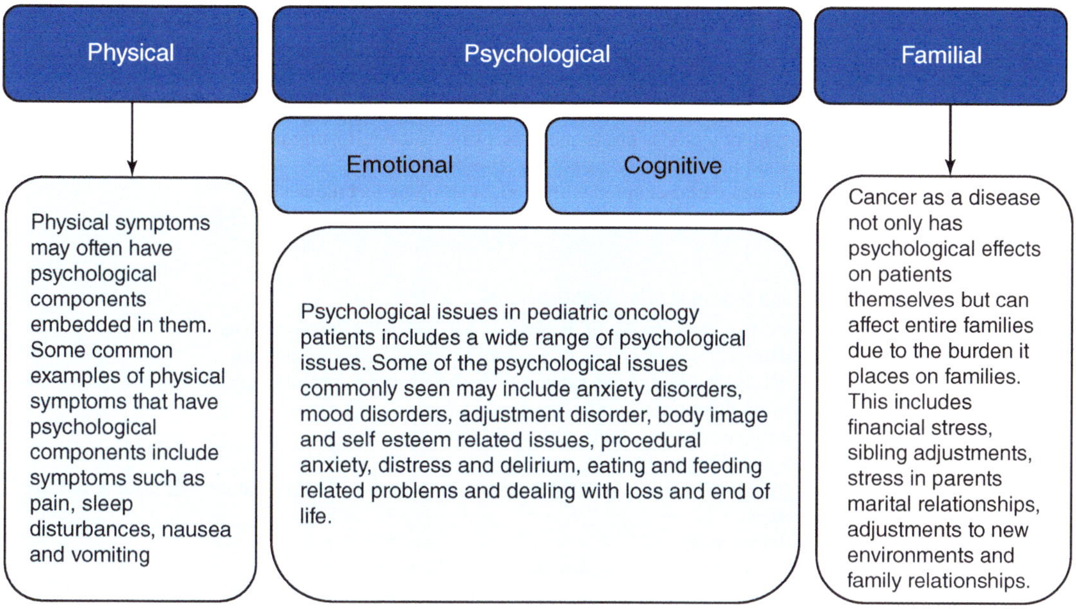

Fig. 22.3 Three main domains that may contribute to psychological issues in children with cancer

pediatric patient experience and decreased distress [97].

22.7 Delirium in Cancer

Delirium is a common neuropsychiatric syndrome seen in cancer patients, particularly in older patients toward end of life. It is associated with increased morbidity and mortality and is a distressing experience for the patient, family, and the staff [36]. The prevalence of delirium is 25% in hospitalized cancer patients [98], 13.3% to 42.3% in palliative care units or hospices [99], and terminal delirium is seen in over 90% at end of life [99–101].

The pathophysiology is multifaceted, involving abnormalities of multiple neurotransmitter systems, inadequate oxidative mechanism, and dysregulation of the circadian rhythm or the hypothalamic-pituitary-adrenal (HPA) axis [102]. Multiple overlapping pathophysiological routes contribute to delirium in cancer, and they vary depending on the etiology [15]. The risk factors for delirium are classified as either "predisposing" or "precipitating" factors [102]. Predisposing factor exists at baseline and increases the person's vulnerability for delirium, whereas the precipitating factor triggers or activates the specific delirium response [102]. The risk factors for delirium are outlined in Table 22.7 [15, 101].

Delirium has a diverse etiology and is multifactorial [103, 104]. In patients with high vulnerability at baseline (elderly, frail with advanced cancer) a less noxious insult (like constipation) can cause delirium, whereas in those with low vulnerability severe noxious insult is needed to cause delirium [15, 103, 105].

The presentation of delirium is varied, and includes a broad range of neurocognitive and motor symptoms [106]. Delirium develops over a short course of time (acute onset), and the symptoms fluctuate over the course of the day [107, 108]. The clinical features are outlined in Table 22.8.

The gold standard for diagnosis is clinician assessment by a trained and competent healthcare professional utilizing standard diagnostic criteria like the Diagnostic and Statistical Manual of Mental Disorders Fifth Edition (DSM-5) or International Classification of Diseases and Related Health Problems, Eleventh Edition (ICD-11) [III, C] [15, 107, 108]. The available evidence

Table 22.4 Commonly seen emotional symptoms in clinical practice along with basic intervention outlines for each of the psychological issues

Emotional symptoms	Psychological interventions for children with cancer
Anxiety	Anxiety management includes one or a combination of the following psychological techniques: – Desensitization to the anxiety provoking stimulus – Positive self-statements coupled with thought stopping techniques – Positive reinforcements to desirable behaviors despite presence of the anxiety provoking stimulus – Distraction – Modeling and rehearsal of desired behaviors
Depression	Based on the severity of diagnosis of depression psychological interventions, psychopharmacological interventions or a combination of both may be used. Psychological interventions that use models of cognitive-behavioral therapy have been found to be effective in treating depression Some CBT techniques that may be used for children are: – Mood diary and thought diary – Identifying the predisposing, precipitating and perpetuating factors that may be leading to the mood disturbances – Coping skills training – Identifying and challenging negative automatic thoughts – Reducing ruminations, monitoring activity and planning activity [77, 78]
Adjustment issues	Adjustment-related problems are common in children with cancer. Often these issues may settle on their own. Some techniques that can be used to help children adjust faster include – Creating a safe, nonthreatening environment – Explaining and preparing children in advance if there are procedures, surgery, admission planned – Allowing space for open and free conversation – Providing children with activity and play while they are in hospital – Allowing a parent or familiar elder to be present at all times as this serves as reassurance and safety for the child – Coping skills training [79]
Body image	Body image issues are especially prevalent in the adolescent age group. Some interventions that are useful for children and adolescents undergoing treatment include: – Preparing children for possible changes they will experience due to treatment such as hair fall, darkening of skin and nails etc. – Emphasizing that these changes are only temporary – coping skills training – Decentralizing the belief that self-esteem, appearance, etc. are dependent on physical characteristics like hair – Encouraging parents and friends to treat the child normally and not focus on the bodily changes that may be occurring – Creating a safe space for children to express their feelings and needs – Support groups for children which help build inclusivity in treatment phase [80, 81]
Temper tantrums and anger	Temper tantrums, anger and other behavioral problems may occur in children undergoing treatment for childhood cancer. It is important to note however that these may sometimes be transient and attributed to medications such as steroids When behavioral problems are psychological in nature, clinicians may use the following interventions: – Behavioral techniques such as reinforcement, punishment and modeling to promote healthy behaviors – "Time outs" as a method of punishment – Pretend play wherein the psychologist creates situations with play figures that may involve undesired behavior and the child themselves suggests how it can be corrected. This method allows for less resistance from children as they may feel less targeted – Ensuring parents are consistent with their behaviors – Encouraging development of healthy problem solving skills so that children are not angered by difficult situations [82, 83]

Table 22.5 Commonly seen physical symptoms in clinical practice along with basic intervention outlines for each of the psychological issues

Physical symptoms	Psychological intervention
Pain	Pain management may often involve a combination of psychological as well as pharmacological methods. Some of the psychological techniques that are used in pain management include the following – Relaxation training using techniques such as diaphragmatic breathing, progressive muscle relaxation, guided imagery [84, 85] – Using cognitive behavioral techniques such as coping skills training, problem solving skills, cognitive restructuring [84, 85] – Managing emotional concomitants of pain – Encouraging physical exercise and activity – Distraction [86]
Poor treatment adherence	Poor adherence to treatment is common in adolescent age groups which greatly affect survival and prognosis. Some interventions that – Patient education – Liaising with teams to understand medication regimen management (using combination pills to reduce the number of pills patients take daily) – Co-management of care where children are also included in decision making – Cognitive behavioral therapies such as motivational interviewing and explaining ABC model – Positive reinforcement to promote adherence [87, 88] – Proper communication skills by treating teams as well as addressing children's information needs helps improve adherence as children feel equally included in treatment
Nausea/vomiting	Nausea and vomiting are common side effects in cancer treatments. Anticipatory nausea vomiting may occur as well and have psychological components that may be addressed by using cognitive and behavioral interventions, especially progressive muscle relaxation training and systematic desensitization [89]
Eating/feeding	Food refusal and negative reactions are common among children undergoing treatment for cancer especially immediately after chemotherapy. However, the importance of good and adequate nutrition during treatment is emphasized abundantly in literature as this greatly influences prognosis and treatment tolerance. Some interventions that are useful in addressing feeding and eating problems on children with cancer include: – Psychoeducation about the importance of adequate and good nutrition for parents and families – Positive reinforcements for maintaining healthy eating practices – Encouraging small, frequent meals – Involving children in decisions about food as often there may be change in taste preferences with treatment (example children prefer warm foods or semisolid foods) [90–92]
Sleep (may be both an emotional symptom or a physical symptom)	Sleep disturbances in children with cancer are common. These may be attributed to a number of factors that are psychological as well as treatment related such as steroidal medications commonly used in treatment. Some common interventions are: – Following sleep hygiene techniques – Creating a safe, comfortable environment for children to sleep in – Maintaining routine – Physical activity – Exposure to adequate sunlight that aids in maintaining circadian rhythms
Distress (may be both an emotional symptom or a physical symptom)	Distress in pediatric oncology may be due to a variety of reasons such as pain, anxiety, discomfort or medical reasons. It is important to explore and understand the cause of the distress by clinical interview. Distress due to medical reasons may best be handled by treating teams. Interventions for distress caused due to psychological reasons such as anxiety require management of the acute distress phase such as relaxation, removing the distress causing stimulus, breathing exercises and psychological support. Addressing the distress causing stimulus long term is an important factor in ensuring that the patient copes better in the future ("COGNITIVE BEHAVIOURAL TREATMENT FOR IMPROVING DISTRESS IN PAEDIATRIC ONCOLOGY," 2019)

Table 22.6 Nonpharmacological age-appropriate methods used by radiation centers to improve the pediatric patient experience and decreased distress during RT

Management of procedural anxiety: Children undergoing RT
Prior to CT simulation/treatment • Prior to the procedure: Meet child and parents for planning and education about RT • Information and education about the treatment process: Booklets and Audio-visual aids, Web-based Education Tools. Videos of other children undergoing simulation and/or RT • Medical play therapy with medical toys (CT, RT machines etc.) by trained professionals • Tour of CT SIM and Treatment Vault • CT SIM Practice Session (Dry run) • Video/Photographs of Radiation Process • Home Practice Session Lying Still • Radiation Activity/Coloring Book • Model CT SIM/Radiation Machine
Management of procedural anxiety for children undergoing RT
• Mask acclimation • Mask modification (cutting out holes for eyes) • Mask decoration • Child practices making mask on object • Child watches mask made for family member
During CT simulation and radiation treatment • Comfort object (stuffed animal, blanket, stress ball, etc.) • Child's choice of music or audio • Anticipatory guidance • Consistent (same) RT Team • Key team member to be present (who has prepared the child): Psycho-oncologist, psychologist, Counselor, MSW or RT team member) • Two-way audio with caregiver and/or team member • No tattoo policy • Video-based distraction therapy (e.g., watch movie or TV show) • Child-friendly décor or environment to be created • Caregiver in CT room (wearing lead shield) • Video between caregiver and child
Additional departmental activities • Children's play room • End-of-treatment celebration • End-of-treatment gift • Child-friendly consultation room • Planned events (e.g., holiday party, art therapy, guests) • Bravery beads (e.g., receive bead after each treatment) • Rides in Mini-Car/Plane to CT Sim/treatment vault • Therapy animal visits • Treasure chest (e.g., child picks toy after each week check) • Treatment countdown chart (e.g., sticker for each treatment)

is insufficient either to recommend for or against the use of any screening, diagnostic, or rating tools for delirium [III, C] [15]. The Confusion Assessment Method (CAM) [103, 109] is a copyrighted instrument used in screening and diagnosis of delirium, Delirium Rating Scale (DRS) [110, 111] and Memorial Delirium Assessment Scale (MDAS) [110, 111] scales used to assess severity of delirium in cancer patients.

The principles of management involve: (a) identification of the predisposing and precipitating factors through comprehensive clinical assessment including detailed history, clinical examination, laboratory testing, and imaging as appropriate (III, A) [15], (b) correction of potentially reversible precipitating causes, (c) tailoring of interventions as per patient and family preferences, and (d) palliative sedation for management of refractory terminal delirium [15].

Table 22.7 Risk factors for delirium [15, 101]

Predisposing factors	Precipitating factors
Old age (>70 years)	Sepsis at any site
Frailty with declining performance status	Electrolyte imbalance
Male gender	Cerebral edema
Sensory impairment (vision, hearing)	Constipation
Cognitive impairment	Dehydration
Neoplastic disease • Primary tumors of the CNS • Brain or meningeal metastasis • Paraneoplastic syndromes	Medications • Anticholinergics or drugs with anticholinergic activity • Opioids • Corticosteroids • Polypharmacy
Cachexia	Radiation to the brain
Malnutrition	Chemotherapy
End-stage organ impairment	
Neurological disorders—stroke, epilepsy	
Substance use disorders	
Past history of delirium	

Table 22.8 Clinical features of delirium

Global disturbances of cognition: • Impaired level of arousal, attention and awareness • Impaired orientation to time, place or person • Disturbances in short-term and long-term memory • Impaired executive functions
Perceptual disturbances • Hallucination • Illusions
Mood disturbances • Increased irritability and anger • Increased fear and anxiety • Apathy or withdrawal • Emotional lability
Altered sleep-wake cycle
Thought disorders • Incoherent and irrelevant speech • Tangentiality • Delusions
Motor disturbances • Agitation • Restlessness, combativeness • Lethargy or listlessness • Enhanced startle reaction
Neurological disturbances • Tremor, myoclonus or frontal release signs
Subtypes based on psychomotor features and arousal level include [23]: • Hypoactive delirium • Hyperactive delirium • Mixed delirium

There is insufficient evidence to support use of antipsychotics in the prevention or treatment of delirium [112]. Pharmacologic interventions are limited to patients with distressing symptoms like hallucinations, or safety concerns. Haloperidol and risperidone provide no symptomatic benefit in treatment of delirium in cancer patients (I, D) [15, 112]. Olanzapine (III, C), quetiapine (V, C), and aripiprazole (IV, C) may offer symptomatic benefit [15]. Methylphenidate may improve cognition in hypoactive delirium (V, C) [15]. Benzodiazepines are deliriogenic and to be used with caution to control severe distress as second-line agents in delirium (II, C) [15]. Recent reviews have explored the role of dexmedetomidine in reducing incidence of delirium, but evidence is limited [113, 114]. The pharmacological and nonpharmacological interventions are outlined in Table 22.9 [15].

Delirium is a clinical emergency but is often under-recognized and under-treated. Early detection, assessment, and management is possible through systematic screening of high-risk individuals. Reversal may not always be possible or the goal, but symptomatic management with nonpharmacological interventions should be available for all patients, and supported with pharmacological interventions.

Table 22.9 Management of delirium

Nonpharmacological interventions • Psychoeducation of and support for the patient (when lucid), family members and healthcare staff • Reduce polypharmacy—Deprescribe • Optimal nutrition, hydration, bowel and bladder function, and • Preservation of sleep—improve exposure to sunlight, reduce daytime nap, minimize disruptions at night • Minimize sensory deficits—encourage use of visual and hearing aids, • Early mobilization, active range of motion exercises, avoid physical restraints • Reorientation—place orientation board, clock or familiar objects in patient room • Encourage cognitive stimulation
Pharmacological interventions Antipsychotics • Haloperidol 0.5–1 mg p.o./sc. stat. (0.5–1 mg prn.) • Chlorpromazine 12.5–25 mg p.o./pr. stat. (12.5–25 mg prn.) • Olanzapine 2.5–5 mg p.o./sc. stat • Quetiapine 25 mg p.o. stat • Risperidone 0.5 mg p.o. stat • Aripiprazole 5 mg p.o. stat Benzodiazepines • Midazolam 2.5 mg s.c./i.v. q1h prn. (up to 5 mg max) • Lorazepam 1 mg s.c./i.v. stat. (up to 2 mg max)

22.8 Role of Psychotropic Medications in Cancer Patients

Premorbid psychiatric conditions or new onset anxiety, depression, delirium, behavioral disturbances are common in people with cancer. In view of vulnerability and risk of drug interaction and side-effects in this patient group, judicious use of psychotropic is of utmost importance.

Antidepressants are required in patients with disabling anxiety and depression and efficacy of this group of medications has been supported by literature [115–117]. SSRIs are often the first choice while treating affective symptoms in cancer. Cardiac and anticholinergic side effects limit the use of older tricyclic antidepressants in cancer. Sedative antidepressants like Mirtazapine may be preferred in patients with insomnia and depressive symptoms. Mirtazapine helps in improving pain and nausea in cancer patients [118, 119]. SSRIs and Venlafaxine are found to be useful in cancer-related pain and hot flush like symptoms after surgical or chemical menopause in cancer [120, 121].

Estrogen receptor positive breast cancer patients, premenopausal or metastatic postmenopausal patients are treated with tamoxifen. Tamoxifen is metabolized in the liver to produce pharmacologically active metabolites of the medicine called Endoxifen. Literature supports cautious use of CYP2D6 inhibitor SSRIs for such patients and encourages using weak CYP2D6 inhibitors like Venlafaxine [122–124].

Benzodiazepines in cancer are used for anxiety disorders and substance withdrawals. Longer acting benzodiazepines like clonazepam and chlordiazepoxide are preferred for short durations of time. Relatively shorter acting benzodiazepines like lorazepam are to be used when the liver function is deranged in order to reduce risk of benzodiazepine accumulation in the body and toxicity [125]. Benzodiazepines may be poorly tolerated by some cancer patients and result in daytime somnolence, risk of fall, anterograde amnesia, and paradoxical agitation [125]. Non-benzodiazepine sedatives like Zolpidem are often preferred due to shorter duration of action, less risk of daytime somnolence and less chance of dependence.

Mood stabilizers like Lithium are often avoided during cancer chemotherapy due to the narrow therapeutic window and sodium valproate is used with caution as it is metabolized in the liver and can lead to drug interaction with chemotherapeutic agents. Carbamazepine is a mood stabilizer with risk of bone marrow toxicity and hence is often avoided with various cancer treatments. Second-generation antipsychotics with mood stabilizing properties like Olanzapine may be used as monotherapy agents during the period of chemotherapy in patients who need mood stabilizer prophylaxis [126].

Antipsychotics are used in cancer for control of agitation in delirium, organic brain syndrome, CNS metastasis or acute psychosis. Parenteral antipsychotics like Injectable Haloperidol are also used for rapid tranquilization [127]. Antipsychotics only reduce behavioral symptoms in organic condition and tapered off once the underlying organic cause has been addressed with interventions like antibiotics, correction of electrolytes etc. Antipsychotics like Haloperidol, Olanzapine are also used for symptomatic relief from nausea, vomiting in cancer patients [128, 129].

22.9 Tobacco Cessation

According to the World Health Organization, annual deaths resulting from tobacco-related diseases stand at eight million globally and tobacco use is a major risk factor for many chronic diseases, including cancer, lung disease, cardiovascular disease and stroke. Tobacco consumption is found to be the leading cause of preventable deaths and is reported to be the cause of 12% of premature deaths globally [130]. India is the second largest consumer and producer of tobacco and therefore bears a heavy disease burden due to tobacco use.

Management of tobacco dependence is important in the general population to prevent cancer. Similarly it is crucial in patients with cancer in India as tobacco-related cancers are common in the subcontinent and many cancer patients struggle to get over the addiction. Additionally they may have guilt associated with the use of tobacco in the past. Tobacco dependence is often managed with a judicious combination of tobacco replacement products as nicotine replacement patch, nicotine chewing gums as well as psychological management of the tobacco dependence. Mainstreaming of specialist tobacco cessation clinics in medical establishments including oncology centers is thus essential. It is evident that tobacco control and cessation must work hand in hand to ensure a complete effectiveness of the efforts being undertaken.

22.10 Long-Term Care of Cancer Survivors: An Outline

Current dispositions of cancer care recognize that many cancer patients are getting cured, entering into complete remission, returning to normal functioning after the completion of treatment and are able to live relatively symptom-free. However, cancer and its treatment can also leave a person with profound emotional and physical symptoms that do not always recede with time [131]. The issues range from hypertension, heart and lung issues, hormonal problems, cognitive difficulties, dental, oral problems, fatigue and emotional difficulties. The clinical management of cancer survivorship care includes not only the long-term medical aspect but also the psychosocial care of the patients. Researches have stated that nearly 31–52% of cancer patients experience fear of cancer recurrence (FCR), and 24–40% of them reported moderate to high levels of need for help in coping with it [132]. While cancer therapies improve survival, it is also found to cause problems in pubertal development, infertility and sexual dysfunction in survivors [133]. The need for a well-integrated standard care of cancer survivors is emphasized, as these long-term side effects of cancer treatment may lead to poor quality of life [134]. This need for a survivorship care plan has paved the way for the development of models of care for postcancer treatment follow-up. The Institute of Medicine (IOM) report states that survivorship care ideally includes (1) prevention of new (primary) and recurrent cancers and other late effects; (2) surveillance for recurrence or new cancers; (3) interventions for illnesses secondary to cancer and cancer treatment; and (4) coordination between specialists and primary care providers to ensure that all the health needs of survivors are met [135]. Nekhlyudov et al. delineated a cancer survivorship care framework which included individual, interpersonal, organizational, community and policy factors [136].

Table 22.10 Cancer survivorship care framework

Domains of care	Prevention and surveillance for recurrence and new cancers
	Surveillance and management of physical effects
	Surveillance and management of psychosocial effects
	Surveillance and management of chronic medical conditions
	Health promotion
	Disease prevention
Contextual domains	Clinical structure
	Communication/decision-making
	Care coordination
	Patient or caregiver experience
Healthcare outcomes	Health-related quality of life
	Healthcare utilization
	Costs of care
	Mortality
Implementation	Implementation barriers
	Implementation facilitators

Models of survivorship care are highly individualized to the institution or setting where they are provided as the components of models, survivor population and target outcomes vary differently across various suggested models of care. All of the established models of survivorship care planning included psychosocial care into their ambit. The specific type of psychological intervention being adopted for a particular patient will be at the discretion of the treating therapist following an assessment.

The paucity of a well-integrated survivorship cancer care plan in our healthcare system needs to be fulfilled [134]. This need for a survivorship care plan has paved the way for the development of models of care for postcancer treatment follow-up, which includes the aspects of the quality of cancer survivorship care framework as defined by Nekhlyudov et al. This model can be broadly classified into four domains, mainly domains of care, contextual domain, healthcare outcomes and implementation [137] (Table 22.10, Fig. 22.1).

22.11 End of Life Care

The burden and mortality due to cancer is rising globally with an estimated 10.0 million deaths in 2020 [138]. Patients and families face multiple and complex challenges when dealing with cancer. Palliative care is a philosophy of care that is aimed at alleviating this health-related suffering [90].

In patients with advanced and metastatic cancer, when cure and prolongation of life is no longer possible, the focus shifts to comfort care [139]. End-of-life care (EOLC) refers to care of people with terminal illnesses during the last few months or weeks of their life [140]. When physical functioning is rapidly declining and death is inevitable, comfort, symptom control, quality of life (QOL), and quality of dying and death (QODD) become the focus of care [140, 141]. Care during end of life (EOL) involves cessation of all cancer-directed therapies and disproportionate interventions, and focuses on alleviating the physical, psychosocial, spiritual, and existential suffering of patients and their families [142]. End of life care is a holistic, person- and family-centered care of the dying, and is an integral component of palliative care [143]. It is provided by multidisciplinary palliative care teams, in hospitals, hospices or in the community, and is tailored to meet the specific needs of the dying patients and their families [140].

22.11.1 Components of Quality End of Life Care

The European Society for Medical Oncology (ESMO) Clinical Practice Guidelines provides the framework for care during EOL for adult patients with advanced cancer [140]. This guideline advocates for a seamless transition into EOLC for patients and their families, with the oncologists recognizing and coordinating the transition and facilitating effective and timely palliative care interventions.

- Prognostication:
  Skilled prognostication is an essential prerequisite for good EOLC, but is a challenging task [144]. Physician estimates often overestimate survival. Prognostic models incorporating physician estimates, objective physical symptoms especially declining performance status, laboratory investigations [I, A], and

routine use of prognostic tools [III, B] improve the accuracy of clinical prediction [145]. Prognostication helps in timely preparation and aids in eliciting patient/family preferences about site and type of EOLC.

- Communication:
  Effective and compassionate communication and shared decision-making are important aspects of quality EOLC (II, A). EOLC discussions involve both the patient and the family and includes eliciting their knowledge about diagnosis and prognosis; preferences around disclosure; honest, compassionate truth-telling; discussions around available treatment options, preferences regarding life-sustaining treatments, and preferred place of care and death [146]. Eliciting important unfinished businesses while acknowledging and addressing emotional concerns is also an important aspect of these discussions [147]. Effective EOLC is anticipatory and strategic and is aligned to the patient's and family's values, preferences, and cultural beliefs (III, A) [140].
- Symptom Management:
  Majority of the patients during EOL experience distressing symptoms like pain, dyspnea, delirium, and respiratory secretions. Proactive management of these symptoms is essential for comfort, especially when patients have difficulty communicating (V, C) [140, 148]. An individualized treatment plan, with regular monitoring, and use of regular and as-needed medications ensure comfort (V, B). Burdensome and nonessential medications are discontinued in EOL. Oral route is preferred, but if patients are unable to swallow either subcutaneous, transdermal, or intravenous routes are used. Refractory symptom in a dying patient is a medical emergency and is best addressed through palliative sedation [149].
- Management of Psycho-socio-spiritual Issues:
  Psychological, existential, and spiritual distress are highly prevalent in patients at EOL. Depression and anxiety are prevalent in about 58% [98], and 73% expressed at least one spiritual need [150]. Depression is an independent risk factor for mortality in cancer patients [151]. Psycho-spiritual distress negatively impacts healthcare outcomes and QOL. Psycho-spiritual domains warrant appropriate assessment and management (I, A) [140]. Addressing emotional and psychosocial factors through pharmacological and nonpharmacological management alongside physical symptom management enhances treatment adherence, less use of aggressive interventions at EOL, and better patient and family satisfaction. Spiritual well-being reduces depression, demoralization and the desire for hastened death and can be addressed by appropriate spiritual care [140].
  Grief and bereavement issues are common in family members of patients nearing EOL. While the majority of caregivers adjust to the loss, a small proportion may develop complicated grief [152]. Caregivers of patients nearing EOL should be screened for psychosocial distress in the pre-death phase. Referral to psycho-oncology for intervention and support should be considered for those at risk.
  The care of patients at EOL is complex and challenging and requires a multidisciplinary and collaborative approach. Proactive, vigilant, and compassionate EOLC addressing the physical, psychosocial, existential, and spiritual concerns will alleviate some of the distress experienced by the terminally-ill patients and their families.

22.12 Conclusion

Identification of psychiatric comorbidities and integrated management of these are important for patients with cancer. Psycho-oncology is a new subspecialty that integrates the care of cancer patients with mental health services. We recommend that all major cancer centers should have a multidisciplinary psycho-oncology team that has consultant psychiatrists and clinical psychologists to help with this aspect of care. This would not only improve the satisfaction of patients and service users but also engage the patient to the treatment procedure and hopefully will lead to a better outcome of the disease.

22.13 Summary (Precepts Box)

1. Cancer patient and survivors face emotional and social challenges which impacts their life in several ways.
2. Pschyco-oncology is the study of psychological aspects of cancer. It needs a multidisciplinary team comprising of oncologists, nurses, psychiatrist, psychologist, and social workers.
3. Accepting diagnosis of cancer is challenging. Patient may experience wide spectrum of symptoms ranging from self-limiting distress and adjustment to suicidal tendency.
4. Nonpharmacologic means to overcome psychosocial impact include education, specific psychological strategies, and facilitating environmental change.
5. Most common and proven psychological interventions include counseling, psychoeducation, facilitating coping, and providing psychological support.
6. Procedural anxiety may be overwhelming for some patients. Table 22.1 summarizes steps to overcome this.
7. Neuropsychological testing in patients with brain tumors is important as it plays an essential role in the assessment and management of central nervous system tumors.
8. Psychological problems in children are diverse; hence, psychological intervention is key in preventing, identifying, treating, and supporting children and their families with emotional and behavioral problems.
9. Psychotropic medications can be used to address certain issues along with tobacco cessation.

Multiple Choice Questions

1. What term was introduced by NCCN in the 1990s to better quantify cancer-related psychological sufferings?
   (a) Adjustment disorders
   (b) Distress
   (c) Anxiety
   (d) Panic attack
2. What psychiatric illness is frequent in cancer patients and often underdiagnosed and misdiagnosed?
   (a) Anger
   (b) Pain
   (c) Delirium
   (d) Panic attacks
3. Nonpharmacological interventions in oncology mainly comprise of
   (a) Education
   (b) Using specific psychological strategies
   (c) Facilitate environmental change
   (d) All the above
4. During which phase of RT patients have maximum distress?
   (a) Prior to staring RT
   (b) At the end of RT
   (c) During simulation
   (d) After completion of RT
5. The acute stress caused by medical procedure is termed as
   (a) Procedural anxiety
   (b) Therapeutic anxiety
   (c) Doctor anxiety
   (d) None of above
6. What are causes of treatment room anxiety?
   (a) Freezing room temperature
   (b) Hard couch
   (c) Noise of machine
   (d) All the above
7. Cognitive screening is assessed by
   (a) Mini mental status examination
   (b) Test battery of attentional performance
   (c) D2 Test
   (d) Trail making test
8. How many subsets are present in Wechsler Memory Scale revised (WMS-r)?
   (a) 11
   (b) 12
   (c) 13
   (d) 14
9. Anxiety management includes
   (a) Desensitization to the anxiety-provoking stimulus
   (b) Positive self-statements coupled with thought stopping techniques

(c) Positive reinforcements to desirable behaviors despite presence of the anxiety provoking stimulus
(d) All the above

Answers

1. b
2. c
3. d
4. b
5. a
6. d
7. a
8. c
9. d

References

1. Mallath MK, Taylor DG, Badwe RA, Rath GK, Shanta V, Pramesh CS, et al. The growing burden of cancer in India: epidemiology and social context. Lancet Oncol. 2014;15(6):e205–12.
2. Mitchell AJ, Chan M, Bhatti H, Halton M, Grassi L, Johansen C, et al. Prevalence of depression, anxiety, and adjustment disorder in oncological, haematological, and palliative-care settings: a meta-analysis of 94 interview-based studies. Lancet Oncol. 2011;12(2):160–74.
3. Hess CB, Chen AM. Measuring psychosocial functioning in the radiation oncology clinic: a systematic review. Psychooncology. 2014;23(8):841–54.
4. Maurer J, Schäfer C, Maurer O, Kölbl O. Anxiety and depression in cancer patients during the course of radiotherapy treatment. Strahlenther Onkol. 2012;188(10):940–5.
5. Peng Y-N, Huang M-L, Kao C-H. Prevalence of depression and anxiety in colorectal cancer patients: a literature review. Int J Environ Res Public Health. 2019;16(3):E411.
6. Alanazi AA. Reducing anxiety in preoperative patients: a systematic review. Br J Nurs. 2014;23(7):387–93.
7. Clover K, Oultram S, Adams C, Cross L, Findlay N, Ponman L. Disruption to radiation therapy sessions due to anxiety among patients receiving radiation therapy to the head and neck area can be predicted using patient self-report measures. Psychooncology. 2011;20(12):1334–41.
8. Chen AM, Jennelle RLS, Grady V, Tovar A, Bowen K, Simonin P, et al. Prospective study of psychosocial distress among patients undergoing radiotherapy for head and neck cancer. Int J Radiat Oncol Biol Phys. 2009;73(1):187–93.
9. Glaser AW, Fraser LK, Corner J, Feltbower R, Morris EJA, Hartwell G, et al. Patient-reported outcomes of cancer survivors in England 1-5 years after diagnosis: a cross-sectional survey. BMJ Open. 2013;3(4):e002317.
10. Harris L, Menzies R, Robinson J. Predictors of panic symptoms during magnetic resonance imaging scans. Int J Behav Med. 2001;8:80–7.
11. Henderson JM, Ord RA. Suicide in head and neck cancer patients. J Oral Maxillofac Surg. 1997;55(11):1217–21; discussion 1221–2.
12. Kam D, Salib A, Gorgy G, Patel TD, Carniol ET, Eloy JA, et al. Incidence of suicide in patients with head and neck cancer. JAMA Otolaryngol Head Neck Surg. 2015;141(12):1075–81.
13. Zeller JL. High suicide risk found for patients with head and neck cancer. JAMA. 2006;296(14):1716–7.
14. McFarland DC, Walsh L, Napolitano S, Morita J, Jaiswal R. Suicide in patients with cancer: identifying the risk factors. Oncology (Williston Park). 2019;33(6):221–6.
15. Bush SH, Lawlor PG, Ryan K, Centeno C, Lucchesi M, Kanji S, et al. Delirium in adult cancer patients: ESMO clinical practice guidelines. Ann Oncol Off J Eur Soc Med Oncol. 2018;29(Suppl 4):iv143–65.
16. Bush SH, Bruera E. The assessment and management of delirium in cancer patients. Oncologist. 2009;14(10):1039–49.
17. Stiefel F, Holland J. Delirium in cancer patients. Int Psychogeriatr. 1991;3(2):333–6.
18. Moussas GI, Papadopoulou AG. Substance abuse and cancer. Psychiatriki. 2017;28(3):234–41.
19. Passik SD, Portenoy RK, Ricketts PL. Substance abuse issues in cancer patients. Part 1: prevalence and diagnosis. Oncology (Williston Park). 1998;12(4):517–21, 524.
20. Yusufov M, Braun IM, Pirl WF. A systematic review of substance use and substance use disorders in patients with cancer. Gen Hosp Psychiatry. 2019;60:128–36.
21. Passik SD, Portenoy RK, Ricketts PL. Substance abuse issues in cancer patients. Part 2: evaluation and treatment. Oncology (Williston Park). 1998;12(5):729–34; discussion 736, 741–2.
22. Male DA, Fergus KD, Cullen K. Sexual identity after breast cancer: sexuality, body image, and relationship repercussions. Curr Opin Support Palliat Care. 2016;10(1):66–74.
23. Katz A, Dizon DS. Sexuality after cancer: a model for Male survivors. J Sex Med. 2016;13(1):70–8.
24. Girgis A, Lambert S, Johnson C, Waller A, Currow D. Physical, psychosocial, relationship, and economic burden of caring for people with cancer: a review. J Oncol Pract. 2013;9(4):197–202.
25. Rhoten BA. Head and neck cancer and sexuality: a review of the literature. Cancer Nurs. 2016;39(4):313–20.

26. Katz MR, Irish JC, Devins GM, Rodin GM, Gullane PJ. Reliability and validity of an observer-rated disfigurement scale for head and neck cancer patients. Head Neck. 2000;22(2):132–41.
27. Katz MR, Irish JC, Devins GM, Rodin GM, Gullane PJ. Psychosocial adjustment in head and neck cancer: the impact of disfigurement, gender and social support. Head Neck. 2003;25(2):103–12.
28. Falk Dahl CA, Reinertsen KV, Nesvold I-L, Fosså SD, Dahl AA. A study of body image in long-term breast cancer survivors. Cancer. 2010;116(15):3549–57.
29. Andersen BL, Yang H-C, Farrar WB, Golden-Kreutz DM, Emery CF, Thornton LM, et al. Psychologic intervention improves survival for breast cancer patients. Cancer. 2008;113(12):3450–8.
30. Johannsen M, Farver-Vestergaard I, Hansen N, Zachariae R. The efficacy of psychosocial intervention for pain in breast cancer patients and survivors: a systematic review and meta-analysis. Breast Cancer Res Treat. 2013;138(3):675–90.
31. Hunter EG, Gibson RW, Arbesman M, D'Amico M. Systematic review of occupational therapy and adult cancer rehabilitation: part 1. Impact of physical activity and symptom management interventions. Am J Occup Ther. 2017;71(2):7102100030p1–7102 100030p11.
32. Grassi L, Spiegel D, Riba M. Advancing psychosocial care in cancer patients. F1000 Res. 2017;6:2083.
33. Classen CC, Kraemer HC, Blasey C, Giese-Davis J, Koopman C, GronskayaPalesh O, et al. Supportive-expressive group therapy for primary breast cancer patients: a randomized prospective multicenter trial. Psychooncology. 2008;17(5):438–47.
34. Johnson JA, Rash JA, Campbell TS, Savard J, Gehrman PR, Perlis M, et al. A systematic review and meta-analysis of randomized controlled trials of cognitive behavior therapy for insomnia (CBT-I) in cancer survivors. Sleep Med Rev. 2016;27:20–8.
35. Kissane DW, Bloch S, Smith GC, Miach P, Clarke DM, Ikin J, et al. Cognitive-existential group psychotherapy for women with primary breast cancer: a randomised controlled trial. Psychooncology. 2003;12(6):532–46.
36. Breitbart W, Rosenfeld B, Pessin H, Applebaum A, Kulikowski J, Lichtenthal WG. Meaning-centered group psychotherapy: an effective intervention for improving psychological well-being in patients with advanced cancer. J Clin Oncol Off J Am Soc Clin Oncol. 2015;33(7):749–54.
37. Carlson LE. Mindfulness-based interventions for coping with cancer. Ann N Y Acad Sci. 2016;1373(1):5–12.
38. Mustian KM, Alfano CM, Heckler C, Kleckner AS, Kleckner IR, Leach CR, et al. Comparison of pharmaceutical, psychological, and exercise treatments for cancer-related fatigue: a meta-analysis. JAMA Oncol. 2017;3(7):961–8.
39. Graves KD. Social cognitive theory and cancer patients' quality of life: a meta-analysis of psychosocial intervention components. Health Psychol. 2003;22(2):210–9.
40. Tamagawa R, Garland S, Vaska M, Carlson LE. Who benefits from psychosocial interventions in oncology? A systematic review of psychological moderators of treatment outcome. J Behav Med. 2012;35(6):658–73.
41. Sharp L, Lewin F, Johansson H, Payne D, Gerhardsson A, Rutqvist LE. Randomized trial on two types of thermoplastic masks for patient immobilization during radiation therapy for head-and-neck cancer. Int J Radiat Oncol Biol Phys. 2005;61(1):250–6. https://doi.org/10.1016/j.ijrobp.2004.04.047.
42. Holmes N, Williamson K. A survey of cancer patients undergoing a radical course of radiotherapy, to establish levels of anxiety and depression. J Radiother Pract. 2008;7(2):89–98.
43. Halkett GKB, Kristjanson LJ, Lobb E, Little J, Shaw T, Taylor M, et al. Information needs and preferences of women as they proceed through radiotherapy for breast cancer. Patient Educ Couns. 2012;86(3):396–404.
44. Mitchell D, Loranzo R. Understanding patient psychosocial issues. Radiat Ther. 2012;21:96–9.
45. Lewis F, Merckaert I, Liénard A, Libert Y, Etienne A-M, Reynaert C, et al. Anxiety and its time courses during radiotherapy for non-metastatic breast cancer: a longitudinal study. Radiother Oncol J Eur Soc Ther Radiol Oncol. 2014;111(2):276–80.
46. Watts S, Leydon G, Birch B, Prescott P, Lai L, Eardley S, et al. Depression and anxiety in prostate cancer: a systematic review and meta-analysis of prevalence rates. BMJ Open. 2014;4(3):e003901.
47. Bristow BJ, McGuffin M, Szumacher E, Fitch M, Di Prospero L, D'Alimonte L. Assessing the psychological impact of daily bowel preparation on prostate patients who receive radiation therapy. J Med Imaging Radiat Sci. 2018;49(1):70–5.
48. Sharma N, Purkayastha A. Prevalence of anxiety and depression in cancer patients during radiotherapy: a rural Indian perspective. J Cancer Res Ther. 2021;17(1):218–24.
49. Kirchheiner K, Czajka-Pepl A, Ponocny-Seliger E, Scharbert G, Wetzel L, Nout RA, et al. Posttraumatic stress disorder after high-dose-rate brachytherapy for cervical cancer with 2 fractions in 1 application under spinal/epidural anesthesia: incidence and risk factors. Int J Radiat Oncol Biol Phys. 2014;89(2):260–7.
50. Rose P, Yates P. Quality of life experienced by patients receiving radiation treatment for cancers of the head and neck. Cancer Nurs. 2001;24(4):255–63.
51. Nixon JL, Cartmill B, Turner J, Pigott AE, Brown E, Wall LR, et al. Exploring the prevalence and experience of mask anxiety for the person with head and neck cancer undergoing radiotherapy. J Med Radiat Sci. 2018;65(4):282–90.
52. Nixon JL, Brown B, Pigott AE, Turner J, Brown E, Bernard A, et al. A prospective examination of mask

anxiety during radiotherapy for head and neck cancer and patient perceptions of management strategies. J Med Radiat Sci. 2019;66(3):184–90.
53. Krauss BS, Krauss BA, Green SM. Managing procedural anxiety in children. N Engl J Med. 2016;374(16):e19.
54. Forbes E, Baker AL, Britton B, Clover K, Skelton E, Oultram S, et al. Non-pharmacological approaches to procedural anxiety reduction for patients undergoing radiotherapy for cancer: systematic review protocol. BMJ Open. 2020;10(10):e035155.
55. Grilo AM, Santos B, Baptista I, Monsanto F. Exploring the cancer patients' experiences during external radiotherapy: a systematic review and thematic synthesis of qualitative and quantitative evidence. Eur J Oncol Nurs Off J Eur Oncol Nurs Soc. 2021;52:101965.
56. Halkett G, O'Connor M, Aranda S, Jefford M, Merchant S, York D, et al. Communication skills training for radiation therapists: preparing patients for radiation therapy. J Med Radiat Sci. 2016;63(4):232–41.
57. Braeken APBM, Kempen GIJM, Eekers D, van Gils FCJM, Houben RMA, Lechner L. The usefulness and feasibility of a screening instrument to identify psychosocial problems in patients receiving curative radiotherapy: a process evaluation. BMC Cancer. 2011;11:479.
58. Jimenez-Jimenez E, Mateos P, Ortiz I, Aymar N, Vidal M, Roncero R, et al. Do patients feel well informed in a radiation oncology service? J Cancer Educ Off J Am Assoc Cancer Educ. 2018;33(2):346–51.
59. Williams K, Blencowe J, Ind M, Willis D. Meeting radiation therapy patients informational needs through educational videos augmented by 3D visualisation software. J Med Radiat Sci. 2017;64(1):35–40.
60. Canil T, Cashell A, Papadakos J, Abdelmutti N, Friedman AJ. Evaluation of the effects of pre-treatment education on self-efficacy and anxiety in patients receiving radiation therapy: a pilot study. J Med Imaging Radiat Sci. 2012;43(4):221–7.
61. Miller C. Radiation oncology: an Irish hospitals approach to supporting patients. Radiography. 2009;15(1):20–5. Available from: https://www.radiographyonline.com/article/S1078-8174(07)00097-1/fulltext.
62. Dong S, Butow PN, Costa DSJ, Dhillon HM, Shields CG. The influence of patient-centered communication during radiotherapy education sessions on post-consultation patient outcomes. Patient Educ Couns. 2014;95(3):305–12.
63. Halkett G, O'Connor M, Jefford M, Aranda S, Merchant S, Spry N, et al. RT prepare: a radiation therapist-delivered intervention reduces psychological distress in women with breast cancer referred for radiotherapy. Br J Cancer. 2018 Jun;118(12):1549–58.
64. PKaradag, Ezgi, Özlem Uğur, and Oğuz Çetinayak. The Effect of Music Listening Intervention Applied during Radiation Therapy on the Anxiety and Comfort Level in Women with Early-Stage Breast Cancer: A Randomized Controlled Trial. European Journal of Integrative Medicine 2019; 27 (April):39–44.
65. Montgomery GH, Sucala M, Dillon MJ, Schnur JB. Cognitive-behavioral therapy plus hypnosis for distress during breast radiotherapy: a randomized trial. Am J Clin Hypn. 2017;60(2):109–22. https://doi.org/10.1080/00029157.2017.1335635.
66. Schnur JB, David D, Kangas M, Green S, Bovbjerg DH, Montgomery GH. A randomized trial of a cognitive-behavioral therapy and hypnosis intervention on positive and negative affect during breast cancer radiotherapy. J Clin Psychol. 2009;65(4):443–55.
67. Ds H. The role of neuropsychology in the assessment and management of CNS tumors. Clin Oncol. 2016;1:3.
68. Dwan TM, Ownsworth T, Chambers S, Walker DG, Shum DHK. Neuropsychological assessment of individuals with brain tumor: comparison of approaches used in the classification of impairment. Front Oncol. 2015;5:56.
69. What is a neuropsychological assessment? [Internet]. Sage Psychology. [cited 2022 Feb 8]. Available from: https://www.sagepsy.com/blog/what-is-a-neuropsychological-assessment.
70. Schroeder RW, Martin PK, Walling A. Neuropsychological evaluations in adults. Am Fam Physician. 2019;99(2):101–8.
71. Marcus J. Psychosocial issues in pediatric oncology. Ochsner J. 2012;12(3):211–5.
72. Sahler OJZ. Psychological aspects of childhood cancer. Am J Dis Child. 1981;135(4):389. Available from: https://jamanetwork.com/journals/jamapediatrics/article-abstract/509868.
73. Van Dongen-Melman JE, Sanders-Woudstra JA. Psychosocial aspects of childhood cancer: a review of the literature. J Child Psychol Psychiatry. 1986;27(2):145–80.
74. Grissom S, Boles J, Bailey K, Cantrell K, Kennedy A, Sykes A, et al. Play-based procedural preparation and support intervention for cranial radiation. Support Care Cancer Off J Multinatl Assoc Support Care Cancer. 2016;24(6):2421–7.
75. Cantrell MA, Ruble K. Multidisciplinary care in pediatric oncology. J Multidiscip Healthc. 2011;4:171–81.
76. Balazy KE, Gutkin PM, Skinner L, von Eyben R, Fowler T, Pinkham DW, Rodriguez S, Maxim PG, Donaldson SS, Loo BW Jr, Bush K, Hiniker SM. Impact of Audiovisual-Assisted Therapeutic Ambience in Radiation Therapy (AVATAR) on anesthesia use, payer charges, and treatment time in pediatric patients. Pract Radiat Oncol. 2020;10(4):e272–9. https://doi.org/10.1016/j.prro.2019.12.009.
77. Gautam M, Tripathi A, Deshmukh D, Gaur M. Cognitive behavioral therapy for depression. Indian J Psychiatry. 2020;62(Suppl 2):S223–9.

78. Lewinsohn PM, Clarke GN. Psychosocial treatments for adolescent depression. Clin Psychol Rev. 1999;19(3):329–42.
79. Aldridge AA, Roesch SC. Coping and adjustment in children with cancer: a meta-analytic study. J Behav Med. 2007;30(2):115–29.
80. Lee M-Y, Mu P-F, Tsay S-F, Chou S-S, Chen Y-C, Wong T-T. Body image of children and adolescents with cancer: a metasynthesis on qualitative research findings. Nurs Health Sci. 2012;14(3):381–90.
81. Pendley JS, Dahlquist LM, Dreyer Z. Body image and psychosocial adjustment in adolescent cancer survivors1. J Pediatr Psychol. 1997;22(1):29–43.
82. Sukhodolsky DG, Smith SD, McCauley SA, Ibrahim K, Piasecka JB. Behavioral interventions for anger, irritability, and aggression in children and adolescents. J Child Adolesc Psychopharmacol. 2016;26(1):58–64.
83. What are ABA interventions for tantrums? [Internet]. Applied Behavior Analysis Programs Guide. [cited 2021 Nov 25]. Available from: https://www.appliedbehavioranalysisprograms.com/faq/interventions-for-tantrums/.
84. Roditi D, Robinson ME. The role of psychological interventions in the management of patients with chronic pain. Psychol Res BehavManag. 2011;4:41–9.
85. Sturgeon JA. Psychological therapies for the management of chronic pain. Psychol Res BehavManag. 2014;7:115–24.
86. American Psychological Association. (2013). Managing chronic pain: how psychologists can help with pain management. Available from: https://www.apa.org/topics/pain/management.
87. Carr E, Rosengarten L. Teenagers and young adults with cancer: an exploration of factors contributing to treatment adherence. J Pediatr Oncol Nurs Off J Assoc Pediatr Oncol Nurses. 2021;38(3):190–204.
88. Kini V, Ho PM. Interventions to improve medication adherence: a review. JAMA. 2018;320(23):2461–73.
89. Figueroa-Moseley C, Jean-Pierre P, Roscoe JA, Ryan JL, Kohli S, Palesh OG, et al. Behavioral interventions in treating anticipatory nausea and vomiting. J Natl Compr Cancer Netw. 2007;5(1):44–50.
90. Braden A, Rhee K, Peterson CB, Rydell SA, Zucker N, Boutelle K. Associations between child emotional eating and general parenting style, feeding practices, and parent psychopathology. Appetite. 2014;80:35–40. https://doi.org/10.1016/j.appet.2014.04.017.
91. Brinksma A, Sulkers E, IJpma I, Burgerhof JGM, Tissing WJE. Eating and feeding problems in children with cancer: prevalence, related factors, and consequences. Clin Nutr. 2020;39(10):3072–9.
92. Scaglioni S, Arrizza C, Vecchi F, Tedeschi S. Determinants of children's eating behavior. Am J Clin Nutr. 2011;94(suppl_6):2006S–11S.
93. Klosky JL, Tyc VL, Tong X, Srivastava DK, Kronenberg M, de Armendi AJ, et al. Predicting pediatric distress during radiation therapy procedures: the role of medical, psychosocial, and demographic factors. Pediatrics. 2007;119(5):e1159–66.
94. Ångström-Brännström C, Engvall G, Mullaney T, Nilsson K, Wickart-Johansson G, Svärd A-M, et al. Children undergoing radiotherapy: Swedish parents' experiences and suggestions for improvement. PLoS One. 2015;10(10):e0141086.
95. Alexander M. Managing patient stress in pediatric radiology. Radiol Technol. 2012 Aug;83(6):549–60.
96. Gårdling J, Törnqvist E, EdwinsonMånsson M, Hallström I. Parents' lived experiences during their children's radiotherapy. J Pediatr Oncol Nurs Off J Assoc Pediatr Oncol Nurses. 2017;34(2):140–7.
97. Holt DE, Hiniker SM, Kalapurakal JA, Breneman JC, Boik N, Cooper BT, et al. Improving the pediatric patient experience during radiation therapy—a Children's Oncology Group Study. Int J Radiat Oncol Biol Phys. 2020;108(3):e216.
98. Massie MJ, Holland J, Glass E. Delirium in terminally ill cancer patients. Am J Psychiatry. 1983 Aug;140(8):1048–50.
99. Hosie A, Davidson PM, Agar M, Sanderson CR, Phillips J. Delirium prevalence, incidence, and implications for screening in specialist palliative care inpatient settings: a systematic review. Palliat Med. 2013;27(6):486–98.
100. Lawlor PG, Gagnon B, Mancini IL, Pereira JL, Hanson J, Suarez-Almazor ME, et al. Occurrence, causes, and outcome of delirium in patients with advanced cancer: a prospective study. Arch Intern Med. 2000;160(6):786–94.
101. Seiler A, Schubert M, Hertler C, Schettle M, Blum D, Guckenberger M, et al. Predisposing and precipitating risk factors for delirium in palliative care patients. Palliat Support Care. 2020 Aug;18(4):437–46.
102. Maldonado JR. Delirium pathophysiology: an updated hypothesis of the etiology of acute brain failure. Int J Geriatr Psychiatry. 2018;33(11):1428–57.
103. Inouye SK, Westendorp RGJ, Saczynski JS. Delirium in elderly people. Lancet Lond Engl. 2014;383(9920):911–22.
104. Fong TG, Tulebaev SR, Inouye SK. Delirium in elderly adults: diagnosis, prevention and treatment. Nat Rev Neurol. 2009;5(4):210–20.
105. Laurila JV, Laakkonen M-L, Laurila JV, Timo SE, Tilvis RS, Reijo TS, et al. Predisposing and precipitating factors for delirium in a frail geriatric population. J Psychosom Res. 2008;65(3):249–54.
106. Meagher DJ, Moran M, Raju B, Gibbons D, Donnelly S, Saunders J, et al. Motor symptoms in 100 patients with delirium versus control subjects: comparison of subtyping methods. Psychosomatics. 2008;49(4):300–8.
107. DSM-5 [Internet]. [cited 2022 Jan 25]. Available from: https://www.psychiatry.org/psychiatrists/practice/dsm.
108. World Health Organization. The ICD-10 classification of mental and behavioural disorders: clinical descriptions and diagnostic guidelines. World Health

Organization; 1992. [cited 2022 Jan 25]. Available from: https://apps.who.int/iris/handle/10665/37958.
109. Shi Q, Warren L, Saposnik G, Macdermid JC. Confusion assessment method: a systematic review and meta-analysis of diagnostic accuracy. Neuropsychiatr Dis Treat. 2013;9:1359–70.
110. Grassi L, Caraceni A, Beltrami E, Borreani C, Zamorani M, Maltoni M, et al. Assessing delirium in cancer patients: the Italian versions of the Delirium Rating Scale and the Memorial Delirium Assessment Scale. J Pain Symptom Manag. 2001;21(1):59–68.
111. Breitbart W, Rosenfeld B, Roth A, Smith MJ, Cohen K, Passik S. The Memorial Delirium Assessment Scale. J Pain Symptom Manag. 1997;13(3):128–37.
112. Agar MR, Lawlor PG, Quinn S, Draper B, Caplan GA, Rowett D, et al. Efficacy of oral risperidone, haloperidol, or placebo for symptoms of delirium among patients in palliative care: a randomized clinical trial. JAMA Intern Med. 2017;177(1):34–42.
113. Ng KT, Shubash CJ, Chong JS. The effect of dexmedetomidine on delirium and agitation in patients in intensive care: systematic review and meta-analysis with trial sequential analysis. Anaesthesia. 2019;74(3):380–92.
114. Flükiger J, Hollinger A, Speich B, Meier V, Tontsch J, Zehnder T, et al. Dexmedetomidine in prevention and treatment of postoperative and intensive care unit delirium: a systematic review and meta-analysis. Ann Intensive Care. 2018;8(1):92.
115. Holland JC, Romano SJ, Heiligenstein JH, Tepner RG, Wilson MG. A controlled trial of fluoxetine and desipramine in depressed women with advanced cancer. Psychooncology. 1998;7(4):291–300.
116. Fisch MJ, Loehrer PJ, Kristeller J, Passik S, Jung S-H, Shen J, et al. Fluoxetine versus placebo in advanced cancer outpatients: a double-blinded trial of the Hoosier Oncology Group. J Clin Oncol Off J Am Soc Clin Oncol. 2003;21(10):1937–43.
117. Strong V, Waters R, Hibberd C, Murray G, Wall L, Walker J, et al. Management of depression for people with cancer (SMaRT oncology 1): a randomised trial. Lancet Lond Engl. 2008;372(9632):40–8.
118. Economos G, Lovell N, Johnston A, Higginson IJ. What is the evidence for mirtazapine in treating cancer-related symptomatology? A systematic review. Support Care Cancer Off J Multinatl Assoc Support Care Cancer. 2020;28(4):1597–606.
119. Kim S-W, Shin I-S, Kim J-M, Kim Y-C, Kim K-S, Kim K-M, et al. Effectiveness of mirtazapine for nausea and insomnia in cancer patients with depression. Psychiatry Clin Neurosci. 2008;62(1):75–83.
120. Rada G, Capurro D, Pantoja T, Corbalán J, Moreno G, Letelier LM, Vera C. Non-hormonal interventions for hot flushes in women with a history of breast cancer. Cochrane Database Syst Rev. 2010;2010(9):CD004923. https://doi.org/10.1002/14651858.CD004923.pub2.
121. Gallagher HC, Gallagher RM, Butler M, Buggy DJ, Henman MC. Venlafaxine for neuropathic pain in adults. Cochrane Database Syst Rev [Internet]. 2015;2015(8):CD011091. https://doi.org/10.1002/14651858.CD011091.pub2.
122. Andrade C. Breast cancer and antidepressant use. J Clin Psychiatry. 2012;73(9):e1156–7.
123. Irarrázaval OME, Gaete GL. Antidepressants agents in breast cancer patients using tamoxifen: review of basic and clinical evidence. Rev Med Chile. 2016;144(10):1326–35.
124. Pritchard KI. Do selective serotonin receptor inhibitor antidepressants reduce tamoxifen's effectiveness and increase the risk of death from breast cancer? Breast Cancer Res. 2010;12(Suppl 4):S18.
125. Mazzocato C, Stiefel F, Buclin T, Berney A. Psychopharmacology in supportive care of cancer: a review for the clinician. II. Neuroleptics. Support Care Cancer Off J Multinatl Assoc Support Care Cancer. 2000;8:89–97.
126. Narasimhan M, Bruce TO, Masand P. Review of olanzapine in the management of bipolar disorders. Neuropsychiatr Dis Treat. 2007;3(5):579–87.
127. Alexander J, Tharyan P, Adams C, John T, Mol C, Philip J. Rapid tranquillisation of violent or agitated patients in a psychiatric emergency setting. Pragmatic randomised trial of intramuscular lorazepam v. haloperidol plus promethazine. Br J Psychiatry J Ment Sci. 2004;185:63–9.
128. Walsh D, Davis M, Ripamonti C, Bruera E, Davies A, Molassiotis A. 2016 Updated MASCC/ESMO consensus recommendations: management of nausea and vomiting in advanced cancer. Support Care Cancer Off J Multinatl Assoc Support Care Cancer. 2017;25(1):333–40.
129. Sutherland A, Naessens K, Plugge E, Ware L, Head K, Burton MJ, et al. Olanzapine for the prevention and treatment of cancer-related nausea and vomiting in adults. Cochrane Database Syst Rev. 2018;9:CD012555.
130. Tobacco [Internet]. [cited 2022 Mar 11]. Available from: https://www.who.int/news-room/fact-sheets/detail/tobacco.
131. Stein KD, Syrjala KL, Andrykowski MA. Physical and psychological long-term and late effects of cancer. Cancer. 2008;112(11 Suppl):2577–92.
132. Sun H, Yang Y, Zhang J, Liu T, Wang H, Garg S, et al. Fear of cancer recurrence, anxiety and depressive symptoms in adolescent and young adult cancer patients. Neuropsychiatr Dis Treat. 2019;15:857–65.
133. Gracia CR. Reproductive health after cancer. Cancer Treat Res. 2010;156:3–9.
134. Ganz PA. Cancer survivorship: today and tomorrow. Springer Science & Business Media; 2007. p. 316.
135. From cancer patient to cancer survivor: lost in transition | The National Academies Press [Internet]. [cited 2022 Apr 7]. Available from: https://nap.nationalacademies.org/catalog/11468/from-cancer-patient-to-cancer-survivor-lost-in-transition.
136. Nekhlyudov L, Mollica MA, Jacobsen PB, Mayer DK, Shulman LN, Geiger AM. Developing a quality of cancer survivorship care framework: implications

for clinical care, research, and policy. J Natl Cancer Inst. 2019;111(11):1120–30.
137. Chan RJ, Crawford-Williams F, Crichton M, Joseph R, Hart NH, Milley K, et al. Effectiveness and implementation of models of cancer survivorship care: an overview of systematic reviews. J Cancer Surviv. 2023;17(1):197–221. https://doi.org/10.1007/s11764-021-01128-1.
138. Ferlay J, Colombet M, Soerjomataram I, Mathers C, Parkin DM, Piñeros M, et al. Estimating the global cancer incidence and mortality in 2018: GLOBOCAN sources and methods. Int J Cancer. 2019;144(8):1941–53.
139. Salins N, Gursahani R, Mathur R, Iyer S, Macaden S, Simha N, et al. Definition of terms used in limitation of treatment and providing palliative Care at the end of life: the Indian Council of Medical Research Commission report. Indian J Crit Care Med Peer-Rev Off Publ Indian Soc Crit Care Med. 2018;22(4):249–62.
140. Crawford GB, Dzierżanowski T, Hauser K, Larkin P, Luque-Blanco AI, Murphy I, et al. Care of the adult cancer patient at the end of life: ESMO Clinical Practice guidelines. ESMO Open. 2021;6(4):100225.
141. Downey L, Curtis JR, Lafferty WE, Herting JR, Engelberg RA. The Quality of Dying and Death Questionnaire (QODD): empirical domains and theoretical perspectives. J Pain Symptom Manag. 2010;39(1):9–22.
142. Steinhauser KE, Clipp EC, McNeilly M, Christakis NA, McIntyre LM, Tulsky JA. In search of a good death: observations of patients, families, and providers. Ann Intern Med. 2000;132(10):825–32.
143. Ellershaw J, Ward C. Care of the dying patient: the last hours or days of life. BMJ. 2003;326(7379):30–4.
144. Glare PA, Sinclair CT. Palliative medicine review: prognostication. J Palliat Med. 2008;11(1):84–103.
145. Simmons CPL, McMillan DC, McWilliams K, Sande TA, Fearon KC, Tuck S, et al. Prognostic tools in patients with advanced cancer: a systematic review. J Pain Symptom Manag. 2017;53(5):962–970.e10.
146. Gilligan T, Coyle N, Frankel RM, Berry DL, Bohlke K, Epstein RM, et al. Patient-clinician communication: American Society of Clinical Oncology Consensus guideline. J Clin Oncol Off J Am Soc Clin Oncol. 2017;35(31):3618–32.
147. Zhang B, Nilsson ME, Prigerson HG. Factors important to patients' quality of life at the end of life. Arch Intern Med. 2012;172(15):1133–42.
148. Lacey J, Sanderson C. The oncologist's role in care of the dying cancer patient. Cancer J Sudbury Mass. 2010;16(5):532–41.
149. Cherny NI. Sedation for the care of patients with advanced cancer. Nat Clin Pract Oncol. 2006;3(9):492–500.
150. Delgado-Guay MO, Chisholm G, Williams J, Frisbee-Hume S, Ferguson AO, Bruera E. Frequency, intensity, and correlates of spiritual pain in advanced cancer patients assessed in a supportive/palliative care clinic. Palliat Support Care. 2016;14(4):341–8.
151. Mols F, Husson O, Roukema J-A, van de Poll-Franse LV. Depressive symptoms are a risk factor for all-cause mortality: results from a prospective population-based study among 3,080 cancer survivors from the PROFILES registry. J Cancer Surviv Res Pract. 2013;7(3):484–92.
152. Aoun SM, Breen LJ, O'Connor M, Rumbold B, Nordstrom C. A public health approach to bereavement support services in palliative care. Aust N Z J Public Health. 2012;36(1):14–6.
153. Becerra IG, Ruíz-Castañeda D, Flujas-Contreras JM, Román AS, de Salazar Arboleas AM, López MD. Cognitive behavioral treatment for improving distress in pediatric oncology: a pilot study. Rev Argent Clín Psicol. 2019;28(5):869–83. https://doi.org/10.24205/03276716.2019.1155.

Online Links

https://www.bing.com/videos/search?q=pscho+oncology%27&docid=603516404879131104&mid=ED1D2603080B1D91D1EAED1D2603080B1D91D1EA&view=detail&FORM=VIRE
http://artofmedicine.in/psycho-oncology/
https://www.youtube.com/watch?v=CfVmjK2OG90
https://www.youtube.com/watch?v=GLFRaxDwLGI

Artificial Intelligence in Radiation Oncology

23

Anusheel Munshi, Indranil Mallick, Rohit Chauhan, and Gaganpreet Singh

23.1 Introduction

Humanity has long been haunted by whether intelligence other than human one can be used for any purpose in our everyday lives. Alan Turing famously developed the "Turing machine" that was instrumental in breaking the German enigma code. Almost instantly, Artificial Intelligence (AI) was born. So, in the truest sense, the use of AI is as old as the use of first computers. AI involves developing and using complex computer algorithms. The intent is to perform and ease tasks that typically require human intelligence. AI includes but is not limited to decision-making, problem solving, visual perception and pattern recognition.

Even before the world became aware of the term Artificial Intelligence, the latter's use had silently entered our lives. Engineering, statistics, stock market, robotics and medicine have witnessed the use of AI. Driverless cars, driverless metros, software-based share market buying and selling, news writing, Alexa and Siri (used commonly with Google and Apple platforms) are examples of AI's use in our lives (Table 23.1).

Table 23.1 Current everyday applications of AI

Speech recognition	Customer service	Computer vision	Recommendation engines	Automated stock trading
Natural language processing (NLP) to process human speech	Online chatbots replacing human agents	Derivation of meaningful info from digital images/videos, etc.	Uses past consumption behavior data	Designed to optimize stock portfolios
Used in mobile devices for voice search, e.g., Siri, Alexa	Answer FAQs, provide personalized advice, e.g., suggesting sizes for users	Applications in social media, radiology imaging in healthcare and self-driving cars	AI algorithms help discover data trends that can be used to develop more effective cross-selling strategies	AI-driven trading platforms make millions of trades per day without human intervention

AI Artificial intelligence, *FAQ* Frequently asked questions

A. Munshi (✉)
Department of Radiation Oncology, IOCI, Fortis Noida, Uttar Pradesh, India

I. Mallick
Department of Radiation Oncology, Tata Medical Center, Kolkata, West Bengal, India

R. Chauhan
Department of Physics, GLA University, Mathura, Uttar Pradesh, India

G. Singh
Department of Radiotherapy, PGIMER, Chandigarh, India

Department of Medical Physics, Apollo Proton Cancer Centre, Chennai, Tamil Nadu, India

A. Munshi et al. (eds.), *Radiation Oncology – Principles, Precepts and Practice*,
https://doi.org/10.1007/978-981-97-8389-2_23

Being a specialty highly dependent on advanced software and employing high-tech hardware, Radiation oncology has posited itself for the use of AI in all the links of the process chain.

23.2 Methods Related to AI and Their Evolution

Broadly, AI refers to an umbrella term for using a computer to perform tasks that typically require human intelligence and thought. A subset of AI that allows the machine to learn from their own experience is called Machine Learning (ML). ML typically refers to a set of computational algorithms. These use input data collection and produce the desired output without being explicitly programmed to do so. Table 23.2 lists the broad features of common ML algorithms.

ML is separated into two broad categories. These include supervised and unsupervised learning.

Supervised learning refers to situations where the input data are structured and labeled with known output which can be in the form of a number or a class (group). The algorithm strives to learn a rule to produce the results that match the output as closely as possible. In the healthcare domain, supervised learning can be used to predict outcomes (e.g., dose coverage, response to treatment, hospitalization, etc.), and classify presence or absence of diagnostic criteria on radiological, pathological or genomic data. As an example, it is currently being used in radiology to detect and characterize suspicious lesions in mammograms and lung cancer imaging [1]. In unsupervised learning, using only the structured input data, the algorithm finds patterns or groups

Table 23.2 Common machine learning algorithms

S. no	Item	Description
1	Linear regression	Relationship is established between independent and dependent variables by fitting them to a line. This line is known as the regression line and represented by a linear eq. $Y = a \times X + b$
2	Logistic regression	Used to estimate discrete values (usually binary values like 0/1) from a set of independent variables. Helps predict the probability of an event by fitting data to a logit function. Used commonly for classification problems
3	Naïve Bayes Algorithm	A Naive Bayes classifier assumes that the presence of a particular feature in a class is unrelated to the presence of any other feature. A commonly used simple classifier
4	Decision tree	Supervised learning algorithm that is used for classification problems. In this algorithm, we split the population into two or more homogenous sets based on the most significant attributes/independent variables.
5	Random forest algorithm	A collective of decision tree is called a Random Forest. To classify a new object based on its attributes, each tree is classified, and the tree "votes" for that class. The forest chooses the classification having the most votes. Can be used for classification and regression problems
6	Gradient boosting algorithm	A collection of decision trees built in a forward stage-wise manner that combines the results of the individual trees along the way. Can be used for classification or regression problems
7	SVM (support vector machine) algorithm	In SVM algorithm, we plot raw data as points in an n-dimensional space (n = no. of features you have). To classify the individual points a hyperplane is calculated in high-dimensional space that gives the best separation between the groups. A commonly used algorithm for classification problems
8	KNN algorithm	This algorithm can be applied to both classification and regression problems. It stores all available cases and classifies any new cases by taking a majority vote of its k neighbors. The case is then assigned to the class with which it has the most in common
9	K-means clustering	This is an unsupervised learning algorithm. Data sets are classified into a particular number of clusters in such a way that all the data points within a cluster are homogenous and heterogeneous from the data in other clusters. Used to create hitherto unknown classes/groups from a set of data

Table 23.2 (continued)

S. no	Item	Description
10	Dimensionality reduction algorithm	In situations where there are many input features, building a model can become computationally expensive. Each feature mathematically represents a dimension in space. Dimensionality reduction techniques are unsupervised learning techniques that mathematically combine the values of a large number of features to a smaller number of features for use in model building or visualization of data. Common examples are principal component analysis (PCA) and t-distributed stochastic neighbor embedding (t-SNE)
11	Artificial neural networks	Artificial neural networks (ANN) are a collection of connected nodes that simulate the connected neurons in the brain. Typically, the nodes are arranged in layers. There is a) An input layer of nodes that take in the input features b) An output layer that generates the output in the form of a value or a class c) One or more hidden layers between the input and output layers The nodes in a layer transmit a value to the next layer that is generated as a non-linear function of the inputs received from the previous layer. The values are modified on an ongoing basis based on feedback from the accuracy of the results generated through the ANN via a process called backpropagation which continually adjusts the weights of the functions in each layer until the best possible result is achieved.

(clusters) that can represent different classes. As an example, genomic data may be used to develop a new molecular classification.

In the health sciences, we often come across a vast volume of data, where only a fraction is labeled or annotated. A technique called semi-supervised learning, which falls between supervised and unsupervised learning, can be used under these circumstances.

The complexity of the machine-learning task depends greatly on the complexity of the input data that is used to build the model. While sometimes the input data consists of a few to a few thousand variables (also called features), at other times we seek to use whole images, genomes or text passages as input to predict classes or outcomes or perform segmentation. A 256 × 256 pixel sectional image has ~65,000 input features and 100 such slices to make a 3D volume will have more than 6.5 million features. These parameters have a spatial relationship (for images) or a temporal relationship (for text passages or music) with each other. In these scenarios, simple ML algorithms are not sufficient. A method is required to create a "representation" or a simpler feature form that can nonetheless preserve the features and the relationships to use in building a model. The sub-field of ML that uses representation learning with artificial neural networks to build models on such complex input data is called Deep Learning (DL) (Fig. 23.1). The neural network architecture that DL most often employs is known as a convolutional neural network (CNN).

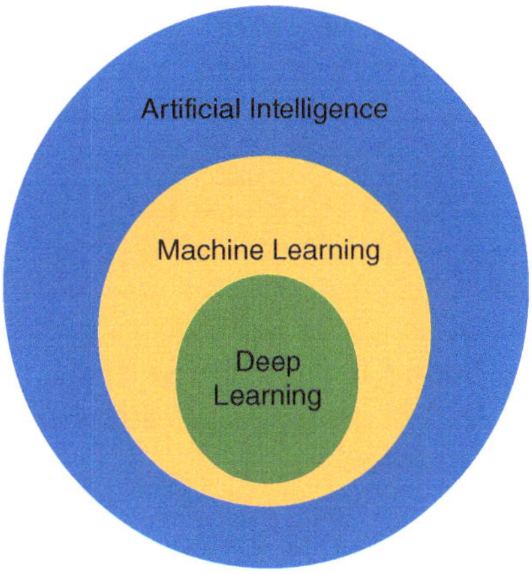

Fig. 23.1 Artificial Intelligence, Machine Learning and Deep Learning

CNNs use techniques called convolution and pooling due derive high-level features from a complex feature matrix, reduce the feature dimension and then perform the function of an ANN (Fig. 23.2). The structure and connections of a DL network is known as its architecture. The architecture of DL networks is based on the type

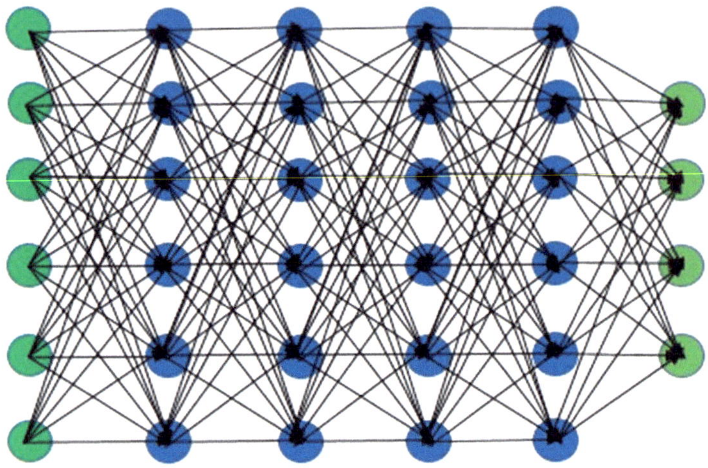

Fig. 23.2 Artificial neural network

of problem it is trying to solve. For example, the U-Net architecture is used commonly for medical image segmentation, while Recurrent Neural Networks (RNNs) or Transformers are commonly used for problems using text data where the sequence of words and sentences is important. Training effective DL models often require substantial computer resources.

23.3 Broad Applications of AI in Oncology and Radiation Oncology

AI has made deep inroads in oncology. This includes oncoradiology, medical oncology, surgical oncology and radiation oncology. Radiation therapy is a crucial cornerstone of cancer treatment and is indicated for 50–60% of all cancer patients. Over the past few decades, radiation therapy has become increasingly complex. This is primarily due to technological advances, resulting in a near-complete reliance on human-machine interactions, including software and hardware.

The use of AI in oncology has already assumed multifaceted proportions. It includes cancer screening, early cancer detection and prognostication. It includes screening where AI can analyze the patients' characteristics and categorize the population into low- and high-risk cohorts. This can enable appropriate modality and frequency of screening. Using AI, the interpretation of mammograms has become 30 times faster than humans and with greater accuracy [2]. An additional gain with AI is to shorten the oncologist's and the diagnostician's time and effort for a particular task.

AI has shown promise in integrating utilization of the traditional parameters like the stage, histopathology characteristics, genetic characteristics of the tumor with newer parameters such as gene scores and help predict appropriate treatment modality and provide a more accurate prognosis [3]. AI has immense potential in radiation oncology, and this shall be dealt in upcoming sections.

23.4 AI in Radiation Oncology

AI has the potential to impact each step of the radiation oncology workflow (Fig. 23.3).

23.4.1 Patient Selection for Radiation Therapy

AI has the potential to determine if a patient is likely to benefit from a particular treatment. Predictive modeling involves learning from a case series of patients where the tumor and

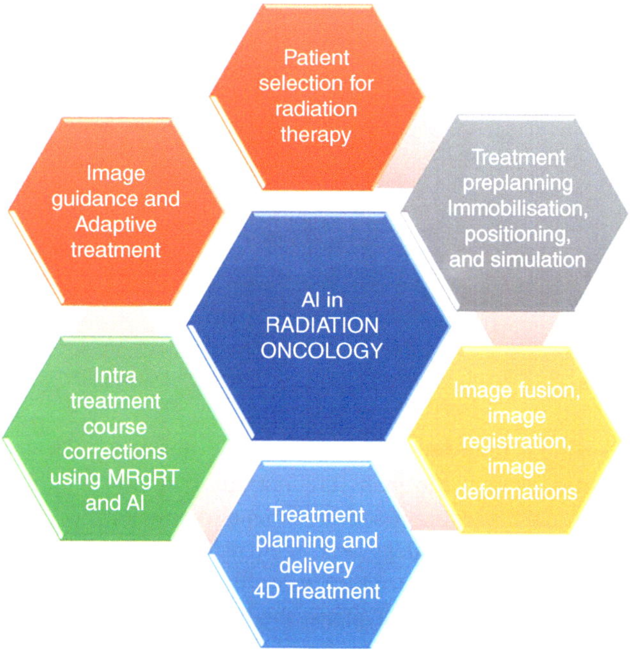

Fig. 23.3 Role of AI in radiation therapy workflow

patient characteristics, treatments offered and outcomes are available and then generating a model that determines the difference in outcomes on the basis of a particular treatment. Traditionally nomograms have been developed to determine such benefits. One example is a nomogram to identify the survival outcomes after stereotactic radiosurgery in brain metastases from non-small cell lung cancer [4]. Recently ML and DL tools are being explored to develop more accurate predictive models or decision-support tools for radiotherapy.

The choice of radiotherapy versus chemoradiotherapy was explored using AI-based tools in the adjuvant therapy of intermediate risk head and neck cancer [5]. Survival outcomes in patients in the validation cohort who received from AI model-recommended treatments were better than those who did not receive AI model-recommended treatments. AI-based models using treatment plan features have been tested to predict patients who are likely to benefit from intensity modulated proton therapy (IMPT) planning vs. photon therapy. One such model could identify those who will not benefit from IMPT and thereby reduce wasted time spent on manual IMPT planning [6].

23.4.2 Treatment Preplanning/Role of AI in Determining the Most Appropriate Machine/Technology (IM)

AI has the potential to help clinicians and planners decide the optimal treatment strategy in terms of the radiation beam and modality (Fig. 23.4). This could be based on the volumes of the target, organs at risk (OARs) and their overlap or anatomical proximity. These features could be used to predict the dose or range of doses that might be delivered to OARs. It could also be based on other clinical factors such as age, stage and prior treatment.

There has been limited work that is currently published on this. A web-based dashboard for library based estimation of doses to the PTV, rectum and bladder has been reported for prostate cancer. The goal is to generate individualized dose-constraints for hypofractionated prostate IMRT, to achieve the best possible plans, rather than stick to generic constraints from QUANTEC [7].

Deciding on the use of advanced beam strategies, e.g., intensity-modulated proton therapy (IMPT), can be challenging. Proton or heavy ion

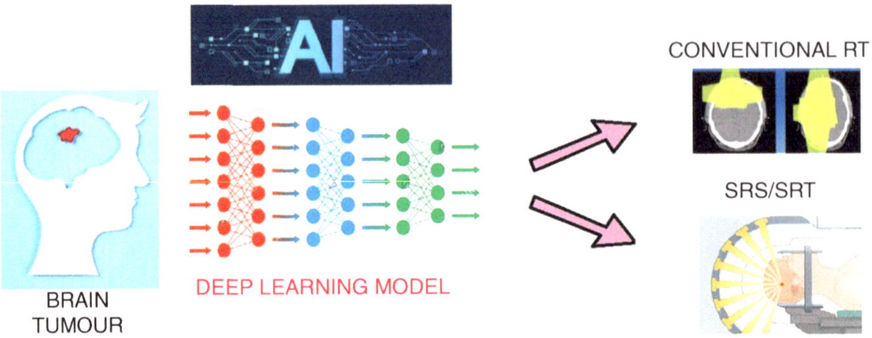

Fig. 23.4 Role of AI in deciding optimal treatment strategy

planning is time consuming and there is time and effort wasted if the plans are not substantially better than photon based treatment. A model based selection strategy has been reported for head and neck cancer where using the structure volumes and the VMAT plans, the improvement with IMPT plans can be predicted without having to go through a treatment planning process [8]. This approach has also been validated using an independent patient cohort.

23.4.3 Immobilization, Positioning and Simulation

Before treatment planning, the process of simulation takes place. In most cases, an immobilization mask or a vacuum cushion is prepared to immobilize the patient. The process can be of varying complexities with the sole objective of appropriate patient immobilization. A common concern is to immobilize in a manner that avoids subsequent gantry collisions, allows free placement of beams from multiple angles. A potential role of AI would be to offer solutions at the immobilization step. Pre-emptive advice regarding arm up versus arms down, neutral neck position versus extended neck, need for shoulder depressor use are some examples where AI can guide at the time of immobilization. This is similar to how AI has been used to expedite treatment planning based on a patient's anatomy [9].

23.4.4 Image Fusion, Image Registration, and Image Deformations

Mono-modality or multimodality rigid/non-rigid (deformable) image registration techniques are frequently used for the accurate identification of the target and OARs before, during and after the radiotherapy treatment. Mono-modality image registration involves registration of the images belonging to the same imaging modality, viz., CT (computed tomography)—CT and MRI (magnetic resonance imaging)—MRI whereas multi-modality image registration involves CT-MRI, MRI-TRUS (trans-rectal ultrasonography), CT-CBCT and X-ray—CBCT imaging modalities. Most of these registration techniques utilize area-based, feature-based or hybrid mode of image registration methods. To mitigate the potential risk of inter-observer delineation and registration errors related to target and OARs definitions, new methods have been developed for image segmentation, rigid and non-rigid image registration. Literature shows AI based rigid and non-rigid registration models superseding the manual task of image registration. The task of performing delineation of the target and OARs for different clinical sites is markedly seen superior with the use of DL-based models as compared to the manual delineation. DL based models have proven itself as the state-of-the-art techniques for image segmentation tasks [10].

The most popular DL architectures for image registration and segmentation problems are AlexNet, VGG, ResNet, DenseNet, SqueezeNet and GoogleNet [11–16]. All these architectures are based upon any of these methods, viz., encoder-decoder CNN like the U-Net, Spatial Transformer Network (STN) or Generative adversarial Network (GAN).

To quantify the accuracy of these algorithms, several metrics have been discussed in various publications. Few of the metrics that compare the ground truth image (X) with the predicted image (Y) are:

Dice similarity coefficient (DSC):

$$DSC(X,Y): \frac{2|X \cap Y|}{|X|+|Y|}$$

DSC is twice the area of overlap between the predicted and the ground truth image divided by the total number of pixels in both image. Its value ranges from 0 to 1. The value of 1 represents the greatest similarity between both of the images.

Jaccard Index or Jaccard Similarity Coefficient (JSC):

$$JSC(X,Y): \frac{|X \cap Y|}{|X \cup Y|}$$

JSC is the area of overlap between the predicted and the ground truth image divided by the area of union between the predicted (Y) and the ground truth image (X). Its value ranges from 0 to 1. The value of 1 represents the maximum overlap between two images.

Structural Similarity Index Measure (SSIM). It measures the similarity between two given images. It extracts 3 key features from an image: luminance, contrast and structure. The combined formula that takes into account all the 3 features is given by:

$$SSIM(x,y): \frac{(2\mu_X\mu_Y + C_1)(2\Sigma_{XY} + C_2)}{(\mu_X^2 + \mu_Y^2 + C_1)(\Sigma_X^2 + \Sigma_Y^2 + C_2)}$$

μ_X and μ_Y are average of the images X and Y respectively and also a measure of luminance of the images, Σ_X and Σ_Y are the variances and also a measure of contrast and Σ_{XY} is the covariance. C_1 and C_2 are regularization constants for the luminance and contrast respectively. Its value ranges from −1 to 1.

To evaluate the performance of these algorithms, metrics parameters, viz., accuracy, sensitivity, specificity, precision, f-measure, and g-mean are commonly used [17–19].

23.4.5 Treatment Planning

Treatment planning requires the planner's skills and computer power to find the solution of the complex mathematical equations used in the different optimization and dose calculation algorithms. Treatment planning involves a number of steps that start with the simulation and end with the completion of the treatment delivery of the patient. In the present era, Treatment planning techniques, intensity-modulated radiation therapy (IMRT), image-guided radiation therapy (IGRT), adaptive radiation therapy (ART), etc., utilize complex computation methods (optimization and dose calculation algorithms) to calculate the required dose distribution which require multiple iterations. This entire process requires the continuous involvement of medical physicists (MPs) and radiation oncologists (ROs) for many hours or sometimes days to get good quality clinical treatment plans. The quality of treatment plans is further evaluated with the help of different evaluation tools that are based on either DVH metric or slice by slice evaluation of the 3D dose distribution. Minimize the planning time and increase the consistency of treatment plans; several approaches have been developed for automatic treatment planning (ATP) [20]. These ATP solutions are based upon following.

23.4.6 Automated Rule Implementation and Reasoning (ARIR)

In this solution, a well-defined logical reasoning scheme follows the steps of planners. It reduces the planner operations in the plan generation process. These solutions are more oriented toward the scripting or template-based functionality in the TPS. The commercial TPS provides application programming interface (API) scripting or template-based support to perform ATP, viz., Eclipse Scripting API (ESAPI) in Eclipse TPS (Varian Medical System, Palo Alto, California, USA), Autoplanning in Pinnacle TPS (Philips Radiation Oncology Systems, Fitchburg, Wisconsin), and scripting solution in RayStation (RaySearch Laboratories, Stockholm, Sweden).

23.4.6.1 Modeling of Prior Knowledge in Clinical Practice

In this solution, prior cases of similar treatment plans are fed into the system to improve the efficiency and quality of the treatment plans. All the plan parameters: beam arrangements, delineated organ geometry and dose volume histogram (DVH) objectives for different OARs, target volumes, etc. were provided to the model to learn from the previous best treatment plans. Table 23.3 shows the brief summary of various approach used for training, testing and validation of the radiotherapy models used for the treatment planning.

23.4.6.2 Multi-Criteria Optimization (MCO)

In this solution, instead of manually iterate through the various steps of the optimization to achieve best possible plan with better OAR sparing without compromising the target coverage [33]. The best solution is achieved with the help of surfing the N-dimensional hyper-surface (Pareto Surface) with the optimal plans following the various dosimetric criteria. Nowadays optimization engine coming up with the MCO features, most of the commercial TPS: Eclipse TPS (Varian Medical System, Palo Alto, California, USA), RayStationTPS (RaySearch Laboratories, Stockholm, Sweden) and Monaco TPS (Elekta, Inc., Stockholm, Sweden). MCO algorithm for IMRT and VMAT developed by RaySearch Laboratories is the most mature and advance algorithm that provides a real-time assessment of the tradeoff between different clinical goals [33, 34].

These ATP solutions have its own advantages and limitations; however, knowledge-based models have tremendous potential for its utility in the clinics. The DL architectures like VGG, U-net, GAN, etc. are mature enough to predict or mimic

Table 23.3 Summary of studies based on the knowledge-based planning

Authors	Site	No. of cases	Planning approach
Yuan et al. [21]	Prostate H&N	64 82	DVH, DTH
Munter et al. [22]	Brain	22	Probabilistic interpretation of DVH
Ahmed [23]	Head and neck	10	Feasibility DVH
Zawadzka [24]	Lung	16	Minimize MLD
Tran [25]	Liver	21	DVH, statistical voxel dose learning
Sheng [26]	Pelvis (prostate)	121	Outlier identification
Zhu [27]	Prostate	198	DVH, DTH
Good [28]	Prostate	132	Similarity matric and deformable registration
Yang [29]	Prostate	21	OVH
Scheibmann [30]	Prostate	83	Mean distance between targets
Zhang [31]	Head and neck	268	DVH, similarity metric, data augmentation
Petit [32]	Lung	89	Dose–distance relationship

OVH Overlap volume histogram, *DTH* distance to target histogram, *MLD* mean lung dose the diagnostic images, contours and dose distribution of the radiotherapy treatment plans. The evaluation of the predicted results and judgment is still remaining in the realm of human intelligence and insight.

23.4.7 Treatment Delivery

Treatment delivery in radiotherapy involves various hardware and software parts that have to be undergone through tighter tolerance criteria before its clinical use. The success of the any radiotherapy treatment delivery depends on the accuracy and performance of entire chain which includes system for data acquisition, MLC movements, gantry rotation, dose rate, image acquisition in different modes (radiography, fluoroscopy and integral film), etc. AI based system have potential to monitor the performance of such complex system over a long period of time and identify the potential errors in the treatment delivery. Many studies have been conducted to explore the role of AI in the quality assurance tests associated with the treatment delivery. Treatment delivery techniques are very complex depending upon the configuration of the machine and technology used for the treatment. Real-time tumor tracking or surface tracking techniques utilize multiple modes of tracking (fiducials, surface, infrared reflector blocks etc.) and monitor the treatment delivery throughout the treatment process. Manual monitoring of dose delivery and accuracy at the time of patient treatment is very tedious task and cannot be performed easily. Table 23.4 shows the brief results of prediction of various factor/parameters that helps in determining the accurate treatment delivery of the complex treatment plan. These various parameters/factors directly or indirectly predict about the machine stability and radiation delivery that assists in taking decisions by the dosimetrists, medical physicists, and radiation oncologist before, during and after the treatment for the next plan of action.

Table 23.4 Summary of publications showing the use of different models to predict the treatment delivery accuracy

Authors	Model	Data used	Prediction
Carlson et al. [35]	Regression, random forest, cubist	74 VMAT plans	MLC leaf positions during delivery
Valdes et al. [36]	Poisson regression	498 IMRT plans	Gamma passing rates
Valdes et al. [37]	Poisson regression	203 IMRT beams	Gamma passing rates
Li and Chan [38]	ANN time series	5 years of daily Linac QA	Linac beam symmetry accurately
Sun et al. [39]	Random Forest, XGBoost, Cubist	1754 Proton fields	Prediction of for double-scatter proton machines
Interian et al. [40]	CNN	498 IMRT plans	Gamma passing rates
Tomori et al. [41, 42]	CNN	60 IMRT plans	Gamma passing rates
Lam et al. [43]	AdaBoost, Random Forest, XGBoost	1497 IMRT beams	Gamma passing rates
Nyflot et al. [44]	CNN	186 IMRT beams	MLC errors
Granville et al. [45]	SVM	1620 VMAT beams	Patient-specific QA using treatment plan complexity and linac performance metrics
Ono et al. [46]	Regression tree, multiple regression, neural network	600 VMAT plans	Dosimetric accuracy using dose difference and gamma index
Li et al. [47]	Poisson Lasso and random forest	255 VMAT beams	Gamma passing rates

Table 23.4 (continued)

Authors	Model	Data used	Prediction
El Naqa et al. [48]	Support vector data description, clustering	119 Images from 8 Linacs	Visualization and failure monitoring as per TG 142 protocol
Wang et al. [49]	Hybrid Model ACLR	576 VMAT beams	Gamma passing rates
Wall and Fontenot [50]	Linear regression, SVM, tree-based, ANN	500 VMAT plans	Gamma passing rates
Hirashima et al. [51]	Hybrid model XGBoost	1255 VMAT plans	Gamma passing rates
Grewal et al. [52]	Gaussian processes, shallow NN	4231 Proton fields	The output factor and Monitor units for uniform SPT
Osman et al. [53]	ANN	400 Machine delivery log files	Deviation in MLC leaf positions
Chuang et al. [54]	Boosted tree outperformed LR	116 IMRT plans, 125 VMAT plans	MLC discrepancies during delivery
Zhao et al. [55]	Multivariate regression (Ridge)	43 Truebeam PDD, Profiles	Modeling of beam data linac commissioning

23.4.8 Image Guidance and Adaptive Treatment

The goal of radiation therapy is to deliver a precise and accurate dose to the tumor while sparing normal tissues. IGRT refers to the use of imaging during radiation therapy to increase dose delivery precision and accuracy. In order to set up the patient in the precise position as during the simulation process, IGRT requires imaging modalities to be connected with therapeutic machines. Cone beam CT (CBCT), electronic portal imaging device (EPID), wall-mounted x-ray devices, 3D-body surface mapping, and MRI are now often used for image guidance.

The addition of AI to IGRT systems can increase the precision and accuracy achieved by IGRT alone. AI-assisted IGRT approaches offer the ability to track tumor movement, reduce treatment uncertainty and increase treatment precision.

Using single view projection radiographs, Shen et al. developed a DL approach to provide more accurate tumor localization as well as measurement of tumor changes in shape, size and position [56]. Isaksson et al. demonstrated the use of a neural network to predict tumor position without using external surrogate indicators [57]. Using 2D kV X-ray pictures acquired from the on-board imager (OBI) system, Zhao et al. present a DL-based technique that eliminates the requirement for marker insertion [58]. Foote et al. use a 2D fluoroscopic projection in real-time to create a CNN model and a patient-specific motion subspace to predict anatomical positions [59]. Satoshi et al. devised a deep convolution neural network (DCNN) strategy for improving CBCT image quality with significant improvements in spatial resolution and signal to noise ratio (PSNR) to improve treatment uncertainty [60]. For more exact online matching and tumor position verification, Matteo Rossi et al. used a convolution neural network (CNN) to construct synthetic CT from CBCT images [61].

Adaptive radiotherapy refers to the modification of a treatment plan in response to significant anatomical changes in the patient during treatment. Adaptive planning is required for a variety of parameters, including tumor progression or shrinkage, patient weight loss and internal organ motions. Radiation-induced toxicities in patients are usually caused by these variables. For adaptive planning, radiation oncologists usually base their decisions on these considerations.

Clinical oncologists can benefit from AI when making decisions for adaptive planning. Guidi et al. showed that a support vector machine tool can identify patients who need adaptive planning and estimate the month's week for adaptive treatment [62]. Clinical oncologists can use this tool to assist them make better judgments, reduce treatment side effects and improve therapeutic gain. Oncologists might decide on dose escalation depending on tumor reduction or patient response to treatment, which can provide additional therapeutic gain with minimal normal tissue toxicities. Tseng HH et al. developed an automated dose escalation approach for non-small cell lung cancer (NSCLC) using a deep reinforcement learning based model [63].

Anatomical changes often occur during therapy, resulting in sometimes under dosed tumors and overdosed normal tissues. Varfalvy et al. employed a Hidden Markov model (HMM) to detect anatomical changes in patients after radiation [64]. Another step toward adaptive planning is AI's role in increasing image quality of CBCT scans for radiation treatment. Harms et al. demonstrated that a GAN-based model can be used to reduce artifacts in CBCT images [65]. As a result, CBCT can be used for dosage computation instead of CT scans in busy radiation departments.

23.4.9 Intratreatment Course Corrections Using MRI-Guided Radiotherapy (MRgRT) and AI

Radiation therapy is not far behind in terms of technological advancements in the realm of healthcare. MR-guided radiation (MRgRT) is a novel addition to technology that combines Linac with on-board magnetic resonance (MR) imaging equipment to boost therapeutic gain even more. MR images are traditionally used in conjunction with CT scans to improve soft tissue contrast and appropriately delineate the tumor and OARs. On-board MR imaging improves soft tissue contrast for tumor delineation while also lowering treatment setup concerns. With the use of continuous MR-cine pictures, MRgRT allows for intra-fraction tumor motion monitoring. It also permits inter-fraction online adaptive planning by taking the daily anatomy of the patient into consideration. At the moment, there are two commercially available systems: MRIdian Linac (ViewRay, Mountain View, California, USA) and Unity (Elekta, Stockholm, Sweden). Many of the treatment processes used so far in MRgRT are manual, time-consuming and prone to human mistake. In MR guided radiotherapy, AI can play a crucial role in streamlining the clinical workflow in MRgRT. Role of AI in synthetic CT (sCT) generation, auto-segmentation, 3D motion management, predictive modeling and auto planning was investigated by many authors. Matteo Maspero et al. used a conditional generative adversarial network (cGAN) for generating a synthetic CT from MRI which can eliminate the idea of using CT acquisition from MRgRT workflow [66]. It can simplify the workflow and can also open the possibility to treat palliative patients in the same simulation session. Yabo Fu et al. demonstrated the efficiency of CNN based AI model in reducing delineation time for online adaptive treatment [67]. In their paper, Paganelli et al. hypothesized that AI might be used to infer the 3D motion trajectory from multi-slice 2D photos, potentially paving the way for real-time volumetric motion management solutions [68]. Isaksson et al. used AI models based on neural networks to predict lung tumor motion during respiration [57]. Terpstra et al. proposed DL-based models to speed up the image reconstruction process for real-time MR acquisitions, which would eliminate the most significant lag in MRgRT treatment delivery [69].

23.4.10 4D Treatment

Tumor motion related to breathing, heartbeats and swallowing is a recognized phenomenon. This motion depends on the patient's respiratory physiology, size, tumor location and the immobilization device used [70]. Several 4D treatment techniques have evolved to counter or account for the tumor motion. Such treatments using 4-D technology can have several benefits, including reducing planning target volume (PTV) size, ensuring the planned dose and delivered dose concurrence, reducing normal tissue dose and the possibility of further dose escalation [71]. Predictive modeling of tumor motion is complex in view of variables such as magnitude, amplitude and frequency and the movement of organs relative to each other. To improve the efficiency and accuracy of the 4D workflow, AI can account for these diverse variables. At the simulation stage, based on the breathing pattern, PFT parameters, information from the diagnostic chest scan, AI could potentially suggest the most appropriate modality for the patient (tracking, gating, breath-hold, MIP, etc.). Further, AI can generate patient-specific dynamic motion-management models that adapt to changes in motion patterns. Currently, most work has focused on using an external surrogate positional marker to predict respiratory motion.

One limitation of tumor tracking is the lag time of a few microseconds between accessing the movement to finally correcting for it (about 0.09 s). This lag is usually corrected with the use of predictive software with an estimated accuracy of about 80% [72].Complex AI algorithms can automatically adjust for complex breathing patterns in real-time. Recently, some researchers have been working on a new predictor for intra- and interfraction data variation, called intra- and interfraction fuzzy deep learning (FDL). When this is integrated with data of breathing patterns, it predicts the movements much more accurately and decreases the computation time for tracking.

23.4.11 Follow-Up Schedule and Investigations

Current standards of post-treatment surveillance are empirical and usually followed uniformly in different cancers without stratification based on risk. AI-based modeling may allow more personalized approaches that may be more time and resource-efficient for hospitals and patients. There is limited published evidence of the use of this method following radiation therapy [73].

23.4.12 Toxicity Management

Prediction of radiation-induced toxicities using machine ML/AI has been a central area of research because of the mitigation of toxicities has been an area of priority in radiation oncology research. Studies on ML-based modeling have been reported since late 2000. While most studies have been based on clinical and dosimetric features, more recent studies have included radiomic and genetic criteria as features. Both traditional ML approaches as well as image-based DL has been used in this evolving field. The current status of research has been summarized in some informative reviews [74]. Table 23.5 summarizes

Table 23.5 Machine learning modeling approaches in different settings

Site	Toxicities	Predictors	Approaches in larger studies	Range of accuracy (studies with more than 200 patients)
Head and neck cancers	Dysphagia and feeding tube dependence Xerostomia Oral mucositis	Clinical Dosimetric Radiomic	Classical ML: logistic regression, random forests, support vector machines	AUC: 0.63–0.83 [75–78]
Breast cancer	Radiation dermatitis Moist desquamation Breast/chest wall pain	Clinical Dosimetric	Classical ML: random forests, decision trees, logistic regression	AUC: 0.77–0.83 [79]
Lung cancer	Pneumonitis Chest wall syndrome	Clinical Dosimetric	Classical ML: logistic regression, random forests, decision trees	AUC: 0.66–0.79 [80–85]
Prostate cancer	Rectal and urinary toxicity	Clinical Dosimetric Radiomic Genomic	Logistic regression. Support vector machines Artificial neural networks	AUC: 0.60–0.76 [86–93]

ML Machine learning

some of the approaches used in different settings, focusing on the larger reported studies. The majority of studies have found moderate success in the methods, although there is still a lack of studies with external validation and reportage of real-life use.

23.4.13 AI in Radiation Therapy Quality Assurance/Risk Management

Quality assurance (QA) is an important component of precision RT. There are two domains of QA that are important—machine-specific QA and individual patient plan QA. ML/AI-based methods have been used for both domains.

Machine-specific QA has different components that require measurement at different intervals. Using this data, predictive models can be developed to optimize QA schedules and predict problems in dose delivery, and verification imaging. Prediction of Individual patient plan QA for IMRT pass rates from the treatment plan characteristics can also make the QA process more efficient leading to faster turnaround for treatment starts.

Table 23.6 summarizes published research looking at the applications of ML/AI in both domains of QA.

23.4.14 AI in Infrastructure/Staffing/Personnel for the Department

Cancer incidence is increasing worldwide, according to GLOBOCAN 2020 reports, regardless of a country's economic condition [94]. This is putting a strain on healthcare facilities all across the world. Around half of all cancer patients will undergo radiotherapy at least once in their lives [95]. The situation is even worse in low- and middle-income countries (LMICs), where infrastructure is lacking and radiation professionals are scarce. The RT practitioners are overworked as a result of the high patient load, resulting in poor quality care. AI can help automate manual repetitious procedures and reduce the workload of radiation experts, especially in LMICs. AI has already been used in clinical contexts for segmentation, treatment planning and adaptive radiotherapy. AI is still in development and research mode in various aspects of the radiotherapy workflow.

Table 23.6 Studies looking at the applications of ML/AI in both domains of QA

Subject studied	Question	References
Machine-specific QA		
Machine dose profiles	Linac beam symmetry over time from prior daily QA data	[38]
	Percentage depth doses from prior measurements	[55]
Leaf motion parameters	Discrepancies of leaf motion parameters with planned dose distributions from delivery log files	[35, 53, 54]
Proton beams	Output factors from prior measurements	[39]
Individual patient plan QA for intensity modulated radiation therapy		
Plan-specific QA	The pass rate of plan-specific QA from carefully designed plan characteristics with or without machine parameters (using ML) or fluence maps (DL)	[40, 43–45, 47, 83]

QA Quality assurance

The International Atomic Energy Agency (IAEA) and the World Health Organization (WHO) offer recommendations for a country's minimum radiotherapy infrastructure and manpower needs. Radiotherapy infrastructure and staffing requirements differ per country, depending on local demographics, geographic location, new cancer incidences, political stability, economic condition, hygiene and habit and other considerations. Policymakers can use an AI-based tool to suggest the best radiotherapy infrastructure for their country. An AI-based model can examine the existing shortage and forecast future infrastructure and staffing requirements by incorporating all conceivable variables that affect infrastructure.

AI is likewise viewed with skepticism by radiotherapy specialists [96–99]. There are concerns about employment losses as a result of AI's implementation in therapeutic settings. However, some specialists believe otherwise. According to them, AI can open up new possibilities and assist clinicians in acquiring confidence in clinical decisions and preserving treatment consistency. Professionals in radiotherapy will have their roles changed. Oncologists who specialize in radiation therapy might concentrate on sophisticated, tailored therapies. Medical physicists and dosimetrists, on the other hand, can refocus their efforts on the development and implementation of novel technologies and high-risk treatments. Radiation therapists and nurses will continue to play a similar role.

23.4.15 Implication of AI for Radiation Oncologists, Medical Physicists and Radiation Technologists

The advent of AI is likely to change the work schedule, timings, and patterns for all personnel in radiation oncology. Routine works are likely to become less routing and vice versa. Table 23.7 presents a snapshot of these likely changes.

23.4.16 AI and BIG Data for Radiation Oncology Research

The volume of data that gets accumulated during radiation therapy treatment is probably the highest in any specialty of medicine. Even in centers lacking paperless facilities, the treatment planning workstations would be storing countless plans, volumes and varied imaging datasets and image guidance datasets. The paperless departments have even more data in their store, including personal parameters, tumor characteristics, treatment details, post-treatment sequelae, survival and quality of life (QOL) outcomes. Much of this crucial information remained unused or is scantily used. This data related to the application of image guidance (radiomics) and biological markers (genomics, proteomics, metabolomics) generated over a treatment period can provide crucial leads in radiation

Table 23.7 Likely rearrangement in wok schedule for radiation oncologists, medical physicists and radiation technologists

Radiation oncologists	Medical physicists	Radiation technologists
Less time needed to be spent on contouring (organs at risk or targets)	Less time spent on planning	Lesser patient treatment times.
More time available for patient consultation, teaching, and research high-value, front-of-house activities of human interaction, such as patient counseling, education, support, and clinical management	Quick calculations, library, DL-based solutions, and multiple plans were quickly generated, providing options for comparison and best plan selection	More slots available for patient treatments
Will need to focus on quality control of AI output and	More time available for special QA procedures	Lesser errors in treatment delivery
Increased standardization of tumor segmentation and reduce unwarranted variation, particularly in under-resourced healthcare environments		
Improved clinical outcomes and quality of care		

QA Quality Assurance

oncology. It carries all the hallmarks "5 V's" (volume, velocity, variety, veracity and value) that characterize big data. Therefore, big data represents an ideal case study for research, quality assessment and clinical care to improve safe radiation delivery and patient treatment outcomes [100].

23.4.17 Challenges and Pitfalls of AI in Radiation Oncology

It is natural that once the data are collected, the next question would be what to do with all the information and data. The robustness of the input, intermediate and output parameters needs to be frequently monitored. Many AI engines would need regular updates based on new research. Making the data useful over short and long terms can be a challenge. Ensuring that it delivers the promise of precision medicine can be even more difficult. Hopefully, it is a matter of time where advanced deep learning soft wares are commonly available to harness the information from this pile of data. This is likely to be an expected challenge on the otherwise glowing horizon of big data in radiation oncology. As discussed in the earlier section, who takes the ethical and legal responsibility for the decision done independent AI tool? The issue of undertrained and overtrained models in the engines of AI and their repercussions is likely to remain a significant one. Since data sharing is not commonly employed in medicine, search engines are likely to behave differently to the same problem, based on the data region represented during the ML and DL process. The radiation oncology community is still undecided on which aspects of care can be completely, partially, or scantily handed over to AI.

Predicting outcomes and complication rates is challenging, particularly in radiation oncology. There is wide heterogeneity in tumor characteristics and normal tissue responses to radiation. Further, there are variations in normal anatomy and the relative position and size of the target volume, radiosensitization with adjuvant therapies, and treatment planning and delivery variability. In addition, there are typically more than a single OAR at risk; each OAR has a variable response to radiation.

Finally, with increasing incorporation of AI in radiation oncology, legal and ethical dilemmas shall come up. Can we leave AI decisions fully unsupervised, relying on the "super intelligence" of AI. In this case, who shall take the legal responsibility of the decision. If we indeed need to supervise each step of the AI process, the very purpose of incorporating AI into the treatment workflow would be defeated. Radiation oncology community needs to be geared up about these fast upcoming issues. Table 23.8 present the SWOT analysis of AI in radiation oncology.

Table 23.8 AI in radiation oncology (strength, weaknesses, opportunities, threats, SWOT analysis)

Strengths	Weaknesses	Opportunities	Threats
Use in nearly all spheres of RO	Variable accessibility to users	Potentially practice transforming	Product fails in unexpected manner
Faster growth trajectory compared to MC or IMRT	Need of robust web based solutions	AI healthcare market likely to grow from 2 bn USD (2018) to 36 bn USD (2025)	Misuse by operator
Capable of handling/ analyzing infinite data	Dependence on high performance computing hardware	Reduction of failures through automated second checks	Mitigation to expected failure not adequate
Standardization of processes	Explainability/need of validation of deep learning tools		Preservation of innovation, creativity and patient safety
	Training of users		Regulation of devices
	Generalization (heterogeneous, fragmented, private nature of healthcare data)		Clinician hesitance to lose control Ethical issues

RO radiation oncology, *USD* United States Dollar, *MC* Monte Carlo, *IMRT* Intensity Modulated Radiotherapy

23.5 Future of AI

As described in the previous sections, AI has tremendous potential to transform the way to practice oncology. There are a plethora of tools on offer, and exploratory studies have shown a potential of benefit in nearly aspects of oncology and radiotherapy practice.

The authors feel that two important aspects will define the role of AI in oncology in the future. The first is the development of multi-institutional imaging and radiotherapy plan archives. The principal limitation of current AI solutions in oncology is the lack of generalizability, where models developed on a limited set of data usually from one or a few institutions fail to work when tried on data or images from other institutions. This issue can be addressed by archives that have images, patient, treatment and outcome details from multiple institutions across the globe. The Cancer Imaging Archive (TCIA) is one such archive based in the USA [101]. In India, the CHAVI archive is being developed at Tata Medical Center and is the only known archive that will bank radiotherapy plan data in addition to clinical data and images [102].

The other main focus on the future will be the design of clinical trials to test the real benefit of AI based tools. Robust study design is essential for high-quality scientific assessment. Guidelines are being developed for both designing and reporting such trials to reduce ambiguity and bias. The SPIRIT-AI extension presents modified guidelines for clinical trial protocols using AI-based treatment interventions [103]. The CONSORT-AI extension has been developed as reporting guidelines for such studies [104]. Only those interventions that are tested and reported following these guidelines can be truly evaluated for their merit.

23.6 Conclusion

AI is making strong inroads into all aspects of life, including in the profession of oncology. Of all the branches of oncology, radiation oncology has the greatest affinity for computational and image-based tasks and therefore has the greatest potential for improvement with AI solutions. Several components of our daily work can be automated or simplified by such tools, and several are in the offing. Much of current scientific research in radiation oncology is focused on AI. Commercial applications are finding their way into the clinic. These have the potential to reduce clinicians' or physicists' time spent in performing repetitive tasks and improve the quality of radiation planning and delivery. Robust studies need to be performed for evaluating these tools to determine if the technical benefit translates into clinical benefit.

23.7 Summary Box for the Entire Chapter

1. Artificial Intelligence (AI) has made its appearance in all aspects of all lives, and medicine is no exception.
2. A subset of AI that allows the machine to learn from their own experience is Machine Learning (ML). Common ML algorithms are listed in Table 23.2.
3. AI has potential in all aspects of the workflow related to radiation oncology, including OPD assessment, simulation, planning, treatment, follow-up and imaging of patients.
4. There is a multitude of data generated in everyday radiation oncology, need to devise means to harness and use this information.
5. AI has tremendous potential to transform radiation practice, with two important aspects being development of multi-institutional imaging and radiotherapy plan archives.
6. Challenges for AI include variation in training platforms, ethical and legal responsibility.
7. Randomized trials are needed to be performed for evaluating if technical benefits of AI translate into clinical benefit.

Multiple Choice Questions

1. Machine Learning typically refers to:
   (a) A set of mathematical calculations
   (b) A set of computational algorithms
   (c) Based on AAA algorithm

   (d) Based on Monte Carlo-based algorithm
2. Deep Learning not employs to
   (a) Convolutional neural network (CNN)
   (b) Cable News Network (CNN)
   (c) Recurrent neural network (RNN)
   (d) long–short-term memory (LSTM)
3. All are nodal layers of CNN except:
   (a) Input layer
   (b) Transmission layer
   (c) An output layer
   (d) Hidden layers
4. All are true of image registration except:
   (a) Literature shows AI based rigid and non-rigid registration models superseding the manual task of image registration.
   (b) The task of performing delineation of the target and OARs for different clinical sites is markedly seen superior with the use of deep learning based models as compared to the manual delineation.
   (c) DL-based models have proven itself as the state-of-the-art techniques for image segmentation tasks
   (d) Manual registration is better than AI
5. All are deep learning architectures for image registration and segmentation problems except:
   (a) AlexNet
   (b) SqueezeNet
   (c) GoogleNet
   (d) YahooNet
6. Advantages of Artificial Intelligence-assisted IGRT approaches are all except:
   (a) Reduce overall treatment time
   (b) Offer the ability to track tumor movement
   (c) Reduce treatment uncertainty
   (d) Increase treatment precision
7. Tseng HH et al. developed an automated dose escalation approach using a deep reinforcement learning-based model for which cancers?
   (a) Head and neck tumors
   (b) Cervical cancers
   (c) Non-small cell lung cancer (NSCLC)
   (d) Breast cancer
8. Which model was employed by Varfalvy et al. to detect anatomical changes in patients after radiation?
   (a) Hidden Markov model (HMM)
   (b) GAN-based model
   (c) Dice Similarity Coefficient (DSC)
   (d) Monte Carlo model
9. Commercially available MRI-guided radiotherapy (MRgRT)-based systems are
   (a) MRIdian Linac (ViewRay, Mountain View, California, USA)
   (b) Unity (Elekta, Stockholm, Sweden)
   (c) Both the above
   (d) None of above
10. All are true about conditional generative adversarial network (cGAN) except:
   (a) Used by Matteo Maspero et al
   (b) Used for generating a synthetic CT from MRI
   (c) Can eliminate the idea of using CT acquisition from MRgRT workflow
   (d) Used in proton beam therapy

Answers

1. b
2. b
3. b
4. d
5. d
6. a
7. c
8. a
9. c
10. d

References

1. Wong KKL, Wang L, Wang D. Recent developments in machine learning for medical imaging applications. Comput Med Imaging Graph. 2017;57:1–3.
2. Patel TA, Puppala M, Ogunti RO, Ensor JE, He T, Shewale JB, et al. Correlating mammographic and pathologic findings in clinical decision support using natural language processing and data mining methods. Cancer. 2017;123:114–21.

3. Burke HB, Goodman PH, Rosen DB, Henson DE, Weinstein JN, Harrell FE, et al. Artificial neural networks improve the accuracy of cancer survival prediction. Cancer. 1997;79:857–62.
4. Zindler JD, Jochems A, Lagerwaard FJ, Beumer R, Troost EGC, Eekers DBP, et al. Individualized early death and long-term survival prediction after stereotactic radiosurgery for brain metastases of non-small cell lung cancer: two externally validated nomograms. Radiother Oncol. 2017;123:189–94.
5. Howard FM, Kochanny S, Koshy M, Spiotto M, Pearson AT. Machine learning-guided adjuvant treatment of head and neck cancer. JAMA Netw Open. 2020;3(11)
6. Kouwenberg J, Penninkhof J, Habraken S, Zindler J, Hoogeman M, Heijmen B. Model based patient pre-selection for intensity-modulated proton therapy (IMPT) using automated treatment planning and machine learning. Radiother Oncol. 2021;158:224–9.
7. Mallick I, Saha S, Arunsingh MA. A web-based dose-volume histogram dashboard for library-based individualized dose-constraints and clinical plan evaluation. J Med Syst. 2021;45(6):62.
8. Tambas M, van der Laan HP, Rutgers W, van den Hoek JGM, Oldehinkel E, Meijer TWH, et al. Development of advanced preselection tools to reduce redundant plan comparisons in model-based selection of head and neck cancer patients for proton therapy. Cancers (Basel). 2022;28(14):681.
9. Nguyen D, Long T, Jia X, Lu W, Gu X, Iqbal Z, et al. A feasibility study for predicting optimal radiation therapy dose distributions of prostate cancer patients from patient anatomy using deep learning. Sci Rep. 2019;9:1076.
10. Litjens G, Kooi T, Bejnordi BE, Setio AAA, Ciompi F, Ghafoorian M, et al. A survey on deep learning in medical image analysis. Med Image Anal. 2017;42:60–88.
11. Krizhevsky A, Sutskever I, Hinton GE. ImageNet classification with deep convolutional neural networks. Commun ACM. 2017;60:84–90.
12. Simonyan K, Zisserman A. Very deep convolutional networks for large-scale image recognition. In 3rd International Conference on Learning Representations, ICLR 2015—Conference Track Proceedings. 2015.
13. He K, Zhang X, Ren S, Sun J. Deep residual learning for image recognition. In Proceedings of the IEEE Computer Society Conference on Computer Vision and Pattern Recognition. 2016. p. 770–8.
14. Huang G, Liu Z, Van Der Maaten L, Weinberger KQ. Densely connected convolutional networks. In Proceedings—30th IEEE Conference on Computer Vision and Pattern Recognition, CVPR 2017. 2017.
15. Iandola FN, Han S, Moskewicz MW, Ashraf K, Dally WJ, Keutzer K. SqueezeNet: AlexNet-level accuracy with 50x fewer parameters and <0.5MB model size [Internet]. 2016. Available from: http://arxiv.org/abs/1602.07360.
16. Szegedy C, Liu W, Jia Y, Sermanet P, Reed S, Anguelov D, et al. Going deeper with convolutions. In Proceedings of the IEEE Computer Society Conference on Computer Vision and Pattern Recognition. 2015. p. 1–9.
17. Powers DM. Evaluation: from precision, recall and F-measure to ROC, informedness, markedness & correlation. J Mach Learn Technol. 2011;2:37–63.
18. Japkowicz N. Why question machine learning evaluation methods? (An illustrative review of the shortcomings of current methods). In AAAI Workshop—Technical Report. 2006. p. 6–11.
19. Olson DL, Delen D. Advanced data mining techniques. Springer; 2008.
20. Hussein M, Heijmen BJM, Verellen D, Nisbet A. Automation in intensity modulated radiotherapy treatment planning-a review of recent innovations. Br J Radiol. 2018;91(1092):20180270.
21. Yuan L, Ge Y, Lee WR, Yin FF, Kirkpatrick JP, Wu QJ. Quantitative analysis of the factors which affect the interpatient organ-at-risk dose sparing variation in IMRT plans. Med Phys. 2012;39:6868–78.
22. Skarpman Munter J, Sjolund J. Dose-volume histogram prediction using density estimation. Phys Med Biol [Internet]. 2015;60(17):6923–36. Available from: http://www.ncbi.nlm.nih.gov/pubmed/26305670.
23. Ahmed S, Nelms B, Gintz D, Caudell J, Zhang G, Moros EG, et al. A method for a priori estimation of best feasible DVH for organs-at-risk: validation for head and neck VMAT planning. Med Phys. 2017;44:5486–97.
24. Zawadzka A, Nesteruk M, Brzozowska B, Kukołowicz PF. Method of predicting the mean lung dose based on a patient's anatomy and dose-volume histograms. Med Dosim. 2017;42:57–62.
25. Tran A, Woods K, Nguyen D, Yu VY, Niu T, Cao M, et al. Predicting liver SBRT eligibility and plan quality for VMAT and 4Π plans. Radiat Oncol. 2017;12(1):70.
26. Sheng Y, Ge Y, Yuan L, Li T, Yin FF, Wu QJ. Outlier identification in radiation therapy knowledge-based planning: a study of pelvic cases. Med Phys. 2017;44:5617–26.
27. Zhu X, Ge Y, Li T, Thongphiew D, Yin FF, Wu QJ. A planning quality evaluation tool for prostate adaptive IMRT based on machine learning. Med Phys. 2011;38:719–26.
28. Good D, Lo J, Lee WR, Wu QJ, Yin FF, Das SK. A knowledge-based approach to improving and homogenizing intensity modulated radiation therapy planning quality among treatment centers: an example application to prostate cancer planning. Int J Radiat Oncol Biol Phys. 2013;87:176–81.
29. Yang Y, Ford EC, Wu B, Pinkawa M, Van Triest B, Campbell P, et al. An overlap-volume-histogram based method for rectal dose prediction and automated treatment planning in the external beam prostate radiotherapy following hydrogel injection. Med Phys. 2013;40:11709.

30. Schreibmann E, Fox T. Prior-knowledge treatment planning for volumetric arc therapy using feature-based database mining. J Appl Clin Med Phys. 2014;15:19–27.
31. Zhang J, Ge Y, Sheng Y, Yin FF, Wu QJ. Modeling of multiple planning target volumes for head and neck treatments in knowledge-based treatment planning. Med Phys. 2019;46:3812–22.
32. Petit SF, van Elmpt W. Accurate prediction of target dose-escalation and organ-at-risk dose levels for non-small cell lung cancer patients. Radiother Oncol. 2015;117:453–8.
33. Craft D, Bortfeld T. How many plans are needed in an IMRT multi-objective plan database? Phys Med Biol. 2008;53:2785–96.
34. Bokrantz R. Multicriteria optimization for volumetric-modulated arc therapy by decomposition into a fluence-based relaxation and a segment weight-based restriction. Med Phys. 2012;2012(39):6712–25.
35. Carlson JNK, Park JM, Park SY, Park JI, Choi Y, Ye SJ. A machine learning approach to the accurate prediction of multi-leaf collimator positional errors. Phys Med Biol. 2016;61:2514–31.
36. Valdes G, Scheuermann R, Hung CY, Olszanski A, Bellerive M, Solberg TD. A mathematical framework for virtual IMRT QA using machine learning. Med Phys. 2016;2016(43):4323.
37. Valdes G, Chan MF, Lim SB, Scheuermann R, Deasy JO, Solberg TD. IMRT QA using machine learning: a multi-institutional validation. J Appl Clin Med Phys. 2017;18:279–84.
38. Li Q, Chan MF. Predictive time-series modeling using artificial neural networks for Linac beam symmetry: an empirical study. Ann N Y Acad Sci. 2017;1387:84–94.
39. Sun B, Lam D, Yang D, Grantham K, Zhang T, Mutic S, et al. A machine learning approach to the accurate prediction of monitor units for a compact proton machine. Med Phys. 2018;45:2243–51.
40. Interian Y, Rideout V, Kearney VP, Gennatas E, Morin O, Cheung J, et al. Deep nets vs expert designed features in medical physics: an IMRT QA case study. Med Phys. 2018;45:2672–80.
41. Tomori S, Kadoya N, Takayama Y, Kajikawa T, Shima K, Narazaki K, et al. A deep learning-based prediction model for gamma evaluation in patient-specific quality assurance. Med Phys. 2018;45:4055–65.
42. Tomori S, Kadoya N, Kajikawa T, Kimura Y, Narazaki K, Ochi T, et al. Systematic method for a deep learning-based prediction model for gamma evaluation in patient-specific quality assurance of volumetric modulated arc therapy. Med Phys. 2021;48:1003–18.
43. Lam D, Zhang X, Li H, Deshan Y, Schott B, Zhao T, et al. Predicting gamma passing rates for portal dosimetry-based IMRT QA using machine learning. Med Phys. 2019;46:4666–75.
44. Nyflot MJ, Thammasorn P, Wootton LS, Ford EC, Chaovalitwongse WA. Deep learning for patient-specific quality assurance: identifying errors in radiotherapy delivery by radiomic analysis of gamma images with convolutional neural networks. Med Phys. 2019;46:456–64.
45. Granville DA, Sutherland JG, Belec JG, La Russa DJ. Predicting VMAT patient-specific QA results using a support vector classifier trained on treatment plan characteristics and linac QC metrics. Phys Med Biol. 2019;64(9):095017.
46. Ono T, Hirashima H, Iramina H, Mukumoto N, Miyabe Y, Nakamura M, et al. Prediction of dosimetric accuracy for VMAT plans using plan complexity parameters via machine learning. Med Phys. 2019;46:3823–32.
47. Li J, Wang L, Zhang X, Liu L, Li J, Chan MF, et al. Machine learning for patient-specific quality assurance of VMAT: prediction and classification accuracy. Int J Radiat Oncol Biol Phys. 2019;2019(105):893–902.
48. El Naqa I, Irrer J, Ritter TA, DeMarco J, Al-Hallaq H, Booth J, et al. Machine learning for automated quality assurance in radiotherapy: a proof of principle using EPID data description. Med Phys. 2019;46:1914–21.
49. Wang L, Li J, Zhang S, Zhang X, Zhang Q, Chan MF, et al. Multi-task autoencoder based classification-regression model for patient-specific VMAT QA. Phys Med Biol. 2020;2020(65):235023.
50. Wall PDH, Fontenot JD. Application and comparison of machine learning models for predicting quality assurance outcomes in radiation therapy treatment planning. Informatics Med Unlocked. 2020;18:100292.
51. Hirashima H, Ono T, Nakamura M, Miyabe Y, Mukumoto N, Iramina H, et al. Improvement of prediction and classification performance for gamma passing rate by using plan complexity and dosiomics features. Radiother Oncol. 2020;153:250–7.
52. Grewal HS, Chacko MS, Ahmad S, Jin H. Prediction of the output factor using machine and deep learning approach in uniform scanning proton therapy. J Appl Clin Med Phys. 2020;21:128–34.
53. Osman AFI, Maalej NM, Jayesh K. Prediction of the individual multileaf collimator positional deviations during dynamic IMRT delivery priori with artificial neural network. Med Phys. 2020;47:1421–30.
54. Chuang KC, Giles W, Adamson J. A tool for patient-specific prediction of delivery discrepancies in machine parameters using trajectory log files. Med Phys. 2021;48:978–90.
55. Zhao W, Patil I, Han B, Yang Y, Xing L, Schüler E. Beam data modeling of linear accelerators (linacs) through machine learning and its potential applications in fast and robust linac commissioning and quality assurance. Radiother Oncol. 2020;153:122–9.
56. Shen L, Zhao W, Xing L. Patient-specific reconstruction of volumetric computed tomography images

from a single projection view via deep learning. Nat Biomed Eng. 2019;3:880–8.
57. Isaksson M, Jalden J, Murphy MJ. On using an adaptive neural network to predict lung tumor motion during respiration for radiotherapy applications. Med Phys. 2005;32:3801–9.
58. Zhao W, Shen L, Han B, Yang Y, Cheng K, Toesca DAS, et al. Markerless pancreatic tumor target localization enabled by deep learning. Int J Radiat Oncol Biol Phys. 2019;105(2):432–9.
59. Foote MD, Zimmerman BE, Sawant A, Joshi SC. Real-Time 2D-3D Deformable Registration with Deep Learning and Application to Lung Radiotherapy Targeting. In Lecture Notes in Computer Science (including subseries Lecture Notes in Artificial Intelligence and Lecture Notes in Bioinformatics). 2019. p. 265–76.
60. Kida S, Nakamoto T, Nakano M, Nawa K, Haga A, Kotoku J, et al. Cone beam computed tomography image quality improvement using a deep convolutional neural network. Cureus. 2018;10(4):e2548.
61. Rossi M, Cerveri P. Comparison of supervised and unsupervised approaches for the generation of synthetic CT from cone-beam CT. Diagnostics. 2021;11(8):1435.
62. Guidi G, Maffei N, Vecchi C, Ciarmatori A, Mistretta GM, Gottardi G, et al. A support vector machine tool for adaptive tomotherapy treatments: prediction of head and neck patients criticalities. Phys Med. 2015;31:442–51.
63. Tseng HH, Luo Y, Cui S, Chien JT, Ten Haken RK, Naqa E. Deep reinforcement learning for automated radiation adaptation in lung cancer. Med Phys. 2017;44:6690–705.
64. Varfalvy N, Piron O, Cyr MF, Dagnault A, Archambault L. Classification of changes occurring in lung patient during radiotherapy using relative γ analysis and hidden Markov models. Med Phys. 2017;44:5043–50.
65. Harms J, Lei Y, Wang T, Zhang R, Zhou J, Tang X, et al. Paired cycle-GAN-based image correction for quantitative cone-beam computed tomography. Med Phys. 2019;46:3998–4009.
66. Maspero M, Savenije MHF, Dinkla AM, Seevinck PR, Intven MPW, Jurgenliemk-Schulz IM, et al. Dose evaluation of fast synthetic-CT generation using a generative adversarial network for general pelvis MR-only radiotherapy. Phys Med Biol. 2018;63(18):185001.
67. Fu Y, Mazur TR, Wu X, Liu S, Chang X, Lu Y, et al. A novel MRI segmentation method using CNN-based correction network for MRI-guided adaptive radiotherapy. Med Phys. 2018;45:5129–37.
68. Paganelli C, Lee D, Kipritidis J, Whelan B, Greer PB, Baroni G, et al. Feasibility study on 3D image reconstruction from 2D orthogonal cine-MRI for MRI-guided radiotherapy. J Med Imaging Radiat Oncol. 2018;62:389–400.
69. Terpstra ML, Maspero M, D'Agata F, Stemkens B, Intven MPW, Lagendijk JJW, et al. Deep learning-based image reconstruction and motion estimation from undersampled radial k-space for real-time MRI-guided radiotherapy. Phys Med Biol. 2020;65(15):155015.
70. Liu HW, Khan R, Nugent Z, Krobutschek K, Dunscombe P, Lau H. Factors influencing intrafractional target shifts in lung stereotactic body radiation therapy. Pract Radiat Oncol. 2014;4(1):e45–51.
71. Sindoni A, Minutoli F, Pontoriero A, Iatì G, Baldari S, Pergolizzi S. Usefulness of four dimensional (4D) PET/CT imaging in the evaluation of thoracic lesions and in radiotherapy planning: review of the literature. Lung Cancer. 2016;96:78–86.
72. Murphy MJ, Pokhrel D. Optimization of an adaptive neural network to predict breathing. Med Phys. 2009;36:40–7.
73. Chen QF, Dai L, Wu Y, Huang Z, Chen M, Zhao M. Surveillance strategy for Barcelona clinic liver cancer B hepatocellular carcinoma achieving complete response: an individualized risk-based machine learning study. Front Bioeng Biotechnol. 2021;9:667641.
74. Isaksson LJ, Pepa M, Zaffaroni M, Marvaso G, Alterio D, Volpe S, et al. Machine learning-based models for prediction of toxicity outcomes in radiotherapy. Front Oncol. 2020;10:790.
75. Wojcieszynski AP, La Cava W, Baumann BC, Lukens JN, Ghiam AF, Urbanowicz RJ, et al. Machine learning to predict toxicity in head and neck cancer patients treated with definitive chemoradiation. Int J Radiat Oncol. 2019;105(1):E139–40.
76. Jiang W, Lakshminarayanan P, Hui X, Han P, Cheng Z, Bowers M, et al. Machine learning methods uncover radiomorphologic dose patterns in salivary glands that predict xerostomia in patients with head and neck cancer. Adv Radiat Oncol. 2019;4:401–12.
77. van Dijk LV, Brouwer CL, van der Schaaf A, Burgerhof JGM, Beukinga RJ, Langendijk JA, et al. CT image biomarkers to improve patient-specific prediction of radiation-induced xerostomia and sticky saliva. Radiother Oncol. 2017;122:185–91.
78. Dean JA, Wong KH, Welsh LC, Jones AB, Schick U, Newbold KL, et al. Normal tissue complication probability (NTCP) modelling using spatial dose metrics and machine learning methods for severe acute oral mucositis resulting from head and neck radiotherapy. Radiother Oncol. 2016;120:21–7.
79. Reddy J, Lindsay WD, Berlind CG, Ahern CA, Smith BD. Applying a machine learning approach to predict acute toxicities during radiation for breast cancer patients. Int J Radiat Oncol. 2018;102:S59.
80. Das SK, Chen S, Deasy JO, Zhou S, Yin FF, Marks LB. Combining multiple models to generate consensus: application to radiation-induced pneumonitis prediction. Med Phys. 2008;35:5098–109.
81. Chen S, Zhou S, Yin FF, Marks LB, Das SK. Investigation of the support vector machine algorithm to predict lung radiation-induced pneumonitis. Med Phys. 2007;34:3808–14.

82. Huang EX, Bradley JD, El Naqa I, Hope AJ, Lindsay PE, Bosch WR, et al. Modeling the risk of radiation-induced acute esophagitis for combined Washington University and RTOG trial 93-11 lung cancer patients. Int J Radiat Oncol Biol Phys. 2012;82:1674–9.
83. Valdes G, Scheuermann R, Hung CY, Olszanski A, Bellerive M, Solberg TD. A mathematical framework for virtual IMRT QA using machine learning. Med Phys. 2016;43:4323–34.
84. Luna JM, Chao HH, Diffenderfer ES, Valdes G, Chinniah C, Ma G, et al. Predicting radiation pneumonitis in locally advanced stage II–III non-small cell lung cancer using machine learning. Radiother Oncol. 2019;133:106–12.
85. Chao HH, Valdes G, Luna JM, Heskel M, Berman AT, Solberg TD, et al. Exploratory analysis using machine learning to predict for chest wall pain in patients with stage I non-small-cell lung cancer treated with stereotactic body radiation therapy. J Appl Clin Med Phys. 2018;19:539–46.
86. Pella A, Cambria R, Riboldi M, Jereczek-Fossa BA, Fodor C, Zerini D, et al. Use of machine learning methods for prediction of acute toxicity in organs at risk following prostate radiotherapy. Med Phys. 2011;38:2859–67.
87. Tomatis S, Rancati T, Fiorino C, Vavassori V, Fellin G, Cagna E, et al. Late rectal bleeding after 3D-CRT for prostate cancer: development of a neural-network-based predictive model. Phys Med Biol. 2012;57:1399–412.
88. Ospina JD, Zhu J, Chira C, Bossi A, Delobel JB, Beckendorf V, et al. Random forests to predict rectal toxicity following prostate cancer radiation therapy. Int J Radiat Oncol Biol Phys. 2014;89:1024–31.
89. Yahya N, Ebert MA, Bulsara M, House MJ, Kennedy A, Joseph DJ, et al. Statistical-learning strategies generate only modestly performing predictive models for urinary symptoms following external beam radiotherapy of the prostate: a comparison of conventional and machine-learning methods. Med Phys. 2016;43:2040–52.
90. Lee S, Kerns S, Ostrer H, Rosenstein B, Deasy JO, Oh JH. Machine learning on a genome-wide association study to predict late genitourinary toxicity after prostate radiation therapy. Int J Radiat Oncol Biol Phys. 2018;101:128–35.
91. Fargeas A, Acosta O, Ospina Arrango JD, Ferhat A, Costet N, Albera L, et al. Independent component analysis for rectal bleeding prediction following prostate cancer radiotherapy. Radiother Oncol. 2018;126:263–9.
92. Carrara M, Massari E, Cicchetti A, Giandini T, Avuzzi B, Palorini F, et al. Development of a ready-to-use graphical tool based on artificial neural network classification: application for the prediction of late fecal incontinence after prostate cancer radiation therapy. Int J Radiat Oncol Biol Phys. 2018;102:1533–42.
93. Rossi L, Bijman R, Schillemans W, Aluwini S, Cavedon C, Witte M, et al. Texture analysis of 3D dose distributions for predictive modelling of toxicity rates in radiotherapy. Radiother Oncol. 2018;129:548–53.
94. Sung H, Ferlay J, Siegel RL, Laversanne M, Soerjomataram I, Jemal A, et al. Global cancer statistics 2020: GLOBOCAN estimates of incidence and mortality worldwide for 36 cancers in 185 countries. CA Cancer J Clin. 2021;71:209–49.
95. Datta NR, Samiei M, Bodis S. Radiation therapy infrastructure and human resources in low- and middle-income countries: present status and projections for 2020. Int J Radiat Oncol Biol Phys. 2014;89:448–57.
96. Ryan ML, O'Donovan T, McNulty JP. Artificial intelligence: the opinions of radiographers and radiation therapists in Ireland. Radiography. 2021;27(Suppl 1):S74–82.
97. Malamateniou C, McFadden S, McQuinlan Y, England A, Woznitza N, Goldsworthy S, et al. Artificial intelligence: guidance for clinical imaging and therapeutic radiography professionals, a summary by the Society of Radiographers AI working group. Radiography. 2021;27(4):1192–202.
98. Fiorino C, Jeraj R, Clark CH, Garibaldi C, Georg D, Muren L, et al. Grand challenges for medical physics in radiation oncology. Radiother Oncol. 2020;153:7–14.
99. French J, Chen L. Preparing for artificial intelligence: systems-level implications for the medical imaging and radiation therapy professions. J Med Imag Radiat Sci. 2019;50(4 Suppl 2):S20–3.
100. Benedict SH, El Naqa I, Klein EE. Introduction to big data in radiation oncology: exploring opportunities for research, quality assessment, and clinical care. Int J Radiat Oncol Biol Phys. 2016;95:871–2.
101. Clark K, Vendt B, Smith K, Freymann J, Kirby J, Koppel P, et al. The cancer imaging archive (TCIA): maintaining and operating a public information repository. J Digit Imaging. 2013;26:1045–57.
102. Kundu S, Chakraborty S, Mukhopadhyay J, Das S, Chatterjee S, Achari RB, et al. Design and development of a medical image databank for assisting studies in radiomics. J Digit Imaging. 2022;35(3):408–23.
103. Cruz Rivera S, Liu X, Chan AW, Denniston AK, Calvert MJ, Ashrafian H, et al. Guidelines for clinical trial protocols for interventions involving artificial intelligence: the SPIRIT-AI extension. Lancet Digital Health. 2020;2:e549–60.
104. Liu X, Cruz Rivera S, Moher D, Calvert MJ, Denniston AK, Ashrafian H, et al. Reporting guidelines for clinical trial reports for interventions involving artificial intelligence: the CONSORT-AI extension. Lancet Digital Health. 2020;2:e537–48.

Online Links for Articles (Including Links for Appropriate Audio/Videos in Free Public Domain), for Further Reading and Other Resources

https://www.youtube.com/watch?v=LVhuj7pOWa0 (Application of AI and ML in Automation of Radiation Oncology Treatment planning).

https://www.youtube.com/watch?v=iivcKrt3Y_s (Machine learning in Radiotherapy).

https://www.youtube.com/watch?v=dS8yJZVggaI&t=1183s (AI in radiation oncology).

https://towardsdatascience.com/a-comprehensive-guide-to-convolutional-neural-networks-the-eli5-way-3bd2b1164a53 (Convolutional neural networks with diagrams).

Setting Up of a New Radiotherapy Department

24

Tharmarnadar Ganesh and Partha Mukhopadhyay

The setting up of a cancer center, of which a radiotherapy facility is an integral part, needs wide coverage. Such a center shall have multidisciplinary cancer care facilities, appropriate infrastructure, human resources, and equipment needed for various services, all of which require in-depth analysis and presentation of various facts. The recent World Health Organization (WHO)–International Atomic Energy Agency (IAEA) joint publications titled "Setting up a cancer center: A WHO-IAEA framework" [1] and "The Comprehensive Cancer Center—Development, Integration and Implementation" by Aljurf et al. [2] form ideal references in this regard. Instead, this chapter will concentrate on how to set up a radiotherapy facility and related issues.

According to a 2015 estimate, worldwide, some 204 million fractions of radiotherapy would be required to treat cancer patients, and in low- and middle-income countries (LMIC), the cost of providing full access to all patients requiring radiotherapy would be close to 100 billion US dollars by the year 2035. It would result in 27 million life-years saved for patients in this region and economic benefits worth more than $280 billion. Only 40–60% of the cancer patients requiring radiotherapy treatment have access to it globally. In LMICs, the situation is worse, with only 10% of the needy patients having access to radiotherapy [3].

The 14,478 radiotherapy machines available globally are distributed unevenly. The high-income countries have a median density of 5.1 machines per million people (range 0.4–11.6), and the low-income countries have 0 (range 0–0.4) [4]. India, with a population of 1400 million, sees about 1.4 million new cancer cases per year, thus requiring 1400 machines going by the WHO recommendation of 1 machine per million population, 3100 machines as per the IAEA recommendation of one machine per 450 patients, and more than 5500 machines based on the ideal figure of 4 machines per million [5–7]. India has approximately 725 machines (as of November 2022), which translates to a mere 0.5 machine per million population—a grossly inadequate figure compared to the recommendations. An additional burden is the uneven geographical distribution of these inadequate radiotherapy facilities.

In developed countries, 50% of the cancer patients would require radiotherapy at some point in the course of their disease management, either alone or in combination with other treatment modalities. The proportion of cancer patients requiring radiotherapy services in low- and middle-income countries (LMICs) would be

T. Ganesh (✉)
Manipal Hospital, New Delhi, Delhi, India

P. Mukhopadhyay
Radiation Oncology, Ruby Cancer Centre, Ruby General Hospital, Kolkata, West Bengal, India

A. Munshi et al. (eds.), *Radiation Oncology – Principles, Precepts and Practice*,
https://doi.org/10.1007/978-981-97-8389-2_24

higher than this figure due to several factors, including late diagnosis, inequality in geographical distribution of the existing radiotherapy facilities, and the nonavailability of other resources [6, 8, 9].

In the recent past, significant advances have taken place in radiation oncology. The field has witnessed an array of technological innovations in the form of newer delivery techniques, treatment machines with far-reaching capabilities, advanced devices for dosimetry, sophisticated systems assisting in patient set-ups, novel treatment accessories, increased integration of imaging modalities into the radiation oncology workflow, total dependence on the information technology (IT) network for every activity in the department, etc. The explosive growth in the technology has made the process of setting up a new radiation oncology facility within a hospital, a highly challenging and demanding task. More than ever before, there is a need for unprecedented level of coordination in the healthcare facility between the involved teams throughout the time of project implementation: right from the top management, the finance team, the human resources team, the project team, the engineering services team (electrical, biomedical, mechanical, civil), the legal team, the construction engineers, nursing services, the quality team, heads of associated disciplines, and of course also the user department team—a spectrum of people and experts must be consulted and involved in the process of establishing radiotherapy services. This includes, but is not necessarily limited to, the choice of equipment, designing of the facility, budgeting, marketing needs, deciding on staff requirements, going through the purchase procedure, interacting with the vendors and getting to know about their requirements, intercompatibility between the vendors involved, deciding on the equipment and accessories, staff recruitment in phases, monitoring adherence to various regulatory requirements, construction, acceptance testing and commissioning, commencing clinical services, and agreeing on maintenance charges, among other things.

24.1 Initial Considerations

The technological advances made feasible the availability of a broad range of machines with new delivery techniques, new features, and new capabilities. However, given their operational complexity, understanding their full potentials, limitations, and clinical implications becomes a herculean task, both for subject experts like radiation oncologists and medical physicists, and even more so for other nontechnical decision makers.

24.2 Assessment of the Needs for a Radiotherapy Facility

Data from the national cancer registry for the concerned region where the radiotherapy facility is proposed, like the number of patients, gender distribution, and tumor site breakdown, shall be carefully considered in assessing the needs for the facility and in determining the projected annual number of patients requiring radiation therapy. The approximate number of new cancer patients per year is estimated from the knowledge of the population of the area from which the institution is likely to get patients and the annual cancer incidence and prevalence for that region. Of these new cancer patients, it can be expected that 50–60% would require radiation therapy, either alone or in combination with other forms of treatment. The admission records of the hospital and patient numbers provided by the hospital's oncologists can prove to be useful data resources. The referral trends of the physicians and the deficiencies in the referral patterns should be well analyzed. Furthermore, information on the distribution of existing radiotherapy facilities in the region, as well as the number and type of machines available, will aid in determining the number of patients likely to be seen in the institution and in selecting the most appropriate equipment [8].

The number of patients seen in the institution in a year with malignancies that are potentially treatable by brachytherapy (both intracavitary and interstitial treatments), along with their stage distribution, shall help to estimate the total number of anticipated brachytherapy treatment courses [8]. This, in turn, would aid in determining whether or not the institution should have a brachytherapy facility. However, most often, the necessity to offer both external beam therapy and brachytherapy services under a single roof will override such calculation-based conclusions.

Consideration should be given as to how the proposed radiotherapy facility is being planned—whether it is part of a comprehensive tertiary hospital or an integral component of a multidisciplinary cancer hospital along with surgical oncology and medical oncology, or a standalone radiotherapy center, which has become increasingly rare these days. Availability and the strengths of other critical supporting services like diagnostic imaging, laboratory services, and pathology should be taken into account [2]. If the facility is planning for teaching, training, and/or research activities, or has plans for these in the future, it will add specific requirements to the planning of the facility.

The cancer center's future requirements for expansion in terms of space, bunker plans, the proposed time scale, and current technology's capability to integrate with the technology proposed for the future should also be considered at the planning stage.

The decision to establish a radiotherapy facility must be accompanied by careful strategic planning. Such a plan should outline the major stages of the project and set out timelines for the same. The project needs continuous coordination and monitoring to ensure that timelines are met and the set goals are achieved at each stage. A special team, called the core implementation team, shall be formed, consisting of professionals from various disciplines, for the purpose [10]. Table 24.1 lists the constituents of this team:

Table 24.1 Constituents of a core implementation team for overseeing a radiotherapy project

Member	Qualifications/expertise
CEO or Director of the facility[a]	Head of the institution
Finance Head	Person who can take decisions on finance matters related to the project
Purchase Head[a]	Person who is familiarized with high-value purchases, global tenders
Projects[a]	Person who has prior experience in handling radiotherapy projects
A team from Engineering services[a]	Electrical engineer, biomedical engineer, mechanical engineer, civil construction engineer
Architect	Qualified architect, specialized in design and construction of RT departments
Head, IT	Person who can advise on cloud-based operations, cloud storage, and establishing network sensitive systems
Head, Legal Affairs	Person who shall be responsible for addressing issues arising from legal point of view
Head of associated disciplines	Heads of Radiological Imaging Services, Nuclear Medicine. If necessary, heads of surgical oncology, medical oncology, anesthesiology, and sterilization may also be consulted
Human Resources Manager	Head of Human Resources preferably experienced in handling staff recruitment for radiotherapy services
Representative from Quality Management department	Expert who specializes in quality management in hospitals
Head of Nursing Services	Head, Nursing services or any senior representative
Project Manager, Equipment Supplier	The Project Lead from the equipment supplier who is well-versed in coordinating RT projects
Site Engineer, Equipment Supplier	Site Engineer from the equipment supplier who will be responsible for ensuring all requirements
Head of Radiation Oncology[a]	Radiation Oncologist heading the department
Radiation Safety Officer (RSO)[a]	If Chief Medical Physicist is not the RSO, then RSO must also be included
Chief Medical Physicist[a]	Chief Medical Physicist of the department

[a] Permanent members who shall be present in every meeting of the core implementation team. Others may be called as and when required depending on the agenda for the meeting

It is not a strict requirement to form the team with the above members only. More members can be added as needed to meet the needs. The team shall meet periodically to assess the progress of the project, form specifications, take appropriate steps to implement decisions (like purchasing, space allocating, coordinating delivery dates, etc.), review regulatory compliance status, determine staffing needs, and time the recruitment of various personnel, among other things. Minutes of every meeting shall be carefully documented, circulated among all members, and preserved as permanent record. Besides overseeing project implementation, the team can be entrusted with the task of making decisions whenever the need arises.

A great deal of thought and brainstorming should go into the formation of the specifications, the tendering, the evaluation, and selection of the treatment machines (EBRT and brachytherapy). The hospital should ideally have the services of a consultant radiation oncologist and a chief or consultant medical physicist in-house *before* framing the specifications of equipment and accessories. Their opinions and suggestions shall carry the maximum weight in this critical task. American Association of Physicists in Medicine (AAPM) Task Group 45 [11] highlights the roles of the radiation oncologist and the medical physicist in the equipment purchase procedure. The report additionally states that in order to realize the full potentials of modern medical linacs, the decision to purchase such a linac must be made concurrently with the purchase of all required dosimetry and treatment planning equipment. Hence, any cost-cutting measure shall not compromise with the purchase of dosimetry, quality assurance and treatment planning equipment. The recent WHO document on technical specifications of radiotherapy equipment for cancer treatment [12] also unambiguously states that radiation oncologists, medical physicists shall be involved in the process of framing technical specifications and in the evaluation of the technical bids. The practice of not involving personnel from the user department in the equipment selection process and relying solely on people who are neither technically competent enough to make equipment specifications, nor aware of the implications of their selections in the operation of clinical services, should be discouraged. Such a practice often leaves the user department to deal with sub-optimal equipment and dosimetry accessories that result in delays in the commissioning of clinical services and, even worse, results in the nonutilization of certain features of the treatment delivery machine(s).

24.3 Basic Essentials for the Functioning of the Clinic

A radiation oncology facility in a comprehensive cancer center or in a multi-specialty tertiary care center shall have at least one external beam treatment unit, one remote afterloading high-dose rate brachytherapy unit, treatment planning system(s), radiation safety instruments, a complete set of dosimetric and quality assurance (QA) equipment mandatorily required for running the clinical services provided, a reliable information technology (IT) network, data backup services, mold room accessories, adequate space, and enough human resources to staff the facility. Ideally, every center shall also have a second external beam therapy unit to provide uninterrupted clinical services when one machine breaks down.

24.4 Technology Acquisition

In India, over the last couple of decades, the number of telecobalt machines has significantly gone down, and the number of new units being added is also down to only two or three units per year. The number of linacs is rapidly growing. There are about 725 teletherapy units in India, with about 600 linacs and 125 telecobalt (Co-60) units as of 2022.

The debate over the choice between Co-60 vs. linac for external beam therapy is still a valid one even after so many decades, especially in LMICs and in a country like India where more cancer hospitals are required in Level-III small towns and at remote locations to increase the accessibility of cancer patients to at least basic, essential radiotherapy services. It is stated that about 25%

Table 24.2 Comparison of telecobalt *vs.* conventional linac

Co-60	Conventional linac
Single energy, 1.25 MeV gamma rays—average energy is 0.9 MeV	Choice for multiple photon and multiple electron energies Choice of flattened and unflattened beam
Simple, robust technology Easy to maintain	Complex technology Requires intense maintenance and strong service backup
Lower initial investment, lower bunker construction cost	Substantial initial investments in terms of machine cost, building cost and running cost
Low annual maintenance cost Annual Maintenance Contract (AMC) is without spare parts and it is roughly at 1% of the total cost of the equipment Comprehensive Maintenance Contract (CMC) is all spare parts inclusive except consumables and it is usually about 5–6% of the total equipment cost	Annual maintenance cost is quite high since AMC and CMC are calculated on the equipment cost
Lack of 3-D CRT, IMRT, IGRT, SGRT capabilities. Special treatment techniques like stereotactic treatments, gated treatments are not possible	All high-end techniques, image guidance techniques, surface guidance technique, latest delivery technologies are available including delivery technologies like tomotherapy and robotic therapy. Special techniques like stereotactic treatment techniques have become standard features
Fixed, low dose rate (maximum possible dose rate is about 250 to 300 cGy per minute)	Variable dose rates that can go as high as 2400 MU/min
Source decays at the rate of 1% every month and needs replacement after 7 years. Timely availability of a replacement source is an issue. Cost can be significant which can offset the gains from lower annual maintenance cost	Does not need such source replacements. But the cost of spare parts can be quite high. If the space parts have to be imported, downtime may be significant. Going in for the AMC instead of CMC is a risky option
Safety and security of the very high activity radioactive source Stringent regulatory requirements mandate safety plan acknowledgement from local police. Needs preplanning for decommissioning and source disposal, including standing fund allocation for the same	No radioactive source and hence no special regulatory conditions
No risk of neutron contamination	At energies 10 MV and above, risk of neutron contamination

of the radical cases could benefit from Co-60 teletherapy with simple conventional treatment approaches [10]. Table 24.2 lists the pros and cons of the two technologies.

24.5 Choice of Machines

24.5.1 Centers with Only One Machine

The choice between the two technologies for a new cancer center with only one teletherapy machine is often heavily skewed in favor of a linac. However, there may be occasions where Co-60 technology may still be suitable, especially for centers with budgetary constraints, and in government setups looking for simple replacements of worn-out machines.

Before deciding on the best technology for the center, the pros and cons of telecobalt and linac should be carefully weighed and debated. Apart from the current cost of the equipment, annual maintenance charges, source replacement cost, cost of spare parts, and life expectancy of the machine should also be considered in making the final choice. For the institutions that are planning to implement new treatment technologies, it will be a worthwhile exercise to go through the questionnaire given in the Appendix-II of the IAEA document [10] before the selection of the technology.

If the choice is telecobalt, the choice should also be made between the options of 80 cm and 100 cm source-to-axis distance (SAD) models. The latter requires a high activity source and thus a corresponding increase in the source replacement cost, but offers the advantage of clearance like a linac. The regulatory body, the Atomic Energy Regulatory Board (AERB), requires a safety plan for a telecobalt unit to be prepared and acknowledged by the local police. In addition to that, it is mandatory for the center to allocate fixed funds upfront at the time of purchase for the decommissioning and disposal of the unit and its source.

If the choice is a single linac, it should ideally be a conventional linac without any field size restrictions. Energy configuration and other features of the linac may vary as per the needs of the facility, and the type of cases seen in the hospital. A Level-III small town may have a basic linac with a single 6 MV energy having only MV image guidance capabilities, whereas a Level-II medium-level city may have a unit with dual-photon energies with four or a maximum of five electron energies ranging from 4 MeV to 20 MeV and IMRT, kV, and MV image guidance capabilities. A linac with advanced features like stereotactic treatment capabilities can be envisaged if it is a Level-I apex center. Brachytherapy should be available in all three levels. There is a recent trend in some centers to go for a closed-ring gantry machine as their choice of the single machine, since the unit has the capability to address heavy patient loads. The machine has the additional capability of kV imaging in addition to MV imaging.

In the multiple photon energy linacs, the number of photon energies should be limited to two or at most three: 6 MV, 10 MV, and a 6 MV unflattened beam. A higher energy like 15 MV or 18 MV would be beneficial only if the department has plans to treat more cases by the 3-D CRT technique than by IMRT and VMAT techniques. For these advanced techniques, 6 MV is sufficient, and energies higher than 10 MV are certainly not used. Restricting the highest energy to 10 MV will have the additional benefit of reduced neutron protection measures and significant savings in the building construction costs [10]. If 15 MV or 18 MV is chosen because there are more cases on 3-D CRT, then it should replace 10 MV. It is not advisable to have 6 MV, 10 MV, and 15 MV (or 18 MV) on a single machine. When the number of beams is limited, less effort is required for beam data collection and commissioning. The burden of periodic quality assurance (QA) tests is also significantly reduced.

It is estimated that the costs of purchasing and maintaining over a 15 year period would come to 1.6 million euros for a cobalt unit, 2.16 million euros for a basic 6 MV linac (without IMRT), and 3.6 million euros for a complex multienergy linac [10].

24.5.2 Centers with Two Machines

An option for the second linac is to have it identical to the first one with matched beams so that in the event of a breakdown of one machine, the other can handle the patient workload without much effort and in less time. Another option is to go for one of the high-end linacs with finer multileaf collimator (MLC) resolution and other advanced features, but they may suffer from severe field size restrictions that may compromise conventional treatments.

24.5.3 Centers with Three or More Machines

Several configurations are possible. The first two linacs can be mid-segment, high-energy, beam matched linacs with IMRT, kV-IGRT capabilities, and the third can be a high-end premium linac with all advanced features, including stereotactic treatment capabilities. The other option for the third unit is to choose a closed-ring gantry linac, a tomotherapy unit, or even a robotic arm type linac. A MR guided linac is another high-end option.

Typical technical specifications for a telecobalt unit, linac, brachytherapy unit, CT-simulator, conventional simulator, treatment planning systems, mold room equipment, immobilization equipment, dosimetry and quality control equipment, radiation safety equipment, and a radiation oncology information system are given in the WHO document [12].

24.6 Maintenance and Lifecycle of Radiotherapy Equipment

The purchase process must include discussions, a cost assessment, and agreement on the terms and conditions of a maintenance contract to cover the lifetime of all the major equipment. Some of the important points to be considered are:

Number of periodical maintenance visits, service hours, and service availability outside the office hours and during the weekends.

Inclusion of all spare parts without exception (only consumable items are excluded). The Comprehensive Maintenance Contract (CMC) is usually 5–6% of the total equipment cost for the first few years after the expiration of the warranty period. It will be higher in the subsequent years. Rates must be worked out for the expected lifetime through a written agreement. A copy of the agreement must be kept in the user department.

Ensure spare parts availability for the expected lifetime of the equipment.

Agreeing on what is included in the upgrades: mandatory, safety related, software (major and minor), and hardware. It shall be unambiguously mentioned in the agreement document.

Remote diagnostics for analyzing machine breakdowns.

Penalty clauses—for uptime and response time guarantees.

If the equipment houses sealed radioactive source, then source replacement cost, frequency, disposal method, responsibility, and cost must all be worked out and agreed upon.

The expected lifetimes of major equipment are given in Table 24.3.

24.7 Staff Requirement

The AERB Safety Code for Radiotherapy Installations [13] stipulates the minimum personnel requirement for a radiation oncology facility as given in Table 24.4. The code is under revision.

Table 24.3 Expected lifetime for various equipment in radiotherapy [12]

Equipment	Expected lifetime
Linac	10–15 years
Telecobalt unit	15 years
CT-Simulator	5–10 years
TPS, OIS	5–10 years depending on hardware/software upgrade
Patient immobilization equipment	Major: 5–10 years Minor: 3–5 years
Dosimetry, QA and radiation safety	10–15 years
Remote afterloading HDR brachytherapy system	10–15 years
Brachytherapy applicators	3–5 years

Table 24.4 Staff requirement as recommended by Atomic Energy Regulatory Board, India

Sr. No.	Category of staff	Requirement
1.	Radiation Oncologists (1) Chief Radiation Oncologist (2) Radiation Oncologist	One per center One additional for every 400 patients treated annually. Each physician can treat up to 40 patients a day
2.	Medical Physicist	One per center treating up to 500 patients in a year. Additional physicist for every 500 patients treated in a year
3.	Radiation Safety Officer	One per center
4.	Chief Radiation Therapy Technologist Radiation Therapy Technologist	One per center Two per teletherapy unit—up to 40 patients treated daily Four per teletherapy unit—up to 80 patients Additionally for every 500 patients simulated, 2 more RTTs and for brachytherapy as per need
5.	Other supporting staff (like nurse, social worker, maintenance engineer, etc.)	As per need

24.8 Space Requirement

The spatial organization of a radiotherapy department must be designed to separate the patient movement areas from the staff movement areas. This would help in smoothening the workflow. For a basic center with one linac, one brachytherapy unit, TPS, etc., the space requirements are given in Table 24.5. The figures given in the table are only tentative and may be altered to some extent. However, major alterations or removal altogether should be carefully considered and discussed. Figure 24.1 shows a typical layout of the facility (not all the spaces listed in the Table 24.5 are shown in the diagram).

High-volume services within the patient service areas should be located near the waiting areas to facilitate easy patient access. Patient areas should be consolidated to control patient access and maintain patient privacy, security, and dignity. A play area for children is also required.

Staff areas should have electronic access control to prevent the entry of patients. Similarly, all controlled areas shall have electronic access control to prevent unauthorized entry. The nursing areas, procedure rooms, clean and soiled utility functions should be close to the patient areas they support. Infrastructure support areas (e.g., electrical, telecommunications, networking/server rooms) should be located in proximity to the radiation therapy suite.

The planning and designing of the radiotherapy facility must be flexible enough to allow for future expansions and the needs for equipment maneuverability. For example, if the existing linac requires any major part replacement, or if the linac itself is replaced with a new one after its lifetime, or if a new linac is procured for installation in a bunker that has been kept empty for years, the design shall allow taking in these equipment without any demolitions or without involving any major civil works. While the facility is under construction, such tasks may not pose issues. However, in a functioning hospital, such requirements may be difficult to deal with if an adequate thought process is not employed as to where to unload the equipment, how to take the equipment to the basement, whether turns will allow maneuvering of long crates, etc. These aspects must be carefully thought over, and the design must be suitably amended.

The radiotherapy department in any hospital is a special area that is frequented by individual patients on a daily basis for several weeks. The design of the department must extend beyond the simple allocation of spaces. There is a need for implementing human-centered design thinking and processes. The designers must work in conjunction with the medical providers and the patients to support person-centered care (PCC). The incidence of anxiety within the cancer patients calls for employing special methods for mediating this emotion through patient empowerment [14]. Design can be a useful tool in the conceptualization and identification of the problem of patient anxiety and its situational triggers

Table 24.5 Typical rooms required in a radiotherapy department and their floor areas

Room/space	No. of rooms	Area of each room (m^2)	Total area required (m^2)
General/office			
Waiting	1	40	40
Bay—wheelchair	1	4	4
Reception (front desk)/clerical, 10 m^2	1	10	10
Toilet—public, 4 m^2 (male and female separate—for exclusive use of patients)	2	4	8
Toilet—accessible	1	6	6
Office—3 person shared	1	15	15
Store—files	1	10	10
Store—files	1	10	10
Office—HOD, radiation oncologist, 18 m^2	1	18	18
Secretary and administration	1	4	4
Office—radiation oncologist, 10 m^2	2	10	20
Office—chief medical physicist	1	12	12
Resident doctors' room	1	20	20
Meeting room—93 m^2	1	93	93
Staff room	1	20	20
Property bay—staff	2	3	6
Toilet—staff (male and female)	2	4	8
Planning zone			
Waiting, 10 m^2	2	10	20
Patient bay—holding, 10 m^2	2	10	20
Toilet—accessible	1	6	6
Change cubicle—accessible	2	4	8
CT-simulator room	1	40	40
CT-Simulator control room	1	15	15
Bay—resuscitation trolley	1	2	2
Mold room	1	22	22
Appliance area			
Clean-up room, 10 m^2	1	10	10
Server room	1	10	10
Medical physics			
Office—workstation (physicist)	1	6	6
Dosimetry/QA room cum physics laboratory	1	24	24
Store—equipment (physics)	1	10	10
Radiotherapy treatment planning room	1	40	40
Radiation treatment			
Change cubicles—patient	2	2	4
Change cubicle—accessible	1	4	4
Toilet–patient	2	4	8
Waiting—near bunker	2	2	4
Radiotherapy bunker room (linear accelerator)	1	150	150
Radiotherapy bunker control room	1	27	27
UPS room	1	20	20
Bay—linen trolley	1	2	2
Store—equipment	1	10	10

(continued)

Table 24.5 (continued)

Room/space	No. of rooms	Area of each room (m^2)	Total area required (m^2)
Brachytherapy suite			
Brachytherapy bunker	1	40	40
Brachytherapy control room	1	8	8
Radioactive source storage room	1	2	2
Scrub-up/gowning, 6 m^2	1	6	6
Anesthetic induction room	1	4	4
Patient bay—holding, 10 m^2	1	10	10
Patient observation and nursing care			
Nursing staff station	1	12	12
Room for keeping drugs/sterile items	1	10	10
Patient bay—holding, 10 m^2	1	10	10
Procedure room, 20 m^2	1	20	20
Bay—linen	1	2	2
Bay—resuscitation trolley	1	2	2
Office—single person, 10 m^2 (nurse manager)	1	10	10
Clean utility	1	12	12
Dirty utility	1	12	12
Disposal room, 8 m^2	1	8	8
Cleaner's room, 5 m^2	1	5	5
Consult rooms	2	14	28
Clean-up room	1	10	10
Toilet—patient	2	4	8
Waiting Area	2	10	20
Total area			1005
Circulation allowance		35%	351.75
Grand total (approximately)			1357

in healthcare experiences by applying new perspectives. Every effort must be made to make every patient's visit to the department a pleasant experience.

24.9 Budget

The radiotherapy department in a hospital needs a hefty budget. The approximate costs of several major equipment are given in Table 24.6.

24.10 Regulatory Process

In the peaceful uses of atomic energy, a wide range of radioactive sources and radiation-generating equipment are used in different areas, including medical applications. Globally, their use is strictly regulated through respective national regulatory agencies, and in India, the Atomic Energy Regulatory Board (AERB) enforces it through a well-structured consenting process for such peaceful uses. AERB ensures that there is a responsible person (designated as the employer or licensee of the institution according to the regulatory regimen) who is or are accountable for all radiation safety-related aspects through this regulatory control.

24.11 Typical Regulatory Steps Involved in the Licensing of Radiotherapy Equipment, from Institution Registration to Licensing in India

Any institution planning to establish a radiotherapy facility in India has to go through a series of steps before acquiring the regulatory consent

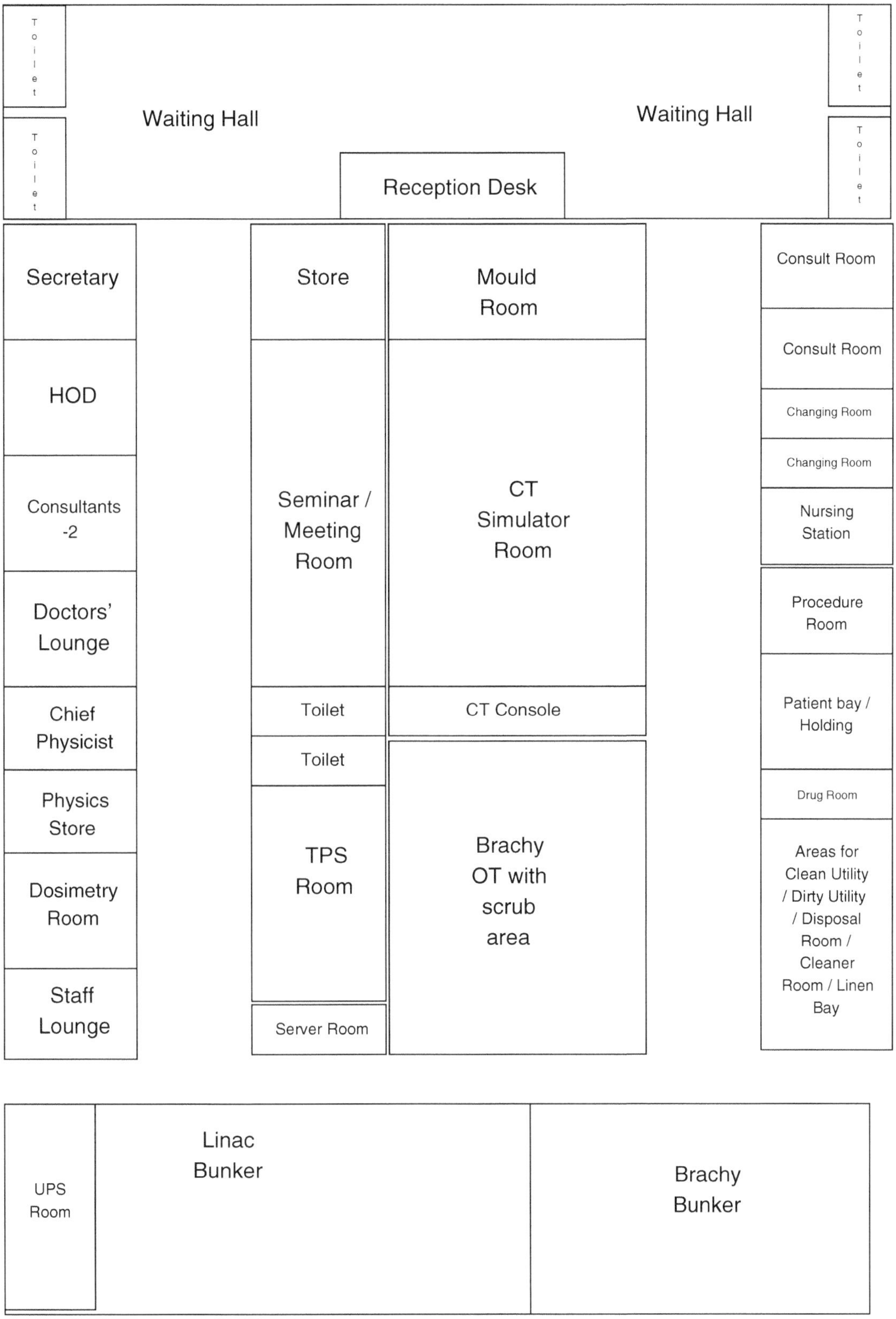

Fig. 24.1 Typical layout of the facility

Table 24.6 Typical costs of various radiotherapy equipment

Machine/building type	Approximate cost
Telecobalt unit	Rs. 5 crores
Telecobalt source cost	Rs. 1–1.2 crores
Single energy (6 MV) linac	Rs. 10 crores
High energy (15 MV plus electron energies)—mid-segment	Rs. 12.5–13 crores
High energy (15 MV plus electron energies)—high-end, premium	Rs. 17–18 crores
Tomotherapy	Rs. 17–18 crores
MR Linac	Rs. 75–85 crores
Dosimetry + QA (up to VMAT)	Rs. 1–1.25 crores
Immobilization accessories	Rs. 25–40 lakhs
Brachytherapy HDR (Iridium 192 based)—10 channels	Rs. 1.8–2 crores
Brachytherapy—Ir-192—per source exchange—once every 120–150 days	Rs. 7 lakhs
Brachytherapy HDR (Cobalt 60 based)—10 channels	Rs. 2.5–3 crores
Bunker for high energy linac	Rs. 1.25–1.5 crores
Bunker for low energy linac	Rs. 1–1.25 crores
Cancer hospital building construction cost depends on size	Rs. 20–50 crores

Customs duty is roughly 29–30% over and above the costs mentioned for the imported equipment (this includes custom clearing charges of 1–2.5% of the total order value)

from AERB to use the equipment for clinical services. It is strongly recommended to go through the regulatory requirements and guidelines for establishing new radiotherapy facility published by AERB. Since the submission of applications, their approvals (or rejections), equipment registrations and the licenses are handled only through the eLORA (e-Licensing of Radiation Applications) online platform, it is also recommended to go through the eLORA guidelines document.

Institute Registration The institution should apply to AERB to register itself as a radiotherapy institution. The following conditions must be met: permission from a local municipal or other appropriate authority to operate as a hospital; employer's identity proof details, including personal valid email and mobile number; and an undertaking by the employer to abide by the Radiation Protection Rules, 2004.

Site and Layout Approval An application for site and layout approval must be submitted with complete details and layout diagrams. The layout plan must conform to the regulatory requirements and be drawn at the recommended scale (1:200), showing all the details: name and address of the institute, all facilities up to 20 m from the external walls of the radiotherapy room, unambiguous indication of the facilities surrounding the installation, and the distance of the plot boundaries from its external walls (shall not be less than 1.5 m). In the main drawing showing the radiotherapy room and its console, all important details like make, model, modes, nominal photon energies, isocenter, central axis, axis of gantry rotation, construction material of each wall, primary barrier width, radioisotope, source position, bed position, all dimensions, conduits, the positions of the gamma zone monitor, the last man out switch, and the closed-circuit-television (CCTV) cameras must be shown. The cross-sectional drawings should be prepared along the length, breadth, and maze of the treatment room. If there are floors above and/or below the installation, they also must be shown in detail. The ducts for central air conditioning must run along the wall of the entrance door without being punctured.

A radiation safety expert and an architect may be hired as consultants during this stage. Standard layout diagrams are available on the AERB website for telecobalt, low-energy and high-energy linacs, and high-dose rate (HDR) brachytherapy installations. If for any reason the make, model, and vendor are to be changed after the approval, it is possible to do so by following the due process. However, the construction process must start only after the site layout plan has been approved.

Authorization to Procure Radiotherapy Equipment The institution must verify whether there is a *valid* type approval certificate for the make and model of the equipment it plans to procure from abroad. If the equipment is among the first few to be imported into the country, it shall have a valid No Objection Certificate. If the equipment is manufactured in India, the procedure is slightly different. The institution shall have an approved radiation safety officer (RSO) and necessary instruments and accessories (either available or proposed) as other prerequisites.

Equipment Receipt Intimation (ERI) Once the equipment is received at the site, the same must be intimated to AERB by filling out the ERI form. This is a critical step, and extreme care shall be taken while filling it, as it would be extremely difficult to correct any mistake, like in the mentioning of energies. Such mistakes would certainly result in project delays.

Commissioning (Beam ON) After the machine is installed, *prior to switching ON the beam*, permission has to be obtained from AERB by declaring in the eLORA portal the details about the availability of manpower, mandatory instruments and accessories, and personnel monitoring services. Further, an undertaking that the bunker has been constructed as per the approved layout and an emergency preparedness plan must also be submitted.

Survey Report With due approval, the beam is switched ON and a detailed radiation survey of the installation shall be carried out. The survey report has to be uploaded in the eLORA portal. If the results of the survey are satisfactory, the survey report is *accepted*. The user institution can go ahead with data collection only after the *acceptance* of the survey report.

License Once the customer acceptance tests are completed, medical physicists shall collect all the data required by AERB, compile them, and submit them in the *required format* to the AERB while applying for the license. The license for clinical use is granted if the submitted data are satisfactory. It should be noted that this process of licensing may take several weeks to scrutinize the submitted data.

Data Collection for Commissioning Treatment Planning System (TPS) Although this is not a regulatory step, the data collection and measurements for TPS are done immediately after applying for the license. This may extend to several weeks depending on the number of beam energies (photon and electron), additional features and capabilities of the machine, requirements of the TPS, etc. The data required for TPS commissioning are different from the data required for license application, though there may be some amount of overlap. A linac is not a fancy electronic gadget that can be taken out of its packaging and put to its intended use immediately thereafter. Medical physicists should never be pressurized into speeding up data collection since it is a highly sensitive work that warrants concentrated attention. Adding pressure will only prolong the process. Everyone involved in the project team must be aware of these facts and acknowledge their importance.

Renewal of License A license for radiotherapy equipment is usually valid for 5 years. Institutions can apply for license renewal 59 days before it expires by submitting the results of QA tests and surveys.

Decommissioning and Disposal Institutions can apply for decommissioning and disposal by submitting an acceptance letter from the suppliers and a concurrence letter from the disposal agency for accepting the disused radioactive source. A separate permission has to be obtained for the transport of the registered source. The process of decommissioning and disposal is completed by intimating the decommissioning and disposal of the source.

The typical steps involved in the licensing process will vary depending on the type of the equipment (whether it is a radiation generator or radioactive source based). Strict compliance to the regulatory requirements at every stage of the licensing process is required. The whole process is done through the online eLORA portal, and the license issued is also in electronic format. As of now, only in the case of authorization for the import/export of radioactive source, a hard copy of the authorization is sent to the user institution to meet the needs of the customs department.

The hospital should keep printed copies of all such authorizations, licenses, permissions, registrations, etc., issued by AERB in a separate file in an organized manner. As per the terms and conditions of the license, a copy of it should be displayed within the premises where the equipment is installed.

The regulatory body makes scheduled visits and can also make surprise visits to user institutions to check on their regulatory compliance status. Any deficiency found during such visits would invoke show-cause notices from them and would get recorded in the institution's compliance history. It is highly recommended to maintain a compliance tracker for the licenses issued (a sample is given in Table 24.7) and another compliance tracker for the calibration status of the dosimetry instruments (a sample is given in Table 24.8) to avoid missing any due dates.

Table 24.8 shows the measuring tools, safety tools, QA tools, monitoring tools declared in the eLORA. User institution can use color coding to attract attention. For example, entries that do not require calibration may be left as such; entries with calibration validity more than a year may be filled in green; entries that require calibration within the next one year may be filled in yellow, and entries that require calibration within the next two months or with expired calibration may be filled in red.

24.12 Site Layout Preparation, Approval, and Construction

The site and layout plans for the radiotherapy facility must be prepared in consultation with the hospital management, radiation oncologists, medical physicists, a radiation safety officer, equipment supplier, and an architect. The shielding drawings shall be prepared by, or under the close supervision of, a qualified radiation safety expert. Any plans for future expansion shall be taken into consideration. A little extra care taken at the time of planning can save substantial costs, result in optimal utilization of space, and avoid teething issues related to engineering, among other benefits.

The radiotherapy facility may either be located in the basement, as is the practice most often, or on the ground floor. The basement option may result in space savings and, to some extent, savings in construction costs, but is riddled with the risks of water seepage, flooding, and lack of natural light. A well-designed ground-floor facility, besides having ample sunlight, will have a psychologically positive impact on the patients while they are waiting for their turn of treatment.

The plans must be well discussed and scrutinized before submission to the AERB for approval, since any alteration to them would involve resubmission and re-approval, requiring justifications and additional time for processing. If such corrections or additions are thought of after the completion of construction, it would not only be expensive to execute them, but also prove to be impossible to carry out. The construction of the facility shall commence only after the final approval of the layout plans by the AERB. There shall not be any deviation or violation of any kind from the approved diagrams, as they would be viewed seriously by the regulatory body. Sometimes, a properly constructed shielding may have to be compromised to accommodate

Table 24.7 The status of various approvals, licenses, no objection certificates issued by Atomic Energy Regulatory Board

Status of AERB licences in e-LORA								
Sr. No.	Name of license	Validity	Issued by	Issued to	Registered in eLORA under entity	eLORA case file number	eLORA document number	eLORA equipment ID
Radiation oncology								
1.	Site and layout approval of medical accelerator installation	From ____ to ____	AERB	Name of the employer/ license	Name of the hospital/ institution	DL-012345-INST	16-SLA-111111	Not applicable
2.	Site and layout approval of remote afterloading brachytherapy facility	From ____ to ____	AERB			DL-012345-INST	17-SLA-111112	Not applicable
3.	NOC for procurement of radiation source—Ir 192—10 Ci	From ____ to ____	AERB			DL-012345-RF-RH-002	17-PROC-111113	Not applicable
4.	License for operation of medical accelerator—TrueBeam SVC with FFF Sr. No. xxxx	From ____ to ____	AERB			DL-012345-RF-RA-004	17-LIC-111114	Equipment ID number in eLORA
5.	License for operation of kV imaging unit—OnBoardImager—Sr. No. xxxx	From ____ to ____	AERB			DL-012345-RF-RV-001	17-LIC-111115	Equipment ID number in eLORA
6.	License for operation of remote afterloading brachytherapy unit—GammaMed Plus iX Sr. No. xxxxxx	From ____ to ____	AERB			DL-012345-RF-RH-002	18-LIC-111116	Equipment ID number in eLORA
Cath lab								
7.	License for operation of medical diagnostic equipment—cathlab—allura clarity FD 10	From ____ to ____	AERB	Name of the employer/ license	Name of the hospital/ institution	DL-012345-RF-XL-016	18-LOP-111124	Equipment ID number in eLORA
	Diagnostic radiology							

(continued)

Table 24.7 (continued)

Status of AERB licences in e-LORA

Sr. No.	Name of license	Validity	Issued by	Issued to	Registered in eLORA under entity	eLORA case file number	eLORA document number	eLORA equipment ID
8.	License for operation of medical diagnostic equipment—CT—ingenuity	From _____ to _____	AERB	Name of the employer/ license	Name of the hospital/ institution	DL-012345-RF-XL-023	18-LOP-111125	Equipment ID number in eLORA
9.	Registration for operation of medical diagnostic equipment—C-Arm—BV Endura	From _____ to _____	AERB			DL-012345-RF-XR-019	18-LOP-111126	Equipment ID number in eLORA
10.	Registration for operation of medical diagnostic equipment—mammography—MammoDiagnost AR	From _____ to _____	AERB			DL-012345-RF-XR-015	18-LOP-111129	Equipment ID number in eLORA
Nuclear medicine								
11.	Site and layout approval of nuclear medicine facility	From _____ to _____	AERB	Name of the employer/ license	Name of the hospital/ institution	DL-012345-INST	17-NMSLA-111111	Beta therapy: 17-NMSLA-111111-beta-therapy-1111 LDT: 17-NMSLA-111112-LDT(low dose therapy)-1112 SPECT-CT: 17-NMSLA-111113-SPECT-CT-1113 Non-imaging techniques: 17-NMSLA-111114-non-imaging techniques-1114 PET-CT: 17-NMSLA-111115-PET-CT-1115
12.	License for operation of NM facility—beta therapy	From _____ to _____	AERB			DL-012345-NM-Equipments-001	18-NMLICENSE-111112	Not applicable

13.	License for operation of NM Facility—SPECT-CT	From _____ to _____	AERB			DL-012345-NM-Equipments-001	18-NMLICENSE-111115	Not applicable
14.	Permission for procurement of PET-CT	From _____ to _____	AERB			AERB/RSD/NM/PET-CT-Proc/DL-12345/20xx/111	Not applicable	Siemens Biograph Horizon CT: Somatom Perspective AERB Type approval: 99-TA-111 130 kVp 345 mA
15.	License for operation of NM Facility—PET-CT	From _____ to _____	AERB			DL-012345-NM-Equipments-001	18-NMLICENSE-111116	Not applicable
16.	NOC for procurement of Sm-153 EDTMP	From _____ to _____	AERB			DL-012345-NM-SourceProcuremnt-006	18-NMSRCPROC-111120	2590 MBq fortnightly
RSO approval								
17.	Approval of RSO -Mr. ______________	From _____ to _____	AERB	Name of the employer/licensee	Name of the hospital/institution	DL-012345-INST	18-RSO-111111	Not applicable

Table 24.8 Instruments declared under eLORA and their calibration validity status

Instrument type	Instrument sub-type	Instrument ID	Calibration valid until	Instrument make	Instrument model	Instrument status	Remarks
Instruments declared under eLORA and their calibration validity status							
Safety tools	Emergency source storage	SE-1111111				Available	
QA tools	RFA phantom	SE-1111112		PTW	MP3-M	Available	
QA tools	QA gadgets	SE-1111120				Available	
Measuring tools	Parallel plate chamber	SE-1111121	15-Jul-25	PTW	Markus Chamber	Available	
Measuring tools	Electrometer	SE-1111122	27-May-25	PTW	UNIDOS E	Available	
Measuring tools	Thimble chamber	SE-1111123	27-May-25	PTW	Farmer Type Chamber	Available	
Monitoring tools	Survey meter	SE-1111126	11-Aug-25	FLUKE BIOMEDICAL	451P- RYR	Available	
Measuring tools	Large volume/ well type ion chamber/dose calibrator	SE-1111127	10-Oct-22	Standard Imaging	HDR 1000 plus	Available	Wrote mail to RSS on 2-1-21
Monitoring tools	Gamma zone monitor	SE-1111128	16-Nov-25	Pulsecho Systems Bombay Pvt Ltd	NZM	Available	
Safety tools	Long forceps	SE-1111129				Available	
Monitoring tools	Pocket dosimeter	SE-1111138	30-Jan-22	Ludlum	23-1	Available	Sent mail to Biomed on 2-1-21
Monitoring tools	Survey meter	SE-1111153	7-Oct-25	G17	DIGIRAD STANDARD	Available	
Measuring tools	Large volume/ well type ion chamber/dose calibrator	SE-1111155	31-Jul-25	CAPINTEC	CRC-55tPET	Available	

unplanned ducts, recessed boxes (for laser lights), or other hardware. These shall be accepted only after careful consideration by the site's radiation safety officer (RSO) and the medical physicists. There shall be periodic checks of the construction to ensure the integrity of the approved barriers. The inspections should evaluate the following: the location and width of primary barriers from isocenter; thickness and density of the concrete walls (density shall not be less than 2.35 gm/cc); checking the density every time a new concrete layer is laid; heating, ventilation, and air conditioning (HVAC) ducts; location and size of the cable conduits for the use by physics; etc. Concrete samples collected for density testing can be dated and saved for future reference. All inspection visits and findings thereof shall be documented and kept as permanent records. Any noncompliance findings shall be brought to the attention of everyone in the project coordination team.

24.13 Power Requirements for Linacs

The power requirement for a high-energy linac is approximately 60 kVA. For a linac with kV X-ray imaging capabilities (X-ray image guidance), the kVA required will be higher. Similarly, the power requirements for closed-ring gantry accelerators, and robotic accelerators are typically around 70–80 kVA. However, for each of these linacs, the recommended power rating for the online uninterrupted power supply (UPS) is 125 kVA. At the time of facility design, special care must be taken to provide uninterrupted power supply to all the critical systems in the department from this UPS, like treatment planning systems (TPS) and all desktop systems, instead of depending on the hospital UPS, which may not be reliable. Electrical wiring has to be done accordingly at the time of construction, as it would be difficult to accommodate it later. A separate space should be provided within the department, preferably near the linac, for keeping such UPS unit(s). The ambient temperature in the area shall be maintained at around 18 °C, preferably with a dedicated split-type air-conditioner, to prevent heat from building up. Brachytherapy systems require about 2–3 kVA, which can be handled by a separate standalone UPS.

The site planning guide or similar document from the manufacturer must be handed over to the hospital project team, which would be responsible for addressing all the needs mentioned therein.

24.14 Grounding of Treatment Machines (Earthing)

For fault-free operation, external beam therapy machines and high-dose rate remote afterloading systems require high-quality grounding. Grounding cables and connections to these systems must be dedicated. They should not share grounding with other pieces of equipment from the hospital, like air-conditioning, air compressors, UPS, etc. Each vendor has specific requirements. Not only must these be addressed, but adequate safeguards must also be in place to prevent thefts of the high-purity copper plates used for this purpose. The project team, especially the electrical engineers, must interact with the vendors to meet all these requirements.

24.15 Electrical Outlets

Everyday workflow in a modern radiotherapy department is completely dependent on a complex network of interconnected desktop workstations stationed at different locations. This includes systems for patient appointments, patient registration, patient scheduling, contouring of volumes on patient images, treatment planning, image guidance, and treatment delivery. These systems and several other accessories require a sufficient number of power sockets to be available in the areas where they are located. Insufficient number of sockets would cause a lot of inconvenience, necessitate the use of electrical extension boxes to connect systems, resulting in power cord clutter and potential electrical hazards.

The linac treatment console area, the linac bunker, the treatment planning system, the CT-simulator console area, the CT-simulator room, the TPS room, brachytherapy bunker, including its console area, are some of the areas that require many sockets. All of these sockets must be powered by the linac UPS. The linac console area must have at least 16 universal sockets that can accept all types of electrical plugs (pins). If, in addition to linac, third-party accessories and equipment are installed, then more sockets would be required. These outlets shall be distributed at the height of the console worktable and also below it. It is critical that the grounding for these outlets be the same as for linac. Power outlets should be installed on all the walls inside the linac bunker every six to ten feet at the floor level. More sockets will be required at 6–7 feet for installing the CCTV cameras and 4–5 feet for the room lasers. If a ceiling laser is planned, then provision must be made for a socket at the ceiling level. In the remote afterloading brachytherapy treatment console area, a minimum of 10 power sockets are required sharing the common grounding. Inside the brachytherapy bunker, power outlets should be provided, like in the linac bunker, on all walls. The TPS room and the QA room will require more power outlets than a normal room. For every workstation planned inside the TPS room, there shall be a minimum of 3 outlets.

Additionally, a few more should be provided for the printers, the scanners, and setting up of systems for temporary use. These requirements shall be discussed with the electrical engineers, documented, and inspected at the time of construction.

Key areas like bunker rooms, mold room, waiting areas, and console rooms shall have an emergency light source, or be backed up through UPS, so that in the event of a disruption in electricity supply, people will have enough light to walk out to a safe place.

24.16 Heating, Ventilation, and Air-Conditioning (HVAC)

For a linac to run smoothly, the room temperature inside a bunker should be between 18 °C and 22 °C, with humidity levels between 30% and 60%. In radiotherapy facilities, most often, downtimes of the treatment units are associated with deficiencies in air-conditioning and humidity control. The problems become acute if the expected workload is heavy, as the continued operation of linac can generate excessive heat within the bunker and also in the console area. The bunker needs air-conditioning capable of six to eight air-exchanges per hour. If the executives in the engineering section of the hospital do not have prior experience in maintaining a radiotherapy department, they may not understand the unique needs of the department. In the initial stages of the project, during project committee meetings, all concerned, including the hospital management, the finance department and the engineering executives, must be made aware of the finer aspects of these requirements and their implications, especially on machine downtimes. It will be worth spending some 10–20% more than the allocated budget for air-conditioning which will more than pay back in the form of reduced downtimes of the machines later. Similarly, the linac's chiller requires specialized care from the engineering division to ensure fault-free operation of the linac. All simulation, treatment, and equipment rooms not required to be under negative pressure shall have positive air pressure with respect to the adjoining areas. This is to help maintain a reduced dust environment with respect to the electronic equipment.

Adequate care must be taken when HVAC services need to penetrate a shielded enclosure. Any penetration in the bunker wall must be authorized by the medical physicist-cum-radiation safety officer, who shall ensure that no regulatory conditions are violated.

24.17 Chiller

Several assemblies and components in the linac need cooling, like the accelerator guide, bend magnet, accelerator solenoid, primary collimator, energy slit, target, klystron, klystron solenoid, circulator, RF driver, RF load, and pulse transformer. The water cooling system of the linac takes the heat from these critical components to a chiller, usually installed on a rooftop, and dissipates it. The chiller should be located as close as possible to the linac bunker so that the water lines are shorter, easier to lay down, and result in better efficiency.

Some linacs, like Tomotherapy, do not a require water chiller and are air-cooled. For such units, manufacturers have specific HVAC requirements that should be discussed with the company's project team.

24.18 Medical Gas

Medical gas outlets should be provided inside the linac bunkers, brachy bunkers, as well as in the stretcher /holding area and procedure room.

24.19 Networking and Information Technology Support

Not long ago, radiotherapy treatment delivery machines were simple standalone units and were manually operated; treatment planning systems were based on huge standalone mainframe systems; imaging modalities were hardly used in radiation oncology; and dosimetry and QA systems were manually operated. But since the early 1990s, computers have made their way into every step in the radiotherapy workflow, like multimodality imaging, advanced treatment planning systems, complex treatment delivery units, fully computerized dosimetry and QA systems, radiation oncology information systems facilitating electronic workflow, etc. In every arena, computers are extensively used, and they are all networked. It can be safely said that if the network is down in the department, no treatment delivery is possible.

The networking needs of a RT department are unique, highly sensitive, and demanding in nature. For uninterrupted patient services, the two departments must work closely together. While the medical physicist is technically competent to spell out the needs of the department, the IT experts are the ones who can ensure those are met. Table 24.9 lists the most important IT needs of the RT department.

The setting up of a radiotherapy facility is a demanding task that calls for unprecedented level of coordination between members from multiple disciplines: administrators, decision makers, professionals like the radiation oncologists, the medical physicists, the engineers, and several others. In the selection of the appropriate technology, besides the administrative support, choice of technical parameters [9], available resources, availability of quality staff, and availability of stable infrastructure and technology servicing are taken into account.

Table 24.10 lists some of the most common errors that happen in clinics and their after effects. Every effort must be made to avoid these common errors.

Table 24.9 List of IT needs of an RT department

Requirement	Explanation
RT department has several client workstations (systems) which are connected to a dedicated server. All the systems are essential for treatment planning, patient scheduling, and treatment delivery. Because of the nature of their functioning, they are called as **medical devices.** An isolated network is required EXCLUSIVELY for interconnecting all these systems since they have to operate in a safe environment, completely protected from outside connectivity and internet. Specifically these systems CANNOT be part of the hospital network.	The dedicated network has to guarantee data exchange between client units and server at faster rates since they are sensitive to time latency. Every treatment delivery is entirely dependent on uninterrupted connectivity between all the systems. None of the systems can have antivirus protection due to their unique nature.
Some of these systems are extremely sensitive since they control radiation generating machines. They are further protected through a network security switch, thus creating a network within a network. None of these systems require internet connectivity EXCEPT the server and two more systems (that sit behind this network security switch).	The internet connectivity is required ONLY to connect with few specific sites the details of which will be shared by the vendors.
IT department must allocate a range of IP addresses (a minimum of 50–100) to this dedicated network. RT department shall have the freedom to allocate IP addresses within this range to any new system added in the department.	RT Department has many permanently installed workstations that are part of the workflow. In addition to these fixed workstations, there are several other mobile QA and dosimetry workstations that are set temporarily for patient related measurements. They shall also have to be in the same network so that data can be shared between the systems.

(continued)

Table 24.9 (continued)

Requirement	Explanation
The server has two network cards. One of the cards is used to connect the server to the rest of the clients through this dedicated network, while the server's other network card can be used to connect it to the network in which the imaging units are functioning.	Images from hospitals' CT scanner, MRI unit, PET-CT unit are required for the treatment planning of cancer patients undergoing radiotherapy. These imaging units are connected to the RT server through this second network card for the transfer of images from these units directly to the RT server. Please note that the imaging units CANNOT transfer images to clients.
A close coordination is required between the service/ installation engineers of CT, MRI, PET-CT units, hospital's IT department, HODs of Radiology, Nuclear Medicine, Radiation Oncology, Medical Physicists from Radiation Oncology and the service engineers of linac, and treatment planning systems. A Task Force shall be created with the above members to ensure smooth communication between the systems.	Data transfer between imaging units and radiotherapy systems is an one-way traffic. Images will be exported from CT/MR/PET-CT to RT Server and/or to RT-Treatment Planning Systems (RT-TPS). No data will be exported by RT Server, RT-TPS to any of the imaging units. However, these RT systems pose stringent requirements: (1) Nodes should be created in the imaging units for data export to RT systems according to specific protocol (AE Title, Port number images, etc.). (2) The exported images should be according to certain specific DICOM format which the service engineers of linac/TPS will provide. Images that are not in this specific format, even if they are in DICOM format, will not be imported in the TPS. The DICOM export configuration in imaging units is a sensitive step and it serves as the source of many teething troubles faced by RT department in many centers. Close coordination and understanding of the unique requirements go a long way in establishing a smooth workflow.
The RT server is the nerve center for all processes in the department. It holds the records of every treatment delivered to each and every patient who was treated in the department from day one. It is the data server of a hospital information system called as the "Radiation Oncology Information System (ROIS)". It has to be kept in the SAFEST PLACE.	The Data Center (Server Room) in a hospital is a well-protected, air-conditioned room with adequate fire-proof measures and with authorized access. It is extremely important that the RT Server has to be located in this Data Center considering the importance of data (patient treatment records) stored in it. During the construction of the hospital, this requirement has to be kept in mind and a suitable conduit has to be laid from the Data Center to the RT Department for laying down cables. The distance has to be the shortest possible. If the hospital does not have a Data Center, then the server must be located in the RT Department. A separate server room must be constructed and strengthened with fire-proof measures and access control.
The RT Server needs backup every day. In the event of any catastrophic failure, if the hospital does not have proper backup solutions, it would lead to the loss of all the treatment records. It will not be possible to continue with the ongoing treatments. This would place the hospital reputability at stake. Legal issues may also arise causing huge losses to hospital. There shall be fool-proof backup solutions in place to get back the data in a matter of hours.	Vendors provide in-built backup solutions. These are in the form of backup tapes in the server. However, these are not adequate since they are in-house. In the event of a catastrophic failure such as fire accident even the backup also would be destroyed. A cloud-based backup solution at a remote location outside of the city must be in place. Identified directories of the server must be backed up on a daily basis during the night to this cloud system. Periodically the integrity of this cloud based solution must be tested. One should download the backed up data from cloud on to a new system and check whether all the data are available. It should be documented. Linac vendors will identify the directories that need to be backed up.

(continued)

Table 24.9 (continued)

Requirement	Explanation
Modern IT solutions are becoming increasingly cloud-based. Radiotherapy vendors are also providing innovative cloud based solutions for the entire RT workflow. In this type of solution, there is no local server and only client systems are provided locally. Everything is stored in a cloud server. Every time a patient has to be treated, patient data is fetched from the cloud, loaded into a local system, treatment is delivered by this system and after the completion of the treatment, the treated data is stored back into the cloud.	The option of going for such a complete cloud-based solution, instead of only cloud backup, should be discussed in length. If one has a good bandwidth, highly reliable internet service with negligible latency period, this is the best solution since there is no fear of loss of data. This option must be explored in depth at the time of setting up a department.
Adequate number of network ports (data ports) shall be provided in the every room in the RT department.	Every room must have at least two data ports—one port in the RT network and another port in the hospital network. The linac control console, brachytherapy console, linac bunker, brachy bunker, TPS room, seminar/meeting room will require more ports. They shall be provided at strategic locations above the floor level and as well as above the worktable height offering freedom to connect the systems in the most optimal way. Linac console, brachy console and TPS room, seminar room will require more number of ports. Ports in TPS room and in seminar/meeting room shall be on all FOUR SIDES OF THE WALLS Linac console—8–10 RT network ports, 4 hospital network ports Brachy console—5 RT network ports, 3 hospital network ports Linac bunker—2 RT network ports, 1 hospital network port Brachy bunker—2 RT network ports, 1 hospital network port TPS Room—10 to 20 RT network ports, 4–6 hospital network port Seminar Room (Meeting Room)—2 RT network ports on each wall, 1 hospital network port on each wall
A network monochrome A3 size printer-cum-color scanner with automatic document feeder shall be provided for departmental use. It will reside in hospital network.	The scanner scans and sends file to a shared folder in hospital computer. Whenever needed, these files can be imported into the Oncology Information System.

Table 24.10 List of common errors and their aftereffects

Improper plans/plans with deficiencies/without mandatory annotations/plans not to the scale are submitted to the AERB.	Will result in rejection of the application leading to inordinate delay.
Purchase decisions are often taken by the management and the finance team.	Neither the management team nor the finance team has the required expertise to take decisions on purchase. They may not be aware of the implications of their decisions on the clinical services and on the quality of healthcare.
Dosimetry items are removed from the list and/or inappropriate items are purchased as a cost-cutting measure.	Without right kind of dosimetry items, data collection and subsequently licensing will be delayed. Certain clinical services cannot be commissioned at all. Dosimetry and QA items shall not be viewed from revenue generation point of view or seen as a financial burden. They should be viewed as commitment to providing quality healthcare.

(continued)

Table 24.10 (continued)

Noncompliance to the eLORA steps.	Designated RSO should check the eLORA portal almost on a daily basis.
Unauthorized changes in the site plan/layout plan.	Will lead to objections from the AERB and significant delays in commissioning.
Missed periodical inspections of the site during the time of construction by qualified personnel.	Risk of poor quality construction; lower concrete density; inadequate shielding; huge cost for rectifying.
Undue pressures on medical physicists after commissioning and acceptance tests by vendors.	Inadequate, insufficient data collection. Possibility for errors.
Choice of too many photon and electron energies.	Does not really serve any purpose. Two photon energies (with FF), one photon energy (FFF) and 4 or 5 electron energies are more than sufficient. More energies mean more time for QA.
Noncompiling of baseline data for all the beams (dosimetric parameters), performance of the imaging systems, TPS, etc.	Baseline data MUST be established at the time of initial commissioning and all subsequent QA results shall be compared against these baseline data.
Inadequate number of electrical sockets, network nodes.	Requires electrical extension boards, installation of hubs leading to cluttering of wires. May go against the hospital policy. Safety hazard.
Hospital's IT department does not understand the unique needs of RT department.	Teething troubles in networking between systems within the department and importing images from different imaging units.
Inadequate attention to HVAC.	Not effective air-conditioning, humid environment—all affecting the functioning of the equipment.
Room lasers should preferably be in a recessed box on the walls; but are not backed up by sufficient lead for providing enough shielding.	If the lasers are protruding outside, chances of bumping into them and losing alignment with isocenter are plenty. However, the recession should be sufficiently compensated by enough lead to compensate for the lost shielding.
Storage space for the transfer tubes in brachytherapy bunker is not provided.	Transfer tubes should ideally be hanging vertically inside a suitably designed box. If they are kept in simple drawers, possibility for kinking.

Precept Box (Take Home Points)

- India's teletherapy machine density stands approximately at 0.5 machine/million population. It is less than the minimum recommended density of 1 machine per million and much less than the ideal figure of 4 machines per million.
- A core implementation team shall be formed with members from different professions to oversee the project implementation and periodically review the progress made.
- The consultant radiation oncologist and the chief/consultant medical physicist shall be the key members in the formation of specifications, tendering, evaluation, and selection of radiotherapy equipment.
- The telecobalt technology cannot be completely ignored, and there may be situations where Co-60 technology is still relevant especially in view of the fact that 25% of the radical cases can benefit from it.
- For the institutions planning to acquire new treatment technologies, it will be a worthwhile exercise to go through the questionnaire given in Appendix II of the IAEA document [10].
- There is no clinical benefit in having all three photon energies in one machine: 6 MV, 10 MV, and 15 MV.

- Comprehensive maintenance contract is approximately 5–6% of the total equipment cost.
- The life expectancy of a linac and an HDR brachytherapy unit is 10–15 years.
- In the design of the facility, care should be taken to keep patient movement areas and staff movement areas separate. Controlled areas should have electronic access control system.
- The regulatory steps involved in the licensing of a radiotherapy equipment are dependent on its type and are part of a long process.
- The hospital shall keep all the authorizations, approvals, permissions, and licenses in an organized manner. Licenses shall be displayed at a suitable location.
- The regulatory body can make surprise visits to the facility. Any deficiency and/or noncompliance found during such visits shall be viewed seriously.
- At the time of construction of bunker, periodic checks shall be carried out to ensure construction conforms to layout approval in every respect. All inspection visits and findings thereof shall be documented and kept as permanent records. Any noncompliance findings shall be brought to the attention of everyone in the project coordination team.
- Power requirement for a linac is around 125 kVA.
- Radiotherapy facilities need many power sockets, especially in the TPS rooms, the control console areas, etc.
- Every radiotherapy department needs a close coordination with the IT department.
- Adequate UPS backup should be there to complete the treatment in the event of a power breakdown.

Multiple Choice Questions

1. India's teletherapy machine density (number of machines per million population) is approximately:
   (a) 0.5
   (b) 1.0
   (c) 2.0
   (d) 4.0
2. The cost of comprehensive maintenance contract (CMC) is approximately about ___ of the total cost of the equipment.
   (a) 1%
   (b) 3%
   (c) 5–6%
   (d) 10–15%
3. The activity of a telecobalt source is approximately around:
   (a) 900 Ci
   (b) 9000 Ci
   (c) 20,000 Ci
   (d) 40,000 Ci
4. The power requirement for a typical linac with kV-imaging capability is approximately:
   (a) 12 kVA
   (b) 25 kVA
   (c) 60 kVA
   (d) 125 kVA
5. The expected lifetime of a linac is:
   (a) 3–5 years
   (b) 5–10 years
   (c) 10–15 years
   (d) 20–25 years
6. The typical cost of a single energy linac (6 MV) is approximately:
   (a) Rs. 1 crore
   (b) Rs. 3 crores
   (c) Rs. 6 crores
   (d) Rs. 10 crores
7. The customs duty for an imported linac is around (in December, 2022):
   (a) 5–10%
   (b) 10–15%
   (c) 15–20%
   (d) 27–29%

8. eLORA is a website portal:
   (a) Owned and maintained by a private company to issue licenses for linacs
   (b) Owned and maintained by a private company to issue licenses for all radiation related applications
   (c) AERB's website portal for licensing of low-energy and high-energy linacs
   (d) AERB's website portal for licensing of all radiation related applications
9. The density of the concrete used in the bunker construction shall *NOT* be less than:
   (a) 1 g/cc
   (b) 1.35 g/cc
   (c) 2 g/cc
   (d) 2.35 g/cc
10. The humidity level inside a linac bunker shall be kept in the range of:
   (a) 1–5%
   (b) 5–10%
   (c) 30–60%
   (d) 70–99%

Answers

1. (a)
2. (c)
3. (b)
4. (d)
5. (c)
6. (d)
7. (d)
8. (d)
9. (d)
10. (c)

References

1. Abdel-Wahab M, Varghese C, editors. Setting up a cancer centre: a WHO–IAEA Framework. Vienna: International Atomic Energy Agency; 2022.
2. Aljurf M, Majhail NS, Koh MB, Kharfan-Dabaja MA, Chao NJ, editors. The comprehensive cancer center: development, integration, and implementation. Cham: Springer; 2022.
3. The ASCO Post. Tackling the global shortfall in radiotherapy. https://ascopost.com/News/33894. Accessed 24 Nov 2022.
4. Abdel-Wahab M, Gondhowiardjo SS, Rosa AA, Lievens Y, El-Haj N, Polo Rubio JA, Prajogi GB, Helgadottir H, Zubizarreta E, Meghzifene A, Ashraf V. Global radiotherapy: current status and future directions—white paper. JCO Glob Oncol. 2021;7:827–42.
5. Laskar SG, Sinha S, Krishnatry R, Grau C, Mehta M, Agarwal JP. Access to radiation therapy: from local to global and equality to equity. JCO Glob Oncol. 2022;8:e2100358.
6. Zubizarreta EH, Fidarova E, Healy B, Rosenblatt E. Need for radiotherapy in low and middle income countries–the silent crisis continues. Clin Oncol. 2015;27(2):107–14.
7. International Atomic Energy Agency. Radiotherapy in palliative cancer care: development and implementation. In: IAEA human health reports No. 2. Vienna: IAEA; 2012. https://www.iaea.org/publications/8128/radiotherapy-in-palliative-cancer-caredevelopment-and-implementation.
8. International Atomic Energy Agency. Setting up a radiotherapy programme, non-serial publications. Vienna: IAEA; 2008.
9. International Atomic Energy Agency. Planning national radiotherapy services: a practical tool. In: IAEA human health series No. 14. Vienna: IAEA; 2010.
10. International Atomic Energy Agency. Selecting megavoltage treatment technologies in external beam radiotherapy. In: IAEA human health reports No. 17. Vienna: IAEA; 2022.
11. Biggs PJ, Ling CC, Purdy JA, van de Geijn J. AAPM code of practice for radiotherapy accelerators: report of AAPM Radiation Therapy Task Group No. 45. Med Phys. 1994;21:1093.
12. World Health Organization. Technical specifications of radiotherapy equipment for cancer treatment. 2021.
13. Atomic Energy Regulatory Board. AERB SAFETY CODE NO. AERB/RF-MED/SC-1 (Rev. 1). Radiation therapy sources, equipment and installations (2011) (The code is under revision).
14. Mullaney T, Pettersson H, Nyholm T, Stolterman E. Thinking beyond the cure: a case for human-centered design in cancer care. Int J Des. 2012;6(3):27–39.

Research, Clinical Trials, and Publications in Radiation Oncology

25

Ajeet Kumar Gandhi, Prashanth Giridhar, Shivakumar Thiagarajan, and Priya Ranganathan

25.1 Introduction

Research is defined as a systematic investigation designed to develop or contribute to generalizable knowledge. Cancer research deals with the studies exploring the various causes of cancer, newer diagnostic methods to identify cancer, and newer treatment modalities (non-surgical and surgical)/drugs/regimens to treat cancer including the various methods of surveillance following treatment for cancer.

Cancer research can be broadly divided as sponsor-driven/initiated (conducted by pharma companies testing investigatory new drug/regimens/device) or investigator-driven/initiated which is typically done at academic institutes. Research is largely funded by government agencies.

A. K. Gandhi (✉)
Department of Radiation Oncology, Dr Ram Manohar Lohia Institute of Medical Sciences, Lucknow, Uttar Pradesh, India

P. Giridhar
Department of Radiation Oncology, MPMMC (A Unit Under Tata Memorial Center), Varanasi, Uttar Pradesh, India

S. Thiagarajan
Department of Surgical Oncology, Tata Memorial Center, Mumbai, Maharashtra, India

P. Ranganathan
Department of Anaesthesiology, Critical Care and Pain, Tata Memorial Center, Mumbai, Maharashtra, India

Types of Research [1]: Medical research can be broadly divided into three types, basic research (analyses the fundamentals of disease and treatment), clinical research (tests new treatments), and epidemiological research (identifies cause and distribution of disease).

Cancer research: Cancer is a major public health problem and a leading cause of death worldwide [2, 3]. Cancer research deals with research that attempts to understand the disease process to develop methods/modalities to prevent cancer, early detection/diagnosis of cancer, and treatment of cancer. Research in oncology is important as it paves way for a better understanding of the disease and for discovering and improving cancer treatment.

25.2 Steps of Conducting Research

25.2.1 Framing a Research Question

The research question should aim to answer an existing uncertainty which on most occasions arises from a clinical scenario that the clinicians have faced in practice. The research question should be framed using the FINER framework (feasibility, interesting, novel, ethical, and relevant) [4, 5]. The research question is the question for which the researcher expects to find an answer after the research. All aspects of the research that

A. Munshi et al. (eds.), *Radiation Oncology – Principles, Precepts and Practice*,
https://doi.org/10.1007/978-981-97-8389-2_25

needs to be carried out in the future need to be connected to this research question. The research question should include the following components, i.e., population (P) of interest, exposure (E) (in observational studies), or intervention (I), comparator (C) outcome of interest (O), and the time frame (e.g., duration of the research)-PICOT framework. Fadino et al. have explained in detail the various aspects within each of the PICOT components that one has to focus on while framing a research question [6].

25.2.2 Review of Literature

It is an exercise of a comprehensive search of the literature on the available up-to-date knowledge on a particular topic of interest. The first step to do after finalizing a research topic is to conduct a thorough search of the literature in search engines and databases such as PubMed, Google Scholar, Science direct, semantic scholar, and others, on the available knowledge on this topic of interest. This search allows one to understand the gaps and controversies in the existing literature and find keywords about the topic of interest that need to be identified. For example, tobacco, smoking, smokeless, and oral cancer. Searching for topics in the various search engines should ideally be done using the keywords with the Boolean operators. Boolean operators are words such as AND, OR, NOT. This helps in obtaining results that are filtered, refined, and focused on the topic of interest largely. The results thus obtained could be articles from journals and/or book chapters predominantly.

The articles or book chapters thus obtained should be further scanned (by reading the abstracts commonly!) to identify if they are relevant or not. If they are deemed pertinent, it is also necessary to look for any further relevant references in the article or chapter's bibliography.

While performing the literature search one needs to identify the existing gaps and controversies and possible themes to conduct research. Also, pay attention to the connections between the different topics, this would help organize the structure of the write-up on literature reviews.

The literature review can be written focusing on various aspects depending on what is found and what we need to emphasize. It can be written chronologically (older to recent publications) or focusing on key themes, or the various methodologies used across studies/publications available so far and it can also focus on the various theories and models.

25.2.3 Finalizing the Research Question

The research question can be discussed and brainstormed with seniors/peers/colleagues or discussed (officially) in department or disease management group meetings to hone and focus the research question. This will not only help hone the research question but also help gain the confidence of colleagues to participate in the study. One should be flexible to suggestions in these meetings. The research question should neither be too broad nor too narrow, both of which might not help one achieve their goals.

A research protocol is a document that has a detailed plan of the proposed project supported by evidence. It is an essential document that would help the research team conduct the study as per the initial plan. This is very useful when there are multiple researchers and centers involved in the study. The research protocol also needs to be submitted to the institute's ethics committee for approval to conduct the study and also for funding agencies to obtain funds to conduct the study.

The research protocol needs to be written with clarity in simple language, paying attention to details about the topic. The protocol has to be written in clear sections which include the background and rationale of the study, aims (which are different from the objectives), objectives (primary and secondary), methodology (includes the study design, eligibility criteria, description of patient population and intervention (if any) and sample size), outcomes and measures, data management, statistical analysis plan, statement regarding study feasibility, budget, references, and appendix (includes quality of life forms, consent forms, etc.) [7, 8]. We should avoid repetition of facts/sentences, with a smooth transition from one section to another. The various study designs are explained in detail in the subsequent sections. Selecting the study design would depend on the research question primarily.

25.2.4 Ethical and Scientific/Peer Review

Once the protocol is written and finalized, it needs to be submitted to the institute's ethics committee (IEC) for approval to conduct the study on-site. Other relevant documents including the case record form and informed consent form along with the protocol need to be submitted for review and approval. The IEC would discuss the protocol including the need and scientific rationale for the study including (but not restricted to) the appropriateness of study design, patient population, eligibility criteria, sample size, ethical concerns if any. Once the committee feels that the protocol is sound with regards to the scientific and ethical aspects approval may be granted for the conduct of the study.

25.3 Medical Ethics and Research

Clinical research eventually aims to improve human health and wellbeing, typically by improving the understanding of the disease, prevention of the disease, and identifying better treatment for the disease. Medical ethics typically attempt to minimize the risks and possibility of exploitation of the participants (patients) and protect the rights and welfare of the subjects. Medical ethics have four pillars, autonomy (respect patients ability to decide), beneficence (do good to the patients), non-maleficence (do not harm the patients), and justice (be fair to the patients) [9].

Concerns of ethics in medical research are as old as medicine itself. The exploitation of participants (patients) has occurred in the past. Various experiments were conducted on the humans in the Nazi concentration camps [10] between 1939 and 1945. These inhuman experiments mainly aimed at improving the rescue and survival of the German troops in the war, testing medical procedures and pharmaceuticals, and confirming the Nazi ideology. Ninety percent of these experiments were criminal. The Nuremberg trial (1946–1947) evaluated the need for such criminal experiments among the available victims-survivors (predominantly European, but not restricted) and perpetrators. Based on the findings of this trial, the 10-point Nuremberg code guides medical research involving human subjects. These are a set (10 points) of research ethics, the first and the most important code being the absolute necessity of obtaining voluntary consent of the participating human subjects [11].

The declaration of Helsinki [12] was adopted by the World Medical Association (WMA) as an important document to be followed while performing research on human subjects. It is reviewed and revised regularly, the last being done in 2013. The declaration of Helsinki remains morally binding for the physicians over and above the local laws and regulations.

Tuskegee experiment [13] was initiated to study the natural course of untreated syphilis in black African-American men. The study was initiated in 1932 and was initially planned for 6 months; however, it was continued until 1972. Around 600 men were enrolled, out of which 399 of them had syphilis and 201 did not have the disease. These men were informed about their disease and no treatment was offered to them despite penicillin becoming the treatment of choice for syphilis in 1947. They were offered free medical care, meals, and burial insurance/stipend. No informed consent was taken from these patients. In 1972, the news of this unethical study was leaked to the press, and there was a subsequently huge public outcry. The national research act (1974) was introduced partly in response to the Tuskegee experiment, which set guidelines to oversee and regulate the use of humans in experimentation medicine. Subsequently, the Belmont report was filed with ethical statements guiding human research in the USA. It reiterated the principles of respect for personal autonomy, beneficence, and justice [14].

There were multiple unethical [15] experiments/studies (22 studies) such as the Willowbrook hepatitis experiment (where children were deliberately infected with hepatitis virus to study the natural history of infectious hepatitis), the Jewish chronic disease hospital cancer experiments (cancer cells were injected into patients body admitted with other chronic debilitating diseases to understand the ability of the body to reject cancer cells) among others that have been done in the past. All these studies describe the various ethical problems related to experimental medicine.

The formation of the international conference on harmonization (ICH) led to the consolidated guidance on good clinical practice (GCP) [16]. This was initiated by the joint initiative by regulators and industry from three regions—the USA, European Union and Japan. The goal of ICH is to decrease the differences in guidelines for research across countries and regulatory authorities. This was essentially done to streamline drug development and regulatory process and to increase the efficiency of clinical research with enforcement of GCP guidelines. The purpose of ICH-GCP is essentially to protect the rights of human participants in clinical trials and to ensure scientific validity and credibility in clinical studies.

25.4 Indian Regulators for Biomedical Research

25.4.1 Schedule Y

It's a law and is a part of drugs and cosmetic acts (schedule A–Y, 1945). Amendments were made subsequently. The purpose of schedule Y is to frame guidelines for the conduct of clinical research and to control regulations for new drugs. The new drug development process is as per schedule Y. It ensures the rights, safety, and well-being of the human participants. It has multiple appendices (I-XIII) that cover an entire array of the process of testing a new drug starting from data submission with application to compensation in case of injury and death.

25.4.2 Indian GCP (Issued by CDSCO) [16]

It has evolved with consideration of the WHO guidelines for research, ICH, US-FDA, and the European GCP guidelines along with ICMR guidelines. It details the prerequisites for conducting clinical trials in India. It clearly states the roles and responsibilities of sponsors, monitors, and investigators, along with record-keeping and data handling. It also ensures quality assurance.

25.4.3 Ethical Guidelines for Biomedical Research (ICMR, 2017) [17]

The Indian council of medical research (ICMR) under the department of health research is the apex body that sets guidelines for biomedical and health research in India. The first guidelines were released in 1980 and subsequently revised in 2000, 2006, and 2017. In addition, there are separate guidelines for research in children and research with stem cells. The ICMR guidelines have twelve sections with a focus on different areas of conduct of research and ethical requirements. It also provides details about the structure of the ethics committee, its composition, the roles, and responsibilities of its members, etc. ICMR guidelines give details regarding the aspects that need to be included in the informed consent documents to promote the voluntary participation of patients in clinical trials. It also describes and identifies the vulnerable population to safeguard the interests of these participants. Apart from these, ICMR also provides guidelines for certain specific types of research to be carried out, such as public health research, social and behavioral science research, etc.

25.4.4 Regulatory Requirements for Research in India

The Central Drugs Standard Control Organization (CDSCO) [18] is the national regulatory authority in India under the Ministry of Health and Family Welfare, Govt. of India. The Drugs Controller General of India (DCGI) is an official of CDSCO and is the final regulatory authority in India for approving the conduct of clinical trials.

The regulators have defined clinical trials, new drugs, and the conduct of clinical trials. As per regulatory requirements, a researcher needs to obtain approval from their respective institutional ethics committee (IEC) which is registered with the central licensing authority (renewed every 3 years). After obtaining approval from IEC, the study needs to be registered with the Clinical Trials Registry of India (CTRI) before the enrollment of the first participant into the study. The investigators must obtain written, informed consent from the study participants. It is important to report serious adverse events that occur in the clinical trial within the timelines mentioned by the regulatory authorities. Also, the investigators should compensate for trial-related death or injury. The preparedness of the research site must be always ensured as the regulatory authorities can inspect the site at any time.

25.4.5 Funding for Research in India [19]

After obtaining the approval for research from the regulatory authorities, it is important to obtain funding for the research to be able to execute the study. Funding is essential for the manpower needed to execute the study and for materials, for example providing new drugs, compensation, etc. Obtaining funds for research is a competitive process. Sources of funding can be intramural (from within the institute) or extramural (from funding agencies). Among the various funding agencies, the thrust areas supported by them are different, which is important to know, to apply for the right funding agencies. This naturally improves the prospect of obtaining the funds. There are various funding agencies in India and abroad, the amount and/or duration of funding differs across these agencies.

25.5 Study Designs for Clinical Trials

25.5.1 Types of Study Designs

Study design refers to the method or strategy used to answer a research question. Primary research involves the collection of data directly through self-conducted research, while secondary research is information derived by synthesis of data gathered from previously conducted studies, as in systematic reviews and meta-analyses. Primary research designs are broadly classified into observational and experimental designs.

25.5.2 Observational Studies

Observational studies "observe" the natural relationship between risk factors (or exposures) and outcomes. There is no active intervention by the researcher. These can be further classified as descriptive or analytical studies

Descriptive studies attempt to describe variables in a sample. They are largely hypothesis-generating. Examples of descriptive studies include case reports, case series and some cross-sectional studies.

Analytical studies attempt to quantify a relationship or association between variables, and usually test a pre-specified hypothesis. Examples of analytical studies include case-control studies, cohort studies and some types of cross-sectional studies

Observational studies may also be classified in other ways:

Based on the direction of enquiry: Forward-directed studies begin by assessing the presence or absence of exposure and then determine the occurrence of the outcome. Backward-directed studies begin by assessing the occurrence (or absence) of outcome, and then determine whether exposure was present.

Based on the timing of the study: In prospective studies, the outcome has not occurred at the time of initiation of the study, and follow-up is needed to determine the outcome. In retrospective studies, the outcome has occurred at the time of initiation of the study.

25.5.2.1 Specific Types of Observational Studies

Case reports and case series: Case reports and case series are descriptive studies that highlight a novel or unusual diagnosis, treatment or outcome in a patient or a group of patients with a similar condition. They have limited value in evidence-based medicine but are useful to draw attention to unique or previously unknown features. Well-known examples of case reports that changed medical practice include the description of Kaposi's sarcoma among individuals infected with human immunodeficiency virus (HIV) or the reports of phocomelia in children of pregnant women who received thalidomide.

Cross-sectional studies: are studies which involve collection of data from a sample (population) at a single timepoint. Depending on how the data is analyzed and presented, cross-sectional studies may be descriptive or analytical. They are useful to describe the prevalence of risk factors or disease in a population, and to generate hypothesis for further research. They are easy to conduct, inexpensive and are a good starting point for early researchers. However, they do not provide evidence of cause and effect, and may be affected by several types of bias. For example, Goula et al. assessed the attitude of health professionals toward radiation protection measures in a cross-sectional study [20] carried out over 3 months with each participant being assessed at a single time-point.

Case control studies: In case control studies, the researcher identifies cases (who have developed the outcome) and matched controls (who have not developed the outcome). Matching is done to control for important prognostic factors. Cases and controls are then assessed to determine the presence or absence of exposure to the risk factor. It is important that the assessor is blinded to whether the individual is a case or a control to avoid ascertainment (information) bias. If a significantly higher proportion of cases show presence of the exposure, then an association between the risk factor and the outcome can be concluded. Case control studies are backward-directed and retrospective. They are useful for rare conditions, need few resources and can be completed within a short time frame. However, they rely on good record keeping, are prone to survival and recall bias, and may have issues with unknown confounders while matching cases to controls [21]. For example, Yoon et al. assessed the impact of mobile phone use (exposure) on the risk of developing glioma (outcome). Two hundred and eighty-five patients with glioma were compared to individually matched controls (without glioma), following which history of mobile phone usage was elicited.

Cohort studies: A cohort refers to a group of people who share a similar characteristic. In cohort studies, the researcher identifies a group of people who have the exposure (risk factor) and another group who do not have the exposure [22]. Both groups are then followed up to determine the occurrence of the outcome. If exposed individuals show a significantly higher incidence of the outcome, then the exposure can be considered a risk factor for the outcome. For example, Morgan et al. compared oncological outcomes of 2 cohorts of patients who received chemotherapy or chemoradiotherapy as neoadjuvant treatment for esophageal cancer. Here, the exposure (chemotherapy or chemoradiotherapy) was decided based on clinical need and the researcher merely noted the outcomes. Cohort studies are forward directed and usually prospective. However, researchers may use established registries or datasets to carry out retrospective cohort studies [22].

25.5.3 Experimental Studies (Clinical Trials)

Experimental studies are prospective studies evaluating the outcomes of interventions in human participants. They involve the manipulation of a variable so that participants are actively assigned to receive a particular intervention. Experimental studies are also called clinical trials. Clinical trials of new drugs in humans must be preceded by pre-clinical studies which establish the safety and toxicity profile of the drug in animals.

Experimental studies are divided into several phases:

Phase 1 clinical trials are the earliest trials of a new drug or device in humans and are carried out to determine the maximum tolerated dose of the drug. These are usually carried out in a small number (usually less than 20) of healthy volunteers [23–30]. However, for oncology drugs, phase 1 studies are carried out in patients with cancer. Phase 1 trials evaluate dose-limiting toxic effects and determine the upper limit of dosing for future larger trials, using various models of dose escalation. A recent concept is phase 0 clinical trials which involves micro-dosing in humans.

Phase 2 clinical trials are "proof of concept studies" to look at early evidence of efficacy of the drug and to gather additional data on safety. Phase 2 studies have sample sizes ranging from 25 to 100 patients. If phase 2 studies show evidence of efficacy, then confirmatory phase 3 trials are planned [31, 32]. Phase 2 studies may be single-arm (without a comparator) or randomized (with a comparator).

Phase 3 clinical trials are considered the gold standard for determining the efficacy of an intervention, and are essential prior to approval of a new drug or device for marketing [33, 34]. Phase 3 trials are randomized trials comparing the study intervention with the current standard of care. They have sample sizes varying from a few hundred to thousands of patients.

Phase 4 clinical trials are post-marketing studies which are done in large samples and provide additional real-world data on safety, quality of life and cost-effectiveness [35].

25.6 Sources of Bias in Clinical Trials

Bias refers to systematic errors in the research process which can skew the results toward a particular outcome, and affect the internal validity of a research study. In a clinical trial, bias can occur at various levels. Some examples of bias include:

Selection bias: This occurs when participants are selectively accrued on the trial, depending on which arm they are likely to be enrolled in. This leads to imbalances between groups and confounding of results. For example, participants who are likely to have worse outcomes may be allocated to the control arm, making the intervention appear more beneficial. Selection bias can be prevented by randomization and allocation concealment (preventing foreknowledge of upcoming allocation).

Performance bias: If researchers are aware of the treatment arm, they may provide better care to participants on the experimental arm, thus influencing the results. Similarly, participants on the experimental arm may believe that they are receiving a superior intervention and report better outcomes.

Detection or Assessment bias: This refers to systematic differences in the way the outcome is measured in the experimental and the control group. This is particularly relevant for subjective outcomes such as nausea or pain scores. Performance and detection bias can be avoided by blinding participants and researchers wherever possible, and by using objective endpoints.

Attrition bias: If participants from one arm are selectively excluded from the study, it can skew the results of the study. First, the exclusion of some participants invalidates the randomization and means that the groups are no longer balanced, and second, the reason for exclusion may be related to study outcomes (e.g., drop-out due to excessive toxicity). Attrition bias is prevented by employing appropriate means of data analysis in the trial.

Phase 3 clinical trials include elements to overcome many of these biases, making them the most reliable method to evaluate the efficacy of an intervention.

25.7 Key Features of Phase 3 Clinical Trials

25.7.1 Randomization

Randomization is a process in which participants are allocated to any of the study groups by chance and not by choice. Randomization ensures that in a clinical trial, groups are balanced with respect to both known and unknown confounders, and any difference in outcomes between the groups is attributable to the effect of the intervention, and not due to baseline differences. Simple randomization may be carried out by techniques such as tossing a coin, or using a computer-generated random number table. Advanced options to maintain prognostic balance between groups include stratified randomization, permuted-block randomization and minimization.

25.7.2 Allocation Concealment

This refers to prevention of foreknowledge of upcoming allocation. Allocation concealment can be done by using sequential numbered opaque sealed envelopes or by using remote randomization techniques such as interactive web or voice response systems.

25.7.3 Blinding or Masking

It is the process of preventing participants, study personnel and assessors from knowing the treatment arm to which the patient has been randomized. Blinding in drug trials is done by using matching placebos which are inert. In surgical and radiotherapy trials, blinding may not always be feasible.

25.7.4 Intention-to-Treat Analysis

In a superiority trial, participants should be analyzed in the group to which they are randomized, irrespective of the treatment that they actually receive. This is known as intention-to-treat (ITT) analysis, and provides an estimate of the "effectiveness" of the experimental treatment in the real-world setting. As opposed to ITT, a per protocol (PP) analysis includes only those participants who strictly adhered to the protocol and provides an estimate of the true "efficacy" of the experimental treatment in a hypothetical situation where there is strict adherence to the protocol.

25.7.5 Control

Every experimental treatment should be concurrently evaluated against a comparator to judge its effectiveness. The comparator may be an active one (current standard of care) or a placebo (if there is no active standard treatment for a particular condition). The use of control allows the researcher to distinguish the effect of the experimental treatment from natural changes in the course of the disease over time, and to overcome the placebo effect. Studies may sometimes use historical data as control; however, this may be flawed due to poor record keeping or changes in ancillary care over time.

25.8 Specific Types of Phase 3 Clinical Trials

Superiority trials: Traditional phase 3 clinical trials aim to show that an intervention is superior to the comparator, [36] which may be an active control or placebo. These are known as superiority trials.

Non-inferiority trials: Aim to show that an intervention is nearly as good as a comparator (not inferior by more than a pre-specified margin) [34, 37], while offering other advantages such as reduced toxicity, easier administration, decreased costs or better compliance.

Equivalence trials: Aim to show that an intervention is not "very different" from the comparator (not better or worse by more than a pre-specified margin). Bioequivalence [38] studies are a subtype of equivalence studies where two drugs or two formulations of the same drug are compared to show similar pharmacokinetics and bioavailability.

Parallel group trials: In parallel-group trials, participants are randomized to receive one of several interventions. Comparisons are made between groups of participants receiving the various interventions [39].

Crossover trials: In crossover trials, participants receive each of the study interventions but in different sequences. Participants are thus randomized to a sequence rather than an individual intervention, and each participant receives all the study interventions. Each participant serves as his or her own control, allowing for comparison between treatments for each participant [40]. For example, Vernia et al. looked at topical butyrate versus saline enemas for the treatment of acute radiation proctitis. Participants were randomly allocated to receive either topical butyrate for the first few weeks, followed by crossover to saline enemas or vice versa.

Factorial design [34]: In factorial designs, two or more interventions are compared simultaneously in the same study, using various combinations of the treatments. Factorial designs allow more than one research question to be answered using the same sample, while also assessing for interactions between the different interventions. For example, the PEACE-2 trial evaluates the benefit of neoadjuvant androgen deprivation therapy (ADT) plus cabazitaxel and pelvic radiotherapy (RT) in the treatment of very high-risk non-metastatic prostate cancer patients. Two levels of randomization (prostate-only versus prostate plus pelvic RT and ADT versus ADT plus cabazitaxel) will result in participants being randomized to one of 4 arms: Arm A : Prostate only RT + ADT, Arm B: Prostate + Pelvic RT + ADT, Arm C: Prostate only RT + ADT + cabazitaxel and Arm D: Prostate + Pelvic RT + ADT + cabazitaxel)

Cluster randomized trials [41, 42]: In cluster randomized trials, the unit of randomization is not the individual, but a group of individuals (e.g., a village, a school or a hospital). These designs are used in situations where it may not be possible to randomize individuals such as health promotion or educational interventions. Examples of cluster randomized trials include an oral cancer screening trial conducted in the community or a hospital-based quality improvement strategy in surgery for rectal cancer.

25.9 Outcomes in Oncology Research

Research outcomes or endpoints refer to measurable variables which are monitored during a research study to assess the impact of the intervention (or exposure) on participants. Most research studies have a primary outcome, which is the main outcome of interest. Occasionally, studies may have co-primary outcomes. Other outcomes may be secondary or exploratory. The sample size for a research study is calculated based on its primary outcome.

In oncology, the most important outcome of any research should be that the patient lives longer or lives better. From this point, patient-centered outcomes such as overall survival (OS) and health-related quality of life (QoL) are the most ideal. However, assessment of OS requires long periods of follow-up and may need large sample sizes to detect small changes in survival. Therefore, tumor-centered outcomes such as progression-free survival (PFS), disease-free survival (DFS), response rates, and biochemical markers are used as surrogates for OS. The use of surrogate endpoints has its limitations which need to be carefully understood while planning or appraising a research study. Researchers may sometimes combine outcomes to define a composite outcome, thereby increasing the statistical efficiency of the research study.

Objective endpoints are those which can be measured quantitatively and without bias. Examples include death, tumor progression, tumor response, and biochemical markers. On the other hand, *subjective endpoints* are those which require interpretation by the participant or investigator, and can lead to a bias. Examples include pain, nausea, and quality of life. Wherever possible, objective endpoints should be used to minimize the risk of bias. If subjective endpoints are used, then techniques like blinding of participant and/or assessor should be considered.

25.10 Data Management in Clinical Research

Every research study should have a data management plan for collection, entry and analysis of data to ensure that the data is reliable and accurate, and to facilitate its analysis. The key steps in research data management include:

- Designing a case report form
- Creating an electronic database
- Data entry, checking and validation
- Data coding and analysis

A case report form (CRF) is an electronic or paper document that is used for recording data in a research study [43]. The CRF is designed based on the variables listed in the protocol, and should be approved by the Institutional Review Board. Some elements of a well-designed CRF include

- The use of clearly worded questions
- The layout is user-friendly
- Data collection is concise and restricted to the minimum needed for the study
- Responses are objective, and free text is minimized.
- Standard formats are used for fields such as dates.

Electronic CRFs take less time to fill, allow direct export of data for analysis, have built-in data validation tools and can be synchronized across sites in a multi-centric study. However, they are technology-dependent, may need training of users and may have expenses related to installation and maintenance. Tools like REDCAP are freely available for electronic data capture.

Data analysis in a research study is done using statistical software. While software like Microsoft Excel allow basic statistical tests to be performed, specialized software like SPSS, STATA, or SAS are needed for more advanced tests.

25.11 Registration of a Clinical Trial [44]

Trial registration involves publishing information about the trial such as the design, methods and administrative aspects on a publicly accessible platform. It is essential to register an interventional study prior to **accrual of the first participant**. For observational research, registration is not mandatory, though several trial registries do accept registration of observational studies. The World Health Organization International Clinical Trials Registry platform compiles data from several independent trial registries, including the Clinical Trials Registry of India. Researchers can register their trials on any of the registries that are included in this platform, or in any other registry approved by the International Council of Medical Journal Editors (ICMJE). Trial registration offers several advantages, some of which include:

- It allows readers, reviewers and editors to ensure that researchers have adhered to the trial protocol.
- It prevents selective reporting and publication of positive results.
- It avoids duplication of research efforts and allows collaboration between researchers.
- It helps to identify studies for inclusion in a systematic review.
- It allows the public to access information about ongoing and completed research.

25.12 Secondary Research: Systematic Reviews and Meta-Analysis

Systematic reviews are a form of secondary research which collates the results of several primary studies. When the results of the studies in a systematic review are statistically combined to give a pooled statistic, it is called a meta-analysis.

Systematic reviews are considered the highest level of research evidence. The steps in a systematic review include [45]

- Clearly stating the research question (in the PICOT format)
- Defining eligibility criteria for including studies in the review
- Defining the search strategy
- Identifying studies for inclusion in the review
- Data extraction
- Pooling of data (if meta-analysis)

Systematic reviews allow combination of data from different studies, thereby increasing the sample size and providing a more precise estimate of treatment effect. However, the quality of a systematic review depends on the quality of the studies included, and the inclusion of studies with poor methodological quality or significant heterogeneity between studies may affect the results of the systematic review. If a systematic review finds adequate number of studies for data extraction, then the results of the individual studies are pooled in a meta-analysis, using different statistical models depending on the degree of heterogeneity between the studies (fixed effects model is there is little heterogeneity and random effects model is there is considerable heterogeneity).

25.13 Scientific Publication

Publishing in peer reviewed journals serves the following purposes:

- Dissemination of scientific information
- Providing a platform for creation of homogenous management in health
- Recognition regarding your work (group's work) in the scientific community opening up pathways for collaboration and further research

Publication ethics: Collection of fundamental values, attitudes and norms considered by most of the population as essential for personal life, life with one another, and life in relation to a society's institution is ethics. Upholding the same principle is necessary for scientific publications. Unethically or improperly conducted research leads to unethical publications. Research misconduct leading to such publications is dangerous especially in healthcare research because it may lead to further harm to patients and society.

25.14 Research Misconduct

Any behavior by a researcher that fails to respect high scientific and ethical standards is research misconduct. Types of research misconduct are:

25.14.1 Fabrication/Falsification

Falsification means changing or omitting research data or its representation in the manuscript after analyses leading to misleading conclusions in the scientific community. Falsification may lead to adoption of wrong practices or when the scientific community is not convinced with the "falsified" results, may lead to further research for the same question and more confusion. Manipulation of images in the manuscript in a manner that distorts the data or "reads too much between the lines" is also falsification [1].

Fabrication is addition or creation of data that was not observed during research process. Adding missing data points without actual observation in order to provide a complete data set is fabrication. The data sets though appearing complete and impressive by such an act lead to wrong results [46].

25.14.2 Inappropriate Claims of Authorship

The ICMJE recommends that authorship be based on the following four criteria [47]:

- Substantial contributions to the conception or design of the work; or the acquisition, analysis, or interpretation of data for the work;
- Drafting the work or revising it critically for important intellectual content;
- Final approval of the version to be published; and
- Agreement to be accountable for all aspects of the work in ensuring that questions related to the accuracy or integrity of any part of the work are appropriately investigated and resolved.

It is often observed that the above mentioned criteria are not adhered to while claiming authorship. Some of the contributing factors include lack of awareness amongst nascent researchers, indirect or direct coercion to include or remove certain researchers, absence of clarity regarding usage of hospital/patient information data and claims to authorship. The corresponding author is the individual who takes primary responsibility for communication with the journal during the manuscript submission, peer-review, and publication process. The corresponding author takes care of various administrative aspects including but not limited to documentation of ethical committee clearance, trial registration details, conflicts of interest of various authors and also documentation of funding resources. The corresponding author is therefore primarily questioned if any research misconduct is spotted in the manuscript. Some large multi-author groups designate authorship by a group name, with or without the names of individuals. When submitting a manuscript authored by a group, the corresponding author should specify the group name and identify the group members who can take credit and responsibility for the work as authors [2].

25.14.3 Plagiarism

Plagiarism is, perhaps, the most common form of research misconduct. Representing work of others as your own work constitutes plagiarism [48]. Plagiarism may be introduced in the research experiment (calling the experiment your own without quoting the original publication) or in manuscript writing wherein statements are copied verbatim from previous publications. Plagiarism though less dangerous than falsification/ fabrication reduces originality of the manuscript and may thus impede future publications from the same research group.

25.14.4 Conflicts of Interest and Disclosures

In a 2003 systematic review of the prevalence of conflicts of interest by Bekelman et al. [49], it was found that a third of biomedical researchers in academic institutions have held conflicts of interest that could introduce a risk of bias. It has been shown in studies that researchers with conflicts of interest tend to use a favorable comparator (one that would help support their bias) and also interpret or present results discordant with actual data. Researchers performing a systematic review with conflicts of interest tend to misrepresent findings. Persons involved in guideline formation may support usage of certain investigations, treatment, etc. even in the absence of unequivocal data. But full disclosure of conflicts of interest also has certain unintended consequences. Users or fellow academicians may not immediately accept results.

25.15 Choosing a Journal for Publication

- Create a list of the journals relevant to your article.
- Verify if journal topics match your area of research. This can be discerned from reading in journal homepage (About the journal) or by screening manuscripts published in recent years
- Check the impact factor and other indices of each journal. It is unlikely that small retrospective studies or small individual patient data analysis will be published in impact factors > 4.
- Check journal requirements

- Check information regarding peer review process as well as indexing
- Read and understand guide for authors
- Check if the journal is invitation-only as some journals will only accept articles after inviting the author.
- Submit paper to only one journal at a time.
- Check journal performance for the review and publication timelines as some journals respond within a week while others take months to complete the initial review process. Delay in publication leads to author frustration as well as reduction in impact of article.

25.16 Indexations, Impact Factors, Citation Metrics

Indices are parameters to evaluate individual researcher or journal productivity.

25.16.1 Researcher-Related Indices

H index: When "*n*" number of publications receives at least "*n*" number of citations, h index of the author is calculated to be "*n*." The biggest advantage of *h* index is simultaneous measurement of quantity and quality of scientific output. *H* index does come with some limitations. *H* index may be falsely increased by self-citation. *H* index also does not provide any weightage to position of author in the author list [5].

M index: M index is *H* index divided by the number of years the researcher has been active. Citations of a researcher will keep on increasing for the number of years of activity. Hence, a person working for many years will have higher *h* index. To nullify this effect, *M* index was created

i10 index: The number of publications with at least 10 citations. It is very easy to calculate, but used by Google Scholar only.

An age-dependent indicator: The AR index: AR index considers of the age of the publication.

25.16.2 Journal-Related Indices

25.16.2.1 Journal Impact Factor

An important predictive measure of research impact is the Journal Impact Factor (JIF). The JIF is the average number of citations per article per year. Impact factors are generally calculated yearly. Journals also present their 5-year impact factors which show how journals have fared over a longer period of time [50].

25.16.2.2 Immediacy Index

Immediacy index measures the average number of times that an article, published in a specific year within a specific journal, is cited over the course of the same year.

25.16.2.3 Eigen Factor

The Eigen factor™ score of a journal is an estimate of the percentage of time that library users spend with that journal [50].

25.17 Preparation of Manuscript for Submission

Manuscript preparation for submission needs to follow certain patterns in order to make reviewing as well as reading easier. Various guidelines including but not limited to PRISMA, CONSORT, STROBE, MOOSE, STARD, and SPIRIT are available.

In recent years, a number of new guidelines have also cropped up for reporting. This often puts editors of journal in doubt regarding what to define as journal manuscript guidelines. The EQUATOR network is an international initiative that collates resources facilitating the reporting of health research, provides education about the available guidelines, and also supports the development of new guidelines [51].

25.18 Referencing

Referencing is an integral part of scientific publications. It is essential that authors quote original references. The most common style recommended in medical research is Vancouver style [52]. A typical Vancouver style referencing is as follows: for a journal article—the sequence includes authors, article title, journal title, date of publication, volume, issue, and location/pagination. Each journal then has its own minor modifications mentioned. It is therefore imperative for the corresponding author to verify all the author guidelines before uploading the manuscript to prevent unnecessary delay in processing. It is difficult to handle a large number of references (e.g., in case of large individual patient data analyses of rare diseases). Certain software like Zotero, Mendeley, RefWorks, Papers, CrossRef and Bibliography help handle references [7]. Explaining individual usage of these software is beyond the scope of this chapter.

25.19 How to Handle Rejections and Revisions?

The response from a journal after manuscript has been uploaded may be the following

1. Desk rejection: The editor finds manuscript unsuitable for the journal. Manuscript has to be uploaded into another journal
2. Major revision: The reviewers and editors have found some major flaws in the study methodology, interpretation, documentation or write up. A major revision is a time taking process. It is appropriate to address all the comments one by one in a methodical manner. Apart from addressing the issues in the manuscript, the corresponding author is required to create a point wise response/answer to the comments as a separate document. The response has to be constructive and polite however irritated the author is. A major revision does NOT guarantee acceptance. Usually in the peer review process, the corrected manuscript is sent again to independent reviewers and decision to reject/accept made after
3. Minor revision: Only minor changes need to be made in manuscript limited usually to grammar, referencing or illustration used. The manuscript is provisionally accepted if only minor revisions are asked for.
4. Accept without changes: This is very rare. Manuscript will then be processed, and proofs sent to you. After final proof reading, manuscript will be published online first.

What do you do if manuscript is rejected?

1. Check appropriateness of journal and choose the next journal wisely
2. Check the impact factor of journal that rejected manuscript. Choosing a higher impact journal next will not be considered wise

Precept Box (Take Home Points)

- Cancer research deals with research that attempts to understand the disease process to develop methods/modalities to prevent cancer, early detection/diagnosis of cancer, and treatment of cancer.
- The first basic step of a health research would be to frame the research question and the FINER framework should be used (feasibility, interesting, novel, ethical, and relevant).
- The next sequential steps are comprehensive review of literature, finalization of research question, ethical, and scientific peer review (Fig. 25.1).
- The core principles underlying medical research are guided by Declaration of Helsinki, Belmont report, and the ICH-GCP guidance and are personal autonomy, beneficence, justice, non-maleficence, and informed voluntary consent (Table 25.1).
- Strict regulatory requirements are to be followed for the conduct of clinical trials, and relevant approvals must be taken before initiation of the clinical trial. In particular, the current guideline mandates the clinical trial to be prospectively registered with CTRI (clinical trial registry of India) before the recruitment of first patient (Table 25.2).
- Study designs in research must be chosen based on the objectives of the study and outcome measures. Phase III ran-

domized clinical trial is generally considered to be a benchmark for evaluation of the efficacy of an intervention and is essential prior to approval of a new drug or device (Fig. 25.2).

- There are several sources of bias in clinical trial like selection bias, performance bias, detection bias, and attrition bias. These biases introduce systematic errors in research process which could negatively affect the outcome and results of the study.
- Phase III clinical trials use several measures to overcome these biases like randomization, allocation concealment, blinding or masking, intention to treat analysis, control, or comparator arm (Table 25.3).
- While the robust outcome measure in oncology research has been survival measures, particularly overall survival, other outcome measures like health-related quality of life, treatment toxicity endpoints, and tumor response/biochemical markers could be equally important depending on the type of trial designs (Fig. 25.3).
- Publication in peer-reviewed journal is the essence of a well conducted research as it not only disseminates the scientific information but also helps other researchers in the field to build upon future research question and help in the advancement of science.
- The EQUATOR network has suggested specific guideline to be followed for the preparation of manuscript as per the type of study designs, and these are valuable for rapid reviewing (Table 25.4).
- Scientific journals mandate specific types of referencing styles to be followed which are specific to the journals and should be followed (Table 25.5).

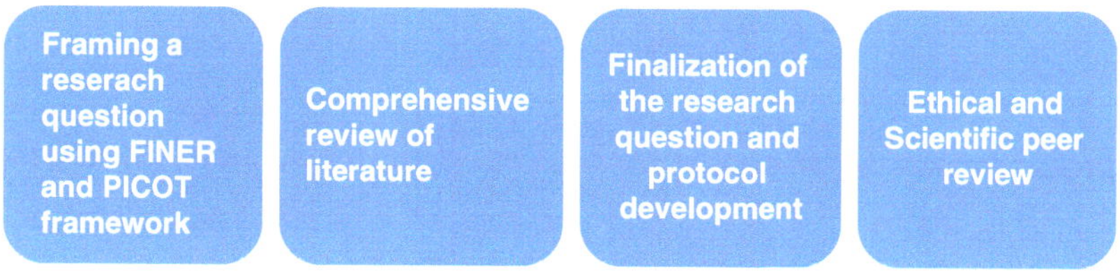

Fig. 25.1 Steps of conducting a research

Table 25.1 Evolution of ethics in medical research

1939–1945	Unethical experiments conducted on humans in the Nazi concentration camps
1946–1947	10-point Nuremberg code to guide medical research. Introduction of voluntary consent of human subjects
1964	Declaration of Helsinki adopted by World Medical Association. Last updated in 2013. Moral code above the local laws and regulations
1979	Belmont report for guiding human research filed with core principles of respect for personal autonomy, beneficence, and justice
1998	International conference on harmonization (ICH) led to the consolidated guidance on good clinical practice (GCP) to protect the rights of human participants in clinical trials and to ensure scientific validity and credibility in clinical studies

Table 25.2 Regulatory requirements for research in India

Any clinical trial	Approval from Institutional Ethics Committee
Any clinical trial	Registration with Clinical Trials Registry of India (CTRI) before the enrollment of the first participant into the study
New drug, dosage, use in new indications, etc. (drug related trials)	The Drug Controller General of India (DCGI) approval required
Recombinant/gene therapy	Department of Biotechnology, Govt. of India approval required
Use of ionizing radiation/ radioisotopes	Bhabha Atomic Research Center approval required
International collaboration/ funding in health research	Health Ministry Screening Committee (HMSC) under Department of Health Research, Indian Council of Medical Research approval required

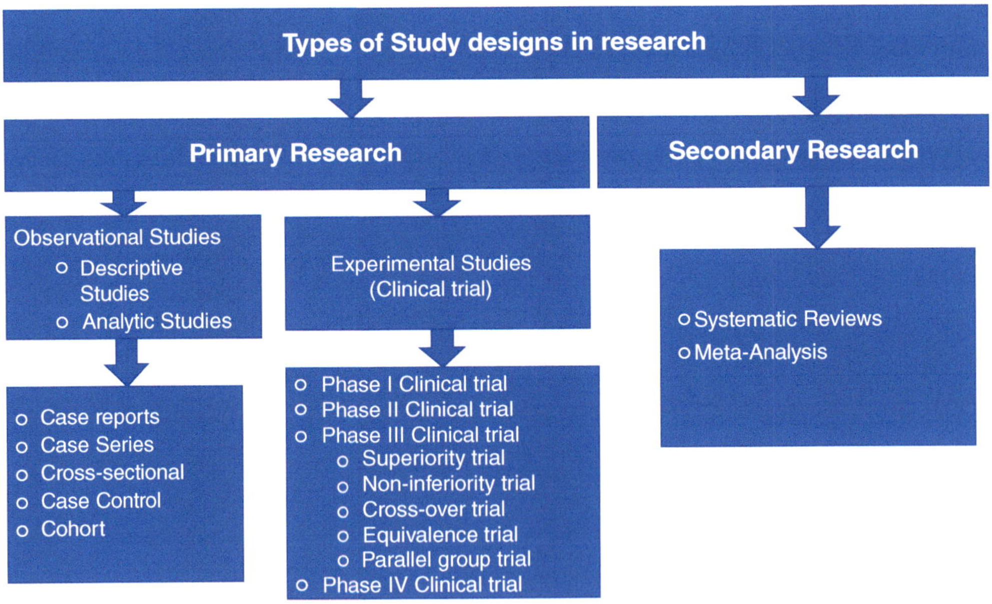

Fig. 25.2 Types of study designs in research

Table 25.3 Important biases in clinical trial and methods to overcome these biases

Types of bias in clinical trial	Methods to overcome bias in clinical trial
Selection bias	Randomization, allocation concealment
Performance bias	Blinding, masking
Detection bias	Blinding, masking
Attrition bias	Intention to treat analysis
Confounder bias	Use of appropriate control/ comparator arm

Table 25.4 Preparation of the manuscript as per the type of study

Name of the guideline [51]	Type of study to be reported
STROBE	Observational studies
CONSORT	Randomized controlled trial
PRISMA	Systematic review and meta analyses in interventional studies
MOOSE	Systematic review and meta analyses in observational studies
STARD	Diagnostic studies
MIAME	Micro-array studies
COREQ	Qualitative studies

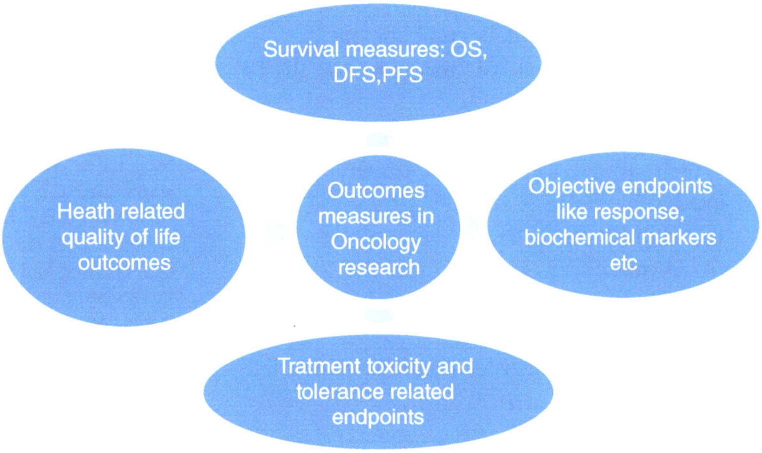

Fig. 25.3 Outcome measures in oncology research

Table 25.5 Common types of referencing styles followed in preparation of manuscript

Type	Particulars	Comments
Vancouver style	For a journal article—the sequence include authors, article title, journal title, date of publication, volume, issue, and location/pagination	Most common style recommended in medical research
Harvard style	Uses references in two places in a piece of writing: in the text and in a reference list at the end. In general, each author name that appears in the text must also appear in the reference list, and every work in the reference list must also be referred to in the main text.	It is also called APA referencing
Chicago style	Chicago style is an "author-date" style, so the citation in the text consists of the author(s) name and year of publication given wholly or partly in round brackets. Use only the surname of the author(s) and the year of publication	

Multiple Choice Questions

1. Selection bias can be overcome by
   (a) Blinding
   (b) Randomization
   (c) Intention to treat analysis
   (d) Use of appropriate control
2. Which of the following constitutes survival measure?
   (a) OS
   (b) Biochemical marker
   (c) Health-related quality of life
   (d) Treatment toxicity

3. ICMJE stands for
    (a) Indian Council of Medical Journal Editors
    (b) International Council of Medical Journal Editors
    (c) International Council of Medical Journal Editions
    (d) International Council of Medical Journal Editions
4. Authorship should be based on what criteria?
    (a) Substantial contributions to the conception or design of the work; or the acquisition, analysis, or interpretation of data for the work
    (b) Drafting the work or revising it critically for important intellectual content
    (c) Final approval of the version to be published
    (d) All the above
5. Which of the following is research-related index?
    (a) H index
    (b) A index
    (c) B index
    (d) C index
6. The average number of citations per article per year is referred a
    (a) Journal impact factor
    (b) Immediacy index
    (c) Eigen factor
    (d) None of above
7. STROBE guidelines are for which type of study?
    (a) Randomized control trial
    (b) Observational study
    (c) Systemic review
    (d) Diagnostic studies
8. Guidelines used for qualitative studies are
    (a) CONSORT
    (b) PRISMA
    (c) MOOSE
    (d) COREQ
9. Which is the most common style of reference recommended in medical research?
    (a) Vancouver style
    (b) Harvard style
    (c) Chicago style
    (d) London style

Answers

1. (b)
2. (a)
3. (b)
4. (d)
5. (a)
6. (a)
7. (b)
8. (d)
9. (a)

References

1. Röhrig B, du Prel JB, Wachtlin D, Blettner M. Types of study in medical research: part 3 of a series on evaluation of scientific publications. Dtsch Arztebl Int. 2009;106(15):262–8.
2. Siegel RL, Miller KD, Fuchs HE, Jemal A. Cancer statistics, 2022. CA Cancer J Clin. 2022;72(1):7–33.
3. Sung H, Ferlay J, Siegel RL, Laversanne M, et al. Global cancer statistics 2020: GLOBOCAN estimates of incidence and mortality worldwide for 36 cancers in 185 countries. CA Cancer J Clin. 2021;71:209–49.
4. Ratan SK, Anand T, Ratan J. Formulation of research question - stepwise approach. J Indian Assoc Pediatr Surg. 2019;24(1):15–20.
5. Fandino W. Formulating a good research question: pearls and pitfalls. Indian J Anaesth. 2019;63:611–6.
6. Aslam S, Emmanuel P. Formulating a researchable question: a critical step for facilitating good clinical research. Indian J Sex Transm Dis. 2010;31:47–50.
7. Chan AW, Tetzlaff JM, Altman DG, Laupacis A, Gøtzsche PC, Krleža-Jerić K, et al. SPIRIT 2013 statement: defining standard protocol items for clinical trials. Ann Intern Med. 2013;158(3):200–7.
8. Chan AW, Tetzlaff JM, Gøtzsche PC, Altman DG, Mann H, Berlin JA, et al. SPIRIT 2013 explanation and elaboration: guidance for protocols of clinical trials. BMJ. 2013;346:e7586.
9. Saxena P, Saxena R. Clinical trials: changing regulations in India. Indian J Community Med. 2014;39(4):197–202.
10. Weindling P, von Villiez A, Loewenau A, Farron N. The victims of unethical human experiments and coerced research under National Socialism. Endeavour. 2016;40(1):1–6.
11. Shuster E. Fifty years later: the significance of the Nuremberg Code. N Engl J Med. 1997;337(20):1436–40.
12. World Medical Association. Declaration of Helsinki. N Engl J Med. 1964;271(9):473–4.
13. Alsan M, Wanamaker M. Tuskegee and the health of black men. Q J Econ. 2018;133(1):407–55.

14. Sims JM. A brief review of the Belmont report. Dimens Crit Care Nurs. 2010;29(4):173–4.
15. Beecher HK. Ethics and clinical research. N Engl J Med. 1966;274(24):1354–60.
16. Imran M, Najmi AK, Rashid MF, Tabrez S, Shah MA. Clinical research regulation in India-history, development, initiatives, challenges, and controversies: still a long way to go. J Pharm Bioallied Sci. 2013;5(1):2–9.
17. ICMR national ethical guidelines for biomedical and health research involving human participants. 2017. Accessed 5 February 2022.
18. Central Drugs Standard Control Organization. https://www.cdscoonline.gov.in/CDSCO/homepage. Accessed 5 February 2022.
19. Srinivasan K, Fredrick J, Gupta R, Singh N. Funding opportunities for health research in India – a technical scan. Indian J Public Health. 2020;64:421–7.
20. Goula A, Chatzis A, Stamouli MA, Kelesi M, Kaba E, Brilakis E. Assessment of health professionals' attitudes on radiation protection measures. Int J Environ Res Public Health. 2021;18(24):13380.
21. Yoon S, Choi JW, Lee E, An H, Choi HD, Kim N. Mobile phone use and risk of glioma: a case-control study in Korea for 2002-2007. Environ Health Toxicol. 2015;30:e2015015. https://doi.org/10.5620/eht.e2015015.
22. Morgan MA, Lewis WG, Crosby TDL, Escofet X, Roberts SA, Brewster E, Harvard TJ, Clark GWB. Prospective cohort comparison of neoadjuvant chemoradiotherapy versus chemotherapy in patients with oesophageal cancer. Br J Surg. 2007;94(12):1509–14.
23. Rachman A, Shatri H, Salamat R. Correlation between higher cumulative dose of cisplatin for concurrent chemoradiation and acute kidney disease incidence among nasopharyngeal carcinoma patients: a comparative study. Int J Gen Med. 2021;14:10527–39.
24. Jastaniyah N, Murtha A, Pervez N, Le D, Roa W, Patel S, Mackenzie M, Fulton D, Field C, Ghosh S, Fallone G, Abdulkarim B. Phase I study of hypofractionated intensity modulated radiation therapy with concurrent and adjuvant temozolomide in patients with glioblastoma multiform. Radiat Oncol. 2013;8:38.
25. O'Rawe M, Wickremesekera AC, Pandey R, Young D, Sim D, FitzJohn T, Burgess C, Kaye AH, Tan ST. Treatment of glioblastoma with re-purposed renin-angiotensin system modulators: results of a phase I clinical trial. J Clin Neurosci. 2022;95:48–54.
26. Mercier C, Claessens M, Buys A, Gryshkevych S, Billiet C, Joye I, Van Laere S, Vermeulen P, Meijnders P, Löfman F, Poortmans P, Dirix L, Verellen D, Dirix P. Stereotactic ablative radiation therapy to all lesions in patients with oligometastatic cancers: a phase 1 dose-escalation trial. Int J Radiat Oncol Biol Phys. 2021;109(5):1195–205.
27. Wei KC, Hsu PW, Tsai HC, Lin YJ, Chen KT, Toh CH, Huang HL, Jung SM, Tseng CK, Ke YX. Safety and tolerability of asunercept plus standard radiotherapy/temozolomide in Asian patients with newly-diagnosed glioblastoma: a phase I study. Sci Rep. 2021;11(1):24067.
28. Maity A, Mick R, Rengan R, Mitchell TC, Amaravadi RK, Schuchter LM, Pryma DA, Patsch DM, Maity AP, Minn AJ, Vonderheide RH, Lukens JN. A stratified phase I dose escalation trial of hypofractionated radiotherapy followed by ipilimumab in metastatic melanoma: long-term follow-up and final outcomes. Onco Targets Ther. 2021;10(1):1863631.
29. Yang KL, Chi MS, Ko HL, Huang YY, Huang SC, Lin YM, Chi KH. Axitinib in combination with radiotherapy for advanced hepatocellular carcinoma: a phase I clinical trial. Radiat Oncol. 2021;16(1):18.
30. Vempati P, Halthore AN, Teckie S, Rana Z, Gogineni E, Antone J, Zhang H, Marrero M, Beadle K, Frank DK, Aziz M, Paul D, Ghaly M. Phase I trial of dose-escalated stereotactic radiosurgery (SRS) boost for unfavorable locally advanced oropharyngeal cancer. Radiat Oncol. 2020;15(1):278.
31. Holyoake DLP, Robinson M, Silva M, Grose D, McIntosh D, Sebag-Montefiore D, Radhakrishna G, Mukherjee S, Hawkins MA. SPARC, a phase-I trial of pre-operative, margin intensified, stereotactic body radiation therapy for pancreatic cancer. Radiother Oncol. 2021;155:278–84.
32. Bananzadeh A, Hafezi AA, Nguyen N, Omidvari S, Mosalaei A, Ahmadloo N, Ansari M, Mohammadianpanah M. Efficacy and safety of sequential neoadjuvant chemotherapy and short-course radiation therapy followed by delayed surgery in locally advanced rectal cancer: a single-arm phase II clinical trial with subgroup analysis between the older and young patients. Radiat Oncol J. 2021;39(4):270–8.
33. Fonteyne V, Dirix P, Van Praet C, Berghen C, Albersen M, Junius S, Liefhooghe N, Noé L, De Meerleer G, Ost P, Villeirs G, Verbeke S, De Maeseneer D, Rammant E, Verghote F, Elhaseen E, De Man K, Decaestecker K. Adjuvant radiotherapy after radical cystectomy for patients with high-risk muscle-invasive bladder cancer: results of a multicentric phase II trial. Eur Urol Focus. 2021;8(5):1238–45.
34. Murthy V, Maitre P, Kannan S, Panigrahi G, Krishnatry R, Bakshi G, Prakash G, Pal M, Menon S, Phurailatpam R, Mokal S, Chaurasiya D, Popat P, Sable N, Agarwal A, Rangarajan V, Joshi A, Noronha V, Prabhash K, Mahantshetty U. Prostate-only versus whole-pelvic radiation therapy in high-risk and very high-risk prostate cancer (POP-RT): outcomes from phase III randomized controlled trial. J Clin Oncol. 2021;39(11):1234–42.
35. Van Heijl M, van Lanschot JJ, Koppert LB, van Berge Henegouwen MI, Muller K, Steyerberg EW, van Dekken H, Wijnhoven BP, Tilanus HW, Richel DJ, Busch OR, Bartelsman JF, Koning CC, Offerhaus GJ, van der Gaast A. Neoadjuvant chemoradiation followed by surgery versus surgery alone for patients with adenocarcinoma or squamous cell carcinoma of the esophagus (CROSS). BMC Surg. 2008;8:21.

36. Ni W, Yu S, Xiao Z, Zhou Z, Chen D, Feng Q, Liang J, Lv J, Gao S, Mao Y, Xue Q, Sun K, Liu X, Fang D, Li J, Wang D, Zhao J, Gao Y. Postoperative adjuvant therapy versus surgery alone for stage IIB-III esophageal squamous cell carcinoma: a phase III randomized controlled trial. Oncologist. 2021;26(12):e2151–60.
37. Halkett GK, Wigley CA, Aoun SM, Portaluri M, Tramacere F, Livi L, Detti B, Arcangeli S, Lund JA, Kristensen A, McFadden N, Grun A, Bydder S, Sackerer I, Greimel E, Spry N, EORTC Quality of Life Group. International validation of the EORTC QLQ-PRT20 module for assessment of quality of life symptoms relating to radiation proctitis: a phase IV study. Radiat Oncol. 2018;13(1):162.
38. Kato K, Ito Y, Nozaki I, Daiko H, Kojima T, Yano M, Ueno M, Nakagawa S, Takagi M, Tsunoda S, Abe T, Nakamura T, Okada M, Toh Y, Shibuya Y, Yamamoto S, Katayama H, Nakamura K, Kitagawa Y, Japan Esophageal Oncology Group of the Japan Clinical Oncology Group. Parallel-group controlled trial of surgery versus chemoradiotherapy in patients with stage I esophageal squamous cell carcinoma. Gastroenterology. 2021;161(6):1878–86.
39. Kim TH, Koh YH, Kim BH, Kim MJ, Lee JH, Park B, Park JW. Proton beam radiotherapy vs. radiofrequency ablation for recurrent hepatocellular carcinoma: a randomized phase III trial. J Hepatol. 2021;74(3):603–12.
40. Veronesi U, et al. Intraoperative radiotherapy versus external radiotherapy for early breast cancer (ELIOT): a randomised controlled equivalence trial. Lancet Oncol. 2013;14(13):1269–77.
41. Sankaranarayan R, et al. Early findings from a community-based, cluster-randomized, controlled oral cancer screening trial in Kerala, India. Cancer. 2000;88(3):664–73.
42. Simunovic M, Coates A, Goldsmith CH, et al. The cluster-randomized Quality Initiative in Rectal Cancer trial: evaluating a quality-improvement strategy in surgery. CMAJ. 2010;182(12):1301–6. https://doi.org/10.1503/cmaj.091883.
43. Bellary S, Krishnankutty B, Latha MS. Basics of case report form designing in clinical research. Perspect Clin Res. 2014;5(4):159–66. https://doi.org/10.4103/2229-3485.140555.
44. Bhaskar SB. Clinical trial registration: a practical perspective. Indian J Anaesth. 2018;62(1):10–5. https://doi.org/10.4103/ija.IJA_761_17.
45. Foerster R, Zwahlen DR, Buchali A, Tang H, Schroeder C, Windisch P, Vu E, Akbaba S, Bostel T, Sprave T, Zamboglou C, Zilli T, Stelmes JJ, Telkhade T, Murthy V. Stereotactic body radiotherapy for high-risk prostate cancer: a systematic review. Cancer. 2021;13(4):759.
46. Khadem-Rezaiyan M, Dadgarmoghaddam M. Research misconduct: a report from a developing country. Iran J Public Health. 2017;46(10):1374–8.
47. International Committee of Medical Journal Editors. Recommendations for conduct, reporting, editing, and publication of scholarly work in medical journals. http://www.icmje.org/recommendations/browse/roles-and-responsibilities/defining-the-role-of-authors-and-contributors.html.
48. Gilbert FJ, Denison AR. Research misconduct. Clin Radiol. 2003;58(7):499–504.
49. Bekelman JE, Li Y, Gross CP. Scope and impact of financial conflicts of interest in biomedical research: a systematic review. JAMA. 2003;289(4):454–65.
50. Satyanarayana K. Impact factor and other indices to assess science, scientists and scientific journals. Indian J Physiol Pharmacol. 2010;54(3):197–212.
51. https://www.equator-network.org/library/guidance-on-scientific-writing/.
52. https://guides.lib.uw.edu/research/citations.

Further Reading

https://www.youtube.com/watch?v=95xRGnHjmY0.
https://www.youtube.com/watch?v=DR2fcn_3ct0.
https://www.youtube.com/watch?v=JO04gjGPOvE.
https://www.youtube.com/watch?v=k2gdzWLY7w8.
https://www.youtube.com/watch?v=TK_VJDLEzuo.

Social Sciences and Economics in Oncology

26

Nidhi Gupta, Shankar Prinja, Bidhu K. Mohanti, Stephen Jan, and Ravi Kannan

26.1 Introduction

Social science in cancer care is gaining recognition, because the cancer care right from prevention and screening to cancer diagnosis and management, and then in situations of terminal and end-of-life care, becomes integral to the sociogeographic, ethno-cultural and financial conditions of the country, region and its citizens.

Globally, the age-standardized incidence rate for all cancers has been reported to be 222 per 100,000 in men and 186 per 100,000 in women [1]. The cancer-specific death rates per 100,000 persons varied from 171 in Eastern Europe to 67 in Central America among males and from 120 in Melanesia to 64 in Central America and Eastern Asia (excluding China) among females [1]. As per the latest global estimates, Asia tops the list with nearly 3.6 and 1.2 million cancer-related deaths in Eastern and South central region, respectively (Fig. 26.1).

On a similar note, the projected incidence of cancer in India for the year 2020 was 94.1 per 100,000 among males and 103 per 100,000 in females. Breast cancer is the commonest cancer in India with an estimated 2.2 million new cases in 2020, followed by lung cancer (Fig. 26.2).

While the age-adjusted mortality rates varied from 153 per 100,000 in Aizawl district to 10 per 100,000 in Osmanabad among males and from 90 per 100,000 in Aizawl district to 8.5 per 100,000 in Aurangabad among females [2]. Thus, we see that in spite of the lower cancer incidence as compared to the global estimates, the mortality due to cancer in India is higher. There are several contextual factors which explain this paradoxical finding for cancer in India. Firstly, there is dearth of treatment facilities for cancer, secondly, these facilities are inequitably located leading to regional disparities [3]. Together, both these reasons lead to poor access to cancer treatment. As a result, by the time patients seek care, they present in advanced stages which results in poorer outcomes. Thirdly, those who seek care, face high out of pocket expenditure (OOPE) [4]. This leads

N. Gupta (✉)
Department of Radiation Oncology, Government Medical College and Hospital, Chandigarh, India
e-mail: nidhiguptaonco@gmail.com

S. Prinja
Department of Community Medicine and School of Public Health, Post Graduate Institute of Medical Education and Research, Chandigarh, India
e-mail: shankarprinja@gmail.com

B. K. Mohanti
Academic and Head of Radiation Oncology, Bagchi Sri Shankara Cancer Centre & Research Institute, Bhubaneswar, Odisha, India

S. Jan
The George Institute for Global Health, Sydney, NSW, Australia
e-mail: sjan@georgeinstitute.org

R. Kannan
Cachar Cancer Hospital, Silchar, Assam, India
e-mail: ravi.kannan@cacharcancerhospital.org

A. Munshi et al. (eds.), *Radiation Oncology – Principles, Precepts and Practice*,
https://doi.org/10.1007/978-981-97-8389-2_26

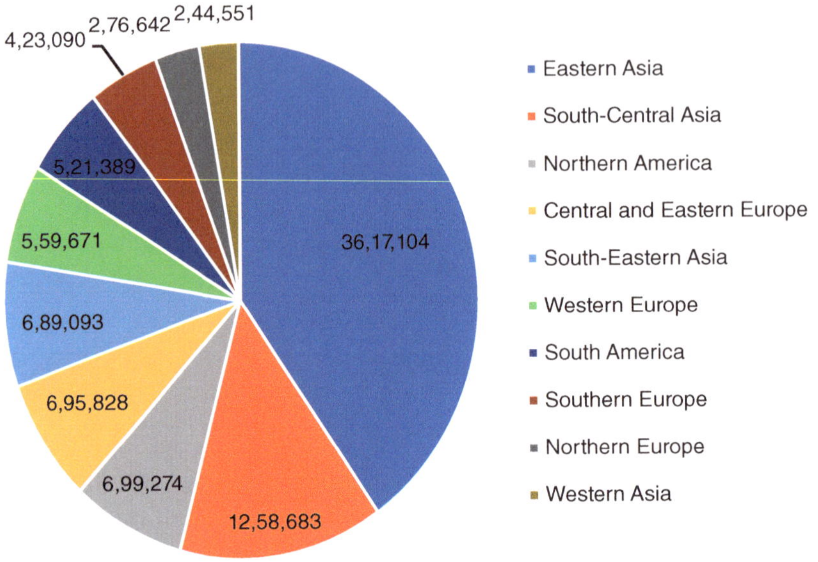

Fig. 26.1 Global deaths due to cancer according to the region (Source: GLOBOCAN 2020)

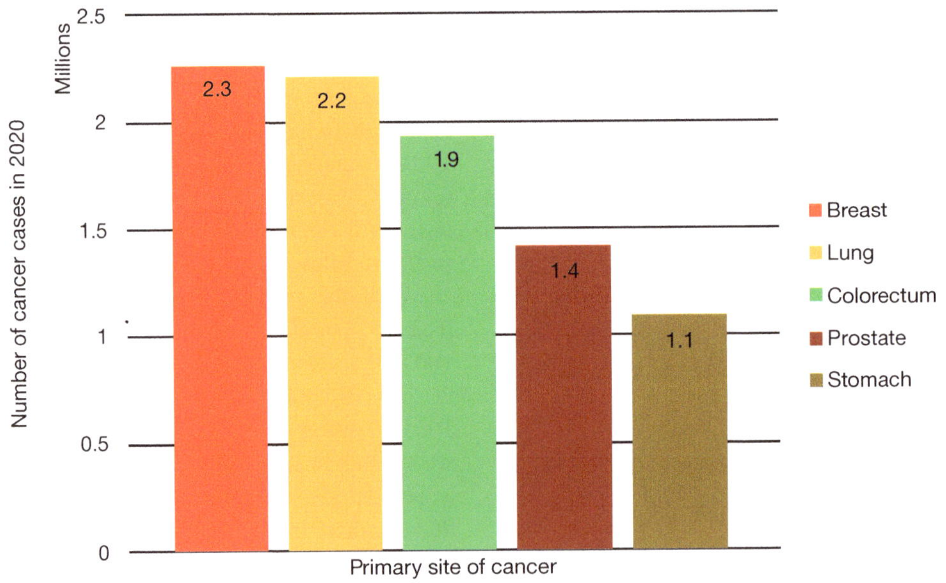

Fig. 26.2 Number of cancer cases in India according to cancer site (Source: GLOBOCAN 2020)

to poor compliance for standard care and financial toxicity.

With a national expenditure of less than 1.3% of the gross domestic product on public healthcare, India has one of the least government spending on healthcare compared to its income. Further, there are large variations in the spending on the health sector across various states [5]. This leads to poor quality, poor functioning and poor access of public healthcare services.

The theme for World Cancer Day 2022 and 2023 was "*Close the Care Gap*." This theme was all about raising awareness of the equity gap that affects almost everyone, predominantly in the LMICs and is costing lives [6]. In this chapter we describe the inequality in access to cancer care treatment; secondly, how the financing of cancer care leads to poor affordability and financial toxicity for the poor. Thirdly, we outline, what can be done both at the physician level and the government level to remove these inequities, make standard treatment both accessible and affordable for the majority of the economically weaker population, thus saving lives and improving outcomes. In the process, communication, social support and research should be encouraged in LMICs for the vast majority of our cancer patients.

26.2 Health Policy for Cancer Management

Since more than two-third of all cancers are diagnosed in LMICs and more than 70% of all cancer deaths occur in these countries, it is recognized that social science research and health system improvements should understand the various causes. The reasons are outlined below.

26.2.1 Inequality in Access to Cancer Care

Access to healthcare in India is determined by various social, demographic and economic factors. The most common barriers to access are financial constraint and geographical divide. Other common barriers include illiteracy, ignorance, fear, social prejudice, gender disparities and faith in alternative therapies [7]. These together are responsible for delay, discontinuation and deferring treatment.

Cancer is one of the most expensive diseases to treat. The inability of the individual and the family to bear the cost of the expensive cancer treatment is one of the major determinants of access to cancer care, in a country like India where majority of costs of cancer care are borne out of pocket. The OOPE for cancer treatment is the highest for any ailment and is about 2.5 times of overall average hospitalization expenditure [8]. It is observed that an average cancer patient may have to spend between 3 and 5 times of the monthly per capita income (MPCI) during the cancer therapy course [9]. Catastrophic expenditure on cancer inpatient treatment is the highest among any non-communicable diseases (NCDs) [4].

The average out of pocket spending on inpatient care in private facilities is about three-times that of public facilities [10]. The more affordable public hospitals have a limited health workforce that includes doctors, nurses and technical staff. The infrastructure is also limited with respect to diagnostic equipment, drug supply, day care facilities, operation theatres, radiotherapy machines, indoor beds, etc. which fails to cater to the increasing demand leading to long waiting periods and subsequent delay in diagnosis and treatment [11]. Limited communication with healthcare workers, failure to understand the treatment options, inability to participate in informed treatment decisions and poor social support limits the access to tertiary healthcare by the cancer patients [12]. It has been observed that the socioeconomic inequities manifest in caste, class and gender differentials; and get further accentuated by inequities in the availability, utilization and affordability of health services. The distribution of the more expensive private sector facilities is even more skewed in the urban affluent regions. The access to private hospitals is also limited by its cost and over prescription of investigations, drugs and procedures [13].

Besides financial barriers, geographical barriers significantly determine the access to cancer care. Overall, at the national level we have a shortfall of the oncology services that exist, for example WHO recommends minimum of 1 teletherapy unit per million populations. As per this recommendation, our national requirement becomes nearly 1350 teletherapy units as against the currently available 545 teletherapy units [3, 14]. This shortfall exists, irrespective of the rising volumes in terms of incidence and prevalence of cancer cases. Further, there is an inequitable regional distribution of these limited treatment facilities. Taking into account the proportion of deaths due to cancer, or proportion of total population, we have higher number of oncology treatment and training facilities in the north, west and south zones of India, compared to the availabilities in the central zone and the eastern zone [15]. Thus, the cancer patients getting timely treatments for disease control or for prolongation of qualitative life is far less in those states of India.

Apart from the regional variation, there is significant intrastate variation that predominantly divides the urban from the rural areas. Nearly 70% of Indian population lives in rural areas [16]: the key determinants and factors which affect can be enumerated as travelling long distances to tertiary care cancer centers or the private cancer centers in the urban areas, arranging travel for frequent visits of multiple family members, arranging stay and food near to the treating hospital, and cultural and language barriers. All these lead to inequities in utilization of cancer care across various geographical areas [17].

From above descriptions, it has been realized that cancer in India and in most LMICs can have one or more of the below mentioned six themes which become the major hurdles [18]:

(a) the journey of patients to access care, (b) the psychological journey of patients, (c) stigma of cancer patients, (d) decision-making and adherence to treatment, (e) economic costs of cancer care and its impact and (f) modifiers to accessing cancer care.

Illiteracy and ignorance also play major barriers in access to healthcare with people at risk being reluctant for cancer screening or vaccination, even when provided at no cost by the government. Gender plays a significant social challenge. Access to healthcare by women remains low, secondary to lack of family support, family responsibilities, financial dependence and social taboos. Similar constraints are encountered by the elderly and children. Ignoring the symptoms of cancer, failure to perceive the severity of the problem and approaching alternative treatment therapies (which are accessible and low cost) leads to patients presenting in advanced stages [7, 12]. This is aggravated by the high costs and poor access to standard of care. Similar challenges are faced by the elderly and children diagnosed with cancer in most of the LMICs [19]. The barriers and delays are seen at three phases, mainly, (1) access to the proper cancer center or specialist team, (2) duration of diagnostic tests and (3) treatment course for the optimal management. The factors are more difficult in low income countries compared to middle and high income countries.

26.2.2 Factors Which Require Attention for Making Changes

The diagram (Fig. 26.3) below, in a summarized manner, depicts the factors which affect the cancer care in India and in most LMICs [20]. These will need adequate social science and health system approaches to mitigate the long-standing hurdles. It is imperative to reduce the inequalities in cancer control and survival between developed and developing countries (within LMICs).

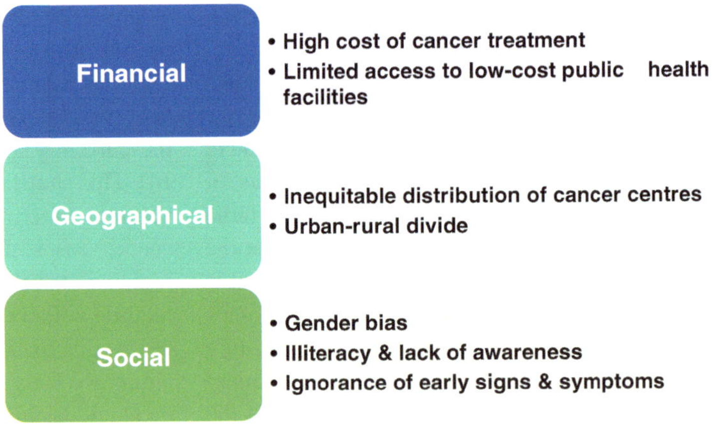

Fig. 26.3 Factors affecting cancer care in India

26.3 Socioeconomic Factors in Cancer Care

Chronic illness like hypertension, diabetes and aging; diseases of neurologic, cardiac, liver, kidney and musculoskeletal systems; and cancer require lifetime healthcare and monitoring in order to achieve disease control and maintain the quality of life. Hence, it is appropriate to invest and spend on the necessary diagnostic tests, procedures, equipment, medicines and ancillary support measures. It is needed for the citizens in India and in LMICs to have the expansion of health insurance/assurance coverage as a vital step to achieve Universal Health Coverage (UHC). As part of its efforts to achieve UHC in India, initiatives like the Ayushman Bharat-Pradhan Mantri Jan Arogya Yojana (AB-PMJAY) by the Govt. of India and similar schemes (e.g., Central Government Health Scheme, CGHS; Employees' State Insurance Scheme, ESIC) provide reasonably comprehensive hospitalization cover to the bottom 40% of the country's population, while another 18% is insured through other government and group health schemes. Retail health insurance penetrates only a meager 3.2% of the 138 crore population, leaving a huge segment of the citizens, which is about 56 crore individuals—unprotected from any sort of health insurance cover, making them prone to financial shocks and healthcare debt. Due to this, India has one of the highest OOPE, close to 62%, making it very difficult and costly for a common man to access quality healthcare [21]. The cancer patients and their families can often spend more than 5 times of the household MPCI in order to comply with the cancer treatments [8].

26.3.1 Financial Toxicity

The average total expenditure on cancer care is estimated to be Rs. 29,066 (US $1715.82 at 2014 purchasing power parity) and Rs. 84,320 (US $4977.56) for the public and the private sectors, respectively [10]. In majority of the developed countries, the government bears the healthcare expenditures of the citizens or the citizens have health insurance to pay for their healthcare expenditures. In contrast, in India, the vast majority of the patients do not have any health insurance and bear the cost of cancer treatment out of their pocket, i.e., OOPE. A meager 18% of the urban population and 14% of the rural population in India have some kind of health insurance to finance their healthcare needs [17, 21–23]. Thus, more than 75% of the cancer treatment in India is financed through OOPE by the patients and their families [8–10]. In a systematic review published in 2021, it was reported that the direct OOPE incurred on the inpatient and outpatient cancer care were Rs. 83,396 and Rs. 2653 respectively [23].

Access to government schemes in India is also low due to poor awareness with limited hospitals at the tertiary care centers, offering the schemes with complex enrolment procedures. Thus, poor expenditure on health by the government coupled with lack of insurance leads to significant out of pocket payments by individuals or their families, for the high cost of cancer treatment, which in turn leads to a substantial financial burden on the patient and their family. As per the NSSO survey, the OOPE is much higher for richer. The total expenditure for highest quintile (Rs. 95,422) is more than twice the average expenditure reported by the poorest quintile (Rs. 44,500). This high OOPE influences the household, predominantly the low socioeconomic status adversely [10, 22, 24].

Cost of cancer treatment includes both direct and indirect costs. Direct and indirect costs associated with cancer treatment were estimated to be ₹25,606 and ₹8772 as per a study conducted among 235 cancer patients in a tertiary level hospital in Kerala, India [25]. Direct costs include outpatient consultations, diagnostic investigations, treatment costs including procedures, medicines and surgery, transportation, food etc. Largest proportion of OOPE constitutes the spending on purchase of medicines. It has been reported that the per episode OOPE on drugs and diagnostics for patients receiving outpatient cancer care was about five times greater than that for patients receiving outpatient care for any other chronic ailments and high-OOPE chronic ailments [26]. Indirect costs include loss of income due to absence of the earning family member from work, who had to frequently accompany the sick patient to the distant cancer centers [27, 28].

The financial burden experienced by patients undergoing cancer treatment and the deleterious effects of such treatments on the financial security of the patient is often referred to as "financial toxicity." This term was first used in 2013, from a study conducted at Duke University Medical Center, USA [29]. As per this study, mainly related to co-payment for the cancer drugs, the patients and caregivers had to reduce spending on food, clothes and leisure and had to exhaust their savings, which raised serious concerns. Patients with chronic illness such as cancer also have a higher risk of incurring catastrophic health expenditure than those with a diagnosis of a communicable disease [8]. The family often finances such healthcare expenditure by cutting down on the consumption levels of other family members. Thus, illness of one family member can lead to impoverishment of the entire household. This leads to catastrophic health expenditure (CHE). According to the WHO, any out-of-pocket medical expenditure in excess of 40% of a household's non-subsistence income is considered to be catastrophic health expenditure [30].

One third of households having individuals with cancer are estimated to spend more than half of their per-capita annual household expenditure on hospitalizations due to the disease [9, 10]. A cancer diagnosis is increasingly responsible for CHE. Incidence of catastrophic expenditure is highly concentrated among poor households. For instance, more than 50% patients from low-income households are reported to be spending more than 10–20% of per capita household expenditure whereas, about 26% of richer households are reported to be spending more than 10–20% of their annual income for a cancer patient in the family [9]. Overall, 36% and 34% of households with cancer patients are spending more than 10% of their annual per capita household expenditure on public and private healthcare respectively. Over 60% of the households who seek care from the private sector incur OOPE in excess of 20% of their annual per capita household expenditure [9].

To pay for this CHE, patients often resort to distress financing that includes using the reserve savings of the family, borrowing from family members and friends, selling or pawning their assets like land or jewelry [8, 26]. About 50% of low-income households raise greater proportion of the cancer treatment expenditure via such distressed means. At national level, more than 40% household resort to distress means as the main source of financing cancer treatment even while seeking treatments in public hospitals. The incidence of distressed financing in rural areas is high with 49% and 58% of households using such means for treatment in public and private hospitals, respectively [9].

It is observed that the poorest are the hardest hit in LMICs [31]. One of the systematic reviews on financial hardships in cancer patients showed that household income, type of health-care facility used, stage of disease, area of residence, age at the time of diagnosis, recurrent cancer, educational status, lack of insurance coverage and treatment modality are the chief determinants of financial toxicity in cancer patients [32]. In other studies from LMICs, a household size of more than four [33] and multiple cycles of chemotherapy [34] were regarded as independent risk factors of financial toxicity. Utilization of private health-care facilities was reported to be associated with escalated cost of treatment across most of the studies, specially so for indoor treatment, and when cancer chemotherapy is delivered [32]. Average hospitalization expenditure in the private sector in India was almost three times that in the public sector [9].

Often patients are unable to undergo the optimal treatment due to paucity of funds, which lead to delay or even deferment of treatment course once their limited finances are exhausted. This explains why people of low socioeconomic status present with advanced disease, receive less aggressive and suboptimal treatment and have a lower 5-year survival when compared to the economically better off patients, even in high-income country like the USA [35]. In all studies evaluating the economic burden and OOPEs

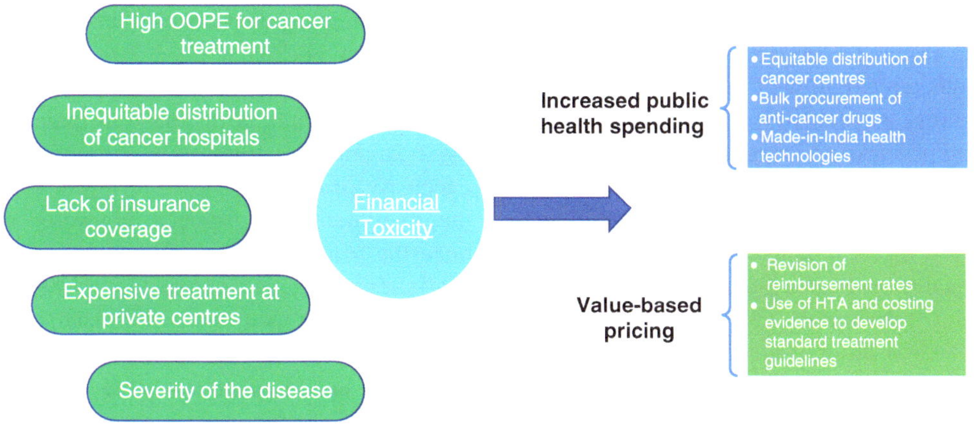

Fig. 26.4 A schematic diagram representing the high cost of cancer care in India

highlight that cancer leads to debilitating financial outcomes in LMICs, even in countries that have achieved some degree of universal health coverage or health insurance protection [8–10, 23, 30–35]. Hence, the health systems in developing countries should address the various issues related to "financial toxicity" in order to protect families of cancer from the financial risks of seeking healthcare [31, 34]. Figure 26.4 depicts the main reasons for the CHE in India and suggests the mitigating measures such as (a) increased public health spending and (b) value-based pricing, which can be equally adapted by most LMICs.

26.4 Public Investment for Affordable Cancer Care

Delivering affordable and equitable cancer care remains a great public health challenge in a diverse and populous country like India. To improve the accessibility to cancer care, we need to bridge the gap between the rural and urban areas and provide economic support to the majority of low socioeconomic strata. However, in the quest to provide accessible equitable cancer care, we must not compromise on the quality of cancer care.

To bring equity in cancer care between the rural and urban areas, the skewed distribution of public tertiary care centers and private centers in the urban areas needs to be corrected. This emphasis on urban-based cancer facility distribution is guided both by logistic reasons and commercial gains. The government needs to regulate the distribution of public cancer centers and provide incentives to the private centers to set up new centers of tertiary cancer care in the rural areas. The recent launch of 7 new cancer centers in the North-eastern part of India is a positive step in the provision of high quality and affordable cancer care to one of the most affected regions of the country [36].

Though there has been considerable expansion of the cancer services in the private sector in the past but utilization has been poor both due to the geographical and financial inaccessibility by majority [3]. The public private balance is essential if India aims to provide universal health coverage for cancer.

The health infrastructure in the public health systems like diagnostic equipment, treatment infrastructure like operation theatres, surgical equipment, radiotherapy equipment, availability of essential chemotherapy drugs, palliative care and increased manpower to make the public health system more efficient and accountable with reduced waiting period can increase the cancer control results.

Another challenge to provision of affordable cancer care is the ever-rising high-cost technologies which has an inflationary effect on cancer treatment. Cancer care is the most evolving field of healthcare, with introduction of multiple therapeutics each day. However, majority of the cancer chemotherapy, immunotherapy, molecular targeted drugs are highly priced and beyond the purchasing power of an average patient [37].

The government can regulate the pricing of the most commonly used cancer drugs by mass drug procurement and subsequent distribution with quality assurance [38]. Tamil Nadu Medical Supplies Corporation (TNMSC) has been a successful early starter example of this form of mass drug procurement for government medical institutions [39]. This example of bulk procurement through negotiated prices for quality drugs has been taken up by most of the state governments in India. The National Cancer Grid [40], a consortium of about 100 cancer hospitals in India, provides a unique opportunity for negotiating drug prices using its economy of scale. National Pharmaceutical Pricing Authority (NPPA) of India can regulate prices of anti-cancer drugs and engage in price negotiations with pharmaceutical companies to significantly reduce the cost of cancer care [41].

Advances in systemic therapy have led to improved survival in many cancers. Inclusion of the cancer drugs in WHO's essential list of medicines (WHO-EML), and in the National list of essential medicines (NLEM), which require to be part of the clinical guidelines has important implications [42]. Equally important is to have affordable access, which needs further action to address the barriers because of quality, differential pricing and stock outs. In a survey conducted by the WHO-EML group, covering 948 oncologists from 82 countries, the most used essential cancer drugs were not universally accessible; risks of significant OOPE for each medicine were reported by 19–58% of oncologists; and the risk of catastrophic expenditure was reported by 58–67% of oncologists for rituximab and trastuzumab [43]. Recently, 42 important anti-cancer medicines which have come under price regulation are being evaluated in a multi-institutional study to generate evidence on cost effectiveness for value-based pricing. The study also aims to assess the broader impact of price regulation of anti-cancer drugs from the perspective of insurer, patient and industry [44].

India has truly proved itself to be the pharmacy of the world, with "made in India" initiative. Another alternative to these essential expensive drugs is the generic formulations of the exorbitantly high-priced innovator drugs, once the patent has expired, which are significantly less costly as compare to innovator molecules. There are many successful examples such as the development of generic imatinib for chronic myeloid Leukemia where in India led the race for affordable cancer medications, with proven randomized study that the cheaper generic imatinib was similar in safety and efficacy when compared with the innovator imatinib [45]. However, the quality of these generic medicines vis-à-vis innovator drugs needs to be regulated by the government [46]. Since cancer drugs, for many cancers, have become integral to cancer management and cure, the health policy makers should intervene when innovator drugs are unaffordable for the patients in LMICs. Differential pricing can become an acceptable policy to increase global affordability and access to the biosimilar highly active innovator drugs for the cancer patients with low socioeconomic status [47].

One major gap in cancer management is the availability of radiotherapy in LMICs. Only 10–40% of the approximately 4.0 million cancer patients annually in LMICs who required radiation therapy were able to access such treatment. With many LMICs having inadequate or, in many cases, no radiation therapy centers, it is projected that about 12,600 radiation therapy machines will be needed globally over the next two to three decades to meet the demand in LMICs [9, 48]. Since the beginning of the twenty-first century, Bhabha Atomic Research Centre, Mumbai, in collaboration with scientists and manufacturers has made efforts to manufacture indigenous low cost radiation therapy equipment like Bhabhatron cobalt-60 machine and Siddhartha indigenous linear accelerator, which can help to make radiotherapy treatment accessible and affordable [49, 50]. These initiatives, though commendable, have not been able to make the anticipated progress in the last 20 years to manufacture and deliver the equipment to prospective cancer facilities in India and LMICs.

Besides installation of cost-effective radiotherapy equipment, the cancer patients in LMICs can benefit by the adoption of hypofractionation schedules for recommended cancer sites and type of neoplasm, which are equally effective with lesser visits to the health facilities [51, 52].

26.5 Communication in Cancer Care

The need to train physicians, nurses and health workers to communicate science and medicine effectively to the patients and public at large is, for the most part, ignored. Even though the traditional medical and nursing curriculum and case-based learning may emphasize on doctor/nurse-patient communication, the structured methods to learn the skills are lacking [53]. It is well recognized that doctors, nurses and health professionals are the trusted sources of information regarding a particular cancer and its management, yet more often than not the patient and the family caregiver is not adequately explained and empowered. At another level, cancer management and cancer research require the involvement of human resources with different qualification, knowledge and experiences. Hence, the registration, diagnostic tests, multi-disciplinary management decision, treatment course, disease prognosis, psycho-social counselling and clarification on financial and morbidity-related matters require to be communicated by different health professionals and support staff within the cancer care team.

Communication in cancer care has gained attention in recent time because on the one side the doctors, nurses and healthcare team should be responsible and accountable toward patients and their family caregivers; and at same time on the other side, the patients (and their caregivers) can benefit from a good communication during the cancer trajectory in many ways [54].

Key Points (as Stated by National Cancer Institute, USA [54])

1. Good communication between patients, family caregivers and the healthcare team is very important in cancer care.
2. Patients with cancer have special communication needs.
3. Some patients and families want a lot of information and choose to make decisions about care.
4. Communication is important at different points during cancer care.
5. End-of life discussions with the healthcare team may lead to fewer procedures and better quality of life.

The learning and training in communication skills for the doctor (and nurse, other healthcare worker) can be approached in a systematic and step-wise manner.

1. Knowing the clinical history, diagnosis, stage and likely treatment options.
2. A multi-disciplinary and logical understanding about the cancer treatment course, its results and likely morbidities.
3. Making the conversation and communication in a professional and empathetic manner.
4. During the communication process, the two key aspects are known as "breaking the news" (about cancer diagnosis, stage, likely anti-cancer therapy course) and "breaking bad news" (about advanced stage, limited survival, palliative care, end-of-life status, pre-bereavement phase).

Several models of cancer communication have evolved. The two relevant models, viz., COMFORT and/or SPIKES (Tables 26.1 and 26.2), in which each cancer doctor should learn and practice in the clinical management, are described briefly. It is imperative that reference sources should be read for more in-depth learning.

It is to recognize within our health systems and in the cancer management scenario of LMICs

Table 26.1 COMFORT communication curriculum [55]

Module	Communication processes
Communication	Understanding the patient's story Recognizing task and relationship practices
Orientation and options	Gauging health-literacy levels Understanding cultural humility
Mindful communication	Engaging in active listening Understanding nonverbal communication Being aware of self-care needs
Family	Observing family communication patterns Recognizing caregiver communication patterns Responding to the varying needs of family caregivers
Openings	Identifying pivotal points in patient/family care Finding common ground with patients/families
Relating	Realizing the multiple goals for patients/families Linking care to quality-of-life domains
Team	Developing team processes Cultivating team structures Distinguishing successful collaboration from group cohesion

Source: The COMFORT Communication Project developed by Elaine Wittenberg, PhD and Joy Goldsmith, PhD (www.CommunicateComfort.com), revised 2016

that communication can vastly improve the cancer diagnosis and treatment compliance and reduce the myths and fears.

26.6 Support of Society and Cancer Advocacy

Supporting those diagnosed and living with cancer (and their family caregivers) becomes an important component for the society outside of the doctors, nurses and medical institutions. For cancer control to be successfully implemented, cancer advocates and support groups can deliver several valuable services at local, regional, national and global levels. The International Union Against Cancer (UICC) has a long-standing history of support and advocacy activities where non-government organizations (NGO) and volunteer groups take leading role in the provision and delivery of various services which are integral to cancer management at any level (https://www.uicc.org/). The key tasks which can be performed by NGOs and volunteers include the following [57]:

1. To support and help in cancer diagnosis and treatments.

Table 26.2 SPIKES model (comprises of six steps) [56]

Module	Components
S (setting up the interview)	• Arrange for some privacy • Attendance of the family
P (assessing the patient's **p**erception)	• Correction of wrong perceptions • Questions that can be answered freely
I (obtaining the patient's **i**nvitation)	• Confirmation of the information that the patient demands • Confirmation of what a patient thinks he/she wants to know
K (giving **k**nowledge and information to the patient)	• Try to use nontechnical words • Periodically check the patient's understanding
E (addressing the patient's **e**motions with empathic response)	• Let the patient know that you understand the emotion • Attitude to understand the emotion of the patient
S (strategy and summary)	• Put together information to be provided and confirm it for the last time • Set an aim that is achievable in the future

2. To provide social support and guidance for transport, stay near cancer facility and financial assistance to the needy.
3. To facilitate the counseling and understanding for the cancer care duration and in subsequent rehabilitation and cancer survivorship.
4. To develop professional relationship and networks with the cancer patient and family caregivers.
5. To carry out specific cancer advocacy tasks—such as public awareness, cancer screening and prevention, improving cancer center facilities and advancing cancer research directly related to local/national cancer types.
6. Support and make significant liaison with government and other regulatory agencies for the cancer control.
7. Support and create networks with doctors, nurses and health workers, mainly in the dispersed and interior areas, to create cancer screening and cancer care trajectories for the community and citizens.

26.7 Value-Based Cancer Care

Newer diagnostic and treatment options are constantly introduced in high income countries and may include newer chemotherapy drugs, immunotherapies, targeted therapies, newer radiation techniques like tomotherapy, proton therapy, use of robotic surgery etc. These may not justify the health benefits corresponding to the cost and their application in resource limited countries like India [58, 59]. Few days or weeks of extra life in advanced cancers can be bought at disproportionate high costs [37]. Equitable cancer care can be achieved only if we follow: (a) value-based guidelines, backed by evidence for resource limited country like India, (b) set priorities on what the country or region would need to spend to achieve qualitative cancer services and to some extent provide the universal health coverage.

When we look at newer expensive anti-cancer molecules like trastuzumab, temozolomide, bevacizumab, sorafenib, etc., and the cancer screening services, these require to be cost effective to provide value-based effectiveness for LMICs like India [60–62]. Implementation of cost-effective treatment strategies should become an important component of affordable, equitable and accessible cancer care in India. The studies also suggest the amount of reduction required either in the reimbursement rates of the health benefit packages under AB PM-JAY or the market rates of anti-cancer drugs which offers useful evidence to the respective stakeholders (Table 26.3).

For the population-based screening for cervical cancer in India, visual inspection with acetic acid (VIA), PAP smear cytology, and, HPV DNA,were assessed. It is learnt that maximum life years would be gained by using HPV DNA testing, VIA and pap smear in decreasing frequency respectively [61]. However, when we look at the cost of implementing these screening strategies, the most cost effective for screening women in the age group of 20–60 years is VIA in low socioeconomic settings, done every 5 years [63]. Hence, VIA should be recommended for screening guidelines in our country, if efficiency is the main criteria. Screening and HPV vaccination are not mutually exclusive interventions for

Table 26.3 Examples of published studies on value-based pricing evaluations in India

Type of cancer	Drugs/regimen	Incremental cost per QALY gained	Whether cost-effective or not?	Value-based price
Advanced/ metastatic cervical cancer [58]	Bevacizumab	₹1,560,840	Not cost-effective	72% reduction in the reimbursement rate
Early breast cancer (HER2+) [60]	Trastuzumab along with adjuvant chemotherapy at 1 year/9 months/6 weeks	₹178,877	Not cost-effective	35% reduction in market prices

QALY quality-adjusted life-years, *HER+* human epidermal growth factor receptor-2 positive the prevention of cervical cancer. However, each of these has a different set of population that has to be targeted. HPV vaccination is a very cost-effective intervention especially when we combine screening with HPV vaccination. When this vaccinated cohort of adolescent girls is also screened later in their life (with VIA every 5 or 10 years), it leads to 69–76% reduction in cancer cases and 71–81% reduction in cancer deaths [64].

Using results and endpoints, like lack of disease progression or radiological and biomarker responses without meaningful benefit to patients should be discouraged for LMICs. The emphasis should be on life years (LYs) gained or quality-adjusted life years (QALYs) gained which should then be weighed against the money spent on these LYs or QALYs to decide the effectiveness and application of newer technology and molecules in cancer care [65]. Evidence generated on the cost effectiveness of all screening, diagnostic and treatment modalities through research in Health Technology Assessment (HTA) for cancer care should be integrated with standard treatment guidelines to ensure equitable cancer care for all [66]. To institutionalize HTA in India, HTA board has been established, the secretariat of which is located in the Department of Health Research. This HTA board is chaired by a member Niti Aayog and several other experts are its members. There are many technical partners and resource centers, for assessment of HTA, which in turn is appraised by technical appraisal committees and decision conveyed back as to whether a particular intervention should be included in the benefit package [67]. Thus, HTA evidence can be used to guide standard treatment guidelines.

Country-specific research into cost-effective procedures for screening, diagnostic procedures, chemotherapy, immunotherapy, targeted therapy, radiotherapy, cancer surgery and palliative care remains limited. The establishment and implementation of country-specific value-based guidelines begins with generation of cost effectiveness evidence for the various oncology settings followed by linkage of the payment systems both in the public and private sector with adherence to these value-based guidelines. Only justifiable healthcare costs should be reimbursed. Policy implications are required for application of these value based standard treatment guidelines to be mandatory followed in both public and private sectors [68]. These recent guidelines by the National Cancer Grid of India have taken initial steps by trying to differentiate between essential, desirable and optional modalities based on cost of treatment.

National Cancer Grid of India, created in 2012, is a partnership of more than 100 cancer centers in India. It aims to provide uniform guidelines for prevention, diagnosis and treatment of cancer across the country, so that distance no more determines the quality or accessibility of cancer care. It also focuses on training personnel, conducting research, generating cost effectiveness evidence and reporting cancer trends across the country [40]. It should further, also aim to integrate the rural centers in its partnership.

Cachar Cancer Hospital and Research Centre is a Department of Science and Industrial Research (DSIR, Government of India) recognized comprehensive cancer centre situated on the outskirts of Silchar in the Barak Valley of Assam on land provided by the Government of Assam. It was established in 1996 and is administered by a registered nonprofit society. Public philanthropy from several organizations and individuals has helped the hospital grow from 20 beds and 23 staff (at beginning) to 146 beds and 450 staff (in 2024), serving a population of about 10 million hailing from the four districts in southern Assam and the neighbouring states of Tripura, Manipur, Meghalaya, and Mizoram. Presently, the hospital treats over 5000 new cancer patients and 25,000 patients on follow-up every year. Eighty percent patients are daily wage earners (often the only earning member in family), agricultural or tea garden workers, and support large families. The hospital follows a multipronged strategy comprising of proactive pro-poor branding, adopting several cost-reduction strategies, fast-forwarding diagnostic workup and treatments, eliminating hidden and indirect costs, and using information technology wisely to ensure equitable and inclusive care. This mission of value-based cancer care is extended to cancer prevention strategies in nearby communities and home-based palliative care. Cachar Cancer Hospital is a shining example of setting up the comprehensive cancer centre in rural India and has become a model of community-centric and pro-poor cancer management in LMIC.

26.8 Social Sciences Research in Cancer

A greater collaboration between cancer care community and social scientists has been emphasized in recent years. In this chapter at the introduction the high burden of new cancer cases and the disproportionately high cancer mortality have been enumerated [1]. The growing cancer burden that low- and middle-income countries face is raising social, political and economic challenges of global cancer care that require interdisciplinary research beyond the traditional biomedical-clinical nexus [69].

There is need to develop affordable, cost-effective cancer treatment delivered locally for rural patients in LMICs. We need to have a more extensive rural cancer registry to justify the establishment of newer cancer centers in the rural areas, in their familiar environment based on incidence and type of cancer. This would also decongest the existing urban cancer centers with long waiting periods.

The financial burden per cancer patient in India is one of the lowest ($641 as compared to $86,758 in the USA), even this is out of reach for majority of cancer patients [70]. Direct economic burden assessment of cancer patients has proven this to be true [9, 10, 15, 17–19]. Social science research for policy changes has to focus on two key challenges faced by the cancer patients in LMIC.

1. To increase the physical and economic accessibility to cancer care, the government needs to find ways to fund the cancer care of the population specially of the economically vulnerable group.
2. For an optimal use of the meager resources in a country like India, health benefits gained at the population level need to be weighed against the money spent. Affordable and equitable cancer care requires cost-effective population-based clinical packages for cancer care.

Public health insurance schemes are the most feasible method to attain universal health coverage in LMICs, provide affordable, effective cancer care, reduce financial toxicity and catastrophic health expenditure and improve access and outcomes of treatment for dreaded disease like cancer [71].

The ambitious Ayushman Bharat Pradhan Mantri Jan Arogya Yojana (AB-PMJAY) scheme was launched by the government of India in 2018 to publicly fund healthcare of up to 500 million people living in India with the aim to provide Universal Health Coverage (UHC) and address the inaccessibility due financial hardship. Under AB-PMJAY all eligible citizens get access to free cancer care with a financial aid of up to 0.5 million Indian rupees (US $6700) per family per year. Eligibility for the scheme is determined based on socioeconomic criteria measured in the 2011 Socio-Economic Caste Census. There is no limit to the number of family members covered, and patients will be eligible, to take cashless benefits from any public or empanelled private hospital across the country [72, 73]. Currently, there are approximately 166 packages for cancer care under AB-PMJAY which includes medical, surgical and radiation oncology.

The Ministry of Health and Family Welfare is charged with overall health policy, including cancer care in India. However, in a federal country like India, constitutionally, the responsibility for implementation of health interventions lies largely on the state governments, with the central government providing policy directions and the financing of national health programs [74]. AB-PMJAY aims to work alongside state-based programs such as those in Andhra Pradesh, Karnataka, Kerala, Odisha, Tamil Nadu, Telangana, Tamil Nadu, West Bengal or a mix of the two (as in Gujarat and Tamil Nadu). Encouraging progress has been made in the delivery of good health insurance coverage, either through government schemes and through private insurance packages.

Other interventions which can help to increase access to cancer care include, the importance of service delivery with good communication skills and compassion, guidance for accommodation, nutrition counseling, travel assistance for patients and caregivers and general assistance in patient navigation. Involvement of the social workers and counsellors from non-governmental organizations (NGOs) is helpful for the patients. Conversations on cost of care often help patients to prepare for the financial burden of their cancer care. Such conversations will enable patients to make well informed and appropriate decisions regarding their treatment, resulting in better management of financial resources [57, 75]. For those cancer patients who complete the cancer treatments, to make follow up visits cost effective, one may consider involving the primary care physician, using tele-consultation or mobile medical units.

Emphasis on cancer prevention helps in improving cancer control and alleviating financial toxicity. Encouraging the vulnerable groups to come forward for screening, control on tobacco consumption, cancer awareness in the primary care setting, and refuting myths and encouraging early diagnosis can save many lives. Early diagnosis is associated with less aggressive treatment with less outpatient visits and fewer diagnostic tests, less hospitalization and better outcomes with improved quality of life. Adopting a healthy diet and a healthy lifestyle is equally important [76]. Resource mobilization for essential requirements like blood donation and recruiting cancer care coordinators at the community level to help with referrals for cancer centers requires a focused action plan. Similarly, awareness about the government sponsored schemes for the cancer patients is important to reduce the OOPE.

Thus, we conclude that if we have to "close the gap for cancer care," there is an urgent need to increase the geographical and economic accessibility for the cancer patients of LMICs. Public funding of cancer care with implementation of value-based guidelines seems to be the way forward for a diverse and huge country like India.

Precept Box (Take Home Points)

- Social science in cancer care is integral to the sociogeographic, ethno-cultural and financial conditions of the country, region and its citizens.
- Cancer health economics refers to application of health economics theory and models for cancer prevention, screening, management and end of life care.
- In India, national expenditure on public health care is less than 1.3% of total GDP. Poor expenditure by government and lack of adequate insurance coverage leads to out-of-pocket expenditure for treatment.
- With rising cost of cancer treatment, out of pocket expenditure leads to poor compliance and financial toxicity.
- Dearth of treatment facilities and the geographical divide of cancer centers have led to poor access to cancer treatment.
- To improve the accessibility to cancer care, we need to bridge the gap between the rural and urban areas and provide economic support to the majority of low socioeconomic strata.
- Pricing of the most commonly used anti-cancer drugs by bulk procurement, collective bargaining and subsequent distribution with quality assurance needs to be organized.
- "Make in India" program should be encouraged to decrease the cost of drugs, manufacture diagnostic tools and radiotherapy equipment.
- Equitable cancer care can be achieved only if we follow value-based guidelines, backed by evidence for resource limited country like India, which needs to set priorities for healthcare spending to achieve universal health coverage.
- Public health insurance schemes like AB-PMJAY and other state government insurance policies are feasible ways to attain universal health coverage.

Multiple Choice Questions

1. What is the global age-standardized incidence rate for all cancers per 100,000?
   (a) 111
   (b) 222
   (c) 333
   (d) 444
2. What is national expenditure of India on public healthcare?
   (a) less than 0.5% of GDP
   (b) less than 1% of GDP
   (c) 1.3% of GDP
   (d) 4% of GDP
3. What was the theme for World Cancer Day (WCD) 2022 and 2023?
   (a) Widen the care gap
   (b) Healthier world for everyone
   (c) Close the care gap
   (d) Year of the nurse and mid-wife
4. Access to healthcare in India is determined by
   (a) Geographic factor
   (b) Social factor
   (c) Financial factor
   (d) All of the above
5. What financial factors become the hurdle for cancer patients in India?
   (a) Gender bias
   (b) Rural–urban divide
   (c) High cost of cancer treatment
   (d) Low-cost diagnosis
6. Which major cancer treatment modality has low availability and poor access in LMIC?
   (a) surgery
   (b) chemotherapy
   (c) radiotherapy
   (d) immunotherapy
7. Which healthcare scheme was launched by government in 2018 to publicly fund healthcare for nearly 500 million person in India?
   (a) Ayushman Bharat Pradhan Mantri Jan Arogya Yojna
   (b) Janani Suraksha Yojna
   (c) Mission Indradhanush
   (d) Mission Antyodaya
8. Value-based cancer care helps to achieve
   (a) Equitable cancer management
   (b) Effective cancer management
   (c) Affordable cancer management
   (d) All of the above
9. Communication is important in cancer care. Doctors and nurses can learn two commonly effective tools, such as
   (a) MOON & MAGIC
   (b) COMFORT & SPIKES
   (c) HEAD & HEART
   (d) Any of above
10. Non-government organization (NGO) and volunteers can make improvements in cancer management by performing which of the following task(s)?
   (a) To support and help in cancer diagnosis and treatment
   (b) To provide social support and guidance for transport, stay near cancer facility and financial assistance to the needy
   (c) To facilitate the counselling and understanding for the cancer care duration and in subsequent rehabilitation and cancer survivorship
   (d) All of the above

Answers Keys

1. (b)
2. (c)
3. (c)
4. (d)
5. (c)
6. (c)
7. (a)
8. (d)
9. (b)
10. (d)

References

1. Sung H, Ferlay J, Siegel RL, Laversanne M, Soerjomataram I, Jemal A, et al. Global cancer statistics 2020: GLOBOCAN estimates of incidence and mortality worldwide for 36 cancers in 185 countries. CA Cancer J Clin. 2021;71(3):209–49.

2. Mathur P, Sathishkumar K, Chaturvedi M, Das P, Sudarshan KL, Santhappan S, et al. Cancer statistics, 2020: report from national cancer registry programme, India. JCO Glob Oncol. 2020;6:1063–75.
3. Munshi A, Ganesh T, Mohanti BK. Radiotherapy in India: History, current scenario and proposed solutions. Indian J Cancer. 2019;56(4):359.
4. Yadav J, Menon GR, John D. Disease-specific out-of-pocket payments, catastrophic health expenditure and impoverishment effects in India: an analysis of national health survey data. Appl Health Econ Health Policy. 2021;19(5):769–82.
5. India - estimated public health expenditure 2017-2020 [Internet]. Statista. [cited 2022 Jun 16]. https://www.statista.com/statistics/684924/india-public-health-expenditure/.
6. Campaign theme I World Cancer Day [Internet]. [cited 2022 Jun 16]. https://www.worldcancerday.org/sites/default/files/styles/max_2600x2600/public/2021-12/WCD22_web%20thumbnail_ENG.png?h=f36f5bad&itok=Iqx2uh3H.
7. Pati S, Hussain MA, Chauhan AS, Mallick D, Nayak S. Patient navigation pathway and barriers to treatment seeking in cancer in India: a qualitative inquiry. Cancer Epidemiol. 2013;37(6):973–8.
8. Joe W. Distressed financing of household out-of-pocket health care payments in India: incidence and correlates. Health Policy Plan. 2015;30(6):728–41.
9. Mohanti BK, Mukhopadhyay A, Sharma K, Das S, Dash S. The economic burden of cancer. Econ Polit Wkly. 2011;43:112–7.
10. Rajpal S, Kumar A, Joe W. Economic burden of cancer in India: Evidence from cross-sectional nationally representative household survey, 2014. PLoS ONE. 2018;13(2):e0193320.
11. Faruqui N, Bernays S, Martiniuk A, Abimbola S, Arora R, Lowe J, et al. Access to care for childhood cancers in India: perspectives of health care providers and the implications for universal health coverage. BMC Public Health. 2020;20(1):1641.
12. Baru R, Acharya A, Acharya S, Kumar AKS, Nagaraj K. Inequities in access to health services in India: caste, class and region. Econ Polit Wkly. 2010;45(38):49–58.
13. Flores G, Krishnakumar J, O'Donnell O, van Doorslaer E. Coping with health-care costs: implications for the measurement of catastrophic expenditures and poverty. Health Econ. 2008;17(12):1393–412.
14. Indicator Metadata Registry Details [Internet]. [cited 2022 Jun 16]. https://www.who.int/data/gho/indicator-metadata-registry/imr-details/2441.
15. Mallath MK, Taylor DG, Badwe RA, Rath GK, Shanta V, Pramesh CS, et al. The growing burden of cancer in India: epidemiology and social context. Lancet Oncol. 2014;15(6):205–12.
16. Socio Economic and Caste Census (SECC) [Internet]. [cited 2022 Jun 16]. https://secc.gov.in/getSeccDataSummaryNationalReport.htm.
17. Banavali SD. Delivery of cancer care in rural India: experiences of establishing a rural comprehensive cancer care facility. Indian J Med Paediatr Oncol. 2015;36(2):128.
18. Datta SS, Ghose S, Ghosh M, Jain A, Mandal S, Chakraborty S, Caduff C. Journeys: understanding access, affordability and disruptions to cancer care in India. Ecancermedicalscience. 2022;16:1342. https://doi.org/10.3332/ecancer.2022.1342.
19. Brand NR, Qu LG, Chao A, Ilbawi AM. Delays and barriers to cancer care in low- and middle-income countries: a systematic review. Oncologist. 2019;24(12):e1371–80. https://doi.org/10.1634/theoncologist.2019-0057.
20. Sirohi B, Chalkidou K, Pramesh CS, Anderson BO, et al. Developing institutions for cancer care in low-income and middle-income countries: from cancer units to comprehensive cancer centres. Lancet Oncol. 2018;19(8):e395–406. https://doi.org/10.1016/S1470-2045(18)30342-5.
21. Jain R. Budget 2022: plugging the health insurance gap for India's Missing Middle. https://www.outlookindia.com/business/budget-2022-plugging-the-health-insurance-gap-for-india-s-missing-middle-news-48117.
22. Household social consumption in India: Health NSS 75th round (July, 2017 – June, 2018) [Internet]. [cited 2022 Jun 16]. https://pib.gov.in/newsite/PrintRelease.aspx?relid=194918.
23. Dhankhar A, Kumari R, Bahurupi YA. Out-of-pocket, catastrophic health expenditure and distress financing on non-communicable diseases in India: a systematic review with meta-analysis. Asian Pac J Cancer Prev. 2021;22(3):671–80.
24. India - Social Consumption: Health, NSS 71st Round: Jan - June 2014 [Internet]. [cited 2020 Jul 10]. http://microdata.gov.in/nada43/index.php/catalog/135.
25. Dinesh TA, Nair P, Abhijath V, Jha V, Aarthy K. Economics of cancer care: a community-based cross-sectional study in Kerala, India. S Asian J Cancer. 2020;9(1):7–12.
26. Goyanka R. Economic and non-economic burden of cancer: a propensity score matched analysis using household health survey data of India. Cancer Res Stat Treat. 2021;4(1):29.
27. Rice DP. Cost of illness studies: what is good about them? Inj Prev J Int Soc Child Adolesc Inj Prev. 2000;6(3):177–9.
28. American Cancer Society Cancer Action Network. The costs of cancer 2020. 2020.
29. Zafar SY, Peppercorn JM, Schrag D, Taylor DH, Goetzinger AM, Zhong X, Abernethy AP. The financial toxicity of cancer treatment: a pilot study assessing out-of-pocket expenses and the insured cancer patient's experience. Oncologist. 2013;18(4):381–90. https://doi.org/10.1634/theoncologist.2012-0279.
30. World Health Organization. Designing health financing systems to reduce catastrophic health expenditure.

Geneva: World Health Organization; 2005. https://apps.who.int/iris/handle/10665/70005.
31. Kruk ME, Goldmann E, Galea S. Borrowing and selling to pay for health care in low- and middle-income countries. Health Aff Proj Hope. 2009;28(4):1056–66.
32. Boby JM, Rajappa S, Mathew A. Financial toxicity in cancer care in India: a systematic review. Lancet Oncol. 2021;22(12):541–9.
33. Piroozi B, Zarei B, Ghaderi B, Safari H, Moradi G, Rezaei S, et al. Catastrophic health expenditure and its determinants in households with gastrointestinal cancer patients: evidence from new health system reform in Iran. Int J Hum Rights Healthc. 2019;12(4):249–57.
34. Bhoo-Pathy N, Ng CW, Lim GCC, Tamin NSI, Sullivan R, Bhoo-Pathy NT, et al. Financial toxicity after cancer in a setting with universal health coverage: a call for urgent action. J Oncol Pract. 2019;15(6):537–46.
35. Du XL, Fang S, Coker AL, Sanderson M, Aragaki C, Cormier JN, et al. Racial disparity and socioeconomic status in association with survival in older men with local/regional stage prostate carcinoma. Cancer. 2006;106(6):1276–85.
36. 7 New cancer facilities to offer affordable care in Ne Region | Guwahati News - Times of India [Internet]. [cited 2022 Jun 16]. https://timesofindia.indiatimes.com/city/guwahati/7-new-cancer-facilities-to-offer-affordable-care-in-ne-region/articleshow/91163875.cms.
37. Sullivan R, Pramesh CS, Booth CM. Cancer patients need better care, not just more technology. Nature. 2017;549(7672):325–8.
38. World Health Organization. Pricing of cancer medicines and its impacts. Geneva: World Health Organization; 2018.
39. Tamil Nadu Medical Services Corporation Limited [Internet]. [cited 2023 Apr 05]. https://tnmsc.tn.gov.in/html/Procurement%20&%20Tender.php.
40. National Cancer Grid-collaboration for cancer care. About Us - NCG [Internet]. [cited 2022 Jun 16]. https://tmc.gov.in/ncg/index.php/overview/about-us.
41. Home | Official Website of National Pharmaceutical Pricing Authority, Ministry of Chemicals and Fertilizers, Government of India [Internet]. [cited 2022 Jun 16]. Available from: https://www.nppaindia.nic.in/en/.
42. World Health Organization. Pricing of cancer medicines and its impacts. 2018. https://apps.who.int/iris/bitstream/handle/10665/277190/9789241515115-eng.pdf?sequence=1&isAllowed=y.
43. Sengar M, Fundytus A, Hopman W, Pramesh CS, et al. Cancer medicines: what is essential and affordable in India? JCO Glob Oncol. 2022;8:e2200060. https://doi.org/10.1200/GO.22.00060.
44. Prinja S, Dixit J, Gupta N, Mehra N, Singh A, Krishnamurthy MN, et al. Development of national cancer database for cost and quality of life (CaDCQoL) in India: a protocol. BMJ Open. 2021;11(7):e048513.
45. Danthala M, Gundeti S, Kuruva SP, Puligundla KC, et al. Generic imatinib in chronic myeloid leukemia: survival of the cheapest. Clin Lymphoma Myeloma Leuk. 2017;17(7):457–62. https://doi.org/10.1016/j.clml.2017.05.006.
46. George T, Baliga MS. Generic anticancer drugs of the Jan Aushadhi Scheme in India and their branded counterparts: the first cost comparison study. Cureus. 2021;13(11):e19231.
47. Ocran Mattila P, Ahmad R, Hasan SS, Babar ZU. Availability, affordability, access, and pricing of anti-cancer medicines in low- and middle-income countries: a systematic review of literature. Front Public Health. 2021;9:628744. https://doi.org/10.3389/fpubh.2021.628744.
48. Ige TA, Jenkins A, Burt G, Angal-Kalinin D, et al. Surveying the challenges to improve linear accelerator-based radiation therapy in Africa: a unique collaborative platform of all 28 African countries offering such treatment. Clin Oncol. 2021;33(12):e521–9. https://doi.org/10.1016/j.clon.2021.05.008.
49. Sahani G, Kumar M, Sharma PD, Sharma DN, et al. Compliance of Bhabhatron-II telecobalt unit with IEC standard - radiation safety. J Appl Clin Med Phys. 2009;10(2):120–30. https://doi.org/10.1120/jacmp.v10i2.2963.
50. https://www.panaceamedical.in/news/panacea-medical-technologies-receives-fda-510k-clearance-for-sbrt-enabled-linear-accelerator-siddharth-ii/. Accessed 4 October 2021.
51. Rodin D, Tawk B, Mohamad O, Grover S, et al. Hypofractionated radiotherapy in the real-world setting: an international ESTRO-GIRO survey. Radiother Oncol. 2021;157:32–9. https://doi.org/10.1016/j.radonc.2021.01.003.
52. Taylor P, Castilho MS, Marta GN. Cost containment analysis and access to treatment associated with adopting hypofractionated radiation therapy from the Brazilian perspective. Lancet Reg Health Am. 2022;13:100292. https://doi.org/10.1016/j.lana.2022.100292.
53. Arora VM, Krishnamoorthi VR, Bloomgarden E, Jain S, et al. Restoring trust in science and medicine: empowering and educating future physicians in science communication with the public. Acad Med. 2023; https://doi.org/10.1097/ACM.0000000000005178.
54. https://www.cancer.gov/about-cancer/coping/adjusting-to-cancer/communication-pdq. Accessed 2 April 2023.
55. Wittenberg E, Goldsmith J. The COMFORT communication project. www.CommunicateComfort.com.
56. Baile WF, Buckman R, Lenzi R, Glober G, Beale EA, Kudelka AP. SPIKES-A six-step protocol for delivering bad news: application to the patient with cancer. Oncologist. 2000;5:302–11. https://doi.org/10.1634/theoncologist.5-4-302.
57. https://www.uicc.org/.
58. Gupta N, Nehra P, Chauhan AS, Mehra N, Singh A, Krishnamurthy MN, et al. Cost effectiveness of bevacizumab plus chemotherapy for the treatment of advanced and metastatic cervical cancer in India—a

model-based economic analysis. JCO Glob Oncol. 2022;8:e2100355.
59. Xia Z, Wang J, Xia J, Wang M, Cheng Z. Inequality in accessibility of proton therapy for cancers and its economic determinants: a cross-sectional study. Front Oncol. 2022;12:876368. https://doi.org/10.3389/fonc.2022.876368.
60. Gupta N, Verma RK, Gupta S, Prinja S. Cost effectiveness of trastuzumab for management of breast cancer in India. JCO Glob Oncol. 2020;6:205–16. https://www.ncbi.nlm.nih.gov/pmc/articles/PMC7051799/.
61. Sefuthi T, Nkonki L. A systematic review of economic evaluations of cervical cancer screening methods. Syst Rev. 2022;11(1):162. https://doi.org/10.1186/s13643-022-02017-z.
62. Prinja S, Jyani G, Gupta N, Rajsekar K. Adapting health technology assessment for drugs, medical devices, and health programs: methodological considerations from the Indian experience. Expert Rev Pharmacoecon Outcomes Res. 2021;21(5):859–68.
63. Bruni L, Serrano B, Roura E, Alemany L, Cowan M, Herrero R, Poljak M, Murillo R, Broutet N, Riley LM, de Sanjose S. Cervical cancer screening programmes and age-specific coverage estimates for 202 countries and territories worldwide: a review and synthetic analysis. Lancet Glob Health. 2022;10(8):e1115–27. https://doi.org/10.1016/S2214-109X(22)00241-8.
64. Chauhan AS, Prinja S, Srinivasan R, Rai B, Malliga JS, Jyani G, et al. Cost effectiveness of strategies for cervical cancer prevention in India. PLoS ONE. 2020;15(9):e0238291.
65. Gupta N, Jyani G, Rajsekar K, Gupta R, Nagar A, Gedam P, et al. Application of health technology assessment for oncology care in India: implications for Ayushman Bharat Pradhan Mantri Jan Aarogya Yojana. Indian J Med Paediatr Oncol. 2021;42(5):479–83.
66. Kim H, Goodall S, Liew D. Health technology assessment challenges in oncology: 20 years of value in health. Value Health. 2019;22(5):593–600. https://doi.org/10.1016/j.jval.2019.01.001.
67. Health Technology Assessment in India (HTAIn) - Health Technology Assessment in India (HTAIn) [Internet]. [cited 2022 Jun 16]. https://htain.icmr.org.in/.
68. National Cancer Grid. National cancer guidelines manual. 2021. https://tmc.gov.in/ncg/images/NCG%20Guidelines_15122021.pdf. Accessed 7 April 2023.
69. Caduff C, Skelton M, Banerjee D, Djordjevic D, et al. Analysis of social science research into cancer care in low- and middle-income countries: improving global cancer control through greater interdisciplinary research. J Glob Oncol. 2018;4:1–9. https://doi.org/10.1200/JGO.18.00045.
70. Goss PE, Strasser-Weippl K, Lee-Bychkovsky BL, Fan L, Li J, Chavarri-Guerra Y, et al. Challenges to effective cancer control in China, India, and Russia. Lancet Oncol. 2014;15(5):489–538.
71. Universal health coverage (UHC) [Internet]. [cited 2022 Jun 16]. https://www.who.int/news-room/fact-sheets/detail/universal-health-coverage-(uhc).
72. Angell BJ, Prinja S, Gupt A, Jha V, Jan S. The Ayushman Bharat Pradhan Mantri Jan Arogya Yojana and the path to universal health coverage in India: overcoming the challenges of stewardship and governance. PLoS Med. 2019;16(3):e1002759.
73. Ayushman Bharat Yojana | National Health Portal Of India [Internet]. [cited 2022 Jun 16]. https://www.nhp.gov.in/ayushman-bharat-yojana_pg.
74. Ministry of Health & Family Welfare. National Health Policy 2015 Draft. 2014. https://www.nhp.gov.in/sites/default/files/pdf/draft_national_health_policy_2015.pdf.
75. Sanadgol A, Doshmangir L, Majdzadeh R, Gordeev VS. Engagement of non-governmental organisations in moving towards universal health coverage: a scoping review. Glob Health. 2021;17(1):129.
76. Pramesh CS, Badwe RA, Bhoo-Pathy N, Booth CM, et al. Priorities for cancer research in low- and middle-income countries: a global perspective. Nat Med. 2022;28(4):649–57. https://doi.org/10.1038/s41591-022-01738-x.

Further Reading

https://www.cancer.gov/about-cancer/managing-care/track-care-costs/financial-toxicity-hp-pdq.

https://www.mdanderson.org/documents/education-training/project-echo/10%2027%2016%20ECHO-PACA%20SPIKES.pdf.

https://www.uicc.org/.

https://www.who.int/activities/controlling-cancer.

Nuclear Medicine in Oncology

27

Dikhra Khan, Rakesh Kumar,
and Nilendu Purandare

Recently, nuclear medicine has been leadingly helping the diagnostics and therapeutics in oncology. This branch explores the functional status of an organ or lesion, which gives it an edge over solitary anatomical imaging, especially in oncology, where we need to comprehend how a tumor behaves. After the advent of PET/CT, a combination of positron-emission tomography (PET) with computed tomography (CT), which integrates anatomical and functional modalities, it has aided in the main objectives of oncological imaging like lesion detection, lesion characterization, evaluation of the extent of the tumor, staging, response assessment and restaging of various cancers.

The most extensively used radiotracer in the field of oncology for PET/CT is ^{18}F-Fluorodeoxyglucose (FDG), a radiolabeled analog of glucose. Malignant cells have increased glucose utilization due to upregulation of hexokinase activity and also exhibit an increased glucose transporter (GLUT) expression [1]. Therefore, FDG is taken up by malignant cells. The uptake of FDG can be graded visually as well as semiquantitatively. Standardized uptake value (SUV) is a semiquantitative measurement of tissue uptake. Multiple factors affect the SUV, including dose calibration accuracy, injection time and imaging time (dose to scan time), patient weight (changes are expected in oncology patients), motion artifacts and blood glucose levels. Other tumor measurement features include metabolic tumor volume (MTV) and total lesion glycolysis (TLG). MTV is defined as the sum of the volume of voxels with SUV surpassing a threshold value in a tumor. TLG is calculated by multiplying the mean SUV and MTV of the desired lesion [1–3]. Recent studies confirm the superiority of MTV and TLG compared with regard to prognostic value in various malignancies like head-and-neck cancer, nonsmall cell lung cancer and epithelial ovarian cancer [2, 3].

For performing FDG-PET/CT, patients should be fasting for 4–6 h before ^{18}F-FDG injection to decrease serum glucose level and to maintain a low insulin level. Patients should also be encouraged to increase hydration and excretion of radiotracer. The glucose level of the patients should be below 200 mg/dL (11.1 mmol/L); the optimal level is below 140 mg/dL (7.8 mmol/L) before injection of FDG-PET/CT. After intravenous injection of the recommended 185–370 MBq (5–10 mCi) dose of FDG, scan of patient is acquired on a dedicated PET/CT scanner after 30–45 min of tracer injection. Whole body image from vertex to mid-thigh is acquired along with additional spot images, after proper history taking and if necessitated after the acquisition of whole body images [1, 2].

D. Khan · R. Kumar (✉)
Division of Diagnostic Nuclear Medicine, Department of Nuclear Medicine, AIIMS, New Delhi, India

N. Purandare
Department of Nuclear Medicine, Tata Memorial Hospital, Mumbai, Maharashtra, India

A. Munshi et al. (eds.), *Radiation Oncology – Principles, Precepts and Practice*,
https://doi.org/10.1007/978-981-97-8389-2_27

The utility of ^{18}F-FDG-PET/CT has been explored and studied in most of the cancers, and therefore, reference of PET/CT in this article would be referring to ^{18}F-FDG-PET/CT unless otherwise specified. Due to the vastness of this topic, in this chapter, we will be mainly focusing on the pivotal role of PET/CT in oncology specifically FDG-PET/CT with short discussion on the scope of ^{68}Ga-DOTANOC and prostate-specific membrane antigen (PSMA) PET/CT in neuroendocrine tumors and prostate cancers, respectively. We will also shine light over the newly developed radiotracers like FAPI PET/CT and also an overall coverage of the established and emerging roles of PET/CT in different domains of various cancers. Emphasizing on all the above areas, each cancer will be covered as a specific section.

27.1 Lung Cancer

Lung cancer is the most common cancer which causes maximum death in people [4]. Almost 80–85% of lung cancers are non small cell lung cancer (NSCLC), while the rest are classified as small cell lung cancer (SCLC) [5]. Earlier for lung, T N M staging, exclusively contrast-enhanced CT (CECT), mediastinoscopy and magnetic resonance imaging (MRI) were used. Nowadays as per National Comprehensive Cancer Network (NCCN) guidelines, ^{18}F-FDG-PET/CT scan is indicated for initial staging in essentially all cases and also before a diagnostic biopsy to be cognizant of the ideal site and tumor viability (Fig. 27.1). It also recommends ^{18}F-FDG-PET/CT for presurgical evaluation (within 60 days). For nodal staging, mediastinoscopy is considered the gold standard and is indicated to confirm histopathologically when ^{18}F-FDG-PET/CT demonstrates positive mediastinal lymph nodes. PET/CT is very effective in detecting distant metastases as it covers the most common metastatic diseases sites of lung cancer, i.e., brain, adrenals, bone, liver, etc., in a typical PET/CT scan from orbit to mid-thigh.

It has been reported in the literature that PET/CT can detect sites which may not be apparent by standard imaging strategies in 10–14% of the

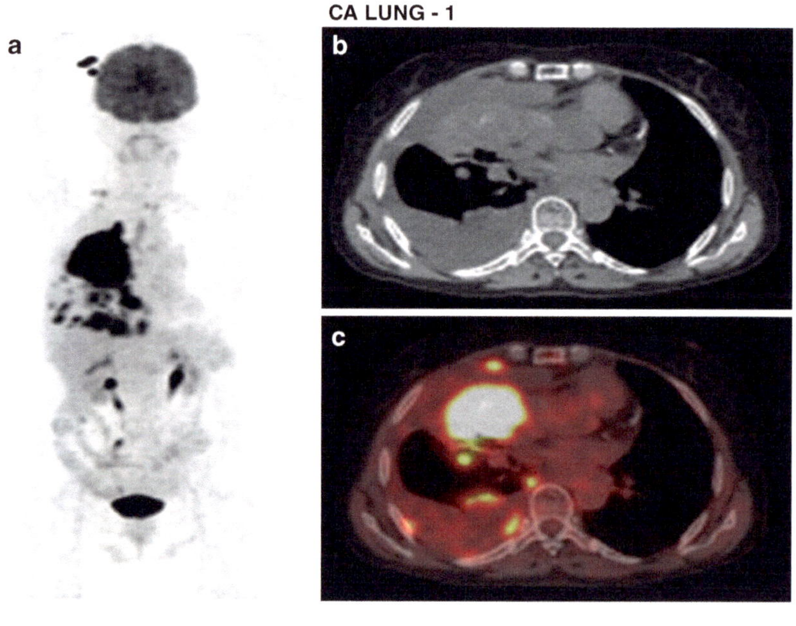

Fig. 27.1 A 74-year-old female patient with a known case of carcinoma lung on erlotinib came for FDG-PET/CT for response assessment. Maximum intensity projection (MIP) (**a**), transaxial CT (**b**) and fused PET/CT (**c**) show soft tissue density lesion with increased FDG uptake noted in right lung upper lobe closely abutting the costal pleura (measuring ~9.7 × 5.0 × 8.0 cm). The mass is encasing the right upper lobe bronchus causing distal non-FDG-avid collapse of the anterior segments along with multiple discrete FDG-avid pleural and parenchymal soft tissue nodules seen in the right lung (Source: Department of Nuclear Medicine, AIIMS, Delhi)

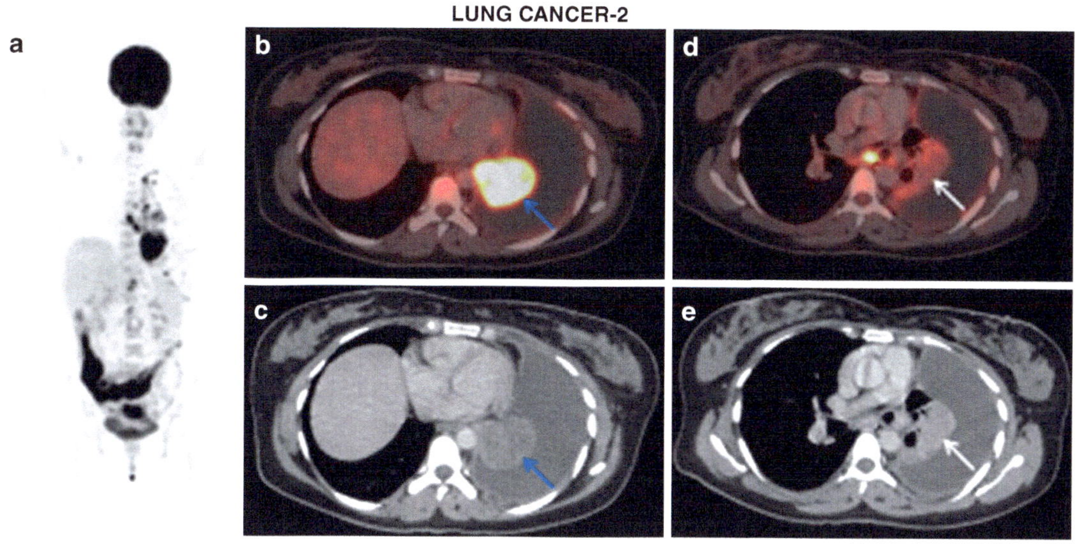

Fig. 27.2 A 39-year-old female patient with a known case of carcinoma lung underwent PET/CT for response assessment. Maximum intensity projection (**a**), fused PET/CT (**b**, **d**) and CT axial (**c**, **e**) sections show FDG-avid left infrahilar mass (blue arrows) measuring 2.9 × 3.5 cm showing encasement of left main bronchus, and subsequent distal segmental collapse (white arrows) along with evidence of left pleural effusion and multiple sub-pleural and parenchymal nodules with mediastinal lymph-adenopathy (Source: Department of Nuclear Medicine, AIIMS Delhi)

lung cancer patients, thus playing a pivotal role in changing the management of the patients [6]. In radiotherapy (RT) planning, ^{18}F-FDG-PET/CT is recommended as per NCCN guidelines and therefore on follow-up to differentiate recurrent disease from post-RT changes in such a scenario (Fig. 27.2). RT target volumes and treatment position can be based on pretreatment PET/CT (4 weeks-Ideal) [7, 8]. In special case context, it is also helpful to assess the tumor tissue in a background of collapsed lung parenchymal tissue with its increased tumor avidity [8]. Other indications for ^{18}F-FDG-PET/CT include re-staging and for surveillance solely in NSCLC category.

Recently, role of ^{18}F-FDG-PET/CT has shown potential to accurately differentiate between active pulmonary tuberculosis (TB) and lung cancer based on radiomics and semantic features as extracted from pretreatment PET/CT [9].

27.2 Breast Cancer

Breast cancer is the most common malignancy in women [10]. The most common imaging technique used to diagnose and screen for breast cancer is mammography but it has limited sensitivity in case of dense breasts and breast implant patients. MR imaging and ultrasonography (USG) carry limited role in differentiating benign abnormalities from malignancies. ^{18}F-FDG-PET/CT is mainly helpful in case of metastatic breast carcinoma and specifically for Stage IIIA /IIIB/IV as per NCCN guidelines in invasive breast cancer. It is a valuable modality in cases of equivocal and suspicious cases, notably in a setting of locally advanced breast carcinoma. However, initial staging and breast tissue characterization of breast cancer patients can be best done by ^{18}F-FDG-PET/MRI. Also ^{18}F-FDG-PET/CT is particularly recommended for all stages of inflammatory breast cancer [11].

Accurate assessment of locoregional axillary nodal status is pivotal for its implication in treatment and prognosis. The 10-year survival rate in breast cancer patients with axillary lymph nodal metastases depends on the number of nodes involved ranging from 30% (>10 nodes) to 70% (1–3 nodes) in contrast to 90% in patients without axillary nodal metastases [12, 13]. ^{18}F-FDG-PET/CT can be used for axillary lymph nodal detection but has a limited role to detect micro-

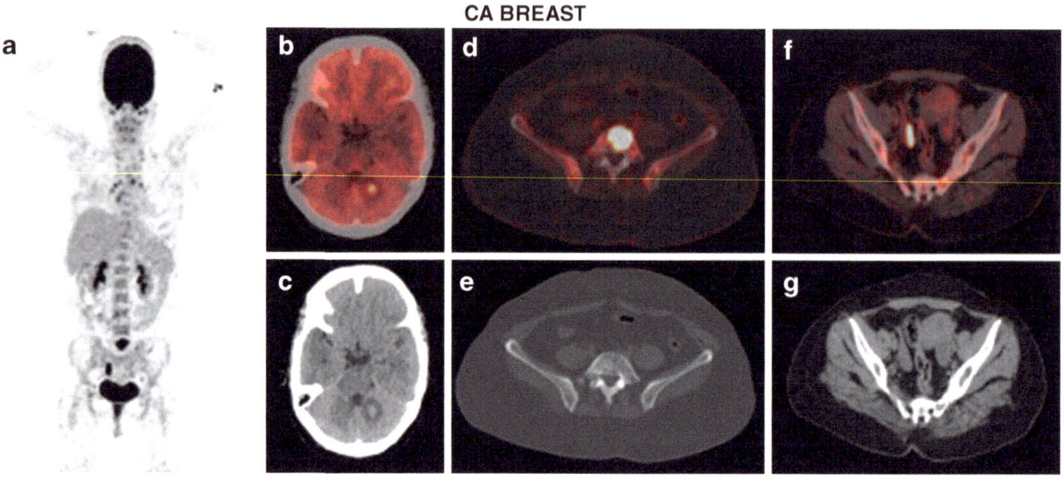

Fig. 27.3 A 33-year-old female with carcinoma right breast who underwent right modified radical mastectomy, 6 cycles of neoadjuvant chemotherapy and radiotherapy. She was on tamoxifen and came for FDG-PET/CT to assess disease status. Maximum intensity projection (MIP) (**a**), transaxial CT (**c**, **e**, **g**) and fused PET/CT (**b**, **d**, **f**) show avid hyperdense lesion (measuring ~1.3 × 0.8 cm) with surrounding perilesional edema noted in the left cerebellum along with lytic-sclerotic lesion with increased FDG uptake noted in the L5 vertebral body and suspicious involvement of right ovary (Source: Department of Nuclear Medicine, AIIMS Delhi)

metastases like the sentinel lymph nodal biopsy (due to limited resolution of the PET/CT). PET/CT has high sensitivity to detect avid level III axillary, internal mammary and supraclavicular lymph nodes [14, 15].

Breast cancer metastases have a predilection to bone, liver and brain depending on the tumor receptor expression and tumor biology. ^{18}F-FDG-PET/CT is therefore an excellent modality in the metastatic workup of the patients (Fig. 27.3). Recently, advances in ^{18}F-FDG-PET/MRI have shown it to be distinctly useful in brain, liver and bone marrow metastases.

Society of Nuclear Medicine & Molecular Imaging (SNMMI) appropriate use criteria recommends ^{18}F-FDG-PET/CT for the purpose of restaging, particularly for detection of local recurrence and distant metastases [16]. It also recommends ^{18}F-FDG-PET/CT for response evaluation too. It has been shown that ^{18}F-FDG-PET/CT can detect metabolic changes within the tumor burden as early as 8 days after chemotherapy initiation. Several studies reported ^{18}F-FDG-PET/CT can differentiate responders from nonresponders just after first cycle of chemotherapy, thus helping in management of the patients. American Society of Clinical Oncology (ASCO) and NCCN guidelines recommend ^{18}F-FDG-PET/CT in posttreatment surveillance only in symptomatic patients not otherwise [11, 17]. Nearly 3.8–8.3% of breast cancer malignancy show contralateral breast involvement; therefore, ^{18}F-FDG-PET/CT can be used for screening. Also ^{18}F-FDG-PET/MRI may play a role in such a setting [18].

Each of the hormonal receptors exhibited by breast cancer tissues has been exploited for imaging purpose in the field of nuclear medicine. Out of the lot, ^{18}F-FES PET/CT is best explored to assess the ER receptor expression and response to hormonal therapies like tamoxifen, fulvestrant, etc. It helps in assessing the receptor blockade [19].

27.3 Lymphoma

Lymphomas, originating from the lymphocytes are mainly divided into two categories namely non-Hodgkin lymphoma (NHL) (90% of cases) and Hodgkin lymphoma (HL) (10%) [20]. ^{18}F-FDG-PET/CT is an essential modality for initial workup in lymphoma especially in Hodgkin's lymphoma and even in non-Hodgkin's lymphoma like diffuse large B-cell lymphoma (DLBCL) and follicular lymphoma. FDG-uptake varies between histological subtypes, with indolent lymphomas generally exhibiting lower SUV_{max}, than aggressive lymphomas, a feature relevant when selecting biopsy site in patients with suspected histologic transformation. The advantage of ^{18}F-FDG-PET when compared with conventional imaging modalities lies in the high tumor-to-background ratio, resulting in high sensitivity and specificity for lymphoma. Accurate anatomical staging of lymphoma is crucial for treatment planning. The Ann Arbor classification remains the standard staging system for noncutaneous lymphomas and separates patients with lymphoma into four disease stages [21, 22].

During the management of Hodgkin's lymphoma, which shows a better survival rate, FDG-PET/CT plays significant role for assessment of disease burden at different stages of the disease. Hutchings et al. demonstrated that the increased sensitivity of FDG-PET/CT for spleen and bone marrow involvement was 83% and 70%, respectively, compared to 36.8% and 13.3% for CT, respectively. Splenic involvement on FDG-PET/CT may either be focal or diffuse. Focal involvement is seen as FDG avid areas in spleen. Splenic uptake 1.5 times higher than the liver uptake, in the absence of diffuse marrow uptake, is the criterion for detection of diffuse spleen involvement. FDG-PET/CT is a one step, noninvasive procedure for staging of nodal and extra-lymphatic disease with a strong prognosis value (Fig. 27.4). It lays aside the need for invasive procedure like bone marrow biopsy in the patients nowadays and it is no longer required for bone marrow staging in Hodgkin lymphoma [23]. ^{18}F-FDG-PET/CT allowed for great improvements in response evaluation when compared to conventional imaging. Particularly two response assessment timings are now documented in the literature, namely end-of-treatment PET/CT after the first line of chemotherapy (±radiotherapy) and also interim PET/CT performed during the first line of chemotherapy, after the first two cycles. The first consensus for interpretation criteria of ^{18}F-FDG-PET/CT was reported in 2007 by the Imaging subcommittee of the International Harmonization Project in which end of treatment PET/CT interpretation was based on the size of the residual mass and thereby compared to the mediastinum background or surrounding background [24]. With use of this criteria, many studies have demonstrated that a fraction of PET positive patients still presented with favorable outcomes. A more specific and better classification was proposed at the Deauville first international workshop, defining the visual Deauville

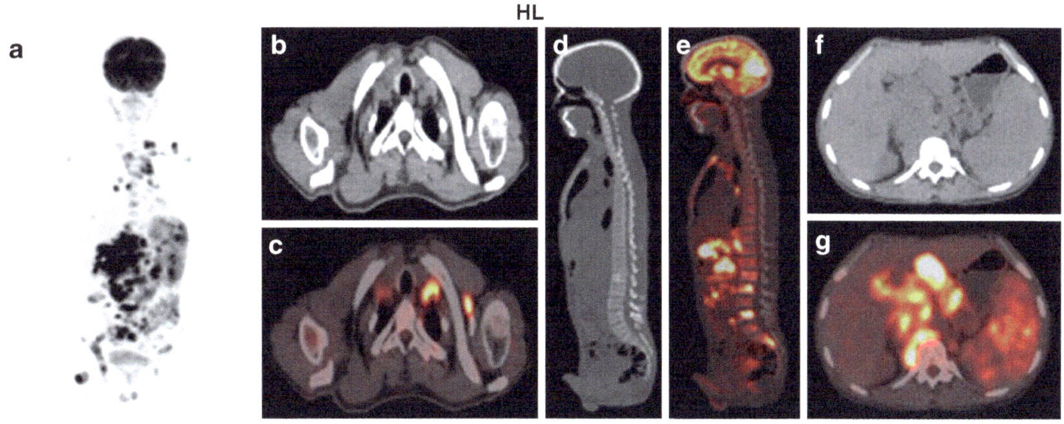

Fig. 27.4 An 18-year-old male patient with a known case of Hodgkin's lymphoma underwent chemotherapy with recent history of abdominal pain and weight loss for a period of two weeks. Maximum intensity projection (**a**) shows increased uptake in the cervical (**b**, **c**), mediastinal, splenic, abdomino-pelvic (**f**, **g**) and bone marrow (**d**, **e**) regions with transaxial-fused PET/CT and CT sections showing lymph nodal, splenic and bone marrow involvement in the above regions (Source: Department of Nuclear Medicine, AIIMS Delhi)

five-point scale for both end of treatment and interim ^{18}F-FDG-PET/CT treatment [25]. A recent consensus modified this Deauville five-point scale in a newer classification called "Lugano Criteria" which depend on the same visual definition of uptake significance [23]. Five-point scale scores 1–3 are defined as Complete Metabolic Response (CMR) while Deauville score of 4 and 5 are considered either no-metabolic-response (NMR), partial response (PR) or progressive disease (PD) based on the baseline PET/CT scan uptake (no change from baseline scan, reduced uptake compared to baseline scan and increased uptake or new lesion, respectively). The current Lugano Classification recommends routine use of FDG-PET/CT in all FDG-avid lymphomas (essentially all subtypes except some indolent non-Hodgkin lymphoma [NHL] subtypes such as small lymphocytic lymphoma and marginal zone lymphomas). It is nowadays used to standardize ^{18}F-FDG-PET/CT reports since it has shown good prognosis discrimination and good interobserver reproducibility (Table 27.1) [26].

Usage of ^{18}F-FDG-PET/CT for interim PET/CT have shown a high prognosis value and this early evaluation leads to an chemo sensitivity evaluation of the of the tumor and, thereby, identifies patients with different outcomes at the first cycle of the first line of treatment. With the use of immune check point inhibitors or the novel immunotherapy in relapsed or refractory Hodgkin lymphoma, PET/CT reporting was challenging. New criteria called LYRIC (lymphoma response to immunomodulatory therapy criteria) were defined. They introduced delayed [^{18}F]FDG-PET/CT evaluation to differentiate pseudo-progression from real disease progression. In pseudo-progression, the imaging findings suggest an increase in size and metabolic activity of the lesions, without any clinical deterioration of the patient. This is called as indeterminate response (IR) in LYRIC criteria and such a response mandates repeat imaging after 12 weeks to differentiate pseudo-progression from true progression of disease [27, 28].

Table 27.1 Deauville 5 point scale definition and Lugano criteria

Deauville S point scale	Definition	Lugano
Score 1	No uptake above the background	CMR
Score 2	Uptake < mediastinum	CMR
Score 3	Uptake > mediastinum but Sliver	CMR
Score 4	Uptake moderately increased compared to the liver at any site	PR/NMR/PD
Score 5	Uptake markedly increased compared to the liver at any site	PR/NMR/PD

CMR complete metabolic response, *PR* partial response, *PD* progressive disease, *NMR* no metabolicresponse

Recently, the total metabolic tumor volume (TMTV) has been introduced as a new functional parameter to predict patient prognosis at baseline. This parameter involves tumor burden quantification on [^{18}F]FDG-PET/CT baseline evaluation. Several publications have shown a significantly worse prognosis in patients who have a high TMTV at baseline in Hodgkin lymphoma [29, 30].

Even in non-Hodgkin's lymphoma, ^{18}F-FDG-PET/CT is now the standard imaging modality for staging and assessment of response in all FDG avid varieties. Non-Hodgkin's lymphomas that are FDG avid, are diffuse large B-cell lymphoma (DLBCL) (Fig. 27.5), Burkitt lymphoma, follicular and mantle cell lymphoma (MCL). However, some indolent NHLs like small lymphocytic lymphoma, chronic lymphocytic leukemia and marginal zone lymphoma show low or variable FDG avidity, and thus, FDG-PET/CT has limited utility [31, 32].

The Lugano classification still recommends a bone marrow biopsy in DLBCL as PET/CT can miss low-volume disease in 10–20% of patients [23]. However in advanced-stage NHL, if the PET/CT is positive for marrow involvement, a bone marrow biopsy may be avoided. If PET/CT is negative for marrow involvement, then a bone marrow biopsy is indicated in DLBCL. In all other NHLs, bone marrow biopsy is indicated [33].

In aggressive NHL, PET/CT is used to assess remission at the end of treatment and to identify patients who have failed the first line treatment

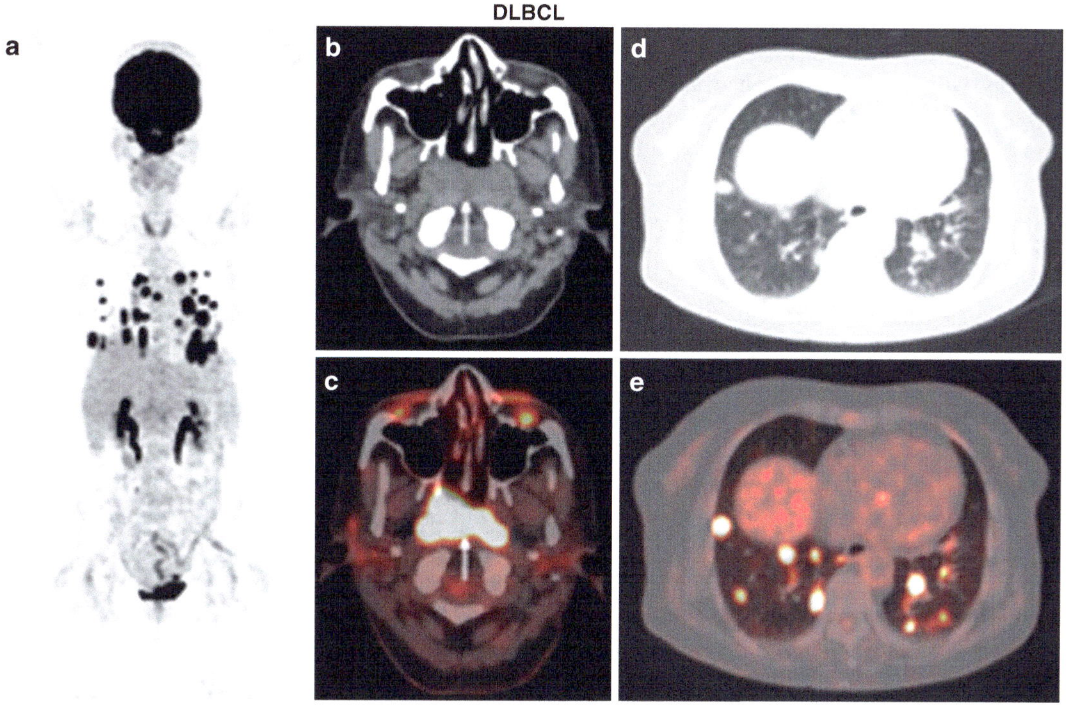

Fig. 27.5 A 63-year-old female with a known case of stage IV DLBCL with primary in the nasopharyngeal region has undergone six cycles of RCHOP chemotherapeutic regimen and underwent PET/CT as part of response evaluation. Maximum intensity projection (**a**), transaxial-fused PET/CT (**c**, **e**) and CT (**b**, **d**) sections show soft tissue thickening (max thickness ~2 cm) in the posterior wall of nasopharynx (white arrows) along with multiple random parenchymal lung nodules in the bilateral lungs (Source: Department of Nuclear Medicine, AIIMS Delhi)

by doing interim PET/CT during the treatment. Interim PET/CT is helpful in assessing the chemosensitivity in aggressive NHL but, unlike Hodgkin's lymphoma, cannot be routinely used for modification of the treatment regime as false positives are common [34]. Deauville criteria are used for response assessment for end of treatment and interim PET/CTs. Most patients with a score of 3 at interim evaluation have good prognosis. In most NHLs a score of 4 or 5 at the end of treatment represents residual disease, even if there is decrease in the SUV value from the baseline. On interim evaluation, if the uptake has reduced from baseline, it is considered as partial metabolic response and that the disease is sensitive to chemotherapy. However, a score of 4 or 5 with no decrease or increase in uptake from baseline and new lymphomatous uptake is considered as failure of treatment at both interim and end of therapy evaluation. In aggressive NHL, end-of-treatment PET/CT has a very high negative predictive value of 80–100% and positive PET/CT at the end of R-CHOP has poorer prognosis [34, 35].

Radiotherapy planning with PET/CT helps in precise localization of disease and also helps in modifying the target volume and gives the exact metabolic and anatomic information needed for it [36]. In relapsed aggressive NHL, second-line, high dose chemotherapy is given to induce remission and accordingly to facilitate autologous stem cell transplant (ASCT). ASCT is a curative method of treatment. Pretransplant PET/CT is an important prognosticator in patients undergoing ASCT, with a progression-free survival and overall survival of 87% and 94% in patients with negative PET/CT [37, 38]. There is no established role of surveillance scan in a treated case of NHL. Similar to Hodgkin's lymphoma, Lyric criteria is used in aggressive NHLs also for interpretation PET/CTs during immune based therapies [27].

Lately, several phase III randomized prospective trials validated the PET-driven therapeutic strategy, which has allowed a new area of personalized medicine and optimized disease control and toxicity for each patient [39].

27.4 Esophageal Cancer

Esophageal cancer, traditionally has had a poor prognosis with an estimated 5-year survival of only 17–34% mainly because of its invasive early spread through the esophageal wall and to adjacent lymph nodes [40]. There are basically two major histological types of esophageal cancer, namely, esophageal squamous cell carcinoma and esophageal adenocarcinoma. Esophageal squamous cell carcinoma can occur anywhere in the esophagus with a predilection to proximal to mid-esophageal, while esophageal adenocarcinoma predominantly occurs in the distal esophagus [41, 42].

Endoscopic ultrasound (EUS) is considered the most accurate procedure for preoperative local T staging due to its ability to determine the depth of penetration of the primary lesion. Endoscopic ultrasound may up-stage the tumor due to peri-tumoral fibrotic changes or down-stage lesions due to microscopic tumor invasion which cannot be overcome even by PET/CT due to its limited resolution. CT is highly specific in the detection of advanced esophageal cancer, especially when an obstructing esophageal mass is present, with invasion or encasement of adjacent structures/vessels with PET/CT providing similar coincidental findings. CT-based N staging based on pathologic size criteria of lymph nodes may result in false-negative downstaging of smaller nodes or upstaging enlarged hyperplastic nodes. In such a context EUS with fine needle aspiration of sonographically detected abnormal lymph nodes is a well-accepted technique. PET/CT is useful mainly in detecting distant metastatic adenopathy, especially in the mediastinum, supraclavicular and celiac regions, as locoregional nodes adjacent to hypermetabolic primary tumors may be obscured by the activity. PET/CT is not found to be useful in the initial evaluation of primary esophageal carcinoma and detection of locoregional disease in patients with superficial tumors (Tis and T1a). Therefore, PET/CT can be used as a complementary technique to EUS/CT [43, 44].

Earlier distant nodes were classified as M category, but recently, it has been included in the N category in American Joint Committee on Cancer (AJCC) classification 8th edition [45]. PET/CT images can be used to delineate the nodal locations and adjacent anatomical landmarks for planning neoadjuvant radiotherapy sessions. PET/CT has greater role for M staging (Fig. 27.6) in case of esophageal cancer particularly in case of metastatic lung involvement.

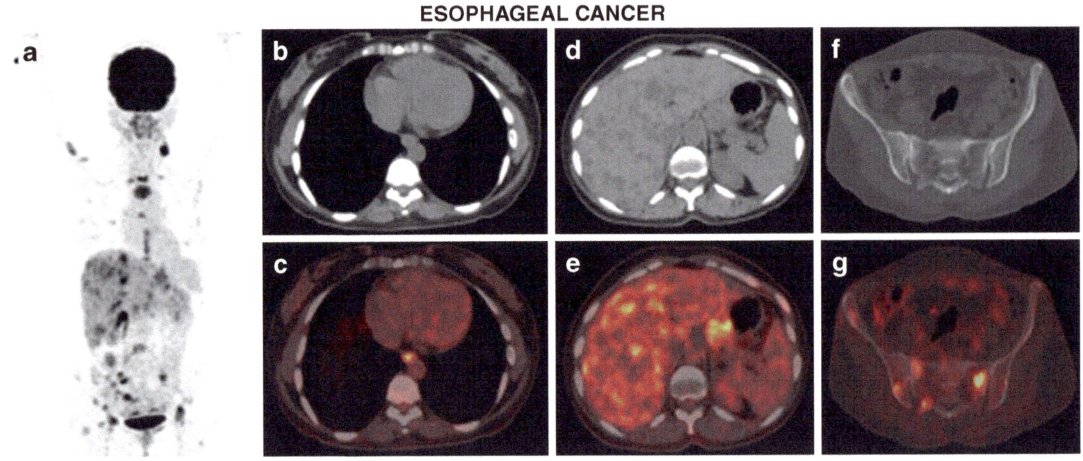

Fig. 27.6 A 68-year male patient underwent ^{18}F-FDG PET/CT as part of response assessment post-concurrent palliative chemoradiotherapy. FDG-avid lesion noted in lower esophagus along with involvement of GE junction with metastases of paraesophageal and left supraclavicular lymph nodes, liver and bone marrow in MIP (**a**), transaxial-fused PET/CT (**c**, **e**, **g**) and CT (**b**, **d**, **f**) sections (Source: Department of Nuclear Medicine, AIIMS Delhi)

PET/CT (usually performed 6 weeks after chemoradiotherapy completion) also helps in the identification of early responders to post- and preoperative neoadjuvant chemoradiotherapy and can prelude esophagectomy in patients with poorer prognosis and other comorbidities [46]. It can also be used in the follow-up of patients post surgery to detect tumor recurrence [47].

Multiple studies have tried to explore the prognostication value of various PET/CT parameters. Elimova et al. showed in a prospective study that PET/CT is a powerful prognostic tool in predicting longer median overall survival in esophageal cancer patients after chemoradiotherapy by using baseline TLG value [41, 48].

27.5 Gastric Cancer

Gastric cancer is the third most common cause of cancer-related death in the world with an estimated 5-year survival of 20–25% [49, 50]. Gastric adenocarcinoma is the most common gastric cancer, which accounts for 95% of all gastric cancers. Most common histologic types of primary gastric adenocarcinoma include intestinal adenocarcinoma (commonly related to chronic autoimmune gastritis or chronic *H. pylori* associated gastritis) and diffuse (mucinous or signet ring) adenocarcinoma. Other gastric malignancies include mucosa-associated lymphoid tumor (MALT) lymphoma, gastrointestinal stromal tumors (GIST), other more aggressive lymphomas and less frequently sarcomas and carcinoid tumors.

The only effective curative treatment of gastric adenocarcinoma is complete resection of the primary tumor and adjacent lymph nodes. According to AJCC 8th edition, as many nodes as possible should be removed at surgery to allow for accurate pathologic staging and prognosis, which is preferably equivalent to 30 nodes compared to the older standard of lesser number of nodes [45]. EUS and CECT remain the most definitive investigative modalities for preoperative T staging of gastric cancer [51]. PET/CT has a limited role in the T staging due to various factors such as low spatial resolution of PET/CT, variable physiologic FDG uptake in the gastric mucosa and gastric motility during scanning [52]. In a nondistended stomach, it is difficult to discern abnormal uptake; therefore, a negative oral contrast agent is often helpful for PET/CT as well as CECT interpretation.

While reporting N stage on PET/CT, one should be aware of the D classification (D1–4) for extent of nodal dissection. D1–2 lymphadenectomy involves dissection of perigastric nodes, celiac axis nodes and nodes along splenic artery, while D3–4 lymphadenectomy is more extensive and refers to dissection of nodes at the root of the mesentery, around the middle colic vessels and para-aortic nodes [53]. Even though PET/CT is not a good modality to assess perigastric lymph nodes (due to relatively low spatial resolution of PET/CT), it does not make a significant difference as D1 dissection is a usual procedure protocol followed during surgical gastric resection. Despite its low sensitivity, PET/CT has high specificity in the detection of N2 or N3 nodal disease, which is more impacting as it may help in decision-making regarding D2 to D4 lymphadenectomy or preventing needless surgery. PET/CT is also superior to CT in the detection of metastatic disease in smaller lymph nodes (Fig. 27.7), not classified as pathologic by CT size criteria, especially near the hepatoduodenal ligament, posterior peripancreatic region, mesenteric, middle colic and para-aortic lymph nodes. This knowledge is important during surgical procedures [54].

PET/CT has a major role in the detection of sites of distant metastases, like the liver, lung, adrenals, bone, peritoneum and ovaries (Krukenberg tumor). Even though usage of PET/CT to predict early response to neoadjuvant chemotherapy (NAC) is promising, it has not sufficiently proven conclusive to guide therapy. Following definitive gastrectomy and NAC, there are mixed results for the utility of PET/CT as a tool for detecting recurrence due to certain limitations like postsurgical inflammatory uptake at the surgical anastomosis, small size of peritoneal metastases, technical difficulty distinguishing vascular uptake from nodal uptake and physiological urinary tract FDG uptake [55, 56]. A meta-analysis that included 500 gastric cancer patients and evaluated the value of ^{18}F-FDG-PET/CT for the detection of recurrence after surgical resection showed a sensitivity, specificity,

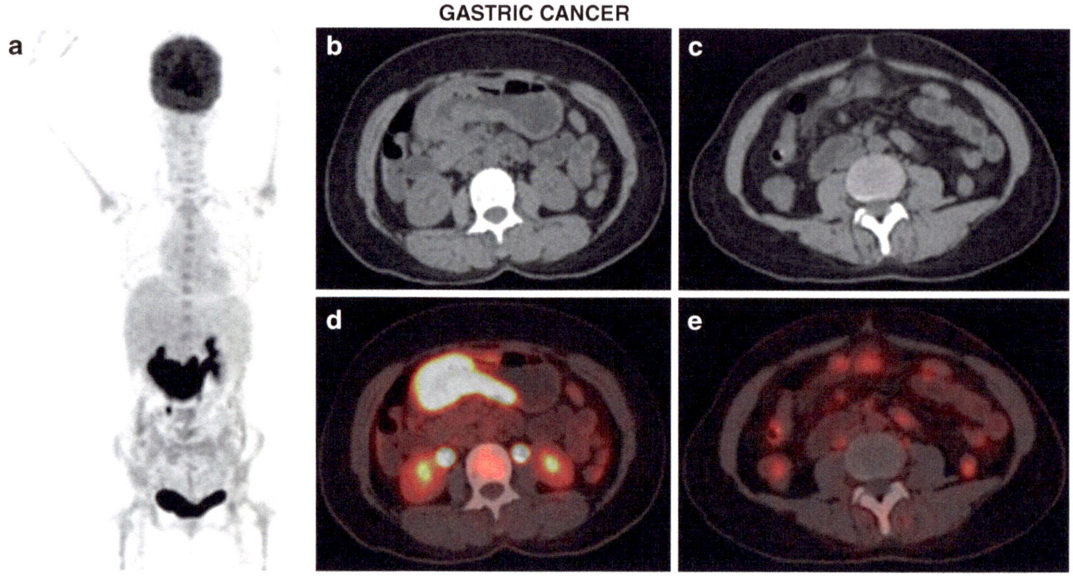

Fig. 27.7 A 65-year-old male patient underwent PET/CT as part of metastatic workup for carcinoma stomach. Maximum intensity projection (**a**), transaxial-fused PET/CT (**d**, **e**) and CT (**b**, **c**) sections show FDG-avid soft tissue circumferential wall thickening involving stomach pylorus along with metastases to peri-pyloric lymph nodes (Source: Department of Nuclear Medicine, AIIMS Delhi)

positive likelihood ratio and negative likelihood ratio of 86.0%, 88.0%, 17.0% and 0.16%, respectively [57]. NCCN guidelines recommend the use of ^{18}F-FDG-PET/CT in the posttreatment assessment and restaging of these patients [58].

PET/CT can provide prognostic information especially the primary tumor uptake. In a study of 321 gastric cancer patients, it was found that a primary tumor SUV_{max} higher than 5.74 was a poor prognostic indicator of progression-free survival (P = 5.034, hazard ratio [HR] 3.59) [59]. The uptake in metastatic lymph nodes can also provide prognostic information. Song and colleagues found that the nodal SUV max measured preoperatively was an independent risk factor for both recurrence-free survival and overall survival [60].

Some other histopathologies like mucinous, signet cell, poorly differentiated adenocarcinomas and gastric adenocarcinoma of intestinal type typically have low FDG uptake. And even diffuse type gastric tumors also have low uptake, mainly due to diffuse infiltration of adenocarcinoma cells resulting in relatively low level homogeneous uptake along with low expression of the glucose transporter 1 (GLUT-1) and an increase in inert mucous content [54, 56]. Even misregistration caused due to diaphragmatic movement can also affect primary gastric uptake.

27.6 Colorectal Cancer

Colorectal carcinoma is the third most common malignancy [40]. Primary tumor resection combined with adjuvant or neoadjuvant therapy in selected patients, is the only proven curative treatment. And recently, chemoablation and radioablation are added in patients with isolated liver or lung metastases for curative intent.

Initial diagnosis and staging of a new colon cancer is widely done with CECT of the chest, abdomen and pelvis. PET/CT has limited role in this setting. However, it is proven to be beneficial for surgical planning in those patients with equivocal CT findings for advanced M1 disease, multiple regional nodes on CT situated outside of the local lymph node drainage and/or significantly increased tumor markers (indicating more likelihood of distant/metastatic disease)

[61]. PET/CT has been more frequently done in rectal cancer compared to colon cancer, in the setting of initial staging and diagnosis, even when EUS and MRI are used for this purpose to assess the local invasion and the anatomic relationship of the tumor and local nodes to the mesorectal fascia. Generally, PET/CT has high specificity for the detection of metastatic lymph nodes and high sensitivity and specificity for liver metastases in patients with colorectal cancer [62, 63].

Recent studies including the study by Petersen et al. have reported that inclusion of PET/CT in the initial staging of colorectal cancer results in the change of treatment in 30% of the patients as compared to conventional CT imaging [64, 65]. Also as per multiple case series the change of treatment is basically due to evidence of liver metastases [66–69]. A meta-analysis by Niekel et al. concluded that, the mean sensitivities of CT, MR and PET/CT were 81.2%, 93.4% and 94.2%, respectively, on a per-patient basis, with the sensitivity of CT being significantly lower than PET/CT [68].

In colon cancer, neoadjuvant chemotherapy is mainly reserved for T4b disease (i.e., invasion of adjacent organs or structures). In this scenario, CECT usually will suggest this invasion, but PET/CT can be performed to have a better assessment of local and distant sites of metastases. The follow-up PET/CT is preferred to be performed 4 weeks after the neoadjuvant chemotherapy for colon carcinoma (or neoadjuvant CRT for rectal carcinoma).

During surgery, if T4 disease, regional or distant nodal involvement, or organ involvement is incidentally found, postoperative imaging including PET/CT is likely performed along with the intention to administer chemotherapy. After reassessment of disease status with PET/CT, if there is complete remission, subsequent follow-up will be mainly done clinically with serial CEA and CECT assessment [70]. In such a framework, serial screening PET/CT is not recommended but on the other hand if there is case of rising CEA, PET/CT should be considered (as it is more sensitive than conventional CT in identifying the site of recurrence) [71]. PET/CT has also proven to be accurate to distinguish between tumor from scar tissue or radiation necrosis, usually in the pelvis and presacral region, after CRT for rectal cancer [72].

Physiologic FDG uptake in colon is common, and due to mucosal, muscular, bacterial or lymphocytic uptake, often most pronounced in the cecum and terminal ileum. Usually, medications such as metformin often concentrate in the colonic mucosa and therefore alter the FDG uptake pattern in the colon [73]. PET/CT may be less sensitive in the detection of mucinous (58%) compared to nonmucinous (92%) colorectal carcinomas [74].

27.7 Pancreatic Cancer

Pancreatic cancer has poor prognosis, with a 5-year survival rate of only 9% and with >80% of patients suffering from unresectable disease together with distant metastases at the time of diagnosis [75]. Therefore, it is highly important for early diagnosis and appropriate treatment care for effective therapy in pancreatic cancer patients. Surgery is the only curative treatment and various imaging procedures like Endoscopic Ultrasounds (EUS), CT, MRI and PET/CT has helped in accurate assessment of the disease status. They have to provide important details on the characterization and diagnosis of the pancreatic mass and anatomical variants, vascular involvement, perineural and lymphatic invasion to the detection of metastases and also most importantly assessment of resectability [75, 76].

^{18}F-FDG-PET/CT has been shown to be superior for the detection of distant metastases by several studies [77–79], and also Heinrich et al. reported a change in the management due to PET/CT in 16% of the cases [80]. Even with this, international guidelines do not recommend the routine use of PET/CT. According to NCCN guidelines, role of PET/CT remains unclear in pancreatic cancer and is not a substitute for high-quality CT. However, PET/CT may be considered after formal pancreatic CT protocol in high-risk patients with borderline resectable disease and noticeably elevated CA 19.9, large primary tumor and large regional nodes to detect extra-pancreatic metastases (Fig. 27.8) [81].

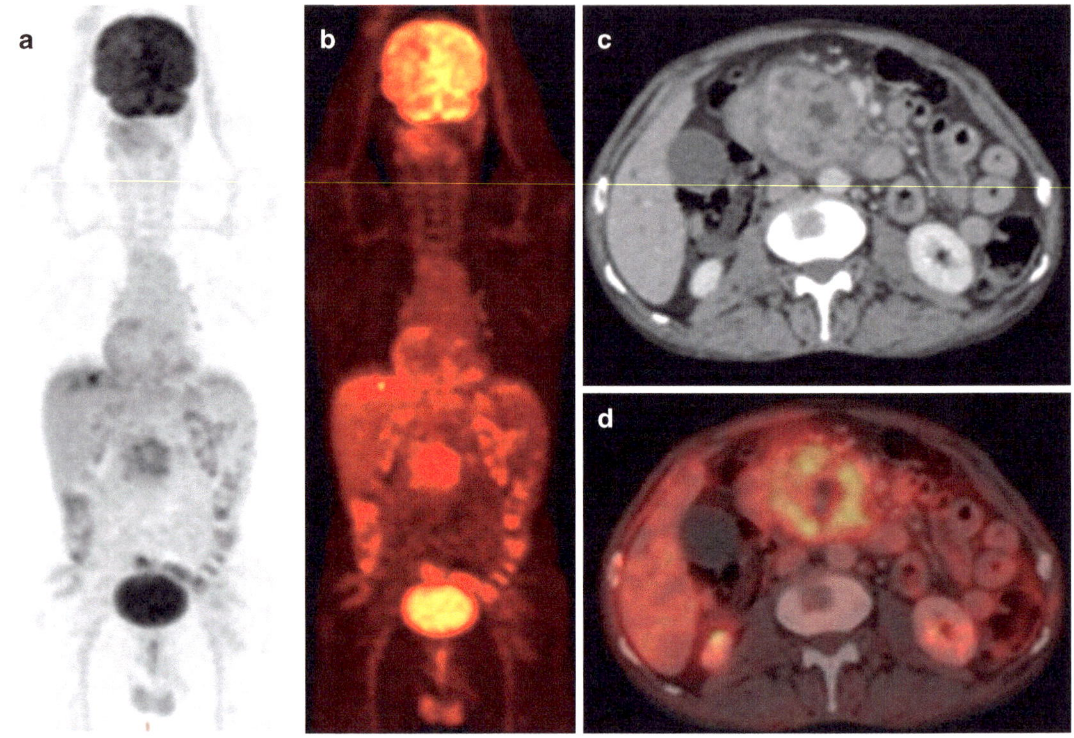

Fig. 27.8 FDG-PET/CT (**a**, **b**) shows metabolically active heterogeneously enhancing soft tissue density mass lesion with areas of necrosis and perilesional fat stranding involving the head and neck and uncinate process of the pancreas (**c**, **d**) with metastatic spread to liver and periportal, portocaval and peripancreatic lymph nodes in a 53-year-old male patient with carcinoma pancreas (Source: Department of Nuclear Medicine, AIIMS Delhi)

Neoadjuvant Chemotherapy (NAC) is used prior to curative resections for increasing the resectability status prior to surgery. PET/CT is widely used to assess local involvement with a sensitivity and specificity of 81–97% and 72–76%, respectively [82, 83] and therefore may have an impact on patient's management prior surgery and/or after NAT, by providing a more accurate preoperative staging and predicting the treatment response to NAT [84].

The capacity of CT, MRI and EUS to properly define tumor borders and local involvement was higher compared to PET/CT. However, some interesting perspectives on the role of ^{18}F-FDG-PET/CT were pointed out by Lai and colleagues. They concluded that PET/CT could likely provide adequate information on tumor size and volume prior to surgery [85, 86]. Also due to the high glucose uptake of pancreatic cells, it can detect occult metastases [87] and there is higher probability of localization of distant metastasis with a sensibility of approximately 90%, higher than CT and MRI [88]. A recent meta-analysis confirmed it, reporting a pooled sensitivity and specificity of 80% (95% CI: 67–89%) and 100% (95% CI: 89–100%), respectively [89].

Even though PET/CT shows higher performance in detecting distant metastasis, its role in properly evaluating nodal (N) stage is debated. Basically due to the high ^{18}F-FDG uptake of the primary tumors, identification of local lymph nodes can be difficult showing sensitivity of 55% (95% CI: 38–72%) and specificity of 94% (95% CI: 81–98%), respectively [89]. However, quantitative imaging PET/CT metrics can help in fulfilling this issue. Pergolini and colleagues identified a patient subgroup (with $SUV_{max} \geq 6.0$ and CA 19.9 $\geq$ 200 U/mL) had lower disease-free survival, after surgery and thereby benefitted from systemic treatment before surgery [90]. These findings were supported by Moon and colleagues who identified CA19.9 (>150 U/mL) along with SUV_{max} (>5.5) as strong predictors of overall survival [91]. In general as reported by a recent meta-analysis the ability of PET/CT to modify clinical management varies between 30% and 44% [89]. Nevertheless the heterogeneity of multiple studies and the different analyses methods (qualitative vs. quantitative) may deeply impact the reported results. Therefore, due to the above-mentioned limitations, Nguyen and colleagues recently reported that PET/CT could be used as initial diagnostic investigation modality to avoid futile examinations in patients with distant metastasis in almost 20% of early-stage cancer patients (resectable or borderline) and who therefore may be down-staged preoperatively [88].

The inability of CT to differentiate between tumor and fibrosis has given a reason to prefer PET/CT for assessment of resectability post neoadjuvant therapy. Yokose and colleagues therefore suggested to adopt PERCIST (Positron Emission Tomography Response Criteria in Solid tumors) criteria over RECIST (Response Evaluation Criteria in Solid tumors) criteria for CT to assess preoperative response to neoadjuvant therapy using PET/CT. They analyzed 22 patients with both CT and PET/CT before and after neoadjuvant chemotherapy and after evaluation with RECIST protocol, partial response and stable disease was defined 9.1% and 90.9% of the cases. However following PERCIST criteria, a higher concordance with pathological analyses was observed with 22.7% of patients showing complete metabolic response, 40.9% partial metabolic response and 36.4% stable metabolic disease. They also depicted that metabolic tumor value (MTV) was associated to prognosis and patients with more than 50% reduction in MTV showed improved 1- and 3-year overall survival (100.0% and 87.5% vs. 90% and 45%, respectively) compared to MTV less than 50% reduction [92]. Another interesting application of PET/CT suggested that metabolic uptake in a PET/CT could be used as a functional biomarker investigation to guide neoadjuvant therapy instead of CA 19.9. CA 19.9 has a limited utility mainly due to multiple confounding factors like jaundice, etc. Conversely, ^{18}F-FDG-PET/CT uptake was more definitive to depict the biology of the tumor and its prognosis [93].

Recently, there are multiple studies to explore the utility of radiomics in pancreatic cancer. Toyama and colleagues assessed the prognostic merit of pretherapy ^{18}F-FDG-PET/CT radiomics analysis in 161 pancreatic cancer patients. They reported 10 features as showing statistical significance for overall survival prediction; nevertheless, multivariate analysis showed that gray-level nonuniformity (GLNU) and gray-level zone length matrix (GLZLM) were the only PET/CT parameters that depict statistical significance. These parameters, reflect tumor heterogeneity by providing a representation of the heterogeneity among the pixels within a specific image. Incidentally, based on the prognostic risk score, the amalgamation of GLNU, GLZLM and total lesion glycolysis helps to differentiate patients into three groups [94]. Lim et al. evaluated 48 patients with pancreatic cancer patients undergoing pretherapy ^{18}F-FDG-PET to establish if gene mutations in SMAD4, TP53, KRAS and CDKN2 genes were correlated to imaging phenotype. KRAS and SMAD4 alterations were significantly associated with ^{18}F-FDG-PET/CT radiomics features. Also SMAD4 mutations depicted significant correlation with long-run emphasis, SUV skewness and high-intensity textural features [95].

A recent meta-analysis on all the available literature on radiomics of the pancreas concluded that radiomics is a promising quantitative imaging biomarker for both focal pancreatic lesions and diffuse changes, and that the utility of radiomics may vary depending on the purpose of their application. Standardization of images and image preprocessing is necessary before considering the use of radiomics in routine clinical practice [96].

Another recent investigative areas are based on the high expression of cancer-associated fibroblasts in pancreatic cancer patients and therefore assessing the utility of ^{68}Ga coupled with a fibroblast-activated protein inhibitor (FAPI). It has increasingly being evaluated for higher specificity with respect to traditional ^{18}F-FDG-PET/CT [97, 98].

27.8 Hepatic Tumors

Hepatic tumors can be classified to be benign and malignant tumors. Hepatocellular carcinoma (HCC) is the most common primary malignancy of the liver and there are many types of benign liver tumors that need to be differentiated from it, to avoid unnecessary interventions for benign lesions. There are also other hypervascular malignant tumors, such as cholangiocarcinoma, cholangiolocellular carcinoma, mixed type tumors and metastatic liver tumors, which also need to be segregated from HCC due to variation in management [99].

The role of ^{18}F-FDG-PET/CT for the liver lesions is limited as CT and MRI can provide an accurate diagnosis of most of the liver tumors. Most of primary and secondary malignant tumors show high FDG uptake but several tumors can show false-negative uptake and the degree of FDG uptake can differ according to tumor differentiation or proliferation rate [100–102]. However, in certain hepatic tumors, the degree of FDG uptake can be used to predict prognosis or therapeutic response [101, 103, 104].

^{18}F-FDG-PET/CT FDG accumulation is mostly high in poorly differentiated HCCs than in well- and moderately differentiated HCCs (Fig. 27.9) [100], and this is well related to the expression of GLUT (glucose transporter) ½ and HK (hexokinase)-II, and G6Pasc (Glucose-6-Phosphatase) activity. Well- to moderately differentiated HCCs has low expression of GLUT1/-2 and high G6Pase activity, while poor- and undifferentiated HCCs show high expression of GLUT-1/-2 and low G6Pase activity [105]. Because of poor detection of well- and moderately differentiated HCCs, there is a limited overall sensitivity for FDG-PET/CT in detecting HCC, with a reported range of 50–65% [100, 101]. Therefore, FDG uptake can serve as a predictor of HCC differentiation grade. After liver transplantation, high FDG uptake can act as an independent predictor of recurrence and poor survival [102] and of poor response to radiation [103] and transarterial chemoembolization [104]. Even though extrahepatic disease is rare in HCC, mainly in well differentiated tumors, FDG-PET/CT has shown to have a high sensitivity for the

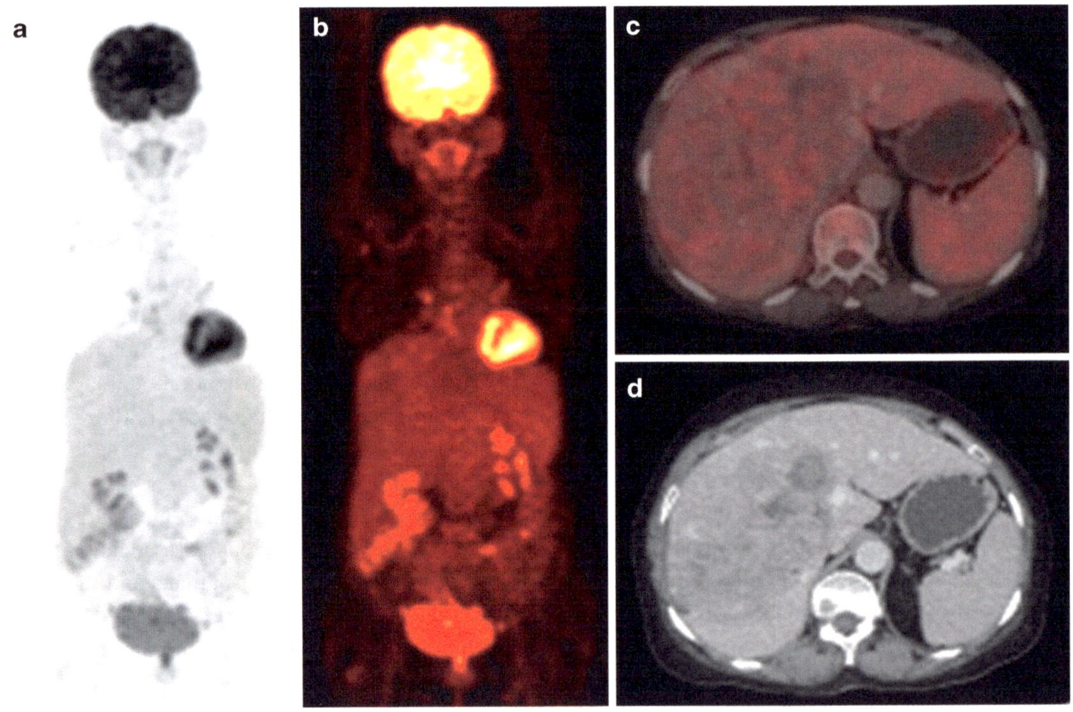

Fig. 27.9 FDG-PET/CT (**a**, **b**) of well-differentiated hepatocellular carcinoma of a 66-year-old female patient demonstrates enhancing mass involving segments I, IVA, VIII and VII of the liver (**c**, **d**) (Source: Department of Nuclear Medicine, AIIMS Delhi)

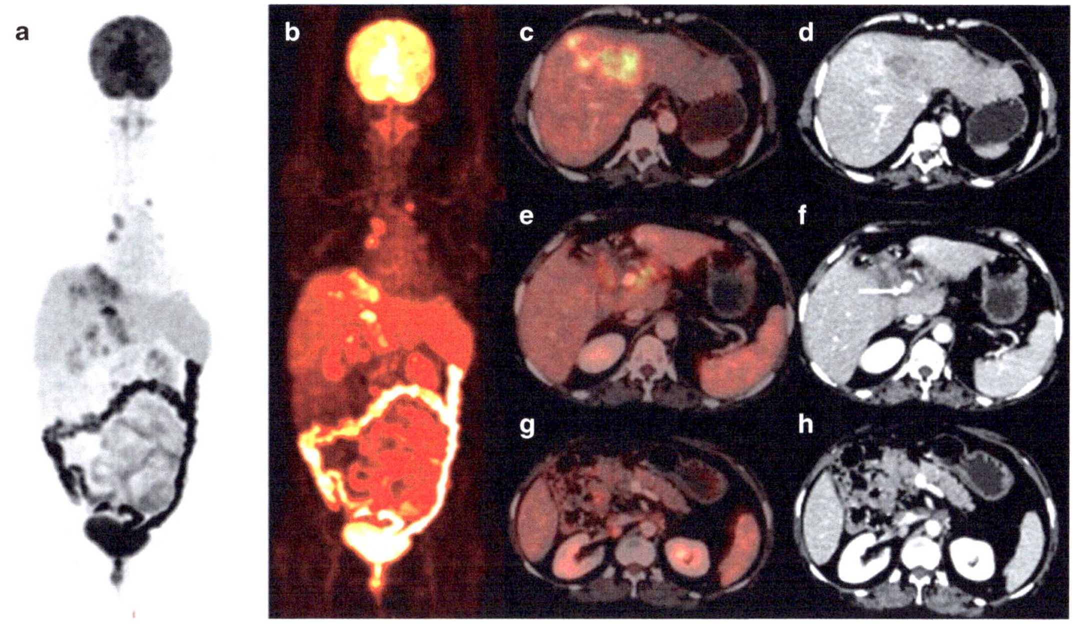

Fig. 27.10 A 53-year-old male with metastatic cholangiocarcinoma presented for posttreatment PET/CT after 5 cycles of chemotherapy cycles. Maximum intensity projection (MIP) and transaxial-fused PET/CT and CT sections (**a**–**d**) show poorly enhancing hypodense lesion in segment IVA/VIII of liver with metastases to abdominal, retroperitoneal (**g**, **h**) and mediastinal lymph nodes and tumor thrombus (**e**, **f**—white arrow) (Source: Department of Nuclear Medicine, AIIMS Delhi)

detection of lymph node and bone. According to NCCN guidelines, PET/CT may be considered in case of equivocal findings.

27.9 Cholangiocarcinoma

Cholangiocarcinoma (CC), arising from the bile duct epithelium, is the second most common primary neoplasm of the liver. CC is further subclassified into mass-forming and infiltrating types, based on the morphology; and into peripheral type (arising from small bile ducts) and perihilar type (arising from large bile ducts) based on the location of the lesion [106]. Lesion location and morphology play a significant role in the sensitivity of FDG-PET/CT for lesion detection with more than 90% of CCs showing high FDG uptake noted predominantly in nodular or mass-forming cholangiocarcinoma located in the periphery showing ring like tracer based on the morphology. While the remaining 10% accounts perihilar and infiltrating type, shows less uptake of FDG in a focal nodular or linear branching fashion. Therefore, FDG-PET/CT is more useful mainly in mass-forming cholangiocarcinoma [107, 108]. In this scenario, FDG-PET/CT imaging has proven beneficial in detecting unsuspected or unknown metastases (Fig. 27.10) than detecting the primary lesion and has shown an impact of 17–30% on the selection of therapy [109, 110].

27.10 Gall Bladder Cancer

Gallbladder carcinoma (GBC) is the most common and aggressive carcinoma derived from biliary cells [111]. They are rapidly progressive causing direct and local hepatic metastases mainly due to absence of a serosal layer separating the gallbladder from the liver, and thereafter followed by lymphatic dissemination and hematogenous spread. Adenocarcinoma is the most common histopathologic subtype of gallbladder tumors [112].

Gallbladder carcinoma result in high mortality due to risk of tumoral spread induced by the manipulation of unknown GBCa, the absence of effective chemotherapy the complexity of the porto-hepatic system, as well as the morbidity and mortality of liver resection [113]. Surgery is the only curative treatment; therefore, accurate evaluation and staging is significant to provide an indication for surgery or to avoid unnecessary interventions [2]. Several investigational modalities have been helpful for this purpose namely ultrasonography (US), computed tomography (CT), magnetic resonance (MR), percutaneous transhepatic cholangiography (PTC) and endoscopic retrograde cholangiopancreatography (ERCP).

There is less literature available highlighting the role and utility of FDG-PET in gallbladder cancer. In 2015, Salvatore et al. conducted a meta-analysis including 13 studies to assess the role of FDG-PET or PET/CT in the evaluation of primary tumor in patients with gallbladder cancer. It revealed a pooled sensitivity of 87% (95% CI: 82–92%) and specificity 78% (95% CI: 68–86%) with PET/CT showing better diagnostic accuracy than PET alone [114]. Another recent meta-analysis by Parida et al. demonstrated a pooled sensitivity and specificity of ^{18}F-FDG-PET/CT for detection of local disease as 96% (95% CI, 90–99%) and 91% (95% CI, 77–98%), respectively, while that for detection of metastatic disease resulted as 95% (95% CI, 88–98%) and 97% (95% CI, 90–100%), respectively. Slightly lower values were observed for nodal disease by FDG-PET/CT, i.e., 75% (95% CI, 53–90%) and 91% (95% CI, 77–98%), respectively [115].

Recently, Jin et al. published a study that investigated the prognostic values of volume-based metabolic parameters (MTV_{total}, TLG) using ^{18}F-FDG-PET/CT in locally advanced and metastatic GBC. They found out that these were useful for the identification of patients with poor prognosis [116]. Shifali Goel et al. explored the role of FDG-PET/CT in preoperative staging of gall bladder cancer patients, and demonstrated that usage of PET/CT preoperatively changed the management plan in approximately 25% of all resectable gall bladder cancer patients and in 33% of locally advanced cases [117].

As per international oncological guidelines like European Society For Medical Oncology (ESMO) and NCCN [118, 119], usage of PET/CT is usually not recommended in biliary tract cancers like gall bladder cancer and cholangiocarcinoma; however, increasing source of evidence suggests that it may be useful for detection of regional nodal metastases and distant metastases (Fig. 27.11) in patients classified as potentially resectable disease, thus changing the management in these cancers [119].

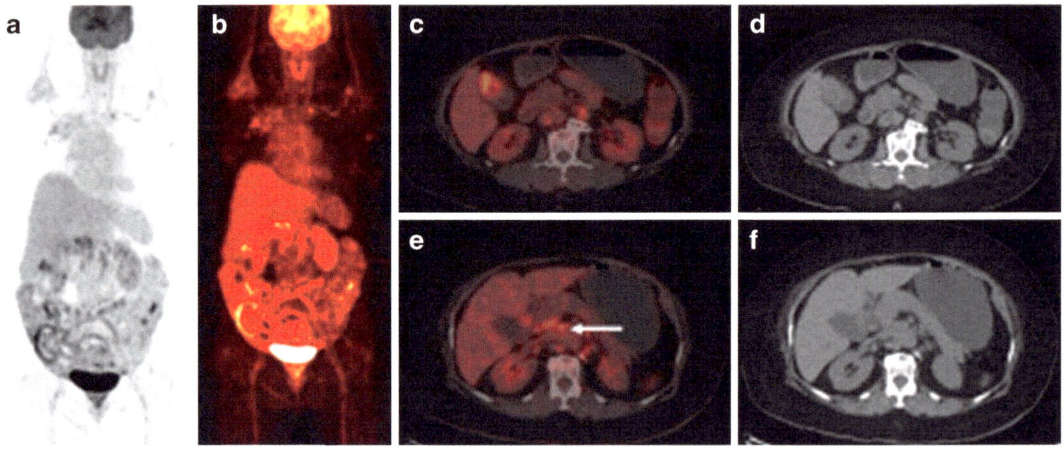

Fig. 27.11 A 53-year-old male with metastatic cholangiocarcinoma presented for posttreatment PET/CT after 5 cycles of chemotherapy cycles. Maximum intensity projection (MIP) and transaxial-fused PET/CT and CT sections (**a**–**d**) show poorly enhancing hypodense lesion in segment IVA/VIII of liver with metastases to abdominal, retroperitoneal and mediastinal lymph nodes and tumor thrombus (**e**, **f**—white arrow) (Source: Department of Nuclear Medicine, AIIMS Delhi)

27.11 Renal Cancer

Renal cancer accounts for 3–4% of all malignancies with an estimated 5-year relative survival rate close to 75% [120, 121]. There are mainly three histological subtypes of renal cell carcinoma (RCC) namely, clear cell (75%), papillary (15–20%) and chromophobe (5%).

Imaging plays a significant role for effective management. Contrast-enhanced CT and MRI have proven to be useful for assessing the primary tumor at initial diagnosis, with detection of locoregional nodal or abdominal visceral metastases. The recommended CT protocol for the initial staging includes both arterial (15–30 s) and nephrographic/portal venous phases (80–90 s) to identify hypervascular tumors and also delineate arterial and venous structures. MRI is generally used when iodinated contrast is contraindicated or when further characterization of soft tissue is needed to determine the disease extent [122]. But these modalities are nonspecific and distinction between benign and malignant lesions is also difficult.

FDG-PET/CT's role in staging and metastatic disease RCC is evolving. Many of the earlier studies have reported less sensitivity and specificity in the detection of RCC by FDG-PET compared to FDG-PET/CT and since the former technology is limited nowadays, such studies won't be discussed in this section on RCC [123, 124]. However, the utility of FDG-PET/CT for detecting primary RCC is limited mainly due to renal excretion of the radiotracer which makes it difficult to differentiate RCC from the normal renal tissue [125]. ^{18}F-FDG-PET/CT is a sensitive modality for detecting distant metastasis in high-risk RCC patients (Fig. 27.12) [126]. A recent meta-analysis revealed a sensitivity of 86% (95% CI 88–93%) and specificity of 88% (95% CI, 84–91%) for the detection of recurrent or metastatic sites of disease [127]. According to national and international guidelines, FDG-PET/CT is not currently recommended for the diagnosis and staging of RCC [128].

Recently, there is growing interest in non-FDG molecular PET imaging agents in RCC. Tumor targets that have been evaluated, includes malignant cell proliferation (^{18}F-FLT), hypoxia (^{18}F-FMISO), aerobic metabolism (^{11}C-acetate), or amino acid transport (^{11}C-methionine). In 2016 another FDA approved PET agent ^{18}F-NaF was introduced which showed more sensitivity in detecting bone metastasis and thereby helped greatly in initial staging and monitoring of bone lesions. Contradictory results

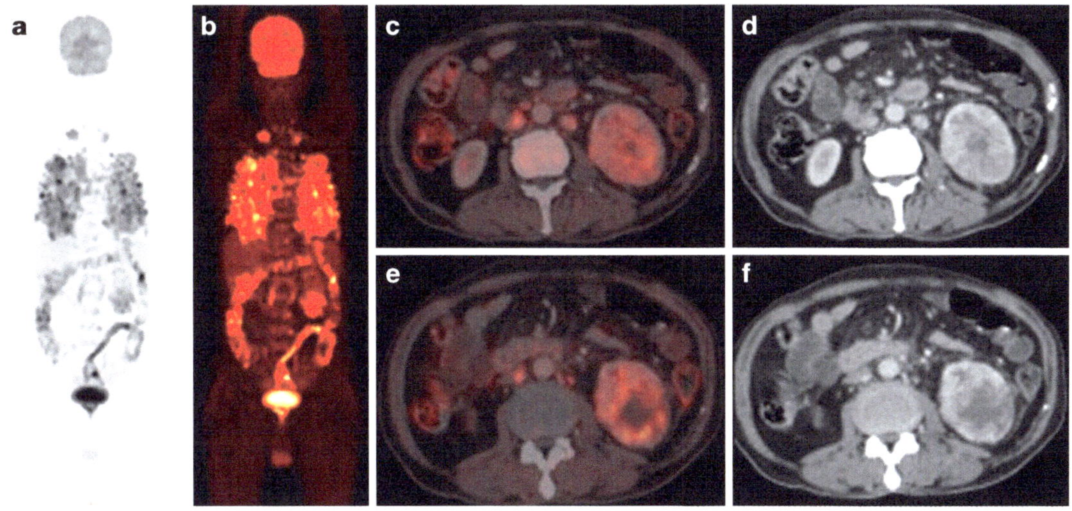

Fig. 27.12 Baseline staging PET/CT (**a**, **b**) of renal carcinoma of a 66-year-old male patient demonstrating a metabolically active soft tissue mass in lower pole of left kidney (**c**, **d**) with metastases to retroperitoneal (**e**, **f**), bilateral pelvic, mediastinal and bilateral cervical lymph nodes and to bilateral lungs (Source: Department of Nuclear Medicine, AIIMS Delhi)

were observed by Shreve et al. and Kotzerke et al. in their studies to assess the role of ^{11}C-acetate in RCC [129, 130]. Another study explored the role of antibody, cG250 (girentuximab), that binds to carbonic anhydrase 9, which is overexpressed in cell membranes of clear cell RCC. In a phase III study, ^{124}I-cG250 PET/CT showed a sensitivity and specificity of 86.2% and 85.9% for the diagnosis of clear cell RCC [131]. Altogether, so far none of these radiotracers could be established in the clinical routine.

Recently, prostate-specific membrane antigen (PSMA) targeting is being researched for the detection of metastatic RCC. The rationale behind this usage deals with the proangiogenic nature of RCC, which is a consequence of overexpression of hypoxia-inducible factor (HIF), due to genetic mutation in VHL gene, commonly clear cell RCC [132]. Rhee et al. investigated ten patients (eight with proven ccRCC) and found that ^{68}Ga-PSMA-HBED-CC PET/CT had greater sensitivity (92% versus 69%) and higher PPV (97% versus 80%) compared to conventional CT in the scenario of detection of metastatic disease [133]. Currently clinical trials are underway to investigate utility of PSMA PET/CT especially in the framework of clear cell RCC [134, 135].

27.12 Urinary Bladder Cancer

Bladder cancer (BC) is the most common urological cancer and the ninth most common cancer worldwide with more than 90% of BCs being urothelial (transitional cell) carcinomas. Approximately 70% of the patients have superficial tumors that tend to recur, whereas the remaining 30% of patients present with a muscle-invasive BC and a high chance of early systemic dissemination [136].

The standard method for evaluation of primary tumor in the urinary bladder (T-staging) is mainly performed with transurethral resection of the bladder (TURB). However, it is an invasive modality. Less invasive modality like CT and MR cannot accurately discriminate between the different layers in the bladder wall, with diagnostic accuracy of CT and MR (for T-staging) being 35–55% and 62–85%, respectively (with MR superior to CT). Nevertheless, these modalities have better performance for detecting extravesical invasion of the tumors [137, 138]. Routinely PET/CT is not used for evaluation of primary tumor in the bladder because of urinary excretion of ^{18}F-FDG. A meta-analysis assessing the diagnostic accuracy of ^{18}F-FDG-PET/CT for detecting bladder lesions reported a pooled sensitivity of 80.0% and pooled specificity of 84.0% [139]. Various methods have been suggested to overcome urinary radioactivity interference, including oral hydration and voiding, early imaging, dual phase imaging, post voiding imaging, bladder catheterization or irrigation and forced diuresis with reported sensitivities ranging from 85% to 96% [140, 141]. Nonetheless, oral hydration followed by voiding has been demonstrated to be superior to standard FDG-PET/CT protocol, and being more patient friendly and less invasive. Lately Eulitt et al. evaluated the role of preoperative FDG-PET/MR in 21 BC patients who were scheduled for cystectomy. FDG-PET/MRI was compared with cystectomy pathology and CT. The results demonstrated that FDG-PET/MR performed like CT for detection of primary tumor in the bladder; however, the determination of lymph node metastases was limited for few patients with true pathologic lymph node involvement [142]. Further larger studies are warranted to evaluate the role for FDG-PET/MRI in the staging of muscle invasive bladder cancer (MIBC).

Nearly 25% of the BC patients (high risk) have lymph node metastases if the primary tumor is classified as T2, and the risk is increased up to 50% in stage T3 [143, 144]. For detection of lymph node metastases, CT and MR are not very effective. They use size criteria for evaluation of lymph node metastases with the present recommendation suggest a cut-off value for suspicious lymph node to be ≥8 mm for pelvic lymph node and ≥1 cm for retroperitoneal lymph node [145]. However in BC patients, lymph nodal metastases often cause little enlargement of the lymph node, resulting in high false-negative rates reported for both CT and MR, with accuracy of CT ranging from 73% to 92%, and for MR from 73% to 90% [146]. In this framework, a modality based on metabolic uptake can play a significant role like FDG-PET/CT, and it has been investigated multiple times. Recently, a large retrospective study in 300 BC patients by Moussa et al. reported FDG-PET/CT as superior compared to CT planned for cystectomy [147]. Various other studies also concluded that FDG-PET/CT is superior to CT for detection of lymph nodal metastases in BC patients [148–151]. According to the literature, FDG/PET/CT detects more malignant lesions than conventional CT/MR in 20–40% of patients [152, 153] and may change the clinical management in up to 68% of the patients. There was only one study which compared FDG-PET/CT and MRI for detection of LNM in MIBC FDG-PET/CT and MR reported sensitivities of 33% and 0%, and specificities of 93% and 80%, respectively [154].

The most common sites of distant metastases in BC patients are the bones, lungs, liver and peritoneum [155]. Multiple studies have evaluated the role of FDG-PET/CT for detection of distant metastases (including LNM) and the reported sensitivities and specificities for detection of distant metastases in the studies are 50–100%, and 86–100%, respectively [156, 157]. Kim et al. depicted a pooled sensitivity of FDG-PET/CT reportedly superior compared to CT (74.9% vs. 43.7%), with pooled specificity of FDG-PET/CT 94.9% vs. 96.6% for CT [158]. Also another study by Bouchelouche et al. compared FDG-PET/CT and MRI in BC patients proving its superiority of former over the latter [137].

FDG-PET/CT can also prognosticate bladder cancer patients based on presence of extravesicular FDG avid lesions and has shown to be an independent variable for reduced survival rate [159]. Currently, there is an ongoing phase II prospective trial (EFFORT-MIBC) looking forward to evaluate the impact of FDG-PET/CT on risk stratification and treatment adaption for BC patients [160].

Few studies have explored the role of FDG-PET/CT in response assessment of neoadjuvant chemotherapy in BC [161, 162]. Van de Putte et al. studied the accuracy of FDG-PET/CT for detection of response following neoadjuvant or induction chemotherapy (before and after FDG-PET/CT) in 37 BC patients. Downstaging of the tumor (partial or complete response) was identified with 83% sensitivity and 80% specificity [162]. Post radical cystectomy, detection of disease recurrence by FDG-PET/CT is rather high, with an overall reported sensitivities and specificities ranging from 87% to 100% and 76% to 100%, respectively [138, 163–165]. It seems to be superior as compared to CT and MR. Zattoni et al. investigated in a multicenter retrospective study (n = 287) the accuracy of FDG-PET/CT in recurrent BC with sensitivities, specificities and accuracies of PET/CT vs. conventional imaging (CT/MR) were 94% vs. 86%, 79% vs. 59% and 91% vs. 81%, respectively [165].

Since urinary excretion of FDG may conceal tumors in the urinary tract and in the bladder, multiple non-FDG radiotracers like ^{11}C-choline, ^{11}C-acetate and ^{11}C-methionine that are not excreted in urine have been studied in small group of patients [166–168].

In conclusion, the most widely accepted indication for PET/CT in BC is particularly in case of metastatic lesions (NM staging) and disease recurrence (Fig. 27.13).

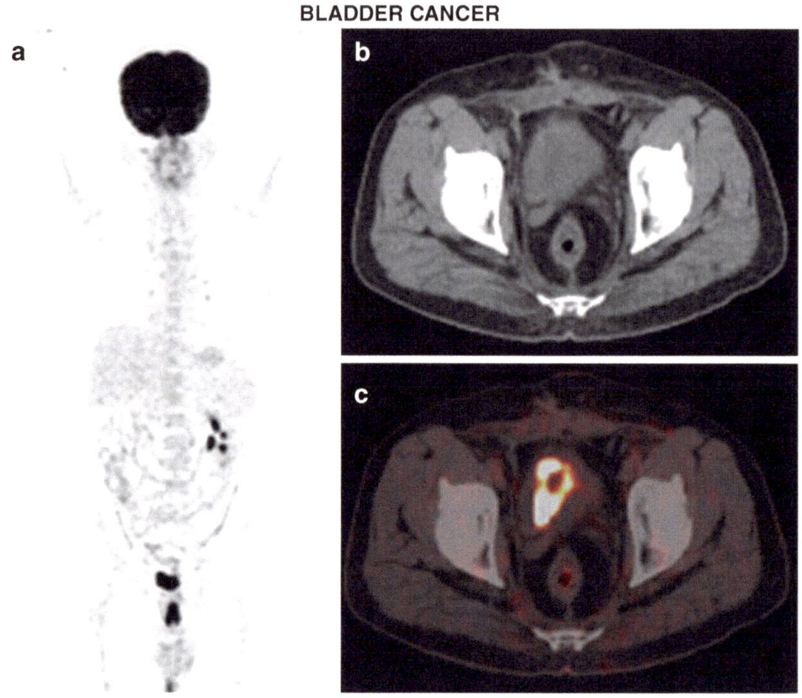

Fig. 27.13 A 74-year-old male patient with a known case of bladder cancer underwent PET/CT to assess the disease status posttransurethral resection of bladder cancer. (a-c) FDG-avid soft tissue lesion is noted in right anterior and anteromedial wall of the bladder visualized in pelvic CT and transaxial-fused PET/CT sections (Source: Department of Nuclear Medicine, AIIMS Delhi)

27.13 Prostate Cancer

Prostate cancer (PCa) is the second most commonly diagnosed cancer [169] with an increasing incidence worldwide [170]. Increased risk of PCa with higher grade and aggressiveness has been associated with germline mutations in BRCA1/2 [171] with most common histological type being acinar adenocarcinoma [172]. Generally PCa is diagnosed using digital rectal examination, serum PSA level assessment and transrectal ultrasonography (TRUS)-guided biopsy. PCa is classified into low-, intermediate- and high-risk, based on the sum of Gleason score, prostate-specific antigen (PSA) level and clinical stage [173]. The five-year survival rate being around 100% for localized PCa, while it drastically drops to 30% in those with distant metastasis [174]. Through the lymphatic system metastasis usually spreads to the pelvic and para-aortic lymph nodes while takes a hematogenous route to the bones. Metastases to other organs (like lungs and liver) are uncommon [175].

In the last decades management of prostate cancer has dramatically changed, grossly due to advancements in diagnostic modalities and also immense development of new therapeutic strategies. Imaging plays a significant role in all the steps of prostate cancer management with multiparametric MRI and molecular imaging being recently most valuable among the lot and found its way to the international guidelines [176]. Multiparametric MRI of the prostate integrates the anatomic information from T1- and T2-weighted sequences (i.e., soft tissue delineation) [177] with functional details from diffusion-weighted imaging (DWI) and dynamic contrast enhancement (DCE) for overall detection and characterization of PCa [178]. It plays a significant role in initial diagnosis of the PCa specially by enabling targeted biopsy in small lesions or patients with negative result on TRUS guided biopsy [179, 180]. However, its sensitivity is suboptimal for LN staging. And also due to limited field of view of MRI in the standard procedure it overlooks metastasis outside the acquisition area [181].

Molecular imaging agents that were explored in PCa included initial monoclonal antibody targets like [^{111}In]In-capromab pendetide (employed by Wynant in 1991) to subsequent agents based on cell proliferation targeting like $^{11}C/^{18}F$-choline and ^{11}C acetate. All of these were FDA approved and has their own advantages and disadvantages. $^{11}C/^{18}F$-choline PET/CT is mainly used for restaging PCa patients in the setting of biochemical recurrence, since it has shown better accuracy than conventional imaging [182, 183]. In a study conducted among 2686 patients, ^{11}C-choline PET/CT had an overall detection rate for recurrent disease of 62% (95% CI, 53–71%) [184]. Even though such overall number could be misleading in the lot, stratifying patients on the basis of PSA levels helps in easy detection with imaging modalities. In the setting of recurrent disease, $^{11}C/^{18}F$-choline PET/CT findings can change clinical decision-making in approximately 50% of cases in which the agent is used [185–187] and it has been incorporated into several international guidelines [188]. Choline PET/CT imaging was found to exhibit excellent diagnostic performance for the detection of bone lesions; as presented by a meta-analysis (compiling total 655), with sensitivity and specificity ranging from 50% to 100% and from 89% to 100%, respectively [189]. Nevertheless, a negative choline PET did not guarantee that bone metastases were not present. One of the main limitations of choline PET/CT is that it cannot distinguish benign prostatic hyperplasia from cancer since uptakes are similar in both cases.

Contrary to the high urinary excretion of radiotracers witnessed in choline PET/CT, another FDA-approved synthetic radiotracer anti--1-amino-3-^{18}F-fluorocyclobutane-1-carboxylic acid (^{18}F-FACBC), also known as ^{18}F-fluciclovine, or Axumin (brand name) was introduced. It is a synthetic amino acid L-leucine analog, that targets amino acid transporters (LAT and ASCT2) on PCa cancer cells, and also manifests negligible activity in the urinary tract within one hour of injection which reduces the possibility of missing small sites of disease relapse within the lower pelvis. It has a favorable dosimetry and biodistribution, that differs from choline PET/CT. Tumor-to-background ratio is highest between 4 and 10 min after injection. Normal biodistribution of fluciclovine activity includes pancreas, marrow, salivary glands, lymphoid tissues and pituitary [190–192].

Multiple trials like LOCATE, EMPIRE-1, etc. have explored the role of ^{18}F-fluciclovine PET/CT in the management of prostate cancer. A meta-analysis with six studies in 251 patients with biochemical recurrence (BCR) reported a pooled sensitivity and specificity for ^{18}F-fluciclovine of 87% (95% CI, 80–92%) and 66% (95% CI, 56–75%), respectively, for biochemical recurrent disease [193]. Likewise, Schuster et al. described a high specificity of 96.7% and positive predictive value (PPV) of 95.7% for extraprostatic detection of disease (e.g., lymph nodes and bones) in recurrent PCa [194]. Generally, ^{18}F-fluciclovine outperforms choline, in terms of biodistribution, image quality and acquisition protocol, with lower physiologic background that improves lesion conspicuity and also detecting a higher number of true positive and true negative lesions in the prostate bed, lymph nodes and bones [195, 196]. Overall, ^{18}F-fluciclovine appears to be a highly sensitive technique for detecting recurrent disease but has low-to-moderate specificity with a relatively high false positive rate that could cause false upstaging of disease. Despite its relatively good performance, the utility of this agent, similar to choline compounds, may need to be considered in the setting of the newer promising PSMA-based PET/CT agents.

To overcome their limitations, multiple studies were conducted. Recently, more focus is provided on the PSMA-based PET/CT agents. PSMA (Glutamate carboxypeptidase II) is a membrane bound glycoprotein, coded by gene FOLH1, catalyzes the hydrolysis of N-acetylaspartylglutamate (NAAG) to glutamate and N-acetylaspartate (NAA) [197–199]. It is an excellent target for diagnostic imaging and therapy of prostate cancer, due to its high expression these cancers, acceptable binding affinity and internalization of PSMA ligands. It is overexpressed in intermediate high-grade prostate tumors and increases in expression with increasing aggressiveness of the disease but shows reduced expression in tumors with neuroendocrine features [200–204]. The first PSMA imaging agent was a radiolabeled monoclonal antibody (mAb), ^{111}Indium-7E11 (^{111}In-capromab;

ProstaScint) that was investigated clinically as a single photon emission computed tomography (SPECT) imaging agent, targeting the intracellular domain of PSMA and, therefore, lacked sensitivity and specificity [205]. The humanized mAb hJ591 that binds to an extracellular domain of PSMA has also been investigated for PET diagnostic imaging in the form of ^{89}Zr-hJ591, and demonstrated higher affinity, and more efficient targeting [206].

The recent development of PSMA radiopharmaceuticals has given attention on small-molecule inhibitors (known as ligands) that target the active substrate recognition site of the PSMA folate hydrolase enzyme. Those ligands, having a urea binding motif, binds with high affinity to the PSMA receptor and also has high tumor-to-background ratios and exhibit rapid plasma clearance. The most commonly used PSMA ligand in clinical studies (PSMA-HBED-CC) is labeled with ^{68}Ga and is known as ^{68}Ga-PSMA-11 (Fig. 27.14) [207]. PSMA has also been labeled with other tracers like ^{68}Ga, ^{18}F, ^{111}In and ^{99}mTc, such as ^{18}F-DCFBC/^{18}F-DCFPyL, ^{18}F-PSMA-1007, ^{68}Ga/^{111}In-PSMA-617 or ^{123}I-MIP-1095 among others.

PSMA PET/CT uptake, measured by maximum standard uptake value, increases with increasing tumor grade demonstrating positive correlation with Gleason score, although considerable overlap exists for any Gleason category [208]. Moreover, PSMA PET/CT can discriminate clinically significant high-grade PCa from benign prostatic hyperplasia (BPH) nodules which usually fail to take up the agent [209, 210]. Eiber et al. assessed single modality and combined ^{68}Ga-PSMA-11 PET/MRI in 53 intermediate- to high-risk prostate cancer patients, and the PET agent demonstrated superior accuracy with the hybrid approach compared to either modality alone, with sensitivities and specificities of 76% and 97% for hybrid ^{68}Ga-PSMA-11 PET/MRI; 58% and 82% for mpMRI alone; and 64% and 94% for ^{68}Ga-PSMA-11 PET alone [211]. Likewise, Zamboglou et al. demonstrated that the combination of PSMA PET with mpMRI increased the accuracy of PET or mpMRI performed alone for primary tumors, showing a sensitivity of 82% and a specificity of 89% [212]. Preoperative assessment of lymph node status is important in intermediate- and high-risk PCa

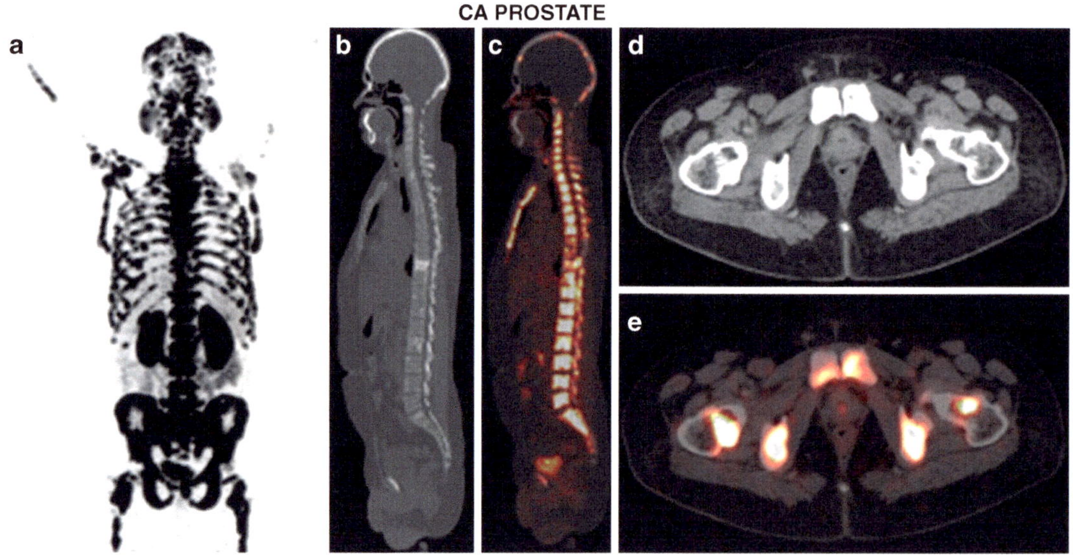

Fig. 27.14 Prostate-specific membrane antigen (PSMA) PET/CT scan in a 56-year-old male patient who is a known case of carcinoma prostate with a rising serum PSA (146 ng/ml) with maximum intensity projection image (MIP, **a**), sagittal & transaxial-fused PET/CT (**c**, **f**) and CT (**b**, **e**) sections showing increased tracer uptake in multiple sclerotic lesions in skull bones, clivus, medial end of right clavicle, sternum, few bilateral ribs, left scapula cervico-dorso-lumbar vertebrae, bilateral pelvic bone, sacrum and bilateral femora (Source: Department of Nuclear Medicine, AIIMS Delhi)

patients, and conventional imaging like CT and mpMRI performs poorly for the detection of pelvic lymph node metastases, almost 80% of which measure less than 8 mm and thus cannot be detected on conventional imaging [213]. Maurer et al. compared ^{68}Ga-PSMA-11 imaging with conventional imaging, for lymph node staging showing that the sensitivity, specificity and accuracy for the detection of nodal metastases was 65.9%, 98.9% and 88.5%, respectively, for ^{68}Ga-PSMA-11, compared with 43.9%, 85.4% and 72.3% for conventional imaging [214]. A recent meta-analysis found a pooled sensitivity and specificity of 71% and 95% for lymph node staging in patients with newly diagnosed intermediate to high-risk PCa; however, such values must be stratified by PSA value [215–217]. ^{68}Ga-PSMA uptake showed higher diagnostic performance in detecting osteolytic and bone marrow metastases than osteoblastic metastases. Data from PSMA PET and CT complement each other for the diagnosis of different types of bone metastases [218]. The routine use of PSMA PET/CT in staging prostate cancer should be reserved for patients with intermediate or high-risk cancers based on PSA and Gleason scores, with a reasonable likelihood of extraprostatic disease.

PSMA PET/CT imaging has been studied extensively in the setting of biochemical recurrence following therapy (PSA of 0.2 ng/mL after radical prostatectomy, or a 2 ng/mL rise above the PSA nadir postradiation). Detection of recurrence sites guides therapy, as it can ultimately help in deciding whether to use aggressive locoregional salvage therapy or systemic palliative therapy. This remains most important decision making at low serum PSA values, when there are potential curative salvage radiation therapy (RT) options, which remain most effective at serum PSA values less than 0.5–1.0 ng/mL [219]. Several reports have demonstrated superiority of PSMA PET/CT to conventional imaging, for detecting sites of recurrent disease. Giesel et al. reported the ability of PSMA PET to detect positivity in two-thirds of patients with nodes with a short axis with BCR [220]. Likewise, in a different study, PSMA-PET was able to detect nodal metastases in 78% of the histologically proven metastatic nodes, whereas morphological imaging was positive in only 26.9% of the same patients [221]. In a recent meta-analysis performed by Perera et al. including data from 16 studies, the overall percentage of positive ^{68}Ga-PSMA-11 PET was 76% and demonstrated the sensitivity of PSMA-PET depends on PSA values. The tumor detection rates for the PSA categories of 0–0.2 ng/mL, 0.2–1.0 ng/mL, 1–2 ng/mL and >2 ng/mL were 42%, 58%, 76% and 95%, respectively; and ^{68}Ga-PSMA PET positivity was associated with shorter PSA doubling time (PSAdt) [222]. The largest retrospective study with 1007 patients reported tumor detection rates for ^{68}Ga-PSMA11 PET/CT of 79.5% [223]. Ceci et al. identified an association between PSA level and PSAdt with a positive ^{68}Ga-PSMA-11 PET/CT in 70 patients with BCR after radical prostatectomy. Receiver operating characteristic analysis showed that a PSAdt of 6.5 months and a PSA of 0.83 ng/mL were optimal cut-off values for ^{68}Ga-PSA PET-positivity, which was observed in 85% of patients with PSAdt 6.5 months [224]. These data imply that positive PSMA scans are associated with clinically significant PSA recurrences. Salvage lymph node resection are also performed in the setting of BCR. In this background, specificities of up to 100% have been reported in presurgical nodal assessment prior to nodal salvage surgery using ^{68}Ga-PSMA [221, 225–227]. PSMA PET imaging has also shown promise for guiding salvage RT [228]. A change in radiotherapy planning occurred in 20–60% of patients after imaging with ^{68}Ga-PSMA PET/CT [229, 230]. PSMA PET can identify sites of disease in patients after external beam radiation, which allows for potential salvage therapy if the recurrence site is outside a prior RT field or it is possible to reirradiate the site of recurrence [231]. In the setting of BCR, a negative PSMA PET scan is associated with favorable RT response compared with patients with a positive scan [232].

Although the term PSMA implies specificity for the prostate it is not entirely prostate-specific, and it is expressed in other neoplasms and in infection and inflammation and is physiologically expressed in normal cells including those in the small intestine, proximal renal tubules and salivary and lacrimal glands. It has shown expression in multiple other cancers [233].

Another radiotracer like ^{18}F-fluoro-5alpha dihydrotestosterone (^{18}F-FDHT) which is a radiolabeled analog of dihydrotestosterone, the androgen receptor (AR) primary activating binding ligand, which offers the possibility to delineate prostate cancer cells with high expression of wild-type AR [234, 235] and in deciding the usage of appropriate chemotherapeutic agents [236].

Few new tracer molecules targeting bombesin or antagonist of the gastrin releasing-peptide (GRP) receptor, labeled with ^{68}Ga, ^{64}Cu and ^{18}F for visualization of GRP receptor-expressing tumors, are being explored in prostate cancer [237, 238]. Also additionally PET tracers that may play a future role in PCa imaging like ^{64}Cu-TP3805 has demonstrated high affinity for the vasoactive intestinal polypeptide receptor 1 receptor, a G-protein coupled receptor that is overexpressed in prostate cancer is being studied [239]. Another cell surface protein, namely STEAP1 has also been identified as a target for castration-resistant prostate cancer. A PET radiopharmaceutical 89Zr-2109A, derived from a fully humanized mAb targeting STEAP1, was tested as an imaging agent to measure changes in STEAP1 expression in a preclinical castration-resistant prostate cancer model. Extensive works in these field are needed to establish their role in the clinical setting of prostate cancer [240].

PSMA PET/CT has also helped in guided targeted PSMA therapies namely 177Lu PSMA and 225Ac PSMA therapies in terminal metastatic patients unresponsive to conventional therapies [241, 242]. The discussion of this broad area is beyond the scope of this article.

27.14 Neuroendocrine Cancers

Neuroendocrine tumors (NET) are heterogeneous slow-growing cancers, occurring in 1–4/100,000 people per year [243, 244]. NET cells originates from the neural crest and belongs to the amine precursor uptake and decarboxylation (APUD) cells system as a result, they can produce different types of substances since they consume, accumulate and thereafter decarboxylate amine precursors such as dihydroxyphenylalanine (DOPA) and hydroxytryptophan. Another typical feature of NET cells is the expression of several receptors in high quantities [245]. The most common sites of NET onset are the bronchus/lungs and gastro-entero-pancreatic (GEP) tract. Less frequent localizations are the skin, the adrenal glands, the thyroid and the genital tract [246]. NET of the gastro-entero-pancreatic tract can be classified depending on their origin in NET of the foregut (pancreas, stomach, duodenum), midgut (ileum, appendix) or hindgut (colon, rectum). European Neuroendocrine Tumor Society (ENETS), taking into account the most recent WHO classifications and the need for a better prognostic assessment of NET and histologic grading, identifies three categories based on the Ki67 levels (G1: Ki67 < 2%, G2: Ki67 between 3% and 20%, G3: Ki67 > 20%) [247, 248]. The diagnostic work up has relied on conventional morphological imaging procedures including computed tomography (CT), ultrasound (US) and magnetic resonance imaging (MRI) combined with gamma camera functional imaging, namely whole-body somatostatin receptor scintigraphy (SRS) [249, 250]. However, although SRS showed a higher accuracy than CT [10] for NET diagnosis at both the primary and metastatic site, the development of novel PET tracers specific for NET has revolutionized the diagnostic approach. In the past decade, several beta emitting tracers have been developed for NET imaging, mainly as a consequence the marginal role of FDG for the assessment of these tumors [251]. In fact, well-differentiated NETs are characterized by a low metabolic rate and therefore low-glucose consumption, rendering FDG not suitable for the evaluation of well-differentiated forms while its role may still be valuable in highly proliferating undifferentiated tumors or in cases of lesions presenting a low expression of somatostatin receptors (e.g., medullary thyroid carcinoma).

^{68}Ga-DOTA-peptides are a group of PET radiotracers that specifically bind to somatostatin receptors that are over-expressed on NET cells. Somatostatin receptors (SSTR) are G-protein-coupled receptors and are internalized after specific ligand-binding. Six different somatostatin receptors have been identified (SSTR 1, 2A, 2B, 3, 4, 5) in humans [252]. ^{68}Ga-DOTA-peptides structure has three parts namely, (a) the active part binding to SSTR (TOC, NOC, TATE), (b) the chelant (DOTA) and (c) the isotope (^{68}Ga). The most relevant difference among these compounds relies in a variable affinity to SSTR subtypes [253]: all can bind to SSTR 2 and SSTR 5, only DOTA-NOC presents a good affinity also for SSTR 3. The observed differences in receptors binding affinity have not yet found a direct clinical correlate and at present there is no indication that such differences are associated with advantages in clinical employment. From a practical point of view, ^{68}Ga-DOTA-peptides presents several advantages. First of all the labeling with ^{68}Ga does not require an onsite cyclotron since ^{68}Ga can be easily eluted from a commercially available Ge-68/Ga-68 generator. The long half-life of parent radionuclide 68 Ge (270.8 days) makes it possible to use the generator for approximately 9–12 months depending on requirement. Approximately 68 Ga ($t1/2 = 68$ min) is a high energy positron emitter with 89% of positron emission and a negligible gamma emission of 3.2% (1077 keV). For labeling, the eluate containing 68 Ga is first concentrated and purified using a micro chromatography method [254]. Radiolabelling yields of >95% can usually be achieved within 15 min. Other advantages include the fact that ^{68}Ga-DOTA-peptides are not dependent on cells metabolism (as compared to ^{18}F-DOPA), have an easy and economic synthesis process and provide relevant information of SSTR expression, with direct therapeutic implications. At present, the most commonly employed DOTA-peptides are ^{68}Ga-DOTA-TOC, ^{68}Ga-DOTA-NOC and ^{68}Ga-DOTA-TATE.

Indications to perform ^{68}Ga-DOTA-peptides studies in NET patients include: staging (Fig. 27.15), re-staging after therapy, identification of the site of the unknown primary tumor in patients with proven NET secondary lesions [255] and selection of cases eligible for therapy

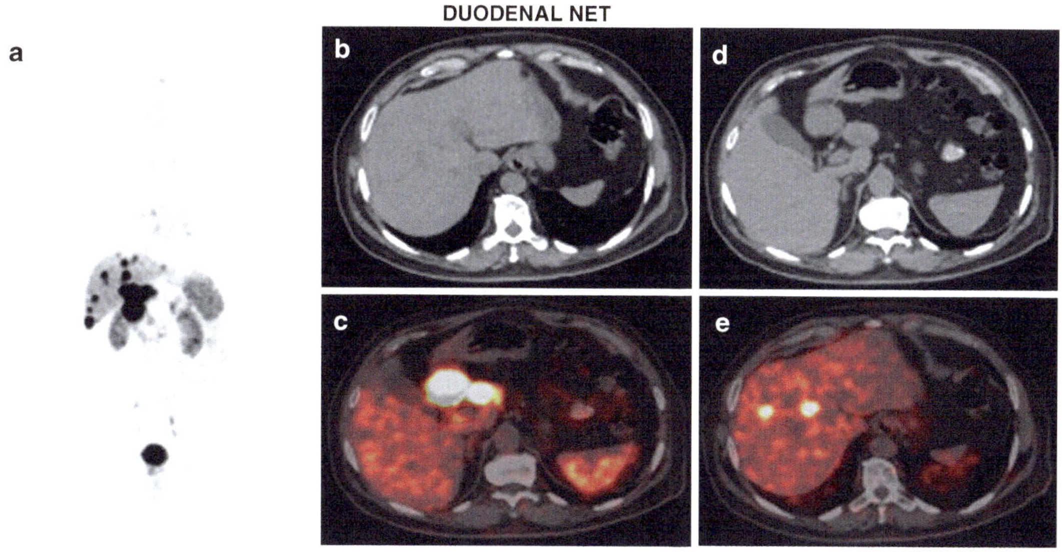

Fig. 27.15 A 54-year-old male patient underwent 68Ga-DOTANOC PET/CT in a case of duodenal NET with rising serum chromogranin. Tracer avid soft tissue lesion was visualized in first part (4.7 × 4.0 cm) of the duodenum with multiple tracer avid isodense lesions noted in segment II, IVA, IVB, V/VI and VI of the liver with largest measuring 1.5 × 1.3 cm (segment VI) seen in MIP (**a**), transaxial-fused PET/CT (**c**, **e**) and CT (**b**, **d**) sections. Also noted tracer avid paraduodenal/portocaval and peripancreatic lymph nodes (Source: Department of Nuclear Medicine, AIIMS Delhi)

with peptide receptor radionuclide therapy (177Lu-DOTATATE/225Ac -DOTATATE) [256, 257].

27.15 Multiple Myeloma

Multiple myeloma is an abnormal plasma cell malignancy characterized by presence of abnormal clonal plasma cells in the bone marrow, with a potential for uncontrolled growth causing destructive bone lesions, kidney damage, anemia and hypercalcemia. Among patients with multiple myeloma, about 73%, 79% and 19% have anemia, osteolytic bone disease and acute kidney injury at the time of presentation [258]. Evaluation of patients with multiple myeloma comprises of assessment of hemoglobin, serum creatinine, serum calcium, serum-free light chain levels and serum protein electrophoresis with immunofixation. Also 24-h urine protein electrophoresis; and full-body skeletal imaging with CT, PET/CT, or MRI have been utilized during the initial workup.

The Revised International Staging System combines details from the serum biomarkers β_2 microglobulin, albumin and lactate dehydrogenase together with malignant plasma cell genomic features detected on fluorescence in situ hybridization—t(4;14), del(17p) and t(14;16) are used to assess approximate progression-free survival and overall survival of multiple myeloma patients. During diagnosis and staging of multiple myeloma, 28% of patients are classified as stage I multiple myeloma according to Revised International Staging, and these patients have a median 5-year survival of 82% [258]. In multiple myeloma patients the standard first-line (induction) therapy consists of a combination of an injectable proteasome inhibitor (i.e., bortezomib), an oral immunomodulatory agent (i.e., lenalidomide) and dexamethasone. They are associated with median progression-free survival of 41 months, compared with historical reports of 8.5 months without therapy. This induction therapy combined with autologous hematopoietic stem cell transplantation followed by maintenance lenalidomide is standard of care for some eligible multiple myeloma patients [259].

For diagnosis, conventional skeletal survey is to be replaced by cross-sectional imaging techniques. The International Myeloma Working Group (IMWG) recommends using cross-sectional imaging in the initial workup like MRI, whole-body low-dose CT, or PET/CT, depending on availability and resources. For the initial evaluation of bone lesions or bone marrow involvement, whole-body low-dose CT and whole-body MRI are the recommended imaging modalities. Two or more focal bone marrow lesions suspicious for myeloma on MRI can presently define symptomatic disease particularly even in the case of intact mineralized bone. Follow-up imaging is not clearly specified so far. Recent guidelines concerning the definitions of minimal residual disease include the assessment of focal lesions before and after treatment using ^{18}F-FDG-PET/CT, with the potential to redefine its role in the diagnosis of multiple myeloma [260].

IMWG recommendations for the use of ^{18}F-FDG-PET/CT in patients with multiple myeloma and other plasma cell disorders are as follows:

Active Multiple Myeloma

(a) ^{18}F-FDG-PET/CT is now the preferred technique for evaluating and monitoring response to therapy (Fig. 27.16) and compared to MRI it provides an earlier metabolic response.
(b) ^{18}F-FDG-PET/CT should be considered as part of the initial investigations in patients with newly diagnosed multiple myeloma as it provides information useful for prognostication and allows careful assessment of the bulk of the disease, particularly extramedullary disease sites; assessing the bulk of the disease with ^{18}F-FDG-PET/CT also applies to patients with relapsed or refractory multiple myeloma.
(c) In patients with newly diagnosed multiple myeloma, with or without extramedullary disease, and more than three focal lesions, ^{18}F-FDG-PET/CT identifies subgroups of patients with unfavorable outcomes; controversies exist about the prognostic role of SUV_{max}.

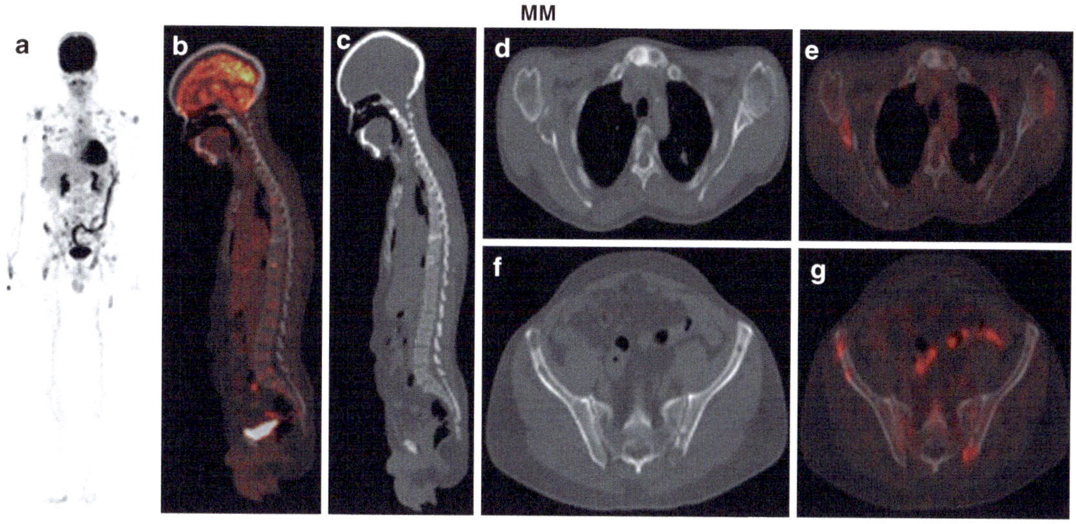

Fig. 27.16 A 42-year-old male patient who is a known treated case of multiple myeloma underwent PET/CT for response assessment. Maximum intensity projection (MIP, **a**), sagittal (**b**, **c**) and axial (**d**–**g**)-fused PET/CT and CT sections showed multiple ill-defined FDG-avid osteolytic lesions with some showing soft tissue component in skull, bilateral clavicles, bilateral humeri, bilateral scapulae, multiple bilateral ribs, sternum, multiple cervico-dorso-lumbar vertebrae, bilateral pelvic bones, bilateral femoral head and shaft of right femur (Source: Department of Nuclear Medicine, AIIMS Delhi)

(d) ^{18}F-FDG-PET/CT should be combined with sensitive bone marrow-based assays as part of minimal residual disease detection inside and outside the bone marrow.

Smoldering Multiple Myeloma

(a) Patients who meet the diagnostic criteria for smoldering multiple myeloma and have one or more lytic lesions on ^{18}F-FDG-PET/CT should be defined as having multiple myeloma that requires immediate therapy
(b) ^{18}F-FDG-PET/CT is recommended to distinguish smoldering multiple myeloma from active multiple myeloma if whole-body X-ray is negative and whole-body MRI is unavailable.

Solitary Plasmacytoma

(a) Patients with focal lesions on PET but without underlying lytic lesions on the CT part of ^{18}F-FDG-PET/CT are at high risk of progression to active multiple myeloma.
(b) Patients with suspected solitary plasmacytoma, either extramedullary plasmacytoma or solitary bone plasmacytoma without symptoms or signs suggestive of cord compression, should receive ^{18}F-FDG-PET/CT to unequivocally confirm the diagnosis, provided whole-body MRI is unavailable.

The advantages of using ^{18}F-FDG-PET/CT at the time of diagnosis are that it provides a whole body evaluation at one session while detecting with relatively high sensitivity and specificity myeloma bone disease and helps in detection of lytic bone lesions (diameter of >5 mm) and extramedullary sites of the disease. It helps in assessing the burden and metabolism of the tumor and also is a valuable tool for predicting the prognosis (particularly, absence or presence of focal lesions, and of extramedullary disease). ^{18}F-FDG-PET/CT is mandatory to confirm the diagnosis of solitary plasmacytoma. After treatment ^{18}F-FDG-PET/CT also helps in evaluating the metabolic response to therapy and also provides earlier assessment of response compared with MRI and can correlates well with biochemical markers of response. It also identifies those patients who are imaging MRD-negative too.

Conversely, the disadvantages of ^{18}F-FDG-PET/CT for initial diagnosis and workup are, that it is a suboptimal investigative modality for detecting diffuse bone marrow plasma-cell infiltration and lytic lesions in the skull and has poor availability and higher costs compared with MRI and whole-body MRI. The radiation dose of ^{18}F-FDG-PET/CT is greater than that given for whole-body X-ray, and whole-body low-dose CT. ^{18}F-FDG-PET/CT can show false positivity due to bone metallic implants, accumulation of the tracer in physiological districts, inflammation, infections, postsurgical areas, concomitant tumors, or recent chemotherapy or radiotherapy or use of growth factors inducing a false diffuse bone marrow pattern. Also false negativity in ^{18}F-FDG-PET/CT can be due to hyperglycemia, recent use of high-dose steroids inducing a transient metabolic suppression, presence of subcentimeter lytic lesions in the skull, or close to the brain. In the setting of using ^{18}F-FDG-PET/CT after treatment, lack of standardization plays a major drawback in its usage on a main stream level [260].

Lately, multiple other radiopharmaceuticals like ^{68}Ga-Pentaxifor, ^{11}C-Choline, ^{18}F-Fluorocholine, etc. have been explored in the scenario of multiple myeloma, but none of them has been studied extensively and standardized like ^{18}F-FDG-PET/CT [261–263].

27.16 Bone Tumors

Bone tumor (BT) is a tumor that occurs in the bone or its subsidiary tissues. BT are classified as benign BT and malignant BT. The most common primary malignant bone tumors are Ewing's sarcoma, osteosarcoma and chondrosarcoma. Disease staging is necessary for determination of the treatment plan as well as for follow-up and also for assessment of response to therapy. This is done using imaging modalities and histopathological confirmation of the suspected entity primary malignant bone tumors [264, 265].

Diagnostic imaging has a principal role in the evaluation and management of bone cancers. Standard imaging modalities such as conventional X-rays, computed tomography (CT) scanning, magnetic resonance imaging (MRI) and skeletal scintigraphy cannot provide the accurate staging of the tumor as they depend mainly on the morphologic features and cannot differentiate between posttreatment changes and recurrence or residual tumor due to change in normal architecture of the regional anatomy by surgery and/or radiation [266–268].

The introduction of the hybrid ^{18}F-FDG-PET/CT has helped in the advancement of the knowledge of pathophysiology of cancers that significantly impacts the evaluation, staging and management of the tumors and thus providing more effective treatment, better quality of life and increased survival [269]. Also, its certain unique advantages, like high sensitivity and long field of view during scanning, provide an opportunity for wider clinical applications, including, total-body dynamic scanning, low administered activity and long acquisition delay [270, 271]. Viable malignant primary bone tumors are usually FDG avid. FDG accumulation may shed light on the tumor characteristics based on its metabolic tumor activity and thus providing better characterization of indeterminate lesions and also as a to guide biopsy of the most metabolically active region within larger heterogeneous tumors especially tumors of mixed grade and/or cell type [2, 272]. FDG-PET/CT is also used for accurate staging of the histologically confirmed tumors (Fig. 27.17), response assessment and also to detect any recurrent lesions. It is a noninvasive method to estimate tumors grade depending on the amount of FDG uptake in the lesions. This indicates its role as a reliable independent prognostic indicator in most of the bone tumors [273, 274]. According to NCCN guidelines, PET/CT is recommended as an optional investigational modality during initial patient workup [275].

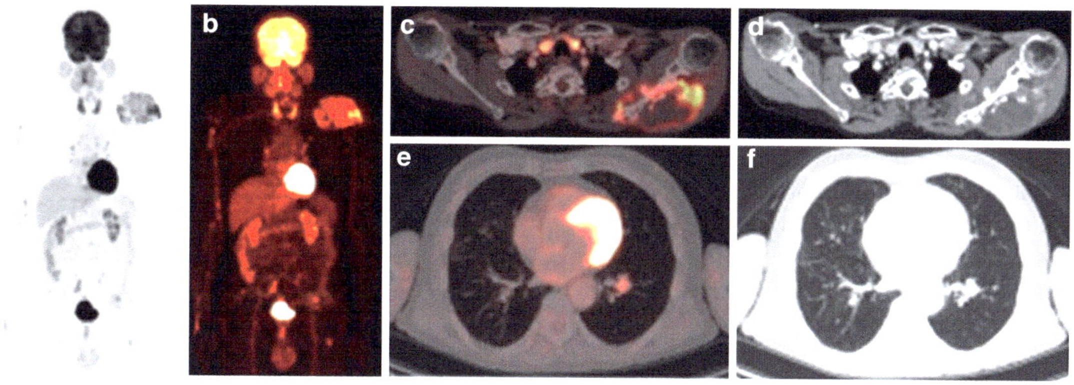

Fig. 27.17 Metabolically active lytic-sclerotic destructive lesion (**a**–**d**) in left scapula with associated soft tissue component—primary (osteosarcoma) with metastatic bilateral pulmonary nodules (**e**, **f**) in FDG-PET/CT of a 65-year-old male patient done to assess the disease status (Source: Department of Nuclear Medicine, AIIMS Delhi)

27.17 Brain Tumors

Primary brain tumors refer to a heterogeneous group of tumors arising from cells within the CNS. Gliomas represent 75% of malignant primary brain tumors in adults [276]. Gliomas are tumors of neuroectodermal origin arising from glial or precursor cells and include astrocytomas, oligodendrogliomas and ependymomas. Their classification has undergone major restructuring in the 2016 version of the WHO classification of CNS tumors [277], with a subsequent effect on diagnostic criteria, diagnostic testing approaches, grading, prognosis and treatment planning.

Patients with primary brain tumors can present with focal (i.e., related to a specific location in the brain) or generalized symptoms over days to weeks, or months to years, depending on the speed of growth and location of the tumor. Tumors can also be found by brain imaging that has been done for unrelated purposes.

MRI is the standard modality for neuroimaging and is used for diagnosis of brain tumors, for achieving stereotactic biopsy and also for surgical planning in neuro-oncology [278]. Recently, hybrid imaging modalities such as PET/CT and PET/MRI, were found to be useful methods for early diagnosis of brain tumors [278, 279].

Multiple PET radiotracers have been used to evaluate brain tumors, including fluorine-18 fluorodeoxyglucose (^{18}F-FDG), fluorine-18 fluoroethyltyrosine (^{18}F-FET), radiolabelled choline (^{11}C-choline or ^{18}F-choline), carbon-11 methionine (^{11}C-methionine), fluorine-18 fluorodihydroxyphenylalanine (^{18}F-FDOPA) and fluorine-18 fluorothymidine (^{18}F-FLT). ^{18}F-FDG is the most common radiotracer that is widely used in oncology and allows for the detection of neoplastic cells due to their frequently increased glucose metabolism. The main concerns about the use of ^{18}F-FDG-PET for evaluating brain lesions are the high background ^{18}F-FDG uptake in the normal brain [279] and the increased 18F-FDG uptake in inflammatory or benign lesions [280]. Therefore, other PET tracers characterized by increased uptake in brain tumors and low uptake in the normal brain have been developed in recent decades [279].

^{11}C-methionine is a radiolabelled amino acid; methionine is used by the cells in two main metabolic functions: protein synthesis and conversion to S-adenosylmethionine. In many cancer cells, there is an increase in protein synthesis, transmethylation and transsulfuration, leading to an increased uptake of ^{11}C-methionine by the malignant cells [279, 281]. Unfortunately, the use of this PET radiotracer is restricted to PET/CT centers with an on-site cyclotron facility because of its shorter half-life compared to ^{18}F (20 min versus 110 min, respectively) [281].

^{18}F-FET is a fluorinated amino acid radiotracer used to detect brain tumors. It is taken up into cancer cells due to their increased amino acid uptake through L-type amino acid transport

(LAT) channels overexpressed in the neoplastic cells, and also it is not incorporated into proteins [278].

^{18}F-FDOPA has been proposed as a useful PET/CT radiotracer for imaging brain tumors. It is transported across the blood-brain barrier by a number of amino acid transporters, which have been shown to be overexpressed in brain tumors. After intracellular uptake through the LAT transporter, ^{18}F-FDOPA is decarboxylated by DOPA decarboxylase to ^{18}F-dopamine, and transported by vesicular monoamine transporters into storage granules and thereby trapped intracellularly [279, 282].

^{18}F-FLT is a radiotracer that acts as an imaging biomarker of cell proliferation, which is particularly increased in neoplastic cells especially during the S phase of the cell cycle. ^{18}F-FLT is phosphorylated by thymidine-kinase-1 enzyme. It is trapped inside the cell but not incorporated into the DNA. It has been reported that the cellular thymidine-kinase-1 activity is proportional to the proliferation activity of the tumor [283].

Lastly, as tumor cells present a high turnover of cellular membranes, radiolabelled choline (using ^{11}C or ^{18}F) may be used to detect brain tumors. The uptake of radiolabelled choline increases in tumor tissue to keep up with the demands of phospholipids synthesis in cellular membranes [279, 284]. Enough literature data already exist about the diagnostic performance and prognostic value of PET with different tracers in brain tumors, especially in a challenging scenario like diagnosing brain tumor recurrence after treatment. MRI has an exceptional structural resolution, significantly needed from the perspective of treatment planning. However, the ability of MRI to detect recurrence is relatively lower, when compared to metabolic imaging. Metabolic imaging is more sensitive to the early changes associated with the recurrence and relatively immune to confounding by treatment related changes. There is no single-step investigation for the diagnosis of recurrence in brain tumors and also the sensitivity of metabolic imaging is not a substitute for the good resolution of the MRI, making a multimodal approach the only way forward in case of brain tumors depending on the indication [285].

27.18 Head and Neck Cancers

Head and neck (H&N) cancer was identified as one of the first clinical indications for PET [286, 287]. FDG-PET/CT is now a key diagnostic tool for H&N cancer. It is an essential examination for the detection of the primary site in patients who present with a malignant neck lymph node and no evidence of cancer on the upper aero-digestive tract mucosa [288–291]. In addition, FDG-PET/CT has been established as one of the main investigations for patients with squamous cell H&N cancer, i.e., mainly in M staging. FDG-PET/CT has a chief role for the staging of patients with T4 cancer of the hypopharynx and nasopharynx and also patients with N3 nodal disease. FDG-PET/CT is indicated for the staging and for the detection of residual and recurrent disease in this group. In head and neck cancer patients with advanced neck nodal disease, who are treated with chemoradiotherapy, there is captivating evidence that patients with no FDG uptake in the neck region in the PET/CT scan, performed 12 weeks following completion of treatment, do not require neck dissection. It also has emerging role in the surveillance of H&N cancer survivors [292].

There is limited evidenced guidance for the use of FDG-PET/CT in non-squamous cell carcinoma, as these tumors are extremely rare. In cases where FDG-PET/CT is done priory, it can be used for pretreatment staging, and subsequently to detect residual/recurrent disease. Most H&N malignant tumors are FDG avid. However, salivary gland cancers and tumors with adenoid cystic, clear cell and mucoepidermoid histology show paucity to FDG uptake, especially when they recur. In addition, perineural invasion cannot be detected reliably with FDG-PET/CT [293].

PET/CT has led to more accurate diagnoses and it has contributed to improved outcomes in H&N cancer patients. According to NCCN guidelines [294], PET/CT is recommended as the imaging modality of choice to detect neck lymph node locoregional metastasis and distant metastasis except for nasopharynx cancer where NCCN recommends chest CT with contrast as an equivalent option.

27.19 Thyroid Cancer

Thyroid cancer remains the most common endocrine malignancy and comprises 2% of all cancers reported in the United States [295]. Differentiated thyroid cancer (DTC), comprising papillary and follicular thyroid cancer, generally carries a good prognosis. About 5–15% of differentiated thyroid cancer and 50% of metastatic differentiated thyroid cancers are refractory to radioactive iodine treatment contributed by progressive dedifferentiation of differentiated thyroid cancer and loss of sodium iodide symporter (NIS) that is required for iodine intake [296–298]. The role of ^{18}F-FDG-PET/CT has been explored particularly in the detection of persistent and recurrent DTCs refractory to RAI treatment.

The role of PET/CT in the management of thyroid cancer is mainly in the evaluation of recurrent or metastatic disease, particularly in poorly differentiated subtypes. Despite its limited role in the surveillance of differentiated carcinomas, there is a definite benefit in patients with negative radioiodine scans. Meta-analysis studies revealed that PET/CT has a sensitivity and specificity of more than 80% in detecting extensive disease [299, 300].

FDG-PET/CT has demonstrated value in the assessment of recurrence, radioiodine refractory disease and prognosis of thyroid malignancy [301–303]. PET/CT using ^{68}Ga-Somatostatin analogs, ^{68}Ga-DOTA-RDG, ^{68}Ga-FAPI and 68 Ga-PSMA, has a promising role in the management of thyroid cancers [304–307]. Furthermore, ^{124}I PET/CT is a sensitive modality for the detection of recurrent disease and dosimetry [308, 309].

27.20 Melanoma

Melanoma, also known as malignant melanoma (MM), is a cancer that develops from melanocytes. It is an aggressive tumor with an early tendency to metastasize to regional lymph nodes followed by distant metastasis mainly into, liver, bone and brain. Over the past 50 years the incidence of melanoma has rapidly increased primarily in Caucasians, despite the increased awareness of the risk of excessive sun exposure and consequent UV damage to DNA. Melanoma mostly affects young and middle aged people unlike other solid tumors [310]. Other than skin (>95%), less than 5% of all primary melanomas arise from meninges, choroidal layer of the eye and mucosal surfaces, that has a common ectodermal origin [311]. High mortality rate in terminal stages calls for early detection at initial stages for effective treatment [312].

PET/CT with its inherent whole body scanning ability is the best imaging modality to search for melanoma metastases with the ability to change treatment plan in as high as 49% [313]. It was understood that among different imaging modalities like sonography, CT, PET and PET/CT, PET/CT had the highest sensitivity and specificity of 86% and 91% for detecting metastases in deep soft tissue, lymph node and also visceral metastases [314]. Importantly FDG-PET/CT can influence treatment decisions and prove to be of great value by locating distant metastases in advanced stages of malignant melanoma (i.e., III–IV) [315–317]. It can save the patients from ineffective therapy and its associated toxicities. One of the main indications of PET/CT in cutaneous melanoma is to detect recurrence or to restage disease following the identification of recurrence, particularly in patients with high-risk melanoma who are candidates for resection with curative intent. There is still lack of agreement on the indications of post-therapy follow-up PET/CT scans for malignant melanoma patients.

27.21 Gynecological Cancers

Gynecological cancers are the cancers of women's reproductive organs and genitals namely cervical, ovarian, uterine, vaginal and vulvar, and even rarely fallopian tube cancer. Earlier role of PET/CT in gynecological cancers was limited, and recently due to the improvement in technical aspects and also due to widespread research in this field, we have better understood specific indications of PET/CT. The specificity of ^{18}F-FDG-

PET/CT is lowered by infective and inflammatory processes, which are seen at the time of presentation of malignancy or manifested secondary to treatment.

In ovarian cancers, PET/CT is superior to CT alone in characterizing adnexal masses but inferior to the combination of MRI and US [318, 319]. In different subtypes of ovarian cancer, clear cell and mucinous subtypes showed significantly lower FDG uptake compared to serous and endometrioid subtypes. Although PET/CT is superior on a lesion-by-lesion basis and in the detection of nodal and distant extra-abdominal metastases (Fig. 27.18) compared to contrast-enhanced CT alone at original staging, this superiority does not translate to significant management change for the patient [320–322]. PET/CT has greatest impact for patients with documented or suspected relapse with a rising CA125 considered for salvage therapy [323]. In recent years, there has been great interest in intratumoral FDG heterogeneity. The heterogeneous uptake of the FDG tracer is hypothesized to be due to a combination of tumor features such as necrosis, angiogenesis and glucose uptake and therefore predictive of disease response to chemo and radiotherapy. Lee et al. retrospectively studied preoperative ^{18}F-FDG-PET/CT studies of 61 patients who had primary surgery in ovarian cancer. They analyzed multiple PET metrics including maximum SUV_{max}, TLG, MTV and intratumoral FDG heterogeneity (IFH). Out of these various PET metrics, high values of IFH were significantly associated with reduced progression-free survival and the authors suggested that patients with high IFH could therefore potentially be managed with more intensive chemotherapy regimens. PET/CT therefore shows promise as a surrogate biomarker to predict chemotherapy response in patients undergoing neoadjuvant chemotherapy [324].

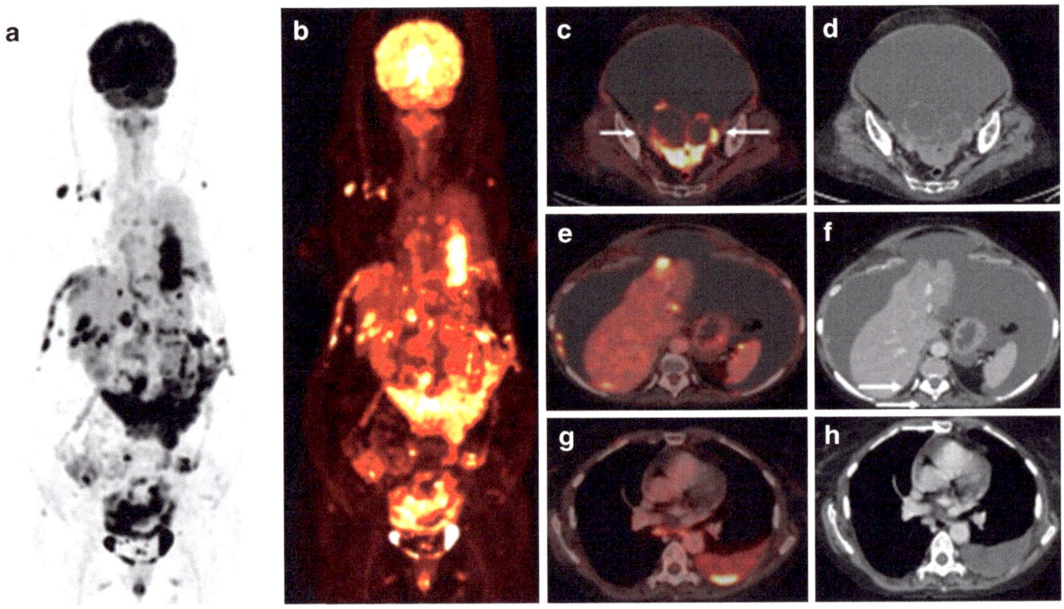

Fig. 27.18 Baseline staging with PET/CT in carcinoma ovary in a 54-year-old female with complex cystic mass lesion in bilateral adnexa (**a**–**d**—white arrows) with metabolically active solid component with thickening of parietal peritoneum, omentum and mesentery with hepatic and splenic subserosal deposits (**e**, **f**) with metabolically active axillary, internal mammary, mediastinal, abdominal and retroperitoneal lymph nodes and left pleural effusion (**g**, **h**) and massive ascites (Source: Department of Nuclear Medicine, AIIMS Delhi)

^{18}F-FDG-PET/CT plays a well-established role in the management of patients with cervical cancer. Although primary tumors are usually metabolically active, ^{18}F-FDG-PET/CT has a very limited role in the locoregional primary staging of cervical cancer as the anatomic details and definition is far superior with MRI [325, 326]. Lymph node status (pelvic and para-aortic nodes) in cervical cancer is one of the most important prognostic indicators, adversely affecting patient prognosis and ^{18}F-FDG-PET/CT has an established and important role in the detection of lymph node and distant metastases where its performance is superior to conventional MRI and contrast-enhanced CT mainly in advanced cervical cancers (Fig. 27.19) [327, 328]. ^{18}F-FDG-PET/CT performs functional assessment of the primary tumor (SUV_{max}) and mean SUV of pelvic nodes provides information that can be used as a biomarker for response, predicting disease recurrence and survival parameters. Pelvic metastatic nodes with higher values of the mean SUV usually correlates with an increased risk of persistent disease at the end of treatment and is an independent predictor of developing pelvic disease recurrence [329]. In the setting of utility of PET/CT for surveillance, PET/CT is indicated in FIGO Stage IB2 with high risk factors and stage II-IV who have received post operative adjuvant radiation or chemoradiation 3–6 months after completion of treatment. It is also recommended in case of symptomatic suspicion of local recurrence when conventional imaging is equivocal or when patients are being considered for salvage surgery [330]. It has been found that the false positive FDG uptake, lowering the specificity and positive predictive value, was seen secondary to radiotherapy induced inflammatory changes. ^{18}F-FDG-PET/CT can also be integrated as an investigative modality in radiotherapy treatment planning for dose escalation and simultaneously boost regimes to achieve increased tumor shrinkage and likeli-

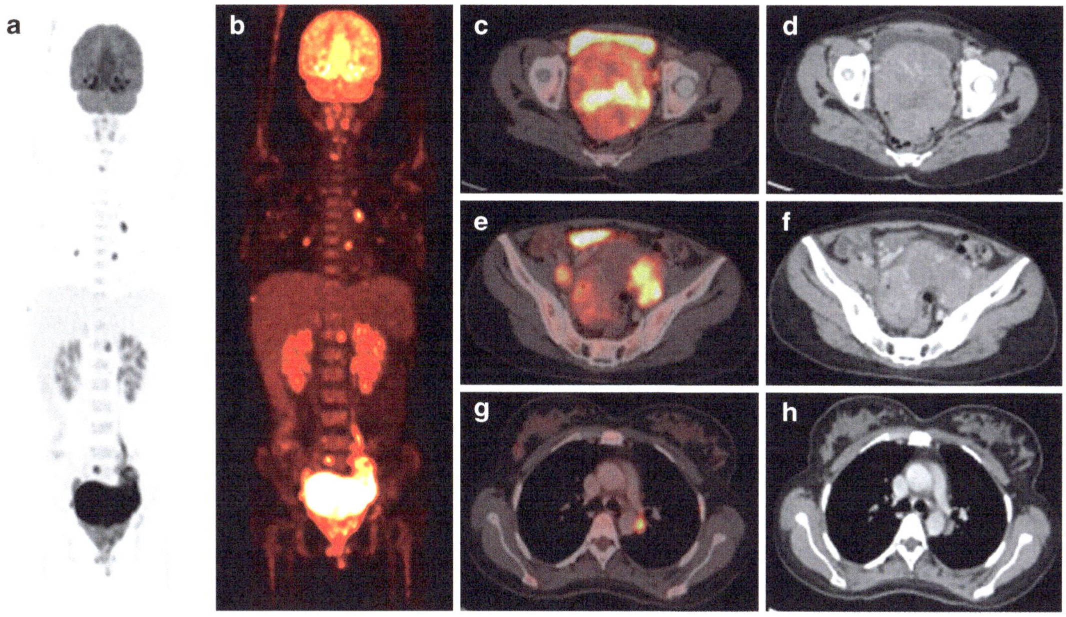

Fig. 27.19 A 55-year-old female with cervical mass as detected on MRI for PET/CT of initial staging. Maximum intensity projection (MIP) and transaxial-fused PET/CT and CT sections (**a**–**d**) show lobulated well-defined measuring 10 × 9.1 × 9.2 cm involving cervix with metastases to pelvic lymph nodes (**e**, **f**) and bilateral lungs (**g**, **h**) (Source: Department of Nuclear Medicine, AIIMS Delhi)

hood of local control [331]. A recent systematic review and meta-analysis of eight studies with 221 patients, found sensitivity of 94% and a specificity of 84% for ^{18}F-FDG-PET/CT on per patient basis for detection of recurrence in cervical cancer. ^{18}F-FDG-PET/CT has an established role in detecting distant recurrent disease and is therefore essential when selecting patients for exenterative surgery [332, 333].

Endometrial cancer is now the most common gynecological cancer in the Western world and the role of ^{18}F-FDG-PET/CT has evolved over the past decade with a wider range of applications in the care of patients with endometrial cancer. ^{18}F-FDG-PET/CT has no established role in the detection and local staging of endometrial cancer. One of the reasons for this limitation is that there is normal physiological endometrial FDG uptake in pre- and peri-menopausal females during their normal ovulatory and menstrual phases of the menstrual cycle. In addition there is lack of spatial resolution of FDG-PET/CT and also the accurate definition of depth of myometrial invasion and tumor extension is difficult. The risk of nodal metastases, that decreases the patient survival, is dependent on the depth of myometrial invasion, cervical tumor involvement, adverse tumor subtype and histological lymphovascular invasion [334]. The main established role for ^{18}F-FDG-PET/CT in endometrial cancer is for extra uterine disease detection. Reported accuracies for preoperative nodal staging of endometrial cancer by ^{18}F-FDG-PET/CT are modestly better than conventional imaging; however, it is dependent on nodal size and still not good enough to replace lymphadenectomy [335]. Bollineni et al. included eight studies on endometrial cancer recurrence in a recent systematic review and meta-analysis and studied lymph node staging and recurrence in endometrial cancer after primary surgical treatment. They demonstrated ^{18}F-FDG-PET/CT had pooled sensitivities and specificities of 95 and 91%, respectively, in the detection of recurrent endometrial cancer [336]. ^{18}F-FDG-PET/CT is excellent in the detection of recurrent disease in endometrial cancer, and also in patients considered for exenterative surgery. However, it is recommended to exclude unsuspected distant metastases. The specific value of ^{18}F-FDG-PET/CT as a predictive biomarker is promising but requires improved standardization and validation of metabolic threshold values.

Data on the role of PET/CT in the management of vulvar and vaginal cancer is sparse. Several studies have suggested that FDG-PET/CT is useful for detecting nodal and distant metastases in vulvar and vaginal cancer [337, 338]. Identification of positive pelvic nodes on PET/CT may allow a patient to avoid extensive surgery, and be treated alternatively with chemoradiation [339]. One study using PET/CT for the assessment of suspected or known disease (primary or recurrent) in vulvar and vaginal cancers reported a change in prognostic impression and intended patient management, in 51 and 36% of studies, respectively [340].

There are specific recognized pitfalls that are related to gynecological cancer imaging. Many changes are noted secondary to physiological variations in the normal menstrual cycle and concurrent infective or post-treatment changes. There is a high chance of coexistence of pelvic inflammatory changes in patients presenting with ovarian and cervical cancer. These limitations should be kept in mind while reporting PET/CT scan in gynecological cancers. Currently, preliminary studies including all gynecological malignancies depict that ^{18}F-FDG-PET MRI is superior to ^{18}F-FDG-PET/CT for primary tumor delineation and eventually may become the preferred imaging modality for staging cervical and endometrial cancers. Although it is a reliable tool in the recognition of extra-abdominal distant metastasis, it does not confer any significant advantage over ^{18}F-FDG-PET/CT in the detection of nodal or distant metastases [341, 342].

27.22 Testicular Cancers

Testicular germ cell tumors are one of the most common neoplasms in young men, affecting men between 15 and 35 years [343]. Testicular cancers are commonly classified into two entities, germ cell tumors and stromal tumors. Nearly 90–95% of testicular cancers are germ cell tumors and they are further classified mainly into seminoma, nonseminoma and mixed germ cell tumors. Nonseminomatous germ cell tumors are further divided into teratoma, embryonal cell carcinoma, yolk sac tumor and choriocarcinoma. Mixed germ cell tumors have components of seminoma and nonseminoma tumor types. Testicular stromal tumors are a rare group and are further classified into Leydig cell tumors and Sertoli cell tumors. Ultrasound of the testis is the main stay imaging modality in the primary diagnosis of testicular tumors with a sensitivity of nearly 100%. However, MRI has an even higher sensitivity and specificity, nearly from 95% to 100%, but it plays a secondary role in this clinical scenario due to its higher costs [344]. Up to this date the role of ^{18}F-FDG-PET/CT for evaluation of scrotal masses is a controversial issue. ^{18}F-FDG-PET/CT can help to differentiate between benign and malignant lesions in primary evaluation of scrotal masses. In case of seminoma, the current European Association of Urology (EAU) Guidelines recommends F-FDG-PET/CT for the follow-up of residual masses >3 cm after chemotherapy. In a meta-analysis including four studies that has 130 patients ^{18}F-FDG-PET/CT was proved to be superior to standard imaging procedures in predicting viable residual tumors [345]. The use of ^{18}F-FDG-PET/CT in workup led to a significant reduction of overtreatment from 72% to 30% and undertreatment was decreased from 14% to 7% [345]. In cases of nonseminomas, the current EAU Guidelines do not recommend ^{18}F-FDG-PET/CT for primary staging or restaging because of its similar sensitivity and specificity to conventional restaging in the follow-up [346, 347].

There are several limitations of the PET/CT in case of imaging of seminomas. False negative results have been described, mainly caused by transitory suppression of tumor cell activity. To reduce this particular effect, imaging is recommended mostly by 6 weeks after chemotherapy [348]. Moreover, lesions less than 10 mm are difficult to detect because of the low spatial resolution of F-FDG-PET/CT. A significant limitation of the F-FDG-PET/CT in nonseminomas is the characterization of residual masses. Also FDG-PET/CT cannot distinguish precisely between necrosis and fibrosis areas as there is a variability of absorption of FDG in teratoma. In addition, false positive findings can be caused by inflammatory or granulomatous tissue [349].

27.23 Penile Cancers

Penile cancer is a rare disease with an incidence of 0.6–2.1 in Western countries [350–352]. Majority of the penile cancers are squamous cell carcinoma, but other histological types and subtypes such as warty squamous cell, basaloid and melanoma can occur and has been reported. Even though all penile cancers are FDG avid, PET/CT is not very useful for primary tumor staging, because of its limited spatial resolution it is unsuitable for evaluating cancer growth in surrounding structures or also to even assess the exact size of the tumor. Furthermore, FDG is excreted via the urine, which therefore can interfere with imaging of the primary tumor. Physical examination and, in selected cases MRI, have shown excellent accuracy for staging of primary penile tumors [353]. It has limited role in assessing inguinal lymph nodes [354]; however, in advanced penile cancers where likelihood of metastatic disease (pelvic metastases and occult metastases) is high, it has role in assessing the metastatic burden (Fig. 27.20) prior to systemic chemotherapy and/or prior to an extensive surgical resection [355, 356]. There is a significant scarcity of literature to establish its role in this field.

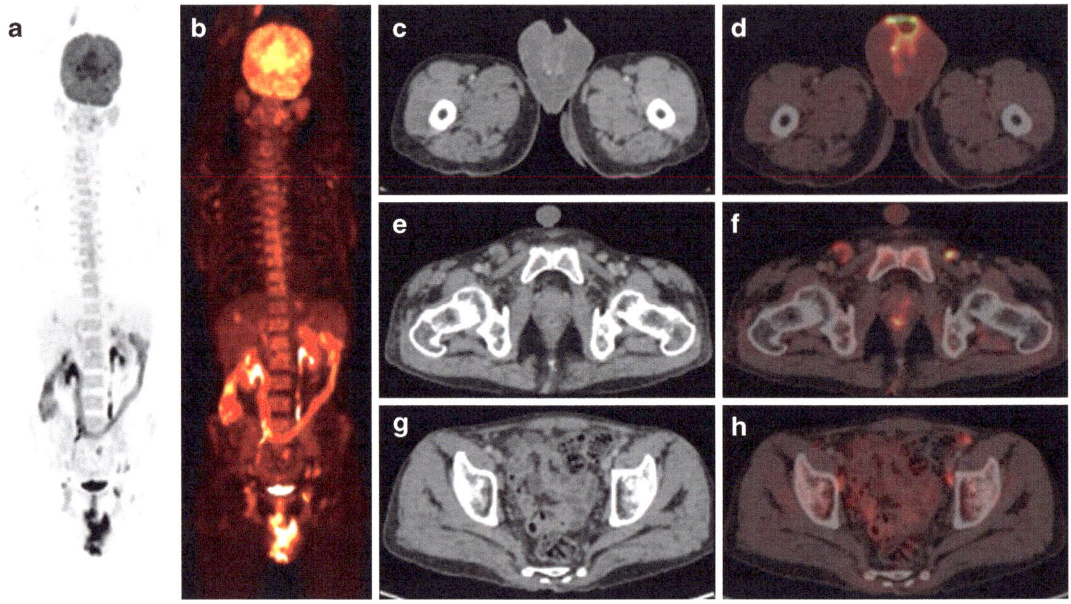

Fig. 27.20 A 39-year-old male with penile cancer underwent PET/CT post four cycles of chemotherapy. Maximum intensity projection (MIP) and transaxial-fused PET/CT and CT sections (**a**–**d**) show metabolically active poorly demarcated soft tissue mass involving root and body of penis that extends into the scrotal sac with metastases to bilateral inguinal (**e**, **f**) and pelvic lymph nodes (**g**, **h**) (Source: Department of Nuclear Medicine, AIIMS Delhi)

27.24 Recent Developments

Lately we have seen a finest radiopharmaceutical like ^{68}Ga-FAPI evolving and overcoming most of the drawbacks of FDG-PET/CT. ^{68}Ga-FAPI is a promising radiotracer that acts on the fibroblast activation protein which is highly expressed in the stroma of several tumor entities, especially in breast, colon, pancreatic carcinomas, etc. [357]. It has shown its undeniable role in detection of the brain metastases in several cancers as there is negligible background uptake. With available research evidence, it can complement FDG-PET/CT and is better than FDG-PET/CT in gastrointestinal, liver, breast and nasopharyngeal cancers [358, 359]. Recently, even therapies based on FAPI have been studied in various cancers like breast and thyroid cancers and it is expected to be the new molecule of the century [360].

27.25 Conclusion

PET/CT is finding its utility in the oncological field and each study, specifically clinical trials, has proved undoubtedly its impeccable role in common cancers like lung cancer, breast cancer, lymphoma, etc. Throughout years, we have seen its evolution from an optional investigative modality to a significant modality useful during the workup and overall management of the oncological patients. With the introduction of hybrid PET/CT, we have seen tremendous rise in its indications for various malignancies, clearly evident from its rising recommendations seen in recent NCCN (summarized in Table 27.2) and other oncological guidelines by various international associations. Lately in areas where PET/CT has limited role, PET/MRI has found its significant role especially in gynecological, hepatic and CNS cancers. This field will undergo

Table 27.2 NCCN recommendations for various malignancies

Cancers	Initial staging	Subsequent staging
CNS tumors	Primary CNS lymphoma • Consider FDG PET/CT for initial staging Brain metastases • Consider FDG PET/CT for diagnosis if primary malignancy is not detected in a case with limited/extensive brain metastases Metastatic spine tumors • Systemic imaging (i.e., PET, CT MRI, bone scan) is recommended to identify primary for the purpose of diagnosis/staging in case of incidentally detected spine tumors in asymptomatic patient	Anaplastic gliomas/glioblastoma • Brain-PET is considered in recurrence disease to differentiate viable tumor from radiation necrosis
Head and neck cancers • Lip • Oral Cavity • Oropharynx • Nasopharynx • Hypopharynx • Larynx • Paranasal tumors (maxillary and ethmoid sinus tumors) • Salivary gland tumors • Mucosal melanoma of the head and neck	• PET/CT scan is indicated in case of occult primary (before EUA for diagnosis) or providing biopsy site for a imaging assessment of primary site • PET/CT is indicated in case of high grade tumour histology or multistation involvement or lower neck nodal involvement • PET/CT is indicated in case of tumour approaching midline- to plan for approach to contralateral neck • PET/CT is indicated to identify involved nodes prior to radiotherapy • FDG PET/CT is indicated in case of distant metastases • For loco-regional advanced disease (<6 months), PET/CT is indicated 3–6 months after radiotherapy and stereotactic radiosurgery for response assessment and assessment of residual disease • For loco-regional advanced disease (>6 months to 5 years post treatment) – PET/CT is most sensitive for detection of recurrence or second primaries	
Non-small cell lung cancer	PET/CT scan is indicated for initial staging in essentially all cases If PET/CT scan is positive in mediastinal lymph node, it requires needs radiological or histopathological confirmation	PET/CT is useful for • Pre-surgical evaluation • Radiotherapy planning • Restaging after induction therapy • Surveillance Other indication PET/CT for metabolic characterization of solid, non-calcified pulmonary nodule >8 mm in size
Small cell lung cancer	• PET/CT is recommended for initial staging of small cell or combined small if limited stage is suspected • PET/CT is suggested for radiation treatment planning. RT target volumes based on pre-treatment PET/CT (4 weeks-ideal, treatment position)	PET/CT not recommended for routine follow-up after initial therapy (restaging)

(continued)

Table 27.2 (continued)

Cancers	Initial staging	Subsequent staging
Breast cancer	Invasive breast cancer • Stage I, II or operable III: PET/CT not recommended • Stage IIIA (T3, N1, M0) or IIIB: FDG PET/CT optional (category 2B) for initial staging • After lumpectomy or mastectomy and surgical axillary staging with >4 positive axillary nodes: optional FDG PET/CT • Recurrent or stage IV: FDG PET/CT optional for staging and restaging Inflammatory breast cancer • Stage T4d, No-N3, M0: FDG PET/CT recommended	
Hodgkin's lymphoma	PET/CT is considered essential during initial work up	Interim or end-of-treatment based on Deauville scores • PET scans with a score of 1, 2, or 3 are considered negative findings • PET scans with a score of 4 and 5 are considered positive findings
Non-Hodgkin's lymphoma	• DLBCL-PET/CT is considered essential during initial work up • Follicular lymphoma – PET/CT scan (including neck) essential if radiotherapy planned for stage I, II disease – PET/CT (CT with contrast of diagnostic quality) if systemic therapy is planned • For the following lymphomas, PET/CT OR CECT is considered "essential" in initial workup – Mantle cell lymphoma (B) – AIDS-related B-cell lymphoma – Peripheral T-cell lymphoma – Extra-nodal NK/T cell lymphoma – Nasal Type PTLD – Castleman disease • CECT or PET/CT has utility only for Richter's transformation in cases of – Chronic lymphocytic leukemia/small lymphocytic lymphoma – Marginal zone lymphomas (MZL) Gastric MALToma NGNC MALToma Nodal MZL Splenic MZL • PET/CT scan considered useful in selected cases only and initiation of treatment is not delayed for PET-CT in the following lymphomas – Burkitt lymphoma (B cell) – Lymphoblastic lymphoma – Adult T-cell leukemia/lymphoma	Subsequent response assessment based on Deauville scores and Lugano response criteria

(continued)

Table 27.2 (continued)

Cancers	Initial staging	Subsequent staging
Neuroendocrine tumor	PET/CT is indicated in • Well differentiated grade 3 NET • Extrapulmonary, poorly differentiated neuroendocrine carcinoma • Large or small cell carcinoma other than in the lungs • Poorly differentiated unknown primary • Adrenal gland tumours • Pheochromocytoma/paraganglioma • MEN 1/MEN 2 syndrome • Merkel cell carcinoma	
Esophagus cancer	PET/CT recommended for staging if no evidence of M1disease (staging)	• PET or PET/CT recommended to assess treatment response following both pre-operativeand definitive chemoradiation, assess >5–8 weeks after therapy
Gastric cancer	PET/CT recommended for staging if no evidence of M1disease (staging)	PET/CT as clinically indicated for • Restaging of unresectable or medically unfit patients following primary treatment • Radiation therapy planning-imaging studies including PET or PET/CT when available should be reviewed • Surveillance in stage II-III treated with neo-adjuvant ± adjuvant therapy
Gastrointestinal stromal tumors	PET/CT is considered for baseline staging in case of definitely unresectable or metastatic disease	PET/CT is considered during follow up of following cases • Marginally resectable or unresectable with risk of considerable morbidity: • Definitely unresectable or metastatic disease PET/CT is also considered for treatment response if – There is a need for rapid readout of activity with PET (after 2–4 weeks of therapy) compared to CT which can be performed only after 3 months – CT or MRI findings are ambiguous to progression or to reassess therapeutic response in that scenario
Hepatocellular carcinoma	PET/CT may be considered when there is an equivocal finding. Even though it is highly specific, it has limited sensitivity	
Gall bladder cancer and cholangiocarcinoma	PET/CT is not routinely recommended, but emerging evidence suggests that it may be useful for detection of regional nodal metastases and distant metastatic disease in patients with otherwise potentially resectable disease	

(continued)

Table 27.2 (continued)

Cancers	Initial staging	Subsequent staging
Pancreatic cancer	PET/CT scan may be considered after formal pancreatic CT protocol in high risk pancreatic adenocarcinoma patients to detect extra-pancreatic metastases. PET/CT can help in radiation therapy treatment planning	
Colorectal cancer	• PET/CT is not indicated in resectable disease • Consider PET/CT scan if – Potentially surgically curable M1 disease i.e., suspected or proven metastatic or synchronous adenocarcinoma – Planned for image guided liver directed therapies for liver metastases (ablation, radio-embolization)	– Consider PET/CT scan if – Potentially resectable, to identify other potential sites of disease which might render patient unresectabe for cure. – If CT and colonoscopy negative with serial CEA elevation (recurrence) PET/CT is not indicated for restaging/follow-up and surveillance
Anal cancer	• Consider PET/CT for initial staging (esp. pelvic node evaluation) of anal or anal marginal squamous cell carcinoma. (Note: PET/CT alone does not replace diagnostic CT) • PET/CT suggested for radiation treatment planning-simulation for anal carcinoma IMRT planning	• PET/CT for restaging for suspected recurrence
Kidney cancer	• PET not recognized in guideline	
Bladder cancer	• PET/CT is not recommended for non-muscle invasive bladder cancer • In case of muscle invasive bladder cancer it is beneficial for staging in selected patients with cT2 disease and may change management in patients with cT3 disease	PET/CT is indicated in following cases • Follow-up of patients, with or without cystectomy, if metastasis is suspected • Follow-up of cT4b disease, where mets is suspected, and for guiding biopsy • When bone mets are suspected (symptomatic patients, high risk, laboratory parameters suggestive), or additional extraosseous sites are expected/documented
Ovarian cancer	PET/CT may be indicated for indeterminate pelvic lesions if results will alter management	• PET/CT is clinically indicated for monitoring/follow-up patients of Stage I–IV (post primary treatment/surgical procedure) • PET/CT is indicated in rising CA-125 with or without usage of chemotherapy and in case of clinical relapse after chemotherapy
Uterine cancers	Endometrial carcinoma • PET/CT is indicated for initial staging if metastases is suspected in selected patients Uterine sarcoma • PET/CT is used only to clarify ambiguous findings	Endometrial carcinoma • PET/CT is indicated if recurrence/metastasis is suspected Uterine sarcoma • PET/CT is considered if metastases is suspected in selected patients

(continued)

Table 27.2 (continued)

Cancers	Initial staging	Subsequent staging
Cervical carcinoma	• PET/CT is indicated for initial work up in stage I (FIGO 1B1–1B3) and II–IVA • PET/CT is indicated for initial work up in small neuro-endocrine cervical carcinoma	• PET/CT is indicated for follow-up/surveillance (3–6 months post completion of therapy) in Stage I (FIGO 1B1-1B3) and II-IVA • PET/CT is indicated for treatment response assessment and surveillance in small neuro-endocrine cervical carcinoma Other indication • PET/CT suggested for radiation treatment planning in patients who are not surgically staged to define nodal volume coverage
Vulval carcinoma	PET/CT indicated for T2 or larger tumours or when metastases is suspected	PET/CT is indicated for follow-up and surveillance in case of recurrence or suspected metastasis
Testicular carcinoma	PET/CT is not indicated for staging of all testicular cancers and in any management level of non seminomatous germ cell tumor	In case of seminoma (stage IIA, IIB, IIC, III), PET/CT is indicated post treatment
Penile carcinoma	PET/CT is considered in case of suspected or inguinal lymph nodal disease	PET/CT to assess disease response and progression in a setting of patients with suspected inguinal lymph nodal disease
Skin cancers	Non-melanoma skin cancers • *Basal and squamous cell skin cancers:* PET not recognized in guideline • *Dermatofibrosarcoma protuberans:* PET not recognized in guideline • *Merkel cell carcinoma* Melanoma • PET/CT is recommended in Stage III (Category 2B recommendation) • PET/CT is recommended in Stage IV for baseline evaluation	Melanoma • PET/CT is recommended in stage IV for staging, recurrence and restaging
Thyroid cancer	• If stimulated Thyroglobin>2–5 ng/mL and I-131 imaging negative, then consider non-radioiodine imaging (e.g. neck ultrasound, neck CT, chest CT, FDG PET/CT) in case of papillary, follicular and Hurthle cell carcinoma • FDG PET/CT scan recommended for initial staging of anaplastic carcinoma	FDG PET/CT scan recommended for assessment of treatment response 3–6 months after initial therapy
Thymic carcinoma	PET/CT is clinically indicated for evaluation (metabolic characterization/staging) of a mediastinal mass	PET/CT is indicated for post chemotherapy preoperative assessment of thymoma/thymic carcinoma

(continued)

Table 27.2 (continued)

Cancers	Initial staging	Subsequent staging
Multiple myeloma	Whole Body PET-CT is indicated for initial staging (replaced skeletal survey). And if it is negative, MRI is used to differentiate smoldering myeloma from multiple myeloma	
Bone cancers	Osteosarcoma • PET scan and/or bone scan for staging Ewing sarcoma • PET scan and/or bone scan for staging Chondrosarcoma • PET not recognized in guideline Chordoma • Consider PET scan or bone scan (if PET negative) for staging	Osteosarcoma • Consider PET scan or bone scan for restaging after chemotherapy; consider PET and/or bone scan for surveillance Ewing sarcoma • Consider PET scan or bone scan for restaging after chemotherapy • Also consider PET scan or bone scan for surveillance
Sarcoma	In case of sarcoma sites in extremities, body wall and head and neck regions • PET may be useful in staging, prognostication and grading Rhabdomyosarcoma • PET scan may be useful for initial staging because of the possibility of nodal metastases and the appearance of unusual sites of initial metastases in adult patients	In case of sarcoma sites in extremities, body wall and head and neck regions • PET may be useful in determining response to systemic therapy Other indications: retroperitoneal/ intra-abdominal sarcomas; desmoid tumors • PET/CT helps in differentiating well differentiated and de-differentiated liposarcoma • PET/CT also helps in determining biopsy sites
Chronic myelogenous leukemia	• PET not recognized in guideline	
Acute myeloid leukemia	• PET/CT is considered when extramedullary involvement and/or lymphadenopathy is suspected	
Carcinoma of unknown primary	• Routine use of PET/CT is not recommended for initial evaluation of carcinoma of unknown primary, but PET/CT may be warranted in some situations, especially when considering local or regional therapy	• PET/CT is considered for remission surveillance • PET/MR better for localizing primary site, PET/CT and PET/ MR comparable for LN and distant metastases
Acute lymphoblastic leukemia	• PET/CT is considered when lymphomatous extramedullary involvement is suspected	• PET/CT is considered when for response assessment • Complete Response- PET is negative n size lesion • Partial response (PR)->50% decrease in sum of product of diameters (SPD) of mediastinal enlargement • Progressive disease (PD)- >25% increase in SPD of mediastinal enlargement • Non responder-failure to qualify for PD/PR

immense changes in the coming years especially when new radiotracers like ^{68}Ga-FAPI comes to main stream by covering up the disadvantages of FDG-PET/CT. The future of molecular imaging and therapies like 177Lu/225Ac-DOTATATE/PSMA are bringing forth our dream of precision oncology. Intensive research and widespread awareness will bring forth best treatment and overall improvement of the patient survival in oncological fields, especially with the introduction of ^{68}Ga-FAPI. More validated effort, time and goal -oriented research is warranted in this field.

Precept Box (Take Home Points)

- Nuclear medicine has both diagnostic and therapeutic roles in oncology. It explores the functional status of an organ or lesion and comprehends on tumor biology.
- Most extensively used radiotracer for PET/CT is ^{18}F-Fluorodeoxyglucose (FDG), a radiolabeled analog of glucose, which is taken by malignant cells due to increased hexokinase activity and increased GLUT expression.
- SUV is measurement of tissue uptake. Factors affecting SUV include dose calibration accuracy, injection time and imaging time (dose to scan time), patient weight (changes are expected in oncology patients), motion artifacts and blood glucose levels.
- Lung cancer—FDG-PET/CT is indicated for initial staging, RT planning and assessment of treatment response.
- Breast cancer—Important for metastatic work up and particularly recommended for all stages of inflammatory breast cancer.
- Lymphoma—Initial work up, response evaluation during and post systemic therapy and for RT planning. Deauville five-point scale is used in response assessment on PET/CT in Lymphoma patients.
- Prostate cancers—PSMA PET is used in initial diagnosis and response assessment and RT planning. Molecular therapies like 177Lu/225Ac-DOTATATE/PSMA are bringing forth our dream of precision oncology.
- Neuroendocrine tumors—^{68}Ga-DOTA-TOC, ^{68}Ga-DOTA-NOC and ^{68}Ga-DOTA-TATE peptides are used for staging and response evaluation.
- PET/CT has seen tremendous rise in its indications for various malignancies, clearly evident from its rising recommendation seen in recent NCCN (summarized in Table 27.2) and other oncological guidelines by various international associations.

Multiple Choice Questions

1. The most extensively used radiotracer for PET/CT is
   (a) ^{68}Ga-DOTA-NOC
   (b) PSMA PET/CT
   (c) ^{18}F-Fluorodeoxyglucose (FDG)
   (d) FAPI PET/CT
2. What is the principle of FDG?
   (a) Normal cells have increased glucose utilization due to upregulation of hexokinase activity and also exhibit an increased glucose transporter (GLUT) expression.
   (b) Malignant cells have increased glucose utilization due to upregulation of hexokinase activity and also exhibit an increased glucose transporter (GLUT) expression.
   (c) Uses electromagnetic fields to concentrate in malignant cells
   (d) None of above
3. SUV stands for
   (a) Standard uptake value
   (b) Standardized uptake value
   (c) Stratification uptake value

   (d) Stanford uptake value
4. TLG is calculated by
   (a) multiplying the mean SUV and MTV of the desired lesion
   (b) multiplying the mean MTV and SUV of the desired lesion
   (c) dividing the mean SUV and MTV of the desired lesion
   (d) dividing the mean MTV and SUV of the desired lesion
5. What radionucleotide is used for neuroendocrine tumors?
   (a) ^{68}Ga-DOTA-NOC
   (b) PSMA PET/CT
   (c) ^{18}F-Fluorodeoxyglucose (FDG)
   (d) FAPI PET/CT
6. What radionucleotide is used for prostate lesions?
   (a) ^{68}Ga-DOTA-NOC
   (b) PSMA PET/CT
   (c) ^{18}F-Fluorodeoxyglucose (FDG)
   (d) FAPI PET/CT
7. PET/CT is routinely recommended for which stages of breast cancer?
   (a) DCIS
   (b) Stage I
   (c) Stage IIA
   (d) Inflammatory breast cancer
8. What PET/CT scoring system is used for treatment response in lymphoma patient?
   (a) Lugano criteria
   (b) Peter's criteria
   (c) Stanford criteria
   (d) Deauville criteria

Answers

1. (c)
2. (b)
3. (b)
4. (a)
5. (a)
6. (b)
7. (d)
8. (d)

Acknowledgments We thank Dr. Sambit Sagar (Senior Resident, AIIMS New Delhi) for providing his encouragement, technical support and academic assistance during the process of creation of this chapter. We also thank Dr. Dharvesh (3rd year Junior Resident, AIIMS New Delhi) for proofreading and providing insights for easy comprehension of the entire chapter.

References

1. de Geus-Oei LF, Vriens D, van Laarhoven HW, van der Graaf WT, Oyen WJ. Monitoring and predicting response to therapy with ^{18}F-FDG PET in colorectal cancer: a systematic review. J Nucl Med. 2009;50(Suppl 1):43S–54S.
2. Hirata K, Tamaki N. Quantitative FDG PET assessment for oncology therapy. Cancer. 2021;13(4):869.
3. Parghane RV, Basu S. PET-CT based quantitative parameters for assessment of treatment response and disease activity in cancer and noncancerous disorders. PET Clin. 2022;17(3):465–78.
4. Doghish AS, Ismail A, Elrebehy MA, Elbadry AM, Mahmoud HH, Farouk SM, Serea GA, Elghany RA, El-Halwany KK, Alsawah AO, Dewidar HI, El-Mahdy HA. A study of miRNAs as cornerstone in lung cancer pathogenesis and therapeutic resistance: a focus on signaling pathways interplay. Pathol Res Pract. 2022;237:154053.
5. Lu C, Onn A, Vaporciyan AA, et al. Chapter 84: cancer of the lung. In: Holland-Frei cancer medicine. 9th ed. Hoboken: Wiley; 2017.
6. Saif MW, Tzannou I, Makrilia N, Syrigos K. Role and cost effectiveness of PET/CT in management of patients with cancer. Yale J Biol Med. 2010;83(2):53–65.
7. Agrawal A, Rangarajan V. Appropriateness criteria of FDG PET/CT in oncology. Indian J Radiol Imaging. 2015;25(2):88–101. https://doi.org/10.4103/0971-3026.155823. PMID: 25969632; PMCID: PMC4419439.
8. Yang RM, Li L, Wei XH, Guo YM, Huang YH, Lai LS, Chen AM, Liu GS, Xiong WF, Luo LP, Jiang XQ. Differentiation of central lung cancer from atelectasis: comparison of diffusion-weighted MRI with PET/CT. PLoS One. 2013;8(4):e60279.
9. Du D, Gu J, Chen X, Lv W, Feng Q, Rahmim A, Wu H, Lu L. Integration of PET/CT radiomics and semantic features for differentiation between active pulmonary tuberculosis and lung cancer. Mol Imaging Biol. 2021;23(2):287–98.
10. Alkabban FM, Ferguson T. Breast cancer. In: StatPearls. Treasure Island: StatPearls Publishing; 2022. https://www.ncbi.nlm.nih.gov/books/NBK482286/.
11. NCCN. National Comprehensive Cancer Network guidelines: breast cancer. 2021. https://jnccn.org/

view/journals/jnccn/19/5/articlep484.xml?ArticleBodyColorStyles=inline%20pdf. Accessed 6 August 2022.
12. Orsaria P, Varvaras D, Vanni G, Pagnani G, Scaggiante J, Frusone F, Granai AV, Petrella G, Buonomo OC. Nodal status assessment in breast cancer: strategies of clinical grounds and quality of life implications. Int J Breast Cancer. 2014;2014:469803.
13. Yang SK, Cho N, Moon WK. The role of PET/CT for evaluating breast cancer. Korean J Radiol. 2007;8(5):429–37.
14. Groheux D, Hindie E. Breast cancer: initial workup and staging with FDG PET/CT. Clin Translat Imaging. 2021;9(3):221–31.
15. Ming Y, Wu N, Qian T, Li X, Wan DQ, Li C, Li Y, Wu Z, Wang X, Liu J, Wu N. Progress and future trends in PET/CT and PET/MRI molecular imaging approaches for breast cancer. Front Oncol. 2020;10:1301.
16. Pennant M, Takwoingi Y, Pennant L, Davenport C, Fry-Smith A, Eisinga A, Andronis L, Arvanitis T, Deeks J, Hyde C. A systematic review of positron emission tomography (PET) and positron emission tomography/computed tomography (PET/CT) for the diagnosis of breast cancer recurrence. Health Technol Assess. 2010;14(50):1–103.
17. Fletcher JW, Djulbegovic B, Soares HP, et al. Recommendations on the use of ^{18}F-FDG PET in oncology. J Nucl Med. 2008;49:480–508.
18. Fowler AM, Strigel RM. Clinical advances in PET-MRI for breast cancer. Lancet Oncol. 2022;23(1):e32–43.
19. Ulaner GA, Jhaveri K, Chandarlapaty S, Hatzoglou V, Riedl CC, Lewis JS, Mauguen A. Head-to-head evaluation of ^{18}F-FES and ^{18}F-FDG PET/CT in metastatic invasive lobular breast cancer. J Nucl Med. 2021;62(3):326–31.
20. The lymphoma guide information for patients and caregivers (PDF). Leukemia and Lymphoma Society. 2013. Archived (PDF) from the original on 14 July 2014. Retrieved 20 June 2014.
21. Lister TA, Crowther D, Sutcliffe SB, et al. Report of a committee convened to discuss the evaluation and staging of patients with Hodgkin's disease: Cotswolds meeting. J Clin Oncol. 1989;7:1630–6.
22. Tubiana M. Report of the committee on Hodgkin's disease staging classification. Cancer Res. 1994;31:1860–1.
23. Cheson BD, Fisher RI, Barrington SF, Cavalli F, Schwartz LH, Zucca E, Lister TA. Recommendations for initial evaluation, staging, and response assessment of Hodgkin and Non-Hodgkin lymphoma: the Lugano classification. J Clin Oncol. 2014;32:3059–67.
24. Juweid ME, Stroobants S, Hoekstra OS, Mottaghy FM, Dietlein M, Guermazi A, Wiseman GA, Kostakoglu L, Scheidhauer K, Buck A, et al. Use of positron emission tomography for response assessment of lymphoma: Consensus of the Imaging Subcommittee of International Harmonization Project in Lymphoma. J Clin Oncol. 2007;25:571–8.
25. Meignan M, Gallamini A, Meignan M, Gallamini A, Haioun C. Report on the first international workshop on interim-PET-scan in lymphoma. Leuk Lymphoma. 2009;50:1257–60.
26. Gallamini A, Barrington SF, Biggi A, Chauvie S, Kostakoglu L, Gregianin M, Meignan M, Mikhaeel GN, Loft A, Zaucha JM, et al. The predictive role of interim positron emission tomography for Hodgkin lymphoma treatment outcome is confirmed using the interpretation criteria of the Deauville five-point scale. Haematologica. 2014;99:1107–13.
27. Bröckelmann PJ, Borchmann P, Engert A. Current and future immunotherapeutic approaches in Hodgkin lymphoma. Leuk Lymphoma. 2016;57:2014–24. https://doi.org/10.1080/10428194.2016.1185789.
28. Cheson BD, Ansell S, Schwartz L, Gordon LI, Advani R, Jacene HA, Hoos A, Barrington SF, Armand P. Refinement of the Lugano classification lymphoma response criteria in the era of immunomodulatory therapy. Blood. 2016;128:2489–96. https://doi.org/10.1182/blood-2016-05-718528.
29. Kanoun S, Rossi C, Berriolo-Riedinger A, Dygai-Cochet I, Cochet A, Humbert O, Toubeau M, Ferrant E, Brunotte F, Casasnovas R-O. Baseline metabolic tumour volume is an independent prognostic factor in Hodgkin lymphoma. Eur J Nucl Med Mol Imaging. 2014;41:1735–43.
30. Cottereau AS, Versari A, Loft A, Casasnovas O, Bellei M, Ricci R, Bardet S, Castagnoli A, Brice P, Raemaekers J, et al. Prognostic value of baseline metabolic tumor volume in early stage Hodgkin's lymphoma in the standard arm of H10 trial. Blood. 2018;131:1456–63.
31. Zelenetz AD, Maragulia J, Horwitz SM. Baseline staging evaluation in lymphoma: the role of FDG PET, CT, and bone marrow biopsy. Blood. 2011;118(21):2640.
32. Weiler-Sagie M, Bushelev O, Epelbaum R, et al. (18) F FDG avidity in lymphoma readdressed: a study of 766 patients. J Nucl Med. 2010;51(1):25–30.
33. Khan AB, Barrington SF, Mikhaeel NG, et al. PET-CT staging of DLBCL accurately identifies and provides new insight into the clinical significance of bone marrow involvement. Blood. 2013;122:61–7.
34. Carr R, Fanti S, Paez D, et al. Prospective international cohort study demonstrates inability of interim PET to predict treatment failure in diffuse large B-cell lymphoma. J Nucl Med. 2014;55:1936–44.
35. Pregno P, Chiapella A, Bello M, et al. Interim 18-FDG-PET/CT failed to predict the outcome in diffuse large B-cell lymphoma patients treated at the diagnosis with rituximab-CHOP. Blood. 2012;119:2066–73.
36. Terezakis SA, Hunt MA, Kowalski A, et al. [^{18}F]FDG-positron emission tomography coregistration with computed tomography scans for radiation treatment planning of lymphoma and hematologic malignancies. Int J Radiat Oncol Biol Phys. 2011;81:615–22.
37. Derenzini E, Musuraca G, Fanti S, Stefoni V, Tani M, Alinari L, Venturini F, Gandolf L, Baccarani M, Zinzani PL. Pretransplantation positron emis-

sion tomography scan is the main predictor of autologous stem cell transplantation outcome in aggressive B-cell non-Hodgkin lymphoma. Cancer. 2008;113(9):2496–503.
38. Sauter CS, Matasar MJ, Meikle J, Schoder H, Ulaner GA, Migliacci JC, Hilden P, Devlin SM, Zelenetz AD, Moskowitz CH. Prognostic value of FDG-PET prior to autologous stem cell transplantation for relapsed and refractory diffuse large B-cell lymphoma. Blood. 2015;125(16):2579–81.
39. Casasnovas RO, Ysebaert L, Thieblemont C, Bachy E, Feugier P, Delmer A, Tricot S, Gabarre J, Andre M, Fruchart C, Mounier N, Delarue R, Meignan M, Berriolo-Riedinger A, Bardet S, Emile JF, Jais JP, Haioun C, Tilly H, Morschhauser F. FDG-PET-driven consolidation strategy in diffuse large B-cell lymphoma: final results of a randomized phase 2 study. Blood. 2017;130(11):1315–26. https://doi.org/10.1182/blood-2017-02-766691. Epub 2017 Jul 12.
40. American Cancer Society. Cancer Facts & Figures 2017. 2017. https://www.cancer.org/content/dam/cancer-org/research/cancer-facts-and-statistics/annual-cancer-facts-and-fgures/2017/cancer-facts-and-fgures-2017.pdf. Accessed 2 September 2019.
41. Elimova E, Wang X, Etchebehere E, Shiozaki H, Shimodaira Y, Wadhwa R, Planjery V, Charalampakis N, Blum MA, Hofstetter W, Lee JH, Weston BR, Bhutani MS, Rogers JE, Maru D, Skinner HD, Macapinlac HA, Ajani JA. 18-fuorodeoxy-glucose positron emission computed tomography as predictive of response after chemoradiation in oesophageal cancer patients. Eur J Cancer. 2015;51(17):2545–52.
42. Reid BJ, Blount PL, Feng Z, Levine DS. Optimizing endoscopic biopsy detection of early cancers in Barrett's high grade dysplasia. Am J Gastroenterol. 2000;95(11):3089–96.
43. Lerut T, Flamen P, Ectors N, Van Cutsem E, Peeters M, Hiele M, De Wever W, Coosemans W, Decker G, De Leyn P, Denefe G, Van Raemdonck D, Mortelmans L. Histopathologic validation of lymph node staging with FDG-PET scan in cancer of the esophagus and gastroesophageal junction: a prospective study based on primary surgery with extensive lymphadenectomy. Ann Surg. 2000;232(6):743–52.
44. Heeren PA, Jager PL, Bongaerts F, van Dullemen H, Sluiter W, Plukker JT. Detection of distant metastases in esophageal cancer with (18)F-FDG PET. J Nucl Med. 2004;45(6):980–7.
45. Amin MB, Edge SB, American Joint Committee on Cancer. AJCC cancer staging manual. Chicago: American Joint Committee on Cancer; 2017.
46. Lordick F. Optimizing neoadjuvant chemotherapy through the use of early response evaluation by positron emission tomography. Recent Results Cancer Res. 2012;196:201–11.
47. Goense L, van Rossum PS, Reitsma JB, Lam MG, Meijer GJ, van Vulpen M, Ruurda JP, van Hillegersberg R. Diagnostic performance of (1)(8) F-FDG PET and PET/CT for the detection of recurrent esophageal cancer after treatment with curative intent: a systematic review and meta-analysis. J Nucl Med. 2015;56(7):995–1002.
48. Weber MA, Bender K, von Gall CC, Stange A, Grunberg K, Ott K, Haberkorn U, Kauczor HU, Zechmann C. Assessment of difusion-weighted MRI and ^{18}F-fuoro-deoxyglucose PET/CT in monitoring early response to neoadjuvant chemotherapy in adenocarcinoma of the esophagogastric junction. J Gastrointestin Liver Dis. 2013;22(1):45–52.
49. Ferlay J, Ervik M, Lam F, Colombet M, Mery L, Piñeros M, et al. Global cancer observatory: cancer today. Lyon: International Agency for Research on Cancer; 2020.
50. De Rafele E, Mirarchi M, Cuicchi D, Lecce F, Cola B. Evolving role of FDG-PET/CT in prognostic evaluation of resectable gastric cancer. World J Gastroenterol. 2017;23(38):6923–6.
51. Van Cutsem E, Sagaert X, Topal B, Haustermans K, Prenen H. Gastric cancer. Lancet. 2016;388(10060):2654–64.
52. Minamimoto R, Senda M, Jinnouchi S, Terauchi T, Yoshida T, Inoue T. Performance profile of a FDG-PET cancer screening program for detecting gastric cancer: results from a nationwide Japanese survey. Jpn J Radiol. 2014;32(5):253–9.
53. Lim JS, Yun MJ, Kim MJ, Hyung WJ, Park MS, Choi JY, Kim TS, Lee JD, Noh SH, Kim KW. CT and PET in stomach cancer: preoperative staging and monitoring of response to therapy. Radiographics. 2006;26(1):143–56.
54. Yun M. Imaging of gastric cancer metabolism using ^{18}F-FDG PET/CT. J Gastric Cancer. 2014;14(1):1–6.
55. Suttie SA, Welch AE, Park KG. Positron emission tomography for monitoring response to neoadjuvant therapy in patients with oesophageal and gastro-oesophageal junction carcinoma. Eur J Surg Oncol. 2009;35(10):1019–29.
56. Cayvarlı H. The role of ^{18}F-FDG PET/CT in the evaluation of gastric cancer recurrence. Mol Imaging Radionucl Ther. 2014;23(3):76–83.
57. Zou H, Zhao Y. 18FDG PET-CT for detecting gastric cancer recurrence after surgical resection: a meta analysis. Surg Oncol. 2013;22(3):162–6.
58. NCCN. National Comprehensive Cancer Network guidelines: gastric cancer. 2021. https://www.nccn.org/professionals/physician_gls/pdf/gastric.pdf. Accessed 6 January 2022.
59. Kim J, Lim ST, Na CJ, et al. Pretreatment F-18 FDG PET/CT parameters to evaluate progression-free survival in gastric cancer. Nucl Med Mol Imaging. 2014;48(1):33–40.
60. Song BI, Kim HW, Won KS, et al. Preoperative standardized uptake value of metastatic lymph nodes measured by ^{18}F-FDG PET/CT improves the prediction of prognosis in gastric cancer. Medicine. 2015;94(26):e1037.
61. Chan BPH, Patel R, Mbuagbaw L, Thabane L, Yaghoobi M. EUS versus magnetic resonance imaging in staging rectal adenocarcinoma: a diagnostic test accuracy meta-analysis. Gastrointest Endosc.

2019;90(2):196–203. https://doi.org/10.1016/j.gie.2019.04.217.
62. Floriani I, Torri V, Rulli E, Garavaglia D, Compagnoni A, Salvolini L, Giovagnoni A. Performance of imaging modalities in diagnosis of liver metastases from colorectal cancer: a systematic review and meta-analysis. J Magn Reson Imaging. 2010;31(1):19–31.
63. Patel S, McCall M, Ohinmaa A, Bigam D, Dryden DM. Positron emission tomography/computed tomographic scans compared to computed tomographic scans for detecting colorectal liver metastases: a systematic review. Ann Surg. 2011;253(4):666–71.
64. Petersen RK, Hess S, Alavi A, Hoilund-Carlsen PF. Clinical impact of FDG-PET/CT on colorectal cancer staging and treatment strategy. Am J Nucl Med Mol Imaging. 2014;4(5):471–82.
65. Park IJ, Kim HC, Yu CS, Ryu MH, Chang HM, Kim JH, Ryu JS, Yeo JS, Kim JC. Efficacy of PET/CT in the accurate evaluation of primary colorectal carcinoma. Eur J Surg Oncol. 2006;32(9):941–7.
66. Venkat SR, Mohan PP, Gandhi RT. Colorectal liver metastasis: overview of treatment paradigm highlighting the role of ablation. AJR Am J Roentgenol. 2018;210(4):883–90.
67. Sacks A, Peller PJ, Surasi DS, Chatburn L, Mercier G, Subramaniam RM. Value of PET/CT in the management of liver metastases, part 1. AJR Am J Roentgenol. 2011;197(2):256–9.
68. Niekel MC, Bipat S, Stoker J. Diagnostic imaging of colorectal liver metastases with CT, MR imaging, FDG PET, and/or FDG PET/CT: a meta-analysis of prospective studies including patients who have not previously undergone treatment. Radiology. 2010;257(3):674–84.
69. Tsunoda Y, Ito M, Fujii H, Kuwano H, Saito N. Preoperative diagnosis of lymph node metastases of colorectal cancer by FDG-PET/CT. Jpn J Clin Oncol. 2008;38(5):347–53.
70. Glazer ES, Beaty K, Abdalla EK, Vauthey JN, Curley SA. Effectiveness of positron emission tomography for predicting chemotherapy response in colorectal cancer liver metastases. Arch Surg. 2010;145(4):340–5. https://doi.org/10.1001/archsurg.2010.41.
71. Metser U, You J, McSweeney S, Freeman M, Hendler A. Assessment of tumor recurrence in patients with colorectal cancer and elevated carcinoembryonic antigen level: FDG PET/CT versus contrast-enhanced 64-MDCT of the chest and abdomen. AJR Am J Roentgenol. 2010;194(3):766–71. https://doi.org/10.2214/ajr.09.3205.
72. Even-Sapir E, Parag Y, Lerman H, Gutman M, Levine C, Rabau M, Figer A, Metser U. Detection of recurrence in patients with rectal cancer: PET/CT after abdominoperineal or anterior resection. Radiology. 2004;232(3):815–22.
73. Tucker GT, Casey C, Phillips PJ, Connor H, Ward JD, Woods HF. Metformin kinetics in healthy subjects and in patients with diabetes mellitus. Br J Clin Pharmacol. 1981;12(2):235–46.
74. Whiteford MH, Whiteford HM, Yee LF, Ogunbiyi OA, Dehdashti F, Siegel BA, Birnbaum EH, Fleshman JW, Kodner IJ, Read TE. Usefulness of FDG-PET scan in the assessment of suspected metastatic or recurrent adenocarcinoma of the colon and rectum. Dis Colon Rectum. 2000;43(6):759–67.
75. Kaur S, Baine MJ, Jain M, Sasson AR, Batra SK. Early diagnosis of pancreatic cancer: challenges and new developments. Biomark Med. 2012;6:597–612.
76. Zakharova OP, Karmazanovsky GG, Egorov VI. Pancreatic adenocarcinoma: outstanding problems. World J Gastrointest Surg. 2012;4:104–13.
77. Asagi A, Ohta K, Nasu J, Tanada M, Nadano S, Nishimura R, Teramoto N, Yamamoto K, Inoue T, Iguchi H. Utility of contrast-enhanced FDG-PET/CT in the clinical management of pancreatic cancer: impact on diagnosis, staging, evaluation of treatment response, and detection of recurrence. Pancreas. 2013;42:11–9.
78. Chang JS, Choi SH, Lee Y, Kim KH, Park JY, Song SY, Cho A, Yun M, Lee JD, Seong J. Clinical usefulness of (1)(8)F-fluorodeoxyglucose-positron emission tomography in patients with locally advanced pancreatic cancer planned to undergo concurrent chemoradiation therapy. Int J Radiat Oncol Biol Phys. 2014;90:126–33.
79. Strobel K, Heinrich S, Bhure U, Soyka J, Veit-Haibach P, Pestalozzi BC, Clavien PA, Hany TF. Contrast-enhanced ^{18}F-FDG PET/CT: 1-stop-shop imaging for assessing the resectability of pancreatic cancer. J Nucl Med. 2008;49:1408–13.
80. Heinrich S, Goerres GW, Schafer M, Sagmeister M, Bauerfeind P, Pestalozzi BC, Hany TF, von Schulthess GK, Clavien PA. Positron emission tomography/computed tomography influences on the management of resectable pancreatic cancer and its cost-effectiveness. Ann Surg. 2005;242:235–43.
81. National Comprehensive Cancer Network. Pancreatic adenocarcinoma guidelines version 2.2021, 2. Plymouth: National Comprehensive Cancer Network; 2021.
82. Katz MH, Fleming JB, Bhosale P, Varadhachary G, Lee JE, Wolff R, Wang H, Abbruzzese J, Pisters PW, Vauthey JN, et al. Response of borderline resectable pancreatic cancer to neoadjuvant therapy is not reflected by radiographic indicators. Cancer. 2012;118:5749–56.
83. Orlando G, Pilone V, Vitiello A, Gervasi R, Lerose MA, Silecchia G, Puzziello A. Gastric cancer following bariatric surgery: a review. Surg Laparosc Endosc Percutan Tech. 2014;24:400–5.
84. Panda A, Garg I, Truty MJ, Kline TL, Johnson MP, Ehman EC, Suman G, Anaam DA, Kemp BJ, Johnson GB, et al. Borderline resectable and locally advanced pancreas cancer: FDG PET/MRI and CT tumor metrics for assessment of neoadjuvant therapy pathologic response and prediction of survival. AJR Am J Roentgenol. 2020;217(3):730–40.
85. Lai JP, Yue Y, Zhang W, Zhou Y, Frishberg D, Jamil LH, Mirocha JM, Guindi M, Balzer B, Bose S, et al.

Comparison of endoscopic ultrasound guided fine needle aspiration and PET/CT in preoperative diagnosis of pancreatic adenocarcinoma. Pancreatology. 2017;17:617–22.
86. Yeh R, Dercle L, Garg I, Wang ZJ, Hough DM, Goenka AH. The role of ^{18}F-FDG PET/CT and PET/MRI in pancreatic ductal adenocarcinoma. Abdom Radiol. 2018;43:415–34.
87. Ghaneh P, Hanson R, Titman A, Lancaster G, Plumpton C, Lloyd-Williams H, Yeo ST, Edwards RT, Johnson C, Abu Hilal M, et al. PET-PANC: multicentre prospective diagnostic accuracy and health economic analysis study of the impact of combined modality ^{18}fluorine-2-fluoro-2-deoxy-d-glucose positron emission tomography with computed tomography scanning in the diagnosis and management of pancreatic cancer. Health Technol Assess. 2018;22:1–114.
88. Nguyen AH, Melstrom LG. Use of imaging as staging and surgical planning for pancreatic surgery. Hepatobiliary Surg Nutr. 2020;9:603–14.
89. Lee JW, Choi M, Choi JY. Impact of F-18 Fluorodeoxyglucose PET/CT and PET/MRI on initial staging and changes in management of pancreatic ductal adenocarcinoma: a systemic review and meta-analysis. Diagnostics. 2020;10:952.
90. Pergolini I, Crippa S, Salgarello M, Belfiori G, Partelli S, Ruffo G, Pucci A, Zamboni G, Falconi M. SUV_{max} after (18)fluoro-deoxyglucose positron emission tomography/computed tomography: a tool to define treatment strategies in pancreatic cancer. Dig Liver Dis. 2018;50:84–90.
91. Moon D, Kim H, Han Y, Byun Y, Choi Y, Kang J, Kwon W, Jang JY. Preoperative carbohydrate antigen 19–9 and standard uptake value of positron emission tomography-computed tomography as prognostic markers in patients with pancreatic ductal adenocarcinoma. J Hepatobiliary Pancreat Sci. 2020;29(10):1133–41.
92. Yokose T, Kitago M, Matsusaka Y, Masugi Y, Shinoda M, Yagi H, Abe Y, Oshima G, Hori S, Endo Y, et al. Usefulness of (18) F-fluorodeoxyglucose positron emission tomography/computed tomography for predicting the prognosis and treatment response of neoadjuvant therapy for pancreatic ductal adenocarcinoma. Cancer Med. 2020;9:4059–68.
93. Zimmermann C, Distler M, Jentsch C, Blum S, Folprecht G, Zophel K, Polster H, Troost EGC, Abolmaali N, Weitz J, et al. Evaluation of response using FDG-PET/CT and diffusion weighted MRI after radiochemotherapy of pancreatic cancer: a non-randomized, monocentric phase II clinical trial-PaCa-DD-041 (Eudra-CT 2009–011968–11). Strahlenther Onkol. 2021;197:19–26.
94. Toyama Y, Hotta M, Motoi F, Takanami K, Minamimoto R, Takase K. Prognostic value of FDG-PET radiomics with machine learning in pancreatic cancer. Sci Rep. 2020;10:17024.
95. Lim CH, Cho YS, Choi JY, Lee KH, Lee JK, Min JH, Hyun SH. Imaging phenotype using (18)F-fluorodeoxyglucose positron emission tomography-based radiomics and genetic alterations of pancreatic ductal adenocarcinoma. Eur J Nucl Med Mol Imaging. 2020;47:2113–22.
96. Abunahel BM, Pontre B, Kumar H, Petrov MS. Pancreas image mining: a systematic review of radiomics. Eur Radiol. 2020;31:3447–67.
97. Taveira M. Comparison of (68)Ga-FAPI versus (18) F-FDG PET/CT for initial cancer staging. Radiol Imaging Cancer. 2021;3:e219007.
98. Rohrich M, Naumann P, Giesel FL, Choyke PL, Staudinger F, Wefers A, Liew DP, Kratochwil C, Rathke H, Liermann J, et al. Impact of (68)Ga-FAPI PET/CT imaging on the therapeutic management of primary and recurrent pancreatic ductal adenocarcinomas. J Nucl Med. 2021;62:779–86.
99. Lee JM, Yoon JH, Joo I, Woo HS. Recent advances in CT and MR imaging for evaluation of hepatocellular carcinoma. Liver Cancer. 2012;1(1):22–40.
100. Torizuka T, Tamaki N, Inokuma T, Magata Y, Sasayama S, Yonekura Y, et al. In vivo assessment of glucose metabolism in hepatocellular carcinoma with FDG-PET. J Nucl Med. 1995;36(10):1811–7.
101. Wudel LJ Jr, Delbeke D, Morris D, Rice M, Washington MK, Shyr Y, et al. The role of [^{18}F] fuorodeoxyglucose positron emission tomography imaging in the evaluation of hepatocellular carcinoma. Am Surg. 2003;69(2):117–26.
102. Garin E, Le Jeune F, Devillers A, Cuggia M, de Lajarte-Thirouard AS, Bouriel C, et al. Predictive value of ^{18}F-FDG PET and somatostatin receptor scintigraphy in patients with metastatic endocrine tumors. J Nucl Med. 2009;50(6):858–64.
103. Lee SD, Kim SH, Kim YK, Kim C, Kim SK, Han SS, et al. (18) F-FDG-PET/CT predicts early tumor recurrence in living donor liver transplantation for hepatocellular carcinoma. Transpl Int. 2013;26(1):50–60.
104. Huang WY, Kao CH, Huang WS, Chen CM, Chang LP, Lee MS, et al. 18F-FDG PET and combined ^{18}F-FDG-contrast CT parameters as predictors of tumor control for hepatocellular carcinoma after stereotactic ablative radiotherapy. J Nucl Med. 2013;54(10):1710–6.
105. Lee JD, Yang WI, Park YN, Kim KS, Choi JS, Yun M, et al. Different glucose uptake and glycolytic mechanisms between hepatocellular carcinoma and intrahepatic mass-forming cholangiocarcinoma with increased (18) F-FDG uptake. J Nucl Med. 2005;46(10):1753–9.
106. Anderson CD, Rice MH, Pinson CW, Chapman WC, Chari RS, Delbeke D. Fluorodeoxyglucose PET imaging in the evaluation of gallbladder carcinoma and cholangiocarcinoma. J Gastrointest Surg. 2004;8(1):90–7.
107. Kim YJ, Yun M, Lee WJ, Kim KS, Lee JD. Usefulness of ^{18}FFDG PET in intrahepatic cholangiocarcinoma. Eur J Nucl Med Mol Imaging. 2003;30(11):1467–72.
108. Suzuki H, Komuta M, Bolog A, Yokobori T, Wada S, Araki K, et al. Relationship between 18-F-fluoro-deoxy-D-glucose uptake and expression of glu-

cose transporter 1 and pyruvate kinase M2 in intrahepatic cholangiocarcinoma. Dig Liver Dis. 2015;47(7):590–6.
109. Jadvar H, Henderson RW, Conti PS. [F-18] fluorodeoxyglucose positron emission tomography and positron emission tomography: computed tomography in recurrent and metastatic cholangiocarcinoma. J Comput Assist Tomogr. 2007;31(2):223–8.
110. Corvera CU, Blumgart LH, Akhurst T, DeMatteo RP, D'Angelica M, Fong Y, et al. 18F-fluorodeoxyglucose positron emission tomography influences management decisions in patients with biliary cancer. J Am Coll Surg. 2008;206(1):57–655.
111. Ramos-Font C, Gómez-Rio M, Rodríguez-Fernández A, JiménezHeffernan A, Sánchez Sánchez R, Llamas-Elvira JM. Ability of FDG-PET/CT in the detection of gallbladder cancer. J Surg Oncol. 2014;109:218–24.
112. Kanthan R, Senger JL, Ahmed S, Kanthan SC. Gallbladder cancer in the 21st century. J Oncol. 2015;2015:967472.
113. Albazaz R, Patel CN, Chowdhury FU, Scarsbrook AF. Clinical impact of FDG PET-CT on management decisions for patients with primary biliary tumours. Insights Imaging. 2013;4:691–700.
114. Annunziata S, Pizzuto DA, Caldarella C, Galiandro F, Sadeghi R, Treglia G. Diagnostic accuracy of fluorine-18-fluorodeoxyglucose positron emission tomography in gallbladder cancer: a meta-analysis. World J Gastroenterol. 2015;21(40):11481–8.
115. Parida GK, Panda RA, Agrawal K. Impact of fluorine-18-fluorodeoxyglucose PET/computed tomography in staging of patients with gallbladder cancer: a systematic review and meta-analysis. Nucl Med Commun. 2021;42(8):846–54.
116. Chun YJ, Jeung HC, Park HS, Park JS, Rha SY, Choi HJ, Lee JH, Jeon TJ. Significance of metabolic tumor volume and total lesion glycolysis measured using ^{18}F-FDG PET/CT in locally advanced and metastatic gallbladder carcinoma. Yonsei Med J. 2019;60(7):604–10. https://doi.org/10.3349/ymj.2019.60.7.604. PMID: 31250573; PMCID: PMC6597472.
117. Goel S, Aggarwal A, Iqbal A, Gupta M, Rao A, Singh S. 18-FDG PET-CT should be included in preoperative staging of gall bladder cancer. Eur J Surg Oncol. 2020;46(9):1711–6.
118. Valle JW, Borbath I, Khan SA, Huguet F, Gruenberger T, Arnold D, ESMO Guidelines Committee. Biliary cancer: ESMO clinical practice guidelines for diagnosis, treatment and follow-up. Ann Oncol. 2016;27(suppl 5):28–37.
119. Benson AB, D'Angelica MI, Abbott DE, Anaya DA, Anders R, Are C, Bachini M, Borad M, Brown D, Burgoyne A, Chahal P, Chang DT, Cloyd J, Covey AM, Glazer ES, Goyal L, Hawkins WG, Iyer R, Jacob R, Kelley RK, Kim R, Levine M, Palta M, Park JO, Raman S, Reddy S, Sahai V, Schefter T, Singh G, Stein S, Vauthey JN, Venook AP, Yopp A, McMillian NR, Hochstetler C, Darlow SD. Hepatobiliary cancers, version 2.2021, NCCN clinical practice guidelines in oncology. J Natl Compr Cancer Netw. 2021;19(5):541–65.
120. Siegel RL, Miller KD, Jemal A. Cancer statistics, 2019. CA Cancer J Clin. 2019;72(1):7–33.
121. Cronin KA, Lake AJ, Scott S, Sherman RL, Noone AM, Howlader N, et al. Annual report to the nation on the status of cancer, part I: national cancer statistics. Cancer. 2018;124(13):2785–800.
122. Ng CS, et al. Renal cell carcinoma: diagnosis, staging, and surveillance. AJR Am J Roentgenol. 2008;191(4):1220–32.
123. Kamel EM, Jichlinski P, Prior JO, Meuwly JY, Delaloye JF, Vaucher L, et al. Forced diuresis improves the diagnostic accuracy of ^{18}F-FDG PET in abdominopelvic malignancies. J Nucl Med. 2006;47(11):1803–7.
124. Wang H-Y, Ding H-J, Chen J-H, Chao C-H, Lu Y-Y, Lin W-Y, et al. Meta-analysis of the diagnostic performance of [^{18}F]FDG-PET and PET/CT in renal cell carcinoma. Cancer Imaging. 2012;12(3):464–74.
125. Fletcher JW, Djulbegovic B, Soares HP, Siegel BA, Lowe VJ, Lyman GH, et al. Recommendations on the use of ^{18}F-FDG PET in oncology. J Nucl Med. 2008;49(3):480–508.
126. Vikram R, et al. ACR appropriateness criteria renal cell carcinoma staging. J Am Coll Radiol. 2016;13(5):518–25.
127. Ma H, Shen G, Liu B, et al. Diagnostic performance of ^{18}F-FDG PET or PET/CT in restaging renal cell carcinoma: a systematic review and meta-analysis. Nucl Med Commun. 2017;38:156–63.
128. Win AZ, Aparici CM. Clinical effectiveness of (18) f-fluorodeoxyglucosepositron emission tomography/computed tomography in management of renal cell carcinoma: a single institution experience. World J Nucl Med. 2015;14(1):36–40.
129. Shreve P, Chiao PC, Humes HD, Schwaiger M, Gross MD. Carbon-11-acetate PET imaging in renal disease. J Nucl Med. 1995;36:1595–601.
130. Kotzerke J, Linne C, Meinhardt M, et al. [1-(11)C] acetate uptake is not increased in renal cell carcinoma. Eur J Nucl Med Mol Imaging. 2007;34:884–8.
131. Divgi CR, Uzzo RG, Gatsonis C, et al. Positron emission tomography/computed tomography identification of clear cell renal cell carcinoma: results from the REDECT trial. J Clin Oncol. 2013;31:187–94.
132. Cohen D, Zhou M. Molecular genetics of familial renal cell carcinoma syndromes. Clin Lab Med. 2005;25:259–77.
133. Rhee H, Blazak J, Tham CM, Ng KL, Shepherd B, Lawson M, Preston J, Vela I, Thomas P, Wood S. Pilot study: use of gallium-68 PSMA PET for detection of metastatic lesions in patientswith renal tumour. EJNMMI Res. 2016;6:76.
134. National Institute of Health ClinicalTrials.gov. STUDY of PSMAtargeted^{18}F-DCFPyL PET/CT in the evaluation of patients with renal cell carcinoma. Accessed 4 Feb 2022.

135. National Institute of Health ClinicalTrials.gov preliminary evaluationof uptake of [^{68}Ga]P16-093 in Metastatic Prostate and Renal Cancer. Accessed 4 Feb 2022.
136. Kaufman DS, Shipley WU, Feldman AS. Bladder cancer. Lancet. 2009;374:239–49.
137. Bouchelouche K, Turkbey B, Choyke PL. PET/CT and MRI in bladder cancer. J Cancer Sci Ther. 2012;14(1):11. https://doi.org/10.4172/1948-5956.S14-001.
138. Einerhand SMH, van Gennep EJ, Mertens LS, et al. ^{18}F-fluoro-2-deoxyD-glucose positron emission tomography/computed tomography in muscle-invasive bladder cancer. Curr Opin Urol. 2020;30:654–64.
139. Wang N, Jiang P, Lu Y. Is fluorine-18 fluorodeoxyglucose positron emission tomography useful for detecting bladder lesions? A meta-analysis of the literature. Urol Int. 2014;92:143–9.
140. Lodde M, Lacombe L, Friede J, Morin F, Saourine A, Fradet Y. Evaluation of fluorodeoxyglucose positron-emission tomography with computed tomography for staging of urothelial carcinoma. BJU Int. 2010;106:658–63.
141. Nayak B, Dogra PN, Naswa N, Kumar R. Diuretic ^{18}F-FDG PET/CT imaging for detection and locoregional staging of urinary bladder cancer: prospective evaluation of a novel technique. Eur J Nucl Med Mol Imaging. 2013;40:386–93.
142. Eulitt PJ, Altun E, Sheikh A, et al. Pilot Study of [(18) F] fluorodeoxyglucose positron emission tomography (FDG-PET)/magnetic resonance imaging (MRI) for staging of muscle-invasive bladder cancer (MIBC). Clin Genitourin Cancer. 2020;18:378–86.
143. Herr HW, Bochner BH, Dalbagni G, et al. Impact of the number of lymph nodes retrieved on outcome in patients with muscle invasive bladder cancer. J Urol. 2002;167:1295–8.
144. Stein JP, Lieskovsky G, Cote R, et al. Radical cystectomy in the treatment of invasive bladder cancer: long-term results in 1054 patients. J Clin Oncol. 2001;19:666–75.
145. Stein JP, Lieskovsky G, Cote R, et al. Radical cystectomy in the treatment of invasive bladder cancer: long-term results in 1,054 patients. J Clin Oncol. 2001;19:666–75.
146. Gurram S, Muthigi A, Egan J, et al. Imaging in localized bladder cancer: can current diagnostic modalities provide accurate local tumor staging? Curr Urol Rep. 2019;20:82.
147. Moussa M, Chakra MA, Saad W, et al. The role of ^{18}F-FDG PET/CT scan compared to CT-scan alone for lymph node staging before radical cystectomy in patients with bladder cancer. Urol Oncol. 2021;39(12):833.
148. Girard A, Rouanne M, Taconet S, et al. Integrated analysis of (18)F-FDG PET/CT improves preoperative lymph node staging for patients with invasive bladder cancer. Eur Radiol. 2019;29:4286–93.
149. Jeong IG, Hong S, You D, et al. FDG PET-CT for lymph node staging of bladder cancer: a prospective study of patients with extended pelvic lymphadenectomy. Ann Surg Oncol. 2015;22:3150–6.
150. Goodfellow H, Viney Z, Hughes P, et al. Role of fluorodeoxyglucose positron emission tomography (FDG PET)-computed tomography (CT) in the staging of bladder cancer. BJU Int. 2014;114:389–95.
151. Hitier-Berthault M, Ansquer C, Branchereau J, et al. 18F-fluorodeoxyglucose positron emission tomography-computed tomography for preoperative lymph node staging in patients undergoing radical cystectomy for bladder cancer: a prospective study. Int J Urol. 2013;20:788–96.
152. Apolo AB, Riches J, Schoder H, et al. Clinical value of fluorine-18 2-fluoro-2-deoxy-D-glucose positron emission tomography/computed tomography in bladder cancer. J Clin Oncol. 2010;28:3973–8.
153. Mertens LS, Fioole-Bruining A, Vegt E, et al. Impact of (18) F-fluorodeoxyglucose (FDG)-positron-emission tomography/computed tomography (PET/CT) on management of patients with carcinoma invading bladder muscle. BJU Int. 2013;112:729–34.
154. Jensen TK, Holt P, Gerke O, et al. Preoperative lymph-node staging of invasive urothelial bladder cancer with ^{18}F-fluorodeoxyglucose positron emission tomography/computed axial tomography and magnetic resonance imaging: correlation with histopathology. Scand J Urol Nephrol. 2011;45:122–8.
155. Shinagare AB, Ramaiya NH, Jagannathan JP, et al. Metastatic pattern of bladder cancer: correlation with the characteristics of the primary tumor. AJR Am J Roentgenol. 2011;196:117–22.
156. Kibel AS, Dehdashti F, Katz MD, et al. Prospective study of [^{18}F]fluorodeoxyglucose positron emission tomography/computed tomography for staging of muscle-invasive bladder carcinoma. J Clin Oncol. 2009;27:4314–20.
157. Soubra A, Hayward D, Dahm P, et al. The diagnostic accuracy of ^{18}F-fluorodeoxyglucose positron emission tomography and computed tomography in staging bladder cancer: a single-institution study and a systematic review with meta-analysis. World J Urol. 2016;34:1229–37.
158. Kim SK. Role of PET/CT in muscle-invasive bladder cancer. Transl Androl Urol. 2020;9:2908–19.
159. Drieskens O, Oyen R, Van Poppel H, et al. FDG-PET for preoperative staging of bladder cancer. Eur J Nucl Med Mol Imaging. 2005;32:1412–7.
160. Verghote F, Poppe L, Verbeke S, et al. Evaluating the impact of ^{18}F-FDG-PET-CT on risk stratification and treatment adaptation for patients with muscle-invasive bladder cancer (EFFORT-MIBC): a phase II prospective trial. BMC Cancer. 2021;21:1113.
161. Mertens LS, Fioole-Bruining A, van Rhijn BW, et al. FDG-positron emission tomography/computerized tomography for monitoring the response of pelvic lymph node metastasis to neoadjuvant chemotherapy for bladder cancer. J Urol. 2013;189:1687–91.

162. van de Putte EEF, Vegt E, Mertens LS, et al. FDG-PET/CT for response evaluation of invasive bladder cancer following neoadjuvant chemotherapy. Int Urol Nephrol. 2017;49:1585–91.
163. Jadvar H, Quan V, Henderson RW, et al. [F-18]-Fluorodeoxyglucose PET and PET-CT in diagnostic imaging evaluation of locally recurrent and metastatic bladder transitional cell carcinoma. Int J Clin Oncol. 2008;13:42–7.
164. Kitajima K, Yamamoto S, Fukushima K, et al. FDG-PET/CT as a posttreatment restaging tool in urothelial carcinoma: Comparison with contrast-enhanced CT. Eur J Radiol. 2016;85:593–8.
165. Zattoni F, Incerti E, Colicchia M, et al. Comparison between the diagnostic accuracies of ^{18}F-fluorodeoxyglucose positron emission tomography/computed tomography and conventional imaging in recurrent urothelial carcinomas: a retrospective, multicenter study. Abdom Radiol. 2018;43:2391–9.
166. Gofrit ON, Mishani E, Orevi M, et al. Contribution of ^{11}C-choline positron emission tomography/computerized tomography to preoperative staging of advanced transitional cell carcinoma. J Urol. 2006;176:940–4.
167. Schoder H, Ong SC, Reuter VE, et al. Initial results with (11)C-acetate positron emission tomography/computed tomography (PET/CT) in the staging of urinary bladder cancer. Mol Imaging Biol. 2012;14:245–51.
168. Letocha H, Ahlstrom H, Malmstrom PU, et al. Positron emission tomography with L-methyl-^{11}C-methionine in the monitoring of therapy response in muscle-invasive transitional cell carcinoma of the urinary bladder. Br J Urol. 1994;74:767–74.
169. Bray F, Ferlay J, Soerjomataram I, Siegel RL, Torre LA, Jemal A. Global cancer statistics 2018: GLOBOCAN estimates of incidence and mortality worldwide for 36 cancers in 185 countries. CA Cancer J Clin. 2018;68:394–424.
170. Wong MCS, Goggins WB, Wang HHX, Fung FDH, Leung C, Wong SYS, Ng CF, Sung JJY. Global incidence and mortality for prostate cancer: analysis of temporal patterns and trends in 36 countries. Eur Urol. 2016;70(5):862–74.
171. Patel VL, Busch EL, Friebel TM, et al. Association of genomic domains in BRCA1 and BRCA2 with prostate cancer risk and aggressiveness. Cancer Res. 2020;80(3):624–38.
172. Humphrey PA. Histopathology of prostate cancer. Cold Spring Harb Perspect Med. 2017;7:a030411.
173. Mottet N, Cornford P, van den Bergh RCN, Briers E, De Santis M, Fanti S, Gillessen S, Grummet J, Henry AM, Lam TB, et al. EAU-ESTRO-ESUR-SIOG guidelines on prostate cancer. https://uroweb.org/guideline/prostate-cancer/. Accessed 28 June 2021.
174. National Cancer Institute Surveillance, Epidemiology, and End Results Program. Cancer stat facts: prostate cancer. https://seer.cancer.gov/statfacts/html/prost.html. Accessed 16 October 2021.
175. Samaržija I. Site-specific and common prostate cancer metastasis genes as suggested by meta-analysis of gene expression data. Lifestyles. 2021;11:636.
176. National Comprehensive Cancer Network. Prostate cancer (version 3.2020). 2020. https://www.nccn.org/professionals/physician_gls/pdf/prostate.pdf.
177. Felker ER, Margolis DJ, Nassiri N, Marks LS. Prostate cancer risk stratification with magnetic resonance imaging. Urol Oncol. 2016;34(7):311–9.
178. Gaur S, Harmon S, Gupta RT, Margolis DJ, Lay N, Mehralivand S, Merino MJ, Wood BJ, Pinto PA, Shih JH, Choyke PL, Turkbey B. A multireader exploratory evaluation of individual pulse sequence cancer detection on prostate multiparametric magnetic resonance imaging (MRI). Acad Radiol. 2019;26(1):5–14.
179. Kasivisvanathan V, Rannikko AS, Borghi M, Panebianco V, Mynderse LA, Vaarala MH, Briganti A, Budaus L, Hellawell G, Hindley RG, Roobol MJ, Eggener S, Ghei M, Villers A, Bladou F, Villeirs GM, Virdi J, Boxler S, Robert G, Singh PB, Venderink W, Hadaschik BA, Ruffion A, Hu JC, Margolis D, Crouzet S, Klotz L, Taneja SS, Pinto P, Gill I, Allen C, Giganti F, Freeman A, Morris S, Punwani S, Williams NR, Brew-Graves C, Deeks J, Takwoingi Y, Emberton M, Moore CM. MRI-targeted or standard biopsy for prostate-cancer diagnosis. N Engl J Med. 2018;378(19):1767–77.
180. Brown LC, Ahmed HU, Faria R, El-Shater Bosaily A, Gabe R, Kaplan RS, Parmar M, Collaco-Moraes Y, Ward K, Hindley RG, Freeman A, Kirkham A, Oldroyd R, Parker C, Bott S, Burns-Cox N, Dudderidge T, Ghei M, Henderson A, Persad R, Rosario DJ, Shergill I, Winkler M, Soares M, Spackman E, Sculpher M, Emberton M. Multiparametric MRI to improve detection of prostate cancer compared with transrectal ultrasound-guided prostate biopsy alone: the PROMIS study. Health Technol Assess. 2018;22(39):1–176.
181. Trabulsi EJ, Rumble RB, Jadvar H, Hope T, Pomper M, Turkbey B, Rosenkrantz A, Verma S, Margolis DJ, Froemming A, et al. Optimum imaging strategies for advanced prostate cancer: ASCO guideline. J Clin Oncol. 2020;38:1963–96.
182. McCarthy M, Siew T, Campbell A, et al. (1)(8) F-Fluoromethylcholine (FCH) PET imaging in patients with castration-resistant prostate cancer: prospective comparison with standard imaging. Eur J Nucl Med Mol Imaging. 2011;38:14–22.
183. Piccardo A, Paparo F, Piccazzo R, et al. Value of fused ^{18}F-Choline PET/MRI to evaluate prostate cancer relapse in patients showing biochemical recurrence after EBRT: preliminary results. Biomed Res Int. 2014;2014:103718.
184. Fanti S, Minozzi S, Castellucci P, et al. PET/CT with (11) C-choline for evaluation of prostate cancer patients with biochemical recurrence: meta-analysis

and critical review of available data. Eur J Nucl Med Mol Imaging. 2016;43(1):55–69.
185. Ceci F, Castellucci P, Mamede M, et al. (11) C-Choline PET/CT in patients with hormone-resistant prostate cancer showing biochemical relapse after radical prostatectomy. Eur J Nucl Med Mol Imaging. 2013;40(2):149–55.
186. Soyka JD, Muster MA, Schmid DT, et al. Clinical impact of ^{18}F-choline PET/CT in patients with recurrent prostate cancer. Eur J Nucl Med Mol Imaging. 2012;39(6):936–43.
187. Goldstein J, Even-Sapir E, Ben-Haim S, et al. Does choline PET/CT Change the management of prostate cancer patients with biochemical failure? Am J Clin Oncol. 2017;40:256–9.
188. Heidenreich A, Bastian PJ, Bellmunt J, et al. EAU guidelines on prostate cancer. Part II: treatment of advanced, relapsing, and castration-resistant prostate cancer. Eur Urol. 2014;65(2):467–79.
189. Guo Y, Wang L, Hu J, Feng D, Xu L. Diagnostic performance of choline PET/CT for the detection of bone metastasis in prostate cancer: a systematic review and meta-analysis. PLoS ONE. 2018;13(9):e0203400.
190. Oka S, Hattori R, Kurosaki F, et al. A preliminary study of anti-1-amino-3-^{18}F-fluorocyclobutyl-1-carboxylic acid for the detection of prostate cancer. J Nucl Med. 2007;48:46–55.
191. Asano Y, Inoue Y, Ikeda Y, et al. Phase I clinical study of NMK36: a new PET tracer with the synthetic amino acid analogue anti-[^{18}F] FACBC. Ann Nucl Med. 2011;25:414–8.
192. Nanni C, Schiavina R, Rubello D, et al. The detection of disease relapse after radical treatment for prostate cancer: is anti-3-^{18}F-FACBC PET/CT a promising option? Nucl Med Commun. 2013;34:831–3.
193. Ren J, Yuan L, Wen G, Yang J. The value of anti-1-amino-3-^{18}F-fluorocyclobutane-1-carboxylic acid PET/CT in the diagnosis of recurrent prostate carcinoma: a meta-analysis. Acta Radiol. 2016;57:487–93.
194. Schuster DM, Nieh PT, Jani AB, et al. Anti-3-[(18)F]FACBC positron emission tomography-computerized tomography and (111)In-capromab pendetide single photon emission computerized tomography computerized tomography for recurrent prostate carcinoma: Results of a prospective clinical trial. J Urol. 2014;191:1446–53.
195. Nanni C, Schiavina R, Brunocilla E, et al. 18F-Fluciclovine PET/CT for the detection of prostate cancer relapse: a comparison to ^{11}C-Choline PET/CT. Clin Nucl Med. 2015;40:e386–91.
196. Schuster DM, Nanni C, Fanti S, et al. Anti-1-amino-3-^{18}F-fluorocyclobutane-1-carboxylic acid: Physiologic uptake patterns, incidental findings, and variants that may simulate disease. J Nucl Med. 2014;55:1986–92.
197. O'Keefe DS, Su SL, Bacich DJ, Horiguchi Y, Luo Y, Powell CT, et al. Mapping, genomic organization and promoter analysis of the human prostate-specific membrane antigen gene. Biochim Biophys Acta. 1998;1443(1–2):113–27.
198. Rojas C, Frazier ST, Flanary J, Slusher BS. Kinetics and inhibition of glutamate carboxypeptidase II using a microplate assay. Anal Biochem. 2002;310(1):50–4.
199. Mesters JR, Barinka C, Li W, Tsukamoto T, Majer P, Slusher BS, et al. Structure of glutamate carboxypeptidase II, a drug target in neuronal damage and prostate cancer. EMBO J. 2006;25(6):1375–84.
200. Ghosh A, Heston WD. Tumor target prostate specific membrane antigen (PSMA) and its regulation in prostate cancer. J Cell Biochem. 2004;91:528–39.
201. Bostwick DG, Pacelli A, Blute M, Roche P, Murphy GP. Prostate specific membrane antigen expression in prostatic intraepithelial neoplasia and adenocarcinoma: a study of 184 cases. Cancer. 1998;82:2256–61.
202. Mannweiler S, Amersdorfer P, Trajanoski S, Terrett JA, King D, Mehes G. Heterogeneity of prostate-specific membrane antigen (PSMA) expression in prostate carcinoma with distant metastasis. Pathol Oncol Res. 2009;15:167–72.
203. Ross JS, Sheehan CE, Fisher HA, et al. Correlation of primary tumor prostate-specific membrane antigen expression with disease recurrence in prostate cancer. Clin Cancer Res. 2003;9:6357–62.
204. Fendler WP, Eiber M, Beheshti M, et al. (68) Ga-PSMA PET/CT: joint EANM and SNMMI procedure guideline for prostate cancer imaging: version 1.0. Eur J Nucl Med Mol Imaging. 2017;44:1014–24.
205. Sodee DB, Malguria N, Faulhaber P, Resnick MI, Albert J, Bakale G. Multicenter ProstaScint imaging findings in 2154 patients with prostate cancer. The ProstaScint Imaging Centers. Urology. 2000;56:988–93.
206. Pandit-Taskar N, O'Donoghue JA, Durack JC, et al. A Phase I/II study for analytic validation of 89Zr-J591 ImmunoPET as a molecular imaging agent for metastatic prostate cancer. Clin Cancer Res. 2015;21:5277–85.
207. Prasad V, Steffen IG, Diederichs G, Makowski MR, Wust P, Brenner W. Biodistribution of [(68)Ga] PSMA-HBED-CC in patients with prostate cancer: characterization of uptake in normal organs and tumour lesions. Mol Imaging Biol. 2016;18:428–36.
208. Uprimny C, Kroiss AS, Decristoforo C, et al. (68) Ga-PSMA-11 PET/CT in primary staging of prostate cancer: PSA and Gleason score predict the intensity of tracer accumulation in the primary tumour. Eur J Nucl Med Mol Imaging. 2017;44:941–9.
209. Rowe SP, Gage KL, Faraj SF, et al. (1)(8)F-DCFBC PET/CT for PSMAbased detection and characterization of primary prostate cancer. J Nucl Med. 2015;56:1003–10.
210. Turkbey B, Mena E, Lindenberg L, et al. 18F-DCFBC prostate-specific membrane antigen-targeted PET/CT imaging in localized prostate cancer: correlation

with multiparametric MRI and histopathology. Clin Nucl Med. 2017;42:735–40.

211. Eiber M, Weirich G, Holzapfel K, et al. Simultaneous (68)Ga-PSMA HBED-CC PET/MRI improves the localization of primary prostate cancer. Eur Urol. 2016;70:829–36.
212. Zamboglou C, Drendel V, Jilg CA, et al. Comparison of (68)Ga-HBEDCC PSMA-PET/CT and multiparametric MRI for gross tumour volume detection in patients with primary prostate cancer based on slice by slice comparison with histopathology. Theranostics. 2017;7:228–37.
213. Hovels AM, Heesakkers RA, Adang EM, et al. The diagnostic accuracy of CT and MRI in the staging of pelvic lymph nodes in patients with prostate cancer: a meta-analysis. Clin Radiol. 2008;63:387–95.
214. Maurer T, Gschwend JE, Rauscher I, et al. Diagnostic Efficacy of (68) Gallium-PSMA positron emission tomography compared to conventional imaging for lymph node staging of 130 consecutive patients with intermediate to high-risk prostate cancer. J Urol. 2016;195:1436–43.
215. Kim SJ, Lee SW, Ha HK. Diagnostic performance of radiolabeled prostate-specific membrane antigen positron emission tomography/computed tomography for primary lymph node staging in newly diagnosed intermediate to high-risk prostate cancer patients: a systematic review and meta-analysis. Urol Int. 2018;102:27–36.
216. Herlemann A, Wenter V, Kretschmer A, et al. (68) Ga-PSMA positron emission tomography/computed tomography provides accurate staging of lymph node regions prior to lymph node dissection in patients with prostate cancer. Eur Urol. 2016;70:553–7.
217. Budaus L, Leyh-Bannurah SR, Salomon G, et al. Initial experience of (68)Ga-PSMA PET/CT imaging in high-risk prostate cancer patients prior to radical prostatectomy. Eur Urol. 2016;69:393–6.
218. Janssen JC, Woythal N, Meissner S, et al. [(68)Ga] PSMA-HBED-CC uptake in osteolytic, osteoblastic, and bone marrow metastases of prostate cancer patients. Mol Imaging Biol. 2017;19:933–43.
219. King CR. The timing of salvage radiotherapy after radical prostatectomy: a systematic review. Int J Radiat Oncol Biol Phys. 2012;84:104–11.
220. Giesel FL, Fiedler H, Stefanova M, et al. PSMA PET/CT with Glu-ureaLys-(Ahx)-[(6)(8)Ga(HBED-CC)] versus 3D CT volumetric lymph node assessment in recurrent prostate cancer. Eur J Nucl Med Mol Imaging. 2015;42:1794–800.
221. Rauscher I, Maurer T, Beer AJ, et al. Value of ^{68}Ga-PSMA HBED-CC PET for the assessment of lymph node metastases in prostate cancer patients with biochemical recurrence: comparison with histopathology after salvage lymphadenectomy. J Nucl Med. 2016;57:1713–9.
222. Perera M, Papa N, Christidis D, et al. Sensitivity, specificity, and predictors of positive (68) Ga-prostate-specific membrane antigen positron emission tomography in advanced prostate cancer: a systematic review and meta-analysis. Eur Urol. 2016;70:926–37.
223. Afshar-Oromieh A, Holland-Letz T, Giesel FL, et al. Diagnostic performance of (68)Ga-PSMA-11 (HBED-CC) PET/CT in patients with recurrent prostate cancer: evaluation in 1007 patients. Eur J Nucl Med Mol Imaging. 2017;44:1258–68.
224. Ceci F, Castellucci P, Graziani T, et al. (68) Ga-PSMA-11 PET/CT in recurrent prostate cancer: efficacy in different clinical stages of PSA failure after radical therapy. Eur J Nucl Med Mol Imaging. 2018;46:31–9.
225. Jilg CA, Drendel V, Rischke HC, et al. Diagnostic accuracy of Ga-68-HBED-CC-PSMA-ligand-PET/CT before salvage lymph node dissection for recurrent prostate cancer. Theranostics. 2017;7:1770–80.
226. Clarebrough E, Duncan C, Christidis D, Lavoipierre A, Lawrentschuk N. PSMA-PET guided hook-wire localization of nodal metastases in prostate cancer: a targeted approach. World J Urol. 2018;37(7):1251–4.
227. Porres D, Pfister D, Thissen A, et al. The role of salvage extended lymph node dissection in patients with rising PSA and PET/CT scan detected nodal recurrence of prostate cancer. Prostate Cancer Prostatic Dis. 2017;20:85–92.
228. Guler OC, Engels B, Onal C, et al. The feasibility of prostate-specific membrane antigen positron emission tomography (PSMA PET/CT)-guided radiotherapy in oligometastatic prostate cancer patients. Clin Transl Oncol. 2018;20:484–90.
229. Bluemel C, Linke F, Herrmann K, et al. Impact of (68)Ga-PSMA PET/CT on salvage radiotherapy planning in patients with prostate cancer and persisting PSA values or biochemical relapse after prostatectomy. EJNMMI Res. 2016;6:78.
230. Calais J, Czernin J, Cao M, et al. (68)Ga-PSMA-11 PET/CT mapping of prostate cancer biochemical recurrence after radical prostatectomy in 270 patients with a PSA level of less than 1.0 ng/mL: impact on salvage radiotherapy planning. J Nucl Med. 2018;59:230–7.
231. Hruby G, Eade T, Kneebone A, et al. Delineating biochemical failure with (68)Ga-PSMA-PET following definitive external beam radiation treatment for prostate cancer. Radiother Oncol. 2017;122:99–102.
232. Emmett L, van Leeuwen PJ, Nandurkar R, et al. Treatment outcomes from (68)Ga-PSMA PET/CT-informed salvage radiation treatment in men with rising PSA after radical prostatectomy: prognostic value of a negative PSMA PET. J Nucl Med. 2017;58:1972–6.
233. Calais J, Fendler WP, Herrmann K, Eiber M, Ceci F. Comparison of (68) Ga-PSMA-11 and (18) F-Fluciclovine PET/CT in a case series of 10 patients with prostate cancer recurrence. J Nucl Med. 2018;59:789–94.
234. Beattie BJ, Smith-Jones PM, Jhanwar YS, et al. Pharmacokinetic assessment of the uptake of ^{16}beta-^{18}F-fluoro-5alpha-dihydrotestosterone (FDHT) in

prostate tumors as measured by PET. J Nucl Med. 2010;51:183–92.
235. Rathkopf DE, Morris MJ, Fox JJ, et al. Phase I study of ARN-509, a novel antiandrogen, in the treatment of castration-resistant prostate cancer. J Clin Oncol. 2013;31:3525–30.
236. Dehdashti F, Picus J, Michalski JM, et al. Positron tomographic assessment of androgen receptors in prostatic carcinoma. Eur J Nucl Med Mol Imaging. 2005;32:344–50.
237. Wieser G, Mansi R, Grosu AL, et al. Positron emission tomography (PET) imaging of prostate cancer with a gastrin releasing peptide receptor antagonist from mice to men. Theranostics. 2014;4:412–9.
238. Sah BR, Burger IA, Schibli R, et al. Dosimetry and first clinical evaluation of the new ^{18}F-radiolabeled bombes in analogue BAY 864367 in patients with prostate cancer. J Nucl Med. 2015;56:372–8.
239. Tripathi S, Trabulsi EJ, Gomella L, et al. VPAC1 targeted (64)CuTP3805 positron emission tomography imaging of prostate cancer: preliminary evaluation in man. Urology. 2016;88:111–8.
240. Doran MG, Watson PA, Cheal SM, et al. Annotating STEAP1 regulation in prostate cancer with 89Zr immuno-PET. J Nucl Med. 2014;55:2045–9.
241. Emmett L, Willowson K, Violet J, Shin J, Blanksby A, Lee J. Lutetium ^{177} PSMA radionuclide therapy for men with prostate cancer: a review of the current literature and discussion of practical aspects of therapy. J Med Radiat Sci. 2017;64(1):52–60.
242. Agrawal S. The role of 225Ac-PSMA-617 in chemotherapy-naive patients with advanced prostate cancer: is it the new beginning. Indian J Urol. 2020;36(1):69–70.
243. Modlin IM, Kidd M, Latich I, Zikusoka MN, Shapiro MD. Current status of gastrointestinal carcinoids. Gastroenterology. 2005;128(6):1717–51.
244. Taal BG, Visser O. Epidemiology of neuroendocrine tumours. Neuroendocrinology. 2004;80(1):3–7.
245. Reubi JC. Neuropeptide receptors in health and disease: the molecular basis for in vivo imaging. J Nucl Med. 1995;36:1825–35.
246. Jensen RT. Endocrine tumors of the gastrointestinal tract and pancreas. In: Kasper DL, Fauci AS, Longo DL, et al., editors. Harrison's principles of internal medicine. 16th ed. New York: McGraw-Hill; 2005.
247. Rindi G, Klöppel G, Alhman H, et al. TNM staging of foregut (neuro)endocrine tumors: a consensus proposal including a grading system. Virchows Arch. 2006;449(4):395–401.
248. Rindi G, Klöppel G, Couvelard A, et al. TNM staging of midgut and hindgut (neuro) endocrine tumors: a consensus proposal including a grading system. Virchows Arch. 2007;451(4):757–62.
249. Sundin A, Garske U, Orlefors H. Nuclear imaging of neuroendocrine tumours. Best Pract Res Clin Endocrinol Metab. 2007;21(1):69–85.
250. Ramage JK, Davies AH, Ardill J, et al. UKNETwork for neuroendocrine tumours, guidelines for the management of gastroenteropancreatic neuroendocrine (including carcinoid) tumours. Gut. 2005;54(Suppl. 4):1–16.
251. Adams S, Baum R, Rink T, Schumm-Dräger PM, Usadel KH, Hör G. Limited value of fluorine-18 fluorodeoxyglucose positron emission tomography for the imaging of neuroendocrine tumors. Eur J Nucl Med. 1998;25:79–83.
252. Cescato R, Schulz S, Waser B, et al. Internalization of sst2, sst3, and sst5 receptors: effects of somatostatin agonists and antagonists. J Nucl Med. 2006;47(3):502–11.
253. Antunes P, Ginj M, Zhang H, et al. Are radiogallium-labelled DOTA-conjugated somatostatin analogues superior to those labelled with other radiometals? Eur J Nucl Med Mol Imaging. 2007;34(7):982–93.
254. Zhernosekov KP, Filosofov DV, Baum RP, et al. Processing of generator produced ^{68}Ga for medical application. J Nucl Med. 2007;48(10):1741–8.
255. Campana D, Ambrosini V, Pezzilli R, et al. Standardized uptake values of (68)GaDOTANOC PET: a promising prognostic tool in neuroendocrine tumors. J Nucl Med. 2010;51(3):353–9.
256. Kim K, Kim SJ. Lu-177-based peptide receptor radionuclide therapy for advanced neuroendocrine tumors. Nucl Med Mol Imaging. 2018;52(3):208–15.
257. Bal C, Ballal S, Yadav M. Long-term outcome of ^{225}Ac-DOTATATE targeted alpha therapy in patients with metastatic gastroenteropancreatic neuroendocrine tumors. JNM. 2021;62(supplement 1):19.
258. Cowan AJ, Green DJ, Kwok M, Lee S, Coffey DG, Holmberg LA, Tuazon S, Gopal AK, Libby EN. Diagnosis and management of multiple myeloma: a review. JAMA. 2022;327(5):464–77.
259. van de Donk NWCJ, Pawlyn C, Yong KL. Multiple myeloma. Lancet. 2021;397(10272):410–27.
260. Cavo M, Terpos E, Nanni C, Moreau P, Lentzsch S, Zweegman S, Hillengass J, Engelhardt M, Usmani SZ, Vesole DH, San-Miguel J, Kumar SK, Richardson PG, Mikhael JR, da Costa FL, Dimopoulos MA, Zingaretti C, Abildgaard N, Goldschmidt H, Orlowski RZ, Chng WJ, Einsele H, Lonial S, Barlogie B, Anderson KC, Rajkumar SV, Durie BGM, Zamagni E. Role of ^{18}F-FDG PET/CT in the diagnosis and management of multiple myeloma and other plasma cell disorders: a consensus statement by the International Myeloma Working Group. Lancet Oncol. 2017;18(4):e206–17.
261. Luo Y, Cao X, Pan Q, Li J, Feng J, Li F. 68Ga-pentixafor PET/CT for imaging of chemokine receptor 4 expression in Waldenström macroglobulinemia/lymphoplasmacytic lymphoma: comparison to ^{18}F-FDG PET/CT. J Nucl Med. 2019;60(12):1724–9.
262. Mesguich C, Hulin C, Lascaux A, Bordenave L, Marit G, Hindié E. Choline PET/CT in multiple myeloma. Cancer. 2020;12(6):1394.
263. Lecouvet FE, Vekemans MC, Van Den Berghe T, Verstraete K, Kirchgesner T, Acid S, Malghem J, Wuts J, Hillengass J, Vandecaveye V, Jamar F, Gheysens O, Vande Berg BC. Imaging of treat-

ment response and minimal residual disease in multiple myeloma: state of the art WB-MRI and PET/CT. Skeletal Radiol. 2022;51(1):59–80.
264. Kumar N, Gupta B. Global incidence of primary malignant bone tumors. Curr Orthop Pract. 2016;27(5):530–4.
265. Kindblom LG. Bone tumors: epidemiology, classification, pathology. In: Davies AM, Sundaram M, James SJ, editors. Imaging of bone tumors and tumor-like lesions: techniques and applications, vol. 1. Berlin: Springer; 2009. p. 1–15.
266. Biermann JS, Chow W, Reed DR, et al. NCCN guidelines insights: bone cancer, version 2.2017. J Natl Compr Cancer Netw. 2017;15(2):155–67.
267. Mittra E, Iagaru A. 18F-FDG-PET and PET/CT for evaluating primary bone tumors. Pet Clin. 2010;5(3):327–39.
268. Behzadi AH, Raza SI, Carrino JA, et al. Applications of PET/CT and PET/MR imaging in primary bone malignancies. PET Clin. 2018;13(4):623–34.
269. Choi YY, Kim JY, Yang SO. PET/CT in benign and malignant musculoskeletal tumors and tumor-like conditions. Semin Musculoskelet Radiol. 2014;18(2):133–48.
270. Tan H, Yusen G, Haojun Y, et al. Total-body PET/CT: current applications and future perspectives. Am J Roentgenol. 2020;215:325–37.
271. Tian R, Su M, Tian Y, et al. Dual-time point PET/CT with F-18 FDG for the differentiation of malignant and benign bone lesions. Skeletal Radiol. 2009;38(5):451–8.
272. Lakkaraju A, Patel CN, Bradley KM, et al. PET/CT in primary musculoskeletal tumours: a step forward. Eur Radiol. 2010;20(12):2959–72.
273. Hillner BE, Siegel BE, Liu D, et al. Impact of positron emission tomography/computed tomography and positron emission tomography (PET) alone on expected management of patients with cancer: initial results from the National Oncologic PET Registry. J Clin Oncol. 2008;26(13):2155–61.
274. Lim HJ, Johnny Ong CA, Tan JWS, et al. Utility of positron emission tomography/computed tomography (PET/CT) imaging in the evaluation of sarcomas: a systematic review. Crit Rev Oncol Hematol. 2019;143:1–13.
275. National Comprehensive Cancer Network. Bone cancer (version 1.2023). 2022. https://www.nccn.org/professionals/physician_gls/pdf/bone.pdf.
276. Ostrom QT, Gittleman H, Liao P, et al. CBTRUS statistical report: primary brain and other central nervous system tumors diagnosed in the United States in 2010–2014. Neuro-Oncology. 2017;19:1–88.
277. Louis DN, Ohgaki H, Wiestler ODCW. WHO classification of tumours of the central nervous system. 4th ed. Lyon: World Health Organization; 2016.
278. Muoio B, Giovanella L, Treglia G. Recent developments of ^{18}F-FET PET in neuro-oncology. Curr Med Chem. 2018;25:3061–73.
279. Treglia G, Sadeghi R, Del Sole A, Giovanella L. Diagnostic performance of PET/CT with tracers other than F-18-FDG in oncology: an evidence-based review. Clin Transl Oncol. 2014;16:770–5.
280. Treglia G. Diagnostic performance of ^{18}F-FDG PET/CT in infectious and inflammatory diseases according to published meta-analyses. Contrast Media Mol Imaging. 2019;2019:3018349.
281. Mattoli MV, Treglia G, Trevisi G, Muoio B, Cason E. Usefulness of ^{11}C-methionine positron emission tomography in differential diagnosis between recurrent tumours and radiation necrosis in patients with glioma: an overview. Open Neurosurg J. 2012;5:8–11.
282. Bell C, Dowson N, Puttick S, Gal Y, Thomas P, Fay M, Smith J, Rose S. Increasing feasibility and utility of (18) F-FDOPA PET for the management of glioma. Nucl Med Biol. 2015;42:788–95.
283. Bertagna F, Biasiotto G, Giubbini R. The role of F-18-fluorothymidine PET in oncology. Clin Translat Imaging. 2013;1:77–97.
284. Treglia G, Giovannini E, Di Franco D, Calcagni ML, Rufini V, Picchio M, Giordano A. The role of positron emission tomography using carbon-11 and fluorine-18 choline in tumors other than prostate cancer: a systematic review. Ann Nucl Med. 2012;26:451–61.
285. Sharma A, Kumar R. Metabolic imaging of brain tumor recurrence. AJR Am J Roentgenol. 2020;215(5):1199–207.
286. Wong WL, Hussain K, Chevretton EB, et al. Validation and clinical application of computer combined CT and PET with FDG Head and neck images. Am J Surg. 1996;172:628–32.
287. Wong WL, Chevretton EB, McGurk M, et al. A prospective study of PET FDG imaging for the assessment of head and neck squamous cell carcinoma. Clin Otolaryngol. 1997;22:209–14.
288. Wong WL, Gharpurhy A, Gollub F. Assessment of patients with occult primary head and neck cancers-An FDG PET/CT study. Clin Oncol. 2012;24:190.
289. Dong MJ, Zhao K, Lin XT, et al. Role of fluorodeoxyglucose-PET versus fluorodeoxyglucose-PET/computed tomography in detection of unknown primary tumor: a meta-analysis of the literature. Nucl Med Commun. 2008;29:791–802.
290. Rusthoven KE, Koshy M, Paulino AC. The role of fluorodeoxyglucose positron emission tomography in cervical lymph node metastases from an unknown primary tumor. Cancer. 2004;101:2641–9.
291. Cancer of the upper aerodigestive tract: assessment and management in patient aged 16 and over NICE guidelines NG36.
292. Lonneux M, Hamoir M, Reychler H, et al. Positron emission tomography with [^{18}F] fluorodeoxyglucose improves staging and patient management in patients with head and neck squamous cell carcinoma: a multicentre prospective study. J Clin Oncol. 2010;28:1190–5.
293. Wong WL. PET-CT for staging and detection of recurrence of head and neck cancer. Semin Nucl Med. 2021;51(1):13–25.

294. National Comprehensive Cancer Network. Head and neck cancer (version 2.2022). 2022. https://www.nccn.org/professionals/physician_gls/pdf/head-and-neck.pdf.
295. Sipos JA, Mazzaferri EL. Thyroid cancer epidemiology and prognostic variables. Clin Oncol. 2010;22(6):395–404.
296. Zarnegar R, Brunaud L, Kanauchi H, Wong M, Fung M, Ginzinger D, Duh Q, Clark OH. Increasing the effectiveness of radioactive iodine therapy in the treatment of thyroid cancer using Trichostatin A, a histone deacetylase inhibitor. Surgery. 2002;132:984–90.
297. Worden F. Treatment strategies for radioactive iodine-refractory differentiated thyroid cancer. Ther Adv Med Oncol. 2014;6:267–79.
298. Gild ML, Topliss DJ, Learoyd D, Parnis F, Tie J, Hughes B, Walsh JP, McLeod DSA, Clifton-Bligh RJ, Robinson BG. Clinical guidance for radioiodine refractory differentiated thyroid cancer. Clin Endocrinol. 2018;88(4):529–37.
299. Leboulleux S, Schroeder PR, Schlumberger M, Ladenson PW. The role of PET in follow-up of patients treated for differentiated epithelial thyroid cancers. Nat Clin Pract Endocrinol Metab. 2007;3(2):112–21.
300. Treglia G, Muoio B, Giovanella L, Salvatori M. The role of positron emission tomography and positron emission tomography/computed tomography in thyroid tumours: an overview. Eur Arch Otorrinolaringol. 2013;270(6):1783–7.
301. Kim SJ, Lee SW, Pak K, Shim SR. Diagnostic performance of PET in thyroid cancer with elevated anti-Tg ab. Endocr Relat Cancer. 2018;25:643–52.
302. Robbins R, Wan Q, Grewal R, Reibke R, Gonen M, Strauss H, et al. Real-time prognosis for metastatic thyroid carcinoma based on 2-[^{18}F]fluoro-2-deoxy-D-glucose-positron emission tomography scanning. J Clin Endocrinol Metab. 2006;91:498–505.
303. Brose MS, Nutting CM, Jarzab B, Elisei R, Siena S, Bastholt L, et al. Sorafenib in radioactive iodine-refractory, locally advanced or metastatic differentiated thyroid cancer: a randomised, double-blind, phase 3 trial. Lancet. 2014;384(9940):319–28.
304. Binse I, Poeppel TD, Ruhlmann M, Ezziddin S, Görges R, Sabet A, et al. 68Ga-DOTATOC PET/CT in patients with iodine- and ^{18}F-FDG-negative differentiated thyroid carcinoma and elevated serum thyroglobulin. J Nucl Med. 2016;57:1512–7.
305. Heitkötter B, Steinestel K, Trautmann M, Grünewald I, Barth P, Gevensleben H, et al. Neovascular PSMA expression is a common feature in malignant neoplasms of the thyroid. Oncotarget. 2018;9:9867–74.
306. Wu HB, Wang QS, Wang MF, Li HS. Utility of ^{11}C-choline imaging as a supplement to F-18 FDG PET imaging for detection of thyroid carcinoma. Clin Nucl Med. 2011;36:91–5.
307. Fu H, Wu J, Huang J, Sun L, Wu H, Guo W, Qiu S, Chen H. 68Ga fibroblast activation protein inhibitor PET/CT in the detection of metastatic thyroid cancer: comparison with ^{18}F-FDG PET/CT. Radiology. 2022;304(2):397–405.
308. Freudenberg LS, Jentzen W, Stahl A, Bockisch A, Rosenbaum-Krumme SJ. Clinical applications of 124I-PET/CT in patients with differentiated thyroid cancer. Eur J Nucl Med Mol Imaging. 2011;38:48–56.
309. Van Nostrand D, Moreau S, Bandaru VV, Atkins F, Chennupati S, Mete M, et al. (124)I positron emission tomography versus (131)I planar imaging in the identification of residual thyroid tissue and/or metastasis in patients who have well-differentiated thyroid cancer. Thyroid. 2010;20:879–83.
310. Rastrelli M, Tropea S, Rossi CR, Alaibac M. Melanoma: epidemiology, risk factors, pathogenesis, diagnosis and classification. In Vivo. 2014;28:1005–11.
311. Thoelke A, Willrodt S, Hauschild A, Schadendorf D. Primary extracutaneous malignant melanoma: a comprehensive review with emphasis on treatment. Onkologie. 2004;27(5):492–9.
312. Papaspyrou G, Garbe C, Schadendorf D, Werner JA, Hauschild A, Egberts F. Mucosal melanomas of the head and neck: new aspects of the clinical outcome, molecular pathology, and treatment with C-kit inhibitors. Melanoma Res. 2011;21(6):475–82.
313. Hess S, Blomberg BA, Zhu HJ, et al. The pivotal role of FDG-PET/CT in modern medicine. Acad Radiol. 2014;21(2):232–49.
314. Krug B, Crott R, Lonneux M, Baurain JF, et al. Role of PET in the initial staging of cutaneous malignant melanoma: systematic review. Radiology. 2008;249:836–44.
315. Perng P, Marcus C, Rathan M. 18F-FDG PET/CT and melanoma: staging, immune modulation and mutation-targeted therapy assessment, and prognosis. AJR. 2015;205:259–70.
316. Petersen H, Holdgaard PC, Madsen PH, et al. FDG PET/CT in cancer: comparison of actual use with literature-based recommendations. Eur J Nucl Med Mol Imaging. 2016;43(4):695–706.
317. Mena E, Taghipour M, Sheikhbahaei S, Mirpour S, Xiao J, Subramaniam RM, et al. 18F-FDG PET/CT and melanoma: value of fourth and subsequent post therapy follow-up scans for patient management. Clin Nucl Med. 2016;41:e403–9.
318. Rieber A, Nüssle K, Stöhr I, Grab D, Fenchel S, Kreienberg R, et al. Preoperative diagnosis of ovarian tumors with MR imaging: comparison with transvaginal sonography, positron emission tomography, and histologic findings. AJR Am J Roentgenol. 2001;177:123–9.
319. Risum S, Høgdall C, Loft A, Berthelsen AK, Høgdall E, Nedergaard L, et al. The diagnostic value of PET/CT for primary ovarian cancer-a prospective study. Gynecol Oncol. 2007;105:145–9.
320. Dauwen H, Van Calster B, Deroose CM, Op de Beeck K, Amant F, Neven P, et al. PET/CT in the staging of patients with a pelvic mass suspicious for ovarian cancer. Gynecol Oncol. 2013;131:694–700.

321. Hynninen J, Kemppainen J, Lavonius M, Virtanen J, Matomäki J, Oksa S, et al. A prospective comparison of integrated FDG-PET/contrast-enhanced CT and contrast-enhanced CT for pretreatment imaging of advanced epithelial ovarian cancer. Gynecol Oncol. 2013;131:389–94.
322. Gu P, Pan LL, WuSQ SL, Huang G. CA 125, PET alone, PET-CT, CT and MRI in diagnosing recurrent ovarian carcinoma: a systematic review and meta-analysis. Eur J Radiol. 2009;71:164–74.
323. Lee M, Lee H, Cheon GJ, Kim HS, Chung HH, Kim JW, et al. Prognostic value of preoperative intratumoral FDG uptake heterogeneity in patients with epithelial ovarian cancer. Eur Radiol. 2017;27:16–23.
324. Sahdev A, Sohaib SA, Wenaden AE, Shepherd JH, Reznek RH. The performance of magnetic resonance imaging in early cervical carcinoma: a long-term experience. Int J Gynecol Cancer. 2007;17:629–36. https://doi.org/10.1111/j.1525-1438.2007.00829.x.
325. Sugawara Y, Eisbruch A, Kosuda S, Recker BE, Kison PV, Wahl RL. Evaluation of FDG PET in patients with cervical cancer. J Nucl Med. 1999;40:1125–31.
326. Sironi S, Buda A, Picchio M, Perego P, Moreni R, Pellegrino A, et al. Lymph node metastasis in patients with clinical early-stage cervical cancer: detection with integrated FDG PET/CT. Radiology. 2006;238:272–9. https://doi.org/10.1148/radiol.2381041799.
327. Choi HJ, Roh JW, Seo SS, Lee S, Kim JY, Kim SK, et al. Comparison of the accuracy of magnetic resonance imaging and positron emission tomography/computed tomography in the presurgical detection of lymph node metastases in patients with uterine cervical carcinoma: a prospective study. Cancer. 2006;106:914–22.
328. Kidd EA, Siegel BA, Dehdashti F, Grigsby PW. The standardized uptake value for F-18 fluorodeoxyglucose is a sensitive predictive biomarker for cervical cancer treatment response and survival. Cancer. 2007;110:1738–44.
329. The National Comprehensive Cancer Network. 2022. https://www.nccn.org/professionals/imaging/. Accessed 9 August 2022.
330. Lazzari R, Cecconi A, Jereczek-Fossa BA, Travaini LL, Dell'Acqua V, Cattani F, et al. The role of [^{18}F] FDG-PET/CT in staging and treatment planning for volumetric modulated rapidarc radiotherapy in cervical cancer: experience of the European Institute of Oncology, Milan, Italy. Ecancermedicalscience. 2014;8:405.
331. Chu Y, Zheng A, Wang F, Lin W, Yang X, Han L, et al. Diagnostic value of ^{18}F-FDGPET or PET-CT in recurrent cervical cancer: a systematic review and meta-analysis. Nucl Med Commun. 2014;35:144–50. https://doi.org/10.1097/MNM.0000000000000026.
332. Burger IA, Vargas HA, Donati OF, Andikyan V, Sala E, Gonen M, et al. The value of ^{18}F-FDG PET/CT in recurrent gynecologic malignancies prior to pelvic exenteration. Gynecol Oncol. 2013;129:586–92.
333. Kakhki VR, Shahriari S, Treglia G, Hasanzadeh M, Zakavi SR, Yousefi Z, et al. Diagnostic performance of fluorine 18 fluorodeoxyglucose positron emission tomography imaging for detection of primary lesion and staging of endometrial cancer patients: systematic review and metaanalysis of the literature. Int J Gynecol Cancer. 2013;23:1536–43.
334. Crivellaro C, Signorelli M, Guerra L, De Ponti E, Pirovano C, Fruscio R, et al. Tailoring systematic lymphadenectomy in high-risk clinical early stage endometrial cancer: the role of ^{18}F-FDG PET/CT. Gynecol Oncol. 2013;130:306–11.
335. Bollineni VR, Ytre-Hauge S, Bollineni-Balabay O, Salvesen HB, Haldorsen IS. High diagnostic value of ^{18}F-FDG PET/CT in endometrial cancer: systematic review and meta-analysis of the literature. J Nucl Med. 2016;57:879–85.
336. Husain A, Akhurst T, Larson S, Alektiar K, Barakat RR, Chi DS. A prospective study of the accuracy of ^{18}Fluorodeoxyglucose positron emission tomography (^{18}FDG PET) in identifying sites of metastasis prior to pelvic exenteration. Gynecol Oncol. 2007;106(1):177–80.
337. Peiró V, Chiva L, González A, Bratos R, Alonso S, Márquez R, Carballo N, Alonso-Farto JC. Utility of the PET/CT in vulvar cancer management. Rev Esp Med Nucl Imagen Mol. 2014;33(2):87–92.
338. Viswanathan C, Kirschner K, Truong M. Multimodality imaging of vulvar cancer: staging, therapeutic response, and complications. AJR. 2013;200(6):1387–400.
339. Robertson NL, Hricak H, Sonoda Y, Sosa RE, Benz M, Lyons G, Abu-Rustum NR, Sala E, Vargas HA. The impact of FDG-PET/CT in the management of patients with vulvar and vaginal cancer. Gynecol Oncol. 2016;140(3):420–4.
340. Queiroz MA, Kubik-Huch RA, Hauser N, Freiwald-Chilla B, von Schulthess G, Froehlich JM, et al. PET/MRI and PET/CT in advanced gynaecological tumours: initial experience and comparison. Eur Radiol. 2015;25:2222–30.
341. Veit-Haibach P, Hauser N, Freiwald-Chilla B, Gvon S, Froehlich J, Kubik-Kubik R. First evaluations of PET/MRI in gynaecological malignancies. J Nucl Med. 2013;54:401.
342. Boccellino M, Vanacore D, Zappavigna S, et al. Testicular cancer from diagnosis to epigenetic factors. Oncotarget. 2017;8:104654–63.
343. Kim W, Rosen MA, Langer JE, et al. US MR imaging correlation in pathologic conditions of the scrotum. Radiographics. 2007;27:1239–53.
344. EAU Guidelines Office, Arnhem, The Netherlands. EAU Guidelines. Presented at the EAU Annual Congress Copenhagen 2018.
345. Müller J, Schrader AJ, Jentzmik F, et al. Assessment of residual tumours after systemic treatment of metastatic seminoma: (1)(8)F-2-fluoro-2-deoxy-D-glucose positron emission tomography - meta-analysis of diagnostic value. Urologe A. 2011;50:322–7.

346. Spermon JR, De Geus-Oei LF, Kiemeney LA, et al. The role of (18)fluoro-2-deoxyglucose positron emission tomography in initial staging and re-staging after chemotherapy for testicular germ cell tumours. BJU Int. 2002;89:549–56.
347. Karapetis CS, Strickland AH, Yip D, et al. Use of fluorodeoxyglucose positron emission tomography scans in patients with advanced germ cell tumour following chemotherapy: single-centre experience with long-term follow up. Intern Med J. 2003;33:427–35.
348. Bachner M, Loriot Y, Gross-Goupil M, et al. 2-(1)(8) fluoro-deoxy-D-glucose positron emission tomography (FDG-PET) for postchemotherapy seminoma residual lesions: a retrospective validation of the SEMPET trial. Ann Oncol. 2012;23:59–64.
349. Oechsle K, Hartmann M, Brenner W, et al. [^{18}F] Fluorodeoxyglucose positron emission tomography in nonseminomatous germ cell tumors after chemotherapy: the German multicenter positron emission tomography study group. J Clin Oncol. 2008;26:5930–5.
350. Kirrander P, Sherif A, Friedrich B, et al. Swedish National Penile Cancer Register: incidence, tumour characteristics, management and survival. BJU Int. 2016;117:287–92.
351. Barnholtz-Sloan JS, Maldonado JL, Pow-Sang J, et al. Incidence trends in primary malignant penile cancer. Urol Oncol. 2007;25:361–7. https://doi.org/10.1016/j.urolonc.2006.08.029.
352. Arya M, Li R, Pegler K, et al. Long-term trends in incidence, survival and mortality of primary penile cancer in England. Cancer Causes Control. 2013;24:2169–76.
353. Lont AP, Besnard AP, Gallee MP, et al. A comparison of physical examination and imaging in determining the extent of primary penile carcinoma. BJU Int. 2003;91:493–5.
354. Sadeghi R, Gholami H, Zakavi SR, et al. Accuracy of ^{18}F-FDG PET/CT for diagnosing inguinal lymph node involvement in penile squamous cell carcinoma: systematic review and meta-analysis of the literature. Clin Nucl Med. 2012;37:436–41.
355. Zhang S, Li W, Liang F. Clinical value of fluorine-18 2-fluoro-2-deoxy-D-glucose positron emission tomography/computed tomography in penile cancer. Oncotarget. 2016;7:48600–6.
356. Barski D, Georgas E, Gerullis H, et al. Metastatic penile carcinoma - an update on the current diagnosis and treatment options. Cent Eur J Urol. 2014;67:126–32.
357. Giesel FL, Kratochwil C, Lindner T, Marschalek MM, Loktev A, Lehnert W, Debus J, Jäger D, Flechsig P, Altmann A, Mier W, Haberkorn U. 68Ga-FAPI PET/CT: biodistribution and preliminary dosimetry estimate of 2 DOTA-containing FAP-targeting agents in patients with various cancers. J Nucl Med. 2019;60(3):386–92. https://doi.org/10.2967/jnumed.118.215913. Epub 2018 Aug 2. PMID: 30072500; PMCID: PMC6424229.
358. Huang R, Pu Y, Huang S, Yang C, Yang F, Pu Y, Li J, Chen L, Huang Y. FAPI-PET/CT in cancer imaging: a potential novel molecule of the century. Front Oncol. 2022;12:854658. https://doi.org/10.3389/fonc.2022.854658. PMID: 35692767; PMCID: PMC9174525.
359. Zhao L, Chen J, Pang Y, Fu K, Shang Q, Wu H, Sun L, Lin Q, Chen H. Fibroblast activation protein-based theranostics in cancer research: a state-of-the-art review. Theranostics. 2022;12(4):1557–69.
360. Ballal S, Yadav MP, Moon ES, Kramer VS, Roesch F, Kumari S, Bal C. First-in-human results on the biodistribution, pharmacokinetics, and dosimetry of [^{177}Lu]Lu-DOTA.SA.FAPi and [^{177}Lu]Lu-DOTAGA.(SA.FAPi)$_2$. Pharmaceuticals. 2021;14(12):1212. https://doi.org/10.3390/ph14121212. PMID: 34959613; PMCID: PMC8707268.

Further Reading

Bouchelouche K. PET/CT in bladder cancer: an update. Semin Nucl Med. 2022;52(4):475–85. https://doi.org/10.1053/j.semnuclmed.2021.12.004.

Bruckmann NM, Morawitz J, Fendler WP, Ruckhäberle E, Bittner AK, Giesel FL, Herrmann K, Antoch G, Umutlu L, Kirchner J. A role of PET/MR in breast cancer? Semin Nucl Med. 2022;52(5):611–8. https://doi.org/10.1053/j.semnuclmed.2022.01.003.

Dendl K, Koerber SA, Tamburini K, Mori Y, Cardinale J, Haberkorn U, Giesel FL. Advancement and future perspective of FAPI PET/CT in gynecological malignancies. Semin Nucl Med. 2022;52(5):628–34. https://doi.org/10.1053/j.semnuclmed.2022.04.002.

Fonti R, Conson M, Del Vecchio S. PET/CT in radiation oncology. Semin Oncol. 2019;46(3):202–9. https://doi.org/10.1053/j.seminoncol.2019.07.001.

Kann BH, Hosny A, Aerts HJWL. Artificial intelligence for clinical oncology. Cancer Cell. 2021;39(7):916–27. https://doi.org/10.1016/j.ccell.2021.04.002.

Langbein T, Weber WA, Eiber M. Future of theranostics: an outlook on precision oncology in nuclear medicine. J Nucl Med. 2019;60(Suppl 2):13S–9S. https://doi.org/10.2967/jnumed.118.220566.

Sharma P, Singh SS, Gayana S. Fibroblast activation protein inhibitor PET/CT: a promising molecular imaging tool. Clin Nucl Med. 2021;46(3):e141–50. https://doi.org/10.1097/RLU.0000000000003489.

Vahidfar N, Eppard E, Farzanehfar S, Yordanova A, Fallahpoor M, Ahmadzadehfar H. An impressive approach in nuclear medicine: theranostics. PET Clin. 2021;16(3):327–40. https://doi.org/10.1016/j.cpet.2021.03.011.

Wang L, Tang G, Hu K, Liu X, Zhou W, Li H, Huang S, Han Y, Chen L, Zhong J, Wu H. Comparison of ^{68}Ga-FAPI and ^{18}F-FDG PET/CT in the evaluation of advanced lung cancer. Radiology. 2022;303(1):191–9. https://doi.org/10.1148/radiol.211424.

Correction to: Principles of Surgical Oncology

Nita Nair, Abhay Kattepur, Shraddha Patkar,
Prasanth Penumadu, Caleb Harris, and S. V. S. Deo

Correction to:
Chapter 8 in: A. Munshi et al. (eds.), *Radiation Oncology – Principles, Precepts and Practice*, https://doi.org/10.1007/978-981-97-8389-2_8

This book was recently published. The authors' name and affiliation in Chap. 8 have been updated and should be changed as below.

Current output:
N. S. Nair (*)
Department of Surgical Oncology, Tata Memorial
Centre, Homi Bhabha National Institute, Mumbai,
Maharashtra, India
Consultant Breast Surgeon, Apollo Hospital,
Navi Mumbai, Maharashtra, India

P. Penumadu
Department of Surgical Oncology, JIPMER,
Pondicherry, India
Department of Surgical Oncology, Sri Venkateswara
Institute of Cancer Care and Advanced Research,
Tirupati, Andhra Pradesh, India

The updated version of this chapter can be found at https://doi.org/10.1007/978-981-97-8389-2_8

A. Munshi et al. (eds.), *Radiation Oncology – Principles, Precepts and Practice*,
https://doi.org/10.1007/978-981-97-8389-2_28

Required Output:

Nita Nair
Consultant Breast Surgeon
Apollo Hospital
Navi Mumbai, India

Prasanth Penumadu
Medical Director and Head of the Department, Surgical Oncology
Sri Venkateswara Institute of Cancer Care and Advanced Research
Tirupati, India

- In FM
- Chapter 8

GPSR Compliance

The European Union's (EU) General Product Safety Regulation (GPSR) is a set of rules that requires consumer products to be safe and our obligations to ensure this.

If you have any concerns about our products, you can contact us on ProductSafety@springernature.com

In case Publisher is established outside the EU, the EU authorized representative is:

Springer Nature Customer Service Center GmbH
Europaplatz 3
69115 Heidelberg, Germany

Batch number: 09336751

Printed by Printforce, the Netherlands